A Journey through the History of Numerical Linear Algebra

A Journey through the History of Numerical Linear Algebra

Claude Brezinski
University of Lille, Lille, France

Gérard Meurant
Paris, France

Michela Redivo-Zaglia
University of Padua, Padua, Italy

siam.
Society for Industrial and Applied Mathematics
Philadelphia

10 9 8 7 6 5 4 3 2 1

Publications Director	Kivmars H. Bowling
Executive Editor	Elizabeth Greenspan
Managing Editor	Kelly Thomas
Production Editor	Ann Manning Allen
Copy Editor	Claudine Dugan
Production Manager	Donna Witzleben
Production Coordinator	Cally A. Shrader
Compositor	Cheryl Hufnagle
Graphic Designer	Doug Smock

Library of Congress Cataloging-in-Publication Data

Names: Brezinski, Claude, 1941- author. | Meurant, Gérard A., author. | Redivo Zaglia, Michela, author.

Title: A journey through the history of numerical linear algebra / Claude Brezinski, Gérard Meurant, and Michela Redivo-Zaglia.

Description: Philadelphia : Society for Industrial and Applied Mathematics, 2022. | Series: Other titles in applied mathematics; 183 | Includes bibliographical references and index. | Summary: "The book describes numerical methods proposed for solving problems in linear algebra from antiquity to the present. Focusing on methods for solving linear systems of equations and eigenvalue problems, the book also describes the interplay between numerical methods and the computing tools available for solving these problems. Biographies of the main contributors to the field are included"-- Provided by publisher.

Identifiers: LCCN 2022029258 (print) | LCCN 2022029259 (ebook) | ISBN 9781611977226 (hardback) | ISBN 9781611977233 (ebook)

Subjects: LCSH: Algebras, Linear--History. | Matrices--History. | Numerical calculations. | Numerical analysis. | AMS: Numerical analysis -- Historical. | Computer science -- Historical.

Classification: LCC QA184.2 .B74 2022 (print) | LCC QA184.2 (ebook) | DDC 518/.43--dc23/eng/20211105

LC record available at *https://lccn.loc.gov/2022029258*

LC ebook record available at *https://lccn.loc.gov/2022029259*

This book is dedicated to the memory of

Gene Howard Golub (1932-2007)

and

Richard Steven Varga (1928-2022).

Study the past if you would divine the future.

– *Confucius*

I do not know what I may appear to the world; but to myself I seem to have been only like a boy playing on the seashore, and diverting myself in now and then finding a smoother pebble or a prettier shell than ordinary, whilst the great ocean of truth lay all undiscovered before me.

– *Isaac Newton, quoted in David Brewster, Memoirs of the Life, Writings, and Discoveries of Sir Isaac Newton (1855)*

Contents

Introduction

In this book we invite the reader to take a journey through the history of numerical linear algebra.

Nowadays numerical linear algebra plays a central role in numerical analysis, applied mathematics and, more generally, scientific computing; see, for instance, [465]. In fact, here we are mainly interested in *matrix computations*, that is, algorithms to solve linear systems of algebraic equations and to compute eigenvalues and eigenvectors of matrices. This also involves considering determinants and matrix factorizations.

We think that the development of the methods and algorithms to solve these problems was intimately linked to the progresses of the computing tools that were available through the ages. This is why we also provide a brief history of computing machines, from the abacus to the supercomputer.

We must warn the reader that the authors of this book are not historians, but applied mathematicians. Even restricted to matrix computations, numerical linear algebra had a rich history. Hence, choices had to be made. Consequently, the history told in this book is biased by the background of the authors. In [3074], the American mathematician Clifford Ambrose Truesdell (1919-2000) wrote

> *Historians of science have often remarked that mathematicians are the most unhistorical of scientists because they tend to regard mathematicians of old as if they were colleagues today. For a mathematician, a mistake is a mistake no matter how old it be, or what great man made it.* [...] *On the other hand, some mathematicians are the most historical of scientists in that they study old sources for what they can learn from them and use.*

We are probably facing the danger that historians will find there is too much mathematics in this book, and that mathematicians will find there is too much history. But we are interested in showing how some of the algorithms that we use today gradually emerged, and we have to explain them, even briefly. Of course, their developments were not linear. There were successes and failures. Many methods that were proposed in the past are now almost forgotten, even though there are sometimes some revivals like, for instance, for the conjugate gradient method for solving symmetric positive definite linear systems. Some methods disappeared because more efficient ones were proposed later, and some others disappeared because the problems they were solving were not of interest any longer or because they were not well suited for new computing machines.

Another danger is to look at the past with the eyes of today. It may sometimes seem anachronistic to use a modern matrix representation to explain results obtained before the introduction of matrices. But this is sometimes easier than providing long explanations in words.

There are several kinds of historical works about mathematics. Some are very detailed studies about a particular topic, like, for instance, the controversy between Camille Jordan (1838-1922) and Leopold Kronecker (1823-1891) in 1874 [421] or a short period of time; see, for instance [1369]. Others try to cover the whole history of mathematics in a few pages, like, for instance, the book *A Concise History of Mathematics* [2930] by Dirk Jan Struik (1894-2000). The publisher added on the cover of the book *Professor Struik has achieved the seemingly impossible task of compressing the history of mathematics into less than three hundred pages*! Another more recent example is *Taming the unknown: A history of algebra from antiquity to the early twentieth century* [1880] by Victor Joseph Katz and Karen Hunger Parshall in 2014. Our work tries to stay in between these two extremes.

But are we really doing history? The historian of mathematics Ivor Owen Grattan-Guinness (1941-2014) [1426] distinguished *history* from *heritage*. He wrote

> *By "history" I refer to the details of the development of* [a notion]*: its prehistory and concurrent developments; the chronology of progress, as far as it can be determined; and maybe also the impact in the immediately following years and decades.* [...] *By "heritage" I refer to the impact of* [a notion] *upon later work, both at the time and afterward, especially the forms which it may take, or be embodied, in later contexts.* [...] *Heritage addresses the question "how did we get here?," and often the answer reads like "the royal road to me."* [...] *History and heritage are twins, each profiting from practices used in the other. In heritage chronology is much less significant: however, mathematicians often regard questions of the type "Who was the first mathematician to ...?" as the prime type of historical question to pose, whereas historians recognize them as often close to meaninglessness when the notion involved is very general or basic.*

Johanna Pejlare and Kajsa Bråting [2476] wrote

> *It is important to point out that the utilization of heritage often results in a modernization of old results in order to show their current place in mathematics, but the historical context is not always taken into consideration.* [...] *one should be aware of that mathematicians of the past based on their definitions within the conceptual framework available at that particular time and not assume that they strived for the modern definitions of today.*

Maybe this book is more on the heritage side, but it is up to the reader to decide. Of interest on these topics are also the essays of Kenneth Ownsworth May (1915-1977) [2172, 2173, 2174].

The way research is done has changed over the ages. Until the 17th century, mathematicians published their work in books or communicated with other researchers by letter. People like Marin Mersenne (1588-1648), a French catholic priest and mathematician, tried to facilitate the communication of information. He had many contacts in the scientific world and acted as a go-between with several mathematicians like Pierre de Fermat (1601-1665), René Descartes (1596-1650), and Blaise Pascal (1623-1662). Some mathematicians, like, for instance, Isaac Newton (1643-1727), were even delaying the publication of their findings. Then, scientific academies were founded: the Royal Society of London in 1660 (see [2563]), followed by the French Academy of Sciences in 1666, the Berlin Academy in 1700, and the Saint Petersburg Academy in 1724. Publications of memoirs were sponsored by the academies, and it was the start of peer review. Later on, mathematical journals were founded, making possible a faster communication and spreading of research findings. Communication is a key to the rapid development of science.

Up until the beginning of the 20th century, research was mostly done by single (male) individuals. As Hans Schneider (1927-2014) wrote in [2705],

> *what might in the nineteenth century have been the work of one man over a period of twenty years, is today the work of four in five years.*

Many papers that are published today have two or even more co-authors.

Things have also changed geographically. For instance, elimination algorithms to solve linear systems originated many centuries ago in China, India, and Mesopotamia. It is not really known if this knowledge spread slowly to the west, or if these methods were found independently elsewhere. Nevertheless, many of the algorithms considered in this book were developed much later in western Europe and in the United States.

Another drastic change is the number of scientific papers produced each year. Only in our area of interest, during the last 50 years the number of journals has dramatically increased and the number of published papers has increased exponentially, not mentioning the rejected papers. Today, it is almost impossible to follow all that is produced regarding our topics, even though the Internet is a great help to obtain information.

We won't enter the old debate: Are mathematics discovered or invented? According to dictionaries *discovery* is the act of finding something that had not been known before, for instance, the discovery of gold in Califormia or the discovery of America by Cristoforo Colombo (Christopher Columbus) (or somebody else before him), whereas *invention* is the process of creating something that has never been made before, for instance, the phone or the steam engine.

It is likely that there are instances of both discoveries and inventions in mathematics. For example, we can say that the positive integers were discovered, but the proof that there is an infinite number of them was invented. Concerning one of the topics of this book, Cornelius Lanczos (1893-1974) invented a method to solve linear equations, but he discovered the properties of the constructed vectors.

An interesting essay [1418] on theses issues was written by Sir William Timothy Gowers (Fields medalist in 1998). He wrote

> *The idea that either all of mathematics is discovered or all of mathematic is invented is ridiculous.*

In this book we tried to avoid as much as possible the words "discovered" and "invented."

We have chosen to try to describe the evolution of numerical methods from the origin to the present days. This may be controversial since for the last years it may be difficult to identify what are the main trends of research, and we may lack hindsight to what will be considered important later. However, there were interesting methods that were developed in the last years, for instance, for eigenvalue computations or for parallel computing. Of course, describing recent methods may look more like a review than a history book, but we thought that it could be interesting for our readers to have at least a few pointers to the recent literature.

Let us now briefly describe the contents of the book, which is in two parts: History and Biographies.

Since matrices are the main mathematical objects we are interested in, Chapter 1 describes how matrices were introduced after 1850 by Arthur Cayley (1821-1895) and James Joseph Sylvester (1814-1897). This can be seen as a consequence of previous studies on quadratic forms, determinants, the solution of linear systems, and of the quaternions of William Rowan Hamilton (1805-1865), that can be considered as an invention. Then, we consider some important properties of matrices, the rank of a matrix, norms, the condition number and the notion of ill-conditioning, the Schur complement, and how matrices were used by physicists working on quantum mechanics in the 1920s, since this contributed to the success of the matrix framework.

Chapter 2 is devoted to the elimination methods to solve systems of linear equations. According to the present sources of information, the story started in Babylonia, Egypt, China, and Greece, as well as India and Persia. Then, we study what happened in Western Europe starting in the Middle Ages and through the next centuries, up to our present days. We also consider how elimination methods were specialized for sparse linear systems, that is, systems with matrices having many zero entries.

Elimination methods for linear systems and also the quest for solutions of polynomial systems led to the introduction of determinants in the 17th century, that is, more than one and a half centuries before matrices appeared on the scene. However, the theory of determinants started to be established at the end of the 18th century and at the the beginning of the 19th century. In Chapter 3 we study how the theory was developed, mainly in France, Germany, and England.

In Chapter 4 we consider several matrix factorizations that helped solve some problems or reveal some properties of the matrix, the QR factorization, the singular value decomposition (SVD), the polar and CS factorizations, as well as the Jordan and Frobenius canonical forms. The use of matrix factorizations was a big step forward in numerical linear algebra.

Iterative methods for solving linear systems are considered in Chapter 5. We start with the classical methods introduced in the 19th century and continue with the alternating direction and Krylov methods that were developed starting in the mid-20th century. This chapter also contains a short description of the history of preconditioning, domain decomposition methods, and multigrid methods.

The computation of eigenvalues and eigenvectors is the topic of Chapter 6. We describe how these problems occurred after the mid-18th century when people were looking for solutions of linear differential equations. Until the 20th century, eigenvalues were obtained as solutions of the (polynomial) characteristic equation. Then, more accurate and efficient methods were developed, like the Lanczos and Arnoldi methods, as well as the QR algorithm.

Many numerical methods for solving linear algebra problems were designed to be well suited for some computing machines. In Chapter 7 we review the tools that were available to speed up the computations, starting from simple devices to mechanical and electric calculators, and to the development of computers in the second half of the 20th century.

After the introduction of digital electronic computers there was a need for efficient programs to solve linear algebra problems. In Chapter 8 we describe the development of some software packages and tools like EISPACK, LINPACK, and LAPACK, as well as MATLAB.[1]

Chapter 9 is devoted to topics that did not fit in the previous chapters: fast solvers, special methods for Toeplitz and Hankel matrices, the computation of functions of matrices, the solution of ill-posed problems, and matrix equations, like the Sylvester and Lyapunov equations.

In a conference given in 1926 and published in [1989], the French physicist Paul Langevin (1872-1946) wrote[2] (our translation)

> *In order to contribute to general culture and to draw from the teaching of science all that it can give for the formation of the mind, nothing can replace the history of past efforts, made alive by contact with the life of the great scientists and the slow evolution of ideas. Only by this means can one prepare those who will continue the work of science, to give them the sense of its perpetual movement and of its human value.* [...] *nothing is better than going to the sources, to be in contact as frequently and as completely as possible with those who have made science and who*

[1]MATLAB is a trademark of The MathWorks Inc.

[2]*Or pour contribuer à la culture générale et tirer de l'enseignement des sciences tout ce qu'il peut donner pour la formation de l'esprit, rien ne saurait remplacer l'histoire des efforts passés, rendue vivante par le contact avec la vie des grands savants et la lente évolution des idées. Par ce moyen seulement on peut préparer ceux qui continueront l'oeuvre de la science, leur donner le sens de son perpétuel mouvement et de sa valeur humaine.* [...] *rien ne vaut d'aller aux sources, de se mettre en contact aussi fréquent et complet que possible avec ceux qui ont fait la science et qui en ont le mieux représenté l'aspect vivant.* [...] *Les exemples précédents montrent bien comment, au point de vue de l'enseignement comme au point de vue de la recherche scientifique, il est indispensable de ne pas oublier l'histoire des idées - et concurremment celle des hommes, puisque c'est par eux qu'on éclaire les idées. Rien de tel que de lire les oeuvres des savants d'autrefois, rien de tel que de vivre avec ceux qui sont contemporains pour pénétrer la pensée intime des uns et des autres.*

> *have best represented its living aspect.* [...] *The preceding examples clearly show how, from the point of view of teaching as well as from the point of view of scientific research, it is essential not to forget the history of ideas - and concomitantly that of men, since it is through them that ideas are illuminated. There is nothing better like reading the works of scholars of the past, nothing better like living with those who are contemporaries to penetrate their intimate thoughts.*

Following what Langevin told us, the second part of the book contains short biographies of researchers who contributed significantly to the field of numerical linear algebra. We consider only people who are, unfortunately, deceased. Of course, doing so, we neglect some people who had a great impact on the development of our field in the last 50 years, like Beresford Neill Parlett, Gilbert Wright (Pete) Stewart, Cleve Barry Moler, and a few others, but, fortunately, these people have a long and fruitful life. We thought that it could be embarrassing for some people to see their lives commented upon in a book. On the use of biographies in the history of science, see [1572] by Thomas Leroy Hankins in 1979.

Full names (with dates of birth and death for deceased people) of the contributors are only given once in each chapter. The subject index only covers Part I.

We must also add that the authors of this book made some contributions to numerical linear algebra. In the text, they are denoted by their initials (C.B., G.M., and M.R.-Z.), but they are listed in the name index. As we said above, we are aware that our own works may have biased us. We apologize for any omission or erroneous attribution of a discovery or an invention. It does not mean that they were less important for the domain. But due to the very large literature, we had to make choices. Moreover, it could be that we simply were not aware of some of them.

Acknowledgments

The authors thanks Jack Dongarra, Iain Duff, and Cleve Moler for their reading of a first version of this book and valuable comments, as well as three anonymous reviewers for their suggestions.

We also thank, in alphabetical order, Tom Altman, Dario Andrea Bini, Yigal Burstein, Andrzej Cegielski, Stefano Cipolla, Richard Farebrother, Catherine Harpham, Tim Hopkins, Yuri Kuznetsov, Jörg Liesen, Sergey A. Matveev, Volker Mehrmann, Howard Phillips, Lucy Shepherd, Jurjen Duintjer Tebbens, Dominique Tournès, and Jens-Peter M. Zemke.

It has been a pleasure working with the SIAM publication staff, in particular Elizabeth Greenspan and Ann Manning Allen, and Cheryl Hufnagle from The PCH Network.

Part I

History

1

Matrices and their properties

The concept of a matrix has become so universal in the meantime that we often forget its great philosophical significance. To call an array of letters by the letter A was much more than a matter of notation. It had the significance that we are no longer interested in the numerical values of the coefficients a_{ij}. The matrix A was thus divested of its arithmetic significance and became an algebraic operator.

– Cornelius Lanczos, *Linear Differential Operators, [1987], p. 100*

1.1 ▪ The birth of matrices

Many people consider papers by James Joseph Sylvester (1814-1897) [2959] in 1851 and by Arthur Cayley (1821-1895) [587] in 1858 as defining the starting point of the history of matrices. However, as quite often is the case, things are more complicated than that. The history of matrices has its roots in what had been done since antiquity to solve systems of linear equations (see Chapter 2), in the development and the theory of determinants (see Chapter 3), and in studies on linear transformations of variables in quadratic and bilinear forms at the beginning of the 19th century. In fact, even though this may seem paradoxical, some properties of matrices were considered or proved before matrices were formally defined by Sylvester and Cayley.

Matrices were introduced independently of the concept of vector space, which appeared gradually through the works of Sir William Rowan Hamilton (1805-1865) [1558] in 1843, Hermann Grassmann (1809-1877) [1424, 1425] in 1844 and 1862 ([1657]), Giuseppe Peano (1858-1932) who gave an explicit axiomatic definition of the concept of vector space over real numbers in 1888, Josiah Willard Gibbs (1839-1903) [1346] in 1901, and Oliver Heaviside (1850-1925); see [781, 1146, 1431, 938, 2270] (ordered by date).

Johann Carl Friedrich Gauss (1777-1855) studied quadratic forms of order 2 and 3 and their linear transformations in 1801. What he wrote about combinations of linear transformations can be now considered as formulas for the product of the corresponding matrices.

At the beginning of the 19th century, several mathematicians considered arrays of numbers arising from linear equations or determinants. In 1812, Jacques Philippe Marie Binet (1786-1857) presented to the French Academy of Sciences a memoir [322, 323] stating what we now call the Cauchy-Binet formula, which shows how to compute the determinant of the product of two rectangular arrays with compatible dimensions, the product being defined as what is now the standard product of matrices. At the same time, Augustin-Louis Cauchy (1789-1857) presented a 83-page memoir [564] (that was finally published in 1815) in which he gave a better proof of the formula. Moreover, he denoted the entries of his arrays with the modern notation $a_{i,j}$, considered the transpose of an array that he called the *conjugate system*, and he proved that the determinant is not changed by this operation. In that paper, Cauchy also gave a solid basis to the theory of determinants; see Chapter 3. In 1829, he proved rigorously that the roots of the characteristic polynomial corresponding to a symmetric determinant are real [569]. In modern terms, the eigenvalues of a real symmetric matrix are real. Cauchy's result was extended in 1855 by Charles Hermite (1822-1901) to arrays with complex entries where $a_{j,i}$ is the conjugate of $a_{i,j}$; see [1651]. The corresponding matrices are now called *Hermitian.*

In 1844, the German mathematician Ferdinand Gotthold Max Eisenstein (1823-1852) studied ternary forms in two variables [1075]. He considered linear substitutions or transformations of the variables that he denoted by a single letter,

$$t_1 = \begin{pmatrix} \alpha, & \beta \\ \gamma, & \delta \end{pmatrix}.$$

He showed how to add and multiply them and also computed their inverses that he called reciprocals; see page 96 of [1075]. Also of interest is his other paper [1076] published in 1844-1845.

In 1850-1851, Sylvester was interested in the intersections of curves and surfaces. In the paper [2955] he introduced the term "matrix" that he considered as a rectangular array of terms from which several different determinants can be constructed. This array was the womb of the determinants, as he explained in [2959] in 1851.

In 1854, Hermite was interested in finding all the substitutions that leave a nonsingular ternary form invariant [1650]. Cayley also gave a solution of this problem [584] in 1855. In that paper, he used matrices that he denoted with vertical bars. Matrices and their inverses and the multiplication rule appeared also in [583]. In that paper, written in French but published in the German journal founded by August Leopold Crelle (1780-1855), one can read[3] (our translation)

> *There will be many things to say about this matrix theory which must, as it seems to me, precede the theory of determinants.*

As we see, at that time, Cayley was thinking in the same way as Sylvester. Matrices also appeared in several papers [582, 584, 585] from Cayley in 1855-1857.

Cayley's paper that is often considered as the start of matrix theory is *A memoir on the theory of matrices*, published in 1858 in the *Philosophical Transactions of the Royal Society of London.* In that paper he used a slightly different notation for matrices; see Section 10.8. He considered matrices as objects, not just as a mere notation, and showed how to add, multiply, and invert them. He noticed that the multiplication is not commutative and that there is no inverse if the determinant vanishes. He stated what he called *a remarkable theorem*, which is now called the Cayley-Hamilton theorem, but he only proved it for 2×2 matrices. He wrote that he had verified the theorem for matrices of order 3, but he did not try to prove the general case. Cayley also considered the problem of characterizing matrices that commute with a given matrix, but his result was not correct, only being true for non-derogatory matrices.

Some people were thinking that the origin of matrices can be found in the theory of *quaternions* developed in the 1840s by Hamilton; see Section 10.31. Peter Guthrie Tait (1831-1901), who was a follower of Hamilton and an advocate of quaternions, wrote in a letter to Cayley in 1872 (see the book by Cargill Gilston Knott (1856-1922) [1919] in 1911),

[3] *Il y aurait bien des choses à dire sur cette théorie de matrices, laquelle doit, il me semble, précéder la théorie des Déterminants.*

> *It is a most singular fact that you seem to have been working simultaneously with Hamilton in 1857-8, just as I found you had been in a very much earlier year* [...] *I have had but time for a hurried glance at your paper on Matrices - and I see that it contains (of course in a very different form) many of Hamilton's properties of the linear and vector function* [...] *I send you a private copy of my little article, by which you will see how closely the adoption of Hamilton's method has led me to anticipate almost every line of your last note.*

The correspondence between Tait and Cayley went on for many years. In an answer to a letter from Tait in 1894 (page 164 of [1919]), Cayley wrote

> *I certainly did not get the notion of a matrix in any way through quaternions: it was either directly from that of a determinant; or as a convenient mode of expression of the equations*
>
> $$x' = ax + by,$$
> $$y' = cx + dy.$$

We observe that Hamilton introduced the term *scalar* as a part of his quaternions.

In 1867, Edmond Nicolas Laguerre (1834-1886) published *Sur le calcul des systèmes linéaires* [1976], at the beginning of which he wrote[4] (our translation)

> *I call, according to the ordinary usage, a linear system the table of coefficients of a system of n linear equations with n unknowns, and, with one exception which I will discuss later, I will always denote it by a single capital letter, reserving the lower case letters to denote especially the entries of the linear system.* [...] *In all what follows, I will consider these capital letters representing linear systems as real quantities, subject to all algebraic operations.*

He denoted these "systems" as arrays of numbers without any bars or parenthesis. He defined the addition, the multiplication, noticing that it is not commutative, and called two systems that commute, "permutables." Unfortunately, he called "inverse" the transpose of an array A and denoted it by A_1. Our matrix inverse multiplied by the determinant is named the "reciprocal" and denoted by A_0. It is the transpose of the matrix of the cofactors, that is, in modern terms, the adjugate. He probably did that because this reciprocal is always defined, even if the determinant is equal to zero. Laguerre defined linear systems as equations $AX = B$, where A, X, and B are arrays of the same order. In fact, for Laguerre, our modern vectors are zero square arrays, except for the first column, as

$$\begin{matrix} x_1 & 0 & \cdots & 0 \\ x_2 & 0 & \cdots & 0 \\ \vdots & \vdots & & \vdots \\ x_n & 0 & \cdots & 0 \end{matrix}$$

denoted as $(\, x_1 \quad x_2 \quad \cdots \quad x_n \,)$, which is counterintuitive for us. Therefore, for him, a quadratic form $f = ax^2 + 2bxy + cy^2$ was equivalent to

$$\begin{matrix} f & 0 \\ 0 & 0 \end{matrix} = \begin{matrix} x & y \\ 0 & 0 \end{matrix} \times \begin{matrix} a & b \\ b & c \end{matrix} \times \begin{matrix} x & 0 \\ y & 0 \end{matrix},$$

[4] *J'appelle, suivant l'usage habituel, système linéaire le tableau des coefficients d'un système de n équations linéaires à n inconnues et, sauf une exception dont je parlerai plus loin, je le représenterai toujours par une seule lettre majuscule, réservant les lettres minuscules pour désigner spécialement les éléments du système linéaire.* [...] *Dans tout ce qui suit, je considérerai ces lettres majuscules représentant les systèmes linéaires comme de véritables quantités, soumises à toutes les opérations algébriques.*

with the right-hand side denoted as X_1AX with (in his notation) $X = (x \quad y)$. He stated the Cayley-Hamilton theorem, but gave a proof only for $n = 2$. He remarked that, from that result, his reciprocal of A is a polynomial in A. On page 226 of [1976], Laguerre stated that the reciprocal of an array with a zero determinant is what we now call a rank one matrix. But note that this is only true if the rank of the matrix is equal to $n - 1$. Laguerre wanted to apply this result to $A - \lambda$, where λ is an eigenvalue (of course, he did not use that name, in fact he did not even give a name to these numbers), and he observed that he cannot do it if there is a polynomial ψ of degree smaller than n such that $\psi(A) = 0$. He also defined what is a symmetric positive definite system, that is, one having positive eigenvalues.

Laguerre applied his "linear systems" to the study of linear transformations of quadratic forms and to Abelian functions. The only mathematicians who are cited in [1976] are Eisenstein, Gauss, Hermite, and Sylvester. Laguerre's paper was published in the *Journal de l'École Impériale Polytechnique* and probably did not have much influence outside France. Nevertheless, his work on this topic was praised in 1898 by the famous Henri Poincaré (1854-1912), who was one of the editors of the Complete Works of Laguerre in two volumes. About Laguerre's 1867 paper, Poincaré wrote[5] (our translation)

> *Linear substitutions have acquired such importance in Analysis that it seems to us today difficult to deal with a single question without them being introduced.* [...] *Since the turn of the century, great efforts have been made to generalize the concept of number; from real quantities, we have risen to imaginary quantities, to complex numbers, to ideals, to quaternions, to Galois imaginaries; Laguerre rises to a point of view from which one can embrace all these horizons at a glance. All these new notions, and in particular the quaternions, are reduced to linear substitutions.* [...] *Undoubtedly, there is only a new notation in all of this; but make no mistake: in the mathematical sciences, a good notation has the same philosophical importance as a good classification in the natural sciences.*

The last sentence shows that Poincaré did not foresee the future of matrices. About Laguerre, see [2609] by Eugène Rouché (1832-1910) in 1887.

In 1878, Ferdinand Georg Frobenius (1849-1917) wrote a monograph [1250] whose title can be translated as *On linear substitutions and bilinear forms*. He represented bilinear forms by capital letters (related to their coefficients) and manipulated them as if they were matrices, including for inverses. The Frobenius (rational) canonical form (of matrices) was introduced in a long paper [1251] published in 1879. Thomas W. Hawkins wrote in [1599] that Frobenius

> ... *brought together for the first time the work on spectral theory of Cauchy, Jacobi, Weierstrass and Kronecker with the symbolical tradition of Eisenstein, Hermite and Cayley.*

Frobenius gave the definition of the rank and of an orthogonal matrix. He also proved the general case of the Cayley-Hamilton theorem, even though his proof was not completely satisfactory. Frobenius started using matrices in 1896 to study forms in [1254]. Note that Frobenius introduced the concept of irreducibility in 1912; see [2704] by Hans Schneider (1927-2014) in 1977.

[5] *Les substitutions linéaires ont acquis dans l'Analyse un telle importance qu'il nous semble aujourd'hui difficile de traiter une seule question sans qu'elles s'y introduisent.* [...] *Depuis le commencement du siècle, de grands efforts ont été faits pour généraliser le concept de grandeur; des quantités réelles, on s'est élevé aux quantités imaginaires, aux nombres complexes, aux idéaux, aux quaternions, aux imaginaires de Galois; Laguerre s'élève à un point de vue d'où l'on peut embrasser d'un coup d'oeil tous ces horizons. Toutes ces notions nouvelles, et en particulier les quaternions, sont ramenées aux substitutions linéaires.* [...] *Sans doute, il n'y a dans tout cela qu'une notation nouvelle; mais qu'on ne s'y trompe pas : dans les Sciences mathématiques, une bonne notation a la même importance philosophique qu'une bonne classification dans les Sciences naturelles.*

The notion of matrix, as developed by Cayley, was not much used, even in England, before 1880-1890. For instance, even though the book [2732] by Robert Forsyth Scott (1849-1933) in 1880 was on determinants, the author spoke about "arrays" and not matrices. Cayley and Sylvester were cited in that book, but not for their works on matrices. Henry John Stephen Smith (1826-1883) was one of a few in England who referred to matrices in 1862. The notion of matrix was more largely adopted in the 1880s; see, for instance, [489, 490] by Arthur Buchheim (1859-1888) in 1884, who published a proof of the Cayley-Hamilton theorem and [1183] by Andrew Russell Forsyth (1858-1942). Sylvester himself did not use matrices from 1853 to 1882. In 1883, Sylvester wrote

> *Much as I owe in the way of fruitful suggestion to Cayley's immortal memoir, the idea of subjecting matrices to the additive process and of their consequent amenability to the laws of functional operation was not taken from it, but occurred to me independently before I had seen the memoir or was acquainted with its contents; and indeed forced itself upon my attention as a means of giving simplicity and generality to my formula for the powers or roots of matrices, published in the Comptes Rendus of the Institute for 1882 (Vol. XCIV. pp. 55, 396). My memoir on Tchebycheff's method concerning the totality of prime numbers within certain limits, was the indirect cause of turning my attention to the subject, as (through the systems of difference-equations therein employed to contract Tchebycheff's limits) I was led to the discovery of the properties of the latent roots of matrices, and had made considerable progress in developing the theory of matrices considered as quantities, when on writing to Prof. Cayley upon the subject he referred me to the memoir in question: all this only proves how far the discovery of the quantitative nature of matrices is removed from being artificial or factitious, but, on the contrary, was bound to be evolved, in the fulness of time, as a necessary sequel to previously acquired cognitions.*

Sylvester coined the term *latent roots* for what we now call "eigenvalues" (see Section 10.67), and he gave a formula for a general function of a matrix whose eigenvalues are distinct. He defined the *nullity* of a matrix and *derogatory* matrices and considered matrix equations. In 1884, in his article [2969], Sylvester called Cayley's memoir on matrix theory from 1858 *a revival of algebra.*

Except in the work of Laguerre under another form, matrices appeared, probably for the first time in mainland Europe with a reference to Cayley, in the work of the Czech mathematician Eduard Weyr (1852-1903) in 1889. In 1885, Weyr published a 4-page note [3221] written in French in the *Comptes Rendus de l'Académie des Sciences de Paris* in which he introduced what is now called the *Weyr canonical form.* This was followed by a 110-page paper [3223] written in Czech in which Weyr cited Cayley and Sylvester. The title can be translated as *On the theory of bilinear forms.* A German version was published in the first issue of the *Monatshefte für Mathematik und Physik* [3224] in 1890.

Matrices started to be used in Germany in the 1890s in relation with bilinear forms, hypercomplex systems, and group theory. The first list of writings on the theory of matrices was probably in the appendix of a paper [2299] by Sir Thomas Muir (1844-1934) in 1898. He listed 50 papers published between 1858 and 1894 by 13 authors: 20 by Sylvester, 7 by Buchheim, 6 by Taber, 5 by Cayley, 3 by William Henry Metzler (1863-1943), 2 by Weyr, 1 by Laguerre, 1 by Benjamin Osgood Peirce (1854-1914), 1 by Charles Sanders Peirce (1839-1914), 1 by William Hugh Spottiswoode (1825-1883), 1 by Frobenius, 1 by A.R. Forsyth, and 1 by Muir. Muir published a second list [2301] in 1929 which contained 172 papers, of which 18 are for the period 1854 to 1894 and 14 are from the period 1895 to 1900. From this, we see that there was an acceleration of the use of matrices at the turn of the century.

Even though he called them *substitutions linéaires*, matrices and some of their properties were studied by Paul Matthieu Hermann Laurent (1841-1908) [2000] in 1896. Note that this is not the Laurent of Laurent series, who was Pierre Alphonse Laurent (1813-1854).

Matrices were considered by Léon César Autonne (1859-1913) [105, 106] in 1902 and [108, 109] in 1903, and by Kurt Hensel (1861-1941) [1648] in 1904. Albert Châtelet (1883-1960) [643] in 1911 used what he called *tableaux* (arrays) or matrices and applied the results to number theory.

For more about the gradual acceptance of the matrix theory by the mathematical community, see [2865] by Marie Stepánová in 2011.

Let us see what was the opinion of some mathematicians about the origin of matrices.

Henry Taber (1860-1936), an American mathematician, wrote in his paper *On the theory of matrices* [2988] published in 1890,

> *Accordingly, Cayley laid down the laws of combination of matrices upon the basis of the combined effect of the matrices as operators of linear transformation upon a set of scalar variables or carriers. The development of the theory, as contained in Cayley's memoir, was the development of the consequences of these primary laws of combination. Before Cayley's memoir appeared, Hamilton had investigated the theory of such a symbol of operation as would convert three vectors into three linear functions of those vectors, which he called a linear vector operator. Such an operator is essentially identical with a matrix as defined by Cayley; and some of the chief points in the theory of matrices were made out by Hamilton and published in his Lectures on Quaternions (1852). [...] Hamilton must be regarded as the originator of the theory of matrices, as he was the first to show that the symbol of a linear transformation might be made the subject-matter of a calculus. Cayley makes no reference to Hamilton, and was of course unaware that results essentially identical with some of his had been obtained by Hamilton; and, on the other hand, Hamilton's results related only to matrices of the third and fourth order, while Cayley's method was absolutely general. The identity of the two theories was explicitly mentioned by Clifford in a passage of his Dynamic, and was virtually recognized elsewhere by himself and by Tait. Sylvester carried the investigation much farther, developing the subject on the same basis as that which Cayley had adopted. Subsequent to Cayley, but previous to Sylvester, the Peirces, especially Charles Peirce, were led to the consideration of matrices from a different point of view; namely, from the investigation of linear associative algebra involving any number of linearly independent units.*

In 1932, Herbert Westren Turnbull (1885-1961) and Alexander Craig Aitken (1895-1967) published a book [3081] on canonical forms of matrices in which they wrote

> *Matrices, considered as arrays of coefficients in homogeneous linear transformations, were of course tacitly in existence long before Cayley in 1857 proposed to develop their properties as a pure algebra of multiple number. But the intrinsic properties of the arrays were not studied for their own sake; only as much information was extracted in passing as would be useful for the application in hand, such as determinants, co-ordinate geometry, and the like. Rectangular arrays, too, had been well known from the time of the Cauchy-Binet theorem [...] and had found applications, for example in the normal equations of Least Squares, where the determinant is the row-by-row square of an array; and they had also been used, when premultiplied by a determinant, to express a set of determinantal equalities.*

> *Hamilton's quaternions (1843) can be regarded* [...] *as matrices of special form. But Cayley may fairly be credited with founding the general theory of matrices, for the same kind of reasons as leads us to credit Vandermonde with the founding of determinants.*

In his book *Lecture on Matrices* [3199] published in 1934 the Scottish mathematician Joseph Henry Maclagan Wedderburn (1882-1948), who taught mainly in the USA, favored Hamilton as the origin of matrices:

> *The calculus of matrices was first used in 1853 by Hamilton (1, p. 559ff, 480ff) under the name of "Linear and vector functions." Cayley used the term matrix in 1854, but merely for a scheme of coefficients, and not in connection with a calculus. In 1858 (2) he developed the basic notions of the algebra of matrices without recognizing the relation of his work to that of Hamilton; in some cases (e.g., the theory of the characteristic equation) Cayley gave merely a verification, whereas Hamilton had already used methods in three and four dimensions which extend immediately to any number of dimensions. The algebra of matrices was rediscovered by Laguerre (9) in 1867, and by Frobenius (18) in 1878.*

He concluded his book with a chronological list of references from 1853 to 1933. This list contains 549 works devoted to matrices and quaternions, of which 149 were for the period up to 1896, 37 from 1897 to 1900, 132 from 1901 to 1914, and 231 from 1915 to 1933. This was supplemented by 129 references in the next editions. We observe that the relations of the quaternions with the matrices of order 3 were considered by Buchheim [490] in 1884.

The American mathematician Cyrus Colton MacDuffee (1895-1961) wrote [2107] in 1943,

> *The theory of matrices had its origin in the theory of determinants, and the latter had its origins in the theory of systems of equations. From Vandermonde and Laplace to Cayley, determinants were cultivated in a purely formal manner. The early algebraists never successfully explained what a determinant was, and indeed they were not interested in exact definitions.*
>
> *It was Cayley who seems first to have noticed that "the idea of matrix precedes that of determinant". More explicitly, we can say that the relation of determinant to matrix is that of the absolute value of a complex number to the complex number itself, and it is no more possible to define determinant without the previous concept of matrix than it is to have the feline grin without the Cheshire cat.*
>
> *In fact, the importance of the concept of determinant has been, and currently is, vastly over-estimated. Systems of equations can be solved as easily and neatly without determinants as with, as is illustrated in Chapter I of this monograph. In fact, perhaps ninety per cent of matric theory can be developed without mentioning a determinant. The concept is necessary in some places, however, and is very useful in many others, so one should not push this point too far.*
>
> *In the middle of the last century matrices were approached from several different points of view. The paper of Hamilton in 1853 on "Linear and vector functions" is considered by Wedderburn to contain the beginnings of the theory. After developing some properties of "linear transformations" in earlier papers, Cayley finally wrote "A Memoir on the Theory of Matrices", in which a matrix is considered as a single mathematical quantity. This paper gives Cayley considerable claim to the honor of introducing the modern concept of matrix, although the name is due to Sylvester (1850).*

> *In 1867 there appeared the beautiful paper of Laguerre entitled "Sur le calcul des systèmes linéaires" in which matrices are treated in almost the modern manner. It attracted little attention at the time of its publication. Frobenius, in his fundamental paper "Ueber lineare Substitutionen und bilinearen Formen" of 1878, approached matric theory through the composition of quadratic forms.*
>
> *In fact, Hamilton, Cayley, Laguerre, and Frobenius seem to have worked without the knowledge of each others' results. Frobenius, however, very soon became aware of these earlier papers and eventually adopted the term "matrix".*

The historian of mathematics Hawkins [1601] wrote in 1977,

> *During the thirty-year period 1852-1882 the idea of matrix algebra was conceived, more or less independently, by no less than five mathematicians: Eisenstein, Cayley, Laguerre, Frobenius and Sylvester. We have seen that there were great differences in what each did with the idea, differences which reflect differences in motivation and in professional training and attitudes. Although it has been argued that in no meaningful historical sense did Cayley's 1858 memoir constitute the foundation stone of the theory of matrices, I have also shown that Cayley occupies a special place in the history of that theory by virtue of his work relating to the Cayley-Hermite problem. During the 19th century that problem was one of the principal reasons for the introduction and development of the symbolical algebra of matrices. Hermite's role was therefore also important. To him we owe the formulation of the problem, and the form of his solution proved conducive to the introduction of matrix algebra. An important role was also played throughout by research activity in the theory of numbers. Thus we have emphasized the influence of Gauss' Disquisitiones arithmeticae and indicated the centrality of number-theoretic considerations to Eisenstein, Hermite and Laguerre. In particular arithmetical research motivated Hermite's formulation of the Cayley-Hermite problem and Bachmann's revival of interest in the problem in the 1870s, through which it was brought to the attention of Frobenius. Finally, Frobenius' contribution must be stressed, for it was in his memoir of 1878 that the two basic components of the theory of matrices, matrix algebra and spectral theory, were united and the advantages of such a union convincingly demonstrated.*

H. Schneider wrote [2705] in 1977,

> *In the 1850s there was an outward turn of matrix theory, and one may speculate that this was due in part to the profound psychological effect of the definition of "matrix" as a separate and independent entity - I am tempted to say matrix theory turned outward at least partly because matrices had been named. In distinction to its previous role, a matrix was no longer merely a "schema" for writing a determinant or linear substitution but a "quantity", if it be allowed that that term is properly applied to whatever is subject of functional operation.*

Ivor Owen Grattan-Guinness (1941-2014) and Walter Ledermann (1911-2009) wrote in their article *Matrix theory* [1427],

> *Cayley's memoir was ignored for many years. It may be that, with a few notable exceptions, mathematicians of his generation, or even one or two subsequent generations, were reluctant to accept mathematical objects that could neither be seen (geometry) nor computed (algebra and analysis). At best, most mathematicians paid lip-service to vectors and matrices by regarding them as convenient abbreviations for certain sets of numbers which must be brought into play if concrete results are to be achieved.*

A few books referring to matrices were published starting at the end of the 19th century:

- The German mathematician Eugen Otto Erwin Netto (1848-1919) published *Vorlesungen über Algebra volume I* (Lectures on Algebra) [2331] in 1896. See also the textbook [2332] in 1904.

- *Determinanti. Teoria ed Applicazioni* (Determinants. Theory and Applications) [2469] by the Italian mathematician Ernesto Pascal (1865-1940) was published in 1897. A German translation by Hermann Leitzmann was published three years later. Pascal also referred to matrices in *Repertorio di Matematiche Superiori: 1: Analisi* [2470] in 1898. This was also translated to German.

- The English mathematician and philosopher Alfred North Whitehead (1861-1947) is the author of *A Treatise on Universal Algebra with Applications* [3226] published in 1898. There is a chapter about matrices, which are the topic of articles 140 to 155, on pages 248-269.

- Some lectures of the German mathematician Leopold Kronecker (1823-1891) were published in 1903 with the title *Vorlesungen über die Theorie der Determinanten* (Lectures on the Theory of Determinants) [1954]. Even though the main topic is determinants, there are two lectures devoted to matrices.

- Matrices are introduced from determinants in the textbook *Lezioni di Algebra Complementare* [2496] by the Italian mathematician Salvatore Pincherle (1853-1936) in 1906.

- The book *Introduction to Higher Algebra* [348] written in 1907 by the American mathematician Maxime Bôcher (1867-1918) was very successful. It had many editions and was translated to German and Russian.

- Probably the first book having the word "matrix" in its title was *Matrices and Determinoids* in two volumes [784, 785] in 1913 and 1918 by Cuthbert Edmund Cullis (1868-1954), who was teaching at the University of Calcutta in India. Contrary to what was done in the previous books in this list, Cullis started from rectangular matrices.

Things started to change in the 1920s and 1930s. Some books still started from determinants and then introduced matrices and some other books did the opposite.

- The American algebraist Leonard Eugene Dickson (1874-1954) published *Modern Algebraic Theories* [895] in 1926. Matrices appeared only in the third chapter, together with bilinear forms and linear equations.

- The German mathematician Rudolf Hans Heinrich Beck (1876-1942), who was the translator of Bôcher's book [348], published *Einführung in die Axiomatik der Algebra* [245] in 1926. This book gave a presentation of the fundamental operations and theorems of algebra from the view point of formal logic. Chapters 2 to 5 define the algebraic operations for point sets, vectors, and matrices. Chapters 6 to 10 treat the theory of linear equations, linear vector forms, linear and quadratic forms, matrices, and determinants.

- The book *Algebra* [2483] in two volumes by Oskar Perron (1880-1975) was published in 1927. The first volume is an introduction to the field. Matrices were considered in the third chapter.

At the end of the 1920s and in the 1930s, there were more and more books with the word "matrices" in the title:

- Turnbull in *The Theory of Determinants, Matrices and Invariants* [3080] published in 1928-1929 had a small introduction about determinants and introduced matrices almost at the beginning. In 1932, he published *An Introduction to the Theory of Canonical Matrices* [3081] with Aitken.

- In 1933, the American mathematician MacDuffee published *The Theory of Matrices* [2106] in which he introduced matrices right at the beginning. This book contains interesting remarks and notes about the history of the topic. There was a second edition in 1946.

- *Lectures on Matrices* [3199] by Wedderburn in 1934 is based almost entirely on matrices. The knowledge of determinants was assumed.

- Aitken's book *Determinants and Matrices* [18] in 1939 could have been called "Matrices and Determinants" since Aitken introduced matrices first.

George Abram Miller (1863-1951), in his short paper [2249] about the history of determinants, wrote in 1930,

> *Many modern writers have based their definitions of a determinant on the existence of a square matrix* [...] *From this point of view a determinant does not exist without its square matrix, and, judging from many of the textbooks on elementary mathematics, it is likely that many students consider the square matrix as an essential part of a determinant, so that the term determinant conveys to them a dual concept composed of a square matrix and a certain polynomial associated therewith. When they speak of the rows and columns of a determinant they naturally are thinking of its matrix and when they speak of the value thereof they are naturally thinking of the polynomial implied by the term determinant.*

Hence, by the end of the 1930s, matrix theory was well established.

Interesting papers or books on the history of matrices are [1599, 1600, 1601, 1603] by Hawkins, [309] by Jindřich Bečvář, and [416, 417, 418, 419, 420, 421, 422, 423, 424] by Frédéric Brechenmacher. The early days of matrix computations were described by Beresford Neill Parlett in [2452].

In the next sections we consider the history of some properties of matrices and how matrices were used by physicists working on quantum mechanics in the 1920s.

The history of methods for eigenvalues and eigenvectors is considered in Chapter 6, and the singular values and the different possible factorizations of a matrix in Chapter 4.

1.2 ▪ Matrix rank

The notion of *rank* was introduced by Frobenius in 1879. In the paper [1251] devoted to the study of bilinear forms, in Section 1, page 148, he wrote[6] (our translation)

> *Given a finite system A of quantities $a_{\alpha,\beta}$, ($\alpha = 1, \ldots, m$, $\beta = 1, \ldots, n$), which are ordered by rows and columns. If in this system all determinants of degree $(\ell + 1)$ vanish, but the ℓth degree are not all zero, then ℓ is called the rank of the system.*

[6] *Gegeben sei ein endliches System A von Grössen $a_{\alpha,\beta}$, ($\alpha = 1, \ldots, m$, $\beta = 1, \ldots, n$), die nach Zeilen und Colonnen geordnet sind. Wenn in demselben alle, Determinanten $(\ell + 1)$ten Grades verschwinden, die ℓten Grades aber nicht sämmtlich Null sind, so heisst ℓ der Rang des Systems.*

According to MacDuffee in [2106], page 10, the notion of rank was implicit in a work [1629] published in 1858 by Ignaz Heger (1824-1880), an Austrian physician and mathematician. A discussion of vanishing minor determinants can also be found in [2956] by Sylvester in 1851.

In 1884, Sylvester defined the *nullity* of a matrix [2970] as the difference between its order and rank, even though he did not use that word. He proved results about the nullity of a product of matrices.

Some of the books we have mentioned above defined the rank of a matrix. In 1896 [2331, p. 185], Netto defined the rank like Frobenius from the minors. In 1898, Whitehead considered the null space [3226, p. 252] with a reference to Sylvester. Bôcher in 1907 defined the rank with the minors [348, p. 22]. He discussed linear dependence of vectors in Chapter III, starting on page 34, and made the connection with vanishing determinants and the rank. Cullis defined the rank from the minors in Chapter IX of [784, p. 265]. He discussed linear independence but did not use that term and spoke of connected and unconnected rows and columns.

In 1905, Autonne defined the rank of a matrix using the minors [110]. In 1913 and 1915, he published papers [112, 113] in which he considered what we now call the *Singular Value Decomposition* (SVD); see Chapter 4. He noticed that if A is of rank r, there are only r nonzero singular values.

In 1911, Frobenius published a paper [1257] in which he studied some properties of the rank and proved results about the rank of product of matrices.

MacDuffee [2106, Theorem 8.1] showed the relation between the rank and the linear dependence of the rows and columns of a matrix in 1933.

With the advent of digital computers, the question arose of computing the rank of a matrix. This is not an easy task since the computations are not done exactly. This problem was discussed by Gilbert Wright Stewart [2874] in 1984. The most reliable way is to compute the singular values and to set the rank as the number of nonzero singular values. However, one has to decide if a "small" singular value is genuine or if it arises from rounding errors and has to be set to zero. Moreover, a small perturbation of the matrix could change the rank. This leads to the notion of *numerical rank*, which is the number of singular values larger than a defined threshold. For instance, in MATLAB,[7] for an $n \times m$ matrix A, the threshold is $\max(n, m)\, \varepsilon(\|A\|)$, where $\varepsilon(x)$ is the distance from $|x|$ to the next larger floating point number. For the history of the computation of the SVD, see Section 4.3.

Since computing the SVD is expensive, some research was done to deal with rank deficiency in the 1970s and 1980s, particularly for least squares problems; see, for instance, [1612] by Michael Thomas Heath in 1982.

An alternative to the SVD is to compute a rank-revealing QR (RRQR) factorization of the matrix A, that is $A\Pi = QR$, where Π is a permutation matrix, Q is orthogonal, and R is upper triangular; for the standard QR factorization, see Section 4.2. These factorizations were introduced by Gene Howard Golub (1932-2007) [1377] in 1965. In RRQR, if A has rank r, Π must be computed such that

$$R = \begin{pmatrix} R_{1,1} & R_{1,2} \\ 0 & R_{2,2} \end{pmatrix},$$

where $R_{1,1}$ is $r \times r$ and the norm of $R_{2,2}$ is small. An algorithm for computing the column permutation was described by Tony Fan-Cheong Chan [613] in 1987. Other algorithms were proposed by Shivkumar Chandrasekaran and Ilse Clara Franziska Ipsen [634] in 1994, and Ming Gu and Stanley Charles Eisenstat (1944-2020) [1473] in 1996.

[7]MATLAB is a trademark of The MathWorks.

The problem of rank determination is even more complicated when the matrix is large and sparse. Algorithms tailored to this type of matrices were developed by Leslie V. Foster and Timothy Alden Davis [1200] in 2013, and Shashanka Ubaru and Yousef Saad [3083] in 2016.

1.3 ▪ Norms

In 1821, Cauchy was interested in proving results about averages of some numbers. In Note II of [566, pp. 455-456], he proved, among other results, that the sum of the dot product of two finite sequences of numbers is less than the product of the square roots of the sums of the squares of the numbers in each sequence. This is what we now call the Cauchy-Schwarz inequality for the ℓ_2 norm of a vector. About the square root of the sum of squares, Cauchy remarked that[8] (our translation)

> *This expression, which is larger than the largest of the numerical values* [of the sequence of numbers], *is what we can call the modulus of the system of numbers,* [...] *The modulus of a system of two numbers a and b would then be nothing other than the modulus of the imaginary expression $a + b\sqrt{-1}$.*

Therefore, Cauchy saw that the square root of the sum of squares can be used to give a magnitude to complex numbers. In 1832, Gauss considered complex numbers with integer real and imaginary parts and defined what he called (in Latin) the *norma* of $a + ib$ as $a^2 + b^2$, that is, the square of the modulus. The same definition and name was used by Hamilton in 1856.

Cauchy's result was extended to what we now call the ℓ_p norm by the British mathematician Leonard James Rogers (1862-1933) [2585] in 1888. The corresponding inequality was also proved by Otto Ludwig Hölder (1859-1937), who was a student of Kronecker, Karl Theodor Wilhelm Weierstrass (1815-1897), and Ernst Eduard Kummer (1810-1893) in Berlin; see [1724] in 1889. The inequality is (improperly) called Hölder's inequality; see [2122] by Lech Maligranda in 1998.

In 1887, in his paper [2474, 2475] about linear differential equations, Peano considered what he called complex variables $(\, x_1 \quad \cdots \quad x_n \,)$ (which are vectors for us) and defined their *modulus*

$$\text{mod. } x = \sqrt{x_1^2 + \cdots + x_n^2}.$$

He also considered matrices R and defined their modulus as

$$\text{mod. } R = \max \frac{\text{mod. } (Rx)}{\text{mod. } x}.$$

In fact, he forgot to say that one must have $x \neq 0$. Nevertheless, he proved the triangle inequality

$$\text{mod. } (R + S) \leq \text{mod. } R + \text{mod. } S,$$

as well as

$$\text{mod. } (RS) \leq \text{mod. } R \cdot \text{mod. } S.$$

He also stated without proof that the square of mod. R is the largest root of $\det(R^T R - \lambda I)$, that is, the modulus of the largest singular value.

In 1908, Erhard Schmidt (1876-1959) defined a positive quantity that can be seen as the Euclidean norm of an infinite vector [2700].

---

[8] *Cette expression, qui surpasse la plus grande des valeurs numériques dont il s'agit, est ce qu'on pourrait appeler le* module *du système des quantités,* [...] *Le module du système de deux quantités a et b ne serait alors autre chose que le module même de l'expression imaginaire $a + b\sqrt{-1}$.*

Bôcher and his student Louis Brand (1885-1971) defined in 1911 a k-tuple of numbers $(a_1 \quad a_2 \quad \cdots \quad a_k)$ as a complex quantity, an inner product of two quantities, and the *norm* as $|a_1|^2 + \cdots + |a_k|^2$ in [350].

Albert Arnold Bennett (1888-1971) discussed extensions of the norm of a complex number [265] in 1921. He referred to *Geometrie der Zahlen*, a posthumous book of Hermann Minkowski (1864-1909) in 1910, and to *Les systèmes d'équations linéaires à une infinité d'inconnues* by the Hungarian mathematician Frigyes Riesz (1880-1956) in 1913.

In his paper [176] in 1922, the Polish mathematician Stefan Banach (1892-1945) considered general sets for which an addition and a multiplication by scalars are defined, and he stated the properties of a general norm on such a set,

- $\|X\| \geq 0$,
- $\|X\| = 0 \Rightarrow X = 0$,
- $\|aX\| = |a| \, \|X\|$,
- $\|X + Y\| \leq \|X\| + \|Y\|$,

where X and Y are elements of the set and a is a scalar. He also proved some properties of the norm and of limits of sequences. In his book [177] in 1932, he considered complete normed vector spaces that we now call Banach spaces.

Results about the Frobenius norm of a product of matrices were published by Wedderburn [3198] in 1925.

In 1932, Turnbull and Aitken considered x^*x for a vector x and wrote that *it is often called the norm of the complex vector* x [3081].

In his 1934 book [3199, p. 125], Wedderburn defined what we call the Frobenius norm as the *absolute value of a matrix* (as in his 1925 paper); he related this quantity to the trace of A^TA, showed that it satisfies the submultiplicative property $\|AB\| \leq \|A\| \, \|B\|$, and observed that this norm is invariant under multiplication by orthogonal matrices. Wedderburn referred to Peano, but it seems he did not realize that his absolute value was different from Peano's modulus.

Curiously enough, in 1943 MacDuffee called "norm" the absolute value of the determinant in [2107]. That same year, the word "norm" appeared in a paper [1732] of the statistician Harold Hotelling (1895-1973). He used the Frobenius norm with a reference to Wedderburn.

Another American statistician Albert Hosmer Bowker (1919-2008) published a paper [389] about matrix norms in 1947. He stated the properties that must be satisfied, including the submultiplicative property. He observed (with what was probably a printing mistake) that the modulus of any eigenvalue is smaller than or equal to a matrix norm satisfying his hypotheses. He considered the two norms

$$R(A) = \max_i \sum_j |a_{i,j}|, \quad C(A) = \max_j \sum_i |a_{i,j}|.$$

As remarked by Alston Scott Householder (1904-1993) in his book [1745] and in [1748] there were not many uses of norms before World War II, even though they were used by Leonid Vitalievitch Kantorovich (1912-1986) in 1939 [1870].

Norms were used in 1947 by Arvid Turner Lonseth (1912-2002) to study the propagation of errors in linear problems [2091].

John von Neumann (1903-1957) and Herman Heine Goldstine (1913-2004) [3158] defined a norm $N(A)$ as the square root of the trace of A^*A [3158] in 1947. They also defined the spectral norm that they denoted as $|A|$.

In 1948, Alan Mathison Turing (1912-1954) introduced several norms in Section 7 of [3079]. What he denoted $N(A)$ is the Frobenius norm, the maximum expansion $B(A)$ is the spectral norm, and $M(A)$ is the max of $|a_{i,j}|$. He also listed some properties of these norms.

In 1952, in their paper [1664] on the conjugate gradient method, Magnus Rudolph Hestenes (1906-1991) and Eduard Ludwig Stiefel (1909-1978) used the square of the A-norm of the error, for a symmetric positive definite matrix A, but they just called it the *error function*, not mentioning that it is a norm.

Matrix norms were studied intensively by Householder and Friedrich Ludwig Bauer (1924-2015) in the 1950s and 1960s; see [1737, 1739, 1740, 227, 229, 231, 219, 1745, 223, 1748] (ordered by date). Of interest is also [2902] by Josef Stoer in 1964.

Bounds for norms of powers of matrices were obtained by Werner Gautschi (1927-1959) [1304, 1305] in 1953. Properties of matrix norms and their relations to eigenvalues were studied by Alexander Markowich Ostrowski (1893-1986) [2398] in 1955.

Another scholar who contributed to the study of matrix norms was Noël Gastinel (1925-1984) in his doctoral thesis [1287] in 1960 and later with [1288, 1289, 1292]. See also the thesis [2580] of François Robert, who was a student of Gastinel, in 1968. Generalized norms of matrices were studied by Miroslav Fiedler (1926-2015) and Vlastimil Pták (1925-1999) [1164] in 1962. For more about Fiedler and Pták, see [1163, 1162, 1550, 3139].

Comparison theorems for supremum norms were obtained by H. Schneider and Gilbert Strang [2707] in 1962. That same year, ratios of matrix norms were considered by Betty Jane Stone (1935-2017), who was a student of George Elmer Forsythe (1917-1972) in Stanford; see [2910]. Also of interest are the works of the Czech mathematicians Jan Mařík (1920-1994) and Pták [2138] in 1960, Fiedler and Pták [1165], and Pták [2522] in 1962. About norms of matrices and operators, see the thesis of Jean-François Maitre [2118] in 1974.

1.4 ▪ Ill-conditioning and condition numbers

Ill-conditioned problems arose once people began solving linear systems. For instance, in 1809 [1299] Gauss calculated the corrections to the elements of the minor planet Pallas (which is the second asteroid that was discovered after Ceres by the astronomer Heinrich Olbers (1758-1840) in 1802). Gauss obtained a linear system with 12 equations in 6 unknowns. However, because some of the observations were not precise enough, Gauss removed the tenth equation, obtaining a set of 11 equations of which he was seeking the least squares solution by solving the normal equations. Gauss' paper was translated to French by Joseph Louis François Bertrand (1822-1900) [299] in 1855. In 1933, Aitken wrote a paper on the solution of linear equations [15] in which he considered as an example Gauss' least squares problem. Aitken wrote

> *At the suggestion of an interested friend the author some time ago applied the routine to the set of six normal equations by which Gauss in 1809 calculated the corrections to the elements of the minor planet Pallas. This particular set of equations has a formidable appearance, owing to the wide range of magnitude of the numbers involved; but the solution was performed, and the results verified by substitution in all six equations - a check well worth applying in any circumstances - in some two hours or less. The results proved to be in considerable disagreement with those of Gauss.*

Using IEEE double precision arithmetic and taking the value of the coefficients given by Gauss in his paper for the 11 equations, the matrix A^TA of order 6 is

$$\begin{pmatrix} 5.9156742361 & 7203.9003559 & -0.09345842 & -2.2851325505 & -0.3466410932 & -0.1819746513 \\ 7203.9003559 & 10834257.177 & -49.0642641 & -3229.7922897 & -198.6393882 & -143.058056 \\ -0.09345842 & -49.0642641 & 0.7191875933 & 1.1338413202 & 0.0586161038 & 0.262837384 \\ -2.2851325505 & -3229.7922897 & 1.1338413202 & 12.003467166 & -0.3713729341 & -0.1204035557 \\ -0.3466410932 & -198.6393882 & 0.0586161038 & -0.3713729341 & 2.2821291327 & -0.3626104294 \\ -0.1819746513 & -143.058056 & 0.262837384 & -0.1204035557 & -0.3626104294 & 5.6246473351 \end{pmatrix}.$$

The right-hand side is

$$(A^T b)^T = \begin{pmatrix} 367.9868069 & 569753.4236 & 116.4417978 & -261.4826751 & 146.7106422 & -32.3809018 \end{pmatrix},$$

and the solution (transposed) is

$$\begin{pmatrix} 8.4003736723 & 0.040746831448 & 207.99744955 & -27.197281769 & 57.587281786 & -11.038112526 \end{pmatrix}.$$

The condition number of the matrix is $1.8896\ 10^7$, showing that we have an ill-conditioned problem. The smallest eigenvalue is 0.57335 and the largest one is $1.0834\ 10^7$. The norm of the residual vector for the previous solution is $1.2790\ 10^{-13}$.

However, this is not the matrix given by Gauss in his paper, which is

$$\begin{pmatrix} 5.91569 & 7203.91 & -0.09344 & -2.28516 & -0.34664 & -0.18194 \\ 7203.91 & 10834225 & -49.06 & -3229.77 & -198.64 & -143.05 \\ -0.09344 & -49.06 & 0.71917 & 1.13382 & 0.064 & 0.26341 \\ -2.28516 & -3229.77 & 1.13382 & 12.0034 & -0.37137 & -0.11762 \\ -0.34664 & -198.64 & 0.064 & -0.37137 & 2.28215 & -0.36136 \\ -0.18194 & -143.05 & 0.26341 & -0.11762 & -0.36136 & 5.62456 \end{pmatrix}.$$

The absolute relative difference with our matrix above is

$$\begin{pmatrix} 2.66477\ 10^{-6} & 1.33873\ 10^{-6} & 1.97093\ 10^{-4} & 1.20122\ 10^{-5} & 3.15369\ 10^{-6} & 1.90418\ 10^{-4} \\ 1.33873\ 10^{-6} & 2.96992\ 10^{-6} & 8.69085\ 10^{-5} & 6.90128\ 10^{-6} & 3.07995\ 10^{-6} & 5.63128\ 10^{-5} \\ 1.97093\ 10^{-4} & 8.69085\ 10^{-5} & 2.44627\ 10^{-5} & 1.88035\ 10^{-5} & 9.18501\ 10^{-2} & 2.17859\ 10^{-3} \\ 1.20122\ 10^{-5} & 6.90128\ 10^{-6} & 1.88035\ 10^{-5} & 5.59557\ 10^{-6} & 7.90068\ 10^{-6} & 2.31186\ 10^{-2} \\ 3.15369\ 10^{-6} & 3.07995\ 10^{-6} & 9.18501\ 10^{-2} & 7.90068\ 10^{-6} & 9.14379\ 10^{-6} & 3.44841\ 10^{-3} \\ 1.90418\ 10^{-4} & 5.63128\ 10^{-5} & 2.17859\ 10^{-3} & 2.31186\ 10^{-2} & 3.44841\ 10^{-3} & 1.55272\ 10^{-2} \end{pmatrix}.$$

Some relative differences are of the order 10^{-2}. So, either Gauss used slightly different values of the initial coefficients or he made some slight mistakes in rounding his results. His right-hand side (transposed) is

$$\begin{pmatrix} 371.09 & 580104 & 113.45 & -268.53 & -94.26 & 31.81 \end{pmatrix}.$$

Again, the maximum relative differences with the components of our right-hand side are of the order of 10^{-2}. Using the Gauss matrix and right-hand side we find the solution (transposed)

$$\begin{pmatrix} -15.681862783 & 0.054012857666 & 219.2003128 & -33.206164113 & -51.763873062 & -7.7636784767 \end{pmatrix}.$$

We observe that this is completely different from our solution, but this is also quite different from the solution that was given by Gauss, which is

$$\begin{pmatrix} -3.06 & 0.054335 & 166.44 & -4.29 & -34.37 & -3.15 \end{pmatrix}.$$

In fact, if we look at the results of the first step of the elimination done by Gauss reported in [1299], the absolute values are almost correct but there are sign mistakes. The signs of his coefficients [ce,1] and [dn,1] are wrong. If we also do these mistakes in our modern elimination, we finally find a solution which is not too far from Gauss' solution, except for the first component. But some partial results of Gauss have some digits which are not fully correct. Because these errors are repeated in Bertrand's translation, he probably did not check the calculation.

In his paper, Aitken wrote that Gauss had made a mistake in his computation by misreading or miscopying the $(1, 2)$ coefficient of the matrix. However, we have seen that the cause of a wrong solution are sign mistakes. The solution given by Aitken (transposed) is

$$\begin{pmatrix} -15.593 & 0.053996 & 218.41 & -33.092 & -51.197 & -7.6995 \end{pmatrix}.$$

It looks like our solution for the matrix and right-hand side from Gauss but not exactly. Aitken probably started from the matrix $A^T A$ given by Gauss and not from the initial 11 equations. Moreover, the right-hand side used by Aitken is not exactly Gauss' right-hand side. Aitken's right-hand side is

$$\begin{pmatrix} 371.09 & 580104 & 113.34 & -268.39 & -94.274 & 31.764 \end{pmatrix},$$

with the last three components slightly different. If we use Gauss' matrix and Aitken's right-hand side, we obtain

$$\begin{pmatrix} -15.592028066 & 0.053994954224 & 218.40825022 & -33.091065634 & -51.196211461 & -7.698857088 \end{pmatrix}.$$

But again, this is slightly different from what Aitken had obtained. So, this example shows that even great mathematicians like Gauss and Aitken made some mistakes in their computations but, as we now know, for an ill-conditioned system small changes in the entries of the matrix or the right-hand side may give large changes in the solution. However, to be fair, the different solutions that we have seen do not make much difference if we consider the norm of the residual $\|b - Ax\|$ for the initial system of 11 equations. With the modern solution of the correct normal equations it is $2.8706\ 10^2$, for the modern solution of Gauss normal equations it is $3.3040\ 10^2$, for Aitken's solution it is $3.3001\ 10^2$, and for Gauss' solution it is $3.3047\ 10^2$. This example shows the difficulties in solving badly conditioned linear systems.

In the first half of the 20th century, several papers were published on the subject of perturbation of solutions of linear systems. They were written by astronomers or statisticians whose linear systems arose from measurements that may be inaccurate.

Forest Ray Moulton (1872-1952), an American astronomer whose name is also associated in mathematics with the Adams-Moulton method for solving ordinary differential equations, wrote the paper [2298] in 1913 in which he tried to relate the sensitivity of the solution to perturbations of the data to the determinant of the matrix. As an example he took a linear system $Ax = b$ of order 3,

$$A = \begin{pmatrix} 0.34622 & 0.35381 & 0.36518 \\ 0.89318 & 0.90274 & 0.91143 \\ 0.22431 & 0.23642 & 0.24375 \end{pmatrix}, \quad b = \begin{pmatrix} 0.24561 \\ 0.62433 \\ 0.17145 \end{pmatrix}.$$

The condition number of A is 705.93905 and its determinant is $5.15302702\ 10^{-5}$. Hence, the matrix is not badly conditioned and the determinant is small, but not too small. The solution given by Moulton is

$$x_M = \begin{pmatrix} -1.027066 & 2.091962 & -0.380515 \end{pmatrix}^T.$$

The norm of the residual vector for that solution is $2.61140\ 10^{-7}$. The double precision solution is

$$x = \begin{pmatrix} -1.027057 & 2.091916 & -0.3804789 \end{pmatrix}^T,$$

with a residual norm equal to $2.305551\ 10^{-16}$. The point stressed by Moulton is that vectors

$$x_{M2} = \begin{pmatrix} -1.022773 \\ 2.084125 \\ -0.376941 \end{pmatrix}, \quad x_{M3} = \begin{pmatrix} -1.031229 \\ 2.099457 \\ -0.383879 \end{pmatrix},$$

also satisfy the equations with the same accuracy as x_M. However, this depends on how one defines the accuracy. In fact, the residual norms of these last two vectors (computed in double precision) are, respectively, $3.149109\ 10^{-5}$ and $3.143103\ 10^{-5}$, which is far less accurate than for x_M. Hence, attributing the problems to the value of the determinant, Moulton misunderstood the problem. Moreover, his explanation was based on using Cramer's formulas.

In 1936, Helmut Wittmeyer [3265] studied the changes to the solution of linear systems due to perturbations of the entries. This was an outcome of the thesis he obtained in Darmstadt in 1935, in which he also considered the changes in the eigenvalues and eigenvectors.

Louis Bryant Tuckerman II (1915-2002) from the National Bureau of Standards published a paper [3077] in 1941 in which he distinguished the "observational errors," that is, the uncertainties on the coefficients from the "computational errors" due, for instance, to the rounding errors. In the end he wrote pessimistically

> *Unfortunately there seem to be no reasonably simple formulae for determining upper bounds of the relative errors that arise in the solution of simultaneous linear equations in more than two variables. This does not absolve the computer from the necessity of ensuring that his computational errors are suitably limited.*

In this quote, the "computer" is the human being performing the computation.

In 1942, Lonseth [2089] was interested in how the solution of a linear system changes when there are errors in the matrix entries and the components of the right-hand side. As a "measure of sensitivity," he proposed $\sum_\mu \sum_\nu |A_{\mu,\nu}|/|A|$, where $A_{\mu,\nu}$ is the cofactor of $a_{\mu,\nu}$ and the vertical bars denote determinants. He was not concerned with any particular method for solving the system; see also [2090] in 1944.

In 1947, he considered the same problem in the more general setting of Banach spaces [2091]. In that work he considered norms of linear maps and realized that his "measure of sensitivity" was nothing other than $\sum_\mu \sum_\nu |A^{-1}_{\mu,\nu}|$.

In 1943, the American statistician Harold Hotelling (1895-1973) was concerned with the accuracy in solving linear systems; see [1732]. He wrote

> *The question how many decimal places should be retained in the various stages of a least-square solution and of other calculations involving linear equations has been a puzzling one. It has not generally been realized how rapidly errors resulting from rounding may accumulate in the successive steps of such procedures.*

However, his answer to this question (which was to dismiss direct methods) was far too pessimistic and he did not realize that the condition number was involved.

In 1944, Franklin E. Satterthwaite, who was working for the Aetna Life Insurance Company, started his paper from an LU factorization of the matrix A written in matrix terms; see [2689]. He used the Frobenius norm denoted as $N(\cdot)$ and gave bounds for the norm of the error matrix in the Doolittle variant of Gaussian elimination, which is described in Chapter 2.

The condition number of a matrix appeared for the first time in a paper by Goldstine and von Neumann [3158] in 1947, although not under that name. They considered symmetric positive definite matrices. The ratio of the extreme eigenvalues, $\lambda_{\max}/\lambda_{\min}$, appears on page 1062 and page 1093, relation (7.5′) of that paper. This ratio was called *the figure of merit*. A thorough analysis of [3158] was done by Joseph Frank Grcar in [1434]. This ratio also appeared in [1374] but that paper was never published, except in the different publications of the collected works of von Neumann. In fact, initially von Neumann defined the condition number as the ratio of the extreme singular values; see Figure 2.6 in [1434] which shows a letter from von Neumann to Goldstine in 1947. What we now call the Frobenius norm is defined on page 1042 of [3158].

In 1971, in his John von Neumann lecture [3252], James Hardy Wilkinson (1919-1986) gave his appraisal of the von Neumann and Goldstine paper:

> *von Neumann and Herman Goldstine [1] were certainly responsible for a paper which has dominated modern error analysis. Since its appearance in 1947 it has been perhaps the most widely quoted work in numerical analysis; indeed I have the impression that authors have sometimes included it in their list of references even when it had no particular relevance, using it, as it were, as a prestige symbol.* [...]
>
> *The commanding position held by this early paper in the field of error analysis led perhaps to one or two unfortunate side effects which cannot in any respect be attributed to weaknesses in the paper itself. It is a very substantial paper, some 80 pages long, and it is not exactly bedside reading, though when the basic plan of the proof is grasped it is not essentially difficult.*

In 1951, Goldstine and von Neumann reconsidered the problem from a probabilistic point of view in [1373]; see also H.P. Mulholland [2303] in 1952.

The term "condition number" apparently appeared for the first time in a paper [3079] by Turing in 1948. Norms of matrices are defined in Section 7, starting on page 296. Our modern Frobenius norm of a matrix A is denoted as $N(A)$. Then Turing defined two quantities he denoted as the *maximum expansion* $B(A)$ and *maximum coefficient* $M(A)$,

$$B(A) = \max_x \frac{\|Ax\|_2}{\|x\|_2}, \quad M(A) = \max_{i,j} |a_{i,j}|,$$

where $\|\cdot\|_2$ is the ℓ_2 norm of a vector. Then he stated inequalities satisfied by $N(A)$, $B(A)$, and $M(A)$, including the equivalences of norms.

In Section 8, he started by writing

> *When we come to make estimates of errors in matrix processes we shall find that the chief factor limiting the accuracy that can be obtained is "ill-conditioning" of the matrices involved. The expression "ill-conditioned" is sometimes used merely as a term of abuse applicable to matrices or equations, but it seems most often to carry a meaning somewhat similar to that defined below.*

Instead of A of order n, he considered matrices $A - S$, where S are random matrices. Averaging over these random matrices, he found that the root mean squares (RMS) satisfy

$$\frac{\text{RMS error on components of the solution}}{\text{RMS components of the solution}} = \frac{1}{n} N(A) N(A^{-1}) \frac{\text{RMS error on entries of } A}{\text{RMS entries of } A}.$$

In reference to the possible ill-conditioning, he called $(N(A)N(A^{-1}))/n$ the N-condition number of A and $nM(A)M(A^{-1})$ the M-condition number. He also observed that

> *the determinant may differ very greatly from the above-defined condition numbers as a measure of conditioning*

and that

> *the best conditioned matrices are the orthogonal ones, which have N-condition numbers of 1.*

Then he studied the error in computing the inverse of A using the Gauss-Jordan method and in the numerical solution given by Gaussian elimination. His bounds involve $M(A^{-1})$ and, in fact, he did not use $N(A)$ any longer.

Concerning the relations of his work with what was done a little before by von Neumann, Turing wrote

> *In the meantime another theoretical investigation was being carried out by J. v. Neumann, who reached conclusions similar to those of this paper for the case of positive definite matrices, and communicated them to the writer at Princeton in January 1947 before the proofs given here were complete.*

We observe that Cholesky is cited in Turing's paper through the paper [266] published by Commandant Benoît in 1924. That paper was the only source for Cholesky's method at that time.

The paper [3049] by John Todd (1911-2007) on condition numbers was written before his other paper [3048] but published later in 1950. In [3049] Todd wrote

> *The phenomenon of ill-condition of a system of equations, or of a matrix, has been known for a long time, but it is only recently that attempts have been made to put it on a quantitative footing.*

As references about ill-conditioning he cited Gauss and a paper of 1901 by Enno Jürgens (1851-1907) [1853] whose title can be translated as *Numerical calculation of determinants.* In that paper there is an example of an integer nonsymmetric matrix of order 4 with a determinant of order 10^{14} (the condition number is 1.8604 but, of course, this is not computed in that paper). Another example of order 4 is given for which, if the entries are computed inaccurately with relative errors of order 10^{-2}, relative errors in the components of the solution are of the same order. The matrices of that paper are not really ill-conditioned.

In his paper, Todd recalled the condition numbers defined by Turing and introduced the P-condition number which is defined as the ratio of the largest modulus of the eigenvalues to the smallest modulus of the eigenvalues (apparently the "P" stands for Princeton in reference to von Neumann). This was seen as a generalization of what was defined by von Neumann and Goldstine for symmetric matrices. He considered the matrix of order n arising from the standard finite difference scheme of the one-dimensional Laplacian. This is a symmetric tridiagonal Toeplitz matrix having -2 on the diagonal and 1 on the first upper and lower diagonals. The eigenvalues of this matrix are known explicitly. From this he obtained that its P-condition number is of the order of $4n^2/\pi^2$ for n large. This means that the matrix is more and more ill-conditioned when n increases, that is, when the mesh size tends to zero. He also considered the asymptotic values of the N and M condition numbers.

In [3048] Todd considered the matrix arising from the five-point finite difference approximation of the two-dimensional Laplacian in a square with homogeneous Dirichlet boundary conditions. The matrix A is a symmetric block tridiagonal matrix with tridiagonal Toeplitz blocks on the block diagonal. Again he derived asymptotic values of the P and N condition numbers.

In 1950, Todd's wife, Olga Taussky (1906-1995) [3005], proved a result that she loosely stated (in our modern notation) as

> *Let A be a real $n \times n$ nonsingular matrix and A^T its transpose. Then AA^T is more "ill-conditioned" than A.*

She proved it rigorously for Turing's N-condition number (using the Frobenius norm). She also considered the P-condition number that was proposed by Todd [3049]. This is not what we consider now as the condition number (which, as we have seen, was introduced by von Neumann), that is, the ratio of the largest to the smallest singular values of A. As we said above, this definition is similar to that of the von Neumann and Goldstine paper for a symmetric matrix. But, she added in a footnote,

> *These authors consider symmetric matrices only, but it is reasonable to apply the definition to the general case.*

In 1956, Ewald Konrad Bodewig (1901-?) published a book [353] in which he recalled the condition numbers defined by Turing and von Neumann and Goldstine but he attributed the problem of ill-conditioning to small determinants.

In 1958, the condition number $\|A\|\,\|A^{-1}\|$ was considered by Householder in [1742]. About $\|A\|\,\|A^{-1}\|$, he wrote

> *If the norm is the spectral norm, this is the ratio of the largest to the smallest of the singular values. The ratio of the proper values themselves is not a suitable measure since a triangular matrix can be poorly conditioned even when all proper values are equal.*

The proper values are the eigenvalues.

In his "Thèse d'Etat" in 1960, Gastinel considered general norms and condition numbers. In Chapter 2, he recalled and proved the properties of vector and matrix norms. He proved that for real matrices there does not exist a norm such that $\|A\| = \|SAS^{-1}\|$ for all nonsingular S. Chapter 3 is devoted to condition numbers. His definition of the condition number is the reciprocal of the usual definition, that is, $1/(\|A\|\,\|A^{-1}\|)$. He showed that the condition number for the ℓ_2 (or Euclidean) norm is the ratio of the smallest and largest singular values (even though he did not refer to singular values). On page 44 he showed that if A tends to a singular matrix, one of his condition numbers tends to zero (that is, infinity for the usual definition). However, he did not consider explicitly the distance to the nearest singular matrix.

Nevertheless, Gastinel was cited in 1966 by William Morton Kahan [1861], who proved that the distance to the nearest singular matrix is $1/\|A^{-1}\|$. In 1987, James Weldon Demmel [867] extended this type of result to other problems than matrix inversion.

For the same type of results componentwise, see Siegfried Michael Rump [2626, 2628] in 1997-1999. For more results on structured matrices, see [2627, 2629].

In 1960, Leon Mirsky (1918-1983) considered the distance to the nearest matrix of a given rank, which can be seen as a generalization of the previous problem; see [2254].

That same year, Bauer and Householder [229] proved that if two vectors x and y satisfy $|y^*x| \leq \|x\|\,\|y\|\cos\varphi$ with $0 \leq \varphi \leq \pi/2$ (where the $*$ denotes the conjugate transpose), then for a matrix A we have

$$|(Ay)^*Ax| \leq \|Ax\|\,\|Ay\|\cos\psi,$$

with $\cot\psi/2 = \|A\|\,\|A^{-1}\|\cot\varphi/2$. They used this result to study projection methods for solving linear systems.

In 1962, Marvin Marcus (1927-2016) in [2137] proved an inequality between the P-condition number of a Hermitian matrix and its determinant. In his notation, if p is the condition number of the matrix A of order n, then

$$\det(A) \geq q^{n-1}\prod_{i=1}^{n} a_{i,i}, \quad q = \frac{4p}{(p+1)^2}.$$

In 1965, Albert W. Marshall and Ingram Olkin (1924-2016) [2146] extended the results of Taussky. If ϕ is a norm and $c_\phi(A) = \phi(A)\phi(A^{-1})$, they proved that $c_\phi(A) \leq c_\phi(AA^*)$ if ϕ is a unitarily invariant norm and A^* is the conjugate transpose of A. In addition, if $\phi(E_{ij}) = 1$

where E_{ij} is the zero matrix except for the (i, j) entry equal to 1, then $c_\phi(A) \geq [c_\phi(AA^*)]^{1/2}$. In 1969, they proved in [2147] that if A and B are positive definite and if $c_\phi(A) \leq c_\phi(B)$ then $c_\phi(A+B) \leq c_\phi(B)$ when ϕ is a monotone norm, that is, $\phi(A) \leq \phi(B)$ if $B - A$ is positive definite. Note that if ϕ is unitarily invariant, ϕ is monotone.

The condition number was heavily used in the works of Wilkinson to study direct methods for solving linear systems, starting at the beginning of the 1960s; see [3244]. His work culminated in his book [3248] in 1965; see Chapter 4.

Using a norm to measure the sensitivity of the solution of a linear system is not always satisfactory if the matrix has some special structure, like, for instance, being sparse, that is, with many zero entries. Then, it is better to introduce elementwise perturbations. This led Bauer [222] in 1966 and Robert David Skeel [2786] in 1979 to introduce $\| \, |A| \, |A^{-1}| \, \|$ as a condition number. It is called the Bauer-Skeel or *componentwise condition number*. Nowadays $\|A\| \, \|A^{-1}\|$ is called the *normwise condition number*.

In the beginning of the studies of condition numbers the notation for the norms was not really standardized. Today almost everybody use the same notation. The most widely used norms are

$$\|A\|_F = \left(\sum_{i,j} |a_{i,j}|^2 \right)^{\frac{1}{2}} \text{ (Frobenius norm)}, \quad \|A\|_1 = \max_j \sum_i |a_{i,j}| \ (\ell_1 \text{ norm}),$$

$$\|A\|_\infty = \max_i \sum_j |a_{i,j}| \ (\ell_\infty \text{ norm}), \quad \|A\|_2 = \frac{\sigma_{\max}}{\sigma_{\min}} \ (\ell_2 \text{ norm}),$$

where $\sigma_{\max}$ and $\sigma_{\min}$ are, respectively, the largest and smallest singular values of A. The ℓ_2 norm is sometimes called the spectral norm or the Euclidean norm since it is the norm subordinate to the Euclidean norm of vectors.

Since the condition numbers involve the inverse matrix they are not cheap to compute. This is why researchers looked for condition number estimates that can be computed cheaply.

In 1979, a technique for computing condition number estimates was developed by Alan Kaylor Cline, Cleve Barry Moler, G.W. Stewart, and Wilkinson [714], to be included in the LINPACK software package [914] for the ℓ_1 norm. They started by considering triangular matrices. For a triangular matrix R one computes [714],

$$R^T x = b, \quad Ry = x.$$

If the vector b is suitably chosen, $\|y\|/\|x\|$ gives an approximation of $1/\sigma_{\min}(R)$, giving an estimate of $\|R^{-1}\|$. Heuristic choices for b are described in [714]. Their aims were to maximize $\|x\|/\|b\|$. For a matrix A such that $A = LU$, replace R by A in the previous technique and solve four triangular linear systems. Moreover, $\|A\|$ can be estimated by using the power method. Note that LINPACK computed estimates of the inverse of the condition number to avoid overflow problems. This justifies the definition of Gastinel for the condition number.

Estimates for the ℓ_1 norm were proposed by Dianne Prost O'Leary [2367] in 1980. In 1982, Cline, Andrew Roger Conn, and Charles Francis Van Loan [713] considered other choices for the vector b in the LINPACK estimator.

In 1984 William Ward Hager [1534] published a new technique for estimating the ℓ_1 condition number. His algorithm amounts to maximize $\|A^{-1}x\|_1$ over the unit ball. This is done by a finite number of steps (at most $2n$) solving $Ay = x$, $A^T z = \xi$, with $\xi_i = 1$ if $y_i \geq 0$ and -1 otherwise, at each step. Even though the number of steps may be small, this can be more expensive than the LINPACK estimate.

Since condition number estimators are based on heuristics they do not work correctly for all examples. Some counter-examples for the LINPACK estimators were exhibited by Cline and Russell Keith Rew [715] in 1983.

In 1987, Nicholas John Higham [1673] did a survey of condition number estimation for triangular matrices. His conclusion was

> *Our tests confirm that the LINPACK condition estimator is very reliable in practice, despite the existence of counter-examples.* [...] *The convex optimisation algorithm* [proposed by Hager] *appears, from our tests, to produce estimates generally sharper than those of the LINPACK algorithm, at a similar computational cost.*

In 1988, Higham described Fortran codes for estimating the ℓ_1 norm of a matrix with application to condition number estimation [1674]. The methods used in these codes are used in LAPACK for estimating the condition number; see [1675].

In 1990, Christian Heinrich Bischof [330] introduced an algorithm for estimating the smallest singular value of a triangular matrix when it is generated one row or one column at a time. This leads to an incremental condition estimator. He considered a lower triangular matrix L which is extended by one row,

$$\begin{pmatrix} L & 0 \\ v^T & \gamma \end{pmatrix}.$$

Given a vector x such that $Lx = d$ with $\|d\| = 1$ and $\sigma_{\min}(L) \approx 1/\|x\|$ one finds $s = \sin\varphi$ and $c = \cos\varphi$ (that is, with $s^2 + c^2 = 1$) such that $\|y\|$ is maximized where y solves

$$\begin{pmatrix} L & 0 \\ v^T & \gamma \end{pmatrix} y = \begin{pmatrix} sd \\ c \end{pmatrix} \Rightarrow y = \begin{pmatrix} sx \\ (c - s\alpha)/\gamma \end{pmatrix},$$

with $\alpha = v^T x$. The optimal solution is

$$\begin{pmatrix} s \\ c \end{pmatrix} = \frac{1}{\sqrt{\mu^2 + 1}} \begin{pmatrix} \mu \\ -1 \end{pmatrix},$$

with $\mu = \eta + \text{sign}(\alpha)\sqrt{\eta^2 + 1}$ and $\eta = (\gamma^2 x^T x + \alpha^2 - 1)/2\alpha$. The new estimate is $1/\|y\|$. This technique was referred to as ICE (Incremental Condition Estimator).

Extension of this method to sparse matrices was considered by Bischof, John Gregg Lewis (1945-2019), and Daniel J. Pierce [331] in 1990. Generalization for the estimation of several extremal singular values was done by Bischof and Ping Tak Peter Tang [333] in 1992. Another extension called ACE (Adaptive Condition Estimator) was proposed by Pierce and Robert James Plemmons [2494] in 1992.

An algorithm using right and left singular vectors was proposed in 2002 by Iain Spencer Duff and Christof Vömel [1012] in 2002. When comparing ICE and the Duff-Vömel approach for condition number estimation, Jurjen Duintjer Tebbens and Miroslav Tůma [1021] found in 2014 that the preferable approach is the algorithm of Duff and Vömel when appropriately applied to both the triangular matrix itself and its inverse.

The condition number of condition numbers was discussed in [1671] by Desmond John Higham in 1995.

1.5 ▪ The Schur complement

In his 1917 paper [2718] (submitted in September 1916), whose title in English is *About power series that are bounded inside the unit circle*, Issai Schur (1875-1941) introduced the notion of what is now named the *Schur complement*. On pages 216-217, he considered the matrix

$$M = \begin{pmatrix} P & Q \\ R & S \end{pmatrix},$$

where P, Q, R, and S are square matrices of dimension n, and P is nonsingular. He gave the formula (his notation)

$$\begin{pmatrix} P^{-1} & 0 \\ -RP^{-1} & E \end{pmatrix} \begin{pmatrix} P & Q \\ R & S \end{pmatrix} = \begin{pmatrix} E & P^{-1}Q \\ 0 & S - RP^{-1}Q \end{pmatrix},$$

where E is the identity matrix I. Schur added that the determinant $|M|$ of M is equal to the determinant $|PS - RQ|$ if P and R commute,

$$|M| = |P| \cdot |S - RP^{-1}Q| = |PS - PRP^{-1}Q| = |PS - RQ|,$$

and that

$$|P^{-1}| \cdot |M| = |S - RP^{-1}Q|.$$

This formula was apparently found independently by the Polish mathematician Tadeusz Banachiewicz (1882-1954) in 1937 [178].

The name "Schur complement" and the notation (M/P) for $S - RP^{-1}Q$ were introduced by Emilie Virginia Haynsworth (1916-1985) [1611] in June 1968. Obviously, Schur complements can also be defined by starting from any corner of the matrix M assuming that the matrix in the corner diagonally opposite is nonsingular.

The Schur complement is related to block Gaussian elimination. Indeed

$$M = \begin{pmatrix} A & B \\ C & D \end{pmatrix} = \begin{pmatrix} I & BD^{-1} \\ 0 & I \end{pmatrix} \begin{pmatrix} (M/D) & 0 \\ C & D \end{pmatrix},$$

with $(M/D) = A - BD^{-1}C$.

Let us mention that Richard William Farebrother [1141, pp. 116-117] recognized the notion of Schur complement in the work of Pierre-Simon de Laplace (1749-1827) [1998, p. 273, 1812 edition] when he used the ratio of two successive leading principal minors of a symmetric determinant. According to David Hilding Carlson [552], it can also be traced back to Sylvester in 1851 [2959]. For a discussion of the appearance of the Schur complement in the 1800s, see Richard Anthony Brualdi and H. Schneider [484] about determinantal identities in the works of Gauss, Schur, Cauchy, Sylvester, Kronecker, Jacobi, Binet, Laplace, Muir, and Cayley.

In 1968, Douglas Edward Crabtree and Haynsworth published the *quotient formula* [755]. Consider

$$M = \begin{pmatrix} A & B \\ C & D \end{pmatrix} \quad \text{with} \quad A = \begin{pmatrix} E & F \\ G & H \end{pmatrix},$$

then, assuming the nonsingularity of (A/E), they showed that

$$(M/A) = \frac{(M/E)}{(A/E)}.$$

Another proof of this formula is due to Ostrowski [2407] in 1971. A simpler proof can be found in [447].

More properties of the Schur complement were given in papers by Richard Warren Cottle [751] in 1974, and [432] by C.B. in 1988. See also [2408] by Diane Valérie Ouellette in 1981 for some properties and applications to statistics.

In [553] by Carlson, Haynsworth, and Thomas Lowell Markham in 1974, the Schur complement has been generalized by replacing the inverse with the Moore-Penrose inverse. These authors extended Sylvester's determinantal formula and the quotient property originally given in [755] to this case. Another extension when the block to be inverted is rectangular and/or singular was given in [2542] by M.R.-Z. by replacing its inverse with its pseudo-inverse. It is the notion of *pseudo-Schur complement.* This generalization is related to the least squares solution of systems of linear equations in partitioned form. Several of its properties were studied, and the proof of the quotient property given in [447] was extended to this case. Pseudo-Schur complements were also defined for matrices partitioned into an arbitrary number of blocks.

Everything you always wanted to know about the Schur complement can be found in the book [3324] edited by Fuzhen Zhang in 2005, in particular its complete history, its mathematical properties, and its applications in eigenvalue and singular value inequalities, block matrix techniques, matrix inequalities, statistics and probability, and numerical analysis in particular, for formal orthogonality, Padé approximation, continued fractions, extrapolation algorithms, bordering methods, projections, preconditioners, domain decomposition methods, triangular recursion schemes, and linear control.

1.6 ▪ Matrix mechanics

Obviously it is not our purpose, in this book, to give even a small overview of quantum mechanics; see, for example, [2096, 3322]. We only want to explain how matrix theory was introduced into this area of physics in the 1920s, but for that, it is necessary to give a minimum of explanations.

Between late 1925 and early 1926, two theories of atomic physics emerged: wave mechanics due to the Austrian physicist Erwin Rudolf Josef Alexander Schrödinger (1887-1987), and matrix mechanics by Werner Karl Heisenberg (1901-1976), a German physicist. Both of them received the Nobel Prize in Physics: Heisenberg in 1932, Schrödinger in 1933.

Schrödinger's approach combined intuitive mathematical notions with physical concepts based on the original ideas of the French physicist Louis Victor Pierre Raymond, 7th Duc de Broglie (1892-1987), who had obtained the Nobel Prize in Physics in 1929. De Broglie's approach, named *mécanique ondulatoire* by its author, was based on the hypothesis that any moving particle or object had an associated wave, an idea in the footsteps of that of Augustin Louis Fresnel (1788-1827) on the duality wave-particle for rays of light, and that of Albert Einstein (1879-1955) about the quanta of light.

Wave mechanics uses operators acting on functions. Each physical quantity is represented by an operator, and the result of a measurement of this quantity can only be one of the eigenvalues of the operator. For example, the energy of a system is represented by Schrödinger's operator whose eigenvalues are the possible energy levels, the operator associated to a coordinate x is the multiplication by x, denoted $x\cdot$, and the operator associated with the component p_x of the momentum is denoted $\frac{ih}{2\pi}\frac{\partial}{\partial x}\cdot$, where i is the imaginary unit and h Planck's constant. Two quantities can be measured simultaneously only if their corresponding operators, say A and B, have a common eigenfunction φ, that is, $A\cdot\varphi = a\varphi$ and $B\cdot\varphi = b\varphi$. A necessary and sufficient condition for that to hold for any eigenvalue of A or B is that they commute. Thus, two coordinates x and y can be measured simultaneously since $x\cdot y = y\cdot x$. On the contrary, this is not true for x and p_x since, by the rule for the derivation of a product, $\frac{\partial}{\partial x}x\cdot\varphi = x\cdot\frac{\partial\varphi}{\partial x}+\varphi$, and it follows that $p_x\cdot x - x\cdot p_x = \frac{ih}{2\pi}$ [884].

Wave mechanics was enthusiastically received not only by the community of conservative physicists, but also by the majority of the young quantum theorists [260].

Spectroscopy developed during the 19th century, and atomic spectroscopy was its first application. An atom can absorb or emit one photon when an electron makes a transition from one stationary state, or energy level, to another one. The law of conservation of energy determines the energy of the photon, and thus the frequency emitted or absorbed. These absorptions and emissions are often referred to as atomic spectral lines. Atoms of different elements have distinct spectra, and thus, atomic spectroscopy allows the identification and quantification of the various components of a combination of atoms or molecules.

The history of matrix mechanics started with research on the spectral lines emitted by chemical elements. Quantum physics tells us that there is only a discrete infinite set of states in which an atom can exist, contrarily to the predictions of classical physics. Attempts to develop a theoretical understanding of the states of the hydrogen atom had an important impact in the history of quantum mechanics. Its spectral lines are characterized by their frequency and their brightness. The four wavelengths discovered by the Swedish physicist Anders Jonas Ångström (1814-1874) in 1855 were known but their values remained unexplained. In 1885, the Swiss physicist and mathematician Johann Jakob Balmer (1825-1898) discovered that the wavelengths of these four lines are given by the empirical formula $\lambda_n = An^2/(n^2-4)$ for $n = 3, 4, 5$, and 6, where A is a physical constant. Subsequent measurements confirmed that the formula remained valid for larger values of n.

The problem was to know if a similar formula could explain the spectral lines of other elements. Balmer's formula was first extended to alkali metals by the Swedish physicist Johannes Robert Rydberg (1854-1919) in 1888, and, in 1903, to all atoms by the Swiss theoretical physicist Walther Heinrich Wilhelm Ritz (1878-1909), also known for the variational method named after him. It is $\lambda_{mn} = Bm^2n^2/(n^2-m^2)$ with $n > m$, and B a constant. Thus, the frequencies of the lines are given by $\nu_{mn} = B(1/m^2 - 1/n^2)$. Remember that the wavelength λ and the frequency ν are related by the formula $c = \lambda\nu$, where c is the speed of light.

Then, in 1913, the Danish Niels Bohr (1885-1962), Nobel Prize in Physics in 1922, came with his planetary model of the atom. Each electron moves on a circular orbit around the nucleus of the atom, and only a discrete set of circular stable orbits, called stationary states, are possible. As long as an electron stays on a stationary orbit, it does not radiate electromagnetic energy. Emission or absorption of a radiation, namely a photon, arises only when an electron jumps, with the probability A_{nm} per unit of time, from one orbit characterized by the energy E_n to another one of energy E_m. Since the energy and the frequency are related by $E = h\nu$, where h is Planck's constant, by Rydberg's formula, the transition energy $E_n - E_m$ between the two states determines the frequency $\nu_{nm} = (E_n - E_m)/h$ of the radiation, and the probability A_{nm} its intensity.

However, Heisenberg, while working on the problem of calculating the spectral lines of hydrogen in 1925, was not satisfied with Bohr's model because the trajectory of an electron in an atom cannot be directly observed. He claimed that concepts which are not observable, such as the circular orbits of the electrons with fixed radii and periods, should not be used in a theoretical model. He decided to use only measurable quantities, that is, the energies of the quantum states of all electrons in an atom, and the frequencies with which they spontaneously jump from one state to another one while emitting a photon.

Bohr's theory described the motion of a particle by a classical orbit, with a well defined position $X(t)$ and momentum $P(t)$ (product of the mass by the velocity), with the restriction that the time integral over one period T of the momentum times the velocity $dX(t)/dt$ must be a positive integer multiple of Planck's constant h. When a classical particle is weakly coupled to a radiation field, it emits a periodic radiation whose frequencies are integer multiples of the orbital frequency T. This property reflects the fact that $X(t)$ is periodic and can be represented by a Fourier series

$$X(t) = \sum_{n=-\infty}^{\infty} X_n e^{2\pi i n t/T},$$

where the coefficients X_n are complex numbers. However, since $X(t)$ is real, $X_{-n} = X_n^*$. But, a quantum particle is not emitting radiation continuously, it only emits photons. Assuming that the quantum particle starts in orbit n, emits a photon, then ends up in orbit m, the energy of the photon is $E_n - E_m$, which means, as we said above, that its frequency is $\nu_{nm} = (E_n - E_m)/h$. For large n and m, but with $n - m$ relatively small, these are the classical frequencies by Bohr's correspondence principle $E_n - E_m \simeq h(n-m)/T$. This principle, formulated by Bohr in 1920, states that the behavior of a system described by the theory of quantum mechanics (or by the old quantum theory) has to reproduce classical physics in the limit of large quantum numbers. In other words, it says that for large orbits and for large energies, quantum calculations must agree with classical calculations. But, when n and m are small, or $n - m$ is large, the frequencies are no longer integer multiples of a single frequency.

After a collaboration with the Dutch physicist Hendrik Anthony "Hans" Kramers (1894-1952), Heisenberg realized that the transition probabilities were not classical quantities, because the only frequencies that must appear in the Fourier series have to be those that are observed in quantum jumps. Since the frequencies emitted by the particle are the same as those in the Fourier series, this suggests that something in the time-dependent description of the particle is oscillating with frequency $(E_n - E_m)/h$. Heisenberg denoted this quantity X_{nm} and imposed that it reduced to the classical Fourier coefficients in the classical limit. For large values of n and m but with $n - m$ relatively small, X_{nm} is the $(n - m)$th Fourier coefficient of the classical motion at orbit n. Since X_{nm} has a frequency opposite to that of X_{mn}, and $X(t)$ is real, $X_{nm} = X^*_{mn}$. Thus X_{nm} only has the frequency $(E_n - E_m)/h$, and its time evolution is simply

$$X_{nm}(t) = e^{2\pi i(E_n - E_m)t/h} X_{nm}(0).$$

Heisenberg started from the relation $\nu_{mn} = (E_m - E_n)/h$, from which it follows that $\nu_{nk} + \nu_{km} = \nu_{nm}$, that $\nu_{nm} = -\nu_{mn}$, and that $\nu_{nn} = 0$. He puts these quantities into a table

$$\begin{matrix} \nu_{00} & \nu_{01} & \nu_{02} & \cdots \\ \nu_{10} & \nu_{11} & \nu_{12} & \cdots \\ \nu_{20} & \nu_{21} & \nu_{22} & \cdots \\ \vdots & \vdots & \vdots & \ddots \end{matrix}$$

The row i contains all the frequencies generated from E_i, while the column j contains all frequencies (in increasing order) that can be generated upon reaching a certain level. He performed mathematical operations on this table, which led him to define new tables corresponding to other dynamical quantities related to the transition from n to m such as the position, the velocity, or the square of the velocity of an electron. The table entries were the so-called transition amplitudes, that is, quantities whose squares specify a transition probability.

Thus, every physical quantity was represented by a matrix. Given two arrays X_{nm} and P_{nm} describing two physical quantities, Heisenberg formed a new array of the same type by convolution of the two corresponding Fourier series, which led him to the multiplication rule

$$(XP)_{mn} = \sum_{k=0}^{\infty} X_{mk} P_{kn}.$$

Under this rule, $XP \neq PX$.

At that time, Heisenberg was in touch with the theoretical physicists Max Born (1882-1970), Ernst Pascual Jordan (1902-1980), and Paul Adrien Maurice Dirac (1902-1984). He soon realized that the operations he applied to these tables were well known to mathematicians. His tables were called *matrices*, and his multiplication rule was named *matrix multiplication*. By the end of the year 1925, Heisenberg's ideas were transformed into a comprehensive and systematic version of quantum mechanics, which is now referred to as *matrix mechanics*.

With the help of the new matrix mechanics, the German physicist Wolfgang Ernst Pauli (1900-1958) managed in the following January to completely solve the calculation of the energy levels of the hydrogen atom. Within a year, Schrödinger himself showed that his wave mechanics and Heisenberg's matrix mechanics were equivalent. In 1927, Heisenberg introduced the *uncertainty principle*, which states that the position of a particle and its momentum cannot be simultaneously measured with precision. Indeed, any measurement, since it implies photons, disturbs the system. This principle is a consequence of the non-commutativity of the matrix product, which means that the results of the measurements of two physical quantities depend on the order they are made.

Up until those times, matrices were seldom used by physicists who considered them as pure mathematics. However, the German physicist Gustav Adolf Feodor Wilhelm Ludwig Mie (1868-1957) had already used them in a paper on electrodynamics in 1912, and Born in his work on the lattices theory of crystals in 1921. Born had learned matrix algebra from Jakob Rosanes (1842-1922) and he also knew Hilbert's theory of integral equations and quadratic forms with an infinite number of variables. E.P. Jordan had been an assistant to Richard Courant (1888-1972) at Göttingen.

In 1926, von Neumann became assistant to David Hilbert (1862-1943). In 1932, he established a rigorous mathematical framework for quantum mechanics, and explained its physics by the mathematical theory of Hilbert spaces and linear operators acting on them.

Max Karl Ernst Planck (1858-1947) in 1918, Einstein in 1921, Dirac in 1933, and Pauli in 1945 were also awarded the Nobel Prize in Physics.

1.7 ▪ Lifetimes

In this section, we show the lifetimes of the main deceased contributors to the introduction of matrices and their properties, starting in 1777. The language of the author is given by the color of the bars and by letters: E (red) for English, G (black) for German, I (green) for Italian, F (blue) for French, and O (magenta) for the others. The contributors are ordered by date of birth.

Matrices main contributors (a)

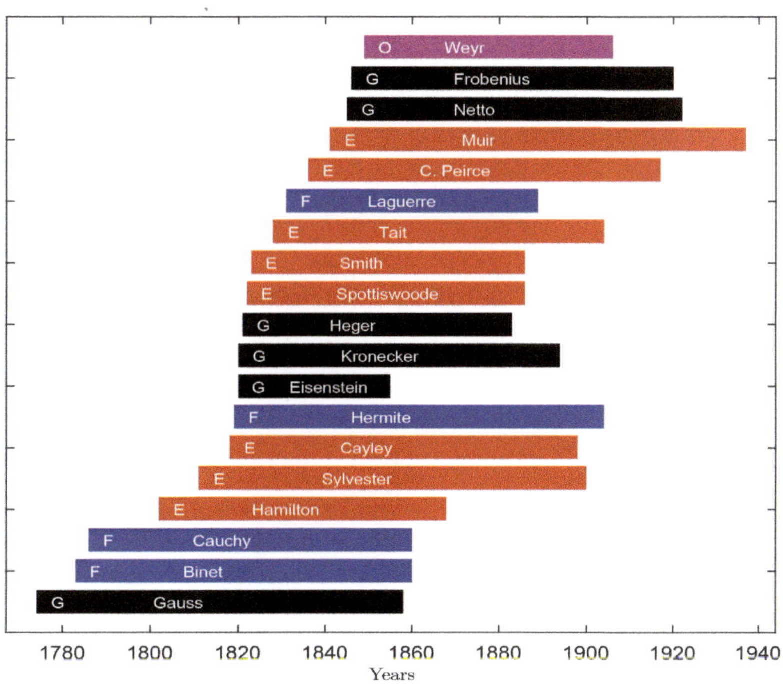

Matrices main contributors (b)

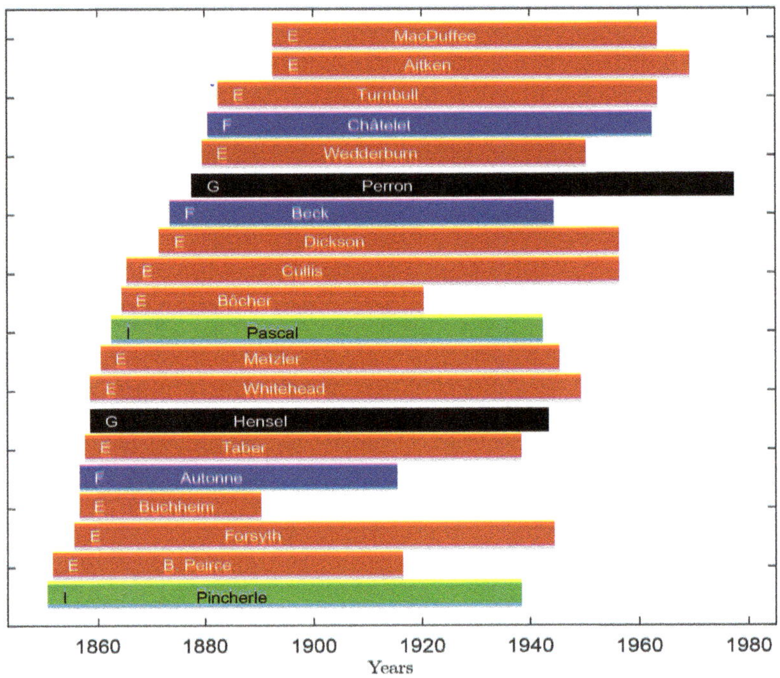

Condition number

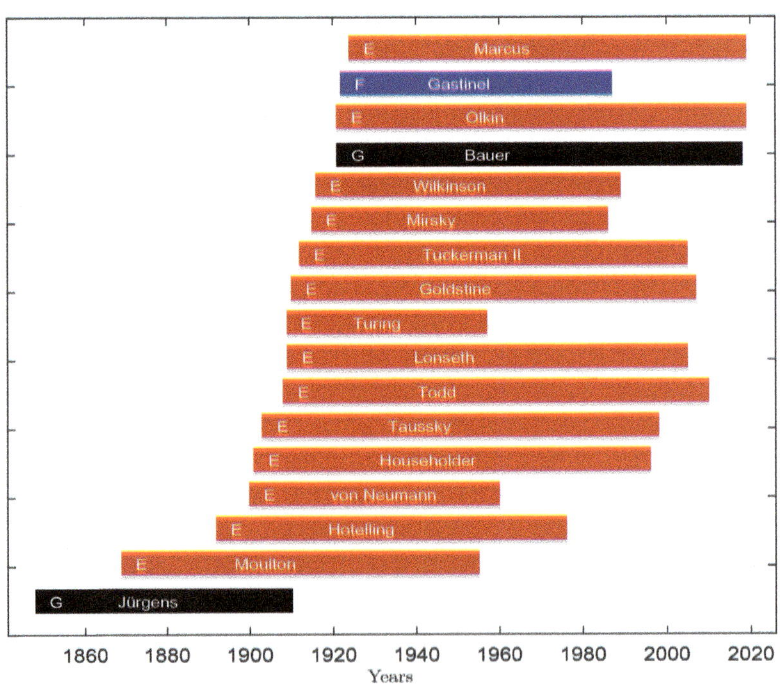

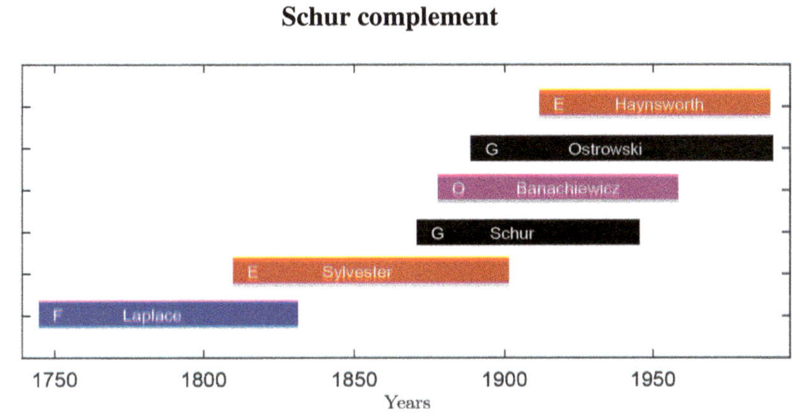
Schur complement
E Haynsworth
G Ostrowski
O Banachiewicz
G Schur
E Sylvester
F Laplace
1750
1800
1850
1900
1950
Years

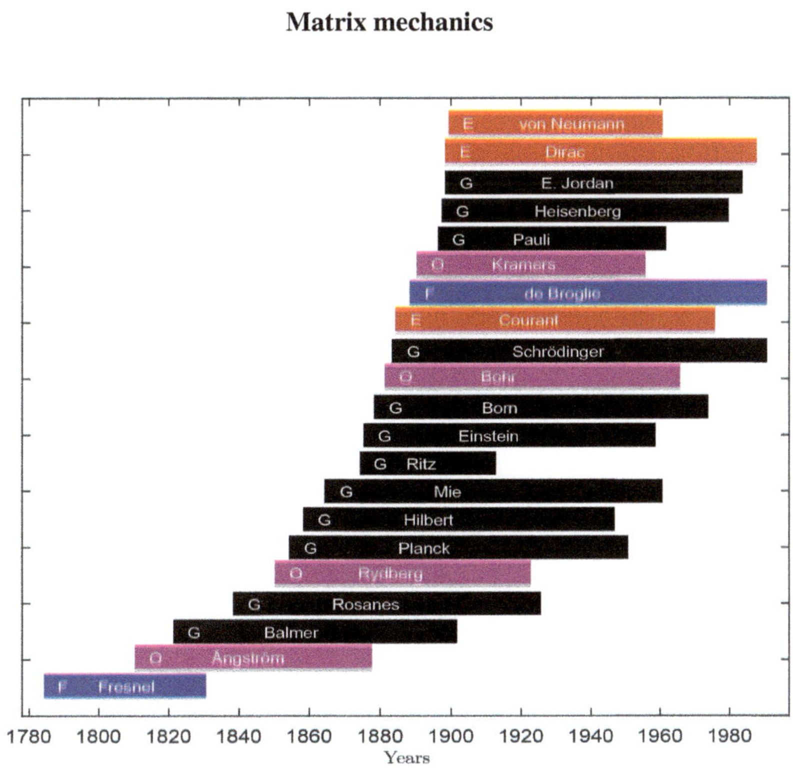
Matrix mechanics
E von Neumann
E Dirac
G E. Jordan
G Heisenberg
G Pauli
O Kramers
F de Broglie
E Courant
G Schrödinger
O Bohr
G Born
G Einstein
G Ritz
G Mie
G Hilbert
G Planck
O Rydberg
G Rosanes
G Balmer
O Angström
F Fresnel
1780
1800
1820
1840
1860
1880
1900
1920
1940
1960
1980
Years

2 Elimination methods for linear systems

The two main classes of rounding error analysis are not, as my audience might imagine, 'backwards' and 'forwards', but rather 'one's own' and 'other people's'. One's own is, of course, a model of lucidity; that of others serves only to obscure the essential simplicity of the matter in hand.
– James Hardy Wilkinson, *NAG 1984 Annual General Meeting*

2.1 ▪ Antiquity

It has been written (see, for instance, [1435, 1436]) that the earliest solutions of linear equations were obtained at the "Old Babylonian" epoch (2000 to 1600 BC) in Mesopotamia, the region between the Tigris and Euphrates rivers, centered more or less about the city of Babylon. The oldest written texts are dated around 3300 BC. They were written in cuneiform writing on clay tablets, their basic language being Babylonian, one of the two main dialects of Akkadian. The cuneiform characters were first decoded around 1840 by the English archeologist Sir Henry Creswicke Rawlinson (1810-1895). Around half a million tablets have been found, of which a few hundred contain mathematical statements and more than one thousand with numerical tables and metrological data. Many of these mathematical tablets were deciphered and translated in the beginning of the 20th century by Otto Eduard Neugebauer (1899-1990), an Austrian-American mathematician and historian of science, and François Thureau-Dangin (1872-1944), a French archeologist, assyriologist, and museum curator. However, new translations and study of other tablets continue to be done; see, for instance, [1243, 1244, 1758, 1760, 1761].

According to Jens Høyrup [1759],

> *Ancient Babylonian [...] mathematics were powerful calculational tools for the solution of scribal tasks - accounting, planning of resources, measurement of land; they were developed and taught for that purpose.*

The Babylonians were using a positional sexagesimal (that is, base 60) number system and fixed point numbers. They represented problems by a step-by-step list of rules whose evaluation was given in words. It turns out that problems on some tablets correspond, when translated in our modern mathematics, to solving linear and nonlinear equations. However, as noted by Høyrup [1760],

> *This is exclusively the result of blunt mapping onto the conceptual grid of present-day mathematics.*

In 1972, the well-known computer scientist Donald Ervin Knuth [1921] saw some of these tablets as describing algorithms. This view was challenged by Høyrup [1760] in 2012,

> *The prescriptions turn out to be neither renderings of algebraic computations as we know them nor mindless rules (or algorithms) to be followed blindly. [...] The so-called "algebra" is indeed best known from texts written between 1800 and 1600 BC. The "linear equations with several unknowns" are (e.g.) word problems dealing with fields of different area and different rent per area unit - problems whose translation into algebraic equations presupposes choices and therefore is not unambiguous. [...] What we know about was a technique, no mathematical theory. It was formulated in words, in a very standardized but not always unambiguous language. [...] To assert that it was an algebra merely because its procedures can be described in modern equations, or to declare that it was not because it did not itself write such equations, is perhaps a bit superficial.*

Hence, seeing the Babylonians as the first people to ever solve linear equations and invent linear algebra may be simply an overstatement. However, see the modern interpretation of the tablet VAT8391 ♯ 2 by Jöran Friberg and Farouk N.H. Al-Rawi in [1244] and a similar discussion of VAT 8389 ♯ 1 and VAT 8391 ♯ 3 by Høyrup [1758]. These tablets are in the Vorderasiatisches Museum in Berlin. For the constitution of collection of mathematical clay tablets, see the thesis [883] of Magali Dessagnes in 2017.

Other examples of linear systems in antiquity are found in the Rhind Papyrus from Egypt. This papyrus from 1550 BC is named after Alexander Henry Rhind (1833-1863), a Scottish antiquarian, who purchased the papyrus in 1858 in Luxor. It was copied by the scribe Ahmes from some older texts. The papyrus was mathematically translated in the late 19th century. It was reproduced and studied by Arnold Buffum Chace (1845-1932), an American businessman and mathematician, in 1929 [607, 608]. Problem 40 (see page 12 of [607] and pages 31-32 and 172 of [608]) is to divide 100 loaves among five men in such a way that the shares received are in arithmetical progression and that $1/7$ of the sum of the largest three shares is equal to the sum of the smallest two. In modern terms, this can be formulated as a linear system with the base x and the increment y as unknowns. The solution is $y = 9 + 1/6$ and $x = 1 + 2/3$. However, this was, in fact, an exercise about arithmetic progressions which was solved by the double false position method (*regula falsi*). The majority of the papyrus is now kept in the British Museum in London.

Other Egyptian mathematical texts are on the Moscow Mathematical Papyrus, also called the Golenishchev Mathematical Papyrus. It was bought by the egyptologist Vladimir Golenishchev (1856-1947) in 1892 or 1893 in Thebes. It is kept in the Pushkin State Museum of Fine Arts in Moscow.

Note that we do not have that many remaining mathematical texts from Egypt, probably because papyrus is not a very strong material, in contrast to the Babylonian clay tablets.

2.2 ▪ Ancient China

Elimination methods were used in ancient China long before our era. The book in which we find these methods is the most important Chinese mathematical classic *Jiuzhang Suanshu* (which has been translated as *Nine Chapters on the Mathematical Art*). This text is one of the earliest surviving mathematical texts from China. It is believed to have been compiled some time between 100 BC and 100 AD, but it is likely that the content of *Nine Chapters* was much older than its date of compilation. In 1984, archeologists discovered a tomb that had been sealed c. 186 BC at Zhangjiashan in the Hubei Province. It contained a text entitled the *Book of Mathematical*

Procedures or *Book on Numbers and Computations* (*Suanshu Shu*, see [813]), which is dated at around 200 BC. Its contents exhibit a resemblance to that of the Jiuzhang Suanshu, including even some identical numerical data. The Suanshu Shu is approximately seven thousand characters in length, written on 190 bamboo strips. This book shows how to solve systems of two equations and two unknowns using the double false position method.

From what is known from some commentaries (see [3314]), it seems that Zhang Cang (256 BC-152 BC), a politician, mathematician, and astronomer, played an important role in composing *Nine Chapters on the Mathematical Art*, and that the current version of the book remains more or less the same as it was in the 2nd century BC, but probably not the same as it had been before the Qin Dynasty. Qin Shi Huang (259 BC-210 BC) was the first emperor of China, whose tomb in Xian is famous for its Terracotta Army. In 213 BC, he ordered most existing books, including *Nine Chapters*, to be burned.

Nine Chapters was later commented by several scholars, the most prominent one being Liu Hui (c. 220-c. 280 AD), an official who lived in the northern Wei kingdom. Not much is known about Liu Hui's life, even though he is considered one of the greatest Chinese mathematicians; see [2917]. No ancient edition of *Nine Chapters* has survived that does not contain the commentaries of Liu Hui in 263 AD and the explanations added by a group of scholars under the supervision of Li Chunfeng (c. 602-c. 670 AD) in the 7th century AD.

There are several editions of *Nine Chapters* in Chinese; see [1481, 2048]. The book was partially translated in the late 19th century. It was fully translated later into English [2752], French [655], as well as German and Russian.

A positional base-10 number system was used in ancient China. Computations were done by manipulating counting rods on a counting board (like a checkerboard) or on the sand. Counting rods were used from 500 BC until approximately 1500 AD when they were gradually replaced with the abacus. Counting rods, which were generally bamboo sticks, were used to represent the digits 1 to 9 (see Figure 2.1), and the arrangement of the rods on a counting board indicated the place value, with an empty space for zero, which was not used in China. Negative and positive numbers were distinguished by the color of the counting rods, generally red for positive and black for negative.

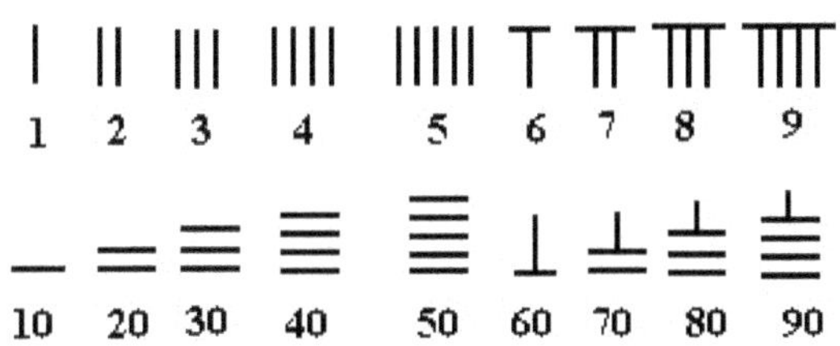

Figure 2.1. *Chinese numbers*

As its title implies, *Nine Chapters* has nine chapters, organized by topic (see [653, 1175, 2752]):

1. Rectangular Fields. This chapter is concerned with arithmetical operations on fractions, land measurement, and formulas for finding areas of fields of several shapes.

2. Millet and Rice. Chapters 2 and 3 contain problems from agriculture, manufacturing, and commerce.

3. Distribution by proportion.

4. Short Width. Computations of square and cubic roots and problems with circles.

5. Construction Consultations. This chapter contains formulas for volumes of various solids.

6. Fair Levies: The problems in this chapter come from taxes and distribution of labor.

7. Excess and Deficit. The rule of double false position for solving linear equations of order 2 is used to solve a variety of problems.

8. Rectangular Arrays. The *fangcheng* procedure is introduced to solve systems of linear equations.

9. Right-angled Triangles. This chapter includes the Gougu Rule, nowadays known as the Pythagorean Theorem.

Overall, there are 246 problems in the book. Of particular interest for us is Chapter 8. The term *fangcheng* is not easy to translate exactly. However, as a first approximation it has been translated as "rectangular arrays," "square arrays," or "measures in a square"; see [2864] by Marie Stepánová.

Chapter 8 contains 18 practical problems which amount to solving nonsingular linear systems of order 2 to 5 and one underdetermined system with 5 equations and 6 unknowns. In fact, it is an anachronism to speak of "equations" since the Chinese did not use that term and did not formulate their problems in terms of equations.

The procedure or rule is described for Problem 1. The problem is first formulated in words. In the book, the solution is given right after the problem is stated. Here is one of the available translations of the problem in [651]:

> *Suppose that 3 bing of high-quality grain, 2 bing of medium-quality grain and 1 bing of low-quality grain produce 39 dou; 2 bing of high-quality grain, 3 bing of medium-quality grain and 1 bing of low-quality grain produce 34 dou; 1 bing of high-quality grain, 2 bing of medium-quality grain and 3 bing of low-quality grain produce 26 dou. Then, how much is produced respectively by one bing of high-, medium- and low-quality grain?*

The bing is a unit of capacity whose relation to the dou, the other unit of capacity used in these statements, is not defined either by *Nine Chapters*, or by Liu Hui. Other documents give 1 bing as equal to 160 dou. One dou is one deciliter.

Then, the data is set up on the counting board. The coefficients and the right-hand side are arranged in columns. Using arabic numbers instead of the Chinese numbers, it looks like

$$\begin{bmatrix} 1 & 2 & 3 \\ 2 & 3 & 2 \\ 3 & 1 & 1 \\ 26 & 34 & 39 \end{bmatrix}.$$

Note that the first row corresponds to high-quality grain, the second row to medium-quality grain and the third row to low-quality grain. The last row contains the right-hand side.

The method of solution proceeds step by step on the columns of the board. First, the second column is multiplied by the term in position $(1, 3)$ (in our modern matrix notation) which is equal to 3. It yields

$$\begin{bmatrix} 1 & 6 & 3 \\ 2 & 9 & 2 \\ 3 & 3 & 1 \\ 26 & 102 & 39 \end{bmatrix}.$$

Then, one subtracts 2 times the third column from the second one. This puts a zero (or a blank for the Chinese book) in position $(1,2)$

$$\begin{bmatrix} 1 & & 3 \\ 2 & 5 & 2 \\ 3 & 1 & 1 \\ 26 & 24 & 39 \end{bmatrix}.$$

The same kind of manipulation is done with columns 1 and 3. Column 1 is multiplied by 3 and a subtraction is done

$$\begin{bmatrix} & & 3 \\ 4 & 5 & 2 \\ 8 & 1 & 1 \\ 39 & 24 & 39 \end{bmatrix}.$$

The same procedure is applied to what is remaining of the first two columns. The first column is multiplied by 5 and four subtractions are done, yielding

$$\begin{bmatrix} & & 3 \\ & 5 & 2 \\ 36 & 1 & 1 \\ 99 & 24 & 39 \end{bmatrix}.$$

Today, we would directly obtain that the value for the low-quality grain is $99/36 = 2.75$ and then proceed backwards. But this is not what is done in the fangcheng procedure, the goal being to compute only with integers as long as possible and to avoid rational numbers, except for the last step.

The term $(4,2)$ is multiplied by the term $(3,1)$, and the term $(3,2)$ is multiplied by the term $(4,1)$

$$\begin{bmatrix} & & 3 \\ & 5 & 2 \\ 36 & 99 & 1 \\ 99 & 864 & 39 \end{bmatrix}.$$

In the second column, the terms, except the diagonal term, are subtracted from the last one, and the result is divided by the diagonal term

$$\begin{bmatrix} & & 3 \\ & & 2 \\ 36 & & 1 \\ 99 & 153 & 39 \end{bmatrix}.$$

For the last column, $(4,3)$ is multiplied by $(3,1)$, $(3,3)$ is multiplied by $(4,1)$, and $(2,3)$ is multiplied by $(4,2)$

$$\begin{bmatrix} & & 3 \\ & & 306 \\ 36 & & 99 \\ 99 & 153 & 1404 \end{bmatrix}.$$

The terms, except the diagonal term, are subtracted from the last one, and the result is divided by the diagonal term

$$\begin{bmatrix} & & \\ & & \\ 36 & & \\ 99 & 153 & 333 \end{bmatrix}.$$

Finally, the last row is divided by $(3, 1) = 36$, and one obtains the solution, whose components are sums of integers and rational numbers,

$$\frac{99}{36} = 2 + \frac{3}{4} = 2.75, \quad \frac{153}{36} = 4 + \frac{1}{4} = 4.25, \quad \frac{333}{36} = 9 + \frac{1}{4} = 9.25.$$

The reader may wonder why the result of the subtractions in a column is divisible by the diagonal term. In modern terms, this can be seen as follows. Let our board be

$$\begin{bmatrix} c_{11} & c_{12} & c_{13} \\ c_{21} & c_{22} & c_{23} \\ c_{31} & c_{32} & c_{33} \\ d_1 & d_2 & d_3 \end{bmatrix}.$$

After two main steps, we have

$$\begin{bmatrix} 0 & 0 & c_{13} \\ c_{21}c_{13} - c_{23}c_{11} & c_{22}c_{13} - c_{12}c_{33} & c_{23} \\ c_{31}c_{13} - c_{33}c_{11} & c_{32}c_{13} - c_{12}c_{33} & c_{33} \\ d_1c_{13} - d_3c_{11} & d_2c_{13} - d_3c_{12} & d_3 \end{bmatrix}.$$

Now, we reduce the first column using the second column. It becomes

$$c^{(2)} = \begin{bmatrix} 0 \\ 0 \\ (c_{31}c_{13} - c_{33}c_{11})(c_{22}c_{13} - c_{12}c_{33}) - (c_{21}c_{13} - c_{23}c_{11})(c_{32}c_{13} - c_{12}c_{33}) \\ (d_1c_{13} - d_3c_{11})(c_{22}c_{13} - c_{12}c_{33}) - (d_2c_{13} - d_3c_{12})(c_{21}c_{13} - c_{23}c_{11}) \end{bmatrix}.$$

We transform the second column by multiplying the last term $d_2c_{13} - d_3c_{12}$ by $c_3^{(2)}$ and subtracting the term $c_{32}c_{13} - c_{12}c_{33}$ multiplied by $c_4^{(2)}$. When computing the resulting value, it turns out that some terms cancel since

$$\begin{aligned} &- (c_{21}c_{13} - c_{23}c_{11})(c_{32}c_{13} - c_{12}c_{33})(d_2c_{13} - d_3c_{12}) \\ &+ (c_{21}c_{13} - c_{23}c_{11})(d_2c_{13} - d_3c_{12})(c_{32}c_{13} - c_{12}c_{33}) = 0. \end{aligned}$$

The remaining terms are

$$\begin{aligned} &+ (c_{31}c_{13} - c_{33}c_{11})(c_{22}c_{13} - c_{12}c_{33})(d_2c_{13} - d_3c_{12}) \\ &- (d_1c_{13} - d_3c_{11})(c_{22}c_{13} - c_{12}c_{33})(c_{32}c_{13} - c_{12}c_{33}), \end{aligned}$$

which are proportional to $c_{22}c_{13} - c_{12}c_{33}$ by which we have to divide. Hence, the bottom term of the second column, which is the only one we have to keep in the procedure, is an integer if all the c_{ij}'s are integers.

Of course, there is no such proof in *Nine Chapters*, even in Liu Hui's comments. What he did was to give details about the procedure and to check its correctness on the given examples, even though he wrote (see Footnote 8 in [651])

> *This procedure is general, but it is difficult to explain in terms of abstract expressions. Thus, in order to eliminate this difficulty, it is deliberately linked to grains.*

On the concept of proof in *Nine Chapters* and its commentaries, see the papers [650, 654] by Karine Chemla.

The book became a classic through the next centuries and had influence not only in China, but also in Korea and Japan. It remained a reference work for practitioners of mathematics in China until at least the 14th century. It is not known if the ideas expressed in the book found their ways to Western Europe.

Clearly, we can see *Nine Chapters* as a collection of algorithms. Details and discussions about *Nine Chapters* can be found in [650, 651, 653, 2752, 2124, 655, 1594, 1436, 1435, 1433, 654, 3314, 1175] (ordered by date). On the history of Chinese mathematics, see [2162, 2785, 2953].

Whether or not the Chinese mathematicians discovered Gaussian elimination can be a subject of debate. What is certain is that they were solving small linear systems with integer coefficients using an elimination method. But their method is not exactly what we now call Gaussian elimination, particularly in the backward phase. Moreover, they did not use pivoting techniques and symbols for the unknowns and did not have the concept of linear equation.

Roger Hart in [1594] claims that determinants can also be found in *Nine Chapters*. This is based on what he calls "cross-multiplications" of elements of the counting board that we have seen above. Although this can be interpreted as a 2×2 determinant and even if Japanese mathematicians were certainly influenced by China, pretending that determinants were discovered or invented in ancient China before they appeared in Japan and Europe seems an overstatement. Reviews of Hart's book were written by Joseph Frank Grcar [1436] and Eberhard Knobloch [1916].

2.3 ▪ Ancient Greece

Even though Greek mathematicians were mostly interested in geometry, there were a few examples of what corresponds to solving linear systems. For instance, Thymaridas of Paros (c. 400-c. 350 BC) was the author of a rule for solving a certain set of n simultaneous simple equations with n unknowns. Another example of work in late antiquity where we can find linear system solves is the *Arithmetica* of Diophantus. Not much is known about Diophantus, not even his birth date. However, it is known that he lived in Alexandria (Egypt), a city founded by Alexander the Great in 331 BC, which was the center of the Hellenistic world.

At the end of the 19th century, Paul Tannery (1843-1904), a French historian, claimed that Diophantus lived in the middle of the third century AD [2997]. This dating is based on an analysis of one of the problems of the Arithmetica. This is problem V.1, which is the only problem giving some data, that is, numbers of amphorae of wine and their prices. It was the prices that Tannery used to obtain a date. However, there are other hypotheses: Wilbur Richard Knorr (1945-1997) [1918] thought that Diophantus lived at the same time as Hero of Alexandria, that is, two centuries earlier. What seems sure so far is that Diophantus was not cited by other authors before Theon of Alexandria at the end of the 4th century [2692].

In Arithmetica, it is written that there are thirteen chapters. Up to 1968, only six of these chapters, written in Greek, were known, and the other chapters were believed to be lost. However, in 1968, an Arabic translation of four of the lost chapters, attributed to Qustā ibn Lūqā (c. 820-c. 912), was found in a library in Iran and later translated and published; see [2538, 2539, 2746].

Several translations of the Greek version were published in the 16th and 17th centuries, in 1463 by Johannes Müller (1436-1476) also known as Regiomontanus, in 1570 by Rafael Bombelli (1526-1572), who was one of the first to use continued fractions in Europe, and in 1575 by Wilhelm Xylander (1532-1576). The most praised translation into Latin was, in 1621 (see Figure 2.2), that of Claude Gaspard Bachet de Méziriac (1581-1638), a French mathematician, linguist, and poet; see [726] for a biography. It was in a copy of that edition that Pierre de Fermat (1607-1665), a French lawyer and mathematician, wrote in the margin that there are no

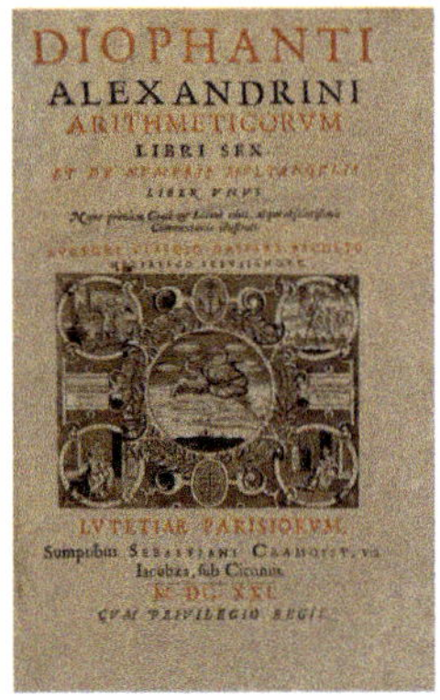
DIOPHANTI
ALEXANDRINI
ARITHMETICORVM
LIBRI SEX.
LVTETIAE PARISIORVM.

Figure 2.2. *Translation by Bachet de Méziriac*

three positive integers x, y, z satisfying the relation $x^n + y^n = z^n$ for an integer $n > 2$. Fermat also wrote that he had a wonderful proof but that it cannot fit in the margin. Bachet's translation was republished in 1670 by Samuel Fermat (1630-1690), the son of Pierre, with the annotations of his father. The theorem was only proved in 1994 by Andrew Wiles, with the help of Richard Taylor.

There were other translations, in fact sometimes translations of translations, in 1585 by Simon Stevin (c. 1548-1620), a Flemish mathematician from Bruges, by Albert Girard (1595-1632) in 1625, by Otto Schultz in 1822, and by Gustav Wertheim (1843-1902) in 1890. It is also interesting to read the book [1617] by Sir Thomas Little Heath (1861-1940), a British mathematician and historian of Greek mathematics, in 1885.

Of interest for us is the first chapter (also called Book I) of the Arithmetica. This chapter contains 25 problems of the first degree, 21 determined with a unique solution and 4 undetermined. Diophantus considered problems with integer coefficients and with integer or rational solutions. The problems are posed in general terms but the solution is given and explained with an example; see [823, 1617]. Let us consider two of these problems.

Problem 1: To divide a given number in two numbers whose difference is given.

In modern terms, this is

$$
\begin{aligned}
x + y &= a, \\
y - x &= d.
\end{aligned}
$$

Clearly, $2y = a + d$, $y = (a + d)/2$ and $x = (a - d)/2$. In Diophantus' example, $a = 100$ and $d = 40$. Even though the solution is given in words, he assumed that x is the smallest of the two numbers. Therefore, they are x and $x + 40$, which yields $2x + 40 = 100$ and $x = 30$, $y = 70$.

Problem 17: Find four numbers which, added three by three, produce given numbers with the condition that one third of the sum of the numbers is greater than any of them.

In modern terms, this is a linear system of order 4,

$$
\begin{aligned}
x + y + z &= a, \\
w + x + y &= b, \\
w + x + z &= c, \\
w + y + z &= d.
\end{aligned}
$$

The trick to solve this problem easily is to consider the sum $s = w + x + y + z$. By summing the four equations, we have $3s = a + b + c + d$ giving the value of s and the equations

$$\begin{aligned} s - w &= a, \\ s - z &= b, \\ s - y &= c, \\ s - x &= d, \end{aligned}$$

a diagonal linear system with the obvious solution

$$w = s - a, \quad z = s - b, \quad y = s - c, \quad x = s - d.$$

The example given by Diophantus is $a = 20, b = 22, c = 24, d = 27$. He considered the sum s of the four numbers, and his reasoning was that the four numbers are $s-20, s-22, s-24, s-27$. Consequently, $4s - 93 = s$ and $s = 31$. We observe that Diophantus reduced the system to only one equation with one unknown. All the determinate linear problems that he considered have this same characteristic.

What is important is that, first of all, the problems are stated in general terms and that, secondly, Diophantus gave the name *arithmos* to one unknown that he abbreviated with the letter ς. This is probably the first occurrence of symbolic unknowns in arithmetic and algebra.

For details on Diophantus and his work, see [2997, 2998, 2999, 2538, 2539, 2746, 674, 2692, 675, 823, 677, 1666, 676] (ordered by date).

2.4 ▪ Ancient India

According to [1620] *the first use of algebra in India is found in the Sulba Sūtras (800-500 BC). This early algebra was geometrical in nature, similar to Babylonian algebra.* A rule for solving a linear equation with one unknown was given in 499 by Aryabhata (476-550 AD) in the Aryabhatīya. He is also known as Aryabhata I or Aryabhata the Elder, to distinguish him from the later mathematician of the same name who lived about 400 years later. His book is a collection of 123 verses, divided into four sections. It is one of the most influential works of Indian mathematics since, even though it is centered on astronomy, it contains many significant topics like arithmetic and geometric methods to compute areas, to extract square and cube roots and arithmetic progressions. It was translated into French by Léon Rodet (1832-1895) in 1879.

In 1881, a manuscript was unearthed by a peasant near the village of Bakhshālī, 80 km away from Peshawar, then part of British India, and now part of Pakistan. This manuscript has 70 "pages" of mathematical formulas and examples written in ink on birch bark in the form of verse rules with a prose commentary. It was studied and deciphered by Rudolf Hoernlé (1841-1918), of German origin but born in India, and later given to the Bodleian Library in Oxford (UK), where it arrived in 1902. Facsimiles were published by George Rusby Kaye (1866-1929) in his edition of the text in 1927 [1881]. The definitive study on this manuscript is the translation and commentary published by Takao Hayashi in 1995 [1607].

There were and still are some disputes about the dating of the manuscript. Hoernlé originally suggested the third or fourth century, Kaye was thinking it was from the 12th century, and Hayashi assigned the date of the commentary's composition to the 7th century, and of the manuscript itself to somewhere between the 8th and the 12th centuries. A carbon dating was done at the Bodleian Library in 2017 on three samples of bark, avoiding areas containing ink. The date ranges found for these pieces are 224-383, 680-779, and 885-993. These findings were reported in an electronic document and a YouTube video with claims that it was "one of the earliest uses of zero... as a placeholder, i.e., the use of zero to indicate orders of magnitude in a number system." This dating and these claims were criticized in [2502].

Nevertheless, the Bakhshālī manuscript contains, among other things, solution of particular determinate and undeterminate linear equations; see [2685, 3315].

In the early 19th century, Henry Thomas Colebrooke (1765-1837), an English judge, botanist, and orientalist, published translations of three classics of Indian mathematics, the *Brāhmasphuta-siddhānta* of Brahmagupta (c. 598-c. 668), an astronomer and mathematician in 628, and the *Līlāvatī* and the *Bijaganita* of Bhāskara II (1114-1185) in 1150; see also [720].

For solving linear equations, Brahmagupta gave the following rule, as stated in [24]:

> *Removing the other unknowns from (the side of) the first unknown and dividing by the coefficient of the first unknown, the value of the first unknown (is obtained). In the case of more (values of the first unknown) two and two (of them) should be considered after reducing them to common denominator. And (so on) repeatedly. If more unknowns remain (in the final equation) the method of pulverizer (should be employed). (Then proceeding) reversely (the values of the other unknowns can be found).*

Bhāskara II was interested in systems of underdetermined systems of linear equations in several variables. Interestingly, he was also the first to give the recurrence relationships for the numerators and the denominators of the successive convergents of continued fractions, 500 years before Western mathematicians.

There were certainly early exchanges between India and China. But there were discussions and disputes about what were the relations between the Greek, Arabic, and Indian mathematics. It is well-established that the positional numbering system with Hindu-Arab numbers was introduced in Western Europe from India through Arab and Persian translations; see [1620, 1623]. Hindu-Arab numbers appeared in the *Codex Vigilanus* written in Spain around 880. Gerbert d'Aurillac (c. 945-1003), a French monk who later became pope under the name Sylvester II, is believed to be one of the people who introduced the Hindu-Arabic numbering system in Europe. He was not really followed on that matter at that time. The decimal positional system reappeared in Italy in 1202 in the *Liber Abaci* by Leonardo Pisano (c. 1170-c. 1235), also known today as Leonardo Fibonacci. This book had much more influence for the adoption of the new numbering system.

For the history of mathematics in India, see [2501] by Kim Plofker and [143] by A.K. Bag.

2.5 ▪ Ancient Persia

The most well-known figure in the realm of ancient Persian mathematics is Muhammad ibn Mūsā al-Khwārizmī (c. 780-c. 850). He was a Persian Muslim who lived in Baghdad (now in Iraq). Not much is known about the life of al-Khwārizmī except that he was a scholar at the House of Wisdom developed by the Caliph al-Ma'mun (786-833), the son of Harun al-Rashid (c. 765-809) who funded it. For more about al-Khwārizmī, see [23, 1757].

Al-Khwārizmī wrote several books, including the famous *Kitāb fī al-jabr wa al-muqābalah*. *Al-jabr* and *al-muqābalah* were later translated as restoration and opposition. The aim of this book was to present solutions of practical problems. It includes rules for finding (an approximation of) the area of a disk, as well as for finding the volume of solid shapes, like spheres and pyramids. It also includes methods for solving quadratic equations algebraically (in words) and geometrically. In the section titled "On legacies" [2595, pp. 86-133]) in Rosen's translation, al-Khwārizmī solved problems which, in modern terms, correspond to one linear equation with one unknown. Even when the problems have several unknowns, they are reduced easily to a single linear equation by the statement of the problem. Al-Khwārizmī did not use any symbols or algebraic notation. Everything was expressed in words or with geometrical figures.

Three Latin translations of al-Khwārizmī's book were done by Robert of Chester around 1145, Gherardo (Gerardo) da Cremona (1114-1187) around 1170, and Guglielmo de Lunis in the first half of the 13th century. The first English translation was done by Friedrich August (or Frederic) Rosen (1805-1837) in 1831 [2595].

Al-Khwārizmī was (wrongly) named "the father of Algebra" by some scholars, but he was not manipulating equations or using numbers or symbols. Albrecht Heeffer [1621] wrote

> *In fact, there are no equations in Arabic algebra as we currently know them. However, some structures in Arabic algebra can be compared with our prevailing notion of equations. There is no separate algebraic entity in al-Khwārizmī's treatise which corresponds with an equation. In early Arabic algebra there are no operations on equations. On the other hand, there are operations on polynomials. Although linear problems are later approached algebraically by al-Karkhī, no rules are formulated for solving linear problems, as common in Hindu algebra.*

It is generally admitted that our word "algebra" comes from transformations of *al-jabr* and our word "algorithm" from *al-Khwārizmī*. Let us quote Heeffer again:

> *Much has been written about the origin of the names al-jabr and al-muqābala, and the etymological discussion is as old as the introduction of algebra into western Europe itself. The jabr operation is commonly interpreted as "adding equal terms to both sides of an equation in order to eliminate negative terms". It appears first in al-Khwārizmī's book in the first problem for the "equation" $x^2 = 40x - 4x^2$. In this interpretation the al-jabr is understood as the addition of $4x^2$ to both parts of the equation. The al-jabr or restoration operation consists of completing the original term $40x$. The second operation, al-muqābala, is generally understood as the addition of homogeneous terms in a polynomial.*

However, al-Khwārizmī's book had a great influence on the development of algebra in Western Europe, but not for solving (systems of) linear equations.

Many Greek and Arabic books were translated to Latin in Western Europe up to the 12th century. For the treatment of algebraic equations in the east and the west until the Middle Ages, see [652] by Chemla.

2.6 ▪ The Middle Ages

Algebra was introduced in medieval Europe through Latin translations of Arabic texts in the 12th century.

As we have written above, the *Liber Abaci* of Fibonacci greatly contributed to the adoption of the Hindu-Arabic number system in Europe. This book, written in medieval Latin, appeared first in 1202 with a second enlarged edition in 1228. Curiously enough, an English translation [2765] was only published in 2002. In the preface, Fibonacci explained how, in his travels, he has found the Hindu number system to be superior to all the other ones. He traveled to several countries around the Mediterranean sea, including North Africa where he was probably in contact with Arabic mathematics. Chapter 12 of the book is mainly devoted to the method of false position for solving one linear equation with 259 examples, and Chapter 13 to double false position, which is called *elchataym* by Fibonacci. Problems with several unknowns are also solved, but by ad hoc techniques and not by a general elimination method. For an analysis of these chapters, see [1574] by John Hannah. In the *Liber Abaci*, problems are stated in sentences and not written symbolically but some methods are described independently of a practical problem. The diffusion of the book in France in the 15th century was studied in [2843] by Maryvonne Spiesser. For the early history of algebra in Italy, see Raffaella Franci [1211] and Jeffrey A. Oaks [2359].

We must remember that when we speak of "equations" about medieval mathematics, this is almost an anachronism because the mathematicians were solving practical problems with given data and, generally, did not give algorithms for solving generic problems. This is why they had to give many examples to convince the reader that problems which look similar can be solved by using the same technique. Before the 17th century, techniques for solving algebraic problems used a single unknown identified as *the thing* or in Latin texts as *res* (or sometimes *radix*), *cosa* in Italian, and *coss* in German. The problems were described in rhetorical form. Then, the given problem was reformulated in terms of the *cosa* and an analytical method was applied to arrive at a value for the unknown. The other unknown quantities, which were not considered as unknowns in the modern sense, can then easily be determined.

In 1484, Nicolas Chuquet (c. 1445-1488), a French mathematician, wrote *Triparty en la Science des Nombres* [695], a book that was never published during his lifetime. Chuquet solved problems corresponding to systems of linear equations, used negative numbers, even negative powers, and notation for exponents. He was cited by people whose work we are going to discuss in a moment, the French Jean Borrel and Guillaume Gosselin as well as the English John Wallis. However, his work was overshadowed by *Larismethique* of Estienne de La Roche (1470-1530) published in 1520 and 1538. De La Roche was accused of plagiarism by Eugène Aristide Marre (1823-1918), a French linguist, when he published the *Triparty* with some comments. He stated that many parts of *Larismethique* were copied from the *Triparty*, but this conclusion was partly challenged by Heeffer in [1626].

One of the problems solved by Chuquet can be stated in modern terms as

$$\begin{aligned} a + 7 &= 5(b + c - 7) + 1, \\ b + 9 &= 6(a + c - 9) + 2, \\ c + 11 &= 7(a + b - 11) + 3. \end{aligned}$$

This problem was already used by Fibonacci, who gave a purely arithmetical solution using the sum $s = a + b + c$. Even though for us this problem has three unknowns a, b, c, Chuquet used only a second unknown in a (complicated) elimination algorithm; see [1626]. This technique was called *Regula quantitatis* or *Rule of Quantity*; see also [2844, 2845] by Spiesser. Slowly, the formulation and the solution of the problems moved from rhetoric algebra to symbolic algebra, which appeared at the beginning of the 17th century [1624].

Johannes Widmann (1462-1498), a German mathematician, published *Behende und Hübsche Rechenung auff allen Kauffmanschafft* (Agile and neat calculation in all trades) in 1489. It is believed to be the first book where the symbols $+$ and $-$ appeared in print. Before, words or abbreviations like $\overline{p}$ and $\overline{m}$ were used for addition and multiplication. These symbols were also used by Heinrich Schreiber, also known as Henricus Grammateus (c. 1495-c. 1525), in *Libellus de Compositione Regularum pro Vasorum Mensuratione. Deque Arte ista tota Theoreticae et Practicae* published in 1518.

Luca Bartolomeo de Pacioli, also known as Luca Pacioli or Luca di Borgo (c. 1447-1517), was a Franciscan monk and mathematician who taught mathematics to his friend Leonardo da Vinci (1452-1519). Da Vinci illustrated Pacioli's book *Divina Proportione*. Pacioli did not introduce any new techniques or methods, but his main work *Summa de Arithmetica Geometria Proportioni et Proportionalita*, written in Italian and originally published in Venice in 1494, greatly contributed to the popularization of algebra in Italy and abroad. The eighth chapter explains contemporary algebra in 78 pages with solutions of linear and nonlinear equations. In the algebra part of the book, Pacioli used $\overline{p}$ and $\overline{m}$ for addition and multiplication. A relation like $3x + 1 = 16$ was written $3\,co\,\overline{p}\,1\,ae\,16$, where co refers to "cosa" and ae to "aequalis" (equal). The part on geometry and some examples were largely inspired (if not copied) from some books by Piero della Francesca (c. 1415-1492) who, besides being a famous painter, was also a mathematician. We observe that Pacioli and della Francesca were born in the same city of Borgo San Sepolcro in Tuscany, and Pacioli is portrayed in one of della Francesca's paintings.

2.7 ▪ The 16th century

Christoff Rudolff (1499-1545), a German mathematician, considered the solution to linear problems by means of the second unknown in his *Behend und Hübsch Rechnung durch die Kunstreichen Regeln Algebre so Gemeincklich die Coss Genent Werden* published in 1525.

In the 15th and 16th centuries, mathematicians were more interested in finding methods for solving cubic and quartic polynomial equations in one unknown than solving systems of linear equations. In 1545, Gerolamo (or Girolamo) Cardano (1501-1576), an Italian physician and mathematician, published a solution for the cubic equation problem and the solution of one of his students, Ludovico Ferrari (1522-1565), for the quartic equation, in his book *Ars Magna* written in Latin. It is interesting to note that Cardano was far from modest or humble, since on the cover of the book, he presented himself as

> *Girolamo Cardano, outstanding mathematician, philosopher and physician, in one book, being the tenth in order of the whole work on arithmetic which is called the perfect book.*

The solution of the cubic equation was the subject of one of the most famous feuds in the history of mathematics. In fact, Scipione Dal Ferro (1465-1526) found an algebraic solution for equations which we write today as $x^3 + bx = c$, where b and c are positive. He did not publish it but communicated it to one of his students, Antonio Maria Del Fióre, and maybe to others. In 1535, while preparing for a problem-solving contest against Fióre, Niccolò Fontana (c. 1499-1557), known as Tartaglia (the stutterer), also found a way to solve Dal Ferro's problem, and maybe $x^3 + ax^2 = d$. In 1539, Cardano approached Tartaglia, and he was finally able to have Tartaglia give him his solution after probably swearing an oath that he would not publish the method. Tartaglia was upset by Cardano's publication, and this led to a 10-year dispute; see [1627].

The first ten chapters of *Ars Magna* deal with linear and quadratic equations and basic algebraic techniques. In Chapters 9 and 10, Cardano solved 2×2 linear systems. All the problems he considered were with specific given integer coefficients. He gave a rule for solving a system of two linear equations that he called *regula de modo*, which is equivalent to Cramer's rule for this particular case. Cardano did not use symbols, but mainly abbreviations like p $(+)$, m $(-)$, $\overline{q}d$ (square), and R (radix).

In 1553, Michael Stifel (1487-1567), a German monk and mathematician, produced a new edition of Rudolff's book, but this was not a simple editing exercise since he more than doubled its length by adding some material of his own. In his *Arithmetica Integra* Stifel demonstrated how the rules of algebra can be derived from the notions of logic stated by Proclus (412-485):

> *1) things which are equal to the same thing are equal to one another, 2) if equals be added to equals the wholes are equal, 3) if equals be subtracted from equals the remainders are equal, 4) the whole is greater than the part, and 5) things which coincide with one another are equal to one another.*

Stifel used the letters A, B, and C and proposed a notation for the powers and products of unknowns.

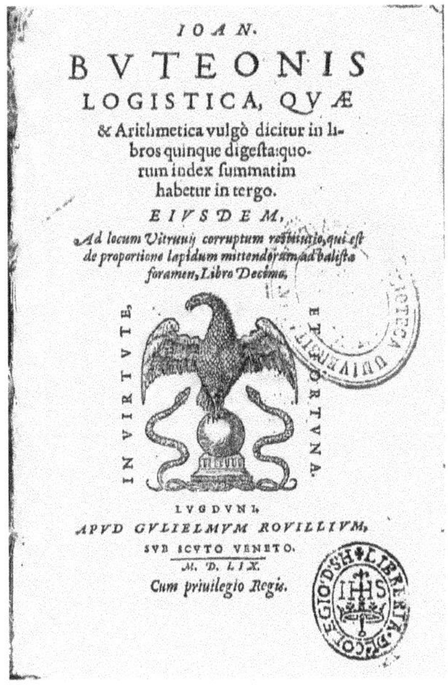

IOAN.
BVTEONIS
LOGISTICA, QVÆ
& Arithmetica vulgò dicitur in libros quinque digesta: quorum index summatim habetur in tergo.
EIVSDEM,
Ad locum Vitruuij corruptum restitutio, qui est de proportione lapidum mittendorum ad balistæ foramen, Libro Decimo.
IN VIRTVTE, ET FORTVNA.
LVGDVNI,
APVD GVLIELMVM ROVILLIVM,
SVB SCVTO VENETO.
M. D. LIX.
Cum priuilegio Regis.

Figure 2.3. *Buteo's Logistica*

Jacques Peletier du Mans (or Pelletier) (1517-1582 or 1583) was a French mathematician, physician, and poet; see [1851, 699, 700]. He wrote in French an algebra book in 1547 with an augmented edition in 1554. He cited Pacioli, Cardano, and Stifel and explained how to manipulate terms in equations. With a reference to Stifel, he used letters to denote several unknowns in linear equations. On pages 107-110 of the 1554 edition of his book, he solved a given linear system of order 3. However, the elimination method is described in words and not using equations.

Johannes Scheubel (1494-1570) published *Algebrae Compendiosa Facilisqve Descriptio* in 1551. It is believed to be the first book published in France in which the $+$ and $-$ signs appeared. However, they were not used with exactly their modern meanings.

The first algebra book in English, *The Whetstone of Witte* [...], was written in 1557 by Robert Recorde (c. 1512-1558), a Welsh physician and mathematician. In this book he introduced the equal sign. It represented not only the arithmetical equivalence of both sides, but symbolized the possible operations on the equation.

Jean Borrel, also known as Johannes Buteo (1492-c. 1570), a French mathematician, published *Logistica quae et Arithmetica Vulgò Dicitur in Libros Quinque Digesta* [515] in 1559 (see Figure 2.3). This is the most important work of Borrel. In this book in five parts, he solved systems of linear equations of order 3 and 4 by elimination. In the third part, under the title *De Regula Quantitatis*, he solved a linear system (see Figure 2.4), which in modern notation, is

$$\begin{aligned} 3x + y + z &= 42, \\ x + 4y + z &= 32, \\ x + y + 5z &= 40. \end{aligned}$$

As we can see in Figure 2.4, Borrel first took the second equation (if we can say so) multiplied by 3 and the first equation to eliminate x. Then, he took the third equation multiplied by 3 and the first equation to eliminate x again. He was left with two equations. Multiplying the second one

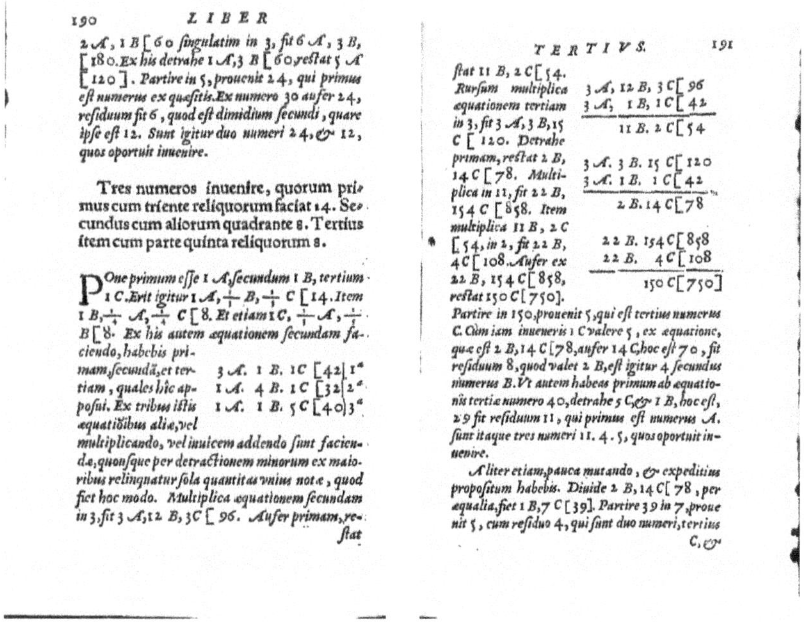

190 LIBER

2 A, 1 B [60 singulatim in 3, fit 6 A, 3 B, [180. Ex his detrahe 1 A, 3 B [60, restat 5 A [120]. Partire in 5, prouenit 24, qui primus est numerus ex quæsitis. Ex numero 30 aufer 24, residuum fit 6, quod est dimidium secundi, quare ipse est 12. Sunt igitur duo numeri 24, & 12, quos oportuit inuenire.

Tres numeros inuenire, quorum primus cum triente reliquorum faciat 14. Secundus cum aliorum quadrante 8. Tertius item cum parte quinta reliquorum 8.

Pone primum esse 1 A, secundum 1 B, tertium 1 C. Erit igitur 1 A, $\frac{1}{3}$ B, $\frac{1}{3}$ C [14. Item 1 B, $\frac{1}{4}$ A, $\frac{1}{4}$ C [8. Et etiam 1 C, $\frac{1}{5}$ A, $\frac{1}{5}$ B [8. Ex his autem æquationem secundam faciendo, habebis primam, secundã, et tertiam, quales hîc apposui. Ex tribus istis æquatiõibus aliæ, vel

3 A. 1 B. 1 C [42 | 1ª
1 A. 4 B. 1 C [32 | 2ª
1 A. 1 B. 5 C [40 | 3ª

multiplicando, vel inuicem addendo sunt faciendæ, quousque per detractionem minorum ex maioribus relinquatur sola quantitas vnius notæ, quod fiet hoc modo. Multiplica æquationem secundam in 3, fit 3 A, 12 B, 3 C [96. Aufer primam, restat

TERTIVS. 191

stat 11 B, 2 C [54. Rursum multiplica æquationem tertiam in 3, fit 3 A, 3 B, 15 C [120. Detrahe primam, restat 2 B, 14 C [78. Multiplica in 11, fit 22 B, 154 C [858. Item multiplica 11 B, 2 C [54, in 2, fit 22 B, 4 C [108. Aufer ex 22 B, 154 C [858, restat 150 C [750].

3 A, 12 B, 3 C [96
3 A, 1 B, 1 C [42
11 B. 2 C [54

3 A. 3 B. 15 C [120
3 A. 1 B. 1 C [42
2 B. 14 C [78

22 B. 154 C [858
22 B. 4 C [108
150 C [750]

Partire in 150, prouenit 5, qui est tertius numerus C. Cũm iam inueneris 1 C valere 5, ex æquatione, quæ est 2 B, 14 C [78, aufer 14 C, hoc est 70, fit residuum 8, quod valet 2 B, est igitur 4 secundus numerus B. Vt autem habeas primum ab æquationis tertiæ numero 40, detrahe 5 C, & 1 B, hoc est, 29 fit residuum 11, qui primus est numerus A. Sunt itaque tres numeri 11. 4. 5, quos oportuit inuenire.

Aliter etiam, pauca mutando, & expeditius propositum habebis. Diuide 2 B, 14 C [78, per æqualia, fiet 1 B, 7 C [39]. Partire 39 in 7, prouenit 5, cum residuo 4, qui sunt duo numeri, tertius C, &

Figure 2.4. *Buteo's Logistica, pages 190-191*

by 11 and the first one by 2, he eliminated y and obtained the solution for z. Clearly, he took the equations in this order to avoid working with negative numbers and he multiplied the equations to avoid rational numbers. Note that there is no sign for the addition which is implicit and just a "[" for the equality. On page 194, he solved a linear system of order 4 by the same elimination method. However, he did not completely eliminate all the unknowns but one. Nevertheless, he found the solution. But we can conclude that he was not completely mastering the elimination method, even though his exposition is quite clear and close to what we are doing today (see page 191 in Figure 2.4). The coefficients of his equations are always given positive integers and not letters.

Inspired by Peletier du Mans, Guillaume Gosselin (death around 1590) published *De Arte Magna, seu de Occulta Parte Numerorum quae et Algebra et Almucabala Vulgo Dicitur Libri Quatuor* [1413] in 1577. Note the reference to *al-muqābalah* in the title. Not much is known about Gosselin, except that he was from Caen (Normandy, France). In 1578, he published *Arithmétique*, a translation of a book by Tartaglia. Gosselin also referred to Stifel, Cardan, Peletier du Mans, Pierre Forcadel (?- c. 1576) who published *L'Arithmeticque* in 1556 and 1573, Borrel, and Pedro Nuñes (1502-1578), a Portuguese mathematician who published *Livro de Algebra en Arithmetica y Geometria* in 1567. Gosselin's algebra is mainly rethorical, even though he introduced abbreviations.

Curiously, in [1413], written in Latin, only every second page is numbered. On page 82, Gosselin solved a linear system of order 4 by elimination, which written in modern notation is

$$
\begin{aligned}
2x + y + z + w &= 34, \\
x + 3y + z + w &= 36, \\
x + y + 4z + w &= 52, \\
x + y + z + 6w &= 78.
\end{aligned}
$$

Note that the corresponding matrix is symmetric positive definite. This system was the same as the one considered and solved incompletely by Borrel [515, p. 194]. Gosselin eliminated what corresponded to x first, then successively y and z, and arrived at the equation $91w = 910$, which gave him $w = 10$. From this, he went backwards, computing z, y, and finally x. The solution is $x = 6$, $y = 4$, $z = 8$, $w = 10$. It is not clear if Gosselin considered what he was computing as unknowns in the modern sense or just as numbers which values he was looking for; see [1940] by Odile Kouteynikoff. Gosselin added that Borrel tried to solve this problem in three different ways without success. But, we have seen that Borrel got the solution. Like Borrel, Gosselin did not use the $+$ and $-$ signs. For an analysis of Gosselin's book, see [377] by Henri Bosmans (1852-1928) in 1906.

Stevin, in his book *L'Arithmétique*, written in French, and published in 1585, drew the conclusion that a number is a continuous quantity:

> *as continuous water corresponds to a continuous humidity, so does a continuous magnitude correspond to a continuous number.*

He stated that there are no *absurd, irrational, irregular, inexplicable or surd numbers*, meaning that one number is not different from any other, even if it is irrational. He also used some symbols in his treatment of polynomials and in *La Disme* in 1585 he introduced decimal fractions.

2.8 ▪ The 17th century

An important step in the history of algebra occurred at the end of the 16th and the beginning of the 17th centuries. This step was the advent of symbolic algebra; see Heeffer [1622]. Even though some people had already used letters to denote known quantities or geometric objects like points or curves and abbreviations for frequently used words like "square" or "cube," the first mathematicians to use letters for known quantities in equations were François Viète and René Descartes. Although they were not interested in solving linear equations, their works are important for the notation that was later used in algebra.

François Viète (1540-1603) was a French lawyer and counselor of French kings (King Henri III and King Henri IV). He did mathematics in his spare time. In 1591, he published *In Artem Analyticem Isagoge* written in Latin [3145], with a second edition in 1631. This means "Introduction to the art of analysis" and was supposed to be an introduction to a new algebra. This book was translated into French in 1631 by A. Vasset (whose real name was probably Claude Hardy (c. 1604-1678)); see Figure 2.5. Viète is credited with introducing the use of letters to represent known quantities in equations rather than solving problems with specific given values of the parameters. He used vowels for unknowns and consonants for known values.

Viète called his new algebra *logistice speciosa* (logistic in "species"). He worked with two kinds of algebra: traditional algebra with numbers, which he called *logistice numerosa*, and his new algebra. However, for Viète, unknowns and coefficients had a dimension, could not be identified only with numbers, and all the terms in an equation had to be homogeneous. There is clearly a geometric foundation for the *logistice speciosa*. Viète stated a formal language whose symbols can be manipulated according to stated rules and whose equations can be interpreted either geometrically or arithmetically. In 1591, in *Zeteticorum Libri Quinque*, Viète solved some of Diophantus' problems to show the efficiency of his "new algebra." Viète was convinced that this could solve all problems and he concluded the *Isagoge* with *nullum non problema solvere*.

Opera Mathematica, a collection of mathematical works by Viète on algebra, arithmetic, and geometry, was published in the Netherlands after his death by Franz van Schooten (1615-1660) in 1646. For analyses of Viète's work, see [1622] by Heeffer and [2360] by Oaks.

Viète had four students or followers: Jacques Aleaume (1562-1627), Marino Ghetaldi (c. 1568-1626) from Republic of Ragusa (now Dubrovnik, Croatia), Jean de Beaugrand

Figure 2.5. *French translation of Viète's book*

(c. 1586-1640), and Alexander Anderson (c. 1582-c. 1620), a Scottish mathematician. They contributed to transmit Viète's works.

Viète did not know the multiplication notation (given by William Oughtred (1574-1660), an English mathematician and clergyman, who published *Clavis Mathematicae* in 1631) or the symbol of equality.

Pierre Hérigone (c. 1580-1643), a French mathematician, popularized and extended Viète's notation in his *Cursus Mathematicus* in six volumes, published between 1634 and 1642.

René Descartes (1596-1650), a French soldier, philosopher, and mathematician, wrote the *Discours de la Méthode* (Discourse on the Method) in 1637. One of the appendices of this book was *La géométrie*. It was written in French and did not have a large circulation before 1664 and 1705. Descartes lived for a long while in the Netherlands. The book was translated to Latin by van Schooten in 1649 and 1659. It was in this essay that Descartes described his idea of a coordinate system, even though he was not using orthogonal axes and a fixed origin. This is considered the birth of analytic geometry (which was also considered by P. de Fermat).

What is important for our topic is the notation Descartes introduced, which is more or less the one we use today. He used the first letters in the alphabet for known parameters or coefficients, letters late in the alphabet for unknowns, and numerical superscripts for powers. Descartes' understanding of his symbolic language is different from Viète's notion of *logistice speciosa*. For Descartes, something like x^2 is a number and not an area. In fact, his goal in *La géométrie* was to solve polynomial equations of high order by intersection of algebraic curves. His book was not intended to solve practical problems of tradespeople, but to solve a mathematical problem. Even though it took almost twenty years for Descartes' ideas to spread, it is known that he had an influence on Isaac Newton. For an analysis of *La géométrie*, see [3180] by André Warusfel (1936-2016) and [2743] by Michel Serfati (1938-2018).

It may be an overstatement, but one can say that Viète put geometry in algebra and Descartes put algebra in geometry.

In between Viète and Descartes, Christophorus Clavius (1538-1612), a German Jesuit, astronomer, and mathematician, wrote an algebra book in 1608. It seems he was the first to use a decimal point. His book was published in Italy and, apparently, this was the first appearance in print of the symbols $+$ and $-$ in that country. However, this book belongs to the old tradition of algebra before Viète.

It is not the same for Girard, a French mathematician who mainly lived in Holland. He published *Invention Nouvelle en l'Algèbre* in 1629. He was a follower of Stevin and was also influenced by Viète. It is maybe in this book that we can see the first statement of the fundamental theorem of algebra (of course, without a proof). He wrote[9] (our translation)

All the algebraic equations have as many solutions as the value of the highest power.

Note that the first author of this result may be Peter Roth (1579-1617) in 1608. Girard also wrote rational powers as well as square roots of negative numbers but said that the corresponding solutions are "impossible."

The book *Artis Analyticæ Praxis* [1592] by Thomas Harriot (1560-1621), an English mathematician and astronomer, was published posthumously in 1631. He mainly used symbolic algebra with letters and symbols $+$, $-$, $=$ to explain how to manipulate polynomials. He is believed to be the first to use the symbols $>$ and $<$ for inequalities (see page 72 of his book).

John Wallis (1616-1703), an English clergyman and mathematician, was Oxford's Savilian Professor of Geometry from 1649 to 1703, and Oxford University archivist. In 1685, Wallis published his last great mathematical work, *A Treatise of Algebra, Historical and Practical: its Original, Progress and Advancement*. This book contains an historical account of the development of algebra. Clearly, Wallis read Pisano, Pacioli, Cardano, Borrel, and Viète, among others. The second edition, published in 1693, which is the second volume of his *Opera Mathematica*, was considerably enlarged. Wallis referred to a linear system solved by Borrel (Buteo) on page 65, and to Viète. He gave his own interpretation of the *logistice speciosa*, which he called *specious arithmetick*. He commented at length on Oughtred, Harriot, and Pell and asserted that Descartes (which he wrote "Des Cartes") had actually plagiarized Harriot's results. But this was demonstrated to be wrong; see [2854, p. 141ff.]. It may be worth mentioning that Wallis, in his book *Aritmetica Infinitorum* of 1655, coined the terminology *continue fractum* for continued fractions.

Johann Heinrich Rahn (1622-1676) was a Swiss mathematician. In 1659, he published *Teutsche Algebra*. He is credited with using the symbols $*$ for multiplication and $\div$ for division. He solved a linear system of order 3 with given numbers as coefficients by elimination on page 86 of his book. The book was translated into English by John Pell (1611-1685), who was sent to Zürich as a political agent by Oliver Cromwell (1599-1658), the Lord Protector of the British Isles. Pell substantially enlarged the book by almost 100 pages. Another translation was done by Thomas Brancker (1633-1676) in 1668.

In 1690, Michel Rolle (1652-1719), a French mathematician, published *Traité d'Algèbre, ou Principes Généraux pour Résoudre les Questions de Mathématique* (A treatise on algebra or general rules to solve mathematical problems). In Chapter IV, which is titled *De la Méthode* (pages 42-55), he described how to solve systems of linear equations. First, he described the several steps of the method in words. To eliminate one unknown he used the word "Dégager" which means to free, extricate, give off. Then, he proposed to organize the computation in two columns. The first one corresponded to our modern forward elimination and the second one to the backward step. Pages 44-47 gave an example of order 4; see Figure 2.6. He described

[9] *Toutes les équations d'algèbre reçoivent autant de solutions que la dénomination de la plus haute quantité le démontre.*

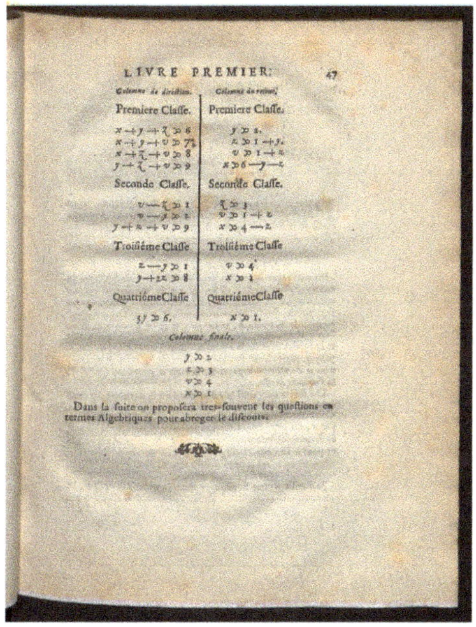

LIVRE PREMIER. 47

Colomne de direction.	Colomne du retour.
Premiere Classe.	Premiere Classe.
x + y + z ⊃ 6	y ⊃ 2.
x + y + v ⊃ 7.	z ⊃ 1 + y.
x + z + v ⊃ 8	v ⊃ 1 + z
y + z + v ⊃ 9	x ⊃ 6 − y − z
Seconde Classe.	Seconde Classe.
v − z ⊃ 1	z ⊃ 3
v − y ⊃ 2	v ⊃ 1 + z
y + z + v ⊃ 9	x ⊃ 4 − z
Troisiéme Classe	Troisiéme Classe
z − y ⊃ 1	v ⊃ 4
y + 2z ⊃ 8	x ⊃ 1
QuatriémeClasse	QuatriémeClasse
3y ⊃ 6.	x ⊃ 1.

Colomne finale.

y ⊃ 2
z ⊃ 3
v ⊃ 4
x ⊃ 1

Dans la suite on proposera tres-souvent les questions en termes Algebriques pour abreger le discours.

Figure 2.6. *Rolle's Traité d'algèbre, page 47*

several variants of the method, considered singular systems on pages 55-59, and discussed Viète's *logistice speciosa* on pages 59-65. On pages 61-62, he used letters for a generic right-hand side of his linear system of order 4, but, unfortunately, not for the coefficients. We observe that Rolle did not use the $=$ symbol.

The main figure of the second half of the 17th century for physics and mathematics in England is, of course, Isaac Newton (1643-1727). The Gregorian calendar was adopted in England in 1752. According to the calendar in use at that time, he was born on Christmas Day in 1642. The corresponding birth date in our present calendar is January 4, 1643. As the Lucasian Professor of Mathematics at the University of Cambridge since 1669, Newton had the duty of providing a manuscript of the ten best lessons of the year to the university library. He honored his duties only in 1684 with a work written in Latin, based on lecture notes on algebra for the period 1673 to 1683. William Whiston (1667-1752), who succeeded Newton as Lucasian Professor in 1702 (after Newton moved to London to become the warden of the Royal Mint), published the manuscript in 1707 with the title *Arithmeticæ Universalis: Sive de Compositione et Resolutione Arithmetica Liber*. Newton was very unhappy with this edition. It has been said that he refused to have his name appear as the author. A second edition, due to John Machin (1686-1751), was published in 1722 with some additions [2340] and still no author's name with the title *Arithmeticæ Universalis: Sive de Compositione et Resolutione Arithmetica Liber. Editio seconda, in qua multa immutantur et emendantur, nonnulla adduntur*, which means that many things have been changed and added. There were many later editions of this book, including seven editions in the 18th century. An English translation [2339] by Joseph Raphson (c. 1668-c. 1712), revised and corrected by Samuel Cunn (after Raphson's death), was published in 1720 under the title *Universal Arithmetick: or a Treatise of Arithmetical Composition and Resolution*, still without the author's name, even though Newton's name appears in the preface, with a second edition in 1728, and a third one in 1769 with notes by Theaker Wilder (1717-1778), an eccentric Anglo-Irish academic with expertise in mathematics and Greek. It is quite difficult to find exact details on Raphson's life. Up until now no documents that record his birth and death dates have been found. In several biographies it is written (c. 1648-c. 1715). We only know for sure that Raphson was proposed for Fellowship of the Royal Society at the meeting of November 27, 1689, by the renowned Edmond Halley (1656-1742), who had himself been elected in 1678 to the Royal Society at the young age of 22. Note that in the 1720 and 1728 editions, after the translation of Newton's work, as indicated in the complete title, there was the addition of *Dr. Halley's Method of Finding the Roots of Aequations Arithmetically* [2339, p. 259ff.]. Following some reviews of Raphson's works in English, and since his name appears in the 1712 list of fellows of the Royal Society but not in the 1713 list, he probably died in 1712. For other information see, for instance, [3032].

We observe that in his book Newton followed Descartes' notation by denoting the unknowns as $x, y, z, \ldots$ The elimination method for systems of equations is found on page 68 and following of the 1722 Latin edition [2340], and page 60 and following of the 1720 and 1728 English editions [2339]. Newton started with a session titled *Of the Transformation of two or more Æquations into one, in order to exterminate the unknown Quantities*. The method is described in words as

> [...] *those Æquations (two by two if there are more than two) are to be so connected, that one of the unknown Quantities may be made to vanish at each of the operations, and so produce a new Æquation.* [...] *And you are to know, that by each Æquation one unknown Quantity may be taken away, and consequently, when there are as many Æquations as unknown Quantities, all may at length be reduced into one, in which there shall be only one Quantity unknown.*

Then, Newton gave two ways to eliminate an unknown: *The extermination of an unknown Quantity by an Equality of its Values* and *The extermination of an unknown Quantity by substituting its Value for it.* The first one corresponds to obtain $x = \ldots$ from two equations and to equate them to obtain a new equation. The second one is to get x from one equation and to replace this value in the other equations. Note that Newton use the word *exterminatio*, that was translated as "extermination," and not "elimination."

However, Newton was mainly interested in polynomial equations, and almost all his examples led to polynomials and not to linear systems. Many of these examples arose from geometry or from mechanics.

Another giant of 17th century mathematics was the German polymath Gottfried Wilhelm Leibniz (1646-1716). In a letter he wrote to Guillaume François Antoine, marquis de l'Hôpital (or Hospital) (1661-1704) in 1693, he described how to use elimination in an overdetermined system of three equations with two unknowns with symbolic coefficients. We consider Leibniz's contributions to the theory of determinants in Chapter 3.

2.9 ▪ The 18th century

The English banker Nathaniel Hammond (?-1776), a follower of Wallis and Newton, published *The Elements of Algebra in a New and Easy Method* in 1742. There is a brief history of algebra in the introduction in which Hammond reproduced the wrong criticisms of Wallis against Descartes. He described an elimination method for two equations with two unknowns on page 142 and following. He considered three equations with three unknowns on page 219 and following, and four equations with four unknowns on page 296 and following, but did not describe a general method.

The *Treatise of Algebra* by Thomas Simpson (1710-1761), an English mathematician, had many editions, starting in 1745. The section starting on page 63 of the third edition in 1767 is titled *Of the Extermination of unknown quantities or the reduction of two or more equations to a single one*. As in many other books, the method is described in words and examples of small linear systems with given coefficients are discussed. However, on page 98, in Problem XXXVII, Simpson solved the problem

$$\begin{aligned} ax + by + cz &= p, \\ dx + ey + fz &= q, \\ gx + hy + kz &= r, \end{aligned}$$

with letters as coefficients. It was the same for the second edition in 1755.

Colin Maclaurin (1698-1746) was a Scottish mathematician who wrote *A Treatise of Algebra* [2111] in 1748. The elimination method for a linear system of order 3 is given in words on page 77, but he wrote that it is obvious to extend the rule to more equations and more unknowns. Theorems giving the formulas with symbolic coefficients for orders 2 and 3 are on pages 82-83. A manuscript dated 1729 contains the formulas for a linear system of three equations; see [1619].

Alexis Claude Clairaut (1713-1765), a French mathematician, astronomer, and geophysicist, wrote a book for beginners in 1746, *Elémens d'Algébre*. His purpose was educational. In the preface he wrote[10] (our translation)

> *I tried to give the rules of algebra in an order that the Inventors could have followed. No truth is presented here in the form of Theorems, all of them seem to be discovered by practicing the Problems that need or curiosity have made them undertake to solve.*

On page 91 and following he explained how to solve by elimination a linear system of order 3 with letters as coefficients. Contrary to many of his predecessors, to eliminate an unknown, say x, he divided by the coefficient of x to equate two values. Saying that the computations can become difficult, he introduced intermediate quantities which are, in fact, determinants of order 2. Clairaut's book had seven editions up to 1801. It was translated into German in 1752, Dutch in 1760, and English in 1766.

The general solution of a nonsingular linear system of any order using ratios of determinants was given in 1750 by the Swiss mathematician Gabriel Cramer (1704-1754) in an appendix of his book [758] about algebraic curves; see Chapter 3. It may seem that this was the end of the problem, but of course we know that it was not so because computing determinants of large order is painful by hand and costly when using any computing device.

At the end of the 18th and in the 19th centuries, many books were written when algebra started to be part of the curriculum. Most of these books explained some variants of elimination on small linear systems. Some described the formula arising from Cramer's rule without explaining the rule or how to compute determinants.

In 1765, the famous Swiss mathematician Leonhard Euler (1707-1783) wrote *Vollständige Anleitung zur Algebra*, a textbook about algebra. This book was first published in Russian in 1768 by the Academy of Sciences of Saint Petersburg, after Euler was back in Russia; see Section 10.19. The German version was published in 1770. It was translated to French by Jean III Bernoulli (1744-1807) in 1774 with notes from Bernoulli and additions by Joseph-Louis Lagrange (1736-1823). A translation into English, from the French version, was published in 1797. For a list of the translations, see [1625].

[10] *J'ai taché d'y donner les règles de l'algèbre dans un ordre que les Inventeurs eussent pû suivre. Nulle vérité n'y est présentée sous forme de Théorèmes, toutes semblent être découvertes en s'exerçant sur les Problèmes que le besoin ou la curiosité ont fait entreprendre de résoudre.*

Chapter IV of Section IV is devoted to the solution of linear systems. Euler started from two equations with two unknowns x and y, in Paragraph 606 and following, that he solved by equating the two values of x and computing y, but he did not relate the solution to determinants. In Paragraph 613, he explained how to solve a linear system of order 3 using an example with given coefficients. He did not write down the general formulas. For the general case, Euler wrote

> *If there were more than three unknown quantities to determine, and as many equations to resolve, we should proceed in the same manner; but the calculations would often prove very tedious.*

The French mathematician Étienne Bézout (1730-1783) is well known for his work on systems of polynomial equations, *Théorie Générale des Équations Algébriques* [311] in 1779, in which he proved that n polynomial equations in n unknowns have, at most, as many solutions as the product of their degrees. On pages 171-181, he described a method, based on determinants, for solving linear systems; see also [310]. For the life and work of Bézout, see [27, 28, 29] by Liliane Alfonsi, and for the history of elimination methods in polynomial systems, see [2478] by Erwan Penchèvre.

Bézout was, for some time, professor in several military schools, and he wrote textbooks for the students. In the second volume of his *Cours de Mathématiques à l'Usage du Corps de l'Artillerie* (Mathematics Course for the Artillery Corps) [313] published after his death in 1788, he explained on page 68 that a linear system may have no solution or an infinite number of solutions (underdetermined system). He solved two equations in two unknowns using the equality of values and gave general formulas on page 71. He considered three unknowns on page 72 using the same method as for two equations. The general method is described in words on page 75. Then, he explained another method only by examples. We will see this method explained later in Lacroix's book. On page 86, he came back again to underdetermined systems. Linear systems were also considered in [312] in 1781.

The *Encyclopédie ou Dictionnaire raisonné des sciences, des arts et des métiers* (Encyclopedia or reasoned dictionary of sciences, arts and crafts) was published from 1751 to 1775 under the supervision of the French philosopher Denis Diderot (1713-1784) and the mathematician Jean Le Rond d'Alembert (1717-1783). However, some people were thinking that since some topics were spread over many different articles, it was difficult to have an overview of a subject. The project of the *Encyclopédie Méthodique, ou par ordre de matières: par une société de gens de lettres, de savans et d'artistes* (Methodical Encyclopedia, or by order of subjects: by a society of writers, scholars and artists), written in a different way, organized by topics, was launched in 1782 by Charles-Joseph Panckoucke (1736-1798), a writer, publisher, and bookseller. In total there were 206 volumes, 125,000 pages, which were published over 50 years. There are three volumes about mathematics published under the supervision of the abbott Charles Bossut (1730-1814). The first volume [380] was published in 1784. In that volume there is an article *Élimination*, written by Bossut, in which we can read[11] (our translation)

> *ELIMINATION. We thus call an operation by which, given a number n of equations which contain a number n of unknowns, we find an equation which contains only one unknown: so that if we can solve this equation, we will know the unknown it contains; and going up, we will know the other unknowns. From this, eliminating a quantity is the same thing as removing it or making it disappear.*

[11] *ÉLIMINATION. On appelle ainsi une opération par laquelle, étant données un nombre n d'équations qui contiennent un nombre n d'inconnues on trouve une équation qui ne contient plus qu'une seule inconnue: de sorte que si l'on peut résoudre cette équation, on connaîtra l'inconnue qu'elle contient; et en remontant, on connaîtra les autres inconnues. De-là, éliminer une quantité signifie la même chose que faire évanouir, faire disparaître cette quantité.*

Examples of order 2 and 3 with letters as coefficients are given, and it is stated that this process can be used for any number of equations. The remaining part of the article is devoted to elimination in polynomial systems.

It is worth noting that Bossut is also the author of an essay on the history of mathematics [379] in two volumes that were published in 1802.

2.10 ▪ The 19th century

Sylvestre-François Lacroix (1765-1843) was a French mathematician whose textbooks had a large influence in the 19th century. In 1799, he published *Elémens d'Algèbre à l'Usage de l'École Centrale des Quatre-Nations*. There were at least 21 editions during the 19th century. It was translated to English by John Farrar (1779-1853) in 1818. In the fourth edition of 1804, Lacroix wrote in the preface, about algebra (our translation):

> *The language of this science bore the imprint of its weakness; the signs, very different from those we are using now, were not fixed; the quantities were sometimes represented by lines; a large number of technical terms adopted then, and so abandoned today that hardly anyone understand them, were taken from the writings of the ancients, and related, for the most part, to Geometry. This first period of the Elémens d'Algèbre produced in France those of Buteo and Jacques Pelletier du Mans. Viète, if not by introducing, at least by extending the use of letters to designate known quantities as well as unknowns, increased the power of calculation, and made science take a great step forward, although he still preserved and that he even created some strange denominations which seemed destined to slow down its progress. Descartes, by the notation of the exponents, completed the set of symbols; and his discoveries, those of Albert Girard and Harriot, on the theory of equations, made of Algebra a body of doctrine capable of being self-sufficient: such was the Algebra of Wallis. Finally, Newton's Universal Arithmetic appeared, which marked a third age much superior to the previous ones; but those who then wrote elementary books, could not achieve all that this work contained, and dragged themselves in the footsteps of Newton's predecessors, until Clairaut shed a new light on the principles of Algebra.*

On page 114, Lacroix described the elimination method in words:

> *One can, as we did in the previous numbers, take in one of the equations the value of one of the unknowns, as if all the others were known, and substitute this value in all the other equations, which will only contain the other unknowns after that. This operation, by which one of the unknowns is chased away, is called "elimination".*

On page 124 and following, he considered linear systems with letters as coefficients. He multiplied the equations by ad hoc values to eliminate one unknown. Then, he described a recursive method due to Bézout. For the linear system,

$$\begin{aligned} ax + by + cz &= d, \\ a'x + b'y + c'z &= d', \\ a''x + b''y + c''z &= d'', \end{aligned}$$

one multiplies the first equation by m, the second one by n, adds them, and subtracts the third one. It yields

$$(am + a'n - a'')x + (bm + b'n - b'')y + (cm + c'n - c'')z = dm + d'n - d''.$$

The coefficients of x and y vanish if we choose

$$am + a'n = a'',$$
$$bm + b'n = b''.$$

Recursively, one knows the solution of this system, and the value of z is

$$z = \frac{dm + d'n - d''}{cm + c'n - c''}.$$

Lacroix wrote down the complete formulas for z, x, and y and used them to introduce determinants.

Another popular textbook, *Éléments d'Algèbre* [385], was written by Pierre Louis Marie Bourdon (1779-1854), a French mathematician, in 1820. In the 9th edition (1843), the solution of linear systems by elimination is described on page 73 and following. This book had many editions and was translated twice into English in the United States, by Edward Coke Ross (1800-1851) [2600] in 1831, and by Charles Davies (1798-1876), a professor in the Military Academy, [815] in 1835. This textbook was used there for the rest of the century. In 1893, it was supplemented by *A Treatise on Algebra* by Charles Smith (1844-1916), an English mathematician.

In 1847, Benjamin Michel Amiot (1806-1878), Professor of Mathematics at the Collège Royal de Rouen in France, discussed all the possible cases for a linear system of order 3 [55].

Let us also mention the *Elémens d'Algèbre* in 1820 by Jean-Guillaume Garnier (1766-1840), a French mathematician who mainly worked in Belgium, and the *Manuel d'Algèbre ou Exposition Élémentaire de cette Science à l'Usage des Personnes Privées des Secours d'un Maître* (Manual of algebra or elementary exposition of this science for the use of persons deprived of the help of a teacher) by Olry Terquem (1782-1862), a French mathematician, in 1827. He explained how to solve linear systems with determinants but without any theoretical result about determinants. Cramer's rule was explained by Terquem [3025] in 1846.

Augustus De Morgan (1806-1871), a well-known British mathematician, wrote a textbook *Elements of Algebra, Preliminary to the Differential Calculus* in 1835. As in some other books, he only considered linear systems of order 2 and 3, explaining how to eliminate unknowns by substitution, subtraction, and equality of values. He also described Bézout's method of indeterminate multipliers that was explained by Lacroix. One can also cite the book [2473] by George Peacock (1791-1858) in 1830.

Isaac Todhunter (1820-1884), an English mathematician and historian of mathematics, published *Algebra for the Use of Colleges and Schools* in 1858. This book contains many exercises for students. He also wrote *Algebra for Beginners* in 1863. An example of a textbook in the USA is *Elements of Algebra* [710] by the Rev. Davis Wasgatt Clark (1812-1871) in 1843.

In Italy, Pietro Cossali (1748-1815) described the mathematical achievements from the emergence of algebra due to Fibonacci to the new research in the 18th century, see [747] in 1799. One can also cite the books [2495] in 1893 and [2496] in 1906 by Salvatore Pincherle (1853-1936).

In Germany, we can cite the elementary textbook *Algebra* by Johann Friedrich Maler (1714-1764) in 1774, [1909] by Diedrich August Klempt in 1880, and [2240] by Friedrich Meyer (1842-1898) in 1885.

We have seen that roughly up to the middle of the 18th century, only very small linear systems with integer coefficients were solved, either as exercises in books or as examples with problems related to tradespeople. As we will now see, this was going to change starting in the second half of that century.

One problem which led to some changes is the reduction of quadratic forms. Let us remind that a quadratic form on $\mathbb{R}^n$ is

$$q(x_1,\ldots,x_n)=\sum_{i,j=1}^{n} a_{i,j}x_ix_j,$$

with $a_{i,j}\in\mathbb{R}$. In modern matrix form, this is x^TAx, where $A=(a_{i,j})$ is symmetric and $x=(x_1,\ldots,x_n)^T$. The reduction of q consists, by successive elimination, in writing q as the weighted sum of the squares of n linear forms L_i in the variables $x_i,\ldots,x_n$, that is,

$$q(x_1,\ldots,x_n)=\sum_{i=1}^{n}\alpha_iL_i^2(x_i,\ldots,x_n).$$

For example,

$$\begin{aligned} q(x_1,x_2,x_3) &= 2x_1^2+2x_1x_2+4x_1x_3-x_2^2+6x_2x_3+2x_3^2,\\ &= 2(x_1+x_2/2+x_3)^2-3(x_2^2-4x_3/3)^2+8x_3^2/3. \end{aligned}$$

This problem was considered by Lagrange in 1759 when he was looking for a method of finding local extrema [1968]. He was concerned with determining sufficient conditions for a stationary point of a function of several variables to be a minimum or a maximum. Lagrange considered the cases of two and three variables and wrote that this can be extended to functions of four or more variables. What Lagrange did was replacing linear combinations of the original variables by new variables, a procedure equivalent, in modern terms, to the construction of a triangular matrix U and a diagonal matrix D such that $A=U^TDU$, from which it follows that $q=(Ux)^TD(Ux)$. This led Gilbert Wright (Pete) Stewart to write in 1991 in the NA-Digest,[12] that maybe we should call Gaussian elimination "Lagrange elimination."

The next important step was the development of the method of least squares, which grew out from problems in astronomy, geodesy, and topography. It is due to Adrien-Marie Legendre (1752-1833) in 1805 (dated 15 ventôse an XIII, that is, March 6, 1805) [2021]. He considered various measurements of the same quantity with errors expressed by $E=a+bx+cy+fz+\cdots$, where $a,b,c,f,\ldots$ are known coefficients and $x,y,z,\ldots$ are variables to be determined such that E is minimum. When having more measurements than variables, the error E can only be minimized. Legendre considered several equations $E,E',E'',\ldots$, and looked for the minimum of $E^2+[E']^2+[E'']^2+\cdots$. Then, he set to zero the partial derivatives of this sum with respect to the variables, which led to a system of linear equations that he solved, as he wrote, by the ordinary methods.[13]

This was the foundation of the method that Legendre called the *Méthode des moindres quarrés* (note that the spelling of the French word meaning "squares" has now changed to "carrés"). He applied his method to the computation of the length of the Paris meridian between Dunkirk and Barcelona, which was measured by Jean-Baptiste Delambre (1749-1822) and Pierre Méchain (1744-1804), starting from 1792, for defining the meter following a decision of the French Assemblée Constituante (Parliament) in 1790.

Robert Adrain (1775-1843), an American mathematician of Irish origin, published a paper [7] in 1808 in which he made use of the method of least squares for a problem of topography. He gave a justification of the principle of least squares and the normal distribution of errors. His work remained unknown in Europe although he later wrote several papers on this method for

[12] NA-Digest, June 30, 1991, volume 91, issue 26.

[13] *Il faudra résoudre ces équations par les méthodes ordinaires.*

the determination of the flattening of the Earth and the calculation of the axis of the terrestrial ellipsoid. However, since Adrain's library contained a copy of Legendre's memoir, in which only the normal distribution was lacking, he cannot be credited for the discovery of least squares.

The interest of Johann Carl Friedrich Gauss (1777-1855) for the method of least squares came from his activities in astronomy and in geodesy. He was asked to map the Kingdom of Hanover, the region in which he was living. The method of *triangulation*, which is used for establishing maps, goes back, at least, to the 16th century. The region to be mapped is covered by triangles. For refinement, one can use networks of smaller and smaller triangles. Surveying needs the measurements of lengths and angles of the triangles. Angles have to be adjusted to make the triangles consistent. One begins by measuring the length of a side of the first triangle (called the basis) and then, from each corner of the adjacent triangles, the other angles are measured. This procedure is easier and less subject to errors than measuring distances. Vertical angles have also to be measured for obtaining a map in a horizontal plane, an operation called *levelling*. Then, the usual trigonometric formulas (spherical trigonometry for long distances) give all the lengths. Thus, the topographer obtains a network of adjacent triangles. Obviously, the three angles of each triangle (instead of two) could be measured for safety, and a second basis also, thus leading to a system of linear equations with more equations than unknowns that has to be solved. This is where the method of least squares comes in. The various sources of errors also have to be taken into consideration, a procedure named *compensation* that is due to Gauss. For more on Gauss and least squares, see [605].

Let us digress about the least squares method for a while and explain this in modern terms. We consider N measurements $l_1, \ldots, l_N$ of n quantities $X_1, \ldots, X_n$ (angles or distances). It is necessary to correct the errors ν_i which affect them, due to the imperfections of the instruments and the experimental flaws. The measurements l_i are related to the $n + N$ unknowns $X_1, \ldots, X_n$ and $\nu_1, \ldots, \nu_N$ by means of nonlinear functions f_i. Let $l_i^* = f_i(X_1^*, \ldots, X_n^*)$ be the measurements computed from approximated values X_i^*, obtained from only some of the l_i's by an adequate procedure. Keeping only the first-order term in the Taylor series expansions of the functions f_i leads to a system of linear equations. Let A be the matrix of the partial derivatives of the f_i's with respect to the X_j^*'s, b the vector with components $b_i = l_i - l_i^*$, x the vector of the *compensations* $x_i = X_i - X_i^*$, and ν the vector of the ν_i. We have $Ax = b + \nu$. This system is solved in the least squares sense, that is, minimizing some norm of the difference of both sides. If the errors ν_i are normal random variables, centered and independent, and each l_i is a random variable with standard deviation σ_i, the most probable solution minimizes $\varepsilon^2 = p_1\nu_1^2 + \cdots + p_N\nu_N^2$, where $p_i = 1/\sigma_i^2$. Setting to zero all the partial derivatives of ε^2 with respect to the x_j's results in the system $A^T P\nu = 0$, where P is the diagonal matrix with diagonal entries p_i. It is a system of n equations with N unknowns. But $\nu = Ax - b$, and thus $A^T PAx = c$ with $c = A^T Pb$. This system is called the *normal equations*, and it only contains the n unknowns x_i. So in the end, one has to solve a linear system with a symmetric positive definite matrix, even though today, the least squares problem is generally solved differently.

Gauss did not use the least squares method only for geodesy. In 1809, he published *Theoria Motus Corporum Coelestium in Sectionibus Conicis Solem Ambientium* [1298] on the calculation of the orbits of celestial bodies. In this work he presented several important mathematical and statistical concepts, such as the method of least squares, the method of maximum likelihood, and the normal distribution; see [2877, 2901].

Denoting by $M, M', M''\ldots$, the measurements of an unknown quantity V, Gauss wanted to find its most probable estimator, that is, the one that maximizes the probability of the product $P = \varphi(M - V) \cdot \varphi(M' - V) \cdot \varphi(M'' - V) \cdots$, where $\varphi(x)$ is the probability law of the measurement errors of magnitude x. But ignoring what the function φ was, he first showed that the arithmetic mean is indeed the best estimator when changing both the probability density and the method of estimation. Starting from these principles, Gauss proved that the only law that rationalizes the choice of the arithmetic mean as an estimator is the normal law of errors $\varphi(x) = e^{-h^2x^2}/\sqrt{\pi}$, where h denotes the precision of the observations. Let us mention that even if Gauss was the first to suggest the normal law, it was Pierre Simon de Laplace (1749-1827) who first settled the problem of aggregating several observations in 1774, although his solution led to the Laplace distribution [1992]. Laplace was also the first to calculate the value of the integral $\int_{-\infty}^{\infty} e^{-t^2}\, dt = \sqrt{\pi}$ in 1782, thus providing the normalization constant for the normal distribution [1994, 1995]. Finally, it was Laplace who, in 1809, proved the central limit theorem, which emphasized the theoretical importance of the normal distribution [1997]; see [2758].

With the normal distribution as the probability law φ, the product P becomes $P \sim \exp(-h^2\Omega)$ with $\Omega = (\nu, \nu)$. As explained in [1490], the procedure used by Gauss is the following. Let

$$u_1 = \frac{1}{2}\frac{\partial\Omega}{\partial x_1}, \quad \Omega_1 = \Omega - u_1^2/r_{11},$$

with $r_{ij} = \sum_k a_{i,k}a_{j,k}$ and r_{11} the coefficient of x_1 in u_1. Then, clearly, $\partial\Omega_1/\partial x_1 = 0$ and so Ω_1 is independent of x_1. For the next step, we have

$$u_2 = \frac{1}{2}\frac{\partial\Omega_1}{\partial x_2}, \quad \Omega_2 = \Omega_1 - u_2^2/r_{22}.$$

Ω_2 is independent of x_1 and x_2, and so on. Thus, Gauss arrived at

$$\Omega = \sum_k u_k^2/r_{kk} + \rho,$$

where u_k is independent of $x_1, \ldots, x_{k-1}$ and ρ is a constant. Then he considered the expression

$$\exp(-h^2\Omega) \sim \exp(-h^2u_1^2/r_{11}) \cdots \exp(-h^2u_m^2/r_{mm})$$

and integrated it with respect to x_1 over the real line. Since the last $m-1$ terms are independent of x_1, they remain unchanged by integration and the first term becomes a constant. Continuing this process, Gauss found that the distribution of x_m is proportional to $\exp(-h^2u_m^2/r_{mm})$ with $u_m = r_{mm} - s_m$ with $s_m = \sum_k a_{m,k}b_k$, and he concluded that the most probable value of x_m, obtained by setting $u_m = 0$ is s_m/r_{mm}, that its precision is $h/r_{mm}{}^{1/2}$, and that the same result holds for the other unknowns since the result does not depend on their position.

At that time, as we have seen above, few mathematicians were using subscripts and superscripts, even though Leibniz had proposed a notation close to the one we use today. Gauss was writing linear systems under the form

$$\begin{aligned} v &= ax + by + cz + \cdots + l \\ v' &= a'x + b'y + c'z + \cdots + l' \\ v'' &= a''x + b''y + c''z + \cdots + l'' \\ &\vdots \end{aligned}$$

In 1810, he solved linear systems by what he called "common elimination" (*eliminationem vulgarem*). He gave the details of his method in [1299]; see [606, Chap. 9] and [1435]. Let us use the notation $\tilde{v} = (v, v', v'', \ldots)^T$, and a similar one for the other coefficients. The system above becomes

$$\tilde{v} = \tilde{a}x + \tilde{b}y + \tilde{c}z + \cdots + \tilde{l}.$$

Denoting by $[uv]$ the dot product (u, v), Gauss wrote the normal equations as

$$
\begin{aligned}
0 = [\tilde{a}\tilde{v}] &= (\tilde{a}, \tilde{a})x + (\tilde{a}, \tilde{b})y + (\tilde{a}, \tilde{c})z + \cdots + (\tilde{a}, \tilde{l}), \\
0 = [\tilde{b}\tilde{v}] &= (\tilde{b}, \tilde{a})x + (\tilde{b}, \tilde{b})y + (\tilde{b}, \tilde{c})z + \cdots + (\tilde{b}, \tilde{l}), \\
0 = [\tilde{c}\tilde{v}] &= (\tilde{c}, \tilde{a})x + (\tilde{c}, \tilde{b})y + (\tilde{c}, \tilde{c})z + \cdots + (\tilde{c}, \tilde{l}). \\
&\vdots
\end{aligned}
$$

Then he defined new quantities for the first step of the elimination method,

$$
\begin{aligned}
[\tilde{b}\tilde{b}, 1] &= [\tilde{b}\tilde{b}] - [\tilde{a}\tilde{b}]^2/[\tilde{a}\tilde{a}], \\
[\tilde{b}\tilde{c}, 1] &= [\tilde{b}\tilde{c}] - [\tilde{a}\tilde{b}][\tilde{a}\tilde{c}]/[\tilde{a}\tilde{a}], \\
&\vdots
\end{aligned}
$$

Dividing the first equation by $(\tilde{a}, \tilde{a})$, multiplying it by $(\tilde{a}, \tilde{b})$ and subtracting it from the second equation, the unknown x disappears, the coefficient of y becomes equal to $[\tilde{b}\tilde{b}, 1]$, that of z is $[\tilde{b}\tilde{c}, 1]$, and $[\tilde{b}\tilde{l}]$ is replaced by $[\tilde{b}\tilde{l}, 1]$. Then x can be eliminated similarly from the other equations. Continuing the same process for the other unknowns, Gauss arrived, in the end, at a triangular system of equations that can be directly solved. This is exactly Gauss' method as we know it today (when we do not have to pivot for stability, but Gauss' system was symmetric positive definite).

Thus, Gauss went beyond Legendre by connecting the method of least squares with probability theory and the normal distribution. He managed to complete Laplace's program of specifying a mathematical form of the probability density for the observations, depending on a finite number of unknown parameters, and defined a method of estimation that minimizes the error of estimation.

Then one of the most famous priority disputes in the history of science began. Ceres is the largest object in the main asteroid belt between the orbits of Mars and Jupiter. It was accidentally discovered on January 1, 1801, by Giuseppe Piazzi (1746-1826), the director of the Palermo observatory in Sicily. After observing it 24 times, Piazzi lost it. Gauss was able to localize it again from three observations, thus validating the method of least squares; see [3016]. He claimed that he already used his method (*Unser Prinzip*) in 1795 but that its publication was delayed by the Napoleonic wars. Although his name was mentioned by Gauss, Legendre was badly offended. In a letter to Gauss dated May 31, 1809, he rightfully stated that priority is only established by publication. Gauss did not answer him.

Gauss improved upon his predecessors, whose works we have reviewed before, by describing the elimination process in very general terms. Contrary to what was done before, he did not rewrite the equations after each elimination step, but just computed the coefficients with his bracket notation. He eliminated preserving the symmetry of the system. Note that the linear systems he had to solve had real coefficients and that their orders were larger than what was considered before. Doing these calculations by hand was cumbersome and prone to error. In 1809, Gauss calculated the corrections to the elements of the minor planet Pallas that has been discovered by the astronomer Heinrich Olbers (1758-1840) in 1802; see [1299]. He had to solve a linear system with 11 equations in 6 unknowns which led to a system of order 6 for the normal equations. Looking at the results of the first step of the elimination done by Gauss, we can observe that the absolute values are almost correct, but unfortunately, there are sign mistakes which are not due to the publisher; for details, see Section 1.4. For details on Gauss' calculations about Pallas, see [2038] by the American astronomer Armin Otto Leuschner (1868-1953) in 1922.

Gauss' least squares method was popularized by Friedrich Wilhelm Bessel (1784-1846), a student of Gauss, and Johann Jacob Baeyer (1794-1885) as soon as 1838 [303, 304], when cartographers adopted his notation, and the method became part of the curriculum of geodesists; see [2318]. In 1888, Wilhelm Jordan (1842-1899) published *Handbuch der Vermessungskunde* (Surveyor's manual), a book [1845] on geodesy in which he showed how to use elimination (that he called *Gauss'sche elimination*), but with a slightly different notation. It is thanks to that book that this method spread [33]. The book had 27 editions in German, published between 1888 and 1948 with additions by Paul Hermann Otto Eggert (1874-1944) and Max Kneissl (1907-1973). An English translation from German of selected portions of its 8th edition, Vol. 1 (Metzler, Stuttgart, 1935), was done by Martha W. Carta of the Army Map Service (Washington, DC) in 1962.

Johann Franz Encke (1791-1865), a German astronomer who became director of the Berlin observatory, published the *Berliner Astronomisches Jahrbuch* (astronomical almanac) for the years 1834, 1835, and 1836 in which there are sections titled *Über die Methode der kleinsten Quadrate* (On the method of least squares). In these three volumes, he explained Gauss' least squares method in great detail, including the method for solving the normal equations and how to organize the computation.

In 1840, Bessel published in the *Astronomische Nachrichten* some formulas that had been communicated to him by Carl Gustav Jacob Jacobi (1804-1851) for the solution of the normal equations in the method of least squares when the number of unknowns amounted only to three. These formulas were based on the use of determinants. A proof was given by Hugo Seeliger (1849-1924) in 1873. As it was explained in 1874 by James Whitbread Lee Glaisher (1848-1928), an English mathematician and astronomer, in the *Monthly notices of the Royal Astronomical Society*, the aim was to reduce the number of operations. For a (brief) tutorial history of least squares, see [2349].

From these last publications, it is clear that at that time, people doing least squares computations were using logarithms and exponentials to avoid doing lengthy multiplications. Hence, it was important to try to minimize the number of logarithms that had to be computed from tables.

Gauss' memoirs on least squares were translated into French by Joseph Louis François Bertrand (1822-1900) in 1855 [299], contributing also to spreading the least squares method. However, some people were looking for other possibilities. In 1835, Augustin-Louis Cauchy (1789-1857) wrote a paper [571] on interpolation that was published in 1837. He was interested in the interpolation problem when the values to be interpolated are known only approximately. He constructed a method for which he claimed that *the largest error is the smallest possible.* In modern terms, Cauchy's method amounts to solve $K^T Ax = K^T b$ by elimination, using a matrix K such that $(K)_{i,j} = \pm 1$.

In 1853, Irénée-Jules Bienaymé (1796-1878), a statistician who was a follower of Laplace and a proponent of least squares (see [317]), started a controversy with Cauchy. In [318], he wrote[14] (our translation)

> *Lately, the attention of several people went to M. Cauchy's interpolation method that was published in 1835 and it seems they regard this method has having the same advantages as the well celebrated method of least squares. It will be annoying that they could be misled by what had been said of the two methods because they differ completely.* [...] *it will be allowed to show that this method* [Cauchy's] *is nothing else than ordinary elimination* [...]

[14]*Depuis quelque temps, l'attention de plusieurs observateurs s'est portée sur une méthode d'interpolation que M. Cauchy a publiée en 1835, et il semble qu'on ait regardé cette méthode comme ayant quelque chose d'analogue aux avantages de la célèbre méthode des moindres carrés. Il serait fâcheux que les observateurs fussent trompés à cet égard par ce qui a pu être dit des deux méthodes, car elles diffèrent complétement.* [...] *il sera permis de montrer que ce procédé* [de M. Cauchy] *n'est qu'une modification de l'élimination ordinaire* [...]

Then, Bienaymé showed, by manipulating the equations, that Cauchy's method is solving $K^T Ax = K^T b$ by elimination and that Cauchy's choice of K cannot minimize the residual norm. We observe that Bienaymé claimed that the number of arithmetical operations for solving a dense linear system of m equations in m unknowns is $(m-1)(2m^2+5m+6)/3$. Another paper [319] about the 1853 controversy was only published in 1867.

William Chauvenet (1820-1870), an American professor of mathematics, astronomy, navigation and geodesy, published a textbook [647] on least squares in 1868, contributing to the popularization of the method in the United States. This was in fact a reprint of an appendix from his book *Manual of Spherical and Practical Astronomy*. Chauvenet's explanation of the elimination method is not much different from Encke's. After his death, the Mathematical Association of America established the Chauvenet Prize in his honor. Also of interest is the book [711] by Colonel Alexander Ross Clarke (1828-1914) in 1880.

The names "Gauss' method," "Gauss' elimination," and "Gaussian elimination" were rapidly adopted by the people using least squares; see, for instance, Chauvenet in 1868, Mansfield Merriman (1848-1925) [2204] in 1877, William Woolsey Johnson (1841-1927) [1836] in 1892, Asaph Hall (1829-1907) [1549] in 1893, and Dana Prescott Bartlett (1863-1935) [211] in 1915.

Since Gauss solved linear systems by elimination in a way clearly different from his predecessors, by considering only the coefficients obtained during the elimination process and by not rewriting the equations at each step of the algorithm, it seems to be fair to name the method we are using today "Gaussian elimination," even though the basic technique is much, much older.

During the second half of the 19th century more and more geodetic surveys were done in different parts of the world. Using least squares, the linear systems to be solved became larger and larger. This led the "human computers" to try to improve the techniques used to perform the computations.

According to the obituary notice in *The Washington Times* of June 28, 1913, Myrick Hascall Doolittle (1830-1913) had been working at the U.S. Coast and Geodetic Survey in Washington, DC, for 35 years. Among other duties, he was in charge of solving by hand the normal equations arising from the least squares method. In the Computing Division, whose acting director was Charles Anthony Schott (1826-1901) in 1878, there were three other "computers" and some temporary assistants.

In the report for the year 1878 [935], published in 1881, Schott wrote on page 93:

> *Paper No. 3 is submitted by special direction of the Superintendent, who desired that a complete account of the method of solution of large numbers of normal equations be presented, and I have accordingly requested Mr. M.H. Doolittle, of the Computing Division, to draw up this paper. This method of solution of linear normal equations is rigorous and direct, and with systematic arrangement merely involves mechanical labor. It may be applied to a far greater number of equations than we have had hitherto occasion to solve (52). Its essential peculiarity consists in first obtaining an approximate solution effected by operating only with a small number of significant figures of the coefficients involved so as to bring into use Crelle's tables. Logarithms are altogether dispensed with.*
>
> *An approximate value for the first unknown quantity having been obtained from the first equation, and expressed in terms of the other unknown quantities contained in that equation, it is substituted in the second equation, and from this an approximate value of the second unknown quantity is found in terms of the remaining unknown quantities. Next follows a substitution in the third equation, and so on to the end. Approximate values of all unknown quantities having thus been obtained, a rigorous substitution in the original equations is made, and the new and necessarily small*

residuals are treated a second time in a manner similar to the first described process, and the small corrections now found are added to the first approximations. It is not likely that a third treatment will ever be needed. To the paper Mr. Doolittle has added some remarks pertinent to the proper solution and arrangement of conditional equations, especially for the case involving small angles. Practically, two computers are employed in the solution of any large number of equations, and they compare their results at the close of each day's work, making the commission of any mistakes, without speedy detection, almost impossible.

From this, we learn that Doolittle had set up a new way to organize the computations, that he computed with only a few decimal digits using Crelle's multiplication tables and that he used *iterative refinement* to improve the approximate solution. August Leopold Crelle (1780-1855) was a German mathematician who was the founder of the *Journal für die reine und angewandte Mathematik*, also known as Crelle's Journal. In 1820, he published multiplication tables [762] in two volumes containing integer products up to 1000×1000. They were later also published in other countries.

Doolittle's Paper No. 3 is only six pages long. He described his technique by using the following example

$$\begin{aligned}
5.4237w + 2.1842x - 4.3856y + 2.3542z - 3.6584 &= 0,\\
2.1842w + 6.9241x - 1.2130z + 2.8563 &= 0,\\
-4.3856w + 12.8242y + 3.4695z + 8.7421 &= 0,\\
2.3542w - 1.2130x + 3.4695y + 7.1243z + 0.6847 &= 0.
\end{aligned}$$

Doolittle managed the computation by using several tables. Tables A and B (see, respectively, Tables 2.1 and 2.2) correspond to a forward elimination. They were probably written on different sheets of paper. For doing the elimination, one has to alternate between the two tables. The first columns of these tables give the number of the line and the second columns give the step number, that is, the order of the operations. In column A.3 we have the negatives of the inverses of the pivots; for instance, -0.184 is the rounded value of $-1/5.424$. Columns A.4 to A.8 correspond to the unknowns w, x, y, z and the negative of the right-hand side. The values of the coefficients, rounded to three significant digits, are written in line A.1. For line A.2, the negative of the inverse of the pivot is computed, put in column 3, and multiplied by the coefficients in columns 5 to 6. Note that this already introduces rounding errors since all the operations are rounded. Line A.2 gives w as a linear function of x, y, z. Now, we move to Table B.

Table 2.1. *Doolittle's Table A*

1	2	3	4	5	6	7	8
			w	x	y	z	
1	1		$+5.424$	$+2.184$	-4.386	$+2.354$	-3.658
2	2	-0.184	$w =$	-0.401	$+0.807$	-0.433	$+0.6731$
3	5			$+6.048$	$+1.759$	-2.157	$+4.323$
4	6	-0.165		$x =$	-0.29	$+0.356$	-0.7133
5	10				$+8.77$	$+5.996$	$+4.538$
6	11	-0.114			$y =$	-0.684	-0.5173
7	16					$+1.236$	$+0.704$
8	17	-0.809				$z =$	-0.57

Table 2.2. *Doolittle's Table B*

1	2	3	4	5	6
		x	y	z	
1	3	+6.924		−1.213	+2.856
2	4	−0.876	+1.759	−0.944	+1.467
3	7		+12.82	+3.470	+8.742
4	8		−3.54	+1.900	−2.950
5	9		−0.51	+0.626	−1.254
6	12			+7.124	+0.685
7	13			−1.019	+1.584
8	14			−0.768	+1.539
9	15			−4.101	−3.104

The coefficients of x, y, z and the constant term in the second equation are written in line B.1. The numbers in line A.1, columns 5 to 8, are multiplied by the number in line A.2, column 5, and the results put in line B.2, columns 3 to 6. Lines B.1 and B.2, columns 3 to 6, are added, and the results are put in Line A.3, columns 5 to 8. This yields a new pivot 6.048 whose inverse rounded and with a negative sign is put in line A.4, column 3.

Then, Doolittle proceeded in the same way as in the first phase. The coefficients in line A.3 are multiplied by the negative of the inverse of the pivot and the results put in columns 6 to 8 of line A.4. The rounded coefficients of y, z and the constant term of the third equation are put in Line B.3, columns 4 to 6. The coefficients in line A.1 (resp., line A.3), columns 6 to 8, are multiplied by the coefficient in line A.2 (resp., line A.4), column 6, and put in line B.4 (resp., line B.5), columns 4 to 6. Then, lines B.3 to B.5 are added, and the results are put in line A.5, columns 6 to 8.

The next phase eliminates y in the same way and, finally, gives the value of $z = -0.57$ at the bottom right of Table A. We observe that when Doolittle had to do multiplications, he multiplied several numbers by the same number. This was nice when using Crelle's tables because all the results were on the same page; the human computer just had to know where to put the decimal point. The numbers in Table B are just the results of auxiliary computations.

If we look at the numbers in lines A.1, A.3, A.5, and A.7, columns 4 to 7, they are (approximately due to the rounding) what is obtained for the matrix U in an LU factorization of the matrix A of the linear system with the lower triangular matrix L having a unit diagonal,

$$A = LU = \begin{pmatrix} 1 & 0 & 0 & 0 \\ 0.40271 & 1 & 0 & 0 \\ -0.8086 & 0.29219 & 1 & 0 \\ 0.43406 & -0.35753 & 0.6853 & 1 \end{pmatrix} \begin{pmatrix} 5.4237 & 2.1842 & -4.3856 & 2.3542 \\ 0 & 6.0445 & 1.7661 & -2.1611 \\ 0 & 0 & 8.762 & 6.0045 \\ 0 & 0 & 0 & 1.2149 \end{pmatrix}.$$

Doolittle used two other Tables C and D (see, respectively, Tables 2.3 and 2.4) for the backward phase of the solve. Table C is obtained from Table A, lines A.2, A.4, and A.6. The line D.1 is obtained from column 8 of Table A, lines 2, 4, 6, 8, corresponding to the right-hand side. The numbers in the last column of Table C are multiplied by the value $z_1 = -0.57$ and the results put in line D.2, columns w, x, and y. The sum of the two numbers in lines D.1 and D.2, column y gives an approximate value $y_1 = -0.127$ put in line D.3, column y. This value is multiplied by the numbers in Table C, column y, and the results are put in line D.3, columns w and x. The sum of the three numbers in column x, lines D.1, D.2, and D.3, which is the approximate value

Table 2.3. *Doolittle's Table C*

		x	y	z
1	-0.184	-0.401	$+0.807$	-0.433
2	-0.165		-0.29	$+0.356$
3	-0.114			-0.684
4	-0.809			

Table 2.4. *Doolittle's Table D*

	w	x	y	z
1	$+0.6731$	-0.7133	-0.5173	-0.57
2	$+0.2468$	-0.2029	$+0.3899$	$= z_1$
3	-0.1024	$+0.0368$	-0.127	$= y_1$
4	$+0.3524$	-0.879		$= x_1$
	$+1.17$			$= w_1$

x_1, is put in line D.4. The next step is to use column x of Table C to compute w_1 by summing the numbers in column w, lines B.1 to B.4.

The approximate solution is $w_1 = 1.17$, $x_1 = -0.879$, $y_1 = -0.127$, and $z_1 = -0.57$. The "exact" solution is $w = 1.2079$, $x = -0.89791$, $y = -0.10743$, and $z = -0.59580$.

Let D_U be a diagonal matrix whose diagonal entries are those of U defined above. Then, what we have in Table C is $-D_U^{-1}U$ (note that because of the symmetry, $D_U^{-1}U = L$) and the first line of Table D is the transpose of $D_U^{-1}L^{-1}b$, where b is the right-hand side in $Au = b$.

Doolittle knew that his approximate solution had to be improved. He used what we now call *iterative refinement.* He computed the residual with more accuracy, writing

> *These values must be substituted in the original equations, and a sufficient number of decimal places must now be used to insure the requisite degree of accuracy.*

He used two other tables E and F to obtain the solution of the system with the residual as the right-hand side. After adding the solution to the previous approximation he obtained $w = 1.205$, $x = -0.8966$, $y = -0.1089$, and $z = -0.5939$. The residuals are, respectively, $1.7628\ 10^{-3}$, $-5.1364\ 10^{-4}$, $-3.6057\ 10^{-4}$, and $-1.3648\ 10^{-4}$. The first value is different from what was given by Doolittle, who, apparently, made a small mistake.

At the end of the description of the method, Doolittle wrote

> *As the multiplication is performed by Crelle's Tables, no multiplier is allowed to extend beyond three significant figures. Other numbers may be extended to four; but it would be a waste of time to extend any number farther, except in the process of substitution for the determination of residuals. By this process, Mr J.G. Porter*[15] *and myself have solved in five and one-half days, or 36 working hours, with far greater than requisite accuracy, 41 equations containing 174 side coefficients counting each but once, or 430 terms in all. Each of us made a complete solution, duplicating the work, and making frequent comparisons in order to avoid errors. For the sake of perspicuity in explanation and convenience in printing, I have here made some slight*

[15]Jermain Gildersleeve Porter (1852-1933) was an American astronomer employed by the U.S. Coast and Geodetic Survey in 1878, where he spent six years in the Computing division, who assisted Doolittle several times, for example, in the solution of numerous normal equations arising from triangulations.

> *departures from actual practice. For instance, in the solution of a large number of equations, it would be inconvenient to pass the eye and hand out to a vertical column of reciprocals; and they are better written in an oblique line near the quantities from which they are derived and with which they are to be employed.*

In the remaining part of the note, Doolittle explained how to add new equations without recomputing everything from scratch. This is a kind of bordering method. He also gave some hints on how to order the equations in survey problems.

Doolittle's method was quickly adopted by geodesists; see, for instance, the book [3275] by Thomas Wallace Wright (1842-1908), who worked as assistant engineer at the U.S. Lake Survey, in 1884. In this book, Wright wrote on page 156

> *Normal equations maybe solved by the ordinary algebraic methods for the elimination of linear equations or by the method of determinants. When, however, they are numerous the methods of substitution and of indirect elimination, both introduced by Gauss, are more suitable.*

He used Gauss's bracket notation. What he called the *indirect elimination* is iterative refinement. Doolittle's technique is explained on pages 167-174. On pages 172-173, Wright wrote

> *The labor involved in solving a series of normal equations, and the consequent time employed, increases enormously with an increase in the number of normal equations. To any one who has never been engaged in such work it will seem out of all reason. Thus Dr. Hügel,*[16] *of Hessen, Germany, states that he has solved 10 normal equations in from 10 to 12 hours, using a log. table, but that 29 equations took him 7 weeks.*
>
> *The following are examples of rapid work: Gen. Baeyer in the Kürstenvermessung (Vorwort p. vii.) mentions that Herr Dase solved 86 normal equations between the first of June and the middle of September; and Mr. Doolittle, of the U.S. Coast Survey, solved 41 normal equations in 5 1/2 days, or 36 working hours.*
>
> *A great deal depends, so far as speed is concerned, on the form of solution and on the mechanical aids used. With a machine or with Crelle's tables much better time can be made than by the logarithmic method, which is by far the most roundabout.*

We see that computing speed was already quite a concern in 1884.

A second edition of Wright's book, with John Fillmore Hayford (1868-1925) as a co-author, was published in 1906 [3276]. Hayford was the successor of Schott as head of the Computing Division of the U.S. Coast and Geodetic Survey. How to solve the normal equations is explained on pages 106-121. Doolittle's example is reproduced on pages 119-120. After Doolittle's retirement, the method continued to be used at the Coast and Geodetic Survey; see, for instance, [5]. It was also used in other areas, for example, in statistics [782, 1717]. A simplified Doolittle method was published by Frederick Vail Waugh (1898-1974) [3194] in 1935. For more about Doolittle's method, see the paper by Paul Sumner Dwyer (1901-1982) [1035] in 1941.

Doolittle's technique was also mentioned in the book [1981] by Charles Jean-Pierre Lallemand (1857-1938), a French geophysicist, in 1912. So, it was not only known in the USA, Doolittle was also cited in the book [2034] published in 1921 by Ora Miner Leland (1876-1962) in 1921. In one appendix of this book, Leland wrote about least squares works

> *Prior to 1805, 22 titles were found. From that time on, averaging by decades, the rate of publication increased steadily from about two per year in 1810 to about ten*

[16]Original footnote in the book: *General-Bericht über die europäischen Gradmessung, 1867, p. 109.*

> *per year in 1870. Altogether, 408 titles were listed up to 1875. Of these, 153 were published in Germany, 78 in France, 56 in Great Britain, and 34 in the United States, the remaining ones being scattered over eight countries. The German language was used in 167 instances, French in 110, and English in 90.*

We observe that none of the works on least squares we have reviewed above used the concept of matrix that was already known at that time since it was introduced by Arthur Cayley (1821-1895) in 1858. But maybe the geodesists were not looking too closely at the works of mathematicians (and vice versa).

In 1888, the abbot Bernard Isidore Clasen (1829-1902), from Luxembourg, proposed a method for solving systems of linear equations that is known under the name of *equal coefficients method* [712].[17] It is followed, in the same journal, by a note [2131] of the Belgium mathematician Paul Mansion (1844-1919), professor at the University of Ghent, who was quite regularly corresponding with Clasen; see [1892].

Clasen considered the linear equations denoted $X_1, Y_1, Z_1, \dots$

$$\begin{aligned} X_1 &= a_1 x + b_1 y + c_1 z + d_1 s + \cdots + m_1 k = 0, \\ Y_1 &= a_2 x + b_2 y + c_2 z + d_2 s + \cdots + m_2 k = 0, \\ Z_1 &= a_3 x + b_3 y + c_3 z + d_3 s + \cdots + m_3 k = 0, \text{ etc.} \end{aligned}$$

Then, he eliminated x and y with a linear combination of the first two equations

$$\begin{aligned} X_2 &= b_2 X_1 - b_1 Y_1 = (a_1 b_1 - a_2 b_1) x + (c_1 b_2 - c_2 b_1) z + \cdots = 0, \\ Y_2 &= a_1 Y_1 - a_2 X_1 = (a_1 b_1 - a_2 b_1) y + (a_1 c_2 - a_2 c_1) z + \cdots = 0. \end{aligned}$$

The coefficient of x in the equation X_2 is the same as the coefficient of y in Y_2, hence the name of the method. Denoting by m this coefficient, multiplying Z_1 by m, Y_2 by $-b_3$, X_2 by $-a_3$ and adding gives $Z_2 = Rz + Ss + \cdots = 0$, and so on until the last equation. In the case of a system of order 3, the system has been transformed into a diagonal one that can be immediately solved. Then, Clasen explained the fundamental principle of his method. He realized the method is, as he wrote,[18] already tedious to solve a system of four equations and impractical for a system of six, and he proposed an abridged variant with a numerical example. The case of a zero coefficient was discussed. Finally, he also showed how his method could be used to compute determinants.

To a modern reader, the abbott's method may seem similar to the Gauss-Jordan elimination devised by the geodesist W. Jordan [1845], but both methods were published the same year; it is likely that Clasen was not aware of Jordan. Clasen's method was considered by Rudolf Mehmke (1857-1944) [2188] in 1930 and also explained in [1368], published in 1953.

2.11 ▪ The 20th and 21st centuries

The progress of mechanical calculators at the end of the 19th century (see Chapter 7) allowed to compute the solution of larger and larger problems for geodetic surveys. Hence, there was a need for efficient methods to solve linear systems. Gaussian elimination was well established as a technique for solving linear systems. What was done later was devising variants of the algorithm to reduce the computing times whatever was the computing engine (Doolittle's method is an early example) and deriving error analyses to show that the method computes a good approximation of the solution.

[17]The paper can be read at the address: https://play.google.com/books/reader?id=qzNOAAAAMAAJ&pg=GBS.RA2-PP14&hl=en (accessed August 2021).

[18]*fastidieuse déjà pour résoudre un système de quatre équations et impraticable pour un système de six.*

In 1907, Otto Toeplitz (1881-1940), a German mathematician who was at that time working in Göttingen, proved, using the language of bilinear forms, that a symmetric positive definite matrix can be factorized as LL^T, where L is a lower triangular matrix; see [3056]. However, this is not so clearly stated in that paper and he did not provide a practical algorithm since the entries of the transformation were given by determinants. This result was cited (wrongly since Toeplitz considered only real matrices) by Cyrus Colton MacDuffee (1895-1961) [2106] on page 80 in 1933. An analysis of this result was given by Olga Taussky (1906-1995) and John Todd (1911-2007) in [3011].

In 1909, Issai Schur (1875-1941) published a paper [2717] in which his aim was to prove results about eigenvalues. In his proofs and using matrices, he reduced a matrix A to lower triangular A form with a unitary matrix P (citing Ludwig Stickelberger (1850-1936), a Swiss mathematician, [2890] in 1877), obtaining $P^*AP = \mathsf{A}$. Then he observed that

$$P^*APP^*A^*P = P^*AA^*P = \mathsf{A}\mathsf{A}^*.$$

Therefore, he obtained the factorization of the symmetric positive definite matrix on the left.

Another method for solving the least squares normal equations was introduced by Cholesky. André Louis Cholesky (1875-1918) was an army officer working for the French Army Geographic Service; see Section 10.12. He had to do surveying and establish triangulations. Thus, he was led to systems of linear equations with more equations than unknowns that he solved by the method of least squares. Cholesky never published his method during his lifetime. It was published by a fellow officer, Major Ernest Benoît (1873-1956), in the *Bulletin Géodésique* [266] in 1924.

In 2005, Cholesky's family gave the documents in its possession to the *École Polytechnique*, where André Louis had been a student from 1895 to 1897. One of his grandsons, Michel Gross, and Claudine Billoux, the archivist of the school, asked C.B. to classify these documents. They rapidly found an unknown and unpublished manuscript titled *Sur la résolution numérique des systèmes d'équations linéaires* (On the numerical solution of systems of linear equations). It contains 8 pages in which Cholesky described his method. It was dated December 2, 1910, and since, contrarily to other Cholesky's manuscripts, very few words were crossed out, it is almost certain that the method was obtained before 1910, probably between 1902 and 1904 when Cholesky was in Northern Africa. The manuscript is entirely reproduced in [439, 441], and in [464], where it is translated into English; see [440, 464] for details.

Cholesky first considered the system (in his notation)

$$\text{(I)}\left\{\begin{array}{lcl} \alpha_1^1\gamma_1 + \alpha_2^1\gamma_2 + \alpha_3^1\gamma_3 + & \cdots & +\alpha_n^1\gamma_n + C_1 = 0 \\ \alpha_1^2\gamma_1 + \alpha_2^2\gamma_2 + \alpha_3^2\gamma_3 + & \cdots & +\alpha_n^2\gamma_n + C_2 = 0 \\ \dots\dots\dots\dots\dots & \dots & \dots\dots\dots\dots \\ \alpha_1^n\gamma_1 + \alpha_2^n\gamma_2 + \alpha_3^n\gamma_3 + & \cdots & +\alpha_n^n\gamma_n + C_n = 0. \end{array}\right.$$

Let Υ be the matrix of this system, $\gamma = (\gamma_1, \dots, \gamma_n)^T$ its solution, and $c = -(C_1, \dots, C_n)^T$ its right-hand side. The system (I) can be written as $\Upsilon\gamma = c$. Then, Cholesky carried out the linear transformation represented by the system

$$\text{(II)}\left\{\begin{array}{l} \gamma_1 = \alpha_1^1\lambda_1 + \alpha_1^2\lambda_2 + \cdots + \alpha_1^n\lambda_n \\ \gamma_2 = \alpha_2^1\lambda_1 + \alpha_2^2\lambda_2 + \cdots + \alpha_2^n\lambda_n \\ \dots\dots\dots\dots\dots\dots\dots\dots \\ \gamma_n = \alpha_n^1\lambda_1 + \alpha_n^2\lambda_2 + \cdots + \alpha_n^n\lambda_n. \end{array}\right.$$

Setting $\lambda = (\lambda_1, \ldots, \lambda_n)^T$, we thus have $\gamma = \Upsilon^T \lambda$. So, the system (I), giving the unknowns γ_i, is replaced by the system (III)

$$(\text{III}) \left\{ \begin{array}{l} A_1^1\lambda_1 + A_2^1\lambda_2 + \cdots + A_n^1\lambda_n + C_1 = 0 \\ A_1^2\lambda_1 + A_2^2\lambda_2 + \cdots + A_n^2\lambda_n + C_2 = 0 \\ \cdots\cdots\cdots\cdots\cdots\cdots\cdots\cdots\cdots\cdots \\ A_1^n\lambda_1 + A_2^n\lambda_2 + \cdots + A_n^n\lambda_n + C_n = 0. \end{array} \right.$$

In matrix terms, this system is $A\lambda = \Upsilon\Upsilon^T\lambda = c$, and it gives the λ_i's, thus allowing the computation of the unknowns γ_i by the system (II). The coefficients of the system (III) are given by the formulas

$$A_p^p = \sum_{k=1}^n (\alpha_k^p)^2, \quad A_p^q = \sum_{k=1}^n \alpha_k^p \alpha_k^q.$$

Cholesky noticed that the matrix of this system is symmetric, but he was probably not aware of the notion of positive definiteness. He remarked that the system (III) could be solved if the γ_i's are easily obtained from the system (I). This is the case if, in (I), the first equation only contains γ_1, if the second equation only contains γ_1 and γ_2, and so on. So, Cholesky proposed to find a system of the form

$$\begin{array}{ll} \alpha_1^1\gamma_1 & +C_1 = 0 \\ \alpha_1^2\gamma_1 + \alpha_2^2\gamma_2 & +C_2 = 0 \\ \alpha_1^3\gamma_1 + \alpha_2^3\gamma_2 + \alpha_3^3\gamma_3 & +C_3 = 0 \\ \cdots\cdots\cdots\cdots\cdots\cdots\cdots\cdots & \cdots\cdots \\ \alpha_1^n\gamma_1 + \alpha_2^n\gamma_2 + \alpha_3^n\gamma_3 + \cdots + \alpha_n^n\gamma_n & +C_n = 0. \end{array}$$

Then, the λ_i's can be directly obtained by

$$\begin{array}{cccccc} \alpha_1^1\lambda_1 + & \alpha_1^2\lambda_2 + & \cdots & \cdots & +\alpha_1^n\lambda_n - \gamma_1 = 0 \\ & \alpha_2^2\lambda_2 + & \alpha_2^3\lambda_3 & \cdots & +\alpha_2^n\lambda_n - \gamma_2 = 0 \\ & & \alpha_3^3\lambda_3 & \cdots & +\alpha_3^n\lambda_n - \gamma_3 = 0 \\ & & \cdots & \cdots & \cdots\cdots\cdots\cdots \\ & & & & \alpha_n^n\lambda_n - \gamma_n = 0 \end{array}$$

starting from the last equation, which gives λ_n. Finally, Cholesky gave the formulas we use today for computing the coefficients α_i^j. Although he used the same notation for the entries of the matrix Υ and for those of these triangular matrices, we clearly see that Cholesky factorized A as $A = LL^T$. Then, he solved $L\gamma = c$, and finally $L^T\lambda = \gamma$, which gave him the solution he was looking for. He also considered the possibility that one triangular matrix is not the transpose of the other. But he showed that the rounding errors are smaller for the LL^T factorization.

Moreover, Cholesky explained how to implement easily his method on the mechanical calculating machine *Dactyle* (see Chapter 7) and how to check that no mistake was made in the application of his method. He did so because the computations were done by soldiers with only a minimal mathematical background.

Cholesky also proposed a method for computing the square root of a number since this is needed to compute the diagonal entries of L, and proved its quadratic convergence rate. It is in fact the standard iterations, $x_{n+1} = (x_n + a/x_n)/2$ for $\sqrt{a}$, a procedure attributed to Hero (or Heron) of Alexandria (first century AD). Cholesky ended his note with numerical examples. He reported solving a system of 10 equations with 5 decimal digits in 4 to 5 hours, and that his method was also successfully used for several systems of dimension 30, and for one of dimension 56.

Cholesky died on the battlefield in the last days of World War I. His method remained unknown outside the community of French topographers until 1924, when Benoît published his paper. Then, a period of 20 years followed in which there were not too many mentions of the work. According to [1433], it was used or rediscovered by Sven Tryggve Salomon Rubin (1874-1946), a Swedish geodesist, in 1926 [2615]. This paper was also cited by George Elmer Forsythe (1917-1972) in [2432]. We will see later that Cholesky's method was brought back to life in the 1940s.

Cholesky's method was later independently rediscovered many times. Let us mention, in particular, the *square root* method of Tadeusz Banachiewicz (1882-1954), a Polish astronomer and mathematician (see Section 10.4), in 1938 [179], that is closely related to Cholesky's. This method is based on the notion of *Cracovian* introduced by Banachiewicz as soon as 1924 [1927]. Banachiewicz introduced Cracovians by using the transpose A^T of A, and multiplying the columns of A^T by the column vector x. This led to the definition of a different type of matrix multiplication that we denote by $\wedge$. Thus $x \wedge A^T = b = Ax$. The Cracovian product of two matrices A and B is defined by $A \wedge B = B^T A$, where the dimensions of B^T and A have to be compatible for the usual matrix product. Since $(AB)^T = B^T A^T$, the products $(A \wedge B) \wedge C$ and $A \wedge (B \wedge C)$ are generally different. Thus, Cracovian multiplication is non-associative. The square Cracovian T, which is called the *transposing Cracovian*, is similar to the identity matrix, satisfying $A \wedge T = A$, but $T \wedge A = A^T$.

For Cracovians, Banachiewicz adopted a column-row convention for denoting individual entries as opposed to the usual row-column notation. This made manual multiplication easier, as one needs to follow two parallel columns, instead of a vertical column and a horizontal row in the matrix notation. It was supposed to speed up computer calculations because the entries of both factors are used in a similar order, which is more compatible with the computing engines of those times. Modern references to Cracovians are in connection with their non-associative multiplication.

According to Forsythe [1188], it seems that the interpretation of the elimination process as a matrix factorization $A = LU$ or $A = LDU$ is also due to Banachiewicz [179] in 1938; see also [2725]. However, this was written in terms of Cracovians and it was not immediately obvious that it corresponded to an LU factorization.

Block elimination for the solution of least squares problems was considered by Hans Boltz (1883-1947) [370] in 1923.

In the 20th century and before World War II, the progresses of physics and engineering as well as other sciences led to new needs in numerical computations. The collection of data by government agencies gave rise to regression analyses, statisticians had to solve normal equations, and geodesists and astronomers were no longer the only ones using these techniques. Developments in aeronautics, structural mechanics and other areas gave new problems to solve. For instance, Lewis Fry Richardson (1881-1953) foresaw the use of discretization methods for partial differential equations and the numerical modeling for weather forecasting; see [2566, 2567].

However, elimination methods for the solution of the normal equations had not already convinced everybody. For instance, in 1924, Edmund Taylor Whittaker (1873-1956) and George Robinson [3227] wrote (on page 239 of the third edition)

> *We may say that the Determinantal method is on the whole the best for the solution of a set of normal equations.*

The future proved that they were wrong.

In 1938, Robert Alexander Frazer (1891-1959), William Jolly Duncan (1894-1960), and Arthur Roderick Collar (1908-1986), three Englishmen who were working in aeronautics, particularly on aircraft flutter, published a book *Elementary Matrices & some Applications to Dynamics and Differential Equations* [1222]. Of interest to us are pages 96-131. The authors were first interested in the inverse (which they called the *reciprocal*) of a matrix that they denoted by a. They multiplied the relation $a^{-1}a = I$ from the right by matrices M_i that they called *postmultipliers*, to obtain $a^{-1}\sigma = M_1M_2\cdots M_s$ with $\sigma = aM_1M_2\cdots M_s$. It means that they were doing operations on the columns of a. They observed that if the inverse of σ can be easily computed, it yields $a^{-1} = M_1M_2\cdots M_s\sigma^{-1}$. Then, they chose the M_i as *elementary matrices* which are the identity matrices, except for the entries of one row to the right of the diagonal. The coefficients were chosen to reduce a to lower triangular form. They showed how to proceed on a general matrix of order 4, finally obtaining an upper triangular matrix with a unit diagonal for $M_1M_2M_3$. They also noticed that, in some cases, one has to use permutations to avoid divisions by zero. However, they did not do the last step, which would have been to realize that they had computed a factorization of a.

Curiously enough, it seems they did not try to compute the inverses of the elementary matrices, which are easily obtained. In fact, they computed the product of the postmultipliers before computing its inverse.

On pages 126-131, they showed how to solve linear systems. They first used postmultipliers after transposing the system. Then they showed how to solve the systems by reducing the matrix to upper triangular form by combinations of rows.

The book by Frazer, Duncan, and Collar was very successful and had many editions. Forty years later, Collar wrote [723] that it was *the first book to treat matrices as a branch of applied mathematics*.

Note that the elementary matrices, as well as their inverses, appeared in the book by Herbert Westren Turnbull (1885-1961) and Alexander Craig Aitken (1895-1967) in 1932 (page 12 of the 1944 edition). The authors attributed this type of matrix to Charles Hermite (1822-1901) without giving a reference.

In 1935, Yue-Kei Wong (1904-1983) described Gaussian elimination on the symmetric positive definite matrix of the normal equations as a transformation to upper triangular form by left multiplication with elementary matrices in [3270]; see Chapter 4.

Prescott Durand Crout (1907-1984) was a professor in the Department of Mathematics of the Massachusetts Institute of Technology, interested in electrical engineering. In 1941, he presented a communication at an AIEE (American Institute of Electrical Engineers) symposium in Toronto (Canada), and a 6-page paper [775] was published that same year. That paper is about a method for solving linear systems with general matrices and, as a by-product, computing determinants. Crout wrote

> *The process is particularly adapted for use with a computing machine, for each element is determined by one continuous machine operation (sum of products with or without a final division).*

Crout described his process in words and gave the formulas in a "Mathematical Appendix." At each step, the algorithm computes first a diagonal entry, say (k,k), the entries (i,k), $i = k+1,\dots,n$ and then, the entries (k,j), $j = k+1,\dots,n$. Crout proved by induction that (provided there is no zero pivot) the algorithm does what it is intended for. But he also briefly considered pivoting. The method was also applied to symmetric matrices and to systems with complex entries.

Crout illustrated the use of the method with a small example

$$A = \begin{pmatrix} 12.1719 & 27.3941 & 1.9827 & 7.3757 \\ 8.1163 & 23.3385 & 9.8397 & 4.9474 \\ 3.0706 & 13.5434 & 15.5973 & 7.5172 \\ 3.0581 & 3.151 & 6.9841 & 13.1984 \end{pmatrix}, \quad b = \begin{pmatrix} 6.6355 \\ 6.1304 \\ 4.6921 \\ 2.5393 \end{pmatrix}.$$

What he called the "auxiliary matrix," obtained from the given formulas, is

$$\begin{pmatrix} 12.1719 & 2.2506 & 0.16289 & 0.60596 & 0.54515 \\ 8.1163 & 5.072 & 1.6793 & 0.0057629 & 0.33632 \\ 3.0706 & 6.6327 & 3.9585 & 1.4193 & 0.19891 \\ 3.0581 & -3.7316 & 12.7526 & -6.7332 & 0.060806 \end{pmatrix}.$$

Even though Crout did not notice or mention this, what he was doing was computing an LU factorization without pivoting of the matrix A, with L lower triangular and U upper triangular with a unit diagonal. If we do the computation with IEEE double precision, we obtain

$$L = \begin{pmatrix} 12.1719 & 0 & 0 & 0 \\ 8.1163 & 5.072 & 0 & 0 \\ 3.0706 & 6.6327 & 3.9585 & 0 \\ 3.0581 & -3.7316 & 12.7526 & -6.7332 \end{pmatrix},$$

$$U = \begin{pmatrix} 1 & 2.2506 & 0.16289 & 0.60596 \\ 0 & 1 & 1.6793 & 0.0057629 \\ 0 & 0 & 1 & 1.4193 \\ 0 & 0 & 0 & 1 \end{pmatrix}.$$

Crout's solution and the IEEE solution are, respectively,

$$x_C = \begin{pmatrix} 0.15942 \\ 0.14687 \\ 0.11261 \\ 0.060806 \end{pmatrix}, \quad x = \begin{pmatrix} 0.159291 \\ 0.146918 \\ 0.112575 \\ 0.0608407 \end{pmatrix}.$$

The norm of the difference is $1.4605\ 10^{-4}$, due to the difference in the accuracy of the computations.

Shortly after the paper appeared, Crout's method became popular, particularly for solving problems in matrix structural analysis. Most of the people using and citing it probably did not realize that Crout's method was just a variant of Gaussian elimination, even though it was well adapted to mechanical calculators. A few early reports citing Crout are [2043] (structural analysis in aeronautics), [2488] (lectures on numerical analysis), and [3177] (aeronautics).

In 1941, Dwyer, an American mathematician interested in statistics, published a paper [1038] titled *The solution of simultaneous equations*. This paper was published in *Psychometrika*, the journal of the Psychometric Society, founded in 1935. In this paper, Dwyer had an ambitious goal. In the abstract, he wrote

> *This paper is an attempt to integrate the various methods which have been developed for the numerical solution of simultaneous linear equations.*

He illustrated the methods with the following small symmetric system

$$\begin{aligned} x_1 + 0.4x_2 + 0.5x_3 + 0.6x_4 &= 0.2, \\ 0.4x_1 + x_2 + 0.3x_3 + 0.4x_4 &= 0.4, \\ 0.5x_1 + 0.3x_2 + x_3 + 0.2x_4 &= 0.6, \\ 0.6x_1 + 0.4x_2 + 0.2x_3 + x_4 &= 0.8. \end{aligned}$$

He mainly considered ten methods plus some block generalizations. In the following, we give Dwyer's attributions of the methods within parenthesis:

1. The method of division (Edward Vermilye Huntington (1874-1952), 1924).
2. The method of single division (this is essentially Crout's version of Gaussian elimination, but Dwyer attributed it to the chemist Horace Grove Deming (1885-1970), 1928, and Aitken [17], 1938).
3. The method of single division-Symmetric.
4. Abbreviated method of single division (this corresponds to delay the update of the coefficients of the reduced matrix).
5. Abbreviated method of single division-Symmetric.
6. Abbreviated Doolittle method (Waugh, 1935).
7. Method of multiplication and subtraction (Paul Reece Rider (1888-1984), 1939, Frank Lunwood Wren (1894-1976), 1938. This is the old schoolbook technique of elimination by considering two equations at a time and adequately multiplying the coefficients before subtracting).
8. Method of multiplication and subtraction-Symmetric.
9. Abbreviated method of multiplication and subtraction.
10. Abbreviated method of multiplication and subtraction-Symmetric.

After reading this paper, we observe two things. First, clearly, Dwyer was thinking of computations done with a mechanical (or electrical) calculator and pencil and paper. Hence, his goal was to find methods minimizing the recording of numbers on paper and being able to use accumulated sums. Second, most of his references were from people working in statistics. His favorite methods were the abbreviated Doolittle method and the abbreviated method of multiplication and subtraction.

Also in 1941, Dwyer gave a proof that Doolittle's method actually solves the normal equations [1035]. He did so by using the formulas describing the method and not by using matrices.

However, in 1944, Dwyer gave a matrix interpretation of Doolittle's method [1039], factoring the matrix as $A = S^T T$ with S and T upper triangular. He also showed that $S = DT$, where D is a diagonal matrix, and he noticed that, since the system is symmetric, it would be more interesting if the matrices S and T were identical in order to perform only half of the computations. For this, it would be enough to take the square roots of the diagonal terms of D, and he noticed that the method obtained in that way is similar to Banachiewicz's. This is indeed Cholesky's method.

In 1945, Waugh and Dwyer [3195] recognized that the methods studied by Dwyer were variations of Gaussian elimination. They also cited Crout's paper and introduced an LU factorization.

In his book [1041] in 1951, Section 6.5, pages 113-118, Dwyer described the square root method. He wrote (with the numbers corresponding to our list of references)

> *The square root method has been worked out independently by a number of different authors. Banachiewicz published it in 1938 in Poland* [180, 181]. *It has more recently been advocated by Dwyer* [1040], *Duncan*[19] *and Kenney*[20] [1025], *and Laderman* [1966] *as an excellent method for studying least squares, correlation, and*

[19]David Beatty Duncan (1916-2005).

[20]John Francis Kenney (1897-1974).

> *regression problems. Articles* [1206, 1654, 1735, 3079] *have traced the method to the earlier work of Cholesky and Schur. It is doubtful that such a simple method was not considered by authors prior to Cholesky. Before the days of computing machines, the method would probably have been discarded as impractical by any author who considered it.*

Dwyer gave no reference for Cholesky and Schur. According to Derrick Shannon Tracy (1933-1998), one of Dwyer's Ph.D. students, he was not willing to credit Cholesky for his procedure since he did not publish it in his lifetime [3066].[21]

In the preceding book's quotation, Dwyer cited Jack Laderman (1914-1984). He was one of people involved in the National Bureau of Standards (NBS) Mathematical Tables Project, a project of the Works Progress Administration, a New Deal agency established by President Roosevelt to alleviate unemployment through public works, that was in operation since 1938. The purpose of the Mathematical Tables Project was to compute tables of higher mathematical functions. In 1943, Laderman published with Arnold Noah Lowan (1898-1962), for instance, the *Table of Fourier coefficients* [2094]. In 1947, George Bernard Dantzig (1914-2005) created the *Simplex method* for linear programming. At the end of that same year, Laderman undertook a computation for the solution on the so-called *Diet problem* by using Dantzig's method. It needed nine clerks using hand-operated desk calculators. The diet problem was one of the first optimization problems, and it was first motivated by the Army's desire to meet the nutritional requirements of soldiers while minimizing the cost. In [810], Dantzig wrote

> *One of the first applications of the simplex algorithm was to the determination of an adequate diet that was of least cost. In the fall of 1947, J. Laderman of the Mathematical Tables Project of the National Bureau of Standards undertook, as a test of the newly proposed simplex method, the first large-scale computation in this field. It was a system with nine equations in seventy-seven unknowns. Using hand-operated desk calculators, approximately 120 man-days were required to obtain a solution.*

Before World War II, people were using tabulating machines and then punched card machines to do numerical computations. It was also the start of the development of digital computers; see Chapter 7. This started the need for some error analyses for the computations, in particular for solving linear equations by elimination methods.

Harold Hotelling (1895-1973) was an American statistician and economist. In 1943, he published a paper [1732] about finding the inverse of a matrix and solving systems of linear equations. It was a revision of a talk given at a symposium in 1941. In his introduction, Hotelling referred to punched-card machines. He cited Doolittle, Dwyer, Crout and Frazer, Duncan, and Collar. Section 3 of this paper is titled *Accuracy of direct solutions of linear equations.* At the beginning of that section, as we have already seen in Chapter 1, Hotelling wrote

> *The question how many decimal places should be retained in the various stages of a least-squares solution and of other calculation involving linear equations has been a puzzling one. It has not generally been realized how rapidly errors resulting from rounding may accumulate in the successive steps of such procedures as, for example, the Doolittle method.*

Hotelling did a short forward analysis of the error in the entries of the upper triangular matrix with some simplifying assumptions, and he continued by writing on page 7:

[21] Personal communication from Richard William Farebrother, The University of Manchester, UK.

> *we find* [...] *the estimated limit of error* $4^{p-1}\epsilon$ [...] *The rapidity with which this increases with p is a caution against relying on the results of the Doolittle method or other similar elimination methods with any number of moderate decimal places when the number of equations and unknowns is at all large.*

Here p was the order of the matrix and ϵ was $10^{-k}/2$, where k is the number of decimal places. These conclusions raised concerns about the possible use of direct methods for solving linear systems on mechanical and digital computers.

In 1942, Arvid Turner Lonseth (1912-2002), an American mathematician of Norwegian origin, studied what happens to the solution of the linear system when the entries of the matrix A and the right-hand side are perturbed. In 1947, he obtained perturbation results for a linear operator in a Banach space using norms. His results are very close to the bounds we use today; see [2089, 2090, 2091].

In 1944, the American statistician Franklin E. Satterthwaite wrote an LDU factorization of a matrix A in the form $A = (R_1 + I)S_1(I + T_1)$, where R_1 (resp., T_1) is "prediagonal" (resp., "postdiagonal"), that is, strictly lower (resp., upper) triangular, and he obtained formulas for the entries of these matrices; see [2689]. Then, he considered perturbations in the Doolittle method using the Frobenius norm. He reached the strange conclusion that the method is accurate provided A is close to the identity matrix, $\|A - I\|_F \leq 0.35$. However, he noticed that if one has an approximation M of A, by solving $MAx = Mb$, this condition can be satisfied for MA. He referred to Dwyer and Hotelling.

In 1944, the Danish geodesist Henry Jensen (1915-1974) published a paper [1816] in which he compared several methods for solving systems of linear equations, including Cholesky's and Banachiewicz's. He recommended that Cholesky's *ought to be more generally used than is the case*, claiming that *Cholesky's method seems to possess all advantages.* In 1946, Todd had to teach a numerical analysis course at King's College in London. Todd and his wife, Olga Taussky, looked into the volumes of *Mathematical Reviews*, and found a review of Jensen's paper by Bodewig. Since the method was clearly explained by Jensen, Taussky and Todd did not try to find the original paper. It is not known if Todd was presenting the method in an algebraic way or if he was using matrix notation.

After Todd's lectures, several of his colleagues and students undertook the study of Cholesky's method. This was done by Leslie Fox (1918-1992), Harry Douglas Huskey (1916-2017), James Hardy Wilkinson (1919-1986), and Alan Mathison Turing (1912-1954); see the report [1207] and the paper [1206] by Fox, Huskey, and Wilkinson in 1948. In this paper the authors wrote that

> *Another method, due to Choleski, for the reciprocation of a symmetric matrix was pointed out to us recently by John Todd. This method is so simple in theory and straightforward in practice that its neglect in this country is surprising.*

Cholesky's name was also mentioned in the paper [1202] by Fox, published in 1954 but written in 1950. In that paper, Fox also clearly stated that the methods of Doolittle, Crout, and Cholesky (in the symmetric case) are variants of the LU factorization. Cholesky's method was adopted by geodesists; see, for instance, the book [2529] by Hume Frederick Rainsford in 1957.

Following Hotelling in 1946, Valentine Bargmann (1908-1989), Deane Montgomery (1909-1992), and John von Neumann (1903-1957) recommended in [193] not to use elimination methods for the solution of a linear system because they were thought to be unstable.

Fortunately, von Neumann changed his mind (if he was responsible for the conclusions of [193]). In 1947, he published a paper [3158] with Herman Heine Goldstine (1913-2004) about the inversion of matrices. On pages 1050-1056 they described the LU factorization of a matrix as well as the LDU factorization with D diagonal and they proved uniqueness. They called the matrices L and U *lower and upper semi-diagonal*, names that did not catch on. Then, they specialized to symmetric positive definite matrices. They used the formalism of fixed point arithmetic and considered what they called *a digital matrix* representable in this format and proved that

$$\|A - U^T DU\|_2 \leq O(n^2 \beta^{-s}),$$

where n is the order of the matrix, β the base of the arithmetic, and s the number of fractional digits. They also gave a long error analysis for computing the inverse X, proving that

$$\|AX - I\|_2 \leq 14.2 n^2 u \ell,$$

where $u = \beta^{-s}/2$ and ℓ is the ratio of the largest to the smallest eigenvalue of A, that is, the condition number (see Section 1.4). For Goldstine's rememberings, see [1370].

In 1971, Wilkinson [3252] gave his appraisal of the von Neumann and Goldstine paper:

> *It is a very substantial paper, some 80 pages long, and it is not exactly bedside reading, though when the basic plan of the proof is grasped it is not essentially difficult.*

A detailed analysis of the von Neumann and Goldstine paper was done by Grcar [1433] in 2011.

In 1947 and 1948, Bodewig wrote a series of papers in which he described several methods for solving linear systems [351]. Following Jensen, he also wrote the LU factorization. This was summarized in his book [353] in 1956 and 1959.

In 1948, a linear system of order 38 was solved by Herbert Francis Mitchell Jr. (1913-2008), on the *Aiken Relay Calculator*, constructed for the Naval Proving Ground, Dahlgren, Virginia, by the staff of the Computation Laboratory of Harvard University [2256]. Mitchell, after serving as a colonel in WW II, was the first Ph.D. student of Howard Hathaway Aiken (1900-1973), a mathematician considered one of the pioneers in the development and application of digital computers; see Chapter 7.

Turing met von Neumann before and during WW II when Goldstine and von Neumann were working on their paper. His interest in linear systems was also triggered by solving a system with L. Fox, Goodwin, and Wilkinson using desk calculators.

In 1948, Turing published a paper titled *Rounding-off errors in matrix processes* [3079] in which he wrote

> *Let us describe a matrix which has zeros above the diagonal as 'lower triangular' and one which has zeros below as 'upper triangular'. If in addition the coefficients on the diagonal are unity the expressions 'unit upper triangular' and 'unit lower triangular' may be used. The resolution is essentially unique, in fact we have the following:*
>
> *Theorem on triangular resolution: If the principal minors of the matrix A are non-singular, then there is a unique unit lower triangular matrix L, a unique diagonal matrix D, with non-zero diagonal elements, and a unique unit upper triangular matrix U such that $A = LDU$. Similarly there are unique L', D', U' such that $A = U'D'L'$.*

More details on this paper are given in Section 10.72. Together with the von Neumann and Goldstine paper, it has been influential in rehabilitating Gaussian elimination.

Clearly, the interpretation of elimination methods for solving linear systems as ways to compute factorizations of the matrix showed that methods that were thought to be different were, in fact, slight variations of the same method.

In the 1950s and 1960s, people became more and more interested in direct elimination methods; see, for example, William Gee Bickley (1893-1969) and John Michael McNamee [316] in 1960. However, there were still people interested in computing inverses of matrices; see, for instance, [2755] by Jack Sherman in 1951, [251] by Gabriel G. Bejarano and Bruce R. Rosenblatt in 1953, as well as [1454] by Donald Greenspan (1926-2010) in 1955.

A forward error analysis for Gaussian elimination for a diagonally dominant tridiagonal linear system was done by Jim Douglas (1927-2016) [948] in 1959.

Errors in digital computations were a concern with the first electronic computers; see, for instance, [1738] by Alston Scott Householder (1904-1993) in 1955.

Important progress in the rounding error analysis was made in the 1950s and 1960s thanks to Wilkinson. After WW II, he was working on the development of the *Pilot ACE* computer at the National Physical Laboratory in the UK. He acquired a very deep knowledge of numerical methods, in particular Gaussian elimination and methods for computing eigenvalues; see Section 10.76. Already in 1948, in a report about the design of the *Pilot ACE*, he gave a program implementing Gaussian elimination with partial pivoting and iterative refinement. His experience with solving linear systems on the *Pilot ACE* in 1953 is described in [3237]; see also [2325]. Later, he developed backward error analysis that was first pioneered by James Wallace Givens (1910-1993). In 1961 Wilkinson published a paper [3244] on the error analysis of direct methods of matrix inversion and then he wrote two books: *Rounding Errors in Algebraic Processes* [3247] in 1964 and *The Algebraic Eigenvalue Problem* [3248] in 1965. Despite the title of this second book, it contains a thorough rounding error analysis of Gaussian elimination. The terms *partial pivoting* (looking for the pivot in the lower part of the present column k) and *complete pivoting* (looking in the current reduced submatrix of order $n-k+1$) were introduced in 1961. See also [2553] by John Ker Reid, written in 1971, when he was at the Mathematics Branch, Theoretical Physics Division, of AERE Harwell, now part of Rutherford Appleton Laboratory (RAL), UK. Iterative refinement using fixed point arithmetic was studied by Wilkinson in [3247] in 1964, and by Cleve Barry Moler [1196] using floating point arithmetic [2259] in 1967.

What Wilkinson essentially proved in [3244, 3247] at the beginning of the 1960s is that Gaussian elimination with partial pivoting for solving $Ax=b$ yields a solution $\tilde{x}$ satisfying

$$(A+\Delta A)\tilde{x}=b, \quad \|A\|_\infty \le n^2 \frac{3nu}{1-3nu}\rho_n \|A\|_\infty,$$

where u is the unit roundoff and ρ_n is the growth factor,

$$\rho_n = \frac{\max_{i,j,k} |a_{i,j}^{(k)}|}{\max_{i,j} |a_{i,j}|},$$

the elements $a_{i,j}^{(k)}$ are those produced at step k of the exact elimination process; see [1681]. With partial pivoting the growth factor is bounded by 2^{n-1} and this bound can be reached for certain matrices. Note that this was already known by Wilkinson in 1953; see [3237]. However, practical problems with an exponential growth are rare. With complete pivoting, he proved that

$$\rho_n \le n^{\frac{1}{2}} (2\, 3^{\frac{1}{2}} \cdots n^{\frac{1}{n-1}})^{\frac{1}{2}}.$$

It was conjectured at the end of the 1960s that ρ_n is smaller than n, but counter-examples were found later. Wilkinson's analysis was rapidly cited, for instance, in the books by L. Fox [1204] in 1964 and Forsythe and Moler [1196] in 1967.

Other pivoting strategies were proposed later, such as, for instance, [660] Yun-Tzong Chen and Reginald Prabhakar Tewarson (1930-2014) in 1972, and rook's pivoting [2511] by George Douglas Poole and (Bill) Larry Neal (1944-1995) from East Tennessee State University in 1992. At the kth step, the algorithm is the following. Let

$$r_1 = \min\{r|\ |a_{r,k}^{(k)}| \geq |a_{i,k}^{(k)}|,\ k \leq i \leq n\}$$

and

$$c_1 = \min\{c|\ |a_{r_1,c}^{(k)}| \geq |a_{r_1,j}^{(k)}|,\ k \leq j \leq n\}.$$

If $c_1 = k$, then $a_{r_1,k}^{(k)}$ is the selected pivot. If $c_1 \neq k$, column c_1 is searched for the entry with maximum modulus. Therefore, rook's pivoting searches for coefficients of maximum modulus in rows, then columns, and then rows and columns, until an entry $a_{r,c}^{(k)}$ satisfies $|a_{r,c}^{(k)}| \geq |a_{i,c}^{(k)}|,\ k \leq i \leq n$ and $|a_{r,c}^{(k)}| \geq |a_{r,j}^{(k)}|,\ k \leq j \leq n$. Other examples of strategies are [2867] by G.W. Stewart in 1974 and [208] by Ian Barrodale and Gordon Francis Stuart in 1977.

Gaussian elimination may be in difficulty when the matrix is symmetric but indefinite, that is, with positive and negative eigenvalues. It is important to preserve symmetry during the elimination and, so, it is not practical to use partial or complete pivoting. An idea, introduced by James Raymond Bunch and Beresford Neill Parlett [504] in 1971 and further developed by Bunch and Linda Kaufman [501] in 1977, was to use diagonal pivoting with either 1×1 or 2×2 pivots; see also [502]. It can be proved that A being nonsingular, it is always possible to find such pivots, and a strategy was devised in [501]. This problem was further studied by Joseph Wai-Hung Liu [2080] in 1987, Iain Spencer Duff, Nicholas Ian Mark Gould, Reid, Jennifer Ann Scott, and Kathryn Turner [991] in 1991, and Cleve Cleveland Ashcraft, Roger G. Grimes, and John Gregg Lewis (1945-2019) [87] in 1998.

Wilkinson was also interested in perturbation analysis, that is, what happens to the solution of a linear system when we perturb the matrix and/or the right-hand side. Let

$$Ax = b, \quad (A + \Delta A)y = b + \Delta b,$$

where the perturbation terms satisfy

$$\|\Delta A\| \leq \alpha\omega, \quad \|\Delta b\| \leq \beta\omega,$$

where ω is given, α (resp., β) is 0 or $\|A\|$ (resp., $\|b\|$) depending on whether A, or b, or both are perturbed. Let

$$\eta_{\mathrm{T}} = \inf\{\omega|\ \omega \geq 0,\ \|\Delta A\| \leq \omega\alpha,\ \|\Delta b\| \leq \omega\beta,\ (A + \Delta A)y = b + \Delta b\},$$

be the *normwise backward error*, where the subscript T refers to Turing. It was proved in 1967 by Jean-Louis Rigal and Jean Gaches [2571], two French mathematicians, that

$$\eta_{\mathrm{T}} = \frac{\|r\|}{\alpha\|y\| + \beta},$$

where $r = b - Ay$. The associated condition number is

$$\kappa_{\mathrm{T}} = \|A^{-1}\| \frac{\alpha\|x\| + \beta}{\|x\|}.$$

If $\alpha = \|A\|$ and $\beta = 0$, it reduces to $\|A^{-1}\|\ \|A\|$; see Section 1.4. It can be shown that the relative forward error is approximately the product of the backward error and the condition number.

Another kind of error analysis is called *componentwise error analysis*. Instead of bounding norms, bounds are obtained for the absolute values (or moduli) of individual entries of the matrices or right-hand sides. In fact, Wilkinson could have developed this type of analysis since most of his bounds in intermediate steps were for entries of matrices before taking norms. This type of analysis was done for perturbations bounds and for rounding errors.

Componentwise analysis was used by Friedrich Ludwig Bauer (1924-2015) in 1966 for linear systems and matrix inversion; see [222]. Bruce A. Chartres and his Ph.D. student James Charles Geuder [641] gave in 1967 an analysis of the Doolittle version of elimination. They derived a backward error result $(A+\Delta A)y = b$, with a componentwise bound on ΔA. Reid [2553] proved in 1971, without assumptions on the pivoting strategy, that $LU = A + \Delta A$ with $|(\Delta A)_{i,j}| \leq 3.01\, u \min(i-1, j) \max_k |a_{i,j}^{(k)}|$, u being the unit roundoff. In 1972, Joseph Stoer [2903] gave some componentwise bounds for the perturbation matrix; see also Stoer and Roland Zdeněk Bulirsch [2906] in 1980. In 1979, Robert David Skeel published a careful componentwise error analysis [2786, 2787] with the aim of finding a good scaling for the linear system and to study iterative refinement.

Componentwise bounds were also obtained in perturbation analysis. The perturbations are such that $|\Delta A| \leq \omega E, \quad |\Delta b| \leq \omega f$. The componentwise backward error is

$$\eta_{\rm BS} = \inf\{\omega |\, \omega \geq 0,\ \|\Delta A\| \leq \omega E,\ \|\Delta b\| \leq \omega f,\ (A + \Delta A)y = b + \Delta b\}.$$

The subscript BS refers to Bauer and Skeel. Note that looking at the perturbations componentwise is useful when the entries of the matrix have different magnitudes and also to account for the zero/nonzero structure. It was shown in 1964 by Werner Oettli (1937-1999) and William Prager (1903-1980) [2361] that

$$\eta_{\rm BS} = \max_i \frac{|(b - Ay)_i|}{(E|y| + f)_i}.$$

See also [2362] with Wilkinson in 1965.

Shivkumar Chandrasekaran and Ilse Clara Franziska Ipsen gave componentwise bounds for the error [634, 635] in 1994-1995. See also the book by G.W. Stewart and Ji-Guang Sun [2888] in 1990.

With some assumptions on the arithmetic, Siegfried Michael Rump and Claude-Pierre Jeannerod [2630] showed in 2014 that, in backward error bounds, the term $nu/(1 - nu)$ can be replaced by nu. That is, they removed the condition $nu < 1$ that may not be satisfied for the very large linear systems that could be solved today and tomorrow.

The bounds given by a backward error analysis are generally far from being tight. This is why some authors tried to obtain probabilistic estimates. Probably the first ones to do this were Goldstine and von Neumann [1373] in 1951. The probabilistic approach led to the rule of thumb that constants in rounding error bounds, depending on the order of the matrix, can be replaced by their square roots.

Another probabilistic error analysis of Gaussian elimination was done by Jesse Louis Barlow and his advisor Erwin Hans Bareiss (1923-2003) [198] in 1985. More recently, in 2019, under the hypothesis of independence of the rounding errors, Nicholas John Higham and Theo Mary [1688] derived probabilistic bounds for the LU factorization; see also [1689] in 2020.

There are classes of matrices for which it is not necessary to pivot for stability. The most well known class is the symmetric positive definite (SPD) matrices for which there exist a Cholesky factorization. If one does an LU factorization of an SPD matrix there is no zero pivot because all the principal minors are nonzero. A fixed point error analysis was done by Wilkinson [3248] in 1965 and a floating point analysis [3251] in 1966. Jean Meinguet gave a componentwise error analysis [2194] in 1983; see also James Weldon Demmel [868] in 1989 and J.-G. Sun [2941] in 1992.

A matrix A is *diagonally dominant by rows* if

$$|a_{i,i}| \geq \sum_{j \neq i} |a_{i,j}|, \quad i = 1, \ldots, n.$$

It is *diagonally dominant by columns* if A^T (or A^*) is diagonally dominant by rows. A matrix diagonally dominant by rows has an LU factorization without pivoting and the growth factor is smaller than or equal to 2. This was proved by Wilkinson [3244] in 1961.

A is an M-matrix if and only if $a_{i,j} \leq 0$ for $i \neq j$ and $A^{-1} \geq 0$ entrywise. A symmetric M-matrix is positive definite and there is no need to pivot. Two Czech mathematicians, Miroslav Fiedler (1926-2015) and Vlastimil Pták (1925-1999), proved in 1962 that an M-matrix has an LU factorization; see [1165]. The proof can also be found in Fiedler's book [1160].

Let us define $M(A)$ as the matrix having entries $m_{i,j}$ such that

$$m_{i,i} = |a_{i,i}|, \quad m_{i,j} = -|a_{i,j}|, \quad \forall\, i, j,\; i \neq j.$$

A is an H-matrix if and only if $M(A)$ is an M-matrix. An H-matrix has an LU factorization; see Abraham Berman and Robert James Plemmons [290] in 1979, Robert Edward Funderlic (1937-2009) and Plemmons [1272] in 1981, and Funderlic, Michael Neumann (1946-2011), and Plemmons [1271] in 1982.

Scaling is a transformation of the linear system to be solved trying to give a better behaved system. Let D_1 and D_2 be two nonsingular diagonal matrices. The system $Ax = b$ is transformed into

$$A'y = (D_1 A D_2)y = D_1 b,$$

and the solution x is recovered as $x = D_2 y$. One of the diagonal matrices can be the identity. Notice that left multiplication by D_1 is a row scaling and right multiplication by D_2 is a column scaling. Scaling has been used for at least two goals: improve the condition number of the matrix and improve the behavior of Gaussian elimination.

On the first goal, see Forsythe and Ernst Gabor Straus (1922-1983) [1198] in 1955, Bauer [220, 224] in 1963 and 1969, Abraham van der Sluis (1928-2004) [3094] in 1969, Charles Alan McCarthy (1936-2018) and Gilbert Strang [2175] in 1973, and Gene Howard Golub (1932-2007) and James Martin Varah [1400] in 1974.

If Gaussian elimination is used with partial pivoting to solve a scaled system, the row scaling influences the choice of the pivot. On these issues, see Forsythe and Moler [1196] in 1967, van der Sluis [3095] in 1970, Alan Raymond Curtis (1922-2008) (who was working at the AERE Harwell Laboratory), Reid [794] in 1972, and G.W. Stewart [2871] in 1977. In 1979, Skeel [2786] showed that a good scaling matrix is obtained by choosing the diagonal elements of D_1 as $d_i = (|A|\,|y|)_i$, where y is the computed solution. Of course, this is impractical, as the solution y depends on the scaling. However, if an approximation c of the solution is known, then A could be scaled by $(|A|\,|c|)_i$.

All these topics were clearly explained and also extended in N.J. Higham's books [1679, 1681] in 1996 and 2002, along with some interesting historical notes; see also [1684, 1686].

We observe that, in 1969, Volker Srassen published a paper [2927] whose title was *Gaussian elimination is not optimal* in which he stated that a linear system can be solved in $O(n^{2.8})$ operations. This paper led to an algorithm for fast matrix-matrix multiplication but not for solving linear systems. On this topic, see also the thesis [850] by Christian de Polignac, defended in

1970 under the supervision of Noël Gastinel (1925-1984). Strassen's result contrasts with those of V.V. Klyuyev (or Klyuev, or Kljuev) and Nikolay Ivanovich Kokovkin-Shcherbak (1922-1995) [1910] in 1965, who showed that Gaussian elimination is optimal if one restricts oneself to operations on rows and columns as a whole. We were not able to find precise information about Klyuyev, but on a Russian website[22] we learned that Kokovkin-Shcherbak was drafted into the Red Army in 1940, and that after the war, he graduated from the physics and mathematics faculty of the Chkalovsk Pedagogical Institute. He worked for 26 years in the Department of Physics and Mathematics of the Pyatigorsk Pharmaceutical Institute.

In 1953, Everett Wayne Purcell (1924-2012) published an orthogonalization method for solving linear systems [2523] whose ith equation is written as

$$a_{i1}x_1 + \cdots + a_{in}x_n + a_{i,n+1}t = 0,$$

where t is a parameter later set to 1. He denoted by V_i the row vector containing $a_{i1} \ldots a_{i,n+1}$, and V_i^j the ith vector orthogonal to the first j vectors of the matrix. Its coordinates satisfy the first j equations. Any linear combination of the V_i^j also satisfies the first j equations, and linear combinations exist that satisfy the other equations.

Purcell outlined a procedure for finding such combinations. Let V_i^0 be the $n+1$ dimensional row vector with all components equal to 0 and the ith equal to 1. The first step of the method is to find the coefficients $C_i^1, i = 1, \ldots, n$ such that $V_i^1 = C_i^1 V_1^0 + V_{i+1}^0$ be orthogonal to V_1. Obviously $C_i^1 = -(V_1, V_{i+1}^0)/(V_1, V_1^0)$. Then, he defined the coefficient C_i^2 so that $V_i^2 = C_i^2 V_1^1 + V_{i+1}^1$ be orthogonal to V_2, that is, $C_i^2 = -(V_2, V_{i+1}^1)/(V_2, V_1^1)$, and so on.

Purcell ended up with a vector orthogonal to all the V_i, which gives the solution of the system. The number of multiplications is $n(n-1)(2n+5)/6$, and the number of additions and divisions is the same as for Crout's method. Then, Purcell showed how his method can be used for evaluating determinants.

In his review of that paper (MR0059065), Forsythe used a more convenient notation. He also noticed that the method, although unpublished, had already been obtained by Theodore Samuel Motzkin (1908-1970), a fact he certainly knew because they were coauthors of other papers. Consequently, the method is sometimes referred to with both names. Jenő Egerváry (1891-1958), a Hungarian mathematician [2536], published several papers on Purcell's method but we only cite [1052], the only one written in English. He showed that it can be given by his rank reduction technique, a detailed exposition of which can be found in [1274].

Properties of Purcell's method were given in [1628] by Csaba J. Hegedüs in 2010, and in the references quoted there. In [3342], Krystian Zorychta proved that for the method to deliver a unique solution of the system it is sufficient that all the principal minors of the matrix be different from zero. He also connected Purcell's method to the matrices of Gaussian factorization, and showed how to avoid a division by zero. Finally, he pointed out that the number of arithmetic operations in both methods is the same, but claimed that Purcell's method is superior to Gauss' since it is a recurrence procedure instead a two-stage method, and that, at each step, only one row of the matrix is required.

Although the method of Purcell is fairly well known, very little is known about his life. However, we were able to find a few details. He earned different degrees in Electrical Engineering, Aerospace Engineering, and Mathematics and Theology at the Universities of Nebraska and Southern California. He served in the battle of Iwo Jima during WW II. He worked as an engineer for Douglas Aircraft, and later, for the Aeronautic Division of the Ford Motor Company. When he retired, he completely changed interests, studying biblical languages and theology because he was interested in reading the Bible in its original languages. He was very active in the creation

[22] https://www.moypolk.ru/soldier/kokovkin-shcherbak-nikolay-ivanovich (accessed on January 12, 2022)

science movement, leading the *Creation Science Association of Orange County* for many years. In 2007 he published a book about the Sign of Jonah and the dates of the celebration of Easter.

Orthogonalization methods were later used to construct preconditioners for iterative methods; see [281] and Chapter 5.

Codes implementing Gaussian elimination started to be published in journals and books in the 1960s. Later on, starting in the 1970s, packages implementing the method became available; see, for instance, the well-known LINPACK and LAPACK packages that we consider in Chapter 8.

We have already seen that algorithms like those of Doolittle, Crout, or Cholesky were designed to use efficiently the computing machines at hand in their time. When coding Gaussian elimination using a high-level programming language, say, FORTRAN, there are three nested loops that can be ordered in different ways. This leads to six possible variants which may have different efficiencies depending on the computer architecture. For simplicity and to avoid pivoting, let us consider a factorization $A = LDL^T$ of a symmetric matrix A. The matrix L is lower triangular with a unit diagonal and D is diagonal.

The first algorithm is called the *outer product algorithm*. Suppose $a_{1,1} \neq 0$,

$$L_1 = \begin{pmatrix} 1 & 0 \\ l_1 & I \end{pmatrix}, \quad D_1 = \begin{pmatrix} a_{1,1} & 0 \\ 0 & A_2 \end{pmatrix},$$

and

$$A = \begin{pmatrix} a_{1,1} & a_1^T \\ a_1 & B_1 \end{pmatrix} = L_1 D_1 L_1^T.$$

By equating blocks, we obtain expressions for l_1 and A_2

$$l_1 = \frac{a_1}{a_{1,1}},$$

$$A_2 = B_1 - \frac{1}{a_{1,1}} a_1 a_1^T = B_1 - a_{1,1} l_1 l_1^T.$$

The matrix A_2 is obviously symmetric. If we suppose that the $(1,1)$ element of A_2 is nonzero, we can proceed further and write

$$A_2 = \begin{pmatrix} a_{2,2}^{(2)} & a_2^T \\ a_2 & B_2 \end{pmatrix} = \begin{pmatrix} 1 & 0 \\ l_2 & I \end{pmatrix} \begin{pmatrix} a_{2,2}^{(2)} & 0 \\ 0 & A_3 \end{pmatrix} \begin{pmatrix} 1 & l_2^T \\ 0 & I \end{pmatrix}.$$

Similarly,

$$l_2 = \frac{a_2}{a_{2,2}^{(2)}},$$

$$A_3 = B_2 - \frac{1}{a_{2,2}^{(2)}} a_2 a_2^T = B_2 - a_{2,2}^{(2)} l_2 l_2^T.$$

If we denote

$$L_2 = \begin{pmatrix} 1 & 0 & 0 \\ 0 & 1 & 0 \\ 0 & l_2 & I \end{pmatrix},$$

then

$$D_1 = \begin{pmatrix} a_{1,1} & 0 \\ 0 & A_2 \end{pmatrix} = L_2 \begin{pmatrix} a_{1,1} & 0 & 0 \\ 0 & a_{2,2}^{(2)} & 0 \\ 0 & 0 & A_3 \end{pmatrix} L_2^T = L_2 D_2 L_2^T.$$

After two steps, we have $A = L_1 L_2 D_2 L_2^T L_1^T$ with

$$L_1 L_2 = \begin{pmatrix} 1 & 0 \\ l_1 & \begin{pmatrix} 1 & 0 \\ l_2 & I \end{pmatrix} \end{pmatrix}.$$

The product of L_1 and L_2 is a lower triangular matrix. If all the pivots are nonzero, we can proceed, and at the last step, we obtain

$$A = A_1 = L_1 L_2 \cdots L_{n-1} D L_{n-1}^T \cdots L_1^T = LDL^T,$$

where L is unit lower triangular and D is diagonal.

The matrix L has been constructed column by column. This method is called the outer product algorithm since, at each step, an outer product aa^T is involved.

The second way to proceed is called the *bordering algorithm*. We partition the matrix A in a different way as

$$A = \begin{pmatrix} C_n & a_n \\ a_n^T & a_{n,n} \end{pmatrix}.$$

Suppose that C_n has already been factored as

$$C_n = L_{n-1} D_{n-1} L_{n-1}^T,$$

L_{n-1} being unit lower triangular and D_{n-1} diagonal. Then,

$$A = \begin{pmatrix} L_{n-1} & 0 \\ l_n^T & 1 \end{pmatrix} \begin{pmatrix} D_{n-1} & 0 \\ 0 & d_{n,n} \end{pmatrix} \begin{pmatrix} L_{n-1}^T & l_n \\ 0 & 1 \end{pmatrix}.$$

By identification,

$$\begin{aligned} l_n &= D_{n-1}^{-1} L_{n-1}^{-1} a_n, \\ d_{n,n} &= a_{n,n} - l_n^T D_{n-1} l_n. \end{aligned}$$

By induction, one can start with the decomposition of the 1×1 matrix $a_{1,1}$, adding one row at a time and obtaining at each step the factorization of an enlarged matrix. The main operation we have to perform at each step is solving a triangular system. In order to proceed to the next step, we need the diagonal entries of D_n to be nonzero. This type of method was used, in a different formulation involving inverses, by Herbert Saul Wilf (1931-2012) [3233] in 1959.

A formula related to the two previous methods is the so-called *Sherman-Morrison formula*, which gives the inverse of a rank-one modification of a given matrix. The Sherman-Morrison formula was published in 1950 in the four-page paper [2757] by two statisticians, Jack Sherman and Winifred J. Morrison, working at the Texas Company Research Laboratory, Beacon, New York; see also [2756]. They were interested in the change of the inverse of a matrix when one of its entries is modified. The formula appeared in the form we use today in a paper by another statistician, Maurice Stevenson Bartlett (1910-2002) [212] in 1951. The extension to matrices instead of vectors appeared in a 1950 report by Max Atkin Woodbury (1917-2010) [3271] at that time at the Statistical Research group at Princeton University, and then at New York University. However, the formula had been published before in some disguise in papers by W.J. Duncan, a professor of aeronautics and fluid mechanics in the University of Glasgow, [1026] in 1944, and Louis Guttman (1916-1987) [1516] in 1946; see [1535] by William Ward Hager in 1989. For the inverse of a sum of matrices, see [1637] by Harold V. Henderson and Shayle Robert Searle in 1981.

For matrices, if A is an $n \times n$ matrix and U and V two $n \times k$ matrices with $k \leq n$, the formula is

$$(A+UV^T)^{-1} = A^{-1} - A^{-1}U(I + V^T A^{-1}U)^{-1}V^T A^{-1}.$$

Of course A and $I + V^T A^{-1}U$ have to be nonsingular.

The third way to program the elimination steps leads to the *inner product algorithm.* It is obtained by writing down the formulas for the matrix product, $A = LDL^T$. Supposing $i \geq j$, we have

$$a_{i,j} = \sum_{k=1}^{j} l_{i,k} l_{j,k} d_{k,k}.$$

If we consider $i = j$ in this formula, then since $l_{i,i} = 1$, we obtain

$$d_{j,j} = a_{j,j} - \sum_{k=1}^{j-1} (l_{j,k})^2 d_{k,k},$$

and for $i > j$,

$$l_{i,j} = \frac{1}{d_{j,j}} (a_{i,j} - \sum_{k=1}^{j-1} l_{i,k} l_{j,k} d_{k,k}).$$

Locally, one has to consider the product of the transpose of a vector times a vector, which is a dot (or inner) product. This is a variant of Cholesky's method.

We see that the data is accessed differently in those three methods, and this has an impact on the performances. These three ways of coding the elimination process were studied, for vector computers, in a paper [922] by Jack J. Dongarra, Fred G. Gustavson, and Alan H. Karp in 1984.

With the rapid development of digital computers in the 1950s and 1960s (see Chapter 7), there was an almost exponential increase in the number of papers concerned with elimination methods for solving linear systems. Therefore, it is impossible to cite all these works. In the following we will just describe what are, in our opinion, the most interesting ones. Most of the developments were done to improve computing codes or to refine the error analyses. Sometimes the improvements were based on theoretical results and sometimes on heuristics.

In the 1950s and early 1960s, people using Gaussian elimination started to be concerned by the fact that, in many applications, some of the entries of the matrix are zero. It was recognized very early that it was not necessary to operate on the zero entries, but it was not a big issue for the small systems that were solved at that time. In the 1950s, people started to be interested in larger problems with many zero entries. Because the computers were slow and had small memories, it became important to be able to exploit the nonzero structure of the matrices. This occurred, for instance, in structural analysis (see John Hadji Argyris (1913-2004) [73] in 1954), linear programming (see Dantzig [812, 810] in 1951), studies of power systems, computational circuit design, and discretization of partial differential equations (see Forsythe and Wolfgang Richard Wasow (1909-1993) [1199] in 1960, Robert Davis Richtmyer (1910-2003) [2570] in 1957).

As a simple example, let us discretize the problem $-\Delta u = f$ in $\Omega = [0,1]^2$, $u|_{\partial\Omega} = 0$ with standard finite differences on a regular Cartesian mesh with $(m+2) \times (m+2)$ mesh points. We obtain a symmetric matrix A of order $n = m^2$, with only at most five nonzero entries in each row. With a standard ordering of the mesh points for left to right and bottom to top, there are only five nonzero diagonals, the outer diagonals being parallel to the principal diagonal and $a_{i,j} = 0$ for $|i-j| > m$. The diagonal entries of A are equal to 4, and the entries on the other nonzero diagonals are -1 or 0 (because of the boundary conditions). The *half-bandwidth* is m and the bandwidth $2m+1$. However, some authors defined the bandwidth to be m or sometimes

$m+1$. This matrix can also be considered as a block tridiagonal matrix, the diagonal blocks corresponding to lines in the mesh. It is what was called a *band matrix* or a *banded matrix*.

Storing these matrices arising for regular finite difference problems by diagonals was a possibility. However, this was not feasible when using elimination methods to solve the linear system. When we use Gaussian elimination, the entries of the matrix at step $k+1$ are modified as follows

$$a_{i,j}^{(k+1)} = a_{i,j}^{(k)} - \frac{a_{i,k}^{(k)} a_{k,j}^{(k)}}{a_{k,k}^{(k)}}, \quad k+1 \le i \le n, \quad k \le j \le n,$$

$$a_{i,j}^{(k+1)} = a_{i,j}^{(k)}, \quad 1 \le i \le k, \quad 1 \le j \le n, \text{ and } k+1 \le i \le n, \quad 1 \le j \le k-1.$$

We immediately see that even if $a_{i,j}^{(k)} = 0$, $a_{i,j}^{(k+1)}$ can be nonzero if $a_{i,k}^{(k)} a_{k,j}^{(k)} \neq 0$. This phenomenon is called *fill-in*. We also see that fill-in entries can create more fill-in later on in the elimination process. When the matrix is banded, the fill-in is located within the band. This was noticed, for instance, by Robert Kenneth Livesley [2086] in 1960, about structural analysis problems:

> *It is sometimes assumed the elimination requires storage of the complete matrix, since elements which are initially zero may become non-zero during the solution process. Consideration quickly shows, however, that not all the zero elements are affected in this way. If the matrix consists of a number of bands parallel to the leading diagonal, then the parts of the matrix lying outside the extreme band will not be affected.*

Algol codes for linear system solves with banded matrices were described by Roger S. Martin and Wilkinson for symmetric positive definite matrices [2155] in 1968, and for nonsymmetric matrices [2156] in 1967. These papers were first published in the journal *Numerische Mathematik* and then in the *Handbook for Automatic Computing* [3256] in 1971; see Section 8.3. These codes stored only the entries within the band (even if they were zero).

Matrices with many zero entries are called *sparse matrices*. It is difficult to give a precise definition of what a sparse matrix is and to say how many zeros must be there or even what percentage of zeros it should have. Special techniques are used to store sparse matrices and special algorithms are defined in order to minimize the storage and the number of operations during Gaussian elimination. Therefore, a definition that has sometimes been given is that a matrix is sparse when it pays (either in computer storage or in computer time) to use these special sparse techniques as opposed to the more traditional dense (or general) algorithms.

We do not know who used the expression "sparse matrix" for the first time, but an Internet search for "sparse matrix" from 1930 to 1960 gave mainly references to papers about geology! The first occurrence we found for a mathematical paper was by William Orchard-Hays (1918-1989) [2383] in 1956. The title of his paper was *An efficient form of inverse for sparse matrices*. At that time Orchard-Hays was working on linear programming codes for the Rand Corporation with Dantzig, the father of the famous simplex algorithm [810]. In the abstract, he referred to Harry Max Markowitz, who received the Nobel Prize in Economic Sciences in 1990. In 1954-1955, Markowitz proposed a method based on the Gauss-Jordan elimination for inverting a matrix with many zero entries [2143, 2144] that he called *the elimination form of the inverse*, but he did not use the expression "sparse matrix." In that paper Markowitz also defined a pivotal strategy to reduce the fill-in but we will come to that later.

It has been proved that the inverse of an irreducible sparse matrix is a full (dense) matrix; see [987, 988]. Hence, it is not a very good idea to compute the inverse of a large sparse matrix. This is why Markowitz kept the inverse as a product of sparse matrices.

Concerning Gaussian elimination for sparse matrices, the situation is different for symmetric matrices, particularly positive definite ones, for which it is not always necessary to pivot and nonsymmetric matrices for which pivoting is mandatory to have some (weak) form of stability. When it is not necessary to pivot, a data structure for storing the triangular factors can be determined and constructed before doing the numerical elimination. These two distinct phases are sometimes called "Analyze" and "Factorize." This decoupling is not always possible for nonsymmetric problems. This is why the research on sparse matrices focused first on symmetric matrices.

Conferences about sparse matrices were organized at the end of the 1960s and in the 1970s; see the proceedings edited by Ralph Arthur Willoughby (1934-2001) [3258] in 1968, Reid [2555] in 1970, Donald James Rose (1944-2015) and Willoughby [2594] in 1971, Bunch and Rose [505] in 1975, Duff and G.W. Stewart [1007] in 1978, and Duff [978] in 1981.

In the 1960s, it was realized that storing all the entries in the band of a sparse matrix was a waste of resources for problems with an irregular structure like, for instance, those arising from power systems or structural mechanics. Hence, new ways of storing sparse matrices were developed. In 1966, Alan Jennings, from the Queen's University in Belfast, proposed a compact storage scheme for symmetric matrices [1810], in which only entries from the leftmost nonzero element to the diagonal are stored for each row in a one-dimensional array. The beginning of each row in the array is given by a pointer. Since, for positive definite matrices, it is not necessary to pivot, the data structure was ready to accommodate the fill-in. Such a scheme could be used also to store the matrix by columns. It has been sometimes called the "variable band", "envelope" or "skyline" (when the entries are stored by columns) scheme.

Let $\beta_i = \min\{i \,|\, a_{i,j} \neq 0\}$. The envelope is defined as $\mathrm{Env}(A) = \{(i,j) \,|\, 0 < i - j \leq \beta_i(A),\, i \geq j\}$ and the profile is $P_r(A) = |\mathrm{Env}(A)| = \sum_{i=1}^{n} \beta_i(A)$. We see that one definition of the bandwidth is $\beta = \max_i\{\beta_i(A),\, 1 \leq i \leq n\}$. We also observe that the fill-in is contained in the envelope; see [1327] by John Alan George and J.W.-H. Liu in 1975. A column j is active in row i if $j > i$ and there is a nonzero entry in column j in some row $k \leq i$. Let w_i be the number of active columns in row i, the wavefront is defined as $W = \max_i w_i$.

The compact storage scheme stores for each row all the elements of the envelope in a vector and uses another vector of integers to point to the start of each row.

Other storage schemes were proposed in the 1960s and 1970s. In 1968, Knuth [1920] described a scheme in which all the matrix entries are stored in a one-dimensional array with two other integer arrays for the row and column numbers, as well as pointers to the next entry in a row or a column (linked lists) and pointers to the start of rows and columns. Clearly, this stores redundant information for the row and column numbers. Curtis and Reid [793] in 1971, and Gustavson [1491] in 1972 used what is now known as the *compressed storage by rows* (CSR). In this scheme, the nonzero entries of a row are stored consecutively in a one-dimensional array with an integer array giving the column numbers. The only additional storage is an integer array giving the start of each row. The drawback of this scheme is that it is difficult to insert new entries.

Sparse techniques were used from problems arising from electric networks by Gary Deane Hachtel, Robert King Brayton, and Gustavson [1521] in 1971; see also [1520] by Hachtel in 1976. An early package for solving sparse linear systems was described by McNamee [2184, 2185] in 1971.

In those years the connection between sparse matrices and graphs was also established. This was considered by David Rosenblatt (1919-2001) [2597] in 1959, by Frank Harary (1921-2005) [1589] in 1959 for the sake of finding a permutation to block triangular form (see also [1590, 1591]), and by Seymour Victor Parter [2466, 2467] in 1960.

A graph can be associated with every matrix. For a general nonsymmetric matrix A of order n, a directed graph (or digraph) is associated. A digraph is a couple $G = (X, E)$, where X is a set of nodes (or vertices) and E is a set of directed edges. There are n nodes and there is a directed edge from i to j if $a_{i,j} \neq 0$. Usually, self-loops corresponding to $a_{i,i} \neq 0$ are not included. For a symmetric matrix, the graph is undirected and the arrows on the edges are dropped since $a_{j,i} = a_{i,j}$. The degree of a node is the number of its neighbors in the graph.

Consider a sequence of graphs where $G^{(1)} = G$, the graph of a sparse matrix A, and $G^{(i+1)}$ is obtained from $G^{(i)}$ by removing the node x_i from the graph as well as all its incident edges and adding edges such that all the remaining neighbors of x_i in $G^{(i)}$ are pairwise connected. Then, proceeding from $G^{(i)}$ to $G^{(i+1)}$ corresponds to the ith step of Gaussian elimination. This was proved by Parter [2467] in 1961; see also Rose [2589, 2590] in 1970. The use of graphs greatly simplified the studies of elimination methods for sparse matrices.

An important tool to study the relations between Gaussian elimination for symmetric matrices and graphs is the elimination tree; see Robert S. Schreiber [2711] in 1982 and J.W.-H. Liu [2079, 2081] in 1986-1990. The elimination tree $T(A)$ of A, a symmetric matrix of order n, is a graph with n nodes such that the node p is the parent of node j, if and only if

$$p = \min\{i \,|\, i > j,\ \ell_{i,j} \neq 0\},$$

where $L = (\ell_{i,j})$ is the Cholesky factor of A. A topological ordering of a rooted tree is defined as a numbering that orders children nodes before their parents. Algorithms for determining the elimination tree structure were given by J.W.-H. Liu [2081]. Let $i > j$, $\ell_{i,j} \neq 0$ if and only if x_j is an ancestor of some x_k in $T(A)$ such that $a_{i,k} \neq 0$. The structure of column j of L is given by

$$\mathrm{Adj}_G(T[x_j]) \cup \{x_j\} = \{x_i |\, \ell_{i,j} \neq 0, i \geq j\},$$

where $T[x_j]$ is the subtree of $T(A)$ rooted at x_j and $\mathrm{Adj}_G(Y)$ is the set of the nodes adjacent to the nodes of Y.

When developing sparse matrix techniques, it became interesting to try to minimize the fill-in since this could reduce the storage and the number of operations. This can be done by changing the ordering of the unknowns. Research on this topic started at the beginning of the 1960s. It first considered problems with symmetric matrices or matrices with a symmetric nonzero structure.

An ordering method was proposed by William Frank Tinney (1921-2019) and his collaborators working on electric power networks at the Bonneville Power Administration in Portland, Oregon; see N. Sato and Tinney [2688] in 1963, and Tinney and J.W. Walker [3045] in 1967 as well as [637] in 1969 and [2364] in 1970. The matrices had a symmetric nonzero structure and, in fact, Tinney and Walker used a symmetric version of the Markowitz strategy (probably without knowing it). Later on, Rose [2591] developed a graph theoretic model of the algorithm, and he renamed Tinney and Walker's algorithm the *minimum degree algorithm* since, at each step of the ordering algorithm, a node of minimum degree is chosen.

Research on efficient implementations of the algorithm was done in the 1970s and 1980s; see Duff and Reid [999] in 1974, Stanley Charles Eisenstat (1944-2020), M.C. Gursky, Martin Harvey Schultz, and Andrew Harry Sherman [1072, 1069, 1074] in 1976-1982, George and J.W.-H. Liu [1329, 1330] in 1980. An important issue was the choice of the tie-breaking strategy when there are several candidate nodes at some step. For a history of the minimum degree algorithm up to 1989, see [1332].

Even if well implemented, the minimum degree is expensive. In 1996 an approximate minimum degree algorithm was proposed by Patrick Amestoy, Timothy Alden Davis, and Duff [47] to reduce the costs.

Early papers proposing algorithms for bandwidth or profile reduction were by G.G. Alway and D.W. Martin [42] in 1965, R. Rosen [2596] in 1968, Elizabeth Hahnemann Cuthill (1923-2011) and James McKee [797] in 1969, and Ilona Arany, Lajos Szóda, and W.F. Smyth [68, 69] in 1971.

Alway and Martin considered all $n!$ possible permutations of the lines and columns of the matrix, making this method too expensive for large problems. Rosen's algorithm is iterative, looking for an interchange of rows and columns to reduce the bandwidth at each iteration. Unless one starts from a good initial guess, the result can be poor. What was proposed by Cuthill and McKee for matrices with a symmetric nonzero structure is a heuristic algorithm which, given a starting vertex, labels successively for $i = 1, \ldots, n$, in order of increasing degree, all the unlabeled adjacent vertices of node i. Even though it was designed as a bandwidth reduction algorithm, it can be used to reduce the profile. Slight improvements of this algorithm were described by Arany, Szóda, and Smyth; see also [2815].

Elizabeth Cuthill obtained her P.h.D. from the University of Minnesota in 1951 and was working for the Carderock Division, David Taylor Naval Ship Research and Development Center in the USA. In 1972, she wrote a review of ordering algorithms [796]. The Cuthill-McKee algorithm (CM) is still popular, particularly in its reverse version (RCM) that was proposed by George in his Ph.D. thesis [1322] in 1971. It consists of using CM and then, labelling the nodes in the reverse order, from n to 1. It was proved by J.W.-H. Liu and A.H. Sherman [2085] in 1976 that the reverse ordering is always at least as good as the original one, and it gives a smaller envelope. A linear time implementation of RCM [629] was proposed in 1980 by Wing-Man Chan and George.

An important issue for the efficiency of CM or RCM was the choice of the starting node. The distance $d(x, y)$ between two nodes x and y of a graph G is the length of the shortest path between x and y. The eccentricity $e(x)$ of a node is $e(x) = \max\{d(x, y) \mid y \in X\}$. The diameter δ of G is $\delta(G) = \max\{e(x) \mid x \in X\}$. A node x is peripheral if $e(x) = \delta(G)$.

A good choice would be to choose a peripheral node as a starting node, that is, one whose eccentricity equals the diameter of the graph. Peripheral nodes are not that easy to find quickly. Therefore, researchers devised heuristics to find *pseudo-peripheral* nodes, e.g., nodes whose eccentricities are close to the diameter of the graph. Such an algorithm was proposed by Norman E. Gibbs (1941-2002), William George Poole, and Paul Kelly Stockmeyer [1347, 1348] in 1976. A modification of this algorithm was suggested by George and J.W.-H. Liu [1328] in 1979; see also [66, 67]. These algorithms used the notion of level structure. Given a node x which is the only element of the level set $L_0(x)$, the first level $L_1(x)$ is defined by the neighbors of x, the second level $L_2(x)$ is the set of the neighbors of the nodes in $L_1(x)$, and so on. Starting from a given node x, the level sets $L_i(x)$ are generated. For the nodes in the last level, the level sets are generated. If there is a node y with more levels than for x, y is the new selected node and one continues until it is not possible to increase the number of levels; see also [66, 67].

Other ordering algorithms were proposed by Richard Arthur Snay [2816] in 1976 (the so-called banker's algorithm), Scott William Sloan (1954-2019) [2799] in 1986, and Gary Kumfert and Alex Pothen [1962] in 1997; see also J.W.-H. Liu [2082] in 1991.

Finding an ordering of the unknowns that minimizes the fill-in is NP-hard. It means that it cannot be found in polynomial time. This was proved in 1981 by Mihalis Yannakakis in a three-page paper [3292]. Therefore, all the algorithms that were devised were based on heuristics. A review of bandwidth reduction methods was done in [2114]. Even though the English of that paper is quite poor, it contains interesting information. Other comparisons of ordering algorithms were published in [843, 1101, 1118]. In 2015, the Brazilian authors of [292] identified 74 heuristic algorithms for profile reduction of symmetric matrices, of which eight were considered "the best."

In 1970, Bruce Moncur Irons (1924-1983), a British engineer and mathematician, described a *frontal method* for solving linear systems with a symmetric positive definite matrix arising from finite element problems [1784]. In finite element methods, the entries of the matrix are sums of contributions from different elements, usually resulting from approximate values of some integrals. This is called the *assembly* of the matrix. In frontal solvers, the assembly and the elimination are intertwined and the global matrix is never stored. This was well suited for out-of-core solutions of large problems on computers with a small memory since only a small frontal matrix had to be stored in memory. A variable is eliminated as soon as it is fully assembled. The efficiency of the frontal method is dependent upon the order in which the elements are assembled. It must keep the size of the frontal matrix as small as possible, that is, the elements need to be ordered such that partially summed variables become fully summed as soon as possible. Since the original work of Irons, the frontal method has been developed and generalized to nonsymmetric systems by a number of authors, including P. Hood [1727] in 1976, Duff [977, 980] in 1981 and 1984, Duff and J.A. Scott [1006] in 1996, and J.A. Scott [2731] in 2006; see Section 8.9; see also Bert Speelpenning [2841] in 1973-1978.

In 1971, R. Levy, working on a preprocessor of the Nastran structural analysis program, labeled vertices successively such that a minimum increase in the wavefront occurs as each vertex is labeled [2042]. Ian P. King's algorithm [1902] in 1970 is similar, but at each step only unlabeled vertices adjacent to already labeled vertices are considered. A more recent paper on wavefront reduction is [1763] by Yifan Hu and J.A. Scott in 2001.

The nested dissection algorithm was introduced by George [1323] in 1973 for finite element problems on a regular mesh with symmetric matrices and then generalized to general sparse matrices; see also Garrett Birkhoff (1911-1996) and George [327] in 1973. It is very close to an old idea used in mechanics called substructuring (see Janusz Stanislaw Przemieniecki (1927-2017) [2520] in 1963, who was also a forerunner of domain decomposition methods).

An example of nested dissection of a regular mesh is shown in Figure 2.7. The nodes labeled 1 are numbered first, and then successively those labeled 2, 3, and 4.

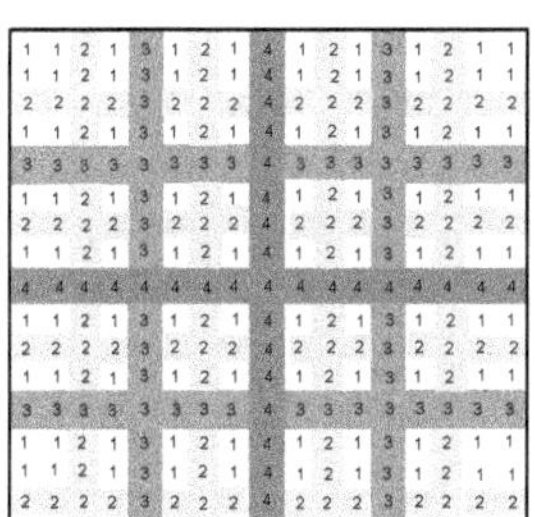

Figure 2.7. *Dissection of a regular mesh*

This strategy was justified later by Rose, Robert Endre Tarjan, and George Schick Lueker [2593] in 1976, and George and J.W.-H. Liu [1331] in 1981. Let $G = (V, E)$ be a graph and $S \subset V$, $x \in V$, $x \notin S$, x is said to be reachable from $y \notin S$ through S if there exists a path $(y, v_1, \dots, v_k, x)$ from y to x in G such that $v_i \in S$, $i = 1, \dots, k$. We define

$$\text{Reach}(y, S) = \{x | x \notin S, x \text{ is reachable from } y \text{ through } S\}.$$

Let $k > j$; there will be a fill between nodes x_j and x_k if and only if $x_k \in \text{Reach}(x_j, \{x_1, \dots, x_{j-1}\})$. Let us consider the example in Figure 2.8. The diagonal from the bottom-left node to the top-right node separates the graph in three pieces 1, 2, 3 as shown in the right part of the

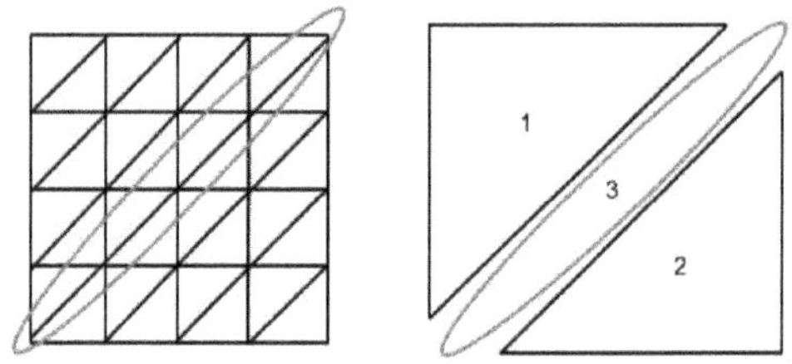

Figure 2.8. *Dissection of a graph*

figure. It is called a *vertex separator*. If we number the nodes in the separator 3 last, there won't be any fill-in between the nodes in 1 and those in 2. Then, we can find recursively separators for the nodes in 1 and for the nodes in 2. When the sets of nodes become small another ordering algorithm is used, for instance, the minimum degree. This explains the name "nested dissection," which is a fill-in reducing algorithm.

The storage schemes that we have seen above are not well suited for nested dissection orderings. In nested dissection, there is a natural block structure that arises, the blocks corresponding to separators or to the nodes that are ordered first.

Duff, Albert Maurice Erisman, and Reid extended the method to irregularly shaped grids and 3D problems [989] in 1976. Nested dissection was generalized to symmetric matrices with planar graphs by Richard Jay Lipton, Rose, and Tarjan [2077] in 1979; see also Lipton and Tarjan [2078].

For a general matrix with a symmetric nonzero structure, what has to be dissected is the graph of the matrix. A method based on level sets was proposed by George and J.W.-H. Liu in their book [1331] in 1981. Of course, one can use heuristic partitioning methods that were devised without relation to linear algebra. Many of these methods compute edge separators but they can be converted to vertex separators; see, for instance, Brian Wilson Kernighan and Shen Lin [1885], who proposed an iterative swapping algorithm in 1970 and Charles M. Fiduccia and Robert M. Matheyses [1156] in 1982 with an improvement of the Kernighan-Lin algorithm.

Other more recent methods used spectral information about the graph. In 1991, Horst D. Simon [2769] relied on the computation of the smallest nonzero eigenvalue of the Laplacian matrix of the graph. The Laplacian $\mathcal{L}$ of the graph is constructed as follows: for row i, $\mathcal{L}_{i,j} = -1$ if node j is a neighbor of node i and the diagonal term is minus the sum of the other entries. From the corresponding eigenvector, a bisection of the graph is obtained by considering the components of the eigenvector larger or smaller than the median value. This eigenvector is called the *Fiedler vector* in reference to Miroslav Fiedler [1158, 1161].

The Fiedler vector was also used by Pothen, Simon, and Kang-Pu Liou [2514] in 1990 and by Stephen T. Barnard, Pothen, and Simon to devise a spectral algorithm for envelope reduction [201] in 1995.

Partitioning large graphs is expensive. Therefore, multilevel methods were introduced where the graph is coarsened several times allowing a partitioning which is less costly; see Bruce Hendrickson and R. Leland [1638] in 1995, and George Karypis and Vipin Kumar [1876, 1877] in 1998. These multilevel methods were implemented in the Chaco and METIS libraries.

When the matrix is nonsymmetric, one can work with the graph of $A + A^T$; see George and Esmond Gee-Ying Ng [1336] in 1988. Another possibility is to use hypergraph partitioning. A hypergraph (V, E) is a set of vertices V and a set of hyperedges E. A hyperedge is a subset of V; see, for instance, U.V. Çatalyürek and C. Aykanat [593] in 1999. Another approach was used by Laura Grigori, Erik Gunnar Boman, Simplice Donfack, and T.A. Davis [1462] in 2010.

The multifrontal method has been introduced for symmetric indefinite matrices by Duff and Reid [1003, 1004] in 1983-1984 as a generalization of the frontal method developed by Irons for finite element problems. The main goal of the multifrontal method is to be able to use dense matrix technology for sparse matrices. Technical details are quite complex and many refinements are necessary to make the method efficient. A nice exposition of the principles of the method was given by J.W.-H. Liu [2083] in 1992. In essence, the method is able to work on small dense matrices which are all independent until the algorithm reaches a point in which some of these matrices need to be combined. For a symmetric matrix, the nodes composing the frontal matrices can be determined by looking at the elimination tree. The constraint is that the computation of a parent front must be done after the computation of all its child fronts is complete. Some contiguous nodes in the tree can also be amalgamated to reduce the size of the tree and increase the order of the frontal matrix. The efficient storage of the frontal matrices and the memory management are also quite complicated. Dense algorithms based on the use of Level 3 BLAS (see Chapter 8) can be used when factoring the frontal matrices. There is parallelism at two levels, some frontal matrices can be processed in parallel and parallelism can also be used when factoring dense frontal matrices. The method was incorporated in the HSL software library as MA41 [1003] (see Chapter 8).

The multifrontal method was applied to nonsymmetric matrices by Duff and Reid [1004] in 1984. The idea was to consider the sparsity pattern of $A + A^T$ to construct the elimination tree. Numerical pivoting takes place within the frontal matrices. This works well if the pattern of A is nearly symmetric. However, the results may be poor if the pattern of A is far from being symmetric. The parallel aspects of the multifrontal method were considered by Duff in [983, 985] in 1986-1989. In the 1990s, T.A. Davis and Duff introduced an other extension of the multifrontal algorithm to nonsymmetric matrices; see [828, 829] as well as Steven M. Hadfield and T.A. Davis [1527, 1528]. This resulted in the UMFPACK package; see [824]. In the nonsymmetric case, the frontal matrices are rectangular and the elimination tree is replaced by a directed acyclic graph (DAG).

The original MA41 was a shared memory parallel code. A working name for MA41 was MUPS (MUltifrontal Parallel Solver). When the team working on this method started the PARASOL project, the distributed memory code was renamed MUMPS (MUltifrontal Massively Parallel Solver); see [48, 49, 44, 50, 2163, 45, 46] (ordered by date). For details on the history of multifrontal methods, see [832].

The *left-looking LU factorization* (also known as the fan-in algorithm) computes L and U one column at a time. At the kth step, it accesses columns 1 to $k-1$ of L and column k of A. In the *right-looking LU factorization* (also known as the fan-out algorithm) at each step, an outer product of the pivot column and the pivot row is subtracted from the lower right submatrix of A. This is the schoolbook Gaussian elimination. These methods are the nonsymmetric equivalents of the different ways of coding the LDL^T factorization we have seen above.

Quite often, in the LU factorization, several columns and rows have the same structure and can be grouped. The *supernodal method* takes advantage of this to save time and storage space. It stores less integer information and it operates on dense submatrices rather than on individual rows and columns. All the nodes in a supernode are eliminated in the same step of the algorithm.

This technique was first developed for symmetric problems. In 1987, Ashcraft, Grimes, J.G. Lewis, Barry W. Peyton, and Simon [88] proposed a left-looking method; see also [2083] by J.W.-H Liu in 1992. A supernodal symbolic factorization to identify the supernodes was described by J.W.-H. Liu, E.G. Ng, and Peyton [2084] in 1993. Edward Rothberg and Anoop Gupta [2603, 2604] showed in 1991 and 1993 that the supernodal method can be implemented as a right-looking method quite similar to the multifrontal method. In 1993, E.G. Ng and Peyton [2342] described a parallel left-looking supernodal algorithm. A parallel version of their algorithm was published by Rothberg and An. Gupta [2605] in 1994. In 2002, Pascal Hénon, Pierre Ramet, and Jean Roman [1639] combined both left- and right-looking approaches. A DAG-based scheduler for a parallel shared memory supernodal Cholesky factorization was presented by Jonathan D. Hogg, Reid, and J.A. Scott [1723] in 2010.

The generalization of the supernodal method to nonsymmetric matrices was far from being trivial. In 1995, Demmel, Eisenstat, John Russell Gilbert, Xiaoye Sherry Li, and J.W.-H. Liu [871] introduced the idea of nonsymmetric supernodes and implemented them in a left-looking code named SuperLU with a paper published in 1999. Demmel, Gilbert, and X.S. Li [872] presented a shared memory version of SuperLU, named SuperLU MT. It is a left-looking method that exploits two levels of parallelism. In 2003, X.S. Li and Demmel [2058] extended SuperLU to distributed memory machines with SuperLU DIST. It uses a static numerical pivoting, but one step of iterative refinement is usually enough for SuperLU DIST to reach a good accuracy; see also [2057, 2059].

We have already seen that there is an additional difficulty in Gaussian elimination for nonsymmetric sparse problems, namely the need to pivot to improve numerical stability. If the pivots are chosen as it is done for dense systems (for instance, partial pivoting), there is no room for preserving sparsity.

In 1966, Tewarson [3026] used the Gauss-Jordan algorithm to compute the *product form of the inverse* (PFI) of a sparse matrix (which was much in use in linear programming [811]), trying to maintain sparsity and avoiding small pivots; see also [3027, 3028, 3029, 3031]. Comparisons with other methods were done by Brayton, Gustavson, and Willoughby [415] in 1970.

For sparse matrices, the constraints for choosing a pivot have to be relaxed. In a right-looking LU factorization the usual strategy is to consider candidate pivots satisfying the inequality

$$|a_{i,j}^{(k)}| \geq w \max_{\ell} |a_{\ell,j}^{(k)}|,$$

where w is a user-defined parameter in $]0,1]$. From these candidates, one is selected that minimizes $(r_i^{(k)}-1)(c_j^{(k)}-1)$, where $r_i^{(k)}$ (resp., $c_j^{(k)}$) is the number of nonzero entries in row i (resp., column j) of the remaining matrix of order $n-k$. This is a relaxed version of the Markowitz criterion [2143]; see the first edition of the book [990] in 1987.

Curtis and Reid [793] provided in 1971 an implementation of Markowitz' pivot strategy. In 1974, Duff and Reid [999] compared this method with four others for selecting a pivot during a right-looking factorization. They recommended Markowitz' strategy. Duff [982] used dynamic data structures in 1985 since the nonzero structure of the LU factors is not known a priori. In 1980, Zahari Zlatev [3337] limited the search to just a few of the sparsest columns. Duff and Reid [1005] used this pivot strategy in 1996 for the code MA48 of the HSL Library, as well as a switch to dense matrix when the number of zeros became too small.

Let us consider a list of papers concerned with the computation of the nonzero structure of the L and U factors. Of course, this list is nonexhaustive and reflects our own preferences, but it shows what kind of research was done on that problem in the last years of the 20th century and later.

- 1976, Tarjan [3001] discussed graph theory and Gaussian elimination.
- 1978, Rose and Tarjan [2592] were the first to consider the symbolic structure of Gaussian elimination for nonsymmetric matrices.
- 1978, A.H. Sherman [2754] proposed a left-looking method with partial pivoting and a dynamic data structure.

- 1982, Jochen A.G. Jess and H.G.M. Kees [1818] described a parallel right-looking LU factorization algorithm for problems with a symmetric nonzero pattern.
- 1985, George and E.G. Ng [1334] showed that if $PA = LU$ and $A = QR$, where P is determined by partial pivoting and Q is orthonormal, then R represents an upper bound on the pattern of U; see also [1335] in 1987.
- 1987, Randolph Edwin Bank and R. Kent Smith [189] described Fortran routines for the LDU factorization of a sparse matrix where the numerical factorization can be carried out without an additional permanent integer data structure for the factors; see also Bank and Rose [188] in 1990.
- 1988, Gilbert and Timothy Peierls [1353] described a left-looking method taking time proportional to the number of floating point operations.
- 1990, George and E.G. Ng [1337] described a parallel algorithm for their method [1337], suitable for shared-memory computers.
- 1993, Gilbert and J.W.-H. Liu [1351] characterized the structure of the triangular factors without pivoting by generalizing the elimination tree for a symmetric matrix to a pair of directed acyclic graphs (DAGs) for a nonsymmetric matrix.
- 1993, Eisenstat and J.W.-H. Liu [1071] generalized the quotient graph representation of George and Liu to the nonsymmetric case.
- 1993, Gilbert and E.G. Ng [1352] showed that the bound with R is tight if A has the strong Hall property. A matrix is strong Hall if it cannot be permuted to block upper triangular form with more than one block.
- 1994, Gilbert [1349] wrote a survey of the use of graph algorithms for symbolic analysis, in particular, for the LU factorization.
- 2001, Gilbert, X.S. Li, E.G. Ng, and Peyton [1350] described an algorithm for computing the row and column counts in the LU factorization.
- 2004, Pothen and Sivan Toledo [2515] wrote a survey of algorithms and data structures for the symbolic analysis of both symmetric and nonsymmetric factorizations.
- 2005, Eisenstat and J.W.-H. Liu introduced a single tree that takes the place of the two dags of Gilbert and Liu [1351].
- 2006, T.A. Davis described an implementation of the Gilbert and Peierls algorithm [1353].
- 2007, Grigori, Michel Cosnard and E.G. Ng [1463] considered the column elimination tree when A is not strong Hall (the so-called row-merge tree).
- 2009, Grigori, Gilbert, and Cosnard [1464] showed that the row-merge tree provides a tight bound on the structure of L and U.

Parallelizing the algorithms for the solution of sparse linear systems was a difficult task, even though there is sometimes more parallelism available in the sparse case than for dense matrices. Many papers have been written on this topic, including a few surveys by Michael Thomas Heath, E.G. Ng, and Peyton [1614] in 1991, Duff and Hendrik Albertus (Henk) van der Vorst [1010] in 1999, and T.A. Davis, Sivasankaran Rajamanickam, and Wissam M. Sid-Lakhdar [832] in 2016; see also Chapter 8.

In some cases it is possible to reorder the matrix to a block triangular form. Then, it is only necessary to factorize the diagonal blocks before a block forward solve. The block triangular form reordering is based on the Dulmage-Mendelsohn decomposition using maximum matching on bipartite graphs; see Andrew Lloyd Dulmage (1917-1989) and Nathan Saul Mendelsohn (1917-2006) [1022, 1023, 1024] in 1958-1963. On this problem, see Gustavson [1492] in 1976, Duff [975] in 1977, Duff and Reid [1001, 1000] in 1978, T.A. Davis [826] in 2006, and Duff and Bora Uçar [1008] in 2010 and [1009] in 2012.

Block LU factorization has also been considered. In these factorizations L and U are block triangular but not necessarily triangular. This was considered, for instance, by Eugene Isaacson (1919-2008) and Herbert Bishop Keller (1925-2008) [1785] in 1966, Varah [3124] in 1972 for block tridiagonal matrices, Robert M.M. Mattheij [2168, 2169] in 1984, and Paul Concus, Golub and G.M. [735] in 1985 for the sake of obtaining preconditioners for block tridiagonal matrices.

Error analyses were done by Demmel and N.J. Higham [875] in 1992, and Demmel, N.J. Higham, and Schreiber [876] in 1995. Block LU factorization (without pivoting) is unstable in general, although it has been found to be stable for matrices that are block diagonally dominant by columns, that is,

$$\|A_{i,i}^{-1}\|^{-1} \geq \sum_{i \neq j} \|A_{i,j}\|,$$

where the $A_{i,j}$ are the blocks of the matrix A with square diagonal blocks. This definition was introduced by David G. Feingold and Richard Steven Varga (1928-2022) [1150] in 1962; see also George and Khakim D. Ikramov [1326] in 2005.

Reviews about Gaussian elimination of sparse matrices were written by Tewarson [3031] in 1970, Willoughby [3259] in 1971, Duff [976, 979, 981, 984, 986] in 1977, 1982, 1984, 1989, and 1998 (see also [1544]), G.M. [2219] in 2000, E.G. Ng [2341] in 2013, and T.A. Davis, Rajamanickam, and Sid-Lakhdar [832] in 2016.

Some books about solving linear systems with direct methods are by Francis Begnaud Hildebrand (1915-2002) [1696] in 1956, Émile Durand (1911-1999) [1030] in 1960-1961, Dmitry Konstantinovich Faddeev (1907-1989) and Vera Nikolaevna Faddeeva (1906-1983) [1127] with an English translation in 1963, Wilkinson [3247, 3248] in 1964-1965, Gastinel [1290] with a translation into English [1291] in 1970, André Korganoff and Monica Pavel-Parvu [1937] in 1967, Forsythe and Moler [1196] in 1967 (see also [1195] in 1977), Wilkinson and Christian Reinsch [3256] in 1971, Jennings [1812] in 1977, George and J.W.-H. Liu [1331] in 1981, Sergio Pissanetsky [2498] in 1984, Duff, Erisman, and Reid [990] in 1986 with a second edition in 2017, Zlatev [3338] in 1991, G.M. [2217] in 1999, Carl Dean Meyer [2239] in 2000, and T.A. Davis [826] in 2006.

There exist collections of test sparse matrices available on the Internet; see the *Matrix Market*[23] (NIST) and the *SuiteSparse* matrix collection[24] [830].

2.12 ▪ Lifetimes

In this section, we show the lifetimes of the main deceased contributors to the elimination methods for solving systems of linear equations, starting in 1440. The language of the author is given by the color of the bars and by letters: E (red) for English, G (black) for German, I (green) for Italian, F (blue) for French, and O (magenta) for the others. The contributors are ordered by date of birth.

[23] https://math.nist.gov/MatrixMarket/
[24] https://sparse.tamu.edu/

Elimination methods for solving linear systems (a)

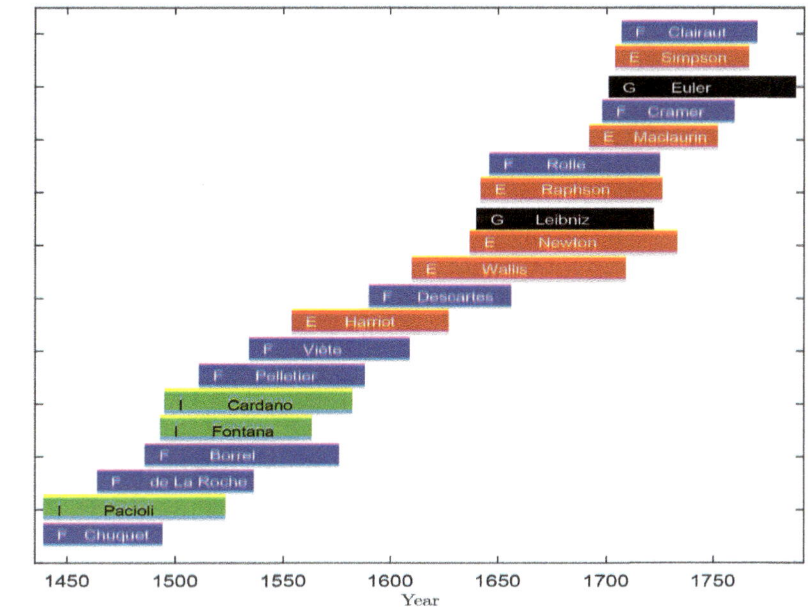

Elimination methods for solving linear systems (b)

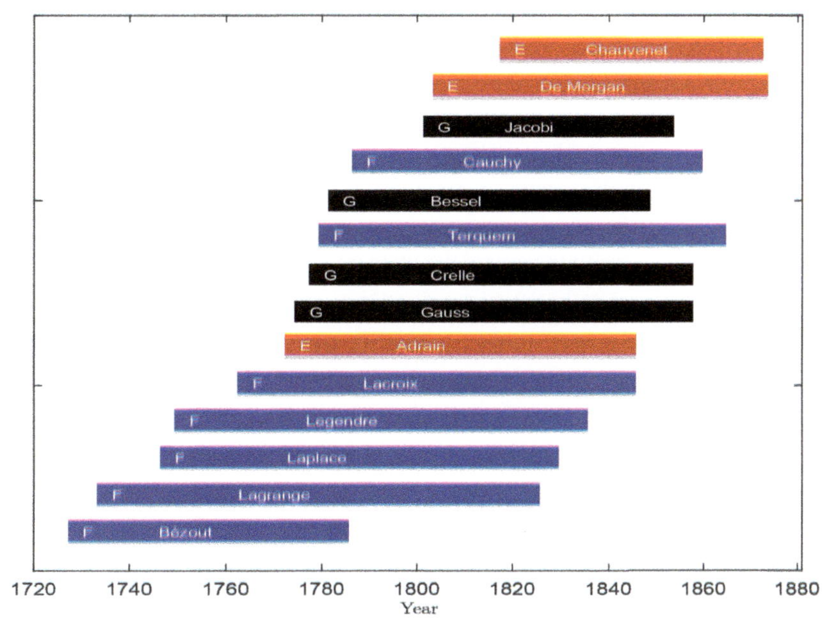

Elimination methods for solving linear systems (c)

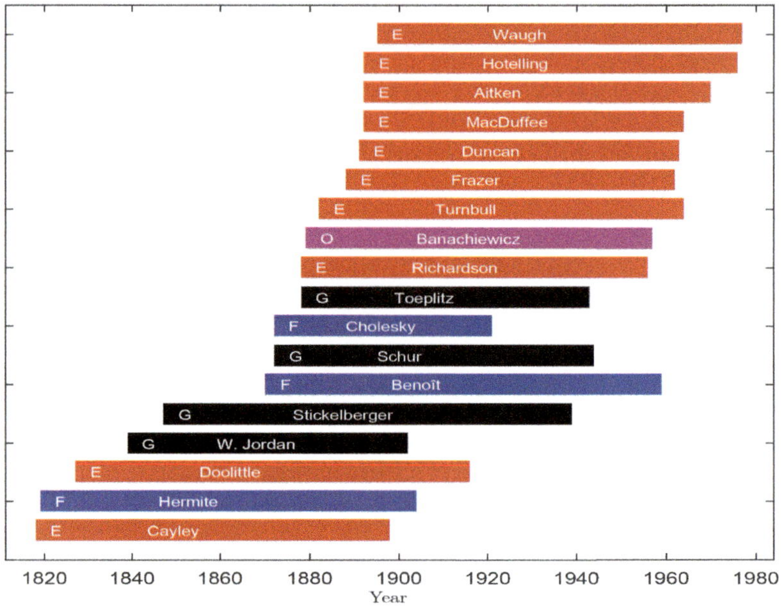

Elimination methods for solving linear systems (d)

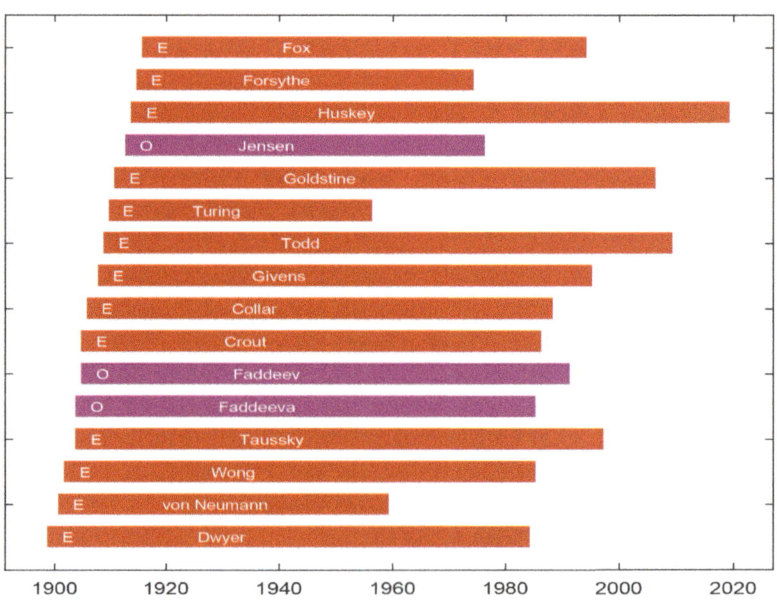

Elimination methods for solving linear systems (e)

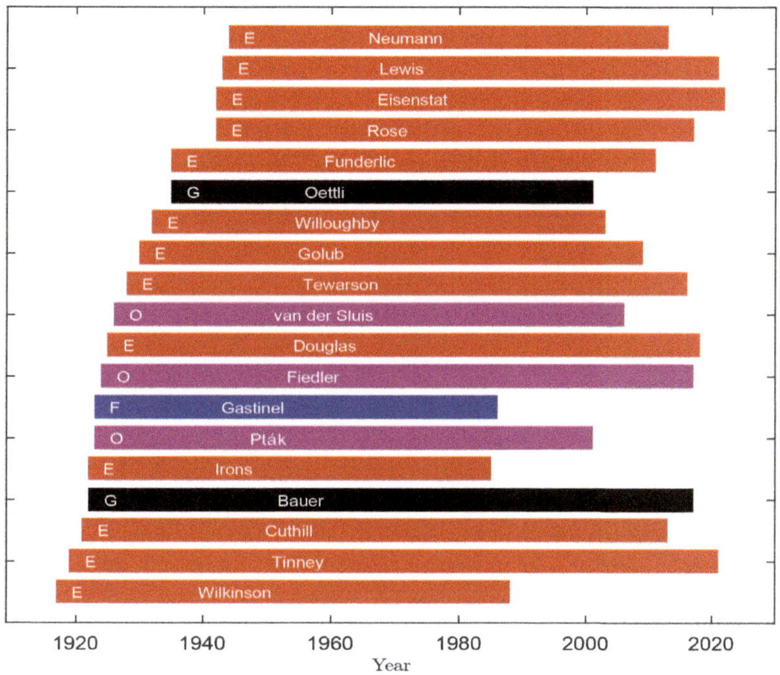

3 Determinants

Je vais maintenant examiner particulièrement une certaine espéce de fonctions symétriques alternées qui s'offrent d'elles-mêmes dans un grand nombres de recherches analytiques. C'est au moyen de ces fonctions qu'on exprime les valeurs générales des inconnues que renferment plusieurs équations du premier degré.

– Augustin-Louis Cauchy, *1815*

The starting point of determinants was mainly the use of elimination methods for solving systems of linear equations. Hence, it is likely that all the people who have tried to solve systems of order 2 or 3 have seen determinants at work without knowing it. This was probably the case in ancient China several hundred years BC. How this could have influenced later developments is, unfortunately, unknown so far.

3.1 ▪ The 17th century

It is generally admitted that the real use of determinants started in the 17th century with two major figures, Seki Kôwa in Japan (see [1608, 2246]) and Gottfried Wilhelm Leibniz (1646-1716), in what is now Germany.

Until recently, not much was known about the life of Seki Kôwa, and certainly not his birthdate; see [2120]. It is thought that he was born in between 1640 and 1645 in Fujioka or Edo (now Tokyo) in Japan. His alias name was Shinzuke and his given name was Takakazu. He was born in the Uchiyama family and adopted by the Seki family. Between 1661 and 1699 he wrote several books and became the prominent Japanese mathematician. He was working for the Tokugawa family as an accountant. When Ienobu Tokugawa (1662-1712) became the sixth Shōgun of the Tokugawa dynasty, Seki Kôwa became chief of a Team of Ceremonies. He died in December 1708. He worked on many different mathematical topics, but here we are interested in his work related to determinants.

He was interested not in linear systems, but in what in our modern notation we call the common solution of two polynomial equations. In the *Kai Fukudai no Hō* (Methods of solving concealed problems) in 1683, Seki gave two eliminating methods for a system of two polynomial equations of degree n. The idea was to combine the equations to obtain a system of n polynomial equations of degree $n-1$. For this system to have a solution, the determinant formed from the coefficients must be zero. Let us illustrate this, in our modern notation, with an example of degree 3

$$a_4x^3 + a_3x^2 + a_2x + a_1 = 0, \tag{3.1}$$
$$b_4x^3 + b_3x^2 + b_2x + b_1 = 0. \tag{3.2}$$

Multiplying the first equation (3.1) by b_4, the second one by a_4, and subtracting, we obtain

$$(a_3b_4 - a_4b_3)x^2 + (a_2b_4 - a_4b_2)x + (a_1b_4 - a_4b_1) = 0. \tag{3.3}$$

Now, we use again (3.1) and eliminate the coefficient of x^2. It yields

$$(a_4b_3 - a_3b_4)x^3 + (a_2b_3 - a_3b_2)x + (a_1b_3 - a_3b1) = 0.$$

To this equation we add equation (3.3) multiplied by x and we obtain

$$(a_2b_4 - a_4b_2)x^2 + [(a_2b_3 - a_3b2) + (a_1b_4 - a_4b1)]x + (a_1b_3 - a_3b_1) = 0. \tag{3.4}$$

This gives us a second equation of degree 2. Continuing in this way, eliminating the constant coefficients in (3.1), we obtain three equations of degree 2. We know (see [1831]) that Seki read some Chinese mathematical books where he probably learned about elimination techniques that the Chinese had mastered for quite a long time.

To compute the determinant (which, of course, does not have this name in his work) Seki considered two methods. The first one is named *chikushiki kōjō* (successive multiplication of each equation by coefficients of other equations). He gave the results only for $n = 2, 3, 4$. It is quite complicated, and Seki introduced another method called *kōshiki* (shuffles) and *shajō'* (oblique multiplications); see [2389]. This method can be seen as a generalization of what we call Sarrus' rule (see below).

In modern terms, shuffles correspond to permutations of the columns of the determinant. The method applies multiplication of entries located on diagonals of several arrays of coefficients obtained from shuffles. The problem was to know which shuffles to use. Seki gave a rule for computing a shuffle for the order n from the shuffle of order $n - 1$. Unfortunately, his result for $n = 5$ was wrong. This was corrected later by Yoshisuke Matsunaga (1693-1744) but his algorithm had other errors. An interpretation of the shuffles using group theory was given in [2389] by Naoki Osada in 2018.

According to Tsuruichi Hayashi (1873-1935) in [1608],

> *The book consists of only a few pages, but they are very difficult to understand, like riddle-books, or secret marks. No book is entirely written in ordinary language. The methods of kōshiki shajō and seikoku which are used to expand a determinant are explained by marks consisting of circles and lines; otherwise only the results are given without explanation.*

Figures 3.1 and 3.2 show modern recreations of Seki's drawings for $n = 3$ and 4. The products of the terms on plain lines are given a $-$ sign and the ones on dotted lines are given a $+$ sign. In the originals the dotted lines were in red. The disks are empty because this process has to be applied to all permutations of columns obtained from the shuffles.

Let us consider the case $n = 3$ for which there is only one permutation of the indices which was chosen as $(3, 2, 1)$ because the Japanese write from right to left. Let $x_{i,j}$, $i, j = 1, 2, 3$ be the entries of the determinant. Filling these entries in the disks of Figure 3.1 we obtain

$$\{-x_{1,1}\, x_{3,2}\, x_{2,3},\ -x_{2,1}\, x_{1,2}\, x_{3,3},\ -x_{3,1}\, x_{2,2}\, x_{1,3}\}$$

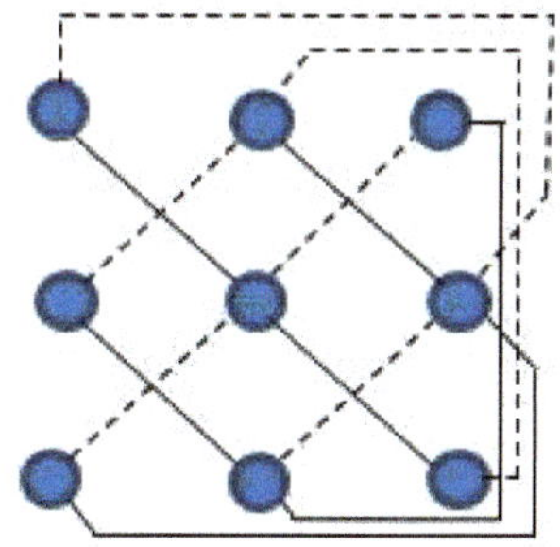

Figure 3.1. *Seki Kôwa's drawing for $n = 3$*

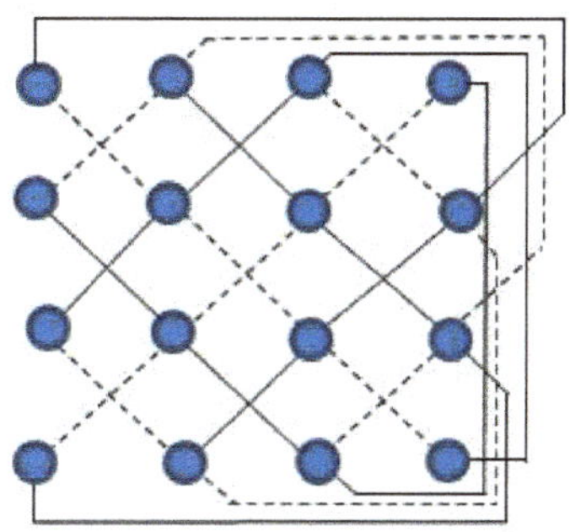

Figure 3.2. *Seki Kôwa's drawing for $n = 4$*

for the plain lines and

$$\{+x_{1,1}\, x_{2,2}\, x_{3,3},\ +x_{2,1}\, x_{3,2}\, x_{1,3},\ +x_{3,1}\, x_{1,2}\, x_{2,3},\}$$

for the dotted lines. When we sum these six quantities we obtain the value of the determinant of order 3. For $n = 4$ we have 24 terms to sum, each term being the product of four entries, because there are three shuffles to consider, $(4, 3, 2, 1)$, $(2, 4, 3, 1)$, and $(2, 2, 4, 1)$, and there are eight products for each shuffle. These three shuffles are obtained by taking the permutations of $(3, 2, 1)$, adding 1 to each number and appending 1 at the end of each tuple.

For $n = 3$, if we copy to the right of the three columns, the two first columns, we obtain

$$\begin{vmatrix} x_{1,1} & x_{1,2} & x_{1,3} & x_{1,1} & x_{1,2} \\ x_{2,1} & x_{2,2} & x_{2,3} & x_{2,1} & x_{2,2} \\ x_{3,1} & x_{3,2} & x_{3,3} & x_{3,1} & x_{3,2} \end{vmatrix}.$$

Now, taking the products along the first three diagonals with three entries (with a $+$ sign) and the products along the last three anti-diagonals (with a $-$ sign), we obtain the same values as before. This is known as Sarrus' rule. It is of course easier to remember than what we have in Figure 3.1. This rule is attributed to Pierre-Frédéric Sarrus (1798-1861), a French mathematician, in 1833; see, for instance, Pierre Joseph Étienne Finck (1797-1870) [1167] in 1846. About Sarrus, see [1700].

Seki's methods continued to be used in Japan until the beginning of the 19th century, when the theory of determinants became gradually neglected there; see Yoshio Mikami (1875-1950) [2246] in 1914. About Japanese mathematics, see [2363, 2806].

In Europe, solving linear equations appeared in print in the 16th century; see Chapter 2. For instance, Girolamo Cardano (1501-1576), in his *Ars Magna* written in Latin and published in 1545, gave formulas for solving some types of cubic equations with one unknown. But, in this book, he also solved 2×2 linear systems in Chapters 9 and 10. All the problems he considered were with specific given integer coefficients.

Later, in the 17th century, Leibniz became interested in the solution of linear equations because he was thinking that the solution of general algebraic equations (that is, polynomial systems) was depending on linear systems. As in many of his other works, he was looking for generality.

Leibniz experimented with many notation systems. He used the indices for the entries and the indices of the diagonal to denote determinants with overbars as

$$\begin{vmatrix} 10 & 11 \\ 20 & 21 \end{vmatrix} = \overline{0.1},$$

where the ij's are indices and not integers. For more on this notation, which was revolutionary in those times, see [2743, 2744] by Michel Serfati. Leibniz already knew what we call Cramer's rule in 1680 for linear systems of order 5; see Eberhard Knobloch [1914]. However, he did not immediately find the correct rule of signs for the terms in the expansion of a determinant. He considered problems of small dimensions and tried to generalize what he found. But his attempts did not always work. So, he struggled with this problem until January 1684, when he found the correct rule and he wrote in French[25] (our translation),

> *Two terms which are distinguished from each other only by an odd number of transpositions of the left or right subscripts have opposite signs. Those that are distinguished from each other by an even number have the same sign.*

Leibniz communicated this rule in a letter to Guillaume François Antoine, marquis de l'Hôpital (or Hospital) (1661-1704) in 1693; see [936, 1915, 1917]. As we said, this letter was written in French; see [2030, pp. 236-241]. A partial English translation can be found in the book [2805] by David Eugene Smith (1860-1944) on pages 267-270. In this letter, Leibniz also wrote

> *For example, consider three simple equations in two unknowns, the object being to eliminate the two unknowns and indeed by a general law. I suppose that*
>
> $$\begin{aligned} 10 + 11x + 12y &= 0, \\ 20 + 21x + 22y &= 0, \\ 30 + 31x - 32y &= 0, \end{aligned}$$
>
> *where, in the pseudo number of two digits, the first tells me the equation in which it is found, the second, the letter to which it belongs.*

Then, Leibniz eliminated y from the two first equations and from the first and third equations to obtain

$$\begin{aligned} 10.22 + 11.22x - 12.20 - 12.21x &= 0, \\ 10.32 + 11.32x - 12.30 - 12.31x &= 0, \end{aligned}$$

where the point represents multiplication. He eliminated x in these two equations, changed notation, and stated that one must have

$$1_0.2_1.3_2 + 1_1.2_2.3_0 + 1_2.2_0.3_1 = 1_0.2_2.3_1 + 1_1.2_0.3_2 + 1_2.2_1.3_0.$$

[25] *Deux termes qui se distinguent l'un de l'autre seulement par un nombre impair de transpositions des indices gauches ou droits ont des signes opposés. Ceux qui se distinguent l'un de l'autre par un nombre pair ont le même signe.*

This is the nullity of the determinant for the three original equations. In his previous notation, it would have been

$$10.21.32 + 11.22.30 + 12.20.31 = 10.22.31 + 11.20.32 + 12.21.30.$$

Being proud of his notation, Leibniz added

> [...] *is the final equation freed from the two unknowns that we wished to eliminate, which carries its own proof along with itself from the harmony observable throughout, and which we should find very troublesome to discover using the letters a, b, c, especially when the number of letters and equations is large. A part of the secret of analysis is the characteristic, rather the art, of using notation well, and you see, Sir, by this little example, that Viète and Descartes did not even know all of its mysteries. Continuing the calculation in this fashion, one will come to a general theorem for any desired numbers of letters and simple equations. Here is what I have found it to be on other occasions:*
>
> *Given any number of equations which is sufficient for eliminating the unknown quantities which do not exceed the first degree: -for the final equation are to be taken, first, all possible combinations of coefficients, in which one coefficient only from each equation is to enter; secondly, those combinations, after they are placed on the same side of the final equation, have different signs if they have as many factors alike as is indicated by the number which is less by one than the number of unknown quantities: the rest have the same sign.*

In this letter the general recipe is given in Latin, even though the rest of the letter is written in French.

The rule is also given in a manuscript, assumed to have been written before 1693, see [2805, p. 269], in which he wrote

> *Make all possible combinations of the coefficients of the letters, in such a way that more than one coefficient of the same unknown and of the same equation never appear together. These combinations, which are to be given signs in accordance with the law which will soon be stated, are placed together, and the result set equal to zero will give an equation lacking all the unknowns. The law of signs is this: To one of the combinations a sign will be arbitrarily assigned, and the other combinations which differ from this one with respect to two, four, six, etc...factors will take the opposite sign: those which differ from it with respect to three, five, seven, etc...factors will of course take its own sign.*

Leibniz was aware that a determinant can be given as sums of products using permutations of indices; see [1915]. Even though he did not write it explicitly, this is what we now call the Leibniz formula,

$$\sum_{\sigma \in S_n} sgn(\sigma) a_{1,\sigma(1)} \cdots a_{n,\sigma(n)},$$

where S_n is the symmetric group of permutations and σ is a permutation. He knew that if two columns or two rows are permuted, the value of the determinant changes sign. He also developed a determinant of order 4 from the entries of a row multiplying by the cofactors. Except for the sign, this is equivalent to the Laplace expansion. Since, in this work, Leibniz was only interested in determinants equal to zero, the difference in sign was not a problem.

However, since Leibniz only noted his ideas on determinants in unpublished notes or in letters, his discoveries did not influence other scholars until much later.

3.2 ▪ The 18th century

Colin Maclaurin (1698-1746), a Scottish mathematician who was a professor at the University of Edinburgh, published *A Treatise of Algebra* in 1748 [2111]. This treatise was elaborated in 1729 but only published in 1748, two years after his death. A hand-written copy by John Russell dated 1729 contains the formulas for a linear system of three equations; see Rev. Bruce A. Hedman [1619].

The elimination method for linear equations is described on page 77 of [2111]. The determinant formula for the solution of a linear system of order 2 is given on page 82 and the formulas for the order 3 on the next page, the coefficients of the equations being denoted by letters. But, these formulas were obtained by elimination and no general rule was given. The rest of the book is devoted to polynomial (system of) equations and applications to geometry.

On pages 59-61 of the book [758] published in 1750 in Geneva, Gabriel Cramer (1704-1754) studied the problem to find a curve in the plane whose equation is (in his notation) $A + By + Cx + Dyy + Exy + xx = 0$ and whose five points are known. He wrote the corresponding five linear equations and he had to solve a linear system. On page 60 there is a footnote[26] in which he said (our translation)

> *I think I found a rule that is easy and general when we have any number of equations and unknowns of which none has a degree larger than one. It can be found in Appendix I.*

This Appendix I is on pages 657-659 of [758]. Cramer used a strange notation with upper case letters and upper indices for the coefficients of the linear system; see Figure 3.3.

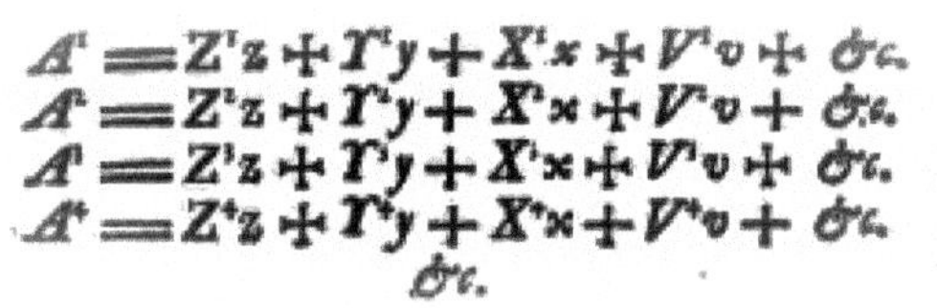

Figure 3.3. *Cramer's notation*

This is

$$
\begin{aligned}
A^1 &= Z^1 z + Y^1 y + X^1 x + V^1 v + \cdots \\
A^2 &= Z^2 z + Y^2 y + X^2 x + V^2 v + \cdots \\
A^3 &= Z^3 z + Y^3 y + X^3 x + V^3 v + \cdots \\
\cdots &= \cdots
\end{aligned}
$$

Cramer's solutions are shown in Figure 3.4.

The general solution for a system of order n is given in rhetoric terms, probably because of a lack of a good notation.

> *An examination of these formulas provides this general rule. The number of equations and of unknowns being n, we will find the value of each unknown by forming n fractions whose common denominator has as many terms as there are various arrangements of n different things. Each term is composed of letters $ZYXV\ldots$,*

[26] *Je crois avoir trouvé pour cela une règle assez commode & générale, lorsqu'on a un nombre quelconque d'équations & d'inconnues dont aucune ne passe le premier degré. On la trouvera dans l'Appendice, No. I.*

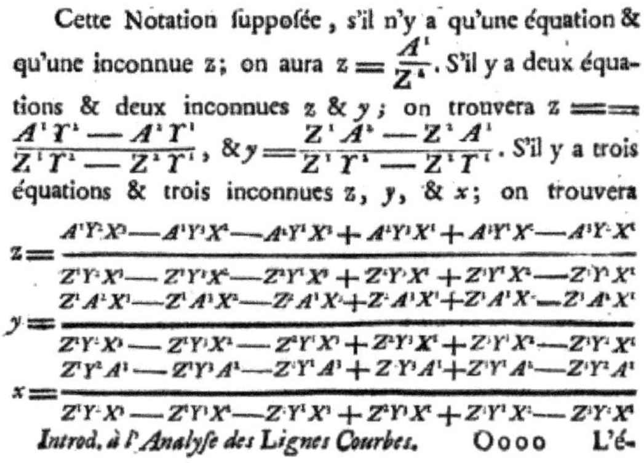

Cette Notation ſuppoſée, s'il n'y a qu'une équation & qu'une inconnue z; on aura $z = \frac{A^1}{Z^1}$. S'il y a deux équations & deux inconnues z & y; on trouvera $z = \frac{A^1Y^2 - A^2Y^1}{Z^1Y^2 - Z^2Y^1}$, & $y = \frac{Z^1A^2 - Z^2A^1}{Z^1Y^2 - Z^2Y^1}$. S'il y a trois équations & trois inconnues z, y, & x; on trouvera

$$z = \frac{A^1Y^2X^3 - A^1Y^3X^2 - A^2Y^1X^3 + A^2Y^3X^1 + A^3Y^1X^2 - A^3Y^2X^1}{Z^1Y^2X^3 - Z^1Y^3X^2 - Z^2Y^1X^3 + Z^2Y^3X^1 + Z^3Y^1X^2 - Z^3Y^2X^1}$$

$$y = \frac{Z^1A^2X^3 - Z^1A^3X^2 - Z^2A^1X^3 + Z^2A^3X^1 + Z^3A^1X^2 - Z^3A^2X^1}{Z^1Y^2X^3 - Z^1Y^3X^2 - Z^2Y^1X^3 + Z^2Y^3X^1 + Z^3Y^1X^2 - Z^3Y^2X^1}$$

$$x = \frac{Z^1Y^2A^3 - Z^1Y^3A^2 - Z^2Y^1A^3 + Z^2Y^3A^1 + Z^3Y^1A^2 - Z^3Y^2A^1}{Z^1Y^2X^3 - Z^1Y^3X^2 - Z^2Y^1X^3 + Z^2Y^3X^1 + Z^3Y^1X^2 - Z^3Y^2X^1}$$

Introd. à l'Analyſe des Lignes Courbes. Oooo L'é-

Figure 3.4. *Cramer's notation*

always written in the same order, but to which one gives as exponents, the n first numbers arranged in all possible ways. Hence, when we have three unknowns, the denominator has [$1\times2\times3 =$] *6 terms, composed of three letters ZYX, which receive successively the exponents* 123, 132, 213, 231, 312, 321. *We give to these terms the $+$ or $-$ sign according to the following Rule. When an exponent is followed in the same term by an exponent smaller than it, I call that a derangement. One counts, for each term, the number of derangements: if it is even or zero, the term has the $+$ sign; if it is odd, the term has the $-$ sign. For instance, in the term $Z^1Y^2V^3$ there is no derangement: this term will have the $+$ sign. The term $Z^3Y^1X^2$ will have the $+$ sign, because it has two derangements,* 3 *before* 1 *and* 3 *before* 2. *But the term $X^3Y^2X^1$, which has three derangements,* 3 *before* 2, 3 *before* 1 *and* 2 *before* 1, *will have the $-$ sign.*

The common denominator thus formed, we will have the value of z by giving to this denominator the numerator which is obtained by changing Z to A in all terms. And the value of y is the fraction with the same denominator and, as a numerator, the quantity obtained from changing Y to A in all terms of the denominator. One finds the other unknowns in a similar way.

Generally speaking, the problem is determined. But, there may be some particular cases, where it is undetermined; and others where it is impossible.

What Cramer called a "derangement" is what we now call a transposition. Then, Cramer discussed the case where the denominator is zero but he did not give any clue about the number of solutions. As we can see, Cramer did not give a proof of the formulas he wrote in his book.

Many proofs of Cramer's rule were published later on. For instance, in 1825 Heinrich Ferdinand Scherk (1798-1885) published a 17-page proof by induction on the number of unknowns. Carl Gustav Jacob Jacobi (1804-1851) proved the rule in 1841. Nicola Trudi (1811-1884) gave a proof in 1862. Some people are still looking for shorter proofs or proofs using different techniques; see, for instance, Maurizio Brunetti [487] in 2014.

There are still some disputes about the attribution of the result to Cramer; see [1619, 1939].

In 1764, Étienne Bézout (1730-1783) investigated the degree of the equation resulting from elimination of unknowns in a system of polynomial equations [310]. This led him to consider determinants because he reduced the existence of solutions to a homogeneous linear system whose determinant has to be zero; see [27, 28, 29] by Liliane Alfonsi. Bézout did not use Cramer's rule but gave his own rule for a homogeneous system (our translation):

> *Let $a, b, c, d, \ldots$ the coefficients of the first equation. Let $a', b', c', d', \ldots$ the coefficients of the second equation and $a'', b'', c'', d'', \ldots$ those of the third equation and so on. [...] Form the two permutations ab and ba and write ab − ba; with these two permutations and the letter c, form all the possible permutations, changing the sign each time c changes places in ab and the same thing for ba. You obtain abc − acb + cab − bac + bca − cba. With these six permutations and the letter d, form all the possible permutations, changing the sign each time d changes places in a same term and so on until you have exhausted all the coefficients of the first equation. Then, keep the first letter of all terms, give to those occupying the second place the marks they have in the second equation, to those occupying the third place the marks they have in the third equation and so on; equal all of that to zero and you obtain the sought condition.*

This looks a little bit like a cooking recipe, but, in fact, it is the development of the determinant using the entries of a row. Bézout's methods for polynomial systems were improved in 1779 [311].

A memoir of Alexandre-Théophile Vandermonde (1735-1796) was read at the French Académie des Sciences in 1771 and published [3121] in 1772. In this work, Vandermonde introduced what we can recognize as determinants, even though he did not name them, and he used them to solve linear systems with formulas equivalent to Cramer's rule.

The notation of an entry of the linear equations is $\genfrac{}{}{0pt}{}{\alpha}{a}$, α being the number of the equation and a being the rank of the coefficient in the equation. This notation is close to that of Leibniz we have seen above. Vandermonde added the notation

$$\begin{array}{c|c}\alpha & \beta \\ \hline a & b\end{array} = \genfrac{}{}{0pt}{}{\alpha}{a} \cdot \genfrac{}{}{0pt}{}{\beta}{b} - \genfrac{}{}{0pt}{}{\alpha}{b} \cdot \genfrac{}{}{0pt}{}{\beta}{a},$$

which, for us, is a determinant of order 2,

$$\begin{vmatrix} a_\alpha & b_\alpha \\ a_\beta & b_\beta \end{vmatrix} = a_\alpha b_\beta - b_\alpha a_\beta,$$

$$\begin{array}{c|c|c}\alpha & \beta & \gamma \\ \hline a & b & c\end{array} = \genfrac{}{}{0pt}{}{\alpha}{a} \cdot \begin{array}{c|c}\beta & \gamma \\ \hline b & c\end{array} + \genfrac{}{}{0pt}{}{\alpha}{b} \cdot \begin{array}{c|c}\beta & \gamma \\ \hline c & a\end{array} + \genfrac{}{}{0pt}{}{\alpha}{c} \cdot \begin{array}{c|c}\beta & \gamma \\ \hline a & b\end{array},$$

and so on with more terms. In modern terms, Vandermonde gave a recursive definition of determinants. It was the first time somebody gave a formal definition of determinants without directly starting from the linear equations, considering them as a mathematical object.

Then, Vandermonde wrote (our translation)

> *I will give for a number n of first degree equations, an elimination formula which is a sort of function of n, whose form is concise and useful. [...] It is clear that $\begin{array}{c|c}\alpha & \beta \\ \hline a & b\end{array}$ represents two terms, one positive, the other negative resulting from all the permutations of a and b; that $\begin{array}{c|c|c}\alpha & \beta & \gamma \\ \hline a & b & c\end{array}$ represents six terms, three positive, the other three negative resulting from all the permutations of a, b and c. [...] Moreover, forming these quantities is such that the unique change that can come from any permutation of the letters of the alphabet is a change of sign.*

In other words, permuting two rows or two columns changes the sign. Vandermonde described the general case but did not prove it, giving only examples for up to four columns in the previous notation. But, he suggested that the general case can be proved by induction. He remarked that if we think of the $\alpha, \beta, \gamma, \ldots$ as exponents, this corresponds to what he did in his *Mémoire sur la résolution des équations* (Vandermonde determinant). This is the only reference (if we exaggerate a bit) to what we call a Vandermonde matrix. For more on the naming or misnaming of this matrix, see [3293]. This remark may have inspired Augustin-Louis Cauchy (1789-1857) for his own definition of determinants. He also noticed that if two letters are the same then the expression is zero.

Vandermonde gave Cramer's rule for linear systems of order 2 and 3. Then, he developed the expressions using expressions of lower order. This corresponds to using the minors for writing down the determinants. Finally, he gave the general Cramer's rule in an obscure way.

A second part of that paper is devoted to polynomial systems. This was the only work of Vandermonde on determinants. In fact, Vandermonde abandoned mathematics to study and teach political economy. Nevertheless, he was called "the founder of the theory of determinants" by Sir Thomas Muir (1844-1934) in [2300], who later wrote a book [2302] about the history of determinants (about Muir, see [2139]). Henri Lebesgue (1875-1941) was less kind in [2020]. He wrote

> *What could have been personal, is the Vandermonde determinant. Yet it is not there, nor anywhere else in Vandermonde's work.*

As remarked by Bernard Ycart in [3293], linear systems with Vandermonde matrices had been written and solved long before Vandermonde, by Isaac Newton (1642-1727) and Abraham de Moivre (1667-1754). The linear equations (resulting from interpolation) are written in Newton's *Methodus Differentialis*, a manuscript that appeared in 1711. The explicit solution was given by de Moivre. A Vandermonde matrix appeared (of course, implicitly) in a paper by Gaspard Clair François Marie Riche, baron de Prony (1755-1839) in 1795.

In 1772, Pierre-Simon de Laplace (1749-1827) published a memoir about the differential equations modeling the movement of planets [1993]. In this memoir he had to solve systems of linear equations. He wrote (our translation)

> *Geometers have given general rules (see Introduction à l'analyse des lignes courbes by M. Cramer and the Mémoires de l'Académie pour l'année 1764); but, since it seems to me that they have been proved so far only by induction and that they are not practical when the number of equation is large; I am going to give some simpler processes than those already known to eliminate in any given number of equations.*

He denoted the coefficients by ${}^1a, {}^1b, \ldots, {}^2a, {}^2b, \ldots$ and the unknowns by $\mu', \mu'', \ldots$ He gave Cramer's rule and recalled the work of Bézout in 1764. Laplace used the word "résultante" for the determinant. He showed that transposing two letters changes the sign. Then, Laplace considered a homogeneous system of order 3. For simplicity, let us denote the coefficients as $a^1, b^1, \ldots$ Laplace showed that the "résultante" is

$$\begin{aligned} &a^1b^2c^3 - a^1c^2b^3 + c^1a^2b^3 - b^1a^2c^3 + b^1c^2a^3 - c^1b^2a^3 \\ &\quad = a^1[b^2c^3 - c^2b^3] + a^2[c^1b^3 - b^1c^3] + a^3[b^1c^2 - c^1b^2] = 0, \end{aligned}$$

that is, up to the sign, the expansion of the determinant using the first column. Then, he showed how to use this for solving a system of order 3 with a nonzero right-hand side. For the order 3, he gave the following recipe: write $+ab$, combine it with the letter c in all possible ways, changing the sign each time c changes places, and you get $abc - acb + cab$; in each term give the index 1 to the first letter, the index 2 to the second, the index 3 to the third, obtaining $a^1b^2c^3 - a^1c^2b^3 + c^1a^2b^3$; then instead of $+a^1b^2c^3$, write $(a^1b^2 - b^1a^2)c^2$ [there was a misprint in the original paper], instead of $-a^1c^2b^3$, write $-(a^1b^3 - b^1a^3)c^2$, and instead of $c^1a^2b^3$, write $(a^2b^3 - b^2a^3)c^1$, and you obtain the result. This is the expansion using the last column.

Then, he used the same process for the orders 4, 5, and 6. He showed that for the order n, he obtained the same number of terms as Cramer and Bézout with the good signs. Then, he described the same rule using another notation. Looking at this paper, we may conclude that, for the general case, it is not a very clear exposition of what we now call the Laplace expansion.

In [1970], published in 1773, Joseph-Louis Lagrange (1736-1813) considered the problem of the rotation of a rigid body. This problem had been already considered by Leonhard Euler (1707-1783) and Jean Le Rond d'Alembert (1717-1783) but Lagrange handled it in a different way. He considered three points in space (with coordinates x, y, z, etc.), and if we denote

$$A = \begin{pmatrix} x & x' & x'' \\ y & y' & y'' \\ z & z' & z'' \end{pmatrix},$$

the equalities given by Lagrange amount to writing $\det(A^T A) = [\det(A)]^2$. In another memoir in the same volume, he linked the expressions he gave to the volume of a tetrahedron with one vertex at the origin.

3.3 ▪ The 19th century

In 1801, Carl Friedrich Gauss (1777-1855) studied quadratic forms of order 2 and 3. He named their discriminants *determinants*. For $ax^2 + 2bxy + cy^2$ this is $b^2 - ac$. Gauss also studied linear transformations of the variables. It amounts to considering the formulas for the product of the corresponding matrices.

In November 1812, Jacques Philippe Marie Binet (1786-1857) presented a memoir [322, 323] at the Institut National des Sciences et des Arts that replaced the Académie des Sciences during 1795-1816, following the French revolution. In this memoir he stated what we now call the Cauchy-Binet formula.

Unfortunately, Binet used a very awkward notation, making his derivation difficult to follow. He denoted the entries of what is going to be the determinant as a', a'', and so on and $\sum a = a' + a'' + a''' + \cdots$ as well as $\sum ab = a'b' + a''b'' + a'''b''' + \cdots$. Even more awkward is $\sum ab' = a'b'' + b'a'' + a'b''' + \cdots$.

Binet named determinants "résultants" as Laplace did. He denoted the resultants with parentheses as $(y', z'') = y'z'' - z'y''$, in which we recognize a 2×2 determinant,

$$(x', y'', z''') = x'y''z''' + y'z''x''' + z'x''y''' - z'y''x''' - y'x''z''' - x'z''y''',$$

a 3×3 determinant, and so on. He stated some relations for the sums above, like

$$\sum ab' = \sum a \sum b - \sum ab,$$

$$\sum ab'c'' = \sum a \sum b \sum c + 2\sum abc - \sum a \sum bc - \sum b \sum ca - \sum c \sum ab,$$

and so on. Binet used these relations in the developments of his determinant of products. He first considered two systems, each constructed with two letters, $\{y, z\}$ and $\{\nu, \zeta\}$ and the sums of all the possible ordered combinations of $(\cdot, \cdot)$ that can be formed and their products,

$$(y', z'')(\nu', \zeta'') + (y', z''')(\nu', \zeta''') + \cdots + (y'', z''')(\nu'', \zeta''') + \cdots.$$

Then, he stated that using the formulas above, this can be proved to be equal to

$$\sum y\nu \sum z\zeta - \sum z\nu \sum y\zeta.$$

He remarked that this is a 2×2 "résultant." In modern terms, this is the statement of the Cauchy-Binet formula for the product of a $2 \times n$ matrix A with rows constructed from y and z by an $n \times 2$ matrix B with columns constructed from ν and ζ. It is given by

$$\det(AB) = \sum_{\mathcal{I}_2} \det(A_{:,\mathcal{I}_2}) \det(B_{\mathcal{I}_2,:}),$$

where $\mathcal{I}_2$ is the set of indices i_1, i_2 with $1 \leq i_1 < i_2 \leq n$ and the matrices of the right-hand side are 2×2. Binet's formula implicitly gives the rule for finding the entries of AB but, of course, nobody was explicitly using matrices in 1812.

Then, he gave the result for the case of "résultant" of order 3 like (x', y'', z'''). More generally, Binet's statement is that *the sum of products of resultants is a single resultant.* Then, he generalized that to *the sum of products of sums of resultants as a sum of resultants.* In [322], Binet added (our translation)

> *At the same session of the Institute, M. Cauchy, engineer of the "Ponts et Chaussées", presented a memoir which contains several of the formulas that we just gave. He obtained them in a different way.*

Binet's proofs are far from being obvious and complete. An analysis of Binet's memoir was given by Muir in [2300] but Muir's explanations are almost as obscure as those of Binet. It must be noted that Binet had already published some results about the multiplication theorem for determinants of symmetric equations in 1811.

Cauchy presented his 83-page memoir to the Institute at the same session as Binet's memoir, on November 30, 1812. This memoir [564] was published in 1815. Cauchy used results from a previous work [563] in which he studied the number of different values that a function depending on several variables may have when they are permuted. In the first part of [564] he considered functions that do not change values when permuting the variables. They are called *symmetric functions.* An example is $a_1^2 + a_2^2 + 4a_1a_2$. But, he was more interested in functions whose absolute values stay the same but of which the sign may change. An example is $\sin(a_1 - a_2)\sin(a_1 - a_3)\sin(a_2 - a_3)$. He called them *alternating symmetric functions.*

For a given function K, Cauchy denoted by $S(K)$ the sum of the different values that the function may take when permuting the m indices that may not be consecutive and by $S^n(K)$ the sum when the indices can be replaced also by indices in $\{1, 2, \dots, n\}$, where n is larger than or equal to the largest of the m indices. Then, he considered the sum of the values obtained by an even number of transpositions of the indices and the sum obtained by an odd number of transpositions. The difference of these two values is an alternating symmetric function denoted by $S(\pm K)$. For example, $S(a_1b_2) = a_1b_2 + a_2b_1$ and $S(\pm a_1b_2) = a_1b_2 - a_2b_1$.

In the second part of his memoir Cauchy studied a particular type of alternating symmetric function. He first considered several variables $a_1, \dots, a_n$ and the function

$$a_1 \cdots a_n \, (a_2 - a_1)(a_3 - a_1) \cdots (a_n - a_1)(a_3 - a_2) \cdots (a_n - a_2) \cdots (a_n - a_{n-1}).$$

This is an alternating symmetric function $S(\pm a_1 a_2^2 a_3^3 \cdots a_n^n)$. Then, Cauchy replaced the upper indices, meaning the power, by a second lower index. For instance, a_i^j becomes $a_{i,j}$. He obtained a new alternating function $S(\pm a_{1,1}a_{2,2} \cdots a_{n,n})$ that he called a *determinant.* As we have seen above, this name had been used previously by Gauss in a slightly different situation. Curiously enough, Cauchy still used the name "résultante" in some of his later publications. Examples are

$$\begin{aligned} S(\pm a_{1,1}a_{2,2}) &= a_{1,1}a_{2,2} - a_{2,1}a_{1,2}, \\ S(\pm a_{1,1}a_{2,2}a_{3,3}) &= a_{1,1}a_{2,2}a_{3,3} + a_{2,1}a_{3,2}a_{1,3} + a_{3,1}a_{1,2}a_{2,3} \\ &\quad - a_{1,1}a_{3,2}a_{2,3} - a_{3,1}a_{2,2}a_{1,3} - a_{2,1}a_{1,2}a_{3,3}. \end{aligned}$$

Cauchy remarked that the entries $a_{i,j}$ can be arranged in an array and that the general $S(\pm a_{1,1} \cdots a_{n,n})$ corresponds to the array

$$\begin{matrix} a_{1,1} & a_{1,2} & \cdots & a_{1,n} \\ a_{2,1} & a_{2,2} & \cdots & a_{2,n} \\ \vdots & \vdots & & \vdots \\ a_{n,1} & a_{n,2} & \cdots & a_{n,n} \end{matrix}.$$

He called the set of these entries a *symmetric system*, even though this is not a symmetric array! The diagonal entries were called *principal terms*; two entries $a_{i,j}$ and $a_{j,i}$ were said to be *conjugate*.

Cauchy added that each term of the determinant is the product of n distinct entries $a_{i,j}$. The second indices can be taken as $1, 2, \ldots, n$ and the first indices are taken as a permutation of these indices. The sign of the term $a_{1,1}a_{2,2} \cdots a_{n,n}$ is positive. The sign of $a_{\alpha,1}a_{\beta,2} \cdots a_{\zeta,n}$ is obtained by considering the permutation

$$\begin{bmatrix} 1, & 2, & 3, & \ldots & n \\ \alpha, & \beta, & \gamma, & \cdots & \zeta \end{bmatrix}.$$

Let g be the number of substitutions, then the sign is positive if $n - g$ is even, and negative otherwise. Cauchy gave an example for $n = 7$ and the term $a_{1,3}a_{3,6}a_{6,1}a_{4,5}a_{5,4}a_{2,2}a_{7,7}$ for which we have to consider

$$\begin{bmatrix} 1, & 3, & 6, & 4, & 5, & 2, & 7 \\ 3, & 6, & 1, & 5, & 4, & 2, & 7 \end{bmatrix}.$$

This permutation is obtained by four circular substitutions

$$\begin{bmatrix} 1, & 3, & 6 \\ 3, & 6, & 1 \end{bmatrix}, \quad \begin{bmatrix} 4, & 5 \\ 5, & 4 \end{bmatrix}, \quad \begin{bmatrix} 2 \\ 2 \end{bmatrix}, \quad \begin{bmatrix} 7 \\ 7 \end{bmatrix}.$$

Hence, $g = 4$ and $n - g = 3$, which means that the sign is negative.

Cauchy also considered the transpose of the array above that he called the *conjugate system*, and he proved that the determinant is not changed. He noticed that if we exchange two rows or two columns of the array, the determinant changes sign. He also wrote the formulas that can be obtained by developing the determinant using the last column and minors of order $n - 1$. He introduced a cofactor array whose entries, up to the sign, are the determinants obtained by deleting a row and a column of the array, and proved, even though he did not use matrices, that the product of the array by the transpose of the cofactor array is zero except for the diagonal entries which are equal to the determinant. In modern terms, it is $A\,\mathrm{Adj}(A) = \det(A)I$, where Adj is the adjugate. On page 127 of the memoir, equation (10), we have the rule for multiplication of two matrices (without any matrix).

Cauchy attributed most of these results to Laplace. The result about the determinant of a product was given in Section 2. On page 138, Cauchy considered two systems of coefficients $a_{i,j}$ and $\alpha_{i,j}$ for $i, j = 1, \ldots, n$ and wrote the n^2 relations, whose results he denoted by $m_{i,j}$, that we can identify as a matrix product for square matrices. Then, he studied the determinant of the $m_{i,j}$ system, replacing each $m_{i,j}$ by its value in terms of $a_{i,j}$ and $\alpha_{i,j}$. Finally, he showed that he can recover all the terms involved in the product of the determinant generated by the $a_{i,j}$'s and $\alpha_{i,j}$'s. Hence, the result is that the determinant of a product is the product of determinants. So far, this can be seen as a particular case of Binet's result, even though Cauchy's proof is easier to follow than the one given by Binet. Many different proofs of the Cauchy-Binet formula had been given since 1812.

Using his result, Cauchy proved that the determinant of the adjugate is the $(n-1)$th power of the determinant of $(a_{i,j})$.

In the third section, Cauchy considered the array constructed with the minors of order p obtained by choosing the entries of the array $a_{i,j}$ from p ordered indices for the rows and p ordered indices for the columns, where p can take any value from 1 to $n-1$. Cauchy called them *derived systems* and obtained relations between the determinants of the derived systems and the determinant of the original system $(a_{i,j})$.

The fourth and last section is concerned with the relations between the derived systems of two systems and the derived systems of the product of the two systems. Cauchy showed that any minor of the determinant of the product can be expressed as the sum of products of minors of the two factors.

A discussion about whether Cauchy can be also considered responsible for the Cauchy-Binet formula was given by Muir in [2300]. Muir wrote about the theory of determinants

> *It is, no doubt, impossible to call him* [Cauchy], *as some have done, the formal founder of the theory. This honour is certainly due to Vandermonde, who, however, erected on the foundation comparatively little of a superstructure. Those who followed Vandermonde contributed, knowingly or unknowingly, only a stone or two, larger or smaller, to the building. Cauchy relaid the foundation, rebuilt the whole, and initiated new enlargements; the result being an edifice which the architects of today may still admire and find worthy of study.*

Cauchy gave a brief exposition of the alternating functions in Chapter III of his textbook [565] in 1821. In 1829, he was looking for minimizing a quadratic form [569]. For doing so he considered what we call the characteristic polynomial of a symmetric matrix, that is, a determinant. In 1840, he used determinants in his memoir about elimination in systems of polynomial equations [572]. See also [575] in 1841 in which he used a rule of signs different from that of 1812. The Cauchy determinant appeared in [574].

Joseph Diez Gergonne (1771-1859), the editor of the *Annales de Mathématiques Pures et Appliquées*, one of the first mathematical journals founded in 1810, gave an expository paper on determinants, mainly based on Laplace's results, in 1813.

An early publication about determinants in England was the book *Elementary Illustrations of the Celestial Mechanics of Laplace* by Thomas Young (1773-1829) in 1821; see its Appendix B.

The third chapter of the treatise [2726] by Franz Ferdinand Schweins (1780-1856) in 1825 is devoted to determinants. Following Bézout, he gave a recursive definition of determinants. He proved results on the transformation of an aggregate of products of pairs of determinants into another aggregate of a similar kind.

In the paper [2954] in 1840, James Joseph Sylvester (1814-1897) introduced functions which are in fact determinants. Apparently, he was unaware of the works of his predecessors. As was usual with Sylvester, he introduced a strange notation. He defined *zeta-ic multiplication* as

$$\zeta(a_1-b_1)(a_1-c_1)=a_2-a_1b_1-a_1c_1+b_1c_1, \quad \zeta(a_1+b_1)^2=a_2+2a_1b_1+b_2,$$

and ζ_{+r} when the indices are increased by r. These expressions are obtained by considering the lower indices as exponents. The product-differences are defined as

$$\begin{aligned}PD(abc) &= (b-a)(c-a)(c-b),\\ PD(abcd) &= (b-a)(c-a)(c-b)(d-a)(d-b)(d-c),\\ PD(0abc) &= abc(b-a)(c-a)(c-b).\end{aligned}$$

As noted by Muir in [2300, p. 228], using this notation a 3×3 determinant is

$$\begin{vmatrix} a_1 & a_2 & a_3 \\ b_1 & b_2 & b_3 \\ c_1 & c_2 & c_3 \end{vmatrix} = \zeta abcPD(abc).$$

After studying the properties of zeta-ic product of differences, Sylvester used his notation to write down the solution of linear systems. Muir [2300] observed that

> *This early paper, one cannot but observe, has all the characteristics afterwards so familiar to readers of Sylvester's writings, fervid imagination, vigorous originality, bold exuberance of diction, hasty if not contemptuous disregard of historical research, the outstripping of demonstration by enunciation, and an infective enthusiasm as to the vistas opened up by his work.*

Jacobi published several papers between 1827 and 1841 in which he used determinants. Of interest to us is his expository paper [1794] published in 1841. This paper was really influential for the later use of determinants. Obviously, Jacobi had read Cauchy's 1815 memoir, but he adopted a different point of view. Jacobi started by studying positive and negative permutations of indices and their properties. From this he derived Cramer's and Cauchy's rules of signs. Then, he defined a determinant as a sum of products with a positive or negative sign according to the sign of the permutation of the indices; see also [1793].

Jacobi considered the derivative of a determinant with respect to the entries of the array considered as independent variables. He gave the Laplace expansion result. He studied minors of the determinant and also the multiplication theorem of Cauchy and Binet. In 1833, Jacobi proved that the determinant of an orthogonal matrix is ± 1. What we call Jacobians appeared in Jacobi's works between 1829 and 1833, but they had been considered by Cauchy, of course not under that name, in 1815 and 1822. The name "Jacobian" was coined by Sylvester in 1853.

Hessians, that is, determinants whose entries are the second-order derivatives of a function, appeared in works by Lagrange in 1773, Gauss in 1801, and Ludwig Otto Hesse (1811-1874), who was a student of Jacobi, in 1842-1844.

Skew-symmetric determinants were considered by Johann Friedrich Pfaff (1765-1825). The term "Pfaffian" was coined by Arthur Cayley (1821-1895). The determinant of such matrices can be written as the square of a polynomial in the matrix entries as proved by Cayley in 1849.

In his first paper [579] in 1841, Cayley considered the problem: How are the distances between five arbitrary placed points in space related? This problem had already been solved by Binet in 1812. Cayley used determinants to solve this problem, introducing the notation with vertical bars that we still use today. His proof is based on the multiplication theorem of Cauchy and Binet. Note that in 1841, Cayley was still a student in Cambridge, since he graduated in 1842.

In 1843, Cayley published a paper on the theory of determinants [580]. However, what is the real subject of this paper is a bilinear form. About his notation for determinants, Cayley wrote

> *In the first section I have denoted a determinant, by simply writing down in the form of a square the different quantities of which it is made up. This is not concise, but it is clearer than any abridged notation.*

In 1849, Charles Hermite (1822-1901) gave an interesting proof of the fact that a 4×4 determinant can be written, up to a multiplying factor, as a 3×3 determinant constructed with 2×2 minors by using the multiplication theorem,

$$\begin{vmatrix} a_1 & b_1 & c_1 & d_1 \\ a_2 & b_2 & c_2 & d_2 \\ a_3 & b_3 & c_3 & d_3 \\ a_4 & b_4 & c_4 & d_4 \end{vmatrix} \begin{vmatrix} \xi_1 & b_1 & 0 & 0 \\ \xi_2 & -a_1 & c_1 & 0 \\ \xi_3 & 0 & -b_1 & d_1 \\ \xi_4 & 0 & 0 & -v_1 \end{vmatrix} =$$

$$\begin{vmatrix} 1 & 0 & 0 & 0 \\ 0 & a_2b_1 - b_2a_1 & b_2c_1 - c_2b_1 & c_2d_1 - d_2c_1 \\ 0 & a_3b_1 - b_3a_1 & b_3c_1 - c_3b_1 & c_3d_1 - d_3c_1 \\ 0 & a_4b_1 - b_4a_1 & b_4c_1 - c_4b_1 & c_4d_1 - d_4c_1 \end{vmatrix},$$

where the entries ξ_i are those of the first column of the inverse of the leftmost matrix. The determinant of the second matrix on the left is $-b_1c_1$. Hence, the determinant of the matrix on the left is $-1/(b_1c_1)$ times the determinant of the 3×3 principal minor on the right; see [1649].

In 1850, Sylvester gave the definition of an rth minor determinant by removing r rows and r columns for a given array; see [2955]. It is in this paper that Sylvester used the word *matrix* for the first time.

In 1851, William Hugh Spottiswoode (1825-1883) published what can be considered the first elementary book entirely devoted to determinants [2846]. He used Cayley's notation for determinants and (i, j) to denote the entries. An enlarged second edition was published in Crelle's journal[27] in 1853.

In 1851, Sylvester [2958] wrote

> *The subjoined theorem, which is one susceptible of great extension and generalization, appears to me, and indeed from use and acquaintance (it having been long in my possession) I know to be so important and fundamental, as to induce me to extract it from a mass of memoranda on the same subject; and as an act of duty to my fellow-labourers in the theory of determinants, more or less forestall time (the sure discoverer of truth) by placing it without further delay on record in the pages of this Magazine.*

In modern notation, the result is the following. Let A be a matrix of order n, $\mathcal{I}$ and $\mathcal{J}$ be two sets of m indices with $m < n$, and $A^{\mathcal{I},\mathcal{J}}$ be the matrix of order $n - m$ obtained by deleting the rows in $\mathcal{I}$ and the columns in $\mathcal{J}$. The matrix $\tilde{A}^{\mathcal{I},\mathcal{J}}$ is defined as

$$[\tilde{A}^{\mathcal{I},\mathcal{J}}]_{i,j} = \det(A^{\tilde{\mathcal{I}},\tilde{\mathcal{J}}}),$$

where $\tilde{\mathcal{I}}$ (resp., $\tilde{\mathcal{J}}$) is obtained by deleting $\mathcal{I}_i$ from $\mathcal{I}$ (resp., $\mathcal{J}_j$ from $\mathcal{J}$). Then,

$$\det(A)\,(\det(A^{\mathcal{I},\mathcal{J}}))^{m-1} = \det(\tilde{A}^{\mathcal{I},\mathcal{J}}).$$

In those years, Sylvester also proved results about bordered determinants. For instance, in 1852, he stated

> *If the determinants represented by two square matrices are to be multiplied together, any number of columns may be cut off from the one matrix, and a corresponding*

[27] *Journal für die reine und angewandte Mathematik*, founded by August Leopold Crelle (1780-1855).

number of columns from the other. Each of the lines in either one of the matrices so reduced in width as aforesaid being then multiplied by each line of the other, and the results of the multiplication arranged as a square matrix and bordered with the two respective sets of columns cut off, arranged symmetrically (the one set parallel to the new columns, the other set parallel to the new lines), the complete determinant represented by the new matrix so bordered (abstraction made of the algebraical sign) will be the product of the two original determinants.

This illustrates how Sylvester was exposing his theorems in words. For 2×2 determinants, this result states that the determinant of the product of the two determinants

$$\begin{vmatrix} a & b \\ c & d \end{vmatrix}, \quad \begin{vmatrix} \alpha & \gamma \\ \beta & \delta \end{vmatrix},$$

satisfies

$$\begin{vmatrix} a\alpha + b\beta & a\gamma + b\delta \\ c\alpha + d\beta & c\gamma + d\delta \end{vmatrix} = \begin{vmatrix} a\alpha & a\gamma & b \\ c\alpha & c\gamma & d \\ \beta & \delta & 0 \end{vmatrix}.$$

Cayley and Sylvester used determinants as tools in many of their papers on invariant theory. Cayley published several generalizations of determinants from 1843 to 1851. Determinants whose elements are determinants were named *compound determinants* by Sylvester in 1850. They had appeared before in works by Lagrange in 1773 (2×2 determinants), and by Cauchy in 1812. An example is the adjugate.

Felice Chiò (1813-1871), an Italian mathematician, physicist, and politician, published a memoir on determinants [662] in 1853. In modern matrix terms, Chiò's result is the following.

Let A be a matrix of order n with $a_{n,n} \neq 0$. In the matrix B obtained by deleting the last row and last column, we replace $a_{i,j}$ by $a_{i,j}a_{n,n} - a_{i,n}a_{n,j}$, that is, a determinant of order 2. Then, $\det(B) = a_{n,n}^{n-2} \det(A)$. More generally, if $a_{i,j} \neq 0$, and E being the matrix obtained by deleting row i and column j and $e_{k,\ell} = (-1)^t(a_{k,\ell}a_{i,j} - a_{k,j}a_{i,\ell})$ with $t = (k-i)(\ell-j)$, then $\det(E) = a_{i,j}^{n-2} \det(A)$.

This process of reducing the order of the determinant is called *condensation*. Chiò's method uses a particular case of Sylvester's result given above. This is also related to results of Hermite around 1850.

This method has been used to compute determinants numerically since it is cheaper than Laplace's expansion; see [1270, 1518, 2047, 4] (ordered by date). For a determinant of order n, the number of operations is proportional to n^3, but Chiò's method is more expensive than Gaussian elimination for computing a determinant.

In 1866, Charles Lutwidge Dodgson (1832-1898), a professor at Oxford better known as Lewis Carroll, published another method of condensation for computing determinants [903]. Dodgson's method is based on a result of Jacobi from 1833. To see the difference with Chiò's method, let us consider a small example taken from [4],

$$\begin{vmatrix} 1 & -2 & 3 & 1 \\ 4 & 2 & -1 & 0 \\ 0 & 2 & 1 & 5 \\ -3 & 3 & 1 & 2 \end{vmatrix}.$$

The first step of Chiò's method yields

$$\left|\begin{array}{cc|cc|cc} 1 & -2 & -2 & 3 & 3 & 1 \\ 4 & 2 & 2 & -1 & -1 & 0 \\ \hline 4 & 2 & 2 & -1 & -1 & 0 \\ 0 & 2 & 2 & 1 & 1 & 5 \\ \hline 0 & 2 & 2 & 1 & 1 & 5 \\ -3 & 3 & 3 & 1 & 1 & 2 \end{array}\right| = \left|\begin{array}{ccc} 10 & -4 & 1 \\ 8 & 4 & -5 \\ 6 & -1 & -3 \end{array}\right|,$$

where each square corresponds to a 2×2 determinant. Then, the same method is applied on the 3×3 determinant. For the same problem, when choosing $a_{4,4} = 2$ as the pivot, the first step of Dodgson's method is

$$\frac{1}{2}\left|\begin{array}{cc|cc|cc} 1 & 1 & -2 & 1 & 3 & 1 \\ -3 & 2 & 3 & 2 & 1 & 2 \\ \hline 4 & 0 & 2 & 0 & -1 & 0 \\ -3 & 2 & 3 & 2 & 1 & 2 \\ \hline 0 & 5 & 2 & 5 & 1 & 5 \\ -3 & 2 & 3 & 2 & 3 & 2 \end{array}\right| = \frac{1}{2}\left|\begin{array}{ccc} 5 & -7 & 5 \\ 8 & 4 & -2 \\ 15 & -11 & -3 \end{array}\right|.$$

The 2×2 determinants are all based on the pivot $a_{4,4}$ appearing in the bottom right corner of the determinants. Hence, the difference from Chiò's method is in the choices of the 2×2 determinants.

In 1854 Francesco Brioschi (1824-1897), an Italian mathematician and politician, published a textbook on the theory of determinants [469]; see also [470]. This book was translated into French and German in the following years. In these two works published in 1854, Brioschi showed what are the eigenvalues of what we call an orthogonal matrix Q by using the determinant $\det(Q - \lambda I)$.

After 1850, we see the appearance of more books on the theory of determinants, meaning that the topic had reached some maturity; see, for instance, the books by Giusto Bellavitis (1803-1880) in 1857, Heinrich Richard Baltzer (1818-1887) [175] also in 1857, George Salmon (1819-1904) [2678] in 1859, Francesco Faà di Bruno (1825-1888) also in 1859, Dodgson in 1867 [904], Georges Dostor in 1877, Robert Forsyth Scott (1849-1933) in 1880, Paul Henry Hanus (1855-1941) in 1886, Alferdo Capelli (1855-1910) and Giovanni Garbieri (1847-1931) [547] in 1886, and Leopold Kronecker (1823-1891) in 1903.

Karl Theodor Wilhelm Weierstrass (1815-1897) gave an axiomatic definition of a determinant in the lessons he delivered in 1886-1887. It was published in the third volume of Weierstrass' Complete Works in 1903. An editorial footnote said (our translation)

> *The above communication, apart from formal changes, is taken from an elaboration made by Paul Günther*[28] *and owned by the Mathematisches Verein der Universität Berlin from the lecture given by Weierstrass in the winter semester 1886-87 on the theory and application of bilinear and quadratic forms. But, already in earlier years, Weierstrass had suggested the derivation of the fundamental theorems of the determinant theory in the mathematical seminar of the university with the help of the three characteristic properties laid down above.*

[28] Paul Günther (1867-1891) studied in Berlin with Kronecker and Weierstrass among others. He received his doctorate in 1889 with Lazarus Immanuel Fuchs (1833-1902) and qualified as a professor in 1890. He became seriously ill in the spring of 1891 and died in the autumn of the same year.

A determinant is defined as a function of n^2 elements that (i) is linear and homogeneous in the elements of each row, (ii) changes only in sign if two rows are interchanged, and (iii) has value 1 for the identity matrix. A similar definition was used by Kronecker in [1954].

An important notion in linear algebra is the rank; see Chapter 1. It was formally defined by Ferdinand Georg Frobenius (1849-1917) in 1875-1879. But it is interesting to look for its origins; see Jean-Luc Dorier [937]. One is the work of Euler on Cramer's paradox about algebraic (that is, bivariate polynomial) curves in the plane. Two curves of degrees m and n without a common component have at most mn intersections (this result was guessed but not yet proved in the 18th century) and a curve of degree n is determined by $n(n+3)/2$ points. For $n \geq 3$, $n^2 \geq n(n+3)/2$, hence two curves of the same order n seemed to have more intersection points than necessary to determine one of the curves. Apparently, Cramer sent this problem to Euler in 1750. This led Euler to look for conditions for a linear system of equations to have a unique solution. Euler's conclusion is that the equations must be different and that none of them is "included" in the others. We can see here the beginnings of the notion of linear independence.

In 1840, Sylvester considered homogeneous linear equations. He stated that for a system of order n in n unknowns the determinant must be zero to have a nonzero solution. He also considered cases with $n-1$ or $n+1$ unknowns. In 1857, Benjamin Michel Amiot (1806-1878) discussed the number of solutions of a linear system of order 3. In 1850, Sylvester studied the minors of order p in an $m \times n$ matrix. He stated that the maximum number of simultaneously nonzero minors cannot be larger than $(n-p+1)(m-p+1)$.

In 1857, Baltzer showed in [175] that if in Cramer's rule the denominator is zero and one of the numerators is zero, then all the numerators are zero. This result was also proved by Trudi in 1862 as well as the fact that one equation can be obtained from the other ones. In 1864, Baltzer reconsidered the problem and proved that in a linear system of order n, if there is at least one nonzero minor of order k and if all minors of order $k+1$ constructed from it are zero, then the system is consistent and all the solutions are obtained from $n-k$ arbitrary variables. In fact, all the minors of order $k+1$ are zero and k is the maximal order of nonzero minors.

In 1867, Dodgson [904] studied all the cases for the solutions of a linear system.

In 1877, Frobenius [1249] studied homogeneous linear systems of m equations in n unknowns with $m < n$ and wanted to characterize the set of solutions. He defined the solutions $x^{(i)}$, $i = 1,\dots,k$ to be independent if, for $j = 1,\dots,n$, $\sum_{i=1}^{k} \alpha_i x_j^{(i)} = 0$ implies $\alpha_i = 0$, $i = 1,\dots,k$. He constructed bases of solutions and showed that all bases have the same number of elements that is $n-m$, and that $n-m+1$ solutions are always dependent. Finally, he showed that if a system with n unknowns has $n-m$ independent solutions, all the minors of order $m+1$ are zero. The rank was formally defined by Frobenius in 1879.

Note also the work of Henry John Stephen Smith (1826-1883) [2807] in 1861, who considered problems with integer entries.

In 1880, Eugène Rouché (1832-1910) proved that a linear system $Ax = b$ with n unknowns has solutions if and only if rank(A)=rank($[A\, b]$), and if the rank is less than n there is an infinite number of solutions; see [2608]. This result was also proved by Alfredo Capelli (1855-1910) in 1888 using determinants, as well as by Kronecker and also Georges Fontené (1848-1923). This theorem is referred to by several combinations of those names: Rouché-Capelli, Rouché-Fontené, Capelli-Kronecker, etc.

Capelli and Garbieri published a book in 1886 in which they defined the rank through what we now call Gaussian elimination. Considering a homogeneous linear system $m \times n$, they reduced it by elimination to a triangular system of order k (where k is the rank of the original system) with nonzero diagonal entries. They also showed that the column rank is equal to the row rank.

The question of linear independence was also considered by the Czech mathematician Eduard Weyr (1852-1903) in 1889; see [3223, 3224].

Today the rank of a matrix is generally computed using the singular value decomposition (SVD) or a rank-revealing QR factorization; see Chapter 4.

If a matrix M is partitioned into blocks as

$$M = \begin{pmatrix} A & B \\ C & D \end{pmatrix},$$

with D square and nonsingular, the Schur complement of D is $S_D = A - BD^{-1}C$ and we have $\det(M) = \det(D)\ \det(S_D)$; see Section 1.5. If A is nonsingular, we also have $\det(M) = \det(A)\ \det(D - CA^{-1}B)$. This was used by Issai Schur (1875-1941) in 1917; see [2718]. A special case was proved by Frobenius [1255] in 1908. The name *Schur complement* was coined by Emilie Virginia Haynsworth (1916-1985) [1610] in 1968; see also [751, 2408].

In 1893, Jacques Hadamard (1865-1963) gave an upper bound for the determinant of a matrix A with vector columns $A_{:,i}$, $i = 1, \dots, n$,

$$\det(A) \leq \prod_{i=1}^{n} \|A_{:,i}\|.$$

If $\det(A) \neq 0$, the inequality is an equality if and only if the columns are orthogonal.

3.4 ▪ The 20th century

The 19th century was the golden age for determinants. The mathematical theory of determinants was more or less complete at the end of the 19th century and the beginning of the 20th century. This was summarized, for instance, in the book [18] by Alexander Craig Aitken (1895-1967) in 1939. However, some books on linear algebra still started from determinants; see, for instance, [2253] by Leon Mirsky (1918-1983) in 1955.

Since then, many people have been looking for analytic expressions for certain determinants; see, for instance, [1944, 1945].

For the history of the notation for determinants, see Florian Cajori (1859-1930) [523, 524], pages 87-102.

Today, we teach our students not to compute determinants numerically for solving linear systems because we have much more efficient methods. However, determinants are useful theoretical tools for some problems; see, for instance, how to find expressions for the GMRES residual norms in [1018] using Cramer's rule and the Cauchy-Binet multiplication theorem. In some cases, determinants of large sparse matrices can be approximated cheaply; see [1386]. Determinants are still of interest because they have applications in algebra, combinatorics, geometry, number theory, and physics. They are also used in extrapolation methods and Padé approximation.

Some people have tried to revive determinants for solving linear systems using variants of Chiò's method; see [1518, 2047]. It shows that there exist perfectly parallel methods for solving linear equations, even though their complexities might be too large.

3.5 ▪ Lifetimes

In this section, we show the lifetimes of the main deceased contributors to determinants, starting in 1640. The language of the author is given by the color of the bars and by letters: E (red) for English, G (black) for German, I (green) for Italian, F (blue) for French, and O (magenta) for the others. The contributors are ordered by date of birth.

Determinants (a)

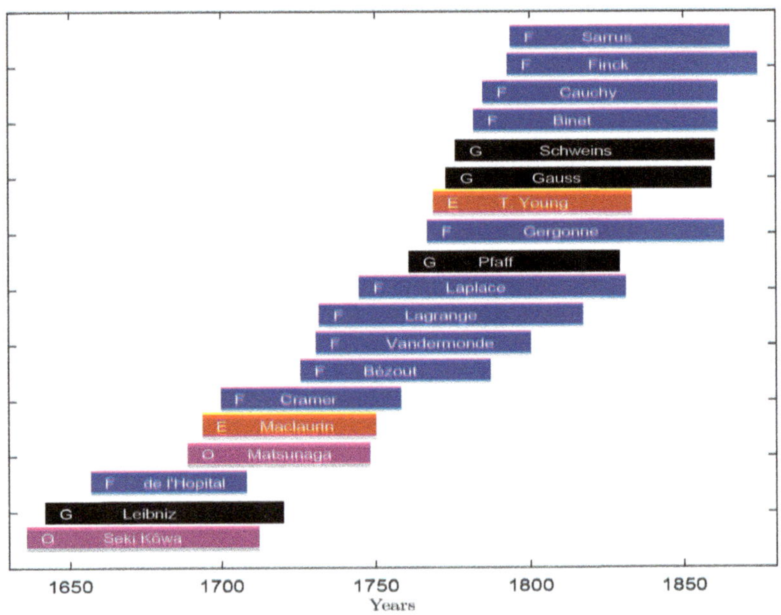

Determinants (b)

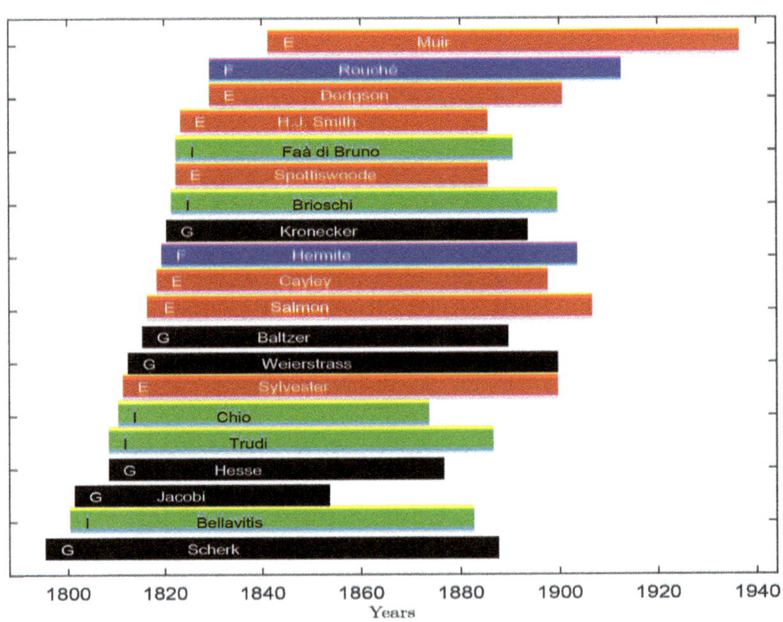

Determinants (c)

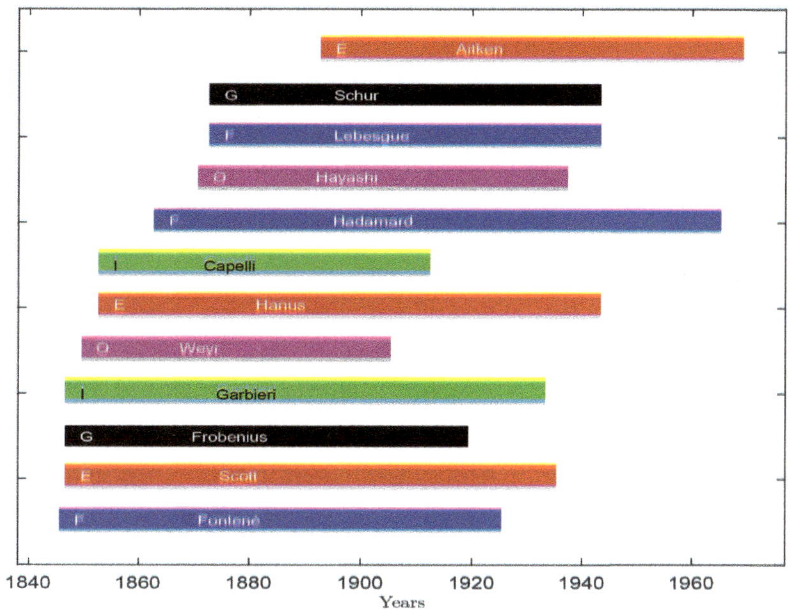

4

Matrix factorizations and canonical forms

> *The underlying principle of the decompositional approach to matrix computation is that it is not the business of the matrix algorithmists to solve particular problems but to construct computational platforms from which a variety of problems can be solved. This approach, which was in full swing by the mid-1960s, has revolutionized matrix computation.*
>
> – Gilbert Wright Stewart, *2000*

A *matrix factorization* expresses a given matrix A as a product of two or more other matrices. It is of interest if the matrices in the product have a special structure, like being upper or lower triangular, or have special properties, like being orthogonal or unitary; see, for instance, G.W. Stewart's book [2878] and [2879].

Some benefits of matrix factorizations are first, that they can be used to solve many problems involving the same matrix, and second, that they can be updated if the matrix is enlarged. For instance, if one wants to solve a linear system $Ax = b$ and if a factorization $A = LU$ is available with L lower triangular and U upper triangular, it can be reused cheaply to solve another system $Ax = c$ with the same matrix, but with a different right-hand side.

In this chapter, we consider the main matrix factorizations. Some of them are also called *canonical forms* because they exhibit intrinsic properties of the matrix.

4.1 ▪ LU factorization

The most widely used matrix factorization is of the form LU, where L is lower triangular and U is upper triangular. Not every square matrix A has an LU factorization; it is necessary that the determinants of the principal matrices of order $\leq n-1$ are nonzero, where n is the order of the matrix A. If the factorization exists and if A is nonsingular, it is unique. However, there is always a permutation matrix P such that $PA = LU$.

The LU factorization was derived from elimination methods for solving linear systems. We discuss elimination methods in Chapter 2.

4.2 ▪ QR factorization

If A is an $m \times n$ real or complex matrix with $m \geq n$, we have

$$A = QR = Q\begin{pmatrix} R_1 \\ 0 \end{pmatrix} = \begin{pmatrix} Q_1 & Q_2 \end{pmatrix}\begin{pmatrix} R_1 \\ 0 \end{pmatrix} = Q_1 R_1,$$

where Q is a unitary matrix of order m, that is, $Q^*Q = I$ (where Q^* is the conjugate transpose of Q) and R_1 is $n \times n$ upper triangular. If A is of full rank and R_1 has positive diagonal entries, Q_1 and R_1 are unique. Moreover, $A^*A = R_1^*R_1$. When A is real, Q is orthonormal with $Q^TQ = I$.

The name *QR factorization* was coined by John Guy Figgis Francis in 1961 in his famous paper [1212] about the QR algorithm for computing eigenvalues; see the section on this algorithm in Chapter 6.

The QR factorization is linked to the notion of *orthogonality*. This word comes from two Greek words, $o\rho\theta o\varsigma$, which means straight, upright, and $\gamma\omega\nu\iota\alpha$, which means corner, angle. There are, of course, examples of orthogonal things in nature and in humans' works. Quite often, when you want to build a wall, you put it at a right angle with the ground. Orthogonality of lines or segments in the plane was considered in Greek geometry. Consider, for instance, Pythagoras' theorem (which was already known many years before Pythagoras) for a right triangle.

The mathematical definition of orthogonality of vectors is linked to the notion of dot product. What we would consider today as a dot product appeared in a work [1971] by Joseph-Louis Lagrange (1736-1813) in 1773. He was considering problems with tetrahedrons (which he called "pyramides triangulaires"). One vertex was assumed to be at the origin, and he was considering the coordinates of the three other vertices. If we think of the coordinates of one point (x, y, z) as a vector, then expressions which are dot products appeared in this paper, even though Lagrange was not really thinking in this way.

The introduction of vectors and vector spaces can be seen as a distant consequence of the introduction of the geometric representation of complex numbers. Its first appearance was in a paper written in Danish in 1797 by Caspar Wessel (1745-1818), a Danish-Norwegian mathematician. That paper did not attract too much attention. This representation was found again later by Jean-Robert Argand (1768-1822), of Swiss origin, in 1806, the abbott Pierre-Louis Buée (1740-1827) in 1806, and Carl Friedrich Gauss (1777-1855) in 1831, even though Gauss considered complex numbers earlier in his proof of the fundamental theorem of algebra.

An important step was the introduction of quaternions by William Rowan Hamilton (1805-1865) in 1843; see Section 10.31. Quaternions are of the form $a + b\mathbf{i} + c\mathbf{j} + d\mathbf{k}$, a, b, c, and d being real numbers, and $\mathbf{i}$, $\mathbf{j}$, and $\mathbf{k}$ being the quaternion units. A quaternion with $a = 0$ is a vector quaternion. The product of two vector quaternions $(0, x)$ and $(0, y)$, where x and y are three-dimensional vectors, is $(-x \cdot y, x \times y)$. The scalar part is the opposite of the dot product and the vector part is the cross product of the two vectors.

In 1878, William Kingdon Clifford (1845-1879), an English mathematician, building on the work of the German Hermann Grassmann (1809-1877), published his *Elements of Dynamic*. In this book he defined the product of two vectors to have a magnitude equal to the area of the corresponding parallelogram and a direction perpendicular to their plane. Orthogonality of vectors was defined using the dot product by Grassmann in the 1862 version of his book [1424, 1425].

The dot and cross products were formally introduced by Josiah Willard Gibbs (1839-1903), an American mathematician, in 1901. The book *Vector Analysis* [1346], written with Edwin Bidwell Wilson (1879-1964) and based on Gibbs' lectures at Yale University, gave the definition of the dot product of two vectors in three-dimensional space as the product of their magnitudes multiplied by the cosine of their angle or as the sum of the product of their components. They introduced the dot notation that gave its name to the product, which is also called an inner product or a scalar product since it produces a scalar from two vectors.

Vector calculus was also used by Oliver Heaviside (1850-1925) in England. He used it to reformulate Maxwell's equations and to give them the form that is still in use today. For more on the history of vector analysis, see the book [781] by Michael J. Crowe.

Today two vectors are said to be orthogonal if their dot product is zero.

Probably, orthogonal transformations appeared in coordinate geometry with the problem of transformation from one set of rectangular axes to another set having the same origin. In fact, when we write how to rotate rectangular axes in the plane, we implicitly construct an orthogonal matrix. This problem was considered in the three-dimensional space by Leonhard Euler (1707-1783) in 1748; see [1108]. Euler came back to this problem in 1770 and 1775 when he introduced what is now known as the Euler angles [1111]. In that paper Euler gave formulas which correspond to the product of rotations around each of the three axes.

In 1773, Lagrange studied the problem of rotation of rigid bodies. In his paper [1970], which we already cited above, he started by establishing some algebraic identities. Even though he did not state things in these terms, on page 580, he first gave a lemma on the square of the determinant of a matrix of order 3, say Q. Then, he gave the formulas corresponding to the product Q^TQ and the solution of the linear system $Q^Tx = (\,\beta \quad 0 \quad 0\,)^T$ which corresponds to a multiple of the first column of the inverse of Q^T. On page 585, he showed that if the three columns of Q correspond to three points in space and if the segments linking them to the origin form an orthogonal frame, then, in modern terms, Q^TQ is diagonal. If the distances of the three points to the origin are equal to 1, what he had shown before proved that the first column of Q^{-1} is the transpose of the first row of Q. In this case we have an orthonormal matrix of order 3. But of course, Lagrange was just thinking in terms of algebra and geometry.

Gauss also worked on this problem in 1818 for his computations related to astronomy [1301]. Among other contributions, let us cite those by Carl Gustav Jacob Jacobi (1804-1851) in 1827 [1790] and Augustin-Louis Cauchy (1789-1857) in 1828 [568]. Jacobi returned several times to this problem.

With the dot product $(\,\cdot\,)$ and a vector x, we can define a norm $\|x\| = \sqrt{(x \cdot x)}$. Then, for two vectors x and y, we have the inequality

$$|(x \cdot y)| \leq \|x\|\,\|y\|.$$

This inequality was proved in 1821 by Cauchy [566] Note II, pp. 455-456. However, Cauchy proved only that the sum of the product of two finite sequences of numbers is less than the product of the square roots of the sums of the squares of the numbers in each sequence. He did not introduce the notion of dot product.

When the dot product of two functions is defined by the integral of their product, the inequality was proved in 1859 by Viktor Yakovlevich Bunyakovsky (1804-1889), a Russian mathematician, and rediscovered in 1888 by Karl Hermann Amandus Schwarz (1843-1921), a German mathematician. Generally, the inequality is called the Cauchy-Schwarz inequality or the Cauchy-Bunyakovsky-Schwarz inequality.

In 1829, Cauchy proved, only using linear equations, that the eigenvalues of a symmetric matrix are real and that, for two distinct eigenvalues, the dot product of the corresponding eigenvectors is zero, that is, they are orthogonal, but he did not use that word; see [569].

The terms *orthogonal transformation* and *orthogonal substitution* appeared in a paper [2960] by James Joseph Sylvester (1814-1897) in 1852. The definition of an orthogonal matrix was given by Ferdinand Georg Frobenius (1849-1917) in 1878 [1250]. For the definition of a unitary matrix (that is, such that $Q^*Q = I$), see [105] by Léon César Autonne (1859-1916) in 1902.

Gram-Schmidt orthogonalization

The QR factorization appeared implicitly before the theory of matrices. It turns out, and this should not come as a surprise for our readers, that its development was closely linked to the method of least squares. An early appearance of orthogonalization is in the work of Pierre-Simon de Laplace (1749-1827) in 1820 in the first appendix of his book *Théorie Analytique des Probabilités* [1998]. Laplace wanted to compute the masses of Jupiter and Saturn from systems of equations that were provided by his assistant Alexis Bouvard (1767-1843). Laplace also wanted to compute the distribution of error in the solution assuming a normal distribution of the noise in the observations. He gave an example for s equations in six unknowns, where s was the number of observations. He used a strange notation where the unknowns are denoted as $z, z', z'', z''', z^{iv}, z^{v}$ and the columns of the linear system as $p, q, r, t, \gamma, \lambda$. Laplace proceeded backwards. In modern terms, he orthogonalized the last column λ against the previous ones. For the first one, his operation is

$$p^{(1)} = \left(I - \frac{\lambda\lambda^T}{\|\lambda\|^2} \right) p.$$

Then, he continued by orthogonalizing the new last column $\gamma^{(1)}$ and so on. In the end, he obtained what can be considered a matrix

$$T = \left(p^{(5)} \quad q^{(4)} \quad r^{(3)} \quad t^{(2)} \quad \gamma^{(1)} \quad \lambda \right),$$

with orthogonal columns. What he implicitly computed is a QL factorization with L lower triangular. Laplace did things in this way because he was only interested in the $(1,1)$ entry of L, which is the norm of $p^{(5)}$ that he computed from the normal equations. He observed that at each step, the remaining columns are orthogonal to the least squares residual, but the word "orthogonal" was never used.

Of course, it would be an anachronism to think that Laplace invented one variant of the Gram-Schmidt algorithm, in fact, what is now known as the modified version of the algorithm. As we said, he was only interested in the L part of the factorization. For a partial translation to English of Laplace's work and some explanations, see the report [1990] by Julien Langou.

In 1843, William Thomson (1824-1907), also known as Lord Kelvin, proved that, if the equations corresponding to $Q^TQ = I$ for Q orthonormal of order 3 are satisfied, then one has $QQ^T = I$.

Jørgen Pedersen Gram (1850-1916), after his doctoral thesis in 1879, which was in Danish, published in 1883 a paper [1423], written in German, whose title can be translated as *On the development of real functions in series using the least squares method.* He was concerned with finding approximations of functions from $\mathbb{R}$ to $\mathbb{R}$ in the least squares sense. First, given a discrete inner product with weights v_x and a finite series

$$y_x = a_1X_1 + \cdots + a_nX_n,$$

where the X_i's are functions of x, Gram wanted to minimize $\sum_x v_x(o_x - y_x)^2$, where o_x denotes the given observations. Denoting

$$y_x^{(m)} = a_{m,1}X_1 + \cdots + a_{m,n}X_m, \; m \le n,$$

and

$$s_i = \sum_x v_xX_io_x, \quad p_{i,k} = p_{k,i} = \sum_x v_xX_iX_k,$$

the minimization problem leads to the normal equations

$$\sum_{i=1}^{m} a_{m,i}\, p_{j,i} = s_j, \quad j = 1, \ldots, m.$$

Gram introduced the difference $y_x^{(m)} - y_x^{(m-1)} = \sum_{i=1}^m b_{m,i} X_i$. It gave him $m-1$ equations for the $b_{m,i}$ if $a_{m,m}$ is known. But

$$a_{m,m} = \frac{\sum_i P_{m,i}^{(m)} s_i}{P^{(m)}},$$

where $P^{(m)}$ is the symmetric determinant of the coefficients $p_{i,j}$ and $P_{i,k}^{(m)}$ is the minor determinant corresponding to the element $p_{i,k}$. Moreover, one has the following relations

$$\frac{b_{m,1}}{P_{m,1}^{(m)}} = \cdots = \frac{b_{m,m-1}}{P_{m,m-1}^{(m)}} = \frac{a_{m,m}}{P_{m,m}^{(m)}} = \frac{\sum_i P_{m,i}^{(m)} s_i}{P^{(m)} P^{(m-1)}}.$$

Therefore,

$$y_x^{(m)} - y_x^{(m-1)} = \frac{\sum_i P_{m,i}^{(m)} s_i}{P^{(m)} P^{(m-1)}} \sum_i P_{m,i}^{(m)} X_i.$$

But $y_x^{(1)} = s_1/P^{(1)}$. It yields $y_x^{(n)}$ by summation. However, Gram used other functions defined as

$$\Phi_n(x) = \sum_i P_{n,i}^{(n)} X_i = \begin{vmatrix} p_{1,1} & \cdots & p_{1,n-1} & X_1 \\ \vdots & & \vdots & \vdots \\ p_{n,1} & \cdots & p_{n,n-1} & X_n \end{vmatrix}. \tag{4.1}$$

If, for the sake of simplicity, $\Phi_n(x) = \sum_{i=1}^n P_{n,i} X_i$, then

$$\sum_x v_x \Phi_m(x) \Phi_r(x) = \sum_{i=1}^m P_{m,i} \sum_x v_x X_i \Phi_r(x)$$

and

$$\sum_x v_x X_i \Phi_r(x) = \sum_{j=1}^r P_{r,j} p_{i,j}.$$

The sum is 0 if $i < r$ and $P^{(r)}$ if $i = r$. Hence $\sum_x v_x \Phi_m(x) \Phi_r(x) = 0$ if $m < r$ and also if $m > r$ by symmetry. Moreover, $\sum_x v_x \Phi_r(x)^2 = P^{(r-1)} P^{(r)}$. Hence, Gram had a set of orthogonal functions with which he expressed the solution of the least squares problem.

In fact, Gram's paper is far from being clear and well written. In particular, the notation is not clearly defined. Nevertheless, if we were thinking of the X_i's as transposes of vectors, the formula (4.1) defines a transposed vector orthogonal to the previous ones.

The remainder of Gram's paper is concerned with the extension to a dot product defined by an integral over a real interval $[\alpha, \beta]$, and to some examples.

Erhard Schmidt (1876-1959), who was a student of David Hilbert (1862-1943), obtained a thesis in Göttingen in 1905 on integral equations. His work was published in a subsequent series of papers [2698, 2699, 2700, 2701]. The title of some of these papers can be translated as *On the theory of linear and nonlinear integral equations*. Following the Swedish mathematician Ivar Fredholm (1866-1927) and Hilbert, E. Schmidt was concerned with studying integral equations and their spectrum. This led him to construct sets of orthogonal functions related to a dot product defined by an integral.

Given functions φ_i defined on $[a,b]$, Schmidt constructed a set of orthogonal functions ψ_i as

$$\psi_1(x) = \frac{\varphi_1(x)}{\sqrt{\int_a^b [\varphi_1(y)]^2 dy}},$$

$$\psi_n(x) = \frac{\varphi_n(x) - \sum_{i=1}^{n-1} \psi_i(x) \int_a^b \varphi_n(z)\psi_i(z)\,dz}{\sqrt{\int_a^b \left[\varphi_n(y) - \sum_{i=1}^{n-1} \psi_i(y) \int_a^b \varphi_n(z)\psi_i(z)\,dz\right]^2 dy}}.$$

Gram is cited on page 437 of Schmidt's paper [2698]. Apparently, Gram and Schmidt were unaware of Laplace's work.

Let us now briefly look for the influence of the works of Gram and Schmidt. Many of the references to their works were related to functions and integral equations.

Maxime Bôcher (1867-1918), an American mathematician, cited Gram and Schmidt on page 53 of his book [349] on integral equations in 1909, in the section about orthogonal functions. Orthogonality was defined with a dot product defined by an integral over a real interval. However, he did not call the orthogonalization method the Gram-Schmidt process. Bôcher considered the problem of the approximation of a function by a finite set of orthogonal functions using least squares. According to Bôcher, the word "orthogonal" for sets of functions was first used by Felix Klein (1849-1925) in 1889.

In 1909, Gerhard Kowalewski (1876-1950), a German mathematician, orthogonalized a set of vectors as it can be seen on pages 191-192 of the 1924 edition of his book [1942], which is a shortened version of the first edition in 1909. Orthogonalization of a set of continous functions appeared on pages 226-227 of [1942]. There is no reference to E. Schmidt but Gram's name appeared in that book just in relation to the Gram determinant. That work is not often cited, probably because Kowalewski later became a member of the Nazi party.

In a footnote on page 266 of his 1950 paper [1984], Cornelius Lanczos (1893-1974) attributed the idea of successive orthogonalization of a set of vectors to Otto Szász (1884-1952) in a paper [2978] written in Hungarian in 1910. This paper was expanded and translated to German in 1917 [2979]. It is also available in Szász's collected mathematical papers [2980]. This attribution is rather strange because Szász cited Gram and E. Schmidt in later papers. Szász and Lanczos had been at Frankfurt University at the same time in the 1930s and much later at the Institute of Numerical Analysis of the National Bureau of Standards in Los Angeles. So, Lanczos probably knew about the works of Gram and Schmidt, but maybe he was thinking that orthogonalizing vectors is slightly different from orthogonalizing functions.

Anyway, Szász's purpose was to give an elementary proof of Hadamard's inequality for determinants. For a positive definite symmetric matrix, the determinant is less than or equal to the product of the diagonal entries. To prove this, Szász orthogonalized the rows of the matrix using what corresponds to the classical Gram-Schmidt algorithm. He was reasoning only with determinants and did not relate the orthogonalization to a matrix factorization. He obtained Hadamard's upper bound by using the relationship between orthogonalization and minimization.

Systems of orthogonal functions were also considered by Louis Brand (1885-1971), an American mathematician [405], in 1912. Schmidt orthogonalization was cited on page 44 of [2304] published by Herman Müntz (1884-1956) in 1913; see also page 177 of the paper [2991] written in 1922 and published in 1924 by Satoru Takenaka about systems of integral equations.

In the famous book by Richard Courant (1888-1972) and Hilbert, *Methoden der Mathematischen Physik* [753], first published in 1924 with a corrected and expanded second edition in 1931, the Gram-Schmidt process is described for vectors and later for functions but without references to Gram or E. Schmidt.

In 1932, Herbert Westren Turnbull (1885-1961) and Alexander Craig Aitken (1895-1967) published a book on canonical forms of matrices. The orthogonalizing process of Schmidt is described on page 95 of the 1944 edition. The QR factorization of a matrix is written (with a different notation) on pages 96-97. We were not able to check if this description is similar in the first edition.

The name "Gram-Schmidt orthogonalization" was already widespread at the beginning of the 1930s since, for instance, Joseph Leonard Walsh (1895-1973), from Harvard University, wrote about *the well known Gram-Schmidt method of orthogonalization* on page 35 of [3175] in 1935. Walsh was concerned with approximation of functions by polynomials.

There was an interesting paper [3270] by Yue-Kei Wong in 1935. The Gram-Schmidt orthogonalization process is cited on page 54, and the classical algorithm is described for a set of vectors on page 57. Wong wanted to minimize (in his notation),

$$\|\sum_{i=1}^{r} p_i A_i - L\|^2,$$

where the A_i are the columns of an $n \times r$ real matrix ($r \le n$) and L is a vector with n components. He orthogonalized the columns A_i and wrote $\sum_{i=1}^{r} p_i A_i = \sum_{i=1}^{r} k_i C_i$ with the C_i's orthonormal. The minimum of the norm is given by $k_i = (C_i, L)$, $i = 1, \dots, r$, where $(\cdot,\cdot)$ is the dot product. It yields the p_i's as

$$\sum_{i=k}^{r} (C_k, A_i) p_i = (C_k, L), \quad i = 1, \dots, k.$$

This is clearly an upper triangular linear system. Hence, Wong had solved the least squares problem by using Gram-Schmidt orthogonalization. Wong observed that the residual vector of the least squares problem must be orthogonal to the columns A_i. He proved that $A_1, \dots, A_r$ are linearly independent if and only if the corresponding Gram matrix of order r with entries (A_i, A_j) is positive definite. He described Gaussian elimination on the symmetric positive definite matrix of the normal equations as a transformation to upper triangular form by left multiplication with elementary matrices of the form $I + \alpha y e_i^T$, where e_i is the ith column of the identity matrix. However, he did not immediately relate that to a matrix factorization since he did not consider the inverses of the elementary matrices.

Wong denoted by α the matrix whose rows are the A_i^T's, and β the matrix obtained from α by orthogonalization of the rows. So, there was a matrix κ such that $\beta = \kappa\alpha$. Because of orthogonality $\beta\alpha^T$ is upper triangular and $\kappa(\alpha\alpha^T) = \beta\alpha^T$. Then, Wong proved that κ is equal to the product of the elementary matrices in his version of Gaussian elimination. In modern terms, if we define the LU factorization, $\alpha\alpha^T = LU$, and set $\kappa = L^{-1}$, we have

$$\beta = L^{-1}\alpha, \quad \beta\beta^T = UL^{-T} = D, \quad \beta\alpha^T = U,$$

with D diagonal, and we also have $\alpha^T = \beta^T L^T$, which is a QR factorization of the matrix whose columns are the A_i's.

Wong had therefore exposed the relationship of Gram-Schmidt to Gaussian elimination on the normal equations. This relationship was rediscovered in the paper by Lyle Eugene Pursell (1926-2015) and Selden Y. Trimble in 1991 [2524]. Pursell was professor of mathematics and statistics at Ohio State University, and from 1967, at the University of Missouri-Rolla (now Missouri University of Science and Technology), and Trimble was also at the Department of Mathematics and Statistics of the University of Missouri.

Let us briefly digress about Wong's life. Yue-Kei Wong (1903-1983) was a student at the University of Chicago, where he received a Ph.D. in 1931. He was the last student of Eliakim

Hastings Moore (1862-1932), an algebraist who is the "Moore" of the Moore-Penrose inverse. His other supervisor was Raymond Walter Barnard (1890-1962); see [3286, p. 297]. The title of the thesis was *Spaces associated with non-modular matrices with applications to reciprocals.* Wong is also mentioned as a draftsman working with Jacob Viner (1892-1970), an economist, professor at the University of Chicago from 1916 to 1917 and from 1919 to 1946. Viner made a famous error in his 1931 paper *Cost Curves and Supply Curves*, and when Wong refused to draw a family of descending U-shaped cost curves with a lower envelope that went through their bottoms, in a footnote of his paper, Viner made a comment[29] about Wong. That footnote made Wong's reputation among economists because his economics, and not only his mathematics, were correct. Viner only acknowledged his error in 1952, in a note added to a reprint of his paper. Wong's address in the paper [3270] in 1935 is given as Academia Sinica, Peiping, China. Peiping was one of the translations of the Chinese name of the capital Beijing. In the fall of 1936, he was the second Chinese mathematician, after Jiang Zehan (1902-1994), to be briefly admitted as a member of the Institute for Advanced Study in Princeton. He was again a member there from 1948 to 1950. In 1944, he was lecturer at the University of North Carolina, and was then appointed as professor in several other American universities. Of the six students of Moore and Barnard during this period, he was the only one to pursue a substantial activity. With Moore, he studied matrices and their reciprocals; in his later research he was concerned with the use of Minkowski-Leontief matrices in economics.

Francis Joseph Murray (1911-1996), an American mathematician who worked with John von Neumann (1903-1957), cited the Gram-Schmidt process [2310, 2311] in 1935 and 1939.

Hence, by the end of the 1930s, the "Gram-Schmidt process" had become a sort of common term for an orthogonalization process.

After World War II, numerical analysts considered the problem of computing orthogonal transformations using digital computers. One of the main applications was solving least squares problems. Probably the first paper in the literature on orthonormalizing a given set of vectors using a computer was by Philip Jacob Davis (1923-2018) and Philip Rabinowitz (1926-2006) in 1954. In [820], they used the classical Gram-Schmidt algorithm; see also [821] for applications to least squares. An Algol implementation named ORTHO by Philip J. Walsh, working at the National Bureau of Standards, included reorthogonalization in 1962 and was much used; see [3176]. This code was later translated to Fortran.

The numerical properties of the Gram-Schmidt algorithms started to be investigated in the 1960s. In 1966, John Rischard Rice [2564] showed experimentally that the classical and modified Gram-Schmidt algorithms (CGS and MGS) have very different properties when executed in finite precision arithmetic, even though they are mathematically equivalent. Let us assume that we want to orthogonalize the columns a_j of a matrix A of order n. In CGS, the orthonormal vectors q_j are computed as $q_1 = a_1/\|a_1\|$ and then, for $j = 2, \dots, n$, the next vector q_j is computed by

$$w = a_j - \sum_{k=1}^{j-1} (q_k, a_j) q_k, \quad q_j = w/\|w\|.$$

In MGS, q_1 is the same, and then, with $w = a_j$ and for $k = 1, \dots, j-1$, we do

$$w = w - (q_k, w) q_k,$$

and we normalize at the end of the loop to obtain $q_j = w/\|w\|$. The difference with CGS is that the dot product is computed with the current vector w instead of the vector a_j.

[29][...] *My instructions to the draftsman were to draw the AC curve so as never to be above any portion of any ac curve. He is a mathematician, however, not an economist, and he saw some mathematical objection to this procedure which I could not succeed in understanding. I could not persuade him to disregard his scruples as a craftsman and to follow my instructions, absurd though they might.* [...]

Both CGS and MGS compute factors Q and R such that $QR \approx A$, but the orthogonality in the columns of Q is much worse with CGS than with MGS. Note that there exist two variants of MGS since R can be computed by rows or by columns.

In 1967, Åke Björck, from Sweden, gave error bounds for the computed orthogonal factor in the modified Gram-Schmidt algorithm in [335]. He essentially proved that for MGS,

$$\|A - QR\| \leq c_1 u \kappa(A), \quad \|I - Q^T Q\| \leq \frac{c_2 u \kappa(A)}{1 - c_2 u \kappa(A)},$$

where u is the roundoff unit, $\kappa(A)$ is the condition number of A, c_1 and c_2 are constants depending only on the dimensions of A, and the norm is the Euclidean norm. It was assumed that $c_2 u \kappa(A) < 1$. In 1968, Björck published two Algol subroutines based on MGS for the solution of linear least squares problems.

A remedy for the potential loss of orthogonality in the computed vectors is reorthogonalization. Since full reorthogonalization against all the previous vectors is expensive, Heinz Rutishauser (1918-1970) [2642] introduced in 1967 a criterion for selective reorthogonalization of the vectors computed by the Gram-Schmidt algorithms.

In 1968, Charles Sheffield (1935-2002), a physicist and award-winning science fiction writer, made the surprising observation that the MGS QR factorization of a matrix A is equivalent to the Householder QR algorithm (see below) applied to A with a square matrix of zeros on top. This equivalence holds also in finite precision arithmetic. In 1992, Björck and Christopher Conway Paige [342] described this equivalence in detail and used it to derive backward stable MGS least squares algorithms. In short, they show that one can solve problems as accurately using MGS as when using Householder or Givens QR factorizations if, instead of using the columns of the matrix computed by MGS, one formulates the problems in terms of the almost orthonormal matrix obtained from the augmented matrix. It turns out that this can be done without computing any extra quantity. For more on this relationship see also Paige's paper [2418]. A good book about least squares problems is [337] by Björck in 1996.

The paper [2620] by Axel Ruhe (1942-2015) in 1983 was about the numerical aspects of Gram-Schmidt orthogonalization. He showed that the classical and modified variants of the Gram-Schmidt algorithm correspond to the Jacobi and Gauss-Seidel iterations for solving the system of normal equations.

Walter Hoffmann (1944-2013) considered iterative algorithms for Gram-Schmidt orthogonalization in [1721] in 1989. He used the classical and the modified Gram-Schmidt algorithms and the iterations allow to obtain a matrix Q that is orthogonal to almost full working precision.

An interesting paper summarizing the many variants of Gram-Schmidt algorithms is by G.W. Stewart [2883] in 2004. For a history of Gram-Schmidt orthogonalization, see [2036].

The potential loss of orthogonality of the Gram-Schmidt algorithms in finite precision arithmetic has been studied intensively; see the paper by Björck and C.C. Paige [342] that we have already cited above and Björck [336] in 1994.

To orthogonalize the columns of a matrix of full numerical column rank, one reorthogonalization step suffices for CGS and MGS to achieve orthogonality to roundoff levels. This is often called the *twice is enough* algorithm. This term was coined by William Morton Kahan in connection with orthogonalizing a Krylov sequence. In fact, one reorthogonalization with CGS is enough and using MGS in this context is not required.

These issues were studied in a series of papers by Luc Giraud and Langou [1359, 1360] at the beginning of the 2000s; see Giraud, Langou, and Miroslav Rozložník [1361], and Giraud, Langou, Rozložník, and Jesper van den Eshof [1362]. Their results were slightly improved in 2006 in the context of studying the backward stability of GMRES by C.C. Paige, Rozložník, and

Zdeněk Strakoš [2424]; see Chapter 5. Robust selective reorthogonalization techniques were described in [1360].

In practical least squares problems, it is often important to be able to add or remove equations. These operations, which are known as *updating* and *downdating*, can be done without recomputing everything from scratch.

Algorithms for updating the QR factorization were described by James Wilson Daniel, William Bryant Gragg (1936-2016), Linda Kaufman, and G.W. Stewart [807] in 1976. Their methods are based on the use of elementary reflection matrices and the Gram-Schmidt process with reorthogonalization.

Downdating algorithms were considered by Björck, Haesun Park, and Lars Eldén [343] in 1994, Kyeongah Yoo and Park [3302] in 1996, and Jesse Louis Barlow, Alicja Smoktunowicz, and Hasan Erbay [200] in 2005. See also the report [2885] by G.W. Stewart in 2010.

To improve the performance of Gram-Schmidt orthogonalization algorithms on modern computers, taking advantage of memory caches, and parallelism, block methods were developed starting in the 1990s. These algorithms use a block of vectors. In 1991, William Jalby and Bernard Philippe proposed and analyzed a block algorithm [1803]. Another block algorithm using a dynamically chosen block size was published in 2000 by Denis Vanderstraeten [3122].

In 2008, in the paper [2884], G.W. Stewart's goal was to devise a block algorithm satisfying

$$\|I - Q^*Q\| \leq c_1\epsilon, \quad \|A - QR\| \leq c_2\epsilon,$$

where ϵ is the rounding unit and c_1, c_2 are constants that depend on the dimension of the problem. He gave a MATLAB implementation of his algorithms.

Block Gram-Schmidt algorithms were published by Barlow and Smoktunowicz [199] in 2013, and new block modified Gram-Schmidt algorithms by Barlow [197] in 2019. In 2015, Barlow considered the problem of downdating block Gram-Schmidt algorithms [196].

Finally, we must mention that Gram-Schmidt algorithms are not only used for least squares problems, they are at the heart of some Krylov iterative methods for solving linear systems or for computing eigenvalues. This is considered in Chapters 5 and 6.

Rotations and reflections

Another way to obtain a QR factorization of a matrix is to use orthogonal or unitary transformations.

Rotations were already used by Jacobi in the middle of the 19th century. His goal in the paper [1795], written at the end of 1844 and published in 1845, whose title can be translated as *A new way of resolving the linear equations that occur in the least squares method*, was to solve the normal equations arising from least squares problems. The linear system has symmetric coefficients and the diagonal coefficients are positive. Jacobi's goal was to apply the iterative method that we now call the Jacobi method. However, he wanted to apply it to a matrix as close as possible to a diagonal matrix. Hence, he had to get rid of the large off-diagonal entries but he wanted to maintain the symmetry of the equations. He denoted the coefficients by (ij), $i, j = 0, 1, \ldots$ and assumed first that the coefficient to eliminate is $(10) = (01)$. What he did corresponds to multiplying from the left and from the right by a rotation matrix and choosing the angle to annihilate the (10) coefficient. The (local) rotation matrix was

$$\begin{pmatrix} \cos\alpha & \sin\alpha \\ \sin\alpha & -\cos\alpha \end{pmatrix},$$

and the condition for the angle was

$$\{(00) - (11)\} \cos\alpha \cdot \sin\alpha = (01)\left\{\cos^2\alpha - \sin^2\alpha\right\}.$$

In 1846, Jacobi published a paper whose title can be translated as *On an easy procedure to numerically solve the equations occurring in the theory of secular disturbances.* In that paper he was interested in computing eigenvalues from a system of equations corresponding to a symmetric matrix. He used a different notation for the coefficients that he denoted by $(a,a), (a,b), \ldots$ At each step he annihilated the coefficient of largest modulus by rotations to iteratively reduce the linear system to diagonal form by a sequence of rotations. The paper of 1846 is generally cited when people want to refer to Jacobi's rotations, but as we have seen, he was using them before that date.

Jacobi's method for computing eigenvalues was rediscovered in 1949 by Herman Heine Goldstine (1913-2004), Murray, and John von Neumann. Goldstine presented their work at a National Bureau of Standards (NBS) Symposium organized at the Institute of Numerical Analysis in Los Angeles in August 1951. However, Alexander Markowich Ostrowski (1893-1986), who attended that symposium, pointed out that this was similar to Jacobi's method. An updated version of the paper [1371] was only published in 1959 after von Neumann's death.

Apart from Jacobi, rotations are also attached to the name of Wallace Givens (1910-1993). Givens' work was first presented at the 1951 NBS Symposium that we cited above [1364]. A detailed account of the method was given in a 116-page report from Oak Ridge National Laboratory [1365] in 1954 in which Givens referred to Jacobi's rotations. In that report, Givens first recalled the Gram-Schmidt process to show that a matrix can be reduced by orthogonal transformations to Hessenberg form, which is tridiagonal when the matrix is symmetric. Then, he gave details on the construction of the plane rotations.

Givens' goals were different from those of Jacobi. Jacobi wanted to obtain an almost diagonal matrix. Using rotations to do so is not possible in a finite number of steps because some entries which have been zeroed are changed by subsequent rotations. So, one has to iterate, hoping that the moduli of the off-diagonal entries are decreasing. Givens used rotations to bring the symmetric matrix to tridiagonal form. This can be done in a finite number of steps since the created zero entries are not changed by the next rotations. The eigenvalues of the tridiagonal matrix were computed using Sturm's sequences.

In [1367] Givens applied rotations to nonsymmetric matrices to obtain an upper triangular matrix. As we shall see below Givens' method was soon superseded by Householder's reflections, which are less costly. However, Givens rotations are still of interest for sparse matrices or matrices of special structures. For instance, rotations are used to transform to upper triangular form the upper Hessenberg matrix which is constructed column by column in the GMRES iterative method for solving nonsymmetric linear systems [2667].

A paper on the reliable computation of Givens rotations [321] was published in 2002 by David Bindel, James Weldon Demmel, Kahan, and Osni Marques.

In 1973, William Morven Gentleman (1942-2018) showed how to avoid square roots when computing a QR factorization with Givens rotations [1319]. The number of operations is also smaller than for standard rotations; see also the paper by Sven Hammarling [1563] in 1974.

Rounding error analyses of Givens rotations were done by James Hardy Wilkinson (1919-1986) [3246] in 1963, and in his book [3248] in 1965 for square matrices. The analysis was extended to tall and thin rectangular matrices by Gentleman [1319] in 1973. He simplified his analysis in 1975; see [1320].

Unitary matrices which are rank-one modifications of the identity matrix were used by Turnbull and Aitken in their book [3081]; see page 103 and also page 109 of the 1944 edition.

They defined

$$\forall z \neq 0,\ Q = \frac{2}{z^*z}zz^* - I, \quad Q^*Q = QQ^* = I, \quad Q^2 = I.$$

They observed that any two nonzero vectors having the same norm can be transformed into each other by unitary transformations. Consequently, a given vector $x \neq 0$ can be transformed to $\|x\|e_1$, where e_1 is the first column of the identity matrix.

Aitken and Turnbull did not give any reference for these transformations. Later on, unitary matrices of the form $I - u\,u^*$ were used in 1951 by William Feller (1906-1970) and George Elmer Forsythe (1917-1972) in [1151]. Their goal was to obtain similarity transformations for computing eigenvalues.

This type of transformation was popularized by Alston Scott Householder (1904-1993). He used them to compute the QR factorization of a matrix. They are now known as *Householder reflections* or *Householder transformations* even though, as we have seen, he was not the first to use them. In [1744], he wrote

> *A method for the inversion of a nonsymmetric matrix, due to J.W. Givens, has been in use at Oak Ridge National Laboratory and has proved to be highly stable numerically but to require a rather large number of arithmetic operations, including a total of $n(n-1)/2$ square roots.* [...] *The triangular form is brought about by means of a sequence of $n(n-1)/2$ plane rotations, whose product is an orthogonal matrix.* [...] *The purpose of the present note is to point out that the same result can be obtained with fewer arithmetic operations, and, in particular, for inverting a square matrix of order n, at most $2(n-1)$ square roots are required, instead of $n(n-1)/2$.*

Then he proved, as Aitken and Turnbull did, that for any vector $a \neq 0$, and any unit vector v, a unit vector u exists such that

$$(I - 2u\,u^*)a = \|a\|\,v.$$

This result was said to be *almost self-evident*. Then, Householder wrote

> *Let a be the first column of A and take $v = e_1$, the first column of the identity. Application of the lemma provides a unitary matrix*
>
> $$U_1 = I - 2u_1\,u_1^*$$
>
> *such that the first column of U_1A is null except in the first element.*

Householder referred to these transformations as *elementary Hermitians*. In the end he obtained $UA = R$ with U unitary and R upper triangular. Householder referred only to his previous paper [1742], to von Neumann and Goldstine [3158], and to a book by Norman Earl Steenrod (1910-1971) on fiber bundles in 1951. The matrices considered in [1742] were rank-one modifications of the identity, that is, $I - \sigma u\,v^T$. Of interest is also [1741] in 1957. In that paper Householder emphasized factorizations like $PA = Q$, where Q is "easily inverted." He wrote

> *Most commonly P is a lower triangle and Q an upper triangle. Other possibilities are to take P a lower triangle and Q a matrix of orthogonal rows, or to take P strictly orthogonal and Q an upper triangle.*

Rotations were also described in that paper; see page 162. Householder wrote

> *The method was programmed at Oak Ridge National Laboratory under the direction of J.W. Givens and has been used very successfully on some matrices of order up to about 200. In actual practice the sines and cosines required for reduction of a given column are computed and stored, and each column is reduced before any rotations are applied to the next.*

The QR factorization was denoted as $A = QP$, where P is a unit upper triangular matrix.

The normwise backward stability of Householder reflections to compute a QR factorization was proved by Wilkinson in [3250] and in his book in 1965. Householder reflections were used in the algorithms that were published in the *Handbook for Automatic Computation* [3256] that appeared in 1971.

A drawback of the Householder reflections for computing the QR factorization is that the matrix Q is not readily available. In 1987, Christian Bischof and Charles Francis Van Loan [332] proposed what they called a *WY representation*. The product Q_r of r successive Householder matrices $I - v_i v_i^T$, $i = 1, \dots, r$, is represented as $I + W_r Y_r^T$ with

$$W_1 = -v_1,\ Y_1 = v_1, \quad W_i = (\, W_{i-1} \quad v_i \,),\ Y_i = (\, Y_{i-1} \quad Q_{i-1}^T v_i \,).$$

See also [2712] by Robert S. Schreiber and Van Loan in 1989. The construction of elementary unitary matrices was considered by Richard Bruno Lehoucq [2026] in 1996.

A columnwise error analysis of the Householder QR factorization was given by Nicholas John Higham in his book [1681, Chapter 19]. A componentwise perturbation analysis was done by Xiao-Wen Chang and C.C. Paige [638] in 2001.

A generalization of Householder reflections to block reflectors was considered by Schreiber and Beresford Neill Parlett [2713] in 1988.

Parallel algorithms

Parallel algorithms for computing a QR factorization started to be developed in the 1980s. The first algorithms used Givens rotations. For references to some early papers, see the paper by Dianne Prost O'Leary and Peter Whitman [2374] in 1990. In that paper the authors partitioned the matrix by blocks of rows and used Householder reflections.

Some methods were devised to construct orthonormal bases for Krylov subspaces; see, for instance, [1098] by Jocelyne Erhel in 1995. She used an algorithm named RODDEC (ROw partitioning Diagonal oriented DEComposition) that was defined in 1994 by Roger Blaise Sidje and Philippe [2763]. They used Householder reflections to factor the blocks of rows and then Givens rotations to merge them to obtain the final factorization.

A QR factorization algorithm based on multifrontal techniques and Householder reflections was proposed by Iain Spencer Duff, Patrick Amestoy, and Chiara Puglisi [51] in 1996. This method used a symbolic factorization of the symmetric matrix $A^T A$ since if $A = QR$ then $A^T A = R^T R$. Multifrontal QR factorizations were also developed by Timothy Alden Davis [827] in 2011 in his *SuiteSparse*[30] collection of programs.

A tiled QR algorithm was proposed in 2008 by Alfredo Buttari, Langou, Jakub Kurzak, and Jack Dongarra [516], based on the algorithm in the paper [1479] by Brian Christopher Gunter and Robert Alexander van de Geijn in 2005. The tiles are small blocks of data.

A partitioning of the matrix in block of rows was also used in the TSQR (Tall and Skinny QR) algorithm described by Demmel, Laura Grigori, Mark Frederick Hoemmen, and Langou [873] in 2012. When the blocks have been factored, they are combined recursively two by two; see also [1719] by Hoemmen in 2011. This was designed for matrices with many more rows than columns, hence the name "tall and skinny."

[30]https://people.engr.tamu.edu/davis/suitesparse.html (accessed January 2022)

4.3 ▪ Singular value factorization

Every $n \times m$ real or complex matrix A has a factorization $A = U\Sigma V^*$, where U is a unitary matrix of order n, Σ is an $n \times m$ rectangular diagonal matrix with positive diagonal entries, and V is a unitary matrix of order m. The diagonal entries of Σ are called the singular values of A. If A is real, U and V are real orthogonal. This factorization is also most often called the *singular value decomposition* (SVD). The SVD has many applications, particularly in statistics, and more recently, in data mining problems. The early history of the SVD was described in the paper [2876] by G.W. Stewart.

In 1873, Eugenio Beltrami (1835-1900), an Italian mathematician, published a paper in a journal whose Italian title can be translated as *Journal of mathematics for use of the students of Italian universities*. In that paper [262] he introduced what can be seen as the SVD of a square nonsingular real matrix having distinct singular values. The paper was translated to English by Daniel Lucius Boley of the University of Minnesota in 1995[31] [358].

Beltrami did not use matrices but bilinear forms,

$$f = \sum_{r,s} c_{r,s} x_r y_s,$$

with x and y, two vectors with n components. Then, he introduced linear changes of variables,

$$x_r = \sum_p a_{r,p}\xi_p, \quad y_s = \sum_q b_{s,q}\eta_q.$$

It yields a new form

$$\varphi = \sum_{p,q} \gamma_{p,q}\xi_p\eta_q,$$

with n^2 relations,

$$\gamma_{p,q} = \sum_{r,s} c_{r,s} a_{r,p} b_{s,q}.$$

Then, Beltrami showed that if the linear substitutions are orthogonal, they can be chosen such that $\gamma_{p,q} = 0$, $p \neq q$ and $\gamma_{p,p} > 0$. The equations written by Beltrami for the diagonal coefficients γ correspond to $\det(C^TC - \gamma^2 I) = 0$ and $\det(CC^T - \gamma^2 I) = 0$. If we write what he did in terms of matrices, he obtained the SVD of a square nonsingular real matrix with distinct singular values; see [2876].

In 1874, the French mathematician Camille Jordan (1838-1922) published a paper [1842] whose title can be translated as *Memoir on bilinear forms*; see also [1843]. He considered the form

$$P = \sum A_{\alpha\beta} x_\alpha y_\beta, \quad (\alpha, \beta = 1, 2, \dots, n),$$

which, despite the title of his paper, he called a "linear polynomial" in x and y. He solved three problems, the first one of which was to bring P to a simple canonical form using orthogonal linear substitutions, one for x and another one for y. He started (without explaining why) by considering an optimization problem: find the maximum and minimum values of P with the constraints

$$x_1^2 + \cdots + x_n^2 = 1, \quad y_1^2 + \cdots + y_n^2 = 1.$$

[31] The translation is available at https://www-users.cse.umn.edu/~boley/publications/ (accessed January 2022).

To obtain the solution, Jordan differentiated P and the constraints but his argument is not very clear. It would have been easier to introduce Lagrange multipliers. Nevertheless, he found that there exist λ and μ such that the following equations (that we write in matrix terms) must be satisfied:

$$Ay = \lambda x, \quad A^T x = \mu y,$$

and that $\mu = \lambda$. He observed that λ is given by the vanishing of the determinant,

$$D = \begin{vmatrix} -\lambda I & A \\ A^T & -\lambda I \end{vmatrix} = 0.$$

The determinant does not change when applying orthogonal transformations and D is a polynomial in λ with only even powers. C. Jordan chose a solution λ_1 of $D = 0$ (in fact, he should have chosen the largest one) and the corresponding vectors x and y. The vectors x and y were the first columns of the orthogonal matrices. In the new system of coordinates the maximum is obtained for $\xi_1 = \eta_1 = 1$ and all the other components equal to zero. Moreover, the first column of the array of coefficients is $(\lambda_1 \ \ 0 \ \ \cdots \ \ 0)^T$. The form P is reduced to $\lambda_1\xi_1\eta_1 + P_1$. Jordan proceeded by applying recursively the same technique to P_1. He also showed that the case of multiple roots of $D = 0$ can be handled.

In 1889, Sylvester, probably ignoring the works of Beltrami and C. Jordan 15 years before, published a three-page note in French in the *Comptes Rendus de l'Académie des Sciences de Paris* [2975], and a paper in English in the *Messenger of Mathematics* [2977] about a canonical form of a bilinear form. The singular values were called *canonical multipliers*. The difference with the papers of Beltrami and Jordan is that Sylvester used matrices. In the note, he considered as an example the form

$$F = 8x\xi - x\eta - 4y\xi + 7y\eta.$$

He wrote the matrix

$$A = \begin{pmatrix} 8 & -1 \\ -4 & 7 \end{pmatrix},$$

such that

$$F = (x \quad y)\, A \begin{pmatrix} \xi \\ \eta \end{pmatrix}.$$

Then, he computed the (equal) eigenvalues of AA^T and A^TA, which are 26 and 104, from which he got the "canonical multipliers" $\sqrt{26}$ and $2\sqrt{26}$. Finally, he obtained the two orthogonal matrices by computing the eigenvectors of A^TA and AA^T.

In the paper [2977], Sylvester used a method he had already used before for quadratic forms in [2976]. To diagonalize quadratic or bilinear forms, he introduced what he called *infinitesimal substitutions* constructed to reduce the sum of squares of nondiagonal elements. However, he considered an "infinite succession of infinitesimal orthogonal substitutions" without any justification. Hence, this paper is far from being a model of rigor. Apparently, Sylvester was not aware of Jacobi's method from 1846.

In his 1907 paper [2698] that we cited above, E. Schmidt introduced an infinite-dimensional version of the SVD. Schmidt was dealing with eigenvalue problems for integral equations with nonsymmetric kernels. He obtained an expansion formula which is the continuous equivalent of the formula

$$A = \sum_i \sigma_i u_i v i^T.$$

Schmidt also showed that the decomposition can be used to obtain approximation of low rank.

Harry Bateman (1882-1946) used the term "singular values" in a paper in 1908; see [213]. Independently, the name "valeurs singulières" (singular values) was introduced in 1909 by Émile Picard (1856-1941), a French mathematician who was the son-in-law of Charles Hermite (1822-1901), on page 1568 of [2493], a paper in which, using the results of E. Schmidt, he proved an existence theorem for the solution of integral equations.

In 1912, Hermann Weyl (1885-1955) proved inequalities for singular values of perturbations of integral equations; see [3219]. Much later, in 1949, he proved inequalities between the squared moduli of eigenvalues of a matrix and its singular values [3220].

In 1915, Autonne published a memoir on square complex hypohermitian and unitary matrices [113]. What he called *une matrice hypohermitienne* is a positive semi-definite Hermitian matrix. A nonsingular hypohermitian is Hermitian. In this memoir, he studied factorizations with hypohermitian and unitary matrices. In Chapter II, page 24, Autonne proved that any complex matrix A of order n and rank $r \leq n$ can be factored as (in his notation) $A = LFM$ with L, M unitary matrices and F diagonal with r nonzero diagonal entries. From this, he also obtained a proof of the polar decomposition; see below. Autonne's results were cited by Edward Tankard Browne (1894-1959) [477] in 1930. Autonne also considered the factorization of complex symmetric matrices. This is generally known as the Autonne-Takagi factorization since it was also proposed by Teiji Takagi (1875-1960) in 1925. For Autonne's works on matrices, see [105, 106, 108, 109, 107, 110, 111, 112].

In 1936, Carl Henry Eckart (1902-1973), an American physicist, and Gale J. Young (1912-1990), an American engineer and physicist, extended the SVD to rectangular real matrices [1047]. They were interested in finding the closest matrix of a given lower rank in the Frobenius norm to solve problems in statistics. For square matrices, they just referred to Sylvester [2976] in 1889, and to the books of Courant and Hilbert (1924 edition) [753], and MacDuffee [2106] in 1933. We saw above that the approximation problem had been solved by E. Schmidt in 1907 for integral equations but Eckart and Young were probably unaware of that. In their 1939 paper [1048] in which they extended their results to complex matrices, they referred to Beltrami, C. Jordan, Autonne, and Browne. In 1960, Leonid Mirsky (1918-1983) solved the approximation problem for an arbitrary unitarily invariant norm in [2254]. In 1987, Gene Howard Golub (1932-2007), Alan Jerome Hoffman (1924-2021), and G.W. Stewart [1381] extended the Eckart-Young approximation result by adding the constraint that a specified set of columns remains fixed.

The SVD for a nonsingular square matrix was mentioned on page 78 of the book by Cyrus Colton MacDuffee (1895-1961) [2106] in 1933. He cited Beltrami, C. Jordan, Sylvester, and Autonne.

The numerical computation of the SVD really started with the advent of digital computers after World War II. With the computers of the 1950s and 1960s, computing A^*A and AA^* accurately was not really feasible and not an option to compute the singular values. In [1929], Ervand George Kogbetliantz (1888-1974) iteratively reduced A to a diagonal form with a Jacobi-like method by using complex rotations to decrease the moduli of the nondiagonal entries. This method was used in 1954 on an *IBM 701* computer [1929]. It took 19 minutes to diagonalize a matrix of order 32. For the convergence of the algorithm, see C.C. Paige and Paul Van Dooren [2429] in 1986.

In 1958, Magnus Rudolph Hestenes (1906-1991) orthogonalized the columns of A with non-increasing norms to obtain $H = AV$, and then normalized the columns of $H = U\Sigma$; see [1660].

In 1960, Forsythe and Peter Karl Henrici (1923-1987) proposed the cyclic Jacobi method for computing eigenvalues and singular values [1193]. As Kogbetliantz, they used rotations but with a predetermined order, to avoid looking for the largest off-diagonal entry. In 1961, Vera Nikolaevna Kublanovskaya (1920-2012) proposed a QR-like method for computing singular values [1959, 1960].

The main algorithm for computing singular values of an $m \times n$ matrix [1382] was published in 1965 by Golub and Kahan. They first reduced A to an upper bidiagonal form J by left and right multiplications with Householder transformations. They also proposed a Lanczos-like algorithm using A and A^* to do so. This is now known as the *Golub-Kahan bidiagonalization*. The zero rows of J can be dropped to compute its singular values. They considered the matrix

$$\tilde{J} = \begin{pmatrix} 0 & J \\ J^* & 0 \end{pmatrix},$$

which was permuted to a tridiagonal form and symmetrized. Its eigenvalues were computed by Sturm sequences. Another possibility was to compute the eigenvalues of J^*J. Golub and Kahan discussed how to compute the singular vectors using deflation. They also described some applications of the SVD like computing the Moore-Penrose inverse. Later on, Golub became a strong advocate of the SVD; see the license plate of his car in Figure 4.1.

Figure 4.1. *Gene H. Golub's license plate*
Courtesy of Cleve Moler (MATLAB blog: Cleve's Corner: Cleve Moler on Mathematics and Computing)

In 1968, Golub [1378] proposed to use Francis' QR algorithm to compute the eigenvalues of $\tilde{J}$. A program implementing this algorithm was published by Peter Arthur Businger and Golub [514] in 1969. A paper [1391] was published in 1970 by Golub and Christian Reinsch in the *Linear Algebra series of the Handbook for Automatic Computation*. An Algol program was described in that paper. It used Householder reflections and the QR algorithm directly on the bidiagonal matrix.

In 1982, Tony Fan-Cheong Chan first reduced A to upper triangular form R using Householder reflections, and then, bidiagonalized R; see [612]. This reduces the number of operations, particularly when the number of rows of A is much larger than the number of columns. This idea was proposed in 1974 for solving least squares problems [2004] by Charles Lawrence Lawson (1931-2015) and Richard Joseph Hanson (1938-2017).

An issue that was investigated in the 1990s was the accurate computation of the small singular values, which was not guaranteed by the algorithms we have seen above.

In 1990, to improve the second phase of the SVD algorithm, Demmel and Kahan [877] presented an algorithm to compute all the singular values of a bidiagonal matrix to a high relative accuracy. Their algorithm is an hybrid of the usual shifted QR algorithm and an implicit zero-shift QR algorithm.

In the paper [874] in 1999, Demmel, Ming Gu, Stanley Charles Eisenstat (1944-2020), Ivan Slapničar, Krešimir Veselić, and Zlatko Drmač analyzed when it is possible to compute the singular values and singular vectors of a matrix with high relative accuracy. They used a rank-revealing decomposition (RRD) $A = XDY^T$, where D of order r, the rank of A, is a diagonal matrix. Then, they did a QR factorization with column pivoting $XD\Pi = QR$ and form $W = R\Pi^T Y$. They computed the SVD of $W = \tilde{U}\Sigma V^T$ using a one-sided Jacobi algorithm and finally obtained $A = (Q\tilde{U})\Sigma V^T$. The authors used Gaussian elimination with complete pivoting to compute the RRD.

In 2002, N.J. Higham used the same algorithm but computed the RRD with a QR factorization with complete pivoting (column pivoting and row sorting or pivoting) [1680].

Proposals to improve the first phase, that is, the reduction to bidiagonal form, were made in the 2000s. The algorithm by Rui Ralha [2530] was fast, but unstable. An algorithm based on Givens rotations [195] was proposed by Barlow in 2002. Other algorithms were considered by Barlow, Nela Bosner and Drmač [1788] in 2005, and Bosner and Barlow [378] in 2007.

Accurate computations of the singular values using (modified) Jacobi iterations were studied by Drmač [954] in 1997 and Drmač and Veselić [956, 957] in 2008; see also [955].

With the advent of parallel computers the question arose of efficiently computing the SVD on these architectures. One of the early attempts at the beginning of the 1980s was the work by Franklin Tai-Cheung Luk [2099], who implemented an SVD algorithm on the *ILLIAC IV*. For details on this computer, see Chapter 7. Luk used Hestenes' 1958 method but iteratively orthogonalized the rows with rotations instead of the columns. For a theoretical implementation on a "systolic array" of processors, see [427] by Richard Peirce Brent and Luk in 1985.

In 1989, Michael W. Berry and Ahmed Sameh [298] used a block Householder reduction to upper triangular form, and then one-sided Jacobi iterations to obtain the SVD. They implemented their algorithms on an *Alliant FX/8* and a *CRAY X-MP/416*.

Berry, Dani Mezher, Philippe, and Sameh [297] wrote a review of some parallel algorithms for computing the SVD in 2005.

Elizabeth Redding Jessup and Danny Chris Sorensen focused on the computation of the SVD of the bidiagonal matrix B using divide and conquer techniques [1819] in 1994. The matrix B is partitioned into two blocks plus a rank-one correction. The SVDs of the blocks can be computed recursively in parallel. To recover the SVD of B from those of the blocks, a secular equation has to be solved. Numerical results were given on an *Alliant FX/8*. Divide and conquer methods had been used before for tridiagonal eigenvalue problems at the beginning of the 1980s, and had been proposed for the SVD by Peter Arbenz in 1989.

In 1995, Gu and Eisenstat [1471] also used divide and conquer techniques. They computed the singular values in the same way as Jessup and Sorensen, but used a different approach for the singular vectors as well as a different technique for deflation.

Mark Gates, Stanimire Tomov, and Dongarra implemented a two-stage bidiagonal reduction (first to banded form and then to bidiagonal form) and divide and conquer on GPUs (Graphics Processing Units) [1294] in 2018.

In 1994, N.J. Higham and Pythagoras Papadimitriou [1691, 1692] introduced a method which differs from many of the other ones. They first computed the polar decomposition (see below) by an iterative method and then computed the spectral decomposition of the polar factor. This method is well suited for parallel computing since the computation of the polar decomposition requires only matrix multiplications and (eventually) computations of inverses. The matrix A is factorized as $A = UH$, where U has orthonormal columns and H is Hermitian positive semidefinite. The matrix H is factorized as $H = VDV^*$ (spectral factorization) and $A = (UV)DV^*$. Numerical experiments were provided on a parallel shared memory computer.

The same technique but with different and more efficient ingredients was used by Yuji Nakatsukasa and N.J. Higham [2322] in 2013. Their algorithms exploited the QR-based dynamically weighted Halley (QDWH) algorithm of Nakatsukasa, Zhaojun Bai, and François Gygi [2319], which computes the polar decomposition using a cubically convergent iteration based on the building blocks of QR factorization and matrix multiplication. So, there is no computation of inverses. The eigendecomposition of H is computed by QDWH-eig, an algorithm for the symmetric eigenproblem described in [2322]. This gives the QDWH-SVD algorithm; see also [2321].

In 2016, Dalal Sukkari, Hatem Ltaief, and David Elliot Keyes [800] described an efficient implementation of the QDWH-SVD algorithm on computers with multicore processors and GPUs. QDWH-SVD was implemented on *CRAY XC* manycore systems in 2019 by Sukkari, Ltaief, Aniello Esposito, and Keyes [2940].

In 2016, Nakatsukasa and Roland W. Freund improved the QDWH-SVD algorithm [2320]. The function used in the QDWH iteration for the unitary polar factor is interpreted as a best rational approximation to the sign function. This type of problem was posed and solved in 1877 by Y.I. Zolotarev.[32] The idea of Nakatsukasa and Freund is to use higher-order Zolotarev rational approximation. The coefficients in Zolotarev's solution are given in terms of Jacobi elliptic functions. With this approximation only two iterations are needed.

Ltaief, Sukkari, Esposito, Nakatsukasa, and Keyes implemented the polar decomposition algorithm using the Zolotarev approximation in parallel [2095] on *CRAY XC* systems with up to 102,400 cores in 2019. Even though its number of floating point operations is larger, the new algorithm outperformed QDWH.

Methods related to Jacobi algorithms were proposed in the 2000s. In 2001, Michiel Erik Hochstenbach published a Jacobi-Davidson type SVD method [1707]. Martin Bečka, Gabriel Okša and Marian Vajteršic published several papers on block-Jacobi methods to compute the SVD [306, 307, 308] in 2002-2015.

We observe that all these algorithms for computing all the singular values and, eventually, the singular vectors were designed for dense matrices. However, some applications only require a small number of singular values and singular vectors. Moreover, the methods we have just described are not well suited for sparse matrices for which methods based on the Golub-Kahan bidiagonalization or on Jacobi methods are to be preferred.

Similar to what had been done for the eigenvalue problem earlier, implicitly restarted bidiagonalization methods were developed in the beginning of the 2000s following an earlier paper by Björck, Eric Grimme, and Van Dooren [340] in 1994. They derived recursion formulas for their restarted bidiagonalization method and applied it to the solution of ill-posed problems. In 2003, Zhongxiao Jia and Datian Niu proposed an implicitly restarted refined bidiagonalization Lanczos method for computing a partial SVD [1826]. Effrosyni Kokiopoulou, Constantine Bekas, and Efstratios Gallopoulos applied the recursion formulas of [340] to a few singular triplets of a large sparse matrix [1931] in 2004.

In 2005, James Baglama and Lothar Reichel proposed a method which is mathematically equivalent but more robust [147]. The essence of the restarted methods is the following. Assume that the ℓ largest singular values and the corresponding singular vectors are sought. Then, $m > \ell$ iterations of the bidiagonalization algorithm are done obtaining matrices B_m, U_m, V_m. A restart is done and a new subspace is constructed as

$$\breve{V}_{\ell+1} = [q_1, q_2, \dots, q_\ell, v_{\ell+1}],$$

with the so-called Ritz vectors $q_j = V_m y_i$, the y_i's being right singular vectors of B_m. The same is done for left singular vectors,

$$\breve{U}_{\ell+1} = [p_1, p_2, \dots, p_\ell, \breve{u}_{\ell+1}],$$

with $p_j = V_m x_i$, the x_i's being left singular vectors of B_m. The vector $\breve{u}_{\ell+1}$ is obtained by the Gram-Schmidt orthogonalization method of Av_{m+1} against the vectors p_i. With this, one

[32] Yegor Ivanovich Zolotarev (1847-1878) was a Russian mathematician who was a student of Pafnuty Lvovich Chebyshev (1821-1894) and Alexander Nikolaevich Korkin (1837-1908) in Saint Petersburg.

obtains a matrix

$$\breve{B}_{\ell+1} = \begin{pmatrix} \sigma_1 & & & & \rho_1 \\ & \sigma_2 & & & \rho_2 \\ & & \ddots & & \vdots \\ & & & \sigma_\ell & \rho_\ell \\ & & & & \alpha_{\ell+1} \end{pmatrix}.$$

Setting $\tilde{v}_{\ell+2} = A^T \breve{u}_{\ell+1} - \alpha_{\ell+1} v_{\ell+1}$, $\breve{\beta}_{\ell+1} = \|\tilde{v}_{\ell+2}\|$ and $v_{\ell+2} = \tilde{v}_{\ell+2}/\breve{\beta}_{\ell+1}$, the following relations are obtained

$$\begin{aligned} A\breve{V}_{\ell+1} &= \breve{U}_{\ell+1}\breve{B}_{\ell+1}, \\ A^T\breve{U}_{\ell+1} &= \breve{V}_{\ell+1}\breve{B}_{\ell+1}^T + \breve{\beta}_{\ell+1}\breve{v}_{\ell+1}e_{\ell+1}^T. \end{aligned}$$

These relations can be extended up to size m by the usual Golub-Kahan recurrences. This algorithm is used iteratively until the obtained approximate singular values satisfy some stopping criteria. Instead of Ritz vectors, harmonic Ritz vectors can be used. A block version of this algorithm was published [148] in 2006.

The drawback of the previous algorithm is that the matrices B_m whose SVD have to be computed are no longer bidiagonal but have spikes due to the restarts. Moreover, it is difficult to implement deflation techniques in the algorithm.

In 2012, Martin Stoll proposed an algorithm that he called a Krylov-Schur approach [2908] based on a slightly more general decomposition. From the Golub-Kahan recurrences and the SVD $B_m = P_m\Sigma_m Q_m^T$, a new decomposition is obtained,

$$\begin{aligned} A\tilde{V}_m &= \tilde{U}_m\Sigma_m, \\ A^T\tilde{U}_m &= \tilde{V}_m\Sigma_m + \beta_{m+1}v_{m+1}p_m^T, \end{aligned}$$

with $\tilde{U}_m = U_m P_m$, $\tilde{V}_m = V_m Q_m$, and $p_m^T = e_m^T P_m$. From this decomposition, deflation can be implemented by just multiplications by permutation matrices. The decomposition can be transformed to the usual form by Householder reflections. The advantage of this method is that the bidiagonal structure is preserved.

In 2010, Jia and Niu proposed a refined harmonic Lanczos bidiagonalization method and an implicitly restarted algorithm to compute the smallest singular values and the corresponding singular vectors [1827].

Recently, the need to use the SVD to compute low-rank approximations of large matrices or to analyze very large data sets leads to the introduction of methods using randomness. Many of these methods use random sampling to identify a subspace that captures most of the action of a given matrix. The input matrix is reduced, explicitly or implicitly, to this subspace, and a low-rank approximation is obtained with a deterministic method. An introduction to this type of method was published in 2011 by Nathan Halko, Per-Gunnar Martinsson, and Joel Aaron Tropp [1548]. For the approximate SVD, the first phase using randomness must produce a matrix Q with (a "small" number of) orthonormal columns such that $A \approx QQ^*A$. The second phase computes $B = Q^*A = \tilde{U}\Sigma V^*$ and finally, $A \approx (Q\tilde{U})\Sigma V^*$.

More recent surveys about randomized methods were written by Michael W. Mahoney [2117] in 2016, Petros Drineas and Mahoney [952] in 2018, and Martinsson and Tropp [2161] in 2020.

Another approach for sampling and a method for computing an approximate SVD [1725] were described by Michael P. Holmes, Alexander G. Gray, and Charles Lee Isbell, Jr. in 2009.

The SVD has been generalized in several directions. The first one is a generalization to a matrix pencil (A, B) for which the two matrices have the same number of columns.

If A is $m \times n$ (with $m \geq n$) and B is $p \times n$, Van Loan proved in 1976 that there exist unitary matrices U and V and a nonsingular matrix X such that

$$U^*AX = \Sigma_A, \quad V^*BX = \Sigma_B,$$

where the diagonal matrix Σ_A is $m \times n$ and the diagonal matrix Σ_B is $p \times n$; see [3110]. The diagonal entries of these matrices are positive. This work was a part of the Ph.D. thesis of Van Loan in 1973, whose advisors were Cleve Barry Moler and George Joseph Fix (1939-2002).

An extension covering all the possible cases was described by C.C. Paige and Michael Alan Saunders in 1981. In [2426], we have the following relations,

$$U^*AQ = \Sigma_A \,(W^*R \quad 0), \quad V^*BQ = \Sigma_B \,(W^*R \quad 0),$$

where U, V, W, Q are unitary matrices, Σ_A $(m \times k)$ and Σ_B $(p \times k)$ are diagonal matrices with 1's, 0's and positive numbers on the diagonals, and with k being the rank of $C^* = (A^*, B^*)$. R is a nonsingular matrix of order k. Other methods were proposed by G.W. Stewart [2873] in 1983 and Van Loan [3112] in 1985.

In 1986, an algorithm to compute the Paige and Saunders decomposition was proposed by C.C. Paige in [2417]. It is an iterative method in the same spirit as what Kogbetliantz did for the computation of the SVD. The method was first developed for square matrices A and B and then extended to the general case. The method was based on transformations of 2×2 matrices which are used to reduce the matrices alternatively to lower and upper triangular matrices. Paige's method is related to the algorithm described in [1613] by Michael Thomas Heath, Alan John Laub, Paige, and Robert C. Ward for computing the singular value decomposition of a product of two matrices [1613] in 1986.

A variant of Paige's algorithm was published by Z. Bai and Demmel [157] in 1993. They added a preprocessing phase, which transforms A and B to upper triangular form, and introduced a new 2×2 generalized SVD algorithm.

The *generalized SVD* (GSVD) of the matrix pencil (A, B) is sometimes called the *quotient SVD* since, when B is square and nonsingular, the SVD of AB^{-1} can be obtained from the GSVD.

The *restricted SVD* (RSVD) was introduced by Hongyuan Zha [3323] and further developed and discussed by Bart De Moor and Golub [847] in 1991. Let A $(m \times n)$, B $(m \times p)$, and C $(q \times n)$ be three given matrices. Then,

$$A = P^{-*}S_AQ^{-1}, \quad B = P^{-*}S_BV_B^*, \quad C = U_CS_CQ^{-1},$$

where P and Q are square nonsingular, V_B and U_C are unitary matrices, and S_A, S_B, S_C are real quasi-diagonal matrices with positive entries. The RSVD of (A, B, I) (resp., (A, I, C)) is the GSVD of (A, B) (resp., (A, C)). De Moor and Golub also described applications of the RSVD.

The RSVD is just one example of generalizations of the SVD for more than two matrices that were considered by De Moor and Zha [849] in 1991. These generalizations were further discussed by De Moor [845, 846] and De Moor and Van Dooren [848].

In 1997, Moody Ten-Chao Chu, Robert Edward Funderlic (1937-2009), and Golub considered a variational formulation of the generalized singular value decomposition [693]. Later, in 2000, Delin Chu and De Moor gave a variational formulation of the QSVD and the RSVD [690] and discussed the non-uniqueness of the factorization factors in the product singular value decomposition [691].

The computation of some of these SVD generalizations was considered by Chu, Lieven De Lathauwer, and De Moor [688, 689]. A Jacobi-Davidson algorithm for computing the GSVD was introduced by Hochstenbach [1708] in 2009.

4.4 ▪ Polar factorization

The *polar decomposition* is a factorization closely linked to the SVD. An $n \times m$ matrix A with $n \geq m$ has a factorization $A = UH$ with U $n \times m$ with orthonormal columns, and H $m \times m$ Hermitian positive semi-definite. The factor H is $(A^*A)^{\frac{1}{2}}$ and the factor U is unique if A has full rank. The factor U is the closest matrix with orthonormal columns to A in the ℓ_2 and the Frobenius norms and in any unitarily invariant norm when $m = n$ (minimization property). As we have seen above, the relation with the SVD is the following: if $A = P\Sigma Q^*$ is the SVD of A, then

$$A = (PQ^*)(Q\Sigma Q^*) = UH,$$

and $Q\Sigma Q^*$ is the spectral decomposition of H. So, the SVD can be computed from the polar decomposition and vice versa.

The polar decomposition can be seen as an extension of the polar form of complex numbers $z = re^{i\theta}$ with $i = \sqrt{-1}$ and $r \geq 0$. With maybe some exaggeration, some people saw its origin in some works on continuum mechanics in the 19th century [2327]; for instance, in the works of Cauchy [573] on the composition of rotations and dilatations in the motion of a body in 1841, and Josef Finger (1841-1925), an Austrian physicist, in 1892 [1168], as well as in the memoir of the brothers Eugène (1866-1931) and François Cosserat (1852-1914) [749] on elasticity theory in 1896. None of these works used matrices, and, of course, they considered only the three-dimensional space.

Anyway, the existence of the polar decomposition for matrices was proved in 1902 by Autonne in Lemma II on pages 124-125 of [105]. He considered only square matrices but his proof is valid for any matrix.

Without any reference to Autonne's work, Aurel Friedrich Wintner (1903-1958) and Francis Dominic Murnaghan (1893-1976) [3262] stated the decomposition for square matrices in 1931 and called it the *polar decomposition*. They proved the uniqueness of the decomposition for nonsingular matrices.

Giuseppe Grioli (1912-2015), an Italian mathematician and physicist, showed the minimization property in the special case of the Frobenius matrix norm and dimension 3 [1467] in 1940; see also [2327]. The minimization property was proved for square matrices and any unitarily invariant norm by Ky Fan (1914-2010) and A.J. Hoffman [1135] in 1955.

Early papers on the computation of the unitary polar factor with iterative methods are by Zdislav Vaclav Kovarik [1941] in 1970 and Björck and Clazett Bowie [338] in 1971.

As we have seen, the polar decomposition can also be computed from the SVD, but other methods for its computation were proposed in the 1980s and 1990s. In 1986, N.J. Higham [1672] considered a Newton-like method to compute the unitary polar factor U of a square matrix A. The sequence

$$X_0 = A, \quad X_{k+1} = \frac{1}{2}(X_k + X_k^{-*}),\ k = 0, 1, 2, \ldots$$

converges to U. The convergence of the method was accelerated by using some scaling factors. At convergence, H is computed as $H = (U^*A + A^*U)/2$. This algorithm was extended to arbitrary matrices A by N.J. Higham and Schreiber [1694] in 1990. They added an initial orthogonal decomposition to extract an appropriate square nonsingular matrix,

$$A = P \begin{pmatrix} R & 0 \\ 0 & 0 \end{pmatrix} Q^*,$$

with P and Q unitary and R nonsingular and upper triangular. They also considered using approximations for the term X_k^{-*} in the Newton iteration. For more on Higham's method, see

[974] by Augustin A. Dubrulle in 1999, as well as the paper [1893] by Andrzej Kiełbasiński and Krystyna Ziętak in 2003.

In 1990, Walter Gander, from Switzerland, proposed several methods to compute the polar decomposition of general real matrices in [1279]. The iterative scheme is

$$X_0 = A, \quad X_{k+1} = X_k\, h(X_k^T X_k),$$

where the function h is suitably chosen. Gander proved that the iteration is of order $p+1$ if and only if h approximates $w(x) = 1/\sqrt{x}$ in such a way that $h^{(i)}(1) = w^{(i)}(1)$, for $i = 1, \ldots, p$. Higham's non-accelerated method of order 2 is given by $h(x) = (1/2)(1 + 1/x)$ and Halley's[33] method of order 3 by $h(x) = (x+3)/(3x+1)$.

The relationships and analogies between the sign function for matrices and the polar decomposition were studied by N.J. Higham [1678] in 1994. This allows him to define new iterative methods for the computation of the polar decomposition. This was used to develop a parallel algorithm by Higham and Papadimitriou [1691, 1692] in 1994 as we have seen above in the SVD subsection.

(John) Alan George and Khakim D. Ikramov addressed the following problem: Is the polar decomposition finitely computable? This means, computable with the four arithmetic operations, extraction of radicals of arbitrary integer degree, and comparisons with zero. They did not succeed [1324] in 1996, but published an addendum [1325] in 1997 in which they proved that the answer to the question is "no" for matrices of order larger than 4.

Ralph Byers (1955-2007) (see [2190]) and Hongguo Xu used Higham's Newton-like method but with scaling factors computed by a scalar iteration in which the initial value depends only on estimates of the extreme singular values of the matrix [519]. This paper was published in 2008, after Byers's death. About this paper, see the comments by Kiełbasiński and Ziętak [1894] in 2010.

As we have seen above in the SVD subsection, in 2010 Nakatsukasa, Z. Bai, and François Gygi [2319] proposed a method to compute the polar decomposition. DWH is a dynamically weighted Halley algorithm using a cubically convergent iteration with weighting and scaling parameters. The iteration is

$$X_0 = A/\alpha, \quad X_{k+1} = X_k\,(a_k I + b_k X_k^* X_k)\,(I + c_k X_k^* X_k)^{-1},$$

where the scalars are positive weight factors. The DWH iteration is implemented through a QR factorization,

$$\begin{pmatrix} \sqrt{c_k} X_k \\ I \end{pmatrix} = \begin{pmatrix} Q_1 \\ Q_2 \end{pmatrix} R, \quad X_{k+1} = \frac{b_k}{c_k} X_k + \frac{1}{\sqrt{c_k}} \left(a_k - \frac{b_k}{c_k} \right) Q_1 Q_2^*.$$

This involves only QR factorizations and matrix multiplications. Hence, this algorithm is more suited to parallel computing than the Newton-like iteration, even though it uses more arithmetic operations per iteration. The QDWH algorithm was improved in 2016 by Nakatsukasa and Freund using a Zolotarev approximation [2320].

In 2014, Johannes Lankeit, Patrizio Neff, and Nakatukasa [1991] proved other optimality properties of the unitary polar factor U of a square matrix $A = UH$. They showed that

$$\min_{Q \in \mathcal{U}(n)} \|\mathrm{Log}\, Q^* A\| = \|\mathrm{Log}\, U^* A\|,$$

[33]Edmond Halley (1656-1741) was an English astronomer, geophysicist, and mathematician who devised a root-finding algorithm for functions with a continuous second derivative.

where $\mathcal{U}(n)$ is the group of unitary matrices of order n. In this relation Log B is any solution of $\exp X = B$. The minimization property also holds for the Hermitian part of the logarithm; see also [2328] by Neff, Nakatsukasa, and Andreas Fischle.

Extensions of the polar decomposition using indefinite scalar products were proposed by Yuri Bolshakov and Boris Reichstein [367] in 1997 and by Bolshakov, Cornelis Victor Maria van der Mee, André C.M. Ran, Reichstein, and Leiba Rodman (1949-2015) [368, 369] in 1997.

In N.J. Higham's book [1683] on functions of matrices, it is shown that every matrix A has a *canonical polar decomposition* $A = UH$, where U is a partial isometry ($UU^*U = U$) and H is Hermitian positive semi-definite. A canonical generalized polar decomposition was described by N.J. Higham, Christian Mehl, and Françoise Tisseur [1690] in 2010.

4.5 ▪ CS factorization

The *CS decomposition* (for cosine-sine) is a factorization of a unitary matrix A. If A is partitioned as

$$A = \begin{pmatrix} A_1 & A_3 \\ A_2 & A_4 \end{pmatrix},$$

with A_1 $m_1 \times n_1$, A_2 $m_2 \times n_1$, A_3 $m_1 \times n_2$, and A_4 $m_2 \times n_2$. Assuming $m_1 \geq n_1$ and $m_1 \geq m_2$, it holds

$$\begin{pmatrix} U_1 & 0 \\ 0 & U_2 \end{pmatrix}^* A \begin{pmatrix} V_1 & 0 \\ 0 & V_2 \end{pmatrix} = \left(\begin{array}{cc|ccc} I_p & 0 & 0 & 0 & 0 \\ 0 & C & -S & 0 & 0 \\ 0 & 0 & 0 & I_{m_1-n_1} & 0 \\ \hline 0 & S & C & 0 & 0 \\ 0 & 0 & 0 & 0 & I_q \end{array}\right),$$

where U_1, U_2, V_1, V_2 are unitary, C, S are diagonal with positive entries, $C^2 + S^2 = I$, $p = \max(0, n_1 - m_2)$, $q = \max(m_2 - n_1)$. If A is $2n \times 2n$, we have

$$A = \begin{pmatrix} U_1 & 0 \\ 0 & U_2 \end{pmatrix} \begin{pmatrix} C & -S \\ S & C \end{pmatrix} \begin{pmatrix} V_1 & 0 \\ 0 & V_2 \end{pmatrix}^*.$$

The early history of the CS decomposition was described by C.C. Paige and Musheng Wei [2430] in 1994. They saw the origin of the decomposition in the work of C. Jordan [1844] in 1875 (see a translation to English and an analysis by G.W. Stewart in [2886]). A part of what Jordan did was related to angles between subspaces. An early contributor to the CS decomposition was Chandler Davis (1926-2022), who was initially interested in finding how the eigenvectors of a Hermitian matrix A must be rotated to get the eigenvectors of $A + H$, where H is a Hermitian perturbation; see [816] in 1963. In 1965, he considered operators in Hilbert spaces [817]. In the paper [818], written in 1969 with Kahan, C. Davis considered angles between finite-dimensional subspaces. They essentially proved a part of the CS decomposition; see the analysis of their work in [2430]. The proofs and the extension to Hilbert spaces were given [819] in 1970.

The work of C. Davis and Kahan was cited and used by Björck and Golub [339] in 1973. They were interested in numerical methods for computing principal (or canonical) angles between two subspaces F and G with $p = \dim(F) \geq q = \dim(G)$. The principal angles $\theta_k \in [0, \pi/2]$ are defined as

$$\cos(\theta_k) = \max_{u \in F} \max_{v \in G} u^* v = u_k^* v_k, \; k = 1, \ldots, q, \quad \|u\| = 1, \; \|v\| = 1,$$

subject to $u_j^* u = 0$, $v_j^* v = 0$, $j = 1, \dots, k-1$. If F and G are the ranges of two matrices A and B and if Q_A and Q_B form unitary bases for the ranges, the cosines of the principal angles are the singular values of the matrix $Q_A^* Q_B$.

Unaware of [818], in 1977 G.W. Stewart, in an appendix of [2870], proved (in his notation) that for a unitary matrix W of order n partitioned as

$$W = \begin{pmatrix} W_{11} & W_{12} \\ W_{21} & W_{22} \end{pmatrix},$$

with W_{11} of order $r \le n/2$, there are unitary matrices U and V such that

$$U^* W V = \begin{pmatrix} \Gamma & -\Sigma & 0 \\ \Sigma & \Gamma & 0 \\ 0 & 0 & I \end{pmatrix},$$

where Γ and Σ are diagonal matrices of order r.

This result was used by Van Loan [3111] in 1979, where the factorization is called the p-SVD. He gave some applications of this result to total least squares, subset selection, and the Riccati equation. The result of G.W. Stewart [2871] was also used, and slightly generalized, by C.C. Paige and Saunders in 1981 in their paper on the generalized singular value decomposition [2426].

The name *CS decomposition* appeared in print for the first time in the title of the paper [2872] by G.W. Stewart in 1982. He considered the computation of the CS decomposition of an $n \times p$ real unitary matrix Q,

$$Q = \begin{pmatrix} Q_1 \\ Q_2 \end{pmatrix},$$

where Q_1 is $k \times p$ and Q_2 is $\ell \times p$, with $k + \ell = n$. Stewart started from the case $k = \ell = p$ for which

$$\begin{pmatrix} U_1^T & 0 \\ 0 & U_2^T \end{pmatrix} \begin{pmatrix} Q_1 \\ Q_2 \end{pmatrix} V = \begin{pmatrix} C \\ S \end{pmatrix},$$

and $U_1^T Q_1 V = C$ is the SVD of Q_1. The straightforward computation of U_2 can be unstable and Stewart used a matrix scaling. The matrices needed for the scaling are computed through Jacobi iterations to diagonalize the matrix $\bar{Q}_2^T \bar{Q}_2$ with $\bar{Q}_2 = Q_2 V$. Then, Stewart extended his algorithm to the general case.

In 1984, C.C. Paige gave an analysis of the sensitivity of the CS singular values in [2416].

The next attempt to compute the CS decomposition was by Van Loan [3112] in 1985. He proposed a stable algorithm which used only SVDs and QR factorizations. A parallel implementation of this algorithm was done by Luk and Sanzheng Qiao [2100] in 1986.

In 2005, Michael Stewart and Van Dooren introduced an hyperbolic CS decomposition for Σ-unitary matrices [2889]. A matrix A is Σ-unitary if

$$A^* \Sigma A = \Sigma, \quad \Sigma = \begin{pmatrix} I_p & 0 \\ 0 & -I_q \end{pmatrix}.$$

Later, Brian D. Sutton published several algorithms for computing CS decompositions. The paper [2942] in 2009 described an algorithm for computing the CS decomposition of a 2×2 block partitioned matrix,

$$X = \begin{pmatrix} X_{11} & X_{12} \\ X_{21} & X_{22} \end{pmatrix}.$$

It is a two-phase algorithm. First, the blocks are simultaneously transformed to bidiagonal form by Householder unitary matrices, lower bidiagonal for X_{11} and X_{21} and upper bidiagonal for X_{12} and X_{22}. Then, the Golub and Reinsch bidiagonal SVD iterative algorithm is applied simultaneously to the four bidiagonal blocks using Givens rotations. Of course, this has to be done carefully to obtain the desired accuracy in finite precision arithmetic. The subsequent paper [2943] in 2012 proved the numerical stability of the reduction to block-bidiagonal form and derived a mathematically equivalent modified algorithm. In 2013, Sutton introduced a divide and conquer algorithm for a unitary matrix partitioned in 2×1 blocks [2944]. An algorithm for reducing the blocks of the matrix X above was presented in 2014 by Kingston Kang, William Lothian, Jessica Sears, and Sutton [1868]. Using block Givens rotations, it generalized the bidiagonalization algorithm of [2942].

A full 2×2 block CS decomposition algorithm was described by Daniela Calvetti, Reichel, and H. Xu [542] in 2015.

In 2016, G.W. Stewart wrote a paper [2887] on a canonical CS representation of a pair of subspaces. Let $\mathcal{X}$ and $\mathcal{Y}$ be subspaces of $\mathbb{R}^n$ with $\dim \mathcal{X} \leq \dim \mathcal{Y}$ and X, Y be orthonormal bases of these subspaces. Stewart described canonical forms of X and Y that exhibit a smallest possible shared subspace of $\mathcal{X}$ and $\mathcal{Y}$ and a smallest possible subspace of $\mathcal{Y}$ that is orthogonal to $\mathcal{X}$.

Evan S. Gawlik, Nakatsukasa, and Sutton [1310] proposed in 2018 to compute the 2×1 CS decomposition via polar decompositions. They first computed the polar decompositions of the two blocks, say, $A_i = W_i H_i$, $i = 1, 2$, using the Zolotarev rational approximation version of QDWH. The eigenvalues of the two matrices H_1 and H_2 are, respectively, the cosines and sines of the principal angles appearing in the CS decomposition, $H_1 = V_1 C V_1^*$, $H_2 = V_1 S V_1^*$. But computing the two spectral decompositions separately is not reliable. Instead, V_1 is computed through the spectral decomposition of $H_2 - H_1$.

4.6 ▪ Jordan canonical form

For every square matrix with complex entries there exists a nonsingular matrix X such that $X^{-1}AX = J$, where J is a block diagonal matrix with blocks of a simple bidiagonal structure. The diagonal blocks are

$$J^{(i,j)} = \begin{pmatrix} \lambda_i & 1 & & \\ & \ddots & \ddots & \\ & & \ddots & 1 \\ & & & \lambda_i \end{pmatrix}, \quad i = 1, \ldots, \ell, \; j = 1, \ldots, g_i, \tag{4.2}$$

where ℓ is the number of distinct eigenvalues λ_i of A, and g_i is the geometric multiplicity of λ_i, that is, the number of linearly independent corresponding eigenvectors. The matrices $J^{(i,j)}$ are called *Jordan blocks*, and have only two nonzero constant diagonals. The factorization $A = XJX^{-1}$ is nowadays called the *Jordan canonical form* (J.c.f.) of A.

Let $d_{i,j}$ be the order of $J^{(i,j)}$. Then, the characteristic (and irreducible) polynomial of $J^{(i,j)}$ is $(\lambda - \lambda_i)^{d_{i,j}}$. There may be several Jordan blocks concerning one eigenvalue, and the block diagonal matrix whose diagonal blocks are all Jordan blocks $J^{(i,j)}$, $j = 1, \ldots, g_i$, associated with the eigenvalue λ_i, is called a *Jordan box* and denoted as $B^{(i)}$ when the concerned Jordan blocks are numbered sequentially. The matrix $B^{(i)}$ is of order m_i (the algebraic multiplicity of λ_i) and contains g_i blocks. The Jordan canonical form is unique up to the order of the diagonal Jordan blocks. A matrix is said to be *defective* if the J.c.f. is not strictly diagonal, and *derogatory* if there is at least one eigenvalue associated with more than one diagonal block.

This decomposition is associated with the name of Camille Jordan and his *Traité des Substitutions et des Équations Algébriques* [1840] in 1870. It would be difficult for today's students to recognize what we now call the Jordan canonical form in what Jordan did. Jordan was concerned with a finite field and linear transformations with integer coefficients taken modulo p, p being a prime. Moreover, Jordan used an awful notation which is quite difficult to understand. Nevertheless, Jordan's aim was to obtain an equivalent form which was general and the simplest possible.

Jordan's work was extended later by people working on group theory at the University of Chicago, notably Leonard Eugene Dickson (1874-1954) [892, 893, 894]. Dickson extended Jordan's canonical form to an arbitrary field in 1902.

The problem of the transformation of couples of bilinear forms (matrix pencils in modern language) was first stated in 1866 in two memoirs by Elwin Bruno Christoffel (1829-1900) and Leopold Kronecker (1823-1891) in Crelle's journal.[34] Their goal was to create a theory of bilinear forms. At that time, Kronecker considered only distinct roots of the characteristic polynomial (that is, eigenvalues). A general solution was given by Karl Theodor Wilhelm Weierstrass (1815-1897) [3202] in 1868. Let $x = \{x_1, \dots, x_n\}$ and $y = \{y_1, \dots, y_n\}$ be two sets of variables and u and v two other sets obtained, respectively, from x and y by nonsingular linear transformations, two forms $pP + qQ$ of x and y and $pP' + qQ'$ coincide if and only if their elementary divisors are the same. The *elementary divisors* of a form (or matrix) A of rank r are obtained from the determinant of $\lambda I - A$, its nonzero minors $D_j(\lambda)$, $j = 1, \dots, r$, and the invariant polynomials

$$i_1(\lambda) = \frac{D_r(\lambda)}{D_{r-1}(\lambda)},\ i_2(\lambda) = \frac{D_{r-1}(\lambda)}{D_{r-2}(\lambda)},\ \dots,\ i_r(\lambda) = \frac{D_1(\lambda)}{D_0(\lambda)},\ D_0(\lambda) \equiv 1.$$

The elementary divisors are the irreducible factors of the invariant polynomials; see Chapter VI of the book [1282] by Felix Ruvimovich Gantmacher (1908-1964) in 1959. The case where the linear transformations can be singular was solved by Kronecker.

In December 1873, C. Jordan published a short note in the *Comptes Rendus de l'Académie des Sciences de Paris* [1841] to introduce the memoir he intended to publish. His goal was to apply his canonical form to three problems: 1) bring a bilinear form to a simple canonical form with orthogonal transformations, 2) bring a bilinear form to a simple canonical form by linear transformations, and 3) bring simultaneously two bilinear forms to a simple canonical form. He wrote that the first problem was new, the second one was solved by Kronecker, and the third one by Weierstrass. But, he added (our translation)

> *The solutions given by the eminent Berlin geometers are not complete because they had left aside some exceptional cases that are of interest. Their analyses are difficult to follow, particularly those of M. Weierstrass. The new methods that we propose are, on the contrary, extremely simple and with no exception.*

This started a feud about the priority for the results between Jordan and Kronecker. Kronecker replied to Jordan in January 1874 by reading a memoir at the Berlin Academy. He wrote about Jordan's note (our translation)

> *The solution of the first problem is not really new; The solution of the second one is not well done and that of the third problem is not sufficiently established. Let me add that this third problem contains the two first as particular cases and that its complete solution was given by M. Weierstrass in 1868 and can be obtained by my additions to his work. Therefore, if I am not mistaken, there are serious reasons for contesting M. Jordan's claims for the original invention of his results, as long as they are correct.*

[34] *Journal für die reine und angewandte Mathematik*, founded by August Leopold Crelle (1780-1855).

Then, there was an exchange of letters between Kronecker and Jordan during the winter of 1874. Kronecker reproached that Jordan published his note without having any previous discussions with him or Weierstrass. Jordan was upset that Kronecker expressed his negative comments publicly. There were communications to the Paris Academy by Kronecker and Jordan in March and April 1874. Kronecker criticized the use of the word "canonical" by Jordan, and they also disagree on what is a "general" and "simple" solution.

For a detailed description and study of this controversy as well as the history of the Jordan canonical form, see the thesis of Frédéric Brechenmacher [416] (in French) in 2005, and the papers [419, 420, 421] in 2007.

The controversy was somehow closed by Frobenius in his 1878 paper [1250], whose title can be translated as *On linear substitutions and bilinear forms*. He wrote

> *If we apply a substitution on only one sequence of variables of a bilinear form, we obtain new coefficients which define a transformed form as if the substitution was a transformation of the form itself.* [...] *These facts led me to consider the transformation of bilinear forms as a composition of linear substitutions.*

See also Frobenius' paper [1251] in 1879. In 1907, Bôcher presented the elementary divisors theory in matrix form in his book [348].

In 1881, Jordan admitted that Kronecker's criticisms were justified; see [416, p. 271].

Probably the first mathematician who published a canonical form involving eigenvalues of a matrix was Eduard Weyr (1852-1903), a Czech mathematician who received a doctorate from the University of Göttingen in 1873. In 1885, Weyr published a four-page note [3221] written in French in the *Comptes-Rendus de l'Académie des Sciences de Paris*. It was followed by papers written in Czech and German [3223, 3224] in 1889-1890.

The *Weyr canonical form* can be described as follows. Every square matrix of order n with p distinct eigenvalues $\lambda_1, \dots, \lambda_p$ is similar to a block diagonal matrix with p diagonal blocks $W(\lambda_1), \dots, W(\lambda_p)$, which are Weyr blocks. A Weyr block associated with an eigenvalue λ is a square block bidiagonal matrix whose diagonal blocks are diagonal, λI_{w_i} where I_{w_i} is the identity matrix of order w_i. The integers w_i, $i = 1, \dots, d$ are computed as follows. Let $r_k(\lambda) = \text{rank}(A - \lambda I)^k$, $r_0(\lambda) = n$. Then,

$$w_k(\lambda) = r_{k-1}(\lambda) - r_k(\lambda), \; k = 1, \dots, d,$$

where d is the index of λ. The sequence $w_1(\lambda), w_2(\lambda), \dots$ is non-increasing and $w(\lambda) = (w_1(\lambda), \dots, w_d(\lambda))$ is the *Weyr characteristic* of A, associated to the eigenvalue λ. Note that Weyr used the nullities of the powers to define the w_i's.

The blocks on the upper block diagonal of a Weyr block are identity matrices, eventually completed by rows of zeros at the bottom. The Jordan form can be obtained from the Weyr form by using permutations. For details on the Weyr canonical form, see the second edition of the book [1729], by Roger Alan Horn and Charles Royal Johnson, whose advisor was Olga Taussky (1906-1995); the paper [2750] by Helene Marian Shapiro, another student of Taussky; or the book by Kevin C. O'Meara, John Clark (1943-2017), and Charles (Chuck) Irvin Vinsonhaler (1942-2020) [2377].

The Weyr canonical form did not receive much attention until the end of the 20th century, although it was cited in papers by William Henry Metzler (1863-1943) [2208] in 1892, Kurt Hensel (1861-1941) [1648] in 1904, and Julius Wellstein (1888-1978) [3214] in 1930. Weyr was mentioned in the book on canonical matrices by Turnbull and Aitken in 1932, but just for the

Weyr characteristics and not for his canonical form. His name is also cited in the book [3199] by Joseph Henry Maclagan Wedderburn (1882-1948) in 1934, but only in the bibliography and not in the text, and in the book by MacDuffee [2106] in 1933.

We have seen that Weyr described a canonical form of a matrix that is equivalent to the Jordan canonical form, but when did the Jordan canonical form of a real or complex matrix as we know it today appear in print?

A matrix which is a Jordan block appeared on page 34 of a memoir by Autonne [110] in 1905. He also defined block matrices, and on page 40 he constructed a Jordan box, and the Jordan form on page 41. Autonne cited Weierstrass and Kronecker but not C. Jordan. He used the Jordan form in 1910 to study the matrices commuting with a given matrix [111]. The Jordan canonical form appeared in matrix form in Bôcher's book [348] in 1907 with its relationship with the elementary divisors. Bôcher did not refer to Jordan but, even though he had a French grandfather, he studied in Germany before returning to the USA. Transposed Jordan blocks also appeared in the paper [851] by Joseph Pierre Arthur de Séguier (1872-1940) in 1908.

The matrix form of the Jordan canonical form appeared in a paper [1598] by Herbert Edwin Hawkes (1872-1943) in 1910; however, see the remark by Turnbull and Aitken [3081, p. 81] about the correctness of Hawkes' proof. One can also see the paper [507] by Horace Thomas Burgess (1881-1939) in 1916.

Hence, the Jordan canonical form of a square matrix was well established in the beginning of the 20th century. It also appeared in books in the 1930s, for instance, Dickson [895] as well as Turnbull and Aitken [3081].

With the advent of digital computers, it was tempting to try to compute the Jordan canonical form of non-diagonalizable matrices, since it reveals the whole structure of the eigensystem. However, in their paper [1406] in 1976, Golub and Wilkinson wrote

> *The principal objective of the remainder of this paper is to show the basic limitations of the J.c.f. from the point of view of practical computation and, indeed, to cast doubt on the advisability of trying to determine it.*

Good reasons for not trying to compute the Jordan canonical form is that first, if the eigenvalues are not known exactly, but computed, it is difficult to decide what is the multiplicity of an eigenvalue, and second, that a small perturbation of the matrix may lead to a large change in the structure of the Jordan form.

Nevertheless, algorithms for computing the Weyr characteristic of a given eigenvalue λ were developed, even by Golub and Wilkinson. The goal was to determine the ranks and nullities of powers of $A - \lambda I$ without explicitly computing the powers [1406].

In 1966, Kublanovskaya used QR factorizations with column permutations in [1961] to compute the Weyr characteristic of the zero eigenvalue of a singular matrix. Kublanovskaya's method was revisited by Ruhe [2619] in 1970 to compute the eigenstructure of the matrix when the eigenvalues are approximated. Ruhe determined which of these approximations can correspond to a numerical multiple eigenvalue and worked with the restriction of the matrix to the subspace corresponding to these eigenvalue approximations. He tried to find a matrix with a true multiple eigenvalue (the mean of the group of eigenvalues) which was close to the original matrix, and determined its eigenvalues and the structure defined by its Weyr characteristics.

Also in 1970, James Martin Varah, who was a former student of Forsythe, considered computing approximate invariant subspaces for numerically defective matrices [3123].

These methods used unitary transformations but their drawback is the complexity, which, for one eigenvalue, is $\mathcal{O}(rn^3)$, where r is the sum of the integers in the Weyr characteristic of the eigenvalue. Note that Golub and Wilkinson used nonorthogonal bases in parts of their algorithm.

In 1980, Bo Kågström and Ruhe [1857] published an algorithm for the numerical computation of the Jordan normal form of a complex matrix.

The goal of the papers that followed on that topic was mainly to reduce the complexity; see the papers by Theo Beelen and Van Dooren [250] in 1990; Nicola Guglielmi, Michael Lockhart Overton, and G.W. Stewart [1476] in 2005, who proposed an algorithm of complexity $\mathcal{O}(n^3)$; and Nicola Mastronardi and Van Dooren [2166] in 2017. In this last paper they first reduced the matrix to Hessenberg form, and then obtained a block upper triangular form that reveals the dimensions of the null spaces of $(A - \lambda I)^i$ at a given eigenvalue via the sizes of the leading diagonal blocks.

We observe that the Jordan canonical form can be computed safely with symbolic mathematical systems when the entries of the matrix are integers or rational numbers. This is also the case when they are in some finite fields. But this is outside the scope of this book.

4.7 ▪ Frobenius canonical form

Let p be a monic polynomial of degree n:

$$p(\lambda) = \lambda^n + \alpha_{n-1}\lambda^{n-1} + \cdots + \alpha_1\lambda + \alpha_0.$$

Associated with p is the *companion matrix*

$$C = \begin{pmatrix} 0 & 0 & \dots & 0 & -\alpha_0 \\ 1 & 0 & \dots & 0 & -\alpha_1 \\ 0 & 1 & \ddots & \vdots & -\alpha_2 \\ \vdots & \ddots & \ddots & 0 & \vdots \\ 0 & \dots & 0 & 1 & -\alpha_{n-1} \end{pmatrix}.$$

The matrix C is square of order n and non-derogatory. The characteristic polynomial of C is equal to $p(\lambda)$ and the roots of p are the eigenvalues of C. Sometimes, the transpose or permutations of C are also called companion matrices.

Every matrix is similar to a block diagonal matrix whose diagonal blocks are companion matrices. This is known as the *Frobenius normal form* or the *rational normal form*. Note that this form is using the coefficients of polynomials whose roots are the eigenvalues and not directly the eigenvalues. If there is only one diagonal block, the matrix is necessarily non-derogatory. In this case, the minimal polynomial is equal to the characteristic polynomial of the matrix.

The Jordan canonical form can be obtained from the Frobenius normal form and vice versa.

The rational normal form was introduced by Frobenius in 1879, when he was a professor in Zürich, in a long paper [1251]. Frobenius did not use matrices at that time. However, already in [1250], he represented bilinear forms by capital letters (related to their coefficients) and manipulated them as if they were matrices, including for inverses. The determinant of a Frobenius companion matrix appears on page 206 of [1250].

Papers related to the Frobenius form were written by Georg Landsberg (1865-1912) [1988] in 1896 and by William Burnside (1852-1927) [512] in 1898, who established the canonical form in its generality.

The Frobenius normal form appeared in the book by Turnbull and Aitken [3081] under the name R.C.F. (for rational canonical form).

As for the Jordan canonical form, the Frobenius form can be computed with symbolic mathematical systems; see, for instance, [2914, 2915].

4.8 ▪ Schur factorization

Today we know that every square matrix A has a factorization $A = QRQ^*$ with R upper triangular, Q unitary. This is called the *Schur factorization* or Schur decomposition. It was proved in 1909 by Issai Schur (1875-1941) in a paper [2717] whose title can be translated as *About the characteristic roots of a linear substitution with an application to the theory of integral equations*.

In fact, in Section I of this paper, Schur proved that $A = QLQ^*$ with L lower triangular. Schur derived this factorization to prove some inequalities on the sum of the moduli of the eigenvalues of A. The factorization helps doing this since the eigenvalues are the diagonal entries of L or R. Schur's proof is similar to those that can be found in modern textbooks. He proved the result by induction, using the fact that any square matrix has at least one eigenvalue, as a consequence of the fundamental theorem of algebra. The inequality is used in other sections of his paper to prove results about integral equations.

Of the main matrix factorizations, the Schur factorization is the only one that was introduced and first proved using matrices.

Of course, a general real square matrix A could have complex eigenvalues. The Schur factorization was extended to cover this case to be able to only use real arithmetic. This was done in 1931 by Murnaghan, an Irish mathematician working in the USA at the Johns Hopkins University in Baltimore, and Wintner, born in Hungary; see [2307]. Their result is that A can be factored as $A = QTQ^T$, where Q is a real orthogonal matrix and T is real and quasi upper triangular (or block upper triangular) with diagonal blocks of order 1 or 2. The 2×2 diagonal blocks correspond to pairs of complex conjugate eigenvalues.

Note that a family of pairwise commuting matrices can be simultaneously triangularized.

The computation of the Schur factorization started in the 1950s. This was motivated by the fact that the eigenvalues of A appear on the diagonal of R.

In 1955, John Lester Greenstadt cited Schur's result and used Jacobi rotations to decrease the subdiagonal entries of the matrix to obtain the Schur form iteratively; see [1455]. He described experiments with small matrices on an *IBM 701* computer.

In 1958 [1367], Givens first reduced the matrix to upper Hessenberg form S (which he called a 1-subtriangular matrix) by right complex rotations. Then, for a given λ, the last column of $S - \lambda I$ is reduced to zero, except for the first component $g(\lambda)$. When, by some method, λ has been determined such that $g(\lambda) = 0$, and that the matrix is left multiplied by the conjugate transpose of the rotations, a matrix is obtained with the last column equal to zero except for the last component, which is λ and a principal submatrix of order $n - 1$, which is upper Hessenberg, that is reduced in the next step of the algorithm.

Also in 1958, Robert Lewis Causey (1927-2021), a former student of Forsythe in Stanford, discussed Greenstadt's method in [577] and showed examples for which the method did not converge. He proposed some modifications and reported numerical experiments on a *UNIVAC 1103A* computer. Greenstadt went back to the problem in 1962 by describing some numerical experiments [1456]. The rotations were chosen to "minimize" the sum of squares of the off-diagonal entries, but this was not working much better than the Jacobi iteration.

In 1985, G.W. Stewart described a method using rotations to compute the Schur form in [2875]. He also discussed its parallel implementation.

Patricia James Wells Eberlein (1923-1998) discussed the Schur decomposition of a matrix for parallel computation [1045] in 1987.

Nowadays, the standard way to compute the Schur factorization is to do a preliminary reduction to Hessenberg form by Householder reflections, and then to use the QR algorithm.

4.9 ▪ Spectral factorization

If in the Jordan canonical form, $A = XJX^{-1}$, all the diagonal blocks of J are of dimension 1, the matrix J is diagonal, the matrix A is diagonalizable, and we have $A = XDX^{-1}$ with D diagonal. This is called the *spectral factorization* of A. Complex normal matrices (that is, such that $A^*A = AA^*$) are unitarily diagonalizable, $A = XDX^*$.

We will discuss the spectral factorization in Chapter 6, devoted to eigenvalue and eigenvector computations.

4.10 ▪ WZ factorization

Several non-standard factorizations of banded matrices were introduced in the 1970s by David John Evans (1928-2005) and Michael Hatzopoulos [1115, 1116] to obtain more parallelism than in the LU factorization. In 1995, with Plamen Y. Yalamov, Evans considered what he called the WZ factorization $A = WZ$ [3287, 3288]. For $n = 5$, the matrices W and Z have the following pattern

$$W = \begin{pmatrix} 1 & 0 & 0 & 0 & 0 \\ w_{2,1} & 1 & 0 & 0 & w_{2,5} \\ w_{3,1} & w_{3,2} & 1 & w_{3,4} & w_{3,5} \\ w_{4,1} & 0 & 0 & 1 & w_{4,5} \\ 0 & 0 & 0 & 0 & 1 \end{pmatrix}, \quad Z = \begin{pmatrix} z_{1,1} & z_{1,2} & z_{1,3} & z_{1,4} & z_{1,5} \\ 0 & z_{2,2} & z_{2,3} & z_{2,4} & 0 \\ 0 & 0 & z_{3,3} & 0 & 0 \\ 0 & z_{4,2} & z_{4,3} & z_{4,4} & 0 \\ z_{5,1} & z_{5,2} & z_{5,3} & z_{5,4} & z_{5,5} \end{pmatrix}.$$

Obvious modifications of the nonzero structure have to be done when n is even.

The existence and uniqueness of this factorization was studied by S. Chandra Sekhara Rao [2535] in 1997. The WZ factorization did not draw too much attention but it was revived in the 2000s by Beata and Jarosław Bylina, working at the Maria Curie-Skłodowska University in Lublin, Poland [520].

4.11 ▪ Lifetimes

In this section, we show the lifetimes of the main deceased contributors to factorizations of matrices, starting in 1700. The language is given by the color of the bars and by letters: E (red) for English, G (black) for German, I (green) for Italian, F (blue) for French, and O (magenta) for the others. The contributors are ordered by date of birth.

Factorizations of matrices (a)

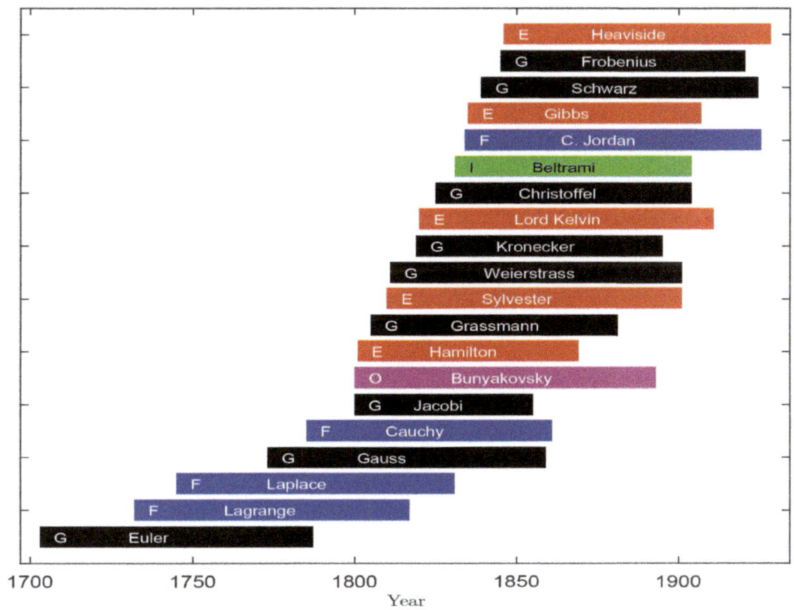

Factorizations of matrices (b)

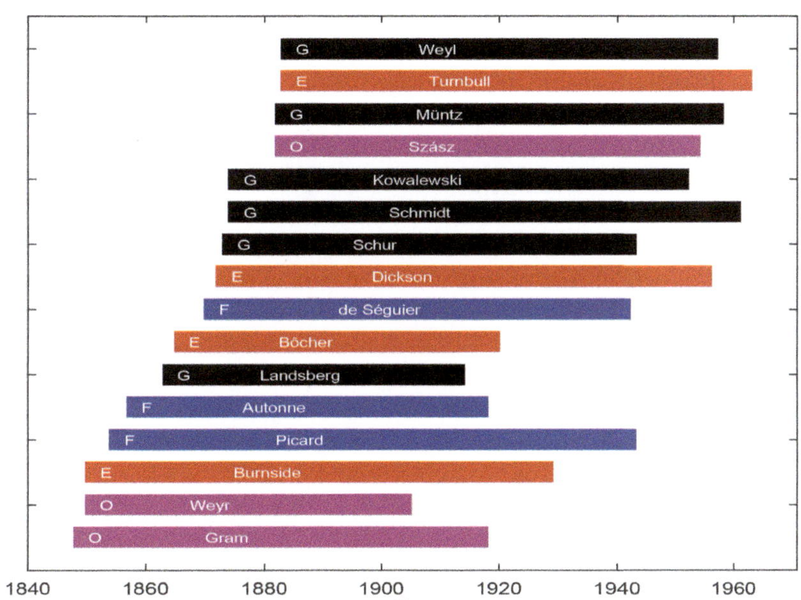

Factorizations of matrices (c)

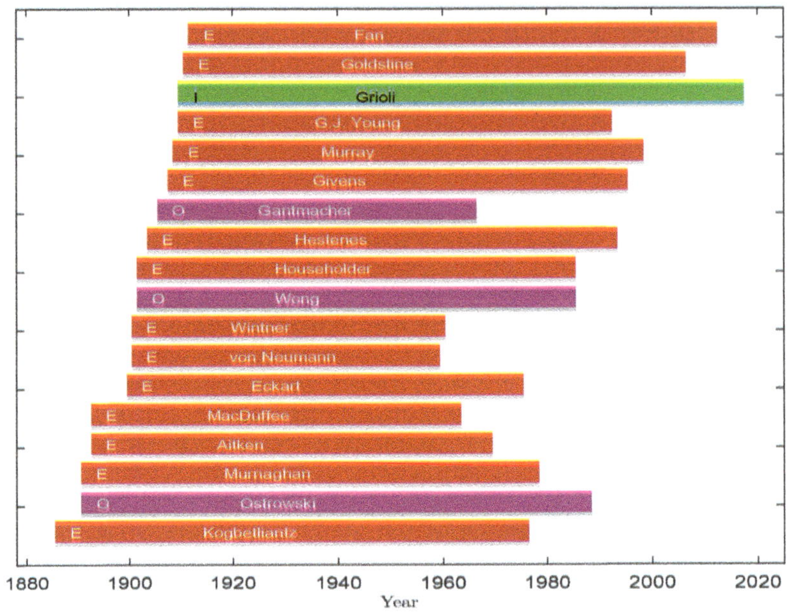

Factorizations of matrices (d)

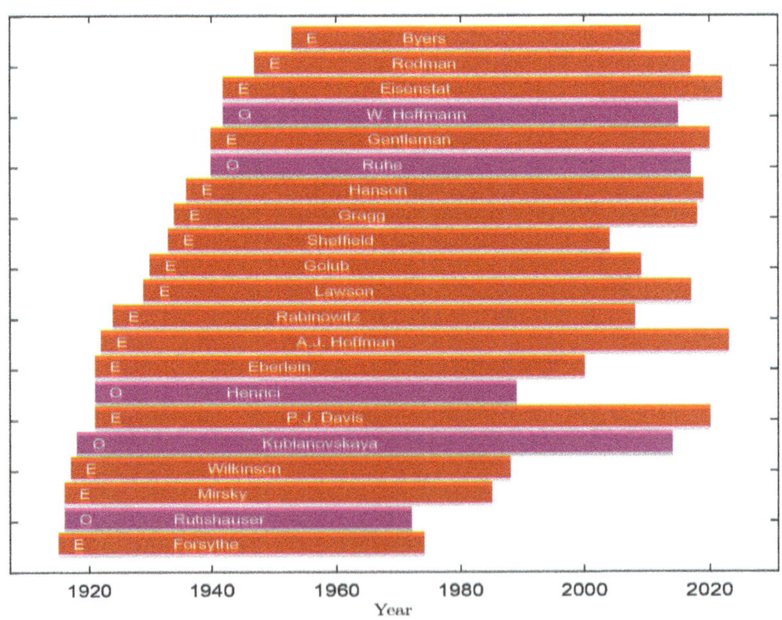

5 Iterative methods for linear systems

> *Few books on iterative methods have appeared since the excellent ones by Varga and later Young. Since then, researchers and practitioners have achieved remarkable progress in the development and use of effective iterative methods. Unfortunately, fewer elegant results have been discovered since the 1950s and 1960s.*
>
> – Yousef Saad, *Iterative Methods for Sparse Linear Systems*

Iterative methods were used a long, long time before people tried to use them for solving linear systems. Remember, for instance, Hero (Heron) of Alexandria (c. 10 AD-c. 70 AD), a Greek mathematician, physicist, and engineer who gave iterative methods for calculating square and cube roots of a number in his *Metrica* (which was lost until 1896), describing how to calculate surfaces and volumes of diverse objects.

Perhaps the most well-known iterative method of all times that is used for solving nonlinear equations is the Newton-Raphson method. The method was described by Joseph Raphson (c. 1668-c. 1712) in his book *Analysis Aequationum Universalis* [2537] published in 1690. It was known to Isaac Newton (1643-1727) in the 1670s since it appeared in *Method of Fluxions*. Although it was written in 1671, it was not published until 1736.

5.1 ▪ Classical methods

Even though, as we have just seen, iterative methods had been used for quite a long time to solve nonlinear problems, it seems they were not used for solving linear systems before Carl Friedrich Gauss (1777-1855), at the beginning of the 1820s. This is understandable since the linear systems that had to be solved before the 19th century were of a small order.

Gauss explained his iterative method in an addition to a long letter he wrote to Gerling, a former student, on December 26, 1823 [1302, letter 163, p. 294]). A partial translation to English was published by George Elmer Forsythe (1917-1972) [1185] in 1951.

Christian Ludwig Gerling (1788-1864) obtained his thesis in 1812 and was professor of mathematics, physics, and astronomy at the University of Marburg from 1817 to 1864. In 1927, Clemens Schaefer (1878-1968) published 388 letters dating from 1810 to 1854, of which 163 are those written by Gauss and 225 by Gerling [1302].

In letter 163, Gauss wrote (our translation)

> *My letter arrived too late at the post office and was returned to me. I am therefore opening it up again in order to enclose practical instructions for elimination. Of course, there are many small local advantages that can only be learned by use.*

As an example, he used a problem from geodesy given by Gerling with six angle measurements that, after some manipulations, he reduced to four. He looked for increments satisfying the following normal equations (obtained by shifting; see [1197])

$$
\begin{aligned}
0 &= +6 + 67a - 13b - 28c - 26d,\\
0 &= -7558 - 13a + 69b - 50c - 6d,\\
0 &= -14604 - 28a - 50b + 156c - 78d,\\
0 &= +22156 - 26a - 6b - 73c + 110d.
\end{aligned}
$$

He noted that the sum of the coefficients in a column or a row is equal to zero. The matrix is singular but positive semi-definite. In fact, Gauss was not looking for a very precise solution. He called his method *indirect elimination*. He wrote

> *Now to eliminate indirectly, I notice that if three of the quantities a, b, c, d are set equal to 0, the fourth one gets the largest value if d is chosen. Of course, each quantity must be determined from its own equation, i.e. d from the fourth. So I set $d = -201$ and substitute this value. The absolute terms then become $+5232$, -6352, $+1074$, $+46$, the other terms remain the same. Now I let b take its turn, find $b = +92$, substitute and find the absolute terms $+4036$, -4, -3526, -506. I continue in this way until there is nothing left to correct.*

Finally, when he stopped, he found the approximation $a = -56$, $b = +90$, $c = +12$, $d = -201$. The residual vector is $(\,26 \quad 14 \quad -14 \quad -26\,)$. He concluded this part of the letter with the following (which has been often quoted):

> *Almost every evening I make a new edition of the tableau, which is always easy to improve. With the monotony of the measuring business, this always makes a pleasant distraction; you also always immediately see if something doubtful has crept in, what still remains to be done, etc. I recommend this method for imitation. You will hardly ever again eliminate directly, at least not if you have more than two unknowns. The indirect method can be performed half asleep or you can think about other things when you use it. [...] With hearty congratulations on the new year. Yours truly.*

Note that a particular solution of the system is $a = 43.96$, $b = 189.84$, $c = 111.58$, $d = -101.55$. If we add a multiple of the null vector to obtain $a = -56$ like Gauss, we find the solution $a = -56$, $b = 89.876$, $c = 11.611$, $d = -201.52$, which is close to what Gauss found.

Gerling published the method in his book [1340] in 1843, whose title can be translated as *The compensation calculations of practical geometry or the method of least squares with its applications to geodetic tasks*. The method was described in words at the end of the book (pages 386-392). Gerling wrote (our translation)

> *Indirect elimination, like all indirect methods, which we use in practice even for ordinary trigonometric tasks, in general assumes that the quantities to be found are very small or almost zero. This prerequisite makes it possible for the elimination work to find an approximate value for one of the quantities, by setting all the others to zero. If one now sets this so found value in all given equations, thus changing the absolute members accordingly, then one can, with all the corrections, which this first quantity still requires, look for an approximate value of a second sought quantity and so on. This procedure brings thus gradually all searched quantities to the solution. We continue until all absolute members have disappeared, or rather until they have become so small that no correction can be found, which exceeds a given limit; for an absolute zero is impossible to be reached because of the incompleteness of the decimal fractions.*

In fact, he explained the method with the help of an example with integer coefficients. Like for Gauss' example, the matrix is singular, the sum of the coefficients by columns being zero,

$$\begin{aligned}0 &= +2.8 + 76x - 30a - 20b - 26c,\\0 &= -4.1 - 30x + 83a - 25b - 28c,\\0 &= -1.9 - 20x - 25a + 89b - 44c,\\0 &= +3.2 - 26x - 28a - 44b + 98c.\end{aligned}$$

Starting from a zero vector, Gerling computed the right-hand sides divided by the diagonal coefficients,

$$x = -\frac{2.8}{76} = -0.03\ldots, \quad a = +\frac{4.1}{83} = +0.04\ldots, \quad b = +\frac{1.9}{89} = +0.02\ldots, \quad c = -\frac{3.2}{98} = -0.03\ldots$$

From this, he chose $a = +0.04$ as the first component to be computed because it had the largest absolute value. He recorded his computation in two columns

The products	The right-hand sides
−1.20	+1.60
+3.32	−0.78
−1.00	−2.90
−1.12	+2.08
−−	−−
0	0

Note that the second right-hand side is not exactly zero because Gerling rounded the value of a to two fractional figures. Then, he proceeded in the same way, obtaining Table 5.1.

Table 5.1. *Iterations*

$a = +0.04$		$b = +0.03$		$a = +0.01$		$x = -0.01$		$c = -0.01$	
−1.20	+1.60	−0.60	+1.00	−0.30	+0.70	−0.76	+0.06	+0.26	+0.20
+3.32	−0.78	−0.75	−1.53	+0.83	−0.70	+0.30	−0.40	+0.28	−0.12
−1.00	−2.90	+2.67	−0.23	−0.25	−0.48	+0.20	−0.28	+0.44	+0.16
−1.12	+2.08	−1.32	+0.76	−0.28	+0.48	+0.26	+0.74	−0.98	−0.24
0	0	0	0	0	0	0	0	0	0

The last row shows the sum of the columns and was used as a check of the computation. He stopped after four corrections because the next one would be $x = -0.002$, which was smaller than the limit of 0.01 he had chosen. Gerling's solution $(\,-0.01 \quad 0.05 \quad 0.03 \quad -0.01\,)$ has a large residual with a norm of 0.37.

In his book, Gerling raised some concerns about the method when the systems did not have integer coefficients and when the sums of the columns are not equal to zero. He also noted that the convergence could sometimes be very slow. All these remarks show that he was not completely mastering the method that, like Gauss, he called *indirect elimination*.

In 1845, Carl Gustav Jacob Jacobi (1804-1851) also used an iterative method [1795] in the solution of the normal equations for least squares problems. He considered diagonally dominant symmetric systems, where the off-diagonal entries are small compared to the diagonal ones. He divided by the diagonal entries to obtain an approximate solution and he iterated the process.

Jacobi did not analyze the convergence of the method. However, in this paper, he also proposed to transform the system before iterating to make it more diagonally dominant. He did so by applying the so-called Jacobi rotations to zero some entries. Doing this symmetrically reintroduced nonzero entries, but by iterating the process, the magnitudes of the off-diagonal entries decrease. See also [1797] in 1846.

Philipp Ludwig Seidel (1821-1896) published his iterative method in 1874 [2740] (see also [2739]), but he used it before in 1862, for a paper [2738] published in 1863.

Seidel started by recalling what is the method of least squares that he already studied in [2735]. He wrote the normal equations using Gauss' bracket notation. He also recalled Gauss' elimination method and Jacobi's method of 1846 for reducing the non-diagonal entries [1797]. Seidel did not use matrices and did not give many details about the derivation of his method. However, it was explained in modern matrix terms by Ewald Konrad Bodewig (1901-?) in his 1959 book [353, pp. 146-147].

Like Bodewig, let us explain what Seidel was doing in matrix terms. Let A be the $m \times n$ matrix of the observations $Ax = b$, which yields the normal equations $A^TAx = A^Tb$ of order n. Given an approximate solution x, Seidel wanted to decrease the square of the norm of the residual $Q = (b - Ax)^T(b - Ax)$. Let us assume that we would like to modify the first component x_1 by changing x to $x + \delta e_1$, where e_1 is the first column of the identity matrix. We plug this in Q and we obtain

$$\begin{aligned}\widetilde{Q} &= b^Tb - 2x^TA^Tb + x^TA^TAx - 2\delta e_1^TA^Tb + 2\delta e_1^TA^TAx + \delta^2 e_1^TAe_1,\\ &= Q - 2\delta e_1^TA^Tb + 2\delta e_1^TA^TAx + \delta^2 e_1^TAe_1.\end{aligned}$$

We choose δ to minimize $\widetilde{Q}$ as

$$\delta = \frac{e_1^TA^T(b - Ax)}{e_1^TA^TAe_1}.$$

Then,

$$\widetilde{Q} = Q - \delta e_1^TA^T(b - Ax) = Q - \frac{[e_1^TA^T(b - Ax)]^2}{e_1^TA^TAe_1}.$$

This decreases the value of Q unless $Ax = b$. The numerator of δ is the first component of the residual of the normal equations, $A^Tb - A^TAx$ and the denominator is the $(1, 1)$ entry of A^TA which is positive. The value of δ means that the new value $x_1 + \delta$ is simply obtained by computing the value of the first component, given the values of the other ones. The new value of x_1 modifies the residuals of the other equations.

The first method proposed by Seidel was to change the components $x_1, x_2, x_3, \ldots$ of the approximation cyclically, but he remarked that any ordering of the components can be used. He suggested to choose at each step the component which reduces Q the most, writing (our translation)

> *Strictly speaking, at every stage of the calculation it would be the most rational to first improve the variable whose correction reduces the sum of the squares of the errors the most,* [...] *It has already been emphasized that it is by no means necessary to correct all the unknowns one after the other, but that one can very well come back to one variable before all the others have been corrected the same number of times. But that one cannot generally arrive at the definitive set of values without having improved all the unknowns, is evident in itself,* [...]

He also remarked that the behavior of the method depended on the properties of the linear system:

> *The advantage of certain convergence, which the normal form (B) of the equations offers for our kind of solution over any other forms, is naturally founded in the distinctive properties of its coefficient system, which are by no means exhausted in the symmetry around the diagonal, and whose most important consists in the fact that every sub-determinant extracted from the system is positive, if it has, as diagonal, a piece of the diagonal of the whole system.*

He noticed that any linear system $Ax = b$ can be modified to the system $A^TAx = A^Tb$, to which his results can be applied. We observe that Seidel did not have the same strategy as Gauss and Gerling for the choice of the component to correct at each step.

In summary, Seidel proved that his method is converging for the normal equations arising from the least squares method, but he did not analyze the rate of convergence.

Gauss and Seidel methods were used by geodesists and astronomers, mainly within the least squares method. An analysis of the convergence was done in 1885 [2329] by the Russian mathematician and Rector of the Imperial University of Moscow Pavel Alekseevich Nekrasov (1853-1924), who was working mainly on probability. The title of his paper, written in Russian, can be translated as *Determination of the unknowns by the method of least squares with a very large number of unknowns*.

Nekrasov wrote

> *Reviewing Seidel's method at the suggestion of the respected astronomer V.K. Cerasky,*[35] *who has to deal with these difficulties, I noticed that in his memoir Seidel does not touch a very important point in practical respect, that is, the question about the speed with which it is possible to approach the searched solutions. To remedy to this shortcoming, I will show that under the right circumstances Seidel's method quite quickly approach the solutions we are looking for, but very often there can be cases when this convergence will be slow, even infinitely slow.*

Nekrasov used the cyclic ordering for the components to be corrected. Moreover, even though he started from least squares problems, later on he did not assume symmetry of the linear system. He wrote the equations

$$\begin{aligned}
(aa)x_{n+1} + (ab)y_n + (ac)z_n + \cdots + (ap)t_n - (aq) &= 0,\\
(ba)x_{n+1} + (bb)y_{n+1} + (bc)z_n + \cdots + (bp)t_n - (bq) &= 0,\\
(ca)x_{n+1} + (cb)y_{n+1} + (cc)z_{n+1} + \cdots + (cp)t_n - (cq) &= 0,\\
\vdots &= \vdots\\
(pa)x_{n+1} + (pb)y_{n+1} + (pc)z_{n+1} + \cdots + (pp)t_{n+1} - (pq) &= 0
\end{aligned}$$

for the components of the approximations to those $x, y, z, \ldots, t$ of the solution of the normal equations. Therefore, the method he used is what we now call (improperly) the Gauss-Seidel method. Then, he wrote equations for the errors

$$\xi_n = x_n - x,\ \eta_n = y_n - y,\ \zeta_n = z_n - z, \ldots,\ \tau_n = t_n - t.$$

[35]Vitold Karlovich Cerasky or Tserasky (1849-1925) was a Russian astronomer who discovered many stars and studied meteors.

These quantities satisfy the equations

$$\begin{aligned}
(aa)\xi_{n+1} + (ab)\eta_n + (ac)\zeta_n + \cdots + (ap)\tau_n &= 0, \\
(ba)\xi_{n+1} + (bb)\eta_{n+1} + (bc)\zeta_n + \cdots + (bp)\tau_n &= 0, \\
(ca)\xi_{n+1} + (cb)\eta_{n+1} + (cc)\zeta_{n+1} + \cdots + (cp)\tau_n &= 0, \\
\vdots &= \vdots \\
(pa)\xi_{n+1} + (pb)\eta_{n+1} + (pc)\zeta_{n+1} + \cdots + (pp)\tau_{n+1} &= 0.
\end{aligned}$$

He looked for solutions of this system,

$$\xi_n = A\alpha^n,\ \eta_n = AB\alpha^n,\ \zeta_n = AC\alpha^n,\ \ldots, \tau_n = AP\alpha^n.$$

The parameter α must satisfy the equation [numbered (7)]

$$\begin{vmatrix}
(aa)\alpha & (ab) & (ac) & \cdots & (ap) \\
(ba)\alpha & (bb)\alpha & (bc) & \cdots & (bp) \\
(ca)\alpha & (cb)\alpha & (cc)\alpha & \cdots & (cp) \\
\vdots & \vdots & \vdots & & \vdots \\
(pa) & (pb) & (pc) & \cdots & (pp)
\end{vmatrix} = 0,$$

and Nekrasov wrote the solution as functions of the nth powers of the roots $\alpha_1, \alpha_2, \ldots$ of this polynomial in α. He concluded

> *Since in the limit at $n \to \infty$ the errors of the quantities $x_n, y_n, z_n, \ldots$ should tend to zero, all roots $\alpha_1, \alpha_2, \ldots$ of equation (7) must have a modulus less than unity.*

We note that if, in modern notation, we write the matrix of the normal equations as $A = D+L+U$ with D diagonal and L (resp., U) strictly lower (resp., upper) triangular, the determinant of equation (7) corresponds to the matrix $\alpha(D+L)+U$ and α is an eigenvalue of $-(D+L)^{-1}U$.

Then, Nekrasov observed that if the roots all have a modulus less than 1, but the root of largest modulus is close to 1, the convergence will be slow. He illustrated his findings with numerical examples. He even exhibited an example for which, if Seidel's strategy for the choice of the order of the components to be corrected is used, the convergence is slow. The conclusions of Nekrasov are correct but his analysis is not very rigorous. Nevertheless, he was the first to relate the convergence to the eigenvalues of the iteration matrix and to analyze the modern Gauss-Seidel method.

Nekrasov published another paper on successive approximations in 1892 [2330]. This publication is not easily accessible. But we have letters sent to and received from Rudolf Mehmke (1857-1944), a German mathematician, who was also looking for sufficient convergence conditions for Seidel's method. Mehmke published in 1892 a paper [2187], written in German, in the Russian journal *Mat. Sbornik* with the title *On Seidel's method to solve linear equations with a very large number of unknowns by successive approximations*. This paper is, in fact, a letter sent to Nekrasov. The letters were published also in 1892 in *Mat. Sbornik* [2189]. In these letters, they compared their "rules," that is, sufficient conditions for convergence. Nekrasov described five rules and Mehmke generalized some of them.

In a letter from the end of March 1892, Nekrasov described three rules. For rule II, he wrote,

> *Rule II. If for $i = 1, 2, \ldots, n$ the value of $|a_{i,i}|$ is greater than the sum of the magnitudes $|a_{i,1}|$, $|a_{i,2}|, \ldots,$ $|a_{i,i-1}|$, $|a_{i,i+1}|, \ldots,$ $|a_{i,n}|$, then Seidel's method should be convergent.*

The $a_{i,j}$'s are the entries of the matrix of the linear system. This condition was also discussed by Mehmke in [2187]. This says that Seidel's method is convergent for a strictly diagonally dominant matrix, a result that was anticipated a bit by Seidel and that was rediscovered many times later on.

In some of the rules, Nekrasov and Mehmke introduced parameters. For instance, Nekrasov's rule III is

> *Rule III. Let for* $i = 1, 2, \ldots, n$,
>
> $$p_i = \left|\frac{a_{i,1}}{a_{i,i}}\right| p_1 + \left|\frac{a_{i,2}}{a_{i,i}}\right| p_2 + \cdots + \left|\frac{a_{i,i-1}}{a_{i,i}}\right| p_{i-1} + \\ + \left|\frac{a_{i,i+1}}{a_{i,i}}\right| + \left|\frac{a_{i,i+2}}{a_{i,i}}\right| + \cdots + \left|\frac{a_{i,n}}{a_{i,i}}\right|.$$
>
> *If the largest of the values* $p_2, p_3, \ldots, p_n$ *is less than* 1, *then Seidel's method should be convergent.*

One of Mehmke's generalizations was

> *Let for* $i = 1, 2, 3, \ldots, n-1$,
>
> $$r_{1,i} = -\frac{a_{i,n}}{a_{n,n}},$$
>
> $$r_{2,i} = -r_{1,i}\frac{a_{n,n-1}}{a_{n-1,n-1}} - \varepsilon_{i,n-1}\frac{a_{i,n-1}}{a_{n-1,n-1}},$$
>
> $$r_{3,i} = -r_{1,i}\frac{a_{n,n-2}}{a_{n-2,n-2}} - r_{2,i}\frac{a_{n-1,n-2}}{a_{n-2,n-2}} - \varepsilon_{i,n-2}\frac{a_{i,n-2}}{a_{n-2,n-2}},$$
>
> $$r_{4,i} = -r_{1,i}\frac{a_{n,n-3}}{a_{n-3,n-3}} - r_{2,i}\frac{a_{n-1,n-3}}{a_{n-3,n-3}} - r_{3,i}\frac{a_{n-2,n-3}}{a_{n-3,n-3}} - \varepsilon_{i,n-3}\frac{a_{i,n-3}}{a_{n-3,n-3}},$$
>
> *and so on, where* $\varepsilon_{i,k} = 1$ *if* $i < k$, *and* $\varepsilon_{i,k} = 0$ *if* $i \geq k$ *further let*
>
> $$q_h = \sum_{k=1}^{n-1} |r_{h,k}|, \quad h = 1, 2, 3, \ldots, n.$$
>
> *If each of the quantities* $q_1, q_3, \ldots, q_n$ *is less than* 1, *then Seidel's method is convergent.*

Note that Nekrasov and Mehmke were not using matrices, but only the linear equations, and it was not easy for them to make their statements in a simple way.

In 1892 [2187], Mehmke described what can be considered a block Seidel's method with 2×2 or 3×3 blocks.

A block Gauss-Seidel-like algorithm was described in 1887 by Paolo Pizzetti (1860-1918), an Italian geodesist, astronomer, geophysicist, and mathematician, in the context of solving least squares problems [2499]. When adding new observations to a previous system, Pizzetti wanted to be able to reuse the work done before. He had several sets of equations and iterated between them until convergence. Pizzetti, on page 288, wrote (our translation)

> *Compensate the observations by taking into account only the first system* (α) *of conditions, consider the compensated values as given directly by the observations, and from the observations, and over them operate a new compensation according*

to (β). Then by means of the system conditions (γ) operate a third compensation on the values already corrected by the previous two operations. And so proceed in such a way as to employ, one after the other, all the partial systems of conditions. After this we will say that we have completed a complete round of compensations. When the first round is completed, the values obtained will not generally satisfy the conditions of the system (α). In this case we will repeat the next compensation by means of the various systems of conditions by performing a second run, and so on until a compensated value system has been obtained, which satisfy all the conditions proposed.

In the 2nd, 3rd, 4th, etc. rounds of compensation, the systems of normal equations to be solved do not differ from the corresponding systems of the first round except for the known terms, so that, for a practical calculator, the calculation of the subsequent rounds, after the first, is very simple and fast.

Pizzetti gave a proof of convergence for two sets of equations that is far from clear. Moreover, because of clumsy notation, this paper is very difficult to read. Pizzetti's paper was cited by Forsythe [1188] in 1953.

In 1910, Lewis Fry Richardson (1881-1953), an English mathematician, physicist, and meteorologist, published a 53-page paper [2566] in which he considered approximate solutions of partial differential equations using finite difference schemes, and an iterative method that will be known later as (first-order) Richardson's method. For finite differences, Richardson referred to George Boole (1815-1864) and John Fletcher Moulton (1844-1921) [374] in 1872, and William Fleetwood Sheppard (1863-1936) [2753] in 1899, even though finite differences had been used also in those times by Carl Runge (1856-1927), Karl Heun (1859-1929), and Martin Wilhelm Kutta (1867-1944).

Richardson did not precisely identify the discrete problem with a matrix. He spoke of "functions" and "tables of numbers." To solve $\mathcal{D}'\phi = 0$, where $\mathcal{D}'$ is a discrete operator, he wrote his iterative method (defined in Section 3.2.1 [2566, p. 319]) as

$$\phi_{m+1} = \phi_m - \alpha_m^{-1}\mathcal{D}'\phi_m,$$

and he wrote

It will be shown that by the judicious choice of $\alpha_1, \alpha_2, \dots, \alpha_t$ it is possible to make ϕ_{t+1} nearer to ϕ_u than ϕ_t was.

Note that for a general problem, $\mathcal{D}'\phi_m$ must be replaced by $\mathcal{D}'\phi_m - f$, where f is the right-hand side. ϕ_u is the exact solution of the discrete problem. To "prove" this, he considered in a loose way the eigenvectors P_k and eigenvalues λ_k^2 of $\mathcal{D}'$ to expand the error "function." He assumed that the eigenvalues are real and all of the same sign. He showed that the choice of the α_i's must be related to the eigenvalues of $\mathcal{D}'$. In fact, he understood that the parameters must be chosen as to make a polynomial, whose roots are related to the eigenvalues, "small." He did not solve this problem but he realized that to find "good" parameters, it is enough to know the interval in which the eigenvalues lie. So, he was probably making a very empirical choice of the α_i's, but he did not fully explain that. He also noticed that the order in which the parameters are used is important.

We observe that, on page 321, he gave a sketch of the power method to compute approximations of the largest eigenvalue. On page 325, there is something amusing. Apparently, he paid human computers to do the calculations, and he wrote

So far I have paid piece rates for the operation [...] of about $n/18$ pence par coordinate point, n being the number of digits. The chief trouble to the computers has

> *been the intermixture of plus and minus signs. As to the rate of working, one of the quickest boys averaged* $2,000$ *operations* [...] *per week, for numbers of three digits, those done wrong being discounted.*

The last and largest part of the paper is devoted to the modelization and the detailed study of a masonry dam, and to solving a biharmonic problem with finite differences. Let us also mention that, in that paper, Richardson introduced the first step of his polynomial extrapolation method, named h^2*-extrapolation* or *deferred approach to the limit.*

The problem of convergence and the choice of the parameters in Richardson's method were studied in the second half of the 20th century when digital computers became available. As we will see below, good parameters were found, but it was also shown that the results were very sensitive to the order in which the parameters are used, the method being somehow unstable. But, finally good orderings were found; for these issues, see Mark Konstantinovich Gavurin (1911-1992) [1309] in 1950, Mikhail Shlemovich Birman (1928-2009) [329] in 1952, David Monaghan Young (1923-2008) [3305] in 1953, Vyacheslav Ivanovich Lebedev (1930-2010) [2017] in 1969, Lebedev and Sergey A. Finogenov [2018] in 1971, Robert Scott Anderssen and Gene Howard Golub (1932-2007) [57] in 1972, Evgenii S. Nikolaev and Alexander Andreevich Samarskii (1919-2008) [2350] in 1972, and Lothar Reichel [2548] in 1991.

In 1918, Karl Otto Heinrich Liebmann (1874-1939), a German mathematician who was then professor in the Technischen Hochschule München, rector of the University of Heidelberg, and dean of the Faculty of mathematics and natural science, published a paper [2063] whose title can be translated as *The approximate determination of harmonic functions and conformal mappings.* His goal was to show practical ways to compute an approximation of the solution of $\Delta u = 0$, with given boundary values, in a square or in more general domains through conformal mapping. He was following some ideas that he attributed to Ludwig Boltzmann (1844-1906), the famous Austrian physicist. Liebmann wrote (our translation)

> *Ludwig Boltzmann, working in Munich in the winter semester 1892/1893, gave a lecture on the mechanical potential, which has been published after the tragic end of the researcher, by his student, the astronomer Hugo Buchholz.*[36] *The editor sees a highlight in the treatment that Boltzmann gave to Dirichlet's principle;* [...] *Boltzmann replaces the differential equation*
>
> $$\Delta u \equiv \frac{\partial^2 u}{\partial x^2} + \frac{\partial^2 u}{\partial x^2} = 0$$
>
> *by the difference equation*
>
> $$D_2 u \equiv u(x-\varepsilon, y) + u(x+\varepsilon, y) + u(x, y-\varepsilon) + u(x, y+\varepsilon) - 4u(x, y) = 0$$

Liebmann used a regular Cartesian mesh for the square. He was interested in the convergence of the approximation but he did not formulate that very clearly, and he did not prove it. He did not use matrices to write the discrete problem. Curiously enough, he started by computing some entries of the inverse matrix of the discrete Laplacian for very small examples. He presented the entries as weights that have to be applied to the right-hand side (arising from the boundary

[36]Hugo Buchholz (1866-1921) received his doctorate in Munich in 1895 and was later Professor of mathematics and Director of the Halle Observatory in Germany; the book is Ludwig Boltzmann, *Vorlesungen über die Principe der Mechanik*, Vol. III, based on lecture notes taken by H. Buchholz, J.A. Barth, Leipzig, 1920. Vol. I and II appeared in 1897 and 1904.

conditions). For example, with nine interior points, some of his results were

$$
\begin{array}{ccccc}
 & 1 & 2 & 1 & \\
1 & . & . & . & 1 \\
2 & . & u_{2,2} & . & 2 \\
1 & . & . & . & 1 \\
 & 1 & 2 & 1 & \\
 & [p] & = & 16 &
\end{array}
\qquad
\begin{array}{ccccc}
 & 11 & 37 & 11 & \\
11 & . & u_{1,2} & . & 11 \\
7 & . & . & . & 7 \\
3 & . & . & . & 3 \\
 & 3 & 5 & 3 & \\
 & [p] & = & 112 &
\end{array}
$$

The locations around the mesh are where the boundary values are to be given. The left diagram corresponds to the fifth row of the inverse (from a numbering from left to right and bottom to top). The coefficients have to be divided by p. In fact, the entries of this row are

$$0.0625 \quad 0.125 \quad 0.0625 \quad 0.125 \quad 0.375 \quad 0.125 \quad 0.0625 \quad 0.125 \quad 0.0625$$

which, multiplied by 16, yield Liebmann's coefficients, except for 0.375, but this coefficient will be multiplied by zero. The right diagram corresponds to the eighth row, whose entries are

$$0.0267857 \; 0.0446429 \; 0.0267857 \; 0.0625 \; 0.125 \; 0.0625 \; 0.0982143 \; 0.330357 \; 0.0982143$$

which, multiplied by 112, yield Liebmann's coefficients. He did not tell how he computed the coefficients, but of course, such results are not too interesting since the coefficients change when the number of points increases. Nevertheless, we observe that for large meshes, Liebmann described a method in which he can use his "inverses" for small matrices, that can be seen as a precursor of domain decomposition techniques. In Section 3 of [2063], Liebmann described an iterative method to compute an approximate solution of the discrete problem that he saw as a modification of Jacobi's method. But in fact, this is not Jacobi's method and not Seidel's method either. He wrote (the mesh is $(n-1) \times (n-1)$)

> *If (for a given n) you first assign some values $u^0_{i,k}$ to the inner mesh points and then the values on the inner lines closest to the boundary with the help of the difference equation itself, i.e. the values $u^0_{i,k}$ for*
>
> $$(i = 1, k = 1, \dots, n-1), \; (i = n-1, k = 1, \dots, n-1),$$
> $$(i = 1, \dots, n-1, k = 1), \; (i = n-1, \dots, n-1, k = 1),$$
>
> *are replaced by*
>
> $$u'_{i,k} = \frac{1}{4}\left(u^0_{i-1,k} + u^0_{i+1,k} + u^0_{i,k-1} + u^0_{i,k+1}\right)$$
>
> *- the boundary values are fixed - the mesh function of the first and last [horizontal and vertical] mesh lines is improved. With these improved values and the values on the third and $(n-3)$th lines, one improves the values in the second and $(n-2)$th lines and so on. When you have reached the innermost square, you start again from the outside and calculate to the center, and so on.*
>
> *We claim that the values improved in this way converge in geometric progression towards the sought values.*

What Liebmann did amounts to use a spiral ordering (see [998]) or to consider the mesh like an onion, for instance, for a mesh with 25 points

$$
\begin{array}{ccccc}
- & - & - & - & - \\
| & - & - & - & | \\
| & | & \cdot & | & | \\
| & - & - & - & | \\
- & - & - & - & -
\end{array}
$$

The computation is first done for the outer ring, with the known values and the boundary values, then on the inner ring using the values just computed for the outer ring, and finally for the center point using the values just computed for the inner ring. Then, the process is repeated until convergence. Liebmann gave a proof of convergence for the Laplace discrete problem.

In modern terms, this method can be seen as a matrix splitting using the spiral ordering. The method is convergent because the spectral radius of the iteration matrix is smaller than 1, according to the Householder-John theorem that we will discuss in a moment.

There is a misunderstanding about Liebmann's method since, in many places, it is said that it is the same as the Gauss-Seidel method; see, for instance, Richard Steven Varga's book [3132, p. 58] (see also D.M. Young's thesis [3304] in 1950). But as a source, Varga cited the paper by Stanley Phillips Frankel (1919-1978) [1217] in 1950. Unfortunately, what is described as Liebmann's method in that paper does not correspond to the method described as above by Liebmann. Frankel cited George H. Shortley (1910-1980), Royal Weller, and Bernard Fried [2760], who probably were the source of the error. In fact, Liebmann was already cited in [2759] by Shortley and Weller in 1938, who used iterative methods to solve Laplace's equation. So, Liebmann did not exactly rediscover Seidel's method. In 1935, Liebmann asked for retirement due to political pressure because of his Jewish ancestry.

In 1921, in his book [2631], Runge explained how to solve linear systems by elimination using small examples. After a chapter on solving nonlinear equations by iteration, he also applied the same methods to solving linear equations. What he explained (pages 70-76) with the aid of a system of order 3 with a strongly diagonally dominant matrix is what we now call Jacobi's method.

Gauss and Seidel methods were cited, when applied to least squares problems in the 1924 book [3227] by Sir Edmund Taylor Whittaker (1873-1956), and George Robinson, a Canadian mathematician who was lecturer in Whittaker's department in Edinburgh,.

Whittaker was an English mathematician, physicist, interested also in astronomy, who in 1896 was elected fellow at Trinity College in Cambridge. There he was an advisor of Godfrey Harold Hardy (1877-1947) and Alexander Craig Aitken (1895-1967), and he also had as a student John Edensor Littlewood (1885-1977). In 1906, he was Professor of Astronomy at the University of Dublin and, from 1912, Professor of Mathematics at the University of Edinburgh, where he remained for the rest of his career.

Kurt Wilhelm Sebastian Hensel (1861-1941) was a German mathematician whose grandmother was Fanny Mendelssohn Hensel (1805-1847), a composer and pianist, and the sister of the famous composer Felix Mendelssohn Bartholdy (1809-1847). In 1826, Hensel published a paper on the convergence of series of matrices (in his notation)

$$\mathfrak{B}(T) = a_0 + a_1 T + a_2 T^2 + \cdots .$$

Denoting by ψ the characteristic polynomial of T, he proved the following result:

A power series $\mathfrak{B}(t)$ diverges for $t = T$ if only one of the r-roots of the main equation $\psi(t) = 0$ for T lies outside the convergence circle $\mathfrak{R}_\rho$; it always converges when all r-roots are inside $\mathfrak{R}_\rho$. For every $\bar{\nu}$ finite that lies on the periphery of $\mathfrak{R}_\rho$, $\mathfrak{B}^{(\bar{\nu}-1)}(\bar{\tau})$ must be convergent if $\bar{\tau}$ is an $\bar{\nu}$-fold root of the main equation of T.

His proof, which used the Jordan canonical form of T (without reference to Jordan), is not very clear. Nevertheless, he reduced the problem to considering only one Jordan block. This technique can be used to characterize the behavior of powers of a matrix which are of interest to study convergence of iterative methods.

Hardy Cross (1885-1959) was an American Professor of Structural Engineering who came to the University of Illinois in 1921, spending there the most productive 16 years of his life. In 1937, he moved to Yale University until his retirement in 1953. He presented a short paper (eight pages) to the American Society of Civil Engineers (ASCE), published in 1930, with the title *Analysis of continuous frames by distributing fixed-end moments*, which is really an application of Jacobi and Seidel methods to structural frames described in a pseudo-physical way more appealing to the engineers of the day. A more refined paper [774] appeared in 1932.

Sufficient conditions for convergence of Jacobi and Gauss-Seidel methods were given in 1929 by Richard Edler von Mises (1883-1953) and Hilda Pollaczek-Geiringer (1893-1973). They were Austrian mathematicians working at that time in Germany. In the paper, published in two parts [3155, 3156], they first considered solving by iteration one nonlinear equation $f(x) = 0$. Then, they somehow applied the method to linear solves. If each of the n linear equations with n unknowns are written as $L_i(x) = 0$, $i = 1, \ldots, n$, the method is given by $x_i^{(\nu+1)} = x_i^{(\nu)} + c_i L_i(x^{(\nu)})$, where the c_i's are parameters. They first showed that, if the c_i's are chosen such that

$$|c_1\, a_{1,i}| + \cdots + |c_{i-1}\, a_{i-1,i}| + |1 + c_i\, a_{i,i}| + |c_{i+1}\, a_{i+1,i}| + \cdots + |c_n\, a_{n,i}| \le \mu < 1,$$

the method converges. Then, they considered the choice $c_i = -1/a_{i,i}$, which cancels the diagonal term in the previous inequality (note that their equation (7), describing the iteration, is not correct since the right-hand side is missing). This corresponds to the Jacobi method, and the sufficient condition is

$$\sum_{i \ne k} \left| \frac{a_{i,k}}{a_{i,i}} \right| \le \mu < 1 \quad k = 1, \ldots, n,$$

that is, strict diagonal dominance by columns. Von Mises and Geiringer named this method *iteration in total steps*. They also tried to give error estimates and gave another sufficient convergence condition

$$\sum_{i,k,\, i \ne k} \left(\frac{a_{i,k}}{a_{i,i}} \right)^2 < 1.$$

It is interesting to note that for the proof of this result they used matrices. In their notation, they wrote the equations as

$$\mathfrak{A}\mathfrak{x} = \mathfrak{r} \quad \text{with } a_{i,i} = 1,\ i = 1, \ldots, n.$$

Then the iteration with $c = -1$ becomes

$$\mathfrak{x}^{(\nu+1)} = \mathfrak{x}^{(\nu)} - (\mathfrak{A}\mathfrak{x}^{(\nu)} - \mathfrak{r})$$

or, if $\mathfrak{z}^{(\nu)} = \mathfrak{x}^{(\nu)} - \mathfrak{x}$,

$$\mathfrak{z}^{(\nu+1)} = \mathfrak{z}^{(\nu)} - \mathfrak{A}\,\mathfrak{z}^{(\nu)} = (\mathfrak{E} - \mathfrak{A})\,\mathfrak{z}^{(\nu)} = \mathfrak{K}\mathfrak{z}^{(\nu)},$$

where $\mathfrak{E}$ is the identity matrix and

$$\mathfrak{K} = \mathfrak{E} - \mathfrak{A},\ k_{i,i} = 0,\ k_{i,\chi} = -a_{i,\chi}\ (i \ne \chi).$$

So,

$$\mathfrak{z}^{(2)} = \mathfrak{K}\mathfrak{z}^{(1)}, \quad \mathfrak{z}^{(3)} = \mathfrak{K}\mathfrak{z}^{(3)} = \mathfrak{K}^2\mathfrak{z}^{(1)}, \ldots.$$

It means they noticed that the powers of the matrix $\mathfrak{K}$ are involved, but they did not relate that to necessary and sufficient convergence conditions. We do not know if they were aware of Nekrasov's paper or of Hensel's result in 1826. However, they related convergence to the eigenvalues in the case that the same parameters $c_i = c$ are used for all equations.

In Section 3, they introduced a method named *iteration in single steps* which corresponds (as they observed) to Seidel's method when $c_i = -1/a_{i,i}$. They considered the symmetric case and positive definite matrices, introducing

$$Q = \frac{1}{2}\sum_{i,j} a_{i,j}\, x_i\, x_j > 0.$$

They aimed at proving that when A is symmetric positive definite, Seidel's iteration converges. To do so, they used the fact that the solution of the linear system $Ax = b$ gives the minimum of

$$F(x) = \frac{1}{2}(Ax, x) - (b, x) = Q - (b, x).$$

Their proof of this fact is not very rigorous. They showed that, under the hypothesis, $F(x^{(\nu)})$ is decreasing when one component of the approximate solution is computed with Seidel's method. They provided an example of order 6 arising from a paper by Walter Ritz (1878-1909) in 1909. The matrix is not symmetric, but Von Mises and Geiringer claimed that it can be symmetrized by right multiplication with a diagonal matrix. Unfortunately, this is not completely true for two of the entries. Nevertheless, the matrix is strictly diagonally dominant and Seidel's method is converging.

Von Mises and Geiringer also defined what they called *iteration in groups* where several components are computed simultaneously by solving a small linear system. They observed that this was already proposed by Seidel. In another part of this paper, they discussed what is now known as the power method to compute eigenvalues; see Chapter 6.

Up to the end of the 1930s, people discussing iterative methods for solving linear systems only used the equations and did not much considered the matrix formulation. This was about to change and to provide more powerful results and insights.

The goal of the paper [3266] by Helmut Wittmeyer in 1936 was to generalize some results of von Mises and Geiringer. Wittmeyer used the matrix framework and some eigenvalue bounds from the dissertation [3264] he obtained from the Technische Universität Darmstadt in 1935, under the supervision of Udo Wegner (1902-1989) and Alwin Oswald Walther (1898-1967). In his notation, he considered the problem $A\mathfrak{x} = \mathfrak{r}$ with

$$A = (a_{i,k}); \quad \det(A) \neq 0; \quad \mathfrak{x} = (x_1, \dots, x_n); \quad \mathfrak{r} = (r_1, \dots, r_n),$$

and the iteration method is

$$\mathfrak{x}^{(i+1)} = (I + DA)\,\mathfrak{x}^{(i)} - D\mathfrak{r}.$$

Wittmeyer stated (our translation)

> *We want to show that the iteration procedure leads to the goal even if D is an arbitrary matrix - i.e. not necessarily a diagonal matrix - as long as the elements of D satisfy certain "convergence conditions".*

Denoting by $|\lambda^B|_{\max}$ the spectral radius of B, he proved that this iterative method is convergent

> *1. if $|\lambda^{I+DA}|_{\max} < 1$, i.e. if the largest absolute value of the characteristic numbers of the matrix $I + DA$ is smaller than one, or*
>
> *2. if the characteristic numbers of DA in the complex plane lie entirely in the circle with center -1 and radius 1.*

Then, he stated without proof that the iteration method of the theorem converges to the solution if the matrix D is "sufficiently close" to the negative inverse of A. He gave another proof of the sufficient conditions for convergence obtained by von Mises and Geiringer and derived bounds for the error. He also discussed a method by August Hertwig (1872-1955) in 1912 which corresponds to choosing D as the negative inverse of

$$\begin{pmatrix} a_{1,1} & a_{1,2} & 0 & 0 & 0 & 0 \\ a_{2,1} & a_{2,2} & 0 & 0 & 0 & 0 \\ 0 & 0 & a_{3,3} & a_{3,4} & 0 & 0 \\ 0 & 0 & a_{4,3} & a_{4,4} & 0 & 0 \\ 0 & 0 & 0 & 0 & a_{5,5} & a_{5,6} \\ 0 & 0 & 0 & 0 & a_{6,5} & a_{6,6} \end{pmatrix}$$

when $n = 6$. Furthermore, he considered a splitting of the matrix $A = B + C$ and the method

$$\mathfrak{x}_k = \mathfrak{x}_{k-1} - B^{-1}A\mathfrak{x}_{k-1},$$

and proved that the method converges if $|\lambda^{I-B^{-1}A}|_{\max} < 1$.

The work of Lamberto Cesari (1910-1990) [600, 601] in 1937 was along the same lines as the work of Wittmeyer, even though the formulation was different. He considered a splitting $B + C = cA$, where B is nonsingular and $c \neq 0$ and proved that the iteration defined (in his notation) by

$$\begin{aligned} BX_0 &= cH - CR, \\ BX_1 &= cH - CX_0, \\ &\vdots \\ BX_{p+1} &= cH - CX_p, \end{aligned}$$

where R is a given vector, converges to the solution of $AX = H$ if all the moduli of the roots of $\det(B + \lambda C) = 0$ are strictly larger than one. Note that, as it was observed by Cesari, this implies that the spectral radius of $B^{-1}C$ is smaller than one. Then, he gave sufficient conditions for the convergence which are similar to those of Wittmeyer. He continued by considering what he called "normal systems," which are, in fact, linear systems with symmetric positive definite (SPD) matrices. For this class of matrices, he considered the method of von Mises and Geiringer, the method of Seidel and its block generalization, and a block method suggested by Mauro Picone (1885-1977), the founder in 1927 of the Istituto per le Applicazioni del Calcolo, where Cesari was working. He proved that what we now call the block Gauss-Seidel method is convergent for SPD matrices.

Cesari also considered polynomial preconditioning to improve the rate of convergence. He observed that the polynomial has to be small at the eigenvalues of A or, at least, on an interval containing the eigenvalues. He derived some polynomials up to order 4. Finally, he gave a numerical example, comparing the different methods.

Relaxation methods started to be developed by Richard Vyne Southwell (1888-1970) in the second half of the 1930s, at first as a pure engineering method for computing stresses in frameworks; see Section 10.64. In the paper [2834] in 1935, Southwell wrote

> *Consider a framework having elastic members and frictionless joints, and imagine that we can at will impose rigid constraints, at any or all of the joints, such that movement of the joints is prevented but the ends of all members are left free to turn. Suppose that initially all joints are constrained in this way and the specified*

> *forces applied. Since the constraints are rigid, they will sustain the whole force at every joint. Next, imagine that one constraint is relaxed, so that one joint is permitted to travel slowly through a specified distance in some specified direction. Then force will be transferred from the constraint to the framework, and strain energy will be stored in the latter: if the initial force on the constraint had a component in the direction of the travel, the force on the constraint will be relieved (as regards this component), and the strain-energy will be stored at the expense of the potential energy of the external forces. All joints but the one being fixed, it is an easy matter to calculate how much force will be transferred as the result of a specified displacement; therefore we can so adjust the displacement that the constraint will be relieved to any desired extent. Now let this constraint be fixed rigidly in its new position, and let some other constraint be relaxed: exactly as before, we can arrange that additional strain-energy is stored in the framework and the force on the second constraint relieved. Proceeding in this way, we shall (in general) continuously increase the amount of the total strain-energy, and by a finite quantity at every step. The increments tend to decrease as the constraints are relieved of load; thus the total (at least, in the final stages) tends asymptotically to its correct value.*

This is the origin of the use of the word "relaxation" in iterative methods. Corrections and details were given in [2835]. The choices of the constraints to be relaxed were based on the intelligence and the insight of the engineer doing the computation. In [679] Sir Derman Guy Christopherson (1915-2000), a research assistant of Southwell and the first to apply the method to the solution of field differential equations, wrote

> *What was new was the replacement of an automatic iterative procedure for the solution of a large number of such equations by a process which, by concentrating on measures of the extent to which the solution had not been reached (the "residual loads" in relaxation parlance), the physical insight of the operator could be employed to shorten the process.*

Then, Southwell realized that he could apply the same techniques to other engineering problems. He summarized his research in a book [2836] published in 1940. In a lecture he gave in 1943, Southwell said

> *Before passing to other types of equation I must deal with another aspect of the relaxation process - whether applied to frameworks or to nets. Is that process always convergent? This is a question for mathematicians, and I, from the standpoint of engineering science, would first propound another: What is an "exact" solution?*

A partial proof of convergence of the method was published by Archibald Niel Black (1912-2001) and Southwell [345] in 1938.

A more mathematical explanation of the relaxation method was given by George Frederick James Temple (1901-1992) [3022] in 1939. Temple wrote

> *During the last few years Southwell and his fellow-workers have developed a new method for the numerical solution of a very general type of problem in mathematical physics and engineering. The method was originally devised for the determination of stresses in frameworks, but it has proved to be directly applicable to any problem which is reducible to the solution of a system of non-homogeneous, linear, simultaneous algebraic equations in a finite number of unknown variables.*

He proved the convergence of the method for symmetric positive definite matrices. However, these authors did not notice that the relaxation method was not much different from the methods proposed by Gauss and also Seidel. The only differences were maybe the order in which the equations were processed and the fact that the modification of one component was not always completely applied.

Temple also described a method of steepest descent for linear operators in Hilbert spaces which can, of course, be applied also to linear systems of finite order. The steepest descent idea was not new. It was used in another context in the 19th century by Augustin-Louis Cauchy (1789-1857) in 1827 to estimate the asymptotic value of some integrals, and formulated for solving general nonlinear equations [576] in 1847, by Bernhard Riemann (1826-1866) in 1863, Nekrasov in 1885, and Peter Debye (1884-1966) in 1909; see [2486]. After the death of his wife in 1980, Temple took monastic vows in the Benedictine order and entered Quarr Abbey on the Isle of Wight, where he remained until his death.

The fact that the powers of a complex matrix A converges to zero if and only if the eigenvalues lie in the unit circle was proved rigorously using the Jordan canonical form by Rufus Oldenburger (1908-1969), an American mathematician and mechanical engineer, in a paper [2365] published in 1940. There is no reference to the work of Hensel in Oldenburger's paper. For powers of matrices and some bounds, see Werner Gautschi (1927-1959) in 1953 [1304, 1305].

Israel Moiseevich Gelfand (1913-2009), a Russian mathematician, proved in 1941 and in a more general setting [1316] that the spectral radius (that is, the largest modulus of the eigenvalues) of A is $\lim_{n\to\infty} \|A^n\|^{\frac{1}{n}}$ for any norm. See also [1879] by Tosio Kato (1917-1999) in 1960.

The necessary and sufficient condition of convergence for the Gauss-Seidel method was also proved, in a complicated way [2702], by Robert J. Schmidt, of Imperial College in the UK, in 1941. Schmidt combined several iterates of the Gauss-Seidel method for obtaining a new approximation of the solution. This new iterate is obtained by solving a linear difference equation, and thus it is given as a ratio of determinants, and can be seen as acceleration of a sequence. He also discussed several numerical examples.

Referring to von Mises and Geiringer, in 1942, Lothar Collatz (1910-1990) considered the *iteration in total steps* (Jacobi) and the *iteration in single steps* (Gauss-Seidel) in [725]. He observed that the methods converge when the matrix is diagonally dominant, and his aim was to find transformations to make a matrix more diagonally dominant and to obtain estimates of the error. He also showed with small examples that the two methods do not have the same convergence conditions. Then, he considered the normal equations $A^TAx = A^Tb$ and called this the *Gaussian transformation*. He noticed that, even though the matrix is now symmetric positive definite, this transformation can worsen the convergence quality. He showed this on a small 2×2 linear system. To find a better method, Collatz referred to Runge in the book [2632] written with Heinz König in 1924, where the goal was to increase the diagonal dominance by combinations of the equations. However, no general method to achieve that goal was given in this book. The authors only discussed a particular system of order 3 with ad hoc transformations.

What is more interesting in Collatz's paper is an estimate of the error for the iteration in single steps, that is, the Gauss-Seidel method. Assuming strict diagonal dominance and defining $b_i = \sum_{k=1,k\neq i}^n |a_{i,k}|$, he proved that

$$|x_i^{(\nu+1)} - x_i| \leq \rho\delta^{(\nu)}, \quad \rho = \max_i \frac{b_i}{|a_{i,i} - b_i|}, \quad \delta^{(\nu)} = \max_i |x_i^{(\nu+1)} - x_i^{(\nu)}|,$$

where x is the exact solution and $x^{(\nu)}$ the iterates.

In 1943, Mark Kormes published a paper [1938] describing how to obtain an approximation of the solution of $\Delta u = 0$ using the different types of punchcard equipment available at that time from the IBM or Remington Rand companies. He used a 5-point finite difference scheme on a regular Cartesian mesh. Multiplying the difference equation by h^2, where h is the mesh size, he used the relation

$$u_{i,k} = \frac{u_{i-1,k} + u_{i,k-1} + u_{i+1,k} + u_{i,k+1}}{4}$$

repeatedly. This is what we now call Jacobi's method (which is convergent for this problem). Having as input cards with the given boundary values and the initial values, the machine punched the results of the first iteration on another set of cards and the operation was repeated as long as it was needed. Kormes reported that a problem with 145 unknowns was solved in 30 minutes and that the 10th approximation gave an accuracy of two decimal places.

Strangely enough, in the following years, some people referred to the method of Jacobi as Kormes' method; see, for instance, D.M. Young's thesis [3304] in 1950, page II.9 of the original manuscript.

In an invited address presented before the American Mathematical Society at New York in March 1944, Howard Wilson Emmons (1912-1998), who was professor of mechanical engineering at Harvard University, discussed the solution of linear systems arising from PDE finite difference approximations [1094]. The first paragraph of the introduction of that paper is as follows:

> *Through a consideration of the fundamental aspects of the universe Newton was led to the invention of fluxions. The physical idea of "rate of change," the geometric idea of "slope of a curve" together with his newly invented mathematics proved to be very powerful in describing and predicting a wide range of phenomena of nature. Since Newton's day, an enormous number of physical phenomena have been described in terms of a few "laws of nature." Very often these laws make use of the calculus, especially when applied to a specific problem. Thus large sections of the phenomena of the physical universe are described by the solutions of differential equations for the appropriate boundary conditions.*

It would be difficult to write something more general!

For solving linear systems iteratively, Emmons was a proponent of relaxation methods. He explained the method and cited the paper by Christopherson and Southwell [680] in 1938. He wrote about "human computers"

> *These "operating instructions" may appear vague. Indeed they are vague. Their vagueness is the source of their great power because the computer may without any effort alter the procedure to attain more rapid approach to the final answer (of no residuals). There is only one way to appreciate fully the meaning of these remarks and that is to do a problem.*

Emmons gave numerical examples on small heat equation problems and also for the problem of the water tube of a boiler. He also discussed how to solve other types of equations. We observe that Emmons cited Liebmann and Kormes as authors of methods he did not want to use. For some insight on how some engineers were thinking about iterative methods in that pre-computer time, let us quote Emmons again:

> *To appreciate fully the power of the flexibility of the relaxation method, one must take pencil and paper and carry out the numerical process in all its uninteresting details. In fact, for the computer (as opposed to those who think only about the logic behind the computation methods) the relaxation method has a spirit lacking entirely from the iteration processes. The former challenges one's intellect at each*

> *step to make the best possible guess, while the latter reduces one to the status of an automatic computing machine (without the advantage of no computational errors). It should not be inferred that the relaxation process requires high intellectual powers. If changes are chosen in a specifiable way it reduces exactly to the iteration process. The computer can then vary from this completely specified process by whatever amount fits his own skill.*

A paper [296] studying convergence of the Gauss-Seidel method was published in 1945 by Clifford Edward Berry (1918-1963). He stated that if $A = A_1 + A_2$, where A_1 (resp., A_2) is the lower (resp., strictly upper) triangular part of A, a necessary and sufficient condition of convergence for a general matrix A with nonzero diagonal entries is $\lim_{m\to\infty}(-A_1^{-1}A_2)^m = 0$. He also observed that a sufficient condition is to have the norm of $A_1^{-1}A_2$ strictly less than 1. Berry referred to von Mises and Geiringer [3155] in 1929 and Hotelling [1732] in 1943, but he did not seem to be aware of the works of Hensel, Cesari, and Oldenburger.

We observe that when he was a graduate student in electrical engineering at Iowa State University, Berry helped John Vincent Atanasoff (1903-1995) to build what is now called the Atanasoff-Berry computer; see Chapter 7. He is considered one of the computer pioneers.

The proposition of the American statistician Harold Hotelling (1895-1973) was to compute matrix inverses to solve linear systems. In the paper [1733] elaborated in 1946 but published in 1949, he used a Newton iteration

$$C_{k+1} = C_k\,(2I - AC_k), \quad k = 1, 2, \ldots$$

for computing $C = A^{-1}$. Of course, we know that this was not a very fruitful idea, to say the least.

Leslie Fox (1918-1992), who worked with Southwell for a while, published a paper on relaxation methods [1201] in 1948. In that paper, he wrote

> *Most of the available literature is due to R.V. Southwell, who, with a small and variable team of research workers, has been engaged since about 1936 with the development of the method and its application to a large number of important problems in engineering and mathematical physics.* [...] *Southwell first used the method in the problem of the determination of stresses in loaded frameworks, and all his subsequent work keeps the vocabulary and notation appropriate to this problem.* [...] *Devices for the acceleration of convergence of the relaxation process come by experience and practice. For example, the largest residual is not the all-important quantity. Quicker convergence is secured if at each stage that residual is liquidated which requires the largest "displacement" for its liquidation.* [...] *"Under-relaxation", in which only a portion of a residual is removed, and "over-relaxation", in which the sign of a residual is deliberately changed, are two useful devices.*

Fox gave some examples of the application of the methods for the solution of problems arising from the finite difference discretizations of the two-dimensional Laplace and biharmonic partial differential equations. Of course, all the calculations were done by hand or with mechanical calculators. In conclusion, Fox added

> *Finally, though the labour of relaxation in three dimensions is prohibitively great, the future use of the new electronic calculating machines in this connexion is a distinct possibility.*

However, this conclusion did not prove to be true and the relaxation methods "à la Southwell" were almost killed by the advent of digital computers.

Philip Bernard Stein (1890-1974) and Reuben Louis Rosenberg (1909-1986) were mathematicians working at the Natal University College in South Africa. In 1948, they published a paper [2862] on the solution of linear simultaneous equations by iteration. They considered a linear system $Ax = b$ that, dividing by the diagonal entries, they reformulated as $x = Cx + d$, where $C = L + R$ is a matrix with a zero diagonal and L (resp., R) is a strictly lower (resp., upper) triangular matrix. They defined two iterative methods

$$x^{(n)} = d + Lx^{(n)} + Rx^{(n-1)},$$

which they called iteration (O) (Gauss-Seidel), and

$$x^{(n)} = d + Cx^{(n)},$$

which they called iteration (S) (Jacobi). Defining the error vector $z^{(n)} = x - x^{(n)}$, they had $z^{(n)} = C^* z^{(n-1)}$ for (O) with $C^* = (I - L)^{-1}R$ and $z^{(n)} = Cz^{(n-1)}$ for (S). They recalled some known sufficient conditions for convergence, but their main result is a comparison of both methods for the case where the matrix C contains no negative entries.

They proved that if λ (resp., μ) is the maximum positive eigenvalue of C (resp., C^*, the conjugate transpose of C), then both methods converge if $\mu < \lambda < 1$ and they both diverge if $1 < \lambda < \mu$. Moreover, if $\lambda = 1$, then $\mu = 1$, and conversely. The proof is based on what is now known as the Perron-Frobenius theorem related to eigenvalues and eigenvectors of non-negative matrices. This theorem was the result of works by Oskar Perron (1880-1975) in 1907 and Ferdinand Georg Frobenius (1849-1917) in 1912.

Olga Taussky (1906-1995) (see Section 10.68) was interested in that paper, and as a result, P.B. Stein was invited to the National Bureau of Standards. There he wrote four papers on iterative methods [2858, 2859, 2860, 2861] in 1951-1952. In [2858], he considered the Gauss-Seidel method for "nearly" symmetric matrices. In [2861], Stein proved that for a square matrix B, $\lim_{n\to\infty} B^n = 0$ if and only if there exist a positive definite Hermitian matrix H such that $H - B^*HB$ is positive definite. Here B^* is the conjugate transpose of B.

Edgar Reich (1927-2009) was a Research Assistant in the Servomechanisms Laboratory at MIT when, in 1949, he published a paper [2546] in which he proved that if A is a real symmetric matrix with a positive diagonal, Seidel's method is convergent if and only if A is positive definite. That this condition is sufficient had already been proved. Reich's contribution is the necessity part. However, his proof is not completely rigorous when the eigenvalues of the iteration matrix are not all distinct.

In 1949, Geiringer proved that the Gauss-Seidel method is convergent for irreducibly diagonally dominant matrices [1314]. A matrix A is irreducible if there does not exist a permutation matrix P such that

$$P^{-1}AP = \begin{pmatrix} A_1 & 0 \\ 0 & A_2 \end{pmatrix},$$

A_1 and A_2 being square matrices. The matrix A is irreducibly diagonally dominant if it is irreducible and diagonally dominant with at least one row i with $|a_{i,i}| > \sum_{j=1, j\neq i}^{n} |a_{i,j}|$. Such a matrix is nonsingular. She also proved the same result as Reich, that is, if the diagonal entries are positive, it is necessary that the matrix be positive definite for Seidel's method to be convergent.

When the matrix is symmetric positive definite, Werner Johannes Schmeidler (1890-1969), a student of Edmund Landau (1877-1938) in Göttingen, proved [2697] in 1949 that the method of Seidel is convergent in the cyclic case, that is, when the unknowns are considered periodically through all indices $1, \ldots, n$. See also Ostrowski [2400].

The name "Gauss-Seidel method" started to be used in the beginning of the 1950s; see, for instance, Hotelling [1733] in 1946, Irving C. Liggett [2072] and Alston Scott Householder (1904-1993) [1735] in 1950.

The idea of a symmetric Gauss-Seidel method was introduced by Aitken in 1950 [19]. The components of the iterates are computed successively in the order $1, \ldots, n$ and then $n, \ldots, 1$. When applied to a symmetric matrix A, the iteration matrix is symmetric and positive definite with real eigenvalues. Aitken did this to be able to apply his Δ^2 extrapolation process to the iterates. The idea of a "symmetric" iteration was further developed by other authors as we will see below.

A landmark in the development of classical iterative methods is the Ph.D. thesis of D.M. Young, done under the supervision of Garrett Birkhoff (1911-1996) at Harvard University; see Section 10.77. It was considered by Golub as the best thesis in numerical linear algebra he had ever seen.

Young was interested in the linear systems arising from finite difference discretizations of linear partial differential equations with Dirichlet boundary condition in two and three dimensions. The first part of the thesis [3304] is devoted to the discretization scheme and results about the convergence of the discrete approximation to the solution of the PDE. The matrices of these linear systems are irreducibly diagonally dominant, and moreover, the off-diagonal entries are negative.

Curiously enough, as we wrote above, Young named the Jacobi method the Kormes' method, and the Gauss-Seidel method the Liebmann's method. We have already seen that Liebmann's method is not exactly the Gauss-Seidel method. Young proved, as was already done by Geiringer, that both methods converge, and he compared their rates of convergence. He did this by considering the iteration matrices. He also introduced a successive overrelaxation (SOR) method,

$$x_i^{(m+1)} = \frac{\omega}{a_{i,i}} \left[b_i - \sum_{j=1}^{i-1} a_{i,j} x_j^{(m+1)} - \sum_{j=i+1}^{n} a_{i,j} x_j^{(m)} \right] + (1-\omega) x_i^{(m)}, \quad i = 1, \ldots, n,$$

where ω is a positive parameter to be chosen. In his thesis, Young introduced the notion of property (A_q). A matrix A of order n has this property if the set of the first n integers can be split into q non empty disjoint subsets $n_1, \ldots, n_q$ such that

$$a_{i,j} = 0 \text{ unless } i = j \text{ or } i \in n_\ell \text{ and } j \in n_{\ell-1} \cup n_{\ell+1}.$$

He proved that if A has this property, it is possible to order the equations in such a way that there is an exact relation between the eigenvalues and eigenvectors of the Jacobi and Gauss-Seidel iteration matrices. A matrix A has property (A_q) if and only if there exist an ordering vector p with integer components assuming q distinct values such that $a_{i,j} \neq 0$ and $i \neq j$ imply $|p_i - p_j| = 1$. An ordering σ is consistent if $p_i > p_j$ implies $i > j$. These definitions were somehow simplified in the 1954 paper [3306] in which Young defined *property (A)*: there exist two disjoint subsets S and T of the set of the first n integers, such that $S \cup T = [1, \ldots, n]$ and if $a_{i,j} \neq 0$ then either $i = j$ or $i \in S$ and $j \in T$ or $i \in T$ and $j \in S$. It corresponds to the previous property (A_2).

The main results of Young's thesis are a relation between the eigenvalues of the iteration matrices of the Jacobi and SOR methods and the derivation of an optimal value (in the sense that it minimizes the spectral radius of the iteration matrix) of the relaxation parameter ω. Assuming property (A) and a consistent ordering, the relation between the eigenvalues μ of the Jacobi iteration matrix and the eigenvalues λ of the SOR iteration matrix is

$$\mu = \omega^{-1}[\lambda^{1/2} + (\omega - 1)\lambda^{-1/2}] \Rightarrow \omega^2\mu^2\lambda = (\lambda + \omega - 1)^2.$$

The optimal parameter ω_b is

$$\omega_b = 1 + \left[\frac{1-(1-\mu_1^2)^{1/2}}{\mu_1}\right]^2 = \frac{2}{1+(1-\mu_1^2)^{1/2}},$$

where μ_1 is the Jacobi spectral radius. Moreover, the spectral radius of the SOR iteration matrix is then $\omega_b - 1$. This dependence on μ_1 is, of course, a problem since in most cases μ_1 is not known and ω_b cannot be computed exactly. This aroused many research works for estimating μ_1.

There was great interest in the SOR method in the 1950s and 1960s because, with the optimal parameter, it gave a large improvement over Gauss-Seidel. With a standard 5-point finite difference scheme for the Poisson problem in a square with mesh size h, the Gauss-Seidel spectral radius of the iteration matrix is $1 - 2\pi^2h^2 + O(h^4)$ whence the optimal SOR spectral radius is $1 - 2\pi h + O(h^2)$. It makes a big difference when h is small.

In [3309] Young gave historical details on how he developed the SOR method. He also explained the difficulties he had in getting his work published in a journal:

> *The original thesis was 150 pages long. (Incidentally, because photocopying was unknown, all formulas had to be filled in by hand on all three copies.) By May 1951, the paper had been condensed to 75 pages and submitted to the Transactions of the American Mathematical Society. The referee, Hilda Geiringer, correctly pointed out that the paper was "far from ready for publication." She made a number of very useful criticisms and suggestions. After much agony and discarding of material, the paper was reduced to 15 pages and resubmitted. Some time later, I was told that I had cut out too much and that some expansion was needed to make the paper intelligible. A final iteration increased the length to 20 pages, and the paper finally appeared in 1954 - four years after the thesis was written. It can truly be said that without Garrett Birkhoff's continued interest and encouragement, the paper would never have seen the light of day!*

Young summarized his research in the book [3307] in 1971. See also his book [1533] co-authored with Louis A. Hageman (1932-2019) in 1981. Hageman earned a Ph.D. in Mathematics from the University of Pittsburgh in 1962 and worked his entire career at Westinghouse Electric Corporation (Bettis Atomic Power Laboratory). The class of *p-cyclic matrices* introduced by Varga, Hageman's thesis advisor, [3130] in 1959 is equivalent with property (A) when $p = 2$. For more on extensions of Young's results, see also [3129] and [1859, 1860] by William Morton Kahan in 1957-1958. Kahan showed that a necessary condition for SOR to converge is $|\omega - 1| < 1$. If ω is real, it implies that $\omega \in]0, 2[$.

The SOR method was also proposed independently by Frankel [1217] in 1950 under the name *extrapolated Liebmann method*. Frankel was a physicist and computer scientist who worked on the Manhattan project in Los Alamos during the war. After losing his clearances after the war, he went to the California Institute of Technology. In [1217] Frankel mainly considered the approximation of the solution of the Laplace equation $\Delta u = 0$ in a rectangle using the linear equations (in his notation)

$$L\phi_{j,k} \equiv \phi_{j-1,k} + \phi_{j+1,k} + \phi_{j,k-1} + \phi_{j,k+1} - 4\phi_{j,k} = 0.$$

He first considered what he called the Richardson method, $\phi_{j,k}^{n+1} = \phi_{j,k}^n + \alpha L\phi_{j,k}^n$, where α was a parameter. Using $\alpha = 1/4$ gave the Jacobi iteration. He analyzed this method using the eigenvectors of L, which are known explicitly. Then he showed, using the extreme eigenvalues of L, that, for this problem, $1/4$ is the optimal value. What he called the extrapolated Liebmann method was

$$\phi_{j,k}^{n+1} = \phi_{j,k}^{n} + \alpha[\phi_{j-1,k}^{n+1} + \phi_{j+1,k}^{n} + \phi_{j,k-1}^{n+1} + \phi_{j,k+1}^{n} - 4\phi_{j,k}^{n}].$$

Frankel derived the optimal value of α for this special case. He also proposed what he named the second-order Richardson method, which was

$$\phi_{j,k}^{n+1} = \phi_{j,k}^{n} + \alpha L\phi_{j,k}^{n} + \beta(\phi_{j,k}^{n} - \phi_{j,k}^{n-1}),$$

where β is a parameter; see also James Daniel Riley [2572] in 1954 and [2573] in 1955.

Even though the work of D.M. Young was more general, since the paper of Frankel was published in 1950 and that of Young only in 1954, the SOR method was often named Frankel's method or the extrapolated Liebmann's method in the 1950s.

Later, in 1958, Göran Kjellberg [1907] showed how to choose the optimal SOR parameter, ω, when the eigenvalues of the Jacobi matrix are complex. After the introduction of the SOR methods, there were many attempts to compute approximate values of the optimal relaxation parameter; see, for instance, [1284] by Paul Roesel Garabedian (1927-2010) in 1956.

Interesting papers at the beginning of the 1950s were [3200] by Wegner in 1951, who gave some remarks on the iterative methods for linear systems of equations, [2687] by Helmut Max Sassenfeld (1920-2014) in 1951, who derived a sufficient convergence criterion and error estimates for the Gauss-Seidel method, and [3210, 3211] by Johannes Weissinger (1913-1995), who wrote on the theory and application of the iteration process and generalizations of Seidel's method in 1952-1953.

In 1954, Alexander Markowich Ostrowski (1893-1986) proved that if $A = D - E - E^*$ is a Hermitian matrix, with D Hermitian and positive definite, and $D - \omega E$ nonsingular for $0 \leq \omega \leq 2$, then the spectral radius of the SOR iteration matrix is strictly less than 1 if and only if A is positive definite and $0 < \omega < 2$; see [2397]. As we have seen, this result was proved by Reich for $\omega = 1$ in 1949. For this reason, this result is often referred to as the Ostrowski-Reich theorem.

It turns out that the Ostrowski-Reich theorem is a corollary of a more general result, which is called the Householder-John theorem. The theorem states that if $A = M - N$ is Hermitian with M nonsingular and such that $M + M^* - A = M^* + N$ is positive definite, then the spectral radius of $M^{-1}N$ is strictly less than 1 if and only if A is positive definite. As reported by Herbert Bishop Keller (1925-2008) in [1883], the sufficiency part was established by Fritz John (1910-1994) [1833, p. 25] in 1956. A similar result was established earlier by Weissinger [3211] in 1953 under the hypothesis that M and $M + N$ are Hermitian. The necessity part was proved by Householder in a report in 1955 and a paper [1740] published in 1956.

The symmetric successive overrelaxation (SSOR) method [2751] was introduced by John Waldo Sheldon (1923-2015) in 1955. This is a generalization to SOR of what Aitken proposed for Gauss-Seidel. The components of the iterates are computed for 1 to n, and then from n to 1. Again, for a symmetric matrix with a positive diagonal, this gives an iteration matrix which is similar to a symmetric positive definite matrix. Computational experiments described by Sheldon indicated that SSOR was superior to SOR. This is not always true for other problems as it was shown by George Joseph Habetler and Eugene L. Wachspress [1517] in 1961. Sufficient conditions for convergence were studied later. There exists an optimal parameter as it was shown in [1517], but its analytical form is not known. However, if the matrix has the block form

$$\begin{pmatrix} I & U \\ L & I \end{pmatrix}$$

(which means that the matrix has property (A)), the optimal relaxation parameter is $\omega = 1$.

A block version of SSOR was proposed by Louis William Ehrlich (1927-2007) in 1964 [1054].

In the 1950s and 1960s, there was great interest in solving linear systems arising from the finite difference discretizations of linear second-order partial differential equations. This generally gives matrices whose diagonal entries are positive and off-diagonal entries negative or zero. This renewed the interest in M-matrices.

The origin of the theory is in the works of Thomas Joannes Stieltjes (1856-1894) [2899] in 1887 and those of Perron and Frobenius on non-negative matrices, but M-matrices and H-matrices were introduced by Ostrowski [2392] in 1937. In fact, this paper is about determinants, but it really contains the idea of M- and H-matrices. The "M" referred to Hermann Minkowski (1864-1909) and the "H" to Jacques Hadamard (1865-1963).

The set of real matrices of order n for which $a_{i,j} \leq 0$ for all $i \neq j$ is often denoted as $\mathbb{Z}^{n\times n}$. Such matrices with $a_{i,i} > 0$ were called L-matrices by D.M. Young. When A is symmetric and positive definite, they are sometimes called Stieltjes matrices; see [3132] page 85. The definition of M-matrix used by Ostrowski was that, for $A \in \mathbb{Z}^{n\times n}$, the principal minors of A are positive. He then showed that $A = sI - B$ with $B \geq 0$ and the scalar s is strictly smaller than the spectral radius of A. This is often taken as a definition of an M-matrix. Moreover, the inverse of A is a positive matrix $A^{-1} \geq 0$ (which means that the entries of A^{-1} are positive). We observe that some people include the fact that A must be nonsingular in the definition of an M-matrix and some people don't.

Papers on M-matrices were published in 1958 by Ky Fan (1914-2010) [1132] and Householder [1743]. The first attempt to characterize M-matrices was done by Miroslav Fiedler (1926-2015) and Vlastimil Pták (1925-1999) [1165] in 1962. One of the properties they essentially showed is that the diagonal entries of A are positive and there exists a positive diagonal matrix D such that AD is strictly diagonally dominant, that is,

$$a_{i,i}d_i > \sum_{i\neq j} |a_{i,j}|\, d_j, \quad i = 1, \dots, n.$$

This is called *generalized strict diagonal dominance*. It can be also stated as there exists a diagonal matrix E, with strictly positive diagonal entries, such that $E^{-1}AE$ is strictly diagonally dominant. For more about Fiedler, see [1550].

A review on M-matrices was done by George Douglas Poole and Thomas Loris Boullion (1940-2012) [2510] in 1974. The relations between many different characterizations of M-matrices were studied by Robert James Plemmons [2500] in 1977. In the 1979 book [290] by Abraham Berman and Plemmons more than 50 characterizations of M-matrices are listed.

Regular splittings were introduced by Varga [3131] in 1960. For real matrices, $A = M - N$ is a regular splitting if M is nonsingular with $M^{-1} \geq 0$ and $N \geq 0$. In the definition of a *weak regular splitting* the condition $N \geq 0$ is replaced by $M^{-1}N \geq 0$. If $A \in \mathbb{Z}^{n\times n}$, A is a nonsingular M-matrix if and only if there exists a regular splitting $A = M - N$ with the spectral radius $\rho(M^{-1}N) < 1$. If A has a regular splitting and $A^{-1} \geq 0$, then

$$\rho(M^{-1}N) = \frac{\rho(A^{-1}N)}{1 + \rho(A^{-1}N)} < 1.$$

Varga gave some comparison theorems for regular splittings. If $A^{-1} \geq 0$ and $A = M_1 - N_1 = M_2 - N_2$ with $N_2 \leq N_1$, then $\rho(M_2^{-1}N_2) \leq \rho(M_1^{-1}N_1) < 1$.

Another landmark in the history of classical iterative methods was the book [3132] by Varga in 1962. It summarized the state of the art at that time and was highly cited.

Let A be a general matrix, the matrix $M(A)$ with entries $m_{i,j}$ is defined by

$$m_{i,i} = |a_{i,i}|, \quad m_{i,j} = -|a_{i,j}|, \; i \neq j.$$

By definition, A is an H-matrix if and only if $M(A)$ is an M-matrix. Hence, A is a nonsingular H-matrix if and only if A is generalized strictly diagonally dominant. H-matrices are interesting because many results concerning iterative methods that hold for M-matrices do also hold for H-matrices. For instance, if A is an H-matrix, then the Gauss-Seidel method converges as well as the block method. Moreover, the SOR method converges for H-matrices if

$$0 < \omega < \frac{2}{1 + \rho(|J(A)|)},$$

where $J(A)$ is the Jacobi iteration matrix. These results were proved by Varga [3134] in 1976. The SSOR method is converging under the same conditions, as was proved by Götz Alefeld and Varga [26] in 1975. For results on domain of convergence, see Arnold Neumaier and Varga [2334] in 1984.

Some of the results about M- and H-matrices and convergence of the classical iterative methods were rediscovered many times, sometimes in a weaker form. See, for instance, the critical and vivid analysis of the 1975 paper [1804] by K.R. James and W. Riha done by Varga in [3134].

Over the years some extensions of SOR and SSOR were proposed; see [1531] by Apostolos Hadjidimos for a review, and [1530]. In the modified SOR (MSOR) method a different ω is associated with each (block) row of the linear system. The idea goes back to the beginning of the 1960s. For 2-cyclic matrices two values of ω can be used; see E. D'Sylva and G.A. Miles [968] in 1964. The unsymmetric SSOR (USSOR) method differs from SSOR by using a different relaxation parameter in the second (backward) part of the SSOR iteration; see [2104]. The accelerated overrelaxation (AOR) method was proposed by Hadjidimos [1529] in 1978. It combines the SOR method and extrapolation, having two parameters to determine. Other splittings involving two parameters were introduced by Miroslav Šisler in the beginning of the 1970s and studied by Wilhelm Niethammer [2347] in 1979.

A Fourier analysis of the SOR method was done by Randall John LeVeque and Lloyd Nicholas Trefethen [2039] in 1989. Fourier analysis was also used by Tony Fan-Cheong Chan and Howard C. Elman for other iterative methods [617] in 1989.

The classical iterative methods are almost never used today because they were superseded by other methods, notably by Krylov methods. However, they are still of interest since they can be used as preconditioners in Krylov methods or as smoothers in multigrid methods.

A review of methods for solving partial differential equations on vector and parallel computer, including iterative methods, was done by James McDonough Ortega and Robert Gary Voigt [2387] in 1985.

5.2 ▪ Alternating direction methods

The ADI methods originated in the beginning of the 1950s for solving linear systems arising from elliptic or parabolic partial differential equations on rectangles. If one uses block methods like block Jacobi or block Gauss-Seidel with a block corresponding to one mesh line, these methods emphasize one space direction over the other. A natural idea was to switch (or alternate) directions. Moreover, at that time, the computer memories were quite small and it was difficult to solve large problems.

ADI methods were first developed by researchers working in reservoir simulation for the oil industry in the USA. They were first interested in solving the heat equation

$$\frac{\partial u}{\partial t} - \Delta u = 0,$$

with suitable (nonzero) initial and boundary conditions using implicit time schemes. These schemes required the solution of a linear system at each time step.

The paper by Donald William Peaceman (1926-2017) and Henry H. Rachford Jr. [2472] was published in 1955. They considered solving the heat equation and also the Laplace equation; see also the paper by Jim Douglas Jr. (1927-2016) [947] in 1955. Let D be a diagonal matrix with $d_{i,i} = 4$ and

$$H = \begin{pmatrix} C & 0 & & & \\ 0 & C & 0 & & \\ & \ddots & \ddots & \ddots & \\ & & \ddots & \ddots & 0 \\ & & & 0 & C \end{pmatrix}, \quad \text{with} \quad C = \begin{pmatrix} 0 & -1 & & & \\ -1 & 0 & -1 & & \\ & \ddots & \ddots & \ddots & \\ & & \ddots & \ddots & -1 \\ & & & -1 & 0 \end{pmatrix},$$

and

$$V = \begin{pmatrix} 0 & -I & & & \\ -I & 0 & -I & & \\ & \ddots & \ddots & \ddots & \\ & & \ddots & \ddots & -I \\ & & & -I & 0 \end{pmatrix}.$$

Then, $A = D + H + V$. The ADI method as defined by Douglas and Rachford [946] in 1956 can be written as

$$\begin{aligned} \left(\frac{1}{2}D + H + \rho_{k+1} I\right) x_{k+\frac{1}{2}} &= b - \left(\frac{1}{2}D + V - \rho_{k+1} I\right) x_k, \\ \left(\frac{1}{2}D + V + \rho'_{k+1} I\right) x_{k+1} &= b - \left(\frac{1}{2}D + H - \rho'_{k+1} I\right) x_{k+\frac{1}{2}}, \end{aligned}$$

where ρ_{k+1} and ρ'_{k+1} are real positive parameters. At the first step one has to solve as many tridiagonal systems as there are rows in the mesh, and at the second, as many tridiagonal systems as there are columns in the mesh. The method that was described by Peaceman and Rachford in [2472] used $\rho_{k+1} = \rho'_{k+1} = \rho$. It was later called the stationary Peaceman-Rachford iterative method. Its convergence was proved by Alefeld [25] in 1976.

The ADI methods were not much in use after the 1970s because it was then necessary to solve more general partial differential problems on non-rectangular domains. But this framework was used in a great variety of other applications.

A method having a structure quite similar to ADI, although it has no geometrical meaning, is the Hermitian and skew-Hermitian splitting (HSS) iteration. It was first proposed by Zhong-Zhi Bai, Golub, and Michael Kwok-Po Ng in 2003 [161] for solving a large class of non-Hermitian linear systems $Ax = b$. Let $H = (A + A^*)/2$ be the Hermitian part of A and $S = (A - A^*)/2$ its skew-Hermitian part, the HSS method consists in the iterations

$$(H + \alpha I)x_{k+1/2} = (\alpha I - S)x_k + b,$$
$$(S + \alpha I)x_{k+1} = (\alpha I - H)x_{k+1/2} + b,$$

where α is a positive parameter. If H is positive definite the method converges to x for any choice of α and x_0. This method immediately attracted much attention and many papers were devoted to its various aspects and applications. In 2009, Benzi [269] proposed a generalization of the method that can outperform the standard HSS method in some situations and can be used as an effective preconditioner for certain linear systems in saddle point form. More references can be found in that paper and on the Internet.

5.3 ▪ Semi-iterative methods

It was noticed quite early with the use of the first digital computers that, in some cases, the convergence of methods like those of Jacobi and Gauss-Seidel, and even SOR, was too slow to be practical. Therefore, the search for methods to accelerate their convergence began. The methods that were proposed were named *semi-iterative methods*, *k-step methods*, or *acceleration methods*. The research went in two main directions: some people were interested in practical algorithms and their computer implementations and others were more interested in the theoretical properties of the methods without paying too much attention to their implementation, in particular how their convergence can be influenced by rounding errors.

One of the most successful techniques was to use Chebyshev polynomials. These polynomials are named after the famous Russian mathematician Pafnuty Lvovich Chebyshev (1821-1894). He was the founder of the Saint Petersburg school of mathematics. His name has been transliterated as Chebysheff, Chebychov, Chebyshov, Tchebychev, Tchebycheff, Tschebyschev, Tschebyschef, Tschebyscheff, and so on. This explains why Chebyshev polynomials are denoted by C_n or, sometimes, T_n.

Chebyshev polynomials were first defined for real intervals and their use was later generalized to the complex plane.

Chebyshev polynomials were used numerically by Cornelius Lanczos (1893-1974) [1983] in 1938 as well as Donald Alexander Flanders (1900-1958) and Shortley [1176] in 1950.

Later, in 1952, Lanczos [1985], pages 42-43, considered a method where the iterates are defined as $x_{m+1} = p_m(A)b$ for a matrix A with real eigenvalues and a polynomial p_m. Therefore, the residual vectors are $r_{m+1} = [I - Ap_m(A)]b$. He wanted to find a polynomial q_{m+1} where $q_{m+1}(x) = 1 - xp_m(x)$ with $q_{m+1}(0) = 1$, and as small as possible on $[0, 1]$ since he normalized the matrix to have a largest eigenvalue equal to 1. He used Chebyshev polynomials (of the first kind). Due to the properties of the Chebyshev polynomials, a short three-term recurrence is obtained. Lanczos referred to the paper of Flanders and Shortley cited above for their use and to Gábor Szegö (1895-1985) [2982] for their definition.

Eduard Ludwig Stiefel (1909-1978) used Chebyshev polynomials and conformal mapping to solve Fredholm integral equations [2895] in 1956.

In 1957, Varga [3128] solved $Au = k$ with A symmetric positive definite. The basic method was $u_{i+1} = Mu_i + k$ and, from the u_i's, he defined another sequence $t_n = \sum_{j=0}^{n} \nu_j^{(n)} u_j$. This is called a *semi-iterative method* because one first computes the u_i's, and then combines them algebraically.

For the Gauss-Seidel method as the basic method, the result was that the successive overrelaxation method converges at least as fast as any semi-iterative method. The successive overrelaxation method with the optimum overrelaxation factor converges at least twice as fast as the Chebyshev semi-iterative method with respect to the Jacobi method, and therefore at least twice as fast as any semi-iterative method with respect to the Jacobi method. For SOR the eigenvalues of the iteration matrix are not all real. This was not really encouraging, even though the optimal SOR relaxation parameter is not easy to obtain.

Stiefel studied kernel polynomials and their numerical applications [2897] in 1958. He used orthogonal polynomials to define iterative methods, including the use of Chebyshev polynomials.

The Ph.D. thesis of Golub [1375] in 1959 was titled *The use of Chebyshev matrix polynomials in the iterative solution of linear equations compared to the method of successive overrelaxation.* Golub's advisor at the University of Illinois was Abraham Haskel Taub (1911-1999). For more about Golub's thesis and his meeting with Varga, see the interview [1543] and Section 10.30.

Golub and Varga wrote a two-part paper [1401, 1402] that was published in 1961. They considered the problem $x = Bx + g$ with $A = I - B$ symmetric positive definite and B a 2-cyclic matrix with $\rho(B) < 1$. A matrix is 2-cyclic if it can be permuted to

$$\begin{pmatrix} 0 & F \\ F^T & 0 \end{pmatrix}.$$

It is primitive if there is no such permutation. This class of matrices was also considered in [1401, 1402].

Golub and Varga studied three methods: Chebyshev, SOR, and second-order Richardson, which are formally the same, except for the Chebyshev coefficients. For primitive matrices, they compared the spectral (ℓ_2) norms driving the convergence. For m iterations, the spectral norms of Chebyshev are the smallest. For cyclic matrices, taking advantage of the structure of the permuted matrix, they defined a cyclic Chebyshev semi-iterative method. Again, the Chebyshev spectral norm is better than that of SOR. They also considered block methods. These results were summarized in Varga's book [3132] in 1962.

The roundoff properties of the second-order Richardson method were first considered by Golub [1376] in 1962.

In 1963, H.E. Wrigley [3277] used Chebyshev polynomials in the complex plane to accelerate the Jacobi method, and an adaptive algorithm to obtain the parameters.

D.M. Young considered second-degree iterative methods in 1972 for the solution of large linear systems [3308]. He recalled the Chebyshev method and a stationary method with fixed parameters for accelerating SSOR. For the problem $u = Gu + k$, the method is defined as

$$u^{(n+1)} = u^{(n)} + d(u^{(n)} - u^{(n-1)}) + e(Gu^{(n)} + k - u^{(n)}),$$

with parameters d and e. Young proposed a "good" parameter for SSOR under some restrictions for $A = I - G$, and then applied the second-order method.

As we have seen above, Anderssen and Golub [57] reviewed in a 1972 report the orderings for the parameters of the second-order Richardson method. In 1977, Henryk Woźniakowski [3272] studied the stability of the Chebyshev method for Hermitian positive definite matrices. His conclusion was that the method is numerically stable.

In 1977-1978, Thomas Albert Manteuffel [2132, 2133] developed a practical algorithm for solving $Ax = b$, using Chebyshev polynomials in the complex plane when the eigenvalues of A are in the right (or left) half-plane. He used a modified power iteration to obtain estimates of the eigenvalues, enclosed them in an ellipse, and then used a Chebyshev iteration. The method converges whenever the spectrum of A can be enclosed in an ellipse that does not contain the origin. He found "optimal" parameters for the ellipse. If the eigenvalues are known, the optimal parameters can be computed. Note that there are several misprints in [2132]; see formula (2.23), which was corrected in [2133]. Chebycode, a computer implementation of Manteuffel's method, was proposed later in 1985 by Steven Flynn Ashby [81].

Other authors proposed using Chebyshev polynomials in the complex plane; see, for instance, Wachspress [3164] in 1962, as we have seen above Wrigley in 1963, Hageman in a 1972 report, and David Ronald Kincaid [1899] in 1974.

The idea of enclosing the spectrum in a bounded region of a complex plane and to use polynomials was used later by several people. In 1986, Howard C. Elman, Yousef Saad, and Paul Edward Saylor (1939-2020) [1085] also used an ellipse enclosing the Ritz values obtained from the Arnoldi process (see Section 6.8) and Chebyshev polynomials. In that same year, Elman and Roy L. Streit [1087] considered the union of convex hulls of subsets of the Ritz values and an L_∞ optimal approximation polynomial. In 1991, Dennis Chester Smolarski and Saylor [2814] enclosed the Ritz values in a polygon and used a least squares polynomial. This was previously used by Saad [2655] in 1987 but his least squares polynomial was computed in a different and more stable way using modified moments.

Linear systems with an indefinite symmetric matrix were considered by Carl-Wilhelm Reinhold de Boor and John Rischard Rice [841] in 1982. This means that the matrix has positive and negative eigenvalues. The spectrum is contained in two intervals with the origin in between $S = [a, b] \cup [c, d]$. The polynomial which is of minimum norm on S is a scaled extremal polynomial of degree n, that is, such that $\|p\|_S = \max_{x \in S} |p(x)| = 1$ and $p(0) = 1/\min\{\|p\|_S,\, p \in \pi_n,\, p(0) = 1\}$. This extremal polynomial is constructed by a Remes (or Remez) iterative algorithm. Evgeny Yakovlevich Remez (1895-1975) was a Russian mathematician working on function theory. His algorithm was published in 1934; see [2558].

The first-order Richardson method was reconsidered by Gerhard Opfer (1935-2020) and Glenn Edward Schober (1938-1991) [2381] in 1984. They showed that when the problem data is real, it can be solved using real arithmetic, even if the coefficients of the method are complex. They gave the explicit solution for the stationary case for all compact sets S containing the spectrum of the matrix.

Let us now turn for a while to the more theoretical side. Niethammer and Varga [2348] used summability theory in 1983 to define k-step stationary methods for the problem $x = Tx + c$. The iterates are defined as

$$y_m = \mu_0(Ty_{m-1} + c) + \mu_1 y_{m-1} + \cdots + \mu_k y_{m-k}, \quad m = k, k+1, \ldots,$$

given $y_0, y_1, \ldots, y_{k-1}$. For $k = 1$ (resp., $k = 2$) this reduces to the first-order (resp., second-order) Richardson method. They discussed how to select the (complex) parameters μ_j. If $\rho(T) < 1$,

$$x = (I - T)^{-1} = \sum_{j=0}^{\infty} T^j c.$$

They used summability theory to obtain a transformation to other polynomials $\nu_j(T) = \sum_{\ell=0}^{j} \pi_{j,\ell} T^\ell$. It corresponds to an infinite triangular transformation matrix $\mathfrak{B}$ with entries $\pi_{j,\ell}$. An Euler function p is meromorphic (a function that is holomorphic on all of an open subset except for a set of isolated points, which are the poles) and univalent in a disk of center the origin and $p(0) = 0, p(1) = 1$. There exists ν such that the series

$$p(\phi) = \sum_{j=1}^{\infty} \pi_{j,1} \phi^j, \quad [p(\phi)]^m = \sum_{j=m}^{\infty} \pi_{j,m} \phi^j,$$

converge for $|\phi| < \nu$. Let p be an Euler function, $\tilde{p} = 1/p$ and D_1 the closed unit disk. If the spectrum $\sigma(T)$ is contained in the exterior of the image of D_1 by $\tilde{p}$, then the expansion

$$(I - p(\phi)T)^{-1} = \sum_{j=0}^{\infty} \nu_j(T)\phi^j$$

is absolutely convergent for $\phi \in D_1$ and conversely. If $p(\phi) = \mu_0\phi/(1 - \mu_1\phi - \cdots - \mu_k\phi^k)$, a finite recursive formula is obtained and the partial sums of the transformed series can be computed with the previous recurrence. This was illustrated for $k = 2$ and real parameters when $\sigma(T)$ is contained in an ellipse. See also [2346].

Michael Eiermann and Niethammer used the same notation in their paper [1061] in 1983. They considered $\sigma(T) \subset U$, U compact, $1 \notin U$ and defined the iterates as $y_m = q_{m-1}(T)c + p_m(T)x_0$ with $q_m(x) = (1 - p_m(x))/(1 - x)$, where $p_m(T)$ is the error polynomial. It led to the problem

$$\min_{p_m(1)=1} \max_{z \in U} |p_m(z)|.$$

In general the solution is not explicitly known, so they solved this problem in an asymptotic sense. When the roots of p_m are uniformly distributed in U, it induces a semi-iterative method which is asymptotically optimal. Of course, there are many different node sets known as being uniformly distributed on U: the Fejér nodes, the Fekete nodes, and the Leja nodes.[37] If the roots do not depend on m, one obtains a first-order Richardson method.

Once again, the same notation was used by Eiermann, Niethammer, and Varga [1062] in 1985. They provided a summary of the previous results and studied the relations between Euler methods in Niethammer and Varga [2348] and asymptotic semi-iterative methods in Eiermann and Niethammer [1061]. They also used subsequences of Fejér nodes and generalized Faber polynomials. These polynomials are named after Georg Faber (1877-1966), a German mathematician. Unfortunately, there were no numerical experiments in these papers, so the real efficiency of these methods is not known.

In the paper [1055] in 1989, Eiermann constructed residual polynomials that are normalized Faber polynomials for Ω, a compact set containing the spectrum of T. He also considered methods whose iteration polynomials are truncated Faber series for $g(z) = 1/(1-z)$. This was illustrated by the construction of four-step methods.

In 1987, William Bryant Gragg (1936-2016) and Reichel were interested in the case where the eigenvalues are contained in the union of compact sets, $S = \cup_{j=1}^{\ell} S_j$; see [1422]. They constructed polynomials orthogonal on the boundary ∂S. The inner product on ∂S was defined by an integral with a weight function ρ.

Another approach was taken by Bradley N. Parsons [2465] in 1987. He considered general k-part stationary iterative solutions to linear systems $(I - B)x = b$. The stationary method was

$$y_m = B\left(\sum_{i=1}^{k} p_i y_{m-i}\right) + \sum_{i=1}^{k} q_i y_{m-i} + \omega b.$$

If one defines $P(z) = \sum_{i=1}^{k} p_i z^{1-i}$, $Q(z) = z - \sum_{i=1}^{k} q_i z^{1-i}$, $\omega = P(1) = Q(1)$, $Q(z) = 0$ implies $P(z) \neq 0$ for $z \neq 0$. Let

$$\Omega(z) = \frac{\sum_{i=1}^{k} q_i z^i}{\sum_{i=1}^{k} p_i z^i}.$$

[37] Lipót Fejér (1880-1959) was a Hungarian mathematician, Michael Fekete (1886-1957) was also born in Hungary but he emigrated to Palestine in 1928, and Franciszek Leja (1885-1979) was a Polish mathematician.

If $\sigma(B)$ is contained in the complement of the image of the closed unit disk by Ω in the complex plane, then the method converges for any initial values. Ω is called the generating function of the method. Parsons studied the relations of his method with the Niethammer-Varga construction. However, he did not give any indication of how to construct Ω.

In 1988, Golub and Michael Lockhart Overton [1390] did a convergence analysis of the Chebyshev and second-order Richardson methods when the matrix is split as $A = M - N$ and the linear systems $Mz_k = r_k$ at each iteration are solved inexactly, for instance, with another iterative method. This was a kind of inner-outer iterative method. They gave results of numerical experiments for the finite difference discretization of non-self-adjoint second-order partial differential equations in the unit square with M corresponding to the discretization of $-\Delta u + cu$. The inner iterations used a preconditioned conjugate gradient algorithm.

In a paper [2547] submitted in 1984, but only published in 1988, Reichel studied some methods using conformal mappings for determining polynomials for the first-order Richardson iteration for complex linear systems. These polynomials can be computed fairly rapidly and give rise to iterative methods with a rate of convergence very close to optimality. Bernd Fischer (1955-2013) and Reichel [1172] in 1989 used the same framework as in [2547], but they changed the ordering of the parameters. They proved the stability of the method and showed numerical results on convection-diffusion problems. This ordering gave better results than those of Eiermann, Niethammer, and Varga. However, it is difficult to implement.

The problem $x = Tx + c$ with spectrum in a compact set Ω was again considered by Eiermann, Xiezhang Li, and Varga [1060] in 1989. They proposed what they called *hybrid semi-iterative methods*. The problem is transformed to $x = \widetilde{T}x + \widetilde{c}$ with $\widetilde{T} = t_n(T)$, where t_n is a polynomial of degree n, and then, asymptotically optimal semi-iterative methods are used to solve the transformed problem. This is a sort of polynomial preconditioning. The set Ω is transformed to $\widetilde{\Omega}$. The aim was to find t_n such that a semi-iterative method with respect to $\widetilde{\Omega}$ can be easily constructed. The given examples are methods with short recurrences.

Martin Hermann Gutknecht published (k,l)-step methods for (eventually nonlinear) problems $x = \Phi x$ [1493] in 1989. They are given by

$$x_m = \frac{1}{\beta_{m,m}} \left[\Phi \left(\sum_{i=0}^{m-1} \gamma_{i,m-1} x_i \right) - \sum_{j=0}^{m-1} \beta_{j,m} x_j \right],$$

with $B = (\beta_{j,m}), C = (\gamma_{j,m})$ matrices with column sums equal to 1. B has upper bandwidth k and C has upper bandwidth $l - 1$. In stationary methods B and C are Toeplitz matrices (that is, with constant diagonals) except for the first row. For the linear problem, $\Phi x = Tx + d$, the residual polynomials of a stationary semi-iterative method are generalized Faber polynomials. These methods were revisited by Gutknecht [1503] in 2015. He gave a simpler matrix interpretation of [1493]. This gave methods with coupled simpler recurrences which are more stable.

In 1991, Reichel came back to the Richardson iteration [2548] with a different approach. He considered the first-order Richardson method and assumed that the eigenvalues of A are contained in a compact subset K of the complex plane. He used weighted Leja points for K. These points are defined in the following way. If $z_0 \in K$ is such that $|z_0|\omega(z_0) = \sup_{z\in K} |z|\omega(z)$, where ω is defined on K and bounded with values on the positive real axis, the points $z_j \in K$ are defined by

$$\prod_{k=0}^{j-1} |z_j - z_k|\, \omega(z_j) = \sup_{z\in K} \prod_{k=0}^{j-1} |z - z_k|\, \omega(z), \quad j = 1, 2, 3, \ldots.$$

When $\omega(z) = 1$, these points were studied by Leja [2032, 2033] in 1957. Leja was probably not aware that a similar definition had been used by Albert Edrei (1914-1998), a mathematician born in Egypt who obtained a Ph.D. from the ETH in Switzerland and emigrated to the USA after World War II. The definition is in [1051], page 78, published in 1940. If ω is the magnitude of an analytic function, the Leja points are on the boundary ∂K. Reichel considered $\omega(z) = |z|$ and $\omega(z) = 1$.

The relaxation parameters of the Richardson iteration are the reciprocals of the weighted Leja points. Reichel proved the stability of the method. In practice, K was replaced by K_m, a finite set of m boundary points of K. The algorithm can also be used to order any set of relaxation parameters.

D.M. Young and Kincaid [3312] solved the problem $u = Gu + k$, assuming that G has real eigenvalues in $[\alpha, \beta]$ with $\beta < 1$, with stationary second-order methods in 1993. They were defined as

$$\begin{aligned} u^{(1)} &= \omega^*\{\gamma(Gu^{(0)} + k) + (1-\gamma)u^{(0)}\} + (1-\omega^*)u^{(0)}, \\ u^{(n+1)} &= \omega\{\gamma(Gu^{(n)} + k) + (1-\gamma)u^{(n)}\} + (1-\omega)u^{(n-1)}. \end{aligned}$$

They derived optimal values for $\gamma = 2/(2-(\alpha+\beta))$ and $\omega = 2/(1+\sqrt{1-\sigma^2})$ with $\sigma = (\beta-\alpha)/(2-(\alpha+\beta))$ and showed that, asymptotically, the convergence rate is independent of ω^*, which was used for the initialization. These results were already given in a 1990 report.

In 1993, Gerhard Starke and Varga [2849] used the Arnoldi process (see Section 6.8), a polygonal domain enclosing the approximate eigenvalues, a conformal map, Faber polynomials, and a Richardson iteration. Several numerical experiments were provided, with comparisons with other iterative methods.

Daniela Calvetti, Golub, and Reichel proposed an adaptive Chebyshev iterative method [525] in 1994. The main difference with other adaptive Chebyshev algorithms enclosing approximate eigenvalues in an ellipse was that these approximations were computed with modified moments. Numerical experiments showed that this can be more efficient than the Manteuffel-Ashby implementation which was using a (modified) power method.

This idea of using modified moments was used again by Calvetti and Reichel [535] in 1996 for solving linear systems with a symmetric positive definite matrix. An interval $[a, b]$ is determined using modified moments and the reciprocals of Leja points for this interval are used in a first-order Richardson iteration. The interval can be updated during the iterations.

An algorithm for linear systems with a symmetric indefinite matrix was proposed by Calvetti and Reichel in [536] in 1996. We observe that *fast Leja points* were introduced by James Baglama, Calvetti, and Reichel [144] in 1998. They are faster and simpler to compute than the Leja points.

The Chebyshev iteration was "revisited" by Gutknecht and Stefan Röllin [1506] in 2002. They observed that the Chebyshev iteration is not seriously affected by roundoff. They used an ellipse with foci $\alpha \pm c$ not containing zero. The standard algorithm with $\beta_{-1} = 0$, $\beta_0 = c^2/(2\alpha)$, $\gamma_0 = -\alpha$, is

$$\beta_{n-1} = \left(\frac{c}{2}\right)^2 \frac{1}{\gamma_{n-1}}, \quad \gamma_n = -(\alpha + \beta_{n-1}),$$

$$\begin{aligned} x_{n+1} &= -(r_n + \alpha x_n + \beta_{n-1}x_{n-1})/\gamma_n, \\ r_{n+1} &= (Ar_n - \alpha r_n - \beta_{n-1}r_{n-1})/\gamma_n. \end{aligned}$$

Their analysis is not completely rigorous because the relations (11b), (11c) in [1443] are not exact in finite precision arithmetic. Other algorithms were proposed. In a second algorithm the residual is computed as $r_n = b - Ax_n$. The third algorithm introduced and updated $\Delta x_n = x_{n+1} - x_n$, $\Delta r_n = r_{n+1} - r_n$. The fourth algorithm updated $\Delta x_n = x_{n+1} - x_n$ and used $r_n = b - Ax_n$. They also introduced coupled two-term recurrences

$$v_n = r_n - \psi_{n-1} v_{n-1}, \quad x_{n+1} = x_n + \omega_n v_n, \quad r_{n+1} = r_n - \omega_n A v_n.$$

These two-term recurrences are generally more stable than the three-term recurrences. In the limit, a second-order Richardson iteration was obtained.

Alicja Smoktunowicz and Iwona Wröbel published a stability analysis of the stationary second-order Richardson method [2813] in 2005 for a symmetric positive definite matrix. The parameter is computed using the smallest and largest eigenvalues.

None of the methods discussed in this section are used much anymore, even though they can be useful on parallel computers. Their main drawback is that approximations of some eigenvalues are needed to compute their parameters. They were replaced by Krylov methods (see Section 5.8) which, in general, do not need approximate eigenvalues. However, some of these polynomial methods can be used as preconditioners in Krylov methods or as a part of hybrid methods.

5.4 ▪ Projection methods

Let us begin this section by recalling what are projection methods for solving a linear system $Ax = b$, and describe some of their properties; see [435, 597, 599, 1273, 2217, 2233, 2662] for more details.

For solving $Ax = b$ we consider an iterative method of the form

$$x_{k+1} = x_k + \lambda_k z_k,$$

where λ_k is a parameter and z_k an arbitrary vector. Obviously, the residuals $r_k = b - Ax_k$ satisfy the recurrence $r_{k+1} = r_k - \lambda_k A z_k$. Let u_k be an arbitrary vector, and let us choose λ_k such that u_k is orthogonal to the new residual $(u_k, r_{k+1}) = u_k^T r_{k+1} = 0$, that is, $\lambda_k = (u_k, r_k)/(u_k, Az_k)$. Such a procedure is a (one-dimensional) *projection method*. Indeed we have

$$x_{k+1} = x_k + \frac{u_k^T r_k}{u_k^T A z_k} z_k, \quad r_{k+1} = r_k - \frac{u_k^T r_k}{u_k^T A z_k} A z_k = \left(I - \frac{A z_k u_k^T}{u_k^T A z_k} \right) r_k. \tag{5.1}$$

Let P_k be the matrix $P_k = I - Az_k u_k^T / u_k^T A z_k$. Then $P_k^2 = P_k$, which shows that it is a projection matrix. More precisely, it is the *oblique* projection on $u_k^\perp$ along Az_k, and it holds $r_{k+1} = P_k r_k$. We also have $x_{k+1} - x = Q_k(x_k - x)$ with $Q_k = I - z_k u_k^T A / u_k^T A z_k = Q_k^2$. Thus Q_k represents the oblique projection on $(A^T u_k)^\perp$ along z_k. If we choose $u_k = Az_k$, then $P_k = P_k^T$, and the projection is *orthogonal*. For the choice $z_k = A^T u_k$, we have $Q_k = Q_k^T$ and x_{k+1} is the orthogonal projection of x_k on $z_k^\perp$, that is on the hyperplane $\{v \mid (u_k, b - Av) = (z_k, x - v) = 0\}$. Obviously, this hyperplane contains the exact solution x. Other choices of λ_k are possible, for example, such that $(x_{k+1} - x, u_k) = 0$, or the one that minimizes $\|r_{k+1}\|_2$, or $\|x_{k+1} - x\|_2$.

It is easy to see that $\|x_{k+1} - x\| \le \|x_k - x\|$ if and only if we are in one of the two following cases

$$(i) \quad 0 \le \lambda_k \le 2\, \frac{(x - x_k, z_k)}{(z_k, z_k)},$$

$$(ii) \quad 2\, \frac{(x - x_k, z_k)}{(z_k, z_k)} \le \lambda_k \le 0.$$

For the particular choice $z_k = A^T u_k$, $\lambda_k = (r_k, u_k)/(A^T u_k, A^T u_k) = (x - x_k, z_k)/(z_k, z_k)$, and the first condition (i) is satisfied. Thus, $\|x_{k+1} - x\|$ is minimum and a well-known property of projection methods is recovered. Similarly, $\|r_{k+1}\| \leq \|r_k\|$ if and only if we are in one of the two following cases

$$(iii) \;\; 0 \leq \lambda_k \leq 2 \, \frac{(r_k, Az_k)}{(Az_k, Az_k)},$$
$$(iv) \;\; 2 \, \frac{(r_k, Az_k)}{(Az_k, Az_k)} \leq \lambda_k \leq 0.$$

Higher-order projection methods are defined as follows. Let $\mathcal{K}_k$ and $\mathcal{L}_k$ be two n_k-dimensional subspaces. The sequence $\{x_k\}$ of iterates is completely defined by the two conditions

$$x_k - x_0 \in \mathcal{K}_k,$$
$$r_k = b - Ax_k \perp \mathcal{L}_k.$$

We denote by U_k and V_k the matrices whose n_k columns form a basis of $\mathcal{K}_k$ and $\mathcal{L}_k$, respectively. The first condition above gives $x_k - x_0 = U_k a$ with $a \in \mathbb{R}^{n_k}$. Setting $W_k = AU_k$, we also have $r_k = r_0 - W_k a$, and the second condition (often named after Petrov and Galerkin) leads to $a = (V_k^T W_k)^{-1} V_k^T r_0$. Thus, the two conditions above completely define the sequence $\{x_k\}$ of iterates if the matrices $V_k^T W_k$ are nonsingular, which holds in two important particular cases: A positive definite and $\mathcal{K}_k = \mathcal{L}_k$, and A nonsingular and $\mathcal{L}_k = A\mathcal{K}_k$. With this assumption, we have

$$x_k = x_0 + U_k (V_k^T W_k)^{-1} V_k^T r_0,$$
$$r_k = (I - P_k) r_0 \quad \text{with } P_k = W_k (V_k^T W_k)^{-1} V_k^T.$$

$I - P_k$ is the projection on $\mathcal{L}_k^\perp$ along $A\mathcal{K}_k$ and P_k is the projection on $A\mathcal{K}_k$ along $\mathcal{K}_k^\perp$. If $\mathcal{L}_k = A\mathcal{K}_k$, these projections are orthogonal. In this case, the orthogonality condition above is called a Galerkin condition. If $\forall k$, $n_k = n$, the dimension of the system, and $\mathcal{K}_k = \mathcal{L}_k = \text{span}(e_1, \ldots, e_n)$ where e_i is the ith column of the identity, the Gauss-Seidel method is recovered. Moreover, it is easy to see that x_k is the solution of the system $A(x_k - x_0) = P_k r_0$, that is, multiplying both sides by P_k, $P_k A x_k = P_k b$.

Let us now describe the history of some methods fitting into this general framework.

The method of Kaczmarz

On June 14, 1937, the Polish astronomer, mathematician, and geodesist Tadeusz Banachiewicz (1882-1954) presented at the Polish Academy of Sciences a note of the Polish mathematician Stefan Kaczmarz (1895-1939) (see Section 10.40) in which the author proposed an iterative method for solving systems of linear equations. Although the method attracted some interest after its publication, and was explained in some university textbooks, it almost disappeared from the research literature for some time because despite the fact that it is always converging, its convergence can be quite slow.

However, the method was rediscovered in the 1970s by Richard Gordon, Robert Bender, and Gabor Tamas Herman [1412], and applied to medical imagining. They named it ART (*Algebraic Reconstruction Technique*). This method is well suited for parallel computations and large-scale problems, since each step only requires one row of the matrix A (or several rows simultaneously in its block version), and no matrix-vector products [1410]. Thus, it immediately had, and still has, a lot of applications in various areas of mathematics as well as many practical applications in physics, in medicine, and in technical sciences, e.g., in computed tomography and in many other modern imaging technologies, including the CAT scan.

The method of Kaczmarz has been widely studied, improved, and cited. It is a method belonging to the class of *alternating projection methods* for finding a point on the intersection of closed subspaces in a Hilbert space; see the book [1104] by René Escalante and Marcos Raydan in 2011. Andrzej Cegielski, from Poland, compiled an impressive list of publications on the Kaczmarz method; see [598].

Let us now look at the original paper of Kaczmarz [1858]. He wrote the linear system as

$$a_{i1}x_1 + a_{i2}x_2 + \cdots + a_{in}x_n + b_i = 0, \quad i = 1, 2, \ldots, n.$$

He assumed, without restricting the generality, that each equation of the system has been normalized such that, for all i, $\sum_{k=1}^{n} a_{ik}^2 = 1$. For his iterative method, he started from a first approximation $x_1^0, \ldots, x_n^0$. Then, he split the following approximations into groups, each of which consisted of n approximate solutions. The solution $x_1^0, \ldots, x_n^0$ forms group zero. The n solutions of the pth group are

$$\begin{gathered} x_1^{(p,1)}, x_2^{(p,1)}, \ldots, x_n^{(p,1)} \\ \vdots \\ x_1^{(p,n)}, x_2^{(p,n)}, \ldots, x_n^{(p,n)}, \end{gathered}$$

and the $(p+1)$th group is given by

$$\begin{aligned} x_i^{(p+1,1)} &= x_i^{(p,n)} - a_{1i}L_1^{(p,n)}, \quad i = 1, \ldots, n, \\ x_i^{(p+1,2)} &= x_i^{(p+1,1)} - a_{2i}L_2^{(p+1,1)}, \quad i = 1, \ldots, n, \\ &\vdots \\ x_i^{(p+1,n)} &= x_i^{(p+1,n-1)} - a_{ni}L_n^{(p+1,n-1)}, \quad i = 1, \ldots, n, \end{aligned}$$

where $L_i^{(r,s)} = \sum_{k=1}^{n} a_{ik}\, x_k^{(r,s)} + b_i$.

Setting $y_{r,s} = \sum_{i=1}^{n} (x_i^{(r,s)} - x_i)^2$, which is a measure of the error, Kaczmarz proved that $y_{r,s+1} = y_{r,s} - (L_{s+1}^{(r,s)})^2$, and he concluded that, since the sequence $(y_{r,s})$ is monotonically decreasing, it has a limit y. Moreover, since the $y_{r,s}$ are bounded from below, a subsequence of $(x_i^{(r,s)})$ tends to a limit $y_1, \ldots, y_n$. In fact, $\lim_{r\to\infty} L_{s+1}^{(r,s)} = 0$, and it follows that $\lim_{r\to\infty} x_i^{(r,s)} = y_i = x_i$. When the system has more than one solution, he stated that $y_1, \ldots, y_n$ is one of them. However, in his proof, some points about the convergence of the subsequence are incomplete.

Kaczmarz noticed that if the a_{ik}^2 do not sum up to 1, but to c_i, all the coefficients of the ith equation and the corresponding b_i have to be divided by c_i to satisfy this condition. Remark also that $c_i = (A^T e_i, A^T e_i)$, where the e_i's are the columns of the identity matrix, and that, since b_i is in the left-hand side of the equations of the system, $L_i^{(r,s)}$ is the usual residual vector corresponding to the approximate solution $x^{(r,s)}$.

Then, Kaczmarz gave a geometric interpretation of his method. The point $x_1^0, \ldots, x_n^0$ is projected orthogonally onto the first hyperplane $L_1 = 0$. This projection is $x_1^{(1,1)}, \ldots, x_n^{(1,1)}$, which is then projected onto the hyperplane $L_2 = 0$, and so on. At the end of this step, the point $x_1^{(1,n)}, \ldots, x_n^{(1,n)}$ is projected onto $L_1 = 0$ and gives $x_1^{(2,1)}, \ldots, x_n^{(2,1)}$, and so on. The paper ends with the sentence *Die Konvergenz des Verfahrens ist geometrisch ohne weiteres einleuchtend* (the convergence of the algorithm is geometric) without further clarification.

Let us now analyze the Kaczmarz method. Obviously, he started from $x_i^{(1,1)} = x_i^0$ for $i = 1, \dots, n$. Let us denote the approximate solutions of the first group by

$$p^{(1,j)} = (x_1^{(1,j)}, x_2^{(1,j)}, \dots, x_n^{(1,j)})^T$$

for $j = 1, \dots, n$. The solutions of the second group are given, for $j = 1, \dots, n$, by

$$p^{(2,j)} = p^{(2,j-1)} - \frac{L_2^{(2,j-1)}}{c_j} A^T e_j,$$

with $p^{(2,0)} = p^{(1,n)}$ and, similarly, $L_2^{(2,0)} = L_1^{(1,n)}$. This means that $p^{(2,1)}$ could have been numbered $p^{(1,n+1)}$, that $p^{(2,2)}$ could have been numbered $p^{(1,n+2)}$, and so on for all vectors of the second group, and similarly for all groups. Thus, the procedure described by Kaczmarz is exactly what is now called the *cyclic Kaczmarz method*, where a cycle is built from n single steps in their natural order,

$$\begin{aligned} p_0 &= x_k, \\ p_j &= p_{j-1} + \frac{(b - Ap_{j-1}, e_j)}{(A^T e_j, A^T e_j)} A^T e_j = p_{j-1} + \frac{b_j - (p_{j-1}, a_j)}{\|a_j\|^2} a_j, \quad j = 1, \dots, n, \\ x_{k+1} &= p_n, \end{aligned}$$

where $a_j = A^T e_j$, i.e., the column vector formed by the jth row of A.

As stated in 2015 by Roman Sznajder in [2983], the Kaczmarz method was forgotten for more than a decade (see also [598]). It resurfaced in the publications of Bodewig [351] in 1948, Charles Brown Tompkins (1912-1971) [3061] in 1949, and Forsythe [1188] in 1953.

In 1957, the Polish mathematician Mieczysław Altman (1916-1997) (see Section 10.2) proposed a new approximate method for linear systems, which *has a very simple geometric sense* [34]. He justified his new method because

> *the geometric sense of Kaczmarz's method is directly connected with considering each separate equation of the system as a certain plane and therefore cannot be formulated in a more abstract form.*

Let X be an n-dimensional linear space with a scalar product, and A be a one-to-one linear transformation of X into itself. Altman considered the linear equation $Ax = b$. Let v_i, $i = 1, \dots, n$, be an arbitrary basis of X, and let P_i be the orthogonal projection on Av_i. Starting from $y_1 = b - P_1 b$, he defined the following infinite sequence

$$y_i = y_{i-1} - P_i y_{i-1} \text{ for } i = 2, \dots, n; \quad y_{n+i} = y_{n+i-1} - P_i y_{n+i-1} \text{ for } i = 1, \dots, n,$$

and then, for $j = 2, 3, \dots$ $y_{jn+i} = y_{jn+i-1} - P_i y_{jn+i-1}, i = 1, \dots, n$. When $X = \mathbb{R}^n$, this method is, in fact, nothing else than the cyclic Kaczmarz method.

Altman gave a complete proof of the convergence of the method, including detailed explanations on the convergence of the subsequence, which was not clearly studied by Kaczmarz. He also mentioned that if A is symmetric positive definite and if the scalar product is defined by $(A^{-1}u, v)$, his method coincides with the Gauss-Seidel method, a property that gives a proof of the convergence of that process. Altman also proved that his process for solving $Ax = b$ is the same, in the general case, as the Gauss-Seidel method applied to $A^*Ax = A^*b$.

It seems that the first matrix interpretation of the Kaczmarz method was given by the French mathematician Noël Gastinel (1925-1984) in 1966 [1290, pp. 168-169]; see also [449]. It is as follows. Let

$$\alpha_j = \frac{A^T e_j}{\|A^T e_j\|^2} = \frac{a_j}{\|a_j\|^2}, \qquad \varrho_j = b - Ap_j,$$
$$P_j = I - A\alpha_j e_j^T, \qquad Q_j = A^{-1}P_jA.$$

We have

$$Q_j = I - \frac{A^T e_j e_j^T A}{\|A^T v_j\|^2} = I - \frac{a_j a_j^T}{\|a_j\|^2} = I - \alpha_j a_j^T.$$

The matrix P_j represents the oblique projection on $e_j^\perp$ along $AA^T e_j$, while Q_j is the rank $n-1$ orthogonal projection on $(A^T e_j)^\perp$ along $A^T e_j$. Thus, for any vector y, $(P_j y, e_j) = 0$, and $(Q_j y, A^T e_j) = 0$.

Defining

$$\begin{aligned} P &= P_n \cdots P_1, \\ Q &= Q_n \cdots Q_1 = A^{-1}P_n \cdots P_1 A = A^{-1}PA, \\ r_k &= b - Ax_k, \end{aligned}$$

it holds

$$x - x_{k+1} = Q(x - x_k), \quad r_{k+1} = Pr_k,$$

and it obviously follows

$$x - x_k = Q^k(x - x_0), \quad r_k = P^k r_0.$$

Finally,

$$x_{k+1} = Qx_k + A^{-1}(I-P)b.$$

Since the Kaczmarz method finally appeared as a renumbering of the single steps after each cycle, it holds $\|x - x_k\| = \mathcal{O}(\rho(Q)^k)$. A rigorous proof of the convergence was given by Gastinel in 1966 [1290, pp. 168-169].

In 1971, Kunio Tanabe studied the behavior of the method when applied to singular systems, and proved its convergence to a limit cycle of vectors [2993]. Two heuristic techniques for accelerating the convergence of Kaczmarz's method were proposed by James S. Dyer in his thesis defended in 1965 under the guidance of Tompkins [1042]. They consisted of the following two sequence transformations

$$x_{k+1}^* = x_k - \frac{x_k^T A^T A(x_{k+1} - x_k)}{\|A(x_{k+1} - x_k)\|_2^2}(x_{k+1} - x_k),$$
$$x_{k+1}^* = x_{k+1} - \frac{(x_{k+1} - x_k, x_{k+1} - x_{k+2})}{(x_{k+2} - 2x_{k+1} + x_k, x_{k+1} - x_{k+2})}(x_{k+1} - x_{k+2}).$$

Obviously, the second method is a generalization of the well-known Aitken's Δ^2 process for accelerating the convergence of scalar sequences. For other acceleration methods, see [449] by C.B. and M.R.-Z. in 2015.

Along the years, improvements and variants of the original Kaczmarz methods were proposed. Let us mention some of them. Firstly, the method has been extended to a mapping in a Hilbert space by Stephen Fahrney McCormick in 1977 [2180]. Then, instead of projecting onto the solution set of a single equation, one can also consider multiple equations as a block and project onto their joint solution set. This was the idea of the *block Kaczmarz method*, which seems to have been first introduced by Tommy Elfving [1081] in 1980. The method was also often used for solving overdetermined consistent systems.

It was observed several times in the literature that using the rows of A in a random order, rather than in their natural order, can greatly improve the rate of convergence. This is the *randomized Kaczmarz method* in which the index i is randomly chosen at each step with a probability proportional to $\|a_i\|_2^2$. It seems that this strategy was first proposed by Boris Teodorovich Polyak [2507] in 2001 in the context of the general convex feasibility problem with a very large (or infinite) number of inequalities. This problem consists in finding a point x in the set $C = \{x \in X, f(x,\omega) \leq 0, \forall \omega \in \Omega\}$, where $X \in \mathbb{R}^n$ is a closed convex set, $f(x,\omega)$ is convex in x for all $\omega \in \Omega$, and Ω is an arbitrary, finite or infinite, set. A convergent version of the method for the infeasible case was also given in the paper.

While this randomized Kaczmarz method is quite appealing for applications, no guarantee of its rate of convergence is known so far. In 2009, Thomas Strohmer and Roman Vershynin proposed the first randomized Kaczmarz method with an exponential expected rate of convergence in [2929]. This rate depends only on a scaled particular condition number of A and not on the number of equations in the system. Moreover, it is not necessary to have access to the whole system, but only to a small random part of it. Thus, their solver outperformed all previously known methods on general extremely overdetermined systems.

In 2015, Robert Mansel Gower and Peter Richtárik developed a versatile randomized iterative method for solving a consistent system of linear equations which includes the randomized Kaczmarz algorithm as a special case [1417]. Of course, after that, randomized block versions of Kaczmarz method appeared in the literature.

The method of Cimmino

In 1938, Gianfranco Cimmino (1908-1989), an Italian mathematician working with Picone at the Istituto per le Applicazioni del Calcolo of the Consiglio Nazionale delle Ricerche (National Council of Research) in Rome, published an iterative method for solving systems of linear equations [701] which is quite close to Kaczmarz's. As pointed out by Michele Benzi in [268], the method dates back, at least, to 1932, but Cimmino, occupied by a work on the generalized Dirichlet problem, did not take the time to publish it until Picone pushed him to do so; see Section 10.13.

As Kaczmarz did, Cimmino explained that the solution x^* of the system $Ax = b$ is the intersection of the n hyperplanes of equation $(a_i, x) = 0, i = 1, \ldots, n$, where $a_i = A^T e_i$ is the vector formed with the ith row of A. Starting from an initial approximation $x^{(0)}$, he projected it on each hyperplane i (this is Kaczmarz's method) and took the symmetric (the reflection) of $x^{(0)}$ with respect to the hyperplane, that is

$$x_i^{(0)} = x^{(0)} - 2\frac{(a_i, x^{(0)}) - b_i}{\|a_i\|_2^2} a_i, \quad i = 1, \ldots, n.$$

Then, choosing n arbitrary positive quantities $m_1, \ldots, m_n$, the next iterate $x^{(1)}$ of the method is the barycenter of the $x_i^{(0)}$'s, that is

$$x^{(1)} = x^{(0)} - \frac{2}{\sum_{i=1}^n m_i} \sum_{i=1}^n m_i \frac{(a_i, x^{(0)}) - b_i}{\|a_i\|_2^2} a_i.$$

Cimmino noticed that $x^{(0)}$ and the $x_i^{(0)}$'s lie on a hypersphere whose center is the solution x^*. Then, the process is repeated and, if the system is compatible and if the rank of A is greater than 1, it converges to x^* as it was proved by Cimmino. Then, the case of incompatibility was discussed, and the method was still proved to converge. If the system is nonsingular, Cimmino showed that

$$\|x^{(k)} - x^*\|_2^2 \leq (1-\varrho)\|x^{(k)} - x^*\|_2^2 \leq (1-\varrho)^k\|x^{(0)} - x^*\|_2^2,$$

where ϱ is the smallest eigenvalue of A. The paper ends with a formal extension of the procedure to the solution of a linear integral equation of the first kind. No references were given in that paper. As reported in [268], Cimmino wrote in [702] that his method *is just a curiosity, being evidently devoid of practical utility*. In fact, the convergence is usually quite slow, slower than that of the method of Kaczmarz.

In matrix form, Cimmino's method can be written as [268]

$$x^{(k+1)} = x^{(k)} - 2A^TD^T(Ax^{(k)} - b)/m,$$

where D is the diagonal matrix with elements $\sqrt{m_i}/\|a_i\|_2$, $i = 1, \ldots, n$, and $m = m_1 + \cdots + m_n$.

At the time when Kaczmarz and Cimmino published their methods, the linear systems to be solved had only moderate dimensions and could be efficiently treated by direct methods such as those of Gauss and Cholesky. Later, other quite effective iterative methods were found, such as Lanczos' method and the conjugate gradient algorithm. Thus, Kaczmarz and Cimmino methods only received few citations in research papers, but they appeared in several university textbooks or review papers such as [353, 1735, 3030]. Later, Cimmino and Kaczmarz methods came out of oblivion because they could be easily parallelized, as well as their block versions.

Let us mention that the connections between both methods were brought to light by Rainer Ansorge in 1984, and that he also proposed a modification of the method of Cimmino which made it as interesting as Kaczmarz's, and even faster in some cases [60].

A block Cimmino algorithm with a conjugate gradient acceleration was described by Mario Arioli, Iain Spencer Duff, Joseph Noailles, and Daniel Ruiz [76] in 1992. They also discussed a method by Chandrika Kamath and Ahmed Sameh [1865].

Acceleration techniques for the block Cimmino method were further considered by Arioli, Duff, Ruiz, and Miloud Sadkane [77] in 1995.

Duff, Ronan Guivarch, Ruiz, and Mohamed Zenadi introduced in 2015 another way of accelerating the convergence of the block Cimmino method by augmenting the matrix so that the subspaces corresponding to the partitions are orthogonal [994].

The method of Altman

Let us give more details on what the Polish mathematician Altman, already introduced in the beginning of this section, did. In 1961, in a series of papers [36, 37, 38, 39, 40, 41], he proposed a method for solving a linear system $Ax = b$ with A Hermitian positive definite, and $\|b\| = 1$, a condition that does not restrict the generality.

He considered the associated problem $Ay = (Ay, b)b$, which, with the help of the orthogonal projector $P = I - bb^*$ onto the orthogonal complement of b, is equivalent to $PAy = 0$. The set of solutions of this projected system is the set of scalar multiples of $A^{-1}b$, and thus, from any non-trivial solution y, a solution of $Ax = b$ is given by $x = y/(Ay, b)$. Setting $r(y) = PAy = -P(b - Ay)$, Altman proposed two iterative methods for computing y, that is x after normalization.

The first method, presented in [40, Eqs. (5) and (6)] and further analyzed and generalized in [37], minimizes the norm of the projected residual

$$y_{k+1} = y_k + \alpha_k r(y_k), \quad \alpha_k = \arg\min\{\|r(y_k + \alpha r(y_k))\|, \alpha \in \mathbb{R}\},$$

starting from an y_0 such that $(y_0, b) \neq 0$.

In the second method, Altman considered a quadratic form [41, Eq. (7)] closely related to the quadratic form (see [2587] for further details)

$$G(y) = \left(\left(A - \frac{bb^*}{(A^{-1}b, b)} \right) y, y \right).$$

It holds $G(y) \geq 0$, and $G(y) = 0$ if and only if y is a solution of the projected system. Then, the so-called generalized steepest descent method introduced in [41, Eqs. (10) and (11)], and further analyzed in [38], is given by

$$y_{k+1} = y_k + \alpha_k r(y_k), \quad \alpha_k = \arg\min\{G(y_k + \alpha r(y_k)), \alpha \in \mathbb{R}\}.$$

Altman obtained the following error estimates ([40, Eq. (16)] and [41, Eq. (12)])

$$\frac{\|r(y_k)\|}{\|r(y_0)\|} \sqrt{\frac{G(y_k)}{G(y_0)}} \leq \frac{2}{q^k + q^{-k}} \leq 2q^k,$$

where $q = (\sqrt{\kappa(\widetilde{A})} - 1)/(\sqrt{\kappa(\widetilde{A})} + 1) < 1$ with $\kappa(\widetilde{A}) = \|\widetilde{A}\| \, \|\widetilde{A}^\dagger\|$ (where $\widetilde{A}^\dagger$ is the pseudo-inverse), and $\widetilde{A} = PAP$. Relaxation approaches for these methods were discussed in [36, 39], and additional projections on other subspaces were investigated in [37, 38].

In [2587], it was showed that both methods may be considered as restarted versions (after one iteration) of generalized Krylov methods. They are mathematically equivalent to the classical algorithms of minimal residuals (MINRES) and conjugate gradients (CG), applied to the Hermitian and semi positive definite matrix $\widetilde{A}$.

In [3030], Reginald Prabhakar Tewarson (1930-2014) mentioned a method whose starting idea is quite close to Altman's but whose iterates look like those of the Kaczmarz method. This method was described in a report written for the Raytheon company from Alexandria in Virginia, which, in 1966-1967, had a contract with the National Aeronautics and Space Administration [2541]. The system was first normalized so that for all i, $b_i = 1$. If for some i, $b_i = 0$, another row can be added to the ith one, and then divided to obtain $b_i = 1$. The row vectors of this new matrix define an $n-1$ dimensional hyperspace. The iterates of this method were

$$x_{k+1} = x_k - \frac{(x_k, x_k - A_i^T)}{\|x_k - A_i^T\|_2^2}(x_k - A_i^T),$$

where A_i is the ith row of the matrix A. In the summary of the report, it was written

> *This procedure may be considered a variant of the Kacmarz and Cimmino procedures, which has some advantages in computation characteristics and in flexibility.*

This method was named *The Hawkins universally convergent indirect method.* It was written that this method was described in the first report (CR-66-198-1) published by the Raytheon company, but it could not be found on the Internet. However, the writer of the second report called it *my method.* Was he Hawkins?

The method of Gastinel

In 1958, Gastinel [1286] proposed a method in which, at each step, the approximate solution is projected on a hyperplane parallel to a face of a convex polyhedron whose center of symmetry is the exact solution x^* of the system.

Writing each equation of the linear system $Ax + b = 0$ as $f_i(x) = 0$ for $i = 1, \ldots, n$, he defined $\Phi(x) = \sum_{i=1}^n |f_i(x)|$. The set of points x such that $\Phi(x) = \Phi(x_0)$, x_0 being the first iterate, is the boundary of a convex polyhedron of $\mathbb{R}^n$ with x^* as its center of symmetry. This boundary consists of faces, one and only one of which contains x_0. The hyperplane parallel to that face and containing x^* satisfies the equation

$$\sum_{i=1}^{n} f_i(x) \frac{|f_i(x_0)|}{f_i(x_0)} = 0.$$

Let $m(x_0)$ be the vector whose components are $|f_i(x_0)|/f_i(x_0) = \text{sign}(f_i(x_0))$. The vector $V(x_0) = A^T m(x_0)$ is orthogonal to that hyperplane. The next iterate of the process is

$$x_1 = x_0 - \frac{\Phi(x_0)}{\|V(x_0\|_2^2} V(x_0) = x_0 - \frac{\Phi(x_0)}{\|A^T m(x_0)\|_2^2} A^T m(x_0).$$

If, for some indices i, $f_i(x) = 0$, the corresponding component of $m(x_0)$ is set to 1 so that x_1 can always be computed. Then, the same process is applied again from x_1. Gastinel proved that

$$\|x^* - x_k\|_2 \leq \|x^* - x_0\|_2 (1 - \lambda_1/n\lambda_n)^{k/2},$$

where λ_1 is the smallest eigenvalue of $A^T A$, and λ_n the largest one, a result that showed that the method always converges if A is nonsingular.

This method was tested numerically on the electronic computer of the University of Grenoble in France, for systems up to the dimension 12 (remember that this was in 1958!). It was noticed that the method was very slow for the Hilbert matrix of dimension 8, an example where the Gauss-Seidel method failed to produce a valid result. For the dimension 12, five exact digits were obtained after 250 iterations.

Gastinel came back to this method, and extended it, in his doctoral thesis [1287] defended at the University of Grenoble on December 17, 1960. He named it the *norm decomposition method.*

A vector norm φ has been *decomposed* if an application $z : x \longmapsto z(x)$ in $\mathbb{R}^n$ is known such that, for all x, $\varphi(x) = (z(x), x)$. Obviously, if $\varphi(x) = \|x\|_1$, then $z(x) = m(x)$ as defined in his 1958 paper [1286] described above. Thus Gastinel proposed the iterative method

$$x_{k+1} = x_k - \frac{\varphi(r(x_k))}{\|A^T z(r(x_k))\|_2^2} A^T z(r(x_k)),$$

where $r(x_k) = Ax_k - b$. It is easy to see that this process consisted in taking $u_k = z_k$ in (5.1) with z_k such that $(z_k, r_k) = \varphi(r_k)$, where the residual vector r_k is now taken as $r_k = Ax_k - b$, that is a change of sign with respect to (5.1).

A norm decomposition is said to be *bounded* if there exists M such that $\forall r_k$, $\varphi(z_k) \leq M$. Gastinel considered the three most used vector norms. If $\varphi \equiv \|\cdot\|_2$, then $z_k = r_k/\|r_k\|_2$. Thus, $\|z_k\|_2 = 1$ and the decomposition is bounded. If $\varphi \equiv \|\cdot\|_\infty$, and if r_k^m is a component of the vector r_k such that $|r_k^m| = \|r_k\|_\infty$, then all the components of z_k are zero except the mth one that is equal to the sign of r_k^m. Thus $\|z_k\|_\infty = 1$. If $\varphi \equiv \|\cdot\|_1$, then z_k is the vector whose components are $+1$ or -1 according to the sign of the corresponding component of r_k. Consequently, $\|z_k\|_1 = n$, the dimension of the system.

Gastinel then proved that if the norm decomposition is bounded, the method converges to the solution of the system for all x_0, and that

$$\|x_k - x\|_2^2 \leq \|x_0 - x\|_2^2 \left(1 - \frac{1}{L^2 \kappa^2(A)}\right)^k,$$

where $\kappa(A) = S_{2\varphi}(A)S_{2\varphi}(A^{-1})$ with $S_{2\varphi}(B) = \max_{u\neq 0}\frac{\|Bu\|_2}{\varphi(u)}$, and L is independent of A, but depends on the chosen decomposition. When $\varphi \equiv \|\cdot\|_2$, one has

$$x_{k+1} = x_k + \frac{\|r_k\|_2^2}{\|A^T r_k\|_2^2} A^T r_k,$$

which shows that if A is symmetric, the steepest descent method is recovered.

Finally, he generalized his method by replacing the orthogonal projections by others parallel to suitably chosen directions, but he did not pursue in this way because of the difficulty of this choice.

Gastinel generalized again his method in 1963, thus leading to the concept of *over-decomposition* [1288]. Let x be an approximation of the exact solution, here denoted by Ω, and let x' be the new approximation given by $x' = x + A^T H_x \Lambda$, where $H_x \in \mathbb{R}^{n\times q}$, $1 \le q \le n$, and where $\Lambda \in \mathbb{R}^q$ is solution of the system $H_x^T AA^T H_x \Lambda = H_x^T(b - Ax)$. The vectors $x - x'$ and $\Omega - x'$ are orthogonal and, thus, $\|\Omega - x'\|_2^2 = \|\Omega - x\|_2^2(1 - L_1)$ with $L_1 = \|x - x'\|_2^2/\|\Omega - x\|_2^2 \le 1$.

For all x, the matrix H_x is defined by

$$x^T H_x (H_x^T AA^T H_x)^{-1} H_x^T x \,\mathrm{tr}(H_x^T AA^T H_x) = \varphi^2(x).$$

The application $x \longmapsto H_x$ is the over-decomposition of order q of the norm φ. It is said to be *bounded* if its elements are bounded. There exist an infinite number of such over-decompositions. The matrix H_x can be chosen such that the over-decomposition is bounded, and its columns are conjugate directions with respect to $x^T AA^T x$. This case was further discussed and an example was given.

The convergence was then discussed in more detail, and the paper ends with a second example which shows that less iterations are needed than with $q = 1$. For more on Gastinel's projection method, see also [1293].

5.5 ▪ Asynchronous methods

Asynchronous iterative methods for solving linear systems were introduced by Daniel Chazan and Willard L. Miranker (1932-2011) [649] in 1969 under the name *chaotic relaxation*. They referred to "free steering" methods by Ostrowski [2399] in 1956 and Samuel Schechter [2694] in 1959, as well as experiments reported by J.L. Rosenfeld [2599] in 1967. For solving $Ax = d$, written as $x = Bx + Cd$, a chaotic iterative scheme is $x_i^{(j+1)} = x_i^{(j)}$ if $i \neq k_{n+1}(j)$ and

$$x_i^{(j+1)} = \sum_{\alpha=1}^{n} b_\alpha^i x_\alpha^{j-k_\alpha(j)} + (c^i, d)$$

otherwise, where the upper indices i refer to the rows of B and C. The iteration is defined by B, C, and S, which is a sequence of vectors $k_1(j), \dots, k_{n+1}(j)$ for $j = 1, \dots$ satisfying

$$0 \le k_i(j) \le s,\ i = 1, \dots, n,\ j = 1, \dots, \quad 1 \le k_{n+1}(j) \le n,\ j = 1, \dots$$

for a given integer s and $k_{n+1}(j) = i$ infinitely often for each i, $s \le i \le n$. The $k_{n+1}(j)$th component is updated while the remaining components are unchanged. Every component is updated infinitely often and no update uses a value which was produced s or more steps previously.

The standard Gauss-Seidel iteration is recovered by taking $k_1(j) = \cdots = k_n(j) = 0$, $k_{n+1}(j) = (j-1)(\mathrm{mod}\ (n))$, and $s = 1$. Chazan and Miranker introduced periodic schemes in which $k_{n+1}(j)$ is a periodic function of j. They gave a convergence theorem for chaotic relaxation, the method converges if there exist a vector v such that $|B| \leq \alpha v$ with $\alpha < 1$. This is satisfied if the spectral radius of $|B|$ is strictly less than 1. If no such vector v exists, there exist sequences S for which the method diverges.

This purely theoretical paper was highly cited; see also [2252] by Miranker in 1971. The Chazan and Miranker results were extended to nonlinear problems by Jean-Claude Miellou [2243] in 1975; see also [2581] by François Robert, Michel Charnay, and François Musy that same year and [2313] by Musy and Charnay in 1974.

Some of the results of Chazan and Miranker for the periodic case were generalized for positive definite matrices by J.D.P. Donnelly [934] in 1971. He named the method he studied *delayed relaxation*.

In the Chazan and Miranker chaotic relaxation, for doing the evaluation of the ith iterate, a process cannot make use of any value of the components of the jth iterate if $j < i - s$. This may cause problems if the results of a process are too much delayed. Therefore, a synchronization is required to avoid this problem. A class of asynchronous methods without this constraint was introduced by Gérard M. Baudet [214] in 1978. The previous methods defined in [649, 2243, 2581] were special cases of Baudet's method. He proved a convergence theorem for a fixed point equation $x = F(x)$, where F is a contracting operator. He gave the results of numerical experiments on *C.mmp* (CMU multiminiprocessor), a parallel computer at Carnegie Mellon University for solving linear systems arising from Poisson's equation in the unit square with asynchronous Jacobi and Gauss-Seidel methods.

Asynchronous iterative methods were studied in the 1980s and 1990s, mainly for nonlinear fixed point equations whose linear systems are a special case. Three different techniques were used to analyze convergence: contracting operators, partial orderings, and nested subspaces. Let us cite a few papers and a book from the 1980s.

Some convergence results were given by Mouhamed Nabih El Tarazi in 1982 [1078], methods for fixed point problems were described by Dimitri Panteli Bertsekas in 1983 [300], asynchronous relaxations for parallel processors were proposed by Debasis Mitra in 1987 [2257], models of parallel chaotic iterations were the subject of Rafael Bru, Ludwig Elsner, and Michael Neumann (1946-2011) in 1988 [480], parallel methods were described by Bertsekas and John Nikólaos Tsitsiklis in 1989 [301], sufficient conditions for the convergence of asynchronous iterations were given by Aydin Üresin and Michel Dubois in 1989 [3092], and a comparison of Jacobi and Gauss-Seidel parallel iterations was done by Tsitsiklis in 1989 [3075].

In the 1990s and the beginning of the 21st century, there were still some papers devoted to convergence, but asynchronous methods were also used combined with domain decomposition techniques. A method of performance evaluation for subdomain decomposition parallel algorithms of three-dimensional elliptic problems was proposed by Miellou, Pierre Spiteri, and G.R. Perrin in 1986 [2245], contractive mappings with maximum norms were studied by Jacques Bahi and Miellou in 1993 [153], asynchronous two-stage iterative methods were described by Andreas Frommer and Daniel Benjamin Szyld in 1994 [1265], effects of asynchronism on the convergence rate of iterative algorithms was the topic of [3093] by A. Üresin and M. Dubois in 1996. Asynchronous weighted additive Schwarz methods were considered by Frommer and Szyld in 1997 [1265]; see also [1263] with H. Schwandt in 1997 and different models of parallel asynchronous iterations with overlapping blocks [2984] by Szyld in 1998. Convergence of asynchronous linear iterations was also studied by Yangfeng Su, Amit Bhaya, Eugenius Kaszkurewicz, and Victor S. Kozyakin in 1998 [2939]. Implementation of asynchronous subdomain methods was discussed by R. Guivarch and Spiteri in 1998 [1478],

Other interesting papers were [1266] by Frommer and Szyld in 2000, [1077] by Didier El Baz, Spiteri, Miellou, and Mohamed Jarraya in 2002. A probabilistic analysis of asynchronous iteration was done by John Charles Strikwerda in 2002 [2928].

A difficult problem in the asynchronous methods we have seen before is how to stop the iterations on parallel computers. Doing that straightforwardly may reintroduce synchronizations in a purely parallel algorithm or it can lead to an early stop, if only local stopping criterion are used. For discussions of this problem, see [898] by Edsger Wybe Dijkstra (1930-2002) and Carel S. Scholten (1925-2009), [636] by Kanianthra Mani Chandy and Leslie Lamport in 1985, [301] by Bertsekas and Tsitsiklis in 1989, [2691] by Serap Ayşe Savari and Bertsekas in 1996, [152] by Bahi, Sylvain Contassot-Vivier, and Raphaël Couturier in 2007, and [2244] by Miellou, Spiteri, and El Baz in 2008.

There was renewed interest in asynchronous methods after 2010 because of the availability of parallel computers with a very large number of processing elements. Let us cite [728] by Tijmen Pieter Collignon and Martin Bastiaan van Gijzen in 2010, [1932] by Giorgios Kollias, Ananth Grama, and Z. Li in 2011, [63] by Hartwig Anzt, Stanimire Tomov, Jack J. Dongarra, and Vincent Heuveline in 2013, [142] by Marc Baboulin, Dongarra, Julien Herrmann, and S. Tomov in 2013, and [2540] by A.S. Rasulov, M.T. Bakoev, and D.R. Akabirhodjaeva in 2014.

Other recent papers studying convergence or discussing implementations are [305] by Iain Bethune, J. Mark Bull, Nicholas J. Dingle, and Nicholas John Higham in 2014, [114] by Haim Avron, Alex Druinsky, and Anshul Gupta in 2015, [3268] by Jordi Wolfson-Pou and Edmond Ten-Fu Chow in 2016 (see also [3269] in 2019), [1728] by James Hook and N.J. Dingle in 2018, [2116] by Frédéric Magoulès and Cédric Venet in 2018, the Ph.D. thesis [721]s of Evan Coleman in 2019, [1815] by Erik J. Jensen, Coleman, and Masha Sosonkina in 2019, and [722] by Coleman, Jensen, and Sosonkina in 2021.

However, the asynchronous methods we have discussed in this section are parallel versions of classical iterative methods like Jacobi and Gauss-Seidel. Even though they can be used as preconditioners in other methods, if using them asynchronously can reduce the computing time, this cannot improve much their convergence, which is rather slow in many cases. This is why some scholars tried to introduce parallelism in Krylov methods; see Sections 5.7 and 5.8.

5.6 ▪ The Institute for Numerical Analysis

What was probably the first applied mathematics institute, the Istituto per le Applicazioni del Calcolo was founded in Italy, as we wrote above, by Picone in 1927. Later on, similar institutes were founded in other countries. In this section, we review the history of the Institute for Numerical Analysis of the National Bureau of Standards of the USA, which played an important role in the early development of numerical linear algebra.

The National Bureau of Standards (NBS) was founded in 1901 to provide standard weights and measures and to be a physical laboratory. NBS had been responsible for the Mathematical Tables Project of the Works Progress Administration since 1938. The group of people involved in this project headed by Arnold Noah Lowan (1898-1962) included Milton Abramowitz (1915-1958), Ida Rhodes (1900-1986), Gertrude Blanch (1897-1996), Herbert Ellis Salzer (1915-2006), and Irene Ann Stegun (1919-2008). The computations were made by a large staff using hand-desk mechanical calculators.

From 1945 to 1951, the Director of the NBS was Edward Uhler Condon (1902-1974). John Hamilton Curtiss (1909-1977) joined the NBS in 1946 as an assistant to the director. He was initially in charge of statistical problems. In 1946, Condon instructed Curtiss to assess the federal needs for computers and for a national computing center. This investigation had its source at the Office of Naval Research (ONR). The ONR was established by the US Navy. In 1946, the Mathematics Branch of ONR was formed with Mina Rees (1902-1997) as its head. It was suggested that NBS and ONR should jointly establish a computing center, to develop and use computers.

Curtiss realized early that the mathematics needed to exploit the new computers also had to be developed. It had been proposed to create four sections: the Institute for Numerical Analysis (INA), a Computation Laboratory (CL), a Statistical Engineering Laboratory (SEL), and a Machine Development Laboratory (MDL). Rees and Curtiss were largely responsible for the formation of INA. In the fall of 1946, Curtiss came to the west coast on a mission for Condon to seek a home for the proposed INA. It was decided to locate the Institute for Numerical Analysis at UCLA, the University of California in Los Angeles. However, INA was temporarily housed at NBS in Washington, DC.

On July 1, 1947, Curtiss was appointed chief of a new division of the NBS called the National Applied Mathematics Laboratories (NAML), also known as Division 11 of the NBS. Later this name was changed to the Applied Mathematics Division (AMD). Curtiss remained at the NBS until mid-1953, except for a leave of one semester in 1952.

The main goal of the INA was to plan and conduct a program of research in pure and applied mathematics directed at developing methods of analysis which will allow the most efficient and general use of high-speed automatic digital computers. The INA was organized on a project basis, with a manager for each project.

Early in 1948, after some problems with computer procurements, a proposal was made for a small machine to be developed at INA by Harry Douglas Huskey (1916-2017). This led to the *Standards Western Automatic Computer* (SWAC) [1769]. This project began in January 1949, and the first Williams tube machine to be completed in the USA was dedicated on April 7, 1950. There was a rather long period of debugging, but the SWAC finally became a reliable machine with many significant accomplishments. When the INA operation closed in 1954, the SWAC was transferred to UCLA and remained in operation until 1967.

The first appointees to INA were Olga Taussky, her husband John Todd (1911-2007), and Albert Cahn (1910-1978). The first two projects were entitled *Characteristic roots of matrices* and *Applications of automatic digital computing machines in algebra and number theory*. The first report about the activities of the NAML was issued in January 1948. Otto Szász (1884-1952) joined the INA in February 1948. The new building for the INA at UCLA was occupied in April 1948. Blanch organized the computational unit. The period April-June 1948 was an organizational period in which the scientific and computational staffs were hired. The table-making tradition of the NBS was also continued at INA.

Douglas Rayner Hartree (1897-1958), a mathematical physicist from Cambridge, UK, was invited to be the first Director of INA. But, he stayed only a short time. After Hartree's leave, Curtiss was acting chief of INA until the fall of 1949. Forsythe was the first regular member of research staff of INA, Lanczos became the second regular member and Wolfgang Richard Wasow (1909-1993) the third regular member of the research staff in 1949. Magnus Rudolph Hestenes (1906-1991), a professor at UCLA, was also strongly involved in the work at INA. He was Assistant Director of Research from mid-1950 and UCLA Liaison Officer.

Several symposia were organized in 1948, with lectures by John von Neumann (1903-1957), Solomon Lefschetz (1884-1972), and George Bernard Dantzig (1914-2005). Taussky and Todd returned to the University of London. A year later they joined the staff of the NAML at NBS in Washington, DC. Todd became Chief of the Computation Laboratory there and Taussky became a mathematics consultant in the Division. They made frequent visits to the INA.

Curtiss was a strong proponent of Monte Carlo methods and he initiated a program of studies of these methods at INA. William Edmund Milne (1905-1981), a numerical analyst from the University of Oregon, undertook the preparation of a monograph on numerical solutions of differential equations. Ostrowski of the University of Basel (Switzerland) became a frequent visitor to the INA. He made several contributions to the Gerschgorin theory and he obtained bounds for eigenvalues of matrices. As we have seen above, he was also concerned with iterative methods for the solution of linear systems.

In April-June 1949, among others, there were Forsythe, Hestenes, Marc Kac (1914-1984), Lanczos, Ostrowski, Marvin Leonard Stein (1924-2015), not to be confused with P. Stein of the Stein-Rosenberg theorem, Szász, Huskey, and Blanch on the staff.

A conference in June 1949 led to the establishment of the *Pacific Journal of Mathematics* under the sponsorship of some west coast universities and INA. Another journal in which NBS took an important part was *Mathematical Tables and other Aids to Computation*, established in 1943. In 1960, its name was changed to *Mathematics of Computation*. The NBS also had its own *Journal of Research*.

In July-September 1949, the staff included Lars Ahlfors (1907-1996) as a consultant, Forsythe, Hestenes, Kac, William Karush (1917-1997) (this is one of the K's of the Karush-Kuhn-Tucker condition; see [752, 1906]), Lanczos, Ostrowski as a consultant, Szász, Wasow, Huskey, and Blanch.

In early September 1949, John Barkley Rosser (1907-1989) became Director. He served for only one year but had a great impact on INA.

Rosser organized a weekly seminar concerned primarily with iterative methods for solving systems of linear equations and the computation of eigenvalues. Forsythe undertook the task of classifying the various known methods for solving linear systems. Hestenes, Karush, and M.L. Stein were responsible for the study of methods for computing eigenvalues with particular emphasis on gradient methods, power methods, and inverse power methods. Lanczos worked on his orthogonalization techniques for computing eigenvalues and for solving linear systems. He also worked on Chebyshev polynomials.

In 1950, Rosser constructed a symmetric matrix of order 8 with integer entries (which is available in MATLAB as `rosser`). Mathematically, this matrix is singular with a zero eigenvalue, another one negative with a large modulus and five eigenvalues close to 1000. It was used to test the methods devised by Lanczos and by Hestenes and Karush. It took about 11 hours to solve the problem on the CPC, the *Card-Programmed Calculator* of IBM, by the gradient method developed by Hestenes and Karush. Forsythe and Rosser proposed improved methods for generating random numbers, Lanczos devised a new method for inverting Laplace transforms and Rosser and Szász studied methods for improving rates of convergence of series.

Rosser returned to Cornell University in September 1950 and F. John, an expert in partial differential equations, became the Director of INA for one year. The dedication of the SWAC occurred on August 17, 1950, followed by a symposium on applications of digital computing machines to scientific problems.

In July-September 1950, the staff included Paul Erdös (1913-1996) as a consultant, the famous physicist Richard Phillips Feynman (1918-1988) as an associate, Forsythe, John, Kac, Karush, Lanczos, Edward James McShane (1904-1989) as an associate, Theodore Samuel Motzkin (1908-1970), Milne, Szász, Wasow, Huskey, and Blanch. In October-December 1950, Huskey and Blanch became assistants to the Director. Forsythe, Karush, Lanczos, Milne, Motzkin, and Wasow were still on the staff.

The year 1951 was to be an important one for the INA. In January 1951, Lowell J. Paige (1919-2010), an algebraist from UCLA, joined the INA. Forsythe made numerical experiments on various schemes for solving systems of linear equations. He found that for ill-conditioned matrices, the optimal gradient method tended to have problems and that an acceleration method was needed. Forsythe and Motzkin suggested one acceleration scheme and Hestenes suggested a simpler one. These acceleration schemes improved the results of the gradient method as well as those of other iterative methods. These experiments led to the formulation of the conjugate gradient method by Hestenes. Motzkin proposed new methods for solving systems of inequalities.

Rosser returned for the summer of 1951 to work on solving linear systems and on the computation of eigenvalues. In that summer, Paul Sumner Dwyer (1901-1982) was at INA as well as Herman Heine Goldstine (1913-2004) and Stiefel from Switzerland. Urs W. Hochstrasser, a Ph.D. student of Stiefel, was at INA as a graduate fellow.

Late in June or early in July, Hestenes devised the conjugate gradient (CG) method for solving a linear system. Theoretically, this method delivers the solution in at most n iterations for a matrix of order n. He gave three versions of his routine. According to Hestenes and Todd [1665],

> *When Stiefel arrived at INA from Switzerland, the librarian gave him a paper describing this routine. Shortly thereafter Stiefel came to Hestenes' office with this paper in hand and said, "Look! This is my talk". It turned out that Stiefel had invented the same algorithm from a different point of view. He looked upon it as a relaxation routine whereas Hestenes viewed it as a gradient routine on conjugate subspaces. The term "conjugate gradient" was coined by Hestenes.*

Stiefel was given a joint appointment with INA and UCLA so that he and Hestenes could write an extensive paper on the conjugate gradient method. Numerical experiments were carried out by Robert Mayo Hayes, Hochstrasser, M.L. Stein, and W. Wilson, who was a member of the computational staff.

Credit should also be given to Rosser, Forsythe, Karush, and L.J. Paige for the development of the conjugate gradient routine because the routine was also an outgrowth of their efforts. Rosser and Stiefel presented the conjugate gradient method at the August 23-25, 1951, *Symposium on Simultaneous Linear Equations and the Determination of Eigenvalues*, which was a part of the Semicentennial Celebration of NBS. The proceedings were edited by L.J. Paige and Taussky [2432]. Two other special publications are [2326, 3006] in 1954 and 1956.

At the same time, Lanczos developed his method of *minimized iterations*. Variations of standard gradient methods with or without accelerations were studied experimentally and theoretically by Forsythe and M.L. Stein. Forsythe started a study of Soviet mathematical progresses which led to the publication of a bibliography of Soviet mathematical books.

In April-June 1951, in addition to the personnel cited above, Shmuel Agmon was at INA as an associate as well as Johannes van der Corput (1890-1975) from the Netherlands. At the beginning of 1952, Derrick Henry Lehmer (1905-1991), initially an expert in number theory, became the Director of INA.

In 1952, Lanczos devised some routines for using the Chebyshev polynomials for solving linear systems. Numerical experimentation on solving linear systems and computing eigenvalues was continued on the SWAC. Tompkins became a regular member of the INA staff. During the fiscal year 1953, Huskey and Lanczos were on leave of absence.

In 1953, methods of preconditioning a symmetric matrix by pre- and post-multiplication by a diagonal matrix were studied by Forsythe and Ernst Gabor Straus (1922-1983) of UCLA. Characterizations of "best conditioned" matrices were obtained. They observed that preconditioning tended to reduce roundoff errors in matrix computations.

In the third quarter of 1953, Tompkins was the acting Director of INA. This period saw the beginning of the troubles because of problems at the NBS linked to the so-called "battery additive scandal." NBS was involved in this from 1948 to 1953. From the early 1920s the NBS had tested a great many commercial battery additives. They never found any to be beneficial. The owner of a small company in California proclaimed that his product, the AD-X2, had a special composition and beneficial effect. This company encouraged satisfied users to flood Congress and the NBS with testimonial letters and they also did some lobbying at the Congress. Some more tests were done at NBS with the same results as before. Due to the pressure put on the federal administration, Allen Varley Astin (1904-1984), who was the Director of NBS since 1951, was asked to resign by the Assistant Secretary of Commerce for Domestic Affairs. All these difficulties created unfortunate consequences for the INA. Because of these problems, the Department of Defense (DoD) stated new rules which made difficult the funding arrangements with the Office of Naval Research. After some committee reports, on August 21, 1953, A.V. Astin was reinstated as Director, but four sections of the NBS were transferred to the DoD. One investigation committee praised the AMD activities, but it questioned the appropriateness of the INA as part of the NBS. After a while it was decided that the NBS had to give up the sponsorship of INA on June 30, 1954.

It was not possible for UCLA to take over all sections of INA, but the university agreed to take over the administration of the research group, the SWAC computer and its maintenance, as well as the library. By the end of June 1954, several members of INA had accepted positions in industry and in various departments of universities.

After his leave of absence, during which he worked for the North American Aviation Company, Lanczos became professor in the School of Theoretical Physics of the Dublin Institute for Advanced Studies in Ireland. Huskey accepted a faculty position at University of California at Berkeley. Forsythe remained with UCLA for a while before going to Stanford University. Wasow remained also with UCLA before joining the University of Wisconsin. Motzkin became a member of the Department of Mathematics at UCLA.

At 12 o'clock on Wednesday, June 30, 1954, INA held a "Gala Party and Luncheon" in honor of the transfer of INA to UCLA, the people leaving INA, and the persons staying at UCLA, and that was the end of an Institute that had produced many interesting results during its brief existence. It was a pity, and probably bad for US research, that such an interesting place was shut down because of administrative problems.

The main sources of information about the history of INA are the report written in 1991 by Hestenes and Todd [1665] and the quarterly reports of activity of the NAML and AMD. Unfortunately, not all the issues are easily available; see also the papers by Dianne Prost O'Leary [2369, 2370, 2371], her joint paper with Golub [1389] and the papers [3051, 3055] by Todd about the NBS. For the history of NBS, see also [355].

5.7 ▪ CG, Lanczos, and other methods

The conjugate gradient method is now the method of choice (particularly in its preconditioned form) for solving linear systems with a symmetric positive definite matrix. As we have seen in Section 5.6, it was developed independently by Hestenes in the USA and Stiefel in Switzerland at the beginning of the 1950s. After meeting at the INA in 1951, they realized that their algorithms were the same, and they wrote a famous and seminal joint paper [1664] which was published in the December 1952 issue of the *Journal of the National Bureau of Standards*. In [1664], page 409, one can read

> *The method of Conjugate Gradients was developed independently by E. Stiefel of the Institute of Applied Mathematics at Zürich and by M.R. Hestenes with the cooperation of J.B. Rosser, G. Forsythe and L. Paige of the Institute for Numerical Analysis, National Bureau of Standards. The present account was prepared jointly by M.R. Hestenes and E. Stiefel during the latter's stay at the National Bureau of Standards. The first papers on this method were given by E. Stiefel [1952] and by M.R. Hestenes [1951]. Reports on this method were given by E. Stiefel and J.B. Rosser at a symposium on August 23-25, 1951. Recently, C. Lanczos [1952] developed a closely related routine based on his earlier paper on eigenvalue problems [1950]. Examples and numerical tests of the method have been by R. Hayes, U. Hoschstrasser and M. Stein.*

Stiefel's paper cited in this quote is [2891], whose title can be translated as *On some relaxation methods*. It was published in 1952 but submitted in mid-July 1951. Stiefel saw relaxation methods in a broad sense as "step-by-step procedures" based on minimizing quadratic forms. He considered methods written as

$$x_{k+1} = x_k + \lambda_k p_k,$$

where p_k are descent directions that can be chosen as we wish. Stiefel reviewed the classical methods, steepest descent and introduced CG where the descent directions are constructed to be mutually A-orthogonal. He gave some numerical examples that were run on Zuse's *Z4* computer installed at the ETH in Zürich (see Chapter 7). The largest example, arising from a problem in structural analysis, the computation of an Airy stress function for the profile of a dam, had 106 equations. Stiefel wrote (our translation)

> *90 computing cycles were carried out. After the ninetieth step, the residuals were so small that the rounding errors began to predominate and could therefore be aborted. The final residuals were approximately the* 10^6*th part of the initial residuals. Due to the limited storage capacity one cycle of the machine took about 2h 20 m.*

He also considered error estimates. When describing CG that he called the n-*Schritt-Verfahren* (the n-step process), he added in a footnote (our translation),

> *After writing the present work, on the occasion of a visit to the Institute for Numerical Analysis (University of California), I discovered that this method had also been developed a little later by a group from this institute. A preliminary internal report for the National Bureau of Standards was prepared by M.R. Hestenes in August 1951 (N.A.M.L. Report, 52-9).*

As Stiefel mentioned, Hestenes' work referenced in [1664] is an NBS report. Hestenes considered CG from the point of view of variational theory and optimization. In that report, Hestenes wrote

> *The methods presented are an outgrowth of discussions with Forsythe, Lanczos, Paige, Rosser, Stein, and others. For the positive Hermitian case it is convenient to separate the methods into two types. The first is a method which is my interpretation of the suggestions made by Forsythe and Rosser. The second is one which grew out of my discussion of the problem with Paige.*

The first method was the three-term recurrence for CG and the second one, the now standard form of CG with coupled two-term recurrences. About the discovery of CG, Hestenes wrote much later in [1662],

> *In June or July 1951, after almost two years of studying algorithms for solving systems of linear equations, we finally hit upon a conjugate-gradient method. I had the privilege of first formulation of this new method. However, it was an outgrowth of my discussions with my colleagues at INA. In particular, my conversations with George Forsythe had a great influence on me. During the month of July 1951, I wrote an INA report on this new development.*

As we already mentioned in Section 5.6, lectures on the method were given by Stiefel and Rosser at the *Symposium on Simultaneous Linear Equations and the Determination of Eigenvalues* held August 23-25, 1951, at INA; see [2432]. Stiefel's lecture was almost taken from his paper [2891] which was only submitted at the time, but not yet published. In [2432], Rosser started by writing

In the fall of 1949 the Institute of Numerical Analysis of the National Bureau of Standards undertook an extended study of methods of solving simultaneous linear equations. Through the use of colloquia and discussion groups, nearly all scientific members of the Institute have made some sort of contribution to the problem. Accordingly, it is impossible to assign complete credit for the results disclosed herein to a single person or a few persons. However, certain members of the staff have given concentrated attention to the problem over an extended period and are primarily responsible for the results acted herein. In alphabetical order, these are G.E. Forsythe, M.R. Hestenes, C. Lanczos, T. Motzkin, L.J. Paige, and J.B. Rosser.

Then, he considered the problem of finding the center of an ellipsoid in n dimensions. A first approach was geometrical but this led to an algebraic problem. Let c be a vector and C be a symmetric matrix. Ellipsoids with center c are given by $(x-c)^T C(x-c) = \phi$ with ϕ given. The method used direction vectors z_i which are C-orthogonal $z_i^T C z_j = 0$, $i \neq j$. Rosser observed that they can be generated by a Gram-Schmidt algorithm, but he preferred using a recurrence relation proposed by Lanczos,

$$z_{i+1} = Cz_i - \alpha_i z_i - \beta_i z_{i-1}.$$

When solving $Ax = k$, $(Ax-k)^T H(Ax-k) = \phi$ defines a family of ellipses with center $A^{-1}k$. If A is symmetric positive definite, choosing $H = A^{-1}$ gives $C = A$ and the z_i's can be generated by a three-term recurrence. The iterates are given by $x_i = x_{i-1} + a_i z_i$ and the z_i's can be computed as $z_{i+1} = r_i + b_i z_i$ with $r_i = k - Ax_i$. The residual vectors are orthogonal and the direction vectors z_i A-orthogonal. Hence, the center of the ellipsoids must be found in at most n steps. Rosser observed that

It is interesting that the iteration outlined above is identical with that derived by Stiefel from an entirely different approach.

He also said that

If at the end of n steps, the r_n is not sufficiently small (either from round off or from mistakes), there is nothing whatever to bring the process to a termination. One can simply let the procedure continue until r_m is sufficiently small, even if this may in some cases require a value of m greater than n.

The proceedings of this symposium were only published in 1953 [2432] after the paper of Hestenes and Stiefel has appeared. In their 28-page paper, Hestenes and Stiefel studied many aspects of CG. Let us briefly review its contents. Hestenes and Stiefel observed at the outset something that was unfortunately forgotten by most of their followers:

At each step the residual $r_i = k - Ax_i$, is computed. Normally this vector can be used as a measure of the "goodness" of the estimate x_i; However, this measure is not a reliable one because, as will be seen in section 18, it is possible to construct cases in which the squared residual $|r_i|^2$ increases at each step (except for the last) while the length of the error vector $h - x_i$ decreases monotonically.

The vector h was the exact solution of the problem and the vertical bars denote a norm. Then, they gave the formulas we use today for CG which are (in their notation) with $p_0 = k - Ax_0$,

$$a_i = \frac{\|r_i\|^2}{(p_i, Ap_i)}, \quad x_{i+1} = x_i + a_i p_i, \quad r_{i+1} = r_i - a_i A p_i,$$

$$b_i = \frac{\|r_{i+1}\|^2}{\|r_i\|^2}, \quad p_{i+1} = r_{i+1} + b_i p_i,$$

and also alternative possibilities for computing the two scalar coefficients. They saw CG as a special case of the method of conjugate directions, observing that this method had also been considered by Fox, Huskey, and James Hardy Wilkinson (1919-1986) [1206] in 1948. The direction vectors denoted as p_i were selected by Hestenes and Stiefel to be mutually conjugate but had no further restrictions. If the p_i's are obtained by A-orthogonalization of the columns of the identity matrix, the method is essentially Gaussian elimination.

Hestenes and Stiefel introduced the *error function* $f(x) = (h - x, A(h - x))$ (which is the square of the A-norm of the error) and showed that x_i minimizes $f(x)$ on the line $x = x_{i-1} + \alpha p_{i-1}$, where α is a scalar, and the $f(x_i)$ are strictly decreasing since A is positive definite. In Section 5, they derived the basic properties of CG, in particular, the relation of p_i with all the previous residual vectors, the orthogonality of the residuals, the A-orthogonality of the descent directions p_i and the fact that the p_i's, as well as the r_i's, also satisfy three-term recurrences. They also showed that the ℓ_2 norm of the error is decreasing. In Section 7, they showed how to compute new approximations $\bar{x}_i$, from the x_i's, having residuals $\bar{r}_i$ whose ℓ_2 norms are decreasing, but the A-norm of the error for $\bar{x}_i$ is larger than for x_i.

In Section 8, they observed that the orthogonality of the residuals and the A-orthogonality of the descent directions can be lost because of the rounding errors on a digital computer. They gave formulas that can be used to check if the computation had gone wrong. They stated informally that the larger the ratios a_i/a_{i-1}, the more rapidly the rounding errors accumulate. This ratio is bounded above by the condition number of A. They proposed modifications of the computation of the coefficients to improve the local orthogonality. To do this one more dot product has to be computed at each iteration. They derived several variants of CG in which the vectors p_i are scaled in different ways.

In Section 10, they considered the case A singular for which a least squares solution is obtained and the case of a general nonsymmetric matrix A. They wrote the algorithm for solving $BAx = Bk$ with BA symmetric and positive definite. An obvious choice was $B = A^T$. The algorithm obtained by applying CG to $A^T Ax = A^T b$ was later named CGNR.

In Sections 14-16, the relations of CG with the theory of orthogonal polynomials and to continued fraction expansions are studied. Hestenes and Stiefel showed that CG implicitly computes orthogonal polynomials corresponding to a dot product defined by a Riemann-Stieltjes integral for a piecewise constant measure m having jumps at the eigenvalues of A. In fact, $r_i = R_i(A)r_0$ and $p_i = P_i(A)r_0$ and these polynomials satisfy

$$\int_a^b R_i(\lambda)R_k(\lambda)\, dm(\lambda) = 0, \quad \int_a^b P_i(\lambda)P_k(\lambda)\, \lambda\, dm(\lambda) = 0,\ i \neq k,$$

where a and b are lower and upper bounds of the smallest and largest eigenvalues of A. The CG coefficients can be calculated as ratios of Riemann-Stieltjes integrals. Unfortunately, the measure m is generally unknown since it depends on the eigenvalues. Using the relation with continued fractions, Hestenes and Stiefel proved that given any two sequences of positive numbers $a_0, \dots, a_{n-1}$ and $b_0, \dots, b_{n-1}$, there exist a symmetric positive definite matrix A and a vector r_0 such that CG applied to this data yields the given numbers a_i and b_i as coefficients.

Numerical experiments are described in Section 19 with small matrices of order at most 6. We observe that Hestenes and Stiefel considered CG as an iterative method but they did not give any bounds for the residual or error norms.

After spending some time at the Boeing Aircraft Company in Seattle in 1946, in January 1949 Lanczos joined the Institute for Numerical Analysis of the National Bureau of Standards in Los Angeles. Lanczos turned his attention to the solution of linear systems and matrix eigenvalue problems in which he had been interested when working in physics and at Boeing. He studied his method of *minimized iterations*, a continuation of some work he began at Boeing. In the paper

[1984] published in 1950, he proposed constructing an orthogonal basis of what we now call a Krylov subspace. He used the algorithm on practical problems: the lateral vibration of a bar, the vibrations of a membrane, and a string.

In the paper [1985] dated September 1951 but published in 1952, he discussed the solution of linear systems. Lanczos started with a general matrix A, but quickly specialized to symmetric positive definite matrices. However, for symmetric matrices, the Lanczos algorithm is related to the Stieltjes procedure that was presented in 1884 by Stieltjes [2898] to compute the coefficients of the three-term recurrence satisfied by orthogonal polynomials with respect to a dot product defined by a Riemann-Stieltjes integral; for more details see [1386].

Lanczos started by refereing to his 1950 paper [1984]:

> *It was pointed out in [14, p. 256] that the inversion of a matrix, and thus the solution of simultaneous systems of linear equations, is contained in the general procedure as a special case. However, in view of the great importance associated with the solution of large systems of linear equations, this problem deserved more than passing attention.*

He wanted to solve the problem $Ay = b_0$. He noticed that the vectors he generated can be written as $b_k = p_k(A)b_0$, where p_k was a polynomial of degree k. However, he also denoted these vectors by p_k. The vectors he introduced satisfied

$$\begin{aligned} p_k &= [A^k - (a_0 + a_1 A + \cdots + A^{k-1})]b_0, \\ p_k^* &= [A^{*k} - (a_0 + a_1 A^* + \cdots + A^{*(k-1)})]b_0^*, \\ \bar{q}_k &= [1 - (\bar{a}_1 A + \bar{a}_2 A^2 + \cdots + \bar{a}_k A^k)]b_0, \\ \bar{q}_k^* &= [1 - (\bar{a}_1 A^* + \bar{a}_2 A^{*2} + \cdots + \bar{a}_k A^{*k})]b_0^*. \end{aligned}$$

The coefficients are determined such that $p_k \cdot p_k^*$ and $\bar{q}_k \cdot \bar{q}_k^*$ are minimized. Then, he normalized the polynomial $\bar{q}_k$ as $q_k = -(1/\bar{a}_k)\bar{q}_k$ such that the highest coefficient is equal to 1. Approximate solutions are defined as

$$\bar{y}_{k-1} = -\frac{q_k(A) - q_k(0)}{q_k(0)A} b_0, \quad y_{k-1} = -\frac{p_k(A) - p_k(0)}{p_k(0)A} b_0,$$

where the division means using the inverse.

He generated the p_k's and the q_k's with the recurrences,

$$p_{k+1} = \rho_k p_k + q_k', \quad p_{k+1}^* = \rho_k p_k^* + q_k^{*\prime},$$

$$q_{k+1} = \sigma_k q_k + p_{k+1}, \quad q_{k+1}^* = \sigma_k q_k^* + p_{k+1}^*,$$

where the prime (resp., $*$ prime) denotes the multiplication by A (resp., A^*) and $p_0 = b_0$. The coefficients of the recurrences are given by

$$\rho_k = -\frac{p_k^* \cdot q_k'}{p_k^* \cdot p_k}, \quad \sigma_k = -\frac{p_k^* \cdot p_{k+1}'}{p_k^* \cdot q_k'}.$$

He also introduced $h_k = p_k^* \cdot p_k$ and $h_k' = p_k^* q_k'$ (which is not a very good notation since the prime does not have the same meaning on both sides of the relation). Then, he considered the symmetric case and showed a small example, the matrix of the one-dimensional Laplacian of order 4. He obtained the solution by writing

$$y = \sum_{i=0}^{n-1} \eta_i q_i.$$

The coefficients are obtained using the relations defining the vectors and the orthogonality relations as

$$\eta_{i+1} = \frac{\eta_i}{\rho_{i+1}}, \quad \eta_0 = -\frac{1}{\rho_0}.$$

He observed that

> *The outstanding feature of the recurrence relations (37) and (42) is the fact that they are two-term relations. This has the following remarkable consequence. We have pointed out before that we can consider the successive stages of our iteration process as a succession of approximations. [...] The expansion (35) into the q-vectors thus imitates the behavior of an orthogonal expansion whose coefficients remain unchanged as we gradually introduce more and more vectors of the function space until finally all dimensions are exhausted.*

He also obtained the relation $Ay_k - b_0 = -\eta_k p_{k+1}$ for the residual vector. Then, he suggested restarting the algorithm (which he called "the block technique") to prevent the growth of rounding errors:

> *The accumulation of rounding errors tends to destroy the orthogonality of the p_i more and more. If we do not want to take recourse to the lengthy process of constant reorthogonalization, we can break our operations in blocks as soon as we notice that the rounding errors have done too much damage to the orthogonality. In that case we evaluate the remaining residual and start the process independently over again. The accumulation of rounding errors is thus avoided, at the price of retarded convergence.*

He noticed that the approximate solution y_k minimizes $(y - y_k) \cdot A(y - y_k)$, which is the square of the A-norm of the error when A is symmetric positive definite:

> *If by some means fast convergence can be enforced, the scheme might terminate in much fewer than n steps. Even if theoretically speaking the last vector vanishes exactly only after n iterations, it is quite possible that it may drop practically below negligible bounds after a relatively few iterations.*

Then, he proposed to accelerate the convergence in the symmetric case by a sort of preconditioner constructed using Chebyshev polynomials. For this, an approximation of the largest eigenvalue of A is needed. Lanczos suggested to use the bound given by the Gerschgorin disks. In Section 6, he discussed the solution of nearly singular systems. Section 7 is devoted to eigenvalue computations. In the end, he added

> *The latest publication of Hestenes [10] and of Stiefel [18] is closely related to the p, q algorithm of the present paper, although developed independently and from different considerations. The present investigation is based on years of research concerning the behavior of linear systems, starting with the author's consulting work for the Physical Research Unit of the Boeing Airplane Company, and continued under the sponsorship of the National Bureau of Standards.*

The 1951 paper [80] by Arnoldi (see below) is cited, as well as papers by Hestenes and Stiefel, but not the Hestenes and Stiefel 1952 paper, since Lanczos' paper was finished in September 1951.

It is fair to say that compared to the Hestenes and Stiefel paper, Lanczos' paper is far from a model of clarity. This may be why it had less impact. We observe that Lanczos' paper was published in July 1952 and that of Hestenes and Stiefel in December 1952 in the same journal.

Stiefel's paper [2891] was published in January 1952. Lanczos used Chebyshev polynomials again [1986] in 1958.

In the paper [2369] in 1996, O'Leary reported a personal communication from Hestenes who said

> *I believe it was done in the following order. 1. Stiefel because he had carried out some large experiments which surely took place more than a month before he came to UCLA. I invented it within a month of his arrival. 2. Hestenes. 3. Lanczos. He is third because he would have been talking about it prior to my invention of the routine. I am sure that when he saw my paper he said to himself, " I knew it all along". The remarkable thing is that it took two years of study of iterative methods at INA before the cg-algorithm was devised*

Computational results for CG were presented by Stiefel [2892] in the proceedings of the 1951 NBS Symposium, but this is essentially similar to his paper [2891]. Joseph W. Fischbach, from the Ballistic Research Laboratories at the Aberdeen Proving Ground in Maryland, computed the solution of linear systems on the *ORDVAC* computer using CG and also the steepest descent method in 1954 [1169].

Stiefel gave a one-hour lecture at the International Congress of Mathematicians held in Amsterdam in September 1954 [2896]. His title was *Recent developments in relaxation techniques*; see also [2894] in 1955. In that paper he discussed the optimality of CG. In 1958, Stiefel published a paper [2897] about kernel polynomials. Once again he described a method in which one filters out error components corresponding to a set of large eigenvalues and then applies CG to remove components corresponding to a few small eigenvalues. Hestenes later published several papers about iterative methods [1658, 1661].

A booklet [1095] was published in 1959 by Max Engeli, Theo Ginsburg (1926-1993), Heinz Rutishauser (1918-1970), and Stiefel from the ETH Zürich. There are five chapters. Chapter I: The self-adjoint boundary value problem by Stiefel; Chapter II: Theory of gradient methods by Rutishauser; Chapter III: Experiments on gradient methods by Ginsburg; Chapter IV: Overrelaxation by Engeli; Chapter V: Conclusions. They considered two problems in elasticity analysis: a plate problem with 70 unknowns and a condition number of 1600, and a bar problem with 44 unknowns and a condition number of $7\ 10^5$. These problems where solved on the *ERMETH*, a computer built at ETH. The plate problem was also solved on an *IBM 704* using a refined grid with 270 unknowns and a condition number of 22000. Careful numerical experiments are described in Chapter III. For the small plate problem on the *ERMETH*, the computing time with CG was around 150 mn to obtain a reduction of 10^7 of the ℓ_2 norm of the residual. One of the conclusions was that

> *Among the gradient methods, CG without smoothing and with recursive residuals seems to be the best one.*

Numerical results were also given in the Ph.D. thesis of Hochstrasser [1713] in 1954.

A generalization of CG was published by Curtiss [795] in 1954 for general matrices A. He used two auxiliary matrices B and T which is symmetric positive definite. Practical cases are $B = I, T = A^{-1}$, which gives the Hestenes and Stiefel algorithm when A is symmetric positive definite; $B = A^T, T = (A^TA)^{-1}$, which is solving $A^TAx = A^Tb$ with CG; and $B = I, T = I$, which is similar to the method described by Edward Joseph Craig (1924-2015) in his Ph.D. thesis [756] in 1954; see also [757] in 1955.

Extension of CG to Hilbert spaces was proposed by R.M. Hayes, a former student of Hestenes, [1609] in 1954, and Altman [35] in 1959.

A textbook description of CG was given in the book [760] by Stephen Harry Crandall (1920-2013), a professor of mechanical engineering at MIT, in 1956.

In the 1960s, people started to solve larger linear systems and CG began to acquire a mixed reputation. Werner Louis Frank [1216] reported slow CG convergence for tridiagonal linear systems of order 50 in 1960. Robert Kenneth Livesley used the *EDSAC II* British computer to solve problems arising from structural analysis in 1960. He tried CG and concluded [2086]

> *Rounding errors show a tendency to build up to such an extent that the solution after N steps is often a worse approximation to the solution than the starting point.*

Therefore, he switched to direct elimination methods. See [2369] for other uses of CG in the 1960s.

D.W. Martin and G.J. Tee wrote a survey [2148] of iterative methods for solving linear systems with symmetric and positive definite matrices in 1961. They reported numerical experiments from [1095, 1216]. Their conclusion was that

> *Although few comparative experiments have been reported to date, these suggest that no single method is to be recommended for universal application.*

In 1962-1963, the Soviet mathematician V.M. Fridman proposed an extension of CG for indefinite symmetric matrices in [1245]. He did not provide any numerical experiments and it turned out that his method is unstable; see [2425, 2904, 2907].

An Algol procedure for CG was published in 1963 by T. Ginsburg in the *Handbook series Linear Algebra* [1355]. Curiously enough, it is the three-term form of CG that was used. A numerical example for a pentadiagonal matrix of order 40 was given. More than 40 iterations were needed to obtain a good approximate solution.

The 1950s and 1960s were the golden age of the relaxation methods. CG was not much used as an iterative method and certainly not as a direct method since there were other more efficient direct methods like Gaussian elimination, Householder reflections, or Givens rotations. Note that CG is not cited in the seminal book about iterative methods [3132] by Varga in 1962. However, a few people were interested in studying CG convergence. The most well-known bound for CG is

$$\frac{\|x - x_k\|_A}{\|x - x_0\|_A} \le 2\left[\left(\frac{\sqrt{\kappa}-1}{\sqrt{\kappa}+1}\right) + \left(\frac{\sqrt{\kappa}+1}{\sqrt{\kappa}-1}\right)\right]^{-1} \le 2\left(\frac{\sqrt{\kappa}-1}{\sqrt{\kappa}+1}\right)^k, \tag{5.2}$$

where $\kappa = \lambda_{\max}/\lambda_{\min}$ is the condition number of A. This bound (obtained for exact arithmetic) is, in most cases, a large overestimate of the A-norm of the error. Nevertheless, it is found in almost every book discussing CG. Its origin can be traced back to results on Chebyshev polynomials; see Engeli, Ginsburg, Rutishauser, and Stiefel [1095] in 1959, a paper [2193] by Günter Meinardus (1926-2007) in 1963, and a paper [1869] by Shmuel Kaniel (1934-2017) in 1966. For the proof, they used an inequality proved by Leonid Vitaliyevich Kantorovich (1912-1986) in 1948 (see [1871] and [1746]). However, these papers were not directly related to CG. Even though the formula above was more or less implicit in the works of Stiefel and Rutishauser [1095], it appeared explicitly in the Ph.D. thesis of James Wilson Daniel in 1965 and the subsequent papers [804, 805] in 1967. These papers contain also the generalization of CG to nonlinear problems; see also Roger Fletcher (1939-2016) and Colin M. Reeves [1179] in 1964. There were errors in the papers by Kaniel and Daniel which were pointed out and corrected in [252, 719, 806, 2412].

Things started to change at the beginning of the 1970s. The first most significant results for explaining the behavior of the 1950 Lanczos algorithm for computing eigenvalues in finite precision arithmetic were obtained by Christopher Conway Paige in his Ph.D. thesis in 1971 [2412] and improved and extended in his subsequent papers in journals [2413, 2414, 2415]. He proved the important result that loss of orthogonality has a close relationship with convergence of Ritz values. Hence, such a method can be used, even in finite precision arithmetic, eventually doing a number of iterations larger than the order of the matrix. Almost at the same time, an equivalent statement was made for CG by John Ker Reid [2554] in the proceedings published in 1971 of a conference organized by the Institute of Mathematics and its Applications held in Oxford (UK) in April 1970; see also [2556] in 1972. Reid stated that CG can be used as an iterative algorithm for large, sparse, well-conditioned linear systems, using much fewer than n iterations. He discussed several computational variants of the algorithm and compared storage requirements, operations counts, and stability. He recommended the method that is standard today. Reid's paper drew the attention of many researchers to the potential of CG as an iterative method for sparse linear systems. It was an incentive for much of the subsequent work on CG.

Another seminal paper that renewed the interest in CG was [734] by Paul Concus, Golub, and O'Leary (who was a Ph.D. student of Golub) published in 1976 in the proceedings of the *Symposium on Sparse Matrix Computations* held at Argonne National Laboratory in September 1975. In that paper they derived what they called a generalized conjugate gradient method which is a three-term recurrence preconditioned CG. This was Golub's favorite form of CG since it formally looked like the second-order Richardson and Chebyshev methods and like an acceleration method of the underlying splitting of the matrix $A = M - N$, where M is the preconditioner. However, they noted that this form of the algorithm was used for expository purposes and that coupled two-term recurrences must be used in practice. They showed the optimality of the algorithm and that approximation of the eigenvalues of $M^{-1}A$ can be easily obtained. They gave numerical examples for problems arising from discretization of partial differential equations and comparisons with the Chebyshev method. This paper was influential to promote the use of preconditioning with CG.

CG was compared to other iterative methods on problems arising from finite element analysis by Erhard G. Schultchen and Celal N. Kostem (1939-2006) [2716] in 1973. Their conclusion was that CG was the method of choice; see also [2715].

In 1980, O'Leary developed the block conjugate gradient algorithm [2366] whose main goal is to solve systems $AX = B$, where B and X are $n \times s$ matrices. The difficulty in this method is that the columns of some $n \times s$ matrices computed during the iterations may become linearly dependent.

In 1987, Roy A. Nicolaides proposed a variant of CG where the residuals verify linear constraints [2345]. The method was formulated for the three-term recurrence form of CG. Let C be a given $n \times m$ matrix ($m < n$) of rank m. The method is constructed such that the residuals r_k satisfy $C^T r_k = 0$ for all k. At each iteration an $m \times m$ linear system must be solved.

After CG was recognized as a method of interest in the beginning of the 1970s, research was resumed to try to analyze and explain CG convergence.

In 1975, Gilbert Wright Stewart studied the case where all the eigenvalues of A are contained in an interval $[\alpha, \beta]$ except for one outlier λ_1; see [2868]. He showed bounds for the component of the error vector lying on the eigenvector associated with λ_1. These bounds implied that when λ_1 is well separated from $[\alpha, \beta]$ this component must decrease rapidly, particularly when $\lambda_1 \gg \beta$.

In 1976, Owe Axelsson (1934-2022), a Swedish mathematician working in the Netherlands, gave bounds for the A-norm of the error when there are large outliers in the eigenvalues [117]. Assume

$$a \leq \lambda_1 \leq \lambda_2 \leq \cdots \leq \lambda_{n-m} \leq b \leq \lambda_{n-m+1} \leq \cdots \leq \lambda_n,$$

and let $k \geq m$, then

$$\frac{\|x - x_k\|_A}{\|x - x_0\|_A} \leq 2 \left(\frac{\sqrt{\frac{b}{a}} - 1}{\sqrt{\frac{b}{a}} + 1} \right)^{k-m}.$$

This was obtained by using another polynomial R_k such that $R_k(\lambda_i) = 0$ for all λ_i, $i = n, n-1, \ldots, n-m+1$. This result is interesting when a few of the largest eigenvalues are well separated from the others. In this case, m is small, so, in theory, one does not lose too much in the exponent and b/a can be much less than $\kappa = \lambda_n/\lambda_1$; see also Axelsson and Gunhild Lindskog [131] in 1986. However, this distribution of eigenvalues does not occur too often in problems arising from the discretization of partial differential equations.

The influence of the eigenvalue spectrum on the convergence rate of CG was also studied by Alan Jennings [1811] in 1977; see also Jennings and G.M. Malik [1814] in 1978.

Hendrik (Henk) Albertus van der Vorst and Abraham van der Sluis (1928-2004), his former advisor, studied the case of separated smallest eigenvalues [3096]. Let

$$0 < \lambda_1 < \lambda_2 < \cdots < \lambda_p \leq a \leq \lambda_{p+1} \leq \cdots \leq \lambda_n \leq b$$

and let l_j be integers such that $l = \sum_{j=1}^{p} l_j \leq k$. Then,

$$\frac{\|x - x_k\|_A}{\|x - x_0\|_A} \leq 2 \left(\frac{1}{\prod_{j=1}^{p} T_{l_j}(\rho_j)} \right) \left(\frac{\sqrt{\frac{b}{a}} - 1}{\sqrt{\frac{b}{a}} + 1} \right)^{k-l},$$

where

$$\rho_j = \frac{b \cos \frac{\pi}{2l_j} + \lambda_j}{b - \lambda_j},$$

and T_{l_j} are Chebyshev polynomials. Van der Sluis and van der Vorst [3096] showed that as soon as one of the Ritz values converges, even moderately, to some eigenvalues of A, then CG behaves as if the component of the error along the corresponding eigenvector does not exist.

However, all these papers assumed exact arithmetic and it was not obvious that these bounds are still valid in finite precision arithmetic. These issues were discussed in a paper by Tomáš Gergelits and Zdeněk Strakoš [1339] in 2014. Their statement is

> *We conclude and numerically demonstrate that in the presence of large outlying eigenvalues such bounds have, apart from simple exceptions, little in common with the finite precision behaviour of the CG method.*

In 1980, Michel Bonnet and G.M. [373] showed that if only local orthogonality is exactly assumed, CG must converge as fast as the steepest descent method. Note that local orthogonality is generally well preserved and thus CG must converge (maybe slowly) in finite precision arithmetic. At the same time, Henryk Woźniakowski [3273] considered a special version of CG and showed that a finite precision implementation converged at least as rapidly as the method of steepest descent. We observe that in 1972, Harlan Crowder and Philip Wolfe (1927-2016) [780] showed that if the initial search direction is chosen incorrectly (that is, $p_0 \neq r_0$) but all other steps of the algorithm are implemented exactly, then convergence may be almost as slow as for the method of steepest descent.

On the foundations provided by C.C. Paige for the Lanczos method for computing eigenvalues, an explanation of the behavior of Lanczos and CG algorithms in finite precision arithmetic was given by Anne Greenbaum [1437, 1439, 1440, 1442], starting in 1981. She showed that the tridiagonal matrix T_k generated at a given iteration k of a perturbed Lanczos recurrence is the same as the one generated by an exact Lanczos recurrence applied to a matrix of larger dimension than A, whose eigenvalues all lie within tiny intervals about the eigenvalues of A, and some special starting vector. These matrix and vector depend on the iteration number k. An analogous statement is valid, within a small inaccuracy, also for the behavior of finite precision CG. This can be considered a backward error analysis in a sense different from the usual one. For instance, for Gaussian elimination, analysis shows that the finite precision computation corresponds to a matrix $A + \Delta A$ and bounds are given on ΔA. Here, we have a larger matrix whose eigenvalues are perturbations around the eigenvalues of A.

As justified in [1439, 1451], the bound (5.2) holds, with a small correction, also for CG computations in finite precision arithmetic. But this is only related to the fact that CG must converge and does not mean that this bound describes the CG behavior accurately.

A parametrized class of matrices for which the rate of convergence of CG varies greatly with the parameter choices was constructed by Strakoš [2920] in 1991. For some choices the orthogonality of the residual vectors is lost very rapidly.

Greenbaum and Strakoš [1451, 2922] demonstrated numerically that the behavior of CG in finite precision arithmetic is very similar to that of the exact algorithm applied to any matrix, say, $\tilde{A}$, which has many eigenvalues spread in tiny intervals about the eigenvalues of A. Note that this is slightly different from Greenbaum's previous results where the matrix depended on the iteration number k. Thus, CG convergence can be understood by considering the behavior of the exact algorithm applied to such matrices. The size of the intervals is a modest multiple of the machine precision. However, as far as we know, this result has not been theoretically proved.

Using results from Greenbaum, Yvan Notay investigated rounding error effects on the convergence of CG [2353] in 1993. He analyzed theoretically and experimentally how finite precision arithmetic affects known bounds on iteration numbers when the spectrum of the system matrix has small or large isolated eigenvalues.

Since the norm of the residual vector $b - Ax_k$ cannot be computed exactly, in 1997, Greenbaum showed how to estimate the maximum attainable accuracy of recursively computed residual methods [1441].

In 2000, Gutknecht and Strakoš studied the accuracy of two three-term and three two-term recurrences for Krylov space solvers [1509]. One of their conclusions was that even though the two-term recurrences are more stable than the three-term recurrences, for CG the difference between two-term and three-term implementations is usually moderate or small. But there may be differences in the maximum attainable accuracy.

The main results about the Lanczos and CG algorithms in finite precision arithmetic were summarized and explained in G.M.'s book [2222] and in the joint paper with Strakoš [2235] in 2006.

Following what was stated by Hestenes and Stiefel, G.M. showed in 2020 how to construct matrices and right-hand sides with prescribed convergence curves for both the A-norm of the error and the residual norm [2231]. This means that one can construct linear systems with a fast convergence of the residual norm and a slow convergence for the A-norm of the error or vice versa. It was also shown that it is not always possible to prescribe both the CG residual norms and the eigenvalues of A.

An interesting line of research about CG was started by Golub in the 1970s. His goal was to find upper and lower bounds for the norm of the error when an approximation ξ of the solution of $Ax = b$ is available. Let $r = b - A\xi$ be the residual and $\mu_m = (A^m r, r)$ be the corresponding moments. The ℓ_2 norm of the error e is such that $\|e\|^2 = (A^{-2}r, r) = \mu_{-2}$ and if A is symmetric and positive definite $\|e\|_A^2 = \mu_{-1}$. In that case, the moments can be written as Riemann-Stieltjes integrals for a piecewise constant measure with jumps at the eigenvalues of A. Upper and lower bounds can be found by using quadrature rules, namely Gauss quadrature and Gauss-Radau[38] or Gauss-Lobatto[39] rules if we know lower and upper bounds for the eigenvalues of A. For the computation of Gauss quadrature rules, see [1405] by Golub and John H. Welsch in 1969. The nodes of the quadrature rules are the roots of the orthogonal polynomials associated with the measure. These polynomials satisfy a three-term recurrence whose coefficients can be computed by the Lanczos algorithm started with the normalized residual vector. This was first stated in a paper by Golub, the Swedish mathematician Germund Dahlquist (1925-2005) and Stanley Charles Eisenstat (1944-2020) [802] in 1972; see also [1379] by Golub in 1973.

This technique was used again in 1978 for symmetric and positive definite matrices by Dahlquist, Golub, and Stephen Gregory Nash [803], but they also showed how an improved solution can be found using CG.

The formula

$$\|\varepsilon_k\|_A^2 = \|\varepsilon_0\|_A^2 - \|r_0\|^2 (T_k^{-1})_{1,1} = \|r_0\|^2 [\,(T_n^{-1})_{1,1} - (T_k^{-1})_{1,1}\,], \tag{5.3}$$

where ε_k is the error vector at iteration k, r_0 is the initial residual, and T_k is the tridiagonal matrix of the Lanczos coefficients that can be obtained from the CG coefficients, was used in [803] and also by Golub and B. Fischer [1171] in 1993. In that paper, $\|\varepsilon_0\|_A$ was computed by means of continued fractions and the error norms $\|\varepsilon_k\|_A$ were not calculated below 10^{-5}.

In 1994, Golub and G.M. [1384] developed methods to obtain bounds or approximations of elements of a matrix $f(A)$, where A is a symmetric positive definite matrix and f is a smooth function. These methods were also based on the use of quadrature rules and the Lanczos algorithm for diagonal entries and the block Lanczos or the nonsymmetric Lanczos algorithms for the non-diagonal entries. Upper and lower bounds for $u^T f(A) v$ were obtained, if the sign of the derivative of f is known, where u and v are given vectors, and also for $W^T f(A) W$, where W is an $n \times m$ matrix, $m \ll n$. Numerical experiments were given with several functions.

The formula (5.3) was also used in [1392] by Golub and Strakoš in 1994 for reconstructing the A-norm of the error in CG. They analyzed the effect of rounding errors in the Gauss quadrature calculations for the distribution function with finite points of increase.

The difficulty of computing bounds for the CG error norm $\|\varepsilon_k\|_A$ using formula (5.3) below a moderate value like $10^{-5} - 10^{-8}$ was overcome in [1385] published by Golub and G.M. in 1997. As in [1392], a positive integer delay d was introduced and the formula which was used is

$$\|\varepsilon_{k-d}\|_A^2 \approx \|r_0\|^2 [\,(T_k^{-1})_{1,1} - (T_{k-d}^{-1})_{1,1}\,], \tag{5.4}$$

but the trick was to turn the difference $(T_k^{-1})_{1,1} - (T_{k-d}^{-1})_{1,1}$ into a sum of positive quantities since the $(1,1)$ entry of the inverse of the tridiagonal matrix T_k can be computed incrementally as one goes through the CG algorithm. This led to the CGQL algorithm which computes bounds using Gauss, Gauss-Radau, and Gauss-Lobatto quadrature rules. For the last two, estimates of the smallest and/or largest eigenvalues of A are needed. The interest of this algorithm is that only a few operations, whose number is independent of the problem size, are needed to compute the bounds for the norm of the error at each iteration. The fact that CGQL gives upper and lower bounds (in exact arithmetic) was proved by G.M. [2216] in 1997. Numerical experiments for the preconditioned CG were given in [2218].

Strakoš and Petr Tichý proved in 2002 that formulas like (5.3) and (5.4) hold also in finite precision arithmetic, up to a small inaccuracy [2923]; see also [2924] in 2005. Hence, the bounds or estimates obtained from them can be used safely.

[38] Jean Charles Rodolphe Radau (1835-1911) was a German astronomer, born in Prussia, who worked mainly in Paris. He obtained French citizenship in 1873.

[39] Rehuel Lobatto (1797-1866) was a Dutch mathematician.

Formulas for computing estimates of the ℓ_2 norm of the error were proposed by G.M. [2221] in 2005.

The work of Golub and G.M. was summarized in the book [1386] published in 2010. This book described also applications of computing bounds for $u^T f(A)v$: computing some entries of $f(A)$, computing estimates of the trace of the inverse and of the determinant of A, the least squares fit of some given data by polynomials, computing the backward error of a least squares solution, total least squares (for total least squares, see [1398] by Golub and Charles Francis Van Loan in 1980) and the determination of the Tikhonov regularization parameter for discrete ill-posed problems.

In the CGQL algorithm, the Lanczos coefficients are computed from the CG coefficients. G.M. and Tichý showed how the bounds for the A-norm of the error can be directly computed from the CG coefficients [2236, 2237] in 2012. The formulas that were obtained are simpler than those in CGQL. In 2019, the same authors considered how to easily approximate the extreme Ritz values and how to compute upper bounds for the A-norm of the error [2238]. Heuristics to set up the value of the delay parameter d adaptively were given in a joint paper [2234] with Jan Papež in 2021.

Computation of bounds for matrix function entries and/or for norms of the error in CG was also considered by Calvetti, Golub, and Reichel [526] in 1999 and Calvetti, Serena Morigi, Reichel, and Fiorella Sgallari [532, 533] in 2000-2001. These publications used anti-Gauss quadrature rules that were introduced by the South African mathematician Dirk Pieter Laurie (1946-2019) [2001] in 1996; see also [539].

Applications of the computation of bounds for $u^T f(A)v$ to network analysis, in particular to identify the most important nodes of a network, were developed by Benzi and Paola Boito [271] in 2010, Benzi, Ernesto Estrada, and Christine Klymko [275] in 2013, Caterina Fenu, D. Martin, Reichel, and Giuseppe Rodriguez [1152] in 2013.

Other papers concerned with computing bounds for the norm of the error in methods related to CG are by C.B. [436] in 1999, whose formulas also provide estimates of the norm and can be applied to any direct or iterative method as well, by Frommer, Karsten Kahl, Thomas Lippert, and Hannah Rittich [1261] in 2013, Ron Estrin, Dominique Orban, and Michael Alan Saunders [1106, 1105] in 2019, and Eric Hallman [1552] in 2020.

The preconditioned CG (PCG) algorithm was very well suited to the vector computers which became available in the 1970s (see Chapter 7). However, it is not the same for parallel computers, particularly for distributed memory computers, since PCG is a very sequential process. The standard Hestenes and Stiefel version of CG is the following

$r_0 = b - Ax_0$, $p_0 = r_0$
for $k = 1, \ldots$ until convergence
$\quad \alpha_{k-1} = \frac{r_{k-1}^T r_{k-1}}{p_{k-1}^T A p_{k-1}}$
$\quad x_k = x_{k-1} + \alpha_{k-1} p_{k-1}$
$\quad r_k = r_{k-1} - \alpha_{k-1} A p_{k-1}$
$\quad \beta_k = \frac{r_k^T r_k}{r_{k-1}^T r_{k-1}}$
$\quad p_k = r_k + \beta_k p_{k-1}$
end for

At each iteration there are two dot products to compute, and there must be a synchronization of all the processing units after each step of the algorithm (except after the computation of x_k). Starting in the 1980s, many people tried to modify the algorithm to increase the degree of parallelism by reducing the data dependencies.

As far as we know, the first proposal was made by John Van Rosendale in a report for the Institute for Computer Applications in Science and Engineering (ICASE) of NASA [3114] in 1983. The author's goal was to remove the data dependencies for the CG coefficients by computing dot products (r_k, r_k) using long recurrences involving dot products $(r_j, A^i r_j)$, $(r_j, A^i p_j)$, and $(p_j, A^i p_j)$ from previous iterations $j < k$ and several powers i. It was proposed also to compute vectors like $A^i r_j$ by recurrences. However, this paper was purely theoretical and, apparently, this method was never implemented. It is likely that it would be highly unstable.

Also in 1983, Lennart Johnson [1837] proposed to compute β_k from α_{k-1} using the formula

$$\beta_k = -1 + \alpha_{k-1} \frac{(Ap_{k-1})^T Ap_{k-1}}{p_{k-1}^T Ap_{k-1}}.$$

This formula is obtained by computing $r_k^T r_k$ recursively and using local orthogonality relations, $r_k^T r_{k-1} = 0$ and $r_{k-1}^T Ap_{k-1} = p_{k-1}^T Ap_{k-1}$. This removes a data dependency at the expense of computing one more dot product $(Ap_{k-1})^T Ap_{k-1}$. As above, this paper did not provide any numerical experiment or roundoff error analysis; see also [1838].

The three-term form of CG is more parallel than the standard algorithm with two-term recurrences. Probably the first attempt to use the three-term form to reduce the number of synchronization points on parallel computers was published by Strakoš [2918] in 1985; see also [2919].

A paper about the use of polynomial preconditioners for CG [2654] was published in 1985 by Saad. Even though this was not the main aim of his paper, Saad considered the application of CG to vector and parallel computing. Like others, he observed that the dot products are a bottleneck of the algorithm and he proposed to use (in our notation)

$$r_k^T r_k = \alpha_{k-1}^2 (Ap_{k-1})^T Ap_{k-1} - r_{k-1}^T r_{k-1} \tag{5.5}$$

to compute β_k. The additional dot product $(Ap_{k-1})^T Ap_{k-1}$ can be computed as the same time as $p_{k-1}^T Ap_{k-1}$. Mathematically, this is equivalent to Johnson's formula. Because of roundoff errors the value of β_k may become negative. The numerical experiments were only concerned with the polynomial preconditioners and not CG modifications.

Another equivalent modification was suggested by David Barkai, Kevin J.M. Moriarty, and Claudio Rebbi [194] in 1985. They used it on the *Control Data Cyber 205* vector computer and seemed satisfied with their results. However, their algorithm is unstable.

Numerical experiments with CG on a two-processor computer *CRAY X-MP/2* (see Chapter 7) were done by G.M. [2210] in 1984. The preconditioned form of relation (5.5) was used to introduce more parallelism as proposed by Saad. However, numerical experiments showed that this algorithm is unstable. To remedy to this problem, it was proposed to use (5.5) to compute β_k and then at the end of the iteration or at the beginning of the next one to recompute the dot product $r_k^T r_k$. Instead of having two dot products in standard CG, one now has four dot products, but they can be computed concurrently. This technique was called the *predictor-corrector* variant of CG. There was no proof of stability but all the numerical experiments were satisfactory. This predictor-corrector algorithm was also used in the report [2211] in 1985 which described numerical experiments on a *CRAY X-MP/48* with four processors; see also [648, 2214]. A paper [2212] reporting these experiments was published in 1985. The algorithm was also described in the book [2217], page 328, in 1999. Unfortunately, this proposal did not receive much attention by people wanting to use CG on parallel computers until much later.

CG was also used on the *CRAY X-MP* by Mark K. Seager [2733] in 1986.

In 1989, Saad published a long paper [2656] about Krylov methods on supercomputers. He did a short error analysis of the formula (5.5) and recalled the predictor-corrector method proposed in [2210].

Another method was described by Anthony Theodore Chronopoulos and Charles William Gear (1935-2022) [684, 683] in 1989. It is called the *s-step conjugate gradient* method. This work was an outcome of Chronopoulos' Ph.D. thesis [681] where several s-step methods were proposed. Roughly speaking, the method uses s (hopefully linearly independent) vectors r_k, Ar_k, $A^2r_k, \dots, A^{s-1}r_k$. The iterates are written as $x_{k+1} = x_k + a_k^1 p_k^{(1)} + \cdots + a_k^s p_k^{(s)}$, where the coefficients are chosen to minimize the A-norm of the error and the directions $p_{k+1}^{(j)}$ are such that $p_{k+1}^{(j)} = A^{j-1}r_{k+1} + b_k^{(j,1)} p_k^{(1)} + \cdots + b_k^{(j,s)} p_k^{(s)}$, where the coefficients are chosen to enforce A-conjugacy. One step like this replaced s steps of standard CG. This reduced the number of global synchronization points of the processors. In exact arithmetic, both methods are equivalent. Implementations on parallel computers were discussed and numerical experiments using an *Alliant FX/8* were described.

In [684] by the same authors, another modification of CG was suggested to increase parallelism. It used only two dot products but the vectors Ar_k are used to compute the coefficients. To avoid another matrix-vector product, they are computed by an additional recurrence.

The s-step CG method was revisited by Erin Claire Carson [556] in 2018. If s is too large, the s-step method can suffer from convergence delay and decreased attainable accuracy relative to the standard CG method. As a remedy, she proposed to use a variable s and she derived estimates allowing to adaptively choose the value of s for each block step.

In the report [837] Eduardo Francisco D'Azevedo and Charles Holland Romine proposed in 1992 to modify the computation of $p_k^T A p_k$. It is computed recursively as $\sigma_k = r_k^T A r_k - \beta_k^2 \sigma_{k-1}$, but to avoid one more matrix-vector product, Ap_k is computed recursively. In this modified algorithm there are only two dot products that can be computed concurrently. Numerical examples were given on a parallel computer for which the gains in computed time were quite modest, ranging from 5% to 13%. Another report [834] in 1999 was co-authored with Victor Eijkhout. This report is mainly a rewriting of [837] with a few modifications involving other formulas to compute $p_k^T A p_k$. The authors claimed they proved the stability of their method showing that it is equivalent to a (more or less) stable variant of the Lanczos method. However, they did so in exact arithmetic, and unfortunately, the rounding errors in their method would not give the same rounding errors as in the standard Lanczos method. So, their "proof" was not fully convincing, even though the numerical experiments seemed to substantiate their claims.

Kevin McManus, Steve Johnson, and Mark Cross [2183] used the same idea as L. Johnson and Saad in 1999 but with the mathematically equivalent formula

$$r_k^T r_k = r_{k-1}^T r_{k-1} + \alpha_{k-1}^2 (Ap_{k-1})^T A p_{k-1} + 2\alpha_{k-1} r_{k-1}^T A p_{k-1}.$$

However, in the paper there is a sign error for the last term on the right-hand side. But nevertheless, their algorithm cannot be stable. The potential speed-up was evaluated with a theoretical model, and it is not clear if any real numerical experiment had been done.

There was not much activity on this domain of research until the beginning of the 2010s. In 2014, Pieter Ghysels and Wim Vanroose proposed a preconditioned pipelined CG method [1345] that has a single non-blocking reduction per iteration. This reduction can be overlapped with the matrix-vector product. They started from the s-step method of Chronopoulos and Gear with $s = 1$; however, the products Ar_k are also computed by recurrence and this needs another recurrence for $A^2 p_k$. The numerical results showed that this algorithm gave a maximum attainable accuracy which is much worse than with the standard CG. A very simple periodic strategy for residual replacement (that is, computation of the residual as $b - Ax_k$) was suggested to solve this problem. They gave numerical examples on a parallel computer.

However, examples for which the Ghysels and Vanroose algorithm not only had a worse maximum attainable accuracy but also a delay in convergence were pointed out by Carson, Miroslav Rozložník, Strakoš, Tichý, and Miroslav Tůma [559] in 2018. They gave a rounding error analysis for the maximum attainable accuracy.

In 2018, Siegfried Cools, Emrullah Fatih Yetkin, Emmanuel Agullo, Luc Giraud, and Vanroose [744] studied the influence of local rounding errors that are introduced by the additional recurrences in pipelined CG; see also [743]. They suggested an automated residual replacement strategy to reduce the loss of maximum attainable accuracy.

Jeffrey Cornelis, Cools, and Vanroose [746] extended pipelined CG to deeper pipelines which used longer recurrences in 2018. In 2019, the same authors presented a redesigned algorithmic variant of the ℓ-length pipelined CG [741] that they claimed to be stable. More numerical experiments on parallel computers were presented in a joint paper with Ghysels [740].

In 2019, Greenbaum, Hexuan Liu, and Tyler Chen [1445] studied the convergence rate of CG variants in finite precision arithmetic. Their goal was to relate the behavior of these variants to Greenbaum's results on Lanczos and CG convergence.

G.M.'s 1984 predictor-corrector idea [2210] was brought back to life more than 35 years later and extended by Carson and T. Chen [658] in 2020. They renamed the technique as *predict-and-recompute*. Recursively updated quantities are first used as a predictor for their true values and then recomputed exactly later in the iteration. They introduced pipelined versions of these algorithms, recomputing some vectors, and did a rounding error analysis for the maximum attainable accuracy, explaining why the technique of recomputing some quantities (either scalars or vectors) works. They also provided extensive numerical experiments.

Another way to enhance parallelism by reducing communication was considered in 2016 by Laura Grigori, Sophie Moufawad, and Frédéric Nataf in [1465]. They enlarged the Krylov subspaces by t vectors at each iteration. They described different methods and discussed numerical experiments.

For the history of CG up to 1976, see the nice paper by Golub and O'Leary [1389] published in 1989. It must be taken into account that solving a linear system with CG is, most of the time, only a part of solving a scientific problem. This issue was discussed in the book [2121] by Josef Màlek and Strakoš in 2015.

Let us now consider methods which are more or less related to the Lanczos and CG methods. As we mentioned above, in [1664] page 418, Hestenes and Stiefel showed that other approximations $\bar{x}_k$ can be obtained from the CG iterates x_k. These $\bar{x}_k$'s minimize the ℓ_2 norm of the residual.

Stiefel showed in his 1955 paper [2894] how to compute the $\bar{x}_k$'s directly in an iterative method without using CG. He used orthogonal polynomials and Riemann-Stieltjes integrals for discrete and continuous measures. It turns out that the residual vectors are A-conjugate and the method is what we now call the *conjugate residual* (CR) method; see also Hestenes [1659] in a 1953 conference whose proceedings were published in 1956. The conjugate residual method was used early by geodesists; see [1013] by Henri Marcel Dufour in 1964.

In principle CG and CR can only be used for symmetric positive (or negative) definite matrices. So there was a need for developing methods for symmetric indefinite matrices with positive and negative eigenvalues.

We have seen above that in 1962-1963, V.M. Fridman published an (unstable) extension of CG for indefinite symmetric matrices [1245]. In 1969, David Gilbert Luenberger proposed the method of hyperbolic pairs [2097]. Unfortunately, this method was also numerically unstable; see also [2098] in 1970.

The starting point of C.C. Paige and Saunders in their 1973 report was the Lanczos algorithm that does not require the matrix A to be definite. CG can be derived from the Lanczos method by using a Cholesky factorization of the tridiagonal matrices T_k which are constructed by the algorithm. The idea of Paige and Saunders was to use instead an L_kQ_k factorization of T_k with L_k lower triangular (with only three nonzero diagonals) and Q_k orthogonal with $Q_k^TQ_k = I$. The method, named SYMMLQ, uses only short recurrences. The corresponding paper [2425] was published in 1975. In that paper, they also derived MINRES, a method that minimizes the ℓ_2 norm of the residual. MINRES is mathematically equivalent to CR.

Fletcher considered methods for indefinite systems in 1975. This was published in the proceedings of the 1975 Dundee Conference on Numerical Analysis [1177] in 1976. This paper described the biconjugate gradient method which is used for nonsymmetric problems. We will come back to it in Section 5.8.

The conjugate residual method was also described in the Ph.D. thesis of Rati Chandra [630] in 1978. This work considered the generation of A^μ-orthogonal vectors p_i, where μ is an integer and a descent method with the p_i's as descent directions. This gives $(r_i, A^{\mu-1}r_j) = 0,\ i \neq j$ for the residual vectors. Bounds for the A^μ-norm of the error were derived and CG (resp., CR) was obtained for $\mu = 1$ (resp., $\mu = 2$). The relations between CG and CR were stated. Preconditioned versions of CG and CR were given, as well as a variant of PCG for 2-cyclic matrices. The SYMMBK method was proposed for indefinite problems. This method is mathematically equivalent to SYMMLQ, but used the Bunch and Kaufman factorization of T_k; see [501]. A modified conjugate residual (MCR) method which is a stabilized version of Luenberger's conjugate residual variant [2098] was also described; see also Chandra, Eisenstat, and Martin Harvey Schultz [631] in 1977.

All these methods were analyzed as polynomial methods for symmetric problems in the nice book [1170] by Fischer.

Augmentation and deflation to improve convergence was first proposed for solving symmetric linear systems with the conjugate gradient (CG) method. Nicolaides [2345] used deflation for CG in 1987. Zdeněk Dostál [940, 941] constructed preconditioners using projectors to eliminate the influence of some eigenvalues in 1988-1990. Deflation was also advocated by Lois Edna Mansfield [2130] in 1988. Augmented CG algorithms were introduced by Jocelyne Erhel and Frédéric Guyomarc'h [1099] in 2000 and Saad, Man-Chung Yeung, Erhel, and Guyomarc'h [2671] in 2000. Deflation preconditioners were used by Cornelis Vuik, Guus Segal, and J.A. Meijerink [3163] in 1999.

Symmetric positive definite systems with multiple right-hand sides were considered by Erhel and Guyomarc'h [1099] in 2000.

5.8 ▪ Krylov methods for nonsymmetric linear systems

A Krylov method is an iterative method which uses in one way or another the so-called *Krylov subspaces*,

$$\mathcal{K}_k(A, v) = \text{span}\{v, Av, A^2v, \ldots, A^{k-1}v\},$$

for a given matrix A and a given vector v. They are named after Alexei Nikolaevich Krylov (1863-1945), a naval engineer and mathematician (see Section 10.42). This is related to a paper [1955] published in 1931, whose title can be translated as *On the numerical solution of the equation by which, in technical matters, frequencies of small oscillations of material systems are determined.* He used what is now called Krylov vectors, but his method was designed to obtain the coefficients of the characteristic polynomial of a matrix to compute eigenvalues, as was common in the beginning of the 20th century. However, Krylov was not the only mathematician of those times to use a sequence $v, Av, A^2v, \ldots$; see, for instance, Herbert Western Turnbull (1885-1961) and Aitken [3081] in 1932, pages 43, 46-51. They used these vectors for theoretical purposes to obtain the Frobenius canonical form and the Jordan canonical form of a matrix (see Chapter 4). The term *Krylov sequence* appeared in a paper by Householder and Friedrich Ludwig Bauer (1924-2015) [1752, p. 33] in 1959.

The conjugate gradient is a Krylov method for symmetric positive definite linear systems. In this section, we more or less use a chronological approach to describe the history of Krylov methods for nonsymmetric problems. Thousands of papers have been written on these methods since the 1950s. A large number of methods have been proposed and many of them are now almost forgotten. Of course, we cannot cite all of them, therefore we only consider what we think are the most important and successful ones.

The 1940s

Some Krylov subspaces and methods are closely linked to Hessenberg matrices. An upper Hessenberg matrix H has entries such that $h_{i,j} = 0$ for $i > 2$, $j = 1, \ldots, i-2$, that is, only the entries of the upper triangular part of H are (possibly) nonzero as well as the entries on the first subdiagonal. A lower Hessenberg matrix is a transposed upper Hessenberg matrix.

Hessenberg matrices are named after Karl Hessenberg (1904-1959), a German mathematician and engineer (see Section 10.35). He was interested in computing eigenvalues of a matrix A and described two methods for obtaining the characteristic polynomial. In one of them, he constructed a basis V of a Krylov subspace and obtained, under some conditions, a relation $AV = VH$, where H is upper Hessenberg. These Hessenberg matrices did not first appear in Hessenberg's thesis as it is often said, but in the report [1655] in 1940, page 23, Equation (58). This report, with handwritten equations, is available on the website http://www.hessenberg.de (accessed January 2022). Hessenberg's method was explained in the book [3343] by Rudolf Zurmühl (1904-1966) in 1950, but even though he showed Hessenberg matrices, he did not give them that name. Zurmühl studied in Darmstadt almost at the same time as did Hessenberg. They had the same advisor, Alwin Oswald Walther (1898-1966). The words "Hessenberg matrix" appeared in a paper [2676] by Edward Aaron Saibel and W.J. Berger in 1953. The characterization of the lower triangular part of inverses of Hessenberg matrices was published by Yasuhiko Ikebe [1772] in 1979 and by Dmitry Konstantinovich Faddeev (1907-1989) in 1981 in Russian [1126] and in 1984 in English.

The 1950s

An algorithm to compute a biorthogonal basis was introduced by Lanczos in 1950 [1984]; see also [1985].

As we have seen above, the Lanczos algorithms were proposed by Lanczos in 1950 to compute eigenvalues [1984] and in 1952 for solving linear systems [1985]. Curiously, concerning orthogonality, Lanczos did not refer to Gram or Schmidt, but, in a footnote, to Szász [2978]:

> *The idea of the successive orthogonalization of a set of vectors was probably first employed by O. Szász, in connection with a determinant theorem of Hadamard.*

Note that Szász was also originally from Hungary, and that, as we have seen above, he worked for a while at the Institute of Numerical Analysis as Lanczos did. But orthogonalization was implicitly used by Pierre-Simon de Laplace (1749-1827) in 1816 (see [1990]) and Jorgen Pedersen

Gram (1850-1916) in 1879-1883 as well as by Erhard Schmidt (1876-1959) in 1908 for sets of functions; see Chapter 4. There is also an interesting footnote on the first page of [1984] with a reference to Krylov:

> *The literature available to the author showed no evidence that the methods and results of the present investigation have been found before. However, A.M. Ostrowski of the University of Basle and the Institute for Numerical Analysis informed the author that his method parallels the earlier work of some Russian scientists;* ... [references to Krylov and Luzin] ... *On the basis of the reviews of these papers in the Zentralblatt, the author believes that the two methods coincide only in the point of departure. The author has not, however, read these Russian papers.*

On this point Lanczos was right since his method is not the same as Krylov's method.

Lanczos constructed a basis of $\mathcal{K}_k(A, v)$, but also a basis of $\mathcal{K}_k(A^T, w)$, where A^T is the transpose of A, with coupled short recurrences that can be used to solve a nonsymmetric linear system.

In 1951, Walter Edwin Arnoldi (1917-1995), an American engineer, published a paper [80] describing what we now call the *Arnoldi process*. Arnoldi earned a degree in mechanical engineering from the Stevens Institute of Technology, Hoboken, New Jersey, in 1937, before achieving a Master of Science degree from Harvard University. From 1939 until his retirement in 1977, Arnoldi worked at the Hamilton Standard Propellers Division of the United Technologies Corporation in Wethersfield, Connecticut, an American multinational conglomerate headquartered in Farmington, Connecticut. His positions there included project engineer 1939-44, system engineer 1944-51, senior technical specialist of advance planning 1951-59, chief advance analyst 1959-60, head product researcher 1962-67, chief division researcher 1967-70, and division technical consultant 1970-1977. His main research interests included modeling vibrations of propellers, engines and aircraft, acoustics, aerodynamics of aircraft propellers, oxygen reclamation problems of space science, and high speed digital computers. He filled several patents about propellers. He was married to Flora (von Weiler) Arnoldi, with whom he had two sons, Douglas and Carl. He lived in West Hartford, Connecticut, from 1950 until his death.

Arnoldi's paper was received in May 1950. The *Proceedings of a Second Symposium on Large-scale Digital Calculating Machinery* organized on September 13-16, 1949, by Howard Hathaway Aiken (1900-1973) at Harvard University show that Arnoldi attended this symposium where Lanczos gave a talk whose title was *An iteration method for the solution of the eigenvalue problem of linear differential and integral operators*. Lanczos' paper in these proceedings is very similar to his 1950 paper [1984] written in September 1949. Arnoldi wanted to find a method somehow similar to Lanczos', but without using A^T. He used the strange notation u for the matrix, but he wrote that he was following the notation of Frazer, Duncan, and Collar [1222]. His paper starts by recalling the Lanczos algorithm but this is written in a more matrix-oriented way than in Lanczos' paper, even though he used row vectors for the second recurrence relation. He described the tridiagonal structure of the resulting matrix in the Lanczos algorithm. It is not before Section 6 of [80] that the Arnoldi process is described. He wrote

> *It will now be shown that a minimized iteration technique, similar in form but slightly different in detail, is also applicable to the solution of the eigenvalue problem by the Galerkin method, and that this method may offer certain computational advantages.*

Arnoldi used a long recurrence based only on the matrix A to generate the orthogonal basis vectors, but they were not normalized as it is done today. The upper Hessenberg structure of the resulting matrix is shown in relation (23). Because of the way he computed the basis vectors, the entries of the first lower diagonal were equal to 1. He wrote the recurrence relation for the characteristic polynomials and wanted to obtain approximations of the eigenvalues from the roots of these polynomials. The algorithm is finally fully stated in Section 8. In his conclusion, Arnoldi wrote

> *The Galerkin variation* [his method] *appears to be of greatest advantage in the eigenvalue problem, requiring substantially only half the number of matrix-column products involved in the Lanczos procedure.*

Apparently, Arnoldi did not realize that he was replacing two short recurrences by a long one whose complexity increases with the number of iterations. However, the matrices that were considered at that time were of small order, and this increase was probably not an issue. We observe that Arnoldi's paper [80], which, essentially, is concerned with computing an orthonormal basis of a Krylov subspace, did not mention Krylov. It only referred to Lanczos.

For a later use, let us just recall that the conjugate residual (CR) method was introduced for symmetric problems by Stiefel [2894] in 1955.

The method known as CGNE [756, 757] proposed by E.J. Craig in 1954-1955 uses CG on $AA^Ty = b$, $x = A^Ty$. It minimizes the norm of the error on the Krylov subspace $\mathcal{K}(AA^T, r_0)$.

The 1960s

A paper by Ishaq Murad Khabaza (1928-2005) [1890] in 1963 was one of the firsts to use iterative minimization techniques for solving nonsymmetric linear systems. The residual given by the method was written recursively as

$$r_k = (I - Ap_m(A))\, r_{k-1},$$

where p_m is a polynomial of degree m with $p_m(0) = 0$. The polynomial coefficients were computed such that the norm $\|r_k\|$ is minimized. This was done by solving the normal equations of the corresponding least squares problem. We observe that the basis which was used is the monomial basis with the vectors A^jr_{k-1}. Therefore, numerically, only small values of m can be used because the vectors A^jv tend to be in the same direction with increasing j, and they are prone to lose their linear independence in finite precision arithmetic. In the (small) numerical examples given by Khabaza, the value $m = 2$ was used. There was no theoretical results about the convergence of the method.

Essentially the same method (without a reference to Khabaza) was proposed by two Russian mathematicians, Gury Ivanovich Marchuk (1925-2013) and his student Yuri Alekseevich Kuznetsov [2135] in 1968. The formulation for the residual was more general than Khabaza's,

$$r_{k+1} = r_k - \left[\sum_{i=1}^{m} \gamma_i^{(k)} (AB)^i\right] r_k,$$

where B is a symmetric positive definite matrix. They studied the convergence of the algorithm (Theorem 1, page 1042). A more detailed 117-page paper [2136] was published in a book in 1974. The method is described in Chapter 4. In Chapter 5, it is stated that solving the normal equations to compute the coefficients of the polynomial using the natural basis is not satisfactory, and it is proposed to use an orthogonal basis for the subspace spanned by the vectors $B(AB)^ir_k$, $i = 1, \ldots$. When $B = I$, this algorithm is essentially the restarted GMRES method (see below) with a different orthogonal basis. The main difference is that the Gram-Schmidt algorithm is used to orthogonalize A^iv_1 and not Av_i. Unfortunately, that paper, published in a book and written in French, remained largely unnoticed.

The 1970s

In 1974, Gragg considered polynomials which turned out to be important later [1420]. He introduced a linear functional on the space of polynomials whose values are given by the moments for a given matrix. He showed the links to formal Laurent series, the Padé table, and the Lanczos nonsymmetric method. He gave a definition of formal orthogonal polynomials (even though he did not refer to them in this way) through determinants. In the conclusion section he wrote

> *The Lanczos polynomials may be generalized to maintain their connection with the Padé numerators and denominators, which are defined even though nontrivial blocks may occur in the Padé table.*

This was an incentive to study the singular case.

Let us just remember that C.C. Paige and Saunders [2425] described the method MINRES that minimizes the residual norm for symmetric indefinite matrices in 1975. The derivation of this method was based on the symmetric Lanczos method.

A method named Orthomin [3147] was proposed in 1976 by Paul Kazmi William Vinsome, a British physicist who was working for the Brunei Shell Petroleum Company. The paper was published in the proceedings of the *Fourth Symposium on Reservoir Simulation organized by the Society of Petroleum Engineers* in Los Angeles. The iterates were defined as $x_{k+1} = x_k + \alpha_k p_k$, and the direction vectors were defined by a truncated (eventually long) recurrence,

$$p_{k+1} = r_{k+1} + \sum_{j=k-d+1}^{k} \beta_j^{(k)} p_j.$$

The vector p_{k+1} is computed to be only $A^T A$-orthogonal to the last d vectors p_j. The convergence of Orthomin(d) when the symmetric part of A is positive definite was proved in [1066, 1082] in the 1980s. This type of methods and their relations with formal orthogonal polynomials and extrapolation methods were also studied in C.B.'s book [430] in 1980; see also [463] with Hassane Sadok in 1993.

The *biconjugate gradient* method (BiCG) was proposed by Fletcher in 1976 in a paper published in the proceedings of the 1975 *Dundee Conference on Numerical Analysis* [1177]. In that paper the method was introduced to solve symmetric indefinite problems. It is based on the nonsymmetric Lanczos process. A direct derivation of BiCG from biorthogonality conditions was done later by Tichý and Jan Zítko [3038] in 1998. We observe that BiCG was used in 2008 by Strakoš and Tichý [2925] to compute approximations of the bilinear form $c^* A^{-1} b$, where b and c are given vectors, without solving the linear system $Ax = b$.

At the end of the 1970s, for a course on numerical analysis, Peter Sonneveld, a Dutch mathematician, was looking for a multidimensional generalization of the secant method; see [2827]. There were already existing methods, but nevertheless, he obtained what can be called the primitive *Induced Dimension Reduction* (IDR) algorithm. The linear system $Ax = b$ is written as $x = Bx + b$ with $B = I - A$. Let r_0 be the initial residual, then $r_1 = Br_0$ and the residual vectors are written as

$$r_{k+1} = B[r_k - \gamma_k(r_k - r_{k-1})], \quad \gamma_k = \frac{p^* r_k}{p^*(r_k - r_{k-1})},$$

where p is a given vector, not orthogonal to any invariant subspace of B. Then, Sonneveld proved that under some mild conditions, $r_{2n} = 0$. The name of the method came from the fact that the dimension of the subspaces to which the residual vectors belong is decreasing. This algorithm was never published since it does not always converge in finite precision arithmetic and it is unstable for some problems.

In 1979, Sonneveld attended the *IUTAM Symposium on Approximation Methods for Navier-Stokes Problems* in Paderborn, Germany. There, he presented the first real IDR algorithm based on what is now called the IDR theorem with a multiplication with a factor $I - \omega_j A$, where ω_j is fixed for two steps and chosen to minimize every second residual norm. In this version of IDR, the residual vectors can be seen as being given by the product of two polynomials in A times the initial residual, one of the polynomials being such that $q_k(\xi) = (1 - \omega_k \xi) q_{k-1}(\xi)$. IDR did not use the transpose of A. This algorithm was published in the proceedings of that conference as a joined contribution with Pieter Wesseling [3218]. It did not receive much attention from the numerical linear algebra community until much later.

The 1980s

These years saw a lot of activity in looking for generalizations of CG to nonsymmetric problems. Many new methods were proposed, and most of the methods which are in heavy use today for nonsymmetric problems were developed in the 1980s and 1990s.

In 1980, Saad [2649] used the Arnoldi process for computing approximations of eigenvalues and proved some convergence results. He introduced the *Incomplete Orthogonalization Method* (IOM) by truncating the Arnoldi relation. This was used later to solve linear systems.

That same year, Axelsson introduced a generalized conjugate gradient algorithm in the paper [118] received in 1978. The work of Vinsome was cited in that paper. A method named GCG-LS was published in 1987 [121] as an extension of the 1980 method. GCG stands for generalized conjugate gradient and LS for least squares. In these methods, the direction vectors are given by short recurrences, but the iterates are defined by long recurrences whose coęfficients are computed to minimize the residual norm. A truncated version was also studied.

Young and his student Kang Chang Jea published the Orthodir and Orthores algorithms in 1980-1983. A paper titled *Generalized conjugate-gradient acceleration of nonsymmetrizable iterative methods* [3310] was published in 1980. Three generalizations of CG, Orthomin, Orthodir, and Orthores, were introduced in that paper. In the simplest version of Orthodir, the directions vectors are computed as

$$p_{k+1} = Ap_k + \sum_{j=0}^{k} \beta_j^{(k)} p_j,$$

where the coefficients are chosen to give a set of $A^T A$-orthogonal vectors. A simple version of Orthores, which is a generalization of the three-term variant of CG, is

$$\alpha_j^{(k)} = \frac{(Ar_k, r_j)}{(r_j, r_j)}, \; j = 0, \dots, k,$$

$$\beta_k = \frac{1}{\sum_{j=0}^{k} \alpha_j^{(k)}}, \quad \gamma_j^{(k)} = \beta_k \alpha_j^{(k)},$$

$$x_{k+1} = \beta_k r_k + \sum_{j=0}^{k} \gamma_j^{(k)} x_j, \quad r_{k+1} = -\beta_k A r_k + \sum_{j=0}^{k} \gamma_j^{(k)} r_j.$$

Simplifications were studied in the 1983 paper [1809]. For details on these methods, see also the Ph.D. thesis of Jea [1808] in 1982.

Block Krylov methods for linear systems were considered by O'Leary [2367] in 1980. She introduced a block BiCG algorithm and derived a block conjugate gradient method from it. She gave a convergence analysis of the block CG method.

In 1981, Saad introduced FOM (*Full Orthogonalization Method*) [2650] under the name *The method of Arnoldi* for solving linear systems. This method used the basis generated from the Arnoldi process. He considered again his previous IOM method which can be seen as a truncated FOM method and proved some convergence results. The name FOM was first used in [2653] in 1984. In that paper, Saad used the LU factorization with pivoting for IOM as well as a QR factorization of the upper Hessenberg matrix constructed by the method.

A *Generalized Conjugate Residual* (GCR) method was considered in the Ph.D. thesis of Elman [1082] in 1982; see Chapter 5 of the thesis. Convergence results were proved for matrices having a positive definite symmetric part and a bound was provided for the norm of the residual vector. Relations between GCR and Orthomin were studied. Parts of this work were summarized in the paper [1066] by Eisenstat, Elman, and Schultz in 1983. Another interesting paper is [2665] published in 1985 by Saad and Schultz.

In 1982 Saad [2651] considered the nonsymmetric Lanczos algorithm as an oblique projection method. He gave an analysis of convergence and feasibility conditions for the Lanczos algorithm and BiCG using moment matrices; see also [2653].

After their previous work on methods with long recurrences [3310], Young and Jea considered Lanczos/Orthodir, Lancos/Orthores, and Lanczos/Orthomin [1809] in 1983. They stated that Lanczos/Orthomin is essentially equivalent to BiCG and briefly review the work of other researchers. They wrote that

> *The Lanczos/Orthomin method converges if and only if the Lanczos/Orthores method converges, and if both converge, then the Lanczos/Orthodir method converges and all three methods are equivalent. From this it would appear that the Lanczos/Orthodir method is the safest of the three.*

This was a statement about the mathematical properties of the methods, but they may not hold in finite precision arithmetic. For this type of method, see also C.B.'s book [430] in 1980.

At the beginning of the 1980s, Sonneveld reconsidered his idea of product of polynomials and applied it to the Lanczos polynomials by squaring them. This gave a method he called *Conjugate Gradient Squared* (CGS). In CGS, the auxiliary polynomial is equal to the Lanczos polynomial. So the factors $I - \omega_j A$ of the first IDR algorithm were abandoned. Compared to BiCG, CGS had the advantage of not using A^T. A report [2823] was written in 1984 and the paper was submitted to the *SIAM Journal on Scientific and Statistical Computing* on April 24, 1984, but a revised version was only accepted on February 2, 1988, and finally published in January 1989 [2824]. When BiCG converges well, the reduction of the residual norm is better with CGS. However, there are many problems for which the CGS residual norms are even more erratic than those of BiCG. This may spoil the accuracy of the computation.

A block Arnoldi algorithm was introduced by Daniel Lucius Boley and Golub [360] in 1984; see also [361].

At the beginning of the 1980s, it was not known if orthogonal bases can be constructed with short recurrences as was done in CG. At the 1981 *Householder Symposium* held in England, Golub offered a $500 prize for the construction of a CG-like descent method for nonsymmetric matrices or a proof that there can be no such method. This was also stated in the report of the meeting given in Volume 16, No. 4, of the SIGNUM Newsletter, on page 7. This question meant that if we have a Hermitian positive definite matrix B, can we have a method defined with a short recurrence such that the B-norm of the error is minimized at each iteration? A negative answer to that question was given by Vance Faber and Manteuffel [1123] in 1984 who got the prize; see also [1124] in 1987. Their result implied that there exists a matrix B leading to a three-term recurrence if and only if the degree of the minimal polynomial of A is less than or equal to 3 or if A has a complete set of eigenvectors and eigenvalues lying on some straight line in the complex plane. A related problem was considered by Valentin Vasilyevich Voevodin (1934-2007) and his student Evgeny (Eugene) Evgen'evich Tyrtyshnikov [3150] in 1981.

A nice exposition of this difficult problem and its solution is given in Chapter 4 of the book by Jörg Liesen and Strakoš [2067] published in 2013; see also [1121] in 2008.

In 1985, Saad and Schultz [2665] presented a general Petrov-Galerkin framework from which several Krylov methods known at that time can be obtained as particular cases.

The *Generalized Minimal Residual* (GMRES) algorithm was presented in the report [2666] in May 1985 by Saad and Schultz. However, the first version of the report was from August 1983. Saad and Schultz saw their method as a generalization of the MINRES method by C.C. Paige and Saunders [2425]. They used the Arnoldi process to generate an orthogonal basis of the Krylov subspace with the modified Gram-Schmidt (MGS) variant and Givens rotations to solve the least squares problems involving Hessenberg matrices that are needed to minimize the norm of the residual. This technique is now considered the standard implementation of GMRES. With $v_1 = r_0/\|r_0\| = (b - Ax_0)/\|b - Ax_0\|$, the basis vector v_{j+1} is mathematically defined as

$$\tilde{v}_j = Av_j - \sum_{i=1}^{j} h_{i,j} v_i, \quad v_{j+1} = \tilde{v}_j / \|\tilde{v}_j\|,$$

with

$$h_{i,j} = (Av_j, v_i),\ i = 1, \dots, j, \quad h_{j+1,j} = \|\tilde{v}_j\|,$$

The $h_{i,j}$ are the entries of an upper Hessenberg matrix. The iterates are computed as $x_k = x_0 + V_k y^{(k)}$, where V_k is an $n \times k$ matrix whose columns are the vectors v_j and $y^{(k)}$ is the solution of the least squares problem $\min_y \| \, \|r_0\| e_1 - \underline{H}_k y\|$, $\underline{H}_k$ being a $(k+1) \times k$ Hessenberg matrix with entries $h_{i,j}$.

Saad and Schultz showed that GMRES is mathematically equivalent to GCR [1082] and Orthodir [3310], even though their implementations are completely different. They also showed that GMRES cannot break down, that, mathematically, it must terminate in at most n iterations and, incidentally, that FOM is equivalent to Orthores [3310]. After describing the algorithm, Saad and Schultz wrote

> *It is clear that we face the same practical difficulties with the above GMRES method as with the Full Orthogonalization Method. When k increases the number of vectors requiring storage increases like k and the number of multiplications like $1/2\, k^2 n$. To remedy this difficulty, we can use the algorithm iteratively, i.e. we can restart the algorithm every m steps, where m is some fixed integer parameter.*

They named this restarted method GMRES(m).

The corresponding paper [2667] was received originally by the editors in November 1983, in revised form in May 1985 and published in July 1986. GMRES is still one of the methods which is used the most, and hundreds of papers have been written about the convergence of FOM and GMRES since Saad and Schultz's paper appeared in 1986. It has been known from the beginning that the FOM residual norms may have an erratic behavior whence GMRES residual norms are monotonically decreasing by construction.

A small example where GMRES stagnates for one iteration was already given in [2667]. A larger example was given in 1992 by Nöel Maurice Nachtigal, Satish Chandra Reddy, and

Trefethen [2316]. It was an orthogonal matrix whose minimal polynomial is $1 - z^n$. Such an example was also mentioned by Peter Norman Brown [474] in 1991.

GMRES implementations different from what was done by Saad and Schultz were proposed over the years. Since, at that time, the backward stability of GMRES-MGS had not been proved yet, Homer Franklin Walker [3168] proposed in 1985 to use Householder reflections to compute the orthonormal basis vectors. This was supposed to be more stable, but it was also more expensive. This was published later in a journal [3169] in 1988; see also [3170].

Algorithms like the Lanczos methods or BiCG may break down for some nonsymmetric problems because some quantities used for division may become zero or tiny. Breakdowns in the nonsymmetric Lanczos process were already considered in Wilkinson's book [3248] in 1965. The proposed remedy was to restart the algorithm. Since then there has been a whole collection of names for the possible breakdowns in Lanczos algorithms and in BiCG. Beresford Neill Parlett, Derek Roy Taylor, and Zhishun A. Liu [2461] in 1985 said we have a *serious breakdown* if $w_k^T v_k = 0$ with the right and left basis vectors v_k and w_k being nonzero. Randolph Edwin Bank and T.F. Chan [183, 184] in 1994 called this a *Lanczos breakdown* and said we have a *pivot breakdown* if the factorization without pivoting of the tridiagonal Lanczos matrix T_k does not exist. Wayne David Joubert [1847] in 1992 defined a *hard breakdown* if $K_k(A^T, \tilde{r}_0)^T A K_k(A, r_0)$, where $K_k(A, r_0)$ is a Krylov matrix, is singular and a *soft breakdown* if $K_k(A^T, \tilde{r}_0)^T K_k(A, r_0)$ is singular. C. B., M.R.-Z., and Sadok [442, 457] in 1994 said we have a *true breakdown* if a formal orthogonal polynomial does not exist and a *ghost breakdown* if it exists but a division by zero occurs in the computation of the recurrence relation under consideration; in that case, a different relation has to be used. Roland W. Freund and Nachtigal [1236] in 1991 said we have a *breakdown of the first kind* in BiCG if $\tilde{p}_k^T A p_k = 0$ with $\tilde{p}_k, p_k \neq 0$ and a *breakdown of the second kind* if $\tilde{r}_k^T r_k = 0$ with $\tilde{r}_k, r_k \neq 0$. The only thing on which all these authors agreed was the *incurable breakdown* which means that there is no way to cure it before we reach the grade of r_0 with respect to A.

Besides restarting the algorithm from scratch, *look-ahead* methods can be used. There was a lot of activity in the 1980s and 1990s concerned with look-ahead techniques. It seems that the name "look-ahead" was coined by Parlett, Taylor, and Liu [2461]; see also the thesis of Taylor [3014], who was a student of Parlett, in 1982. In [2461] the authors used 2×2 diagonal blocks when needed to factorize the tridiagonal Lanczos matrix. Other more elaborated techniques were developed in the 1990s.

John Richard Wallis, Richard P. Kendall, and Todd E. Little [3174] showed in 1985 how to introduce constraints on the residual vector in the GCR algorithm, following ideas from James W. Watts [3193] in 1973 for the line successive overrelaxation method. Their method was used for oil reservoir simulations in two and three dimensions. The residual vectors r_k were constrained to satisfy $C^T r_k = 0$ for a given $n \times m$ matrix C with $m \ll n$. The matrix C was constructed initially from a vector computed using iterations of the power method for $I - (AM^{-1})^T$, where M was the preconditioner.

Smoothing techniques were described in the book by Willi Schönauer [2709] about vector computing in 1987. In smoothing methods, several iterates are combined together to give a smoother residual convergence curve.

IGMRES, a truncated GMRES method, was published by Brown and Alan Carleton Hindmarsh [475] in 1989. Unfortunately, all the basis vectors were needed to compute the solution at convergence. This constraint was relaxed in DQGMRES by Saad and Kesheng Wu [2670] in 1996. In this method the approximate solution is computed recursively. A restarted version was considered by Wu [3278] in 1997.

In 1989, Chronopoulos and Sangbak Ma [686] proposed squaring methods based on the nonsymmetric Lanczos algorithm in the same spirit as CGS.

The 1990s

More new methods were proposed in the 1990s as well as attempts to remedy to some flaws of previous methods, particularly breakdowns in Lanczos-like methods.

The notion of pseudospectrum[40] was used to study GMRES convergence. The ϵ-pseudospectrum of a matrix A is a set in the complex plane, which is defined as

$$\Lambda_\epsilon(A) = \{z \in \mathbb{C} \mid \|(zI - A)^{-1}\| \geq \epsilon^{-1}\}.$$

The use of the pseudospectrum was essentially promoted by Trefethen. Bounds for the GMRES residual norms using the pseudospectrum were given in [3067] in 1990. The computation of the pseudospectrum was discussed in [3068] in 1999. For applications of the pseudospectrum, see the book [3070] co-authored with Mark Embree in 2005.

A collaboration between Sonneveld and van der Vorst (who was also working in Delft at that time) led in 1990 to a report [3102] whose title was *CGSTAB, a more smoothly converging variant of CG-S*. The resulting method was presented at the *Householder Symposium* in Tylösand, Sweden, in June 1990 by van der Vorst.

In Sonneveld's comments [2827] on the problems with CGS, one can read

> *To overcome this disadvantage Henk van der Vorst suggested to replace one of the polynomials ϕ_n in the square ϕ_n^2 by a polynomial that could be used for damping out unwished behaviour of CGS. This lead to the development of BiCGSTAB, a method meant as a stable variant of CGS, although many people consider it as a stabilization of BiCG. In fact BiCGSTAB was mathematically equivalent to IDR, only the recurrence formulae differ completely, since they stem from the α's and β's in the (Bi)-CG method. Both the author and van der Vorst were well aware of that. But apparently the BiCGSTAB algorithm had a significantly better numerical stability, for which reason the IDR-interpretation was buried! This, after all, appears to be wrong. It was not the theoretical basis (IDR versus BiCG) that caused IDR instability, but a not so lucky implementation of the old IDR algorithm.*

In an interview in 1992, van der Vorst said

> *Early ideas by Sonneveld (1984) for improvements in the bi-Conjugate Gradient (Bi-CG) method, for the solution of unsymmetric linear systems, intrigued me for a long time. Sonneveld had a brilliant idea for doubling the speed of convergence of Bi-CG for virtually the same computational costs: CGS. He also published a rather obscure method under the name of IDR. I doubt whether that paper got more than two or three citations altogether. The eventual understanding of that method and the reformulation of it, so that rounding errors had much less bad influence on its speed of convergence, led to the so frequently cited Bi-CGSTAB paper.*

Van der Vorst joined the Mathematical Institute of the University of Utrecht in 1990. Finally, he published a report [3099] in 1990 and a paper whose title is *Bi-CGSTAB: A fast and smoothly converging variant of Bi-CG for the solution of nonsymmetric linear systems*. The paper was

[40]The pseudospectrum was introduced independently starting in the 1960s by J.M. Varah,, H. Landau, S.K. Godunov, L.N. Trefethen, D. Hinrichsen, A.J. Pritchard, and E.B. Davies.

submitted to *SISSC* on May 21, 1990, accepted for publication in revised form on February 18, 1991, and published in March 1992 [3100].

For real linear systems, the algorithm is the following, with $r_0 = b - Ax_0$, $p_0 = r_0$, $\tilde{r}_0$ arbitrary, for $k = 0, 1, \dots$

$$\begin{aligned}
\alpha_k &= \frac{(r_k, \tilde{r}_0)}{(Ap_k, \tilde{r}_0)}, \\
s_k &= r_k - \alpha_k A p_k, \\
\omega_k &= \frac{(As_k, s_k)}{(As_k, As_k)}, \\
x_{k+1} &= x_k + \alpha_k p_k + \omega_k s_k, \quad r_{k+1} = s_k - \omega_k A s_k, \\
\beta_{k+1} &= \frac{(r_{k+1}, \tilde{r}_0)}{(r_k, \tilde{r}_0)} \frac{\alpha_k}{\omega_k}, \\
p_{k+1} &= r_{k+1} + \beta_{k+1}(p_k - \omega_k A p_k).
\end{aligned}$$

Many researchers soon realized that the new methods CGS and BiCGStab were based on residual polynomials which are products of auxiliary polynomials and the Lanczos polynomials. Gutknecht coined the term *Lanczos-type product method* (LTPM) for this type of method. In the 1990s and even later, many researchers tried to improve these LTPM methods and developed new ones. This explains the plethora of Krylov subspace methods that were published until the end of the century and even up to now.

The block GMRES algorithm was probably described for the first time in the Ph.D. thesis of Brigitte Vital (in French) [3149] in 1990.

A switch between Lanczos/Orthomin and Lanczos/Orthodir or the use of the Parlett, Taylor, and Liu algorithm to avoid breakdowns was proposed by Joubert [1846, 1847] in 1990.

The relation between FOM (called the Arnoldi algorithm in that paper) and GMRES residual norms and the fact that when GMRES mathematically stagnates the Hessenberg matrices H_k are singular was established by Brown [474] in 1991. This phenomenon is called the *peak-plateau behavior* because when there is a plateau in GMRES convergence, there is a peak in FOM convergence. It was also later studied in a more general setting by Jane Grace Kehoe Cullum [786, 787] in 1995-1996 and in a joint paper with Greenbaum [789] in 1996.

The behavior in finite precision arithmetic of different variants of GMRES has been a concern since the introduction of the algorithm in 1986. It was the motivation to introduce the Householder implementation which was assumed to be more stable than those based on Gram-Schmidt algorithms. A report [1874] by R. Karlson in 1991, based on forward estimates and experimental observations, studied rounding error effects in GMRES.

To introduce more parallelism in GCR solvers, the s-step GCR method was proposed by Chronopoulos [682] in 1991; see also the joint papers with Charles D. Swanson [2945, 687] in 1992 and 1996. This kind of methods reduces the number of synchronization points on parallel computers.

The convergence of the BiCG residual norm is sometimes very oscillatory. The QMR (*Quasi-Minimal Residual*) method was introduced to obtain smooth convergence curves, but with short recurrences, contrary to GMRES. A first paper [1236] in 1991 was by Freund and Nachtigal. The method is based on the three-term Lanczos nonsymmetric process. The basis is not orthogonal and, instead of minimizing the residual norm, the method minimizes $\| \|r_0\| e_1 - \underline{T}_k y\|$, where $\underline{T}_k$ is a $(k+1) \times k$ tridiagonal matrix constructed by the Lanczos process. A joint paper with

Gutknecht [1230] was published in 1993. These papers were following a few technical reports [1235, 1228, 1229, 1238, 1237] (ordered by date) from 1990 to 1992.

In the 1990s, two groups of people were interested in look-ahead techniques to solve breakdown problems in methods implicitly or explicitly based on the Lanczos process. The first group was composed of Gutknecht, Freund, and Nachtigal. A first paper on QMR with look-ahead was [1236] in 1991 but the corresponding preprint from NASA RIACS [1228] appeared in 1990; see also Nachtigal's thesis [2315] in 1991. This was based on some theoretical work by Gutknecht [1496, 1498] published later in 1992 and 1994 on the nonsymmetric Lanczos algorithm. A paper about the implementation of the algorithm [1230] was published in 1993. A two-term recurrence version of QMR with look-ahead [1239] appeared in 1994. The details of its implementation were given in [1237]. QMRPACK, a package of Fortran subroutines implementing these algorithms, is still available in Netlib.

The second group was composed of C.B., M.R.-Z. and Sadok. They advocated the use of the formal orthogonal polynomial (FOP) framework. This type of polynomial was considered in the thesis of Herman van Rossum (1918-2006) [3115] in 1953, but not under that name. They also appeared in a paper by Peter Wynn (1931-2017) [3284] in 1967 and a paper by C.B. [429] in 1979. According to André Draux's thesis [951] in 1983, it was C.B. who suggested the name *formal orthogonal polynomials* instead of generalized orthogonal polynomials as they were sometimes named before; see [430, 438]. A first paper [454] was published in 1991, but the beginning of the work of this group on look-ahead techniques was really in [456], which was only published in 1992. The authors introduced the MRZ algorithm whose aim was to cure exact breakdowns in the Lanczos algorithm. Contrary to ordinary orthogonal polynomials, some FOPs may be non-existing and to jump to the next existing polynomial, they used the relations that were exhibited by Draux [1003] in the context of Padé approximation. The near-breakdowns were considered in the 1991 paper. Several variants, BMRZ, SMRZ, and BSMRZ, were proposed, and some numerical results on small problems were given; see also [455]. Unfortunately, these methods used the monomial basis for $\mathcal{K}_k(A^T, \tilde{r}_0)$ and therefore did not give good results on large problems. Another discussion of these methods was given in [457] in 1997. In 1999, other bases for $\mathcal{K}_k(A^T, \tilde{r}_0)$ were discussed in [458]. These methods were named MRZ-stab, HMRZ, and HMRZ-stab. Successive multiplications by A^T were still used but just to find the lengths of the jumps, which are in general small. HMRZ-stab gave much better numerical results than, for instance, MRZ. A review of formal orthogonality in Lanczos-based methods [459] was published in 2002.

A systematic study of the relations that can be used for the FOPs and the adjacent family was done in 1994 in the thesis of Carole Baheux [150] who was a student of C.B.; see also [151].

There were a few other papers dealing with the breakdown problems. An interesting paper by Parlett about reduction to tridiagonal form and, in particular, dealing with incurable breakdowns is [2448] in 1992. In 1995, Thomas Kilian Huckle [1764] used a rank-one modification of the matrix A to cure a serious breakdown. This approach can be used with many different iterative methods.

For detecting near-breakdowns in look-ahead algorithms the user usually needs to define some thresholds, which may be problem dependent, to be able to correctly detect the lengths of the jumps that have to be done. To avoid this, in 1996 Jean-Marie Chesneaux and Ana Cristina Matos [661] used stochastic arithmetic on the code BSMRZS of C.B. and M.R.-Z. They were able to obtain good results without using any thresholds, but stochastic arithmetic is much more expensive than the usual double precision finite precision arithmetic.

Then, people started applying look-ahead techniques to product-type methods. This was mainly done by those who had been working on look-ahead for the methods derived from the Lanczos process or BiCG. In 1991, C.B. and Sadok [462] showed how to avoid breakdowns in

Sonneveld's CGS algorithm. Near-breakdowns were handled using FOPs in [444] in 1994 by C.B. and M.R.-Z. A Fortran code for the algorithm BSMRZS is still available in Netlib. Look-ahead for BiCGStab was considered in [445] in 1995. Later, in 1998, these authors considered look-ahead techniques for transpose-free Lanczos-type algorithms [446].

Given a set of initial moments or modified moments, the problem of computing indefinite weights for a discrete inner product and the associated orthogonal polynomials was considered in 1991 by Boley, Sylvan Elhay, Golub, and Gutknecht [359]. They described a modified nonsymmetric Lanczos process which recovers when there is a breakdown by enforcing the biorthogonality in a blockwise sense.

Parallel versions of s-step nonsymmetric Lanczos algorithms [1897, 1898] were proposed in 1991-1992 by Sun Kyung Kim and Chronopoulos. At the same time, the first versions of parallel QMR algorithms [1231, 1232] were published by Freund and Marlis Hochbruck. Freund and Theodore Szeto proposed the QMRS algorithm obtained by squaring the QMR polynomial. Two reports [1240, 1241] appeared in 1991-1992 and a paper was published in the proceedings of a conference [1242] in 1992.

The modified Gram-Schmidt (MGS) algorithm which is used by the Arnoldi process in GMRES is not very well suited for parallel computers. Therefore, some people tried to modify the algorithm to introduce more parallelism. Newton bases were considered by Zhaojun Bai, Dan Yu Hu, and Reichel in 1992, see [159, 160], as well as Calvetti, Johnny Petersen, and Reichel [534] in 1993. The basis vectors are constructed as $v_1 = r_0/\|r_0\|$, $v_{j+1} = [(A - \xi_j I)v_j]/\eta_j$ with shifts ξ_j and normalizing factors η_j. The hope was that the shifts can be chosen such that the basis is better conditioned than with the natural basis when $\xi_j = 0$. The difficult problem was, of course, the choice of the shifts. One possibility is to use sets of points obtained from approximate eigenvalues and to order them as (fast) Leja points; see [144]. When some basis vectors are obtained, they are orthogonalized, and another issue is the QR factorization of these basis vectors, which must be parallelized. These problems were also addressed later by Erhel [1098, 3166] in 1995 and 2013, Roger Blaise Sidje and Bernard Philippe [2763] in 1994, and Sidje [2761] in 1997. A more recent work on Newton bases is by Philippe and Reichel [2489] in 2012; see also, Eric de Sturler and van der Vorst [856] in 1995. A study of the influence of the orthogonalization scheme on the parallel performance of GMRES was done in [1220] by Valérie Frayssé, Giraud, and Hatim Kharraz-Aroussi in 1998.

As we said above, Joubert characterized the breakdown situations in Lanczos algorithms in 1992. He also studied the likelihood of breakdowns using results about zeros of polynomials. His conclusion was that if A is a complex matrix, breakdowns cannot occur for the three Lanczos algorithms (Orthomin, Orthodir, and Orthores) except for a set of initial residual vectors r_0 which is of measure zero in $\mathbb{C}^n$. For the same result to be true when A is a real matrix, it is sufficient that A has at least one real eigenvalue. In particular, this is the case if the order of the matrix is odd.

That same year, Nachtigal, Reichel, and Trefethen [2317] described a hybrid method. They ran GMRES until the residual norm is small enough, and then re-applied the polynomial implicitly constructed by GMRES in a Richardson-type iteration which is fully parallel. They explicitly computed the coefficients of the polynomial, factored the polynomial, and used the Richardson iteration with a Leja ordering of the roots. Of course, it is simpler to directly compute the harmonic Ritz values which are the roots of the GMRES residual polynomial, but this was not known at that time.

A different approach was advocated by Starke and Varga [2849] in 1993. They used the Arnoldi process, enclosed the approximate eigenvalues in a polygon, and, with a conformal map, used Faber polynomials in a Richardson iteration.

In 1993, van der Vorst and Cornelis (Kees) Vuik [3103] proved bounds for the GMRES residual norms. Their aim was to explain the superlinear convergence behavior of GMRES that is sometimes observed. They attributed this phenomenon to the convergence of the Ritz values to the eigenvalues of the matrix. However, we will see below that one can construct examples with a prescribed (superlinear) convergence curve and prescribed eigenvalues and Ritz values. Hence, this link between Ritz values and GMRES convergence cannot always be true.

A variant of BiCG using 2×2 blocks in case of pivot breakdown was introduced by Bank and T.F. Chan [183, 184] in 1993-1994. The resulting algorithm was named the *Composite Step Biconjugate Gradient* (CSBCG).

Gutknecht [1497] derived a new variant of BiCGStab. He pointed out that the damping coefficients ω_j used in BiCGStab are real and this may cause problems since nonsymmetric matrices generally have complex eigenvalues. He modified BiCGStab to allow the auxiliary polynomial q_k to have complex zeros. This method was named BiCGStab2. The paper was submitted to the *SIAM Journal on Scientific Computing* (SISC) on September 9, 1991, but only published in September 1993.

Another method to improve over BiCGStab was proposed in 1993 by Gerard L.G. Sleijpen and Diederik R. Fokkema [2790]. BiCGStab(ℓ) used a polynomial of degree ℓ instead of a first-order polynomial $1 - \omega_j \xi$ in BiCGStab. The minimization of the residual norm occurs in a subspace of dimension ℓ. This paper was received by the electronic journal *ETNA* on March 24, 1993, and published in September 1993. Note that BiCGStab(2) is not equivalent to BiCGStab2.

The TFQMR algorithm from Freund [1226] used the CGS basis vectors in a clever way to compute iterates satisfying a quasi-minimal residual property. The letters TF stand for "transpose free" since the method does not use A^T. The paper was submitted in September 1991 and published in 1993. The TFQMR and the QMR algorithm of Freund and Nachtigal which uses A^T are not mathematically equivalent.

A *Flexible GMRES* method named FGMRES was introduced in 1993 by Saad [2659]. The method is said to be flexible in the sense that a different preconditioner can be used at each iteration. This method is not a Krylov method stricto sensu, but it is very close to them.

Axelsson published a book [122] *Iterative Solution Methods* in 1994. He described some linear algebra theory, classical iterative methods, and preconditioned generalized conjugate gradient methods.

Wolfgang Hackbusch published *Iterative Solution of Large Sparse Systems of Equations* [1524] in 1994, a book in which he described conjugate gradient methods, multigrid methods, and domain decomposition techniques. He also gave implementations in the Pascal language.

GMRESR was derived in 1994 by van der Vorst and Vuik [3104]. The idea was to precondition the linear system with an approximation of the inverse. GMRESR used iterations of GMRES itself as a preconditioner, hence the name GMRESR, the R standing for "recursive." The algorithm is in fact based on GCR rather than on GMRES, but these two methods are mathematically equivalent. A comparison with FGMRES was published by Vuik [3162] in 1995.

The GCRO method was proposed by de Sturler and Fokkema [854, 852] in 1993 and 1996. Let U_k and C_k be two $n \times k$ matrices such that

$$AU_k = C_k, \quad C_k^T C_k = I,$$

with columns u_j and c_j. Let $x_0 = 0$ and $x_k \in \text{range}(U_k)$ such that the norm of the residual is minimized. Then

$$r_k = b - AU_k y_k = b - C_k y_k,$$

with

$$y_k = \text{argmin}_y \|b - C_k y\| = C_k^T b.$$

The iterates are $x_k = U_k C_k^T b$ and $r_k = (I - C_k C_k^T)b$. But one has to compute u_{k+1} and c_{k+1} for the next iteration. In GMRESR, u_{k+1} was chosen to be an approximation of the error vector by running a given number of GMRES iterations starting from r_k. The improvement given by GCRO was to require that the inner iteration maintains orthogonality to the columns of C_k. The storage for GCRO increases with each outer iteration. A truncated version named GCROT was derived in [853] in 1999. The truncation strategy was based on determining which subspace in range(C_k) is the most important for convergence in the inner GMRES iteration and throwing away the rest of C_k. A simplified strategy was considered by Jason E. Hicken and David W. Zingg [1670] in 2010.

The *ideal Arnoldi* and *ideal GMRES* matrix approximation problems were formulated by Greenbaum and Trefethen [1453] in 1994. Their goal was to remove the influence of the right-hand side of the linear system in the bounds. The GMRES residual vectors satisfy

$$\begin{aligned} \frac{\|r_k\|}{\|r_0\|} &= \min_{p \in \pi_k, p(0)=1} \frac{\|p(A) r_0\|}{\|r_0\|}, \\ &\leq \min_{p \in \pi_k, p(0)=1} \|p(A)\|, \end{aligned}$$

where π_k is the set of polynomials of degree k. The minimum problem on the second line is the ideal GMRES approximation problem. Note that in the first line one has to minimize the norm of a vector whence in the ideal GMRES case one has to minimize the norm of a matrix. Greenbaum and Trefethen proved the uniqueness of the solution of the ideal GMRES approximation problem for general nonsingular matrices. See the paper [2070] by Liesen and Tichý in 2009 for extensions and a better proof of uniqueness.

Unfortunately, the solution of the ideal GMRES problem is not explicitly known and this problem does not always explain GMRES convergence.

In the beginning of the 1990s, it was already known that the distribution of the eigenvalues alone cannot explain GMRES convergence for non-normal matrices A. This was first clearly shown the paper [1452] by Greenbaum and Strakoš in 1994. They studied the matrices B that generate the same Krylov residual space as the one given by the pair (A, b). Moreover, it was shown that the spectrum of B can consist of arbitrary nonzero values. The paper [1447] with Pták in 1996 extended these results by proving that any non-increasing sequence of residual norms can be generated by GMRES. For more about Pták, see [1162, 1163]. The paper [79] by Arioli, Pták, and Strakoš in 1998 closed this series of papers. The main result was a full parametrization of the class of matrices and right-hand sides giving a prescribed residual norm convergence history while the system matrix has a prescribed nonzero spectrum; see also [2226] and [2067], Section 5.7. As we will see, this line of research was continued by other authors in the 2010s.

To avoid the triangularization of the upper Hessenberg matrix in GMRES by Givens rotations, H.F. Walker and Lu Zhou [3172] introduced the simpler GMRES method in a paper received in 1992 and published in 1994. This method used a basis W_k of $A\mathcal{K}_k(A, r_0)$. A basis of $\mathcal{K}_k(A, r_0)$ was obtained as $V_k = (r_0/\|r_0\|, W_{k-1})$, satisfying $AV_k = W_k T_k$, where T_k is an upper triangular matrix. The iterates were defined, as in GMRES, by $x_k = x_0 + V_k y^{(k)}$, where $y^{(k)}$ is obtained by solving a linear system with the matrix T_k. There is no Hessenberg matrix and no Givens rotations anymore. The residual norms are minimized as in GMRES. Unfortunately, as it was shown later, this method is prone to instability.

In 1994, Qiang Ye [3295] proposed a new way to solve the breakdown problem in Lanczos algorithms. When one occurs, the length of the recurrence using A^T is increased in such a way that biorthogonality relations are maintained. The matrix that is obtained is no longer tridiagonal but banded with a bandwidth increasing at each breakdown.

Computational variants of the CGS and BiCGstab methods were considered by Eijkhout [1064] in a 1994 technical report from the University of Tennessee. That same year, motivated by TFQMR, T.F. Chan, Efstratios Gallopoulos, Valeria Simoncini, Szeto, and Charles Hok-Shun Tong [618] proposed QMRCGStab, a quasi-minimum residual extension of BiCGStab. In the same vein, a family of transpose-free quasi-minimal residual algorithms QMRCGStab(k) was introduced in 1994 by Tong [3062]. Numerical results for QMRCGStab(2) and QMRCGStab(4) were presented.

In 1994 and 1996, T.F. Chan and Szeto [623, 624] applied the Bank and Chan composite step technique [183, 184], using 1×1 and 2×2 pivots, to BiCGStab and BiCGStab2. This gave the two algorithms CS-BiCGStab and CS-BiCGStab2. But this did not cure all the possible breakdowns.

The convergence of GMRES(m), the restarted version of GMRES, was studied by Joubert [1848] in 1994. He considered the impact of the restart frequency m on the convergence. A good choice can lead to a reduced solution time whence a bad choice may hinder convergence. There must be a balance between efficiency and cost. Unfortunately, it was noticed later by Eiermann, Oliver G. Ernst, and Olaf Schneider [1059] in 2000 and by Embree [1090] in 2003 that increasing m does not always improve GMRES(m) convergence. Strategies for choosing "good" values of m for GMRES were proposed by Joubert [1848], Masha Sosonkina, Layne Terry Watson, Rakesh Kumar Kapania, and H.F. Walker [2831] in 1998, Mitsuru Habu and Takashi Nodera [1519] in 2000, Linjie Zhang and Nodera [3327] in 2005, Kentaro Moriya and Nodera [2285] in 2007 (based on a paper by Naoto Tsuno and Nodera [3076] in 1999), Allison Hoat Baker, Elizabeth Redding Jessup, and Tzanio Valentinov Kolev [166] in 2009, and Rolando Cuevas, Christian E. Schaerer, and Amit Bhaya [783] in 2010.

Two QMR algorithms for solving singular systems with applications to Markov chain problems [1233] were published in 1994 by Freund and Hochbruck.

L. Zhou and H.F. Walker considered residual smoothing techniques for iterative methods [3331, 3171] in 1994-1995. They showed that QMR can be obtained from BiCG by residual smoothing. They also introduced a quasi-minimal residual smoothing. In 1994, M.R.-Z. and C.B. proposed hybrid procedures [443] which are closely linked to residual smoothing since it consisted of combining linearly two arbitrary approximate solutions with coefficients summing up to one, the parameter value being chosen in order to minimize the norm of the residual vector obtained by the hybrid procedure.

The methods we have seen so far were based on orthogonality relations or on the minimization of a residual norm. It was tempting to look for methods minimizing a norm of the error $x - x_k$. We have seen that this was done in CG for symmetric positive definite problems and the A-norm of the error, but it was much more difficult for nonsymmetric problems. As we have seen above, an early example is Craig's method, which is solving $AA^T y = b$ using CG.

Error-minimizing Krylov subspace methods were introduced and studied by Rüdiger Weiss (1954-1999) [3206] in 1994. Unfortunately, Weiss passed away a few years later after a bicycle accident. He had considered what he called generalized CG methods [3205, 3207]. GMERR was a generalization of an (unstable) method proposed by Fridman for symmetric, positive definite matrices. Weiss' method used A^T, the transposed matrix, the iterates being such that $x_k \in x_0 + A^T \mathcal{K}_k(A^T, r_0)$ and such that the ℓ_2 norm of the error is minimized. Two bases were constructed: one is an orthonormal basis for $A^T \mathcal{K}_k(A^T, r_0)$, and the other one is an AA^T-orthonormal basis for $\mathcal{K}_k(A^T, r_0)$. This was done with the classical Gram-Schmidt algorithm.

A problem for GMERR was the control of the convergence because the error norms decrease but they cannot be computed. The norm of the residuals may increase or oscillate even if the errors decrease. Moreover, an additional matrix-vector multiplication was needed if one wants to compute the residuals. It was shown in [1053] by Rainald Ehrig and Peter Deuflhard (1944-2019) that GMERR without restart converges for every right-hand side if and only if the matrix is normal. A more stable variant of GMERR, using Householder transformations, was proposed by Rozložník and Weiss [2614] in 1998. Unfortunately, the error-minimizing methods for nonsymmetric problems have not been very successful up to now. A theoretical overview of Krylov subspace methods [3208] was written in 1995 by Weiss. He defined a general framework that was able to describe many of the methods known so far; see also his book [3209] published in 1996.

The numerical stability of the Householder implementation of GMRES, proposed by H.F. Walker [3168] in 1985, was considered in the paper [953] by Jitka Drkošová, Greenbaum, Rozložník, and Strakoš in 1995. They analyzed the relation between the true and computed residual in finite precision arithmetic. Under some restrictions on the condition number of A, the backward stability of the Householder implementation of GMRES was proved. For details, see the Ph.D. thesis of Rozložník [2610] in 1996, Chapters 3 and 4. The fact that the Arnoldi basis vectors for the modified Gram-Schmidt GMRES implementation lose their linear independence only after the GMRES residual norm has been reduced to its final level of accuracy was also proved. The conclusion of the authors was that the MGS implementation of GMRES can be used safely.

In the mid-1990s, attempts were made to improve the accuracy of BiCGStab and BiCGStab(ℓ). In 1995, Sleijpen and van der Vorst [2794] proposed approaches that help to reduce the effects of local rounding errors on the BiCG iteration coefficients that are used in methods like CGS or BiCGStab. They proposed heuristics to vary ℓ for improving the accuracy. Another paper [2793] was devoted to the computation of the BiCG coefficients and can be viewed as a complement of [2794]. In 1996, the same authors discussed strategies to maintain the computed residuals close to the true residuals $b - Ax_k$; see [2795]. This has a beneficial effect on the maximum attainable accuracy; see also van der Vorst and Q. Ye [3105] in 2000. As we have seen above, estimating the attainable accuracy of recursively computed residual methods was studied by Greenbaum [1441] in 1997.

A common belief in the 1990s was that GMRES convergence can be hampered by "bad" eigenvalue distributions. This is not completely wrong (particularly for normal matrices) but, as we have seen above, not completely true either. Nevertheless, methods were developed to decrease the influence of some eigenvalues.

In a series of papers, Ronald Benjamin Morgan considered augmenting the Krylov subspaces in GMRES with approximate eigenvectors. In the beginning, he used the vectors corresponding to some Ritz values (the eigenvalues of the Hessenberg matrices H_k), the so-called Ritz vectors. He proposed different implementations of this idea, GMRES-E [2274] in 1995, GMRES-IR [2276] in 2000 based on the implicitly restarted Arnoldi algorithm (see Chapter 6) and GMRES-DR [2277] in 2002, which was a generalization to nonsymmetric problems of the idea of thick restarting by Wu and Horst D. Simon [3279] in 2000.

Another implementation of GMRES-IR was proposed by Caroline Le Calvez and Brigida Molina [2007] in 1999. Augmentation with a combination of harmonic Ritz vectors and approximations of the error vector can be found in [2352] by Qiang Niu and Linzhang Lu in 2012.

Algorithms for linear systems with multiple right-hand sides were considered by Simoncini and Gallopoulos [2776] in 1995. They did an analysis of the convergence of block GMRES [2777] based on matrix polynomials in 1996; see also [2778].

The worst-case GMRES residual bound is

$$\frac{\|r_k\|}{\|r_0\|} \leq \max_{\|v\|=1} \min_{p \in \pi_k, p(0)=1} \|p(A)v\|.$$

It is the best bound independent of the initial residual, and it describes the worst possible situation when changing the right-hand side or the initial approximation x_0.

In 1996, Faber, Joubert, Emanuel Knill, and Manteuffel [1120] constructed a small example for which the worst-case GMRES bound was different from the ideal GMRES bound that we have seen above. Kim-Chuan Toh [3058] provided a simpler example in 1997. These examples showed that the ideal GMRES bound is not always descriptive even of the worst possible situation. In 1998, Toh and Trefethen [3060] described an algorithm to compute the polynomial that gives the ideal GMRES bound $\|p(A)\|$. These polynomials are called the Chebyshev polynomials of a matrix.

Stephen LaVern Campbell, Ilse Clara Franziska Ipsen, Carl Timothy Kelley, and Carl Dean Meyer [543] considered in 1996 problems where the matrix A has its eigenvalues in one or several clusters with a few outliers. They gave some bounds of the GMRES residual norms for this situation. Their model said that in the case of a single cluster, the asymptotic rate of convergence of GMRES is proportional to the size of the cluster with a constant reflecting the non-normality of A and the distance of the cluster from the outliers. They extended these results to the case of several clusters.

That same year, Arioli and Claudia Fassino [78] did a forward rounding error analysis of the Householder implementations of FOM and GMRES. The influence of orthogonality of the basis vectors in the GMRES method was studied by means of a numerical example by Rozložník, Strakoš, and Tůma [2613]. On this topic, see also [1450] by Greenbaum, Rozložník, and Strakoš in 1997 who considered the linear independence of the MGS-based basis vectors.

The problem of constructing linear systems with a prescribed BiCG convergence was partly considered by Cullum [787] in 1996. Given the residual norms obtained with BiCG on $Ax = b$, she proved that one can construct a matrix B with the same eigenvalues as A and a right-hand side c such that the FOM residual norms when solving $Bx = c$ are the same as those obtained in the BiCG computation.

In 1996, Fokkema, Sleijpen, and van der Vorst [1180] introduced a class of methods named *Generalized CGS* (GCGS) methods, of which CGS and BiCGStab are special cases. The disadvantages of squaring the iteration polynomial were discussed and two new methods were introduced, CGS2 using another BiCG polynomial (with a different starting shadow vector) and shifted CGS using an approximation of the eigenvalue of largest modulus. In some numerical examples, CGS2 gives a smoother convergence behavior than CGS.

Properties of bases obtained by truncation of Krylov subspaces bases (for instance, in the IOM method) were studied by Zhongxiao Jia [1822, 1824] in 1996-1998.

James Weldon Demmel published *Applied Numerical Linear Algebra* [869] in 1997, a book in which he described numerical methods for solving systems of linear equations and computing eigenvalues.

In [2661], published in 1997, the FGMRES framework was used by Saad to develop Krylov methods where the Krylov subspaces are augmented with some other vectors. This was further studied by Andrew Chapman and Saad in [640]. Augmented GMRES methods were considered by Baglama and Reichel [149] in 1998. This was also studied later by Eiermann, Ernst, and O. Schneider [1059] in 2000 and Simoncini and Szyld [2782] in 2007.

Peter Russell Graves-Morris [1429] proposed in 1997 a technique he called *look-around Lanczos* to avoid breakdowns. In his own words,

> *The algorithms proposed use a method based on selecting well-conditioned pairs of neighbouring polynomials (in the associated Padé table), and the method is equivalent to going round the blocks instead of going across them, as is done in the well-known look-ahead methods.*

Hence, the name of the method.

A parallel variant of BiCG was published [492, 493] by Hans Martin Bücker and Manfred Sauren in 1997 and 1999. Their goal was to reduce the data dependencies of BiCG and to be able to overlap computation and communication by introducing new vectors in the algorithm. A small variation of the same method [3291, 3290] was proposed under the name *improved BiCG* by Laurence T. Yang and Richard Peirce Brent in 2003. The presentation of this algorithm contains some mistakes that can fortunately be easily corrected.

CGS and BiCGStab were combined in 1997 by T.F. Chan and Q. Ye [628]. At each iteration the algorithm chose either a CGS step or a BiCGStab step without restarting, depending on some criterion.

Generalized product-type methods were published in 1997 by Shao-Liang Zhang [3328]. The auxiliary polynomials q_k were chosen to satisfy a general three-term recurrence relation. This paper was submitted in August 1995. This class of methods was further discussed by the same author in 2002 [3329].

A 127-page review of Lanczos-type solvers for nonsymmetric problems was published by Gutknecht in *Acta Numerica* in 1997 [1499]. This nice paper described many variants of Lanczos-like methods.

In 1997, Zhi-Hao Cao [544] presented breakdown-free BiCGStab and BiCGStab2 algorithms, using the theory of formal orthogonal polynomials. This paper is purely theoretical, without any numerical experiments.

A QMR version of block BiCG was proposed in 1997 by Simoncini [2770]. Freund and Manish Malhotra published a block QMR algorithm including deflation [1234] for linear systems with multiple right-hand sides, based on the Lanczos-type process developed by José I. Aliaga, Boley, Freund, and Vicente Hernández [30] in 1996, with the paper published in 2000. T.F. Chan and Wing Lok Wan did an analysis of projection methods for solving linear systems with multiple right-hand sides [627].

Brown and Walker studied the properties of GMRES on (nearly) singular systems [476] in 1997.

In 1997, Greenbaum published a book [1442] whose title was *Iterative Methods for Solving Linear Systems*. She described the main Krylov methods and paid attention to their behavior in finite precision arithmetic.

In 1998, Hochbruck and Christian Lubich [1705] gave upper bounds of the error norm for FOM, GMRES, BiCG, and QMR in terms of projections. This small paper has an amusing title: *Error analysis of Krylov methods in a nutshell*. They also studied the relations between the residual norms of these methods. In the report [1779], Ipsen showed that the GMRES residual norm is large as long as the Krylov basis is well conditioned. She obtained this result by expressing the minimal residual norm at iteration k in terms of the pseudo-inverse of the Krylov matrix at iteration $k + 1$. Examples of exact expressions for residual norms were given for scaled Jordan blocks. The factorization of the Krylov matrix constructed from a normal matrix was used to derive upper and lower bounds for the residual norms. These bounds involved a subset of the eigenvalues of A. Some of these results were published later in the paper [1780] in 2000.

To avoid breakdowns in Lanczos-type methods, El Hassan Ayachour, a student of C.B. [139], introduced in 1998-1999 new biorthogonal polynomials satisfying a three-term recurrence relation instead of jumping over non-existing blocks in the Padé table; see also Ayachour's thesis [138]. He made use of the intermediate biorthogonal polynomials to compute the regular orthogonal polynomials. This technique was also applied to Padé approximation and extrapolation methods.

T.F. Chan, Lisette De Pillis, and van der Vorst squared the Lanczos polynomials to obtain transpose-free implementations of the QMR and the BiCG methods. They named these methods TFiQMR and TFiBiCG; see [615, 616]. However, three matrix-vector products were needed per iteration instead of two in BiCG. Even though there was a preliminary report in 1991, the paper was published in 1998.

Following the approach of L. Zhou and H.F. Walker [3331], QMR smoothing for product-type methods was considered in 1998 by Klaus J. Ressel and Gutknecht [2560]. Mohammed Heyouni and Sadok published a variable smoothing procedure for Krylov subspace methods [1669].

Fortran implementations of FGMRES from Frayssé, Giraud, and Serge Gratton were made available at CERFACS (Toulouse, France) in 1998 [1218]. Frommer and Uwe Glässner introduced restarted GMRES for shifted linear systems [1259]. A survey of properties of GMRES applied to consistent linear systems with a singular matrix was provided by Ipsen and C.D. Meyer [1781].

Two methods, named USYMLQ and USYMQR, were proposed in [2690] by Saunders, Simon, and Elizabeth Lingfoon Yip in 1988. Stricto sensu, they are not Krylov methods, but they are nevertheless interesting. However, they were not much used.

Three tools were used in the 1990s to study GMRES convergence: the field of values, the polynomial numerical hull, and the pseudospectrum. The field of values of a matrix A is a convex subset of the complex plane that is defined as

$$F(A) = \{(Ax, x) \mid x \in \mathbb{C}^n,\ \|x\| = 1\}.$$

It was used to study the convergence properties of some iterative methods (such as the semi-iterative Chebyshev method) in 1993 by Eiermann [1056]. This tool was also used by Starke [2848] in 1997. Results for GMRES using the field of values are summarized in the Habilitation thesis of Ernst [1102] in 2000 and in the nice paper by Eiermann and Ernst [1057] in 2001 (page 47). In 2012, Liesen and Tichý [2071] gave a simple and direct proof that the bounds using the field of values also holds for the ideal GMRES approximation.

The polynomial numerical hull is defined as

$$\mathcal{H}_k(A) = \{z \in \mathbb{C} \mid \|p(A)\| \geq |p(z)|,\ \forall p \in \pi_k\},$$

where π_k is the set of polynomials of order k. It has been shown that $\mathcal{H}_1(A) = F(A)$. It was introduced by Olavi Nevanlinna [2336] in 1993 and mainly used by Greenbaum; see [1443, 1444].

The successes and failures of these three tools for non-normal matrices were discussed by Embree in the report [1089] in 1999. This was illustrated with several examples. Unfortunately, none of these tools was able to succeed in all the examples, showing the limits of these approaches.

Serge Goossens and Dirk Roose [1409] proved in 1999 that the harmonic Ritz values are the roots of the GMRES residual polynomials. For GMRES, the harmonic Ritz values at iteration k are the eigenvalues θ_j given by $H_k^T H_k x = \theta H_k^T x$ when H_k is nonsingular. This was introduced by Freund [1225] in 1992.

It does not seem that the Hessenberg basis obtained by an LU factorization of the Krylov matrices attracted much attention until Sadok used it to define a new method named CMRH in the paper [2673] in 1999. The key ingredient to obtain a viable method was to use partial pivoting in the LU factorization to compute the basis. CMRH does not minimize the norm of the residual because the basis is not orthogonal, but the norm of the quasi-residual. This paper was submitted in June 1998. The corresponding quasi-orthogonal method, named the *Hessenberg method*, was proposed in 1998 in the paper [1669] by Heyouni and Sadok. One of the advantages of CMRH is that there is no dot product to compute, except maybe for computing the norm of the residual. However, the pivoting phase is not very well suited to parallel computing.

An interesting method related to BiCGStab was proposed in 1999 by Yeung and T.F. Chan [3301]. They named their method ML(k)BiCGStab. It can be seen as a transpose-free extension of a method ML(k)BiCG which used k shadow vectors instead of just one in BiCG. The derivation of this method was quite involved. This may be the reason why it did not receive much attention at the time. We note that ML(k)BiCGStab was re-derived in a simpler way by Yeung [3299] in 2012; see also [3300]. This method is very close to some IDR methods that were developed later.

The BiCGStab damping polynomials q_k were introduced in a Lanczos/Orthomin algorithm by Graves-Morris and Ahmed Salam [1430], who used the Graves-Morris' look-around strategy to avoid breakdowns. This paper was published in 1999.

A global version of GMRES was described by Khalide Jbilou, Abderrahim Messaoudi, and Sadok [1807] in 1999 for solving linear systems with several right-hand sides. The idea was to use a matrix dot product. If Y and Z are two $n \times s$ real matrices, their dot product is $< Y, Z >_F =$ trace$(Y^T Z)$. The subindex F stands for Frobenius since $< Y, Y >_F$ is equal to the square of the Frobenius norm of Y. The matrix residual vectors are constructed to be F-orthogonal. A global CMRH algorithm was considered in 2001 by Heyouni [1668].

Laurent Smoch published results about GMRES in the singular case [2811] in 1999; see also his Ph.D. thesis (in French) [2810] and another paper [2812] on the same topic in 2007.

G.M. published a book *Computer Solution of Large Linear Systems* [2217]. This book covered direct and iterative methods for solving linear systems. It described many of the iterative methods known in 1999.

The 2000s

After the intense activity in developing new methods in the 1980s and 1990s, it was probably difficult to find fruitful new ideas, and many scholars put more emphasis on trying to find theoretical explanations for the behavior of the methods that seemed to work well and on trying to improve those with deficiencies.

The convergence of restarted Krylov methods, with a particular emphasis on GMRES, was investigated by Simoncini [2771] in 2000. She related the convergence to the singular values of A. Sufficient conditions for restarted or augmented GMRES convergence were published by Zítko [3332, 3333, 3334, 3335] in 2000-2008 and Zítko and David Nádhera [3336] in 2010.

In 2000, Liesen introduced some computable residual norm bounds for GMRES in [2064]. Convergence of GMRES was also studied for particular classes of linear systems. Problems arising from the discretization of convection-diffusion equations were considered by Ernst [1103] in 2000. This problem was studied later by Liesen and Strakoš [2066] in 2005; see also Liesen and Tichý [2068] in 2004.

Most of the papers we have seen so far concentrated on obtaining bounds for the residual norms. However, exact expressions for these norms were found starting in 2000. In 2000, Ipsen gave in [1780, Theorem 4.1] an exact expression for the GMRES residual norm for normal matrices using a minimization problem over $k+1$ distinct eigenvalues.

In 1994, the study of GMRES convergence using unitary matrices was started by a paper of Greenbaum and Strakoš [1452], who were looking for linear systems having the same GMRES convergence curves. The authors wrote

> *If, for each vector b, we can find a matrix B of the given form, for which we can analyze the behavior of the GMRES method applied to B, then we can also analyze the behavior of the GMRES method applied to A.*

They suggested to find an equivalent matrix B which is unitary because the eigenvalues of such a matrix are located on the unit circle in the complex plane. This was done by Liesen [2064] in 2000. He used an orthogonal basis of $A\mathcal{K}_k(A, r_0)$ and constructed matrices that are GMRES-equivalent. Bounds, which depend on the initial guess, were obtained for the residual norms when the matrix is unitary. These bounds involve the gaps of the eigenvalues on the unit circle. A large gap may induce fast convergence. One bound is computable during the iterations if the simpler GMRES algorithm of H.F. Walker and L. Zhou [3172] is used.

In 2000, Leonid Aronovich Knizhnerman [1912] proved an inverse result, namely that fast GMRES convergence implies a large gap in the spectrum of the unitary matrix $\mathcal{Q}$ in a certain RQ factorization of the upper Hessenberg matrix obtained from GMRES. These issues were further investigated in the paper [1018] by Jurjen Duintjer Tebbens, G.M., Sadok, and Strakoš in 2014. The authors studied to what extent GMRES convergence can be explained using unitary GMRES (A, b)-equivalent pairs, eventually involving different right-hand sides. They characterized the equivalent matrices B in terms of orthonormal bases for the sequence of Krylov subspaces $\mathcal{K}_k(A, b)$ and Krylov residual subspaces $A\mathcal{K}_k(A, b)$, $k = 1, 2, \dots, n$. This showed that linking the spectral properties of unitary GMRES (A, b)-equivalent matrices, which influence GMRES convergence behavior, to some simple properties of A would be, in general, rather difficult. Exact expressions giving the GMRES residual norms for normal matrices as functions of the eigenvalues were given, and it was shown that they can, for some particular eigenvalue distributions, explain the acceleration of convergence observed after some number of iterations.

In 2000, Gutknecht and Ressel [1505] studied look-ahead procedures for general Lanczos-type product methods based on three-term recurrences. They also discussed different look-ahead strategies. We observe that after the year 2000, there was almost no paper considering the treatment of (near-) breakdowns in Krylov methods. This field of research seemed abandoned. This is surprising since the problems of deciding when we have to cure a near-breakdown or of maintaining a sufficient level of linear independence in the basis vectors have not been completely solved yet.

C.H. Tong and Q. Ye [3063] gave an analysis of BiCG in finite precision arithmetic in 2000. The effect of changing the rounding mode in BiCGStab computations was studied experimentally using the CADNA software (which was a library for estimating round-off error propagation; see [1820]) by Marc Montagnac and Chesneaux [2269] in 2000. They presented a dynamic strategy which allows one to detect breakdowns and near-breakdowns and to choose between a look-ahead technique and a restart.

It was recognized very early that Krylov methods using long recurrences were too costly (in computing time and storage) when solving difficult problems; see, for instance, Saad and

Schultz [2667]. Restarting and truncation were proposed as remedies. General papers about these methods are [1059] by Eiermann, Ernst, and O. Schneider in 2000, [2782] by Simoncini and Szyld in 2007, and [1502] by Gutknecht in 2012.

A technique to save some computer time in iterative solution of linear systems using Krylov methods is to do the matrix-vector products inexactly. This was called a *relaxation technique* (even though it has nothing to do with relaxation methods for solving linear systems). It was first considered by Amina Bouras and Frayssé in a 2000 CERFACS technical report [383]. Unfortunately, the corresponding paper [384] was only published in 2005, after some other people published results about this technique. Of course, the problem is to know by how much the matrix-vector product can be "relaxed" for the method to be still converging. This is called the relaxation strategy.

Following this technical report, some theoretical results about these relaxation techniques were published in 2003 by Simoncini and Szyld [2780]; see also the paper [2781] published in 2005. Relaxation strategies were discussed by Sleijpen, Jasper van den Eshof, and Martin Bastiaan van Gijzen in 2004 [2792].

Calvetti, Bryan Lewis, and Reichel investigated under which conditions on the singular matrix A and the right-hand side b GMRES iterates give a least squares solution of $Ax = b$ [529]. They also derived RRGMRES, a range restricted variant of GMRES that, under some conditions, produces the minimum norm least squares solution. The regularizing properties of GMRES for ill-posed problems were studied in [531] in 2002; see also [530].

As we said above, in 2001, Eiermann and Ernst [1057] published a very interesting paper in which they showed that, essentially, any Krylov method can be considered a quasi-orthogonal (Q-OR) or quasi-minimum (Q-MR) method. In fact, by changing the dot product, they can even be considered true orthogonal and minimum residual methods. If we have a basis whose vectors are the columns of matrices (not necessarily orthonormal) V_k, $k = 1, 2, \dots, n$ and Arnoldi-like relations $AV_k = V_{k+1}\underline{H}_k$ (where $\underline{H}_k$ is a $(k+1) \times k$ upper Hessenberg matrix), the iterates are defined as $x_k = x_0 + V_k y^{(k)}$. For a Q-OR method, $H_k y^{(k)} = \|r_0\| e_1$ (e_1 being the first column of the identity matrix) and for a Q-MR method, $y^{(k)}$ minimizes $\| \|r_0\| e_1 - \underline{H}_k y \|$.

Following the same idea as Saad for GMRES, Szyld and Julie A. Vogel published a flexible quasi-minimal residual method [2986] in 2001; see also flexible BiCG and BiCGStab methods derived by Vogel [3153]. Flexible methods were studied in 2001 by Simoncini and Szyld [2779].

We have seen above that smoothing techniques had been proposed to obtain smooth residual convergence curves for methods having an erratic residual norm behavior. In 2001, Gutknecht and Rozložník studied if these smoothing techniques improve the maximum attainable accuracy of iterative solvers [1508]. Unfortunately, the answer was that they don't; see also [1507].

In the paper [2065] in 2002, Liesen, Rozložník, and Strakoš presented identities and bounds for the residual norm of overdetermined least squares problems and applied these results to study minimal residual norm Krylov methods. They stressed that the choice of basis is fundamental for the numerical stability of the implementation and proved that the simpler GMRES method of L. Zhou and H.F. Walker is less numerically stable than the standard GMRES implementation using Givens rotations.

A paper by Röllin and Gutknecht [2588] in 2002 proposed variations of Zhang's Lanczos-type product method that are mathematically equivalent to GPBiCG in [3328]. Their methods were intended to have a better maximum attainable accuracy. A look-ahead version was considered in Röllin's diploma thesis.

Many papers were written on product-type methods by Japanese researchers. Let us cite only a few of them. In 2002, Seiji Fujino published a method named GPBiCG(m, ℓ) [1267] which is a hybrid of BiCGStab and GPBiCG methods. In 2005, Fujino, Maki Fujiwara, and Masahiro Yoshida proposed the BiCGSafe method [1269]. In 2010, Masaaki Tanio and Masaaki Sugihara published GBi-CGSTAB(s, L) [2996]. In 2012, Takashi Sekimoto and Fujino proposed variants of BiCGSafe [2742].

The complete stagnation of GMRES was studied by Ilya Zavorin, O'Leary, and Elman [3317] in 2003; see also Zavorin's Ph.D. thesis [3316] in 2001. They used a factorization of the Krylov matrix $K_k = XD_c\mathcal{V}_k$ that was originally presented by Ipsen in [1780], where X is the eigenvector matrix of A, D_c a diagonal matrix, and $\mathcal{V}_k$ the Vandermonde matrix constructed from the eigenvalues. They stated that GMRES completely stagnates if and only if $\mathcal{V}_k^* D_c^* W y = e_1$, where $W = X^*X$ and $y = X^{-1}b$. They also proved that there are no normal matrices with real eigenvalues for which GMRES could stagnate, but there could be stagnating normal matrices with complex eigenvalues. In 2008, Simoncini and Szyld [2783] gave conditions for GMRES non-stagnation involving the symmetric (or Hermitian) part and the skew-symmetric parts of A. These results were somehow extended by Simoncini [2773] in 2010. Necessary and sufficient conditions for GMRES complete and partial stagnation were given by G.M. in the paper [2227] in 2014. A complete characterization of those linear systems for which we have complete or partial stagnation was given. Moreover, it was shown that it is easy to construct examples of linear systems for which GMRES stagnates. Stagnation of block GMRES and its relationship to block FOM was considered in 2017 by Kirk M. Soodhalter [2829].

A different implementation of GMRES was done by Ayachour [140] in 2003. He used the MGS Arnoldi process as in the standard implementation, but the least squares problem at each iteration is solved by a partitioning of the Hessenberg matrix. The number of operations per iteration is smaller than for the standard GMRES implementation using Givens rotations. But the stability of this approach has not been studied theoretically.

Ahmed El Guennouni, Jbilou, and Sadok derived a block version of BiCGStab for linear systems with multiple right-hand sides [1474] and a block Lanczos method [1475] in 2004.

Van der Vorst published a book [3101] about iterative Krylov methods with an emphasis on GMRES, BiCG, and BiCGStab.

The worst-case GMRES problem was studied again by Liesen and Tichý in 2004 [2069], and in joint papers with Faber [3037] in 2007 and [1122] in 2013. They showed that for a Jordan block of order n ideal and worst-case GMRES are identical at steps k and $n-k$ such that k divides n. They also derived explicit expressions for the $(n-k)$th ideal GMRES approximation. An unpublished report by Arioli [74] in 2009 also studied the worst-case problem.

We have seen above that one of the tools used to study GMRES convergence was the polynomial numerical hull. It can be considered a generalization of the field of values. In [1443] in 2002, Greenbaum studied the properties and equivalent definitions of the polynomial numerical hull. In 2003, Faber, Greenbaum, and Donald Eddy Marshall characterized the polynomial numerical hulls of Jordan blocks [1119]. In 2004, James Vincent Burke and Greenbaum gave six characterizations of the polynomial numerical hull of a matrix and proved results for polynomial numerical hulls of Toeplitz matrices; see [508, 509]. In 2004, Greenbaum derived lower bounds on the norms of functions of Jordan blocks and triangular Toeplitz matrices [1444]. She also obtained estimates of the convergence rate of ideal GMRES applied to a Jordan block. In 2015, Daeshik Choi and Greenbaum examined the relations of Crouzeix's conjecture with GMRES convergence; see [663]. This conjecture was stated by the French mathematician Michel Crouzeix [776] in 2004. For what concerns us, his conjecture said that

$$\|p(A)\| \leq 2 \sup_{z \in F(A)} |p(z)|$$

for any polynomial p, where $F(A)$ is the field of values. The initial value of the constant on the right-hand side for which Crouzeix gave a proof [777] in 2007 was 11.08. He improved the value of the constant to $1 + \sqrt{2}$ in a joint paper with Cesar Palencia [778] in 2017. Note that 2 is the best possible constant since there are examples for which this bound is attained. A numerical investigation of Crouzeix's conjecture [1446] was done in 2018 by Greenbaum and Overton, from which it is likely that the conjecture is true.

A class of matrices that has received much attention are tridiagonal Toeplitz matrices, that is, matrices with constant diagonals. The behavior of GMRES for these particular linear systems systems was studied in Wei Zhang's Ph.D. thesis [3330] in 2007. These results were published in the paper [2054] by Ren-Cang Li and W. Zhang in 2009. Simpler formulas were obtained for right-hand sides $b = e_1$ and $b = e_n$ by the same authors [2055] in 2009 using Chebyshev polynomials of the second kind. Symmetric and normal tridiagonal Toeplitz matrices were considered by R.-C. Li in [2053].

In [2069] in 2004, Liesen and Tichý gave an exact formula for the GMRES residual norms for normal matrices at the next to last iteration $k = n - 1$ (in exact arithmetic).

In 2005, Bernhard Beckermann, Serguei Anatolievich Goreinov, and Tyrtyshnikov [247] improved Elman's bound for the GMRES residual norm. Elman's bound was, in fact, derived for GCR, and is valid only if the origin is not in the field of values of A which corresponds to the symmetric part of the matrix being positive definite; see also [246].

In 2005, Simoncini and Szyld [2781] studied the so-called superlinear convergence of GMRES using spectral projectors. They essentially showed that when the subspace $A\mathcal{K}_k(A, r_0)$ has captured some linearly independent vectors close to vectors in Q, an invariant subspace of A, GMRES behaves almost like another GMRES process applied to an initial vector $(I - P_Q)r_k$ with no components in the invariant subspace where P_Q is the spectral projector onto the range of Q.

Sadok [2674] established expressions for the GMRES residual norms involving the singular values of the Hessenberg matrices H_k which is of order k and $\underline{H}_k$ which is $(k+1) \times k$. Fortran codes implementing GMRES were provided by Frayssé, Giraud, Gratton, and Langou [1219] in 2005.

An augmented method named LGMRES (meaning "loose" GMRES) was proposed by A.H. Baker, Jessup, and Manteuffel [167] in 2005; see also Baker's Ph.D. thesis [162] in 2003. The Krylov subspace for GMRES is augmented by rough approximations of the error vector. Another implementation was published by Baker, John M. Dennis, and Jessup [163] in 2003; see also [164]. In 2016, this method was rediscovered with a different implementation by Akira Imakura, R.-C. Li, and S.-L. Zhang [1774], who proposed two methods: the locally optimal GMRES (LOGMRES) and the heavy ball GMRES (HBGMRES). Related papers were by Imakura, Tomohiro Sogabe, and S.-L. Zhang [1775, 1776] in 2012 and 2018.

Reichel and Q. Ye discussed properties of GMRES solutions at breakdown for singular matrices and presented a modification of GMRES to overcome breakdowns [2551] in 2005. The thesis of O. Schneider was devoted to Krylov subspace methods and their generalizations for solving singular linear operator equations, with applications to continuous time Markov chains [2708]. In 2007, Smoch showed that the singularity of the Hessenberg matrix produced by GMRES applied to a singular system is linked to the spectral properties of the matrix A [2812].

The backward stability of GMRES-MGS was finally proved by C.C. Paige, Rozložník, and Strakoš [2424] in 2006. This was based on previous results on the stability of MGS but the last steps to obtain the backward stability were far from being trivial. This is one of the main results obtained about GMRES. It showed that the modified Gram-Schmidt implementation of GMRES can be used safely, even though the Householder implementation could sometimes give a better maximum attainable accuracy.

Relaxed GMRES was again considered in [1357] by Giraud, Gratton, and Langou in 2007. The inaccuracy of the matrix-vector product was modeled by a perturbation of the original matrix. They proved the convergence of GMRES when the perturbation size is proportional to the inverse of the computed residual norm; this implies that the accuracy can be "relaxed" as the method proceeds.

GMRES for matrices with a skew-symmetric part of low rank (nearly symmetric matrices) was studied by Beckermann and Reichel in 2008 [249]. This method was studied and improved by Embree, Josef Aaron Sifuentes, Soodhalter, Szyld, and Fei Xue [1093] in 2012.

The influence of the choice of the basis on the stability of variants of GMRES was considered in a paper by Pavel Jiránek, Rozložník, and Gutknecht [1830] published in 2008; see also the Ph.D. thesis of Jiránek [1828]. They introduced a generalized simpler approach of which the simpler GMRES of L. Zhou and H.F. Walker is a special case as well as the *Residual-Based Simpler GMRES* (RB-SGMRES). They analyzed the maximum attainable accuracies of these methods in finite precision arithmetic. They showed that RB-SGMRES is conditionally backward stable. An adaptive method mixing simpler GMRES and RB-SGMRES was proposed later by Jiránek and Rozložník [1829].

In 2006, Jens-Peter M. Zemke sent a message to Sonneveld asking what happened to the IDR method. On this occasion, Sonneveld revisited his ideas from the 1970s and realized that the IDR framework can be used with more than one shadow vector. The outcome of this rethinking was the first IDR(s) algorithm and the paper [2828] with van Gijzen in 2008; see [2825, 2827].

However, the idea of using several shadow vectors was not new; remember the work of Yeung and T.F. Chan [3301] in 1999. Block Krylov methods have also been developed for quite some time, mainly to solve problems with several right-hand sides. As we have seen above, in 2000, Aliaga, Boley, Freund, and Hernández [30] had published a nonsymmetric Lanczos-type method which generates two sequences of biorthogonal basis vectors and can handle the case of left and right starting blocks of different sizes.

Numerical experiments with the first IDR(s) algorithm, particularly on linear systems arising from the discretization of convection-diffusion equations, showed that it was not very stable. Increasing the number s of shadow vectors improved the convergence but the maximum attainable accuracy was getting worse; see, for instance, Figure 6.1 in [2828] where the maximum attainable accuracy is larger by more than two orders of magnitude when s is increased from 1 to 4.

Contrary to the IDR paper of 1980, the publication of IDR(s) in 2008 drew the attention of many researchers. In particular, several papers were published by Japanese researchers who were already involved in the study of Lanczos-type product methods; see Yusuke Onoue, Fujino, and Norimasa Nakashima [2378, 2379, 2380] in 2008-2009.

In 2008, Tokushi Ito and Ken Hayami used preconditioned GMRES methods for least squares problems [1786].

In 2009, Arioli and Duff [75] used FGMRES to obtain a backward stability result for the triangular factorization of a matrix in single precision. FGMRES was used for the iterative refinement in double precision.

Based on the *Biconjugate Residual* (BiCR) method of Sogabe, Sugihara, and S.-L. Zhang [2820], in 2010, Kuniyoshi Abe and Sleijpen [2] introduced new product-type variants, replacing the "BiCG part" of those methods by BiCR. These methods were named BiCRStab, GPBiCR, and BiCRstab(2). On some examples they converged faster than their BiCG counterparts. See also [3] in 2012.

Also in 2009, Lakhdar Elbouyahyaoui, Messaoudi, and Sadok [1079] studied the algebraic properties of block GMRES and block Arnoldi methods; see also [386].

The 2010s

Using efficiently the Krylov methods that had been proposed in the 1980s and 1990s on parallel computers was not always easy. The bottleneck of the computation was often the communication between the processing units. We have already seen this problem for CG in the symmetric case. So the researchers were willing to modify the algorithms to solve these problems. The methods trying to avoid communications or to hide them with computations started to be called *communication-avoiding methods*, even though communications cannot be completely avoided when solving linear systems or when computing eigenvalues and eigenvectors (except in very particular cases).

Communication-avoiding variants of the nonsymmetric Lanczos and BiCG methods were briefly described in 2010 by Mark Frederick Hoemmen in [1718]. Later, CA-BiCG was derived from CA-Lanczos [174] in 2014 by Grey Ballard, Carson, Demmel, Hoemmen, Nicholas Knight, and Oded Schwartz. A direct derivation of CA-BiCG [555] was done in the Ph.D. thesis of Carson in 2015.

Residual replacement (that is, the computation of $b - Ax_k$) was considered a remedy for the potential instability of parallel variants of Krylov methods like CA-BiCG or CA-BiCGStab; see [1718, 557, 555] (ordered by date).

To introduce more parallelism, s-step versions of block Krylov methods were proposed by Chronopoulos and Andrey B. Kucherov [685] in 2010.

In 2010, Sleijpen, Sonneveld, and van Gijzen [2791] showed that BiCGStab can be considered an induced dimension reduction method. This was not too surprising since BiCGStab was derived as an improvement of CGS and the first IDR algorithm. A paper by Sleijpen and van Gijzen [2798] proposed the method IDR(s,ℓ). They exploited the idea used in BiCGStab(ℓ), applying a higher-order stabilizing polynomial instead of the linear factors $I - \omega_j A$. As we have seen above, a similar idea was used independently by Tanio and Sugihara with the GBi-CGSTAB(s,L) method [2996], which is IDR(s) with higher-order stabilization polynomials; see also Kensuke Aihara, Abe, and Emiko Ishiwata [10].

Gutknecht published in 2010 an interesting expository paper [1501] whose title is *IDR explained* in which he gave details about the IDR algorithms, relating them to other Krylov methods and summarizing the early history of IDR methods. That same year, Simoncini and Szyld [2784] showed how IDR(s) can be interpreted as a Petrov-Galerkin method. They also proposed a new variant of IDR using some Ritz values to compute the parameters ω_j. This was named the Ritz-IDR variant.

Other interesting papers in 2010 were [1358] by Giraud, Gratton, Xavier Pinel, and Xavier Vasseur, who used deflated restarting with flexible GMRES; [3298] by Yeung, who considered the solution of singular systems with Krylov subspace methods; and [1606] by Hayami, Jun-Feng Yin, and Ito, who used GMRES methods for least squares problems.

In 2011, G.M.'s paper [2224] provided formulas for the FOM and GMRES residual norms, involving a triangular submatrix of the Hessenberg matrix H_k. Lower and upper bounds using the singular values of this submatrix were given. Techniques to compute estimates of the norm of the error for FOM and GMRES during the iterations were explained in the paper [2223]. This was done using the same techniques as for symmetric matrices and CG; see [802, 803, 1171,

1384, 1392, 2216, 2218, 2221]. Their computations are more expensive than in the symmetric case, but these estimates can provide a more reliable stopping criterion than the one based on the residual norms. We observe that other methods to obtain estimates of the norm of the error were described by Giles Auchmuty [99] in 1992 and C.B. [436] in 1999; see also [460, 461] in 2008-2009 by C.B., Giuseppe Rodriguez, and Sebastiano Seatzu (1941-2018).

To improve the stability of the first IDR algorithm, in 2011, van Gijzen and Sonneveld proposed another variant of IDR using biorthogonality properties [3107]. Collignon and van Gijzen studied how to minimize the number of synchronization points in IDR(s) for an implementation on parallel computers [729]. They reformulated the variant with orthogonalization to have only one global synchronization point per step. Collignon, Sleijpen, and van Gijzen showed how to interpret IDR(s) as a deflation process in [727].

Lei Du, Sogabe, and S.-L. Zhang applied residual smoothing techniques to IDR(s) in 2011 [971]. A block IDR(s) method was proposed by L. Du, Sogabe, Bo Yu, Yusaku Yamamoto, and S.-L. Zhang to solve linear systems with multiple right-hand sides [970].

A flexible generalized conjugate residual method was derived by Luiz Mariano Carvalho, Gratton, Rafael F. Lago, and Vasseur [560].

A geometric view of Krylov subspace methods for singular systems was published by Hayami and Sugihara [1604] in 2011; see some corrections in 2014 [1605].

In [1014] by Duintjer Tebbens and G.M. in 2012, a parametrization was given of the class of matrices and right-hand sides generating, in addition to prescribed GMRES residual norms and eigenvalues, prescribed Ritz values (that is, approximations of the eigenvalues) in all iterations. This parametrization was different from what was done by Arioli, Pták, and Strakoš [79] in 1998. Later, in the paper [969] with Kui Du in 2017, it was shown that, instead of the Ritz values, one can prescribe the harmonic Ritz values which are the roots of the GMRES residual polynomials. Prescribing the behavior of early terminating GMRES was studied in the paper [1015] in 2014. A study of the role eigenvalues play in forming GMRES residual norms with non-normal matrices [2232] was published in 2015. Some of these results were extended to block GMRES in 2019 by Marie Kubínová and Soodhalter [1957].

In 2012, Sadok and Szyld [2675] published theoretical comparisons of the residual norms in GMRES and CMRH. Even though CMRH does not minimize the residual norm, the convergence of the two methods is not too different in many cases.

To introduce parallelism, a reordered BiCGStab method [1943] was proposed by Boris Krasnopolsky.

In 2012, Sonneveld published a paper on the convergence behavior of IDR(s) [2826] in which he used statistical arguments to show how close the IDR residual norms are to the GMRES residual norms. However, he made some hypotheses like s being very large and $s > n$, where n is the order of the matrix.

A theoretical study of methods using a deflated matrix was published by Gutknecht [1502] in 2012. A general framework for deflation and augmentation techniques for accelerating the convergence of Krylov subspace methods that satisfy a Galerkin condition was introduced by André Gaul, Gutknecht, Liesen, and Reinhard Nabben [1295] in 2013. Parallel versions of deflated restarted GMRES were proposed by Désiré Nuentsa Wakam and Erhel [3166] and Wakam, Erhel, and William Douglas Gropp [3167] in 2013.

Hiding global communication latency in GMRES was considered by Ghysels, Thomas J. Ashby, Karl Meerbergen, and Vanroose [1343] in 2013. Unfortunately, the authors did not pay too much attention to the stability of the method and to the maximum attainable accuracy. This was also the case in many of the papers in which the authors tried to introduce more parallelism in existing iterative methods.

Olaf Rendel, Anisa Rizvanolli, and Zemke [2559] summarized what was the knowledge about IDR algorithms by describing in detail the generalized Hessenberg relations corresponding to different variants of IDR. This allowed to transfer techniques known for classical Krylov subspace methods to IDR-based methods, in particular, to develop eigenvalue or Q-MR algorithms.

Keichi Morikuni and Hayami studied inner-outer iteration Krylov subspace methods for overdetermined least squares problems [2282]; see also their paper [2283] in 2015 about rank-deficient least squares problems.

A very interesting book [2067] about Krylov methods was published in 2013 by Liesen and Strakoš. They discussed in depth many aspects of these methods.

Bounds for the GMRES residual norms involving the initial residual for diagonalizable matrices were given in 2014 in the paper [3047] by David Titley-Peloquin, Jennifer Pestana, and Andrew John Wathen. This was quite different from what had been done by many previous authors who wanted to get rid of the influence of the initial residual.

In 2014, as we have seen above, Duintjer Tebbens, G.M., Sadok, and Strakoš [1018] gave exact expressions of the GMRES residual norms for normal matrices. These expressions involve the eigenvalues, the eigenvector matrix X, and the projections $X^* r_0$ of the initial residual on the eigenvectors. The proofs used Cramer's rule and the Cauchy-Binet formula to obtain these expressions. This idea was originally from Sadok. In 2015, G. M. and Duintjer Tebbens [2232] extended these results to diagonalizable matrices.

That same year, C.C. Paige, Ivo Panayotov, and Zemke [2422] published an "augmented" rounding error analysis of the nonsymmetric Lanczos process. They stressed the difficulty of obtaining good bounds because, in case of (near-) breakdown, some quantities grow unbounded.

In 2015, van Gijzen, Sleijpen and Zemke [3106] proposed flexible and multi-shift quasi-minimal IDR algorithms, flexibility meaning, as for GMRES, that a different preconditioner can be used in each iteration of the algorithm.

Mohammed Bellalij (1956-2015), Reichel, and Sadok studied properties of range restricted GMRES methods [257].

Duintjer Tebbens and G.M. [1016] showed in 2016 that any residual norm history with finite residual norms is possible for the BiCG method with any nonzero eigenvalues of the matrix A. This holds also for the QMR method. The approach was constructive but prescribing infinite residual norms in BiCG is, as far as we know, still an open problem.

In 2016, Marcel Schweitzer [2727] proved that any residual norms can be generated in the restarted FOM method. It had been proved in 2011 by Eugene Vecharynski and Langou [3140] that any history of decreasing residual norms at the last iteration of every cycle of GMRES(m) is possible. G.M. and Duintjer Tebbens extended these results in 2019 by showing that prescribing the residual norms as well as the (harmonic) Ritz values inside every cycle of GMRES(m) is possible provided there is no stagnation step at the very end of the cycle [1017].

In 2017, exact expressions for the coefficients of the FOM and GMRES residual polynomials as functions of the eigenvalues and eigenvectors were given in the paper [2229] by G.M. An optimal quasi-orthogonal residual Krylov subspace method which minimizes the residual norm was proposed in [2230]. This method used a non-orthogonal basis of the Krylov subspace.

Its construction is based on the fact that the Hessenberg matrices can be factored as $H_k = U_k C^{(k)} U_k^{-1}$, where $C^{(k)}$ is a companion matrix and U_k is an upper triangular matrix and the reciprocals of the Q-OR residual norms are the moduli of the entries of the first row of U_k^{-1}. This method seems to be mathematically equivalent to RB-SGMRES, but the implementation is completely different.

Weighted dot products for GMRES were considered by Embree, Morgan, and Huy V. Nguyen [1092] in 2017.

Zemke [3321] proposed variants of IDR in which some blocks of "basis" vectors are orthogonalized. This was called *partial orthogonalization*. The IDR family of methods was still an active area of research in the end of the 2010s.

A communication-hiding pipelined variant of BiCGStab [742] was proposed in 2016 by Cools and Vanroose; see also [739, 741].

Adaptive multilevel Krylov methods were considered in 2018 by René Kehl, Nabben, and Szyld [1882].

In 2020, G.M. and Duintjer Tebbens published a book [2233] about Krylov methods for nonsymmetric linear systems. They described many of the methods we have seen above.

Reviews about iterative methods were written by Freund, Golub, and Nachtigal [1227] in 1992, Golub and van der Vorst [1396] in 1997, Saad and van der Vorst [2669] in 2000, and Axelsson [124] in 2010. Saad wrote an essay [2664] on the history of iterative methods in 2019. Also of interest is the book [204] by Richard Barrett et al.

5.9 ▪ The method of moments

The method of moments of Y.V. Vorobyev, published in his book [3160], is at the crossing of important topics. It has applications in orthogonal polynomials, continued fractions, quadrature, methods for solving systems of linear and nonlinear equations, control theory, extrapolation methods, and differential and integral equations.

This method is due to Yuri Vasilievich (or Vasilevich) Vorobyev. As mentioned in [2067], very little information is available on this important researcher. Vorobyev is a very common name in Russia. However, we found some additional data on a man who is probably "our" Vorobyev by looking at the site of the Saint Petersburg State University (in Russian), devoted to the front-line university student participants in World War II. According to what is written there, Vorobyev was born in Leningrad on August 5, 1922. He graduated from the 102nd high school. In 1939, he entered the Faculty of Mathematics and Mechanics. He fought from 1941 to 1943, and participated in the Battle of Stalingrad as a tank driver. He was awarded two Orders of the Patriotic War and medals. He participated in the Victory Parade on June 24, 1945. He graduated from the Faculty of Mathematics and Mechanics in 1946 with a degree in mechanics. After graduating, he was sent to the State Optical Institute in Saint Petersburg, named after the physicist Sergey Ivanovich Vavilov (1891-1951), and worked there all his life. He was a Doctor of Physical and Mathematical Sciences. The date of his death is unknown to us. Let us now describe his method of moments.

Let $(u_0, \ldots, u_{k-1})$ and $(v_0, \ldots, v_{k-1})$ be two sets of linearly independent vectors in $E = \mathbb{R}^p$ and let $E_k = \text{span}(u_0, \ldots, u_{k-1})$ and $F_k = \text{span}(v_0, \ldots, v_{k-1})$. Let P_k be the oblique projection operator on E_k along $F_k^{\perp}$, that is, the projection on E_k orthogonal to F_k. Let U_k and V_k be the $p \times k$ matrices whose columns are $u_0, \ldots, u_{k-1}$ and $v_0, \ldots, v_{k-1}$, respectively. P_k exists if and only if the $k \times k$ matrix $V_k^T U_k$ is nonsingular. This is true with our hypothesis. If the matrix $V_k^T U_k$ is singular and if $u_0, \ldots, u_{k-1}$ (resp., $v_0, \ldots, v_{k-1}$) are linearly independent, then $v_0, \dots, v_{k-1}$ (resp., $u_0, \dots, u_{k-1}$) are dependent. If $u_0, \dots, u_{k-1}$ (resp., $v_0, \dots, v_{k-1}$) are independent and if $V_k^T U_k$ is nonsingular, then $v_0, \dots, v_{k-1}$ (resp., $u_0, \dots, u_{k-1}$) are also independent.

Identifying the projection operator P_k with the matrix representing it, we have

$$\mathrm{P}_k = U_k \left(V_k^T U_k\right)^{-1} V_k^T, \text{ and } \mathrm{P}_k = \mathrm{P}_k^2.$$

If y is an arbitrary vector, then $\mathrm{P}_k y = a_0 u_0 + \cdots + a_{k-1} u_{k-1}$ and $(v_i, y - \mathrm{P}_k y) = 0$ for $i = 0, \dots, k-1$. P_k is an orthogonal projector if and only if $\mathrm{P}_k = \mathrm{P}_k^T$, that is, if and only if $v_i = u_i$ for $i = 0, \dots, k-1$ or, equivalently, $E_k = F_k$. In that case, the matrix P_k is symmetric positive semidefinite.

Let us now explain Vorobyev's method of moments in a Hilbert space E. The method can be extended to a general vector space where $u_i \in E$, $v_i \in E^*$, the algebraic dual space of E, and the dot product is replaced by the duality product; see [434, 435] by C.B.

Given $u_0, \dots, u_k \in E$ and $v_0, \dots, v_{k-1} \in E$, the method of moments consists of constructing the linear mapping A_k on E_k such that

$$\begin{aligned}
u_1 &= A_k u_0, \\
u_2 &= A_k u_1 = A_k^2 u_0, \\
&\cdots\cdots\cdots\cdots\cdots\cdots \\
u_{k-1} &= A_k u_{k-2} = A_k^{k-1} u_0, \\
\mathrm{P}_k u_k &= A_k u_{k-1} = A_k^k u_0,
\end{aligned}$$

where P_k is the projection operator defined above.

These relations completely determine the mapping A_k. Indeed, any $u \in E_k$ can be written as

$$u = c_0 u_0 + \cdots + c_{k-1} u_{k-1}.$$

Thus

$$\begin{aligned}
A_k u &= c_0 A_k u_0 + \cdots + c_{k-2} A_k u_{k-2} + c_{k-1} A_k u_{k-1} \\
&= c_0 u_1 + \cdots + c_{k-2} u_{k-1} + c_{k-1} \mathrm{P}_k u_k \in E_k.
\end{aligned}$$

Since $\mathrm{P}_k u_k \in E_k$, we can write

$$\mathrm{P}_k u_k = -\beta_0 u_0 - \cdots - \beta_{k-1} u_{k-1},$$

that is,

$$\beta_0 u_0 + \cdots + \beta_{k-1} u_{k-1} + \mathrm{P}_k u_k = (\beta_0 + \beta_1 A_k + \cdots + \beta_{k-1} A_k^{k-1} + A_k^k) u_0 = 0.$$

But we have

$$(v_i, u_k - \mathrm{P}_k u_k) = 0 \quad \text{for } i = 0, \dots, k-1,$$

that is, for $i = 0, \dots, k-1$,

$$\beta_0 (v_i, u_0) + \cdots + \beta_{k-1} (v_i, u_{k-1}) + (v_i, u_k) = 0.$$

Thus, if we set

$$\widetilde{P}_k(\xi) = \beta_0 + \cdots + \beta_{k-1} \xi^{k-1} + \xi^k,$$

then

$$\widetilde{P}_k(A_k)u_0 = 0,$$

which shows that $\widetilde{P}_k$ is an annihilating polynomial of A_k for the vector u_0.

Let λ be an eigenvalue of A_k. Let us look for the eigenelements u of A_k belonging to E_k. Such a vector u can be written as

$$u = a_0u_0 + \cdots + a_{k-1}u_{k-1}.$$

Then

$$\begin{aligned} A_k u &= a_0 A_k u_0 + \cdots + a_{k-2}A_k u_{k-2} + a_{k-1}A_k u_{k-1} \\ &= a_0 u_1 + \cdots + a_{k-2}u_{k-1} + a_{k-1}\mathrm{P}_k u_k \\ &= a_0 u_1 + \cdots + a_{k-2}u_{k-1} + a_{k-1}(-\beta_0 u_0 - \cdots - \beta_{k-1}u_{k-1}) \\ &= -\beta_0 a_{k-1}u_0 + (a_0 - \beta_1 a_{k-1})u_1 + \cdots + (a_{k-2} - \beta_{k-1}a_{k-1})u_{k-1} \\ &= a_0\lambda u_0 + \cdots + a_{k-1}\lambda u_{k-1}. \end{aligned}$$

Since $u_0, \ldots, u_{k-1}$ are linearly independent in E_k, we must have

$$\begin{aligned} -\beta_0 a_{k-1} &= a_0\lambda, \\ a_i - \beta_{i+1}a_{k-1} &= a_{i+1}\lambda \qquad \text{for } i = 0, \ldots, k-2, \end{aligned}$$

that is, in matrix form,

$$\begin{pmatrix} -\lambda & 0 & 0 & \cdots & 0 & 0 & -\beta_0 \\ 1 & -\lambda & 0 & \cdots & 0 & 0 & -\beta_1 \\ \vdots & \vdots & \vdots & & \vdots & \vdots & \vdots \\ 0 & 0 & 0 & \cdots & 1 & -\lambda & -\beta_{k-2} \\ 0 & 0 & 0 & \cdots & 0 & 1 & -\beta_{k-1} - \lambda \end{pmatrix} \begin{pmatrix} a_0 \\ a_1 \\ \vdots \\ a_{k-2} \\ a_{k-1} \end{pmatrix} = 0.$$

This system has a nonzero solution, hence its determinant must be zero, that is, $\widetilde{P}_k(\lambda) = 0$. Moreover, we must have $a_{k-1} \neq 0$ since, otherwise, all the a_i's would be zero. Since an eigenelement is defined up to a multiplying factor, we can choose $a_{k-1} = 1$ and we have

$$\begin{aligned} a_{k-2} &= \beta_{k-1} + \lambda, \\ a_i &= \beta_{i+1} + a_{i+1}\lambda \qquad \text{for } i = k-3, \ldots, 0. \end{aligned}$$

This polynomial $\widetilde{P}_k$ is the characteristic polynomial of the $k \times k$ matrix $\widetilde{A}_k$ which represents the mapping A_k in E_k. Indeed, as seen above, if

$$u = c_0u_0 + \cdots + c_{k-1}u_{k-1},$$

then

$$A_k u = -\beta_0 c_{k-1}u_0 + (c_0 - c_{k-1}\beta_1)u_1 + \cdots + (c_{k-2} - c_{k-1}\beta_{k-1})u_{k-1}.$$

The transformation mapping the coordinates $c_0, \ldots, c_{k-1}$ of u in the basis formed by the elements $u_0, \ldots, u_{k-1}$ into the coordinates of $A_k u$ in the same basis is given by

$$\begin{pmatrix} 0 & \cdots & 0 & 0 & -\beta_0 \\ 1 & \cdots & 0 & 0 & -\beta_1 \\ \vdots & & \vdots & \vdots & \vdots \\ 0 & \cdots & 0 & 1 & -\beta_{k-1} \end{pmatrix} \begin{pmatrix} c_0 \\ c_1 \\ \vdots \\ c_{k-1} \end{pmatrix} = \begin{pmatrix} -\beta_0 c_{k-1} \\ c_0 - c_{k-1}\beta_1 \\ \vdots \\ c_{k-2} - c_{k-1}\beta_{k-1} \end{pmatrix}.$$

Thus, we see that the characteristic polynomial of the matrix $\widetilde{A}_k$ is $\widetilde{P}_k$. Consequently, $\widetilde{A}_k$ is regular if and only if $\beta_0 \neq 0$, and the rank of A_k is equal to the rank of $\widetilde{A}_k$.

In the case where $u_i = A^i u_0$, $i = 0, 1, \ldots$, it is possible to obtain an expression for A_k. Let $u = c_0 u_0 + \cdots + c_{k-1} u_{k-1}$ be an arbitrary element of E_k. Then

$$\begin{aligned} Au &= c_0 A u_0 + \cdots + c_{k-2} A u_{k-2} + c_{k-1} A u_{k-1} \\ &= c_0 A u_0 + \cdots + c_{k-2} A^{k-1} u_0 + c_{k-1} A^k u_0 \\ &= c_0 A_k u_0 + \cdots + c_{k-2} A_k^{k-1} u_0 + c_{k-1} A^k u_0, \end{aligned}$$

and it follows that

$$\begin{aligned} \mathrm{P}_k A u &= c_0 A_k u_0 + \cdots + c_{k-2} A_k^{k-1} u_0 + c_{k-1} \mathrm{P}_k u_k \\ &= c_0 A_k u_0 + \cdots + c_{k-2} A_k^{k-1} u_0 + c_{k-1} A_k^k u_0 \\ &= A_k (c_0 u_0 + \cdots + c_{k-1} u_{k-1}) = A_k u, \end{aligned}$$

which shows that $A_k = \mathrm{P}_k A$ on E_k. Since if $u \in E_k$, $\mathrm{P}_k u \in E_k$, the domain of A_k can be extended to the whole space E by setting

$$A_k = \mathrm{P}_k A \mathrm{P}_k.$$

Let $P_k(\xi) = \widetilde{P}_k(\xi) / \widetilde{P}_k(0)$. Then

$$P_k(\xi) = \begin{vmatrix} 1 & \cdots & \xi^k \\ (v_0, u_0) & \cdots & (v_0, u_k) \\ \vdots & & \vdots \\ (v_{k-1}, u_0) & \cdots & (v_{k-1}, u_k) \end{vmatrix} \Bigg/ \begin{vmatrix} (v_0, u_1) & \cdots & (v_0, u_k) \\ \vdots & & \vdots \\ (v_{k-1}, u_1) & \cdots & (v_{k-1}, u_k) \end{vmatrix}.$$

Let $Ax = b$ be a system of linear equations, x_0 an arbitrary vector, set $u_0 = r_0 = b - Ax_0$, and consider the sequence of vectors (x_k) defined as the solution of $A_k(x_k - x_0) = r_0$. Then $x_k - x_0 = Q_{k-1}(A_k) r_0$, where the polynomial Q_{k-1} is such that $1 - P_k(\xi) = \xi Q_{k-1}(\xi)$. The polynomial Q_{k-1} has degree $k-1$ at most since $P_k(0) = 1$, then $Q_{k-1}(A_k) = Q_{k-1}(A)$ and thus $x_k = x_0 + Q_{k-1}(A) r_0$. Multiplying both sides by A and subtracting b leads to $r_k = r_0 - A Q_{k-1}(A) r_0$, where $r_k = b - Ax_k$. That is, $r_k = P_k(A) r_0$.

By construction, we have $(v_i, P_k(A) r_0) = (v_i, r_k) = 0$ for $i = 0, \ldots, k-1$, which is the definition of a Lanczos-type method for solving a system of linear equations. Moreover, by the preceding formula for P_k, the vectors r_k and x_k can be written as ratios of determinants. For more details about Vorobyev's method and its connection with Lanczos methods, see [434] and [435, pp. 154-164] by C.B. If $v_i = (A^*)^i v_0$, the original Lanczos' method is recovered. For a block version of Vorobyev's and Lanczos' methods, see [437]. For the connection of Vorobyev's work with the concept of matching moment model reduction, see Strakoš [2921].

5.10 ▪ Preconditioning

In our context, *preconditioning* means multiplying the matrix A by another nonsingular matrix to change its properties. For a linear system $Ax = b$, this can be done in different ways. Given a nonsingular matrix M, if

$$Ax = b \to M^{-1} A x = M^{-1} b,$$

we speak of left preconditioning, and if

$$Ax = b \to A M^{-1} (M x) = b,$$

we speak of right preconditioning. When A is symmetric, it is sometimes important to maintain symmetry and to use

$$Ax = b \to M^{-\frac{1}{2}}AM^{-\frac{1}{2}}(M^{\frac{1}{2}}x) = M^{-\frac{1}{2}}b,$$

with M symmetric positive definite. Note that when using this kind of preconditioning with CG, it is not necessary to compute $M^{-\frac{1}{2}}$. With a change of variable, one has only to consider M^{-1}. Another possibility is to use $M = LL^T$ with L lower triangular and

$$Ax = b \to L^{-1}AL^{-T}(L^Tx) = L^{-1}b.$$

One can also combine left and right preconditioning. This is sometimes called split preconditioning. The matrix M is known as the *preconditioning matrix* or the *preconditioner*. In Krylov subspace iterative methods, it is generally necessary to solve a linear system $Mz = r$ at each iteration.

The goal of the preconditioner can be to improve the condition number or to obtain a distribution of the eigenvalues more favorable to the convergence of some iterative methods. However, in most cases, it is difficult to assess theoretically what is the effect of a preconditioner. The construction of many preconditioners was based on intuition, for instance, that it would be nice if M "looks like" A^{-1}. However, the sense in which $M^{-1}A$ should be close to the identity matrix depends on the iterative method to be used, and it has been said that devising preconditioners is more an art than a science.

There are many ways to construct preconditioners. Some people considered this as a purely algebraic problem, given a matrix A, find a preconditioner M such that CG or GMRES, or any other iterative method has a fast convergence. It is a "matrix given" approach. Another possibility is to consider where the linear system is coming from, for instance, from the discretization of a PDE, and to use this knowledge to build a more specific preconditioner. It is a "problem given" approach.

There is a lot of freedom for constructing a preconditioner, but there are also some constraints. The preconditioner itself M or M^{-1} must be easy to construct. If too much time is spent in constructing the precondtioner, there may be no benefit to use it, except if several linear systems with the same matrix are to be solved. If A is a sparse matrix, M must be sparse and not using too much storage. A linear system with M must be easy to solve because this has to be done at each iteration. If A is symmetric positive definite, M must be symmetric and positive definite. Of course, the most difficult thing is to find a preconditioner that improves the convergence of the given iterative method.

Thousands of papers have been written on preconditioning since the 1960s. Therefore, we will consider only what we feel were the main trends.

Even though they were not using matrices, we may think that the idea of preconditioning goes back to Gauss and Jacobi in the 19th century since, after all, formulating the normal equations $A^TAx = A^Tb$ is using a form of preconditioning. However, getting the normal equations was imposed by the problem since this is how the solution of the least squares problem is characterized. It was realized much later that solving the normal equations is not the best way to solve a least squares problem. However, we observe that in 1845, Jacobi used rotations to improve the diagonal dominance of the linear system he was considering. This can be interpreted as preconditioning.

Probably the first scholar to use preconditioning purposely was Cesari in Italy in 1937. In [601], page 27 and following, he considered transforming a linear system that he wrote $AX = H$ with A symmetric positive definite into another system $A'X = H'$,

> *having the same unknowns, and such that the successive approximation methods result in a more rapid convergence and therefore be of practical applicability.*

He chose to left multiply A by a polynomial he denoted by $f'(A)$. He knew that if η_i are the eigenvalues of A, the eigenvalues of $f(A) = f'(A)A$ are $f(\eta_i)$. He wanted to construct a polynomial f such that $f(0) = 0$ and such that the ratio of the largest value of f to the smallest value of f on the interval containing the eigenvalues was the smallest possible. He computed the coefficients of polynomials of degree 2, 3, and 4. Finally, he gave a numerical example with a matrix of order 3, a polynomial of order 2, and the Gauss-Seidel method. What he did was preconditioning by constructing M^{-1} directly.

The word "preconditioning" appeared on page 299 of [3079] by Alan Mathison Turing (1912-1954) in 1948. However, it does not exactly correspond to the definition we gave above.

One may also consider that preconditioning was already implicitly included in the 1952 paper of Hestenes and Stiefel [1664], page 424. As we have seen in a previous section, they saw the multiplication of the linear system by another matrix as a generalization of the normal equations:

> *There is a slight generalization of the system (10:2) that is worthy of note. This generalization consists of selecting a matrix B such that BA is positive definite and symmetric. The matrix B is necessarily of the form A^*H, where H is positive definite and symmetric. We can apply the cg-algorithm to the system $BAx = Bk$.*

They gave the CG formulas for this last system. But the only choices they proposed were $B = A^*$ or $B = I$.

The best preconditioner is, of course, $M = A$, but this is clearly infeasible. The simplest preconditioner one can think of is a diagonal matrix $M = D$. This is sometimes called the Jacobi preconditioner in reference to the classical iterative method. But, how to choose the diagonal entries of D? If one multiplies from the left, it is a scaling of the rows of A, and multiplying from the right gives a scaling of the columns; one can, of course, do both. This problem was considered independently of iterative methods as it is also sometimes relevant for direct methods. This operation is called *scaling* or *equilibration*.

A more general problem was studied by Forsythe and Straus in [1198]. This work was presented at the *International Congress of Mathematicians* in 1954, and the paper published in 1955. They considered a Hermitian positive definite matrix A and a class $\mathcal{T}$ of regular transformations $A_{\mathcal{T}} = T^*AT$. They said A is best conditioned with respect to $\mathcal{T}$ if the condition number of $A_{\mathcal{T}}$ is larger than or equal to the condition number of A for all $T \in \mathcal{T}$. They gave sufficient conditions for A to be or not to be best conditioned. They used the word "preconditioning" on page 342. However, they did not tell how to choose or compute T.

Preconditioning with diagonal matrices for the computation of eigenvalues was proposed by Elmer Edwin Osborne, a student of L.J. Paige [2390], in 1960.

The optimal diagonal scaling of matrices was considered by Bauer [220] in 1963. He showed that the problem can be solved for the maximum norm condition number and also for Hölder norms if the matrices A and A^{-1} have some blockwise checkerboard sign distribution.

In 1969, Van der Sluis [3094, 3095] characterized the best scaling for different condition numbers, $\|A\|_\infty \|A^{-1}\|_*$ with the $*$ denoting any Hölder norm or the Frobenius norm for the left scaling and $\|A\|_1 \|A^{-1}\|_*$ for the right scaling. A Hölder p-norm is defined for a vector as $\|x\|_p = (\sum_i |x_i|^p)^{\frac{1}{p}}$. The most well-cited result of that paper is for the ℓ_2 condition number

$$\kappa(A) \le m \min_D \kappa(D^*AD)$$

if all diagonal entries of the Hermitian positive definite matrix A are equal, where m is the maximum number of entries in rows of A. Van der Sluis wrote

> *Symmetric scaling for equal diagonal elements can be considered as reasonably optimal.*

In 1973, Charles Alan McCarthy (1936-2018) and Gilbert Strang [2175] defined the optimal condition number c of a matrix A as the minimum, over all diagonal matrices D, of $\|DA\| \, \|(DA)^{-1}\|$. They established that, in the ℓ_2 norm, the inequality $b \leq c \leq 2b$ holds, and disproved the conjecture that always $b = c$, where b is the supremum of $\|A^{-1}UA\|$ over all diagonal unitary matrices U.

Golub and James Martin Varah [1400] considered the characterization of the best ℓ_2 scaling of a matrix in 1974. For a nonsingular matrix A, the problem is

$$\min_{D,E} \|DAE\| \, \|(DAE)^{-1}\| = \min_{D,E} \frac{\sigma_{\max}(DAE)}{\sigma_{\min}(DAE)},$$

where D and E are diagonal matrices and $\sigma_{\max} = \sigma_1$ (resp., $\sigma_{\min} = \sigma_n$) is the maximum (resp., minimum) singular value. Let $A = U\Sigma V^T$ be the singular value factorization of A (see Chapter 4) and $u^{(j)}$ (resp., $v^{(j)}$) be the columns of U (resp., V). Golub and Varah defined the EMC property as $|u_i^{(1)}| = |u_i^{(n)}|$ and $|v_i^{(1)}| = |v_i^{(n)}|$ for $i = 1, \dots, n$. They showed that for matrices having simple $\sigma_{\max}$ and $\sigma_{\min}$, EMC is a necessary and sufficient condition for A to be best scaled. They also discussed the extension of their results to rectangular matrices.

In 1989, Greenbaum and Garry Hector Rodrigue [1449] considered the problem of finding a symmetric positive definite preconditioner M of a given nonzero structure which minimizes the condition number of $M^{-1}A$ for a given symmetric positive definite matrix A. In particular, they used an optimization code to numerically compute the optimal diagonal preconditioners for model problems. For the 5-point finite difference discretization of the Laplacian matrix on a regular Cartesian mesh, the code converged to the diagonal of A as was expected. This same idea of minimizing the condition number was pursued much earlier for a certain class of preconditioners in 1973 by Paul Concus and Golub [733]. They considered one-dimensional model problems and used an optimization code to find an optimal diagonal scaling of the Laplacian.

The next step for constructing simple preconditioners is to use block diagonal matrices. Van der Sluis' theorem was generalized to those preconditioners, but m is now the number of nonzero blocks in a block row. This result was a consequence of theorems by Demmel [865, 866] in 1982-1983 and Eisenstat, John W. Lewis, and Schultz [1070] in 1982, who generalized results by Forsythe and Straus. The equivalence of their results was shown by Ludwig Elsner [1088] in 1984.

The real development of preconditioners started in the 1960s. Let us first concentrate on incomplete factorizations of a matrix A. When doing the LU (or Cholesky in the symmetric case) factorization of a sparse matrix, the fill-in phenomenon makes some zero entries in A to become nonzero in L or U; see Chapter 4. The idea of incomplete factorization is to neglect some or all of these fill-ins during the elimination process.

One of the pioneers of incomplete factorization was the Russian mathematician Nikolai Ivanovich Buleev (1922-1984) (or sometimes transliterated as Buleyev) who published in 1960 a paper [498] written in 1958. A translation of the title is *A numerical method for the solution of two-dimensional and three-dimensional equations of diffusion*. For the two-dimensional problem, Buleev considered a 5-point finite difference approximation of the PDE. He assumed the matrix to be diagonally dominant and solved the linear system by successive approximations. His derivation of the incomplete factorization is far from crystal clear since he did not use matrices but only the difference equations. Nevertheless, his method corresponds to modified incomplete factorizations that we will consider in a moment. Buleev published another paper [499] on the same topic in 1970. Concerning the history of this work, Valery Pavlovich Il'in wrote in his book [1773]

> *Fortunately, the author in his younger years witnessed the beginning of a new trend in computational algebra - methods of incomplete factorization. It goes back to the end of the 50s at the Institute of Physics and Energy, Department of Mathematics headed by G.I. Marchuk, in the town of Obninsk near Moscow, which used to be one of the secret towns, where the first atomic power station in the world was constructed. Prof. N.I. Buleev, head of the Laboratory of Numerical Methods in Hydrodynamics and the author of one of the original turbulent theories was keen on calculations and he created his own algorithm for iterative computations of two- and three-dimensional flows, which were successfully used on URAL-1 computer, with the capacity of 100 operations per second. The method was described in the "close"* [classified] *research reports and was first to appear in the book "Methods of Calculation of Nuclear Reactors" by G.I. Marchuk in 1958. Though the new class of algorithms was practically effective and presented generalizations, it did not draw much attention. First, it was due to the absence of an adequate theory and forced purely experimental grounds. The second reason was due to the preference of then popular methods of optimal relaxation and alternating directions which had attractive theoretical investigations.*

Another pioneer was Varga in 1960. In the paper [3131], he considered the solution of two-dimensional diffusion equations with finite difference methods which yield Stieltjes matrices, that is, irreducible symmetric positive definite matrices with non-positive off-diagonal entries. In Section 6, he proposed to write $A = T^T T - C$, where T is a sparse upper triangular matrix. His choice was to have off-diagonal nonzero entries in T that correspond to the right and top neighbors of a given mesh point. He proved that doing so he obtained a regular splitting of A. He also made proposals for 9-point finite difference approximations.

As noted by Varga, a similar method was proposed by Thomas A. Oliphant Jr., who was working at the Los Alamos Scientific Laboratory, in 1958, with a paper [2375] published in 1962. As did Varga, Oliphant considered a 5-point finite difference scheme for discretizing a diffusion equation on a regular mesh. It yields a block tridiagonal matrix with five nonzero diagonals. Oliphant used $A = D + L + U$ with D diagonal, L strictly lower triangular and U strictly upper triangular and wrote the linear system to be solved as $Cx = b - (1-k)Ux$ with $C = D + L + kU$. He looked for a factorization $C = WV - H$, where W (resp., V) has the same nonzero structure as $D + L$ (resp., $D + U$) and V has a unit diagonal. He noticed that H has only two nonzero diagonals, next to the outer diagonals of A. By identification, he wrote the formulas to compute the nonzero entries of W and V. Then, he proposed to use the iteration $WVx_{i+1} = b + [H - (1 - kU)]x_i$. He also used an extrapolated method and gave a numerical example for the Laplacian on a 10×10 grid; see also [2376].

A clear exposition and study of the methods of Buleev and Oliphant for M-matrices was done in a report [236] by Robert Beauwens in 1973. He also considered a less well-known method proposed by Zbigniew Ignacy Woźnicki (1937-2008), a Polish mathematician, in his Ph.D. thesis [3274] in 1973. Beauwens introduced also block versions of Buleev and Oliphant methods.

In 1968, for solving a diffusion equation in a two-dimensional region discretized with a 5-point finite difference scheme leading to $Ax = b$, Todd F. Dupont, Kendall, and Rachford [1029] used as an iterative method $(A + B)x_{i+1} = (A + B)x_i - \omega(Ax_i - b)$ with $A + B = M = LL^T$, where L is lower triangular with no more than three nonzero entries in a row. In fact, the matrix B is equal to $R + \delta D$, where D is the diagonal of A and R has nonzero entries in two adjacent diagonals to the outer diagonals of A and inside the band in addition to the main diagonal and such that the sums of the entries of each row is equal to zero. The parameter δ was taken proportional to the square of the mesh size h multiplied with a positive constant. The authors essentially showed that under mild conditions, the smallest eigenvalue of $M^{-1}A$ is

$\mathcal{O}(h^{-1})$ as well as the condition number. This shows an improvement of the condition number since for these problems, $\kappa(A) = \mathcal{O}(h^{-2})$. The authors cited Buleev, but not Varga or Oliphant. Their preconditioner was later known as DKR. Its spirit is close to what was done by Buleev. See also Dupont [1028] for a generalization to the 9-point finite difference scheme.

In the same issue of the *SIAM Journal on Numerical Analysis*, there was a paper [2911] by Herbert L. Stone, who was working for the Esso Production Research Company. This paper was following a 1966 report. He used the same technique as Dupont, Kendall, and Rachford, but with an incomplete LU factorization using a modification of the diagonal entries. It is a nonsymmetric procedure, even for a symmetric matrix. Stone's approach was more intuitive and less mathematically rigorous than in [1029]. He did a Fourier analysis to study convergence of his method that he named the *strongly implicit method*. He gave numerical examples with a 31×31 grid and did comparisons with the Jacobi, SOR, and ADI methods. On diffusion problems with discontinuous coefficients, Stone's method was faster than the others. Stone cited Buleev, Oliphant and Dupont, Kendall, and Rachford. Stone's algorithm was studied in 1973 by Amnon Bracha-Barak and Saylor in [390].

A problem with the methods of Buleev, Oliphant, and Stone was that they contained several parameters whose selection was not so obvious. This is probably why these methods were not much used.

In 1973, A.D. Tuff and Jennings described in [3078] a method called *partial elimination* in which they discarded some entries based on their values. It was intended for linear systems arising from structural analysis; see also Jennings and Malik [1813] in 1977.

Approximate (or incomplete) factorizations of a matrix A may break down, with a division by zero or by a small quantity, depending on the properties of the matrix. Therefore, sufficient conditions were looked for that could guarantee the feasibility of the factorization. This was done, for example, by Beauwens [236] in 1973, as well as Beauwens and Lena Quenon [243] in 1976. They considered the Buleev and Oliphant methods as well as a generalization of Stone's method. They derived feasibility sufficient conditions for M- and H-matrices. They also considered the corresponding block methods.

An approximate factorization procedure based on the block Cholesky decomposition to be used with the conjugate gradient method was described by Richard Underwood in a 1976 report [3091].

In 1977, Meijerink and van der Vorst described a general incomplete factorization method [2191] for M-matrices, that is, nonsingular matrices A such that $a_{i,j} \leq 0, i \neq j$ and $A^{-1} \geq 0$. During the usual LU factorization, entries are neglected in the L and U matrices at chosen off-diagonal places. This can be done since the matrix that arises from an M-matrix after one elimination step is again an M-matrix (as was proved by Fan [1133] in 1960) and zeroing off-diagonal entries of an M-matrix also gives an M-matrix. For the symmetric case, the authors proposed to use these preconditoners with the conjugate gradient method. They named this method ICCG, IC standing for Incomplete Cholesky. They showed some examples for two-dimensional diffusion PDEs and a 5-point finite difference scheme. The incomplete factorization without fill-in was named IC(0), but this was changed later to IC(1,1) because there are only two nonzero diagonals in addition to the main diagonal in the factors L and U. For these matrices, in IC(1,1), it is just necessary to compute the diagonal entries and the other ones are equal to the corresponding entries of A. The only previous paper on this topic which is cited is Stone's. The paper of Meijerink and van der Vorst was really influential in the following years and it was highly cited. This paper was complemented by [2192] in 1981. Many of the preconditioners that were proposed later were only slight variations around the same ideas.

David S. Kershaw from the Lawrence Livermore Laboratory was a proponent of incomplete Cholesky and LU factorizations; see [1886]. In 1978-1980, he studied the problem of small pivots that could arise, depending on the properties of the matrix A. The prescription in [1886] was to set the pivot to some arbitrary nonzero value. A better fix was proposed in [1887, 1888] which, in some sense, minimized the modifications induced by the fix. The 1978 paper was also highly cited, probably because it was published in a journal devoted to computational physics.

In the 1977 report [1448], Greenbaum and Rodrigue showed how to implement IC(1,1) for block tridiagonal matrices on the *Control Data STAR-100* vector computer that was used at the Lawrence Livermore Laboratory. The part that was difficult to vectorize was the solution of $Mz_k = LDL^T z_k = r_k$ at each iteration. They wrote L in block form and used recursive doubling [973, 2912] to solve the bidiagonal systems corresponding to each diagonal block.

Chandra [630] proved in 1978 that for the model problem arising from the 5-point finite difference discretization of the two-dimensional Laplacian, the LL^T incomplete factorization IC(1,1) of A is such that

$$\frac{1}{17}\kappa(A) \leq \kappa(L^{-1}AL^{-T}) \leq 17\kappa(A).$$

In fact, the constant in the upper bound is 16.219. This result showed that contrary to DKR, regarded as a function of the mesh size, the condition number of $M^{-1}A$ is of the same order as for A. However, the distribution of the eigenvalues is more favorable and it explains the improvement in convergence.

A class of first-order factorization methods for the solution of sparse systems of linear equations [1488, 1489] was introduced by Ivar Gustafsson, a student of Axelsson, in 1978. He gave asymptotic results for the computational complexity and numerical experiments. This preconditioner with diagonal modifications is known as MIC (Modified Incomplete Cholesky).

In 1979, Beauwens revisited the methods that he called OBV, in relation to Oliphant, Buleev, and Varga [238]. He gave sufficient feasibility conditions for M-matrices and derived bound for the eigenvalues of the preconditioned matrix in the symmetric case. He also related his results to those of Axelsson [115, 116, 117] and Gustafsson [1488] when using a diagonal modification; see also [239] in 1985. A conditioning analysis of approximate factorizations for positive definite matrices was done by Beauwens and Renaud Wilmet [244] in 1989.

Incomplete block factorization for finite difference matrices was also considered by Robert G. Steinke [2863] in 1979.

Manteuffel considered the incomplete Cholesky preconditioner for symmetric positive definite matrices [2134] in 1980. He proved that whatever is the choice of the off-diagonal entries to be neglected, the incomplete factorization is feasible if A is an H-matrix with positive diagonal entries. We observe that this was proved before by Beauwens for slightly different factorizations; see [237]. The factorization proposed by Meijerink and van der Vorst could fail if the matrix does not have more stringent properties than being symmetric positive definite. To handle these problems, Manteuffel proposed to introduce a shift α and, if $A = D - B$ with D being the diagonal of A, to use the incomplete factorization of $D - B/(1+\alpha)$ as a preconditioner. If α is large enough, the factorization is feasible. He studied the asymptotic behavior of the factorization when $\alpha \to \infty$. However, the value of α which minimized the number of iterations or the computing time, or even the condition number of the preconditioned matrix, was unknown. He discussed some numerical experiments for a three-dimensional structural mechanics problem with 18,000 unknowns discretized by a finite element method. This problem cannot be solved by direct methods with the computers available at that time, *CDC 6600* and *7600* (see Chapter 7), and he also gave results for some smaller problems that allow to consider many more values of α. Most of the experiments were done by neglecting the fill-in out of the nonzero structure of A. Manteuffel also suggested to keep the fill-in corresponding to unknowns having a common neighbor in the graph of A.

Varga, Edward Barry Saff, and Volker Mehrmann [3138] characterized the matrices that are incompletely factorizable in 1980. Let $\mathcal{F}_n$ be this set and $\mathcal{H}_n$ be the set of nonsingular H-matrices. It was shown that $\mathcal{H}_n \subset \mathcal{F}_n$. Let

$$\begin{aligned}\Omega(A) &= \{B|\ M(B) = M(A)\},\\ \Omega^d(A) &= \{B|\ |b_{i,i}| = |a_{i,i}|,\ |b_{i,j}| \le |a_{i,j}|, i \ne j\},\\ \mathcal{F}_n^d &= \{A \in \mathcal{F}_n, \Omega^d(A) \subseteq \mathcal{F}_n\}.\end{aligned}$$

The authors showed that $\mathcal{H}_n = \mathcal{F}_n^d$. Moreover, if

$$\mathcal{F}_n^c = \{A \in \mathcal{F}_n,\ \Omega(A) \subseteq \mathcal{F}_n\},$$

then $\mathcal{H}_n$ is strictly contained in $\mathcal{F}_n^c$ which is also strictly contained in $\mathcal{F}_n$. This proved that there exist matrices that can be incompletely factorized which are not H-matrices.

In 1980, Axelsson and Gustafsson [127] considered symmetric matrices arising from 5-point finite difference schemes on two-dimensional grids. The grid points (nodes) can be ordered with the so-called red/black ordering (also called the checkerboard ordering) in which red (resp., black) nodes are only connected to black (resp., red) nodes. They used a symmetric diagonal scaling, eliminated the red (or black) unknowns, and applied a modified incomplete factorization to the reduced matrix.

A clever way of implementing an LDL^T incomplete factorization in the conjugate gradient algorithm allowing to decrease the number of operations was described by Eisenstat [1065] in 1981. This was later known as *Eisenstat's trick*.

Yves Robert [2582] gave another proof of the fact that H-matrices can be incompletely factorizable in 1982. He also defined a general factorization which, for symmetric positive definite matrices A, is the following. Let

$$A = \begin{pmatrix} a_{1,1} & a_1^T \\ a_1 & B_1 \end{pmatrix}, \quad a_1 = b_1 - r_1.$$

The remainder is constructed as follows,

$$R_1 = \begin{pmatrix} r_{1,1}^1 & r_1^T \\ r_1 & D_{R_1} \end{pmatrix},$$

with D_{R_1} diagonal such that $(D_{R_1})_{j,j} = |(r_1)_j|$ and $r_{1,1}^1 = \sum_{j=1}^{n-1} |(r_1)_j| > 0$. The splitting of the matrix A is

$$A = A_1 = \begin{pmatrix} a_{1,1} + r_{1,1}^1 & b_1^T \\ b_1 & B_1 + D_{R_1} \end{pmatrix} - \begin{pmatrix} r_{1,1}^1 & r_1^T \\ r_1 & D_{R_1} \end{pmatrix} = M_1 - R_1.$$

Note that something is added to the (positive) diagonal entry of A. Then, M_1 is factorized,

$$M_1 = \begin{pmatrix} 1 & 0 \\ \ell_1 & I \end{pmatrix} \begin{pmatrix} a_{1,1} + r_{1,1}^1 & 0 \\ 0 & A_2 \end{pmatrix} \begin{pmatrix} 1 & \ell_1^T \\ 0 & I \end{pmatrix}.$$

By identification,

$$\ell_1 = \frac{b_1}{a_{1,1} + r_{1,1}^1}, \quad A_2 = B_1 + D_{R_1} - \frac{1}{a_{1,1} + r_{1,1}^1} b_1 b_1^T.$$

Robert showed that A_2 is positive definite, and therefore, the factorization can continue in the same way and symmetric positive definite matrices can be factorized incompletely whatever is the choice of the off-diagonal entries that are neglected. With this factorization the eigenvalues of $M^{-1}A$ are smaller than 1.

As we wrote above, it was noticed that there can be troubles with small pivots in incomplete factorizations of nonsymmetric problems. Elman [1083] studied the stability of the factorization and of the triangular solves. He considered the following two-dimensional model problem in the unit square with Dirichlet boundary conditions

$$-\Delta u + 2P_1 \frac{\partial u}{\partial x} + 2P_2 \frac{\partial u}{\partial y} = f,$$

which is approximated with finite difference schemes. Elman did not directly study ILU and MILU but the constant coefficient variants where the diagonal entries are replaced by their limits. The factorizations are not stable for all values of P_1 and P_2. There were regions of stability in the (P_1, P_2) plane. For instance, for centered finite differences and ILU when $P_1 \geq 0, P_2 \geq 0$, the condition for stability is $p_1 p_2 \leq 1$, where $p_i = P_i h, i = 1, 2$, h being the mesh size. The numerical experiments showed a high correlation between these stability results and the behavior of those preconditioners for several iterative methods.

In 1991, T.F. Chan studied the so-called *relaxed incomplete factorizations* (RIC and RILU) [614], using Fourier analysis. This is quite similar to modified incomplete factorizations, but the values that are added to the diagonal are multiplied by a relaxation parameter ω, $0 \leq \omega \leq 1$.

We have said that the more one knows about the problem, the easier it is to construct "good" preconditioners. For example, Clemens W. Brand considered the Laplacian matrix arising from a 5-point finite difference scheme in the unit square [404] in 1992. A red/black ordering allows to eliminate about half of the unknowns. Brand's idea was to drop some of the generated fill-ins, obtaining equations for which the red/black ordering can be applied again. This technique is applied recursively for several levels giving the repeated red/black (RRB) preconditioner. This was further studied by Patrick Ciarlet [698] in 1994. Some of these preconditioners gave a condition number of $M^{-1}A$ which is independent of the mesh size, but they are not as general as some other incomplete factorizations.

In 1994, Axelsson published a book about iterative methods [122] in which he described and studied some preconditioners. Another book about preconditioners was written by Are Magnus Bruaset [485] in 1995.

Many variants of incomplete factorizations were proposed over the years, for instance, Miron Tismenetsky [3046] constructed a method based on the description of the triangular factorization of a matrix as a product of elementary matrices in 1991.

In 1994, Saad considered several techniques in [2660]. One was to keep fill-ins according to their level. A level is recursively attributed to each fill-in entry from the level of fill-in of its parents. Each fill-in whose level exceeds a given threshold was dropped. Another possibility was to drop fill-ins according to their value, but to keep only a given number of the largest ones in modulus to control the storage for the preconditioner. There are two parameters in this ILUT preconditioner: the threshold and the number of entries to keep for each row.

Since drop tolerance–based preconditioners may be difficult to use in a black-box fashion, another variant that does not require the selection of a drop tolerance and has predictable storage requirements was proposed by Mark T. Jones and Paul Eugene Plassmann [1839] in 1995. In their approach, a fixed number m_k of nonzero entries was allowed in the kth row of the incomplete triangular factor L. Originally, it was the number of nonzero entries in the lower triangular part of A, but only the nonzeros of largest magnitude were kept. This strategy had the advantage of being a black box method, but convergence was not always very good. An extension, proposed by Chih-Jen Lin and Jorge Jesus Moré [521] in 1999, was to keep $m_k + p$ nonzero entries where p was defined by the user. Compared to Saad's proposal, there was no user-defined threshold.

Incomplete factorizations were still a topic of research in the 2000s, even though the basic principles and the main ideas were known for quite some time. A few examples will follow. Preconditioners based on a decomposition $U^T U + U^T R + R^T U$ of A where U is an upper triangular matrix and R is strictly upper triangular were proposed in 1998 by Igor Evgen'evich Kaporin [1872]. In 2001, Matthias Bollhöfer presented an ILU factorization with a refined dropping strategy that monitored the growth of the inverse factors of L and U; see [362] and also [363]. Na Li, Saad, and Chow [2050] used in 2003 an implementation of ILU based on the Crout factorization (see Chapter 2). A Crout version with pivoting was investigated by Li and Saad [2049] in 2005. Rafael Bru, José Marín, José Mas, and Tůma introduced *balanced incomplete factorizations* [481] in 2008. They computed the triangular factors and their inverses at the same time. Xiaoye Sherry Li and Meiyue Shao used supernodes (see Chapter 2) for an incomplete LU factorization with partial pivoting [2060] in 2011.

When the matrix has a natural block structure, efficient incomplete factorizations that take advantage of this structure can be developed. This was done by several authors in the 1980s and 1990s. Following the work of Underwood, Concus, Golub, and G.M. [735] proposed several block incomplete factorization methods for block tridiagonal matrices in 1985. This paper is an abridged version of a report which was presented at the *SIAM 30th Anniversary Meeting* held in Stanford University in 1982. Let

$$A = \begin{pmatrix} D_1 & A_2^T & & & \\ A_2 & D_2 & A_3^T & & \\ & \ddots & \ddots & \ddots & \\ & & A_{m-1} & D_{m-1} & A_m^T \\ & & & A_m & D_m \end{pmatrix},$$

each block being of order m, $n = m^2$. Let L be the block lower triangular part of A. The exact block factorization of A can be written as $A = (\Sigma + L)\Sigma^{-1}(\Sigma + L^T)$, where Σ is a block diagonal matrix whose diagonal blocks are denoted by Σ_i. By identification,

$$\Sigma_1 = D_1, \quad \Sigma_i = D_i - A_i(\Sigma_{i-1})^{-1}A_i^T, \; i = 2, \ldots, m.$$

The idea used in [735] to obtain an incomplete decomposition is to replace the inverses by sparse approximations. The authors defined

$$M = (\Delta + L)\Delta^{-1}(\Delta + L^T),$$

where Δ is a block diagonal matrix whose elements are computed generically as

$$\begin{aligned} \Delta_1 &= D_1, \\ \Delta_i &= D_i - A_i \,\text{approx}(\Delta_{i-1}^{-1})A_i^T, \end{aligned}$$

where approx(Δ_{i-1}^{-1}) is a sparse approximation of Δ_{i-1}^{-1}. Several approximations were proposed in [735]. One of the most efficient, denoted by INV, just considered tridiagonal approximations of the inverses. A modified block method called MINV was also considered in which the row sums of the remainder are equal to zero. INV and MINV were shown to be feasible for H-matrices. For the computation of these preconditioners, see also [2209] in 1984 and [736] by Concus and G.M. in 1986.

Block preconditioners were analyzed using Fourier analysis by T.F. Chan and G.M. [622] in 1990. This allowed the investigation of the condition numbers and the eigenvalue distributions. It is amusing to note that when submitted, this paper was rejected, and that one referee wrote "this is not mathematics"!

Another proponent of block factorizations was Axelsson, who published many papers on that topic; see [125] with Sjaak Brinkkemper and Il'in in 1984 and [119] in 1985. The idea of block incomplete factorization for block tridiagonal matrices can be easily generalized to any block structure. This was proposed by Axelsson [120] in 1986. However, there are not that many practical situations where this was used in practice. For general sparse matrices, it is not so easy to exhibit a natural block structure except if the ordering of the unknowns has been devised to do so. For other proposals of block factorizations, see Axelsson and Ben Polman [133] in 1986. Approximate block factorizations were also proposed by Beauwens and Mustapha Ben Bouzid [241, 242] in 1987-1988.

In 1995, T.F. Chan and Panayot Spirov Vassilevski [626] described an extension of the block incomplete factorization algorithm. They introduced matrices that are restriction operators transforming vectors to a lower-dimensional space to construct corrections of the factorization. For problems arising from discretization of PDEs this can be seen to be analog to multigrid methods that we will consider in the next sections; see also [620] by T.F. Chan, Svetozar Dimitrov Margenov, and Vassilevski in 1997.

A preconditioner, the so-called *nested factorization*, was derived at the beginning of the 1980s by J.R. Appleyard, I.M. Cheshire, and R.K. Pollard [65] for block tridiagonal matrices arising from finite difference approximations of three-dimensional oil reservoir problems. They used a modified form of this preconditioner; see also [64] by Appleyard and Cheshire in 1983.

Another type of preconditioner was obtained by using the classical iterative methods like Gauss-Seidel, SOR, and SSOR as preconditioners. We have already seen the diagonal preconditioner which corresponds to the Jacobi iterative method.

Using the SSOR iteration matrix as a preconditioner for symmetric positive definite matrices was proposed by David John Evans (1928-2005) [1112] in 1968; see also the book [1114]. SSOR was also considered by Axelsson [116] in 1974. He proved that there exists a value of the SSOR parameter ω that minimizes the condition number of $M^{-1}A$. For two-dimensional finite difference model problems this decreases the condition number from $\mathcal{O}(1/h^2)$ to $\mathcal{O}(1/h)$, where h is the mesh size. However, the optimal value of ω is generally unknown, and what is often used is a symmetric Gauss-Seidel preconditioner, that is, SSOR with $\omega = 1$. See also the paper [117] in 1976 and the book [122]. Note that with the SSOR preconditioner, two triangular systems have to be solved at each iteration as in incomplete factorizations, but there is no cost to compute the entries of the factors since they are directly obtained from those of the matrix A. Another possibility is to use a block SSOR preconditioner.

SOR was used by Michael A. DeLong and Ortega [863] in 1996; see also [864] in 1998. A Gauss-Seidel preconditioner for the Euler equations was considered by Arnold Reusken [2561] in 2003.

Incomplete factorizations and SSOR preconditioners of type LL^T, where L is lower triangular or block lower triangular, were not very well suited to vector or parallel computers that were becoming common in the 1980s. A possibility to somehow increase the level of parallelism was to use different orderings of the unknowns like, for instance, nested dissection (see Chapter 2). It was also thought that using different orderings could increase the efficiency if the number of neglected fill-ins could be smaller. Hence, some people tried to use ordering algorithms that were already used for complete factorizations for reducing the fill-in. However, it turned out that this was not always working nicely.

The effect of the ordering of the unknowns on incomplete factorizations used with the conjugate gradient method was investigated experimentally by Duff and G.M. [998] in 1989, using a 5-point finite difference scheme on two-dimensional model problems. They examined 17 ordering methods including nested dissection, minimum degree, and red-black and considered preconditioners without fill-in. Some orderings, like the minimum degree or nested dissection gave results much worse than others like the Cuthill-McKee ordering. The number of conjugate gradient iterations was not related to the number of fill-ins that were dropped, but almost directly related to the norm of the remainder matrix R in $A = M - R$. Orderings like the minimum degree with very few fill-ins but with a "large" R matrix gave a large number of iterations, and orderings, like a spiral ordering, with a large number of fill-ins but with a "small" R gave a number of iterations comparable to the one given by the standard row ordering. It was also interesting to see that, if some fill-in is allowed to the incomplete factorization, then the relative merits of the orderings differed.

These experimental results have been explained theoretically to some extent by Eijkhout [1063], Shun Doi [906], and Doi and Alain Lichnewsky [908, 909, 2062] in 1991. Eijkhout considered a two-dimensional diffusion PDE in a square with constant diffusion coefficients. He showed that there is a difference between orderings that eliminate nodes between two uneliminated nodes and those that don't. This was formalized by the following definition: a node is naturally ordered if it does not have two neighbors in one direction that both have a larger number. Orderings with all nodes naturally ordered gave better results. Doi and Lichnewsky introduced the notion of compatibility of orderings to explain the differences in performance.

Other papers published later showed that it may be beneficial to color the graph of A with a large number of colors and to order the unknowns by color; see Eugene Lee Poole and Ortega [2509] in 1987, Fujino and Doi [1268], Doi and Atsushi Hoshi [907] in 1992, and Doi and Takumi Washio [910] in 1999. Orderings depending on the values of the entries were proposed by D'Azevedo, Peter A. Forsyth, and Wei-Pai Tang [835, 836] in 1992. Orderings for incomplete factorization preconditioning of nonsymmetric problems were considered later by Benzi, Szyld, and Arno Van Duin [282] in 1999. Many preconditioners were described and studied in the book [2217] by G.M. in 1999.

As we wrote above, incomplete Cholesky or LU factorizations were not well suited for the vector computers of the 1970s and 1980s and even less well suited for parallel computers. The bottleneck is often the triangular solves that have to be done at each iteration. This triggered some early efforts to modify the methods to enhance their parallelism.

After the introduction of the IC(1,1) incomplete Cholesky factorization at the end of the 1970s for block tridiagonal matrices, van der Vorst [3097] proposed in 1982 to use truncated Neumann series for each block solve in the block recurrence for solving systems with L and L^T. He gave numerical results on the *CRAY-1* vector computer. Rodrigue and Donald Wolitzer used an incomplete block cyclic reduction for symmetric block tridiagonal matrices [2584] and gave results on the *CRAY-1* in 1984. In 1986, Axelsson and Polman proposed block incomplete factorizations [133], including an inverse-free method, that can be used on vector computers.

We have seen above that Poole and Ortega [2509] used multicolor orderings in 1987. They described results on the *CDC Cyber 205* vector computer. Cleve Cleveland Ashcraft and Roger G. Grimes [85] described a vector implementation of the factorization and the forward and backward solves of a family of modified incomplete factorization for matrices arising from 5-point (in 2D) and 7-point (in 3D) finite difference schemes. They did so by analyzing the data dependencies and showed results on the *CRAY X-MP*.

In 1989, Axelsson and Eijkhout [126] considered vectorizable preconditioners for block tridiagonal matrices arising from three-dimensional problems. In these problems the diagonal blocks are themselves block tridiagonal matrices. In 1989-1990, G.M. reviewed some techniques for improving the efficiency of conjugate gradient methods on vector and parallel supercomputers [2213, 2214]. This was illustrated by numerical results on *CRAY* and *ETA* computers.

Later papers were more interested in parallelization when parallel computers became commercially available. In 1989, Elman used approximate Schur complement preconditioners for matrices arising from 9-point finite difference schemes on parallel computers [1084]. In 1999, Benzi, Joubert, and Gabriel Mateescu reported the results of numerical experiments with parallel orderings for ILU preconditioners [279]. Their conclusion was that preconditioners derived from finite difference matrices reordered with multicoloring strategies can outperform those derived for the naturally ordered system in a significant number of cases, particularly when some fill-in is retained.

David Hysom and Alex Pothen derived a scalable parallel algorithm for incomplete factorization [1770] using graph partitioning and a two-level ordering strategy in 2001. They reported results with up to 216 processors for a problem with about 20 million unknowns.

More recently, in 2015, Chow and Aftab Patel [670] described a parallel ILU algorithm which was very different from existing approaches. The ILU factorization is seen as the solution of a set of bilinear equations which can be solved using fine-grained asynchronous parallelism. The nonlinear equations are solved using fixed point iterations that are performed in parallel. The authors reported that only a few sweeps were needed. This method was used as a part of the ParILUT algorithm proposed in 2018 by Anzt, Chow, and Dongarra [61]. The main idea is to combine a fixed point iteration from [670] for approximating the incomplete factors for a given sparsity pattern with a process that adaptively changes the sparsity pattern. Nonzeros are added and removed from the sparsity pattern in each adaptive step.

When looking for parallel preconditioners, it was tempting to directly define M^{-1} since then there are no triangular solves to do. We recall that this is what was done in 1937 by Cesari [601], who used a low degree polynomial. Using polynomial preconditioners with CG may seem strange since, for symmetric positive definite matrices, we know that the polynomial generated by CG is, in some sense, optimal. Therefore, applying m iterations of CG to $P_k(A)A$ will generate a polynomial of degree $k+m$ that will be less efficient than $k+m$ CG iterations for reducing the A-norm of the error. But with the preconditioner, there are fewer dot products that are a bottleneck on certain computer architectures.

In 1979, Paul F. Dubois, Greenbaum, and Rodrigue [972] introduced a simple polynomial preconditioner to be used on vector computers. They used a Neumann expansion of the inverse. If $A = D - L - L^T$ with D a positive diagonal matrix and L strictly lower triangular, one has $A^{-1} = D^{-\frac{1}{2}}(I - D^{-\frac{1}{2}}(L + L^T)D^{-\frac{1}{2}})^{-1}D^{-\frac{1}{2}}$. If the spectral radius of $I - D^{-1}A$ is strictly less than 1, the inverse of $I - D^{-\frac{1}{2}}(L + L^T)D^{-\frac{1}{2}}$ can be expanded in Neumann series. The proposed preconditioner was to use a few terms of the Neumann series. The simplest one is $M^{-1} = 2D^{-1} - D^{-1}AD^{-1}$. The authors showed that a Neumann polynomial of odd degree d is more efficient than the Neumann polynomial of degree $d+1$. This simple preconditioner was used on a vector computer.

Since we have $M^{-1}A = p_k(A)A$, it was natural to consider the polynomial q_k such that $q_{k+1}(\lambda) = \lambda p_k(\lambda)$. Note there was $q_k(0) = 0$ as a constraint. The ideal situation would be to have $q_k(\lambda) \equiv 1\ \forall\lambda$. Unfortunately, this is not possible because of the constraint at the origin.

In 1983, Olin Glynn Johnson, Charles Anthony Micchelli, and George Paul [1835] defined a generalized condition number. Let a and b such that the eigenvalues are such that $\lambda_i(A) \in [a, b]$ $\forall i$, the "condition number" is

$$cond(q) = \frac{\max_{\lambda\in[a,b]} q(\lambda)}{\min_{\lambda\subset[a,b]} q(\lambda)}.$$

Let $\mathcal{Q}_k = \{\text{polynomials } q_k \mid \forall\lambda \in [a, b], q_k(\lambda) > 0, q_k(0) = 0\}$. The first constraint gives a positive definite preconditioner. Then, they looked for the solution of the problem

$$\text{find } q_k \in \mathcal{Q}_k \text{ such that } \forall q \in \mathcal{Q}_k, cond(q_k) \leq cond(q).$$

The solution can be written in terms of shifted and scaled Chebyshev polynomials; see also S.F. Ashby [83] in 1991.

Saad [2654] looked for the polynomial p_k of degree k that minimizes

$$\int_a^b (1 - \lambda q(\lambda))^2 w(\lambda)\, d\lambda, \quad q \in \mathcal{Q}_k,$$

where $w(\lambda)$ is a given positive weight. A common choice are the Jacobi weights,

$$w(\lambda) = (b - \lambda)^\alpha (\lambda - a)^\beta, \quad \alpha \geq \beta \geq -\frac{1}{2},$$

because the orthogonal polynomials associated with these weights are explicitly known. Then, it is easy to compute the least squares polynomial. Interesting choices are two special cases of the Jacobi weights: the Chebyshev weight $\alpha = \beta = -\frac{1}{2}$ and the Legendre weight $\alpha = \beta = 0$. Analytic bounds of the generalized condition number, which are functions of b/a, were obtained for these polynomials by Olivier Perlot [2481] in 1995. He reported also numerical experiments on *Connection Machines* (see Chapter 7).

Another polynomial for CG was proposed by O'Leary [2368] in 1991. She used the polynomial defined by the conjugate gradient iteration as a preconditioner in an adaptive recursive procedure on an already preconditioned system.

We have seen above that polynomials for symmetric indefinite matrices were considered by de Boor and Rice [841] in 1982; see also Fischer [1170] in 1996.

Polynomials for nonsymmetric matrices were constructed for Richardson methods, but they can be used as preconditioners for Krylov methods. We recall the works of Manteuffel [2132, 2133] in 1977, Elman, Saad, and Saylor [1085], Elman and Streit [1087] in 1986, Saad [2655] in 1987, and Smolarski and Saylor [2814] in 1991; See also the Ph.D. thesis of S.F. Ashby [82] in 1987 and [84] by S.F. Ashby, Manteuffel, and James S. Otto in 1992.

It appeared that the polynomial preconditioners developed in the 1980s were not so efficient to reduce the number of iterations in many cases. Some researchers tried to investigate other ways to compute approximate inverses.

One idea that was exploited was to use norm minimization. Since what is sought is a matrix $P = M^{-1}$ such that $PA \approx I$ or $AP \approx I$, a possibility is to try to minimize the norm of the difference. Generally, the Frobenius norm was chosen since it can be easily computed. This was considered by Marcus J. Grote and Simon [1469] in 1993. They looked for minimizing $\|AP - I\|_F$ for a matrix P of a given a priori sparsity structure. They noticed that this problem reduced to n independent minimization problems for the columns of P that could be solved in parallel using the normal equations. They gave results on *Connection Machines CM-2* with a banded approximate inverse.

One problem was that P is not necessarily symmetric when A is symmetric. Liliya Yurievna Kolotilina and Alex Yu. Yeremin [1933] looked for P in factored form $P = G^T G$, where G is a sparse lower triangular matrix. Their method was called FSAI and further developed [1934] in 1995. In 1998, Yeremin, Kolotilina, and Andy A. Nikishin [3296] constructed FSAI preconditioners iteratively; see also [3297] by Yeremin and Nikishin in 2004.

In a 1994 report, Huckle and Grote proposed an incremental method for choosing the sparsity patterns. They started from a set of indices G_k^0 (usually corresponding to a diagonal P or to the structure of the matrix A), they solved the least squares problems, and then, iteratively, they enlarged the sets of indices and solved again the least squares problems until some criterion was satisfied. Hence, the sparsity pattern of P was not fixed and chosen a priori. A paper [1468] was published in 1997. This method was named SPAI. Further details were given by Huckle [1764, 1765] in 1995 and 1998 and [1766] in 1999. Some improvements of the method of Huckle and Grote were described by Nicholas Ian Mark Gould and Jennifer Ann Scott [1415] in 1995; see also the paper [1416] in 1998.

In 1994-1998, Chow and Saad [671, 672, 673] proposed a few methods for computing sparse inverses. One of these methods was to solve approximately $Ap_j = e_j$, where p_j (resp., e_j) is the jth column of P (resp., the identity), by using an iterative method. Of course, this problem was as hard to solve as the problem one would like to precondition, therefore only a crude solution was sought. Some of the elements of this crude solution are dropped to preserve a given sparsity structure of the approximate inverse.

Another approach was chosen by Benzi and his co-authors. For a symmetric positive definite matrix A, they remarked that if $Z = [z_1, z_2, \dots, z_n]$ is a set of conjugate directions for A, one has $Z^T AZ = D$, a diagonal matrix with diagonal elements $d_i = (z_i, Az_i)$. This gives an expression for the inverse $A^{-1} = ZD^{-1}Z^T$. A set of A-conjugate directions can be constructed by a Gram-Schmidt orthogonalization-like algorithm applied to a set of linearly independent vectors $v_1, v_2, \dots, v_n$. If $V = [v_1, v_2, \dots, v_n] = I$, then the matrix Z is upper triangular. Some entries are dropped during the process according to a given threshold; see Benzi, Meyer, and Tůma [281] in 1996. This preconditioner was named AINV.

This was extended to nonsymmetric matrices by Benzi and Tůma [284] in 1998. Here, two matrices Z and W are constructed. The authors showed that this approximate factorization is feasible if A is an H-matrix. Numerical experiments and comparisons with other approximate inverses and preconditioners were published [283, 285] in 1998-1999.

In 2000, a variant of the AINV factorized sparse approximate inverse algorithm which is applicable to any symmetric positive definite matrix was presented by Benzi, Cullum, and Tůma in [273]. Their method was named SAINV.

Benzi, John Courtney Haws, and Tůma [278] experimented with nonsymmetric permutations and scalings aimed at placing large entries on the diagonal to construct approximate inverses for highly indefinite and nonsymmetric matrices in 2000. That same year, the effect of different orderings of the unknowns on the AINV approximate inverse was investigated by Benzi and Tůma [286]. Numerical experiments with problems in solid and structural mechanics were described by Benzi, Reijo Kouhia, and Tůma [280] in 2001. A review of preconditioners and comparisons with approximate inverses were done in [287] in 2003. Much later, in 2016, Jiri Kopal, Rozložník, and Tůma [1936] proposed factorized approximate inverses with adaptive dropping.

More than 20 years after their first appearance, approximate inverses are still of interest, particularly on parallel computers; see, for instance, [62] by Anzt, Huckle, Jūrgen Bräckle, and Dongarra in 2018. The authors combined incomplete LU factorizations with approximate inverses for the triangular factors.

Many other ideas were used over the years to construct preconditioners. An Internet search reveals that from 2000 to 2021 there were about 2000 papers with the word "preconditioner" in the title. Let us consider only a few early papers using ideas different from what we have described above.

When computing an approximate solution of a PDE like $-\Delta u + Lu = f$, with a first-order operator L, one can use the discretization of $-\Delta$ as a preconditioner. This is what was done by Concus and Golub [733] in 1973. They used a fast solver to apply the preconditioner. This type of preconditioner was studied by Gergelits, Kent André Mardal, Bjørn Fredrik Nielsen, and Strakoš [1338] in 2019.

Alternating direction (ADI) preconditioners were considered in Chandra's thesis [630]. A DKR-like ADI preconditioner appeared in [619] by T.F. Chan, Kenneth R. Jackson, and Beren Zhu in 1983.

In the context of finite element methods, element by element preconditioners were introduced by Thomas Joseph Robert Hughes, Itzhak Levit, and James Winget [1768] in 1983.

Wavelets were proposed as sparse approximate inverse preconditioners by T.F. Chan, W.-P. Tang, and Wing Lok Wan [625] in 1997.

In 1997, François Alouges and Philippe Loreaux [32] considered preconditioners of the form

$$M^{-1} = \prod_{i=1}^{\ell}(I - E_i)D\left[\prod_{i=1}^{\ell}(I - E_i)\right]^T,$$

where D is diagonal and the matrices E_i are strictly lower triangular with at most one nonzero entry per row.

Support-graph preconditioning is a technique for constructing and analyzing preconditioners. Predecessors can be found in the work by Beauwens [240] and his collaborators in the 1980s in which graph-theoretic notions were used in the analysis of preconditioners. This was extended by Pravin Moreshwar Vaidya, who used them to study spanning tree preconditioners. He described his work in a talk in 1991 but never published a paper on this topic. See [371] by Erik Gunnar Boman and Bruce Hendrickson, [291] by Marshall Bern, John Russell Gilbert, Hendrickson, Nhat Nguyen, and Sivan Toledo in 2006, [2842] by Daniel Alan Spielman and Shang-Hua Teng, and [2480] by Danil Valer'evich Perevozkin and Gulzira Alimovna Omarova [2480] in 2021.

In 2003, Bru, Juana Cerdán, Marín, and Mas [479] showed how the matrix $A_0^{-1} - A^{-1}$, where A_0 is a nonsingular matrix whose inverse is known or easy to compute, can be factorized in the form $U\Omega V^T$ using the Sherman-Morrison formula. When this factorization process is done incompletely, an approximate factorization is obtained.

Preconditioners for solving sequences of nonsymmetric linear systems were constructed by Duintjer Tebbens and Tůma [1019, 1020] in 2007-2010.

In 2007-2009, Axelsson and János Karátson [128, 129, 130] used an equivalent operator approach to construct preconditioners for problems arising from PDEs. One first approximates the given differential operator by some simpler differential operator, and then chooses as preconditioner the discretization of this operator for the same mesh. This is an extension of the idea used in 1973 by Concus and Golub.

Matrices that attracted a lot of interest arise from saddle point problems because they occur in a wide variety of applications, for instance, mixed finite elements methods in fluid and solid mechanics as well as in optimization. These matrices have a 2×2 block structure. In the simplest case, the matrix is positive semidefinite of the form

$$\mathcal{A} = \begin{pmatrix} A & B^T \\ B & 0 \end{pmatrix},$$

where A is a symmetric $n \times n$ matrix and B is $m \times n$. The matrix $\mathcal{A}$ is nonsingular if B^T has full column rank. One of the problems that was investigated was to find "good" preconditioners for saddle point matrices.

Block diagonal preconditioners have been studied in the form

$$\begin{pmatrix} A & 0 \\ 0 & -S \end{pmatrix},$$

where $S = -BA^{-1}B^T$ is a Schur complement; see Malcolm F. Murphy, Golub, and Wathen [2308] in 2000. The left preconditioned matrix has only three distinct eigenvalues. However, in practice, A and S have to be replaced by approximations, and this property does not hold any longer; see de Sturler and Liesen [855] in 2005.

Another possibility was to use block triangular matrices; see Simoncini [2772] in 2004. Three papers on preconditioning for saddle point matrices were [2612] by Rozložník and Simoncini in 2002, [2775] by Simoncini and Benzi, and [276] by Benzi and Golub in 2004.

Benzi, Golub, and Liesen wrote a 137-page review article about saddle point problems and their numerical solution [277] in 2005.

More recent papers are [2745] by Debora Sesana and Simoncini in 2013 and [2484] by Pestana and Wathen in 2015. Of interest for these problems are also the books [1086] by Elman, David James Silvester, and Wathen in 2014 and [2611] by Rozložník in 2018.

Other efficient preconditioners that were used during the last years arose from multigrid methods and domain decomposition methods. They will be considered in the next sections.

Interesting documents about preconditioning are the review article by Benzi [267] in 2002 and the book [657] by Ke Chen in 2005. Finally, it must be noted that preconditioning and iterative methods are only a part of the solution process, as discussed in the book [2121] by Màlek and Strakoš in 2015.

5.11 ▪ Domain decomposition methods

As the name suggests, domain decomposition (DD) methods were initially proposed for solving partial differential equations on a domain which is partitioned into subdomains. Subproblems are solved on each subdomain and their solutions are glued together to obtain the global solution.

It is generally admitted that the first ever DD method was due to Hermann Amandus Schwarz (1843-1921) in August 1870. It is described in the paper *Über einen grenzübergang durch alternierendes verfahren* [2720] published in the *Vierteljahrsschrift der Naturforschenden Gesellschaft in Zürich* (*Quarterly Journal of the Natural Sciences Society*), a publication founded in 1856.

As nicely explained in the historical paper [1278] by Martin Jakob Gander and Gerhard Wanner, Schwarz's goal was to prove the existence of a solution to Laplace's equation $\Delta u = 0$ with Dirichlet boundary conditions $u = g$ on complicated domains. This was to close a discussion between Bernhard Riemann (1826-1866), who had assumed the existence of such a solution, Karl Theodor Wilhelm Weierstrass (1815-1895), and Hermann von Helmholtz (1821-1894). Schwarz considered a two-dimensional domain which was the union of a disk and an overlapping rectangle; see Figure 10.5 in Section 10.61. It was known that solutions of Laplace's equation existed for the disk and the rectangle since solutions had been constructed, respectively, by Siméon Denis Poisson (1781-1840) in 1815 and Jean-Baptiste Joseph Fourier (1768-1830) in 1807. Schwarz constructed an iterative method by solving alternatively subproblems in the disk using a boundary condition from the solution in the rectangle and then, in the rectangle using a boundary condition from the solution in the disk and so on by iterating the process. He proved the convergence of the method using the maximum principle, even though his argument lacked some rigor. This is known as the *Schwarz alternating method*.

The existence of solutions to Laplace's equation was established later using other techniques and Schwarz's method was almost forgotten, except in a few papers. In 1936, Sergei Lvovich Sobolev (1908-1989) gave a variational convergence proof for the case of elasticity [2819]. Solomon Grigorevich Mikhlin (1908-1990) gave a proof of convergence for general elliptic operators [2247] in 1951. Keith Miller [2250] made the following remark in 1965:

> *Schwarz's method presents some intriguing possibilities for numerical methods. Firstly, quite simple explicit solutions by classical methods are often known for*

simple regions such as rectangles or circles. Also, better numerical solutions, from the standpoint of the computational work involved, are often known for certain types of regions than for others. By Schwarz's method we may be able to extend these classical results and these computational advantages to more complicated regions.

Even without going back to Schwarz, DD had been in use for quite a long time by engineers, mainly in structural mechanics under the name *substructuring*; see, for instance, [2520, 2521] by Janusz Stanisław Przemieniecki (1927-2017) in 1963 and 1968. Figure 5.1, which is from [2520], shows the substructures for a delta wing airplane.

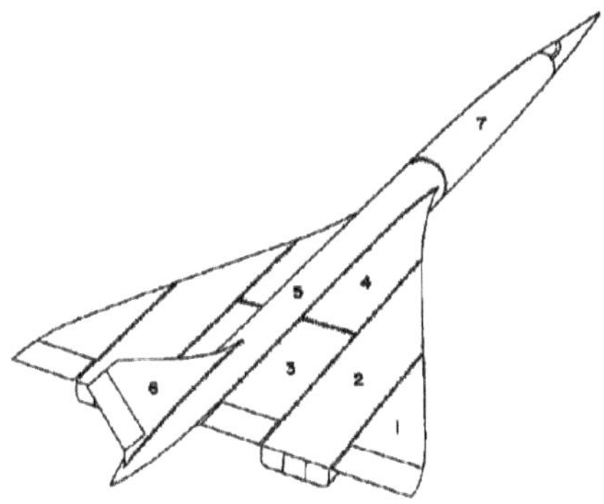

Figure 5.1. *Figure in Przemieniecki's 1963 paper*

DD methods were also used by Russian applied mathematicians in the USSR in the 1960s and 1970s because they had some advantages for solving large problems on computers with small memories.

Another forerunner of DD methods was the *capacitance matrix* method. It seems it was originally proposed by Roger W. Hockney (1929-1999), who was inspired by potential theory in 1970. This method was designed to solve second-order partial differential equations on complex domains by imbedding the domain in a rectangle or, if possible, by splitting the domain into rectangles. It computes the solution by changing some rows of the matrix to those of a matrix whose linear systems can be solved by fast direct solvers. For an application to the discrete Poisson equation, see [517, 518] by Bill L. Buzbee, Fred W. Dorr, J. Alan George, and Golub in 1971. Variants of this method were used for the Helmholtz equation by Włodzimierz Proskurowski and Olof B. Widlund [2519] in 1976 and by Maksymilian Dryja [964] in 1982 to solve the Poisson equation on union of rectangles.

Domain decomposition methods came back to life in the 1980s, probably because of the accessibility of parallel computers. DD methods come in two flavors, with or without overlapping of the subdomains. The main example of overlapping method is Schwarz's, but note that in that method there is no parallelism since the algorithm alternates from one domain to the other. In non-overlapping methods, in general, the unknowns interior to the subdomains are eliminated resulting in Schur complements. Generally, the Schur complement is a nonlocal operator and approximations had to be found. Quite rapidly DD methods became used as preconditioners in Krylov iterative methods. Another potential interest of DD methods was to allow the coupling of different physical models or the use of non-matching meshes at the interfaces between subdomains for which the continuity of the solution is enforced by Lagrange multipliers, the so-called *mortar method*.

Some work was done during the 1980s by Roland Glowinski (1937-2022), Jacques Périaux, and their co-workers; see, for instance, [899] with Quang V. Dinh in 1980. We should also mention the works of Widlund, Petter Bjørstad, and Dryja.

All this activity led to the organization of the first Domain Decomposition Conference in Paris in 1987. The conference was held at the École Nationale des Ponts et Chaussées, which was at that time located in the center of Paris. It was organized by Glowinski, Golub, G.M., and Périaux. The conference was attended by 160 people including 24 speakers. The first paper in the proceedings [2074] was by Pierre-Louis Lions (Fields medalist in 1994), who gave a nice proof of convergence of the alternating Schwarz method, considering it a sequence of projections in a Hilbert space. He also proposed a parallel variant; see also his two other papers on Schwarz's method [2075, 2076] in 1988 and 1989. Later, algebraic versions of Schwarz's method were introduced by using restriction operators corresponding to the "subdomains," that is, subsets of unknowns. Moreover, more sophisticated interface boundary conditions were proposed. About the history of Schwarz methods and for a summary of some variants, see [1276] by M.J. Gander in 2008.

Since 1987, the DD conference has been organized regularly in different countries. The last one so far (in 2021) was the 25th, organized in Canada in 2018. The proceedings of these conferences are one of the best sources of information about DD methods. They are freely available on the Internet at the address www.ddm.org (accessed January 2022). The committee has adopted an idealization of the figure in Schwarz's paper as the logo of the conferences. A study of Schwarz's method on this logo was done by Gabriele Ciaramella and M.J. Gander [696] in 2020.

Domain decomposition has been applied to more and more difficult problems, starting from elliptic model problems in square domains in the 1980s to complex three-dimensional industrial problems at the end of the 1990s. At some point, it became clear that the complexity of the methods was not optimal and that a global process of transfer of information has to be added. This resulted in the addition of the solution of coarse problems to those solved on the subdomains to enhance convergence. This was done in the same spirit as in multigrid methods to be described in the next section.

Thousands of papers were written about the class of DD methods. Let us just mention a few of the early ones.

In [1383], Golub and David Mayers considered in 1984 how to approximate the Schur complements obtained on the interfaces after elimination of the unknowns in the subdomains.

A Schwarz alternating method in a subspace [2167] was proposed by Aleksandr Mikhailovich Matsokin and Sergei Vladimirovich Nepomnyaschikh in 1985. Iterative methods for the solution of elliptic problems on regions partitioned into substructures [344] were published by Bjørstad and Widlund in 1986. Most of the early methods used partitioning of the domain in slices. DD methods with cross points for elliptic finite element problems [965] were considered by Dryja, Proskurowski, and Widlund in 1986.

A series of papers [398, 397, 399, 400, 401] was written by James Henry Bramble, Joseph Edward Pasciak, and Alfred Harry Schatz in 1986-1989. Convergence estimates for product iterative methods with applications to domain decomposition [403] were given by Bramble, Pasciak, Jun Ping Wang, and Jinchao Xu in 1991.

An additive variant of the Schwarz DD alternating method with many subdomains [3228] was studied by Widlund and Dryja in 1987; see also [966] in 1989 and [967] in 1990. Algorithms for implicit difference schemes [1964] were described by Kuznetsov in 1988. Block preconditioning and domain decomposition methods were combined[134] by Axelsson and Polman in 1988.

A method named FETI, which means finite element tearing and interconnecting, [1143] was proposed by Charbel Farhat and François-Xavier Roux for finite element problems in 1991; see also [1142] with Jan Mandel in 1994. Generalized Schwarz splittings [2995] were defined by W.-P. Tang in 1992. Mortar element methods with non-matching meshes at the interfaces [293] were studied by Christine Bernardi (1955-2018), Yvon Maday, and Anthony Tyr Patera in 1993.

A review of DD algorithms [621] was done by T.F. Chan and Tarek P. Mathew in 1994. A DD method for the Helmholtz equation [263] was proposed by Jean-David Benamou and Bruno Després in 1997.

Different techniques were used to precondition the Schur complements for the interfaces. For instance, a Neumann-Neumann algorithm for solving plate and shell problems [2012] was used by Patrick Le Tallec, Mandel, and Marina Vidrascu in 1998.

A restricted additive Schwarz preconditioner [522] was proposed by Xiao-Chuan Cai and Marcus Sarkis in 1999. The convergence of a balancing DD method by constraints and energy minimization [2126] was studied by Mandel and Clark R. Dohrmann in 2003.

Besides the conference proceedings, only a few books were devoted to DD methods, *Domain Decomposition: Parallel Multilevel Methods for Elliptic Partial Differential Equations* [2804] by Barry F. Smith, Bjørstad, and Gropp in 1996, *Domain Decomposition Methods for Partial Differential Equations* [2527] by Alfio Quarteroni and Alberto Valli in 1999, *Domain Decomposition Methods - Algorithms and Theory* [3065] by Andrea Toselli and Widlund in 2004, *An Introduction to Domain Decomposition Methods: Algorithms, Theory, and Parallel Implementation* [911] by Victorita Dolean, Pierre Jolivet, and Frédéric Nataf in 2015, and *Iterative Methods and Preconditioners for Systems of Linear Equations* [697] by Ciaramella and M.J. Gander.

5.12 ▪ Multigrid and multilevel methods

Originally, multigrid methods were developed in the USSR in the 1960s for solving linear systems arising from discretization of partial differential equations. This was first considered for finite difference methods and elliptic problems. These methods used a hierarchy of grids, the goal being to reduce the error to the level of the truncation error of the discretization method and to have, if possible, a complexity proportional to the number of unknowns.

Let us briefly explain the method by first considering only two grids. Let Ω be a two-dimensional domain, say the unit square, with a regular Cartesian mesh of stepsize $h = 1/(m+1)$, m being odd and $A_h u_h = b_h$ be the linear system obtained from the 5-point standard finite difference scheme. Assume that we have an approximation u_k of the solution u_h of $A_h u_h = b_h$ and let $r_k = b_h - A_h u_k$. The idea behind multigrid is to compute an approximation v_k of $A_h^{-1} r_k$ on a coarser grid Ω_H consisting (for instance) of every other point in each direction. The fine grid Ω_h has m^2 points, the coarse grid Ω_H has p^2 points, where $p = (m-1)/2$. We need a process to go from Ω_h to Ω_H so we define a linear *restriction* (projection) operator R,

$$R : \Omega_h \to \Omega_H, \quad (\mathbb{R}^{m^2} \to \mathbb{R}^{p^2}).$$

It is represented by a rectangular matrix. When we have solved the problem on Ω_H, we need to go back to Ω_h, so a linear *prolongation* (interpolation) operator P is defined,

$$P : \Omega_H \to \Omega_h, \quad (\mathbb{R}^{p^2} \to \mathbb{R}^{m^2}).$$

Quite often, one has $P = R^T$. The problem to be solved on the coarse grid can be defined in different ways. The most common one is to define $A_H = RA_hP$ and to solve a linear system with this matrix. When we are back to the fine grid, the next iterate is $u_{k+1} = v_k + u_k$, where v_k is the result of the prolongation of the solution on the coarse grid. A multigrid method is defined by using the two-grid method recursively, that is, the coarse problem itself is solved by the same method until we reach the coarsest level where the problem is solved, say, with a direct method.

However, such a method does not converge. A heuristic explanation for the non-convergence of the two-grid (or multigrid) algorithm is the following. Since A is symmetric, the eigenvectors span a basis of $\mathbb{R}^n$. This basis corresponds to (or converges towards when $h \to 0$) the eigenfunctions of the operator of the PDE. Some of these eigenfunctions vary rapidly (high frequencies), some others are smooth (low frequencies). It is intuitive that components of any vector on the high frequency eigenvectors cannot be well approximated on the coarse grid. Hence, to make the method work, we need smooth residuals such that the restricted $(r_H)_k$ could be a good approximation of r_k. A solution is to do a few iterations of an iterative method which gives smooth residuals (for instance, Gauss-Seidel) before computing the residual and eventually after being back from the coarse grid. The iterative method that is used is called a *smoother*.

What we have described so far is called a V-cycle because it starts on the finest level and traverses all the grids, one at a time, until it reaches the coarsest. Then it traverses them all again in reverse order until it reaches the finest. What is called a W-cycle is similar except that it does two corrections per level. There are also some other possibilities about how to navigate between the levels. It is clear that many possibilities exist for choosing the components of a multigrid method.

The first researcher to describe such a method was Radii Petrovich Fedorenko (1930-2009) in the USSR; see Section 10.20. He published three papers on this topic [1147, 1148, 1149] in 1961, 1964, and 1973. His 1961 and 1964 papers described in detail a multigrid method for solving Poisson's equation in a square with a uniform grid using a standard finite difference scheme. He did a Fourier-like analysis of convergence. The method was devised in the course of solving two-dimensional fluid dynamics equations on a spherical surface as applied to numerical weather prediction. In 1964, he proved that the convergence rate does not decrease when the mesh is refined for Poisson's equation in a rectangle.

The method was extended to general, second-order, variable coefficient elliptic problems on a square domain with a uniform mesh [170] by Nikolai Sergeevitch Bakhvalov (1934-2005) in 1966 and to a finite element discretization [94] by Gennady Petrovich Astrakhantsev in 1971.

Multigrid methods received little attention and applications until they were popularized by the Israeli mathematician Achi Brandt, who was one of the first to recognize their potential in the 1970s by solving many different and difficult problems; see [406, 407, 408, 409, 410] (ordered by date). Brandt did a local-mode (Fourier-like) analysis to estimate the convergence rate.

The golden age of multigrid was the 1980s. More theory was done, the method was extended to hyperbolic and parabolic PDEs, and many more applications used the method. Four sets of multigrid conferences began during the 1980s: the European multigrid conferences, the Copper Mountain conferences, the GAMM workshops, and the Oberwolfach workshops.

Many papers were written on multigrid in the 1980s and 1990s. Some of these papers proposed new techniques and others tried to establish a theoretical basis of multigrid convergence for general problems. Let us just cite a few of them.

Multigrid convergence for indefinite problems [2343] was studied by Nicolaides in 1978; see also [2344] in 1979. Convergence of multigrid iterations applied to difference equations was the main topic of a paper [1522] by Hackbusch in 1980.

In 1981, Bank and Dupont proposed an optimal order process for solving elliptic finite element equations [186]. The contraction number of a multigrid method for solving the Poisson equation [391] was the subject of Dietrich Braess in 1981. An introduction to multigrid methods [1634] was written by Pieter Wilhelm Hemker in 1981.

Klaus Stüben and Ulrich Trottenberg published in 1982 a study of multigrid methods [2935]. The theoretical and practical aspects of a multigrid method [3216] and a robust and efficient multigrid method [3215] were written by Wesseling in 1982. A black box multigrid [880] was proposed by Joel Eugene Dendy in 1982.

A new convergence proof for the multigrid method including the V-cycle [392] was given by Braess and Hackbusch in 1983. An analysis of a multigrid method as an iterative technique for solving linear systems [1438] was published by Greenbaum in 1984. Several papers on multigrid were published by Craig Carl Douglas; see, for instance, [942] in 1984. A convergence theory in a variational framework [2119] was done by Jean-François Maitre and François Musy in 1984; see also [2181] by McCormick in 1985.

Sharp estimates for multigrid rates of convergence with general smoothing and acceleration [185] were obtained by Bank and C.C. Douglas in 1985. An algebraic study of multigrid methods for symmetric and definite problems [2125] was published by Mandel in 1988; see also [2127] by Mandel, McCormick, and John W. Ruge that same year.

A hierarchical basis multigrid method was the topic of the paper [187] by Bank, Dupont, and Harry Yserentant in 1988. Prolongations and restrictions were studied [1635] by Hemker in 1990. The F-cycle [2128] was introduced by Mandel and Seymour V. Parter in 1990.

Convergence estimates for multigrid algorithms without regularity assumptions were given by Bramble, Pasciak, Jun Ping Wang, and J. Xu [402] in 1991. An analysis of smoothers [396] was done by Bramble and Pasciak in 1992. Convergence proofs for multigrid methods were reviewed and improved [3313] by Yserentant in 1993. A unified convergence theory for multigrid or multilevel algorithms [945] was proposed by C.C. Douglas in 1993; see also [943, 944] by C.C. Douglas in 1996.

During these years parallel versions of multigrid were introduced. However, an important step was the introduction of *algebraic multigrid* in 1985 by Ruge and Stüben [2616, 2617, 2934]. What we described above is called *geometric multigrid* because the method was tailored for linear systems arising from discretization of PDEs on some meshes. The goal of algebraic multigrid was to be able to apply multigrid techniques to any linear system by considering only the matrix and replacing the finest grid by the graph of the matrix. Then, coarsening techniques have to be used to construct the coarse levels.

A few books were devoted to multigrid in the 1980s and 1990s: *Multigrid Methods and Applications* [1523] by Hackbusch in 1985, *A Multigrid Tutorial* [466] by William Lambert Briggs in 1987, an updated and extended version [467] was published by Briggs, Van Emden Henson, and McCormick in 2000, *Multigrid Methods* [2182] edited by McCormick in 1987, *An Introduction to Multigrid Methods* [3217] by Wesseling in 1992, *Multigrid Methods* [395] by Bramble in 1993, and *Multigrid* [3073] by Trottenberg, Cornelius W. Oosterlee, and Anton Schüller in 2000.

PLTMG, a software package for solving elliptic partial differential equations in general two-dimensional regions using adaptive refinement and multigrid was developed by Bank, starting in the 1970s. The current version (2022) is 13.0. Multigrid solvers were also included in software packages, for instance, Hypre [165] (see Chapter 7).

More and more, a few steps of multigrid methods were used as preconditioners in Krylov subspace iterative methods. There was also in the 1980s and 1990s the development of algebraic multilevel preconditioners which were often based on approximate Schur complements. They can be seen as using a generalization of the multigrid idea.

Let us cite a few papers on multilevel preconditioners. Algebraic multilevel preconditioning methods [135, 137] were studied in 1989 and 1990 by Axelsson and Vassilevski; see also a survey [136] by the same authors in 1989.

A domain-based multilevel block ILUT preconditioner for general sparse matrices [2672] was proposed by Saad and Jun Zhang in 1999. A multilevel block incomplete factorization preconditioning [2354] was introduced by Notay that same year.

In 2000, J. Zhang studied a sparse approximate inverse and multilevel block ILU preconditioning techniques for general sparse matrices [3326]. The same author considered the relation of preconditioning Schur complement and Schur complement preconditioning [3325] in 2000.

Robust parameter-free algebraic multilevel preconditioners [2355] were proposed by Notay in 2002.

An algebraic recursive multilevel solver (ARMS) for general sparse linear systems [2668] was published by Saad and Brian Suchomel in 2002. The relationships of algebraic multilevel methods and sparse approximate inverses [364] were studied by Bollhöfer and Mehrmann in 2002.

A multilevel AINV preconditioner [2220] was defined by G.M. in 2002. Multilevel preconditioners which are optimal with respect to both problem and discretization parameters [132] were proposed by Axelsson and Margenov in 2003; see also the survey [123] by Axelsson that same year.

A parallel version of the algebraic recursive multilevel solver (pARMS) [2061] was published by Zhongze Li, Saad, and Sosonkina in 2003; see also [2663] by Saad in 2005. A parallel multistage ILU factorization based on a hierarchical graph decomposition [1640] was proposed by Pascal Hénon and Saad in 2006.

Aggregation-based algebraic multilevel preconditioners [2357] were studied by Notay in 2006. Multilevel preconditioners constructed from inverse-based ILU factorizations [366] were proposed by Bollhöfer and Saad in 2006.

The construction of multilevel preconditioners is still an active area of research.

5.13 ▪ Lifetimes

In this section, we show the lifetimes of the main deceased contributors to the iterative methods for solving linear systems, starting in 1777. The language is given by the color of the bars and by letters: E (red) for English, G (black) for German, I (green) for Italian, F (blue) for French, and O (magenta) for the others. The contributors are ordered by date of birth.

Iterative methods for solving linear systems (a)

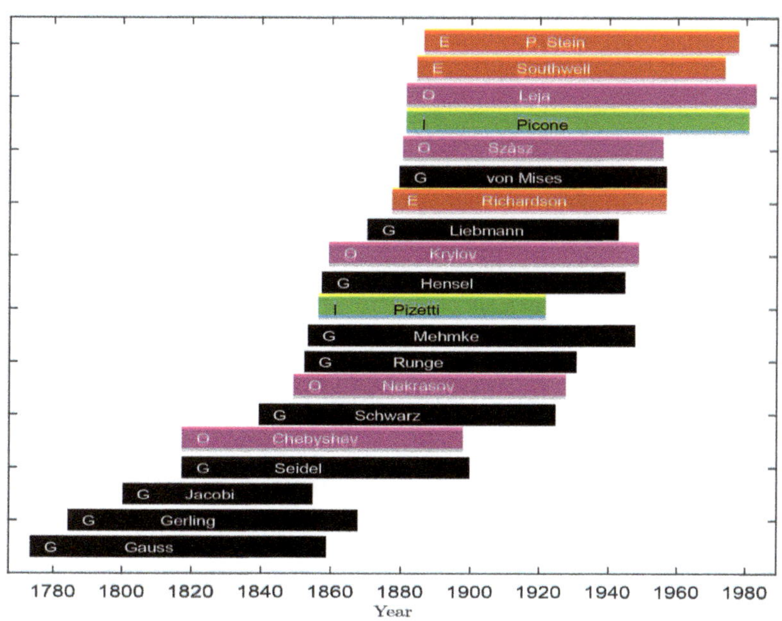

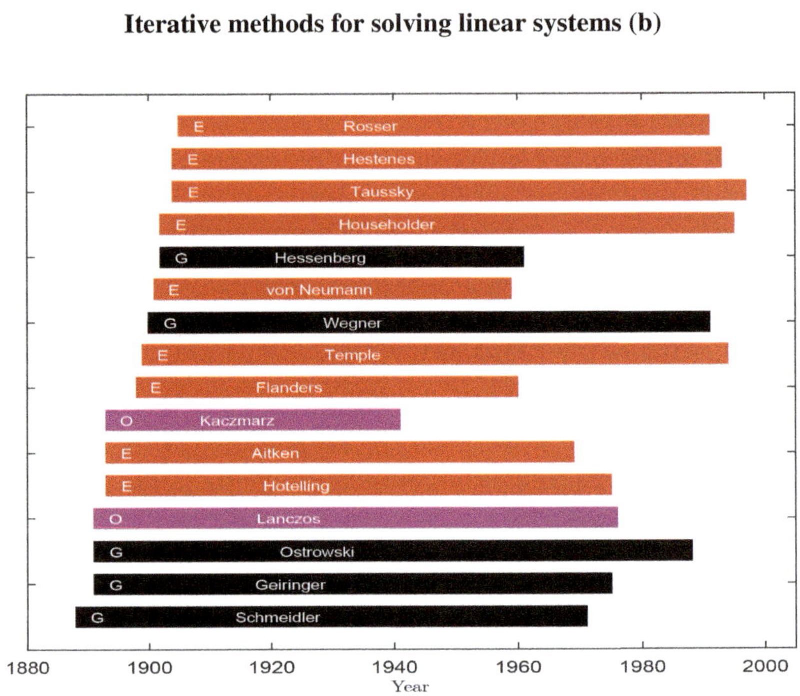
Iterative methods for solving linear systems (b)
E Rosser
E Hestenes
E Taussky
E Householder
G Hessenberg
E von Neumann
G Wegner
E Temple
E Flanders
O Kaczmarz
E Aitken
E Hotelling
O Lanczos
G Ostrowski
G Geiringer
G Schmeidler
1880 1900 1920 1940 1960 1980 2000
Year

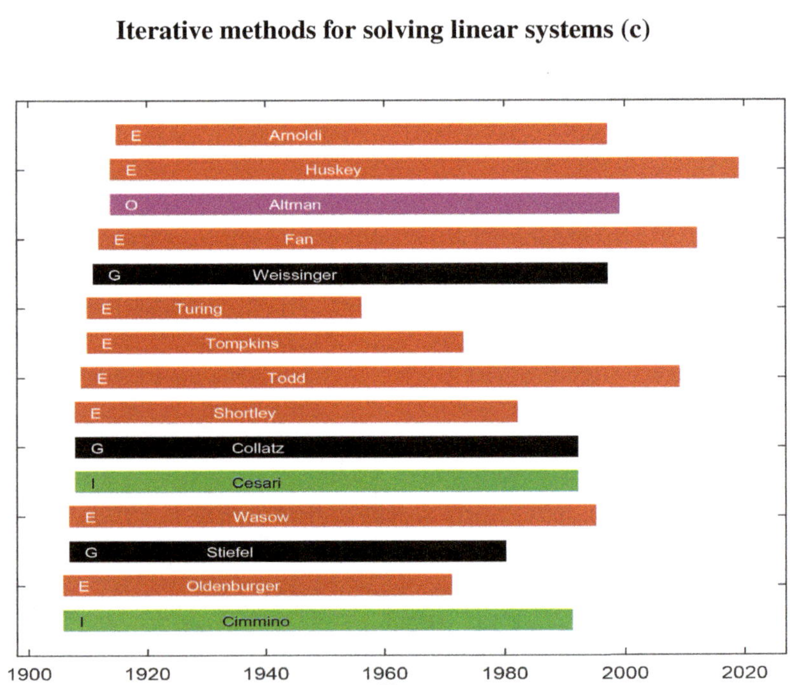
Iterative methods for solving linear systems (c)
E Arnoldi
E Huskey
O Altman
E Fan
G Weissinger
E Turing
E Tompkins
E Todd
E Shortley
G Collatz
I Cesari
E Wasow
G Stiefel
E Oldenburger
I Cimmino
1900 1920 1940 1960 1980 2000 2020
Year

Iterative methods for solving linear systems (d)

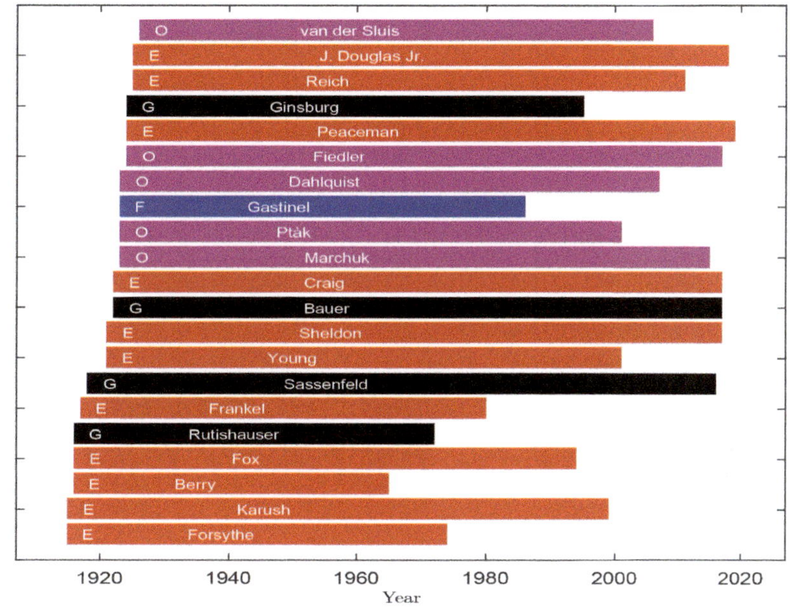

Iterative methods for solving linear systems (e)

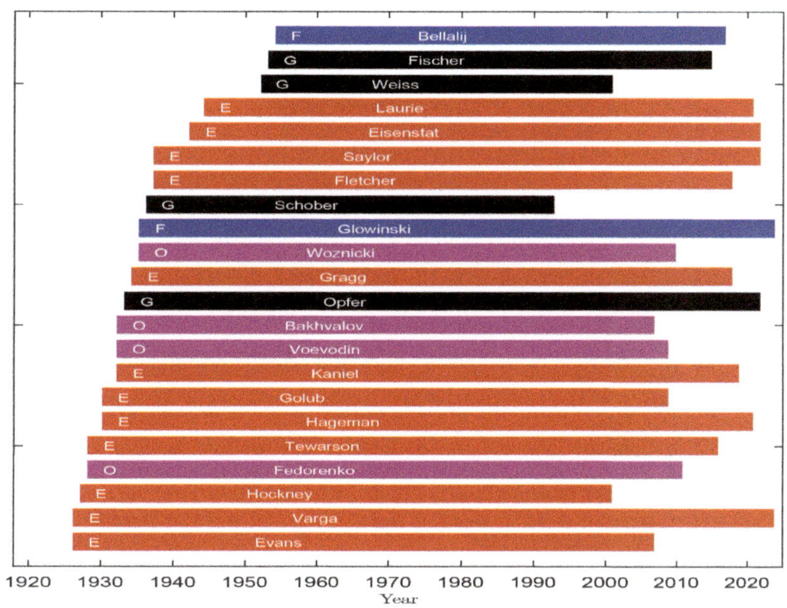

6 Eigenvalues and eigenvectors

It will be convenient to introduce here a notion (which plays a conspicuous part in my new theory of multiple algebra), namely that of the latent roots of a matrix - latent in a somewhat similar sense as vapour may be said to be latent in water or smoke in a tobacco-leaf.

– James Joseph Sylvester, *1883*

Contrary to the solution of linear systems which goes back to antiquity, the definition, properties, and computation of eigenvalues and eigenvectors (also known as spectral theory) is much younger. Its origin can be seen in works about mechanical problems whose solutions were attempted in the 18th century. For the history of spectral theory, see Lynn Arthur Steen [2856] and Thomas W. Hawkins [1600].

6.1 ▪ The early years

Let us consider mainly two examples: the discrete mechanical problem of a weightless string loaded with a finite number of masses with one end fixed and the problem of rotating rigid bodies.

How to describe the motion of a string was studied by Daniel Bernoulli (1700-1782), Jean Le Rond d'Alembert (1717-1783), and Leonhard Euler (1707-1783). For more about Euler's life and works, see Walter Gautschi [1303].

In his book *Traité de Dynamique* [2008] (see Figure 6.1), D'Alembert studied the following problem (Problem V, Section 98, pages 95ff, of the 1743 edition):

Un fil C m M chargé de deux poids m, M, étant infiniment peu éloigné de la verticale CO, trouver la durée des oscillations de ce fil.

It means that he considered a vertical, weightless string with one end fixed, loaded with two masses m and M with weights p and P, and looked for its motion when it is slightly moved away from the vertical.

Even though the notation he used is quite intricate (there is the same notation for points and weights), with the hypothesis he made, d'Alembert was led to two coupled ordinary differential

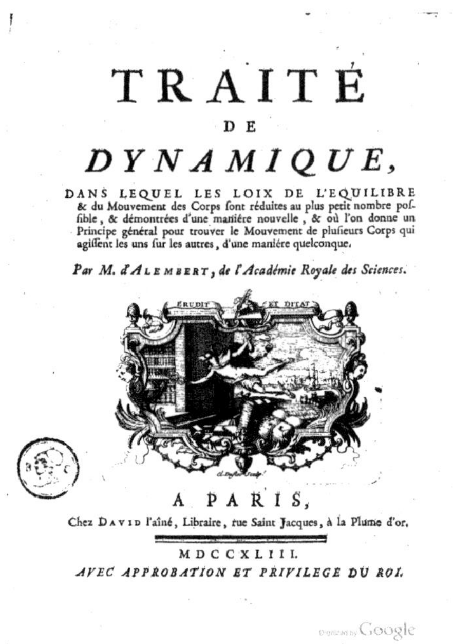

TRAITÉ
DE
DYNAMIQUE,
DANS LEQUEL LES LOIX DE L'EQUILIBRE & du Mouvement des Corps ſont réduites au plus petit nombre poſſible, & démontrées d'une maniére nouvelle, & où l'on donne un Principe général pour trouver le Mouvement de pluſieurs Corps qui agiſſent les uns ſur les autres, d'une maniére quelconque.

Par M. d'ALEMBERT, de l'Académie Royale des Sciences.

A PARIS,
Chez DAVID l'aîné, Libraire, rue Saint Jacques, à la Plume d'or.
M D C C X L I I I.
AVEC APPROBATION ET PRIVILEGE DU ROI.

Figure 6.1. *D'Alembert, Traité de Dynamique, 1743*

equations (ODEs) with constant coefficients,

$$\frac{d^2x}{dt^2} = -\left[\frac{px}{\ell} - \frac{MP}{m}\left(\frac{y}{L} - \frac{x}{\ell}\right)\right],$$
$$\frac{d^2y}{dt^2} = -\left[\frac{yP}{L}\frac{M+m}{m} - \frac{x}{\ell}\left(p + \frac{MP}{m}\right)\right],$$

where ℓ and L are known coefficients.

He remarked that this problem was also considered by D. Bernoulli and Euler. At that time, d'Alembert was unable to solve this problem in full generality. He considered a particular case with $P = p$, $M = m$, and $L = \ell$. Then, the equations simplified to

$$\frac{d^2x}{dt^2} = -\alpha(2x - y),$$
$$\frac{d^2y}{dt^2} = -\alpha(2y - 2x),$$

where α is a constant. He combined the two ODEs to obtain a single equation and derive a "solution" with an intricate reasoning mixing mathematics and physics. He then said that his reasoning can be extended to the general case and to a problem with more masses on a string, but he only gave details for the problem with three masses, which gave him a system with three ODEs.

In the equations he manipulated, he had to find roots. On page 111, he wrote that there can be troubles if some of the roots are not distinct or imaginary. He tried to prove that there is always a real root, but his reasoning was based on experimental physics.

D'Alembert did better in the second edition of the book [2011] in 1758, pages 139 and following, even though he claimed again that if the roots were not real, the solution would be non-physical. To solve the simplified equations, he referred to a paper that was published in 1748

[2009] at the Berlin Academy; see also [2010], published in 1750, in which he considered the solution of systems of linear ODEs.

He did a clever change of variables,

$$u = x + \frac{y}{\sqrt{2}}, \quad u' = x - \frac{y}{\sqrt{2}},$$

and remarked that this gave him two decoupled ODEs

$$\frac{d^2u}{dt^2} = -\alpha(2 - \sqrt{2})u,$$
$$\frac{d^2u'}{dt^2} = -\alpha(2 + \sqrt{2})u'.$$

In modern matrix terms, the system of ODEs can be written as

$$\frac{d^2}{dt^2}\begin{pmatrix} x \\ y \end{pmatrix} = \alpha A \begin{pmatrix} x \\ y \end{pmatrix}, \quad A = \begin{pmatrix} -2 & 1 \\ 2 & -2 \end{pmatrix}.$$

The matrix A has two real eigenvalues $-2 \pm \sqrt{2}$ and the normalized eigenvectors are

$$\frac{1}{\sqrt{3}}\begin{pmatrix} 1 \\ \sqrt{2} \end{pmatrix}, \quad \frac{1}{\sqrt{3}}\begin{pmatrix} -1 \\ \sqrt{2} \end{pmatrix}.$$

The inverse of $\sqrt{3}$ times the matrix of the eigenvectors is

$$\frac{1}{2}\begin{pmatrix} 1 & \frac{1}{\sqrt{2}} \\ -1 & \frac{1}{\sqrt{2}} \end{pmatrix}.$$

Up to a scaling and a change of signs, this matrix corresponds to the change of variables done by d'Alembert. Hence, without knowing it, he computed the eigenvalues and (unnormalized) eigenvectors of the matrix of order 2 to diagonalize his system of ODEs. This was probably one of the first (implicit) appearances of eigenvalues and eigenvectors.

In the late 1740s, there was a long dispute between d'Alembert, Euler, and D. Bernoulli about the problem of vibrating strings. They argued about what is the proper solution of the wave equation; for details, see Gerald F. Wheeler and William P. Crummett [3225] in 1987, Alexandre Guilbaud and Guillaume Jouve [1477] in 2009, Vanja Hug and Thomas Steiner [1767] in 2015, and Jouve [1849] in 2017.

In his memoir *Solution de différents problèmes de calcul intégral* [1969], published in 1766, Joseph-Louis Lagrange (1736-1813) was interested in the motion of a system of interacting bodies having small movements around their equilibrium points. This led him to consider the system of ODEs (our notation),

$$\frac{d^2y_i}{dt^2} = \sum_{j=1}^{n} a_{i,j}y_j = 0, \; j = 1, \dots, n.$$

with arbitrary initial conditions and with constant coefficients. He multiplied the ith equation by $\lambda_i e^{\rho t}$, where $\lambda_i, = 1, \dots, n$ and ρ are so far undetermined, summed the equations, integrated and zeroed some terms, to obtain the equation

$$e^{\rho t} \sum_{j=1}^{n} \left(\lambda_j \frac{dy_j}{dt} - \rho y_j \right) = cste,$$

and a linear system that we can write in modern terms as

$$A^T \lambda + \rho^2 \lambda = 0,$$

where A is the real matrix with entries $a_{i,j}$ and λ has components λ_i. Clearly, $-\rho^2$ must be an eigenvalue of A. Curiously enough, Lagrange did not use determinants, but he noticed that ρ^2 must satisfy a polynomial equation of degree n and therefore, there are only at most n possible values of ρ^2 that he denoted as ρ_i^2, $i = 1, \dots, n$. From the assumed knowledge of the eigenvalues and eigenvectors of A^T and the initial conditions, he explained how to obtain the solution y_i, $i = 1, \dots, n$. To say the least, his discussion of the computation of the eigenvectors is far from crystal clear.

Then, Lagrange discussed the nature of the roots of the monic polynomial equation $P = 0$ for ρ^2. Distinct and negative roots posed no problem since the solutions y_i are sums of sines and cosines. Lagrange considered the case of a double real eigenvalue. For complex eigenvalues, he stated that they occur in complex conjugate pairs. Finally, he gave an a posteriori physical justification:[41]

> *So, for the solution to be mechanically correct, it is necessary* [...] *that the expressions* [of the y_i] *do not contain any term increasing to infinity with time; consequently, it is necessary that the roots of the equation $P = 0$ are all real, negative and different* [...]

Lagrange applied his results to the problem that was considered earlier by d'Alembert starting on page 534. He looked at the equation $P = 0$ and, once again, he relied on the physics of the problem[42],

> *Even though it is difficult, may be impossible, to generally find the roots of the equation $P = 0$, we can be sure, because of the nature of the problem, that the roots are necessarily all real, different and negative, because, if not, the values of y', y'', y''', ... could increase to infinity, something that would be nonsensical.*

The solution of this problem by Lagrange and his followers was discussed in detail by Frédéric Brechenmacher [422] in 2007.

Lagrange also worked on celestial mechanics; see [1972] in 1774. To justify his solutions to the first order ODEs of the model, Lagrange invoked the stability of the solar system. He went back to the problem in 1781-1782; see [1973, 1974]. He studied the system of four planets, Mars, Earth, Venus, and Mercury. He solved the resulting fourth degree equation and, using astronomical data, found numerically four real and distinct roots. However, he was aware that this conclusion depended on the accuracy of the data and that his computation did not give a "proof" that the roots are real and distinct.

Pierre-Simon de Laplace (1749-1827) read Lagrange's memoir. He thought that it would be difficult to show that the roots are real and distinct by studying the polynomial. However, Laplace was the first to realize that the symmetry of the coefficients of the ODEs plays a role.

Another problem in which eigenvalues and eigenvectors appeared (of course, not under that name) is the rotational motion of a rigid body.

[41] *Ainsi il faudra pour que la solution soit bonne mécaniquement:* [...] *que les expressions* [des y_i] *ne contiennent aucun terme qui augmente à l'infini avec le temps t; par conséquent il faudra que les racines de l'équation $P = 0$ soient toutes réelles, négatives et inégales* [...]

[42] *Au reste, quoiqu'il soit difficile, peut-être impossible, de déterminer en général les racines de l'équation $P = 0$, on peut cependant s'assurer, par la nature même du problème, que ces racines sont nécessairement toutes réelles inégales et négatives car sans cela les valeurs de y', y'', y''', ... pourraient croître à l'infini, ce qui serait absurde.*

The angular momentum can be written in matrix form with a symmetric matrix of order 3 (called the tensor of inertia). Its three eigenvectors give the directions of the three principal axes and the three eigenvalues give the moments of inertia with respect to each of these axes. In the coordinate system of the principal axes the tensor of inertia is diagonal.

The term *principal axes* was introduced by Euler in his investigation of the mechanics of rotating bodies [1109, 1110] in 1765. He defined moments of inertia with respect to an axis and the center of inertia, which may be different from the center of gravity. Euler postulated the existence of the principal axes as a starting point. Their characterization was then established.

Lagrange considered this problem [1970] in 1773. In the introduction he wrote[43] (our translation)

> *This problem, one of the most curious and difficult in Mechanics, has already been solved by M. Euler in the Memoirs of this Academy for the year 1758, and in volume III of his Mechanics. M. d'Alembert has also solved it in his Opuscules. The solution methods of these two great Geometers are very different, but they are both based on the mechanical consideration of the rotation of the body around a mobile axis, and they suppose that one knows the position of its three axes of uniform rotation; which requires the solution of a cubic equation. However, considering the problem in itself, it seems that it should be possible to solve it directly and independently of the properties of the axes of rotation, properties which are rather difficult to demonstrate, and which should moreover be consequences of the solution itself rather than the foundations of this solution,* [...] *I therefore thought that it would be an interesting work for the progress of both these sciences to seek a completely direct and purely analytical solution of the question at issue; This is the object I have proposed to fulfill in this Memoir; the difficulties it presents have stopped me for a long time, but finally I have found a way to overcome them by a rather uncommon and entirely new method, which seems to me worthy of the attention of Geometers.* [...] *Moreover, my research has nothing in common with theirs but the problem which is its object; and it is always a contribution to the advancement of Mathematics to show how one can solve the same questions and arrive at the same results by very different ways; the methods lend themselves by this means a mutual light and often acquire a greater degree of evidence and generality.*

He also presented his solution in his book *Méchanique analitique* [1975] in 1788, second part, section VI, pages 337-389. Lagrange wrote differential equations describing the motion of the body and then, in his development he had to reduce a quadratic form with three variables which is transformed to a sum of squares through a change of variable with what we now call an

[43] *Ce Problème, l'un des plus curieux et des plus difficiles de la Mécanique, a déjà été résolu par M. Euler dans les Mémoires de cette Académie pour l'année 1758, et dans le tome III de sa Mécanique. M. d'Alembert l'a résolu aussi dans ses Opuscules. Les solutions de ces deux grands Géomètres sont fort différentes quant à la méthode, mais elles sont fondées l'une et l'autre sur la considération mécanique de la rotation du corps autour d'un axe mobile, et elles supposent qu'on connaisse la position de ses trois axes de rotation uniforme; ce qui exige la résolution d'une équation cubique. Cependant, à considérer le Problème en lui-même, il semble qu'on devrait pouvoir le résoudre directement et indépendamment des propriétés des axes de rotation, propriétés dont la démonstration est assez difficile, et qui devraient d'ailleurs être plutôt des conséquences de la solution même que les fondements de cette solution,* [...] *J'ai donc cru que ce serait un travail avantageux aux progrès de l'une et de l'autre de ces deux sciences que de chercher une solution tout à fait directe et purement analytique de la question dont il s'agit; c'est l'objet que je me suis proposé de remplir dans ce Mémoire; les difficultés qu'il présente m'ont arrêté longtemps, mais enfin j'ai trouvé moyen de les surmonter par une méthode assez singulière et entièrement nouvelle, qui me paraît digne de l'attention des Géomètres.* [...] *D'ailleurs mes recherches n'ont rien de commun avec les leurs que le Problème qui en fait l'objet; et c'est toujours contribuer à l'avancement des Mathématiques que de montrer comment on peut résoudre les mêmes questions et parvenir aux mêmes résultats par des voies très différentes; les méthodes se prêtent par ce moyen un jour mutuel et en acquièrent souvent un plus grand degré d'évidence et de généralité.*

orthonormal transformation. In this problem, he was able to prove that the roots are real because he had a cubic polynomial.

In his *Leçons sur les applications du calcul infinitésimal à la géométrie* [567] Augustin-Louis Cauchy (1789-1857) considered in 1826 general quadratic forms in three variables on pages 248ff. He reduced them to sums of squares by considering the 3×3 characteristic determinant and he proved that the roots are real. In 1828, he studied the principal axes of second degree surfaces in [568].

In 1829, Cauchy published *Sur l'équation à l'aide de laquelle on détermine les inégalités séculaires des mouvements des planètes* [569]. He was looking for the maximum and minimum values of a quadratic form that we write today as (Ax, x) with A symmetric and $\|x\| = 1$. He wrote the derivatives and had to solve $Ax = sx$, where A was written as an array and s was a scalar. This was, maybe, the first explicit appearance of an eigenvalue. He showed that if the roots are distinct, the eigenvectors are orthogonal, and using determinants, he proved that the roots (that is, the eigenvalues) are real and that the quadratic form can be reduced to a sum of squares.

Despite its title, there is nothing directly related to astronomy in Cauchy's paper. In fact, as suggested by Hawkins [1600], Cauchy probably noticed the interest of his results for secular perturbations of planets from Jacques Charles François Sturm (1803-1855). Cauchy was mainly interested in extending Lagrange's results to n variables and in a more rigorous way.

It is also in that paper that he stated his famous interlacing theorem for the eigenvalues, but he proved it only for $n = 3$ and $n = 4$. Cauchy's interlacing results were extended several times; see, for instance, refined results for tridiagonal matrices in [1697] by R.O. Hill Jr. and Beresford Neill Parlett in 1992 and [190] by Ilan Bar-On in 1996.

In a short note [570] in 1830, Cauchy considered the applications of his results to the moments of inertia of rotating bodies. He wrote[44] (our translation)

> *It is known that the determination of the axes of a surface of the second degree, or of the principal axes and moments of inertia of a solid body, depend on an equation of the third degree, whose three roots are necessarily real. However, geometers have only succeeded in demonstrating the reality of the three roots by indirect means, for example by resorting to a transformation of coordinates in space, in order to reduce the equation in question to another equation of the second degree only, or by showing that one would arrive at absurd conclusions if one assumed two imaginary roots. The question that I proposed to myself consists in directly establishing the reality of the three roots, whatever the values of the six coefficients in the given equation. The solution, which deserves to be noticed because of its simplicity, is included in a theorem that I will state.*

According to [1600], the term "characteristic value" (*valeur caractéristique* since Cauchy was not writing in English) was coined by Cauchy in 1839.

In May 1829, a memoir by Sturm was read at the Academy of Sciences in Paris which contained his famous theorem on the number of real roots of a polynomial. This memoir was only

[44] *On sait que la détermination des axes d'une surface du second degré, ou des axes principaux et des moments d'inertie d'un corps solide dépend d'une équation du troisième degré, dont les trois racines sont nécessairement réelles. Toutefois, les géomètres ne sont parvenus à démontrer la réalité des trois racines qu'à l'aide de moyens indirects, par exemple en avant recours à une transformation de coordonnées dans l'espace, afin de réduire l'équation dont il s'agit à une autre équation qui soit du second degré seulement, ou en faisant voir que l'on arriverait à des conclusions absurdes si l'on supposait deux racines imaginaires. La question que je me suis proposée consiste à établir directement la réalité des trois racines, quelles que soient les valeurs des six coefficients enfermés dans l'équation donnée. La solution qui mérite d'être remarquée à cause de sa simplicité, se trouve comprise dans un théorème que je vais énoncer.*

published in 1835, but an excerpt was printed [2936] in 1829 in the 12th volume of the *Bulletin des Sciences Mathématiques, Physiques et Chimiques* that was published by André Étienne Justin Pascal Joseph François d'Audebert de Férussac (1786-1836), a politician and naturalist; see also [2938]. That same year, Sturm published a review [2937, 2938] of Cauchy's *Exercices de Mathématiques*. Therefore, Sturm was aware of Cauchy's works, but he said that his results were developed independently. Sturm was also influenced by results of Jean-Baptiste Joseph Fourier (1768-1830) who generalized Descartes' rule. Sturm applied his results to prove the reality of the roots, but his results were not as general as Cauchy's.

In 1837, Victor Amédée Le Besgue (1791-1875) considered the reduction of a quadratic form with n variables to a sum of squares [302] using determinants. He showed the reality of the roots and referred to Cauchy and Sturm.

The problem of the nature of the roots was solved independently by Karl Theodor Wilhelm Weierstrass (1815-1897) and Camille Jordan (1838-1922); see Chapter 4.

Carl Gustav Jacob Jacobi (1804-1851) was already interested in quadric surfaces in 1837, a topic on which he wrote several papers. After that he became aware of Cauchy's work and published in 1834 a 69-page paper [1791] written in Latin. As it is stated in [1600], he reworked Cauchy's results and apply them to the transformation of multiple integrals in n variables.

In 1846, Jacobi computed the eigenvalues of real symmetric matrices by rotating the matrix (the array of coefficients) to a strongly diagonally dominant one but he did not use the method to full convergence [1796, 1797].

As we said in Chapter 4, Jacobi's iterative technique to diagonalize a matrix was rediscovered in 1949 by John von Neumann (1903-1957), Herman Heine Goldstine (1913-2004), and Francis Joseph Murray (1911-1996) and published in a manuscript. After Alexander Markowich Ostrowski (1893-1986) pointed out that this was actually a rediscovery of Jacobi's method, the revised manuscript [1371] was published only in 1959. The method was used in 1953 by Robert Todd Gregory (1920-1984) on the computer *ILLIAC I* of the University of Illinois (not to be confused with the parallel computer *ILLIAC IV* developed much later) [1457].

The quadratic convergence for the cyclic Jacobi algorithm (using the rotations iteratively in the same order) was proved, under various assumptions, by Peter Karl Henrici (1923-1987) in 1958 [1641], Arnold Schönhage in 1961 [2710], James Hardy Wilkinson (1919-1986) in 1962 [3245, 3248], and Huub P.M. van Kempen in 1966 [3109, 3108].

Extensions of Jacobi's method to nonsymmetric matrices became a topic of research starting in the 1950s. John Lester Greenstadt was one of the first to compute the Schur form in this way in 1955 [1455, 1456].

In 1966, Heinz Rutishauser (1918-1970) wrote an Algol 60 implementation of Jacobi's iteration [2641] which was later published in the *Handbook for Automatic Computation* [3256] in 1971. In the 1960s, the popularity of the method declined. However, variants and extensions were proposed in the 1970s and later. Let us cite Axel Ruhe (1942-2015) [2618] in 1968, Patricia James Wells Eberlein (1923-1998) [1046, 1044] in 1968-1970, Ahmed Hamdy Sameh [2680] in 1971 who discussed an implementation for the *ILLIAC IV* parallel computer, and Michael Hubertus Cornelius Paardekooper [2410] in 1971 for skew-symmetric matrices.

An interesting feature of Jacobi's iteration is its good accuracy; see James Waldon Demmel and Krešimir Veselić [878] in 1992.

6.2 ▪ The Rayleigh quotient

Let A be a Hermitian matrix and x a nonzero vector. The *Rayleigh quotient* is defined as

$$R(A, x) = \frac{x^* A x}{x^* x}.$$

Since A is Hermitian, it is diagonalizable with only real eigenvalues, $\lambda_{\min} \leq \cdots \leq R(A,x) \leq \cdots \leq \lambda_{\max}$. If λ_i is an eigenvalue of A and v_i is the corresponding eigenvector, $R(A, v_i) = \lambda_i$.

In 1877, John William Strutt, third Baron Rayleigh (1842-1919), published the first volume of his book *The Theory of Sound* [2931] in which the Rayleigh quotient is defined. It has been analyzed by Parlett [2444] in 1973 and Ulf Kristiansen [1951] in 2017 in modern mathematical terms. In this section, we will first explain this in physical terms, and then we will see how Rayleigh himself proceeded.

As it was written in 1956 by George Frederick James Temple (1901-1992) and William Gee Bickley (1893-1969) in [3024], Rayleigh's principle can be stated as follows:

> *In the fundamental mode of vibration of an elastic system, the distribution of kinetic and potential energies is such as to make the frequency a minimum.*

See also Temple's paper [3023] in 1952.

As an example, consider a particle of mass m supported by an elastic string such that when the particle is displaced a distance x from its equilibrium position, the restoring force is λx. The equation of motion is $m\ddot{x} = -\lambda x$. At time t, we have $x = a \sin \omega t$, where a is the amplitude and ω the pulsatance (2π times the frequency) given by $\omega^2 = \lambda/m$. The kinetic energy T is given by

$$T = \frac{1}{2} m\dot{x}^2 = \frac{1}{2} m\omega^2 a^2 \cos^2 \omega t = \frac{1}{2} \lambda a^2 \cos^2 \omega t$$

and the potential energy V is

$$V = \frac{1}{2} \lambda x^2 = \frac{1}{2} \lambda a^2 \sin^2 \omega t.$$

Integrating over $[0, \pi]$ shows that the mean kinetic and potential energies, averaged over a period $2\pi/\omega$, are equal. This equality not only holds for systems with one degree of freedom, but also for those with many degrees of freedom, and even for continuous systems. This property allows to compute the frequency of any simply periodic motion if the amplitude of vibration of every part of the system is known.

Let us now see how Rayleigh introduced this quotient. His presentation is not easy to follow since it extends on many pages, and in two editions of his book (in the second one, additions are put into square brackets), and is often hidden into other considerations. We based our exposition on the second edition in 1894. On pages 91-92, one can read[45]

> §70 *The main problem of Acoustics consists in the investigation of the vibrations of a system about a position of stable equilibrium, but it will be convenient to commence with the statical part of the subject. By the Principle of Virtual Velocities, if we reckon the coordinates ψ_1, ψ_2, &c. from the configuration of equilibrium, the potential energy of any other configuration will be a homogeneous quadratic function of the co-ordinates, provided that the displacement be sufficiently small. This quantity is called V, and represents the work that may be gained in passing from the actual to the equilibrium configuration. We may write*
>
> $$V = \frac{1}{2} c_{11}\psi_1^2 + \frac{1}{2} c_{22}\psi_2^2 + \ldots + c_{12}\psi_1\psi_2 + c_{23}\psi_2\psi_3 + \ldots \qquad (1).$$
>
> *Since by supposition the equilibrium is thoroughly stable, the quantities c_{11}, c_{22}, c_{12}, &c. must be such that V is positive for all real values of the co-ordinates.*

[45] The equation numbers are those in Rayleigh's work.

Then, in Section 77, page 97, Rayleigh assumed that the kinetic energy T of the system can be expressed as a homogeneous quadratic function of the generalized coordinates, that is,

$$T = \frac{1}{2}a_{11}\dot{\psi}_1^2 + \frac{1}{2}a_{22}\dot{\psi}_2^2 + \cdots + a_{12}\dot{\psi}_1\dot{\psi}_2 + a_{23}\dot{\psi}_2\dot{\psi}_3 + \cdots,$$

and he added

> *In the present theory the initial kinetic energy T bears to the velocities and impulses the same relations as in the former V bears to the displacements and forces respectively. In one respect the theory of initial motions is the more complete, inasmuch as T is exactly, while V is in general only approximately, a homogeneous quadratic function of the variables.*

Then, setting $D = d/dt$ and $e_{rs} = a_{rs}D^2 + c_{rs}$ if there are no frictional forces, it follows that $e_{rs} = e_{sr}$, and Rayleigh arrived, in Section 84, at the result $\nabla\psi = 0$, where (with an abuse of notation) ∇ is the determinant

$$\nabla = \begin{vmatrix} e_{11} & e_{12} & e_{13} & \cdots \\ e_{21} & e_{22} & e_{23} & \cdots \\ e_{31} & e_{32} & e_{33} & \cdots \\ \vdots & \vdots & \vdots & \ddots \end{vmatrix} \qquad (3).$$

Let $\pm\lambda_1, \ldots, \pm\lambda_m$ be the roots of $\nabla = 0$ considered as an even function of the symbol D of degree $2m$. Then, by the theory of differential equations, the most general solution has the form

$$\psi = Ae^{\lambda_1 t} + A'e^{-\lambda_1 t} + Be^{\lambda_2 t} + B'e^{-\lambda_2 t} + \cdots,$$

where $A, A', B, B', \ldots$ are arbitrary constants. After comments on the various forms of the solution, Rayleigh reduced T and V to sums of squares,

$$T = \frac{1}{2}a_1\dot{\phi}_1^2 + \frac{1}{2}a_2\dot{\phi}_2^2 + \cdots,$$
$$V = \frac{1}{2}c_1\phi_1^2 + \frac{1}{2}c_2\phi_2^2 + \cdots,$$

where the coefficients are necessarily positive if the equilibrium is stable. Lagrange's equations become $a_i\ddot{\phi}_i + c_i\phi_i = 0$, whose solutions are $\phi_i = A_i\cos(n_i t - \alpha_i)$ with $n_i^2 = c_i/a_i$. It follows that the total energy $T + V$ is

$$T + V = \frac{1}{2}c_1A_1^2 + \frac{1}{2}c_2A_2^2 + \cdots.$$

Setting $\phi_i = A_i\theta$, the expressions for T and V become

$$T = \left(\frac{1}{2}a_1A_1^2 + \frac{1}{2}a_2A_2^2 + \cdots\right)\dot{\theta}^2,$$
$$V = \left(\frac{1}{2}c_1A_1^2 + \frac{1}{2}c_2A_2^2 + \cdots\right)\theta^2,$$

whence, if θ varies as $\cos pt$,

$$p^2 = \frac{c_1A_1^2 + \cdots + c_mA_m^2}{a_1A_1^2 + \cdots + a_mA_m^2}.$$

Then, on page 110 of the first volume of the 1894 edition of his book, he wrote

> *The stationary properties of the roots of Lagrange's determinant (3) §84, suggests a general method of approximating to their values. Beginning with assumed rough approximations to the ratios $A_1 : A_2 : A_3 \dots$ we may calculate a first approximation to p^2 from*
>
> $$p^2 = \frac{\frac{1}{2}c_{11}A_1^2 + \frac{1}{2}c_{22}A_2^2 + \cdots + c_{12}A_1A_2 + \cdots}{\frac{1}{2}a_{11}A_1^2 + \frac{1}{2}a_{22}A_2^2 + \cdots + a_{12}A_1A_2 + \cdots} \qquad (3).$$
>
> *With this value of p^2 we may recalculate the ratios $A_1 : A_2 \dots$ from any $(m-1)$ of equations (5) §84, then again by application of (3)* [above] *determine an improved value of p^2, and so on.*

This is the *Rayleigh quotient.* Equation (5), Section 84, was (in modern notation)

$$A \begin{pmatrix} A_1 \\ A_2 \\ \vdots \end{pmatrix} = 0,$$

where A is the matrix corresponding to the determinant ∇.

Then, in Section 106 (Chapter V, *Vibrating systems in general, continued*), page 149, Rayleigh explained that, denoting by p'^2 the value of p^2 *natural to the system when vibrating under the restraint defined by the ratios* $A_1 : A_2 \dots A_r : A_{r+1} : \dots A_m$, p^2 lies within the range of all the possible values of p'^2. This is the fundamental property of his quotient.

At the end of volume I (Dover edition, 1945), there is a *Historical introduction* by the American physicist Robert Bruce Lindsay (1900-1985) in which, after a biographical sketch of the life of Rayleigh, his contributions to acoustics are analyzed. The author wrote

> *The simplest case* [oscillations of a system with one degree of freedom] *is followed by two chapters on the general theory of vibrations of a system of n degrees of freedom, largely a development of his 1873 paper mentioned just above. It was here he emphasized the value of the method of obtaining an approximation to the lowest frequency of vibration of a complicated system in which the direct solution of the differential equations is impracticable. This procedure, which makes use of the expressions for the maximum potential and kinetic energies, was later generalized by Ritz and is now usually known as the Rayleigh-Ritz method; It was proved of value in handling not only all sorts of involved vibration problems but also problems in quantum mechanics.*

In 1908 and 1909, the Swiss theoretical physicist Walther Heinrich Wilhelm Ritz (1878-1909) published two papers in which he gave a procedure for solving numerically boundary value and eigenvalue problems. In his first paper [2578], he laid out the method, and gave its underlying concepts and some applications.

Ritz's second paper [2579] is devoted to new results on the vibrations of a completely free square plate. As explained by Martin Jakob Gander and Gerhard Wanner in [1277], in order to compute Chladni's figures, which correspond to eigenpairs of the biharmonic operator $\Delta^2\omega = \lambda\omega$ in $(-1,1)^2$ with free boundary conditions, Ritz used the principle of energy minimization derived from this problem, instead of solving directly the partial differential eigenvalue problem.

Ernst Florens Friedrich Chladni (1756-1827) was a German physicist doing research on vibrating plates. He invented a technique to visualize the different modes of vibration of a plate. The figures obtained are called Chladni's figures.

Ritz's idea was to search for an approximate solution in a certain given subspace. He pointed out that his method was completely general and could be applied to plates of arbitrary shapes. Two years after Ritz's second paper (that is, after Ritz's death), Rayleigh published a paper [2933], where, after complimenting Ritz, he wrote

> *Ritz rather implies that I had overlooked the necessity of the first two terms in the expressions of an arbitrary function. It would have been better to have mentioned them explicitly; but I do not think any reader of my book could have been misled. In §168 the inclusion of* all *particular solutions is postulated, and in §175 a reference is made to zero values of the frequency.*

Ritz's method was soon widely adopted and, as explained by Arthur W. Leissa in [2031], it was maybe due to Rayleigh's claim, which is now erroneously named the *Rayleigh-Ritz method.* Let us mention that the method of Galerkin is issued from Ritz's, as well as is Arnoldi's method when the subspace is a Krylov subspace.

Bounds of the Rayleigh quotient were given by Temple and Bickley [3024] in 1933, Tosio Kato (1917-1999) [1878] in 1949, Hans Felix Weinberger (1928-2017) [3203] in 1960, and Chandler Davis (1926-2022) and William Morton Kahan [819] in 1970.

For a Hermitian matrix, the minimum and the maximum values of the Rayleigh quotient give, respectively, the smallest and largest eigenvalues. This was extended in 1905 by Ernst Sigismund Fischer (1875-1954) who gave the min-max characterization of the eigenvalues [1174]. It was generalized to infinite-dimensional operators by Richard Courant (1888-1972) and published in the famous book [753] by Courant and David Hilbert (1862-1943). This result is now known as the Courant-Fischer theorem.

6.3 ▪ Localization of eigenvalues and the field of values

The Rayleigh quotient is related to the notion of *field of values* (FOV), which is also called the *numerical range*. It is a set in the complex plane which is defined as

$$\mathcal{F}(A) = \{x^*Ax \mid x^*x = 1, x \in \mathbb{C}^n\},$$

where x^* is the conjugate transpose of x. So, the FOV is the range of the Rayleigh quotient. The *numerical radius* is defined as the maximum of the moduli of the elements in $\mathcal{F}(A)$. It is clear that the eigenvalues of A are in the FOV, but a better localization of the eigenvalues was and still is an interesting topic of research. Therefore, this section is devoted to the localization of eigenvalues in the complex plane. Interesting works on the history of this topic are [1726] and [3136].

The first localization results were given for polynomials and analytic functions. The most famous one is a theorem obtained in 1862 by the French mathematician Eugène Rouché (1832-1910) [2607, Thm. III]. In modern terms, it says that two functions f and $g = f+h$, holomorphic on and inside a simple closed curve in the complex plane, have the same number of zeros inside it if, on the curve, $|h| < |f|$ and $f \neq 0$. Results due to Augustin-Louis Cauchy (1789-1857) in 1829 and Auguste Claude Éliacin Pellet (1848-1935) in 1881 [2477] allowed to localize the zeros of polynomials to certain disks and exclude them from certain annuli (for generalizations, see the paper by Aaron Melman [2200], as well as [2198, 2199, 2201]). Pellet, a professor at the Faculty of Sciences of Clermont (France), and then its Dean, was described by the various Rectors as *A young mathematician, somewhat stiff-looking, gauche and clumsy in speech, lost and absorbed in his mathematical preoccupations. It is an original, very detached, I believe, from problems other than those of pure science* [...] *Mr. Pellet must be a good mathematician, since he was awarded honors. He is a most ordinary teacher, devoid of both clarity of ideas and neatness of expression* [...] *His mind is confused, more original than useful.*[46]

Semyon Aranovich Gerschgorin (1901-1933) published his famous localization theorem in 1931 [1341] (see the paper by Olga Taussky (1906-1995) [3007] for another proof). However, several other scholars before him came quite close to proving it, but stopped before its complete proof. For example, the basis of Gerschgorin's theorem is a nonsingularity test based on diagonal dominance, and the 1881 paper [2041] by Lucien Lévy (1892-1965), which is cited by Gerschgorin, already used a version of diagonal dominance. However, Lévy, who was an electrical engineer and during WW I, the head of the laboratory of the Centre radiotélégraphique militaire, that is, the Eiffel Tower transmitter, only used the concept to prove that a specific matrix arising in electrostatics does have an inverse. Gerschgorin's theorem was rediscovered in 1946 by Alfred Theodor Brauer (1894-1985) [412] who, a year later, introduced Cassini's ovals[47] that are better than Gerschgorin's disks [413]. These ovals were replaced by lemniscates by Richard Anthony Brualdi in 1982 [482]; see also [483] by Brualdi and Stephen Mellendorf in 1994.

A long time before, the first general result for strictly diagonally dominant complex matrices was obtained by Jules Desplanques (1867-1939) in 1887, at that time a student at the École préparatoire de Sainte-Barbe in the class of Désiré André (1840-1918) [882], who proved as an extension of Levy's results that such matrices are nonsingular. His result was independently rediscovered by Jacques Salomon Hadamard (1864-1963) in his book of 1903 [1526], and it is often referred to as "Hadamard's theorem."

Hermann Minkowski (1865-1909) also used a version of diagonal dominance for real matrices with all negative entries except for a positive diagonal, in a lemma of his 1900 article on algebraic number theory [2251]. However, he was only concerned with showing that a particular matrix has a nonzero determinant. The general equivalence between eigenvalue inclusion and nonsingularity results appeared in the 1931 paper of Hans Rohrbach (1903-1993) [2586]. The simultaneous use of row and column sums to obtain better nonsingularity results leading to inclusion results are due to Ostrowski in 1951 [2393], Ky Fan (1914-2010) and Alan Jerome Hoffmann (1924-2021) in 1954 [1134], and the culminating result of Fan in 1958 [1131] whose proof makes use of the Perron-Frobenius theorem on non-negative irreducible matrices and leads to the best Gerschgorin's circles.

Others who came close to Gerschgorin's theorem approached it from a different point of view. The usual proof, that of Gerschgorin himself, is to consider the equation $Ax = \lambda x$, take absolute values, and reason about the sizes of the terms. In contrast to this simplicity, previous localization results were full of restrictive assumptions. Bounds for the real and imaginary parts of the eigenvalues obtained from the eigenvalues of the Hermitian and skew-Hermitian parts of A were given by Ivar Otto Bendixson (1861-1935) [264] and Arthur Hirsch (1866-1948) [1701] in 1902. Bendixson's and Hirsch's results were discussed by Edward Tankard Browne (1894-1959) [477, 478] in 1930 and 1939. These results were proved by writing out the linear system $(A - \lambda I)x = 0$ but they made use of algebraic tricks and the theory of quadratic forms. They clearly foreshadowed Gerschgorin's theorem but the results were more complicated to state and to prove, and not even as useful.

In 1942, Lothar Collatz (1910-1990) described some inclusion sets for eigenvalues [724]. The use of matrix permutations to obtain Gerschgorin-type eigenvalue inclusion results for ma-

[46] *Jeune mathématicien un peu raide d'allure, gauche et maladroit de parole, perdu et absorbé dans ses préoccupations mathématiques. C'est un original, très détaché, je crois, des problèmes autres que ceux de la science pure* [...] *M. Pellet doit être un bon mathématicien, puisqu'on l'a décoré. C'est un professeur des plus ordinaires, dépourvu à la fois de clarté dans les idées et de netteté dans l'expression* [...] *Son esprit est confus, plus original qu'utile.*

[47] Giovanni Domenico Cassini (1625-1712) was an Italian mathematician, astronomer, and engineer who studied these curves in 1680. He was the first director of the Paris Observatory.

trices, involving the closed exteriors of the disks, was independently done by Maurice Parodi (1907-1992)[48] in 1952 [2462] and Hans Schneider (1927-2014) in 1954 [2703]. This idea was later extended by Richard Steven Varga (1928-2022) to Brauer's Cassini ovals and Brualdi's lemniscates [3136].

Exclusion regions for the eigenvalues were obtained by Miroslav Fiedler (1926-2015) in 1960 by establishing the nonsingularity of the matrix through comparisons with M-matrices [1157]. The approach of David G. Feingold and Varga in 1962 [1150], although not fundamentally different, established the nonsingularity of the matrix by generalizing the concept of a diagonally dominant matrix. These exclusion regions may give significant improvements over the usual Gerschgorin's circles in providing bounds for the eigenvalues of the matrix.

Localization results for Hermitian matrices refining Cauchy's result were derived by Nikolaus Joachim Lehmann (1921-1998) in 1949 and 1963-1966 [2022, 2023, 2024]. A clear exposition of these results is given in Parlett's book [2446]. Lehmann's results were somehow extended to nonsymmetric and non-normal matrices by Christopher Beattie and Ilse Clara Franziska Ipsen [235] in 2003.

In 2008, Varga, Ljiljana Cvetković, and Vladimir Kostić explained in [3137] how to compute the minimal Gerschgorin sets that were defined by Varga [3133] in 1968. An algorithm for counting the number of eigenvalues in a region of the complex plane surrounded by a closed curve was described in 2013 by Emmanuel Kamgnia and Bernard Philippe [1866].

Concerning the FOV, Otto Toeplitz (1881-1940) proved in 1918 that the convex hull of the spectrum of A is contained in the FOV, that they are equal when the matrix is normal, and that the boundary of the FOV is a convex curve (even though his proof is not very clear); see [3057]. Toeplitz expressed almost everything in the language of bilinear forms, even though he used them as if they were matrices. For the proof he used the Schur factorization. Toeplitz considered also a generalization of the field of values. In 1919, using the same technique as Toeplitz, that is, decomposing the matrix A with its Hermitian and skew-Hermitian parts, Felix Hausdorff (1868-1942) proved in [1597] that the FOV is a convex set by showing that every segment joining two points of the FOV is contained in it.

The FOV was also studied by the Irish mathematician Francis Dominic Murnaghan (1893-1976) in 1932 [2306], Albert B. Farnell [1144] in 1945, W.V. Parker [2437] in 1951, as well as by Rudolf Kippenhahn (1926-2020) in his thesis in 1951 and in the paper [1904], by investigating the geometric properties of the FOV using the notion of a boundary generating curve, by Wallace Givens (1910-1993) [1363] in 1952, and by Fiedler [1159] in 1981.

The FOV has many applications. For instance, Henrici related in 1962 the distance of a point of the FOV to the convex hull of the eigenvalues with a measure of the non-normality of the matrix [1643]. The FOV has also been used to study the convergence of some iterative methods, see Michael Eiermann [1056] in 1993, Gerhard Starke [2848] in 1997, and Jorg Liesen and Petr Tichý [2071] in 2012. For the uses of the FOV, see also [270] by Michele Benzi in 2021.

Of course, since the FOV is convex, it was interesting to draw its boundary. An algorithm to do so was proposed by Charles Royal Johnson [1834] in 1978 after some other tentatives using Gerschgorin's sets. Points on the boundary of the FOV are obtained by computing eigenvalues of the Hermitian part of rotated matrices. More recent and faster methods were proposed by Sébastien Loisel and Peter Maxwell [2088] in 2018 and by Frank Uhlig [3088] in 2020.

[48] Maurice Parodi was born in Paris on February 7, 1907. After a doctorate in physics in 1938, he participated in WW II and worked on the detection of magnetic mines. In 1944, he entered CNRS (French National Center for Research), and published several notes and memoirs on wave propagation, operational calculus, and the Laplace transform. In 1951, he obtained the chair of *Mathématiques appliquées à l'art de l'ingénieur* at the Conservatoire National des Arts et Métiers. He also held various other teaching positions. In 1967, he was elected as a corresponding member of the French Academy of Sciences. Parodi wrote 11 books, 55 papers, and 158 notes at the Academy of Sciences. He died in Nice on February 4, 1992.

Many generalizations of the FOV were proposed over the years like the one by Friedrich Ludwig Bauer (1924-2015) [219] in 1962 and, for instance, the q-numerical range

$$W_q(A) = \{y^*Ax \mid x^*x = 1,\ y^*y = 1,\ y^*x = q\}$$

for a given complex number q and the joint numerical range with several matrices,

$$W(A_1, \dots, A_k) = \{(x^*A_1x, \dots, x^*A_kx),\ x^*x = 1\},$$

see [2046] by Chi-Kwong Li and Yiu-Tung Poon in 2000.

An interesting inverse problem is to find a nonzero vector x such that $x^*Ax = \mu$, where μ is a given complex number. There is no solution if μ is outside of the FOV of A. This problem can be reduced to the case $\mu = 0$ and x is then called an *isotropic vector*. Algorithms to compute such vectors were given by Uhlig [3085] in 2008, Russell Carden [548] in 2009, Christos Chorianopoulos, Panayiotis Psarrakos, and Uhlig [668] in 2010, and G.M. [2225] in 2012.

6.4 ▪ Where does the name "eigenvalue" come from?

In [1553], Paul Richard Halmos (1916-2006) wrote

> *Almost every combination of the adjectives proper, latent, characteristic, eigen and secular, with the nouns root, number and value, has been used in the literature for what we call a proper value.*

As we have seen above, the name *équation caractéristique* was used by Cauchy in 1839. It gave *valeur caractéristique* and later *polynôme caractéristique* for the eigenvalue and the characteristic polynomial. The adjective *characteristic* was also used by Ferdinand Georg Frobenius (1849-1917) [1250] in 1878. Later on, the French scholars used *valeur propre* for eigenvalue but kept *polynôme caractéristique* for the polynomial.

The name *characteristic value* spread into English publications; see, for instance, [328] by George David Birkhoff (1884-1944) in 1908 and [1699] by Harold Hilton (1876-1974) in 1913.

The poetic term *latent root* for the eigenvalue was coined by James Joseph Sylvester (1814-1897) [2966, 2967] in 1883.

The German names *eigenfunktion* and *eigenwert* (which means proper value) were used by Hilbert in his work [1695] on integral equations in 1904. This terminology was then used in the 1920s by German-speaking physicists working in quantum mechanics.

It seems that the prefix *eigen* was first used in English by physicists, mixing German and English. Paul Adrien Maurice Dirac (1902-1984) used the name *eigenfunction* [901] in 1926. The name *eigenvector* was used by Richard Brauer (1901-1977) and Hermann Weyl (1885-1955) [414] in 1935.

In [1554] Halmos wrote in 1967,

> *For many years I have battled for proper values, and against the one and a half times translated German-English hybrid that is often used to refer to them. I have now become convinced that the war is over, and eigenvalues have won it; in this book I use them.*

As we can see now, Halmos was, unfortunately, right. Some of the information given in this section comes from John Aldrich from the University of Southampton in UK.

6.5 ▪ Using the characteristic polynomial

Polynomials of the form $p(\lambda) = |A - \lambda I|$ are now called *characteristic polynomials*. This is the only remaining reference to "characteristic values" as eigenvalues were called in the 19th century. In the 18th century, $p(\lambda) = 0$ was thought to be the result of the elimination of the components of x in $Ax = \lambda x$, but it began to be considered a determinant after Cauchy's work.

For quite some time, up to the first half of the 20th century, computing the roots of the characteristic polynomial was considered a method to compute eigenvalues. Therefore, methods for computing the coefficients of this polynomial were investigated.

By the mid-19th century, it was known that roots of polynomials of degree larger than five cannot be obtained explicitly using only the four basic operations and radicals (except in some special cases) as a result of the works of Niels Henrik Abel (1802-1829) and Évariste Galois (1811-1832). Hence, iterative methods have to be used to compute roots. Newton's method could be used, but convergence is not guaranteed, depending on the starting point. In 1880, Edmond Nicolas Laguerre (1834-1886) proposed a method to compute roots of polynomials [1977, 1978]. His method almost always converged whatever was the starting point. Laguerre's method was used by Parlett in 1964 to compute eigenvalues after reduction of the matrix to upper Hessenberg form [2439].

A method for computing the coefficients of the characteristic polynomial of a matrix based on Newton's identities (they relate the coefficients of a polynomial with sums of products of the roots) was first given by the French astronomer Urbain Jean Joseph Le Verrier (1811-1877) in 1840 [2013]. The method goes as follows in modern terms. Let A be a matrix of dimension n and $A_0 = A$, $c_n = 1$. The iterates for $k = 1, \dots, n-1$ are

$$A_k = AA_{k-1} + c_{n-k+1}I, \quad c_{n-k} = -\mathrm{tr}(AA_k)/k,$$

where "tr" is the trace operator. The characteristic polynomial of A is $P_n(x) = c_0 + c_1 x + \cdots + c_n x^n$. For more about Le Verrier, see [1852]. This method was rediscovered several times, by the statistician Paul Horst [1730] in 1935, by Jean-Marie Souriau (1922-2012) [2833] in 1948 (see also his book [2832] in 1959), by Dmitry Konstantinovich Faddeev (1907-1989) and Ilya Samuilovich Sominsky (1900-1962) [1128] in 1949, by James Sutherland Frame (1907-1997) [1210] in 1949, by Udo Wegner (1902-1989) [3201] in 1953, by the Japanese theoretical physicist Hyôitirô Takeno (1910-?) [2992] in 1954, and probably others.

After studies in secondary schools in several French cities from 1932 to 1942, Souriau entered the École Normale Supérieure in 1942. After that, he first worked at the Centre National de la Recherche Scientifique (National Center for Scientific Research), and then as an engineer at the Office National d'Études et de Recherches Aérospatiales (an institute for studies in aeronautics and space). After a Doctorate in 1952, he was professor at the Institut des Hautes Études in Tunis (Tunisia), and at the University of Aix-Marseille (France). From 1978 to 1985, he was the Director of the Centre de Physique Théorique in Marseilles.

This method is now known as the Le Verrier-Souriau or Faddeev-Le Verrier method. It is unstable and cannot be used reliably on some problems. However, the method still attracts some researchers who proposed proofs and/or extensions; see, for instance, Stephen Barnett [202] in 1989 and [203] in 1996, Gilbert Helmberg, Peter Wagner, and Gerhard Veltkamp [1633] in 1993, Mordechai Lewin [2045] in 1994, and Shui-Hung Hou [1734] in 1998.

Properties of the characteristic polynomial of a symmetric matrix and a new proof of the reality of its roots were obtained, using Sturm's functions, by Jacobi's student, Carl Wilhelm Borchardt (1817-1880) [375] in 1846. Eigenvalues of other types of matrices began to be studied; for instance, Francesco Brioschi (1824-1897) showed in 1854 what are the eigenvalues of what we call an orthogonal matrix Q by using the determinant $\det(Q - \lambda I)$ in [469, 470].

In [2960] Sylvester gave in 1852 another proof that the roots of the characteristic polynomial $\det(\lambda I - A) = 0$ are real for A symmetric. He wrote

> *M. Cauchy has somewhere given a proof of the theorem, that the roots of A in the above equation must necessarily always be real; but the annexed demonstration is, I believe, new ; and being very simple, and reposing upon a theorem of interest in itself, and capable no doubt of many other applications, will, I think, be interesting to the mathematical readers of this Magazine.*

As we said above, Sylvester also coined the term *latent roots* for eigenvalues; see [2966, 2967]. This naming was mainly used in the UK.

The minimal polynomial and the rational normal form were introduced by Frobenius in a long paper [1251] published in 1878. Even though at that time Frobenius did not use matrices but bilinear forms, the rational normal form stated that every matrix is similar to a block diagonal matrix whose diagonal blocks are companion matrices. If there is only one diagonal block, the matrix is necessarily non-derogatory, that is, its minimal polynomial is equal to its characteristic polynomial; see also [1254]. If p is a monic polynomial of degree n,

$$p(\lambda) = \lambda^n + \alpha_{n-1}\lambda^{n-1} + \cdots + \alpha_1\lambda + \alpha_0,$$

the *companion matrix* associated to p is

$$C = \begin{pmatrix} 0 & 0 & \dots & 0 & -\alpha_0 \\ 1 & 0 & \dots & 0 & -\alpha_1 \\ 0 & 1 & \dots & 0 & -\alpha_2 \\ \vdots & \ddots & \ddots & \vdots & \vdots \\ 0 & \dots & 0 & 1 & -\alpha_{n-1} \end{pmatrix}. \tag{6.1}$$

The matrix C is square of order n and non-derogatory. The characteristic polynomial of C is equal to $p(\lambda)$ and the roots of p are the eigenvalues of C.

In fact, the companion matrices used (implicitly) by Frobenius were permutation of matrices like C. Apparently, the name *companion matrix* was introduced by Cyrus Colton MacDuffee (1895-1961) [2106] in 1933 as a translation of the German "begleitmatrix," a name used by Alfred Loewy (1873-1935) in 1918 [2087].

Between 1926 and 1936, Alexander Craig Aitken (1895-1967) published three interesting papers [12, 14, 16] on the method of D. Bernoulli for computing the dominant zero λ_1 of a polynomial [294] in 1728, and its use in the eigenvalue problem.

Consider the polynomial $p_n(x) = a_0 + a_1x + \cdots + a_nx^n$, and the difference equation $a_0u_k + a_1u_{k+1} + \cdots + a_nu_{k+n} = 0$. Starting from initial given values for $u_0, \dots, u_{n-1}$, the sequence (u_k) can then be computed. Under some assumptions on the zeros of p_n and on the initializations, the ratio $s_k = u_{k+1}/u_k$ converges to the dominant zero λ_1 of p_n with a speed given by $\mathcal{O}(|\lambda_2/\lambda_1|^k)$. In the first part of [12], Aitken considered the ratios of determinants $H_m^{(k-m+2)}/H_m^{(k-m+1)}$, where $H_m^{(k)}$ is the Hankel determinant

$$H_m^{(k)} = \begin{vmatrix} s_k & s_{k+1} & \cdots & s_{k+m-1} \\ s_{k+1} & s_{k+2} & \cdots & s_{k+m} \\ \vdots & \vdots & & \vdots \\ s_{k+m-1} & s_{k+m} & \cdots & s_{k+2m-2} \end{vmatrix},$$

and proved that when k goes to infinity, this ratio converges to the product of the m zeros of p_n of greatest modulus in their respective order of magnitude. He used Sylvester's determinantal identity for the recursive computation of these Hankel determinants. Then, Aitken proposed a method for accelerating the convergence of the sequence (s_k): it was his famous Δ^2 *process*, which consists in building the new sequence

$$t_k = s_k - (\Delta s_k)^2/\Delta^2 s_k,$$

with $\Delta s_k = s_{k+1} - s_k$ and $\Delta^2 s_k = s_{k+2} - 2s_{k+1} + s_k$, and the speed of convergence satisfies $|t_k - \lambda_1| = \mathcal{O}(|\lambda_3/\lambda_1|^k)$, where λ_3 is the third zero of p_n.

In [14], Aitken applied in 1930 the process to the determination of the eigenvalues of a matrix, knowing the coefficients of its characteristic polynomial. His paper [16] in 1936 was a continuation of the two preceding ones. He considered the computation of left and right eigenvectors. Aitken came back to his Δ^2 process (denoted δ^2) and wrote that for practical computation, it can be remembered by the *memoria technica: product of outers minus square of middle, divided by sum of outers minus double of middle*. He also studied the gain brought by the process. For determining the other eigenelements, he proposed two procedures: deflation, which consists in removing from the matrix the part due to the first eigenelements, and λ-differencing.

In 1931, Alexei Nikolaevich Krylov (1863-1945) published a method [1955] to obtain the characteristic polynomial avoiding the computation of $\det(A - \lambda I)$; see Section 10.42. Let v be a vector of grade n with respect to A of order n. Krylov wrote

$$\begin{pmatrix} v & Av & \cdots & A^{n-1}v & A^n v \\ 1 & \lambda & \cdots & \lambda^{n-1} & \lambda^n - p(\lambda) \end{pmatrix} \begin{pmatrix} -\alpha_0 \\ \vdots \\ -\alpha_{n-1} \\ 1 \end{pmatrix} = \begin{pmatrix} 0 \\ \vdots \\ 0 \\ 0 \end{pmatrix},$$

where the α_j's are the coefficients of the characteristic polynomial p. The first row of this matrix relation comes from the Cayley-Hamilton theorem and the second row from the definition of the characteristic polynomial. To obtain a nonzero solution, the determinant of the matrix on the left must be zero, and with a few manipulations one obtains

$$p(\lambda) = \frac{\det \begin{pmatrix} v & Av & \cdots & A^{n-1}v & A^n v \\ 1 & \lambda & \cdots & \lambda^{n-1} & \lambda^n \end{pmatrix}}{\det \begin{pmatrix} v & Av & \cdots & A^{n-1}v \end{pmatrix}}.$$

The interest of this formulation is that λ appears only in the last row of the matrix in the numerator rather than in every row as in $\lambda I - A$.

Krylov's method was discussed in the USSR by Nikolai Nikolaevich Luzin (1883-1950) [2101, 2102] in 1931-1932 and Igor Nikolaevich Khlodovskii [1891] in 1933.

In 1937, Aleksandr Mikhailovich Danilevsky reduced a matrix to its companion form by using the same elementary matrices as in the Gauss-Jordan elimination method [809]. Unfortunately, pivoting cannot be used since this would change the structure of the matrix and the method is not stable. Moreover, a matrix is similar to a single companion matrix only if it is non-derogatory. Therefore, this method can be in trouble for derogatory matrices. The method was explained by Alston Scott Householder (1904-1993) and Bauer [215, 1752] in 1959 as well as by Wilkinson in [3248], page 409, in 1965. Nevertheless, some people tried to introduce some sort of pivoting in the method; see [1575] by Eldon Robert Hansen in 1963.

Danilevsky was born in 1906. He worked at the Kharkiv Electrotechnical University (which is now known as Kharkiv Polytechnic Institute), where he was a docent. He died from starvation during the occupation of the town after its invasion by the German army on October 24, 1941.

The German engineer Karl Hessenberg (1904-1959) described two methods for obtaining the characteristic polynomial in a 37-page report [1655] issued in 1940 and not in his thesis

[1656], as is often written. The first method, using the Cayley-Hamilton theorem and starting from a vector z_0, simply constructed the Krylov vectors $z_i = Az_{i-1}$, the matrix $K = (z_0, \ z_1, \ \ldots \ z_{n-1})$ and solved $Ka = -z_n$ to obtain the vector of coefficients of the characteristic polynomial. Of course, the matrix had to be non-derogatory. In the second method, by using an elimination method, he constructed a basis V of the Krylov subspace and obtained, under some conditions, a relation $AV = VH$, where H is upper Hessenberg. Then, he remarked that if V has n columns, A and H have the same eigenvalues, and he added that the eigenvalues of H can be obtained by *any of the known methods.* He derived a recurrence for computing the determinant of $\lambda I - H$. He also explained how to compute the eigenvectors and gave numerical examples.

Hessenberg's method was explained in the book [3343] by Rudolf Zurmühl (1904-1966) in 1950. It was cited in the book by Ewald Konrad Bodewig (1901-?) [353] in 1959 and in Wilkinson's book [3248] in 1965.

Paul Anthony Samuelson (1915-2009) was an American economist who was awarded the Nobel Prize in Economic Sciences in 1970. In 1942, he published a method to compute the characteristic polynomial [2683]. His method was inspired by the reduction of an nth-order differential equation to a system of first-order equations. An expanded matrix is reduced using the Crout variant of Gaussian elimination. However, Samuelson's method was implicitly using, as Krylov did, powers of the matrix A. This method was discussed by Harold Wayland [3196] in 1945 and Hsin Chu [692] in 1968.

Edward Aaron Saibel (1903-1989) and W.J. Berger reconsidered Hessenberg's second method [2676] in 1953 and discussed its implementation.

Hans Rudolf Schwarz was interested in the stability of linear differential equations $y' = Ay$ with A constant and real [2721] in 1955 and complex [2722] in 1956. He determined the sign of the real parts of the eigenvalues without computing the characteristic polynomial. He first used Hessenberg's results (as described by Zurmühl) to transform the matrix to upper Hessenberg form with subdiagonal entries equal to -1 and then used elementary matrices (doing linear combinations of rows and columns) to put zeros above the first upper diagonal. The diagonal is zero (resp., imaginary numbers) except for the bottom right entry in the real (resp., complex) case. The signs of the eigenvalues were determined using the signs of products of the nonzero entries of the transformed matrix.

This was also considered at length in Schwarz's thesis [2723] under the supervision of Eduard Ludwig Stiefel (1909-1978) at the ETH in Zürich. It was observed in [1752] that this method can also be used to compute the characteristic polynomial.

A stable recursive method for computing the determinant of $H - \lambda I$ where H is an upper Hessenberg matrix was presented by Morton Allan Hyman[49] at the 12th National Meeting of the Association for Computing Machinery in Houston, Texas, in 1957. It was cited in Wilkinson's book [3248] in 1965, page 426; see also Nicholas John Higham's book [1681], page 280. Hyman's method was used by Pradeep Misra, Enrique S. Quintana, and Paul Van Dooren [2255] in 1995, who provided a backward error bound for the characteristic polynomial.

In 1958, Werner Louis Frank [1215] reported the results of numerical experiments for the computation of eigenvalues with Danilevsky's method and other ways of computing the characteristic polynomial and the eigenvalues. He gave examples on the *UNIVAC 1103* computer for which Danilevsky's method was unable to compute any single correct digit of the eigenvalues.

[49] M.A. Hyman was an American mathematician who obtained a Ph.D. in Delft (The Netherlands) in 1953. He worked for Remington Rand in Philadelphia and then for IBM in Yorktown Heights.

Methods for computing the characteristic polynomial were reviewed and discussed in 1959 by Householder and Bauer [1752]; see also [1747]. By the beginning of the 1960s, it was clear that computing eigenvalues from the characteristic polynomial was not reliable due to the sensitivity of the polynomial roots to small changes in the coefficients, and that other methods must be used.

Computation of roots of polynomials was considered in a very interesting and well-written paper [3254] by Wilkinson titled *The perfidious polynomial* in 1984. In that paper, Wilkinson reported his early experiments with polynomials on the *Pilot ACE* computer in the 1950s. About the characteristic polynomial, he wrote

> *Now it was perfectly natural for mathematicians to approach this problem by devising algorithms which would give coefficients of the explicit polynomial corresponding to* $\det(\lambda I - A)$*, ie., the characteristic polynomial of A. This had the advantage of reducing the volume of data from the* n^2 *elements of A to the* n *elements of the characteristic polynomial. However, the real incentive to the development of such algorithms was that it reduced the problem to the solution of an explicit polynomial equation, and this was felt to be a highly desirable transformation, Although attempts were made to analyse the effect of the rounding errors made in the algorithm on the accuracy of the computed coefficients of the explicit polynomial form, the desirability of this form does not seem to have been questioned. Almost all of the algorithms developed before the 1950s for dealing with the unsymmetric eigenvalue problem (i.e., problems in which A is unsymmetric) were based on some device for computing the explicit polynomial equations.*

Wilkinson received the Chauvenet Prize of the Mathematical Association of America for that paper.

Up until the 1950s, characteristic polynomials were used to compute eigenvalues. In the 1970s, things were turned upside down and eigenvalues were used to compute roots of polynomials. This was done by Cleve Barry Moler for the sake of computing roots of polynomials in MATLAB; see Chapter 8. A companion matrix was set up from the polynomial coefficients, the matrix was balanced and its eigenvalues computed by the QR algorithm (see below) to obtain the roots. This strategy was somehow justified by Kim-Chuan Toh and Lloyd Nicholas Trefethen [3059] in 1994 and Alan Edelman and Hiroshi Murakami [1049] in 1995.

Algorithms to compute the eigenvalues exploiting the structure of the companion matrix, which is the sum of a unitary matrix and a rank-one matrix, were proposed by Shiv Chandrasekaran, Ming Gu, Jianlin Xia, and Jiang Zhu [633] in 2007, Dario Andrea Bini, Paola Boito, Yuli Eidelman, Luca Gemignani, and Israel Gohberg (1928-2002) [324] in 2010, and Jared L. Aurentz, Thomas Mach, Raf Vandebril, and David Scott Watkins [104, 103] in 2015.

Although it is not a good idea to compute eigenvalues from characteristic polynomials, they are sometimes useful in physics, for instance, in quantum physics. Perturbation bounds for the coefficients of the characteristic polynomial of a complex matrix when it is perturbed were derived by Ipsen and Rizwana Rehman [1782] in 2008. In 2011, the same authors considered the computation of the coefficients of the characteristic polynomial from the eigenvalues [2545]; see also Rehman's thesis [2544].

6.6 ▪ The power and inverse iterations

Starting from a vector u_0, the power method consists in the iterations

$$u_k = Au_{k-1}/\|u_{k-1}\|_2, \quad \mu_k = (u_k, Au_k)/(u_k, u_k), \quad k = 1, 2, \dots.$$

If the complex matrix A has an eigenvalue λ_1 of modulus greater than the moduli of the other eigenvalues, and if the vector u_0 has a nonzero component in the direction of the eigenvector v_1 associated with λ_1, then the sequence (u_k) converges to v_1, and the sequence (μ_k) converges to λ_1 with a speed given by $|\mu_k - \lambda_1| = \mathcal{O}(|\lambda_2/\lambda_1|^k)$.

Looking at the expression for μ_k, the method is related to the Rayleigh quotient. If the assumptions on A and u_0 are not satisfied, variants of the power method can be used. Under some assumptions, the convergence of the power method can be accelerated by Aitken's Δ^2 process [12] and applying the scalar ε-algorithm of Peter Wynn (1931-2017) [3282] to the sequence (μ_k) can lead to sequences converging to the other eigenvalues of A; see [428] by C.B. in 1975.

The origin of the power method is not completely clear, and it has been rediscovered and used by several scholars. Some references attribute it to the German mathematician Gerhard Kowalewski (1876-1950) in 1909. Although some ratios of dot products related to quadratic forms look like the ratio for μ_k in his book [1942, pp. 191ff], it seems difficult to claim that he invented the method.

In fact, on page 321 of a famous paper published in 1910 [2566], Lewis Fry Richardson (1881-1953) gave a sketch of the power method to compute approximations of the largest eigenvalue. But he did not put too much emphasis on that.

According to others, a method for computing eigenvalues is due to Chaim (Herman) Müntz (1884-1956) in 1913 by computing ratios of minors of the matrices A^k [2304, 2305]. As stated by Eduardo L. Ortiz and Alan Pinkus in [2388],

> *It is very possible that Müntz was the first to develop an iterative procedure for the determination of the smallest eigenvalue of a positive definite matrix. It certainly predates the more generally quoted result of R. von Mises of 1929.*

Clearly, the authors of [2388] overlooked Richardson's paper whose main goal was not computing eigenvalues. Moreover, Müntz denoted the eigenvalues as $1/\lambda$ and thus his method converged to the largest eigenvalue and not the smallest one as stated in [2388].

Nevertheless, the method can be found explicitly in a paper by the Austrian scientist and mathematician Richard Edler von Mises (1883-1953) and his wife (they were later married in 1943) Hilda Pollaczeck-Geiringer (1893-1973) [3156] in 1929.

They considered the sequence $(z^{(\nu)})$ given by $z^{(\nu+1)} = \mu^{(\nu)} A z^{(\nu)}$ with $z^{(0)}$ arbitrary, and where the numbers $\mu^{(\nu)}$ are also left arbitrary. They wrote that considerations will be given below for their choice in the practical implementation of the procedure. Then, they claimed that for any choice of the component $z_i^{(\nu)}$ of the vector $z^{(\nu)}$,

$$\lambda_1 = \lim_{\nu \to \infty} \mu^{(\nu)} \frac{z_i^{(\nu)}}{z_i^{(\nu+1)}},$$

where, for practical purposes, the $\mu^{(\nu)}$ are to be chosen as decadic units, a denomination difficult to understand. However, the procedure works with $\mu^{(\nu)} = 1$ for all ν.

This work was, in fact, also anticipated, in a completely different domain, by the Russian mathematician and economist Georg von Charasoff (1877-1931) in his book [3154] of 1910 where he criticized and reconstructed Karl Marx's price theory. His iterations, analyzed by Kenji Mori in [2280] and named *production series*, were the same as those of von Mises/Pollaczeck-Geiringer. He showed that except for a factor, they converge to an eigenvector and that the associated eigenvalue is obtained as the limit of the quotient of one of the components of two successive vectors. He also discussed the invariance of the limit with respect to the choice of what he called the initial competition.

Although some sources mention that the power method was used by the statistician Harold Hotelling (1895-1973) for obtaining the numerical results of his 1933 paper [1731], it is not so easy to recognize the method in that paper.

It is interesting to see how the power method and its acceleration by Aitken's Δ^2 process were programmed by Wilkinson in 1953; see [3237]. We also observe that some problems with the power method were pointed out in 1954 by Bodewig in [352] who seemed to favor computing the characteristic polynomial. However, his conclusion was *the whole eigenproblem must be considered anew*. But see the remarks [425] about Bodewig's example by Joel Lee Brenner (1917-1997) and George Walter Reitwiesner (1918-1993), who were working at the Aberdeen Proving Ground in Maryland.

Let us mention that the power method was instrumental in the PageRank algorithm for classifying the pages on the Internet [357, 468], and that, as we said above, it can be accelerated by Aitken's process [448, 1867].

Note that applying the power method with A^{-1} gives approximations of the eigenvalue smallest in magnitude, but a linear system has to be solved at each iteration.

The *inverse iteration* aims at computing the eigenvector corresponding to a given eigenvalue μ (or of an approximation). Starting from an arbitrary vector u_0, the method generates the sequence

$$u_{k+1} = (A - \mu I)^{-1} u_k / c_k,$$

where c_k is a number usually chosen as $c_k = \|(A - \mu I)^{-1}\|$. The shift μ can be modified at each iteration and replaced by $\mu_k = (u_k, Au_k)/(u_k, u_k)$, a process known as *Rayleigh quotient iteration* (RQI). Note that this is not the method used by Lord Rayleigh; the name refers only to the use of the Rayleigh quotient.

It seems that the inverse iteration was first introduced in 1943-1944 by the German mathematician Helmut Wielandt (1910-2001), who published several reports on that topic, but only one journal paper [3229] in 1944. The convergence of the method was studied by Ostrowski in a series of papers [2401, 2402, 2403, 2404, 2405, 2406] in 1958-1959. Wielandt's method was explained in Bodewig's book [353] in 1956 and numerical experiments were reported by Frank [1215] in 1958. Wielandt also proved some inclusion results for eigenvalues [3230] in 1953.

Inverse iteration was used and studied by Wilkinson, as early as 1958 for tridiagonal matrices [3239], described in his book [3248] in 1965 and used in the software package EISPACK and later in LAPACK; see Chapter 8. These methods were rediscovered several times; variants of them were also proposed and they received various names, like for instance, shift-and-invert. See Ipsen [1783] in 1996, Rita Meyer-Spasche [2242] in 2017, and Richard Alfred Tapia, John Emory Dennis Jr., and Jan Peter Schäfermeyer [3000] in 2018 for a full account and historical details.

In 1951, Stephen Harry Crandall (1920-2013) proposed a method based on Southwell's relaxation for solving the generalized eigenvalue problem $Ax = \lambda Bx$ [759] for real symmetric matrices. Starting from an initial vector u_0, he computed $\mu_{k+1} = (u_k, Au_k)/(u_k, Bu_k)$ and u_{k+1} is the solution of the linear system $(A - \mu_{k+1}B)u_{k+1} = r$, where r is a residual. He mentioned that this method was already used in 1949 by Walter Kohn (1923-2016), who won the Nobel Prize in Chemistry in 1998, in [1930]. Then Crandall gave an iterative procedure with a cubic convergence rate for modifying r at each iteration by taking $r_{k+1} = Bu_k$. Crandall did not cite Wielandt.

What can be seen as a generalization of the power method was the *Treppeniteration* (staircase iteration) of Bauer [216] in 1957. If T_0 is an $n \times m$ unit lower trapezoidal matrix, the iterations were

$$X_{k+1} = AT_k, \quad T_{k+1}R_{k+1} = X_{k+1},$$

where R_{k+1} is an $m \times m$ upper triangular matrix. If the dominant m eigenvalues have distinct moduli the sequence T_k converges to T, where T is a part of the trapezoidal decomposition of the matrix of the m dominant eigenvectors. In any case, T_k tends to a basis of an invariant subspace of A.

Bauer's method was a prototype for a class of methods named *simultaneous iteration*. Bauer's method was studied by Rutishauser [2643] in 1969. Simultaneous iteration methods were used by Maurice Clint and Alan Kellerman Jennings for symmetric matrices [716] in 1970 and for nonsymmetric matrices [717] in 1971. Starting from an $n \times m$ matrix S_0, one computes the products $A^k S_0$ recursively. However, there can be severe linear dependence problems that can be solved by maintaining an orthonormal basis of the range of $A^k S_0$. Another problem is the extraction of approximations to eigenvectors. Gilbert Wright Stewart [2869] handled this problem in 1976 by using Schur vectors and Rayleigh-Ritz approximations.

6.7 ▪ Morris' escalator method

The *escalator method* is due to Joseph T. Morris, who published several papers [2288, 2289, 2290, 2291, 2295, 2292] (ordered by date) from 1936 to 1946 and a book [2293] in 1947.

Working on aeronautical problems, Morris was interested in solving linear systems and computing eigenvalues. In his first paper [2288], he explained on a small linear system a method which is nothing other than the Gauss-Seidel iteration. He showed how to organize the computation with the help of a table and using differences of consecutive current values of the iterates. Morris came back to this problem in [2290], citing Southwell on relaxation and discussing convergence. Apparently, he was not aware of what had been published previously on the convergence of relaxation methods. The papers [2289] and [2291] do not bring anything new, even though in [2291] he referred to Gauss and Seidel. Note that these papers were published in aeronautical journals whose readers may not have been aware of relaxation methods.

In 1946, he proposed a direct method for solving linear systems [2292] which is a kind of bordering method, obtaining the solution of a system of order n for the solution of the system obtained by removing the last row and last column.

The name "escalator" was introduced in [2294] and then in [2295], both written with J.W. Head. The method, inspired by the bordering method for linear systems, was intended to compute the eigenvalues of a matrix. The eigenvalues of the matrix of order n are obtained by solving a nonlinear equation which is obtained using the eigenvectors of the matrix of order $n-1$.

The authors wrote

> *In consequence of these difficulties* [the drop of the accuracy of the method proposed in [1027]] *and other considerations the authors devised the "Escalator" process when they were confronted with a practical problem in vibration which involved the numerical solution of a twelfth-order Lagrangian frequency equation.*
>
> *The method is essentially based on the successive introduction or elimination of each of the variables involves by definite self-contained stages in which the roots and modes of say fourth-order equation are obtained in terms of those of the preceding third-order equation and* vice versa [mathematical description of the method.]
>
> *We have proved that, given the solution of a set of Lagrangian frequency equations of any order, we can write down the latent root equation for a succeeding order, derived by bordering the original equations by an additional row and column; and in addition we have obtained expressions for the modes of the succeeding order. We shall call the process "escalation" and the particular form of equation (23) the "escalator" equation.*

The work of Morris on this method culminated in the publication of his book [2293]. There were several reviews of this book. In [315], it is pointed out that Morris *had a long struggle for recognition, but now that this recognition has come it is being given to him in well-deserved and full measure*. The second review pointed out that he worked for many years as a consultant in the Structural and Mechanical Engineering department of the Royal Aircraft Establishment, probably in Farnborough in the UK, where he was perhaps faced with a question of secrecy.

Almost nothing is known about Morris' life. According to Sven Hammarling,[50]

> *It turns out that he had been a Captain in the RAF, so Captain Joseph Morris. Born in 1888, he served in the first world war. He wrote at least two technical books:* The Strength of Shafts in Vibration, C. Lockwood and Son, 1929, *and* The Escalator Method in Engineering Vibration Problems, Chapman and Hall, 1947 (Published by John Wiley and Sons in the USA) *as well as* The German Air Raids on Great Britain 1914-1918. *It was first published in about 1925, but seems to have been republished several times, the latest I can see is 2007 by Nonsuch. I have not yet found when he died.*

In fact, Joseph Morris died in 1946.

The escalator method for linear systems was discussed by Robert Alexander Frazer (1891-1959) in 1947, where the notation and the presentation are in matrix form [1221]. For a modern account, see [1127, pp. 168-171].

6.8 ▪ The methods of Lanczos and Arnoldi

For symmetric matrices, the Lanczos algorithm is related to the Stieltjes procedure that was presented in 1884 by the Dutch mathematician Thomas Joannes Stieltjes (1856-1894) [2898] to compute the coefficients of the three-term recurrence satisfied by polynomials orthogonal with respect to a dot product defined by a Riemann-Stieltjes integral; for details see [1386]. However, Lanczos was probably not inspired by that paper.

The paper [1984] by Cornelius Lanczos (1893-1974) on the computation of the eigenvalues of a matrix (that he called the latent roots in the abstract, using Sylvester's terminology) was published in October 1950. As the title of the paper, *An iteration method for the solution of the eigenvalue problem of linear differential and integral operators*, suggests, Lanczos was interested in the solution of integral equations of Fredholm type and the solution of the eigenvalue problem of linear differential and integral operators. However, this paper is mostly remembered for the methods described for the computation of eigenvalues of symmetric and nonsymmetric matrices.

The problem Lanczos wanted to solve is $y - \lambda A y = b$. He defined what we now call the Krylov vectors b_i. He remarked that

> *We first notice that the iterated vectors $b_0, b_1, b_2, \ldots$ cannot be linearly independent of each other beyond a certain definite b_k. All these vectors find their place within the n-dimensional space of the matrix A, hence not more than n of them can be linearly independent.*

Hence, there is a relation

$$b_m + g_1 b_{m-1} + \cdots + g_m b_0 = 0$$

for some $m \leq n$. This relation defines a polynomial $G(x) = x^m + g_1 x^{m-1} + \cdots + g_m$ and also what he called the "inverted polynomial" $S_m(\lambda) = 1 + g_1\lambda + \cdots + g_m\lambda^m$. He considered the

[50]https://blogs.mathworks.com/cleve/2018/08/20/reviving-wilsons-matrix/ (accessed August 2021)

partial sums $S_j(\lambda)$ for $j = 0, \dots, m-1$ and obtained the relation

$$\frac{S_m(\lambda) - \lambda^m G(x)}{1 - \lambda x} = S_{m-1}(\lambda) + S_{m-2}(\lambda)\,\lambda x + \cdots + S_0\,\lambda^{m-1} x^{m-1}.$$

Replacing x by A and applying this to b_0 yields the formula

$$y = (I - \lambda A) b_0 = \frac{S_{m-1}(\lambda) b_0 + S_{m-2}(\lambda)\lambda b_1 + \cdots + S_0 \lambda^{m-1} b_{m-1}}{S_m(\lambda)}.$$

For the eigenvalue problem ($b = 0$) Lanczos obtained $S_m(\lambda) = 0$. The eigenvalues are $\mu = 1/\lambda$.

In Section V, Lanczos defined the Krylov vectors for A^*, which he denoted as b_i^* (note that Lanczos' notation was not so good because A^* is the conjugate transpose of A, but b_i^* is a given vector and not the conjugate transpose of b_i) and (in his notation) scalars $c_{k+i} = b_i \cdot b_k^* = b_k \cdot b_i^*$. He observed that $b_{k-1} \cdot b_{k+1}^* = b_{k-1}^* \cdot b_{k+1} = b_k \cdot b_k^*$ and discussed the case of defective matrices. Then, he showed how to obtain the coefficients of the polynomial G from the c_j's, which are entries of the moment matrix. If we have a polynomial $F(x) = x^n + \eta_{n-1} x^{n-1} + \cdots + \eta_0$ and if all the eigenvalues are distinct, from $F(\mu_i) = 0$, one obtains the relations

$$\begin{aligned} c_0 \eta_0 + c_1 \eta_1 + \cdots + c_{n-1} \eta_{n-1} + c_n &= 0, \\ \cdots &= 0, \\ c_n \eta_0 + c_{n+1} \eta_1 + \cdots + c_{2n-1} \eta_{n-1} + c_{2n} &= 0. \end{aligned}$$

If there is an early stop, we have to replace n by m. Lanczos observed that his method cannot tell the multiplicity of the eigenvalues.

In Section VI, Lanczos described a "progressive" algorithm to compute the coefficients of $G(x)$. For $k < m$, he considered the systems

$$\begin{aligned} c_0 \eta_0^{(k)} + c_1 \eta_1^{(k)} + \cdots + c_{k-1} \eta_{k-1}^{(k)} + c_k &= 0, \\ c_1 \eta_0^{(k)} + c_2 \eta_1^{(k)} + \cdots + c_k \eta_{k-1}^{(k)} + c_{k+1} &= 0, \\ \cdots &= 0, \\ c_k \eta_0^{(k)} + c_{k+1} \eta_1^{(k)} + \cdots + c_{2k-1} \eta_{k-1}^{(k)} + c_{2k} &= h_k \end{aligned}$$

for given h_k. Assuming the solution for a given k is known, Lanczos showed how to obtain the solution for $k+1$. He considered an additional system where the first equation is removed and a new one is added at the bottom

$$\begin{aligned} c_1 \bar{\eta}_0^{(k)} + c_2 \bar{\eta}_1^{(k)} + \cdots + c_k \bar{\eta}_{k-1}^{(k)} + c_{k+1} &= 0, \\ \cdots &= 0, \\ c_{k+1} \bar{\eta}_0^{(k)} + c_{k+2} \bar{\eta}_1^{(k)} + \cdots + c_{2k} \bar{\eta}_{k-1}^{(k)} + c_{2k+1} &= \bar{h}_{k+1}. \end{aligned}$$

Multiplying the first system by $q_k = -\bar{h}_{k+1}/h_k$ and adding the second system, one finds

$$\begin{aligned} c_0 \eta_0^{(k+1)} + c_1 \eta_1^{(k+1)} + \cdots + c_{k+1} &= 0, \\ \cdots &= 0, \\ c_k \eta_0^{(k+1)} + c_{k+1} \eta_1^{(k+1)} + \cdots + c_{2k+1} &= 0, \end{aligned}$$

provided that

$$\eta_0^{(k+1)} = q_k \eta_0^{(k)},\ \eta_1^{(k+1)} = q_k \eta_1^{(k)} + \bar{\eta}_1^{(k)}, \dots, \eta_k^{(k+1)} = q_k + \bar{\eta}_k^{(k)}.$$

Then, $h_{k+1} = c_{k+1}\eta_0^{(k+1)} + c_{k+2}\eta_1^{(k+1)} + \cdots + c_{2k+2}$. Also, $\bar{\eta}_i^{(k+1)}$ can be obtained from $\bar{\eta}_i^{(k)}$ and $\eta_i^{(k+1)}$ by using a multiplier $\bar{q}_k = -h_{k+1}/\bar{h}_{k+1}$.

Lanczos illustrated the algorithm with a 3×3 example, starting from $b_0 = b_0^* = e_1$, the first column of the identity matrix, and $h_1 = 1$. This algorithm yields the coefficient of the minimal polynomial.

In Section VII, Lanczos discussed the shortcomings of this algorithm for "large" and ill-conditioned matrices. Then, he introduced the "minimized iterations" modification, starting with the symmetric case. He introduced the notation $b' = Ab$ for a given vector b. In the first step he chose $b_1 = b_0' - \alpha_0 b_0$ and the scalar α_0 which minimizes the norm of b_1, that is, $\alpha_0 = b_0' \cdot b_0 / b_0 \cdot b_0$. He noticed that $b_1 \cdot b_0 = 0$.

For the next step $b_2 = b_1' - \alpha_1 b_1 - \beta_0 b_0$. Minimizing yields $\alpha_1 = b_1' \cdot b_1 / b_1 \cdot b_1$ and $\beta_0 = b_1' \cdot b_0 / b_0 \cdot b_0$. Note that the vectors are not normalized. For the third step, Lanczos remarked that because of the orthogonality relations, it is not necessary to add a term with b_0. This gives the three-term relation of the symmetric Lanczos method.

When the matrix is nonsymmetric, Lanczos obtained biorthogonal vectors by using A and A^*. In both cases, he wrote the three-term relation for the polynomials $p_{k+1}(x) = (x - \alpha_k)p_k(x) - \beta_{k-1}p_{k-1}(x)$ saying that, in the end, we obtain the minimal polynomial. Today, we know that this is only true without rounding errors in the computation. But Lanczos added

> *Although the new scheme goes basically through the same steps as the previously discussed "progressive algorithm", it is in an incomparably stronger position concerning rounding errors. Apart from the fact that the rounding errors do not accumulate, we can effectively counteract their influence by constantly checking the mutual orthogonality of the gradually evolving vectors b_i and b_i^*. Any lack of orthogonality, caused by rounding errors, can immediately be corrected by the addition of a small correction term.*

Lanczos obtained the eigenvalues as roots of the polynomial and the eigenvectors by using the values of the polynomials p_k at the eigenvalues. We observe that Lanczos did not notice that he was building tridiagonal matrices and that approximations of the eigenvalues can be obtained from these matrices.

Then, he showed some applications of his method. The first one is computing the lateral vibrations of a bar, a one-dimensional problem. The discretization of the partial differential equation gave him a symmetric matrix of order 12. For the iterations he used the inverse of the matrix. Probably by looking at the results, he noticed that his method can be used as an iterative method and that, in some cases, one does not have to continue the steps up to the end. For the bar problem the iterations were done up to $m = 6$ with a sort of selective reorthogonalization.

In the next sections Lanczos studied the eigenvalue problem for linear integral operators and for linear differential operators. The paper ends with a summary and a note:

> *The present investigation contains the results of years of research in the fields of network analysis, flutter problems, vibration of antennas, solution of systems of linear equations, encountered by the author in his consulting and research work for the Boeing Airplane Co., Seattle, Wash. The final conclusions were reached since the author's stay with the Institute for Numerical Analysis, of the National Bureau of Standards. The author expresses his heartfelt thanks to C.K. Stedman, head of the Physical Research Unit of the Boeing Airplane Co. and to J.H. Curtiss, Acting Director of the Institute for Numerical Analysis, for the generous support of his scientific endeavors.*

In 1951, Lanczos was a co-author of a paper [2601] published in the *Journal of the National Bureau of Standards* whose other authors were John Barkley Rosser (1907-1989), Magnus

Rudolph Hestenes (1906-1991), and William Karush (1917-1997). The goal of this paper was to compare the methods of Lanczos and Hestenes and Karush [1663, 1875] for computing eigenvalues of a singular 8×8 matrix with close eigenvalues. We observe that this matrix is now available in MATLAB under the name `rosser`. In the paper one can read

> *The method of Lanczos (see footnote 2 and appendix 2) seems best adapted for use by a hand computer using a desk computing machine. In the present case, the computation according to Lanczos' method was carried out by a hand computer, and required of the order of 100 hours computing time.*

The method of Hestenes and Karush [1663] was a gradient algorithm which was used on an *IBM card-programmed electronic calculator*. For the Lanczos method it was observed that it leads to a tridiagonal matrix and that, by normalizing the Lanczos vectors, one can obtain a symmetric tridiagonal matrix. The matrix was scaled by 1000 and the initial vector was e_2, the second column of the identity matrix. Full reorthogonalization was also used. The eigenvalues were computed as the roots of the characteristic polynomial.

In Lanczos's paper [1985] devoted to the solution of linear systems in 1952, he advocated the preliminary use of a *purification* process with Chebyshev polynomials to concentrate on a part of the spectrum. He wrote

> *The preliminary purification of b_0 served the purpose of increasing the convergence of the final algorithm by properly preparing the vector on which it operates. We were able to effectively eliminate all components of the original vector except those associated with the small eigenvalues.*

This type of technique was also used by Donald Alexander Flanders (1900-1958) and George H. Shortley (1910-1980) [1176] in 1950.

Contrary to what was done later, at that time, Lanczos did not consider his method as generating tridiagonal matrices whose eigenvalues are approximations of the eigenvalues of A. As was still quite common in the 1950s, he was thinking in terms of equations and polynomials. As we said above, the connection to tridiagonal matrices was noticed in 1951, in a paper [2601] co-authored with Rosser, Hestenes, and Karush.

It has been known since the beginning that the properties of the algorithm in finite precision arithmetic are far from the theoretical ones. In particular, as a consequence of rounding errors, the Lanczos basis vectors do not stay orthogonal as they should. Another annoying problem (also a consequence of the rounding errors) is the appearance of multiple copies of approximations of some eigenvalues of A in the set of converged Ritz values (that is, the eigenvalues of the tridiagonal matrices).

Since computing the characteristic polynomial is not a reliable way of computing the eigenvalues, Lanczos' method was afterwards considered an algorithm to reduce a matrix to tridiagonal form. But soon after the method was published, Givens and Householder proposed more efficient and cheaper methods to compute triangular or tridiagonal forms of a matrix using reliable orthogonal transformations.

In his 1965 authoritative book on the algebraic eigenvalue problem [3248], Wilkinson described the Lanczos nonsymmetric algorithm, insisting on the fact that there can be breakdowns and loss of (bi-)orthogonality. He suggested to use reorthogonalization but pointed out that this was more expensive than using Householder's method.

It has been written in many papers that Lanczos' methods were neglected in the 1950s and 1960s because of their sensitivities to rounding errors. Well, this is only partly true. After their publication, Lanczos' papers received some attention in those times. For instance, Lanczos' method is described in Sections 3-5 of the paper by Arnoldi [80] that we will discuss below.

Some papers or books, published before 1970, referring to Lanczos' papers are by Householder [1736] in 1953, Rutishauser [2634] in 1953, Ralph Anthony Brooker (1925-2019) and Frank Hall Sumner (19331-2013) [472] in 1956, Gregory [1458] in 1958, Householder and Bauer [1752] in 1959, Wilkinson [3240] in 1959, Max Engeli, Theo Ginsburg (1926-1993), Rutishauser, and Stiefel [1095] in 1959, Robert Lewis Causey and Gregory [578] in 1961, Wilkinson [3248] in 1965, and Tetsuro Yamamoto [3289] in 1968.

Bounds for the distances of the Ritz values obtained from the Lanczos algorithm to the eigenvalues in exact arithmetic and in the symmetric case were given by Shmuel Kaniel (1934-2007) [1869] in 1966, corrected by Christopher Conway Paige [2412] in 1971, and extended by Yousef Saad [2648] in 1980. The nice thing with these bounds is that they just depend on the distribution of the eigenvalues of A and the initial vector. Saad's result was

$$|\theta_i^{(k)} - \lambda_i| \le (\lambda_n - \lambda_1) \left[\kappa_i^{(k)} \frac{\tan \angle(q_i, v_1)}{C_{k-1}(\lambda_i; \lambda_{i+1}, \lambda_n)} \right]^2 ,$$

where $\theta_i^{(k)}$ are the Ritz values, the vectors q_i are the eigenvectors, the λ_i the eigenvalues (in increasing order from 1 to n), v_1 is the initial vector, and the C_k are shifted and normalized Chebyshev polynomials. Moreover,

$$\kappa_i^{(k)} = \prod_{j<i} \frac{\theta_j^{(k)} - \lambda_n}{\theta_j^{(k)} - \lambda_i}.$$

At the end of the 1960s there was no analysis of the rounding error properties of the Lanczos algorithms even though Wilkinson did a rough analysis in 1956 but did not publish it. At the beginning of the 1970s, Paige started working on a rounding error analysis of the Lanczos algorithm for computing eigenvalues of symmetric matrices. This culminated in his famous thesis [2412] at the University of London in 1971; see also [2411] in 1970. In the abstract of the thesis one can read

> *For the symmetric matrix case one of the most elegant algorithms theoretically is the method of minimized iterations developed by Lanczos in 1950. This method reduces the original matrix to tri-diagonal form from which the eigensystem can easily be found. The method can be used iteratively, and here the convergence properties and different possible eigenvalue intervals are first considered assuming infinite precision computation. Next rounding error analyses are given for the method both with and without reorthogonalization. It is shown that the method has been unjustly neglected, in fact a particular computation algorithm for the method without reorthogonalization is shown to have remarkably good error properties.*

Paige proved the important result that loss of orthogonality has a close relationship with convergence of Ritz values. He derived a matrix equation describing the propagation of the rounding errors in the algorithm. One of the consequences of this equation is to give a relation for the dot product of the new Lanczos vector and the Ritz vectors showing that the new vector stays approximately orthogonal to all unconverged Ritz vectors. This relation also shows that orthogonality is lost with a Ritz vector if there is convergence of the corresponding Ritz value to an eigenvalue of A. It can be said that Paige was the first to show that despite the rounding error problems, the Lanczos algorithm can be used to compute accurate approximations of some eigenvalues of A, and that this method is particularly useful for large sparse matrices.

For a summary of Paige's work and of the results of other people later on, see G.M.'s book [2222] and a joint paper [2235] with Zdeněk Strakoš in 2006.

Paige came back several times to the problem of the behavior of the Lanczos' algorithms in finite precision arithmetic; see [2413, 2414, 2415, 2418, 2419, 2420] and [639, 2421, 2422, 2431] with some co-authors. In [2419] in 2010 and in [639] with Xiao-Wen Chang and David Titley-Peloquin in 2021, Paige proved that after k iterations of the Lanczos algorithm in finite precision arithmetic, the results are the same as for the exact algorithm for the matrix

$$\mathcal{A}_k = \begin{pmatrix} T_k & 0 \\ 0 & A \end{pmatrix} + H^{(k)}, \quad \|H^{(k)}\| \leq \mathcal{O}(\epsilon)\|A\|,$$

where T_k is the Lanczos finite precision tridiagonal matrix and ϵ is the machine precision. This is a kind of extended backward stability. In [2419] Paige studied the convergence and the rate of convergence and showed how the Lanczos algorithm makes available backward stable eigenpairs of A for those eigenvectors that are represented in the initial vector v_1.

There were two schools about what to do with the Lanczos algorithm in finite precision arithmetic when one wants to compute all (or a large number of) the eigenvalues of A. The first possibility is to accept doing more, and sometimes many more, iterations than n, the order of the matrix. This was particularly advocated by Jane Grace Kehoe Cullum and Ralph Arthur Willoughby (1934-2001) in their two-volume book [790, 791]. The other possibility is to use some forms of reorthogonalization to maintain orthogonality between the Lanczos vectors, at least in a weak way.

Paige's work was emphasized in the seminal book [2446] by Parlett in 1980; see also David St Clair Scott's Ph.D. thesis [2729] in 1978, [2730] in 1979, and [2459] with Parlett in 1979 with a proposal to remove the delaying effects of rounding errors on convergence by a *selective orthogonalization* process for the Lanczos vectors.

There were a few other works about Lanczos' method in finite precision arithmetic a little later.

The Ph.D. thesis [1432] of Joseph Frank Grcar in 1981 attempted a more classical forward analysis of the Lanczos algorithm. He attributed the behavior of the Lanczos algorithm, particularly the loss of orthogonality, to a growth, through the use of the three-term recurrence, of the local rounding errors which are of the order of the unit roundoff.

An interesting work about the finite precision Lanczos algorithm is the Ph.D. thesis [2766] of Horst D. Simon, who was in 1982 a student of Parlett at U.C. Berkeley. He derived a recurrence for the level of orthogonality which measures how orthogonal the Lanczos vectors are, and used the fact that, to obtain a behavior of the Lanczos algorithm close to the exact one, it is enough to maintain semi-orthogonality that is at the level of the square root of the unit roundoff. His algorithm was called *partial reorthogonalization*; see [2767, 2768].

As we have said above, Cullum and Willoughby advocated iterating without reorthogonalization. The Lanczos algorithm produces several copies of the same approximate eigenvalue. One has to decide if they are good approximations of close eigenvalues or just artifacts. The identification test for eigenvalues proposed in [790] is the following: any Ritz value of T_k which is also an eigenvalue of $T_{2,k}$ (the matrix obtained from T_k by deleting the first row and the first column) is labeled as "spurious" and discarded from the list of computed eigenvalues.

On the foundations provided by Paige, a model and an explanation of the behavior of the Lanczos and the conjugate gradient (CG) algorithms in finite precision arithmetic were published by Anne Greenbaum in her Ph.D. thesis [1437] in 1981, in two papers [1439] in 1989, [1440] in 1994, and in her book [1442] in 1997. She studied perturbed Lanczos recurrence like

$$AV_k = V_k T_k + w^k (e^k)^T + F_k.$$

But she used a different view point than in other works since she showed that the matrix T_k generated at step k of a perturbed Lanczos recurrence is the same as the one generated by an exact Lanczos recurrence applied to a matrix of larger dimension than A whose eigenvalues are all close to those of A. This can be considered a backward error analysis in a sense different from the usual one. The proofs of these results were far from being easy. Greenbaum wrote

the details of the theorems and proofs are gory even though the basic ideas are not.

These theses and papers have inspired many other works. More recently, results in the same direction were published by Greenbaum and Strakoš [1451] in 1992, Jens-Peter M. Zemke [3318] in 2003, [3319] in 2006, [3320] in 2007, and Wolfgang Wülling [3280] in 2005 and [3281] in 2006.

The nonsymmetric Lanczos process in finite precision arithmetic was studied by Zhaojun Bai [154] in 1994. His work was in the same spirit as Paige's work on the symmetric Lanczos algorithm [2412].

A block Lanczos method for computing eigenvalues was introduced in 1975 by Richard Underwood in his Ph.D. thesis [3090]; see also the joint paper with his advisor Gene Howard Golub (1932-2007) [1394] in 1977. This was also considered by Cullum and William E. Donath [788] in 1974. See also [2366] by Dianne Prost O'Leary, another student of Golub, in 1980.

Z. Bai, David Minot Day, and Qiang Ye introduced *ABLE*, an adaptive block Lanczos method for non-Hermitian eigenvalue problems [155], in 1999.

In 1951, the American engineer Walter Edwin Arnoldi (1917-1995) published a paper [80] describing what is now called the *Arnoldi process*. As we said in Chapter 5, he attended a meeting at Harvard University in September 1949 and probably listened to the lecture Lanczos gave there. Arnoldi's paper starts with recalling Lanczos' algorithm but it is written in a more matrix-oriented way than Lanczos' paper, describing the tridiagonal structure of the generated matrix. In Section 6 of [80] he described his method based also on orthogonalization, but using only the matrix and not its transpose. It is not known if Arnoldi knew about the Gram-Schmidt process, but this is what he used for the Krylov subspace, replacing $A^k v_1$ by Av_i. The matrix relation for the basis vectors v_i is then

$$AV_k = V_k H_k + h_{k+1,k} v_{k+1} e_k^T,$$

where H_k is an upper Hessenberg matrix with entries $h_{i,j}$ and e_k is the kth column of the identity matrix. In fact, in Arnoldi's paper the entries of the first lower diagonal of H_k were equal to 1 because of the way he computed the basis vectors. Apparently, he did not realize that he replaced short recurrences in Lanczos' method with long ones and that this can be an issue for very large problems. Arnoldi obtained the approximate eigenvalues as roots of the characteristic polynomials computed with the entries of the successive matrices H_k.

Because, an explicit orthogonalization is done against all the previous basis vectors, Arnoldi's method is less prone to problems associated with rounding errors than Lanczos'.

Arnoldi's paper is cited in Lanczos' 1952 paper [1985] on solution of linear systems. Otherwise there were not many references to Arnoldi's paper in the 1960s and 1970s, even though it was cited and commented (almost negatively) in Wilkinson's book [3248] in 1965, pages 382-385.

The convergence of Arnoldi's method was investigated by Saad [2649] in 1980. In fact, he considered more general projection methods with Krylov subspaces. Let $\mathcal{P}_k$ be the orthogonal projector onto the Krylov subspace $\mathcal{K}_k(A, v_1)$, which can be written as $K_k(K_k^* K_k)^{-1} K_k^*$, where K_k is the Krylov matrix or $V_k V_k^*$ depending on the basis one considers. Then, the approximate eigenvalue problem amounts to solving

$$\mathcal{P}_k(Ay - \lambda y) = 0, \quad y \in \mathcal{K}_k(A, v_1),$$

or in operator form,

$$\mathcal{P}_k A \mathcal{P}_k y = \lambda y.$$

If one defines $A_k \equiv \mathcal{P}_k A \mathcal{P}_k$ (note that $A_k = V_k H_k V_k^*$), Saad's result stated how accurate the exact eigenpair is with respect to the approximate problem. This is given in terms of the distance of the exact eigenvector x to the Krylov subspace,

$$\|(A_k - \lambda I)\mathcal{P}_k x\| \leq \gamma_k \|(I - \mathcal{P}_k)x\|,$$

$$\|(A_k - \lambda I)x\| \leq \sqrt{|\lambda|^2 + \gamma_k^2}\, \|(I - \mathcal{P}_k)x\|,$$

with $\gamma_k = \|\mathcal{P}_k A(I - \mathcal{P}_k)\|$. Since $\gamma_k \leq \|A\|$, the problem is to bound the term $\|(I - \mathcal{P}_k)x\|$, which is the distance of the eigenvector to its projection onto the Krylov subspace. If the initial vector v_1 can be written as

$$v_1 = \sum_{i=1}^{n} \alpha_i x_i,$$

where the x_i's are the eigenvectors, then

$$\|(I - \mathcal{P}_k)x_i\| \leq \xi_i \min_{p \in P^{(i)}} \max_{j \neq i} |p(\lambda_j)|,$$

where $P^{(i)}$ is the set of polynomials of degree $k - 1$ such that $p(\lambda_i) = 1$ and

$$\xi_i = \sum_{j=1, j \neq i}^{n} |\alpha_j|/|\alpha_i|.$$

Unfortunately, the solution of the min-max problem is, in general, not explicitly known. These results were also stated in Saad's book [2658] in 1992. The second edition of the book in 2011 corrected a small mistake about bounds for the min-max problem in the first edition. Bounds were also given by Zhongxiao Jia for defective and non-derogatory matrices [1821] in 1995.

Upper bounds for $\|(I - \mathcal{P}_k)x_i\|$ were given by Mohammed Bellalij (1956-2015), Saad, and Hassane Sadok [258] in 2007 and [259] in 2010. However, it must be noted that a closed-form expression for $\|(I - \mathcal{P}_k)x_i\|$ was obtained in 2016 by Bellalij, G.M., and Sadok [256]. This distance is given in terms of the eigenvalues and eigenvectors of A as well as the components of v_1 in the eigenvector basis. Unfortunately, the formulas are quite intricate and difficult to interpret. They only simplify when A is a normal matrix. In that case, for the eigenvector x_1, the formula is

$$\|(I - \mathcal{P}_k)\, x_1\|^2 = \frac{N}{D},$$

with

$$N = \sum_{\hat{I}_{k+1}} \left[\prod_{j=2}^{k+1} |\alpha_{i_j}|^2 \right] \prod_{\substack{1 < i_2 \leq i_p < i_q \leq i_{k+1} \\ i_p,\, i_q \in \hat{I}_{k+1}}} |\lambda_{i_q} - \lambda_{i_p}|^2,$$

$$D = \sum_{I_k} \left[\prod_{j=1}^{k} |\alpha_{i_j}|^2 \right] \prod_{\substack{i_1 \leq i_p < i_q \leq i_k \\ i_p,\, i_q \in I_k}} |\lambda_{i_q} - \lambda_{i_p}|^2,$$

where I_k is a set of indices $\{i_1, i_2, \ldots, i_k\}$ such that $1 \leq i_1 < \cdots < i_k \leq n$ and the summation with $\hat{I}_{k+1}$ is over all sets of indices $\{i_2, \ldots, i_{k+1}\}$ such that $1 < i_2 < \cdots < i_{k+1} \leq n$. Moreover, $\alpha = X^* v_1$, where X is the matrix of the eigenvectors.

In [2658] Saad gave a bound for the sine of the angle between an eigenvector and a Ritz vector in the Hermitian case. This was extended to eigenspaces of non-Hermitian matrices by G.W. Stewart [2880] in 2001.

However, the Arnoldi algorithm does not always deliver some eigenvalues before the last iteration n since one can construct matrices and starting vectors with prescribed eigenvalues and prescribed Ritz values at all iterations 1 to $n-1$. This was considered by Paul Smit in a 1995 technical report [2800]. For some work on GMRES convergence and following ideas from Mario Arioli, Vlastimil Pták (1925-1999), and Strakoš [79] in 1998, Jurjen Duintjer Tebbens and G.M. described a parametrization of the class of matrices and right-hand sides generating, in addition to prescribed GMRES residual norms and eigenvalues, prescribed Ritz values in all iterations; see also [1015] in 2014 and [969] with Kui Du in 2017. This problem was also considered with a different approach by Parlett and Gilbert Strang [2460] in 2008.

In [2649] Saad also noticed that the cost of one Arnoldi iteration increases with the iteration number and he suggested to restart the algorithm periodically with v_1 replaced by a Ritz vector or a combination of Ritz vectors.

A good choice for the initial vector v_1 in Arnoldi's method would be to take a vector close to being in an invariant subspace generated by the eigenvectors corresponding to the wanted eigenvalues. The choice of initial vector would be $p_m(A)v_1$ where p_m is a shifted and normalized Chebyshev polynomial. As we have seen above, this was already proposed in 1950-1952 by Lanczos. But the idea was brought back to life by Saad [2652] in 1984. The Chebyshev and Arnoldi algorithm can be used in combination. This was called *explicit restarting*; see the second edition of [2658]. This hybrid method was also investigated by Diem Ho, Françoise Chatelin (1941-2020) and Maria Bennani [1703] in 1990.

A breakthrough for computing a few eigenvalues of large sparse matrices was done in 1992 by Danny Chris Sorensen, who published what is now called the *Implicitly Restarted Arnoldi* (IRA) method [2830]. This method was described as follows. Let

$$AV_{k+p} = V_{k+p}H_{k+p} + r_{k+p}r_{k+p}^T$$

be the result of $k+p$ Arnoldi iterations with k and p two given integers. Forgetting about the sub-indices for a moment, let μ be a shift and $H-\mu I = QR$ be a QR factorization of the shifted upper Hessenberg matrix. Then,

$$(A-\mu I)V - V(H-\mu I) = (A-\mu I)V - VQR = r\, r_{k+p}^T.$$

It yields

$$AV^+ - V^+H^+ = re_{k+p}^T Q, \quad V^+ = VQ, \quad H^+ = RQ + \mu I.$$

This has the effect of doing one iteration of the shifted QR algorithm (see below). Sorensen extended this idea to the use of p shifts for which one obtains

$$AV_{k+p}^+ = (V_{k+p}^+, v_{k+p+1}) \begin{pmatrix} H_{k+p}^+ \\ \beta_{k+p}e_{k+p}^T Q \end{pmatrix},$$

with $V_{k+p}^+ = V_{k+p}\hat{Q}$, $H_{k+p}^+ = \hat{Q}^T H_{k+p}\hat{Q}$, $\hat{Q} = Q_1Q_2\cdots Q_p$, where Q_j is the unitary matrix associated with the shift μ_j. Equating the k first columns, a legitimate Anoldi relation is obtained,

$$AV_k^+ = V_k^+H_k^+ + r_k^+e_k^T,$$

from which p new Arnoldi iterations can be done. Generally, the shifts are chosen from the unwanted Ritz values. The $k+p$ Ritz values of H_{k+p} are computed and then p shifted QR iterations are done with the p unwanted Ritz values as shifts. This method, which can be seen as a polynomial filtering method, was designed to iteratively produce "good" starting vectors.

The practical aspects, particularly the deflation rules, and the implementation of the IRA method were considered by Richard Bruno Lehoucq, who was a student of Sorensen, in his Ph.D. thesis [2025] in 1995 and in the papers [2028] in 1996 and [2027] in 2001. This led to the software ARPACK that was and still is extensively used to compute a few eigenvalues of nonsymmetric sparse matrices; see the user's guide [2029] by Lehoucq, Sorensen, and Chao Yang. A version for parallel computers was also developed.

Implicit restarting was used for the Lanczos algorithm in the symmetric case by Daniela Calvetti, Lothar Reichel, and Sorensen [541] in 1994. Later on, James Baglama, Calvetti, and Reichel developed IRBL, an implicitly restarted block-Lanczos method [145, 146].

A way of restarting the Lanczos algorithm for symmetric matrices that was simpler than IRA was described by Kesheng Wu and Simon [3279] in 2000. It was called the *thick-restart Lanczos* method. This thick-restart idea was already used, for instance, for the Davidson method by Andreas Stathopoulos, Saad, and Wu [2853] in 1998. If a restart is needed at iteration k, p (wanted) Ritz vectors are computed, constructing a matrix Y and used for the restart. Let $\hat{V}_p = V_k Y$ and $\hat{T}_p = Y^T T_k T$, which satisfy the relation

$$A\hat{V}_p = \hat{V}_p \hat{T}_p + \beta_k \hat{v}_{p+1} s^T,$$

with $\hat{v}_{p+1} = v_{k+1}$ and $s = Y^T e_k$. Then, this Lanczos-like relation is extended by maintaining orthogonality,

$$\hat{\beta}_{p+1}\hat{v}_{p+2} = (I - \hat{v}_{p+1}\hat{v}_{p+1}^T)A\hat{v}_{p+1} - \beta_k \hat{V}_p s,$$

and the next vectors are computed with the usual three-term recurrence. The matrix which is obtained is tridiagonal except for the upper left block of order $p+1$ which is an arrrowhead matrix that is diagonal except for the dense last row and last column.

Thick-restarting is equivalent to implicit restarting, as shown by Ronald Benjamin Morgan in [2275]. A thick-restart strategy was used by Morgan [2277] in 2002 for the Arnoldi process to be used in a restarted GMRES algorithm for solving linear systems; see Chapter 5. About restarting the Arnoldi algorithm, see also [2279] by Morgan and Min Zeng in 2006.

Ruipeng Li, Yuanzhe Xi, Eugene Vecharynski, C. Yang, and Saad developed a thick-restart Lanczos algorithm with polynomial filtering for Hermitian eigenvalue problems [2052] in 2016.

To solve some problems when using deflation in IRA, G.W. Stewart introduced in 2002 the so-called Krylov-Schur algorithm which uses a Schur factorization of the matrices replacing the Hessenberg matrices [2882]. It can be seen to be equivalent to an extension of the thick-restart technique for non-Hermitian matrices.

Other filtering procedures were proposed or used by Haw-Ren Fang and Saad [1136] in 2012 and by R. Li, Y. Xi, Lucas Erlandson, and Saad [2051] in 2019.

The convergence of polynomial restart for Krylov methods was studied by Christopher Andrew Beattie, Mark Embree, and John Rossi [233] in 2004 as well as well as by Beattie, Embree, and Sorensen [234] in 2005. However, in 2009, Embree [1091] constructed examples for which the implicitly restarted Arnoldi iteration failed, even in exact arithmetic.

The question of convergence of the Ritz vectors in Arnoldi's method was investigated by Jia [1823] in 1997. In order to improve upon the convergence of Ritz vectors he introduced *refined Ritz vectors*. The refined Ritz vector $\tilde{x}_i$ associated with the Ritz value θ_i at iteration k is defined as

$$\|(A - \theta_i I)\tilde{x}_i\| = \min_{x \in \mathcal{K}_k(A, v_1)} \|(A - \theta_i I)\tilde{x}\|.$$

The vector $\tilde{x}_i$ can be computed by using a small singular value decomposition.

The *harmonic Ritz values* were used by Morgan [2272] in 1991, but not under that name, for computing interior eigenvalues. They were named and defined in 1995 by Paige, Parlett, and Hendrik (Henk) Albertus van der Vorst [2423] for symmetric matrices as the reciprocals of Ritz values for A^{-1} obtained from $A\mathcal{K}_k$. They are weighted harmonic means of the eigenvalues of A, hence their name. This was later extended to nonsymmetric matrices. When used with the Arnoldi process, assuming that the upper Hessenberg matrix H_k is nonsingular, the harmonic Ritz values are given by the generalized eigenvalue problem

$$\underline{H}_k^* \underline{H}_k x = \theta H_k x,$$

where $\underline{H}_k$ is a $(k+1) \times k$ matrix

$$\underline{H}_k = \begin{pmatrix} H_k \\ h_{k+1,k} e_k^T \end{pmatrix}.$$

The convergence of harmonic Ritz values, harmonic Ritz vectors, and refined harmonic Ritz vectors was discussed by Jia [1825] in 2004 and also in [656] with Guizhi Chen.

The Ritz values obtained from Arnoldi's method are contained in the field of values (FOV) of the matrix A in the complex plane. Except for the first iteration, it is not at all obvious that any point in the FOV can be a Ritz value at some iteration and for some starting vector. The inverse problem described in Carden's Ph.D. thesis [549] in 2011 and the paper [551] about normal matrices with Derek J. Hansen in 2013 was, given a matrix A and complex values $\theta_1, \ldots, \theta_k$, to know if there is a subspace of dimension k such that the θ_i's are the corresponding Ritz values. If we restrict ourselves to Krylov subspaces and the Arnoldi process, this amounts to knowing if there is a unit vector v such that the values θ_i are Ritz values for the Krylov subspace; see also [550] with Embree in 2012. A closely related problem was considered for normal matrices by Zvonimir Bujanović [497] in 2013. He was interested in knowing the location of the other Ritz values if one fixes some of the Ritz values in the FOV of A. He gave a necessary and sufficient condition that tells if a set of k complex values can be the Ritz values of a given normal matrix. In 2015, G.M. [2228] gave a necessary and sufficient condition for a set of complex values $\theta_1, \ldots, \theta_k$ to be the Arnoldi Ritz values at iteration k for a given general diagonalizable matrix A. For real normal matrices and real starting vectors and for the case $k = 2$, a characterization was given of the boundary of the region in the complex plane contained in the FOV where pairs of complex conjugate Ritz values are located.

Rational Krylov methods for solving generalized eigenproblems $(A - \lambda B)x = 0$ were developed by Ruhe starting in the 1980s. These methods are similar to shifted and inverted Arnoldi in that they start with a random unit vector v_1 and construct an orthonormal basis V_k, one column at a time, multiplying at each step with a shifted and inverted matrix and orthogonalizing the resulting vector to the already computed part of the basis by means of Gram-Schmidt orthogonalization. They differ from shifted and inverted Arnoldi in that they may use different shifts in different iterations; see [2621, 2622, 2623, 2624] from 1984 to 1998 and the chapter written in [158] in 2000.

6.9 ▪ The methods of Givens and Householder

In 1954, Givens published a 116-page report from Oak Ridge National Laboratory [1365]. He used plane rotations to reduce a symmetric matrix to tridiagonal form. This was different from Jacobi's method, which amounts to iteratively reducing the matrix to a diagonal form. The tridiagonal form is obtained in a finite number of steps. Givens computed the eigenvalues of this tridiagonal matrix T using Sturm sequences. If $T - \lambda I$ has an LU factorization, the product of the diagonal entries $u_{1,1} \cdots u_{k,k}$, which is a polynomial $p_k(\lambda)$, is the determinant of the leading principal minor. The polynomials p_k are called a Sturm sequence and can be used to locate the eigenvalues by bisection; see also [1366] in 1957. Givens' work was first presented at the 1951 National Bureau of Standards Symposium. He used rotations to reduce a general matrix to upper triangular form [1367] in 1958. Givens' method was analyzed by Wilkinson [3249] in 1965.

A method using orthogonal transformations was proposed by Mark Lotkin [2093] in 1956.

In 1958, Householder [1744] used elementary matrices of the form $I - \alpha uu^*$, where u is a given vector to reduce matrices to upper triangular or upper Hessenberg forms. The method was discussed by Wilkinson [3243] in 1960. A rounding error analysis of the symmetric case was done in 1963 by James McDonough Ortega [2385].

6.10 ▪ The qd and LR algorithms

In his Doctoral thesis [1525], defended in 1892, Hadamard considered a rational function with n poles $|\lambda_1| \geq \cdots \geq |\lambda_n|$ represented by the series $f(z) = \sum_{\nu=0}^{\infty} c_\nu / z^{\nu+1}$, and proved that, for $1 \leq k < n$, the Hankel determinant

$$H_k^{(\nu)} = \begin{vmatrix} c_\nu & c_{\nu+1} & \cdots & c_{\nu+k-1} \\ c_{\nu+1} & c_{\nu+2} & \cdots & c_{\nu+k} \\ \vdots & \vdots & & \vdots \\ c_{\nu+k-1} & c_{\nu+k} & \cdots & c_{\nu+2k-2} \end{vmatrix}$$

satisfies

$$H_k^{(\nu)} = C(\lambda_1 \cdots \lambda_k)^\nu (1 + \mathcal{O}(\lambda/|\lambda_k|)^\nu), \qquad \nu \to \infty,$$

with $|\lambda_{k+1}| < \lambda < |\lambda_k|$ if $k < n$, and $\lambda < |\lambda_n|$ if $k = n$. Thus, if $|\lambda_k| > |\lambda_{k+1}|$, $\lim_{\nu\to\infty} H_k^{(\nu+1)}/H_k^{(\nu)} = \lambda_1 \cdots \lambda_k$, a result also proved later by Aitken in 1926 [12], but without assumptions on the poles. Obviously, if $|\lambda_{k-1}| > |\lambda_k| > |\lambda_{k+1}|$, then the quantity

$$q_k^{(\nu)} = \frac{H_k^{(\nu+1)}}{H_k^{(\nu)}} \frac{H_{k-1}^{(\nu)}}{H_{k-1}^{(\nu+1)}}$$

converges to λ_k when ν goes to infinity. When $k = 1$, Bernoulli's method for computing the root of largest modulus of a polynomial is recovered.

In 1949, Rutishauser was hired by Stiefel as a research associate at the Institute for Applied Mathematics at the Eidgenössische Technische Hochschule Zürich (ETHZ). He first worked on the development of a compiler for the Swiss computer *ERMETH*, and on the programming language Algol. Then, thanks to Stiefel's influence, he became interested in numerical linear algebra, in particular in Lanczos's biorthogonalization method [1984] that is related to orthogonal polynomials satisfying a three-term recurrence relationship and to Jacobi matrices [2634] (for more on these formal orthogonal polynomials, see [430] by C.B.).

In the second half of 1952 or early in 1953, Rutishauser considered the function

$$f(z) = ((zI - A)^{-1}x_0, y_0) = \sum_{\nu=0}^{\infty} c_\nu / z^{\nu+1},$$

with $c_\nu = (A^\nu x_0, y_0)$ [2636]. He showed how to compute the ratios $q_k^{(\nu)}$ defined above without explicitly computing the Hankel determinants defining them. As explained in [1504], defining the quantities $e_k^{(\nu)}$ as

$$e_k^{(\nu)} = \frac{H_{k-1}^{(\nu+1)} H_{k+1}^{(\nu)}}{H_k^{(\nu)} H_k^{\nu+1}},$$

he obtained the so-called *qd-algorithm* (qd stands for "Quotienten-Differenzen"), whose rules are

$$\begin{aligned} e_k^{(\nu)} &= e_{k-1}^{(\nu+1)} + q_k^{(\nu+1)} - q_k^{(\nu)}, \\ q_{k+1}^{(\nu)} &= q_k^{(\nu+1)} e_k^{(\nu+1)} / e_k^{(\nu)}, \end{aligned}$$

with $e_0^{(\nu)} = 0$ and $q_1^{(\nu)} = c_{\nu+1}/c_\nu$ for all ν [2636]. From these initializations, the first rule allows to compute the $e_1^{(\nu)}$, the second rule gives us the $q_2^{(\nu)}$, and so on. Usually, these quantities are displayed in a double entry table as follows, and the rules of the algorithm relate quantities located at the four corners of a rhombus:

$$\begin{array}{cccccc}
 & q_1^{(0)} & & & & \\
e_0^{(1)} & & e_1^{(0)} & & & \\
 & q_1^{(1)} & & q_2^{(0)} & & \\
e_0^{(2)} & & e_1^{(1)} & & e_2^{(0)} & \\
 & q_1^{(2)} & & q_2^{(1)} & & \ddots \\
e_0^{(3)} & & e_1^{(2)} & & e_2^{(1)} & \\
 & q_1^{(3)} & & q_2^{(2)} & & \ddots \\
e_0^{(4)} & \vdots & e_1^{(3)} & & e_2^{(2)} & \\
\vdots & & \vdots & q_2^{(3)} & \vdots & \ddots \\
 & & & \vdots & &
\end{array}$$

Rutishauser did not clearly state how he obtained these rules, but one can conjecture that he used Sylvester's determinantal identity for Hankel determinants, thus following Aitken. Then, Rutishauser showed how the coefficients $e_k^{(\nu)}$ and $q_k^{(\nu)}$ appear in the continued fractions associated with and corresponding to the series f. He also introduced adjacent families of polynomials satisfying recurrence relationships whose coefficients are expressed in terms of the $e_k^{(\nu)}$ and $q_k^{(\nu)}$. The links with Lanczos's method and the conjugate gradient algorithm were given.

At the beginning of the paper, he mentioned that it should be noted that excerpts from this work were reported by Stiefel at the GAMM 1953 conference (April 21-24 in Aachen, Germany) [2893], but the title he gave for this reference is wrong.

This first paper of Rutishauser was immediately followed by two others [2635, 2637] in 1954-1955, where the results of the first one were deepened. In particular, in [2635], he gave the progressive form of the *qd*-algorithm, which consists in building the table from its first descending diagonal by the rules

$$q_k^{(\nu+1)} = e_{k-1}^{(\nu+1)} - e_k^{(\nu)} - q_k^{(\nu)},$$
$$e_{k+1}^{(\nu+1)} = q_k^{(\nu+1)} e_k^{(\nu)} / q_k^{(\nu+1)}.$$

In [2637], Rutishauser considered the tridiagonal matrix (notation of [2638])

$$J_k = \begin{pmatrix} \alpha_1^{(k)} & \beta_1^{(k)} & & & \\ 1 & \alpha_2^{(k)} & \beta_2^{(k)} & & \\ & 1 & \ddots & \ddots & \\ & & \ddots & \alpha_{n-1}^{(k)} & \beta_{n-1}^{(k)} \\ & & & 1 & \alpha_n^{(k)} \end{pmatrix}$$

with $\alpha_k^{(k+1)} = q_k^{(k)} + e_k^{(k)}$ and $\beta_k^{(k+1)} = q_{k+1}^{(k)} e_k^{(k)}$. Then, in [2638], he showed that if J_k is factorized as $J_k = L_k R_k$ with

$$L_k = \begin{pmatrix} q_1^{(k)} & & & & \\ 1 & q_2^{(k)} & & & \\ & 1 & \ddots & & \\ & & \ddots & q_{n-1}^{(k)} & \\ & & & 1 & q_n^{(k)} \end{pmatrix}, \quad R_k = \begin{pmatrix} 1 & e_1^{(k)} & & & \\ & 1 & e_2^{(k)} & & \\ & & \ddots & \ddots & \\ & & & 1 & e_{n-1}^{(k)} \\ & & & & 1 \end{pmatrix},$$

then the matrix J_{k+1} is obtained, through the rules of the qd-algorithm, as the product $J_{k+1} = R_k L_k$. He wrote that under certain assumptions, it can be proved that for $i = 1, \ldots, n$, $\lim_{k\to\infty} \alpha_i^{(k)} = \lambda_i$, the eigenvalue of any tridiagonal Jacobi matrix J_0.

But the most important result of this paper was yet to come. Rutishauser started from a matrix $A = A_0$ and decomposed it into the product $A_0 = L_0 R_0$, where L_0 is a lower triangular matrix and R_0 is an upper triangular one. Then, he constructed the matrix $A_1 = R_0 L_0$ and decomposed it into the product $A_1 = L_1 R_1$ with L_1 lower triangular and R_1 upper triangular. Again A_2 is obtained as $A_2 = R_1 L_1$, and so on. This is the LR *algorithm*.

If A is real and symmetric, if its eigenvalues satisfy $\lambda_1 > \cdots > \lambda_n \geq 0$, and if all the matrices A_k, which are similar, are decomposed such that $R_k = L_k^T$, then the sequence (A_k) converges to a diagonal matrix with $\lambda_1, \ldots, \lambda_n$ on its diagonal, in this order. The convergence is geometric, but it could be quadratic by using the progressive form of the algorithm; see [2646]. In [2640], published in 1958, Rutishauser introduced a shift in the method to speed up the convergence; see also [2647] with Schwarz in 1963.

In quite a short time, the LR algorithm gave rise to a bunch of publications about it and around it; see [1504] and [2645] for a list of Rutishauser's papers on these topics. Parlett studied the development and use of methods of LR type [2438] in 1964. In [3249], Wilkinson discussed the existing proofs of convergence of the LR algorithm which were based on rather sophisticated determinantal identities, and he gave elementary and complete proofs of the convergence.

As already mentioned, the qd-algorithm is related to the works of Lanczos, Hestenes, and Stiefel. But it is also related to continued fractions, and, by continued fractions, to Padé approximation (see [1749]), and, by Padé approximation, to the sequence transformation introduced by Daniel Shanks (1917-1996) for accelerating sequences (quoted by Rutishauser in [2639]) [2748, 2749], and to the ε-algorithm of Wynn for its implementation [3282]. Bauer [216, 217, 218], Wynn's advisor, and Henrici [1642], a student of Stiefel, were also involved in this historical background and its numerous related connections. This was a period of intense development of all these topics, with many contacts and collaborations between the actors, although the Internet was yet to be invented. An account of this intellectual atmosphere could be grasped in [450, 451] by C.B. and M.R.-Z.

The qd-algorithm has been used and extended over the years, mainly in relation with eigenvalues of tridiagonal matrices; see Kurukula Vince Fernando and Parlett [1153] in 1994, Parlett [2449] in 1995, Inderjit Singh Dhillon and Parlett [886] in 2000, or for solving inverse problems, see Dirk Pieter Laurie (1946-2019) [2002] in 1999.

About the LR algorithm, Wilkinson [3248, p. 485] commented

> *In my opinion its development is the most significant advance which has been made in connection with the eigenvalue problem since the advent of automatic computers. The QR algorithm, which was developed later by Francis, is closely related to the LR algorithm but is based on the use of unitary transformations. In many respects this has proved to be the most effective of known methods for the solution of the general eigenvalue problem.*

6.11 ▪ The QR algorithm

For symmetric matrices, Rutishauser's original LR algorithm [2638] was based on the decomposition given by André Louis Cholesky (1875-1918) for solving a system of linear equations. For a general matrix, it was related to the LU factorization. But Rutishauser's idea can be generalized. If $A = XY$, with X invertible, then the new matrix $C = YX$ is similar to A, because $C = YX = X^{-1}AX$ and, therefore, C has the same eigenvalues as A.

The idea of replacing the LU factorization by a QR factorization with Q orthogonal or unitary and R upper triangular was a real breakthrough. In [2451] Parlett wrote

> *What makes the experts in matrix computations happy is that this algorithm is a genuinely new contribution to the field of numerical analysis and not just a refinement of ideas given by Newton, Gauss, Hadamard, or Schur.*

The basic algorithm, starting from $A_1 = A$, is

$$A_k = Q_k R_k, \quad A_{k+1} = R_k Q_k, \; k = 1, 2, \ldots.$$

If the eigenvalues have distinct moduli and under mild conditions on the eigenvectors, the matrices A_k converge to an upper triangular form with the eigenvalues of A in monotone decreasing order of absolute value down the diagonal as shown by Wilkinson [3249] in 1965. The basic iteration converges linearly.

The QR algorithm was introduced by a young British computer scientist John Guy Figgis Francis in two papers [1212, 1213] published in 1961-1962. The first paper was submitted on October 29, 1959, to the *Computer Journal*. In 1956 and the following years, Francis was working at the National Research and Development Corporation (NRDC) as an assistant (programmer) to the computer scientist Christopher Strachey (1916-1975). The interesting story of Francis was told in the paper [1393] by Golub and Uhlig in 2009. Francis left NRDC in 1961 and worked in several companies as a computer scientist. In fact, what is amazing is that when Gene Golub was able to contact Francis in 2007, he was entirely unaware that there had been many references to and extensions of his early work and that his QR algorithm was considered one of the ten most important algorithms of the 20th century; see also [3086] by Uhlig in 2009. Francis was then invited to give a talk at the 23rd Biennial Conference on Numerical Analysis at Strathclyde

University in Glasgow, June 23-25, 2009. He also received an honorary doctorate from the University of Sussex in 2015; see [3087]. It is Francis who coined the term QR for the factorization of a matrix as the product of a unitary matrix and an upper triangular matrix.

The basic QR algorithm was proposed independently by Vera Nikolaevna Kublanovskaya (1920-2012) in the USSR. She published several papers [1959, 1960] in 1960-1962. In fact, she was using an LQ factorization, with L lower triangular and Q orthonormal, and she gave a proof of convergence using determinants. Apparently, the method was not really used in the USSR at that time.

A QR algorithm for symmetric tridiagonal matrices without using square roots was proposed in 1963 by Ortega and Henry Felix Kaiser (1927-1992) [2386]. The convergence of the tridiagonal QL algorithm (with L lower triangular) [1722] was studied by Walter Hoffmann (1944-2013) and Parlett in 1978.

In [1212] Francis proved the convergence of the basic QR algorithm under the condition that the eigenvalues have distinct moduli. In fact, Francis' algorithm was more elaborate than the basic QR algorithm. He made several useful observations and introduced new ideas:

- A QR step applied to an upper Hessenberg matrix yields an upper Hessenberg matrix. This was shown in his first paper [1212]. This reduced a cost proportional to n^3 floating point operations for a full matrix to $12n^2$ for one step. Hence, it was useful to do a preprocessing reduction of a general matrix A to upper Hessenberg form by unitary similarities in a stable way.

- Use of shifts can improve convergence. This was considered in his second paper [1213]. With a shift σ and a matrix H, a QR step is

$$H - \sigma I = QR, \quad \hat{H} = RQ + \sigma I.$$

- Deflation can save floating point operations. As soon as the last lower off-diagonal entry is sufficiently small, it is declared zero, and the algorithm proceeds with a smaller matrix of order decreased by 1 (or 2 in the case of complex shifts).

A good shift would be an eigenvalue λ of A because then, after one step, the last row of $\hat{H}$ would be λe_n^T and deflation could take place. However, this cannot be done since the eigenvalues are unknown. When using a single shift, one can choose $\sigma = h_{n,n} = e_n^T H e_n$, the bottom right entry of H. However, Francis remarked that complex shifts must be used to enhance convergence since nonsymmetric real matrices may have complex conjugate eigenvalues. He proposed to use the eigenvalues of the 2×2 bottom right submatrix of H which are, in general, complex conjugate numbers (this is sometimes called *Wilkinson's shifts*, even though it was proposed by Francis when these eigenvalues are complex conjugate). Moreover, he wanted to stay with real arithmetic; in [1213] he devised a clever way to do that. Two consecutive QR steps with μ and its complex conjugate $\bar{\mu}$ as shifts can be done in real arithmetic. One computes

$$M = H^2 - (\mu + \bar{\mu})H + \mu\bar{\mu}\, I, \quad M = QR, \quad H_2 = Q^T H Q,$$

and the matrix H_2 is real upper Hessenberg. But it is expensive to compute H^2 and Francis proved and used the *implicit Q theorem*. There are several versions of this theorem, but roughly speaking it says that if $Q^T A Q = H$ is upper Hessenberg and unreduced (which means $h_{i+1,i} \neq 0$), columns 2 to n of Q are determined uniquely, up to the signs, by the first column of Q. This means that it is enough to consider the first column of M, which is cheap to compute. This first column, Me_1 has only three nonzero components at the top. Let P_0 be a Householder transformation such that it makes Me_1 proportional to e_1. The matrix $P_0 H P_0$ is no longer

Hessenberg because it has a 3×3 top left dense block. which is called the *bulge*. The Hessenberg structure is restored by using Householder transformations that push the bulge down along the subdiagonal. This is called *chasing the bulge*.

We see that Francis' algorithm was much more sophisticated than the basic QR algorithm. He programmed the algorithm for the *Ferranti Pegasus* computer. Numerical results and programs were given in [1213]. As it is written at the end of the paper, Wilkinson read the paper at an early stage and he described the method in his book [3248] in 1965.

The QR algorithm simplifies when the matrix is symmetric or Hermitian since the Hessenberg matrices are tridiagonal, and because the eigenvalues are real, single shifts are sufficient.

Algol codes implementing Francis' algorithm for real upper Hessenberg matrices were described by Roger S. Martin, G. Peters, and Wilkinson [2152] in 1970 and later published in the *Handbook for Automatic Computation, Volume II, Linear Algebra* in 1971. It was noticed in that paper that the algorithm can fail by giving the example of a companion matrix of order 5 with the last column equal to e_1. The shifts are both zero and the matrix is invariant for the algorithm without shift. It was suggested that after some iterations without progress, one does a change to another shift for one iteration to restore convergence.

Convergence proofs under some conditions were given in the following years after the publication of the algorithm. Parlett gave another proof of convergence of the basic QR algorithm [2440] in 1965; see also the correction [2441] in 1967. He considered the convergence of the basic QR algorithm for Hessenberg matrices [2442] in 1968. In 1973, with William George Poole Jr., he gave a geometric theory for the QR, LU, and power iterations [2456] showing the relation between these methods. See also [2443] in 1973.

A simpler proof of convergence of the basic QR algorithm was given in 1968 by Colette Lebaud (1937-2018) using the Jordan canonical form [2015]. She considered the convergence of a QR algorithm with a double shift [2014, 2016] in 1970 and showed that for any unreduced Hessenberg matrix it is safe to switch from the basic to the double shift strategy ultimately, when certain conditions have been fulfilled.

The QR algorithm for real symmetric tridiagonal matrices was shown to be backward stable by Wilkinson [3248]. This was perfectly fine for computing eigenvalues. However, Parlett and Jian Le [2457] showed cases of forward instability in 1993. Moreover, Demmel and Veselić [878] showed in 1992 that Jacobi's method (with a proper stopping criterion) computes small eigenvalues of symmetric positive definite matrices with a uniformly better relative accuracy bound than QR, or any algorithm which first involves tridiagonalizing the matrix.

In 1975, Jean Della Dora (1946-2009) proved a convergence theorem [861] for algorithms based on subgroup decompositions. His theorem could be applied to the QR, LR, and other algorithms, but it was limited to the unshifted case and dealt only with matrices whose eigenvalues have distinct moduli.

Another contributor to the QR algorithm convergence theory is Watkins. In 1982, he published a review paper [3181] about the basic method in which he wrote

> *the QR algorithm is neither more nor less than a clever implementation of simultaneous iteration, which is itself a natural, easily understood extension of the power method.* [...] *it provides a framework within which the rapid convergence associated with shifts of origin may be explained.*

With Ludwig Elsner, he considered in 1991 [3191, 3192] what they called *GR algorithms* (but this naming was already used by Della Dora in [861]) in which the transformation matrices for the similarity transformations $A_i = G_i^{-1} A_{i-1} G_i$ are obtained from a *GR decomposition*

$p_i(A_{i-1}) = G_i R_i$ with an upper triangular matrix R_i and a polynomial p_i whose degree is called the multiplicity of the ith iteration. Its roots are the shifts. This framework covers, in particular, the LR and QR algorithms. They showed that every GR algorithm is a form of nested subspace iteration in which a change of coordinate system is made at each iteration. They gave a convergence theory for GR algorithms based on results for subspace iteration.

Other contributions of Watkins include [3182, 3183, 3184, 3185, 3187, 3189, 3190]. He published a book *The Matrix Eigenvalue Problem: GR and Krylov Subspace Methods* [3188] in 2007.

Improvements and extensions of Francis' QR algorithm have been proposed over the years, building on the accumulated experience with the method. Some were for improving the convergence of the method or specializing to some types of matrices (for example unitary matrices or companion matrices) and others for increasing the degree of parallelism to adapt the method to modern parallel computers.

Increasing the parallelism and the ability to use Level 2 and Level 3 BLAS (see Chapter 8) were the primary goals of Z. Bai and Demmel for introducing a *multishift* QR algorithm [156] in 1989. In this method m shifts (generally chosen as eigenvalues of the trailing principal submatrix of order m) are used simultaneously, resulting in an $m \times m$ bulge which is chased down p columns at a time.

Unfortunately, the multishift algorithm showed serious convergence problems caused by roundoff errors when large values of m were used. These problems (called *shift blurring*) were analyzed and explained [3185] by Watkins in 1996; see also [3183, 3184].

The problems of the multishift QR algorithm with large bulges were addressed in 2002 by Karen Braman, Ralph Byers (1955-2007), and Roy Mathias [393]. The large bulge in a QR step was replaced with a chain of many small (in fact 3×3) bulges which are chased simultaneously. This allowed the use of Level 3 BLAS with matrix-matrix multiplies.

Traditionally, the deflation criterion was to look at the size of the subdiagonal entries of the Hessenberg matrices and set them to zero if they were too small. The eigenvalue problem then decouples into smaller problems. To save some computational work a sophisticated *aggressive early delation* strategy (which allows to deflate sooner than the standard criterion) was defined in [394] by Braman, Byers, and Mathias.

Daniel Kressner related the shift blurring problems observed with large bulges to the notoriously ill-conditioned pole placement problem [1949] in 2005. Nevertheless, his conclusion was that using slightly larger bulges (like 5×5) can still have a positive effect on the performance of the QR algorithm. In [1950], published in 2008, he showed that the early deflation strategy is equivalent to extracting converged Ritz vectors from certain Krylov subspaces involving H^*. He also obtained improved convergence bounds.

In 2011, Vandebril presented a procedure [3117] for transforming a matrix via unitary similarity transformations to a condensed format using products of rotations. He used that to propose a generalized QR algorithm which could use condensed formats different from Hessenberg matrices, for instance, inverses of Hessenberg matrices. This was extended to a multishift algorithm in a joint paper [3120] with Watkins in 2012 in which a convergence theory based on rational Krylov subspaces was developed. The same techniques were used for computing the roots of polynomials using QR-like algorithms on the companion matrices; see [103] by Aurentz, Mach, Vandebril, and Watkins in 2015 and [102] by the same authors and Leonardo Robol.

The QZ algorithm, an analog of the QR algorithm for the generalized eigenvalue problem $Ax = \lambda Bx$, is due to Moler and Stewart [2266] in 1973. For details, see Stewart's book [2881] in 2001 and [3186] by Watkins in 2000.

6.12 ▪ Tridiagonal matrices

We have seen that in the 1950s, Givens solved the eigenvalue problem for a symmetric tridiagonal matrix by using the Sturm sequence property for successive subdeterminants. The corresponding eigenvectors were obtained by inverse iteration. Later on, people used a symmetric version of the QR algorithm. This problem was discussed in Parlett's book [2446] in 1980.

Starting in the 1980s, other methods were developed to solve the eigenvalue problem for symmetric tridiagonal matrices.

In 1981, Johannes Josephus Maria Cuppen, who was a student of Theodorus Jozef Dekker (1927-2021) in Amsterdam, proposed a divide and conquer algorithm [792]. The symmetric tridiagonal matrix T with subdiagonal entries β_j is written as

$$T = \begin{pmatrix} T_1 & 0 \\ 0 & T_2 \end{pmatrix} + \beta_m \begin{pmatrix} e_k \\ e_1 \end{pmatrix} \begin{pmatrix} e_k^T & e_1^T \end{pmatrix},$$

where T_1 is of order k, that is, T is the sum of a block diagonal matrix with tridiagonal diagonal blocks plus a rank-one matrix. Assuming one knows the spectral factorizations of T_1 and T_2, the eigenvalues of T can be recovered by computing those of $D + \beta_k zz^T$, where D is a diagonal matrix whose diagonal entries are the eigenvalues of T_1 and T_2 and z is computed from the eigenvectors of those matrices. Then, the method is applied recursively to T_1 and T_2 until the order is small enough and the QR algorithm is applied to stop the recursion.

The eigenvalues of $D + \beta_k zz^T$ are the roots of the rational function

$$1 - \beta_k \sum_{i=1}^{n} \frac{z_i^2}{\lambda - d_i}.$$

Equating this rational function to zero is known as the *secular equation*. For comparisons of methods for the solution of secular equations, see [2196, 2197] by Aaron Melman in 1997.

In 1987, Jack Joseph Dongarra and Sorensen [930] proposed a parallel implementation of Cuppen's method with a new deflation strategy. For solving the secular equation they used the method of James Raymond Bunch, Christopher P. Nielsen, and Sorensen [503] published in 1978.

In a paper [1472] submitted in 1992 but published only in 1995, Gu and Stanley Charles Eisenstat (1944-2020) proposed another method that they called an *arrowhead divide-and-conquer* algorithm. They used a division strategy close to Cuppen's (with a different notation),

$$T = \begin{pmatrix} T_1 & \beta_{m+1} e_m & 0 \\ \beta_{m+1} e_m^T & \alpha_{m+1} & \beta_{m+2} e_1^T \\ 0 & \beta_{m+1} e_1 & T_2 \end{pmatrix}.$$

From the spectral factorizations of T_1 and T_2 ($T_i = Q_i D_i Q_i^T$) they constructed an orthogonal matrix Q such that $T = QHQ^T$, where

$$H = \begin{pmatrix} \alpha & z^T \\ z & D \end{pmatrix},$$

with $\alpha = \alpha_{m+1}$ a scalar, z a vector, and D a diagonal matrix,

$$D = \begin{pmatrix} D_1 & 0 \\ 0 & D_2 \end{pmatrix}.$$

H is what is called an *arrowhead matrix*. The goal of Gu and Eisenstat was to compute accurate orthogonal eigenvectors. They introduced a new algorithm for computing the eigenpairs of an arrowhead matrix. It was based on a previous paper [1470] by the same authors. The computation of the eigenvalues of H required the solution of a secular equation which has to be solved with enough accuracy. They proved some stability results and discussed the deflation strategy.

We observe that the divide and conquer concept was used by Peter Arbenz for band matrices [70] in 1992.

A new method for computing the eigenvalues and eigenvectors of a symmetric tridiagonal matrix was introduced in the 1997 Ph.D. thesis [885] of Dhillon who was, at that time, a student of Parlett at U.C. Berkeley. The goal was to obtain the eigenvalues and eigenvectors with a high relative accuracy.

It was known that the singular values of a bidiagonal matrix can be computed to high relative accuracy with respect to small relative perturbations in the entries; see Demmel and Kahan [877] in 1990 as well as Fernando and Parlett [1153] in 1994 using the (shifted) differential qd-algorithm.

The Parlett and Dhillon method was named MRRR for *Multiple Relatively Robust Representation*. It is a sophisticated algorithm based on variants of the qd -algorithm and *twisted factorizations* of tridiagonal matrices. A twisted factorization (also called a BABE - burn at both ends - factorization) is obtained from two Cholesky-like factorizations, one started from the top and the other started from the bottom. For the use of these factorizations in computing the eigenvectors, see [1154] in 1995 and [2454] by Fernando, Parlett, and Dhillon in 1997; see also [2450].

After Dhillon's Ph.D. thesis, the ingredients of the algorithm were presented in joint papers with Parlett, [2454] in 1997, [886] in 2000, [887] in 2003, and [888] in 2004. MRRR was implemented in the package LAPACK.

However, the MRRR algorithm can fail in extreme cases when some eigenvalues agree to working accuracy and the algorithm cannot compute orthogonal eigenvectors for these eigenvalues. Dhillon, Parlett, and Christof Vömel [889] described and analyzed these failures and described various remedies in 2005. The design and implementation of the MRRR algorithm was explained by these authors in [890] in 2006.

New representations for bidiagonal and twisted factorizations were proposed by Paul R. Willems and Bruno Lang [3257] in 2012.

Inverse problems, that is, construction of tridiagonal matrices from spectral data, was considered by Carl-Wilhelm Rheinhold de Boor and Golub [840] in 1978, William Bryant Gragg (1936-2016) and William Joseph Harrod [1421] in 1984, and Laurie [2002] in 1999. For the eigenvectors, see [2455] by Parlett, Froilán M. Dopico, and Carla Ferreira in 2016.

6.13 ▪ The Jacobi-Davidson method

In 1975, Ernest Roy Davidson, an American chemist working in quantum chemistry, introduced a projection method for computing the lowest eigenvalues and corresponding eigenvectors of large real symmetric matrices [814]. Davidson wanted to compute the lowest energy levels and the corresponding wave functions of the Schrödinger operator.

The matrices in his applications were (strongly) diagonally dominant. Starting from a given initial vector, Davidson computed the Ritz pairs of interest (the eigenpairs of the projected matrix $V_k^T A V_k$), the residual $r = (A - \theta I)v$ for a pair of interest (θ, v), and expanded the current subspace V_k with the vector $(D - \theta I)^{-1} r$ after its orthogonalization with respect to the columns of V_k. The matrix D is a diagonal matrix with the same diagonal as A. For diagonally dominant matrices this approximates, in some sense, inverse iteration using Rayleigh quotients. The Davidson method was efficient in computing eigenvalues of diagonally dominant matrices and became quite popular for applications in chemistry and physics. Note that because of the orthogonalization, the amount of work per iteration is increasing, just like in Arnoldi's method. We observe that another symmetric matrix can be used to replace the diagonal matrix D. This was interpreted as preconditioning in [2278].

Generalizations of Davidson's method for computing eigenvalues of sparse symmetric matrices were studied by Morgan and D.S. Scott [2278] in 1986; see also [2273] by Morgan in 1992. A block generalization of Davidson's method was analyzed by Michel Crouzeix, B. Philippe, and Miloud Sadkane [779] in 1994. Dynamic thick restarting of the method was proposed by Stathopoulos, Saad, and Wu [2853] in 1998.

In his paper [1797] in 1846, Jacobi, besides his method for increasing the diagonal dominance of the problem he was considering, introduced a much less well known technique that was later called *Jacobi's orthogonal component correction* (JOCC). The original Davidson method with diagonal scaling is related to JOCC. In 1996, Gerard L.G. Sleijpen and van der Vorst used ideas from Davidson's method and JOCC to derive the Jacobi-Davidson method [2795, 2796] for computing a few eigenvalues and eigenvectors of a general matrix A.

The iteration $k+1$ of the method is as follows. Assume that one has an approximate eigenpair (θ, u), an orthogonal matrix V_k, and a residual vector $r = Au - \theta u$. The correction equation

$$(I - uu^*)(A - \theta I)(I - uu^*)t = -r$$

is solved approximately using a few iterations of an iterative method (for instance, GMRES) for the linear system with $A - \theta I$, the result t is orthogonalized against the columns of V_k to give the $(k+1)$st column of V_{k+1}; this corresponds to using JOCC. The matrix $H_{k+1} = V_{k+1}^* A V_{k+1}$ is obtained by computing its last row and last column and a wanted eigenpair (θ, s) is computed from which the new approximate eigenvector $u = V_{k+1}s$ is obtained, as well as a new residual.

This process can be restarted to control the storage. An exact solution of the correction equation, or an approximate solution of high accuracy, leads to cubic convergence for a properly selected sequence of θ's if A is symmetric, and to quadratic convergence in the nonsymmetric case. Restarting techniques for the Jacobi-Davidson method were presented by Stathopoulos and Saad [2852] in 1998.

The method was extended to generalized eigenvalue problems; see [2789] by Sleijpen, Albert G.L. Booten, Diederik R. Fokkema, and van der Vorst in 1996, [2797] by Sleijpen, van der Vorst, and Ellen Meijerink in 1998, and [1181] by Fokkema, Sleijpen, and van der Vorst in 1998.

Another analysis of Davidson's method, as well as comparisons with the Jacobi-Davidson method, was done by Yvan Notay [2356] in 2005. A review of the method was done in a joint paper [1709] with Michiel Erik Hochstenbach in 2006. The application of Jacobi-Davidson to polynomial eigenvalue problems was proposed by Hochstenbach and Notay [1710] in 2007. In [1711], published in 2009, they proposed a strategy to control the inner iterations of the solver used in each Jacobi-Davidson iteration by using the relation between the residual norm of the inner linear system and the residual norm of the eigenvalue problem.

This is an example of inner-outer iterations. Ways to stop the inner iterations for eigenvalue problems were considered for other methods, like the Rayleigh quotient iteration and inverse iteration; see, for instance, [1224, 1407, 2625, 2801, 2850, 2851].

A Fortran software JADAMILU (JAcobi-DAvidson method with Multilevel ILU preconditioning) was written in 2007 by Matthias Bollhöfer and Notay and described in [365]. Davidson and Jacobi-Davidson methods are also part of Trilinos and PETSc; see the end of Chapter 8.

6.14 ▪ Other methods

A trace minimization method for the generalized symmetric eigenvalue problem $Ax = \lambda Bx$ was proposed by Sameh and John A. Wisniewski [2682] in 1982. Let X_k be the current approximation to the eigenvectors corresponding to the p smallest eigenvalues where $X_k^T B X_k = I$. The idea of the trace minimization algorithm is to find a correction term Δ_k that is B-orthogonal to X_k such that

$$\mathrm{tr}\{(X_k - \Delta_k)^T A(X_k - \Delta_k)\} < \mathrm{tr}(X_k^T A X_k).$$

This algorithm was revisited in 2000 by Sameh and Zhanye Tong [2681], who discussed the convergence theory, acceleration techniques, some implementation details, and generalizations.

In the 1980s and 1990s, there had been some attempts to use multigrid techniques (see Chapter 5) to solve eigenvalue problems; see, for instance, [411] by Achi Brandt, Stephen McCormick, and John Ruge in 1983 and [750] by Sorin Costiner and Shlomo Ta'asan in 1995.

Two recent methods, aimed at computing eigenvalues of Hermitian matrices not necessarily at the ends of the spectrum, are based on the idea of filtering, either with polynomials or with rational functions.

A method named FEAST was proposed by Eric Polizzi [2505] in 2009. It was designed to be used in electronic structure calculations, taking its inspiration from the density matrix representation and contour integration in quantum mechanics. It solved the generalized eigenvalue problem $Ax = \lambda Bx$, with A and B Hermitian and B positive definite, for eigenvalues in a given interval of the real line. It used Gauss-Legendre quadrature rules. Linear systems with several right-hand sides had to be solved at each iteration. The description of the method was very much physics-oriented.

A more mathematical derivation and analysis of the method was given in the paper [2994] published in 2014 by Ping Tak Peter Tang and Polizzi. They showed that FEAST is a subspace iteration, used with a Rayleigh-Ritz procedure, accelerated by a rational filter $\rho(M)$, where $M = B^{-1}A$, exploiting the Cauchy integral formula that approximates the spectral projector to the searched invariant eigenspace. FEAST uses subspaces spanned by sets of the form $\rho(M)V$, where V consists of p vectors chosen randomly. The resulting enhancements upon [2505] included estimation of the number of eigenvalues in the interval of interest, and evaluation of whether the dimension chosen for the subspaces is appropriate. The authors gave a convergence analysis and proposed ideas to extend FEAST to non-Hermitian problems.

A non-Hermitian solver was described by the same authors and James Kestyn [1889] in 2016. Improved rational approximants, based on the work of Zolotarev (see the footnote page 139), leading to FEAST variants with faster convergence, were proposed by Stefan Güttel, Polizzi, Tang, and Gautier Viaud [1514] in 2015.

Methods for the estimation of eigenvalue counts in an interval were considered [891] by Edoardo Di Napoli, Polizzi, and Saad in 2016.

The use of Krylov solvers in FEAST was discussed in [1307] by Brendan Gavin and Polizzi in 2018.

Extensions of FEAST to polynomial nonlinear eigenvalue problems were proposed by Gavin, Agnieszka Międlar, and Polizzi [1306] in 2018. For a review of nonlinear eigenvalue problems, see [1515] by Güttel and Françoise Tisseur in 2017.

A method using contour integration and subspace iteration was introduced a few years before FEAST by Tetsuya Sakurai and Hiroshi Sugiura [2677].

As we have seen above, polynomial filtering was used by R. Li, Y. Xi, Vecharynski, C. Yang, and Saad [2052] in 2016. The paper [2051] by Li, Xi, Erlandson, and Saad described in 2019 a software package called EVSL (EigenValues Slicing Library) for solving large sparse real symmetric standard and generalized eigenvalue problems. Eigenvalue slicing is a strategy which divides the spectrum into a number of subintervals and extract eigenpairs from each subinterval independently. EVSL uses polynomial and rational filtering techniques and Krylov subspace methods as well as subspace iteration. It implements Cauchy integral and also least squares rational filters. The authors also gave a review of available packages for computing eigenvalues.

Preconditioned eigensolvers were surveyed by Andrew V. Knyazev [1924] in 1998 with a special attention to the Russian literature; see also [2186]. In 2001, he introduced the *Locally Optimal Block Preconditioned Conjugate Gradient* (LOBPCG) method [1925] for solving symmetric generalized eigenproblems $(B - \mu A)x = 0$. The iterates were defined as

$$\begin{aligned} x_{k+1} &= w_k + \tau_k x_k + \gamma_k p_k, \\ w_k &= M(Bx_k - \mu(x_k)Ax_k), \\ p_{k+1} &= w_k + \gamma_k p_k, \quad p_0 = 0, \end{aligned}$$

where τ_k and γ_k are computed to maximize the Rayleigh quotient $\mu(x_{k+1})$ by using the Rayleigh-Ritz method and $\gamma_0 = 0$. The matrix M is the preconditioner. An extra orthogonalization was used for ill-conditioned problems. This algorithm was used in block form with m starting vectors to obtain the m largest eigenvalues.

A corresponding software package was described by Knyazev, Merico E. Argentati, Ilya Lashuk, and Evgueni E. Ovtchinnikov [1926] in 2007. A convergence theory was given in [72] by Argentati, Knyazev, Klaus Neymeyr, Ovtchinnikov, and Ming Zhou in 2015.

Preconditioned methods for nonlinear eigenvalue problems were studied by Daniel B. Szyld and Fei Xue [2987] as well as Szyld, Vecharynski, and Xue [2985] in 2015-2016.

Since Krylov methods can also be used to compute approximations to eigenvalues of a matrix A, it was tempting to do the same with IDR (*Induced Dimension Reduction*) algorithms; see Chapter 5. This was started in 2013 by Martin Hermann Gutknecht and Zemke [1510]. They considered the matrix pencils arising from the generalized Hessenberg relations satisfied by IDR algorithms. However, there are some difficulties due to the polynomial factors $1 - \omega_j \xi$ arising from the damping factors. They used purification and deflation to construct smaller matrix pencils and to obtain only the Ritz values that are wanted. On this topic, see also [95, 96] by Reinaldo Astudillo and Martin Bastiaan van Gizjen in 2016.

One of the new trends in numerical linear algebra in the 2010s was the introduction of randomized algorithms. Eigenvalue problems did not escape from this trend. Let us just cite [2312] by Cameron Musco and Christopher Musco in 2015 and [2323] by Yuji Nakatsukasa and Joel A. Tropp and the references therein.

6.15 ▪ Books

Several interesting books were written after World War II about solving eigenvalue problems. For quite some time, the "bible" on this topic was Wilkinson's book *The Algebraic Eigenvalue Problem* [3248] published in 1965. Even though it is now a bit outdated, it is still worth reading. Another classic is Parlett's *The Symmetric Eigenvalue Problem* [2446], which was published in 1980 and reprinted by SIAM in 1998. A popular and interesting book is *Numerical Methods for Large Eigenvalue Problems* [2658] by Saad, published first in 1992, with a revised and enlarged edition in 2011. The eigenvalue problem is considered in *Matrix Computations* [1399] by Golub and Charles Francis Van Loan, a book which had four different editions.

It is worth mentioning the two books by Chatelin [644, 645] in 1983 and 1993, *Numerical Methods for General and Structured Eigenvalue Problems* [1948] by Kressner in 2005, *The Matrix Eigenvalue Problem: GR and Krylov Subspace Methods* [3188] by Watkins in 2007, and *Core-Chasing Algorithms for the Eigenvalue Problem* by Aurentz, Mach, Robol, Vandebril, and Watkins [101] in 2018. A review of Watkins' and Kressner's books was written by Parlett [2453].

Also of interest is *Templates for the Solution of Algebraic Eigenvalue Problems: A Practical Guide*, edited by Z. Bai, Demmel, Dongarra, Ruhe, and van der Vorst and published by SIAM in 2000.

Reviews of a historical nature about eigenvalue problems were written by Golub and van der Vorst [1395, 1397] in 1997.

6.16 ▪ Lifetimes

In this section, we show the lifetimes of the main deceased contributors to the methods for computing eigenvalues and eigenvectors, starting in 1700. The language is given by the color of the bars and by letters: E (red) for English, G (black) for German, I (green) for Italian, F (blue) for French, and O (magenta) for the others. The contributors are ordered by date of birth.

Eigenvalues and eigenvectors (a)

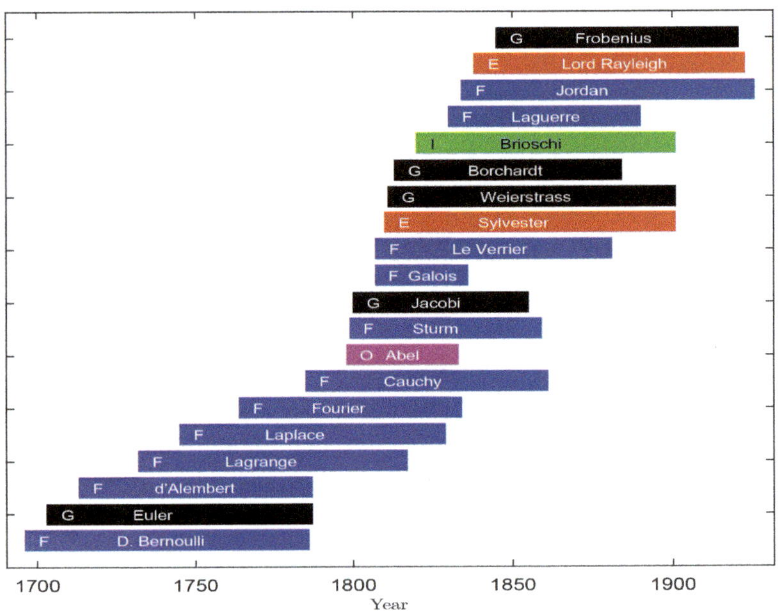

Eigenvalues and eigenvectors (b)

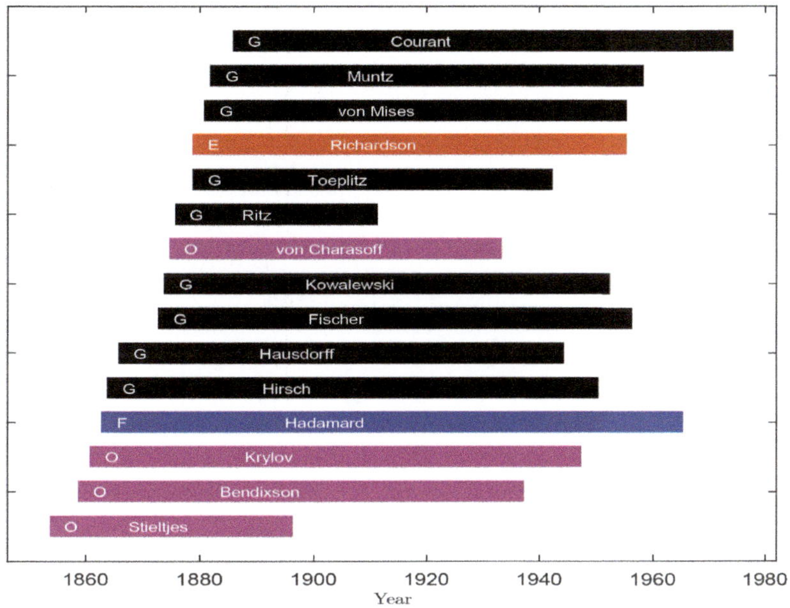

Eigenvalues and eigenvectors (c)

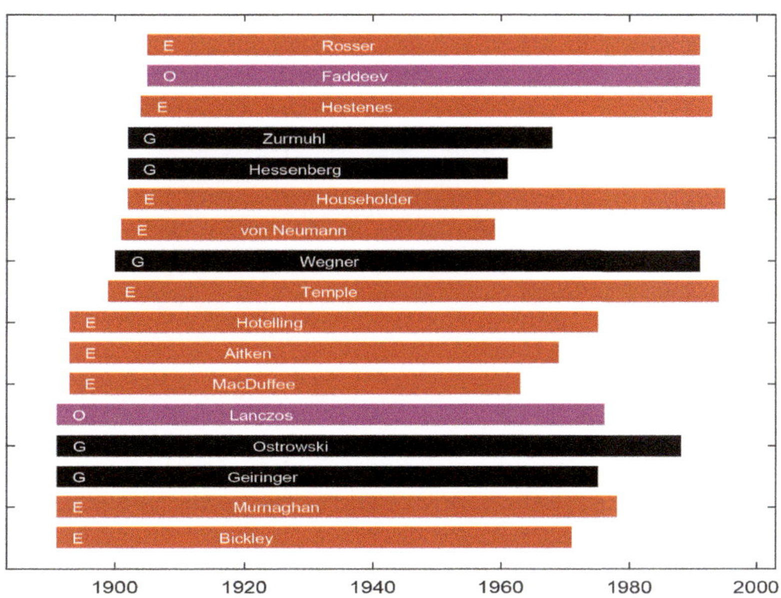

Eigenvalues and eigenvectors (d)

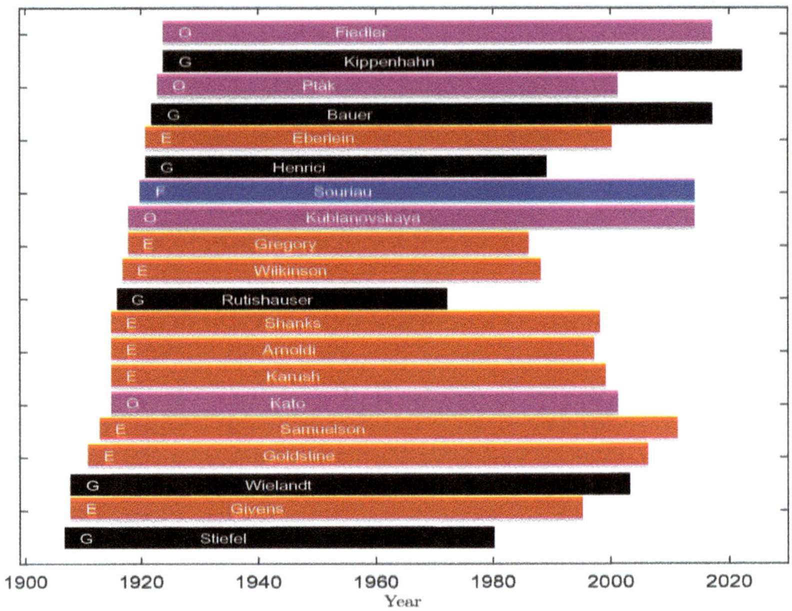

Eigenvalues and eigenvectors (e)

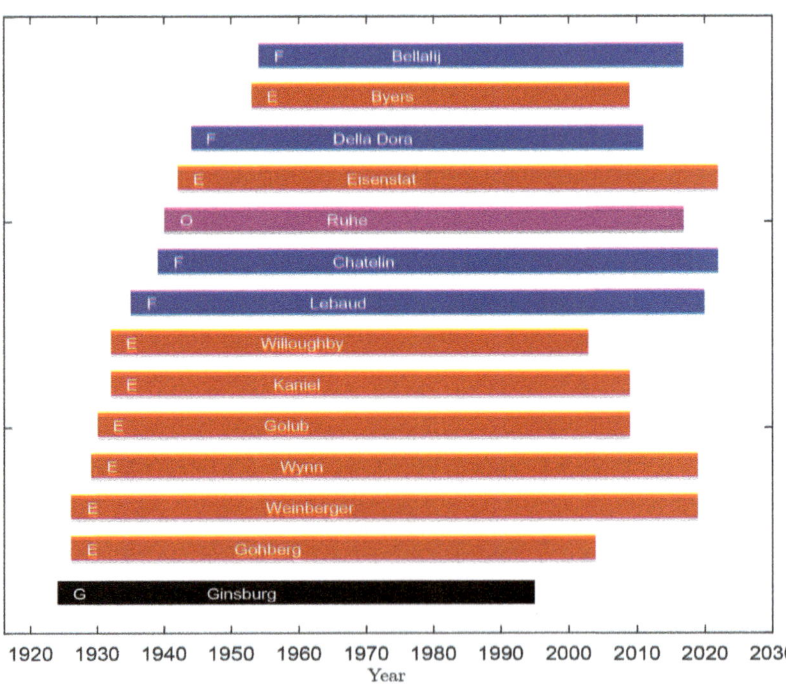

7 Computing machines

> *A major concern which is frequently voiced in connection with very fast computing machines, particularly in view of the extremely high speeds which may now be hoped for, is that they will do themselves out of business rapidly; that is, that they will out-run the planning and coding which they require and, therefore, run out of work. I do not believe that this objection will prove to be valid in actual fact.*
>
> – John von Neumann, *IBM Seminar, 1949*

Our aim in this book is not to write an exhaustive history of the computing machines that have been invented but to give a sketch of their development to be able to understand their influence on the research for the numerical methods we are interested in. We will see that, with only very few exceptions, the algorithm developers did not have much influence on the design of the machines that they had to use for their computations, and they had to spent a lot of time to adapt the methods and algorithms to the peculiarities of the computing machines.

For quite some time, at least until the end of the 19th century, a "computer" meant a human computer, somebody who carried out calculations, mainly series of the four operations $(+, -, \times, \div)$ and maybe square or cubic roots. In antiquity, pebbles were used to perform the basic operations. Calculus is a Latin word meaning pebbles. The word "computer" comes from the Latin "computare" which literally means "reckoning together." It had been used in English at least since the 17th and 18th centuries.

7.1 ▪ The beginnings

The first thing used to compute was probably the human hand(s); see [2035] and [1308]. Devices that have been used to record numbers or to do additions are wooden sticks or bones and pebbles.

Through the studies of remaining clay tablets, there is evidence that between 2000 BC and around 300 BC the Babylonians (in the region which is now Iraq) did some sophisticated (for the time) mathematics, were able to solve some nonlinear equations, and knew what is now known as the Pythagoras theorem, a thousand years before Pythagoras; see, for instance, [1321, 1921] and the book [2333]. The Babylonians were using a sexagesimal numbering system (base 60) for their mathematics. They also used tables with columns of numbers.

This idea of reckoning boards is perhaps one of the origins of the abacus. The etymology of the word is not clear. Some people think that it comes from words meaning "sand" or "dust." Abaci were used in ancient Egypt, in Persia near 600 BC, in Greece near 500 BC (see the Salamis tablet (discovered in 1846), a marble board used with pebbles around 300 BC), in China near 200 BC, in Rome, in India, in Japan, in Korea, and in Russia; see [2693] and Figure 7.1. Gerbert d'Aurillac (born between 945 and 950 and died in 1003), a French monk, is believed to be one of the people who introduced the Hindu-Arabic numbering system in western Europe. He probably learned about it when he traveled to Spain. He was the originator of a type of abacus where multiple beads were replaced by one bead labeled with a decimal number. He also wrote several treatises about arithmetic and became pope under the name Sylvester II.

Figure 7.1. *Abacus*

One of the earliest mechanisms that we know of, used for what we would call scientific purposes, is the Antikythera mechanism discovered in the sea close to the Greek island of Antikythera in 1901; see [1223]. This device was used to predict astronomical positions. It is believed to have been constructed around 200-100 BC. It is a highly sophisticated mechanism. The fine technology used to build it was lost afterwards for many centuries. We could also mention Archimedes' (c. 287-c. 212 BC) device around 200 BC for predicting the movements of planets. But we only know about it from later writings of Greek and Roman authors.

After the re-discovery in 1965 of a manuscript (named Codex I) written in 1493 by Leonardo da Vinci (1452-1519), it was believed by some people that one of his drawings (see Figure 7.2) represented a calculating machine. The controversial Italian engineer Roberto A. Guatelli (1904-1993) built a hypothetical replica of the machine in 1968 that was shown in an IBM exhibition. However, after examination by historians and engineers from MIT, the conclusion was that Da Vinci was only showing a ratio machine and that Guatelli's replica could not have been built with 15th century technology.

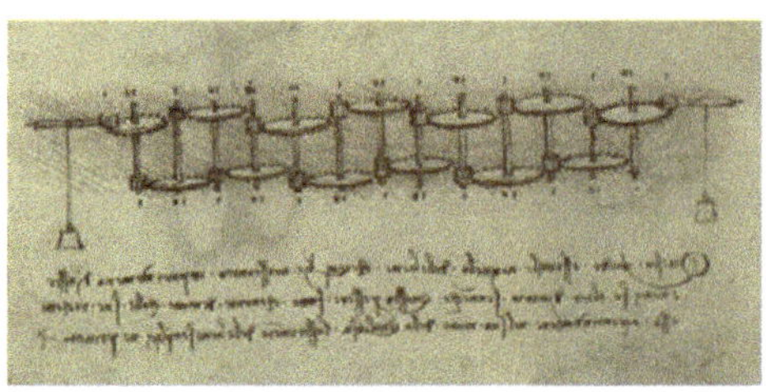

Figure 7.2. *Leonardo da Vinci's drawing*

7.2 ▪ Logarithms and slide rules

John Napier (1550-1617), a Scottish mathematician, physicist, and astronomer, is best known for his work on logarithms in 1614. But he is also known for Napier's bones, a device to do multiplications and divisions by just having to do additions and subtractions (see Figure 7.3).

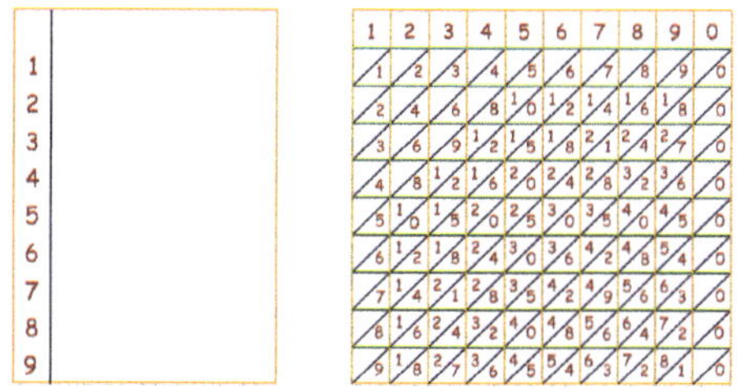

Figure 7.3. *Napier's bones*

The device is made of sticks or bones that are placed in a board whose left edge is divided into nine squares, $1, 2, \dots, 9$. Each vertical stick has nine squares. The one on the top has only one decimal number, the others have two decimal numbers encoding the multiplication table. Let us take a small example: 42×6. We put the sticks 4 and 2 in the box and we look at row 6 for which we have $2/4$ and $1/2$ and we sum the entries by diagonals to get $2, 4+1 = 5, 2$ and we obtain the result 252. Of course, this is not a very fast calculating machine since, when doing a series of multiplications, one has to move the sticks in and out of the board. A variant of Napier's bones was designed in 1891 by Henri Genaille (1856-1903), an engineer in the French railways who invented a number of arithmetic constructions, and the mathematician Édouard Lucas (1842-1891).

Galileo Galilei (1564-1642), the famous Italian astronomer and physicist, popularized the sector at the end of the 16th century. The sector is an instrument, consisting of two graduated rulers of equal length joined by a hinge, that uses trigonometric formulas and a caliper to calculate squares, cubes, reciprocals, and tangents of numbers.

After the introduction of logarithms by Napier, the slide rule was invented by William Oughtred (1574-1660), an English mathematician and clergyman, around 1622, following some logarithmic scales made by Edmund Gunter (1581-1636), an English astronomer, in 1620. Oughtred placed two logarithmic scales side by side to do multiplications and divisions. He also developed a circular slide rule. Slide rules were used all over the world up until the 1970s, when the first electronic pocket calculators appeared; see [3034].

7.3 ▪ Mechanical calculators

Thousands of mechanical calculators to perform some or all of the four elementary operations have been built since the beginning of the 17th century. Many of them are described in the books by Philibert Maurice d'Ocagne (1862-1938) [902] in 1905, and Ernst Martin [2149] in 1925; see also [642] by George Clinton Chase (1884-1973) in 1952. Let us consider the most important ones.

Wilhelm Schickard (1592-1635), a German Lutheran minister and friend of Johannes Kepler (1571-1630), invented a calculating clock for adding and subtracting 6-digit decimal numbers around 1620. There are just three documents about this machine: two letters to Kepler and a sketch of the machine. Schickard's letters mentioned that the original machine was destroyed in a fire while still incomplete. It is not known if the machine had ever been functioning correctly.

Blaise Pascal (1623-1662), a French mathematician, physicist, and philosopher, invented a calculating machine named the *Pascaline* (see Figure 7.4). It was done in 1642 to help Pascal's father, who was in charge of the collection of taxes in the French province of Normandy. This computer was able to add or subtract two numbers and to multiply by repetition in a complicated way. The first machine was presented in 1645. Pascal had around 20 machines built in the next

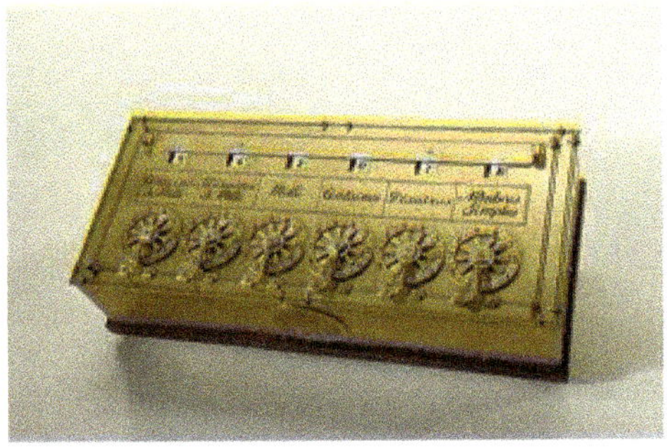

Figure 7.4. *Pascaline*

10 years; nine of these machines are remaining in museums. Pascal gave up this activity in 1654, concentrating on his interest in religion and philosophy.

Gottfried Wilhelm von Leibniz (1646-1716) is one of the fathers, together with Isaac Newton (1643-1727), of the differential and integral calculus. In 1671, he invented a calculating machine, doing all four arithmetic operations, which was a major advance in mechanical calculating; see Figure 7.5. Leibniz's machine incorporated a new mechanical feature, the *stepped drum*, a cylinder bearing nine teeth of different lengths which increase in equal amounts around the drum. Although the Leibniz calculator was not developed for commercial production, the stepped drum principle was used in the next 300 years in many later calculating machines. Leibniz also studied the binary system (see Figure 7.6), which had been considered before by Juan Caramuel y Lobkowitz (1606-1682), a Spanish priest, philosopher, and mathematician in *Mathesis biceps vitus et nova* in 1670; for more details see [1917]. In his paper of 1679, Leibniz was also speaking of a hypothetical machine that can compute using binary numbers.

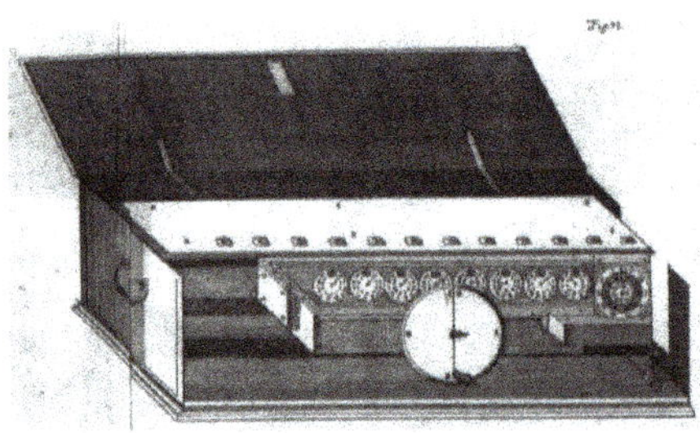

Figure 7.5. *Leibniz's machine as in Brevis descriptio Machinae Arithmeticae, 1710*

Here is an incomplete list of people who proposed calculating machines from the mid-1600s to the mid-1700s: the Italian philosopher Tito Livio Burattini (1617-1681) in 1659, who was probably inspired by the Pascaline; the English diplomat and inventor Samuel Morland (1625-1695) between 1662 and 1666; the French mechanic and watchmaker René Grillet in 1673; the French Huguenot César Caze (1641-1720) in 1696; Giovanni Poleni (1683-1761), who founded the *Teatro di filosofia sperimentale* in Padua, which was the first physics laboratory created in an Italian university, with the first calculating machine with gears with a variable number of teeth in 1709; the German mechanic, constructor, and optician Anton Braun (1686-1728) in 1727; the German scientist Christian Ludwig Gersten (1701-1762) in 1722; the French artisan Lépine, whose machine was described in 1735; and the French mathematician and inventor Jean-Baptiste Laurent de Hillerin de La Touche de Boistissandeau (1704-1779), whose machine was also described in 1735.

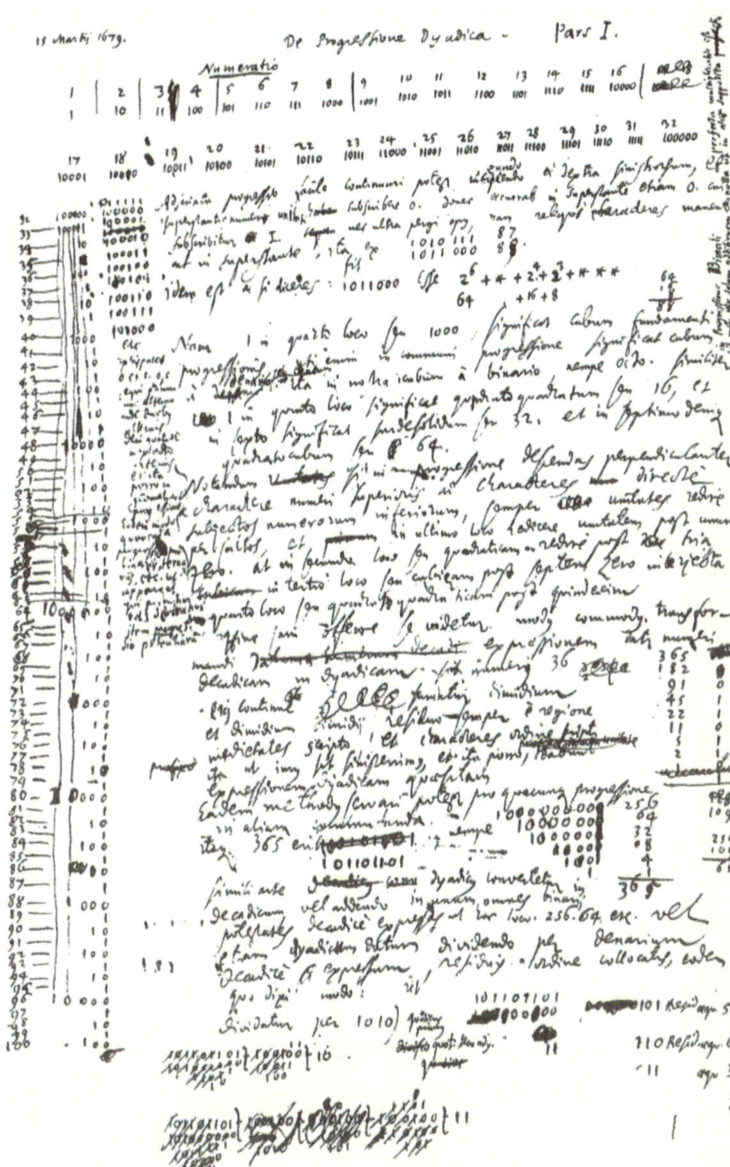

Figure 7.6. *Leibniz, De Progressione Dyadica, March 1679*

In 1735, the French Académie des Sciences asked the army engineer Jean-Gaffin Gallon (1706-1775) to collect descriptions of the calculating machines that were approved by the Academy. Six volumes were published between 1735 and 1754, plus an additional one in 1777. In Germany, Jacob Leupold (1674-1727) wrote a seven-volume book, the *Theatrum Machinarum Generale*, whose publication started in 1724. A part of the last volume is concerned with calculating machines like those of Grillet, Leibniz, and Poleni.

Philipp Matthäus Hahn (1739-1790) constructed a fully functional cylindrical machine doing the four arithmetic operations in 1773. He used Leibniz's stepped drum. His machine became popular in Germany. He also constructed adding machines. Johann Helfrich von Müller (1746-1830) built an improved version of Hahn's machine in 1784. Jacob Auch (1765-1842), a former apprentice of Hahn and clockmaker, built adding machines around 1790. Johann Christoph Reichold (1753-1798) built a machine about which not much is known.

One of the most successful and influential machines was the *Arithmomètre* of Charles-Xavier Thomas (known as Thomas de Colmar (1785-1870)); see Section 10.69 for more details on his life and work. Thomas filed a patent for his machine in 1820. When Thomas passed away 900 Arithmometers had been produced. An unsigned paper published in the *Nouvelles annales de mathématiques* 1re série, tome 13 (1854), pp. 257-259, tells us that the multiplication of two numbers with 8 decimal digits took 18 seconds, and division of a number with 16 decimal digits by a number with 8 decimal digits took 24 seconds. It is interesting and amusing to read what was written about Thomas' machine in that paper[51] (our translation):

> *The arithmometer of M. Thomas, of Colmar, as he has just perfected it, is indeed the most ingenious, the most convenient and the most portable instrument known of its kind. It is very useful to operate on numbers which exceed the limits of the logarithmic tables and especially to verify any operation on large numbers; moreover, additions and subtractions cannot be done by the logarithm tables. Leibnitz searched for such a machine all his life, and spent more than twenty-four thousand "ecus" on it. Inorganic agents have the great advantage of not getting tired.*

After 1840, David Roth (1808-1885), a French medical doctor of Hungarian origin, designed several adders and multipliers using pinwheel mechanisms. Heinrich Gotthelf Kummer (1809-1880), a German musician, devised an adding machine in 1846, while he was in Russia, improving on a previous one by Caze. Kummer's device was mass produced in Russia up to the 1970s.

Another interesting machine is the *Arithmaurel* of François Timoléon Maurel (1819-1879) and Jean Jayet (1796-1856). Their machine was presented at the Paris "Exposition nationale" in 1849, where it won the gold medal, while Thomas obtained only a silver medal. This machine was particularly fast for multiplications, but it was not as reliable as Thomas' Arithmometer and more difficult to build with 19th century technology.

Willgodt Theophil Odhner (1845-1905), an engineer from Sweden living in Saint Petersburg, Russia, designed a pinwheel machine in 1877 when he was working in the factory of Ludvig Nobel (1831-1888), an older brother of Alfred Nobel (1833-1896), the founder of the Nobel prize; see Section 10.54 and Figure 7.7. A pinwheel is a variable-toothed cylinder with sets of nine radial pegs that protrude or retract. Then, he founded his own factory and became very successful. When the factory closed in 1918, 29,000 machines had been made. Franz Trinks (1852-1931) of the German firm Grimme, Natalis & Co. acquired the patents of Odhner's invention in 1892. Trinks manufactured this machine in an improved form and put it on the market under the name *Brunsviga*. The Brunsviga machines were continually improved and soon became the most popular calculating machines in Europe. In 1912, more than 20,000 machines had been sold. It was produced until 1958. In France, Odhner-like machines were sold under the name *Dactyle*, manufactured by Chateau Frères, who, later on, sold the machine under their name. This type of machine was used by André-Louis Cholesky (1875-1918) for solving linear systems arising from least squares problems with his algorithm; see Section 10.12.

[51] *L'arithmomètre de M. Thomas, de Colmar, tel qu'il vient de le perfectionner, est bien l'instrument le plus ingénieux, le plus commode et le plus portatif qu'on connaisse en ce genre. Il est d'une grande utilité pour opérer sur des nombres qui dépassent les bornes des Tables logarithmiques et surtout pour vérifier toute opération sur de grands nombres; d'ailleurs, les additions et les soustractions ne peuvent se faire par les Tables de logarithmes. Leibnitz a recherché une telle machine toute sa vie, et y a dépensé plus de vingt-quatre mille écus. Les agents inorganiques possèdent l'immense avantage de ne pas se fatiguer.*

Figure 7.7. *One of Odhner's early machines*

In the USA, Frank Stephen Baldwin (1838-1925) designed a machine in 1873 with a pinwheel as in Odhner's machine. He filed his patent in the USA before Odhner, but in fact, this system was probably invented by Leibniz and was used in machines by Poleni and Roth. Between 1873 and 1912, Baldwin made several models based on the pinwheel. The commercial success came with his association with Jay Randolph Monroe (1883-1937), started in New Jersey with the Monroe Calculating Machine Company in 1912.

In 1885, Dorr Eugene Felt (1862-1930) constructed a wooden prototype of the *Comptometer*, from rubber bands, meat skewers, string, staples, and a macaroni box. The first practical model was ready in 1886. It was the first machine with a sort of numeric keyboard to input the numbers.

It is probably of interest to our readers to know that the famous Russian mathematician Pafnuty Lvovich Chebyshev (1821-1894) was also interested in mechanical engineering. In the 1870s he designed several calculating machines. In 1876 he presented a ten decimal place adding machine with a continuous carry mechanism. He showed an improved machine able to also do multiplications and divisions in 1881 in Paris; see Figure 7.8.

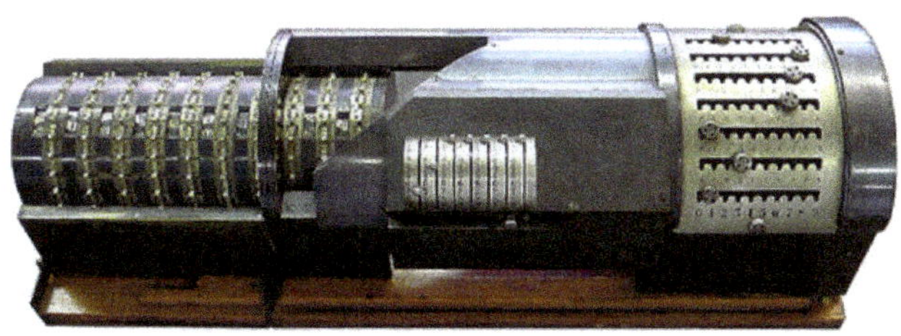

Figure 7.8. *One of Chebyshev's machines*

William Seward Burroughs (1857-1898) was an American inventor and businessman who, in 1888, patented a commercially successful adding machine. Unfortunately, the success came only after the death of Burroughs. Burroughs' machines incorporated a full keyboard, and for the first time, a printing device to record numbers and totals.

Léon-Auguste-Antoine Bollée (1870-1913), a well-known French automobile manufacturer, invented a machine which was the first successful direct-multiplying calculator when most of the previous machines used repeated addition for multiplication. It won a gold medal at the 1889 Paris Exhibition, but the machine was never a commercial product. Bollée was only 18 years old when he designed his machines. He is also known as the founder of the world-renowned car race 24 Hours of Le Mans. This type of multiplication hardware was also used in the *Millionaire* machine (1893).

7.4 ▪ Electric calculators

All the machines that we have mentioned above were put in operation mechanically, most of the time by turning a crank. The availability of electric power changed the landscape. In 1831, electricity became viable for use in technology when Michael Faraday (1791-1867) created the electric dynamo (a crude power generator), which solved the problem of generating electric current in a practical way. Small DC power systems were available in the 1870s and long distance distribution of electricity started in the 1880s-1890s.

Soon, it became obvious that electricity can be used for calculating machines. This started at the end of the 1800s and the early 1900s; see, for instance, the *Mercedes-Euklid* model 7 that was driven by an electric motor (1905) or the *Ensign* (1907) in the USA. Since 1915, the machine manufactured by Marchant in the USA was provided with an electric drive as well as in Monroe machines which originated from Baldwin's design. The Marchant Calculating Machine Company, which was founded in 1911 by Rodney and Alfred Marchant, built mechanical and electromechanical calculators up to 1958.

The Spaniard Leonardo Torres y Quevedo (1852-1936) showed his *Arithmomètre Electrico-Mécanique* at the Paris Exhibition in 1920. The machine was linked to a typewriter by electric wires. The operation to be done was typed on the typewriter and the result was later printed with it. Besides calculating machines, Torres y Quevedo invented many other things: a new type of dirigible in 1902, a radio guided small boat in 1906, a chess-playing automaton in 1910, and the Spanish Aero Car, a cable car over the whirlpool near Niagara Falls, in Canada, in 1914.

The Friden Calculating Machine Company was founded in 1934 by Carl Friden (1891-1945), a former employee of Marchant. This company built robust electromechanical calculators. In 1963, Friden introduced the first fully transistorized desktop electronic calculator.

7.5 ▪ The beginning of automatic computers

The machines that we have considered so far were capable of doing all or some of the four basic arithmetic operations and sometimes of extracting square or cubic roots. But in the meantime, there were some inventors who had more ambitious goals. Unfortunately, they were not always able to build what they designed.

Today Charles Babbage (1791-1871) is credited to have foreseen what we now call a computer must be; see Section 10.3 for more about his life and work. He designed several mechanical machines, but the one that foreshadowed the modern computer was the *Analytical Engine*, on which he worked for many years. Unfortunately, his machines were never completed, partly because they were too complex and too ahead of their time, and partly because of funding problems.

A machine based on Babbage's difference engine was built by the Swedish engineer Pehr Georg Scheutz (1785-1873) in 1843. The design of the machine was later improved by Martin Wiberg (1826-1905). This machine was used to print logarithmic tables.

Percy Edwin Ludgate (1883-1922), an Irish clerk, published a paper in the *Scientific Proceedings of the Royal Dublin Society* in 1909 describing a mechanical analytical machine whose design was very different from Babbage's. Unfortunately Ludgate's drawings have not been found so far; see [2533].

Torres y Quevedo, mentioned above, studied Babbage designs. He was thinking that a machine similar to the *Analytical Engine* could be built with electromechanical relays. He described a machine to compute automatically $a \times (y - z)^2$ [3064] in 1913 in Spanish and 1915 in French; see also [2533]. He advised using electromechanical devices instead of purely mechanical ones for computations. Torres also described what can be considered a formal definition of decimal floating point numbers on page 610 of [3064].

Electromagnetic relays are electrically operated switching mechanisms that were invented in the early 1800s. They were first used in telegraph applications and later in telephone networks and in the early digital computers.

In 1879, Sir William Thomson, also known as Lord Kelvin (1824-1907), the well known physicist, published a paper [3035] describing a machine for solving systems of linear equations. This was probably an outcome of some of his other inventions, notably a tide predicting machine. Thomson's machine is what we would call an analog computer using cables and pulleys. It is also described in the book with Peter Guthrie Tait (1831-1901) [3036]. In this book one can read

> *The actual construction of a practically useful machine for calculating as many as eight or ten or more of unknowns from the same number of linear equations does not promise to be either difficult or over-elaborate. A fair approximation having been found by a first application of the machine, a very moderate amount of straightforward arithmetical work (added very advantageously by Crelle's multiplication tables) suffices to calculate the residual errors, and allow the machines (with the setting of the pulleys unchanged) to be re-applied to calculate the corrections.*

This is clearly what we now call *iterative refinement.*

In 1934, John Benson Wilbur (1904-1996), at MIT in the USA, made some modifications to Thomson's design and built a prototype for solving up to nine linear equations; see [3232]. This was announced in *Nature*, December 1934, and also in *The Tech* newspaper at MIT, October 30, 1934:

> *Completion of an experimental model of what is believed to be the first mechanical calculating machine for the solution of simultaneous equations was announced at the Institute last Saturday. The device was designed by Dr. John B. Wilbur of the department of civil engineering under the direction of Dr. Vannevar Bush, vice-president of the Institute. It represents a new step in Technology's program of developing mechanical devices for the solution of mathematical problems. The significance of Dr. Wilbur's machine is indicated by the fact that the labor involved in the solution of large numbers of simultaneous equations has stood in the way of engineers in their analysis of many important problems. The rapid solution of such equations will provide engineers with a practical tool for the solution of many complex problems of design holds prospects of facilitating important research in several fields. Complicated in appearance, the new machine reproduces mechanically, through a maze of pulleys and steel tapes, the mathematical conditions of the equations to be solved. Some idea of the intricacy of the device may be obtained from the fact that a second proposed model, already designed, calls for almost 1,000 ball bearing pulleys and over 500 feet of steel tape, although it will be only two feet wide, two and a half feet high, and seven feet long. A total of 110 vernier scales enable the setting of the various coefficients and constants of 10 equations, while 10 more angular verniers enable the operator to read directly the solutions of the equations.* [...] *A graduate of the Institute in 1926, Dr. Wilbur received his master's degree*

> *in 1928 and his Doctor of Science degree in 1933. He has served as assistant and instructor in the department of Civil Engineering and last June his appointment to the position of Assistant Professor of Civil Engineering was announced.*

In [3232] Wilbur wrote

> *The total time required to solve nine simultaneous equations on the machine, to an accuracy of three significant figures, is estimated at from one to three hours, depending upon the equations themselves. This time will undoubtedly be reduced when more experience has been acquired in the technique of operation. The solution of similar equations employing a keyboard calculator might be expected to require in the neighborhood of eight hours.*

The order of the systems that can be solved with such a machine is limited by the friction on the pulleys. The accuracy is also not very good. Its assessment has been done in [2840]. A pirate copy of the machine was made at Tokyo University's Aviation Laboratory in Japan in 1944. For the sake of education, Thomas Püttmann [2526] has constructed analog calculators using Thomson's principles for solving systems of order 2 and 4, using Fischertechnik, a German educational toy.

Wilhelm Veltmann (1832-1902), a German mathematician and physicist, described a hydro-mechanic device for solving linear equations [3142] in 1886. It used a parallelepiped tank and cylinders filled with water and a lever. Whether or not this machine was really built is not known. For an analysis of this device, see [3141].

Wilhelm Cauer (1900-1945), a German mathematician working in network and system theory, devised an analog electromechanical machine for solving linear equations in the end of the 1920s; see [2487]. At the beginning of the 1930s, Cauer started to build a prototype to solve a system of order 3 but he never completed the project due to funding problems.

Rawlyn Richard Manconchy Mallock (1885-1959), from Cambridge University in the UK, built electric analog machines for solving linear systems in 1931 and 1933; see [2123]. Alexander Craig Aitken (1895-1967) reported in *Nature*, volume 60, that the setting of the switches and the plugging of the wires for the calculation of a system with six equations took half an hour.

In 1948, Willis Alfred Adkock (1922-2003) designed an analog machine for solving linear systems [6]. His design used resistive voltage dividers to represent coefficients and voltages to represent the variables. Some machines for solving algebraic equations were described in [1209] in 1945.

Even in the 1950s, some people were still constructing electric analog machines for solving systems of linear equations; see, for instance, the proposal by Dušan Mitrović (1912-1962) [2258] in 1952 and the references therein.

7.6 ▪ Tabulating machines

Clearly, from what we have seen above, up to World War II, most scientists wanting to solve systems of linear equations did not have many tools available besides their brain, pencil and paper, an electromechanical desktop calculator, and maybe also a slide rule and logarithmic and multiplication tables. One can look, for instance, at the nice photo of Cornelius Lanczos (see Section 10.46) on the cover of the book [1318]. This photo was taken in 1944 when he was working at Boeing. On the right-hand side of the picture we clearly see a mechanical calculator.

The progress in computing machines was made possible by the progress in some electric devices, like vacuum tubes. Even though there were already developments in the 19th century, the simplest vacuum tube was invented by an Englishman, John Ambrose Fleming (1849-1945), in 1904.

Before moving to electronic calculators and computers, it is interesting to note that an application which led to progress in electromechanical and later on electronic machines is data processing; see the book [1166].

Herman Hollerith (1860-1929), an American inventor and businessman, built a punched card tabulating machine patented in 1884. He was the founder of the Tabulating Machine Company that was merged (via stock acquisition) in 1911 with three other companies to form a fifth company, the Computing-Tabulating-Recording Company (C-T-R), which was renamed International Business Machines (IBM) in 1924. Soon, Thomas John Watson Sr. (1874-1956) became the president of the company. When Watson died, the company had 72,500 employees. With Hollerith's machines, data represented as holes in specified locations on a card, arranged in rows and columns, could be counted or sorted electromechanically. In 1884, Hollerith began working for the United States Census Bureau and the 1890 census. He invented the first automatic card-feed mechanism and the first manual numeric keypunch that was patented in 1901. Hollerith's first *Electrical Tabulating Machine* had a card reader in which the cards had to be inserted one at a time. It was not a calculator but simply a counting machine that kept a running count of the number of cards with a hole in a particular position. It had 40 counters to read the results. In the 1910s C-T-R produced 80 million cards a month. Initially, Hollerith used cards with round holes. The capacity and the format of the cards changed over the years. In the 1920s, it was 45 columns and 10 rows. The holes in the card were sensed by pins dropping through the hole to make contact on a drum. This completed an electrical circuit. Depending on the position of the drum, a counter in the tabulating machine would add one number. Later on, the IBM standard format became 80 columns, 12 rows and rectangular holes. A copper brush instead of dropping pins detected the hole. Some tabulating machines also used paper tapes for input.

In 1923, C-T-R introduced the 011 electric keypunch. In 1929, with the introduction of the rectangular hole and the 80-column card, IBM brought out a new device called the 016 motor-driven electric keypunch. In 1949, they introduced the *024 Card Punch* and the *026 Printing Card Punch* that some of our older readers may remember since they were used up until the 1970s. The last IBM keypunch machine was announced in 1971.

A competitor to IBM was the Powers Accounting Machine Company founded in 1911 by James Legrand Powers (1871-1927). It was bought by Remington Rand in 1928.

The tabulating machines had attached and/or removable plug boards that were used as a primitive way of programming these machines for different tasks. In 1931, IBM produced the 601. This electromechanical machine could read cards and add, subtract, or multiply fields on the card and punch the result in the same card. The 602, which could also divide, was announced in 1946. In that same year, IBM began experimenting with the use of machines based on vacuum tubes. They produced an electronic multiplier called the 603. It was followed in 1948 by the *604 Electronic Calculating Punch* that could do the four arithmetic operations and was programmable with a plug board. In 1949, IBM announced the *605 Card Programmed Electronic Calculator* (CPC), with printing capabilities.

Even though the tabulating machines were designed for business applications, they were sometimes used for scientific calculations. For instance, such a machine was used in 1928 by Leslie John Comrie (1893-1950), an astronomer, to compute some tables by interpolation [730, 731, 732]. It was also used to solve a linear system of order 8 in 1950 at the Institute of Numerical Analysis of the National Bureau of Standards; see Section 5.6. Methods for solving linear equations using desk calculating machines or punched card equipment were described in [1207] in 1948; see also [3144] in 1949.

7.7 ▪ The early digital computers

Let us now consider some of the early computers in the modern sense of the word. But before describing these machines, let us briefly discuss the representation of the real numbers. Being mechanical, electro-mechanical, or electronic, computers cannot represent or store all the real numbers. Two ways were used to represent the real numbers: fixed point and floating point formats.

To simplify, let us use decimal arithmetic. Fixed points numbers are of the form $\pm d_0.d_1 \cdots d_t$ with t decimal digits after the decimal point. Generally, numbers to be represented are assumed to be in a fixed interval, say $[-1, 1]$ in which case $d_0 = 0$. Hence, the quantity that are manipulated have to be scaled. In binary arithmetic the digits are 0 or 1.

Decimal floating point numbers are of the form $\pm m\ 10^{\pm e}$, where m $(m < 1)$ is known as the mantissa or significand and e is the exponent. In a binary machine, the digits of the mantissa and the exponent are 0 or 1. The advantage of the floating point format is that it allows representation of very large real numbers, and no scaling is required. When the magnitude of a number is larger (resp., smaller) than what can be represented we have an overflow (resp., underflow).

The larger the number of digits, the better the accuracy of elementary operations, if the hardware implementation is correct. Of course, when two numbers are added or multiplied the result may not fit in the data format. Then, it has to be rounded and there exist different ways of rounding numbers. A computer designer has also to decide what to do with operations like $x/0$ or $0/0$ and with overflows or underflows. This is called exception handling.

Starting in 1927, Vannevar Bush (1890-1974), a professor at MIT, built a *Differential Analyzer*, an electrical and mechanical analog machine to solve differential equations with up to 18 independent variables. It is less well known that from 1936 to 1942, Bush and some of his students were working on the *Rapid Arithmetical Machine* project [2533]. This work stopped in 1942. Claude Elwood Shannon (1916-2001), a student of Bush, worked on the application of Boolean algebra to electronic circuits in his master's thesis, *A Symbolic Analysis of Relay and Switching Circuits* in 1937. Later on, Shannon became well known for his work on information theory.

In 1937, George Robert Stibitz (1904-1995), a researcher working at Bell Labs, constructed in his kitchen, using phone relays, a device that could sum two single-digit binary inputs. This was called the "K" computer (for Kitchen). Later, in 1940, at the request of Bell Labs, with the help of telephone engineer Samuel Williams, he constructed the *Complex Number Computer* (CNC), whose purpose was to do calculations on complex numbers. It used 450 relays and stored intermediate results in 10 crossbar switches. Input was entered and output received via a teletype writer. At an American Mathematical Society meeting in September 1940 on the Dartmouth College campus, Stibitz demonstrated the CNC, performing calculations in real time via a teletype connection to the machine located in New York. It was the first ever remote use of a computer. But Stibitz's machine was not a general purpose computer; it was more a predecessor of the pocket calculators. Later on, during World War II, several special purpose relay-based machines were built for the army under Stibitz's guidance.

Also in 1937, Howard Hathaway Aiken (1900-1973), at Harvard University, wrote a proposal for a computing machine. He approached IBM, in particular the chief engineer James Wares Bryce (1880-1949), about building the machine. Encouraged by Bryce, he submitted a formal proposal to IBM that got the approval of Watson, the IBM president. The *Automatic Sequence Controlled Calculator* (ASCC, later known as the *Harvard Mark I*) was built at IBM's factory in Endicott, New York. In February 1944, the machine was shipped to Harvard. It was a huge machine using electromagnetic relays with a length of 16 meters and a height of 2.4 meters. It contained about 760,000 electromechanical parts. It was a fixed point decimal machine with 24-digit words. It did 3 additions or subtractions in a second, a multiplication took 6 seconds, and a division took 15.3 seconds. The program was read from a paper tape. John von Neumann (1903-1957) did some calculations for the Manhattan Project on this machine. The *Harvard Mark I* was used until 1959. Aiken built three other computers, the *Mark II* relay computer in 1947, the *Mark III*, using some electronic components and a magnetic drum memory in 1949, and the electronic *Mark IV* with a magnetic core memory in 1952.

The IBM *Selective Sequence Electronic Calculator* (SSEC) was constructed after the ASCC. The machine was ready in 1947. It was a mix of vacuum tubes and electromechanical relays with 12,500 vacuum tubes and 21,400 relays. It was faster than the *Harvard Mark I*. The SSEC was dismantled in 1952.

In 1936, Konrad Zuse (1910-1995), a German engineer, started building mechanical general purpose computers; see Section 10.78 for details. He successively built the *Z1*, *Z2*, *Z3*, and *Z4* machines. What is remarkable is that his machines used binary floating point arithmetic. But because of the war, he worked in isolation. What he did didn't have any impact abroad up until the 1950s, when the *Z4* was installed at the Swiss Federal Institute of Technology (ETH) in Zürich in 1950; see Section 10.66.

John Vincent Atanasoff (1903-1995), an American physicist, was looking between 1936 and 1937 for devices which could help him in his long time-consuming calculations. Finally, he studied electronics and decided to build his own vacuum tube-based binary computer with a regenerative capacitor memory with rotating drums. The design was completed in 1938; see [97, 98]. With a grant of $650 from Iowa State College, the machine was built with the help of Clifford Edward Berry (1918-1963), a graduate student in electrical engineering, and was ready at the end of 1939. It is known as the *Atanasoff-Berry Computer* (ABC). They worked on improving the machine until 1941, when the work was stopped because of the war. This machine, which was not a stored-program computer, was mainly designed for solving systems of linear equations. It was able to solve systems of order up to 29 by successive elimination of variables. It used 50-digit 2's complement integer arithmetic. The machine was disassembled in 1948, but a working replica was built in 1997 by people at Iowa State University and Ames Laboratory; see Figure 7.9 and [1485, 1486]. It is important to note that John William Mauchly (1907-1980) became aware of Atanasoff's project and visited him for a few days in 1941, discussing the details of the machine.

Figure 7.9. *Atanasoff-Berry computer replica (Durham Center, Iowa State University)*

7.8 ▪ The 1940s

In the UK during World War II, at the request of Maxwell Herman Alexander Newman (1897-1994), Thomas (Tommy) Harold Flowers (1905-1998), an electrical engineer, and a team of 50 people spent 11 months designing and building *Colossus* at the Post Office Research Station in North West London. This machine was intended to decipher the German messages coming from the Lorenz SZ cipher machine. After some tests, *Colossus Mark I* was delivered to Bletchley Park in late December 1943 and was working in early February 1944. *Colossus* was an electronic digital machine with programmability using 1,600 vacuum tubes. An improved design, the *Mark II*, with 2,400 vacuum tubes, was ready in June 1944. The machine was programmed with switches. Twelve machines were built under the supervision of Allen William Mark Coombs (1911-1995). *Colossus* was a secret project. Machines 1 to 10 were dismantled after the war and the documentation destroyed. The two other machines were still used after the war for decyphering and were dismantled at the end of the 1960s. The existence of the machine, which was a special purpose computer, was not disclosed until the mid-1970s; see [2531, 2532, 2534]. It took nearly 15 years to rebuild the *Mark II Colossus* computer which was completed in 2008. The replica is exhibited at the National Museum of Computing in Bletchley Park, UK.

In 1943, at the request of Herman Heine Goldstine (1913-2004), the US army gave the University of Pennsylvania a development contract for a computing machine. The ENIAC (*Electronic Numerical Integrator And Computer*) was designed to compute artillery firing tables. It was completed at the end of 1945 and announced in February 1946. It was a thousand times faster than the electromechanical computers of the time. ENIAC was designed by Mauchly and John Adam Presper Eckert (1919-1995) at the University of Pennsylvania. Already in August 1942, Mauchly wrote a report on *The use of high speed vacuum tube devices for calculating*. The cost of ENIAC was around $500,000 (in 1946 dollars, which corresponds to around $7 million today). It was in operation at the Aberdeen Proving Ground in Maryland from 1947 until October 1955. Initially, the machine was programmed by plugging cables like in plug boards of tabulating machines, but after 1948, it was updated as a stored program computer. In the end, the machine had 20,000 vacuum tubes, occupied 167 m^2, and consumed 150 kW of electricity. A 100-word magnetic core memory was added in 1953. The ENIAC was a decimal fixed point arithmetic machine with a basic cycle of 200 μs, a 10- by 10-digit multiplication took $2,800$ μs. There were 10 "accumulators" (registers) holding numbers with 10 decimal digits. It was reported that there were failures of some vacuum tubes every day.

Eckert and Mauchly resigned from the Moore Engineering School of the University of Pennsylvania and founded their own corporation, the Eckert and Mauchly Computer Corporation, in 1947, where they developed the *UNIVAC 1*, a business computer. For more about using this machine to solve Laplace's equation, see [2818] by Frances Elizabeth Snyder (Holberton) (1917-2001) and Hubert M. Livingstone in 1949. In 1950, Eckert and Mauchly sold their company and their computer patents to Remington Rand, which merged later with Sperry Corporation in 1955 to become Sperry Rand.

Over the years there has been a debate about who invented the electronic computer or which was the first electronic machine? See, for instance, [510]. Zuse's first machines were mechanical, and moreover, they were not well known, Stibitz used relays and his machine was a special purpose computer. For many years, the favorites were Eckert, Mauchly, and the ENIAC, since this machine was a general purpose computer.

However, in 1967 Sperry Rand sued Honeywell for patent infringement, but, maybe in retaliation, Honeywell sued Sperry Rand for fraud, asking for the invalidation of ENIAC patents. The trial started in 1971 in Minneapolis. Among many other people, Atanasoff was asked to testify during this trial. He told about his meetings with Mauchly and showed their correspondance.

The judgment of Earl Richard Larson (1911-2001) was published on October 19, 1973. One of his many conclusions was that Mauchly's basic ENIAC ideas were derived from Atanasoff, and the inventions claimed in ENIAC were also derived from Atanasoff. Of course, one can debate about the abilities of judge Larson in computer architecture and technology. But this was not the end of the story.

Another important episode occurred in the mid-1970s when, due to the efforts of Brian Randell from Newcastle University, the British government revealed the existence of *Colossus*. On July 2, 1975, Randell had a meeting with British officials who showed him photos of *Colossus* and an explanatory document. He was also allowed to interview people who participated in the design and building of *Colossus*, particularly Flowers and Coombs. Randell wrote a paper that was presented at the *International Conference on the History of Computing*, which was held in Los Alamos in June 1976.[52] Zuse and Mauchly attended this meeting. Robert William (Bob) Bemer (1920-2004), one of the inventors of the ASCII code, who was also attending this meeting, described Mauchly's reaction (see [2534]):

> *I looked at Mauchly, who had thought up until that moment that he was involved in inventing the world's first electronic computer. I have heard the expression many times about jaws dropping, but I had really never seen it happen before. And Zuse - with a facial expression that could have been anguish.*

Of course, the question "Who invented the electronic computer?" does not make too much sense. It really depends on the chosen criteria. Should it have only electronic parts? Should it be generally programmable? Should it be a stored-program machine? And so on... These early machines are in fact the consequences of what was done before and also of the state of the technology at that time. Clearly just before and during World War II, the idea of building electronic computers was in the air.

For more on the role of WW II and of the military in the development of computers and applied mathematics, see [288, 801].

Let us continue our brief review of the computer research and industry. We give details only for the machines that we think had an importance in the evolution of computers for scientific computations. The computational performance of these machines is generally given in "flops" which does not mean a failure (even though some computers were flops) but floating point operations per second. Normally we should write flop/s, but for simplicity we will use flops. So, one can speak of Kflops (Kilo flops = 10^3 flops), Mflops (Mega flops = 10^6 flops), GFlops (Giga flops = 10^9 flops), Tflops (Tera flops = 10^{12} flops), and Pflops (Peta flops = 10^{15} flops). Computers to appear in the beginning of 2020s are designed to reach the Eflops (Exa flops = 10^{18} flops).

The *Small-Scale Experimental Machine* (SSEM), also known as the *Manchester Baby*, was a binary stored program computer built in 1948 at Manchester University (UK) to test the Williams tube, a random-access memory device using a cathode-ray tube (CRT). It had a 32-bit word length and a memory of 32 words and used 550 vacuum tubes. Most arithmetic operations were implemented in software. The program was separated from the data. This machine had only seven instructions. It was built at the request of Newman by Frederic Calland Williams (1911-1977), Tom Kilburn (1921-2001), and Geoffrey Colin Tootill (1922-2017). It was followed by the *Manchester Mark I* in 1949, a binary fixed point arithmetic machine. The word length was 40 bits, with one 40-bit number or two 20-bit instructions. A distinctive feature of this machine was that it had index registers. The memory used CRT and a magnetic drum. The machine had 30 instructions. It was turned into a commercial product by the Ferranti company.

Another early computer in the UK was the *Electronic Delay Storage Automatic Calculator* (EDSAC). It was developed by Maurice Vincent Wilkes (1913-2010) at the Cambridge University Mathematical Laboratory; see [3234, 3235]. It first ran in May 1949 and was used for nine years. The machine had 512 18-bit words of memory, later extended to 1,024. It was a fixed point arithmetic binary machine. It was superseded by EDSAC 2, built by the same team.

[52] The fact that *Colossus* was the first ever programmable computer was challenged by Thomas Haigh and Mark Priestley in *Colossus and programmability, IEEE Annals of the History of Computing, 40 (4) (2018, pp. 5-27).*

In 1949, the *Electronic Discrete Variable Automatic Computer* (EDVAC) was built by Eckert and Mauchly. This machine was influential mainly because of the report *First draft of a report on the EDVAC* [3157] written by von Neumann in 1945 describing a computer architecture in which the data and the program are both stored in the computer's memory in the same address space; see also [511] in 1946. The EDVAC was a binary serial computer with addition, subtraction, multiplication, and programmed division, with a memory capacity of 1,000 44-bit words. This binary fixed point arithmetic machine had a magnetic tape reader-recorder, a control unit with an oscilloscope, a dispatcher unit to receive instructions from the control and memory and direct them to other units, a computational unit for arithmetic operations on a pair of numbers at a time and for sending results to memory after checking, and a dual memory unit consisting of two sets of 64 mercury acoustic delay lines with 8 words on each line. The addition time was 864 μs and the multiplication time was 2900 μs. The computer had almost 6,000 vacuum tubes and 12,000 diodes, and consumed 56 kW of power. EDVAC was delivered to the Ballistics Research Laboratory in August 1949 and began its operation in 1951. The mean time between failure was eight hours. It ran until 1961.

7.9 ▪ The 1950s

Many machines were developed in the beginning of the 1950s in universities or for government agencies and laboratories. A lot of them had names ending with "AC" for Automatic Computer: BINAC, CSIRAC, SEAC, SWAC, ORDVAC, ILLIAC, MANIAC, AVIDAC, FLAC, JOHNNIAC, MIDAC, RAYDAC, WEIZAC, ...

In the UK, the *Pilot ACE*, resulting from a more ambitious project led by Alan Mathison Turing (1912-1954), became operational at the National Physical Laboratory in 1950; for details see Sections 10.72 and 10.76 as well as [3238, 3255]. It was a stored-program binary computer with 32-bit words. There was no hardware for multiplication and division which were done in software. A follow-on was the *Deuce* computer built in 1955 by English Electric that sold 33 machines up to 1964.

The IBM *Naval Ordnance Research Calculator* (NORC) was started in 1950 and delivered to the US Navy in 1954. It was the fastest machine until 1960. The machine had 9,800 vacuum tubes. The internal clock speed was one μs and it was capable of 15,000 operations per second with decimal arithmetic. Each word could store 16 decimal digits. Initially it had a Williams tube memory holding 3,600 words. Later on this was replaced by 20,000 words of magnetic core memory.

The *IAS* computer was built at the Institute for Advanced Study in Princeton (USA) from 1945 to 1951, implementing the ideas of von Neumann; see [3179]. It was a binary computer with fixed point arithmetic and 1,024 40-bit words of memory able to store two 20-bit instructions in each word. It had two general-purpose registers. The addition time was 62 μs and the multiplication time was 713 μs. Several machines were later developed on the same principles. One of them was the *IBM 701*, the first IBM commercial scientific computer in 1952. It was followed by the *IBM 704* in 1954 with floating point arithmetic and a magnetic core memory. Floating point numbers had a 27-bit mantissa and an 8-bit exponent.

Besides the USA and the UK, computers were also built in other countries, for instance, in the USSR. Examples of soviet computers are *MESM* (1948-1951), an electronic binary stored-program computer designed by Sergey Alekseevich Lebedev (1902-1974), followed by *BESM* machines (1950s-1960s), including the BESM-6 in 1967, *Strela* (1953), *Ural* (1956-1964), *MINSK-2* (1962), *Elbrus 2* (1977), a 10-processor computer, and *Elbrus 3* (1986) a 16-processor computer. At some point the Soviet government decided to stop these homemade developments and to copy the IBM/360 architecture. This resulted in the stagnation of the Soviet computer industry in the 1980s.

Some examples of early European computers are Bull *Gamma 3* in France (1952), *BESK* in Sweden (1953), *ERMETH* in Switzerland (1955), and *G1* (1952), *G2* (1955), *G3* (1961) developed by Heinz Billing (1914-2017), as well as *PERM* in 1952-1956 in Germany, *CEP* and *Olivetti* in 1957-1960 in Italy (see [372]). Some early machines were also developed in Japan, like the *FUJIC* computer in 1956.

A real improvement was given by the invention of the transistor. Following works by Julius Edgar Lilienfeld (1882-1963) in 1925 and Oskar Heil (1908-1994) in 1935, John Bardeen (1908-1991), Walter Houser Brattain (1902-1987), and William Bradford Shockley (1910-1989) developed the first working transistors at Bell Labs in 1947. In 1956, they were awarded the Nobel Prize in Physics for their research. A little later, Shockley invented the bipolar junction transistor. Transistors started to be mass produced in the beginning of the 1950s. Most early transistors were made of germanium. A progress was the use of silicon by Mohamed Attala (1924-2009) at Bell Labs. Later on, he developed the metal-oxide-semiconductor field-effect transistor (MOSFET).

Similar to what we have seen above for the first electronic computer, there are still some disputes about which was the first transistor computer. A computer using transistors was built in Manchester, UK, in 1953 and 1955, but it was also using vacuum tubes. The *TRADIC* computer built at Bell Labs in 1955 was almost fully transistorized, but it used one vacuum tube. The Harwell *CADET*, built at AERE Harwell, UK, was the first fully transistorised computer in Europe in 1955 (and, depending on the definition of what is a transistor computer, maybe the first in the world). Another contender is the *Mailüfterl* computer that was built from May 1956 to May 1958 at the Vienna University of Technology in Austria by Heinz Zemanek (1920-2014). It had 3,000 transistors.

Other early machines using transistors were the *Bull Gamma 60* (1958), the *UNIVAC Solid State* (1958), the *IBM 7090* (1959), the Digital Equipment Corporation (DEC) *PDP-1* (1959), the *UNIVAC LARC* (1960), Zuse's *Z23* (1961), the *IBM 7094* (1962), the *Telefunken TR4* (1962), and the English Electric *KDF9* (1964). The Control Data Corporation (CDC) *1604*, designed by Seymour Roger Cray (1925-1996) (see Section 10.15) in 1960, was one of the first commercially successful transistorized computers.

7.10 ▪ The 1960s

An interesting machine ahead of its time was the *IBM 7030* (Stretch), introduced in 1961. It was the fastest computer in the world from 1961 until the first *CDC 6600* became operational in 1964. It had more than 150,000 transistors. The word length was 64 bits with a 48-bit mantissa and the memory was up to 256 Kwords. The floating point addition time was 1.38-1.5 μs, the multiplication time was 2.48-2.70 μs (around 0.3 Mflops), and the division time was 9.00-9.90 μs. Even though this computer was not commercially successful, with only eight machines being sold, mainly to nuclear weapons laboratories, it introduced many new technologies that were used in latter machines. Multiprogramming, memory protection, generalized interrupts, and the 8-bit byte were all concepts later incorporated in the IBM System 360 line of computers. Instruction pipelining, prefetch and decoding, and memory interleaving were used in later supercomputer designs.

In the following "cp" is the abbreviation for "clock period." This is the smallest time that sequences the computer. A nanosecond (ns) is 10^{-9} seconds. One nanosecond corresponds to a frequency of 1 GHz.

The chief designer of the *Control Data 6600* was Cray. This machine, first introduced in 1964, had a cp of 100 ns, its speed was 0.5-3 Mflops, and it had 128 60-bit Kwords of memory. Floating point numbers had a 48-bit mantissa. It was built with silicon-based transistors as opposed to germanium transistors for many previous machines. It had several functional units for doing the elementary operations that can work in parallel and was freon cooled. Another distinctive feature was that it had peripheral processors handling the Input/Output operations. The performance was 0.48 Mflops on the Linpack 100 benchmark. It was the fastest computer in the world from 1964 to 1969. The *CDC 6400* in 1966 was a downsized version of the 6600.

The Linpack benchmark was a side effect of the tests done during and after the development of the LINPACK software for solving linear systems with Gaussian elimination; see Chapter 8. Initially, it solves a linear system of order 100 because of the small memories of the machines of that time. Later on, it was extended to a system of order 1000 and the computer companies were also allowed to solve the largest system they could. This is known as the HPL (high performance LINPACK) benchmark [927]. It is used for ranking machines in the TOP500 list published every six months, see www.top500.org.

The *CDC 7600*, first introduced in 1969, was also designed by Cray. The cp was 27.5 ns with 40 Mflops peak, the memory was 128 60-bit Kwords plus 512 Kwords of extended memory. This computer had pipelined functional units. This means that they were segmented into several stages. When a stage is completed, it can process new data. Therefore several operations take place simultaneously. Unfortunately, the compilers of the time were not really able to take advantage of these hardware capabilities. The performance was 3.3 Mflops on the Linpack 100 benchmark.

7.11 ▪ The 1970s

Even though changes in the architecture were starting with the *CDC 7600*, the machines that we have considered above can be seen as scalar machines, doing the elementary operations ($+$, $-$, $\times$, $\div$) on two floating point numbers sequentially. So, a good measure of the complexity of a scientific program was the number of floating point operations. This was about to change with the introduction of the so-called vector computers that were able to efficiently perform the elementary operations on vectors, that is, sets of independent floating point numbers. Vector computers were mainly manufactured in the USA: Cray (*CRAY 1S*) and Control Data (*Cyber 205*) and to a small extent by IBM (*VF machines*) and in Japan by NEC, Fujitsu, and Hitachi.

The Control Data Corporation *STAR-100*, designed by James E. Thornton (1925-2005) and introduced in 1974, had some vector capabilities. The goal was to obtain 100 Mflops. But, the vectors were fetched from consecutive addresses in memory and the result written back to memory, and this was not very efficient. The cp was 40 ns and the machine had deep pipelines. All this resulted in the performances being lower than expected. The word length was 64 bits. The *STAR-100* was considered a failure and only five machines were sold.

The *Texas Instruments Advanced Scientific Computer* (ASC) was another early vector computer. The machine was designed and built between 1966 and 1973. As in the *STAR-100*, the vector operations were from memory to memory. The cp was 60 ns. Only seven machines were built before Texas Instruments abandoned the supercomputer market.

The next Cray machine, after he left CDC, started the era of efficient vector computers. The main new feature of Cray, IBM, and Japanese machines was the addition of vector registers. Therefore, if one wants to do (say) the addition of two vectors, those are segmented into chunks by the compiler, loaded in the vector registers, and then the addition pipelined functional unit can be used at full speed provided there is a large enough bandwidth between the memory and the vector registers. With these computers (and good vectorizing compilers) one can deliver at least one result each clock period and more if there are several independent functional units; there were two of them for the *CRAY 1S*. Using these vector machines meant that most of the time, the algorithms had to be rethought or at least reprogrammed in terms of vectors. Early vectorizing compilers were not really efficient. This is why a standardization of low-level vector operations was done, leading to the Basic Linear Algebra Subroutines (BLAS); see Chapter 8. These routines can be implemented efficiently in assembly language. Use of vector computers provided improved performances but added some difficulties for algorithm designers and programmers compared to previous scalar machines.

The *CRAY 1* was introduced in 1976 and extended in 1979 with more memory to the *CRAY 1S*, which had a cp of 12.5 ns, a 160 Mflops peak performance, and 2 64-bit Mwords of memory. This machine was the first available efficient vector computer and was a big commercial success since over 80 machines were sold. The performance was 27 Mflops on the Linpack 100 benchmark and 110 Mflops on Linpack 1000. The machine was freon cooled. The cooling system represented a large part of the patents for the *CRAY 1*. Cray used to say that it was the part of the machine he was the most proud of. The shape of the machine (C-like) was chosen to minimize the length of the cables; see the figure in Section 10.15.

7.12 ▪ The 1980s

At the end of the 1970s the company Floating Point Systems (FPS) started producing so-called *Array Processors*, which were SIMD (Single Instruction Multiple Data) machines attached to minicomputers or mainframes. The *FPS 164 MAX* in 1985 had a maximum performance of 341 Mflops. They can be considered the poor man's supercomputers.

After the *CRAY 1*, Cray developed the *CRAY 2*, which was introduced in 1985. This was a multiprocessor machine with two or four processors. The cp was 4.1 ns. At the time the unique features of the machine were a large memory with 256 Mwords and the fact that the electronic circuits were immersed in the cooling fluid; see Section 10.15. One processor of the *CRAY 2S* reached 384 Mflops on Linpack 1000. When developing the *CRAY 3*, Seymour Cray left Cray Research, but he never fully completed the *CRAY 3* or the *CRAY 4*.

In parallel to Cray's new designs, Cray Research developed another line of vector machines derived from the *CRAY 1*. The *CRAY XMP-4*, first introduced in 1985, had four vector processors, a cp of 8.5 ns, a 940 Mflops peak performance, 16 Mwords of memory with a better memory bandwidth than the *CRAY 1*. There were also vector gather/scatter operations for handling indirect addressing in vectors. It was a parallel computer with a shared memory. The performance of one processor was 218 Mflops on the Linpack 1000. The project leader of the XMP series was Steve Chen, who left the company in 1987.

It was followed in 1988 by the *CRAY YMP* with up to 8 vector processors with a shared memory up to 128 Mwords. The cp was 6 ns and the peak performance was 2.6 Gflops. Using the 8 processors the performance was 2.14 Gflops on Linpack 1000. Real applications can reach more than a Gflops on this machine. The next machine in this line was the *CRAY C90* in 1991, with a 4.1 ns cp and up to 16 processors. The performances were improved with the *CRAY T90* in 1995 which could have between 4 and 32 processors. The cp was 2.2 ns and the peak performance per processor was 1.8 Gflops. The machine implemented the IEEE 754 floating point arithmetic (see below). As in the *CRAY 2*, the circuits were immersed in the fluorinert cooling fluid. The performance on Linpack 1000 with 32 processors was 29.4 Gflops. At the same time, Cray Research manufactured the *CRAY J90* which used CMOS processors and was air-cooled. More than 400 systems were sold.

Cray Research Inc. was bought by Silicon Graphics (SGI) for $740 million in February 1996. Being a division within SGI, Cray introduced the *CRAY SV1*. On March 2, 2000, Cray was sold to the Tera Computer Company, a company founded in 1987 by Burton Smith (1941-2018), which was renamed Cray Inc. On September 25, 2019, Hewlett Packard Enterprise (HPE) acquired Cray Inc.

When designing the floating point arithmetic of the *CDC 6600* and the *CRAY 1*, the goal of Cray was to obtain the maximum performance he could. Therefore, he sometimes traded accuracy for speed. For instance, there was no division on Cray computers. It was obtained by multiplication with the inverse which was obtained by Newton iterations. A more serious problem was the lack of a guard digit when doing subtractions. Cray's arithmetic idiosyncracies were mainly exhibited by William Morton Kahan [1862]. IEEE 754 floating point arithmetic was only adopted on the T90.

In the USA, the competitors of Cray Research on the vector computer market were CDC and to a lesser extent IBM. The Control Data *Cyber 205* in 1980 was an improved version of the *STAR-100* architecture. A machine with four vector pipes had a peak performance of 400 Mflops on 64-bit data but this speed was not attained on real applications. A *Cyber 205* with four pipelines obtained 195 Mflops on the Linpack 1000 benchmark. In 1983, CDC spun off its supercomputer division which became ETA systems. The *ETA-10*, which was announced in 1986, was a follow-on of the *Cyber 205* with up to 8 processors. The cp was 7 ns using CMOS circuits and the high-end models were liquid nitrogen-cooled. A single processor of the *ETA-10G* achieved 93 Mflops on the Linpack 100 benchmark and 496 Mflops on Linpack 1000. Around 30 machines were sold before ETA Systems was reincorporated into CDC in 1989.

In 1985, IBM announced an optional *Vector Facility* for the model 3090. There were 16 vector registers with a number of words, from 8 to 512, depending on the model. One processor delivered 16 Mflops on Linpack 100 benchmark and 97 Mflops on Linpack 1000. An *IBM 3090/600J VF* with 6 processors had a performance of 540 Mflops on Linpack 1000.

Starting in the 1980s, the main competitors of Cray Research for vector processors were Japanese companies. They were highly supported by the Japanese government. The Ministry of International Trade and Industry (MITI) started a project on high-speed calculation for science and engineering, known as the *Supercomputer Project* in 1981. For nine years, this project developed basic technologies for building supercomputers. The three main companies involved in supercomputing were Fujitsu, Hitachi, and NEC.

The Fujitsu *FACOM 230-75 APU* (for Array Processing Unit) was Japan's first vector computer in 1977. Using this technology, in 1982, Fujitsu introduced the *FACOM VP 100*, the *VP 200* with a peak performance of 571 Mflops, and the *VP 400* with a performance of 1 Gflops. These machines were air cooled, they had multiple vector functional units and a separate scalar unit as well as a reconfigurable vector register storage. The cp was 7.5 ns for the vector part and 15 ns for the scalar part. They were followed by the *VP2000* series announced in December 1988.

The Hitachi *S-810* in 1982 had a cp of 14 ns, 12 floating point arithmetic pipelines for the model 20, and a peak performance of 630 Mflops. It was an air cooled machine.

In 1983, NEC announced the *SX-1* and *SX-2* machines. The *SX-2* had a cp of 6 ns, 8 vector pipelines, and a peak performance of 1.333 Gflops. It was water cooled. It was followed by the *SX-3* in 1984 with up to 16 vector pipelines. The peak performance of the multiprocessor high-end model was 22 Gflops. The performance on the Linpack 1000 was 13.4 Gflops.

7.13 ▪ The 1990s

We have already discussed Cray machines above. Let us concentrate on Japanese computers.

The National Aerospace Laboratory of Japan developed the *Numerical Wind Tunnel* (NWT) parallel supercomputer in partnership with Fujitsu. The NWT, introduced in January 1993, was in the top position of the TOP500 list from 1993 to 1995. On this basis Fujitsu developed the *VPP500* commercial supercomputer.

NEC introduced the *SX-4* series in November 1994. The maximum vector performance of one machine with 32 processors was 64 Gflops. It reached 31 Gflops on the Linpack 1000. A full configuration was obtained by linking 16 machines or nodes for a total of of 512 processors and 1 Tflops of peak performance.

The next machine was the *SX-5* in June 1998. Each node had 16 processors with a shared-memory architecture. The peak performance of each node was 128 Gflops. The performance on the Linpack 1000 was 45 Gflops. The peak performance of the maximum configuration with 32 nodes was 4 Tflops.

7.14 ▪ The 2000s

The *SX-6* appeared in 2001. Its processor was a single chip implementation with a vector processor unit and a scalar processor whence the previous machines had a multi-chip implementation. Each node had 8 vector processors. The cp was 1.77 ns. The peak performance of one processor was 9 Gflops. A system with 8 processors obtained 46.3 Gflops on the Linpack 1000.

The *Earth Simulator computer*, developed for simulations with global climate models in 2002, was built by NEC and based on *SX-6* nodes. It had 640 nodes with 9 processors. Its performance on HPL for solving a linear system with a little more than one million unknowns was 35.6 Tflops. It was the fastest machine in the world up until the end of 2004.

The *SX-8* series ranges from the single-CPU *SX-8b* to the *SX-8/4096M512* with 512 nodes and 4,096 processors.

The NEC *SX-9* was introduced in 2008. The vector part had a cp of 0.31 ns. The peak performance of one processor was 102.4 Gflops. Up to 16 processors and one TB of memory may be used in a single node. The maximum was 512 nodes.

There was no *SX-10*. The next generation in 2013 was named *SX-ACE*. It used the first 4-core vector system on a chip.

NEC is still manufacturing vector computers. In 2017-2019 they introduced the *Vector Engine Processor* which integrates eight vector cores and 48 GB of high bandwidth memory providing a peak performance of up to 2.45 Tflops. A single Vector Engine can execute 32 double precision floating point operations per clock cycle with its vector registers holding 256 floating point values. With three fused-multiply-add units, each core has a peak performance up to 307.2 Gflops.

7.15 ▪ The rise of microprocessors

Since some early computers adopted a floating point representation of numbers, many data formats have been used. The base, the numbers of bits in the mantissa (significand), and the exponent were different from one machine to the others. Therefore, what were single and double precisions had no real meaning. The treatment of the underflows and overflows was different, as well as the treatment of cases like $0/0$ and the rounding methods were also different. Moreover, some machines gave wrong results on a few operations. This was not due to hardware failures but to the design of the arithmetic. Bases (or radix) that have been used are 2, 8, 10, 16. One Russian computer, the *Setun*, used base 3. Hence, it was a real mess and it was very difficult for algorithm designers and programmers to develop accurate and portable software. The situation was improved at the end of the 1980s by the definition of a standard for floating point arithmetic; see below.

The rise of microprocessors was the demise of vector processors, which were proprietary and dedicated designs of their manufacturers. Intel Corporation and Advanced Micro Devices (AMD) were selling microprocessors in high volumes, allowing them to sustain large research and development efforts to improve their microprocessors. Therefore, at some point, it was becoming cheaper to build supercomputers with off-the-shelf commodity microprocessors than to develop costly proprietary designs. However, in many cases, these microprocessors were not designed for scientific computing.

Early competitors in the microprocessor business were Motorola (founded in 1928), Texas Instruments (founded in 1951), Fairchild Semiconductors (founded in 1957), Intel (founded in 1968), and AMD (founded in 1969).

The metal-oxide-semiconductor field-effect transistor (MOSFET) originated in 1960 at Bell Labs in the USA. MOS integrated circuit chips appeared in the early 1960s. Large-scale integration (LSI) with hundreds of transistors on a single MOS chip was available by the late 1960s. This led to the integration of CPU functions on a few MOS LSI chips. The first commercially available single-chip microprocessor was the 4-bit Intel 4004 in 1971 followed by the 8-bit microprocessor Intel 8008 in 1972. All this led to the first personal computers in the beginning of the 1970s. The Motorola 8-bit 6800 was announced in 1975 followed by 16-bit designs in the mid-1970s. Intel introduced the 16-bit 8086 in 1978 and the 8088 in 1979, which was used in the first IBM PC.

In 1980, Intel introduced the 8087 coprocessor with hardware floating point arithmetic. The development of the 8087 led to the IEEE 754 standard for floating point arithmetic in 1985. John Palmer, the manager of the 8087 project, who wanted to have the "best" arithmetic, had Kahan (who was a professor at the University of California at Berkeley) as a consultant; see [1545]. A draft of the standard written with his student Jerome Coonen and Harold Stone was published in October 1979 in the SIGNUM Newsletter. The standard for binary floating point arithmetic was adopted in 1985. In 1989, Kahan received the Turing Award for his work. The first standard considered binary arithmetic. This was extended to decimal arithmetic and revised in 2008 and 2019.

The standard defined the arithmetic formats: sets of floating-point data, signed zeros and subnormal numbers, infinities, and special "not a number" values (NaNs), interchange formats to convert from and to decimal input/output data, rounding rules, arithmetic, and other operations such as elementary functions and exception handling.

The adoption of the IEEE standard by most computer manufacturers was a blessing for people developing algorithms and for programmers. In particular a standard model could be used in the rounding error analysis of linear algebra algorithms; see, for instance, [1681].

The 32-bit Motorola processor 68020 was introduced in 1984. It was used in the *Apple Macintosh II* and *Sun* workstations. The 32-bit Intel 80386 appeared in 1985. Starting with the 80486 (also named i486) in 1989, the later Intel processors did not use a separate floating point coprocessor. During the development of microprocessors, the computing speed was increasing faster than the memory access times were decreasing. Hence, the access to memory became the bottleneck of the computations. Since the number of components that was available on a chip was increasing, the designers started to add caches, that is, fast memories on the chip. The data was moving from the (slow) main memory to the (faster) cache before being used by the processor. If there was some locality in the data to be used, this was speeding up the computations. For instance, the 80486 had an 8KB on-chip L1 cache. Later on, several level of caches were added, with different sizes and access times.

The introduction of caches put new constraints on the algorithm designers. To obtain good performance on these processors it became necessary to favor data locality, that is, to reuse as much as possible the data that was in the cache(s). This led to the design and use of block algorithms.

The Intel x86 architecture was licensed to other companies that became second sources, for instance AMD. There was no i586 because it was renamed *Pentium*, a name for which Intel filed a trademark application. Unfortunately, the name Pentium became also (in)famous worldwide because of the "Pentium bug" that affected the 32-bit P5 in 1994. It was an error in a table used by the floating point divide operation. In the end Intel offered to replace the processors that had this bug. AMD introduced the 64-bit *Opteron* microprocessor in 2003 and Intel introduced the 64-bit *Xeon* in 2004.

In 1965, Gordon Earle Moore, who was the co-founder of Fairchild Semiconductor and Intel, stated that the number of transistors in integrated circuits doubles about every year. This was later known as *Moore's law*, even though it was not a law of physics, but just based on observations. In the mid-1970s, Moore revised his "law" to doubling every two years. Unfortunately, after some time, in the beginning of the 2000s the rate of increase started to slow down. For technological reasons, it became difficult to increase again the frequency (which means decreasing the clock period), the microprocessor companies started to put several processing units on one chip. This induced a change of terminology; these processing units that would have been called processors before are now known as *cores*. A processor has several or many cores becoming a parallel computer. The manufacturers proposed successively chips with 2, 4, 8 cores and even more now. This is most certainly the case in your own PC if it is not too old. There are also now multicore chips with tens of cores that are used in supercomputers.

The mid-1980s and 1990s saw the introduction of the first reduced instruction set (RISC) microprocessors. Remember that probably the first commercially available RISC computer was the *CDC 6600* in 1964, even though the name RISC was not used at that time! The first 32-bit RISC microprocessor was introduced by MIPS Computer Systems in 1984. They released a 64-bit RISC microprocessor in 1991. Other companies proposing RISC architectures were ARM, DEC, HP, IBM, Intel, Motorola, and SUN.

The IBM Power architecture appeared at the end of the 1980s and was used in servers and RS/6000 workstations. The PowerPC architecture was the result of an alliance between Apple, IBM, and Motorola that ended at the beginning of the 2000s.

There were also some failures in the microprocessor industry. For instance, the Intel *iAPX 432*, a 32-bit processor introduced in 1981 and discontinued in 1986. Another example is the Intel i860 RISC microprocessor in 1989, which never achieved a commercial success. It was used in the Intel *iPSC/860* parallel computer.

7.16 ▪ Parallel computers

We have seen that in the 1980s there were some multiprocessor computers like the *CRAY 2* or the *CRAY X-MP/Y-MP* with a few processors. The 1990s saw the development of what had been called *massively parallel processors* (MPP). Most of these machines were based on the use of commercially available microprocessors.

Basically there are three kinds of machines depending on how the memory is accessed. In shared memory computers (like the *CRAY 2* or *CRAY X-MP*) there is a single address space and all processors have access to the memory. In distributed memory computers each processor has its own memory that cannot be directly accessed by the other processors. The data has to be explicitly exchanged through the communication network that links the processors. The third type is a mix between the two previous ones. The machine is made of shared-memory nodes with several processors and the nodes are connected through a communication network. These machines are qualified as NUMA (non-uniform memory access) since the times to access memory are smaller within a node than across nodes.

Machines are also classified as being SIMD or MIMD. In SIMD (single instruction–multiple data) computers all the processors execute the same instruction on different data as in a vector instruction. In MIMD (multiple instruction–multiple data) computers each processor executes its own program asynchronously from the others. Of interest is the book [1716] by Roger Willis Hockney (1929-1999) and Chris R. Jesshope in 1988.

In fact, there were some forerunners like the *ILLIAC IV* that was designed at the University of Illinois in the late 1960s. The chief engineer was Daniel Leonid Slotnick (1931-1985) who had been interested in parallel computing since the 1950s. Due to financial problems only one quarter of the original design was built by Burroughs in 1970. The machine had one central processing unit (CU) and 64 processing elements (PE) which were floating point units with a 64-bit word length. Each PE had its own memory but the CU could see the whole address space. It could be seen as an SIMD machine. The peak performance was 50 Mflops. The machine was moved to the NASA Ames Research Center in California in 1972. It was switched off in 1981.

Another early machine was the *Parallel Element Processing Ensemble* designed by Burroughs in the mid-1960s for the US Army.

In the UK, International Computers Limited introduced the ICL *Distributed Array Processor* (DAP) in 1972-1974. It was an SIMD machine with 4,096 (64×64) single-bit processing elements with 4K bits of memory each. It can be considered an array processor hosted by an *ICL 2980* computer.

The Goodyear *Massively Parallel Processor* SIMD machine was built for NASA from 1983 to 1991. It was an SIMD computer with a 128×128 two-dimensional array of 1-bit-wide processing elements.

After the mid-1980s, Intel introduced the series of *Intel Personal Super Computer* (iPSC). These were MIMD distributed memory machines with microprocessors connected by a hypercube communication network. Data was exchanged between the processors using a proprietary message passing software. The first model was in 1985, the *iPSC/1* had from 32 up to 128 nodes linked by ethernet, with each node made of an 80286 processor and a 80287 coprocessor. With 32 nodes it was a five-dimensional cube. If the node addresses are binary-coded, a node is connected only to the nodes whose addresses differ by one bit from its own address. In 1987, the *iPSC/2* used 80386 and 80387 processors linked by a proprietary network. The *iPSC/860* was announced in 1990. It used the i860 microprocessor.

Following the *Touchstone Delta* project at Caltech, Intel introduced the *Paragon* computer with up to 4,096 i860 microprocessors connected with a 2D communication network. In 1993, the Sandia National Laboratories installed an Intel *XP/S 140 Paragon* computer with 3,680 processors. This system held the first position on the June 1994 TOP500 list. The Intel *Paragon XP/S 150* supercomputer at ORNL in 1995 had 1,024 nodes, with each node having two compute processors and one processor for handling the message passing.

Thinking Machines Corporation (TMC) was founded in 1983 by William Daniel Hillis. Hillis published a book [1698] in 1986 derived from his Ph.D. thesis at MIT. The machines were initially targeted for artificial intelligence applications. The *CM-1* in 1985 was an SIMD machine with 64K 1-bit processors linked by an hypercube network, but this machine was not able to do floating point operations. The *CM-2* in 1987 had more memory and Weitek floating point coprocessors, 1 for 32 of the 1-bit processors. This machine had an interesting look, being a small black cube (made of eight subcubes) with many flickering red LED lights. The *CM-5* in 1991 had a different architecture. It was an MIMD machine with Sparc processors and a fat tree network. The *CM-5* had also an appealing look. As for the *CM-2*, it drew much attention from the press. A replica of the *CM-5* appeared in Steven Spielberg's *Jurassic Park* movie. With 1,024 processors, the performance of the *CM-5* on HPL for solving a linear system of order $53,224$ was 59.7 Gflops. For a while the company was heavily financially supported by government agencies, but when the federal funding was decreasing, TMC filed for bankruptcy in 1994.

Departing from its previous policy, in the 1990s Cray Research developed parallel computers using off-the-shelf microprocessors. The *CRAY T3D* in 1993 used the DEC Alpha EV4 RISC microprocessor. It had 32 to 2,048 processors linked by a 3D torus network and was hosted by a *CRAY YMP-E*. With 1,024 processors, it reached 100 Gflops on HPL for a system of order 81,920.

It was followed by the *CRAY T3E* in 1996 with 64 up to 2,048 processors. The processors were initially DEC alpha EV5 and later EV56. With 16 processors the performance was 3.2 Gflops on Linpack 1000. With 1,488 processors, the performance on HPL for a system of order 148,800 was 1.13 Tflops. This machine was considered at the time a massively parallel computer. Then, Cray was bought by Silicon Graphics and the DEC Alpha was discontinued because a part of DEC was bought by Compaq.

During all these years many computer manufacturers went into bankruptcy, were bought by more successful companies, or stopped manufacturing scientific computers. Some examples (in alphabetical order) are Alliant, Amdahl, BBN Technologies, Compaq, Control Data, Convex, Cray Research, Cray Corporation, Cydrome, DEC, Denelcor, ELXSI, Encore Computers, ETA, FPS, Kendall Square, Maspar, Meiko, Multiflow, nCUBE, Parsytec, Pyramid, Stardent, Sun Microsystems, Supercomputer Systems, Suprenum, Thinking Machines, and Trilogy Systems, to name just a few.

Parallel computers were also developed in Japan in the 1990s. The Fujitsu *VPP500* in 1992 was a parallel vector computer with 222 processing units. The peak performance was 355 Gflops. It was followed in 1996 by the *VPP700* with 512 processors and a peak performance of 1.13 Tflops and the *VPP5000* in 1999. NEC announced the *Cenju-3* in 1993, a small machine with up to 256 microprocessors linked by a multi-stage interconnection network. The peak performance was 12.8 Gflops. It was followed by the *Cenju-4* in 1997 with up to 1,024 nodes.

In 1996, the Tsukuba University developed the *CP-PACS* computer in collaboration with Hitachi. It had 2,048 processing units connected with a 3D network. The performance of this system on HPL was 0.368 Tflops. It was the fastest computer in the world for six months. This university built several other massively parallel computers in the following years.

The NEC's *Earth Simulator* was the fastest computer in the world from 2002 to 2004. It was based on the NEC *SX-6* architecture, having 640 nodes with eight vector processors each.

IBM introduced the SP series based on POWER RISC microprocessors in the 1990s. The *PS 2* was released in 1993. Later systems were based on PowerPC chips. The IBM *SP ASCI* "Blue Pacific" was a PowerPC 604-based system with a peak performance of 3.9 Tflops. ASCI, which meant Accelerated Strategic Computing Initiative, was a program of the Department of Defense to boost the US supercomputer industry. It was followed by the Advanced Simulation and Computing Program (ASC). The machine was installed at the Lawrence Livermore National Laboratory (LLNL) in 1998. It obtained a performance of 2.14 Tflops with 5,808 processors on HPL for a system of order 431,344. ASCI "White" was a 512-node system with 8,192 processors with a peak performance of 12.3 Tflops. It was installed at the LLNL in 2001.

Other computers for the ASCI program were the ASCI "Red" made by Intel, installed at the Sandia National Laboratory (SNL) in 1997, and the ASCI "Blue Mountain," an SGI machine with 6,144 MIPS microprocessors installed at the Los Alamos National Laboratory (LANL) in 1998.

IBM started a new line of massively parallel computers in the 2000s. It was the result of the *Blue Gene* project started in 1999 and initially targeted to biology problems. In November 2004, a *Blue Gene/L* system with 16,384 compute nodes obtained the first place in the TOP500 list, with a performance of 70.72 Tflops. The machine used PowerPC processors with floating point accelerators and a three-dimensional torus interconnect with auxiliary networks for global communications. A *Blue Gene/L* system installed at LLNL achieved 478 Tflops on HPL and 596 Tflops peak.

In June 2007, IBM announced *Blue Gene/P* with improved performance. The first installation was at the Forschungszentrum Jülich (Germany) in 2007. It was expanded in 2009 to 73,728 nodes with a total of 294,912 cores and achieved a peak performance of 1 Pflops. A system with 163,840 cores was installed at the Argonne National Laboratory (ANL).

The third generation in 2011, *Blue Gene/Q* used IBM 18-core chips with 1.47 billion transistors. The peak performance was 20 Pflops. It reached 17 Pflops in the TOP500 list. The *Blue Gene/Q* "Sequoia" machine at LLNL had 98,304 compute nodes with 1.6 million cores. A smaller system was installed at ANL in 2012.

In the 2000s, Cray introduced a series of massively parallel computers. The *CRAY X1* in 2003 had up to 4,096 processors with 1,024 shared-memory nodes connected in a 2-dimensional torus network. It was updated to the *X1E* in 2005. The *XT3* "Red Storm" at SNL in 2004 had 10,880 AMD Opteron processors. This machine was updated two times with a final peak performance of 284 Tflops in 2008. The *XT4* in 2006 was an updated version of the *XT3*. The *XT5* in 2007 used AMD Opteron 4-core processors. The "Jaguar" *XT5* system delivered to ORNL had around 150,000 cores and was upgraded in 2009 to 224,256 cores with a peak performance of 1.75 Pflops. The *XT6* in 2009 was an updated version of the *XT5* with 8- or 12-core Opteron processors. The *XE6* in 2010 used the same processors but a different 3D-torus interconnect between the nodes.

In Japan, the RIKEN Advanced Institute for Computational Science chose Fujitsu to develop the *K* computer. Its peak performance was 10.51 Pflops using 88,128 processors with 8 cores each with a total of 705,024 cores. It was the fastest computer in the world in 2011. It was decommissioned in August 2019.

We have seen that there were early parallel computers in the 1960s and 1970s, but they became more common in the 1980s and 1990s. This put more constraints on numerical analysts and programmers. It seems obvious that to obtain good performance on a parallel computer, the program implementing an algorithm must have as many operations as possible that can be executed in parallel. This is not so easy to achieve, for instance, in algorithms for solving linear systems. By definition, some or all the unknowns of a linear system are coupled, except if the matrix is diagonal, in which case the solution is trivially obtained. In distributed memory computers these algorithms have to exchange data between the processors' memories. This is often the bottleneck of the computation. Quite often it is also necessary to compute the dot product of two vectors. If the multiplication of the components is parallel, the summation is a reduction operation that can be costly. This is why in recent years, researchers have been looking for algorithms minimizing the communications. They are sometimes called communication-avoiding algorithms even though some communications are always needed when solving linear systems. It must also be noted that, for instance, in iterative algorithms, introducing more parallelism can lead to a slowing down of convergence. There is often a tradeoff between the performance in terms of flops and the algorithmic efficiency.

Since the individual performances of the available cores were not progressing too much anymore, in the 2000s the increase of the computer peak performance was obtained by increasing the number of cores. Doing this, it was possible to obtain peak performances in excess of 1 Pflops. However, to increase the performance again in reach of the Exaflops, computer designers turned to another type of computing engine known as general-purpose graphics processing units (GPGPU).

A graphics processing unit (GPU) is a type of specialized processor for the fast creation and manipulation of images for output on a display device. They were first developed in the 1970s and 1980s. They were in particular used for video games. With time, they became more sophisticated and efficient, also being able to do floating point arithmetic. The GPUs use a form of SIMD parallelism which may somehow be assimilated to vectorization. They are now used in many supercomputers which become hybrid machines with traditional multicore processors and GPUs. Programming these devices is done with special programming interfaces like CUDA (Compute Unified Device Architecture) or OpenCL (Open Computing Language).

The CRAY *XK6* in 2011 used AMD processors and Nvidia GPUs. Each node had a 16-core AMD Opteron processor and an Nvidia Tesla X2090 GPGPU. It was followed by the *XK7* in 2012 (the "Titan" machine at ORNL). The *XC30* in 2014 had Intel Xeon processors and Nvidia GPUs or Xeon Phi manycore processors. It was followed by the *XC40* and the *XC50* in 2018. A CRAY *XC50* at the Swiss National Supercomputing Center obtained 21.230 Pflops on HPL with a linear system of order $3,743,232$.

Even though they are not GPUs, NEC's vector engines have more or less the same functionalities (and depending on the program, a better efficiency). The National Institute for Fusion Science supercomputer has 4,320 vector engines in 540 2U-8VE server units. This was the largest system NEC had ever supplied up to 2021. The peak performance of the system is 10.5 Pflops.

In 2018, the IBM "Summit" machine was installed at ORNL. It had 4,608 nodes with two IBM POWER9 processors and six Nvidia Tesla GPUs per node. The nodes were connected in a fat-tree topology using an InfiniBand interconnect. The peak performance was 200 Pflops and the machine reached 148.6 Pflops on HPL for solving a dense linear system of order $16,473,600$. With these results we see that the time for solving the system was $2.0057\ 10^4$ seconds, that is, almost 5 hours and 35 minutes. A somewhat similar machine named "Sierra" was installed at LLNL.

Before the 2000s, China did not have (at least officially) any machine that could be considered as a supercomputer. But then they developed programs to catch up. First, they used off-the-shelf western processors, but facing some import restrictions from the USA, they also developed their own processors.

Dawning Information Industry Company Limited (Sugon) developed computers starting in the 1990s. In 2001, the model *3000* had 280 processors. Later, the *4000A* used 2,560 AMD Opteron processors. In 2008, the model *5000A* had 7,680 AMD Opteron 4-core processors, and in 2011 the *6000* had 3,000 8-core Godson (Loongson) 3B processors designed in China.

In 2002, the Legend Group's *DeepComp 1800* computer had 526 Intel Xeon processors with a performance of 1.07 Tflops on HPL. Then, Legend Group changed its brand name to Lenovo. In 2003, Lenovo's *DeepComp 6800* with 1,024 processors obtained 4.18 Tflops on HPL.

The *Tianhe-1* computer built in 2009 by the National University of Defense Technology (NUDT) had 4,096 Intel Xeon E5540 processors and 1,024 Intel Xeon E5450 processors as well as 5,120 AMD GPUs. The peak performance was 1.2 Pflops and 0.563 Pflops on HPL. It was upgraded in 2010 to *Tianhe-1A* with 14,336 Xeon X5670 processors and 7,168 Nvidia Tesla M2050 GPUs. This machine had a proprietary interconnect and a peak performance of 4.7 Pflops.

The *Sunway BlueLight* in 2011 of the National Research Center of Parallel Computer Engineering & Technology (NRCPC) had 8,704 ShenWei SW1600 processors with a peak performance of 1.07 Pflops and 0.796 Pflops on HPL.

In 2013, the *Tianhe-2*, built by NUDT, had 16,000 nodes. Each node had two Intel Xeon processors and three Xeon Phi coprocessors. The peak performance was 54.9 Pflops. In 2014, US government agencies stopped Intel from selling microprocessors to Chinese supercomputer companies. Therefore, China accelerated the development of its own processors. The machine was upgraded to *Tianhe-2A* replacing the Xeon Phi with Chinese processors. The performance was 33.86 Pflops on HPL. This was improved later to 61.44 Pflops.

The NRCPC introduced the *Sunway TaihuLight* in 2018 with 40,960 Chinese-designed SW26010 manycore 64-bit RISC processors. It reached 93.01 Pflops on HPL.

There were also some tentatives to build parallel computers in Europe, even though some of the projects cannot be qualified as supercomputers.

INMOS was a British semiconductor company founded in 1978. The *Transputer* microprocessor in 1985 was mainly intended for parallel processing. INMOS successively designed 16-bit and 32-bit processors. In April 1989, INMOS was sold to SGS-Thomson, which is now STMicroelectronics, and the Transputer development was stopped.

Meiko was founded in 1985 by some INMOS engineers. A system based on 32-bit T414 Transputers was announced in 1986. After the Transputer was discontinued, the main product was the *CS-2*, using Sparc processors and optionally Fujitsu vector processors. This machine used a proprietary network. The largest *CS-2* system built was a 224-processor system. The company had financial problems in the mid 1990s and disappeared in a merger with an Italian company.

In 1985-1992, *Suprenum*, a project from research institutes, universities, and industrial companies in Germany, designed and built a 256-node MIMD machine announced in 1990. It used a 32-bit Motorola 68020 microprocessor, floating-point units Motorola 68882 to execute scalar floating-point arithmetic and vector floating-point unit with Weitek WTL2264/2265 chips. The peak performance was 5 Gflops. Five machines were built and the project was stopped in the mid-1990s.

Parsytec was founded in 1985. In the beginning of the 1990s the company produced the *Gigacluster* with Transputers and Motorola chips. Machines with 1,024 processors were sold in Germany.

The French company Bull was founded (under a different name) in 1931 to manufacture punched-card equipments using patents from a Norwegian engineer Fredrik Rosing Bull (1882-1925). After many changes and partnerships, Bull was nationalized in 1982 but re-privatized in 1994. In 2014, the control of Bull was taken by Atos, a French information technology company. Up to the 1980s Bull was mainly manufacturing business computers and small systems. In 1984, Bull started a project of supercomputer named *Isis*, sponsored by the French Ministry of Defense. Even though some parts of the machine were built, the project was late, facing some technical difficulties, and it was finally cancelled.

Bull entered the supercomputer arena in the 2000s. In 2006, Bull delivered a machine named *Tera-10* to the Commissariat à l'Énergie Atomique (CEA), the French atomic energy commission. The final machine had 9,968 cores with Intel microprocessors and obtained 52.8 Tflops on HPL. It was ranked fifth in the TOP500 list. It was followed in 2010 by *Tera-1000* with 17,296 Intel cores and 1.25 Pflops on HPL. The third machine *Tera-1000-2* with 561,408 cores had a 25 Pflops peak performance and 11.96 Pflops on HPL. The next step *Exa1* was initially scheduled for 2020.

In the first half of the 1990s there was another project of a French supercomputer named *ACRI*, but it failed in 1995 because of financial problems.

The European Union supported many projects and initiatives concerning supercomputing. For instance, the mission of PRACE (Partnership for Advanced Computing in Europe) was to enable high-impact scientific discovery and engineering research by allowing access to several supercomputers within Europe. More recently, the European Commission supported the EuroHPC project and the European Processor Initiative (EPI) which gathers 27 partners from

10 European countries to develop processors that are going to be used on future exascale machines.

Other countries had also supercomputer programs.

The Russian government had a renewed interest in supercomputers in the 2000s after a 15-year hiatus. In 2007-2010, Russian and Belarusian scientists jointly created the *SKIF* supercomputer series. Another supercomputer, the *AL-100*, was announced in 2008. Its peak performance was 14.3 Tflops using Intel Xeon processors. The *Lomonosov-1* computer had a performance of 902 Tflops on HPL with more than 12,000 Intel CPUs, and 2,130 Nvidia Tesla GPUs. In 2014, the 64,384 core *Lomonosov-2* supercomputer in Moscow State University obtained 2.5 Pflops on HPL. Another Russian supercomputer named *Christofari* has 99,600 cores and a 6.67-Pflops performance on HPL. There are also computers for military applications on which we do not have too many details. Russia is also developing homemade processors like, for instance, the Elbrus-8CV, an 8-core processor.

India started a supercomputer program at the end of the 1980s. The *Chipps* computer using INMOS Transputers was developed at the beginning of the 1990s at the Center for Development of Advanced Computing (C-DAC). Then, a series of computers named *PARAM* was designed in the 1990s and 2000s. In 1998, the performance of the *PARAM 10000* with 160 processors was around 100 Gflops. In 2003, the *PARAM Padma* had a peak performance of 1 Tflops. The *PARAM Yuva II* obtained 361 Tflops on HPL in 2013. In 2019, the C-DAC-built supercomputer, *PARAM Shivay*, was installed at the Indian Institute of Technology. It has a peak performance of 837 Tflops. The Indian government has the project of installing more supercomputers and also developing processors.

In 2020, there were several projects of supercomputers aimed at obtaining a peak performance higher than 1 Eflops in the USA, Japan, China, and Europe.

In December 2019, Fujitsu announced that it began shipping the supercomputer *Fugaku* jointly developed with RIKEN with more than 150,000 high-performance processing units A64FX (with vector extensions) connected by a high-speed network Tofu Interconnect. The aim was to achieve up to 100 times the application performance of the *K* computer with approximately only 3 times the power consumption. In November 2021, the Fugaku computer obtained 442 Pflops on the HPL benchmark. It reached 2 Exaflops on HPL-AI, a benchmark used to rank supercomputers for tasks used in artificial intelligence applications in half precision.

Fujitsu announced the *PRIME HPC FX1000* and *PRIME HPC FX700* models which utilize the technology of the Fugaku supercomputer. The shipment was scheduled for March 2020. It is hoped to have systems with a performance in excess of 1.3 Eflops.

A partnership was announced in November 2019 between Cray (now part of HPE) and Fujitsu. The new Cray system will use Fujitsu's A64FX.

There are three projects for Eflops machines in the USA. The ANL "Aurora" computer is expected for 2022. It is a partnership between Cray and Intel using the next generation of Intel Xeon Scalable processors. The ORNL "Frontier" system will use the Cray architecture with AMD Epyc processors and Radeon Instinct GPU technology. It was delivered in 2021 and is expected to be in operation in 2022. The "El Capitan" system at LLNL, scheduled for 2022-2023, is expected to exceed a 2-Eflop peak performance.

China also has at least three exascale projects with homemade processors and high-speed interconnects at the National University of Defense Technology (NUDT), the National Research Center of Parallel Computer (NRCPC), and Sugon company. They all had working prototypes in the beginning of 2020.

As we have seen above, in Europe, Atos-Bull also has a project of an Exascale machine.

It must be noticed that a big problem that supercomputer designers are facing to increase the performance is power consumption. The fastest supercomputers used very large electrical powers that are in the tens of Mwatts (MW). At the beginning of the 2000s the *Earth Simulator* in Japan needed 12 MW. The cost of electric power was $9.6 million/year. In 2019, the IBM Summit machine needed 10.096 MW which resulted in a performance of 14.719 Gflops/W. As another example, the Piz Daint CRAY *XC50* in Switzerland with a performance of 21.2 Pflops needed 2.324 MW, that is, 8.9 Gflops/W. Hence, to obtain Exascale machines with a "reasonable" power consumption, it is needed to use low power components.

Another big issue is data storage. Today's supercomputers are generating huge amounts of data that need to be stored for analysis by scientists, sometimes for several months or years. The computer centers also need efficient hardware and software to manage the data and to visualize the results.

To illustrate the increase of performance of supercomputers with time, Table 7.1 shows the computational speed obtained on the HPL benchmark as a function of time. We use the computer which had the first place in the TOP500 list. The column labeled "order" is the order of the dense matrix of the linear system to be solved.

It is interesting to note that in the June 2019 TOP500 HPL list, China had 219 entries, whereas the United States had 116. So, according to this criterion, there are more fast computers in China than in the USA. Of course, the populations of the two countries are not the same, even though not all of the inhabitants use supercomputers!

Figure 7.10 displays the performance (in Tflops) of the fastest computer in the HPL list. On the x-axis, one unit corresponds to six months. Note that we have a logarithmic scale on the y-axis.

The HPL benchmark was used for ranking supercomputers for historical reasons since, after the LINPACK software was written (see Chapter 8), Dongarra started collecting performances with the LINPACK linear solver on many computers [926].

However, in HPL, the manufacturers are allowed to choose the order of the dense linear system as they wish. Moreover, they can tune the BLAS routines that are used. Hence, this benchmark is not really representative of the performance that can be obtained with a real application. Not everybody would like to solve a dense system of order 16 10^6 everyday. This is why the HPCG benchmark [924] was developed. It solves a sparse linear system with the conjugate

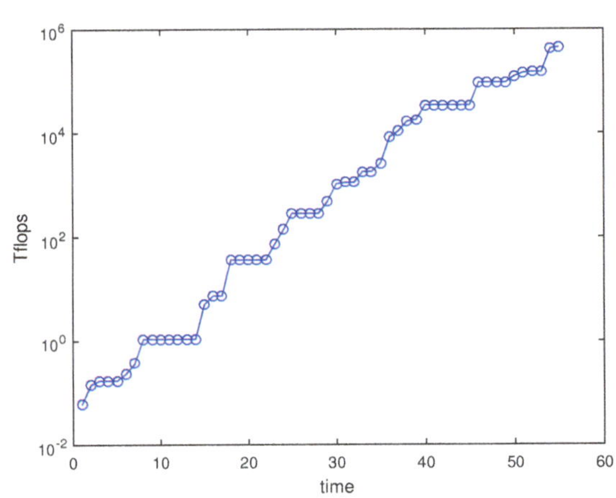

Figure 7.10. *Performance of the fastest computer in the TOP500 HPL list*

Table 7.1. *TOP500 HPL benchmark, first place*

Date	Country	System	nb cores	Tflops	order
November 1993	US	CM-5	1024	0.0597	52,224
June 1994	US	Intel XP/S 140 Paragon	3,680	0.1434	95,000 (?)
November 1994	Japan	Numerical Wind Tunnel	167 vect	0.170	200,848 (?)
June 1995	Japan	same	same	same	same
November 1995	Japan	same	same	same	same
June 1996	Japan	Hitachi SR2201	1,024	0.2324	155,520
November 1996	Japan	CP-PACS (Hitachi)	2,048	0.3682	103,680
June 1997	US	Intel ASCI Red	7,264	1.068	215,000
November 1997	US	same	same	same	same
June 1998	US	same	same	same	same
November 1998	US	same	same	same	same
June 1999	US	same	same	same	same
November 1999	US	same	same	same	same
June 2000	US	same	same	same	same
November 2000	US	ASCI White	8,192	4.9	?
June 2001	US	same	same	7.2	518,096
November 2001	US	same	same	same	same
June 2002	Japan	Earth Simulator	5,120	35.86	1,075,200
November 2002	Japan	same	same	same	same
June 2003	Japan	same	same	same	same
November 2003	Japan	same	same	same	same
June 2004	Japan	same	same	same	same
November 2004	US	BlueGene/L	32,768	70.7	933,887
June 2005	US	same	65,536	136.8	1,277,951
November 2005	US	same	131,072	280.6	1,769,471
June 2006	US	same	same	same	same
November 2006	US	same	same	same	same
June 2007	US	same	same	same	same
November 2007	US	BlueGene/L	212,992	478.2	2,456,063
June 2008	US	IBM Roadrunner	122,400	1,026	2,236,927
November 2008	US	same	129,600	1,105	2,329,599
June 2009	US	same	same	same	same
November 2009	US	Cray Jaguar	224,162	1,759	5,474,272
June 2010	US	same	same	same	same
November 2010	China	Tianhe-1A	186,368	2,566	3,600,000
June 2011	Japan	K computer	548,352	8,162	10,725,120
November 2011	Japan	same	705,024	10,510	11,870,208
June 2012	US	BlueGene/Q, Sequoia	1,572,864	16,324.8	12,681,215
November 2012	US	Cray Titan	560,640 + 261,632	17,590	?
June 2013	China	Tianhe-2A	3,120,000 + 2,736,000	33,862.7	9,960,000
November 2013	China	same	same	same	same
June 2014	China	same	same	same	same
November 2014	China	same	same	same	same
June 2015	China	same	same	same	same
November 2015	China	same	same	same	same
June 2016	China	Sunway TaihuLight	10,649,600	93,014.6	12,288,000
November 2016	China	same	same	same	same
June 2017	China	same	same	same	same
November 2017	China	same	same	same	same
June 2018	US	IBM Summit	2,282,544 + 2,090,880	122,300	13,989,888
November 2018	US	same	2,397,824 + 2,196,480	143,500	16,693,248
June 2019	US	same	2,414,592 + 2,211,840	148,600	16,473,600
November 2019	US	same	same	same	same
June 2020	Japan	Fugaku	7,299,072	415,530	21,288,960
November 2020	Japan	same	7,630,848	442,010	same
June 2021	Japan	same	same	same	same

gradient algorithm using a multigrid preconditioner with a symmetric Gauss-Seidel smoother; see Chapter 5. This is more representative of a real life problem, even though the authors of the benchmark did not use the most parallel version of the conjugate gradient algorithm. The algorithm they used is communication-bounded. We can see in Table 7.2 that this application obtained only around 2-3% of the HPL performance.

Table 7.2. *TOP500 HPCG benchmark, first place*

Date	Country	System	nb cores	Tflops	HPL	ratio
June 2014	China	Tianhe-2	3,120,000	580	33,860	0.017
November 2014	China	same	same	623	same	0.018
June 2015	China	same	same	580	same	0.017
November 2015	China	same	same	580	same	same
June 2016	China	same	same	580	same	same
November 2016	Japan	K computer	705,024	602.7	10,510	0.057
June 2017	Japan	same	same	same	same	same
November 2017	Japan	same	same	same	same	same
June 2018	US	IBM Summit	2,282,544	2,925.75	122,300	0.023
November 2018	US	same	2,397,824	2,925.75	143,500	0.020
June 2019	US	same	2,414,592	2,925.75	148,600	0.020
November 2019	US	same	same	same	same	same
June 2020	Japan	Fugaku	7,299,072	13400.00	415,530	0.032
November 2020	Japan	same	7,630,848	16004.50	442,010	0.036
June 2021	Japan	same	same	same	same	same

There is an interesting thing to observe in the development of computers from the early 1950s up to today: the space in m^2 needed to install the machines. The first electronic computers were huge machines since they used vacuum tubes. Then, with time and the use of transistors, the computers became smaller and smaller. This culminated in the *CRAY*s 2 and 3 which were very compact machines. But then with the increase in the number of processors, the size of the parallel computers became larger and larger again with a huge number of cabinets. For instance, the size of the ORNL *Titan* computer was roughly 400 m^2, and the Chinese *Tianhe-2* was even larger with 700 m^2. We may wonder how much floor space will be needed for the next generations of machines.

Another interesting fact is that, up to the end of the 20th century, computers could be identified by the names of their designers, but this is no longer the case.

During all the years since World War II, researchers in numerical linear algebra had to face many challenges like fixed point arithmetic, floating point arithmetic, different and sometimes weird arithmetics, rounding errors, portability issues, vectorization of algorithms, cache-aware algorithms, parallelism, communication-avoiding algorithms, and so on. But we must note that except for the definition of IEEE 754 standard for floating point arithmetic, numerical analysts did not have much influence on the architecture and design of the computers. Most of the time, they had to use tools designed by people who did not have much knowledge about the algorithms and codes which were going to run on their machines or who were targeting markets other than scientific computing. This is unfortunate, and it is probably going to be worse in the coming years.

New challenges will have to be faced. First of all, since communication is often a bottleneck in modern parallel computers, some people had the "brilliant" idea of reducing the precision of the computation using 16-bit arithmetic to decrease the amount of data that have to be sent and received. Of course, there are cases in which this can be done if we are careful; see, for instance, [558]. But this can be very dangerous in calculations related to safety issues if not done properly.

Another issue is that some people would like to get rid of the IEEE 754 floating point arithmetic, arguing that they can obtain the same or even better accuracy while storing a smaller amounts of bits. For instance, see the logarithmic arithmetic [1903] in the 1970s and tapered floating point arithmetic [2296] also in the 1970s, a format similar to floating point, but with variable-sized entries for the mantissa and exponent. The last avatar is called unums and posits (which are unums of type III); see [1487]. Whether this is good or bad is up for debate. For an analysis of the positive and negative aspects of posits, see [842]. What is sure is that if this is adopted by computer manufacturers, we will have to rethink the error analysis of all the numerical linear algebra algorithms if this is doable!

8
Software for numerical linear algebra

The library is intended to provide a uniform set of subroutines to solve the most common linear algebra problems and to run efficiently on a wide range of architectures. This library, which will be freely accessible via computer network, not only will ease code development, make codes more portable among machines of different architectures, and increase efficiency.
– *Prospectus for the development of LAPACK, 1987*

In this chapter we consider a brief history of the software that has been written for solving linear systems, computing eigenvalues and eigenvectors, computing some matrix factorizations, and solving linear least squares problems. Many experimental codes had been written to check new algorithms or to assess their performance, but we mainly consider the software packages (or libraries) with public access. We do not include the proprietary libraries that were proposed by computer manufacturers for their customers; see, for instance, the IBM Scientific Subroutine Package that was available on the *IBM 7094*. Computer user groups had organized software repositories in the 1960s. These collections contained software which was not always of high quality. Here we are interested in the efforts to write software combining quality and wide distribution.

8.1 ▪ Introduction

If we look in a dictionary, for instance, the Merriam-Webster, one of the definitions of "programming" is *the process of instructing or learning by means of an instructional program*, and a "program" is *a plan or system under which action may be taken toward a goal.* "Program" comes from a Latin word which comes itself from Greek, a mix of two words meaning "before" and "writing."

If you set your alarm clock to wake you up at a certain time, you are programming a device. More seriously and closer to our purposes, in 1801, Joseph-Marie Jacquard (1752-1834), a French weaver, invented a mechanical loom driven by a series of punched cards that simplified the process of manufacturing textiles with complicated patterns. This idea was not entirely new since some years before, in 1725, paper tapes were used by Basile Bouchon to control a loom. Mechanical organs driven by barrels or music boxes are other examples of early programming.

In the 19th century, Charles Babbage (1791-1871), an English mathematician, philosopher, and inventor, designed several computing machines; see Chapter 7 and Section 10.3. The first ones were difference engines for computing polynomial functions. A more sophisticated machine, the *Analytical Engine*, was supposed to be programmed with punched cards. Ada Lovelace (1815-1852), the daughter of the poet Lord Byron (1788-1824), who was corresponding with Babbage, coded an algorithm for the *Analytical Engine* to compute Bernoulli numbers; see Section 10.51. Unfortunately, because of financial problems, only small parts of this machine were really built.

Another example of early programming is the Turing machine, an abstract machine manipulating symbols written on an infinite tape according to some rules. It was published in 1936 by Alan Mathison Turing (1912-1954), a British mathematician and computer scientist.

In the early times of software production, algorithms and codes were published in journals and books. Examples are the *Communications of the ACM* and later on the *Transactions on Mathematical Software* and the *Handbook for Automatic Computing* to be described below. Nowadays, open access software is available on the Internet.

Before we can describe the development of software libraries for linear algebra problems, we have to make a digression with a brief history of high-level programming languages which are used to write the codes.

8.2 ▪ Programming languages

On the first digital computers, coding was done directly using the machine instruction set. This was obviously machine-dependent. Some coding was sometimes even done before the machines were actually built; see, for instance, Herman Heine Goldstine (1913-2004) and John von Neumann (1903-1957) [1372] in 1947-48 and the book by Maurice Vincent Wilkes (1913-2010), David John Wheeler (1927-2004), and Stanley Gill (1926-1975) [3236] in 1951.

Some of the early machines were coded by setting switches or wiring cables on boards. Then, came paper tapes and punched cards to input the code and the data. In some cases only fixed point additions were available and multiplications and divisions were done by software. This was also done on some machines for floating point arithmetic. Fixed point arithmetic meant that programmers had to be very careful about the scaling of variables in their codes.

Coding was also needed for input-output conversions and computation of transcendental functions. Slowly, direct coding with machine instructions evolved to more sophisticated systems which were the early assembly languages with symbols and relative addresses.

So, the first libraries were simply sets of punched cards and written in assembly language. With time passing, people started thinking that high-level programming languages were needed for writing algorithms, even though many people were thinking that this cannot be as efficient as writing in machine code or with an assembly language.

In 1945, Konrad Zuse (1910-1995), a German engineer, was refuged with his *Z4* computer in the small village of Hinterstein in the bavarian Alps. There, unable to work on his machine, he started designing an algorithmic language, which he called *Plankalkül* (PK), as a theoretical study without concerns about its future implementation on real machines. He even wrote programs for floating point arithmetic, for sorting and for playing chess. Unfortunately his manuscript was not published, except for small excerpts, before 1972; see [232, 1923]. Therefore, it did not have much influence on the development of programming languages.

Around the same time Goldstine and von Neumann were looking for a way to represent algorithms in a precise way. They proposed a pictorial representation with boxes joined by arrows and called it a *flow diagram*; see [1372]. This work was not published in journals, but largely distributed to people working with computers. So, contrary to Zuse's work, this had an impact on the future of computer programming.

According to [1923], one or even the first "high-level" language to be implemented was Short Code, suggested by John William Mauchly (1907-1980) in 1949 and coded by William F. Schmitt in 1950 for the *UNIVAC* computer. This system was an algebraic interpreter.

In 1950, Arthur Walter Burks (1915-2008) proposed an *Intermediate Programming Language*.

From 1949 until 1951, Heinz Rutishauser (1918-1970) at ETH in Zürich developed an algebraic language with simple formulas and loop control ("for" statement) for a hypothetical computer and flowcharts for two compilers for that language.

A compiler was written in 1951 by Corrado Böhm (1923-2017), an Italian citizen, for his Ph.D. thesis in 1954 at ETH. It was written in French and its title was *Calculatrices digitales du déchiffrage de formules logico-mathématiques par la machine même dans la conception du programme*. Eduard Stiefel (1909-1978) was a reviewer of that thesis. Böhm developed a hypothetical machine and a language to instruct that machine. For the first time, however, the translator itself was written in its own language. There were no subscripted variables and no loop control.

The first implemented compiler was written by Grace Murray Hopper (1906-1992), who also coined the term "compiler," referring to her A-0 system which functioned as a loader or linker, not the modern notion of a compiler. The first compiler in the modern sense was developed by Alick Edwards Glennie (1925-2003) in 1952 for the *Mark I* computer at the University of Manchester, UK. Glennie called his system "Autocode." His language was still machine-oriented. Glennie's papers had not been published at that time. Another "Autocode" for the *Mark I* was developed by Ralph Anthony Brooker (1925-2019) in 1955.

An important contribution to the development of algebraic languages was done by J. Halcombe Laning Jr. (1920-2012) and Neal Zierler in 1954 for the *Whirlwind* computer at MIT.

In 1955-56 a language called IT (Internal Translator) was developed by Alan Jay Perlis (1922-1990) and Joseph W. Smith (1921-2010), first at Purdue University and then at the Carnegie Institute of Technology for the *IBM 650*. One particularity was the translation to an intermediate language that was translated afterward to a code for a specific machine.

Now, let us consider the two high-level languages that were important for the development of numerical algebra libraries.

The FORTRAN (FORmula TRANslator) team led by John Warner Backus (1924-2007) at IBM introduced the first commercially available compiler in 1957. Their work started in 1954, after Backus was aware of the work of Laning and Zierler. Backus received the ACM Turing Award in 1977. He also developed the Backus Normal Form (also known as the Backus-Naur Form, or BNF), a formal notation able to describe any context-free programming language.

FORTRAN was originally intended to program the *IBM 704*. FORTRAN was widely and quickly adopted by programmers. It used a fixed input format for the punched cards of 80 columns. FORTRAN II in 1958 allowed user-written subroutines as well as COMMON statements (which may not have been the wisest decision!). FORTRAN IV was developed from 1961 to 1966. It added some data types and the logical IF statement. After a while, in the 1960s, other computer manufacturers than IBM started proposing FORTRAN compilers to their customers.

The American National Standards Institute (ANSI), which was then the American Standards Association, defined a standard for FORTRAN in 1966. A new enriched standard, FORTRAN 77, was approved in 1978. It introduced the ELSE statement, DO loops extensions, and the CHARACTER data type. FORTRAN 77 remained the standard for a long time. The next standard, Fortran 90 (note the move to lower case letters), was adopted only in 1991-92 with many new features like free-form source input, recursive procedures, modules, array features, dynamic memory allocation, and pointers. Fortran 95 was a minor revision adopted in 1997. The next standard, Fortran 2003, incorporated object-oriented programming support as well as floating point exception handling. Fortran 2008 introduced constructions for parallel computing. Up until 2021, the last standard was Fortran 2018. The standards are now defined by the working group WG5 of the International Organization for Standardization (ISO), which is a non-governmental, worldwide federation of some 160 national standards institutions. Fortran is the oldest high-level language still in use, but today's Fortran does not have much to do with the initial versions.

A very interesting language is Algol. Unfortunately, it did not have much success, particularly in the USA. But it was influential in further developments of programming languages. For the complicated history of Algol, see [2642] and, more recently, [839]. Note that Algol is the name of a star in the constellation of Perseus.

In October 1955, an international symposium on automatic computing was held in Darmstadt, Germany. Several speakers stressed the need for focusing attention on unification, that is, on one universal, machine-independent algorithmic language [2642]. A GAMM (Gesellschaft für Angewandte Mathematik und Mechanik, which means Society of Applied Mathematics and Mechanics) subcommittee for programming languages was set up after the meeting. A letter was sent to the president of the ACM suggesting that a joint conference with GAMM and ACM be held to define a common algorithmic language. ACM also formed a committee. In April 1958, Friedrich Ludwig Bauer (1924-2015) presented the GAMM proposal at a meeting of the ACM committee. A meeting in Zürich on May 27-June 2, 1958, gathered Bauer, Hermann Bottenbruch (1928-2019), Rutishauser, Klaus Samelson (1918-1980) for GAMM, and Backus, Charles Katz, Perlis, and Joseph Henry Wegstein (1922-1985) for ACM. They agreed on an abstract representation called a reference language. The outcome was a report describing IAL or Algol 58. In Europe, an Algol Bulletin was issued by Peter Naur (1928-2016) from Denmark.

As we have seen above, in 1956, Perlis and Smith had developed a primitive algebraic language named IT at the Carnegie Institute of Technology. At UNIVAC, Katz, in the group of Hopper, had been working on a language called MATH-MATIC.

In 1958, why was FORTRAN not considered as satisfying when Backus was participating in the Zürich meeting? The same question arose for some of the other languages. First, at that time, these languages were machine dependent, not satisfying the need for universality. Second, as we have seen above, FORTRAN was an IBM creation, and IBM became the dominant computer manufacturer. Some people on the committee did not want to become IBM-dependent. Therefore, no existing language was chosen as the candidate by the European (mostly German) and American committees.

An algorithm section was created in the *Communications of the ACM* journal, with algorithms described in Algol. Another meeting took place in Paris in November 1959. At this conference Backus presented "The syntax and semantics of the proposed international algebraic language of the Zürich ACM-GAMM Conference" which led to the BNF. People were selected and met again in Paris in January 1960. They were Backus, Bauer, J. Green, Katz, John McCarthy (1927-2011), Naur, Perlis, Rutishauser, Samelson, Bernard Vauquois (1929-1985), Wegstein, Adriaan van Wijngaarden (1916-1987), and Michael Woodger. A redesign of the language resulted in Algol 60. Backus' notation was used (with some modifications) by Naur for the Algol 60 report. Unfortunately, complex numbers were not considered.

Later some inconsistencies were discovered. Another meeting took place in Rome in April 1962. There was a transfer of responsibility for Algol to IFIP Working Group 2.1. Unfortunately, there were conflicts within the committee. Given the ambiguities in the 1960 report, not all implemented features could be guaranteed to work in the same way on every implementation. The lack of input-output procedures pushed implementors to add extensions to the language. This led to a revised Algol report. Restricted Algol dialects started to appear and the IFIP Working group defined subsets of Algol.

In 1964, WG 2.1 started a new project, developing a new version of Algol called Algol X. A lot of proposals for modifications or extensions were made here and there. In 1965, three proposals for Algol X were done by Niklaus Wirth and Charles Anthony Richard Hoare, Gerhard Seegmüller, and van Wijngaarden. A subcommittee with the four authors of the proposals was formed. But unfortunately, again they were not able to agree on just one final proposal. Wirth and

Hoare published their proposal in CACM. Wirth started implementing the language (known as Algol W) on an *IBM 360* at Stanford. Van Wijngaarden set up another proposal in October 1966. A final version was not ready before 1968. But this proposal created a lot of negative reactions. This was known as Algol 68. Half of the members of the committee produced a minority report stating that the report on Algol 68 was a failure. That was more or less the end of the Algol effort, even though there were still some developments later on. Compilers for Algol W under the Linux operating system still exist; see www.algol60.org.

In the 1960s, Algol 60 was, especially in the academic world, an influential language. After 1960, Algol was used to describe algorithms in several journals: *Communications of the ACM*, the *Computer Journal*, the *Computer Bulletin*, *Numerische Mathematik*, *BIT*, *Algorytmy*. Algol 60 was able to gain more support in Europe than in the USA since Algol was not so popular in the industry, and IBM, the most influential manufacturer at that time, was pushing FORTRAN.

Regarding other languages in use today for developing numerical linear algebra software, C was originally developed at Bell Labs in the beginning of the 1970s by Dennis Ritchie (1941-2011) and Kenneth Thompson. A book about this language was written in 1978 by Brian Wilson Kernighan and Ritchie. The language C++, with object-oriented features, was started in 1979 by Bjarne Stroustrup from Denmark. He published a book in 1985.

8.3 ▪ The Handbook for Automatic Computing

Starting in the mid-1960s, James Hardy Wilkinson (1919-1986) and some colleagues published a series of papers about algorithms for linear algebra problems in the journal *Numerische Mathematik*. This series, under the general editorship of Bauer, was called the *Linear Algebra series of the Handbook for Automatic Computation*. Each paper contains a short description of the algorithm and its properties, the code in Algol 60, and some (small) examples with numerical results that are carefully analyzed. The algorithms are for solving linear systems and computing eigenvalues and/or eigenvectors. The matrices are dense, that is, all the n^2 entries are stored in a two-dimensional array or banded. Since complex numbers were not allowed by Algol 60, the complex matrices were stored as two matrices for the real and imaginary parts.

Let us list these papers since this gives a good idea of the state of numerical linear algebra algorithms at that time:

- *The conjugate gradient method* by Theo Ginsburg (1926-1993) [1355] in 1963,
- *Linear least squares solutions by Householder transformations* by Peter Arthur Businger and Gene Howard Golub (1932-2007) [513] in 1965,
- *Elimination with weighted row combinations for solving linear equations and least squares problems* by Bauer [221] in 1965,
- *Symmetric decomposition of positive definite band matrices* by Roger S. Martin and Wilkinson [2155] in 1965,
- *Symmetric decomposition of a positive definite matrix* by Martin, G. Peters and Wilkinson [2150] in 1965,
- *Iterative refinement of the solution of a positive definite system of equations* by Martin, Peters, and Wilkinson [2151] in 1966,
- *Solution of real and complex systems of linear equations* by Hilary J. Bowdler, Martin, Peters, and Wilkinson [387] in 1966,

- *The Jacobi method for real symmetric matrices* by Rutishauser [2641] in 1966,
- *Solution of symmetric and unsymmetric band equations and the calculations of eigenvectors of band matrices* by Martin and Wilkinson [2156] in 1967,
- *Calculation of the eigenvalues of a symmetric tridiagonal matrix by the method of bisection* by W. Barth, Martin, and Wilkinson [210] in 1967,
- *Solution to the eigenproblem by a norm reducing Jacobi type method* by Patricia James Wells Eberlein (1923-1998) and John Boothroyd [1046] in 1968,
- *Reduction of the symmetric eigenproblem $Ax = \lambda Bx$ and related problems to standard form* by Martin and Wilkinson [2159] in 1968,
- *Householder's tridiagonalization of a symmetric matrix* by Martin, Christian Reinsch, and Wilkinson [2153] in 1968,
- *Tridiagonalization of a symmetric band matrix* by Hans Rudolf Schwarz [2724] in 1968,
- *Rational QR transformation with Newton shift for symmetric tridiagonal matrices* by Reinsch and Bauer [2557] in 1968; see C. Reinsch, Mathematics of Computation, v 25, n 115 (1971), and R.A. Sack, Numerische Mathematik, v 18 (1971), for improvements,
- *The QR and QL algorithms for symmetric matrices* by Bowdler, Martin, Reinsch, and Wilkinson [388] in 1968,
- *Similarity reduction of a general matrix to Hessenberg form* by Martin and Wilkinson [2160] in 1968,
- *The modified LR algorithm for complex Hessenberg matrices* by Martin and Wilkinson [2158] in 1968,
- *The implicit QL algorithm* by Martin and Wilkinson [2157] in 1968 (plus Augustin Dubrulle in the book),
- *Balancing a matrix for calculation of eigenvalues and eigenvectors* by Beresford Neill Parlett and Reinsch [2458] in 1969,
- *The QR algorithm for real Hessenberg matrices* by Martin, Peters, and Wilkinson [2152] in 1970,
- *Solution to the complex eigenproblem by a norm reducing Jacobi type method* by Eberlein [1044] in 1970,
- *Singular value decomposition and least squares solutions* by Golub and Reinsch [1391] in 1970,
- *The QR algorithm for band symmetric matrices* by Martin, Reinsch, and Wilkinson [2154] in 1970,
- *Eigenvectors of real and complex matrices* by LR and QR triangularizations by Peters and Wilkinson [2485] in 1970,
- *Simultaneous iteration method for symmetric matrices* by Rutishauser [2644] in 1970.

These papers, sometimes with slight modifications, were collected in a book titled *Handbook for Automatic Computation, Volume II, Linear Algebra* [3256], edited by Wilkinson and Reinsch and published by Springer in 1971 with three additional papers:

- *A realization of the simplex method based on triangular decompositions* by Richard Harold Bartels, Joseph Stoer, and Christoph Zenger,
- *Inversion of positive definite matrices by the Gauss-Jordan method* by Bauer and Reinsch,
- *The calculation of specified eigenvectors by inverse iteration* by Peters and Wilkinson.

The paper *Implicit QL algorithm* in the book is a merger of a half-page paper by Augustin Dubrulle (1935-2016) and the earlier contributions by Martin and Wilkinson.

The numerical results were obtained on different computers, mainly those available at that time in the institutions of the authors, generally using single precision. These machines were the *KDF9* at the National Physical Laboratory (NPL) in the UK, the *Telefunken TR4* in Munich, the *Elliott 503* at the University of Tasmania in Australia, the *Burroughs B5000* at Stanford University, the *UNIVAC 1108* at Case Western Reserve University in Cleveland, the *CDC 1604* in Zürich, the *CDC 6400* at the University of California in Berkeley, and the *CDC 6600* at the University of Texas at Austin.

One can imagine the difficulties that were faced, since all these machines have different data formats and different floating point arithmetics.

The *KDF9* was a British computer manufactured by English Electric with floating point numbers having a 39-binary digits mantissa. The maximum configuration incorporated 32K 48-bit words of storage. It was a multi-programming machine with time-sharing. The *KDF9* was one of the first computers to accept Algol 60.

The *Telefunken TR4* was, at that time, the largest digital computer developed in Europe. It was installed at the computing center of the Bavarian Academy of Sciences in Munich, Germany. Its operating frequency was 2 MHz, and it had a main storage of about 0.5 MByte and an additional 550 Mbytes of disk storage.

The *Elliott 503* was a transistorized computer introduced by Elliott Brothers (UK) in 1963. The basic configuration had 8192 words of 39 bits each for the main memory. The system operated at a system clock speed of 6.7 Mhz.

The *Burroughs B5000* had computer words of 48 bits with a 39-bit mantissa. This machine was designed to run Algol 60 programs.

The *UNIVAC 1108* was a binary machine with a data format of 36 bits with a sign bit, 7-bit exponent, 27-bit mantissa in single precision and 72 bits: sign bit, 10-bit exponent, 60-bit mantissa in double precision.

The Control Data Corporation *CDC 1604* was a 48-bit computer designed by Seymour Roger Cray (1925-1996). This machine had 32K 48-bit words of magnetic core memory and floating point numbers had a 36-bit mantissa.

The *CDC 6400* and *6600* were also designed by Cray. They had 60-bit words. The *6600* was the fastest machine of its time with multiple functional units that can operate in parallel.

The Springer book, known as the *Handbook*, was really influential and cited many times. It set up some good rules for program design and testing, programming practices, and documentation standards.

Around the same time, papers in a special function series for the *Handbook of Automatic Computation* were also published in *Numerische Mathematik.*

In the 1960s and 1970s, there were also some efforts for developing software in other countries. For instance, in France, two books were published by CNRS (Centre National de la Recherche Scientifique, that is, National Center for Scientific Research) with Algol procedures

Procédures Algol en Analyse Numérique. This was written by teams of numerical analysis researchers from several French universities, Besançon, Clermont-Ferrand, Grenoble, Lille, Nancy, Paris Institut Blaise Pascal, and Toulouse, with prefaces by Jean Kuntzmann (1912-1992) and Noël Gastinel (1925-1984).

Algol 60 procedures for linear systems were also published in The Netherlands by Theodorus Jozef Dekker (1927-2021) and Walter Hoffmann (1944-2013) in 1968-1975; see [857, 858, 859].

The book *Computer Solution of Linear Algebraic Systems* by George Elmer Forsythe (1917-1972) and Cleve Barry Moler [1196] published in 1967 contains programs written in Algol, FORTRAN, and PL/1; see also [1864].

Meetings that were influential for the development of linear algebra software were organized by John Rischard Rice at the end of the 1960s [2565, 2747]. Other meetings were organized in the 1970s by Donald James Rose (1944-2015) and Ralf Arthur Willoughby (1924-2001) [2594], James Raymond Bunch and Rose [505], Wayne Cowell (1926-2021) [754], Iain Spencer Duff, and Gilbert Wright Stewart [1007]. Other meetings about numerical software were organized in the UK: Loughborough [1113] in 1973 and Sussex [1799] in 1977.

The status and the future of mathematical software was discussed by Ronald F. Boisvert [354] in 2000.

8.4 ▪ BLAS

In this section we consider the *Basic Linear Algebra Subprograms*, better known as BLAS. This is a set of subroutines implementing low-level operations that frequently occur in linear algebra algorithms.

Level 1 BLAS

The first BLAS routines originated at the Jet Propulsion Laboratory in California. It is essentially the work of Frederick (Fred) Thomas Krogh, Charles Lawrence (Chuck) Lawson (1931-2015), and Richard Joseph Hanson (1938-2017).

The idea was to identify a small set of useful low-level operations and to provide routines for them. These routines can be implemented in assembly language on different computers to achieve efficiency and portability. Originally, this was the idea of Krogh [1952] in 1972 who wrote some test programs and ran some tests on the *UNIVAC 1108* to see if it was faster to call an assembly language routine than to have the routine written in Fortran. Then Krogh, Lawson, and Hanson formulated the slightly bigger package idea that was described in a JPL report [1588] in 1973.

This project was discussed in the annual meetings of the IFIP Working Group 2.5. The ACM *SIGNUM Newsletter* was also used to communicate this proposal widely and to report its subsequent evolution. A meeting was held at the Math Software II Conference at Purdue University in May 1974 to discuss and modify the proposal. Another meeting was the National Computer Conference in Anaheim in May 1975.

David Ronald Kincaid at the University of Texas at Austin became involved in the project. He was interested in seeing a good version of the BLAS on the Control Data computer there. He programmed an assembly version for the *CDC 6600*. He also suggested other things to do with the package and was included as a co-author.

Hanson had moved to the Sandia Laboratories in Albuquerque. The final version was published as a Sandia report [2005] in 1977. A paper was published in the *Transactions on Mathematical Software* (TOMS) [2006] in 1979.

Assembly versions of some BLAS subprograms had been written by other people for the *UNIVAC 1108*, *IBM 360/67*, and *CDC 6600*. Accuracy testing of all 38 subprograms was done on the *Honeywell 6000*, *PDP10*, *CDC 7600*, *IBM 370/145*, and *Burroughs 6700*.

Meanwhile, the authors of the Level 1 BLAS were in contact with the people writing LINPACK who decided to use the BLAS; see Section 8.6. Jack J. Dongarra had some involvement with the BLAS before it finally came out. He wrote the Fortran software that appears in [2006] and added loop unrolling into the code to improve performance; see [912, 925]. He worked with 40 separate computing facilities that ran timing tests for him.

The routines that were provided in this first version of BLAS were implementing dot product (sdot), vector plus a scalar times a vector (saxpy), Givens transformation (srotg, srot), modified Givens transformation (srotmg, srotm), copy (scopy), swap (sswap), Euclidean norm (snrm2), sum of magnitudes (sasum), multiplying a scalar times a vector (sscal), and locating an element of largest magnitude in a vector (samax). These names were for the single precision version. The routines had an increment parameter, so the vector components were not necessarily contiguous. When looking at the test results, the gains using assembly language depended on the machine, but they were not spectacular.

Level 2 BLAS

At the end of the 1970s, vector supercomputers appeared. To obtain good performance on these machines it was necessary to optimize at the level of matrix-vector operations. Hence, it was natural to try to extend the BLAS to include matrix-vector operations occurring frequently in linear algebra algorithms. This resulted in the Level 2 BLAS, also known as BLAS2. From then on, the original BLAS was referred to as Level 1 BLAS or BLAS1.

A first proposal was issued in 1984 and there were discussions at several meetings in 1984-85. A proposal appeared in the ACM *SIGNUM Newsletter* 20 in 1985 and in an Argonne National Laboratory report in 1986. The paper describing Level 2 BLAS by Dongarra, Jeremy Du Croz, Sven Hammarling, and Hanson was published in TOMS [917, 916] in 1988.

The operations which are in BLAS2 are matrix-vector products like $y = \alpha Bx + \beta y$ with $B = A, A^T, \overline{A}^T$, the product of an upper or lower triangular matrix with a vector, rank-one and rank-two updates of a matrix, that is, $A + \alpha xy^T$, $H + \alpha x\overline{y}^T + \overline{\alpha}y\overline{x}^T$, where H is Hermitian, and the solution of a linear system with a triangular matrix. The matrices are dense, banded or packed (for symmetric and triangular matrices), symmetric or not. The naming conventions are similar to those of LINPACK. For instance, $y = \alpha Bx + \beta y$ for a general matrix is named SGEMV or DGEMV depending on the precision; the fact that it uses the matrix or its transpose or its conjugate transpose is given by the value of an argument of the subroutine. There is no report of tests or performance results in the TOMS paper.

Level 3 BLAS

As time goes on, the architecture of computers is changing. It turned out that BLAS2 was not well suited for computers with a hierarchy of memory, like several cache levels, where locality of the data is important to obtain a good performance. This can be obtained by using matrix-matrix operations with matrices partitioned into blocks. This resulted in the Level 3 BLAS or BLAS3.

A first proposal was issued in 1987. The paper describing BLAS3 by Dongarra, Du Croz, Duff, and Hammarling was published in TOMS [915] in 1990.

The operations which are in BLAS3 are matrix-matrix products like $C = \alpha AB + \beta C$, rank-k and rank-2k updates, multiplication of a matrix by a triangular matrix, and solutions of triangular systems with multiple right-hand sides. The same naming conventions as in BLAS2 were adopted. For instance, $C = \alpha AB + \beta C$ is SGEMM or DGEMM depending on the precision. Again, only dense matrices were targeted.

We observe that to make an effective use of BLAS2 and BLAS3, it was necessary to make major changes to the usual algorithms for the solution of linear equations and eigenvalue algorithms, expressing the algorithms in terms of vectors or partitioning the matrices into blocks. There are no matrix factorization algorithms in BLAS3 because this was supposed to be done at a higher level. For instance, the BLAS3 was used in LAPACK; see Section 8.7.

Other extensions to the BLAS

The introduction of the three BLAS called the attention of many researchers. This led to many proposals of extensions and/or implementations, complying more or less to what had been defined. Of course, this was going away from the original idea, which was to have a *small* set of routines that can be implemented efficiently. There was a need for control of the process, and the BLAS Technical Forum was started in 1995; see www.netlib.org/blas/blast-forum. Many more people were involved, and it was probably not so easy to reach a consensus. The BLAST report appeared in August 2001. It defined extensions: Sparse BLAS, Extended and mixed precision BLAS, interval BLAS; see also [347].

The extensions to the dense BLAS1-2-3 were vector norms, maximum of a vector, Householder transform, combined axpy and dot product, sort of a vector, permutation of a vector, matrix norms, diagonal scaling of a matrix, matrix copy, matrix transpose, and permutation of a matrix. There were column-based and row-based storage, packed storage for symmetric and triangular matrices, and also band storage.

The original BLAS2-3 considered only dense matrices. Sparse BLAS has been discussed for many years. An early proposal was done by David Scott Dodson, Roger G. Grimes, and John Gregg Lewis (1945-2019) from Boeing [905] in 1991 considering level 1 operations. A result of the BLAST was a paper by Duff, Michael Allen Heroux, and Roldan Pozo [995] in 2002 for unstructured sparse matrices. The problem with sparse matrices is that the structure of the matrix is known only when it has been constructed. The Sparse BLAS standard allows complete freedom for the implementor to select the data access pattern. The interface addresses these problems via a generic handle-based representation (pointer to an already created sparse matrix). The storage scheme of the matrix is not specified in the argument list of the routines. This differs with the early proposals done in 1991 [905], 1992 [996], 1994, 1996, 1997 [997], 2000, and 2001. Three levels are defined as for the BLAS. There is support for dense and sparse vectors in packed form. Note that in matrix-matrix operations, only one matrix can be sparse. There are routines to create a handle, to input the values of the entries, to state the properties of the matrix, and to close the construction and destruct the matrix.

Some of the routines of Sparse BLAS are:

- Level 1: USDOT sparse dot product, USAXPY sparse vector update, USGA sparse gather, USGZ sparse gather and zero, USSC sparse scatter,
- Level 2: USMV matrix-vector multiply, USSV matrix-vector triangular solve,
- Level 3: USMM matrix-matrix multiply, USSM matrix-matrix triangular solve.

An implementation in Fortran 95 was done by Duff and Christof Vömel [1011] in 2002.

ATLAS was a research effort started by R. Clint Whaley and Dongarra at the Innovative Computer Laboratory of the University of Tennessee around 2000. It was focusing on applying empirical techniques in order to provide portable performance and to automatically tune the BLAS to a given machine; see [718].

GotoBLAS is an open source implementation of the BLAS with many hand-tuned optimizations for specific processors. GotoBLAS was developed by Kazushige Goto at the Texas Advanced Computing Center at the beginning of the 2000s. OpenBLAS is an open source implementation developed in China and derived from GotoBLAS2 with many optimizations for specific processor types. Optimized implementations were also provided in the Intel MKL library.

We will consider parallel versions of BLAS in Section 8.10.

8.5 ▪ EISPACK

EISPACK was started at the beginning of the 1970s at the initiative of James Christopher Thomas Pool, who was the Assistant Director of the Applied Mathematics Division at the Argonne National Laboratory (ANL), west of Chicago. It was a part of a project called NATS, for *National Activity to Test Software*. Later on, NATS started to unofficially mean NSF, Argonne, Texas, and Stanford because of the institutions and people that were involved in that project. This project was funded in 1971 by the Department of Energy (DOE, maybe under another name at that time) and the National Science Foundation (NSF). The official goal was not to produce software but to study procedures for testing software.

EISPACK was supposed to implement algorithms for computing eigenvalues and eigenvectors. After a start that was not judged satisfactory, it was decided to carefully translate to FORTRAN the Algol procedures of the *Handbook for Automatic Computation* (see Section 8.3), preserving the structure and the numerical properties.

There was a summer visitor program in ANL, with people like Wilkinson visiting for several weeks. He provided advice for the development of EISPACK. The software was mainly written by ANL employees. The people involved were Brian T. Smith, James M. Boyle, Dongarra, and Burton S. Garbow. The participants from outside the ANL were Yasu Ikebe (University of Texas), Virginia C. Klema (ANL and then Northwestern University), and Moler (Stanford); see [2261]. According to Moler [1539], Ikebe was not much involved in the work for the final product. The unofficial team leader was Smith.

The FORTRAN codes were sent to about 20 test sites, universities, industrial companies, and government laboratories with different computers and operating systems; see [929]. The package was released for public use in May 1972. This release contained some of the algorithms in the *Handbook*, that is, 38 routines. The documentation was ready in 1974.

Then, the package was extended, including the SVD algorithm of Golub and Kahan and the newly developed QZ algorithm of Moler and G.W. Stewart [2266] for solving the generalized eigenvalue problem, $Ax = \lambda Bx$ without inverting one of the matrices. The second release of EISPACK in 1976 contained 70 routines. A third version was released in 1983 with minor corrections and 78 routines. It was estimated that the total cost of EISPACK was \$1 million; see [929].

Two books describing EISPACK and its extensions were published by Springer [1285, 2803, 2802]. Execution times on many computers are given in [2802]. Dongarra was not an author of the first edition because he was just arriving at ANL at that time.

For more details on the development of the package, see [1539, 1540, 1541].

8.6 ▪ LINPACK

LINPACK is a package of mathematical software for solving problems in linear algebra, mainly dense and banded linear systems and factorizations.

After EISPACK was near completion, there were discussions at ANL in 1974 about writing another package for solving linear systems with direct methods. The participants were ANL people, visiting scientists, and various consultants. At first, the reactions were negative, but the idea for this package was pushed by Pool, and a year later the opinions had evolved. Again, it was found necessary to involve the summer visitors to ANL. Cowell and G.W. Stewart wrote a prospectus for a LINPACK project in 1975. Stewart wrote a grant proposal that was submitted to the National Science Foundation and the Department of Energy. NSF agreed to fund the project for three years, starting January 1976.

The main developers of LINPACK were Bunch (University of California at San Diego), Moler (University of New Mexico; see [2263]), and Stewart (University of Maryland) with the help and support of Dongarra (ANL). Bunch worked on the routines related to symmetric indefinite systems, Moler and Dongarra on factorization and linear solvers, and Stewart was responsible for the QR decomposition, least squares, and singular value decomposition. Dongarra was also developing a framework for the testing of the software. Wilkinson came to ANL every summer and was sort of an advisor for the project.

The four people developed the software at their own institutions and met in the summer to coordinate their work. There were animated discussions in the summer of 1976 about the naming conventions and the possibility of using or not using the BLAS (see Section 8.4). Finally, it was decided to use the BLAS, and there were discussions and interactions with the ANL people implementing it. The developers used Boyle's TAMPR system to reformat the code, trying to reveal its structure. Using this tool only one version for complex matrices was written, and TAMPR automatically produced the versions for single- and double-precision real matrices. It also cleaned up the codes.

In 1977, the software, together with test programs, was sent to 26 test sites. Using the feedback from the test sites in mid-1978, some changes were incorporated, and by the end of 1978 the codes were ready to be released through the National Energy Software Center at ANL and also through IMSL (see Section 8.9). A first version of the Users' Guide [913] was ready in 1977, and the final version was published by SIAM [914] in 1979. For a short presentation of LINPACK, see the papers [928, 931] by Dongarra and Stewart.

LINPACK was a software project that used well-known algorithms with a solid theoretical background. However, a few new things were introduced, like solvers for indefinite symmetric systems [502, 504] (but these algorithms were improved later on), algorithms to update and downdate the Cholesky factorization, and condition number estimates [714].

In the *Bulletin of the American Mathematical Society*, vol. 2, no. 3, May 1980, Garrett Birkhoff (1911-1996) wrote a review of the LINPACK Users' Guide that is amusing to read:

> *Written for computing professionals, it has the same sparkle as a manual explaining to expert repair mechanics the workings of an automobile engine. Yet it is an important and meaty document, giving an authoritative picture of current mathematical software technology. Its programs have been exhaustively tested at a number of computing centers on a variety of machines, and can be certified as optimal (in the present state of the art) for solving many problems of linear algebra. Many millions of dollars of computing and programming time will be saved by using them!*

During the development of LINPACK, the test sites were asked not only to provide their feedback, but also to measure the time required for two subroutines in the package, DGEFA and DGESL, to solve a dense linear system of order 100. Appendix B of the LINPACK Users' Guide shows the timing results collected by Dongarra. The order 100 was chosen because it was sufficient to obtain meaningful timing results and fit within the primary memory of the computers of that time. These measurements were the beginning of LINPACK as a benchmark. Dongarra continued to collect performance data on many more computers and published the results from time to time [926]. The original LINPACK routines had to be used, only the BLAS can be optimized, and the results had to be checked.

Then, larger linear systems were allowed and parallelism came into play. This is now called High Performance Linpack (HPL). But the same algorithms are still used to rank computers; see www.top500.org. Of course, solving dense linear systems is not really representative of the performances that can be reached on real scientific computing problems. Recently, some other algorithms, like the conjugate gradient algorithm for solving sparse linear systems, had been proposed as test programs [924].

Although this is not directly comparable, in 1977, on the *CRAY 1* computer at the National Center for Atmospheric Research, a performance of 14 Mflops (a Mflops is 10^6 floating point operations per second) for a dense linear system of order $n = 100$ was obtained. On the November 2019 TOP500 list, the fastest machine, which was the IBM Summit computer at the Oak Ridge National Laboratory (with more than two million cores), obtained a performance of 148 Pflops (10^{15} flops) for a linear system of order $n = 16,473,600$.

8.7 ▪ LAPACK

As we have said above, computer architecture changed, and there were also some improvements to known algorithms and even new algorithms that appeared, mainly for eigenvalue problems. Hence, at the end of the 1980s, EISPACK and LINPACK were becoming more or less obsolete. It became obvious that a replacement was needed, particularly for using BLAS2 and BLAS3. It was to be done by restructuring some of the algorithms to a block form in terms of matrix-vector and matrix-matrix products. If the BLAS routines are optimized, this could give a much better performance on computers with several levels of memory. It was also tempting to be able to use parallelism on shared memory computers.

Originally in Fortran 77, LAPACK is now written in Fortran 90 and contains routines for solving systems of linear equations, least-squares solutions of linear systems of equations, eigenvalue problems, and singular value problems, as well as factorizations of matrices. As in EISPACK and LINPACK, sparse matrices are not supported in LAPACK.

A *Prospectus for the development of a linear algebra library* [870] was written in September 1987 by James Weldon Demmel (Courant Institute at that time), Dongarra (ANL at that time), Du Croz (NAG), Anne Greenbaum (Courant Institute at that time), Hammarling (NAG), and Danny Chris Sorensen (ANL at that time). The project was initially funded by grants from the National Science Foundation and the Department of Energy. The principal investigators were Demmel and Dongarra.

In 1989, Dongarra moved from ANL to the University of Tennessee at Knoxville (UTK) and Oak Ridge National Laboratory (ORNL). In 1990, Demmel moved to the University of California at Berkeley. In 1998, Greenbaum moved to the University of Washington. At the beginning of the 1990s Sorensen moved to Rice University.

Contrary to EISPACK and LINPACK, the development of LAPACK involved many people. The authors of the first edition of the Users' Guide published by SIAM [56] were Edward Anderson (UTK), Zhaojun Bai (University of California at Davis), Christian Bischof (ANL and Technical University Aachen, Germany, now in Darmstadt), Susan L. Blackford (UTK), Demmel, Dongarra and Du Croz, Greenbaum, Hammarling, Alan McKenney (UTK), and Sorensen.

Over the years many more people were involved in the development of LAPACK. A long list of collaborators is published on the LAPACK website www.netlib.org/lapack.

The first version 1.0 was released in February 1992, version 2.0 in September 1994, and version 3.0 in June 1999. The latest version (up to 2021) is 3.9.0 released in November 2019. A Fortran 95 interface to LAPACK was published in 2001 [933].

The algorithms and codes continued to be improved. As an example, some of the improvements that have been introduced since version 3.0 are as follows:

- A Hessenberg QR algorithm with a small bulge multi-shift QR algorithm together with aggressive early deflation,
- Improvements of the Hessenberg reduction subroutines,
- New multiple relatively robust representations (MRRR) for eigenvalue algorithms,
- New fast and accurate Jacobi SVD algorithm,
- Recursive QR factorization,
- Recursive Cholesky factorization.

Hence, LAPACK still contains the best known algorithms for the type of matrices that are considered.

8.8 ▪ Other libraries

As we said in the introduction to this chapter, we mainly consider the libraries that were of wide distribution. But it is worth mentioning BCSLIB, which was started in the mid-1960s at Boeing, SLATEC developed by US government laboratories in the 1970s, and MATH77 developed at the Jet Propulsion Laboratory by Lawson, Krogh, and W. Van Snyder.

Software associated with papers published in ACM journals and in the *Transactions on Mathematical Software* (TOMS) were collected. They are still available through Netlib.[53] One can also see the NUMERALGO library from papers published in *Numerical Algorithms*.[54]

The software libraries that we described in the previous sections were targeting dense and banded matrices. In the 1970s and 1980s, the interest of researchers turned to sparse matrices because they appear frequently in problems arising from the discretization of partial differential equations and other areas of scientific computing like optimization. Storage schemes for storing only the nonzero entries of the matrix were devised, and research was done to characterize the fill-in that occurs in the Cholesky or L, U factors, that is, matrix entries which become nonzero during the elimination process. Heuristic orderings of the unknowns, like the minimum degree algorithm [1332] or the Cuthill-McKee algorithm [797], were devised to try to minimize the fill-in; see Chapter 2. Many codes were written at that time to experiment with these techniques, but they did not always produce libraries with public access.

For a review of the situation for sparse matrices almost 40 years ago, see the paper by Duff [979] in 1982 and the software catalog [2384]. Considering the complexity of writing software for sparse direct methods, most people who have sparse matrix problems to solve would find it impossible to write their own sparse solvers. This is why libraries dealing with sparse matrices became so important. Let us briefly describe what, in our opinion, were the main contributions. A survey of direct methods for sparse linear systems has been done by Timothy Alden Davis, Sivasankaran Rajamanickam, and Wissam M. Sid-Lakhdar [832] in 2016.

A package named YSMP (Yale Sparse Matrix Package) was written in the Department of Mathematics at Yale University by Stanley Charles Eisenstat (1944-2020), M.C. Gursky, Martin Harvey Schultz, and Andrew Harry Sherman. A report was issued in 1975 and a paper [1069] was published in 1982. They considered symmetric positive definite (SPD) systems, used a

[53] www.netlib.org

[54] https://www.netlib.org/numeralgo/

minimum degree ordering and a compressed sparse storage by rows (CSR). The listing of the routines was given in the Appendix of the report. A users' guide by Sherman appeared in 1975. Extensions to nonsymmetric matrices were described in a report [1068] in 1977. A new package [1067] was released in 1984. Software for limited core storage was described [1073] in 1979.

In 1978, Sherman introduced NSPIV [2754], a code for solving linear systems with a row-oriented Gaussian elimination with partial pivoting and column interchanges. The name was later changed to NSPFAC with threshold partial pivoting.

SPARSPAK (without a C before the K) was the result of the work of J. Alan George and Joseph Wai-Hung Liu at the University of Waterloo, Canada. A book describing SPARSPAK [1331] was published in 1981. The Fortran package was for SPD matrices. Many orderings of the unknowns were available: reverse Cuthill-McKee (RCMK), one-way dissection, refined quotient tree, nested dissection, and minimum degree. There was an easy user interface. The package was available at a small cost (at that time). Later on checkpoint/restart facilities and least squares solutions were added in a 1984 release. This was done with Esmond Gee-Ying Ng, who was a student of George; see [1333].

Zahari Zlatev published several codes at the end of the 1970s: SIRSM in 1977, SSLEST in 1978, and Y12M in 1981 with Jerzy Wasniewski and Kjeld Schaumburg [3339]. Y12M implemented Gaussian elimination with iterative refinement; see also [2391].

The SLMATH code from Fred Gustavson at IBM was released in 1976. It used a switch to a dense code when the reduced matrix in Gaussian elimination becomes full enough.

In 1987, Randolph Edwin Bank and R. Kent Smith published a Gaussian elimination code, known as BSMP, with minimal integer storage requirements [189]. The listing of the code was in the paper; it is available at the address www.netlib.org/linalg.

In 1988, John Russell Gilbert and Timothy Peierls published a paper [1353] in which they described a left-looking LU factorization algorithm with sparse partial pivoting in time proportional to arithmetic operations. This is known as GPLU. This technique was later used in other codes, such as KLU from Davis [831], and generalized to supernodal techniques.

DSCPACK is a code for symmetric sparse linear systems developed by Padma Raghavan at the University of Pennsylvania [2528] at the beginning of the 2000s; see also the papers about the CAPSS code with Michael Thomas Heath [1615, 1616] in 1995-97.

SPOOLES (SParse Object Oriented Linear Equations Solver) is a C code using MPI or POSIX threads developed by Cleve Cleveland Ashcraft, Roger G. Grimes at Boeing at the end of the 1990s; see [86]. It solves linear systems with a symmetric structure using LDL^T or LDU factorizations. The package also provides several ordering algorithms, a multifrontal QR factorization, and some Krylov iterative solvers. The package is available at www.netlib.org/linalg.

TAUCS by Sivan Toledo, Vladimir Rotkin, and Doron Chen is a library of sparse linear solvers written in C at the beginning of the 2000s at Tel-Aviv University; see [2606]. It uses multifrontal and supernodal techniques and provides out-of-core capabilities. It also computes several preconditioners.

SuperLU is a library for the direct solution of large, sparse, nonsymmetric systems of linear equations, written in C at the Lawrence Berkeley National Laboratory (LBNL). The LU factorization uses the supernodal technique. The development started in the end of the 1990s with a sequential version; see [2056, 871, 2059, 2057] (ordered by date). SuperLU was first released in the beginning of 1997. Parallel versions of SuperLU are described in Section 8.10.

MUMPS (MUltifrontal Massively Parallel sparse direct Solver) is a parallel solver described in Section 8.10, but there is a sequential version.

In the mid-1990s, Davis and Duff proposed a nonsymmetric multifrontal algorithm for the LU factorization [828]. They used rectangular frontal matrices and an approximate degree update algorithm. An assembly direct acyclic graph (DAG) was constructed during the analyze-factorize phase. Frontal matrices were factored using BLAS3. The code was originally written in Fortran. Later they combined (uni)frontal and multifrontal methods to reduce data movement; see [829]. This led to UMFPACK 2.2 and to the code MA38 in the HSL library described in Section 8.9. Then Davis [824, 825] incorporated a symbolic pre-ordering and analysis, using a column ordering to reduce the fill-in. The next versions of the code were written in C. MATLAB relies on UMFPACK in $x = A\backslash b$ when A is sparse and either unsymmetric or symmetric but not positive definite.

CSparse is a sparse Cholesky factorization package written in C by Davis for the book [826]. Other codes by Davis are also available: CHOLMOD, a supernodal Cholesky solver [659], KLU, a sparse LU factorization specially designed for sparse matrices arising from circuit simulation [831], and SPQR, a multifrontal QR factorization [827].

All the packages that we cited above use direct methods which are variations around the LU factorization of the matrix for solving linear systems. There were many fewer packages for iterative methods. This is probably because iterative methods may not converge, depending on the properties of the matrix. Moreover, efficient iterative methods for nonsymmetric problems and good preconditioners were only developed starting in the 1980s.

ITPACK was developed in the 1970s at the Center for Numerical Analysis, the University of Texas at Austin. It is a package for solving large sparse linear systems by adaptive accelerated iterative algorithms: Jacobi method, the SOR method, the SSOR method with Chebyshev or conjugate gradient acceleration. The package is better suited for symmetric positive definite or mildly nonsymmetric matrices; see [1466, 3311]. The package NSPCG (Nonsymmetric Preconditioned Conjugate Gradient), developed in the 1980s, contains ITPACK, but also Krylov methods like Orthomin, Orthores, Orthodir, GCR, GMRES, and (Bi)CGS, and some preconditioners like the incomplete Cholesky and polynomial preconditioners; see [1900, 2382]. This package was targeting vector computers.

On his website, Yousef Saad provides several old and new packages for iterative methods for linear systems and solvers for eigenvalue problems. SPARSKIT is a Fortran 77 package of subroutines for working with sparse matrices. It includes general sparse matrix manipulation routines as well as a few iterative solvers, GMRES, FGMRES, CG, CGNR, BiCG, BiCGStab, TFQMR, and some incomplete LU (ILU) preconditoners. ITSOL, written in C, contains many variants of ILU preconditoners. The package pARMS (parallel Algebraic Recursive Multilevel Solvers), started in the mid-1990s, contains distributed memory parallel solvers for linear equations based on recursive multi-level ILU factorization; see [2061]. The more recent EVSL (EigenValues Slicing Library) package contains routines for computing eigenvalues of real symmetric standard or generalized eigenvalue problems located in a given interval and their associated eigenvectors; see [2051].

Krylov iterative solvers for linear systems are provided in the packages PeTSC, Hypre, and Trilinos described in Section 8.10 below.

Concerning eigenvalue and eigenvector computations, two books [790, 791] devoted to Lanczos algorithms for symmetric problems were published by Jane Grace Kehoe Cullum and Ralph Arthur Willoughby (1934-2001) in 1985. Volume II contains Fortran codes that are available at www.netlib.org/lanczos. Other packages for computing eigenvalues with Lanczos methods are also available in Netlib.

ARPACK is a Fortran package, written by Sorensen and Richard Bruno Lehoucq at Rice University in the 1990s, implementing the implicitly restarted Arnoldi algorithm for computing a few eigenvalues and eigenvectors of a general matrix. The users' guide was published by SIAM [2029] in 1998. ARPACK uses reverse communication for the matrix-vector product.

This package was used by MATLAB for eigenvalue computations with sparse general matrices. The package was not maintained, and a community project, ARPACK-ng (for new generation), was set up by some volunteers from companies using ARPACK. The code is available on GitHub.

Any package that provides a QR factorization of a matrix A can be used to solve a linear least squares problem $\min_x \|b - Ax\|_2$. The iterative method LSQR was proposed by Christopher Conway Paige and Michael Alan Saunders [2428, 2427] in the beginning of the 1980s. Fortran 77, 90, C and C++ codes are available at web.stanford.edu/group/SOL/software/lsqr/ as well as codes for the more recent methods LSMR [1182] developed by Saunders and David Fong, LSRN [2202] by Xiangrui Meng, Saunders, and Michael W. Mahoney, and MATLAB codes for LSLQ [1107] and LNLQ [1107] developed by Saunders with Ron Estrin and Dominique Orban.

An important issue in the production of reliable software is the testing of the code. In the beginning of digital computers, only small problems were solved. Programmers often used small dense matrices like the Hilbert and Cauchy matrices or the Rosser matrix. Later on, people started to build collections of matrices to be used as test cases. An example is the book *A collection of matrices for testing computational algorithms* by Robert Todd Gregory (1920-1984) and David L. Karney [1459] in 1969. In his review of the book, Wilkinson (*The Computer Journal*, Volume 13, Issue 4, November 1970, Page 391) wrote

> *In recent years considerable efforts have been made to develop algorithms of wide application in the field of linear algebra. It is important that such algorithms should be subjected to very stringent tests and for this purpose a well designed set of test matrices is required.*

Packages like EISPACK, LINPACK, and LAPACK were thoroughly tested; see, for instance, [2145] for the testing of LAPACK eigenvalue routines.

With the advent of the Internet it became much easier to set up collections of test matrices. We must mention the Harwell-Boeing collection set up by Duff, Grimes, and Lewis [992, 993], the Matrix Market collection [356], and the University of Florida collection of Davis and Y. Hu [830]. The last one is now known as the SuiteSparse collection at the Texas A&M University https://sparse.tamu.edu. This collection is still growing. In July 2021, it contained 2,893 matrices. The largest matrix, arising from a graph problem, was of order $226,196,185$ with $480,047,894$ nonzero entries.

There are also test matrix collections for MATLAB; see, for instance, [1676].

8.9 ▪ Commercial libraries

In the UK, one of the main centers for research on linear algebra was located at the Harwell Laboratory, west of Didcot and south of Oxford. There were reorganizations of the laboratory, and the group of people working there is now part of the STFC Rutherford Appleton Laboratory (RAL). The main contributors to the linear algebra software were John Ker Reid, Duff, and Jennifer Ann Scott.

HSL (formerly the Harwell Subroutine Library) is a collection of packages for large-scale scientific computation written and developed by the Computational Mathematics Group at RAL. The goal of HSL was to offer robust and efficient numerical software. Even though the strengths were originally more in optimization, the library is best known for its packages for the solution of sparse linear systems of equations and sparse eigenvalue problems. HSL is a commercial product but access is free for academic use.

The library was started in 1963, with Mike Hopper and Michael James David Powell (1936-2015), who were the main organizers. It was originally used at the AERE Harwell Laboratory on IBM computers. Over the years, the library has evolved and has been extensively used on a wide range of computers with adaptation to modern computer architecture. Fortran 90 and 95 versions were developed later.

The library contains routines for manipulating and modifying matrices named as MCxx. Let us briefly consider some of the codes for solving linear systems:

- MA28 by Duff in 1977 and modified in 1983, for general matrices, a follower of MA18. It uses the Markowitz criterion and threshold pivoting as well as a switch to full code when the reduced matrix becomes sufficiently dense; see [1002].

- MA27 by Duff and Reid in 1982-83, for symmetric indefinite matrix. It uses an elimination tree, a fast minimum degree ordering, and an indefinite matrix multifrontal algorithm; see [1003].

- MA32 by Duff in 1980, a nonsymmetric frontal code with out of core facilities. The frontal method was proposed by Bruce Irons (1924-1983) in 1970 for symmetric finite element problems [1784]. It was generalized to nonsymmetric problems by Paul Hood [1727] in 1976.

- MA37 by Duff and Reid in 1984 for matrices with a symmetric or nearly symmetric structure; see [1004].

- MA42 by Duff and Scott in 1992 is a redesign of MA32 with an out of core frontal method (HSL_MA42 is a Fortran 90 version by Scott).

- MA48 by Duff and Reid in 1993 is a replacement for MA28 (an updated version is HSL_MA48 in 2001).

- MA41 by Duff and Patrick Amestoy in 1995 is a multifrontal code for nonsymmetric linear systems.

- MA49 by Amestoy (ENSEEIHT), Duff, and Chiara Puglisi (CERFACS) in 1998 is a multifrontal QR code for solving least squares problems.

- MA57 by Duff in 2004 is a multifrontal code for symmetric matrices with a modified version in 2016.

- MA77 by Reid and Scott in 2006 is an out of core solver for symmetric matrices and MA78 in 2007 is for nonsymmetric matrices.

- MA97 by Jonathan Hogg and Scott in 2011 is a parallel multifrontal code for symmetric definite or indefinite matrices.

The library also contains routines for computing eigenvalues and eigenvectors, mainly using the QR algorithm and also some Lanczos codes, but this is obviously not the main strength of the package.

The *Numerical Algorithms Group* (NAG) is a software and services company in the UK. NAG was founded by Brian Ford, Joan Eileen Walsh (1932-2017), and others in May 1970 under the name Nottingham Algorithms Group as a collaboration between the universities of Birmingham, Leeds, Manchester, Nottingham, and Oxford and the Rutherford Appleton Laboratory. Walsh from Manchester, Shirley Lill from Leeds, and Linda Hayes from Oxford were the early collaborators. It was originally intended to provide software for the *ICL 1906A* computer with Algol and Fortran libraries, but it rapidly expanded with broader goals. The first version of the library was ready in September 1971. In 1972, Ford traveled to the USA with Wilkinson, who was a kind of advisor for NAG. He had contacts with the EISPACK group at Argonne. Nottingham University decided to stop the project there at the end of 1972 and NAG moved to Oxford in August 1973. By the end of 1973, the library was in use on most computers in the vast majority of universities in the UK. NAG was funded by the Computer Board for Universities and Research Councils which was in charge of providing computers to British universities. NAG became a not for profit company in 1976 and financially independent of any government funding in 1980. An American branch, NAG Inc., was established in 1978 and one in Germany in 1990. The Algol 60 library was stopped in 1981 but an Algol 68 library was developed. For details on NAG history, see [1536].

We recall that Du Croz and Hammarling from NAG participated in the LAPACK project. The NAG library contains routines coming from or based on that project for dense and banded matrices. It also uses SuperLU for sparse matrices. Some iterative methods for nonsymmetric matrices are also available, as well as codes to compute functions of matrices.

IMSL (International Mathematics and Statistics Library) was founded in 1970 by Edward L. Battiste (?-2001), who was a statistician working for IBM on the scientific subroutine library in Houston at that time, and Charles W. Johnson (1929-2012). Johnson, whose father owned a construction company, provided most of the initial funding. The two of them had met in 1959 in Milwaukee when they were working for IBM. The company started with four or five people in Houston. The first release of the library was delivered to IBM customers in 1972. Then versions for UNIVAC, CDC, and later Honeywell were released. At the beginning the company was losing money; it took about five or six years to break even. Battiste stepped down as president in 1976 and left the company in 1977. His successor was Walt Gregory. After leaving IMSL, Battiste moved to North Carolina and founded the company C.Abaci, which had a focus on supporting individual users with desktop products.

IMSL had a prerelease version of EISPACK. They incorporated some codes from EISPACK and LINPACK in the IMSL library. There was a major rewrite of the library in 1987.

The name of the company was changed to Visual Numerics in 1992 when IMSL bought a Boulder-based company, Precision Visuals, Inc., which specialized in image and signal processing with a product named PV Wave. There were some layoffs after the merge. In 2009, Rogue Wave Software acquired Visual Numerics; see [11, 1542, 1538].

8.10 ▪ Libraries for parallel computers

Even though there were experimental parallel computers before that, starting in the 1970s and 1980s, parallel computers appeared on the market as an alternative to serial or vector supercomputers.

In the beginning there were basically two types of parallel computer depending on the kind of memory. In shared memory computers, there is a single address space that each processor can access. Of course, care must be taken when two processors want to write to the same memory location. In distributed memory computers each processor has its own memory and the processors have to exchange data through a communication network linking the processors.

Traditional programming languages did not offer ways to express parallelism. Programming a parallel computer became more or less manufacturer dependent. For instance, there were many proprietary message passing systems and interfaces. On these machines, it also became more difficult for end users to write efficient codes, even for simple tasks such as computing the Cholesky factorization of a dense matrix.

Soon, people realized that standards were needed. Two important steps were the advent of PVM and MPI for distributed memory computers.

The Parallel Virtual Machine (PVM) software was primarily written in the 1990s by people at ORNL and UTK. It is a package that allows a heterogeneous collection of Unix and/or Windows computers linked together by a network to be used as a single large parallel computer; see [1315]. The development of PVM started in the summer of 1989 at ORNL involving Vaidy Sunderam at Emory University, Al Geist at ORNL, Robert Manchek at UTK, Adam Beguelin at Carnegie Mellon University and Pittsburgh Supercomputer Center, Weicheng Jiang at UTK, Jim Kohl, Phil Papadopoulos, June Donato, and Honbo Zhou at ORNL, and Dongarra at ORNL and UTK. PVM was used by NAG [923].

The Message Passing Interface (MPI) is a standardized and portable message passing standard. MPI started at the beginning of the 1990s and many people and organizations were involved in its definition and design; see [932, 2817]. The first standard MPI 1.0 appeared in May 1994. The MPI Forum is the standardization forum for the evolution of MPI. MPI 2.0 was defined in 1997 and MPI 3.1 was approved in 2015. This standard is still evolving. The draft of MPI 4.0 had 919 pages! It includes constructions for threads, fault-tolerance, interfaces to debuggers, etc.

OpenMP (Open Multi-Processing) is a programming interface that supports shared memory multiprocessing programming. To use OpenMP the programmer inserts compiler directives in the code and/or calls to subroutines. OpenMP implements multithreading, and directives are provided to manage, synchronize, and assign work to the threads. This has to be supported by the compiler. OpenMP 1.0 for Fortran was released in October 1997. Later OpenMP was extended to support other programming languages and also to describe task dependencies and to support accelerators. The last version so far is OpenMP 5.2 in 2021.

OpenACC (Open Accelerators) is another standard of compiler directives for parallel programming on heterogeneous CPU/GPU systems. It was first released in 2011. Of course, it does not make too much sense to have two "'standards" and this will have to be solved.

Parallel BLAS (known as PBLAS or PB-BLAS) is an implementation of the BLAS intended for distributed memory machines. It was described [664, 666] in 1995-96. A new C++ implementation BLAS++ was proposed in [1963].

Another child in the BLAS growing family was the Batched BLAS [918], whose development started in 2016. In modern parallel algorithms for linear algebra and many other problems, a large problem is decomposed into batches containing thousands of smaller problems, which can be solved independently by BLAS routines. The goal of Batched BLAS is to define an interface to enable users to perform thousands of small BLAS operations in parallel while making efficient use of manycore processors or graphics processing units (GPUs).

ReproBLAS stands for Reproducible BLAS. Its aim was to guarantee the reproducibility of a parallel computation using the BLAS, regardless of the number of processors, of the data partitioning, of the way reductions are scheduled, and of the order in which sums are computed.[55]

ScaLAPACK is a library of high-performance linear algebra routines for parallel distributed memory machines. It was intended to be a parallel version of a subset of LAPACK. It was mainly written in Fortran at UTK and used a block cyclic data distribution for dense matrices as well as block-partitioned algorithms. The development started at the beginning of the 1990s and it was first released in 1995 [665]. The version 2.1.0 was from November 2019. It relies on the PBLAS, LAPACK, MPI, and BLACS (Basic Linear Algebra Communication Subprograms) libraries. A ScaLAPACK users' guide [346] was published in 1997. ScaLAPACK was almost 450,000 lines of code, excluding comments. However, ScaLAPACK is not well suited for today's supercomputers using multi-core processors and/or GPU accelerators.

PLAPACK was a library developed at the University of Texas. Its aim was to provide the functionalities of LAPACK on parallel computers. Its main developer was Robert A. van de

[55] bebop.cs.berkeley.edu/reproblas (accessed December 2021)

Geijn. The development of the PLAPACK project ceased around 2000. It was followed by the development of the FLAME project and the libFLAME library. A somewhat related project was Elemental. Elemental [2517] was a C++ library for distributed memory dense and sparse direct linear algebra. Its main developer was Jack Poulson. The library was initially released in 2011. Elemental has not been maintained since 2016. However, Hydrogen is a fork of Elemental used by the Livermore Big Artificial Neural Network Toolkit.

The PLASMA (Parallel Linear Algebra Software for Multicore Architectures) package from UTK implemented a set of fundamental linear algebra routines using OpenMP.

MAGMA (Matrix Algebra on GPU and Multicore Architectures), also from UTK, was a collection of linear algebra libraries for heterogeneous architectures. It provides dense matrix factorizations and solvers and eigen/singular-value problem solvers.

PaRSEC (Parallel Runtime Scheduling and Execution Controller) was a framework for managing the scheduling and execution of tasks on distributed many-core heterogeneous architectures.

Those libraries were not as successful as LAPACK and ScalaPACK; see [1]. So UTK started a new project named SLATE (Software for Linear Algebra Targeting Exascale). Its goal was to replace ScaLAPACK with a better performance potential and maximum scalability on modern, many-node high-performance computers with large numbers of cores and multiple hardware accelerators per node. It is developed in C++ as part of the Exascale Computing Project (ECP), which is a joint project of the U.S. Department of Energy Office of Science and National Nuclear Security Administration (NNSA). A survey of the needs and the definition of a road map were done in 2017 [1]. SLATE uses a new storage scheme for dense matrices as a collection of individually allocated tiles with a global indexing of tiles. It relies on multi-level scheduling with OpenMP, use of batched BLAS for obtaining node-level performance, and use of MPI. SLATE also handles multiple precisions by C++ templating.

PARDISO is a project, started in the beginning of the 2000s, whose main contributor was Olaf Schenk (Universita della Svizzera Italiana, Switzerland). Its aim was to provide tools for solving large sparse symmetric and nonsymmetric linear systems on shared memory and distributed memory architectures; see [2696, 2695]. It is written in Fortran and C and based on OpenMP and MPI. The solver uses a combination of left- and right-looking BLAS3 supernode techniques. PARDISO may also compute incomplete factorizations and use iterative refinement. The last version is 7.2 in 2021. PSPIKE is a parallel implementation, using PARDISO, of the SPIKE algorithm of Ahmed Sameh [2506, 2129] for solving banded systems of linear equations.

PaStiX (Parallel Sparse matriX package) [1639] was developed at the Université de Bordeaux and Inria in France by Mathieu Faverge, Pascal Hénon, Pierre Ramet, Jean Roman, and others starting in the beginnings of the 2000s. It uses shared memory using POSIX threads for multicore architectures and low-rank compression techniques to reduce the memory footprint.

WSMP (Watson Sparse Matrix Package) was written by Anshul Gupta, who was working at the IBM T.J. Watson Research Center, for solving sparse nonsymmetric linear equations at the beginning of the 2000s; see [1482, 1483]. It used a nonsymmetric multifrontal factorization. A parallel version was described [1484] in 2007.

STRUMPACK (STRUctured Matrix PACKage) is a software library, developed at the Lawrence Berkeley National Laboratory (LBNL), that provides linear algebra routines for sparse matrices and for dense rank-structured matrices; see [1344]. The main investigators are Xiaoye Sherry Li and Pieter Ghysels. It is written in C++ using OpenMP and MPI. STRUMPACK has two main components: a distributed-memory dense matrix computations package (for dense matrices that have a hierarchical structure) and a distributed memory sparse general solver and preconditioners.

NLAFET (Parallel Numerical Linear Algebra for Extreme Scale Systems) was a Horizon 2020 FET-HPC project funded by the European Union. The partners in this project were Umeå University in Sweden, the University of Manchester in United Kingdom, Institut National de Recherche en Informatique et en Automatique (INRIA) in Paris, and the Science and Technology Facilities Council (STFC) in the United Kingdom. Umeå University coordinated NLAFET. The project aimed at providing a collection of libraries with dense matrix factorizations and solvers, solvers and tools for standard and generalized dense eigenvalue problems, sparse direct factorizations and solvers, communication optimal algorithms for iterative methods. A first complete release was delivered in 2019.

We have considered the package SuperLU above. Parallel versions were developed at the end of the 1990s: SuperLU_MT [872] for shared memory computers with Pthreads or OpenMP interfaces and SuperLU_Dist [2058] for distributed memory computers using MPI, OpenMP, and CUDA for computing on GPUs. Both libraries are still maintained. Another solver developed at LBNL is SymPACK for sparse symmetric positive definite matrices for distributed memory computers.

MUMPS (MUltifrontal Massively Parallel sparse direct Solver) was initially funded by the European ESPRIT IV project PARASOL that started in 1996; see [48, 49, 54, 44]. It was developed by Rutherford Appleton Laboratory (UK), CERFACS, and ENSEEIHT in Toulouse (France) with some other collaborations later in Lyon and Bordeaux. The main developer is Amestoy. It uses a multifrontal approach with dynamic pivoting for stability based on the work of Duff and Reid. It has been improved and enriched over the years. MUMPS also has out of core solvers and it is possible to use iterative refinement and backward error analysis. Several reorderings are available, as well as the computation of selected entries in the inverse of a matrix. For some literature about MUMPS, see [51, 52, 53]. Recently, block low rank techniques have been introduced to reduced the memory usage; see [45, 46, 2163, 3204]. MUMPS uses MPI and shared memory parallelism (OpenMP, multithreaded BLAS) within each MPI process. A dynamic distributed scheduling is used to accommodate numerical fill-in and for load balancing. MUMPS is widely used all over the world and it has also been included in many open-source or commercial codes.

PARPACK is a parallel version of the well-known software ARPACK implementing the implicitly restarted Arnoldi algorithm for computing a few eigenvalues and eigenvectors of a general matrix; see [2164, 2029].

PETsC (Portable, Extensible Toolkit for Scientific Computation) is a suite of data structures and routines for the parallel solution of scientific problems modeled by partial differential equations developed at Argonne National Laboratory, primarily funded by the DOE; see [172]. The first version was released in 1995. It includes direct and iterative Krylov solvers for linear systems as well as nonlinear and optimization solvers. It also relies on external packages like SuperLU, MUMPS, UMFPACK, and others. The latest version so far is 3.17 in 2022.

Hypre is a package developed at the Lawrence Livermore National Laboratory for solving linear systems. This project started in the late 1990s. In Hypre, the emphasis was put on algebraic multigrid solvers and preconditioners to be used with iterative Krylov solvers; see [669, 1130, 1129, 165] (ordered by date). Hypre also gives access to solvers for the eigenvalue problem.

Trilinos is a project of Sandia National Laboratory started in the beginning of the 2000s; see [1652]. It is a collection of software packages written in C++. Each package is self-contained. There exist packages with direct (sparse and dense) and Krylov iterative linear solvers as well as for solving eigenvalue problems. Trilinos also provides numerous types of preconditioners, including block ILU and multilevel ones.

An important issue that future libraries will be facing is the fault tolerance problem. In long computations with millions of computing engines it is likely that some of them may have troubles or breakdowns. Then it will be necessary to have algorithms and software that can recover from these breakdowns.

8.11 ▪ Netlib

Netlib (www.netlib.org) is a repository of public-domain software for scientific computing. It also holds the NA-Digest electronic newsletter. The creation of these two facilities was largely due to the efforts of Golub, a professor at Stanford University, Dongarra at Argonne National Laboratory, and Eric Grosse at Bell Labs at the beginning of the 1980s. Originally, there were two repositories, one at Bells Labs and the other at ANL. They started to be used in 1984. The ANL server was moved to Oak Ridge National Laboratory (ORNL) in 1989, when Dongarra moved to the University of Tennessee (UTK). The ORNL server was retired in 1997 and Netlib is hosted at UTK as well at some mirror sites in Europe.

Initially the users' requests were done via e-mail and the files were also obtained by e-mail. At the time Netlib was started the distribution of software was mainly done by sending magnetic tapes through the postal service. This was slow and painful since tape formats can be different depending on the computer. The number of e-mail users was growing and this contributed to the success of Netlib. Moreover, only quality software was included in the repository. The initial collection included EISPACK, LINPACK, and other "PACK"s. Codes from the journal *Transactions on Mathematical Software* (TOMS) were added as well as BLAS routines. Netlib started to be accessed through the Internet and a graphical interface at the beginning of the 1990s. Netlib HTTP servers appeared in 1993. For papers on Netlib, see [920, 921]. For the history of Netlib, see [919].

As of July 3, 2021, there were $1,276,897,587$ requests since 1985. The software most downloaded from Netlib is LAPACK.

In the early 1980s, Golub at Stanford was maintaining a list of e-mail addresses of people in the numerical linear algebra community. He started sending news to people in his list. This was the beginning of NA-Digest.

Golub was the original editor of NA-Digest. In July of 1987, he began a sabbatical leave from Stanford and Moler began editing the digest. With only occasional absences, Cleve continued to edit the digest until September of 2005. Tamara Kolda from Sandia National Laboratories took the position up until October 2010. Since then, the editor has been Daniel M. Dunlavy from Sandia.

Usually, NA-Digest gives announcements of conferences, workshops, jobs, software releases, changes of address, new books, journal contents, and notices about community members. An archive of the past digests is kept in Netlib, which is of interest for historians. The oldest digest in the archive is from February 1987.

8.12 ▪ MATLAB

MATLAB[56] is perhaps the world's most commercially successful piece of numerical analysis software.

MATLAB was started at the end of the 1970s by Moler, who was then a professor at the University of New Mexico; see [1546, 2044]. Moler was one of the people involved in writing EISPACK and LINPACK. He was teaching a course on linear algebra and wanted his students to be able to have access to EISPACK and LINPACK without having to write Fortran programs. To do this he wrote an interactive matrix calculator (MATrix LABoratory). At that time there was no possibility to write functions and no graphics. It was just possible to input matrices, which

[56] MATLAB is a trademark of The MathWorks Inc.

were the only data type, perform operations on them, and print the results. The program was written in Fortran. The underlying routines were from LINPACK and EISPACK. For the syntax and to write the parser of the language, Moler was inspired by a book by Niklaus Emil Wirth [3263] which described a simple language PL0 and a parser for it.

Copies of the software were sent for just the price of the tape to people working on numerical linear algebra, and MATLAB started to become popular.

In the academic year 1979-1980, Moler was teaching a course on matrix computations in Stanford University using MATLAB. It drew the attention of students in electrical engineering. MATLAB was sent to Systems Control Technology (SCT), a company in Palo Alto, near Stanford. One of the students had a friend there, John (Jack) N. Little, who started using MATLAB. Little had a master's degree in electrical engineering from Stanford. Little, with one of his colleagues, Steve Bangert, found some errors in Moler's parser that had to be corrected.

In 1983, Moler was at a conference in Boulder to demonstrate MATLAB on an IBM PC. He was approached by Little, who proposed to start a company based on MATLAB. After Moler agreed, Little quit SCT, bought a Compaq PC, went to his home in the Portola Valley in the Stanford hills, and started writing the commercial version of MATLAB. He was helped by Steve Bangert. They threw away the Fortran code and wrote the program in C starting from scratch. They added M-files, toolboxes, and some control structures.

The MathWorks company was incorporated in California in 1984. The owners were Little, Moler, and Bangert without outside investment. The first and only employee was Little. The first ten copies of the commercial MATLAB were sold to MIT in February 1985. The number of employees doubled every year for the next seven years. After a year, Little moved the company to Sherborn, Massachusetts. Bangert stayed in California, and Moler left New Mexico to work for Intel in Oregon. After a while the company moved again to Natick, Massachusetts, where they still are.

After working for Intel, Moler was working for Ardent, a computer manufacturer. This company was not successful. In 1989, Moler went to work full time for The MathWorks; see Moler's book [2260] and [2262, 2264, 2265].

Over the years MATLAB has expanded with new functionalities, many new toolboxes, and sophisticated graphics. They also have another product called Simulink for simulating dynamical systems.

The MathWorks is based in Natick but has other US offices in Michigan and California as well as offices in the UK, France, China, Singapore, India, Ireland, and other countries. In October 2021 there were 6,000 employees worldwide. Around 60% of the sales are from international customers. The MathWorks has more than four million users of MATLAB worldwide, and there are books about MATLAB in 27 languages.

If you look at the papers that were written in the area of numerical linear algebra with numerical experiments in the last 30 years, many of them used MATLAB to prototype their algorithms.

A free software that is mostly compatible with MATLAB is GNU Octave. It was started by John W. Eaton in 1992. Other alternatives are Scilab, which was created at INRIA in France in the 1990s, and FreeMat. Another software which is not MATLAB-compatible but whose use is growing is Julia.

9
Miscellaneous topics

9.1 ▪ Tridiagonal matrices

Tridiagonal matrices, that is, matrices T for which the entries $t_{i,j}$ of row i are zero except for columns $j = i-1, i, i+1$ with obvious modifications for the first and last rows, arise in many problems. For instance, we may see them in linear systems coming from the discretization of one-dimensional partial differential equations or in block Jacobi or Gauss-Seidel and ADI iterative methods as well as in spline interpolation.

Note that the determinant of a tridiagonal matrix can be computed by a three-term recurrence relation. Tridiagonal matrices are related to orthogonal polynomials because they also satisfy a three-term recurrence relation. Orthogonal polynomials were introduced at the end of the 19th century in studies about continued fractions in works by Pafnuty Lvovich Chebyshev (1821-1894) and his students and later by Thomas Joannes Stieltjes (1856-1894), who was interested in solving moment problems.

Tridiagonal arrays appeared explicitly in the work of James Joseph Sylvester (1814-1897) in 1853 because of the relation of their determinants with the numerators and denominators of the convergents of continued fractions; see the postscript of his paper [2962], page 616 of volume I of his Collected Mathematical Papers and also [2961, 2964]. The determinants of tridiagonal arrays were named *continuants*.

Something similar to what Sylvester did was published by William Hugh Spottiswoode (1825-1883) [2847] in 1856. In those years, continuants were also studied by Francesco Brioschi (1824-1897) [471] in 1856, Arthur Cayley (1821-1895) [586] in 1857, and Louis Félix Painvin (1826-1875) [2433] to solve a problem stated by Joseph Liouville (1809-1882).

For the history of continuants, see Thomas Muir (1844-1934) [2300], page 413 and following. The name "continuants" was used for quite some time since it still appeared, for instance, in [18] by Alexander Craig Aitken (1895-1967) in 1939; see page 126 of the 1944 edition.

Symmetric tridiagonal matrices are sometimes called *Jacobi matrices*, even though some authors use that name only when the off-diagonal entries are positive. It seems that the origin of that name is in a paper [3056] by Otto Toeplitz (1881-1940), from 1907. He considered bilinear forms

$$a_1x_1^2 + a_2x_2^2 + \cdots + 2b_1x_1x_2 + 2b_2x_2x_3 + \cdots,$$

which he called a *Jacobi form*. Clearly, this corresponds to (Tx, x), where T is a tridiagonal matrix whose row i is given by (b_{i-1}, a_i, b_i). However, Toeplitz did not use this relation to a tridiagonal matrix and he was interested in what corresponds to semi-infinite tridiagonal matrices. The name referred to what was done by Carl Gustav Jacob Jacobi (1804-1851) in 1848.

The name appeared again in the paper [1632] by Ersnt David Hellinger (1883-1950) and Toeplitz in 1914 and in the paper [1946] by Mark Grigorievich Krein (1907-1989) in 1933.

The name *Jacobi matrix* can be found in the book [2913] by Marshall Harvey Stone (1903-1989) in 1932 (page 282) and in the title of the paper [1557] by Hans Ludwig Hamburger (1889-1956) in 1944. Note that an irreducible tridiagonal matrix is similar to a symmetric tridiagonal matrix.

It is particularly easy to obtain factorizations or inverses of tridiagonal matrices T. Let us first consider the LU factorization of T. Even though T is a sparse matrix, there is no fill-in in this factorization and L and U are bidiagonal matrices, having only one nonzero diagonal next to the main diagonal. The factorization can be written in different ways, as LU, LDU, or $LD^{-1}U$, where D is a diagonal matrix. Let us consider $LD^{-1}U$; here the main diagonals of L and U are equal to the diagonal of D and the other nonzero entries are equal to those of T. If the nonzero entries of row j of T are $(\beta_j, \alpha_j, \gamma_j)$, the diagonal entries d_j of D are given by

$$d_1 = \alpha_1, \quad d_j = \alpha_j - \frac{\beta_j \gamma_{j-1}}{d_{j-1}}, \; j = 2, \ldots, n.$$

Of course for this factorization to be feasible, the denominators d_{j-1} have to be different from zero. Even if this condition is satisfied, the factorization may not be stable, depending on the properties of T. If T is symmetric positive definite, the solution exists and is a Cholesky-like factorization with

$$d_j = \alpha_j - \frac{\beta_j^2}{d_{j-1}}, \; j = 2, \ldots, n.$$

The other factorizations are obtained by scaling.

Apparently, the LU form of the factorization where the diagonal of L is equal to the diagonal of D and the diagonal entries of U are all equal to 1 was described in 1949 in a report [3033] by Llewellyn Hilleth Thomas (1903-1992), who was a British physicist known for the Thomas-Fermi electron gas model. From 1946 to 1968 he was a member of the staff of the IBM Watson Scientific Computing Laboratory at Columbia University. It is not easy to have access to this report, but the method was used and described in a paper [486] published in 1953 by G.H. Bruce, Donald William Peaceman (1926-2017), Henry H. Rachford, and J.D. Rice. They were doing calculations of a one-dimensional unsteady-state gas flow through porous media discretized with a time implicit scheme using an *IBM Card-Programmed Calculator*. Their tridiagonal matrices were symmetric and strictly diagonally dominant, and therefore positive definite. They attributed the factorization to Thomas but they remarked that this was nothing other than Gaussian elimination. Later on, this type of factorization was called the *Thomas algorithm* by many researchers.

Inverses of tridiagonal matrices have been extensively studied, although it seems that most of the results that have been obtained were unrelated and that many of the authors did not know each others' results. Some of the results are special cases of those obtained for banded matrices or Hessenberg matrices.

In the 1950 expanded edition of the book [1283] by Felix Ruvimovich Gantmacher (1908-1964) and Krein formulas for the inverse of a Jacobi matrix were given on pages 117-118. We were not able to check if it was already in the first Russian edition of 1941.

An early paper considering the explicit inverse of tridiagonal matrices is by David Moskovitz [2297] in 1944, in which analytic expressions were given for the 1D and 2D Poisson model problems. Moskovitz's paper was cited by Paul Charles Rosenbloom (1920-2005) [2598] in 1948 and Fred W. Dorr [939] in 1970.

An important paper for the inverses of band matrices was the seminal work by Edgar Asplund (1931-1974) [90] in 1959; see also the paper [91] by Sven Olof Asplund (1902-1984), a structural engineer in 1959. In 1969, the physicist Charlotte Froese Fischer and her student Riaz Ahmad Usmani (?-1995) gave a general analytical formula for inverses of symmetric Toeplitz tridiagonal matrices [1173].

In 1971, Jacques Baranger and Marc Duc-Jacquet (1941-2019) considered symmetric factorizable matrices (whose elements are $a_i b_j$ for $i \leq j$) and proved that the inverse is tridiagonal (this is Asplund's result) and conversely [191]; see also [205] by Wayne Walton Barrett in 1979, who introduced the *triangle property*: a matrix R has this property if $r_{i,j} = (r_{i,k} r_{k,j})/r_{k,k}$. A matrix having the triangle property and nonzero diagonal entries has a tridiagonal inverse and vice versa. Barrett's result is not strictly equivalent to Asplund's result. In one theorem there is a restriction on the diagonal entries, and in the other there is a restriction on the nondiagonal entries of the inverse.

In 1981, Barrett and Philip Joel Feinsilver established a correspondence between the vanishing of a certain set of minors of a matrix and the vanishing of a related set of minors of the inverse [206]. This gave a characterization of inverses of banded matrices; for tridiagonal matrices it reduces to Barrett's triangle property. A generalization was published in 1984 by Barrett and Charles Royal Johnson [207].

Many papers have been written on inverses of tridiagonal matrices. Closed form explicit formulas for elements of the inverse can only be given for special matrices like Toeplitz tridiagonal matrices. A review on the inverse of symmetric tridiagonal and block tridiagonal matrices was published by G.M. in 1992 [2215]. In that paper it is shown that the entries of the inverse can be computed with LU and UL factorizations of the matrix when they are feasible. Nonsymmetric tridiagonal matrices were considered by Reinhard Nabben in 1999 [2314].

Results on inverses of Hessenberg matrices were obtained by Yasuhiko Ikebe [1772] in 1979 and by Dmitry Konstantinovich Faddeev (1907-1989) in 1981 in Russian [1126]. Specialization of Ikebe's result to tridiagonal matrices gave Asplund's results.

9.2 ▪ Fast solvers

Fast solvers were developed in the 1960s and 1970s for solving separable partial differential equations (PDEs) in rectangular domains with various boundary conditions. These equations can be written as

$$a(x)\frac{\partial^2 u}{\partial x^2} + b(x)\frac{\partial u}{\partial x} + c(x)u + d(y)\frac{\partial^2 u}{\partial y^2} + e(y)\frac{\partial u}{\partial y} + f(y)u = g(x,y).$$

These problems are said to be separable since the coefficients of the partial derivatives with respect to x (resp., y) depend only on x (resp., y). This allows to use separation of variables. The model problem is $-\Delta u + \sigma u = f$ with Dirichlet boundary conditions (that is, the values of u are given on the boundary). Discretizing this problem when $\sigma = 0$ with a standard 5-point finite difference scheme, one obtains a linear system $Ax = b$ with a symmetric block tridiagonal matrix,

$$A = \begin{pmatrix} T & -I & & & \\ -I & T & -I & & \\ & \ddots & \ddots & \ddots & \\ & & -I & T & -I \\ & & & -I & T \end{pmatrix},$$

with

$$T = \begin{pmatrix} 4 & -1 & & & \\ -1 & 4 & -1 & & \\ & \ddots & \ddots & \ddots & \\ & & -1 & 4 & -1 \\ & & & -1 & 4 \end{pmatrix}.$$

For the Poisson model problem we explicitly know the eigenvalues of the matrix A. Since A is symmetric, one can use the spectral factorization $A = Q\Lambda Q^T$, where Λ is a diagonal matrix and Q is orthogonal. The solution is written as $x = Q\Lambda^{-1}Q^T b$. Looking at the values of Λ and Q for the model problem, computing x in this way is called *Fourier analysis*, sometimes denoted as DFA, *Double Fourier Analysis*, as it corresponds to a Fourier decomposition in each direction. For the sums involved in the computation of x, when applying Q^T and Q, the *Fast Fourier Transform* (FFT) can be used.

The introduction of the FFT by James William Cooley (1926-2016) and John Wilder Tukey (1915-2000) in 1965 [738] was a breakthrough in the area of numerical computation. It allowed the development of many applications. For a description of the different versions of the FFT, see the book by Charles Francis Van Loan [3113] in 1992. Some improvements and generalizations were introduced in the 1980s by Clive Temperton [3019, 3020, 3021].

The origin of the FFT can be traced back to an unpublished paper of Carl Friedrich Gauss (1777-1855) in 1805. A method similar to the FFT was published in 1942 by Gordon Charles Danielson (1912-1983) and Cornelius Lanczos (1893-1974) [808]. The history of the FFT was discussed by Michael T. Heideman, Don H. Johnson, and C. Sidney Burrus [1630] in 1985, as well as Cooley [737] in 1990.

For a modern computer implementation, see the FFTW by Steven G. Johnson and Matteo Frigo [1246, 1247] at the end of the 1990s.

Another possibility for solving the model problem is to use the FFT in one direction, and to solve the resulting independent tridiagonal systems. This method was used by Roger Willis Hockney (1929-1999) [1714] in 1965; see also [1715] in 1970.

Another popular fast solver in the 1970s was the *cyclic reduction* method which was developed at the suggestion of Gene Howard Golub (1932-2007); see [1280] with Walter Gander in 1997. This method can be applied to block tridiagonal matrices, but let us briefly explain it on a simple tridiagonal system

$$\begin{pmatrix} a & -1 & & & \\ -1 & a & -1 & & \\ & \ddots & \ddots & \ddots & \\ & & -1 & a & -1 \\ & & & -1 & a \end{pmatrix} x = b,$$

of order $n = 2^{l+1} - 1$. The idea is to first eliminate the odd unknowns. We have

$$\begin{array}{rcrcrcrcrcl} -x_{i-2} & + & ax_{i-1} & - & x_i & & & & & = & b_{i-1}, \\ & - & x_{i-1} & + & ax_i & - & x_{i+1} & & & = & b_i, \\ & & & - & x_i & + & ax_{i+1} & - & x_{i+2} & = & b_{i+1}. \end{array}$$

Multiplying the middle equation by a and summing, the odd unknowns x_{i-1}, x_{i+1} are eliminated, and we obtain

$$-x_{i-2} + (a^2 - 2)x_i - x_{i+2} = b_{i-1} + ab_i + b_{i+1}.$$

The new system has the form

$$\begin{pmatrix} a^2-2 & -1 & & & \\ -1 & a^2-2 & -1 & & \\ & \ddots & \ddots & \ddots & \\ & & -1 & a^2-2 & -1 \\ & & & -1 & a^2-2 \end{pmatrix} \begin{pmatrix} x_2 \\ x_4 \\ \vdots \\ \vdots \\ x_{2^{l+1}-2} \end{pmatrix} = \begin{pmatrix} b_2^{(1)} \\ b_4^{(1)} \\ \vdots \\ \vdots \\ b_{2^{l+1}-2} \end{pmatrix},$$

and the matrix is now of order $2^l - 1$. So, we can apply the same reduction in a recursive way; see the paper [518] by Bill L. Buzbee, Golub, and Clair W. Nielson in 1970.

Cyclic reduction was generalized in the 1970s by Paul Noble Swarztrauber (1936-2011) and Roland Andrew Sweet (1940-2019) in a series of papers [2948, 2946, 2951, 2952, 2949] (ordered by date) leading to the distribution of the FISHPACK package (for our readers who do not speak French, the translation of "poisson" is "fish"). The original cyclic reduction algorithm was prone to instabilities and the stable computation of the right-hand side was proposed by the computational physicist Oscar Buneman (1913-1993) [506] in 1969 even though this important work was never published in a journal. Note that cyclic reduction is a special case of (block) Gaussian elimination without pivoting.

Fourier analysis and cyclic reduction were combined leading to the *FACR(ℓ)* method. It used ℓ steps of cyclic reduction and solved the resulting reduced system by the Fourier/tridiagonal method. For practical dimensions the number of operations is almost proportional to the total number of unknowns. For this method, see Temperton [3017, 3018] in 1979-1980 and Swarztrauber [2947] in 1977.

The fast direct solvers described above were more or less abandoned in the 21st century since people wanted to solve problems in other domains than rectangles or cubes, and other efficient methods, like multigrid, were developed.

9.3 ▪ Hankel and Toeplitz matrices

Let (c_i) be a sequence of numbers, where i is a signed integer. A Hankel matrix $H_n^{(k)}, n \geq 1$ is a matrix of order n whose elements are $a_{ij} = c_{k+i+j-2}$. Thus, each ascending skew-diagonal from left to right is constant. These matrices are named after Hermann Hankel (1839-1873), who studied their determinant in his dissertation [1571] in 1861.

Hankel matrices are encountered in (formal) orthogonal polynomials, Padé approximation, the moment problem, some extrapolation methods, partial realization, coding theory, filtering, and tridiagonalization process; see, for example, [430] by C.B.

A Toeplitz matrix $T_n^{(k)}, n \geq 1$ is such that its elements are $a_{ij} = c_{k+i-j}$. That is, each descending diagonal from left to right is constant. These matrices are named after Otto Toeplitz. They have applications in time series analysis, image processing, Markov chains and queuing theory, control theory, solution of certain partial differential and integral equations, the computation of spline functions, and polynomial, rational, and exponential approximation; see, for example, [500, 1419]. A survey on the use of Hankel and Toeplitz matrices in numerical analysis is given in [141, 2525].

About the names attributed to these matrices, Nicolaï Nikolski wrote [2351]

> *There are legends connecting the appearance of the theory of Toeplitz operators with an article by Otto Toeplitz in 1911 and that of Hankel matrices with the works of Hermann Hankel in the 1860s. In fact, neither the first nor the second of these attributions stands up to close examination.*

But we have seen that he was not completely right.

Hankel and Toeplitz matrices only depend on $2k - 1$, usually complex, numbers, and thus, specially well adapted algorithms for their numerical treatment have been designed. They are related by $H_n^{(k)} = T_n^{(k)} J_n$, where J_n is the exchange matrix with ones on the antidiagonal and all other elements zero. Both matrices have the same eigenvalues up to their signs. As proved by Ke Ye and Lek-Heng Lim [3294], every $n \times n$ matrix is generically a product of $\lfloor n/2 \rfloor + 1$ Toeplitz matrices and always a product of at most $2n + 5$ Toeplitz matrices. The same result holds for Hankel matrices and the generic bound $\lfloor n/2 \rfloor + 1$ is sharp. For properties of Hankel and Toeplitz matrices and forms, see [382, 1778].

An impressive enumeration of the early works on Toeplitz matrices was given in the Ph.D. thesis [2950] of Douglas Robin Sweet in 1982. We will mostly follow him but the reading of his work is recommended. According to this work, the American mathematician Norman Levinson (1912-1975) was the first to solve a Toeplitz linear system in $\mathcal{O}(n^2)$ operations in 1946 [2040]. His algorithm was improved twelve years later by James Durbin (1923-2012), a British statistician and econometrician [1031]. A method consisting in eliminating down the diagonals was proposed by Erwin Hans Bareiss (1922-2003) in 1969 [192]. In 1965, the operation-count of Levinson's algorithm was improved by Ralph A. Wiggins and Enders Anthony Robinson, an American geophysicist who developed the first digital signal filtering methods to process seismic records used in oil exploration [3231]. Their algorithm was re-derived in 1973 by John D. Markel and Augustine Heard "Steen" Gray Jr. (1936-2019) using orthogonal polynomials [2140].

In 1964, William F. Trench (1931-2016) gave an algorithm for inverting a Toeplitz matrix [3071], and, one year later, another one for Hankel matrices [3072]. A detailed proof about this algorithm for non-Hermitian Toeplitz matrices was given in 1969 by Shalhav Zohar (1927-2020) [3340], a radar astronomer, who also proved that the only necessary condition for insuring the validity of the algorithm is that all principal minors be nonzero. He also proposed an algorithm needing less arithmetic operations [3341].

A different algorithm, based on the polynomials due to Gabor Szegö (1895-1985) that are orthogonal on the unit circle [2981], was given by James L. Justice [1854, 1855] in 1972 and 1974. The propagation of errors in various algorithms was studied by George Cybenko [798, 799] in 1979 and 1980.

Between 1968 and 1971, Elwyn Ralph Berkelamp (1940-2019) [289], James Lee Massey (1934-2013) [2165], and Jorma Johannes Rissanen (1932-2020) [2575] developed elimination methods to reduce Toeplitz and Hankel systems to triangular forms. These methods still work when a leading minor is zero, in contrast to other methods. Rissanen also published other papers [2576, 2577] on these topics.

In 1971, James L. Phillips gave an algorithm for the factorization of a Hankel matrix as $R^* D R$, where R is unit upper triangular and D is diagonal [2492]. He pointed out its connection with the Lanczos algorithm and polynomials orthogonal with respect to a bilinear form, that is, formal orthogonal polynomials.

The first algorithm for the inversion of block Toeplitz matrices was proposed by the Japanese statistician Hirotugu Akaike (1927-2009) in 1973 [21].

The complexity of solving a Toeplitz system was reduced from $3n^2$ to $2n^2 + 8n \log n$ by J.R. Jain in 1979 [1802], who also incorporated an iterative improvement needing $8n \log n$ operations by iteration. In 1980, there was a major advance with Toeplitz solvers having an asymptotic complexity of $\mathcal{O}(n \log^2 n)$ by Robert R. Bitmead and Brian David Outram Anderson [334], and by Richard Peirce Brent, Fred G. Gustavson, and David Y.Y. Yun [426]. However, Brent mentioned to D.R. Sweet that n has to be quite large for the new algorithm to be faster that $\mathcal{O}(n^2)$.

A cooperation between the USA and the USSR led, in 1979, to a Toeplitz software package in the style of LINPACK [3151].

The thesis of D.R. Sweet [2950] in 1982 was devoted to Bareiss' algorithm, which is related to Trench's, to its improvement, and to the study of the propagation of rounding errors. The QR factorization and the SVD decomposition of a Toeplitz matrix were also studied.

Fast algorithms for Hankel and Toeplitz systems were given in [3082] by Evgeny (Eugene) Evgen'evich Tyrtyshnikov in 1989. Later, so-called superfast algorithms were developed; see, for instance, [632].

The case of Toeplitz-plus-Hankel matrices was treated, for example, in [2203]. On this topic, see also [2926] by Gilbert Strang and Shev MacNamara in 2014.

It is not possible to refer to all the publications related to Hankel and Toeplitz matrices. Recent methods, with many references, are presented in [1631]. For other structured matrices, see [325, 2434, 3118, 3119].

9.4 ▪ Functions of matrices

For the early history of functions of matrices, we rely on [9] by Sydney Naphtali Afriat in 1959, [1667] by Konrad John Heuvers and Daniel Moak in 1987, and the book on functions of matrices [1683] by Nicholas John Higham in 2008 (see pages 26-27). Note that here a function f of a matrix A is not a matrix whose entries are $f(a_{i,j})$. Let us start with the definitions of matrix functions.

When matrices were formally introduced in 1858 [587], the aim was to be able to define for them the same basic operations that could be done for the real numbers, sum, product, and so on. Therefore, after having defined powers of matrices, Arthur Cayley (1821-1895) considered the square root of a matrix. As was usual for Cayley, he did not consider the general case, but he derived formulas for the square root of a matrix of order 2. In fact, he considered obtaining the square root as an application of the "fundamental theorem" that we now call the Cayley-Hamilton theorem. He came back to this problem in 1872 for a matrix of order 3 [589].

In 1867, Edmond Nicolas Laguerre (1834-1886) defined the exponential of a matrix in Section V of [1976] by its power series. He needed it for using matrices in his study of Abelian functions. The exponential of a matrix was defined in the same way by Giuseppe Peano (1858-1932) in 1887. His paper, written in Italian, was translated to French, slightly modified [2475], and published in a German journal in 1888. Peano's goal was to write the solution of a system of linear ordinary differential equations using the exponential of the matrix of the coefficients.

Sylvester briefly gave a general definition of a function of a matrix in [2967] in 1883 using a Lagrange interpolating polynomial and the eigenvalues (that he called the latent roots). However, his formula cannot be used when the eigenvalues are not distinct, and he wrote, without more details,

> *When any of the latent roots are equal, the formula must be replaced by another obtained from it by the usual method of infinitesimal variation.*

Sylvester's formula was considered by Arthur Buchheim (1859-1888) [490] in 1884. He extended it to the case of multiple eigenvalues using Hermite interpolation [491] in 1886.

Even though he did not explicitly mention matrices in the paper [3222] in 1887, Eduard Weyr (1852-1903) defined functions by power series and gave conditions of convergence, that is, the eigenvalues must lie inside the circle of convergence of the series; see also Kurt Hensel (1861-1941) [1647], who gave necessary and sufficient conditions for convergence in 1926. Henry Taber (1860-1936) defined trigonometric functions of a matrix [2988, 2990] in 1890-1891. Emmanuel Carvallo (1856-1945) defined e^A [561] in 1891. In 1892, William Henry Metzler (1863-1943) defined e^A, $\log(A)$, $\sin(A)$, $\arcsin(A)$ with power series in [2208].

Ferdinand Georg Frobenius (1849-1917) [1253] stated that if the function f is analytic, $f(A)$ is the sum of residues of $(\lambda I - A)^{-1} f(\lambda)$ at the eigenvalues of A.

Henri Poincaré (1854-1912) gave the Cauchy integral definition [2504] in 1900. Other relevant papers are by Henry Bayard Phillips (1881-1973) [2491] in 1918 and by Luigi Fantappié (1901-1956) [1137] in 1928.

Giovanni Giorgi (1871-1950) used the Jordan canonical form to define matrix functions [1356] in 1928; see also Michele Cipolla (1880-1947) [709] in 1932. A small book about matrix functions was published by Hans Wilhelm Eduard Schwerdtfeger (1902-1990) [2728] in 1938.

Matrix functions were studied by Hans Richter (1912-1978) [2569] in 1950. In 1955, Robert F. Rinehart (1907-1985) wrote a survey paper and proved the equivalence of definitions [2574].

Functions of matrices appeared in books about matrices in the 1930s. See Section 6.6 in the book by Herbert Westren Turnbull (1885-1961) and Alexander Craig Aitken (1895-1967) [3081] in 1932, Chapter 9 in the book by Cyrus Colton MacDuffee (1895-1961) [2106] in 1933, and Chapter 8 in the book by Joseph Henry Maclagan Wedderburn (1882-1948) [3199] in 1934.

Robert Alexander Frazer (1891-1959), William Jolly Duncan (1894-1960), and Arthur Roderick Collar (1908-1986) used the exponential function in their 1938 book [1222] for solving ordinary differential equations with constant coefficients. However, as we have seen, this was done before by Peano. Functions of matrices were also discussed in the book [261] by Richard Ernest Bellman (1920-1984) in 1960.

There exist many numerical methods for computing functions of matrices. The choice of the method depends on what is needed. One may want all the entries of $f(A)$, but this can in general be done only for "small" matrices. If only a few of the entries are needed, this can be seen as a particular case of computing the scalar $u^T f(A) v$, where u and v are given vectors. Sometimes, it is only needed to compute the vector $f(A)b$ with b given. Different methods are used in these three cases.

A reference book for computing all the entries of $f(A)$ is *Functions of matrices: Theory and computation* [1683] by N.J. Higham, in which many methods are described and analyzed; see also the papers he published on that topic since 2008. A review paper [1264] on matrix functions was written by Andreas Frommer and Valeria Simoncini in 2008.

For general functions f, a basic method is to use Taylor series, but there are constraints when the radius of convergence is not infinite. One can also replace the function by a polynomial or rational approximation. If the matrix A is diagonalizable $A = X \Lambda X^{-1}$, then $f(A) = X f(\Lambda) X^{-1}$ and $f(\Lambda)$ can be trivially computed since Λ is diagonal. When A is not diagonalizable, one can use the Schur factorization of the matrix (see Section 4.8), then the function of a triangular matrix has to be computed; see [2445] by Beresford Neill Parlett, who derived a recurrence among the entries of functions of triangular matrices in 1976.

As we have seen above, the first function of matrix that was ever considered was the square root. Note that the square root of A is not unique and one has to specify which one has to be computed; see [1683]. The Schur factorization was used by Åke Björck and Sven Hammarling [341] in 1983.

Iterative methods can also be used. In 1937, Cornelis Visser (1910-2001) used a Richardson-like iteration

$$X_{k+1} = X_k + \alpha(A - X_k^2), \quad X_0 = \frac{1}{2\alpha} I.$$

A Newton iteration

$$X_{k+1} = \frac{1}{2}(X_k + X_k^{-1} A), \quad X_0 = A,$$

modeled after the method for computing the square root of a real number was considered by Pentti Laasonen (1916-2000) [1965] in 1958. He noticed the potential instability of the method. Other iterative methods were proposed by Eugene Dale (Gene) Denman (1928-2013) and Alex N. Beavers Jr. [881] in 1976, Bruno Iannazzo [1771] in 2003, and Beatrice Meini [2195] in 2004.

These methods can somehow be generalized for the pth root of a matrix; see [1683].

As we have seen above, the exponential can be defined by its Taylor series. The computation of the matrix exponential was the subject of many papers over the years. In 1978, Cleve Barry Moler and Van Loan reviewed nineteen methods for computing e^A [2267]. They wrote

> *In this survey we try to describe all the methods that appear to be practical, classify them into five broad categories, and assess their relative effectiveness.* [...] *In assessing the effectiveness of various algorithms we will be concerned with the following attributes, listed in decreasing order of importance, generality, reliability, stability, accuracy, efficiency, storage requirements, ease of use, and simplicity.*

They also provided an interesting list of references. An extended version of the paper [2268] was published in 2003.

Igor Moret and Paolo Novati used Faber polynomials to obtain an interpolatory approximation of the matrix exponential [2271] in 2001.

Estimates of condition numbers for matrix functions were obtained by Charles Kenney and Alan John Laub [1884] in 1989. How to compute Fréchet derivatives of matrix functions was studied in 2010 by N.J. Higham and Awad Al-Mohy in [1687]; see also [1693] with Samuel Relton in 2014.

In 2011, Hongguo Xu gave formulas for functions of A in terms of Krylov matrices of A [3285].

Let us now turn to the computation of $f(A)b$. Starting in the 1990s, Krylov methods (see Chapter 5) were used to compute such a product, particularly when the matrix A is sparse and large. Many papers were considering the exponential function.

An early attempt of using Krylov methods and matrix functions was done by Tae Jun Park and J.C. Light [2435] for the exponential in 1986 for studying the time evolution for wave packets and by Henk van der Vorst in 1987 for general functions. In fact, in [3098], he was interested in solving $f(A)x = b$; see also Chapter 11 of his book [3101] in 2003.

At iteration k, Krylov methods provide a matrix V_k of basis vectors and an upper Hessenberg matrix H_k (which can be tridiagonal in some methods). Then, one can compute (if $V_k e_1 = b/\|b\|$)

$$f(A)b \approx \|b\| \, V f(H_k) e_1,$$

where e_1 is the first column of the identity matrix. Generally, H_k is not too large and $f(H_k)$ can be computed with the methods we have discussed before.

In 1989, Vladimir L. Druskin and Leonid Aronovich Knizhnerman used Chebyshev series and the Lanczos algorithm for computing functions of symmetric matrices [959].

Computing functions of matrices using Krylov subspace methods was considered by Thomas Ericsson from Sweden in a 1990 technical report [1100] which, apparently, was never published in a journal.

The Lanczos algorithm was again used for computing functions of symmetric matrices [962] by Druskin and Knizhnerman in 1991. Using Arnoldi's method for nonsymmetric matrices was done by Knizhnerman [1911] in 1991; see also [960, 961] by Druskin and Knizhnerman in 1995 and 1998, and a joint paper [958] with Anne Greenbaum in 1998.

A Krylov method and rational approximations of the exponential were used by Efstratios Gallopoulos and Yousef Saad [1275] in 1992 to obtain approximate solutions of parabolic equations. A theoretical analysis of some Krylov subspace approximations to the matrix exponential operation $\exp(A)v$ [2657] was published by Saad in 1992.

David Edward Stewart and Teresa S. Leyk developed an upper estimate of the error for the Krylov subspace approximations to the exponential [2866] in 1996.

A class of time integration methods for large systems of nonlinear differential equations which use Krylov approximations to the exponential function of the Jacobian instead of solving linear or nonlinear systems of equations in every time step was proposed by Marlis Hochbruck and Christian Lubich [1704] in 1997; see also the joint paper [1706] with Hubert Selhofer in 1998.

Expokit, a software package for computing matrix exponentials [2762], was described by Roger Blaise Sidje in 1998.

The cost per iteration of the Arnoldi process increases with the iteration number, as well as the storage. In 2006, Michael Eiermann and Oliver Gerhard Ernst showed how to restart the process when computing matrix functions [1058]. On this restarting technique, see also [8] by Martin Afanasjew, Eiermann, Ernst, and Stefan Güttel in 2013 and [1260] by Frommer, Güttel, and Marcel Schweitzer in 2014.

Rational Krylov methods for solving eigenproblems were essentially developed by Axel Ruhe (1942-2015) starting in the 1980s. The use of rational Krylov spaces for computing $f(A)b$ was studied in the Ph.D. thesis [1511] of Güttel in 2010; see also [1512] in 2013 and [248] with Bernhard Beckermann in 2012. Note that in rational Krylov methods one has to solve linear systems with shifted matrices $A - \xi_j I$.

A special case of rational Krylov subspace is the so-called extended Krylov subspace which was studied by Carl Jagels and Lothar Reichel [1800] in 2009; see also their study of recursion relations [1801] in 2011. In 2010, Knizhnerman and Simoncini [1913] did an investigation of the extended Krylov subspace method for matrix function evaluations. Thomas Mach, Miroslav S. Pranić, and Raf Vandebril showed how to compute approximate extended Krylov subspaces without explicit inversion [2109, 2110] in 2013.

More recent papers on computing $f(A)b$ are [1262] by Frommer, Kathryn Lund, Schweitzer, and Daniel B. Szyld in 2017, [100] by Jared L. Aurentz, Anthony P. Austin, Michele Benzi, and Vassilis Kalantzis, [381] by Mikhail A. Botchev and Knizhnerman, and [1513] by Güttel, Daniel Kressner, and Lund in 2020.

A literature survey of matrix functions for data science [2909] was done by Martin Stoll in 2020.

Methods for computing bounds or approximations of $u^T f(A)v$ for a symmetric matrix A were proposed and studied by Golub and his collaborators in the 1990s. The idea was to use the spectral factorization $A = X\Lambda X^T$, which yields $u^T f(A)v = u^T X f(\Lambda) X^T v$. This can be written as a Riemann-Stieltjes integral for a piecewise constant measure having jumps at the eigenvalues of A. Then, Gauss or Gauss-Radau quadrature rules are used to obtain approximations or bounds if the derivative of f has a constant sign on the interval of integration. It turns out that the approximations can be obtained by running the Lanczos algorithm when $u = v$ or the nonsymmetric Lanczos algorithm otherwise, and by computing entries of $f(T_k)$, where T_k is the tridiagonal matrix generated by the Lanczos algorithms.

This was developed by Golub and G.M. [1384] in 1994; see also the book [1386] in 2010.

Daniela Calvetti, Sun-Mi Kim, and Reichel developed quadrature rules based on the Arnoldi process when the matrix is not symmetric [528] in 2005. Zdeněk Strakoš and Petr Tichý [2925] used the biconjugate gradient algorithm to approximate bilinear forms $c^* A^{-1} b$ when A is non-symmetric in 2008.

Reichel and his many collaborators proposed other quadrature rules for the approximation of functionals; see, for instance, [1967] in 2008, [1152] in 2013, [2550] in 2016, and [2549] in 2021.

Functionals $u^T f(A) v$ were used to analyze communicability in networks, which has become a fashionable topic since the rise of social networks. This was done by Benzi and Paola Boito [271] in 2010, Benzi, Ernesto Estrada, and Christine Klymko [275] in 2013, and Benzi and Boito [272] in 2020. It was also studied by Caterina Fenu, David Martin, Reichel, and Giuseppe Rodriguez [1152] in 2013 and Omar De la Cruz Cabrera, Mona Matar, and Reichel [844] in 2019.

9.5 ▪ Ill-posed problems

Problems are generally defined as ill-posed when the solution is not unique or does not depend continuously on the data. In practice, problems are defined to be *ill-posed* when a small change in the data may cause a large change in the solution. Problems whose solutions are sensitive to perturbations of the data are called *Discrete Ill-posed Problems* (DIPs). This is typically what may happen in the solution of least squares problems. Ill-posed problems are a vast area in applied mathematics. In this section, we only consider the numerical solution of linear problems.

The solution of ill-posed linear systems arises in many areas of scientific computing; see, for instance, [3126] by James Martin Varah, [3041] edited by Andrey Nikolayevich Tikhonov (1906-1993) in 1992, [1096] by Heinz Werner Engl in 1993, [1569] by Martin Hanke and Per Christian Hansen in 1993, the book [1566] by Hanke in 1995, the book [1097] by Engl, Hanke, Andreas Neubauer in 1996 and the books by Hansen [1579] in 1998, [1581] in 2010, and [1584] in 2006 with James Gerard Nagy and Dianne Prost O'Leary.

Typical problems leading to ill-posed problems are discretizations of integral equations of the first kind,

$$\int_a^b k(s,t) f(t)\, dt = g(s),$$

where the unknown is the function f. The potential instability comes from the decay to zero of the singular values of the kernel k.

Problems also happen when one has to solve an overdetermined linear system $Ax \approx c = \bar{c} - e$, where A is an $m \times n$, $m \geq n$ matrix and the right-hand side $\bar{c}$ is contaminated by a (generally) unknown noise vector e. The standard solution of the least squares problem $\min \|c - Ax\|$ (even using backward stable methods with the QR factorization) may give a vector x severely contaminated by noise.

One of the remedies to obtain a meaningful solution is to use *Tikhonov regularization*, which for our case is solving

$$\min_x \{\|c - Ax\|^2 + \mu \|x\|^2\},$$

where $\mu > 0$ is the regularization parameter.

This kind of method was introduced by Tikhonov, who studied these problems in the 1940s; see [3039] in 1943, [3040] in 1963, and the book [3042] with Vasilii Yakovlevich Arsenin (1915-1990) in 1976. Regularization was also advocated by David L. Phillips [2490] in 1962 for integral equations of the first kind. The technique is sometimes called the *Tikhonov-Phillips regularization*.

Ill-posed problems were intensively studied by Soviet researchers; see, in particular, [1787] by Valentin Konstantinovich Ivanov (1908-1992) in 1962, [171] by Anatolii Borisovich Bakushinskii, and the books [3043] by Tikhonov and Aleksandr Vladimirovich Goncharsky in 1987, [3044] by Tikhonov, Goncharsky, V.V. Stepanov, and Anatolii Grigor'evich Yagola in 1990.

Of course, the main problem is to find a "good" value of the regularization parameter. An overview of methods was published by Golub and Urs von Matt [1404] in 1997. Let us review a few of them.

Vladimir Alekseevich Morozov introduced the *discrepancy principle* [2286] in 1966, assuming that the level of noise is known; see also the book [2287] in 1993. The *quasi-optimality* technique was described by Aleksandr Sergeevich Leonov [2037] in 1978. A method which also needs the knowledge of the norm of the noise vector is the *Gfrerer/Raus* method; see Helmut Gfrerer [1342] in 1987 and [1570] by Hanke and Toomas Raus in 1996.

Charles Lawrence Lawson (1931-2015) and Richard Joseph Hanson (1938-2017) [2004] observed in the 1970s that a "good" way to see how the regularized solution x_μ depends on the parameter μ is to plot the curve $(\|x_\mu\|, \|c - Ax_\mu\|)$ obtained by varying the value of $\mu \in [0, \infty)$. This curve is known as the *L-curve* since it is, in many circumstances, shaped like the letter L. In fact, it is even more illuminating to look at this curve in a log-log scale as suggested by Hansen and O'Leary [1585] in 1993. Lawson and Hanson proposed to choose the value μ_L corresponding to the "vertex" or the "corner" of the L-curve, that is, the point with maximal curvature.

For the study and/or the use of the L-curve, see [1577] by Hansen in 1992, [2543] by Teresa Regínska in 1996, [1567] by Hanke in 1996, [3152] by Curtis R. Vogel in 1996, [527] by Calvetti, Hansen, and Reichel in 2002, [538] by Calvetti and Reichel in 2003, [540] by Calvetti, Reichel, and Abdallah Shuibi in 2005, and [1583] by Hansen, Toke Koldborg Jensen, and Rodriguez in 2007.

Cross-validation and generalized cross-validation (GCV) are techniques for model fitting for given data and model evaluation. The available data can be split into two sets: one for fitting and one for evaluation. The idea of cross-validation, introduced by the statistician David M. Allen [31] for linear regression in 1974, was to recycle the data. GCV was considered by Peter Craven and Grace Wahba [761] in 1979. If we write the regularized least squares problem as

$$\min_x \{\|c - Ax\|^2 + m\mu\|x\|^2\},$$

where A is an $m \times n$ matrix, the GCV estimate of the parameter μ is the minimizer of the function

$$G(\mu) = \frac{\frac{1}{m}\|(I - A(A^TA + m\mu I)^{-1}A^T)c\|^2}{\left[\frac{1}{m}\text{tr}(I - A(A^TA + m\mu I)^{-1}A^T)\right]^2},$$

where "tr" is the trace; see [1380] by Golub, Michael Thomas Heath, and Wahba in 1979. The computation of the GCV parameter using Gauss quadrature and stochastic trace estimators was considered by Golub and von Matt [1403, 1404] in 1997; see also Chapter 15 of the book [1386] by Golub and G.M. in 2010 for a summary and some improvements of the method. A weighted GCV method was proposed by Julianne Chung, Nagy, and O'Leary [694] in 2008.

The determination of the parameter by minimization of the error norm using estimates was proposed by C.B., Rodriguez and Sebastiano Seatzu (1941-2018) [460] in 2008; see also [461] in 2009. Computing solutions for several values of the regularization parameter and extrapolating to zero was considered by C.B., M.R.-Z., Rodriguez, and Seatzu [452] in 1998. In 2003, the same authors used multi-parameter regularization techniques [453]. Multiple regularization parameters in a generalized L-curve framework was also proposed by Murat Belge, Misha Elena Kilmer, and Eric Lawrence Miller [253] in 2002. For a more recent work, see Silvia Gazzola and Reichel [1313] in 2016.

Early in the 1950s, people turned to iterative methods to solve large ill-posed problems since most of them only require the ability to compute the product of the matrix with a vector. Iterative methods can be used in two ways: either to solve the regularized system (for a given or several values of μ) or as regularizing procedures by themselves.

For ill-posed problems, iterative methods can be *semiconvergent*, that is, the residual norms are decreasing up to some iteration and then they start increasing. The iterations must be stopped at that point, which is the most regularizing one.

Conjugate gradient type methods were used, but when solving least squares problems $\min_x \|c-Ax\|$, it was usually preferred to use the LSQR algorithm of Christopher Conway Paige and Michael Alan Saunders [2428] in 1982.

Iterations and regularization of the projected problem were considered by O'Leary and John A. Simmons [2373] in 1981. The MINRES iterative algorithm of Paige and Saunders was used by Kilmer and Gilbert Wright Stewart [1896] in 1999. In 2001, Kilmer and O'Leary studied how to choose regularization parameters in iterative methods for ill-posed problems [1895]. Lanczos-based methods were also used by Hanke [1568] in 2001.

The regularizing properties of GMRES (see Section 5.8) were studied by Calvetti, Bryan Lewis, and Reichel [531] in 2002. Serena Morigi, Reichel, Fiorella Sgallari, and Fabiana Zama considered how to choose members of semiconvergent sequences in iterative methods [2281] in 2006.

Iterative regularization with minimal residual methods was the topic of a paper [1817] by Hansen and Jensen in 2007. In 2008, they studied the noise propagation in regularizing iterations for image deblurring applications [1582].

The Golub-Kahan bidiagonalization algorithm was used in 2010 by Michiel Erik Hochstenbach and Reichel for Tikhonov regularization with a general linear regularization operator [1712].

Lars Eldén and Simoncini solved ill-posed linear systems with GMRES and a singular preconditioner [1080] in 2012.

Implementations of range restricted iterative methods for linear DIPs were published by Arthur Neuman, Reichel, and Hassane Sadok [2335] in 2012. Using FGMRES for DIPs was proposed by Keiichi Morikuni, Reichel, and Ken Hayami [2284] in 2014.

Some properties of the Arnoldi-based methods for linear ill-posed problems were studied by Novati [2358] in 2017.

GMRES does not always perform well when applied to ill-posed problems. The paper [1312] by Gazzola, Silvia Noschese, Novati, and Reichel in 2019 looked for reasons explaining the poor performance of GMRES in certain cases and discussed some remedies based on specific kinds of preconditioning.

MATLAB codes for regularization of ill-posed problems were made available by Hansen; see [1578] in 1994 and [1580] in 2007. A more recent package *IR Tools* was described in 2018 by Gazzola, Hansen, and Nagy [1311].

Another method to obtain meaningful solutions of ill-posed problems is the *truncated SVD* (TSVD), using the singular value decomposition (see Section 4.3). The SVD is computed, the solution is written with the singular values and vectors, and the terms corresponding with the smallest singular values are neglected. Unfortunately, this can only be used for small problems because the SVD computation is expensive.

The truncation of the SVD was explicitly used by Hanson for solving Fredholm integral equations of the first kind [1587] in 1971. Early given references on integral equations were the book [2809] by Frank Smithies (1912-2002) for the use of the SVD in integral equations in 1958 and the paper [168] by Christopher Thomas Hale Baker (1939-2017), Leslie Fox (1918-1992), David Francis Mayers, and K. Wright in 1964.

Numerical methods for linear DIPs, including the truncated SVD, were compared by Varah [3125] in 1979; see also [3126] in 1983. The truncated SVD was also considered a method for regularization by Hansen [1576] in 1987; see also [1586] with Takashi Sekii and Hiromoto Shibahashi in 1992.

The TSVD was also used by statisticians; see, for instance, Jerald Franklin Lawless and P. Wang [2003] in 1976, Arthur Pentland Dempster, Martin Schatzoff, and Nanny Wermuth [879] in 1977, as well as scientists from other areas, such as T. Sekii [2741] in 1991 and Douglas Gough [1414] in 1996.

9.6 ▪ Matrix equations

A *matrix equation* is an equation whose unknown is a matrix. The simplest matrix equations are the linear ones such as $AX = I$ or $AX = B$, which are related to the computation of the inverse or the pseudo-inverse of a matrix or the solution of a system of linear equations. They can be solved by any of the direct or iterative methods described in the preceding chapters.

Among other matrix equations that have been studied, we find $XX^T = A$ [2171], $AXB = C$ [1850], $X^2 - EX = F$ [1480], or $AX^2 = I$ [1965], and so on. Since all equations and contributions cannot be mentioned and analyzed, we restrict ourselves to the equations named after Lyapunov, Riccati, Stein, and Sylvester.

The Sylvester equation is $AX - XB = C$, where A, B, C are given matrices. It is named after the English mathematician J.J. Sylvester; see Section 10.67. If $B = A^*$, and $C = C^*$, this equation is named after Lyapunov, since it is related to the differential equations on the stability of an equilibrium studied by Aleksandr Mikhailovich Lyapunov (1857-1918) in his thesis in 1884 [2103].

The equation $X - AXB = C$ is called the Stein equation. It has a unique solution if and only if $\lambda\mu \neq 1$ for any pair λ, μ of eigenvalues of A and B, respectively.

The algebraic Riccati equation is $XA + A^TX - XBX + C = 0$, where B and C are symmetric positive definite. Its name comes from the fact that it is a difference equation analogous to the differential equation studied by Jacopo Francesco, Count Riccati (1676-1754) between 1719 and 1724 [2562].[57]

An interesting account with historical references is given in Chapter VIII (titled "Matric equations") of the book by MacDuffee [2106].

Let us mention that equations of the form $f(X) = A \in \mathbb{C}^{n\times n}$, where f is a complex holomorphic function, are considered in [1117]. We will not discuss the solution of nonlinear matrix equations since they involve iterative methods mostly based on the form of each equation.

Sylvester equation

As we have seen above, the Sylvester equation is

$$AX - XB = C, \quad A \in \mathbb{C}^{n\times n}, B \in \mathbb{C}^{m\times m}.$$

It plays an important role in matrix eigen-decompositions, control theory, model reduction, numerical solution of the matrix differential Riccati and the algebraic Riccati equations, image processing, etc. A survey of theoretical results for this equation in finite or infinite-dimensional spaces in given in [314].

Sylvester's equation has a unique solution $X \in \mathbb{C}^{n\times m}$ for any $C \in \mathbb{C}^{n\times m}$ if and only if A and B have no common eigenvalue. If this condition is not satisfied, the equation can have

[57] See http://www.17centurymaths.com/contents/euler/rictr.pdf (accessed August 2021), for an English translation

several solutions. If $A = B$, the trace of C is equal to that of $AX - XA$, which is zero since AX and XA have equal traces. Thus, a solution can exist only if the trace of C is 0, and, therefore $AX - XA = I$ has no solution. The matrix C and one of the two matrices A or B can be singular (but not both, since in that case, they will have the zero eigenvalue in common). The uniqueness result was first proved by Sylvester in the case of square matrices with $C = 0$ in 1884 in a paper [2973] written in French. He considered the equation $px = xq$, where p and q are two matrices of order ω. This equation leads to ω^2 homogeneous linear equations between the ω^2 elements of x and those of p and q. Thus, the equation has a solution if the elements of p and q are related by one and only one relation. Sylvester wrote that since the identical equation in p (*l'équation identique en* p), that is, the Cayley-Hamilton theorem, is written under the form

$$p^\omega + Bp^{\omega-1} + Cp^{\omega-2} + \cdots + L = 0,$$

then, since, apparently, $p = xqx^{-1}$, it follows

$$xq^\omega x^{-1} + Bxq^{\omega-1}x^{-1} + Cxq^{\omega-2}x^{-1} + \cdots + L = 0,$$

that is,

$$q^\omega + Bq^{\omega-1} + Cq^{\omega-2} + \cdots + L = 0.$$

Thus the roots (the eigenvalues) of q are identical to those of p and, instead of one single equation, we apparently obtain at least ω equations between the elements of p and q. To make this paradox disappear, the only assumption to be made is that x is a void matrix, and thus x^{-1} does not exist and $p = xqx^{-1}$ cannot hold. Then, Sylvester gave an example of dimension two and explained that the result can be extended to the general case. He wrote[58] (our translation)

> *We only have to prove that if one of the latent roots of p is equal to one of q, the equation $px = xq$ is solvable and moreover, without this condition being satisfied, the equation is irresolvable.*

Note that if the eigenvalues are not equal, $x = 0$ is the only solution. His proof made use of the resultant R of the polynomials in p and q, and the conclusion is that the condition $R = 0$ is not only necessary, but also sufficient for the equation to be solvable.

In [2971, 2972, 2974], Sylvester discussed the more general equation

$$A_1XB_1 + \cdots + A_kXB_k = C,$$

where all matrices are square of order n. Again, this equation can be viewed as a system of n^2 equations in the n^2 elements of X. He considered the matrix $A_1 \otimes B_1 + \cdots + A_k \otimes B_k$ that he named *nivellateur*, although he did not recognized it as a sum of Kronecker products. The coefficients of the characteristic polynomial of this "nivellateur" were obtained in 1922 by Frank Lauren Hitchcock (1875-1957), an American mathematician and physicist known for his formulation of the transportation problem, thus allowing to solve Sylvester's general equation [1702]. A finite series solution based on the "nivellateur," after transforming A and B to Jordan canonical forms, was given by Er-Chieh Ma in 1966 [2105].

In 1885, Cayley discussed the solution of the 2×2 matrix equation $pX - Xq = 0$ in [590], noticing that the problem is equivalent to that for quaternions that he already treated in [591]. The case $AX = XA$ was considered by Aurel Edmund Voss (1845-1931) in 1889 [3161]. In 1891, the American mathematician Taber gave the solution involving the Jordan canonical form of A [2989]. This equation was also discussed by Georg Landsberg (1865-1912) in 1896 [1988],

[58] *On n'a qu'à démontrer que si une des racines latentes de p est égale à une de q, l'équation $px = xq$ est résoluble et de plus, sans que cette condition soit satisfaite, l'équation est irrésoluble.*

later by Rowland Wilson (1895-1980) [3260, 3261] in 1930-1933, and in Chapter X of the book by Turnbull and Aitken [3081] in 1932.

An infinite series solution was given in 1904 by Wedderburn [3197]. He considered the equation $\chi - \phi\chi\psi = \theta$. Replacing χ by $\theta + \phi\chi\psi$ leads to $\chi - \phi^2\chi\psi^2 = \theta + \phi\theta\psi$. Repeating the process n times, he obtained

$$\chi - \phi^{n+1}\chi\psi^{n+1} = \theta + \phi\theta\psi + \phi^2\theta\psi^2 + \cdots + \phi^n\theta\psi^n = \theta_n,$$

and he claimed that the series, when convergent, is a solution of the matrix equation.

The number of linearly independent solutions of $AX = XB$ was studied successively by Frobenius in 1878 [1250], Ludwig Maurer (1859-1927) in his dissertation in 1887 [2170], Voss in 1889 [3161], Landsberg in 1896 [1988], Hensel in 1904 [1648], and Ugo Amaldi (1875-1957) in 1912 [43]. Explicit solutions of this equation were given by Henri Kreis [1947], a student of Adolf Hurwitz (1859-1919), and by Léon César Autonne (1859-1916) [111].

In 1909, the Italian mathematician Francesco Cecioni (1884-1968) proved that a necessary condition that $AX = B$ and $XC = D$ have a common solution is that $AD = BC$ since $AXC = BC = AD$. If either A or B is nonsingular, the condition is also sufficient [595]. He also studied the number of linearly independent solutions in his Habilitation thesis [596] in 1910. The fact that $AX = BX$ has a nonzero solution if and only if A and B have a common eigenvalue seems to have been independently discovered by Cecioni in this thesis, and by Frobenius who also gave a characterization of the number of linearly independent solutions [1256].

Cecioni [596] in 1910, Daniel Edwin Rutherford (1906-1966) [2633] in 1932, and Roland Weitzenböck (1885-1955) [3212], that same year, discussed the case $AX + XB = C$. Let us mention that this last author also studied the equation $X^2 = A$ [3213]. According to Mark Hoyt Ingraham (1896-1982), a mathematician who established the first academic computing center in the United States on the campus of the University of Wisconsin in Madison, and Harold Callander Trimble (1914-1991), one of his students,

> *D.E. Rutherford considered this equation getting a complete solution, but his method involves the necessity of using the characteristic values of A and B. This equation was also studied by R. Weitzenböck, who showed the existence of what he calls a "reine" solution.*

They considered the equation $TA = BT + C$, where T, the unknown, and C are $m \times n$ matrices, A and B are square of dimension n and m respectively, and they gave a rational construction for the complete rational solution; see [1777] in 1941.

In 1950, William Vann Parker (1901-1987) gave the solution X of the equation $AX = XB$, where A is taken in rational canonical form in terms of parameters [2436]. In 1952, William Edward Roth proved that a necessary and sufficient condition that $AX - XB = C$ has a solution is that the matrices

$$\begin{pmatrix} A & C \\ 0 & B \end{pmatrix} \text{ and } \begin{pmatrix} A & 0 \\ 0 & B \end{pmatrix}$$

be similar; see [2602]. This result is known as *Roth's removal rule*. One can check that

$$\begin{pmatrix} I_n & X \\ 0 & I_m \end{pmatrix} \begin{pmatrix} A & C \\ 0 & B \end{pmatrix} \begin{pmatrix} I_n & -X \\ 0 & I_m \end{pmatrix} = \begin{pmatrix} A & 0 \\ 0 & B \end{pmatrix},$$

where I_k is the identity matrix of order k. It seems that the extension to operators of Sylvester's basic result was known to Krein, who apparently lectured on that topic in the late 1940s. It was later proved in the 1950s, but this is outside the scope of this book.

Let us now discuss numerical methods for solving Sylvester equations. Obviously, the first idea consists in transforming the equation into the system of mn linear equations

$$(I_n \otimes A - B^T \otimes I_m)\text{vec}(X) = \text{vec}(C),$$

where $\otimes$ is the Kronecker product and the vectorization operator *vec* transforms a matrix into the vector obtained by stacking the columns of the matrix on top of one another. This approach is too expensive unless the dimensions are small, and, moreover, the matrix of this linear system can be ill-conditioned [1677]. When the matrices A and B are diagonalizable, Guy Antony Jameson proposed, in 1968, a method for obtaining the solution by inversion of an $m \times m$ or an $n \times n$ matrix without recourse to diagonalization [1805].

The Sylvester equation can be written under the following equivalent form

$$(qI_m + B)X(qI_n - A) - (qI_m - B)X(qI_n + A) = -2qC,$$

where q is a scalar, a relation that can be easily checked. Multiplying on the left by $(qI_m + B)^{-1}$ and by $(qI_n - A)^{-1}$ on the right, assuming the nonsingularity of these matrices, gives $X - UXV = W$ with

$$U = (qI_m + B)^{-1}(qI_m - b), \quad V = (qI_n + A)(qI_n - A)^{-1}$$

and $W = -2q(qI_m + B)^{-1}C(qI_n - A)^{-1}$. Thus $X = \sum_{i=0}^{\infty} U^i W V^i$. This series converges if the spectral radius of U and V are strictly less than 1. The eigenvalues of U and V are, respectively, $\lambda_U = (q - \lambda_B)/(q + \lambda_B)$ and $\lambda_V = (q + \lambda_A)/(q - \lambda_A)$. It follows that, if $q > 0$, these eigenvalues have a modulus strictly less than 1 if the real parts of λ_A and λ_B are negative, and vice versa if $q < 0$. In general, the convergence of the preceding series is slow and it is replaced by the iterations $Y_{k+1} = Y_k + U^{2^k} Y_k V^{2^k}$ with $Y_0 = W$. Obviously $U^{2^{k+1}}$ is obtained by squaring U^{2^k} and similarly for the powers of V. The matrix Y_k is the sum of the 2^k first terms of the series and the sequence converges quadratically. This method was proposed and studied by R.A. Smith in 1968 [2808].

In 1972, Richard Harold Bartels and G.W. Stewart gave the first numerically stable method in the paper [209], submitted in 1970. They considered the Schur decompositions of $A = URU^*$ and $B = VSV^*$, where U and V are unitary matrices, and R and S are upper triangular. Premultiplying the Sylvester equation by U^*, and postmultiplying it by V, gives $RY - YS = D$, with $D = U^*CV$ and $Y = U^*XV$. This is a Sylvester equation with upper triangular coefficient matrices. The ith column y_i of Y is the solution of

$$(R - s_{ii}I)y_i = d_i - \sum_{j=1}^{i-1} s_{ji}y_j, \quad i = 1, \ldots, n.$$

Then $X = UYV^*$. Obviously, the matrices $R - s_{ii}I$ have to be nonsingular for all i, a condition holding if $r_{ii} \neq s_{jj}$ for all i and j, that is, if A and B have no common eigenvalue. Using the QR algorithm for the Schur decompositions, the computational cost of the method is $\mathcal{O}(m^3 + n^3)$.

A variant, where only A is reduced to a Hessenberg form, was proposed by Golub, Stephen Gregory Nash, and Van Loan [1388] in 1979. The method is known as the *Hessenberg-Schur algorithm*, and it requires $(5/3)m^3$ flops, compared to the $10m^3$ flops needed for the Schur decomposition of A. The method is numerically stable.

For large equations, the storage and cost of these methods become prohibitive. Iterative methods have to be used. These include projection methods, which use Krylov subspace iterations (see, for example, [1762]), methods based on the alternating direction implicit (ADI) iteration (see, for example, [3165]), and hybridizations that involve both projection and ADI (see, for example, [963]). Iterative methods can also be used to directly construct low rank approximations of the solution; see [1806]. Iterative methods using matrix inversion are described in [2248, 2583]. The literature on these methods is extensive and cannot be described here; see the 2016 survey [2774] by Simoncini, which already contained 272 references.

Lyapunov equation

The Lyapunov equation occurs in control theory. It is related to the stability of the differential system

$$\dot{x}(t) = Ax(t), \quad x(0) = x_0.$$

The equilibrium state $x(t) = 0$ is said to be stable if, $\forall \varepsilon > 0$, $\exists \delta > 0$ such that $\|x_0\| < \delta$ implies $\|x(t)\| < \varepsilon, \forall t$. Moreover, if $\lim_{t\to\infty} \|x(t)\| = 0$, the equilibrium state $x(t) = 0$ is called asymptotically stable. The asymptotic stability arises if all the eigenvalues of A have strictly negative real parts. The stability is guaranteed if the real parts of the eigenvalues of A are less than or equal to zero, and if those with a real part equal to zero are simple.

Lyapunov showed in [2103] that the asymptotic stability is equivalent to

$$A^T X + XA = -Q,$$

where Q is negative definite and X is positive definite. It is a necessary and sufficient condition. This equation is known as the *continuous Lyapunov equation*. The *discrete Lyapunov equation* is

$$AXA^T - X = -Q.$$

It has a solution if and only if the discrete system $x_{t+1} = Ax_t$ is asymptotically stable. These equations can be solved by vectorization. Furthermore, if the matrix A is stable, the solution can be expressed as an integral (continuous time case) or as an infinite sum (discrete time case). In the case where A has eigenvalues with negative real parts and Q is positive semidefinite, a method directly computing the Cholesky factor of the solution was proposed by Hammarling [1564] in 1982.

In 1968, Edward J. Davison and F. Man proposed an iterative algorithm based on the Crank-Nicolson method for solving the continuous equation [833]. They computed iterates of the same form as those of Smith but with U and V replaced by $\Gamma = (I - hA/2 + h^2A^2/12)^{-1}(I + hA/2 + h^2A^2/12)$, and with the initialization hC.

The continuous Lyapunov equation with $Q = I$ was studied in 1969 by James Lucien Howland and John Albert Senez [1756]. When A is an upper Hessenberg matrix, they obtained the solution as a linear combination of n linearly independent symmetric matrices. Another method of the same type was given by Rita Meyer-Spasche in 1971 [2241]. The method of Bartels and Stewart [209] can also be used. A survey of methods available up until 1969 was given in [1532].

The discrete Lyapunov equation was studied by Genshiro Kitagawa in 1976 [1905]. He transformed A into a lower Hessenberg matrix and then to a triangular one by the QR algorithm.

Algebraic Riccati equation

There are two types of algebraic Riccati equations that appear in infinite-horizon optimal control problems and in optimal filter design. The first type is

$$XDX + XA + A^*X + C = 0,$$

where C and D are Hermitian matrices. For physical reasons, it is usually required that X also be Hermitian. Such an equation arises when looking for solutions of the differential system $\dot{x}(t) = Ax(t) + Bu(t)$ that are maximizing some quadratic form in x and u. That is why this equation is named the *Continuous Algebraic Riccati Equation* (CARE). An analogous problem holds for difference equations of the form $x_{k+1} = Ax_k + Bu_k$, which leads to the matrix equation

$$X = X^*XA + Q - A^*XB(B^*XB + R)^{-1}B^*XA,$$

where A and Q have the same dimension $n \times n$ as X, but R may have the dimension $m \times m$, in which case B is $n \times m$. Its name is the *Discrete Algebraic Riccati Equation* (DARE). The role linear algebra could play in the solution of Riccati equations was pointed out by James Emerson Potter (1937-2005) in 1966 [2516] and by David Lee Kleinman [1908] in 1968, as well as the way to numerical methods for their solution. Such a method, based on Schur decomposition, was proposed by Alan John Laub in 1979 [1999].

Let us quote the interesting story of the work of Potter given in [1460, pp. 238-239]:

> *The most reliable and numerical stable implementations of the Kalman filter are collectively called* square-root filters. *The fundamental concept for all these methods is to reformulate the matrix Riccati equation in terms of certain factors of the covariance matrix P of estimation uncertainty.* [...]
>
> *The idea of transforming the Riccati equation in this way originated with J.E. Potter in 1962, when he was a graduate student in mathematics at MIT and working part-time at the MIT Instrumentation Laboratory on the Apollo Project to take Americans to the moon and back. The space navigation problem for Apollo was the first real-world application of the Kalman filter, and there was a serious problem with its intended implementation on the Apollo computer. The critical problem was poor numerical stability of the Riccati equation solution, especially the measurement updates.*
>
> *Potter took the problem home with him one Friday afternoon and returned Monday morning with the solution. He had recast the measurement update equations of the Riccati equation in terms of Cholesky factors and* elementary matrices *of the type introduced by Householder* [1744]. *Potter found that he could factor an elementary matrix into a product of its square roots, and that led to a formula for the measurement update of a Cholesky factor of the covariance matrix. The key factoring formula in Potter's derivation involves taking square roots of elementary matrices, and Potter's implementation came to be called* square-root filtering. *This innovation would contribute enormously to the success of the first major Kalman filtering application. Potter's square-root filter was implemented on the Apollo computer (in 14-bit arithmetic) and operated successfully on all Apollo missions.*

See the book [1982] by Peter Lancaster and Leiba Rodman (1949-2015) for the theory of Riccati equations, and the book [326] by Dario Andrea Bini, Iannazzo, and Meini for the numerical methods.

Part II

Biographies

10

Lives and works

Avant tout, je pense que, si l'on veut faire des progrés en mathématiques, il faut étudier les maitres et non les éléves.
– Niels Henrik Abel, *Paris notebook dated August 6, 1826*

10.1 ▪ Alexander C. Aitken

Alexander Craig Aitken (1915)
Courtesy of Auckland War Memorial Museum

Life

Alexander Craig Aitken was born in Dunedin, New Zealand, on April 1, 1895. He was the eldest of seven children. His father had a grocery in Dunedin.

Aitken attended Otago Boys' High School from 1908 to 1912, where he was not particularly brilliant. But at the age of 15, he realized that he had a real power in mental calculations, and that his memory was extraordinary. He was able to recite the first 1000 decimals of π, and to multiply two numbers of nine digits in a few seconds. He knew Virgil's *Aeneid* by heart. He was also very good at several sports and began to play the violin.

In 1913, he won a Junior University Scholarship to Otago University, where he studied mathematics, French, and Latin. It seems that the professor of mathematics there, David James Richards, a "temperamental, eccentric Welshman," who taught there from 1907 to 1917, was unable to communicate his knowledge to the students, and Aitken's interest in mathematics lowered. But World War I forced him to interrupt his studies at the university.

Aitken enlisted as a soldier in the New Zealand Expeditionary Force in April 1915 and he took part in the Gallipoli landing in the Dardanelles and in the campaign in Egypt as a lieutenant. Then, he was commissioned in the north of France and was wounded in the shoulder and foot during the first battle on the river Somme.

After a stay in a London hospital, he was back to New Zealand in 1917. He spent one year of recovery in Dunedin, where he wrote a first version of his memoirs published later [20]. Then, Aitken resumed his studies at Otago University, and graduated with first class honors in languages, but only with second ones in mathematics.

He married Mary Winifred Betts in 1920. They had two children, Margaret Winifred and George Craig. Aitken became a school teacher at his old Otago Boys' High School. Richards' successor in the Chair of Mathematics at the university was Robert John Tainsh Bell (1876-1963). When Bell required an assistant, he called on Aitken. He encouraged him to apply for a scholarship for studying with Sir Edmund Taylor Whittaker (1873-1956) in Edinburgh.

Aitken left New Zealand for the United Kingdom in 1923. There, he worked on an actuarially motivated problem of fitting a curve to data which were polluted by statistical errors. His Ph.D., completed in 1925, was considered so outstanding that he was awarded a D.Sc. for it instead of a Ph.D. That same year, Aitken was appointed as a Lecturer at the University of Edinburgh, where he stayed for the rest of his life.

However, the effort he expended to obtain his degree led him to a first severe breakdown in 1927, and then he was periodically affected by such crises. They were certainly in part due to his fantastic memory, which did not fade with time, and he was always remembering the horrors he saw during the war.

In 1936, Aitken became a Reader in statistics and he was elected a Fellow of the Royal Society. In 1946, he was appointed to Whittaker's Chair in Mathematics. In 1956, he received the prestigious Gunning Victoria Jubilee Prize of the Royal Society of Edinburgh. He retired from the Chair of Mathematics at the University of Edinburgh on September 30, 1965. He passed away on November 3, 1967, in Edinburgh.

Aitken was fond of long walks, he wrote poetry throughout his life, and he was elected to the Royal Society of Literature in 1964. He also played the violin and composed music to a very high standard.

Work

Aitken's mathematical works cover statistics, numerical analysis, and algebra. In numerical analysis, he introduced the Δ^2 process for accelerating the convergence of a sequence x_n by computing a new sequence

$$\tilde{x}_n = x_n - \frac{(\Delta x_n)^2}{\Delta^2 x_n},$$

where

$$\Delta x_n = x_{n+1} - x_n, \quad \Delta^2 x_n = \Delta x_{n+1} - \Delta x_n.$$

Under some circumstances, the sequence $\tilde{x}_n$ converges faster to the limit.

He is also well known for the Neville-Aitken scheme for the recursive computation of polynomial interpolants. In algebra he made contributions to the theory of determinants, and revived the use of determinantal identities. He also saw clearly how invariant theory fitted into the theory of groups.

Aitken published papers on linear algebraic equations and matrices as early as 1927 [12, 14] and on eigenvalue problems in 1928 [13].

He was also interested in computations. In 1933 he published a paper titled *On the evaluation of determinants, the formation of their adjugates, and the practical solution of simultaneous linear equations* in which he explained how to compute the determinant of a nonsymmetric matrix using elimination with pivoting as well as solving a corresponding linear system [15]. He wrote

> *It seems to us, after considerable numerical practice, that if we are granted the use of a machine capable of forming and adding or subtracting binary products, both positive and negative, and of dividing the resulting sums by arbitrary divisors, then the elementary and old-established process of reduced elimination can be so arranged as to provide almost everything that is desired. The eliminations, which in practice are simply cross-multiplications supplemented by divisions at the proper stage, lead to successively condensed systems, the elements of which are signed minors of the original system, ...*

In fact, the method used by Aitken is close to Chió's method published in 1853; see [4, 662].

In 1932, he published a book *An introduction to the theory of canonical matrices* [3081] with Herbert Westren Turnbull (1885-1961). In this book the authors used a sequence of vectors $v, Av, A^2v, \ldots$, pages 43, 46-51, to obtain the Frobenius canonical form and the Jordan canonical form of a matrix. These vectors were called Krylov vectors much later on. In [3081], Turnbull and Aitken also studied matrix pencils. This book was quite influential for the dissemination of matrix methods; see [1139].

In 1935, Aitken introduced the concept of generalized least squares, together with the now standard vector-matrix notation for linear regression. He considered the computation of eigenvalues (called *latent roots* following Sylvester's terminology) [16] in 1936.

Aitken published a textbook [18] on determinants and matrices in 1939. Later, in 1950, he considered the iterative solution of linear equations [19].

10.2 ▪ Mieczysław Altman

Life

The following biography of M. Altman was sent to the authors of the paper [2587] about Altman's methods by his son, Tom Altman, a professor at the Computer Science and Engineering Department, University of Colorado at Denver.

Mieczysław Altman was born in Kutno, Poland, 50 km north of Łódź, on December 2, 1916. After finishing his secondary school in Łodz in 1935, he studied mathematics at the Warsaw University from 1937 until the outbreak of World War II in 1939. In 1940, he enrolled at Lwów University and worked directly under the tutelage of Stefan Banach (1892-1945) for two years. He was Banach's last student.

After the Nazi invasion of the USSR in 1941, he was forced to flee again, eventually settling at the University of Sverdlovsk (now Yekaterinburg), 1700 km east of Moscow, in 1941-1942, and then in Tashkent, USSR (now Uzbekistan). There, he first obtained his Master in Mathematics in 1944 and he finished his Ph.D. in Mathematics in 1948.

Mieczysław Altman (CalTech, 1959)
Courtesy of Tom Altman

After his return to Poland in 1949, he learned that his entire family, including seven brothers and sisters, perished in the Łódź Jewish ghetto during the war. He held a position at the Institute of Mathematics of the Polish Academy of Sciences in Warsaw from 1949 to 1969, first as an Assistant Professor, 1949-1957, then as an Associate Professor, 1957-1958, and finally as a Full Professor and Director of the Numerical Analysis Department from 1958 to 1969.

In 1953, he married Wanda Kusal, M.D., and they had two children, Barbara (in 1956) and Tom (in 1958). For two years (1959-1960) Altman took visiting positions at the California Institute of Technology in Pasadena and the Courant Institute in New York. He received Poland's Banach Prize in Functional Analysis in 1958 and was Vice President of the Polish Mathematical Society in 1962-1963. Due to political pressures, in 1969 the Altman family left Poland and eventually settled in Baton Rouge, Louisiana, where Altman worked as a Professor of Mathematics at Louisiana State University, from 1970 until his retirement in 1987. He was a visiting professor at the Instituto per le Applicazioni del Calcolo, Consiglio Nazionale delle Ricerche, Rome, Italy, 1969-1970, and Newcastle University, Australia, in 1973.

Mieczysław Altman died in Pittsburgh, on December 14, 1997.

Work

In a series of papers [36, 37, 38, 39, 40, 41] published in the 1960s, M. Altman described and studied methods for solving linear systems. His methods are explained in Section 5.4 in Part I of this book.

Altman's 1977 book *Contractors and Contractor Directions - Theory and Applications* (Marcel Dekker, New York, 1977) received international acclaim and recognition among mathematicians as the most encompassing theory for solving equations by analytical means. To honor the memory of his relatives, friends, and countrymen, he dedicated his 1986 book *A Unified Theory of Nonlinear Operator and Evolution Equations with Applications - A New Approach to Nonlinear Partial Differential Equations* (Marcel Dekker, New York, 1986) to the victims of the Holocaust.

Even after his retirement, Altman remained professionally active, publishing several journal papers and another book *A Theory of Optimization and Optimal Control for Nonlinear Evolution and Singular Equations with Applications to Nonlinear Partial Differential Equations* (World Scientific, Singapore, 1990).

Altman is the author of over 200 research papers in pure and applied mathematics, including functional and numerical analysis, mathematical programming, general optimization and optimal control theory, nonlinear differential and integral equations, and Banach algebras. A man of many talents, he published his mathematical papers in a number of different languages, including

Russian, Polish, French, German, and English. He even coined the French word "contracteurs" (*Contracteurs dans les algèbres de Banach*, C.R. Acad. Sci. Paris, 274 (1972), 399-400).

10.3 ▪ Charles Babbage

Charles Babbage

Life

Charles Babbage was born on September 26, 1791, in London. His father was Benjamin Babbage (1753-1827), a merchant and banker, and his mother was Elizabeth Plumleigh Teape (1759-1844).

Babbage was educated at home, then in Enfield near London, and later in a small school near Cambridge. At the age of 16, he returned to live with his parents in Devon. In 1808, the family moved to East Teignmouth.

Babbage entered Trinity College, Cambridge, in October 1810. At this time he was already quite good at mathematics. There he became friends with John Frederick William Herschel (1792-1871), George Peacock (1791-1858), with whom he founded the *Analytical Society*, and Edward Ryan (1793-1875). The purpose of the Society was to introduce continental mathematical methods into the conservative Cambridge. In 1812, Babbage was transferred to Peterhouse College from which he received a degree without examination in 1814. In 1816, Babbage, Herschel, and Peacock published a translation from French of the lectures of Sylvestre-François Lacroix (1765-1843).

On July 25, 1814, Babbage married Georgiana Whitmore (1792-1827) with whom he had eight children: seven sons and one daughter. Unfortunately, five of them died at a young age. The couple moved to London and Babbage applied for several positions. Despite the fact that he was elected as a Fellow of the Royal Society in 1816, he did not get the positions and was still dependent on his father's money.

In 1819, Babbage traveled to Paris to visit French scientists. In 1820, he was elected a fellow of the Royal Society of Edinburgh, and was influential in founding the Royal Astronomical Society. After his father's death in 1827, Babbage inherited a large estate and he became wealthy.

That same year his wife passed away after giving birth to a son who also died. At the end of 1827 Babbage made a one-year trip to Europe.

In 1828, he was elected Lucasian Professor of Mathematics at Cambridge, the chair that was previously held by Isaac Newton. He held this position for 12 years, even though he never gave a lecture. From 1829 to 1834, Babbage was interested in politics. In 1830, he published a book, *Reflections on the Decline of Science in England*, and in 1832, *On the Economy of Machinery and Manufactures*. In the 1830s, he became interested in developing the efficiency of rail transport in England.

His autobiography, *Passages from the Life of a Philosopher*, was published in 1864 by Longman, Green, Longman Roberts, and Green in London, with a dedication to Victor Emmanuel II, King of Italy. Babbage died on October 18, 1871, in London. He is buried in Kensal Green cemetery, London.

Work

John Napier (1550-1617) introduced logarithms in 1614. In 1617, Henry Briggs (1561-1630) published the first table of logarithms. Many of the tables that were published over the years contained many errors.

In his autobiography Babbage wrote that in 1812 or 1813,

> *I was sitting in the rooms of the Analytical Society, at Cambridge, my head leaning forward on the table in a kind of dreamy mood, with a table of logarithms lying open before me. Another member, coming into the room, and seeing me half asleep, called out, "Well, Babbage, what are you dreaming about?" to which I replied "I am thinking that all these tables" (pointing to the logarithms) "might be calculated by machinery."*

In 1820 or 1821, Babbage started to work on a calculating machine called the *Difference Engine*. The goal was to be able to tabulate functions and to print the results for producing mathematical tables, for instance, for logarithms. The idea was to approximate the function by polynomials and to compute them using differences. A polynomial of degree n has a constant n^{th} difference. The example given by Babbage was $f(x) = x^2 + x + 41$. The first difference is $D(x) = f(x+1) - f(x) = 2x + 2$ and the second difference D^2 is constant, equal to 2. So, if we write the values in a table (Table 10.1),

Table 10.1. $f(x) = x^2 + x + 1$

D^2	D	f	x
		41	0
	2	43	1
2	4	47	2
2	6	53	3
⋮	⋮	⋮	⋮

we just have to sum D to the previous value of f to obtain the result.

In 1823, Babbage obtained that the government would grant some funding for a project which was supposed to take three years. In 1824, he received £1500 and he began to construct the *Difference Engine* for which he hired Joseph Clement (1779-1844), an engineer. By 1828 Babbage had spent £6000 and the work went slowly. The plans were completed in 1830 and thousands of parts had been manufactured. In the meantime Babbage had some quarrels with Clement. In 1834, what was done of the machine was transferred to Babbage but the government had already spent £17,000 and decided to stop the funding. At that time Babbage was interested in another machine and the *Difference Engine* project was never completed.

Difference engines were built later on by Per Georg Scheutz (1785-1873) and Martin Wiberg (1826-1905) in Sweden, as well as George Barnard Grant (1849-1917) in the USA.

After the work on the *Difference Engine* was stopped, or a little before, Babbage started working on another more ambitious project, the *Analytical Engine*, whose first sketch is from September 1834. He wanted to construct a machine capable of doing the four arithmetic operations but also having input from punched cards, printed output, memory, and a processing unit called the *Mill*, with control and synchronization mechanisms. Note that punched cards were used in Jacquard's loom at the beginning of the 1800s.

In 1840, Babbage gave a series of seminars on the *Analytical Engine* in Italy. The Italian engineer Luigi Federico Menabrea (1809-1896) wrote an account of these lectures that was later translated into English by Ada Lovelace (1815-1852), the daughter of Lord Byron, who appended some notes about programming the machine, prepared under Babbage's guidance; see Section 10.51.

A part of the Analytical Engine

Over the years Babbage refined the design of many parts of the machine, for instance, the multiplication unit. He was planning to work with 40-digit numbers. Multiplication by repetition of addition would have been very costly. He was also trying to include some parallelism when several independent operations have to be chained. The basic design of the machine was done by the end of 1837.

After 1847, Babbage worked for a while on another *Difference Engine*, but in 1857, he returned to the design of the *Analytical Engine*, which he simplified. During his lifetime Babbage only constructed parts of the *Analytical Engine*. When he passed away in 1871 he had a simple mill and a printing mechanism near completion. His work was somehow continued by his son, Henry Prevost Babbage (1824-1918), who demonstrated parts of the machine.

In 2011, researchers in the UK started a project, called "Plan 28," to construct Babbage's *Analytical Engine*. They were hoping to complete it in 2021, for the 150th anniversary of Babbage's death. Apparently, in 2022, they are still struggling with the classification and study of Babbage's papers.[59]

[59] See https://plan28.org (accessed November 2021)

10.4 ▪ Tadeusz Banachiewicz

Tadeusz Banachiewicz
Courtesy of Professor Adam Strzałkowski of the Jagiellonian University

Life

Tadeusz Julian Banachiewicz was born on February 13, 1882, in Warsaw, which was at that time in the Russian Empire, but is now in Poland.

His father owned an estate at Cychry, near Warsaw, and Tadeusz spent his early years there. He had an older brother (who died in the concentration camp of Dachau in 1940) and an older sister. Already at an early age, Tadeusz showed an outstanding intellectual ability, with a special interest for mathematics and astronomy. He attended the Fifth Gymnasium, a high school in Warsaw, graduating in 1900 with a silver medal for outstanding academic achievement. That same year he entered the Faculty of Mathematics and Physics at the University of Warsaw, where his major subject was astronomy.

In 1903, he published a paper in the journal *Astronomische Nachrichten* on a mathematical description of an eclipse of a star by the planet Jupiter based on observations. In 1904, he defended a thesis titled *Studies into reduction constants of the Repsold's heliometer in the Pulkovo Observatory* and graduated with a degree of candidate in mathematical and physical sciences. This dissertation was awarded a gold medal by the Warsaw University Senate.

Banachiewicz continued to undertake research at the University. In 1905, he started to develop a method for computing the orbits of comets. He published a paper on a particular case of the n-body problem, and a second one on the three-body problem applied to small bodies near Jupiter.

To understand Banachiewicz's life, it is necessary to have a look at the history of Poland at that time. Poland had ceased to be an independent country in 1795. It was divided between Russia, Prussia, and Austria. As Warsaw was controlled by the Russians, Warsaw University became a Russian university and a new campus was built. But the new authorities were not accepted by the population, in particular by the students. After the beginnings of the Russian revolution and unrest among Polish students, the University was closed. The following years, the authorities were thinking about moving Warsaw University to Russia. They finally decided to keep it in Warsaw but with an even stronger Russian character. Teaching had to be given in Russian and only Russian professors were to be employed. Banachiewicz, a firm Polish patriot, grew up in this atmosphere of Russian domination and Polish nationalism.

In 1906, Banachiewiz published two important papers in the *Comptes Rendus de l'Académie des Sciences*, Paris. Banachiewicz left Warsaw and went to Göttingen in Germany to study astronomy with Karl Schwarzschild (1873-1916), the director of the University Observatory. After one year, he went to the Astronomical Observatory in Pulkovo, 19 km south of Saint Petersburg, as an assistant to Oscar Backlund (1846-1916), and worked on celestial mechanics.

In 1908, the Russian authorities reopened Warsaw University. In 1909, Banachiewicz returned to Warsaw where he was appointed as junior assistant. However, after a year, despite his scientific success, he did not obtain the job he was looking for at the university because of the death of Professor Alexander Vasilevich Krasnov (1866-1907), who had previously supported him. Thus, he went back to his home in Cychry, passed two exams in Warsaw, and earned his master's degree in Moscow.

In 1910, he started to work as an assistant at the Engelhardt Astronomical Observatory in Kazan, where he spent five years conducting heliometric observations of the Moon and taking part in a scientific expedition to the Volga River basin to measure the terrestrial gravity. In 1915, he submitted a thesis for the magister degree to Kazan University. However, the work was not published, and thus it was not considered as a degree thesis. Thus, he moved to the Astronomical observatory at Yuryev University in Dorpat (now Tartu, Estonia) where, in November 1915, he defended his habilitation thesis *Three Essays on Refraction Theory*, and became a docent at Dorpat University. Two years later he again defended a thesis at Dorpat, this time on the Gauss equation $\sin(z - q) = m \sin^4 z$, where z is close to q, and gave tables facilitating its numerical solution. He obtained a position of assistant professor. In March 1918, he was promoted to full professor and became director of the Dorpat Observatory.

Following the 1917 Russian Revolution, Estonia declared itself an independent country. However, the Russians tried to keep control of it and appointed a puppet Communist government. In February 1918, German troops entered the country, which lost its independence and went under German domination. The Russians moved Dorpat University to Voronezh in western Russia in the summer of 1918, and Banachiewicz was invited as a professor of astronomy there. However, he decided to return to Poland.

After a short stay in Warsaw (October 1918-March 1919), where he worked as assistant professor of geodesy at the Technical University of Warsaw, he moved to Kraków in 1919 to take up the chair of astronomy at the Jagiellonian University. He was also appointed the director of the Astronomical Observatory of the University. There, he was very active as a professor, a researcher, and an administrator. But since the observing conditions were not good in Kraków, in 1922, Banachiewicz set up an observing station at Mount Lubomir south of Kraków. In 1936, he organized a series of expeditions to Japan, Greece, and Siberia to make solar observations.

On March 17, 1931, Banachiewicz married Laura (or Larysa) Solohub Dikyj, a Ukrainian poet who died on May 28, 1945. They had no children. After the German invasion of Poland in 1939, he was arrested by the Nazis, along with other members of the Jagiellonian University, on November 6, 1939. He was taken to the Sachsenhausen concentration camp near Berlin, where he spent three months before being allowed to return to Kraków. In 1945, after the reestablishment of the University, he also accepted the position of Professor of higher geodesy and astronomy at the Kraków University of Mining and Metallurgy. As the astronomical station at Mount Lubomir that he set up before the war had been burnt down by the Germans in 1944, in 1953 Banachiewicz obtained permission to set up a new observing station at Fort Skala, a former military site. Later on it became the first site of a Polish radio telescope.

Banachiewicz was president of the Polish Astronomical Society from 1923 to 1933. In 1928, the University of Warsaw endowed him with a *Honoris Causa doctorate* in philosophy, and ten years later the same title was bestowed upon him by the University of Poznań. In 1948, he received an honorary doctorate from the University of Sofia. In the years 1932-1938, Banachiewicz was vice-president of the International Astronomical Union, and in 1938 he was elected president of the IAU Commission 17 (movements and figure of the Moon), which he chaired until 1952. In 1939, he became a member of the Academy of Padua in Italy, and was nominated in 1946 as a correspondent member of the Royal Astronomical Society in London. In 1953, by virtue of the law, he received a doctorate in mathematics at the Jagiellonian University.

Banachiewicz lived and worked in Kraków until his death. He passed away on November 17, 1954, of pneumonia, a complication after a surgery. He is buried in the crypt of distinguished Poles, Church on the Rock (Skałka), in Kraków.

For some pictures related to Banachiewicz, see [496].

Work

Banachiewicz was very interested in the application of mathematics to the numerical solution of problems in astronomy and physics. His main achievement is the theory of the so-called *Cracovians* that he developed when he was in Russia in 1917. He first used them in his lectures at Dorpat University. The scientific basis of the theory was announced in 1923 in the paper *On a certain mathematical notion* published in the *Bulletin of the Polish Academy of Arts and Sciences*.

Cracovians were introduced to simplify the solution of systems of linear equations by hand and on mechanical calculators like arithmometers. Relating Banachiewicz's matrix product, denoted by $\wedge$, to the usual matrix product, we have $A \wedge B = B^T A$. Hence, $x \wedge A^T = Ax$.

For solving the system $Ax = b$, Banachiewicz used A^T instead of A, and the columns of A^T are multiplied by x. This made manual multiplication easier, as one needed to follow two parallel columns, instead of a vertical column and a horizontal row in the matrix notation. It also sped up computer calculations, because both factors' elements are used in a similar order, which is more compatible with the sequential access to memory in computers of those times. However, this definition makes the multiplication of Cracovians non-associative.

For solving the system $LL^T x = b$, where L is a lower triangular matrix, Banachiewicz's algorithm starts from the upper left corner of the matrix L and proceeds to calculate the matrix row by row. Cholesky's method starts from the upper left corner of the matrix L and proceeds to calculate the matrix column by column. It seems that Banachiewicz's method was known in France before Cholesky's method was widely used; see [2916].

Banachiewicz used Cracovians on several occasions, in particular for solving the normal equations in the least squares method, and for least squares interpolation [182]; see also [178, 180, 181]. His monograph *Cracovians Calculus and its Applications* (in Polish) was published in 1959, after his death.

The authors of [495] wrote

> *To simplify matrix computations, Banachiewicz introduced the Krakowian calculus, which greatly simplified and improved computations on calculating machines, which was his principal goal. Theoretically, Krakowians are not as rich in properties as matrices, since they do not involve the associative law. The results obtained by Krakowians can also be obtained by matrices. The Krakowian notation provides formulas that are very easy to remember, which in the conventional form are unwieldy and difficult to memorize. In computations, the Krakowian method makes it possible to obtain results faster and to control the succession of computations. Krakowians reached their zenith of popularity at the start of the 1950s, when calculating machines began to flourish. They found applications in many fields of the natural sciences and technology and simplified many algorithms, freeing formulas from logarithmic ballast.*

10.5 ▪ Friedrich L. Bauer

Friedrich Ludwig Bauer
Courtesy of Manfred Broy

Life

Friedrich Ludwig (Fritz) Bauer was born on June 10, 1924, in the Bavarian city of Regensburg, Germany. His father was an accountant and a bookkeeper in Thaldorf, where the family lived until 1932, near Kelheim and Regensburg. His mother was Elizabeth Hedwig Scheuermeyer.

The family moved to Pfarrkirchen in lower Bavaria, near Passau. After studying in this city, Bauer was sent by his parents to a boarding school in Munich in 1932. He earned his *Abitur* in 1942.

After being in the German Labour service for almost a year, from 1943 to 1945 he served as a soldier in the Wehrmacht during World War II. After a while he was sent to an officer's school and then to Russia. Back to Germany, he was wounded in 1945, captured by the Americans, and sent to a prisoner's hospital in Normandy. After recovering, he was sent to a prisoners' camp and finally he went back home in September 1945.

From 1946 to 1950, he studied mathematics, theoretical physics, astronomy, and logic at the Ludwig-Maximilian University in Munich. Two of his teachers were Oskar Perron (1880-1975), well known for the Perron-Frobenius theorem, and Heinrich Tietze (1880-1964).

After two years as a teacher in a grammar school, in 1951 he came back to the university as a teaching assistant to Professor Friedrich Bopp (1909-1987), a German theoretical physicist who contributed to nuclear physics and quantum field theory. He received his Ph.D. in 1952 on *Group-theoretic investigations of the theory of spin wave equations*.

Then, Bauer became interested in computers and numerical methods. He went to the *Technische Hochschule* in Munich, where he served as a teaching assistant to Professor Robert Sauer (1898-1970). He published his first paper, co-authored with his friend Klaus Samelson (1918-1980), in 1953. He got his Habilitation in 1954 with a work entitled *On quadratically convergent iteration methods for solving algebraic equations and eigenvalue problems*, and he obtained a position of *Privatdozent*. In Munich, Bauer was involved in the early development of computers in Germany including *STANISLAUS*, a relay-based computer, in 1951, and *PERM*, a stored program electronic computer, in 1952-56.

In 1958, he moved to the University of Mainz as an Associate Professor. In 1962, he returned to the *Technische Hochschule* of Munich as a Full Professor, a position he held until his retirement in 1989. He was the doctoral advisor of 39 students, including Josef Stoer and Peter Wynn (1932-2017).

Bauer was married in 1949 to Irene Maria Theresia Laimer (deceased in 1973), and then to Hildegard Vogg. Bauer was then the brother-in-law of Alston Scott Householder (1904-1993), with whom he wrote several papers. He was the father of three sons and two daughters. He died on March 26, 2015.

For an interview with Bauer in 1987, see [92] and [1595] in 2004; see also [1145].

Work

Although he was a physicist at the origin, Bauer soon became fascinated by the use of computers in science, and then by computers in general. He covered almost all the aspects of this area: algorithms, arithmetical operation, propagation of rounding errors, computer architecture, etc.

He was one of the developers of the programming language Algol, and, with Samelson, he invented the *stack* in 1955, the conceptually simplest way of saving information in a temporary storage location. Stacks are the way to handle recursive function calls and the dynamic runtime behavior of computer programs. Any modern microprocessor incorporates them.

Wilkinson's research on algorithms for matrix eigenvalues was published in *Numerische Mathematik* in Algol. Equally important, Algol led directly to Niklaus Wirth's pedagogical programming language PL/0, which was used in the early design of MATLAB. In 1968, Bauer also coined the term *software engineering*. He was most influential in establishing computer science as an independent domain in German universities.

In numerical analysis, Bauer's contributions included matrix algorithms for linear systems and eigenvalue problems, continued fractions, nonlinear sequence transformations, and rounding error analysis. With Householder, he was a pioneer in the use of norms in matrix analysis (see [227, 228, 229, 231]), thus opening the way to Richard Steven Varga (1928-2022) in the USA, and Noël Gastinel (1925-1984) in France. Bauer also introduced the concept of the field of values subordinate to a norm, a notion that also turned out to be useful in functional analysis. The Bauer-Fike theorem is a standard result in the perturbation theory of the eigenvalues of a complex-valued diagonalizable matrix. For some of Bauer's works, see [215, 216, 217, 218, 219, 220, 221, 222, 223, 224, 1753].

After 1975, Bauer devoted most of his efforts to computer science. He was also interested in its history and, more generally, in the history of mathematics.

10.6 ▪ Eugenio Beltrami

Life

Eugenio Beltrami was born in Cremona in the north of Italy, at that time under the Austrian Empire, on November 16, 1835. His grandfather, Giovanni Beltrami (1770-1854), was an engraver of precious stones, especially cameos, who found a patron in the Prince Eugène Rose de Beauharnais (1781-1824), Viceroy of the Kingdom of Italy under his stepfather Napoléon Bonaparte. His father too was an artist, a painter passionate about miniatures. During a stay in Venice, he met and married Elisa Barozzi di Rozzano (1818-1909), of a noble family, a composer and lover of opera music (in 1842, she also sang at the Fenice opera house in Venice) and pupil of the famous opera singer Giuditta Pasta (1797-1865). Thus, it is not a surprise that Beltrami was

Eugenio Beltrami

also passionate about music. Thanks to his mother and also to the composer Amilcare Ponchielli (1834-1886), he became a talented pianist. In several of his letters he also showed a great interest in the relationship between mathematics and music.

His father, after the patriotic uprisings of 1848, was forced to take refuge first in Piedmont and then in France, from where he never returned. Thus, Eugenio Beltrami was educated by his mother and his grandfather.

Eugenio attended elementary, secondary, and high schools in Cremona, apart from the year 1848-49, where he studied in Venice. Beltrami began studying mathematics at the Ghisleri College of the University of Pavia in November 1853. Unfortunately, during the third year of studies he was expelled with others from the College, accused of having promoted riots against the rector, the Abbott Antonio Leonardi. Due to economic hardship, he had to leave his studies. He was forced to take up an administrative job in Verona in November 1856, where, thanks to his uncle Niccolò Barozzi (1826-1906), he became the private secretary of the engineer Diday, director of the Lombardy-Veneto railways. In January 1859, due to the suspicions of the Austrian police, he was fired for political reasons. But after only a few months Lombardy was freed and Diday moved to Milan, taking Beltrami with him again as his private secretary.

Anyway, his vocation towards mathematical studies was always present. Asking for advice from Francesco Brioschi (1824-1897), who had been his teacher during his third year of studies in Pavia, he resumed studying mathematics independently from scratch.

In Milan, in 1860, he also met Luigi Cremona (1830-1903), another student of Brioschi, and become acquainted with him. He also tried to get a teaching position in a secondary school and participated in three competitions for positions of second lieutenant of the military engineers. But his lack of a university degree did not allow him to obtain them. But he did not resign himself and continued to study.

Two of his memoirs appeared in the *Annali di Matematica Pura ed Applicata*, published by Barnaba Tortolini (1808-1847) in Rome, and were greatly appreciated by Brioschi. Therefore, on October 1862, Beltrami was appointed extraordinary professor of complementary algebra and analytical geometry at the University of Bologna, the city to which he moved with his mother, who never left him.

In 1863, on the proposal of Enrico Betti (1823-1892), he was offered the professorship of theoretical geodesy at the University of Pisa. Due to modesty, he was tempted to refuse, but in the end he accepted. There, he formed a lifelong friendship with Betti and was able to meet Bernhard Riemann (1826-1866) who, for health reasons, made long stays in that city from 1863 to 1865 to take advantage of the Mediterranean climate.

Ulisse Dini (1845-1918), an Italian mathematician and politician, said about Beltrami, during the commemoration held at the Italian Senate after his death (our translation),

> *Chosen intelligence, he placed it all, together with his good and meek soul, at the disposal of young people who wanted to start the sacred fire of science. His brilliant and very high works, in which to the importance of the treated material was added a clarity and elegance of exposure that made us want to read and meditate on them, soon made him famous here and outside; and these and the love that he knew how to instill in everyone for mathematics, his goodness, his so much modesty, his kindness of manner inspired in many and many the cult of science.*

On September 1866, after only three years in Pisa, due to the climate not suitable for the health of his beloved mother, he obtained a chair in Bologna, as professor of rational mechanics. It was during this period, in February 1868, that Beltrami married Amalia Pedrocchi, of Venetian origin.

In 1871, Rome became the new capital of the Italian kingdom. The Minister of Education, Antonio Scialoja (1817-1877), who wanted to increase the reputation of the Roman university, sought valid scientists to fill vacant professorship chairs. Beltrami was one of them. In October 1873, he was convinced to move to Rome with his wife, and he obtained the chair of rational mechanics there. He also taught advanced analysis in that university, and was professor at the Application School for engineers.

But in Rome, Beltrami did not find that this position suited him. There had been a reform and he feared that the new situation would not allow him enough leisure for studying and doing research. He was also concerned by his wife's health, and he began to consider the proposals that came to him from other universities. This is the reason why, after three years in Rome, Beltrami moved to Pavia in October 1876, taking the chair of mathematical physics there. In Pavia, Beltrami did not find a better climatic situation, but certainly a more peaceful environment in order to be able to cultivate his studies. He also made new friends, including Felice Casorati (1835-1890), with whom he was very close until the latter's death. This tragic event left Beltrami sad and isolated. He therefore decided to accept the requests from the University of Rome, and returned there in 1891, as professor of physics, acclaimed by students and colleagues. He spent his last years in Rome.

All the documents and testimonies show that Beltrami was a man who entirely dedicated his life to study and had no ambitions for prestigious positions. He only agreed to be a member of the Higher Education Council. But thanks to its scientific merits and his publications, numerous Italian and European academies tried to have him among their associates or members (Accademie delle scienze di Bologna, Accademia delle scienze di Torino, Società italiana delle scienze, Istituto lombardo di scienze e lettere di Milano, Società reale di Napoli, Accademia di scienze, lettere ed arti di Modena, Accademia pontaniana di Napoli, Akademie der Wissenschaften zu Göttingen, German and French Academy of Sciences, and the London Mathematical Society as an honorary member). Beltrami was a corresponding member the Accademia dei Lincei in Rome since 1871.

When Brioschi passed away, there was a unanimous vote in favor of Beltrami to replace him in the role of President of the Lincei Academy (February 10, 1898). Beltrami committed himself with great dedication to this role. Another great recognition came directly from King Umberto, who appointed him senator of the Kingdom of Italy, communicating this appointment to him in person on June 4, 1899, during a solemn meeting of the Accademia dei Lincei.

Beltrami was already suffering from a severe stomach disease and only three days after a serious surgical operation he passed away on February 18, 1900. There were countless published commemorations and writings to honor his memory [488, 763, 900, 1556, 2092, 2115, 2471, 2822].

Work

Eugenio Beltrami published more than 100 papers (at that time called "memoirs"). In 1900, the Faculty of Sciences of the University of Rome decided to bring them together in one new and complete edition of his scientific writings. In May 1902, the first of the four planned volumes was published. It included the work from 1861 to 1868. The other three volumes appeared in 1904, 1911, and 1920.

Beltrami worked, of course, in pure mathematics: analysis, analytical, and infinitesimal geometry. But due to the different chairs obtained in his career (geodesy, rational mechanics, physics, and mathematical physics), he also worked on fluid mechanics, general mechanics, theory of potential, attraction, electricity, physical optics, conduction of heat and thermodynamics, magnetism, electromagnetism, elasticity, and acoustics. All his works were appreciated for clarity and an elegant exposition style.

Beltrami is known worldwide for some of his results in non-Euclidean geometry. In 1868, he presented his *Saggio d'interpretazione della Geometria non Euclidea*, which is considered one of his masterpieces. In this work, he examined the properties of the surface of constant negative curvature, to which he gave the name of pseudosphere. The geometry of such a surface was found to be identical with the geometry described independently by János Bolyai (1802-1860) and Nikolai Ivanovich Lobachevsky (1792-1856).

The reception of these ideas, a novelty at that time, was not enthusiastic. After the publication of this memoir, Hermann Helmholtz (1821-1894), Felix Klein (1849-1925), and Angelo Genocchi (1817-1889) expressed serious doubts on certain points of Beltrami's reasoning and mathematics. In his article, Beltrami did not explicitly say that he demonstrated the consistency of non-Euclidean geometry as independent of the axiom of parallels, but emphasizes the link with the theory developed by Bolyai and Lobachevsky. His proof of the independence of the postulate of parallel lines was underlined by Guillaume Jules Hoüel (1868-1881) in his French translation of the works of Lobachevsky, Bolyai, and Beltrami.

Beltrami also examined how the gravitational potential as given by Newton would have to be modified in a space of negative curvature. He gave a generalized form of the Laplace operator. This is now known as the Laplace-Beltrami operator. He used differential parameters in giving a generalization of Green's theorem. His work indirectly influenced the development of tensor analysis by providing a basis for the ideas of Gregorio Ricci-Curbastro (1853-1925) and Tullio Levi-Civita (1873-1941) that were used by Albert Einstein (1879-1955) when developing his theory of general relativity.

Beltrami's paper on the singular value decomposition [262] appeared in 1873 in the journal *Giornale di matematiche ad uso degli studenti delle università italiane*, founded in 1863 by Giuseppe Battaglini (1826-1894) together with his colleagues Vincenzo Janni (1819-1891) and Nicola Trudi (1811-1884), and directed by him until 1893. From 1894, after the death of Battaglini, its name was changed to *Giornale di matematiche di Battaglini*. The intentions of this journal were to address "mainly young scholars of Italian universities, so that they serve as a link between university lectures and other academic issues [...]", and it was an important point of reference for the diffusion of non-Euclidean geometries in Italy.

10.7 ▪ Augustin-Louis Cauchy

Augustin-Louis Cauchy

Life

Augustin-Louis Cauchy was born on August 21, 1789, in Paris. He was the eldest son of Louis François Cauchy (1760-1848) and Marie-Madeleine Desestre (1767-1839). Augustin had two younger brothers, Alexandre Laurent (1792-1857) and Eugène François (1802-1877).

Cauchy's father was the deputy head of the Parisian police. He lost his position when the French revolution started in July 1789, but he soon obtained some other administrative positions. When his former boss in the police was sentenced to death, Cauchy's family escaped from Paris to its suburbs that were supposed to be safer. But then the family was lacking financial resources. After the death of Maximilien Robespierre (1758-1794) they returned to Paris. After Napoléon Bonaparte's (1769-1821) coup in 1799, Cauchy's father became the secretary general of the Senate where he was in touch with Pierre-Simon de Laplace (1749-1827) and Joseph-Louis Lagrange (1736-1813).

Following the advice of Lagrange, Louis François Cauchy put his son in the École Centrale du Panthéon, a college in Paris where Augustin-Louis learned literature, Latin, and Greek from 1802 to 1804. However, he was more interested in science and went to the Lycée Napoléon to prepare for the examination to enter the École Polytechnique. He was ranked second on the exam and studied in École Polytechnique from 1805 to 1807. His teachers were Sylvestre-François Lacroix (1765-1843), Jean-Nicolas-Pierre Hachette (1769-1834), Gaspard de Prony (1755-1839), and André-Marie Ampère (1775-1836). Then, Cauchy went to the École des Ponts et Chaussées where he graduated in civil engineering.

In January 1810, Cauchy was assigned, as a junior engineer, to the building of a harbor in Cherbourg, Normandy. Building this harbor was a part of Napoléon's plans to invade England. During his stay in Cherbourg, Cauchy published his first mathematical papers. In 1812, he became sick and came back to Paris. He obtained a leave of absence until February 1813.

He was looking for a teaching position that he did not obtain and was assigned as an engineer to the Ourcq canal. After the death of Lagrange in April 1813, Cauchy applied for one of the positions that were held by Lagrange in the Bureau des Longitudes. Although he was ranked first by the committee, he did not get the position.

After the defeat and resignation of Napoléon in 1815, the Bourbon royal family came back to France and Louis XVIII (1755-1824), the brother of the late Louis XVI, became king of France. The Académie des Sciences was reestablished and some people that were close to Napoléon's regime, like Lazare Carnot (1753-1823) and Gaspard Monge (1746-1818), were dismissed. Louis XVIII appointed Cauchy to the Académie but this was considered as an outrage by many scientists who became Cauchy's enemies.

In 1815, Cauchy became the substitute of Louis Poinsot (1777-1859) at the École Polytechnique. In 1818, Cauchy married Aloïse de Bure (1795-1863). They had two daughters: Marie Françoise Alicia in 1819 and Marie Mathilde in 1823.

Louis XVIII was succeeded in 1824 by his brother Charles X (1757-1836). During those years, Cauchy obtained positions in the Collége de France and in Paris University. In July 1830 there was a revolution and Charles X fled the country. He resigned in favor of his grandson Henri but he was replaced by Louis-Philippe Ier (1773-1850), who was from a different branch of the royal family. He was not the king of France but "king of the French." Cauchy was a devout Catholic and zealous partisan of the Bourbon kings and he refused allegiance to the new regime. Consequently, he lost his positions, except for the Académie. He left the country, leaving his family behind.

Cauchy went first to Switzerland and then to Turin in Italy where he obtained a chair in mathematical physics. He taught there in 1832 and 1833. In 1833, at the invitation of Charles X in exile, Cauchy accepted to become the science teacher of Henri d'Artois (1844-1883) and he left Turin for Prague. But he was not very successful as a teacher and his pupil disliked mathematics. Cauchy's family joined him in 1834.

Cauchy came back to Paris in 1838 but still refused to swear an oath of allegiance. He was elected to the Bureau des Longitudes in 1839 but not approved by the king. In those years Cauchy helped the Catholic Church to set up some schools like the Institut Catholique, which was a kind of university.

In 1848, there was yet another revolution and the king fled to England. A republic was installed and the oath of allegiance removed. In March 1849, Cauchy obtained the chair of mathematical astronomy to succeed Urbain Le Verrier (1811-1877) who obtained another position.

In 1852, Louis Napoléon Bonaparte (1808-1873), a nephew of Napoléon, became emperor under the name Napoléon III. An oath of allegiance was reestablished but Cauchy was exempted from swearing.

Cauchy remained professor at the university until his death on May 23, 1857, in Sceaux, near Paris. He is buried in the cemetery of that city.

For more details on Cauchy's life, see the biography [254] by Bruno Belhoste. For some aspects of his work, see [2714] by Gert Schubring in 2015, pages 427-480.

Work

Cauchy was one of the most prolific mathematicians of all time and one of the greatest mathematicians of the 19th century. His collected works were published in 27 volumes. He worked in many different areas of mathematics and is known for using more mathematical rigor than his predecessors, for instance, in his definitions of limits, the continuity of functions, and convergence of series. Perhaps his main achievement is that he was at the start of complex analysis with Cauchy's integral theorem in 1825. One of his masterpieces is the *Cours d'Analyse de l'École Royale Polytechnique* [565] that he wrote for students.

Concerning the topics of our book, the main relevant publications are his two memoirs published in 1815, but read at the Académie in 1812, *Mémoire sur le nombre de valeurs qu'une fonction peut acquérir, lorsqu'on y permute de toutes les manières possibles les quantités qu'elle renferme* [563] and *Mémoire sur les fonctions qui ne peuvent obtenir que deux valeurs égales et de signes contraires par suite des transpositions opérées entre les variables qu'elles renferment* [564], in which he gave a firm basis to the theory of determinants; see Chapter 3. In the second one he gave what is now called the Cauchy-Binet formula for the determinant of a product of matrices.

Probably motivated by some papers by Carl Gustav Jakob Jacobi (1804-1851), he came back to that topic in 1841; see [574, 575].

Another point of interest for us is Cauchy's contribution to spectral theory, even though speaking of spectral theory at that time is an anachronism. The problem came from the search for solutions of linear differential equations arising in mechanical or astronomical problems by considering (in modern terms) the eigenvalues of the matrix of coefficients. However, the solutions have to be real. This problem was considered by Jean Le Rond d'Alembert (1717-1783) and Joseph-Louis Lagrange (1736-1813), who based their reasonings about the eigenvalues on physics. The first to realize that the fact for the eigenvalues to be real is linked to the symmetry of the coefficients was Pierre-Simon de Laplace (1749-1827). In 1829, Cauchy proved rigorously that the roots of the characteristic polynomial corresponding to a symmetric determinant are real [569]. The title of the paper was *Sur l'équation a l'aide de laquelle on détermine les inegalités séculaires des mouvements des planètes*, but this paper has almost nothing to do with astronomy, except for the title. Cauchy was motivated by Lagrange's work on the principal axes of a rotating rigid body, but he probably noticed the connection with astronomy from Jacques Charles François Sturm (1803-1855).

Cauchy wrote down the linear equations corresponding to $Ax = sx$ with a scalar s and A real and symmetric and considered the corresponding determinant. He showed that for two distinct roots, the corresponding eigenvectors are orthogonal, but he did not use that word. The proof that the eigenvalues are real is by induction and contradiction using determinants. He also gave bounds for the eigenvalues and an interlacing theorem. For details about Cauchy and spectral theory, see [1600].

10.8 ▪ Arthur Cayley

Arthur Cayley

Life

Arthur Cayley was born on August 16, 1821, in Richmond, Surrey County, UK. He was the third child of a family with five children: Sophia (1816-1889), William Henry (1818-1819), Arthur, Charles Bagot (1823-1883), and Henrietta Caroline (1828-1886). His parents were Henry Cayley (1768-1850) and Maria Antonia Doughty (1794-1875).

The origins of his father's family were in Norfolk County and later on from Yorskhire. Arthur's great grandfather, Cornelius Cayley, was a lawyer in Kingston-upon-Hull, a harbor on the east cost of Great Britain doing commerce with the Baltic cities as part of the Hanseatic League. Two of Cornelius' sons were doing trading with Russia and they settled in Saint Petersburg. One of the two brothers was John Cayley (1730-1795), Arthur's grandfather, who was British consul in Saint Petersburg. Arthur's father was working for the trading company Thornton, Melville, and Cayley. Arthur's mother was the daughter of William Doughty (1729-1789) and Wilhelmina Johanna Doughty. The family was living in Saint Petersburg, and Arthur was born during a visit of his parents to England. The family returned permanently to England in 1828 when Arthur was seven.

At the beginning, Henry's children were taught by instructors. In a notice about Arthur [2679], George Salmon (1819-1904) wrote

> *At a very early age Arthur gave the usual indication by which mathematical ability is wont first to show itself, namely, great liking and aptitude for arithmetical calculations. A lady, who was one of his first instructors, has told that he used to ask for sums in Long Division to do while the other little boys were at play.*

After a while the family settled in Blackheath near London. For four years Arthur was sent to a private anglican school whose director was Rev. G.B. Potticary. In 1835, Arthur was sent to King's College School in London and entered the Senior Department. King's College was preparing the students to enter universities like Oxford or Cambridge. The chair of mathematics was held by Thomas Grainger Hall (1803-1881), who was a fellow and tutor of Magdalene College in Cambridge. Hall wrote several books for students including *The Elements of Algebra.* Hugh Rose, the principal of King's College, was impressed by Arthur's mathematical abilities and advised Arthur's father to send him to university rather than bringing him back to his own business.

Therefore, Arthur entered Trinity College in Cambridge in 1838 at the age of 17. His first year tutor was George Peacock (1791-1858). Peacock was the author of a *Treatise on Algebra* (1830) and the founder with Charles Babbage (1791-1871) and John Frederick William Herschel (1792-1871) of the Analytic Society in 1815.

The mathematical teaching in Cambridge first cycle was organized around the *tripos* system with an exam after three or four years. The name tripos comes from a stool with three feet which was, in the 16th century, the seat for the dean who was questioning the students. After several weeks of examinations the students were ranked. The first of the list was known as the *Senior Wrangler.* "Wrangle," which means dispute, refers to the oral examinations at the origin of the system. Since the courses taught during the year were not enough to hope for a good ranking, the students also used private tutors. One of the most famous, William Hopkins (1793-1866), was the private tutor of Cayley and also of George Gabriel Stokes (1819-1903), William Thomson (Lord Kelvin) (1824-1907), James Clerk Maxwell (1831-1879), etc. He was known as a Senior Wrangler maker.

Arthur was Senior Wrangler in 1842. The examination started on Wednesday, January 5, and ended the next Tuesday. Being Senior Wrangler was a prestigious position. In [2679], George Salmon wrote

> *Stories were current in Cambridge at the time of the equanimity with which he received the news of his success. The best authenticated one is that he was on the*

> *top of the coach on a night journey from London to Cambridge when the tripos list was put into his hands; he quietly put it into his pocket, resigning himself very contentedly to the necessity of waiting till the morning light for a knowledge of its contents.*

Cayley started publishing papers when he was an undergraduate. His first paper from 1841, entitled *On a theorem in the geometry of position*, was published in Volume 2 of the *Cambridge Mathematical Journal*, whose first editor was Duncan Farquharson Gregory (1813-1844), a fifth Wrangler in 1837.

Cayley passed an examination in September 1842 and was elected a Fellow of Trinity College. However, this fellowship could only be held for seven years since Arthur refused to take Holy Orders, which was the condition to be able to stay in Cambridge. In 1843, Arthur traveled to the continent with his friend Edmund Venables (1819-1895) visiting Switzerland and Italy. In July 1845, Arthur become Master of Arts and was elected senior member of Trinity College. In August of that year he traveled to the nordic countries and then to Berlin, where he met several German mathematicians including Carl Gustav Jacob Jacobi (1804-1851).

In April 1846, Arthur Cayley left Cambridge because he did not want to become a clergyman. He turned to law instead and entered Lincoln's Inn in London where he was the pupil of Jonathan Henry Christie (1793-1876), a conveyancer. He was called to the Bar on May 3, 1849. After that he was still working for Christie doing conveyancing work. However, he was sharing his time between his work and mathematics and continued to publish many papers. During that period he produced more than 200 papers. His most productive years were 1856-1878.

It is not known exactly when Arthur met James Joseph Sylvester (1814-1897) but the first letter from Sylvester to Cayley is dated November 24, 1847. There are around 500 letters and notes from Sylvester to Cayley and, unfortunately, only 60 known letters from Cayley to Sylvester; see [764] for a list. From then on, they exchanged ideas during all their lives but they never wrote a joint paper since this was not a common practice in those times.

Cayley met his friend George Salmon (1819-1904), an Irish mathematician and theologian, in 1848. Cayley was elected Fellow of the Royal Society in 1852. He married Susan Moline (1831-1923) in 1863. They had two children: Henry (1870-1950) and Mary (1872-1950). Arthur was elected to the newly established Sadleirian chair of mathematics at Cambridge in 1863. His duties were "to explain and teach the principles of pure mathematics and to apply himself to the advancement of that science." Cayley held this chair until his ultimate death.

Cayley was much involved in his country mathematical life. He was president of several societies: London Mathematical Society (1868), Cambridge Philosophical Society (1869-1870), Royal Astronomical Society (1872-1874), and British Society for the Advancement of Science (1883). He was also much in favor of the access of women to higher education and universities, something which was not so common in those times. He was for some years chairman of the council of Newnham College for Women in Cambridge. He had as a student Charlotte Angas Scott (1858-1931), who was the first woman to receive a doctorate in Great Britain in 1885 from the University of London since Cambridge was not delivering degrees to women. This was changed only in 1948.

He received the Copley medal of the Royal Society in 1882 (two years after Sylvester) for "his numerous profound and comprehensive researches in pure mathematics" and the De Morgan medal of the London Mathematical Society in 1884 as well as many awards and honorary degrees from foreign universities.

At the invitation of Sylvester he gave lectures on elliptic and Abelian functions at Johns Hopkins University in Baltimore, USA, from January to May 1882.

Cayley passed away on January 26, 1895, after several years of illness. He was buried in Mill Road Cemetery, Cambridge. The monument for Arthur Cayley, in the parish of St. Mary the Great, was a headstone that reportedly disappeared sometime in the 1980s. However the grave register for St. Mary the Great places the grave site east of the east path. A photograph of his grave (in bad shape already) can be seen in [764].

Cayley's collected mathematical papers were published by Cambridge University Press. The first volume is dated 1889 with a preface and notes by Cayley. Volumes 1 to 7 contain notes and references by Cayley. The other volumes were prepared after Cayley's death by Andrew Russell Forsyth (1858-1942), his successor to the Sadleirian chair. The 13 volumes contain 967 papers; see Table 10.2.

Table 10.2. *Cayley's collected papers*

Vol	Pub. year	No. papers	Years
1	1889	100	1841-1851
2	1889	58	1851-1859
3	1890	64	1857-1862
4	1891	77	1856-1862
5	1892	84	1861-1866
6	1893	33	1865-1872
7	1894	69	1866-1872
8	1895	70	1871-1873
9	1896	74	1874-1877
10	1896	76	1876-1880
11	1896	93	1878-1883
12	1897	89	1883-1889
13	1897	80	1889-1895

Work

Cayley worked on many aspects of mathematics. Here we are mostly interested in his work related to linear algebra, that is, with determinants and matrices.

In his first published paper [579] in 1841, Cayley considered the problem: How are the distances between five arbitrary placed points in space related? It had already been solved by Jacques Philippe Marie Binet (1786-1856) in 1812. Cayley used determinantal identities. He gave a solution for the three-dimensional space. This paper is interesting because his solution is

$$\begin{vmatrix} 0 & d_{1,2}^2 & d_{1,3}^2 & d_{1,4}^2 & d_{1,5}^2 & 1 \\ d_{2,1}^2 & 0 & d_{2,3}^2 & d_{2,4}^2 & d_{2,5}^2 & 1 \\ d_{3,1}^2 & d_{3,2}^2 & 0 & d_{3,4}^2 & d_{3,5}^2 & 1 \\ d_{4,1}^2 & d_{4,2}^2 & d_{4,3}^2 & 0 & d_{4,5}^2 & 1 \\ d_{5,1}^2 & d_{5,2}^2 & d_{5,3}^2 & d_{5,4}^2 & 0 & 1 \\ 1 & 1 & 1 & 1 & 1 & 0 \end{vmatrix} = 0,$$

and we can see that Cayley used today's (i, j) notation for the entries. He used the notation $\overline{ij}^2$ for $d_{i,j}^2$, where $d_{i,j}$ is the distance between the points i and j. Of course, the matrix within bars is symmetric but this was not obvious with his notation. In his other papers involving matrices he did not use this positional notation, except for some determinants. In 1843, he studied determinants related to bilinear forms and generalized determinants to higher dimensional arrays. In 1845, he introduced "hyperdeterminants" in developing methods to find invariants.

In 1852, a letter of Sylvester to Cayley showed multiplication of rows of matrices [764]. In 1854, in Cayley's paper [581] about the theory of groups, matrices are mentioned.

In [583], written in French, Cayley used a notation with vertical bars for matrices in 1855. He considered the inverse of a matrix and multiplication of two matrices and wrote[60] (our translation)

> *There will be many things to say about this matrix theory which must, as it seems to me, precede the theory of determinants.*

Matrices also appeared in [582, 584, 585]. In [582], whose title is *Recherches sur les matrices dont les termes sont des fonctions linéaires d'une seule indéterminée*, Cayley was only interested in the determinant and the minors of any order of a matrix whose entries are assumed to be linear functions of a parameter s. His goal was to find, if the determinant vanishes, the exponents α such that $(s-a)^\alpha$ is a factor of a minor of any order when $s-a$ is a factor of the determinant. He gave the rule to compute those integers α. Cayley added that these results were partly due to Sylvester in a paper from 1851.

Somehow, the interest of Cayley for matrices was motivated by what Thomas Hawkins [1599] called the Cayley-Hermite problem: Find all linear substitutions of variables of a nonsingular quadratic form which leave the form invariant.

His most well known and cited paper about matrices is the paper *A memoir on the theory of matrices* [587] in 1858. In that paper Cayley used a strange notation for matrices and vectors that he had already introduced in 1855 [583], with vertical bars and parentheses for the first row. He wrote

A set of quantities arranged in the form a square, e.g.

$$\begin{array}{|ccc|} \multicolumn{3}{c}{(\; a, \quad b, \quad c \;)} \\ a', & b', & c' \\ a'', & b'', & c'' \end{array}$$

is said to be a matrix.

Then, he said that a linear system of order 3 with unknowns $(\, x \quad y \quad z\,)^T$ and right-hand side $(\, X \quad Y \quad Z\,)^T$ can be written as (in his notation)

$$(\; X \quad Y \quad Z \;) \quad = \quad \begin{array}{|ccc|} \multicolumn{3}{c}{(\; a, \quad b, \quad c \;)} \\ a', & b', & c' \\ a'', & b'', & c'' \end{array} \!\!(\, x \quad y \quad z\,) \qquad .$$

So, he wrote the unknowns and the right-hand side as row vectors.

He added, as a summary of the contents of the paper,

> *It will be seen that matrices (attending only to those of the same order) comport themselves as single quantities; they may be added, multiplied or compounded together, &c. : the law of the addition of matrices is precisely similar to that for the addition of ordinary algebraical quantities; as regards their multiplication (or composition), there is the peculiarity that matrices are not in general convertible; it is nevertheless possible to form the powers (positive or negative, integral or fractional) of a matrix, and thence to arrive at the notion of a rational and integral function, or generally of any algebraical function, of a matrix. I obtain the remarkable theorem that any matrix whatever satisfies an algebraical equation of its own order, the coefficient of the highest power being unity, and those of the other powers functions of the terms of the matrix, the last coefficient being in fact the determinant; the rule for the formation of this equation may be stated in the following condensed form, which will be intelligible after a perusal of the memoir, viz. the determinant, formed out of the matrix diminished by the matrix considered as a single quantity involving the matrix unity, will be equal to zero.*

[60] *Il y aurait bien des choses à dire sur cette théorie de matrices, laquelle doit, il me semble, précéder la théorie des Déterminants.*

As it is usual for Cayley, he considered only small examples, 3×3 matrices in this case, but he added

> *For conciseness, the matrices written down at full length will in general be of the order 3, but it is to be understood that the definitions, reasonings, and conclusions apply to matrices of any degree whatever.*

Then, he defined the zero matrix and the identity matrix that he called *the matrix unity* and said *The matrix zero may for the most part be represented simply by 0, and the matrix unity by 1.* He defined the addition of 3×3 matrices and remarked that the addition is commutative. He also defined the multiplication by a scalar and the multiplication of two matrices but his notation for the entries of the product is strange (and not logical). If the first row of the left matrix is $(\, a \quad b \quad c\,)$ and the first column of the right matrix is $(\, \alpha \quad \alpha' \quad \alpha''\,)^T$ he wrote the $(1,1)$ entry of the product as

$$(\, a \quad b \quad c\,)(\, \alpha \quad \alpha' \quad \alpha''\,),$$

which means the product of the first row with the first column. He noticed that the matrix product is not commutative. From the multiplication he defined the powers of a matrix and the inverse of a matrix L by $LL^{-1} = L^{-1}L = 1$. He wrote the inverse of his 3×3 example using what we obtain from Cramer's rule when solving the linear system. He denoted the determinant of the matrix by ∇ and the $(1,1)$ entry of the inverse by A. Then, he wrote

$$A = \frac{1}{\nabla} \begin{vmatrix} 1, & 0, & 0 \\ 0, & b', & c' \\ 0, & b'', & c'' \end{vmatrix}.$$

Then, to write down the whole inverse he used derivatives of the determinant ∇ with respect to the entries of the matrix,

$$\begin{pmatrix} a, & b, & c \\ a', & b', & c' \\ a'', & b'', & c'' \end{pmatrix}^{-1} = \frac{1}{\nabla} \begin{pmatrix} \partial_a \nabla, & \partial_b \nabla, & \partial_c \nabla \\ \partial_{a'} \nabla, & \partial_{b'} \nabla, & \partial_{c'} \nabla \\ \partial_{a''} \nabla, & \partial_{b''} \nabla, & \partial_{c''} \nabla \end{pmatrix}.$$

Looking at this formula, he was concerned about what today we call the singular matrix:

> *The formula shows, what is indeed clear a priori, that the notion of the inverse or reciprocal matrix fails altogether when the determinant vanishes: the matrix is in this case said to be indeterminate, and it must be understood that in the absence of express mention, the particular case in question is frequently excluded from consideration. It may be added that the matrix zero is indeterminate; and that the product of two matrices may be zero, without either of the factors being zero, if only the matrices are one or both of them indeterminate.*

For what is now known as the Cayley-Hamilton theorem, Cayley considered 2×2 matrices

$$M = \begin{pmatrix} a, & b \\ c, & d \end{pmatrix}.$$

In Cayley's explanations about this theorem, there is confusion between matrices and determinants. He made an abuse of notation and considered the determinant (or matrix?)

$$\begin{vmatrix} a - M, & b \\ c, & d - M \end{vmatrix}.$$

This expression does not make too much sense since the terms do not have the same dimension. It must be considered a notation. Of course, the correct way is to replace M by a scalar λ, to compute the determinant and then to replace λ by the matrix M. Nevertheless, Cayley said that the developed expression for the determinant is

$$M^2 - (a+d)M + (ad - bc)M^0.$$

His sentence is also wrong since a determinant which is a scalar cannot be equal to a matrix. He computed M^2 and check that the resulting matrix is indeed the zero matrix.

He added, *I have verified the theorem, in the next simplest case of a matrix of the order 3* and he gave the result for the characteristic polynomial. He then concluded *I have not thought it necessary to undertake the labour of a formal proof of the theorem in the general case of a matrix of any degree.*

Cayley also used his result on 2×2 matrices to compute a square root of M, that is, a matrix L such that $L^2 = M$. After this, he considered the roots of the characteristic polynomial $\lambda^2 - (a+d)\lambda + ad - bc = (\lambda - \lambda_1)(\lambda - \lambda_2) = 0$. But since he had replaced λ by M before, he was a little bit embarrassed by the factorization he obtained. But he correctly concluded that the two matrices $M - \lambda_1 I$ and $M - \lambda_2 I$ are singular or indeterminate in his language.

Cayley considered the problem: When do two matrices commute (are "convertible" in his terms), that is, $LM = ML$? Cayley wrote that the general solutions are polynomials of degree less than the order of the matrix. But this result is not correct, as was shown later. He introduced the transpose of a matrix and symmetric (which he called symmetrical) as well as shew-symmetric matrices and noticed that any matrix can be written as the sum of its symmetric and skew-symmetric parts. He also remarked that when multiplying a matrix with its transpose one obtains a symmetric matrix. All these properties were only proved for 2×2 matrices. Finally, he considered rectangular matrices.

The general case of the Cayley-Hamilton theorem was proved by Ferdinand Georg Frobenius (1849-1917) in 1878. He was using the formalism of bilinear forms and proved that the minimal polynomial of a matrix divides its characteristic polynomial. The theorem was also proved by Arthur Buchheim (1859-1888), a British mathematician, in the third note of [489], saying that his proof was an adaptation of the quaternion proof for the case of $n = 3$ given by Professor Tait's *Elementary Treatise on Quaternions*, p. 81. Peter Guthrie Tait (1831-1901) was a Scottish physicist and mathematician. On this topic see Section 10.67 about Sylvester.

In the paper [588] titled *A supplementary memoir on the theory of matrices*, Cayley considered matrices as single objects and he stated that

$$(A+B)(C+D) = AC + AD + BC + BD \text{ and } (ABCD)^{-1} = D^{-1}C^{-1}B^{-1}A^{-1}.$$

He used his results to give solutions to a problem considered by Charles Hermite (1822-1901) about the properties of the matrix for the automorphic linear transformation of the bipartite quadric function $xw' + yz' - zy' - wx'$.

A large part of Cayley's papers is related to the theory of invariants for algebraic forms. His interest in this topic was triggered by George Boole (1815-1864). The general problem is to consider homogeneous polynomials of degree n in k variables and to find the quantities functions of the coefficients that are invariant under linear transformations of determinant 1. In Cayley's times what was most studied is the case $k = 2$. As an example, let us consider $n = 2, k = 2$, to the equation $ax^2 + 2bx + c = 0$ is associated the homogeneous polynomial $ax^2 + 2bxy + cy^2$. In modern notation this can be written as

$$\begin{pmatrix} x & y \end{pmatrix} \begin{pmatrix} a & b \\ b & c \end{pmatrix} \begin{pmatrix} x \\ y \end{pmatrix}.$$

Then, one considers a linear transformation of the coordinates given by a nonsymmetric matrix and we have

$$(\tilde{x} \quad \tilde{y})\begin{pmatrix} \alpha & \gamma \\ \beta & \delta \end{pmatrix}\begin{pmatrix} a & b \\ b & c \end{pmatrix}\begin{pmatrix} \alpha & \gamma \\ \beta & \delta \end{pmatrix}\begin{pmatrix} \tilde{x} \\ \tilde{y} \end{pmatrix}.$$

If we compute the determinants of the matrices involved, they are $-(b^2 - ac)$ for the first one and the same thing for the second one if the matrix of the transformation has a determinant equal to 1. Hence, in this simple case, an invariant is the discriminant of the polynomial. The aim of Cayley and Sylvester was to compute invariants of homogeneous polynomials as well as relations between those invariants. Cayley proposed a few general methods to do so, the most useful one using partial differential equations, but this was leading to very large computations. For instance, an invariant for $n = 5$ has 58 terms. After 1860 the influence of Cayley's methods started to decline and the lead was taken over by the German school with Alfred Clebsch (1833-1872), Paul Gordan (1837-1912), who proved that one of Cayley's conjectures was false, and also David Hilbert (1842-1943); see [766, 767].

For more details on Cayley's life and work, see [765, 768, 769, 771, 772, 773].

10.9 ▪ Lamberto Cesari

Lamberto Cesari

Life

Lamberto Cesari was born on September 23, 1910, in Bologna, Italy, where he completed his secondary studies at the scientific high school "Righi." In 1929, he won a competition to enter the prestigious Scuola Normale Superiore in Pisa, and enrolled in the course in Mathematics at the university. In 1933 he received his degree *cum laude* from the University of Pisa, with a thesis on Fourier series, under the supervision of Leonida Tonelli (1885-1946), one of the leading Italian analysts of the 20th century, professor at the University, but who also gave courses and seminars at Scuola Normale Superiore, and who is considered the founder of the publication *Annali della Scuola Normale Superiore*. It was Cesari himself who dedicated a commemoration to his teacher in his memoir [604]. Silvio Cinquini (1906-1998), a student who had Tonelli as supervisor in

Bologna, and who followed him in Pisa, remembered in [708] that he was charged by Tonelli to review, with Cesari, the manuscript of his thesis. In the afternoon they stopped reading, and the next day, Cinquini realized that several pages of the manuscript were changed. Cesari said that he worked all night and was up until 5 am making several modifications to improve it.

In 1933-1934 he spent an academic year in Munich, where he studied with the famous Constantin Carathéodory (1973-1950). At this occasion he met, and after, he married, Isotta Hornauer (1913-2003). The next year, he won a scholarship for a post-graduate course at Scuola Normale Superiore and also the competition for being a titular in upper secondary schools. His first position was, in 1935, at the Istituto per le Applicazioni del Calcolo in Rome, at that time run by Mauro Picone (1885-1977). He stayed there up to the end of 1938. During this period he also taught at the University of Rome. At the beginning of 1939, he returned to Pisa. During the academic year 1939-1940, he taught at the Scuola Normale Superiore and, in the next years, he was an assistant professor at the universities of Pisa and Bologna. He moved to Bologna in 1942, and became full professor at the University of Bologna in 1947.

In 1948, he resumed his contacts with international mathematical researchers, thanks to his important research carried out during WW II, since he was solicited by many professors from North American universities to agree to go there as a visiting professor. Among these colleagues there was Tibor Radó (1895-1965) who had much worked on the same kind of problems. Cesari accepted to go to the USA with his wife, and he spent a year at the Institute for Advanced Study in Princeton. Over the next four years, he held visiting positions at the University of California at Berkeley and at the University of Wisconsin in Madison. In 1950 he accepted a professorship at Purdue University where he developed a strong research group and trained students. In 1958-1960, Cesari was one of the first visitors at the Research Institute of Advanced Studies (RIAS) in Baltimore, Maryland, at that time the world's largest group of mathematicians employed in research on nonlinear differential equations, and directed by Solomon Lefschetz (1884-1972) and Joseph Pierre LaSalle (1916-1983).

In 1958 he gave up his professorship in Italy and shortly after, in 1960, he became a professor of Analysis at the University of Michigan in Ann Arbor, where he stayed until his retirement in May 1981. In Michigan, in 1975, he was named the first *Raymond L. Wilder Distinguished Professor of Mathematics* and he also held in 1976 the position of *Henry Russel Lecturer*, the highest title bestowed on senior faculty at the University. In 1976, he became an American citizen.

Cesari spent several periods as a visiting professor at the University of Perugia, where he also formed a strong research group devoted to problems of optimality in multidimensional problems. He greatly contributed to the development of the University of Perugia School of Mathematics, to the point that in 1976 he was awarded the *laurea honoris causa* in Mathematics, for having honored the Italian school of mathematics. The lecture titled *Mathematics in the Mediterranean: Today's View*, together with the one for the Russel Lecture, were published as a book only in 1990 by the University of Perugia, thanks to the collaboration of Cesari's wife. It contains 60 pictures found by Cesari himself in some antique bookshops in Paris. In February 1995, on the initiative of Calogero Vinti (1926-1997), an Italian mathematician, the same university established the Centro Studi Interfacoltà "Lamberto Cesari" (now Centro di Ricerca Interdipartimentale), dedicated to the memory of the distinguished mathematician, with the aim of promoting and disseminating interdisciplinary scientific research in Applied Mathematics.

Cesari was also a member of the Unione Matematica Italiana, and was elected to the Accademia dei Lincei in 1982. He was a corresponding member of the Academies of Science of Bologna, Modena, and Milan, a member of the AMS, the Mathematical Association of America, and of the Society for Industrial and Applied Mathematics; he also belonged to several Italian mathematical societies, and he was member of the editorial board of many international journals.

The importance of his research was also recognized by several invitations to speak at the *International Congress of Mathematicians (ICM)* (Cambridge, Massachusetts, in 1950, Amsterdam in 1954, Nice in 1970), and plenty of other international meetings. He was also invited to teach courses in Italy by the Centro Internazionale Matematico Estivo (C.I.M.E.) (Varenna in 1954, Bressanone in 1967 and 1975). It was in Varenna that Vinti met Cesari for the first time. In his commemoration [3148], Vinti remembered the lively, polemic, and stimulating discussions at the end of the seminars between Cesari and Renato Caccioppoli (1904-1959), who was also invited to give a course, that often saw them taking conflicting positions.

In 1980, for his 70th birthday, special honors were dedicated to him during the *International Conference on Nonlinear Phenomena*, held at the University of Arlington, and in 1982, in a conference on nonlinear analysis and optimization at the University of Bologna.

During his entire career, Cesari devoted a special interest to young people, his students (more than 40), and other researchers, encouraging them and trying to create a relaxed and friendly working environment.

After his retirement, in 1981, he was named *Raymond L. Wilder Distinguished Professor Emeritus of Mathematics*, and he continued to be very active. In 1988 in Bologna, for the celebration of the ninth centenary of the University, he gave a colloquium titled *Scuola Bolognese di Matematica nella storia della Cultura*, reporting in a lively way the mathematical legacy of this University. In 1989, he gave several conferences on nonlinear analysis problems in Sicily, in the framework of a program between the universities of Palermo, Messina, and Catania. Following these seminars, for about six months, he worked day and night in the drafting of the texts of the conferences he gave, that have to be published by the Accademia dei Lincei. Four days before his death he called Vinci, announcing to him that he had finally finished writing the manuscript (400 pages) and that he had given it to his typist. Unfortunately two chapters have been lost, and this work will probably never be published.

Lamberto Cesari died in Ann Arbor, Michigan, on March 12, 1990. On April 27, 1990, a memorial meeting in the Alumni Center of the University of Michigan, dedicated to the life and work of Lamberto Cesari, took place with more than 150 people coming from all around the world. In addition, several commemorations have been published since he passed away [58, 594, 708, 1547, 3148], and most of the reported information in this biography is from them.

From the point of view of the human and scientific side, Cesari is remembered as always smiling, controlled, with a winking and sly air, with an impressive scientific dignity in his profound interventions, characterized by detached elegance. In his talks, both in the university courses and in the conferences, he was appreciated for the clarity, the perfection, and the enthusiasm he showed. He rejected schemes and was always looking to reinvent theories to apply them to concrete problems. He also had an unmatched dexterity in handling mathematical algorithms and theorems. But overall, he was a tireless researcher, trying to find the simplest and most direct way to present his work.

Work

Cesari was known worldwide for his research in a variety of areas. The main fields, among several, are the theory of surface area, the study of problems in the calculus of variations, and differential equations. He also did a great deal of work in optimal control.

During his long career, Cesari wrote about 250 papers (see [594] for a list) and three books. The problem of area of surfaces in parametric form was taken up by Cesari and by Radó, who, independently of each other, arrived, between 1942 and 1946, at a comprehensive solution of it.

On this problem and on those related to the calculus of variations, during his stay in Purdue, he wrote his book, *Surface Area* (1956), which is considered a classic. He also published one of the first modern books on differential equations, *Asymptotic Behavior and Stability Problems in Ordinary Differential Equations* (1959). The third one, published when he was already retired, *Optimization - Theory and Applications. Problems with ordinary differential equations* (the first of three volumes planned), appeared in 1983.

Cesari published his first two works at the age of 22, while he was still a student. These works, even if the treatment was elementary, already showed a profound scientific aptitude.

When he passed away at the age of 79, he was still working on problems of plasticity, in collaboration with a mechanical engineer of the University of Michigan for the application aspect, and with his colleagues of the University of Perugia for the theory.

In numerical analysis, Cesari is known for the paper *Sulla risoluzione dei sistemi di equazioni lineari per approssimazioni successive* (On solving systems of linear equations by successive approximations), written when he was at the Istituto Nazionale per le Applicazioni del Calcolo in Rome. Certainly it was Mauro Picone that adressed him to look at this problem. This work, where the idea of "polynomial preconditioning" was probably presented for the first time, was published in different documents [600, 601, 602, 603], more or less detailed.

All these publications begin by an almost similar introduction by Picone who wrote[61] (our translation)

> *The present work of Dr. Lamberto Cesari, coadjutor of the Direction of the Istituto per le Applicazioni del Calcolo, makes a notable contribution to problem of the numerical solution of systems of algebraic linear equations. The resolution of this problem has always been the subject of study by the Institute, being it is an essential condition for the practical applicability of the integration methods - of the classics and of the new ones, devised by the Institute itself - of the ordinary and partial derivatives differential equations, which translate the problems of physics and high technique. With the contributions brought by this work, the methods of numerical solution of systems of linear algebraic equations, receive an arrangement very advanced, in which, as it is reasonable to assume, the Institute will always be able to find what it needs for difficult definitive numerical evaluations, with a sufficient degree of approximation, relating to the arduous problems assigned to him from Science and Technology.*

The first one is probably [600], since, in its introduction, Cesari wrote that he summarized the results that will be published in a forthcoming memoir, and referred to [602]. This last paper was also reprinted in [601], and referred in [603]. Cesari's method is not widely quoted in numerical analysis textbooks. We found it in Householder's *Principles of Numerical Analysis* [1736]. It is mentioned in some review papers [267, 268, 2669], and by Guido Cassinis (1885-1964) who used it in [562]. Let us open a parenthesis. Cassinis studied in Rome. In 1907 he graduated in Engineering and worked as assistant to Vincenzo Reina (1862-1919). Until 1924 he taught at the School of Engineering in Rome, then (1924-1932) he taught Topography in Pisa. He discovered

[61] *Il presente lavoro del Dott. Lamberto Cesari, coadiutore della Direzione dell'Istituto per le Applicazioni del Calcolo, apporta un notevolissimo contributo al problema della risoluzione numerica dei sistemi di equazioni lineari algebriche. La risoluzione di tale problema è stata sempre oggetto di studio da parte dell'Istituto, essendo essa condizione essenziale per la pratica applicabilità dei metodi di integrazione - dei classici e dei nuovi, escogitati dall'Istituto stesso - delle equazioni differenziali ordinarie e a derivate parziali, che traducono i problemi della fisica e dell'alta tecnica. Con gli apporti recati dal presente lavoro, i metodi di risoluzione numerica dei sistemi di equazioni lineari algebriche, ricevono una sistemazione assai progredita, nella quale, come è ben lecito presumere, l'Istituto potrà sempre trovare quanto occorre per le difficili definitive valutazioni numeriche, con un sufficiente grado di approssimazione, relative agli ardui problemi che gli sono assegnati dalla Scienza e dalla Tecnica.*

the "normal gravity formula" which was adopted internationally in 1930 by the International Union of Geodesy and Geophysics as international gravity formula along with Hayford ellipsoid. Cassinis made an important contribution to the development of photogrammetry in Italy. He also was an Italian Democratic Socialist Party politician, and the mayor of Milan, where he was born.

As we wrote in the biography of Cimmino, Picone complains that Cesari does not consider, in his published work, Cimmino's method.

10.10 ▪ Françoise Chatelin

Françoise Chatelin
Courtesy of Cerfacs

Life

Françoise Chatelin was born in Grenoble, France, on September 21, 1941. Her father, Jean Laborde (1912-1997), was a mathematician and member of the team that founded the Institut d'Informatique et Mathématiques Appliquées de Grenoble (IMAG) in 1960, around Jean Kuntzmann (1912-1992) and Noël Gastinel (1925-1984). Françoise had a brother and a sister.

After her *baccalauréat*, she entered at Lycée Champollion to prepare the entrance examination of superior schools, and after three years at École Normale Supérieure de Sèvres, she obtained her *agrégation* at the University of Paris.

From 1963 to 1969, she taught at IMAG as an assistant professor. She began her doctoral thesis under the supervision of Gastinel with a title that can be translated as *Numerical methods for calculating eigenvalues and eigenvectors of a linear operator*, with a defense in 1971.

She remained in Grenoble as an associate professor until 1974, and then as a professor from 1975 to 1983. Then, she moved to the University of Paris-Dauphine and held consulting positions in several industrial research laboratories (IBM-France, CERFACS, Thomson-CSF, ONERA, etc.).

She was also associated with research centers in the USA and other countries. In 1997, she was promoted to full professor at the University of Toulouse, where she stayed until her retirement.

Yousef Saad was among her 30 doctoral students. Outside the scientific domain, she was interested in the history of science, but also politics, literature, philosophy, and especially that of spirituality.

She was married twice, first to Philippe Chatelin and then to Gregory Chaitin, and divorced twice. She had two children from her first marriage.

She died on May 14, 2020, in Toulouse. She is buried in Veules-les-Roses, a village on the Normandy coast.

Work

Françoise Chatelin is well known for her work in numerical algebra, particularly eigenvalue computations, on which she wrote numerous papers. She also wrote some books in French that were translated into English: *Spectral Approximation of Linear Operators* [644] in 1983 and *Eigenvalues of Matrices* [645] in 1993. These two books have been reprinted by SIAM.

Françoise was an expert in finite precision computations. She wrote a book *Lectures on Finite Precision Computations* with Valérie Frayssé [609] in 1996, and she was at the root of the PRECISE (PRecision Estimation and Control In Scientific and Engineering computing) software to assess the quality of numerical software in industrial environments; see [610] with Elizabeth Traviesas-Cassan in 2005.

In 2012, she published another book, *Qualitative Computing, A Computational Journey into Nonlinearity* [646].

10.11 ▪ Pafnuty L. Chebyshev

Pafnuty Lvovich Chebyshev

Life

One of nine children, Pafnuty Lvovich Chebyshev was born into a wealthy family on May 14, 1821,[62] in Okatovo, Kaluga Region, a small town in western Russia, southwest of Moscow. His father was a former army officer who had fought against Napoléon's invading army.

[62]The dates of his birth and death vary according to different sources.

The young Pafnuty was first educated at home by his mother and a cousin. He learned writing and reading from his mother, French and arithmetic from his cousin, and music with a teacher. Since he had one leg longer than the other and was affected by a limp, he dedicated more time than usual to his studies.

In 1832, the family moved to Moscow. At home, Chebyshev was tutored in mathematics and physics by Platon Nikolaevich Pogorelski (1800-1852), then considered the best teacher in the town who, for example, had taught the writer Ivan Sergeevich Turgenev (1818-1883). In the summer of 1837, Chebyshev passed the registration examinations, and in September of that year, he entered the second philosophical department of Moscow University. The person who influenced him the most was the Czech mathematician Nikolai Dmetrievich Brashman (1796-1866), professor of applied mathematics at the university since 1834. Brashman was particularly interested in mechanics, but in addition to courses on mechanical engineering and hydraulics, he taught his students the theory of integration of algebraic functions and probability.

The department in which Chebyshev studied announced a prize competition for the year 1840-1841. Chebyshev submitted a work on the calculation of roots of equations in which he solved the equation $y = f(x)$ by using a series expansion of the inverse function f^{-1}. The paper was awarded a silver medal, not published at that time, but only in the 1950s. That same year, he obtained his first academic degree.

But Chebyshev's financial situation changed drastically. There was famine in Russia, and his parents were forced to leave Moscow. They were no longer able to support him. Nevertheless, he decided to continue his mathematical studies for his master's degree under Brashman's supervision. Chebyshev passed the final examination in October 1843. In 1846, he defended his master's thesis *An Essay on the elementary analysis of the theory of probability*, in which he examined Poisson's weak law of large numbers.

At that time, Chebyshev was already an internationally recognized mathematician. His first paper, in French that he perfectly mastered, was on multiple integrals and had been published in the *Journal de mathématiques pures et appliquées* in 1843. A second paper, also in French, on the convergence of Taylor's series, had been accepted by the *Journal für die reine und angewandte Mathematik* (also known as Crelle's journal) in 1844. However, no suitable position was offered to him at Moscow University. But in 1847, after submitting his thesis *On integration by means of logarithms*, he was appointed to the University of Saint Petersburg.

In 1848, he submitted his work *The theory of congruences* for a doctorate, which he defended on May 27, 1849. That same year, he published his work on the determination of the number of primes not exceeding a given number. The proof of this result was only completed two years after his death by Jacques Hadamard (1865-1963) and, independently, by Charles de la Vallée Poussin (1866-1962).

In 1850, Chebyshev proved the conjecture made in 1845 by Joseph Louis François Bertrand (1822-1900) that there is always at least one prime between n and $2n$ for $n > 3$. He collaborated with Viktor Yakovlevich Bunyakovsky (1804-1889) in editing the 99 number theory papers of Leonhard Euler (1707-1783), which they published in two volumes in 1849.

Chebyshev was promoted extraordinary professor at Saint Petersburg University in 1850. Simultaneously, he also taught practical mechanics at the Alexander Lyceum in Tsarskoe Selo (now Pushkin), a southern suburb of Saint Petersburg.

Between July and November 1852, Chebyshev visited France, London, and Germany. He was very much interested in windmills that he visited near Lille in the north of France, and in various steam engines and their mechanisms. He had discussions with French mathematicians, including Liouville, Bienaymé, Hermite, Serret, and Poncelet, and English mathematicians, including Cayley and Sylvester. In Berlin, he met Dirichlet. This visit had a great influence on Chebyshev's work. His interest in the theory of mechanisms and in the theory of approximation stem from this trip, after which he set up the foundations of the Russian school of approximation theory.

During his life, Chebyshev made many western European visits, in particular in France to give talks at the congresses of the *Association pour l'Avancement des Sciences* between 1873 and 1882. He attended the celebrations to honor Eugène Charles Catalan's retirement at the University of Liège, Belgium, on December 7, 1881, and made other visits in 1884 and 1893, probably in a number of European universities. In addition to the mathematicians he met during his trips, he also had contacts with other European mathematicians such as Lucas, Borchardt, Kronecker, and Weierstrass.

In 1856, Chebyshev was elected junior member of the Imperial Academy of Sciences, then an extraordinary member, and, in 1858, an ordinary member. The same year, he became an honorary member of Moscow University. Chebyshev was promoted to ordinary professor in 1860.

After 25 years of lectureship, Chebyshev became emeritus professor in 1872. In 1882, he left the university and devoted his life to research. The French Académie des Sciences elected him a corresponding member in 1860, and a full foreign member in 1874. In 1893, he was elected honorable member of the Saint Petersburg Mathematical Society, which had been founded three years earlier. In addition, every Russian university elected him to an honorary position. He also became an honorary member of the Saint Petersburg Artillery Academy, and he was awarded the French *Légion d'Honneur.*

Chebyshev never married and lived in a vast and well decorated house. He had a daughter that he never recognized but who was raised by his sister. He supported her financially even after she married a colonel.

Among Chebyshev's students were Yegor Ivanovich Zolotarev (1847-1878) in 1867, Konstantin Posse (1847-1928) in 1882, Andrei Markov (1856-1922) in 1884, and Alexandr Mikhailovich Lyapunov (1857-1918) in 1885.

Chebyshev died in Saint Petersburg on November 26, 1894. His works and more details on his life can be found in *Œuvres de P.L. Tchebychef, publiées par les soins de MM. A. Markoff et N. Sonin*, Académie Impériale des Sciences, Saint Petersburg, 1899-1907.

Work

The works of Chebyshev cover probability, statistics, number theory, numerical analysis, mechanics, continued fractions, theory of integrals, construction of maps, calculation of geometric volumes, and, in the 1870s, construction of calculating machines (see Chapter 7), etc.

He wrote many papers on his mechanical inventions, and models and drawings of some of them were exhibited at the Conservatoire National des Arts et Métiers in Paris. In 1893, seven of his mechanical inventions were presented at the World's Exposition in Chicago organized for the 400th anniversary of the discovery of America by Christopher Columbus, including his invention of a special bicycle for women.

A direct consequence of his 1852 trip to western Europe was the publication in 1854 of *Théorie des mécanismes connus sous le nom de parallélogrammes*. It was in this work that the famous Chebyshev polynomials appeared for the first time. He showed that using their zeros as interpolation points reduces the error of the interpolation polynomial. Only few families of orthogonal polynomials were known at that time, namely those of Legendre and Hermite. But, it was Chebyshev who saw the possibility of a general theory and its applications. His work arose out of the theory of least squares approximation and probability. He applied his results to interpolation, approximate quadrature, and other topics. He was also already aware of the Christoffel-Darboux formula. His work on best approximation, named after him, is well known.

The Russian name of Chebyshev has been transliterated in many different ways, depending on the language: Chebysheff, Chebychov, Chebyshov or Tchebychev, Tchebycheff (French) as well as Tschebyschev, Tschebyschef, Tschebyscheff (German). This explains why the Chebyshev polynomials are denoted as C_k or T_k.

In 1867, he published a paper *On mean values*, in which he used an inequality due to Irénée-Jules Bienaymé (1796-1878) to give a generalized law of large numbers. This inequality is now known as the Bienaymé-Chebyshev inequality. Twenty years later, in the paper *Sur deux théorèmes relatifs aux probabilités*, he gave the basis for applying the theory of probability to statistical data, generalizing the central limit theorem of de Moivre and Laplace.

10.12 ▪ André L. Cholesky

André Louis Cholesky
Courtesy of the Cholesky family

Life

André Louis Cholesky was born on October 15, 1875, in Montguyon, a village located 35 km north-east from Bordeaux, France. He was the son of André Cholesky, head waiter, and Marie Garnier.

Cholesky spent his childhood in his native village, and then went to the "Lycée" (secondary school) in Saint-Jean-D'Angély. He obtained the first part of his "Baccalauréat" in Bordeaux in 1892, and the second part a year after, as usual.

On October 15 1895, he was admitted as a student to the prestigious École Polytechnique (a military school) in Paris, and he had to sign a three-year contract with the Army. After two years at École Polytechnique, he entered the École d'Application de l'Artillerie et du Génie in Fontainebleau, near Paris, where he stayed for two more years. There, he was trained, among other topics, in topography.

Then, Cholesky began his army officer's career in October 1899 as a Lieutenant. He was sent to Tunisia from January to June 1902, and again from November 1902 to May 1903. From December 1903 to June 1904, he was in Algeria. On June 24, 1905, he was appointed to the Geographical Service of the Army.

At that time, a new triangulation of France had been decided. Cholesky was sent to Dauphiné, a part of the Alps near Grenoble, for measuring the length of the meridian passing through Lyon.

The engineers were faced with the problem of finding a simple, fast, and precise method for correcting the measuring and the instrumental errors. A system of linear equations had to be solved in the least squares sense, and it certainly was at this occasion that Cholesky had the idea of his method.

On April 22, 1907, Cholesky married his first cousin Anne Henriette Brunet. They had four children, the youngest of whom was born after Cholesky's death. From November 1907 to June 1908, Cholesky was sent to Crete, then occupied by international troops. His work consisted in the triangulation of the French and British parts of the island.

In March 1909, Cholesky was promoted to Capitaine (Captain), but he had to leave the Geographical Service for a 2-year training period as the head of an artillery battery. In September 1911, he was appointed to the artillery's headquarters. He received the commandment of the leveling operations in Algeria and Tunisia. He also spent some time in Morocco and Sahara. In May 1913, he became the head of the Topographical Service in Tunis. He stayed there until the declaration of war on August 2, 1914.

Back in France, he was in charge of a battery for some months. From September 1916 to February 1918, Cholesky was sent to Romania as the head of the Romanian Geographical Service. He completely reorganized this Service, showing his great capabilities as a leader. He was promoted to Chef d'Escadron, that is, Commandant (Commander), on July 6, 1917. Back in France, his regiment was involved in the second battle of the Somme region.

On August 31, 1918, Commandant Cholesky died at 5:00 am in a quarry, north of the village of Bagneux (10 km north of Soissons), from wounds received on the battlefield. He is burried in the military cemetery of Cuts, 10 km southeast of Noyon, tomb 348, square A.

Work

Cholesky presented his method in an unpublished manuscript entitled *Sur la résolution numérique des systèmes d'équations linéaires* (On the numerical solution of systems of linear equations). This manuscript had been unknown for many years after its writing. For details on this work, see Chapter 2.

Cholesky's method remained unknown outside of the circle of French topographers until 1924, when another French officer, the Commandant Ernest Benoît (1873-1956), published a paper explaining his fellow's method [266].

Cholesky's method became better known after the 1940s through the works of the Danish geodesist Henry Jensen (1915-1974) and Olga Taussky (1906-1995) and her husband John Todd (1911-2007). This method was later rediscovered many times.

From at least December 1909 to January 1914, Cholesky taught in an engineering school by correspondence, the École Spéciale des Travaux Publics, du Bâtiment et de l'Industrie, founded in 1891 by Léon Eyrolles. He wrote many lecture notes, and published a book of 442 pages that had at least 7 editions and was still in print in 1937 [667], that is, almost 20 years after his death.

Cholesky's archives at École Polytechnique also contain the manuscripts of two other books: *Complément de Topographie* (239 pages), and *Cours de Calcul Graphique* (83 pages) which is reproduced (for the first time) and analyzed in [464].

These archives contain many other scientific papers by Cholesky,

- 3 pages with the title *Sur la détermination des fractions de secondes de temps.*
- A 15-page manuscript entitled *Instructions pour l'exécution des nivellements de précision.*
- A 8-page manuscript entitled *Équation de l'ellipsoïde terrestre rapportée à Ox tangente au parallèle vers l'Est, Oy tangente au méridien vers le Nord, Oz verticale vers le zénith.*
- 16 pages on *Étude du développement conique conforme de la carte de Roumanie.*

- A 3-page manuscript with the title *Instructions sur l'héliotrope-alidade (modèle d'étude 1905).*

- 3 pages and 3 maps on the construction of railroads.

- 3 pages with the title *Remarque au sujet du calcul de correction de mire.*

For a detailed account of Cholesky's life, his works, and the mathematical environment, see [439] by C.B.

10.13 ▪ Gianfranco Cimmino

Gianfranco Cimmino

Life

Gianfranco Luigi Giuseppe Cimmino was an Italian mathematician born on March 12, 1908, in Naples.

Gianfranco's father, Francesco, was a historian, translator from Indian and Persian, and poet, from Naples, who taught history in a lyceum and then Sanskrit at the University of Naples. Cimmino's mother, Olimpia, was from the aristocratic family Gibellini Tornielli Boniperti, from Novara in the Piedmont region of northwestern Italy; she died prematurely in 1917.

In 1923, when he was fifteen years old, Cimmino graduated from high school with a classical diploma. He entered the University of Naples and, in 1927, at the young age of 19, he graduated with a laurea (cum laude) in Mathematics. His thesis, under the guidance of Mauro Picone (1885-1977), who had moved to the University of Naples in 1925, was on methods for approximating the heat equation in two dimensions. In 1928, he was immediately appointed an assistant to Picone who held the chair of Analytical Geometry. He held this position during the academic year 1927-28.

Then, having published two papers, Cimmino was awarded a scholarship to study abroad. He spent the academic year 1930-31 in Germany at the universities of Munich and Göttingen. In 1931, as a *Libero docente*, he was in charge of the courses of Higher Analysis between 1932 and 1935 and Analytic Geometry between 1935 and 1938 in Naples. He also began to work at the Istituto di Calcolo per l'Analisi Numerica, founded in Naples by Picone in 1927 (perhaps the first applied mathematics institute ever founded), and renamed Istituto Centrale di Calcolo in 1931.

Cimmino was one of the four assistants to Picone, together with Renato Caccioppoli (1904-1959), Carlo Miranda (1912-1983), and Giuseppe Scorza Dragoni (1908-1996), who wrote an obituary of Cimmino with a description of his works [950]. They were jokingly referred to as *"Mauro Picone's four musketeers"* by Scorza Dragoni [949].

From the very beginning of his scientific activity Cimmino showed originality of thought and calculation ability [2497]. Picone left Naples for Rome in 1932, where he took the chair vacant after the expulsion of Vito Volterra (1860-1940) who refused to sign the new mandatory oath of loyalty. Picone transferred the Institute in Rome and chaired the new Istituto Centrale di Calcoli Tecnici, named after Istituto Nazionale per le Applicazioni del Calcolo (I.N.A.C.) of the Consiglio Nazionale delle Ricerche (National Council of Research). Miranda followed him to Rome, becoming vice director of this new institute. Cimmino and Caccioppoli remained in Naples. Let us mention that in 1938, Caccioppoli, with his future wife, gave a speech against fascism in a tavern in Naples, and it seems that he perhaps sang the French national anthem. He was arrested, but to avoid imprisonment, his family, in agreement with the academic authorities, mentioned hypothetical mental problems, managing to have him interned in a house of care (instead of prison). Disappointed by politics and his wife's desertion, he committed suicide in 1959.

In 1938, Cimmino obtained the chair of Analysis at the University of Cagliari in Sardinia, but he remained there only for one year. At the end of 1939, he was appointed professor of Mathematical Analysis at the University of Bologna. There he was promoted to a full professorship on January 1, 1942, a position he held until his retirement on November 1, 1978.

In 1940, he married Maria Rosaria Martinez (1914-1998). They had seven children. That same year, after the death of Cimmino's father, his uncle Francesco Gibellini Tornielli Boniperti legally adopted him together with his sister Eugenia, transferring his three surnames to them.

In Bologna, and at the same time at the University of Modena, Cimmino taught Higher Analysis, Mathematical Analysis, Theory of Functions, and Topology. His role in training engineers was notable because he taught Mathematical Analysis at the Faculty of Engineering from 1939 to 1974.

Cimmino was also director of the Institute of Mathematics "Salvatore Pincherle" from November 1, 1950, to October 31, 1952; dean of the Faculty of Sciences of the University of Bologna from 1965 to 1972; and director of the Institute of Astronomy of the University of Bologna from February 22, 1972, to May 31, 1974. From 1966 to 1973, he was the editor-in-chief of the Bulletin of the *Italian Mathematical Union* (UMI). He was appointed extraordinary commissioner of the Istituto Nazionale di Alta Matematica (National Institute of High Mathematics) from 1973 to 1977, and, after his retirement, professor and director of this institute.

In 1980, he was elected a national member of the Accademia Nazionale dei Lincei. He was also a member of the Pontaniana Academy of Naples and of the Academy of Sciences of Bologna. He was awarded the Gold Medal for Meritorious of the School of Culture and Arts. He also won the "Gualtiero Sacchetti" Prize and was elevated to Commander of the Order of Merit of the Italian Republic. In 1984, he was appointed emeritus professor of the University of Bologna. Among Cimmino's students in Bologna were Lamberto Cattabriga (1930-1989) and Bruno Pini (1918-2007).

Cimmino died on May 30, 1989, in Bologna.

Cimmino was an aristocrat, a little detached, but understanding and helpful, as Pini wrote [2497], but certainly he was a man of culture. He knew French, English, German, Russian, and also Latin and Greek. In particular, he was a fervent admirer of Dante Alighieri (1265-1321). One of his last conferences was entitled "Dante and Mathematics," and published [705].

His nephew Diego Vasdeki wrote

> *My uncle Gianfranco Cimmino was a first-rate mathematician, and also a lover of languages and music, much appreciated by his students not only for the clarity of his explanations, but also for an innate kindness of soul. Among his passions were Dante and the Divine Comedy (which he knew entirely by heart): in many everyday situations he found a Dante connection and quoted the triplets, enriching them with explanations and comments that fascinated those who listened to him.*

Work

Cimmino worked, in particular, on the solution of linear equations and systems of linear differential equations of the first and second order (and related limit problems) with functional correspondences that arise from these equations, and on the theory of distributions. In the theory of elliptic partial differential equations, he was the first mathematician to study the Dirichlet problem with generalized boundary conditions, and published some of his most remarkable papers on this theory in 1937-38. Besides his works on partial differential equations and the solution of linear equations, Cimmino contributed to the calculus of variations, differential geometry, conformal and quasi-conformal mappings, topological vector spaces, and the theory of distributions. He published 56 scientific papers, all written alone, except for two of them, and seven didactic books. His work is described in [2497, 950], and, concerning numerical analysis, it has been analyzed in [268]. A book containing selected works of Cimmino, edited by Sbordone and Trombetti, was published [707] in 2002.

In numerical analysis, Cimmino is well known for the paper *Calcolo approssimato per le soluzioni dei sistemi lineari* (Approximate computation of the solutions of linear systems) [701], an important paper written under Picone's insistence, and published in 1938. In that paper, he described a method he found at least in 1932, and that, strangely, is not included in [707]. This paper begins by an introduction by Picone explaining that (our translation)

> *Professor Gianfranco Cimmino can be considered, among other things, as one of the supporters of the Istituto per le Applicazioni del Calcolo, to which he lent his continuing and productive assistance during the embryonic stages of the Institute itself, in Naples, in the laboratory annexed to that university's Calculus chair, from July 1928 to October 1932. Towards the end of that period, Professor Cimmino devised a numerical method for the approximate solution of systems of linear equations that he reminded me of in these days, following the recent publication by Dr. Cesari* [reference], *which provides a systematic treatment of the above mentioned computing methods but which, however, does not consider the one by Cimmino, a method which, in my opinion, is most worthy of consideration in the applications because of its generality, its efficiency and, finally, because of its guaranteed convergence which can make the method practicable in many cases. Therefore, I consider it useful to publish in this journal Professor Cimmino's note on the above mentioned method, a note that he has agreed to write on my insistent invitation.*

Cesari's paper was [600, 601]. As explained in [3143], it was probably under the influence of Picone's ideas and the papers of Lamberto Cesari (1910-1990) on the numerical solution of linear systems that Cimmino became interested in this topic. The method is a projection method on hyperplanes with averaging operations. It can be related to Kaczmarz's method [1411].

In 1967, Cimmino came back to the numerical solution of linear systems proposing a Monte Carlo method [702]. Finally, in 1986, in [703, 704] and [706], he showed that suitable hyperspherical means and their ratios can replace Cramer's determinants and formulas in solving systems of linear algebraic equations. He was convinced that such methods deserved to supplant the use of Cramer's rule in applications, so much so that he recommended his son Giovanni, an engineer, to have these expressions experimented numerically in the real number set [950].

10.14 ▪ Gabriel Cramer

Gabriel Cramer
Bibliothèque de Genève

Life

Gabriel Cramer was born in Geneva on July 31, 1704. In those times Geneva was an independent republic (it joined Switzerland in 1815). His father Jean-Isaac Cramer (1674-1751) was a physician who was also elected to the councils ruling the republic. His mother was Anne Mallet. Gabriel had two brothers. Cramer's family was of German origin.

Gabriel studied in the Collège de Genève and then, starting on May 1719, he entered the Académie de Genève where he studied philosophy and mathematics. On August 31, 1722, he defended a thesis whose title was *Dissertatio physico-mathematica de sono*. He was the student of Étienne Jallabert (1658-1723) who was professor of philosophy.

When Jallabert passed away, there were three candidates for his position: a 26-year old parson named Amédée De la Rive and two younger scientists, Gabriel Cramer and Jean-Louis Calandrini (1703-1758). De la Rive got the position on May 3, 1724. But the council was so impressed by Cramer and Calandrini that the chair of Mathematics was re-established and jointly given to Cramer and Calandrini. They had to share the duties and the salary. There was a condition that they must alternately travel to foreign countries to improve their mathematical knowledge.

Cramer started his journey in 1727. He attended the lectures of Jean I Bernoulli (1667-1748) and of his nephew Nicolas Bernoulli (1687-1759) in Basel between May and October 1727. Then, he traveled to England with a stop in Paris. He spent five months in Cambridge and four months in London where he met James Stirling (1692-1770). In England he became familiar with Newton's works. After his long stay in England he moved to Leiden (Holland) where he met Willem Jacob 's Gravesande (1688-1742). He was in Paris from December 1728 until May 1729 where he had the opportunity to meet many scientists, in particular Bernard de Fontenelle (1657-1757), Pierre Louis Moreau de Maupertuis (1698-1759), Georges-Louis Leclerc, comte de Buffon (1707-1788), and Alexis Claude Clairaut (1713-1765). He corresponded with many of these people during the rest of his life.

Back in Geneva, he resumed his duties as a professor. He received some recognition by being elected to the Conseil des Deux Cents (council of the two hundred, which was a sort of parliament without too much power) in 1734 and to the Conseil des Soixante (council of the sixty) in 1751. But, he could not have a political career because his elder brother was already involved in the government of the republic and there were rules preventing too many members of the same family from holding important political positions.

In 1743, he was elected as a foreign member of the Société royale des sciences de Montpellier (France). In that same year he was elected to the Academy of Sciences of the Bologna Institute. He obtained a more prestigious membership when he was elected to the Berlin Academy in 1746. In 1749, he was elected Fellow of the London Royal Society.

In 1747-1748, he traveled to Paris as a tutor of the young Prince of Saxe-Gotha. During his stay he resumed his friendships with French scientists and met new ones like Jean Le Rond d'Alembert ((1717-1783). Unfortunately, he failed twice to be elected to the French Académie des Sciences.

In 1750, he obtained the chair of philosophy in Geneva. His health started to deteriorate in 1751. In December 1751, he asked for a leave of absence because it was recommended to him to travel to the south of France to recover in better and warmer weather. Unfortunately, things got worse, and he passed away in Bagnols-sur-Cèze on January 4, 1754.

Work

Gabriel Cramer did not publish many papers. The most we know about his scientific research is from his letters to other scientists. He is remembered for his book *Introduction à l'Analyse des Lignes Courbes Algébriques* (Introduction to the analysis of algebraic curves) which was published in 1750. However, there is some evidence that he worked on this book since 1740; for a detailed analysis of this book, see [1832].

It is in a three-page appendix to this book that he described what we now call Cramer's rule; see Chapter 3. But he used elimination and probably knew this result as early as 1744 because in a letter to Clairaut in 1744 about a memoir that is, unfortunately, lost, he wrote (our translation)

> *I cannot enough admire my imprudence to send you this big memoir in which I studied the number of intersections of two algebraic curves. I titled it "Evanouissement des inconnues".*

In 1750, he wrote to Leonhard Euler (1707-1783),

> *What I inserted in the appendix of my book is an excerpt of a longer and more detailed memoir that I sent to Clairaut several years ago. He wanted to read it at the Academy but the notation was difficult to express when speaking.*

Cramer spent a lot of time from 1740 to 1745 editing and annotating the complete works of Jean and Jacques Bernoulli. He was also busy publishing one of Euler's manuscripts.

It is sad that he is mostly remembered for only three pages of his book.

10.15 ▪ Seymour R. Cray

Life

Seymour Roger Cray was born September 28, 1925, in Chippewa Falls, Wisconsin, USA. His father was a civil engineer who worked for the Northern States Power Company and then became the town engineer in Chippewa Falls.

Seymour graduated from high school in 1943. At an early age Cray became very interested in science and engineering, particularly in electrical engineering and electronics. In high school, he spent much of his time in the electrical engineering laboratory.

Seymour Roger Cray

He was drafted into the army as a radio operator. After D-Day in 1944, he was sent to Europe where he saw the Battle of the Bulge (Ardennes) and then to the Philippines and the Pacific. Returning home, he resumed his studies and obtained a B.S. in Electrical Engineering in 1950 and an M.S. in Applied Mathematics in 1951 from the University of Minnesota.

Cray married Verene Voll in 1947. They knew each other since their childhood. After the wedding, they went to Madison and the University of Wisconsin for about a year. But then they went back to the University of Minnesota. Cray had two daughters, Susan and Carolyn and a son, Steven.

After getting his diploma he joined Engineering Research Associates (ERA) in Saint Paul, Minnesota. ERA was created in January 1946 at the unofficial request of the US Navy that wanted to keep together a group of scientists and engineers who had been working on machines for code breaking during the war. The main people involved were William Charles Norris (1911-2006), Howard T. Engstrom (1913-1962), and John Parker as an investor. In the beginning the Navy was the only customer of the company, which was located in a former glider factory that was used to build wooden gliders for the D-Day landing. The company designed custom electronic machines to break a specific code, but then it was decided to construct more general purpose computers that can be re-programmed. This led to a machine named Atlas that was marketed as the *ERA 1101*. In 1952, ERA was bought by Remington Rand which merged with Sperry Corporation to become Sperry Rand in 1955. Later Sperry Rand was bought by Burroughs, and this led to the creation of Unisys. The new company was now more interested in computers aimed at business applications. Being dissatisfied with that and more interested in scientific computers, Cray left ERA in 1957, as did several other employees.

He joined Control Data Corporation (CDC), which was created in September 1957. About this move Cray said: *I had a clear idea of what I wanted to do which was to build large scientific computers*. The CEO of CDC was Norris and Cray became the chief designer. When at CDC, Cray designed scientific computers that were the fastest of their time. CDC became very successful and, after a while, to be able to work in a quiet environment, away from marketing and management activities, Cray asked Norris to move his research and development facility to his home town, Chippewa Falls.

Cray was not always an easygoing man with its management. Norris once recalled that when he asked Cray to write five-year and one-year plans for CDC, his response was: *Five-year goal: Build the biggest computer in the world. One-year goal: Achieve one-fifth of the above.*

At the beginning of the 1970s, there were competing computer projects within CDC. In 1972, after the project of *CDC 8600* on which he was working encountered some reliability problems and was dismissed, Cray left CDC and founded Cray Research Inc. Initially, he took half a dozen people from CDC with him.

The *CRAY 1*, his first machine for Cray Research, first delivered in 1976, was a big commercial success. Cray's laboratory and the manufacturing were, once again, in Chippewa Falls. The headquarters were in Mendota Heights Road, in Minneapolis-Saint Paul. An important man at Cray Research was Lester (Les) T. Davis, who took care of the engineering details of Cray's designs.

In 1975, while he was developing the *CRAY 1*, Cray was divorced from his first wife, Verene. A year later, he met Geri M. Harrand, whom he later married.

The next project of Cray, the *CRAY 2*, was less successful. It took six years of development to complete the project. At the same time, there was an internal competition with the X-MP line. Cray left the CEO position at Cray Research in 1980 to become an independent contractor. The project leader of the XMP series was Steve Chen. In an interview, Cray described his (lack of) relationship with Chen:

> *That's kind of fascinating because we had no relationship at all. I think I only met him half a dozen times. This was a relationship more or less generated by John Rollwagen, at that time CEO of Cray Research because he wanted to have a protege and so he created one. I'm sure Steve has a lot of skills but they didn't relate to me. We had our competitive project but it was very much arm's length. He was doing his project and I was doing mine and there was no interaction.*

After Cray stepped down as CEO, his *CRAY 3* project ran into technological problems and Cray moved to Colorado Springs in 1988. He left Cray Research in 1989 to form the Cray Computer Corporation which, after troubles, sought Chapter 11 bankruptcy protection in March 1995. There was not a single *CRAY 3* sold. Cray began working on the follow-on to the *CRAY 3*, the *CRAY 4*, a machine that would have 64 processors.

In those times, the market for supercomputers was changing. The end of the cold war meant declining government budgets for purchasing machines, and the arrival of cheap and powerful microprocessor chips changed the deal.

After Cray Computer ran out of business, Cray set up a new company, SRC Computers, and started the design of his own massively parallel machine. Note that S.R.C. are Cray's initials.

Unfortunately, on September 22, 1996, on the I-25 highway in Colorado, Cray had a serious car accident. He suffered a broken neck, broken ribs, and severe head injuries. He was taken to Penrose Community Hospital where he remained in critical and unstable condition. He passed away on October 5.

John A. Rollwagen, a former chairman and CEO of Cray Research Inc. said after Cray's death,

> *He had a profound effect on the computer industry. He was always on the leading edge.*

Work

At ERA, Cray started working on the 1100 series computers. Cray's first significant work was to design from scratch the control system of the *ERA 1103* computer, basing his design on simplicity. About his way of working, he said

> *For the last 30 or 40 years I've been getting my insights from the customers who bought the Cray computers. They would tell me what's wrong with it. I'd address*

those issues and we'd go another generation. It's been very evolutionary. Essentially every machine I've designed since that first day have been clear descendants of one another in their structure. [sic]

His next big machine was the *CDC 1604* for which he was using rejected transistors because they were much cheaper. He designed the circuits to be tolerant to their lower performances. The 1604 was very successful in the scientific market. It was the first large transistorized scientific computer.

The next supercomputer Cray designed was the *CDC 6600* in 1964. It was the fastest computer of its time. The 6600 was superseded by Cray's next machine, the *CDC 7600*.

Cray began to work on the *CDC 8600* in 1968, and realized that an improved clock speed would not allow to reach the performance goals. He designed the *8600* with four processors all sharing one memory. However, this project was stopped by the management.

After leaving CDC, Cray designed what can be considered as his best machine, the *CRAY 1*. He used new technology, integrated circuits, and vector register technology. The machine had a C-shape so the wires could be on the short inner surfaces of the modules; see Figure 10.1. It was the second vector machine on the market. The first one was the *CDC Star*, which was less efficient. The *CRAY 1* used the COS operating system. There was also a vectorizing Fortran compiler. *CRAY 1* serial number 001 went to Los Alamos National Laboratory. The first full system was sold to the National Center for Atmospheric Research (NCAR) in Boulder, Colorado. More than 80 machines were sold.

Figure 10.1. *Cray 1S*

The *CRAY 2* (Figure 10.2) was Cray's first multiprocessor computer with two or four vector processors. What was most innovative was that the circuits were immersed in a liquid (Fluorinert) that was injected and then cooled in heat exchangers outside of the machine. Another distinctive feature was the large (at the time) memory of 256 Mwords. There were around 240,000 chips, of which 75,000 were for memory. The operating system was UNICOS, the Cray flavor of Unix.

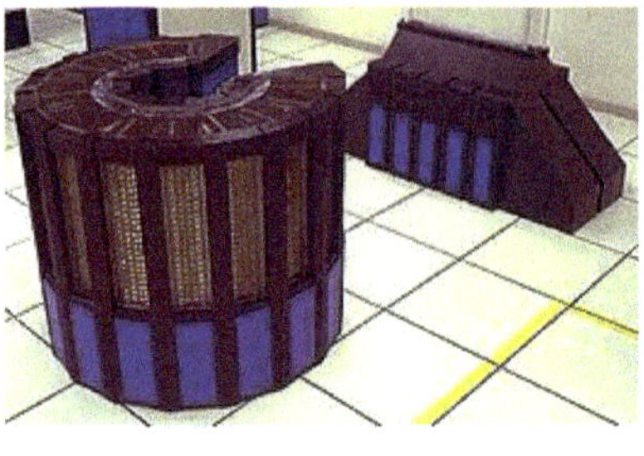

Figure 10.2. *Cray 2*

The first machine went to the Lawrence Livermore National Laboratory. Around 25 machines were sold.

For his *CRAY 3*, Seymour wanted to use gallium arsenide semiconductors to build a machine with 16 processors. Using GaAs was probably Cray's second big mistake, the first one being the too ambitious design of the *CDC 8600*. The first machine was finally only ready in 1993. The only *CRAY 3* made, configured with four processors and 128 MWords of memory, was delivered to the National Center for Atmospheric Research in 1993, but was soon taken out of service because of reliability problems.

Cray started working on the *CRAY 4* in 1994 but Cray Computer filed for bankruptcy in 1995 and the project was stopped.

Let us quote Seymour Cray in 1995:

> *The perception is that microprocessors are so powerful there's no need for anything else. Anything else that needs to be done can be done with microprocessors. So we're seeing great efforts being made to accomplish that goal, massively parallel arrays of microprocessors. Perhaps it will succeed but it presents a lot of software difficulties that we're struggling with today. It's not clear to me that in the long run that it will turn out the best way to do it.*

Interesting documents about Cray's life and work are covered in the interview by David Allison at the Smithsonian Institution and the book [2309] by F. Murray, even though it does not contain too many technical details about the machines.

10.16 ▪ Prescott D. Crout

Prescott Durand Crout
Courtesy of the Antiochiana collection at the Olive Kettering Library of Antioch University

Life

Prescott Durand Crout was born on July 28, 1907, in Ohio. His father was Ray Durant Crout (1878-1954) and his mother was Maryetta Sharp (born 1885). He had a sister and a brother.

Crout lived in different places in Massachusetts. In 1929, he graduated from Massachusetts Institute of Technology (MIT). In 1930, he defended his Ph.D. thesis under the supervision of George Rutledge (1881-1940). It was entitled *The approximation of functions and integrals by a linear combination of functions*.

On January 2, 1933, he married Charlotte Louise Zander (1909-2000). They had four children.

Crout was a member of the MIT Faculty of Mathematics from 1934 to 1973 and emeritus from 1973 up until his ultimate death. He also belonged to the Radiation Laboratory staff from 1941 to 1945. He had seven doctoral students at MIT.

Crout died on September 25, 1984, in Lexington, Massachusetts.

Work

Crout is famous for the method for solving systems of linear equations described in a six-page paper [775] published in 1941. Unaware of the earlier work of André Louis Cholesky (1875-1918), Crout reorganized Gaussian elimination to accumulate sums of products; see more details in Chapter 2. The goal was to obtain a method suited to the hand calculators of the time. The method was also publicized by the manufacturer of calculating machines Marchant. They published a manual with detailed instructions for using Crout's method.

Crout's method became popular in the 1950s, particularly among engineers. It is explained in detail in a book by Francis Begnaud Hildebrand (1915-2002) [1696], a former student of Crout.

10.17 ▪ Myrick H. Doolittle

Myrick Hascall Doolittle circa 1862
Image courtesy of Antiochiana, Antioch College

Life

Myrick Hascall Doolittle was born on March 17, 1830 in Addison, Vermont, USA. His parents were Jared Doolittle (1797-1869) and Phoebe Doolittle (1798-1840). He had a brother Mose Bartholomew (1833-1913) and a sister Lydia Jemina (1837-1915).

He lacked the time or the money or the inclination to enter college. He began by instructing the children of friends and neighbors. At the age of 26, he accepted a position at the New Jersey Normal College, and stayed there until the beginning of the American Civil War in 1861. Deciding it was time for him to get a degree, he entered Antioch College, Ohio, a small religious and abolitionist college. In 1862, Doolittle was awarded a B.A. degree. He volunteered for an

Ohio regiment in the Union army, but his physical condition prevented enlistment. He became a professor of mathematics at Antioch for the following year.

Then, Doolittle went to Harvard College to study mathematics with Benjamin Peirce (1809-1880) who was to become the director of the U.S. Coast Survey from 1867 to 1874. No details about his life in Cambridge are known, but it seems likely that he spent some time in the Nautical Almanach Office with the computers, and not more than 12 months with Peirce. He left Harvard in 1864 without a degree, and joined his wife, Lucy Salisbury, who was a volunteer nurse in the army hospitals in Washington. He may have served a short time in the army sanitary commission.

Then, with the help of Peirce, who gave him a recommendation letter, he obtained a position as an observer and a computer at the Naval Observatory, where he remained for six years (1864-1870). His job required him to record the positions of stars during the night, and to reduce observations by day. Wanting to live a less taxing life, he resigned his position and spent three years (1870-1873) in the Patent Office examining patents on steam boilers.

On April 14, 1873, he was appointed computer in the United States Coast and Geodetic Survey, whose director was Peirce. He stayed there until his retirement, and, among other works, he experimented with binary arithmetic. Doolittle soon became a major figure of the Computing Division, although he never was its formal leader. He had an innate grasp of computation, and was able to design concise and rapid computing plans, that is, algorithms. In 1874, he became interested in least squares, which led him to a simple procedure to handle the computations.

He was for many years an active member of the Philosophical Society of Washington. Further details about his life were given in a letter published by his wife in the January 1895 issue of *The Antiochian*:

> *In 1864 he came to Washington to enter upon Astronomical work in the Naval Observatory. His special work was observing Meridian Zenith Distances. He found the night work very wearing and after six years resigned and went into the Patent Office, where he stayed three years, examining applications for patents for steam boilers. In 1873, he entered upon work in the Coast Survey Office. This was more congenial, and consisted of the least square adjustments of Triangulations. In this department he still remains.*

Let us quote an anecdote about Doolittle, reported by Commander Harry A. Seran (1887-1962):

> *About 8:00 o'clock in the evening of June 30, 1907, direct from a fresh water college in Ohio, I reached Washington, D.C., for the first time. An appointment as Aid in the Coast and Geodetic Survey and instructions to report for duty on July 1st were carefully folded in my pocket.*
>
> *Upon reporting to the Assistant Superintendent and taking the oath of office, I was ordered to duty in the Computing Division. The Computing Division in those days was an entirely different division from what it is to-day. Among others was Mr. Doolittle, a famous computer and the originator of the Doolittle method of solving equations. He was at this time quite old and in order to know when to stop work in the afternoon, he had an alarm clock in his office which sounded off at 4:30. Needless to say, this alarm could be heard throughout the Division. It may have been the forerunner of the signal which is given on the Auto-call today, which tells those who have not already done so to put on their hats, coats and galoshes preparatory to trying to work their various ways home through the jam of traffic, red and green lights, busses, street-cars and trucks. Mr. Doolittle always wore carpet slippers in the office. That type of footwear must be extinct as the dodo bird to-day for it is never seen except in fancy dress costume.*

Doolittle resigned from the Geodetic Survey on September 25, 1911, because of bad health. He died at his home on June 27, 1913, in Linden, Maryland, after a short illness. After his death the Survey sent an account of his life and work to the newspapers where it was written

> *In the details of the work as well as in general principles he displayed a rare talent which has left a lasting impression on the work of the office and made his name well known among computers the world over.*

For more about Doolittle's life, see also [1138] by Richard William Farebrother in 1987.

Work

On November 9, 1878, Doolittle, then at the Computing Division of the U.S. Coast and Geodetic Survey in Washington, presented a method for solving the normal equations arising from problems of triangulation [935]. His method consisted to zero step by step the elements of the matrix in order to transform it into an upper triangular matrix; see Chapter 2 for details. It is equivalent in modern terms to decompose A into a product $A = LU$ by a succession of n intermediate steps, where n is the order of the system, with $L = L_1 + \cdots + L_n$ and $U = U_1 + \cdots + U_n$. The matrix L is lower triangular and U is upper triangular with a unit diagonal, in contrast to Gauss' method.

Doolittle also showed how to solve an enlarged system, obtained by adding new equations and new unknowns, without having to repeat all calculations. This technique is therefore similar to the *bordering method* which is used for solving the normal equations by enlarging the system. According to him, this was one of the main advantages of his method. Doolittle had no calculating machine at his disposal and he only used pencil, paper, and multiplication tables. He said that he solved, with the help of J.G. Porter, a system of 41 equations containing 174 side coefficients in five and a half days [935].

Doolittle's method was quite successful and it was used, with variations, for many years in geodesy. For example, Thomas Wallace Wright (1842-1908), a professor of Applied Mathematics and Chairman of the Department of Physics from 1895 to 1904 at Union College (Schenectady, N.Y.), described it in his 1884 book based on the methods he had developed in his survey works during the decade 1873-1883 he spent as a civil engineer with the U.S. Survey [2821], in the Section entitled *Combination of the direct and indirect methods of solution* [3275, pp. 167-174]. In the second edition of this book, written in 1906 with the cooperation of John Fillmore Hayford (1868-1925), Chief of the Computing Division and Inspector of Geodetic Work, U.S. Coast and Geodetic Survey, the authors wrote [3276, p. 114]

> *In it* [Doolittle's method] *there is a combination of improvements on the Gaussian method of substitution. Its advantage lies mainly in the arrangement of the work in the most convenient form for the computer. This makes the solution more rapid than by the other method, the gain in speed being the more marked the greater the number of equations.*

These two books were standard references for many years. Doolittle's method was also explained in 1912 in the book [1981] by Charles Jean-Pierre Lallemand (1857-1938), then a member of the French Academy of Sciences and head of the Service du Nivellement Général de la France (General Levelling Service of France). Cholesky had a copy of this book which was, however, published after he discovered his own method.

Doolittle's method seems to have been the standard method used by the engineers of the U.S. Coast and Geodetic Survey for a number of years, as exemplified in the small booklet [5] by Oscar Sherman Adams (1874-1962) in 1915. This method was independently rediscovered by Henry Agard Wallace (1888-1965), the 33rd Vice President of the United States. A farmer at the origin, he became a self-taught practicing statistician, and published an influential paper with George Waddel Snedecor (1881-1974) on computational methods for correlations and regressions. Let us quote [1140]:

> *Henry Agard Wallace (1888-1965) served as U.S. Secretary of Agriculture 1929-1932 and as Vice President 1941-1945. (Indeed, he might have succeeded Franklin Delano Roosevelt as President later in 1945 but for the fact that his colleagues thought him too liberal, so he was replaced by Harry Truman when Roosevelt decided to run for a fourth term.)* [...]
>
> *However, he* [Wallace] *was a practicing statistician who published an influential monograph coauthored by George Waddel Snedecor [3173], which has not been properly acknowledged. This pamphlet contained a very useful computational technique which was known to geodesists but not to statisticians at the time, and which apparently Wallace had rediscovered independently* [...]
>
> *At first sight, it seems unlikely that Wallace and other practicing statisticians would not have known of a computational procedure that had been in the public domain for 44 years. But Doolittle's paper was published in an unfamiliar journal in 1881, four years too late for inclusion in the extensive list of writings on the Method of Least Squares [2205] compiled by Mansfield Merriman (1848-1925) in 1877* [...]
>
> *Thus, if we can fully credit the recollections of the 72-year-old Wallace,* [...] *then it is clear that Wallace's contribution to the 1925 monograph was prepared independently of Doolittle's work.*

For more on these developments, see [1461, pp. 159-166].

10.18 ▪ Paul S. Dwyer

Paul Sumner Dwyer
Courtesy of Professor Ingram Olkin

Life

Paul Sumner Dwyer was born on December 8, 1901, in Chester, Pennsylvania, USA.

He received a bachelor's degree from Allegheny College, Pennsylvania, in 1921 and a master's degree from Pennsylvania State University in 1923. He was an instructor in mathematics at Penn State during 1921-26. Then, he went to Antioch College in Ohio as an assistant professor in 1926. He became an associate professor in 1929 and a professor in 1933.

Dwyer's association with the University of Michigan began when he was a graduate student in the summers of 1927, 1930, 1933, and 1934 and the academic year of 1935-1936, at the end of which he received his Ph.D. under the supervision of Harry Clide Carver (1890-1977), a pioneer and a leader in the development of mathematical statistics in the United States. His dissertation *Combined expansions of products of symmetric power sums and of sums of symmetric power products with applications to sampling* was published in 1938 in the *Annals of Statistics* [1034].

He left Antioch College in 1937 to become a research assistant at the University of Michigan. He was promoted to associate professor in 1942 and to professor in 1946. Also in 1946, he joined Cecil Calvert Craig (1898-1995) in founding the Statistical Research Laboratory. In 1942, he was a research associate at Princeton University. Thirteen students completed their Ph.D.'s under his direction. He retired in 1972 as Professor Emeritus.

He was also active in professional societies, particularly in the Institute of Mathematical Statistics. In Canada, he helped building a graduate statistics program at the University of Windsor, which awarded him an honorary Doctor of Science degree in 1971.

In addition to mathematics, Dwyer's interest after his retirement was vegetable gardening. He supplied fresh vegetables to his neighbors. He was a modest and friendly person, and a gifted teacher whose lessons were clear.

Dwyer died on September 17, 1982, at his summer house in Mackinaw City, Michigan.

Work

The first of the two principal areas of Dwyer's work was symmetric functions and finite sampling. The second one included the handling of systems of linear equations of which he became a leading expert. The four papers he published in 1941 show his insight and describe how to save some work in solving linear systems, particularly when the matrix is symmetric, as it is for normal equations. His square root method, an abbreviated Doolittle method, solves a linear system of order n using $4n - 1$ rows instead of $(n^2 + 9n - 4)/2$ rows needed in the Doolittle technique [1038, 1036, 1037, 1035].

In 1944, he gave the matrix interpretation of the method of Doolittle [1039] and, in 1951, he wrote a book [1041] on numerical linear algebra. This book was authoritative and unique. His aim was

> *designing appropriate calculational techniques through the use of better methods that feature less recording, more adequate checking to control mistakes, better control of errors, and relative ease of computation.*

Dwyer published papers with Frederick Vail Waugh (1898-1974) of the United States Department of Agriculture, on a compact method for the inversion of matrices with a study of its errors. He also made numerous contributions to linear regression and multivariate analysis. He did work on linear programming and provided a solution to the transportation problem. Eighteen of his papers were devoted to educational investigations.

Matrix derivatives were another topic on which he published several papers. From 1951 to 1953, he was a consultant to the Personnel Research Section of the Department of the Army, which led to several papers on the group assembly of personnel.

10.19 ▪ Leonhard Euler

Leonhard Euler
Portrait by Emanuel Handmann,1753, Kunstmuseum Basel

Life

Since so many documents on Euler exist, let us only give a brief account of his life (see [1303] for a complete biography).

Leonhard Euler was born on April 15, 1707, in Basel, Switzerland. His father was a friend of the Bernoulli family. The family moved to Riehen, not far from Basel, when he was one year old. Around the age of eight, he was sent to a Latin school in Basel where he lived with his maternal grandmother.

In 1720, at the age of thirteen, he enrolled at the University of Basel, and in 1723, he obtained a Master in philosophy. At the same time, he was receiving lessons from Jean I Bernoulli (1667-1748), then considered as Europe's foremost mathematician, who quickly discovered Leonhard's incredible talent for mathematics.

Euler completed his studies at the University of Basel in 1726 with a dissertation on the propagation of sound. Then, he unsuccessfully applied for a position at the University of Basel before accepting a call from the Academy of Sciences in Saint Petersburg after the death of Nicolas II Bernoulli (1695-1726). He left Basel on April 5, 1727.

Euler first served as a medical lieutenant in the Russian navy from 1727 to 1730. In Saint Petersburg he stayed with Daniel Bernoulli (1700-1782). Then, he became professor of physics at the Academy in 1730 and a full member. When, in 1733, Daniel Bernoulli left his senior chair in mathematics at the Academy, Euler replaced him. The financial improvement that followed allowed him to marry Katharina Gsell (1707-1773), the daughter of a painter from the Saint Petersburg Gymnasium, on January 7, 1734. They had 13 children but only 5 survived their childhood. Euler's health problems began in 1735, and his eyesight problems in 1738, due to his cartographic work.

By 1740, Euler was very highly regarded, having won the Grand Prize of the Paris Academy in 1738 and 1740. In Saint Petersburg, the political turmoil made the position of foreigners particularly difficult and Euler accepted the invitation of Frederick II the Great (1712-1786) to come to Berlin, where an Academy of Science was planned to replace the Society of Sciences. He left Saint Petersburg on June 19, 1741.

Euler spent 25 years in Berlin, where he wrote around 380 papers and published his most famous books *Introductio in Analysin Infinitorum*, a text on functions published in 1748, and *Institutiones Calculi Differentialis*, on differential calculus in 1755.

In addition, he tutored Friederike Charlotte of Brandenburg-Schwedt (1745-1808), who was Princess of Anhalt-Dessau and Frederick's niece. He wrote her over 200 letters, which were later published under the title *Letters of Euler on Different Subjects in Natural Philosophy Addressed to a German Princess*. This work contained a presentation of various scientific topics as well as insights into his personality and religious beliefs. This book became more widely read than any of his mathematical works and was published all over Europe and in the United States. In 1759, Pierre Louis Moreau de Maupertuis (1698-1759), the president of the Berlin Academy, died, and Euler assumed the leadership of the Berlin Academy, without the title of President. The king was, in fact, its real one.

The relationship with Frederick II was not easy. It was *the Age of Enlightenment*, and the monarch had a large circle of intellectuals in his court. But Euler was not a courtier; he was only interested in science. In many ways, he was the opposite of the French philosopher Voltaire (1694-1778), who was in Berlin from 1750 to 1753. Euler was not a skilled debater and he was the frequent target of Voltaire's wit. Moreover, Euler had scientific controversies with Jean Le Rond d'Alembert (1717-1783) and he was unhappy when Frederick offered him the presidency of the Academy in 1763. D'Alembert refused to move to Berlin, but Frederick continued to interfere with the running of the Academy, and Euler decided to leave Berlin.

In 1766, he was invited by Catherine II the Great (1729-1796) to come back to the Saint Petersburg Academy. His conditions were quite exorbitant but all of his requests were accepted. He spent the rest of his life in Russia. However, his second stay in the country was quite tragic. A fire in Saint Petersburg in 1771 destroyed his home, and in 1773, he lost his wife Katharina after 40 years of marriage.

His eyesight had deteriorated during his stay in Germany. He developed a cataract in his left eye and a failed surgical restoration in 1771 rendered him almost totally blind. However, because of his remarkable memory, he was able to continue with his work on optics, algebra, and lunar motion, and it was after his return to Saint Petersburg that he produced almost half of his total works. Three years after his wife's death, Euler married her half-sister, Salome Abigail Gsell (1723-1783). This marriage lasted until his death. On September 18, 1783, after a lunch with his family, Euler was discussing the newly discovered planet Uranus and its orbit with a fellow academician when he collapsed from a brain hemorrhage. He died a few hours later.

Work

Euler's contribution to mathematics and physics is immense. He was the most prolific writer of mathematics of all time. He produced major advances in the study of modern analytic geometry and trigonometry, where he was the first to consider the trigonometric functions as so, instead of chords as Ptolemy. He also introduced the β and Γ functions.

He played a major role in the development of geometry, calculus, and number theory. He assembled Leibniz's differential calculus and Newton's method of fluxions into mathematical analysis. He proposed a method for integrating differential equations. He studied continuum mechanics, lunar theory, the three body problem, elasticity, acoustics, the wave theory of light, hydraulics, and music.

Euler laid the foundation of analytical mechanics, in particular in his *Theoria Motus Corporum Rigidorum* (Theory of the motions of rigid bodies) published in 1765. He came back to this subject in 1776 (Novi Comment. Petrop. XX, 189-207 (1776)). It was at this occasion that orthogonal matrices implicitly appeared. For Euler, any finite rigid motion is equivalent to a translation and a rotation. The translation moves an arbitrary reference point in the rigid body by a rectilinear displacement while the rotation turns the body by three component angles around three axes intersecting at the reference point. Geometrically, Euler's result states that, in three-dimensional space, any displacement of a rigid body such that a point of the rigid body remains fixed, is equivalent to a single rotation about some axis that runs through the fixed point. He proved that by means of spherical geometry. This result also means that the composition of two rotations is also a rotation. Therefore the set of rotations has a group structure, known as a rotation group.

In terms of linear algebra, the theorem states that any two Cartesian coordinate systems with a common origin are related by a rotation about some fixed axis. This means that the product of two rotation matrices is a rotation matrix and that for a non-identity rotation matrix, one eigenvalue is 1 and the two others are both complex, or both equal to -1. The eigenvector corresponding to this eigenvalue is the axis of rotation connecting the two systems; see Chapter 6.

10.20 ▪ Radii P. Fedorenko

Radii Petrovich Fedorenko
Russian Academy of Sciences

Life

Radii Petrovich Fedorenko was born on March 11, 1930, in Voronezh in western USSR. His father was an engineer who became director of a military aircraft factory and his mother was a chemistry engineer.

He graduated from the Rostov State University in 1953 and defended a Ph.D. thesis in 1965. He spent all his career at the Keldysh Institute of Applied Mathematics (IAM). He started in the department headed by Israel Moiseevich Gelfand (1913-2009) and he later headed a department of this institute.

He taught courses on numerical methods in the Moscow Institute of Physics and Technology and wrote a book *Introduction to Computational Physics* in 1994.

Fedorenko died on September 13, 2009. For an obituary, see [22]. According to his colleagues, Fedorenko was a modest and kind person.

Work

In his first years at IAM, Fedorenko worked on problems of mathematical physics related to nuclear weapons using the first Soviet computers. His first publication, which appeared in 1958, was concerned with solving the one-dimensional Magneto-Hydrodynamics (MHD) equations for controlled thermonuclear fusion. In 1960, he worked on finite difference schemes for the Stefan problem.

Fedorenko's first paper on multigrid [1147] was published in 1961. The method was used to solve the Poisson equation as a part of solving two-dimensional fluid dynamics equations on a spherical surface. In 1964, he proved that the convergence rate does not decrease with the mesh refinement for Poisson's equation in a rectangle [1148]. For the development of the multigrid method Fedorenko received the State Prize of the Russian Federation in science and technology.

In 2001, Fedorenko wrote a few lines about the multigrid method[63] that were later translated by Mike Botchev:

> *In my program, a* 48×40 *grid was used so that the unknown grid function and the right-hand side vector occupied almost all operational memory. I started working with the only iterative method familiar to me then, the relaxation method, and soon became convinced of its poor efficiency. If at that time I would have been aware of the ADM (Alternated Direction Method) with optimally chosen parameters (by that time it was already known), or, at least of the overrelaxation (SOR) method, then I would probably not have looked for something else. But "fortunately" I was sufficiently ignorant on this topic. I was trying to understand what caused the slow convergence by looking at the residual evolution in the course of iterations and easily discovered the well known fact: first, the nonsmooth residual decreased fast and became smooth. After this the decrease became desperately slow. How the idea to formulate the correction equation as a problem on a coarse grid with the residual at the right hand side crossed my mind is difficult to recall now. Apparently, a certain hint was given by the Newton method which, for linear equations, leads to the same problem. This problem can be eased by switching to a coarse grid and the smoothness of the right-hand side (the residual) justifies such an approach.*

Later on, Fedorenko worked on fluid mechanics and optimal control problems.

10.21 ▪ Roger Fletcher

Life

Roger Fletcher was born on January 29, 1939, in Huddersfield, Yorkshire, England. He was the only child of the family. His father was a painter and decorator who was killed in North Africa during World War II. His mother remarried another painter and decorator when Roger was about seven years old.

Roger attended a very good grammar school, where he played chess, even on top board for the England junior side on one occasion. He won a state scholarship and went to Cambridge, where, in 1960, he graduated in theoretical physics. As it is told by G. A. Watson,[64] on the

[63] www.keldysh.ru/departments/dpt_2/multigrid.html (accessed August 2021)

[64] G.A. Watson, Roger Fletcher, 29 January 1939 - 15 July 2016.

Roger Fletcher
Courtesy of University of Dundee

evening of his degree presentation, he sleepwalked out of a third floor window of the College where he was staying, broke his back, and suffered paralysis for some time.

Fletcher was very determined, and he went to the University of Leeds with the aid of crutches to begin a Ph.D. under the guidance of Colin M. Reeves. His work involved the development of methods for computing molecular wave functions. At that time, Leeds was one of the first universities to have a computer, which was a *Ferranti Pegasus*. Fletcher used it and became interested in numerical analysis. He defended his doctorate in 1963, stayed in Leeds as a lecturer, and got married.

In 1969, Fletcher moved to the Atomic Energy Research Establishment (AERE) in Harwell as Principal Research Fellow. In 1971, he became a Principal Scientific Officer. His work involved research, teaching, and consultancy, and he was responsible for setting up an advisory service for the Harwell Subroutine Library. But he found his task too commercial, and he left to join the numerical analysis group at the University of Dundee in 1973 as a Senior Research Fellow. He became a Professor in Dundee in 1984, and also took up some administration duties, including acting as Head of Department.

In an interview,[65] Fletcher was asked what were the most important spirits for doing scientific research. He answered

> *Famous Grouse?! Alternatively, treat everything you read about with some skepticism, and be prepared to follow your own intuition. But be willing to change your mind when it becomes clear that other ideas have been demonstrated to be superior. And in Numerical Analysis, watch for what the numbers are telling you.*

The quality of Fletcher's work was recognized by several international honors. He obtained the Dantzig Prize in 1997 and the Lagrange Prize in 2006, both by the Mathematical Optimization Society. He was elected a Fellow of the Royal Society of Edinburgh in 1988, a Fellow of the Royal Society of London in 2003, and was awarded a Royal Medal by the Royal Society of Edinburgh in 2011.

[65] http://www-optima.amp.i.kyoto-u.ac.jp/ORB/issue22/flectcher_interview.html (accessed November 2020)

Fletcher was an enthusiastic swimmer, an accomplished bridge player, and he enjoyed walking in the Scottish hills. He was a fine numerical analyst, a musician, a very educated, kind, and charming man, and a gentleman.

It was during a walk from his holiday accommodation near Dornie on the west coast of Scotland that he went missing on June 5, 2016. His body was only found some six weeks later, and the details of what caused his death are still unclear. He was declared dead on July 15, 2016.

Work

Roger Fletcher is known worldwide for his work on nonlinear optimization and linear algebra. In particular, with William Cooper Davidon (1927-2013), an American physicist, mathematician, and peace activist, and Michael James David Powell (1936-2015), a professor at the University of Cambridge, as co-authors, he proposed in 1959 a formula, known as the DFP formula, which was the first quasi-Newton method generalizing the secant method to several dimensions.

In fact, Reeves, who was Fletcher's advisor, gave him a technical report from Argonne National Laboratory, written in 1959 by Davidon, on a method for a class of optimization problems. Roger programmed the method and realized its potential. The report also went to Michael Powell at the AERE Harwell. By chance, Powell was about to give a seminar in Leeds, and he changed his title at the last moment to describe his own experiences with Davidon's method. Powell presented the essential features of the method in a rather unusual way, and discovered that Fletcher was also working on the method. They collaborated, added some theory, and the result was published in the *Computer Journal* in 1963 [1178]. The method became known as the Davidon-Fletcher-Powell or DFP method. It was a significant achievement since the method allowed to treat problems with hundreds of variables instead of tens.

Around that time, Reeves was writing lecture notes on the conjugate gradient method. The solution of quadratic optimization problems leads to a system of linear equations, and he realized that the line search of the DFP method could be used for non-quadratic optimization problems, that is, for systems of nonlinear equations. Since Fletcher had a line search code, this idea was followed up, leading to the Fletcher-Reeves nonlinear conjugate gradient method, which they jointly published in 1964 [1179].

Fletcher is also one of the authors of the BFGS algorithm, an iterative method for solving unconstrained nonlinear optimization problems, due to Charles George Broyden (1933-2011), Fletcher, Donald Goldfarb, and David Francis Shanno (1938-2019) in 1970.

For problems with constraints, Fletcher introduced the first exact differentiable penalty function which forces convergence even when the initial state is not well chosen. He is also known for having obtained the biconjugate gradient method, the first iterative algorithm for implementing Lanczos's method for solving nonsymmetric linear systems. He presented this work at the *Dundee Numerical Analysis Conference* held in Dundee in 1975; see [1177]. In fact, Lanczos already used the method, but for computing eigenvalues. Fletcher adapted it to linear systems. In addition to the theoretical development of methods, Fletcher also often produced the corresponding computer programs that are widely used today.

Fletcher collaborated widely, and supervised many students and research fellows. In particular, with his former post-doctoral student Sven Leyffer, as stated in the citation for the 2016 Farkas Prize awarded to Leyffer, they

> *proposed a new class of methods for constrained nonlinear programs, which employ the innovative concept of a filter for balancing between feasibility and optimality, and which builds on the concept of domination from multi-objective optimization. This new approach, which has proven to be highly effective, has had a profound influence on nonlinear programming, and has sparked much research among optimizers. Originally proposed within the context of a trust-region sequential quadratic*

> *programming (SQP) algorithm for continuous optimization, it has been extended to non-SQP algorithms, and has also sparked the development of new algorithms and software for non-smooth optimization, derivative-free optimization and nonlinear equations.*

The filter SQP method is a radically different approach to solving large and complex nonlinear optimization problems typical of those faced by the industry.

10.22 ▪ George E. Forsythe

George Elmer Forsythe

Life

George Elmer Forsythe was born on January 8, 1917, in State College, Pennsylvania, USA. His father was a medical doctor who later ran a health service at the University of Michigan. When George was still a small boy the family moved to Ann Arbor, Michigan.

He did his undergraduate studies at Swarthmore College (in the Philadelphia area), where he majored in Mathematics. He obtained a B.S. in 1937. His experience there had a strong influence on his life since it was there that he met his future wife.

His graduate study was in Mathematics at Brown University, Rhode Island, where he received his M.S. in 1938 and his Ph.D. in 1941. The title of his thesis was *Riesz summabilitly methods of order r, for $R(r) < 0$, Cesaro summability of independent random variables*. His two advisors were Jacob David Tamarkin (1888-1945) and William Feller (1906-1970).

In fact, his wedding and Ph.D. defense occurred on the same day, June 14, 1941. George married Alexandra W. Illmer (1918-1980). They had two children. Their son, Warren L. Forsythe, was born June 7, 1944, in Alexandria, Virginia. He died in 2017. Their daughter, Diana E. Forsythe (1947-1997), was a leading researcher in anthropology and a key figure in the field of science and technology studies. She died during a hiking accident in northern Alaska.

George got a position as an instructor at Stanford University, but the war began and he was called up by the Army in 1942. He was sent to a meteorology service at UCLA in Los Angeles.

He took a meteorology course and was kept there as an instructor. Later on he wrote a book, *Dynamic Meteorology*, with Jörgen Holmboe (1902-1979) and William Sharp Gustin as co-authors that was published by Wiley in 1945. George was sent to Washington, working in the Pentagon, in the summer of 1944. Then, the weather service was transferred from the Army to the Air Force, and they moved the headquarters from the Pentagon down to Asheville, North Carolina, and again, a little later, to Langley Field, Virginia.

When out of the Army, George went to work for Boeing in Seattle. But then, in 1947, he got an offer from J. Holmboe who was the head of the meteorology department at UCLA. He was not very pleased with his stay in this department. But the National Bureau of Standards (NBS) started the Institute of Numerical Analysis (INA) on the UCLA campus, which was set up to build what was called the SWAC (*Standards Western Automatic Computer*) machine designed by Harry Douglas Huskey (1916-2017). Forsythe was hired in 1948 by John Hamilton Curtiss (1909-1977) who was acting chief of the INA. According to his wife (see [2177]),

> *He was absolutely thrilled at the idea of that computer. He was just like a little boy. He was so excited about the fact that they were actually going to build one there and he was going to do all sorts of things.*

The INA was gathering many distinguished people who had a great impact on the development of numerical linear algebra. To name just a few: Olga Taussky, John Todd, Cornelius Lanczos, Magnus Hestenes, Wolfgang Richard Wasow, Gertrude Blanch, and Theodore Samuel Motzkin. For more on the work done at INA, see Section 5.6.

The Institute for Numerical Analysis ceased to exist at the end of the quarter, April-June 1954. This was the consequence of the shortage of funding from the national agencies after a scandal that arose at the NBS about a battery additive.

George was transferred to the UCLA Mathematics department, but according to his wife,

> *He never felt that the math department was really sympathetic with applied mathematicians. About that time Stanford began exploring the possibility of George coming to Stanford.*

In 1952, Albert Hosmer Bowker (1916-2008), who was head of the Applied Mathematics and Statistics Laboratory, and Frederick Emmons Terman (1900-1982) who was Dean of Engineering, decided that it was time Stanford got into the computer age. Several machines were installed successively: an *IBM CPC* (card-programmed calculator) in 1953, an *IBM 650* in 1956, a *Burroughs 220* in 1958, and then an *IBM 7094* and a *Burroughs B5500*.

Forsythe came to Stanford in September 1957 in the Mathematics department. His wife said

> *Once he started on the computing, I don't think he ever faltered in his feeling that that was the wave of the future, and of course he was helped along as things developed fairly rapidly.*

Asked why he hired Forsythe, Bowker said [2176]

> *Well, he seemed to have a lot of the qualities, both personal and professional, we wanted. Of course, he was also very acceptable to the Mathematics Department which was where he was housed originally. He was one of the leaders in the field.* [...] *I had a very high opinion of him. He and his wife were very pleasant and attractive people. He worked very hard; he was dedicated to not only his own work but the building of Stanford. He did a splendid job.*

In 1961, John George Herriot (1916-2003) and Forsythe founded the Computer Science Division of the Mathematics Department and started hiring computer science faculty [1653]. Among the people who joined the division were John McCarthy (1927-2011), who coined the term "Artificial Intelligence," and Niklaus Wirth, an expert in programming languages. Gene Howard Golub (1932-2007) came to Stanford in August 1962. The Computer Science Department was established as a separate department on January 1, 1965. Forsythe was its first chairman.

Forsythe was President of the Association for Computing Machinary (ACM) in 1964-1966.

About Forsythe, Gene Golub said [2178]

> *Forsythe went around and tried to get as much support throughout the university for funds for a computer. [...] He was very even tempered, very direct, reserved. [...] He was a very good man to work with. Well, a compulsive note-taker and letter-writer. But you knew things were going to be accomplished.*

Forsythe had 17 Ph.D. students, including Eldon Robert Hansen, James McDonaugh Ortega, Beresford Neill Parlett, Cleve Barry Moler, Roger W. Hockney (1929-1999), James Martin Varah, Richard Peirce Brent, and John Alan George.

He died of pancreatic cancer April 9, 1972, in the Stanford Hospital. For more about Forsythe, see [1750].

Work

Forsythe did not invent any new algorithm, but concerning mathematics, he was mainly interested in linear algebra and had a broad knowledge of the field, both in the USA and abroad. He also stimulated the research of his students and other researchers. On that point, Golub [2178] said

> *He didn't necessarily provide new ideas, but he would suggest areas that one might want to work in. In particular, once Ben Rosen gave a talk on doing a pseudo-inverse calculation and at the end of that lecture he said, "Well, will somebody please work on figuring out how to compute the pseudo-inverse of a matrix?" and later on that's one of the things that I worked on, and actually that particular piece of work is one of the most well-known pieces of work that I ever did.*

He was a visionary about the use of computers in science and, more generally, in society. He was a sort of evangelist or missionary about the development of computer science. Quoting J.G. Herriot [2179],

> *I think that George was one of the big pushers, one of the big influences in the fact that computer science came of age, that it came to be recognized as a discipline.*

Donald Ervin Knuth, one of the most well known figures of the Stanford Computer Science department, wrote [1922]

> *It is generally agreed that he, more than any other man, is responsible for the rapid development of computer science in the world's colleges and universities.*

Forsythe used the term "Computer Science," probably for the first time, in 1961. He wrote around 80 papers (including one with his wife) and several books that became popular:

- *Finite-Difference Methods for Partial Differential Equations*, with W. Wasow (1909-1993), published in 1960 [1199],

- *Computer Solution of Linear Algebraic Systems*, with C.B. Moler, published in 1967 [1196].

The Forsythe and Moler book evolved from notes for the second quarter of a two-quarter basic numerical analysis course that Forsythe taught in the early 1960s and that Moler taught in 1965.

In the paper *Solving linear algebraic equations can be interesting* [1188] in 1953, Forsythe wrote an interesting review of the methods known at that time:

> *It is remarkable how little is really understood about most of the methods for solving* $Ax = b$*. They are being used nevertheless, and yield answers. This disparity between theory and practice appears to be typical of the gulf between the science and the art of numerical analysis.*

In that paper, he attributed the conjugate gradient algorithm to Hestenes, Lanczos, and Stiefel. The word "preconditioning" appears in that article on page 318. There are 131 references out of the 500 that he had in his files.

In the paper *Today's computational methods of linear algebra* in 1967 [1190], he compared the situation in 1966 to what it was in 1953. He wrote

> *For dense, stored matrices we knew Gaussian elimination, of course. We knew that it sometimes produced quite poor results. We weren't always sure why. We debated endlessly about how to pick pivots for the elimination, without settling it. The debate still continues, but now mainly among persons who don't understand that the main lines of the answer have been settled.* [...] *We were not quite aware of the extent of problems of ill conditioning of matrices.* [...] *For the eigenvalue problems, things were in much worse state. We had the power method with matrix deflation.* [...] *For nonsymmetric matrices, things were ghastly. If the power method wouldn't work, we had practically no alternatives.*

About Gaussian elimination for dense matrices of order n, he also wrote

> *The case* $n = 100$ *is now easy and costs around $11 on an IBM 7094. The case* $n = 10,000$ *is likely not to be accessible for a long time, and it would take over 2000 hours now on an IBM 7094. There is beginning to be serious consideration of computers with a substantial amount of parallel operation, so that perhaps much of the solution of a linear system could be done simultaneously.*

About backward error analysis (called *inverse analysis* by him), he wrote

> *Inverse error analysis turns out to be extremely well adapted to the analysis of algorithms of a marching type which continually introduce new data. Both the solution of linear equations and the evaluation of polynomials are of this type.*

About eigenvalue computations for nonsymmetric matrices, he wrote

> *The area of greatest activity in the past decade of research on computational linear algebra has been the eigenvalue problem for unsymmetric matrices. Only one method from before the computer era is still in use -the power method- and it has only limited applications today. Most methods in use today were unheard of 15 years ago.* [...] *For most methods of attacking the eigenvalue problem, the first step is to condense the data, to save time and storage in further work The now universally accepted condensed form is the Hessenberg matrix.* [...] *The LR-algorithm of Rutishauser was an important, development.* [...] *Francis in England, and Kublanovskaja in the Soviet Union devised the very interesting QR-algorithm. This is now widely considered the most satisfactory eigenvalue algorithm for dense, stored unsymmetric matrices.* [...] *Most research goes into the invention of origin shifts when some of the eigenvalues are complex and of equal modulus.*

In *What to do till the computer scientist comes* in 1968 [1191], he expressed his thoughts about what computer science is and how it must be taught in universities. There is an interesting sentence about the place of numerical analysts:

> *In the past fifteen years many numerical analysts have progressed from being queer people in mathematics departments to being queer people in computer science departments!*

Here "queer" is used with the meaning of weird or strange. Other interesting papers are [1194, 1184, 1197, 1186, 1187, 1189, 1192] (ordered by date).

10.23 ▪ Leslie Fox

Leslie Fox

Life

Leslie Fox was born on September 30, 1918, in Dewsbury, Yorkshire, UK. His father, Job Senior Fox, was a coal miner and his mother was Annie Vincent.

The family was quite poor as described by Leslie's brother, Roy Fox, in [2409]. At the age of 10 Leslie won a scholarship to the Dewsbury Wheelwright Grammar School in which he did quite well in almost all subjects, winning several prizes. He was also very good at sports, especially football, cricket, and tennis.

In 1936, he won a scholarship to study mathematics at Christ Church and entered Oxford University. After his undergraduate studies he joined Richard Southwell's group in the Engineering Department where he worked on his Ph.D. that he obtained in 1942. Leslie worked on relaxation methods for the biharmonic equation. This was wartime, and some of his work was classified, which means that the papers were published only after the end of the war.

In 1943, Fox joined the Admiralty Computing Service (ACS) where he stayed for three years, solving differential equations and tabulating functions as was common in those days. He married Pauline Helen Dennis on July 1943.

In 1945, Leslie quit the ACS and joined the newly founded Mathematics Division of the National Physical Laboratory (NPL) in Teddington.

Fox left the NPL in 1956 and was an associate professor in the University of California at Berkeley, USA, for one year. Back in Oxford in 1957, he became director of the Computing Laboratory. In 1963, Leslie became Professor of Numerical Analysis in Oxford University, and was elected to a Fellowship at Balliol College. After a while, the university decided to split the Computing Laboratory into computing services on one hand and teaching and research on the other hand, reinforced with applied mathematicians.

Fox pushed also projects of building libraries of mathematical routines. This led to the funding of the Nottingham Algorithms group, later renamed as the Numerical Algorithms Group (NAG).

In 1970, Leslie had a one-year visiting professorship at the recently founded Open University for which he prepared courses and television programmes. He was also very much involved in the mathematical society, the Institute of Mathematics and its Applications (IMA).

Leslie Fox divorced his first wife in 1973 and married Clemency Clements that same year. After a heart attack in 1981, he retired in 1983. He died from heart problems August 1, 1992, in Oxford.

A Leslie Fox Prize is awarded yearly by the IMA.

Work

In the 1940s, Leslie Fox worked on relaxation methods with Richard Vyne Southwell (1888-1970).

About Fox on relaxation methods, Charles Clenshaw (1926-2004) remembered of his early days at NPL,

> *Given an elliptic partial differential equation with its boundary conditions, Leslie would estimate the value of its solution at each internal grid point, and calculate, sometimes mentally, sometimes with the help of a Brunsviga, the value of the local residual. Then the crawling process would begin; the most significant residuals were attacked by altering the values of the estimated solution in their vicinity. 'Most significant' did not necessarily mean largest in magnitude, because a group of moderate residuals with the same sign could be more significant than an isolated larger one. Moreover, the residuals attacked were not necessarily reduced to zero; an intelligent observation of the surrounding picture might cause one to over-relax or under-relax. Leslie was a master at this work. In the first place he had made profound contributions to the general approach as it had been suggested by Southwell.*

In 1948, Fox published an influential paper [1206] on solving linear systems with Harry Douglas Huskey (1916-2017) and James Hardy Wilkinson (1919-1986); see also [1207].

A favorite area of research for Leslie was finite differences and boundary value problems. He was interested in the difference-correction method and the method of deferred approach to the limit. He published several books as an editor or as an author; see [1203, 1204, 1208].

10.24 ▪ Ferdinand G. Frobenius

Life

Ferdinand Georg Frobenius was born in Charlottenburg, Prussia (now Berlin, Germany) on October 26, 1849. His father, Christian Ferdinand, was a Protestant parson.

From 1860 to 1867, he studied at the Joachimsthalsches Gymnasium. Then he went to the University of Göttingen, but he only stayed there for one semester before going back to Berlin.

Ferdinand Georg Frobenius

At the University of Berlin, he attended lectures by Leopold Kronecker (1823-1891), Ernst Eduard Kummer (1810-1893), and Karl Theodor Wilhelm Weierstrass (1815-1897), and continued to study there for his doctorate, attending the seminars of Kummer and Weierstrass. On July 28, 1870, he was awarded a doctorate with distinction, supervised by Weierstrass and Kummer. The title was *De functionum analyticarum unius variabilis per series infinitas repraesentatione*. Among other topics, it contained developments on continued fractions, orthogonal polynomials, and their generating series, the Christoffel-Darboux identity and its variants.

Then, Frobenius taught at secondary school level, first at the Joachimsthalsches Gymnasium, then at the Sophienrealschule. Having published many papers (a method for finding an infinite series solution of a differential equation at a regular single point, on Abel's problem in the convergence of series, and on Pfaff's problem in differential equations), he earned a fair reputation and, in 1874, due to the strong support of Weierstrass, he was appointed to the University of Berlin as an extraordinary professor of mathematics. However, one year later, he left and went to the Eidgenössische Polytechnikum in Zürich, Switzerland, as an ordinary professor, a position he kept for 17 years, between 1875 and 1892.

He married Augusta Sophia Lehmann (1852-1903) in 1876 and brought up a family. In Zürich, Frobenius did many important works in widely differing areas of mathematics. For a substantial part of his time, he was head of the mathematical seminar, and when he left for Berlin in 1892, the school board noted in its minutes that Frobenius contributed significantly to securing and increasing the scientific reputation of the School during his tenure.

When Kronecker died in Berlin on December 29, 1891, his chair at the Friedrich-Wilhelms-Universität in Berlin became vacant. Weierstrass, strongly believing that Frobenius was the right person to keep Berlin at the forefront of mathematics, used his considerable influence to have Frobenius appointed. However, Frobenius was *occasionally choleric, quarrelsome, and given to invectives* [320], and he had a high opinion of his rank and of his duty to teach pure mathematics. For him, applied mathematics had only to be taught in technical colleges, not in universities, and *the aversion of Frobenius to Felix Klein and Sophus Lie knew no limits* [320]. Due to this attitude, his nomination in Berlin was not as beneficial as it should have been, and it contributed to the relative decline of Berlin in favor of Göttingen. Frobenius was elected to the Prussian Academy of Sciences in 1893.

He died in Berlin on August 3, 1917.

Work

Frobenius is remembered for his formulation of the concept of abstract group. He also made important contributions to the theory of elliptic functions, to the solution of differential equations, and to quaternions.

In 1878, Frobenius introduced the minimal polynomial of a matrix. He stated that it is formed from the factors of the characteristic polynomial and is unique (a result only proved by Kurt Hensel (1861-1941) in 1904 [1648]). He also gave the first general proof of the Cayley-Hamilton theorem, and of the fact that the only finite-dimensional division algebras over the real numbers are the reals, the complex numbers, and the quaternions.

One year later, he introduced the notion of rank of a matrix through the connection with determinants. In linear algebra, his name is also attached to the Perron-Frobenius theorem on the spectral radius of an irreducible nonnegative matrix, due to Oskar Perron (1880-1975) [2482] and him [1258].

The rational normal form was introduced by Frobenius in 1879 in a long paper [1251]. He did not use matrices at that time but the framework of bilinear forms. However, he represented the bilinear forms by capital letters (related to their coefficients) and manipulated them as if they were matrices. The determinant of what is now known as a Frobenius companion matrix appears on page 206 of [1250].

In 1881, following Augustin-Louis Cauchy (1789-1857) [565, pp. 525-528], Carl Gustav Jacob Jacobi (1804-1851) [1796], and Leopold Kronecker [1953], Frobenius studied the rational approximants to a formal power series that are now called *Padé approximants*. He gave the relations linking the denominators or the numerators of three adjacent approximants [1252]. The so-called *missing Frobenius identity* was obtained by Peter Wynn (1931-2017) in 1966 [3283].

Among Frobenius' students, the names Edmund Landau (1877-1938), Konrad Knopp (1882-1957), Walter Schnee (1885-1958), and Issai Schur (1875-1941) have to be mentioned.

10.25 ▪ Noël Gastinel

Noël Gastinel

Life

Noël Gastinel was born in Le Muy, a small town in the southern part of France, on December 25, 1925 (notice that in French, Noël means Christmas).

He was brilliant in mathematics and was preparing in Toulon the common competitive entrance examination to be admitted in a French engineering *Grande École*. But it was World War II, his town was bombed, and his house *fell down over him and he lost most of his hair*, as he often said. He stopped his studies, and spent the last months of the war in clandestinity in the small village of Aups in the Provence Alps.

After the war, he went to the University of Marseilles where, in addition to mathematics, he was very much interested in astronomy. Later, in his house, he computed, constructed, and installed a sundial. He obtained the *Agrégation de Mathématiques* and became a high school teacher in Toulon for several years.

After the war, Jean Kuntzmann (1912-1992), a professor at the University of Grenoble, was developing a course of applied analysis for engineers with practical work on office calculating machines. This course was known under the name of *Techniques Mathématiques de la Physique*, and it was followed by many students in several French universities. In 1951, Kuntzmann founded, at the Institut Polytechnique de Grenoble, the first Laboratoire de Calcul (Computing Laboratory), and it was equipped with the first computer in a French university in 1957. That same year, Kuntzmann asked Gastinel to join him in Grenoble. In 1960, he defended his *Thèse de Doctorat d'État es Sciences Mathématiques* at the University of Grenoble under the guidance of Kuntzmann. He stayed in Grenoble until his death on September 11, 1984, after a long and very painful tongue cancer.

As stated in his biography [862]

> *At this epoch-making stage of numerical linear algebra, N. Gastinel was very interested in and impressed by the pioneering work of A.S. Householder, R.S. Varga, J.H. Wilkinson, F.L. Bauer, A.M. Ostrowski, O. Taussky, and others. He felt that this kind of concrete mathematics had a better future with the advent of computers than the abstract mathematics often studied in France during that period.*

Gastinel was much interested in the advancement of computers and computing techniques. Under his leadership, the Centre Interuniversitaire de Calcul de Grenoble became one of the most important such centers in France, and it received the most modern equipment at that time.

He was a humble man, very open to others, very helpful, but with often a strict judgment on colleagues. Maybe it is for that reason that he never fully participated in the national scientific and academic life. Although Gastinel only published a few papers, he had a considerable influence on the development of numerical analysis in France.

Work

The scientific works of Gastinel were entirely devoted to what he called *les mathématiques du calcul* (computational mathematics). The first part of Gastinel's thesis was devoted to second degree matrices, that is, matrices whose minimal polynomial is of degree 2. The second part treats general norms on finite dimensional spaces, a subject he was particularly fond of (he supervised the thesis of several students on this topic). A general notion of condition number was defined in the following part. The essential problem was to know if the fact that a norm of the error is small implies that the approximate solution is close to the exact one. Particular cases were discussed. Then, he studied the error in the solution of linear systems by elimination, and a method using orthogonalization.

He invented a new projection method for the iterative solution of systems of linear equations, the *méthode de décomposition de norme*; see Section 5.4 for details. It was presented at the IFIP congress in Munich in 1962.

Unfortunately, the norm decomposition method converges slowly. The last chapter of the thesis is devoted to eigenelements of matrices.

It was in his book *Analyse Numérique Linéaire*, translated into English as *Linear Numerical Analysis* [1290], that the first complete proof of the convergence of Kaczmarz's method was given.

Apart from his interest in numerical linear algebra, Gastinel was also fond of functional analysis. But almost all topics in numerical analysis attracted him. This can be seen from the list of theses he supervised (among them that of C.B.) and from the detailed analysis of his work in [862].

10.26 ▪ Johann C.F. Gauss

Johann Carl Friedrich Gauss

Life

Johann Carl Friedrich Gauss was born on April 30, 1777, in Brunswick (Braunschweig), in the Duchy of Brunswick-Wolfenbüttel (now in Germany). He was the only child of his parents.

Gauss' father worked in various low-profit jobs and his mother was illiterate. Despite these difficulties, Gauss entered primary school in 1784 where he soon revealed himself as a child prodigy. There are numerous anecdotes on his potential that was soon recognized by his teachers, in particular Martin Bartels (1769-1836), who later became professor at the University of Kazan. In 1788, Gauss was admitted to a secondary school where he learned High German and Latin.

In 1791, after receiving a stipend from the Duke Karl Wilhelm Ferdinand von Brunswick-Wolfenbüttel (1735-1806), who was impressed by his talents, he went to the Collegium Carolinum (now Braunschweig University of Technology), where he studied from 1792 to 1795. The library there was quite good and Gauss used his time reading the classical mathematical literature. He was then able to do independent research.

From 1795 to 1798, he attended the University of Göttingen against the wish of the Duke who wanted him to study at the local university of Helmstedt. At that time, in German universities, students were free to attend the lectures they wanted; there were no regulations, no tutors to work with, and no examinations. Gauss was attracted to Göttingen by the reputation of its library. He mostly studied alone and left the university without any diploma. During his stay there, it seems that the only acquaintance Gauss had was Farkas (Wolfgang) Bolyai (1775-1856) who was studying philosophy and mathematics.

After he returned to Brunswick in 1798, the Duke agreed to continue Gauss' stipend but requested he submit his doctoral dissertation to the University of Helmstedt. It was defended *in absentia* under the supervision of Johann Friedrich Pfaff (1765-1825) on June 16, 1799, and was entitled *Demonstratio nova theorematis omnem functionem algebraicam rationalem integram unius variabilis in factores reales primi vel secundi gradus resolvi posse* (New proof of the theorem that every integral algebraic function in one variable can be resolved into real factors of the first or second degree). In that thesis, Gauss gave a proof of the fundamental theorem of algebra that states that every non-constant polynomial in one variable with complex coefficients has at least one complex root. Gauss' dissertation contains a critique of the work of Jean le Rond d'Alembert (1717-1783) who gave a wrong proof of the theorem. Ironically, Gauss' own proof is not nowadays acceptable due to the implicit use of the Jordan curve theorem. However, he subsequently produced three other proofs, the last one in 1849 being generally considered rigorous.

Gauss was 21 years old. He wanted to secure an independent life for himself, so he left his parents' house. He enormously expended his scientific interests including mathematical astronomy and the computation of orbits of celestial bodies from inaccurate and scarce measurements. He was in contact with the Hungarian astronomer Franz Xaver Freiherr von Zach (1754-1832).

Gauss' success on computing the position of the newly discovered asteroid Ceres in 1801 (see below) strongly established his reputation and brought him several honors and invitations. He began to travel and visited Heinrich Olbers (1758-1840), who discovered Pallas (the second small planet), in June 1802, and he worked on its orbit. He met Zach in 1803. He also began to correspond with Friedrich Wilhelm Bessel (1784-1846) and Sophie Germain (1776-1831). She was a self-educated young French lady passionate about mathematics, in particular, number theory. She wrote to Adrien-Marie Legendre (1752-1833) who included some of her discoveries in a supplement to the second edition of his *Éssai sur la Théorie des Nombres*. Then, between 1804 and 1809, she wrote a dozen of letters to Gauss under the pseudonym of "M. Le Blanc" fearing to be ignored because she was a woman. Her true identity was revealed to Gauss only after the French occupation of Braunschweig in 1806 when, fearing for Gauss, she asked the French general Joseph Marie de Pernety (1766-1856), who was a friend of her family, to look out for Gauss' safety. When Gauss learned that the intervention was due to Germain, who was also "M. Le Blanc," he gave her even more praise. Among her work done during this period is the study of a particular case of Fermat's last theorem, and a theorem which bears her name.

On October 9, 1805, Gauss married Johanna Osthoff (1780-1809), a tanner's daughter. They had two sons and a daughter. Johanna died on October 11, 1809, as did his most recent child, Louis, the following year. Gauss plunged into a depression from which he never fully recovered. After being injured at the battle of Auerstaedt against the Napoleonic Army, the Duke of Brunswick died on November 10, 1806, and Gauss lost his protection.

In Brunswick, the astronomical instruments at Gauss's disposal were imperfect. In 1807, he accepted the position of director of the Göttingen observatory.

On August 4, 1810, Gauss married Friederica Wilhelmine (Minna) Waldeck (1788-1831), the best friend of his first wife. They had three children, Eugene (1811-1896), Wilhelm (1813-1879), and Therese (1816-1864). Gauss was never quite the same after the loss of his first wife, and just like his father did with him, he had conflicts with his sons. In 1832, Eugene emigrated to the United States where he finally became a businessman. He had seven children. Wilhelm also moved to America in 1837. After Minna's death, Therese took care of her father for the rest of his life. There are still many Gauss descendants in the United States. According to one of Eugene's sons (Charles Henry)

> *Grandfather did not want any of his sons to attempt mathematics, for he said he did not think any of them would surpass him and he did not want the name lowered. Probably he felt the same in a measure of any other line of scientific study.*

From 1845 to 1851, Gauss was involved in the analysis of the pension fund for the widows of the professors of the University of Göttingen. He was still interested in mathematics, and was one of the first European mathematicians to understand and appreciate the work of Nikolaï Lobachevsky (1792-1856) on non-Euclidean geometry. He was also involved in Bernhard Riemann's doctoral thesis. His jubilee was celebrated in 1849.

In January 1849, a dilatation of his heart was diagnosed. Gauss died early on the morning of February 23, 1855. His grave is in the Albani Cemetary in Göttingen.

Work

In 1898, a mathematical diary by Gauss was found. It contains entries from 1796 to 1814, and is the most valuable source to date his discoveries accurately. The first entry concerns the constructibility of the 17-gon by compass and ruler. It was the most major advance in this field since the time of Greek mathematics. Gauss gave a sufficient condition for this construction and stated without proof that this condition was also necessary, but he never published his proof. A full proof of necessity was given by Pierre-Laurent Wantzel (1814-1848) in 1837 [3178]. Other entries are on the expansion of series into continued fractions, on the decomposition of numbers into sums of their squares, on the division of the circle and the lemniscate, on the law of quadratic reciprocity, on the summation of certain series and integrals, on the parallelogram of forces, on the movements of planets and comets, on the foundations of geometry, on the arithmetic-geometric mean, on elliptic integrals, and there is even a formula for determining the date of Easter up to 1999.

In 1801, Gauss published his important work on number theory, *Disquisitiones Arithmeticae* (Investigations in arithmetic) [1297]. Before this work, number theory consisted in a collection of isolated theorems and conjectures. Gauss gathered the works of his predecessors, and added his own original results into a systematic framework. He filled in the gaps, corrected the unsound proofs, and extended the topic in numerous ways. Gauss also illustrated many results with numerical examples. The logical structure of the work (theorem followed by a proof and by corollaries) set a standard for later texts. It was the starting point for the research of many 19th century European mathematicians.

On January 1, 1801, the Italian astronomer Giuseppe Piazzi (1746-1826) discovered the new small planet Ceres, which he tracked for more than one month before it disappeared behind the Sun on February 11. Its orbit was computed by Zach in June 1801, and Ceres was expected to reappear at the end of 1801 or early in 1802. However, when Ceres should have reappeared, Piazzi could not locate it. Another prediction, greatly expanding the search area, was made by Gauss, and on the night of December 31, 1801-January 1, 1802, Zach (and, independently one night later, Olbers from Bremen) found Ceres very close to the position predicted by Gauss.

In 1809, Gauss published his second book *Theoria Motus Corporum Coelestium in Sectionibus Conicis Solem Ambientium* (Theory of the motion of the heavenly bodies surrounding the sun in conic sections[66]). It is a major two-volume treatise on the motion of celestial bodies. In the first volume, he discussed differential equations, conic sections, and elliptic orbits, while in the second volume, the main part of the work, he showed how to estimate and then to refine the estimation of a planet orbit. Section 3 of this second part contains the method of least

[66]Translated into English by Gilbert Wright Stewart, SIAM, Philadelphia, 1995.

squares. The method is described as (Section 186) *the principle that the sum of the squares of the differences between the observed and the computed quantities must be minimum* [494, p. 85]. Gauss claimed that he already used this method in his calculations for Ceres, and this remark was the starting point of his dispute with Legendre about who must be credited for the discovery of the least squares method. Let us mention that because of the Napoleonic occupation, there was a decree that all scientific publications must be written in Latin. It took Gauss several years to prepare his work on the least squares method in Latin, and, therefore, it delayed its publication.

His elimination method for solving linear systems was published in 1810 [1299]. He named it *eliminationem vulgarem* (common elimination) and gave the details of the process; see Chapter 2. Gauss also produced various mathematical works on the hypergeometric function, elliptic integrals, and the arithmetic-geometric mean which gave him access to the theory of modular forms.

On September 16, 1814, Gauss presented his paper on Gaussian quadrature rules [1300] at the Göttingen Society (the volume dates 1815). He was looking for formulas for the numerical computation of definite integrals that achieve the highest accuracy theoretically possible. He made use of the continued fraction expansion of hypergeometric functions [3146]; see [2684] by J.M. Sanz-Serna. Gauss' method was reinterpreted in 1826 by Carl Gustav Jacob Jacobi (1804-1851) on the basis of orthogonal polynomials [1789].

In addition to his activity in astronomy, Gauss was interested in geodesy and topography. The period 1818-1832 was largely dominated by the surveying of the Kingdom of Hanover. He again made use of the method of least squares for the compensation of geodetic networks. However, when the number of measurements is too large, his direct method for solving a system of linear equations became impracticable. This is why, in a letter to his student Christian Ludwig Gerling (1788-1864) dated December 26, 1823, he proposed the iterative procedure[67] that, with modifications, was later known as the Gauss-Seidel method; see Chapter 5. This work on compensation also led Gauss to the theory of conformal mappings, and reactivated his interest in the foundations of geometry and in differential geometry.

For his solution of the problem of the projection, preserving similarity in the smallest parts, of two curved surfaces on each other, he received the Copenhagen prize essay in 1822. His generalization of the theory of curved surfaces culminated with the publication, in 1828, of *Disquisitiones generales circa superficies curva* (General investigations of curved surfaces). This paper also includes his famous *Theorema egregium* (The important theorem) which states that the curvature of a surface can be determined entirely by measuring angles and distances on that surface. Although he never published any original paper on non-Euclidean geometry, Gauss was also interested in this topic and corresponded with János Bolyai (1802-1860), the son of his friend.

In 1831, Wilhelm Weber (1804-1891) arrived in Göttingen as a physics professor. During the six years of his collaboration with Gauss, they investigated the theory of terrestrial magnetism, including Poisson's ideas, the absolute measure for magnetic force, and gave an empirical definition of terrestrial magnetism. Dirichlet's principle was mentioned without proof.

Gauss showed that there can only be two poles in the globe, and he studied the determination of the intensity of the horizontal component of the magnetic force along with the angle of inclination. He used the Laplace equation in his calculations, and specified a location for the magnetic South pole. With Weber, he discovered Kirchhoff's laws, and they even built a primitive telegraph. But, in 1837, Weber had to leave Göttingen for political reasons, and the activity of Gauss decreased. He was still corresponding with many scientists, often claiming that he had known for years the methods they were using, but had never felt the need to publish.

[67]This letter was (partly) translated to English by George Elmer Forsythe [1185]

A contemporary testimony

Wolfgang Sartorius Freiherr von Waltershausen (1809-1876) was a German geologist and a mineralogist, professor at the University of Göttingen. He was a close friend of Gauss. In 1856, one year after Gauss' death, he published a booklet entitled *Gauss zum Gedächtnis*[68] (Gauss memorial). This biography is a first-hand very personal testimony, and this is why it is of value. It is also quite reliable since its author was in possession of original source material and corrected erroneous rumors, which had been given publicity. This 90-page text was translated into English in 1966 by Helen Worthington Gauss, a great-granddaughter of Gauss [3159]. It is quite impossible to relate all the facts it contains. Let us only cite some interesting information and anecdotes that cannot be found elsewhere.

After a foreword from which we learn that Gauss had blue eyes, and a description of his family and his childhood, the author related his school years, his support from Bartels and the Duke, and his entering Göttingen University. According to him, Gauss discovered the method of least squares in 1795 and, on March 30, 1796, the theory of the division of the circle, involving the construction of the 17-angled polygon.

During his stay at Göttingen, Gauss only had a few friends, the jurist Wilhelm Arnold Eschenburg (1778-1861) whom he knew since 1789 when they were boys at school together, F.W. Bolyai, Ide (?) from Braunschweig, and the astronomer, geologist, and physicist Johann Friedrich Benzenberg (1777-1846). Then, when in Helmstedt, he became acquainted with Pfaff; he lived in his house and often went for walks with him. According to von Waltershausen, Gauss did not come to Helmstedt to study under Pfaff, and he was not led by him into his own mathematical researches since *he himself possessed such a completely original mind that he did not need to be led by another* [...].

The author of the booklet told readers about the genesis of the *Disquisitiones Arithmeticae*, the least squares, the discovery of Ceres by Piazzi, the discovery of Pallas by Olbers, and Gauss' visit to him. Then, von Waltershausen described the influence of the Napoleonic wars on Gauss, in particular with the death of the Duke of Braunschweig. He also came back to the method of least squares, the publication of *Theoria motus*, and the dispute with Legendre, and wrote

> *Since Gauss had communicated this important discovery which he made in 1795 in the following year to his friend Bolyai, Bolyai is the only man now living who can give both scientific and contemporary testimony.*

Gauss' works on the triangulation of the Kingdom of Hanover, geodesy, and the invention of the heliotrope are described with many details and anecdotes about the necessary travels around the country. Von Waltershausen wrote

> *Following this discovery Gauss threw out the question, half in earnest, half in jest, of whether the moon might be inhabited by a more intelligent race. Admitting this was not very probable in view of his observations of our nearer planets, he suggested that the heliotrope might help to establish safe telegraphic communication between the two worlds, and without exorbitant cost. He is said even to have calculated the size of the mirrors needed and to have reached a very satisfactory result. "This would be a discovery even greater than that of America" , he said , "if we could get in touch with our neighbors on the moon".*

In the following years, Gauss' interests extended from terrestrial magnetism, dioptrics, and the electromagnetic telegraph to the pension fund for the widows of the professors of the University of Göttingen and the building and the management of railroads. Von Waltershausen

[68] https://archive.org/details/gauss00waltgoog (accessed December 2020)

described in detail the last years of Gauss, his ultimate days, his death, and the following ceremonies. The last part of the booklet described the character and the manner of mind of Gauss. His talent for widely different languages has to be noted. He was familiar with ancient Greek and Latin, and he could read most of the European modern languages. He tried Sanskrit, but after some time, turned to Russian, which he managed to speak with a wholly Russian accent. In literature, he admired Walter Scott, and liked the English historians. On the political side, he was a conservative, and thought that a firmly unified Germany was essential. He was a man of strong character, who only respected men with strong characters. Let us end by quoting von Waltershausen who wrote that *Applied science was for him of secondary importance, though he did not underrate it.*

10.27 ▪ Hilda Geiringer

Hilda Geiringer

© Private possession and courtesy of Geiringer's daughter, Magda Tisza (Boston)

Life

Hilda Geiringer was born into a Jewish family in Vienna, Austria, on September 28, 1893. Her Jewish origin had important consequences on her life. Her father Ludwig Geiringer, originally from Hungary, met her mother Martha Wertheimer when working in Vienna as a textile manufacturer. Hilda was the second of four children.

Already at the Gymnasium, Hilda showed a great mathematical ability and a prodigious memory. In 1913, her parents supported her study of mathematics at the University of Vienna.

In 1917, she defended her doctorate on Fourier series in two variables under the guidance of Wilhelm Wirtinger (1865-1945). The following two years she assisted Leon Lichtenstein (1878-1933) in editing the mathematics review journal *Jahrbuchs über die Fortschritte der Mathematik.*

In 1921, Geiringer moved to Berlin to become an assistant to Richard Edler von Mises (1883-1953), an Austrian Jewish scientist and mathematician, at the Institute of Applied Mathematics. That same year, she married Félix Pollaczek (1892-1981) who, like her, was born in Vienna into a Jewish family. He had been a student of Issai Schur (1875-1941), and he is known, among other things, for the orthogonal polynomials bearing his name. Hilda and Félix had a daughter, Magda, in 1922, but their marriage broke up in 1925, even though the divorce was only finalized in 1932. She continued to work with von Mises, and raised her child alone.

Her mathematical contributions were noticed by Albert Einstein (1879-1955) with whom she corresponded over many years on science topics and her immigration to the United States.

On July 18, 1925, she applied for Habilitation to the University of Berlin. The faculty chose von Mises and Ludwig Bieberbach (1886-1882) as referees. They found a mistake in her work. Bieberbach was very rude since he wrote that he had gotten a *truly shattering impression* of Geiringer's *purely mathematical abilities and achievements.* Moreover, underlying social and religious problems existed between her and Bierberbach, and applied mathematics was not yet well accepted in German academic circles at that time. Two additional referees, Schur and Erhard Schmidt (1876-1959), were nominated, and they confirmed the incorrectness of Geiringer's result, but favored the creation of a special field for Habilitation called *Applied Mathematics*. Finally, after von Mises had found the mathematical condition rescuing the result, and the writing of an addendum to her work, Geiringer got her degree [2764].

After Hitler took power, Jews were no longer allowed to teach in German universities following the Civil Service Law of April 7, 1933. Dismissed from the University of Berlin, Geiringer and her daughter went to Brussels in Belgium where she obtained a position at the Institute of Mechanics for one year. She began to work on the theory of vibrations.

Although he had converted to Catholicism, von Mises left Germany at the very end of 1933 to take up a newly founded chair of mathematics in Istanbul, Turkey. In 1934, Geiringer joined him. There, she was appointed professor of mathematics. She had to learn Turkish to be able to give her lectures. She continued to work on plasticity, statistics, and probability. She wrote or published about 18 papers and a book, in Turkish, based on her lecture notes on introductory calculus for chemistry students. She also became interested in the basic principles of genetics as formulated by Gregor Mendel (1822-1884). She was a pioneer in this emerging discipline, but her work was not fully recognized because it was done in Istanbul and published in Turkish journals [1050].

After the death of Mustafa Kemal Atatürk (c.1881-1938), the founder and first president of the Republic of Turkey, the Jews who had fled Germany worried for their safety, and von Mises emigrated to the United States in 1939. Geiringer and her daughter also succeeded in obtaining visas after she got a temporary lecturer position at Bryn Mawr College in Pennsylvania. Again, she had to learn another language in order to teach, and, as she said, to adjust to *the American form of teaching*.

Simultaneously, as part of the war effort, she did classified work for the United States National Research Council. During 1942, she gave an advanced summer course in mechanics at Brown University in Providence, Rhode Island. She wrote up her outstanding series of lectures on the geometrical foundations of mechanics and, although they were never properly published, these were widely disseminated and used in the United States for many years. However, although Brown University never offered Geiringer a permanent position, the university took full birthplace credit for these *mimeographed notes*.

In 1943, Geiringer and von Mises, who was working at Harvard in Cambridge, Massachusetts, at that time, got married. To be closer to him, she left Bryn Mawr College and accepted a professorship and the position of chairwoman of the Mathematics Department at Wheaton College, a woman's college in Norton, Massachusetts. It was her first permanent position in the USA. During the week, she taught at the college, but she traveled to Cambridge every weekend to be with von Mises.

She became an American citizen in 1945. Geiringer applied for positions at several universities where research was more developed than in her place but, after discrimination against Jews, she experimented discrimination against women. However, she never gave up her research while at Wheaton College. She is one of the founders of the field of applied mechanics, and she is known, in particular, as the creator of the Geiringer equations in plasticity theory.

Von Mises died in 1953 and Geiringer, while pursuing her job at Wheaton College, began to work at Harvard editing her husband's unpublished papers. For that purpose, she received a grant from the Office of Naval Research, and then Harvard offered her a temporary position as a Research Fellow in Mathematics. In 1956, the University of Berlin elected her Professor Emeritus with full salary. She was elected a Fellow of the American Academy of Arts and Sciences in 1959, and retired from Wheaton College, which honored her with an honorary Doctorate of Science. She was also a Fellow of the Institute of Mathematical Statistics. In 1964, she succeeded in editing and complementing the book of von Mises *Mathematical Theory of Probability and Statistics*.

Hilda Geiringer died of pneumonia on March 22, 1973, during a visit to Santa Barbara, California.

Work

We have seen that Hilda Geiringer worked in different areas of pure and applied mathematics as well as in mechanics.

With von Mises, she gave sufficient conditions for convergence of Jacobi and Gauss-Seidel methods in 1929. In the paper, published in two parts [3155, 3156], they first considered solving by iteration one nonlinear equation $f(x) = 0$. Then, they somehow generalized the method to linear solves. It was in this paper that they introduced the power iterations, which converge to the leading eigenvector of a matrix; see Chapters 5 and 6.

In 1949, Hilda Geiringer proved that the Gauss-Seidel method is convergent for irreducibly diagonally dominant matrices [1314].

10.28 ▪ Semyon A. Gerschgorin

Semyon Aranovich Gerschgorin

Life

Semyon Aranovich Gerschgorin (sometimes written Gershgorin or Geršgorin) was born on August 24, 1901, in Pruzhany, Russian Empire (now Belarus).

He studied at Petrograd Technological Institute from 1923, and defended an outstanding thesis submitted to the Division of Mechanics. The papers he published at that time are all in Russian: *Instrument for the integration of the Laplace equation* in 1925, *On a method of integration of ordinary differential equations* in 1925, *On the description of an instrument for the integration of the Laplace equation* in 1926, and *On mechanisms for the construction of functions of a complex variable* in 1926. In these papers, he described original and intricate linkage mechanisms for these problems, thus becoming a pioneer in the construction of analog devices by applying complex variables to the theory of mechanisms. Later, he even designed devices for modeling airfoil profiles.

Gerschgorin obtained a professorship at the Institute of Mechanical Engineering in Leningrad in 1930, and from 1930 he worked in the Leningrad Mechanical Engineering Institute on different topics: algebra, theory of functions of complex variables, numerical methods, and differential equations. He became head of the Division of Mechanics at the Turbine Institute, and taught innovative courses at Leningrad State University. He soon became one of the leading figures in Soviet Mechanics and Applied Mathematics, and he gained a worldwide reputation. He worked on the theory of elasticity, the theory of vibrations, the theory of mechanisms, methods of approximate numerical integration of differential equations and on other areas of mechanics and applied mathematics. He even built a mechanical device for drawing ellipses which can be seen at the Deutsches Museum in Munich.

Gerschgorin was appreciated for the novelty of his methods for approaching a problem, combined with the power and clarity of his analysis. He died on May 30, 1933, in Leningrad, USSR (now Saint Petersburg, Russia), from an accidental illness due to his stressful job.

Work

Finite-difference methods for the approximate solution of partial differential equations were introduced in 1910 by Lewis Fry Richardson (1881-1953). In 1927, Gerschgorin used them for the two-dimensional Poisson equation in a region of the plane with the solution prescribed on the boundary. In 1929, he proposed to solve finite-difference approximations by modeling them with electrical networks. In 1930, he published a seminal paper in which he treated the convergence of finite-difference approximations to the solution of Laplace-type equations.

Gerschgorin's famous paper *Über die Abgrenzung der Eigenwerte einer Matrix* (On the limits of eigenvalues in a matrix) [1341] on the localization of the eigenvalues of a matrix appeared in 1931. It is reproduced in [3136]. The fact that this paper was written in German made it known in the West during the 1940s. However, the first result of this paper was not correct, and his assumption was cleared up by Olga Taussky (1906-1995) in 1949 as explained in [3136]; see also Section 10.68.

Gerschgorin's last paper, submitted on October 9, 1932 but only published in 1933, maybe posthumously, treated numerical conformal maps [3015]. Using Nyström's method, he reduced the problem to the solution of a Fredholm integral equation.

Gerschgorin's result on eigenvalues had been heavily cited and refined many times, for instance, by Alfred Theodor Brauer (1894-1985) [413] in 1947. For more details, see Richard S. Varga's book [3136] in 2004.

Gerschgorin's disks and its generalizations have been used to locate zeros of polynomials by applying them to companion matrices; see papers by Aaron Melman [2198, 2199, 2201].

10.29 ▪ Wallace Givens

James Wallace Givens
© SIAM

Life

James Wallace Givens, Jr., was born on December 14, 1910, in Alberene, Virginia, USA (a small town near Charlottesville). He graduated from high school at the age of 14, and obtained his bachelor's degree from Lynchburg College in 1928 at the age of 17.

In 1931, after a one-year fellowship at the University of Kentucky, he obtained a master's degree from the University of Virginia under Ben Zion Linfield (born 1897), a former student of George David Birkhoff (1884-1944) who defended his Ph.D. *On the theory of discrete varieties* in 1923.[69]

Givens obtained his Ph.D. from Princeton University in 1936 under the guidance of Oswald Veblen (1880-1960). The title was *Tensor Coordinates of Linear Spaces*. He was an assistant to Veblen at the Institute for Advanced Study during his doctoral work from 1935 to 1937. He developed a wide interest in projective geometry that he kept all his life. He followed a series of lectures by John von Neumann (1903-1957) which aroused his interest in computers and numerical methods.

Givens first taught at Cornell University as an instructor from 1937 to 1941. He then spent several years teaching at the Northwestern University (1941-1946 and 1960) in Evanston, Illinois, the University of Tennessee (1947-1956) in Knoxville, and Wayne State University (1956-1960) in Detroit, Michigan. His professional career included work with the earliest vacuum tube computers, first on the *UNIVAC I* at the Courant Institute of New York University (NYU), and later with *ORACLE* at Oak Ridge National Laboratory.

While chair of the mathematics department, in 1957 he organized a workshop on dense matrix computations which can be considered the *Gatlinburg 0* meeting, subsequently organized by Alston Scott Householder (1904-1993).

[69]In July 1925, Linfield defended a doctoral thesis at the University of Strasbourg (France) with the title *Espace discret paramétrique et non paramétrique* with Maurice Fréchet as the president of the committee. One of his students wrote *The late Professor Ben Zion Linfield, at the University of Virginia, was a man who demanded much of his students. Whenever I would show him some proof he would compliment me but then show me that the proof could be made better, more elegant. I learned from him never to be really satisfied upon proving a result until I analyzed the details and found out what really made the proof work, and how to get a simpler or better proof and generalize the result.*

In 1960, Givens joined the staff of the Argonne National Laboratory near Chicago, where he served until his retirement in 1975. In 1963, he became a senior scientist there and, from 1964 to 1970, director of the Division of Applied Mathematics. He established a group of applied mathematicians, computers scientists, and engineers. The group's interests included reactor mathematics, quantum physics, automated theorem proving, programming languages, and image processing. He had a remarkable skill for finding talented individuals and stimulating their interest in computer science and computation; see [3303].

He advocated for the implementation of state-of-the-art algorithms freely available. With James Hardy Wilkinson (1919-1986), he initiated a project for translating the Algol programs for linear algebra published in *Numerische Mathematik* into FORTRAN; see Chapter 8. However, tired of fighting administrative battles, and wanting to devote himself to research, he resigned from his directorship in 1970 [2508].

Givens was honored by the Alexander von Humboldt Foundation in Germany and he visited Bonn in 1974. He served (1969-1970) as president of the Society of Industrial and Applied Mathematics (SIAM), and retired in 1979 as professor emeritus from Northwestern University. He died on March 5, 1993, in El Cerrito, California.

Work

Even though they were already used by Carl Gustav Jacob Jacobi (1804-1851) in the 19th century, rotation matrices are attached to Givens' name because he used them in the 1950s to reduce matrices to tridiagonal form and to compute eigenvalues. Givens' work was first presented at the 1951 *National Bureau of Standards Symposium* [1364]. A detailed description of the method was given in a 116-page report from Oak Ridge National Laboratory [1365] in 1954. Givens first recalled the Gram-Schmidt process to show that a matrix can be reduced by orthogonal transformations to Hessenberg form, which is tridiagonal when the matrix is symmetric. Then, he gave details on the construction of the plane rotations. This reduction can be done in a finite number of steps since the created zero entries are not changed by the next rotations. In [1367] Givens applied rotations to nonsymmetric matrices to transform them to upper triangular form.

Givens was a pioneer of backward error analysis in his 1954 report. It was never published in a journal, but it influenced other researchers, particularly Wilkinson, who developed the technique and applied it to many problems.

10.30 ▪ Gene H. Golub

Gene Howard Golub
Courtesy of Gérard Meurant

Life

Gene Howard Golub was born in Chicago on February 29, 1932. His parents Nathan Golub and Bernice Gelman emigrated, respectively, from Ukraine and Latvia to the United States in 1923, where they met and married. In the USA, Nathan sold bread and Bernice sewed baseball caps, and the family was far from wealthy. Gene had an older brother Alvin.

Gene did his graduate studies first at the University of Chicago and then, from 1953 to 1959, at the University of Illinois at Urbana-Champaign. There, he studied matrix theory and statistics. He had a part time job in the computer center and learned how to program one of the first parallel computers, the *ILLIAC IV*. For a while he was considering doing a Ph.D. in statistics but he became a student of Professor Abraham Haskel Taub (1911-1999), an applied mathematician, who oriented him toward the study of Chebyshev polynomials for solving linear systems, starting from John von Neumann's works. In 1959, Taub invited Richard Steven Varga (1928-2022) to the University of Illinois and Gene discovered that Varga was working on the same topic. This lead to the writing of a seminal joint paper that was published in 1961. About this Golub said (see [1543])

> *The lucky break I had is I was working on something, and Taub invited a man by the name of Richard S. Varga to come. I told Varga what I had done, and Varga said, "Oh, I've done a similar thing. Let's write a paper together." So I had sort of mixed feelings. I like the idea of working with somebody like that. I suspect Taub deliberately called him in just to check me out. Then I finished up. Varga came in February and I left at the beginning of April and I went to England.*

After his Ph.D. thesis Gene received an NSF fellowship that allowed him to spend 15 months in Cambridge, UK, where he met for the first time James (Jim) Hardy Wilkinson (1919-1986), who was working at the National Physical Laboratory; see Section 10.76. Jim was a prominent expert in rounding error analysis in numerical linear algebra algorithms and he became a frequent visitor to Stanford University. Gene met also Cornelius Lanczos (1893-1974) in the UK.

When he was back in the USA, Gene worked for several industrial companies and then decided to go back to academics. In 1962, George Elmer Forsythe (1917-1972) offered him an assistant professorship in Stanford and soon a permanent position. It was in those times that Gene met some of Forsythe's students that were about to become famous like Cleve Barry Moler, the founder of MATLAB, and Beresford Neill Parlett. The Stanford University Computer Science Department was founded by Forsythe in 1966 and Gene naturally got a position in the department together with John McCarthy (1927-2011), Donald Ervin Knuth, and other well-known researchers. Gene was chairman of the department in 1982.

Gene was very much concerned about the applied mathematics community. He was president of SIAM and founded two SIAM journals: *SIAM Journal on Scientific Computing* (SISC, the original name was *Journal on Scientific and Statistical Computing*) and *SIAM Journal on Matrix Analysis and Application* (SIMAX). He was a member of the editorial boards of many journals. He was also a driving force at the beginning of the *International Council for Industrial and Applied Mathematics* (ICIAM). He was closely involved in the organization of the first conference on domain decomposition methods which was held in Paris in 1987. He received more than 10 diplomas and awards from universities around the world. He was elected to the National Academy of Engineering in 1990 and the US National Academy of Sciences in 1993.

Gene traveled around the world to participate in the many conferences to which he was invited and to meet his many collaborators. He also spent a lot of time sending e-mail and answering the tens of messages he received a day. He was a pioneer in this area, and so interested in it that he founded NA-net and the NA-Digest to help the distribution of information through the community. Besides being an exceptional applied mathematician Gene was also a remarkable man. During meetings and conferences he probably spent more time with young students than with distinguished old professors. Many people remember his *Hello, I'm Gene Golub, who are you and what are you doing?*. He had a lot of influence on the careers and lives of many people. He was also very generous to his many friends around the world. He was a great scientist and a very good friend, even though he could be quite depressed on some occasions.

Gene passed away suddenly from the consequences of acute myeloid leukemia in Stanford Hospital on November 16, 2007. An obituary was published in *The New York Times* on December 10, 2007. It is also interesting to read references [2372] and [3069]. An interview with Gene is available at http://history.siam.org/oralhistories/golub.htm (accessed August 2021).

Work

During his brilliant career, Gene authored or co-authored more than 180 papers published in the best journals and many contributed or invited papers to an uncountable number of international conferences.

It is almost impossible to summarize all of his many important contributions to applied mathematics. Perhaps the most well-known ones are the use of QR factorizations to solve least squares problems and the invention of direct and iterative algorithms for the computation of the Singular Value Decomposition (SVD) together with William Morton Kahan. The SVD was a great passion of Gene's and he used it in many applications. He was so in love with SVD that the license plate of his last car was "Pr. SVD."

Noticeable also was the discovery of the cyclic reduction algorithm for solving certain structured linear systems leading to the construction of fast Poisson solvers.

Gene was one of the most active evangelists for the preconditioned conjugate gradient algorithm through his joint papers with Paul Concus and Dianne Prost O'Leary. Later, he worked on algorithms for computing bounds or estimates of bilinear forms $u^T f(A)v$, where u and v are vectors, A is a square symmetric matrix, and f is a smooth function; see [1384] in 1994. The techniques he developed rely on subtle relationships between moments, Gauss quadrature, orthogonal polynomials, and Lanczos algorithms. Applications of these methods are found in approximation of error norms in the solution of linear systems, computation of parameters in Tikhonov regularization of ill-posed problems, as well as the generalized cross-validation algorithms and computing approximations of the determinant of large sparse matrices; see the book [1386].

Gene Golub is also the author of several well-known books, the most famous of which is *Matrix Computations* [1399], written in collaboration with Charles Francis Van Loan. This book is a masterpiece and can be considered the bible of modern matrix computations. Three editions were published starting in 1983 and a fourth one was published posthumously. If one types "Matrix Computations" in Google Scholar, more than 70,000 citations are listed (in 2021). Reference [611] contains reprints of several of Gene's papers, chosen by him as being the most representative of his work, and commented upon by some of his colleagues. It also contains an interesting detailed biography.

10.31 ▪ William R. Hamilton

Life

William Rowan Hamilton was born an August 4, 1805, in Dublin, Ireland. He was the fourth child of a family of nine. His father, Archibald Hamilton (1778-1819), was a solicitor, and his mother was Sarah Hutton (1780-1817).

Sir William Rowan Hamilton

Hamilton showed great skills at a very early age, particularly in foreign languages. Robert Perceval Graves (1810-1893)) wrote in the Dublin University Magazine in 1842,

> *At the very earliest age indications were perceived of W.R.H.'s possession of extraordinary intellectual powers, in consequence of which, his father, unable from professional occupation to superintend their development himself, and recognising with a laudable promptitude their extent and value, consigned him when less than three years old to the care of the Rev. James Hamilton, the uncle of the young genius.* [...] *At the age of four he had made some progress in Hebrew: in the two succeeding years he had acquired the elements of Greek and Latin; and when thirteen years old was in different degrees acquainted with thirteen languages, besides the vernacular - Syriac, Persian, Arabic, Sanscrit, Hindoostanee, Malay, French, Italian, Spanish and German; and we are not sure that this list is a complete one.*

At the age of 10 he started reading Euclid's *Elements*. At 15 years old he began to read Newton's *Principia* and Laplace's *Mécanique Céleste*. In 1822, he met John Brinkley (1763-1835), who he was going to succeed. The paper *Theory of systems of rays* was presented in 1824 by Brinkley to the Royal Irish Academy and finally published in 1828 in the *Transactions of the Royal Irish Academy*.

He entered Trinity College in Dublin in 1823 where he received the best rankings in science, literature, and languages.

In 1826, Brinkley was appointed Bishop of Cloyne (Cork county) and the chair of astronomy became vacant. Hamilton was thought to be a good candidate for the chair. However, at first he did not want to apply, still being an undergraduate. But his tutor urged him to apply, assuring him of the favourable disposition of the Board towards his candidature. On June 1827, he was unanimously elected to the chair and also became Royal Astronomer of Ireland. The observatory was located in Dunsink where he stayed for the rest of his life. However, astronomy was not really one of his best skills. According to Sir Robert Stawell Ball (1840-1913) in [173]

> *To the practical working of the Observatory Sir William Hamilton, the bias of whose genius is undoubtedly to pure mathematics, is not naturally so adapted as to the other departments of scientific labour.*

In 1824, Hamilton fell in love with Catherine Disney (1806-1853) but he did not declare his feelings. Unfortunately, in 1825 he learned that she was to marry somebody else. This was a big disappointment for him. In 1831, he met Ellen De Vere, the sister of one of his friends. But she did not want to move from the place she was living. In the end, in April 1833, he married Helen Maria Baily (1804-1869). They had three children, William Edwin (1834-1902), Archibald Henry (1835-1914), and Helen Eliza Amelia (1840-1870).

Hamilton received the Cunningham Medal of the Royal Irish Academy in 1834 and again in 1848. He also received the Royal Medal of the Royal Society in 1835. Hamilton was knighted in 1835 and became Sir Hamilton. In 1837, he was elected president of the Royal Irish Academy.

In 1841, before leaving England for the United States, James Joseph Sylvester (1814-1897) wrote to Hamilton

> *Believe me, Sir, it is not the least of my regrets in quitting this empire to feel that I forego the casual occasion of meeting those masters of my art, yourself chief amongst the number, whose acquaintance, whose conversation, or even notice, have in themselves the power to inspire, and almost to impart fresh vigour to the understanding, and the courage and faith without which the efforts of invention are in vain. The golden moments I enjoyed under your hospitable roof at Dunsink, or moments such as they were, may probably never again fall to my lot.*

Working on the theory of quaternions occupied a large part of Hamilton's scientific life. At the end of his life he wrote the following to his son about the discovery (or invention) of quaternions:

> *Indeed I happen to be able to put the finger of memory upon the year and month - October, 1843 - when, having recently returned from visits to Cork and Parsonstown, connected with a meeting of the British Association, the desire to discover the laws of multiplication referred to, regained with me a certain strength and earnestness which had for years been dormant, but was then on the point of being gratified, and was occasionally talked of with you. Every morning in the early part of the above-cited month, on my coming down to breakfast, your (then) little brother, William Edwin, and yourself, used to ask me, 'Well, papa, can you multiply triplets?' Whereto I was always obliged to reply, with a sad shake of the head: 'No, I can only add and subtract them.'*
>
> *But on the 16th day of the same month - which happened to be Monday, and a Council day of the Royal Irish Academy - I was walking in to attend and preside, and your mother was walking with me along the Royal Canal, to which she had perhaps driven; and although she talked with me now and then, yet an undercurrent of thought was going on in my mind which gave at last a result, whereof it is not too much to say that I felt at once the importance. An electric circuit seemed to close; and a spark flashed forth the herald (as I foresaw immediately) of many long years to come of definitely directed thought and work by myself, if spared, and, at all events, on the part of others if I should even be allowed to live long enough distinctly to communicate the discovery. Nor could I resist the impulse - unphilosophical as it may have been - to cut with a knife on a stone of Brougham Bridge, as we passed it, the fundamental formula which contains the Solution of the Problem, but, of course, the inscription has long since mouldered away. A more durable notice remains, however, on the Council Books of the Academy for that day (October 16, 1843), which records the fact that I then asked for and obtained leave to read a Paper on 'Quaternions,' at the First General Meeting of the Session; which reading took place accordingly, on Monday, the 13th of November following.*

Brougham Bridge was the old name of the bridge mentioned above. The new name is Broom Bridge. On June 16, there is an annual walk from Dunsink Observatory to the bridge which has been attended by some well-known mathematicians.

Hamilton was also fond of poetry. According to Ball [173],

> *It was the practice of Hamilton to produce a sonnet on almost every occasion which admitted of poetical treatment, and it was his delight to communicate his verses to his friends all round.*

About Hamilton's death, Ball wrote

> *It was on the 9th of May, 1865, that Hamilton was in Dublin for the last time. A few days later he had a violent attack of gout, and on the 4th of June he became alarmingly ill, and on the next day had an attack of epileptic convulsions.*

Hamilton passed away on September 2, 1865, probably from gout and bronchitis. He is buried in Mount Jerome Cemetery in Dublin.

It is written in several books or sketches about Hamilton's life that he was suffering from severe alcohol problems; see [1428] and [1573]. But these views were challenged in [3116].

Work

Hamilton contributed to several areas of mathematics but, of course, what he considered his main achievement was the theory of quaternions, which were thought an extension of the complex numbers. We have seen the circumstances of this discovery above. On November 12, 1843, he presented a paper, *On a new species of imaginary quantities connected with a theory of quaternions* [1558], at a meeting of the Royal Irish Academy, and another paper in 1844. He published many papers on this topic and the book *Lectures on Quaternions* [1559] in 1853. He was still working on the book *Elements of Quaternions* when he passed away. This book was published with the help of his son William in 1866.

He is also famous for Hamiltonian mechanics, a reformulation of classical mechanics proposed by Hamilton in 1833 following some works of Joseph-Louis Lagrange (1736-1813).

For this book, our interest is in the proof of what is now known as the Cayley-Hamilton theorem for $n = 4$. This was done, of course, in the language of quaternions, in three short papers [1560, 1561, 1562]. In the first paper [1560] there is no proof but Hamilton wrote

As early as the year 1846, I was led to perceive the existence of a certain symbolic and cubic equation, of the form

$$0 = m - m'\phi + m''\phi^2 - \phi^3,$$

in which ϕ is used as a symbol of linear and vector operation on a vector, so that $\phi\rho$ denotes a vector depending on ρ, such that $\phi(\rho + \rho') = \phi\rho + \phi\rho'$.

Then, he showed a similar formula for quaternions,

$$0 = n - n'f + n''f^2 - n'''f^3 + f^4,$$

where f is a function satisfying $f(q + q') = fq + fq'$, q and q' being quaternions. A proof was given in [1562], an inverse of f being assumed to exist. This corresponds to a proof for real matrices of order 4.

10.32 ▪ Richard J. Hanson

Life

Richard ("Dick") Joseph Hanson was born in 1938 in Portland, Clackamas County, Oregon, USA. He obtained a bachelor's degree and, in 1962, a master's degree under Robert Eugene

Gaskell (1912-2005) from Oregon State University and received his Ph.D. in mathematics in 1964 from the University of Wisconsin in Madison under Wolfgang Richard Wasow (1909-1993).

He held numerous positions during his career at Rice University, at Sandia National Laboratories in New Mexico, at Washington State University Pullman, at the Jet Propulsion Laboratory in Pasadena in California, and in several industrial companies. Hanson was active professionally until the onset of his illness in 2015.

He was married to Karen H. Haskell with whom he wrote several papers. He served as Algorithms Editor for ACM *TOMS* for eight years, and was a long time member of SIAM, ACM, and of the IFIP Working Group 2.5 on Numerical Software.

He died at age 78 in Albuquerque, New Mexico, on December 4, 2016, after battling against brain cancer for a year and a half.

Richard Hanson loved outdoor activities, walking in the Albuquerque open space almost every day, and traveling throughout the world. He was a quiet and very thoughtful person.

Work

Hanson published numerous papers, including coauthoring the first BLAS paper [2005], and two books: *Solving Least Squares Problems* [2004] with Charles Lawrence ("Chuck") Lawson (1931-2015) and *Numerical Computing with Modern Fortran* with Tim Hopkins.

10.33 ▪ Emilie V. Haynsworth

Emilie Virginia Haynsworth

Life

Emilie Virginia Haynsworth was born on June 1, 1916 in Sumter, South Carolina, USA. She was the daughter of Hugh Charles Haynsworth (1875-1944) and Emilie Edgeworth Beattie Haynsworth (1878-1960). She had two older brothers and a younger sister.

Her talent in mathematics was evident in junior high school, where she was the winner of an annual contest. She graduated in 1937 with a bachelor's degree in mathematics from Coker College in Hartsville, South Carolina. Then, she obtained a master's degree in 1939 from Columbia University in New York, and became a high school mathematics teacher, first at Olympia High School in Columbia, South Carolina, and then at St. Timothy's School in Catonsville, Maryland.

During the war years, she worked at the Aberdeen Proving Ground, Maryland. We don't know what she did there, but during World War II, the technological contributions of this military installation to the war effort included one of the first digital computers (the *Electronic Numerical Integrator and Calculator*, or ENIAC), the first man-portable antitank weapons system (the bazooka) and the first system-wide practical applications of Statistical Quality Control.

After the war, Haynsworth became a lecturer at an extension program of the University of Illinois in Galesburg, Illinois, and in 1948, she registered at Columbia University, in New York, to begin her doctoral studies, but soon transferred to the University of North Carolina at Chapel Hill. Her college teaching career continued in 1951 when she became an instructor at Wilson College in Chambersburg, Pennsylvania.

In 1952, she defended her Ph.D. with the title *Bounds for determinants with dominant main diagonal* at the The University of North Carolina under the guidance of Alfred Theodor Brauer (1894-1985), a German-American mathematician working mainly in number theory. Brauer had been a student of Issai Schur (1875-1941) and Erhard Schmidt (1876-1959) in Berlin. In 1955, Haynsworth moved to the National Bureau of Standards as a mathematician. These were the golden years for matrix theory at the Bureau, and she met many scholars there.

A faculty position was vacant at Auburn University in Alabama. The department chair, William Vann Parker (1901-1987), came to Washington to interview her. She used to say

> *Dr. Parker came to Washington to interview me for a job at Auburn. I met him at his hotel. He said, "I've got this problem in matrix theory...," so we worked on his problem until he had to leave. As we were finishing, he told me that I had the job at Auburn.*

She stayed there until her retirement in 1983. She had 17 doctoral students, and was named a research professor in 1965.

Emilie Haynsworth died on May 4, 1985, in her home town. She is buried in the Holy Cross Cemetery Stateburg, Sumter County, South Carolina.

As stated in [554],

> *To those of us who knew Emilie, no recollection would be complete without acknowledging her marvelous sense of humor and her warm humanity. Her positive, witty, kindly, and loving outlook on life was an inspiration to all of us.*

After her death, Alexander Markowich Ostrowski (1893-1986) wrote

> *I lost a very good, life-long friend and mathematics an excellent scientist. I remember how on many occasions I had to admire the way in which she found a formulation of absolute originality.*

Her biography, with personal recollections and the lists of her publications and students, can be found in [554]. More reminiscences are given in [822, pp. 145-149].

Work

Haynsworth's research concerned the determinants of diagonally dominant matrices, and variants of the Gerschgorin circle theorem for bounding the locations of the eigenvalues of matrices.

In 1968, she produced her most well known work about the Schur complement [1611]. She named it in 1968, but it was a notion she had already been using in her own work since 1959. She is also known for the *Haynsworth inertia additivity formula* about the number of positive, negative, and zero eigenvalues of a block partitioned Hermitian matrix; see [1610].

Her later work involved cones of matrices, and inertia of partitioned Hermitian matrices.

10.34 ▪ Peter Henrici

Peter Henrici

Life

Peter Karl Henrici was born in Basel, Switzerland, on September 13, 1923, and was educated in high schools there. His father, Hermann Henrici, was a lawyer and his mother was Erika Müller.

In 1942, he entered the University of Basel to study law for two years. After World War II, he began to study at the Eidgenössische Technische Hochschule Zürich (ETH) and he obtained a diploma in electrical engineering in 1948[70] and another one in mathematics.

In 1952, he was awarded a doctorate in mathematical sciences[71] under the "formal" (as he said) supervision of Eduard Ludwig Stiefel (1909-1978). Even before the publication of his thesis in 1953, he already had four papers in print, one of them on the computation of the eigenvalues of matrices with punched card machines.

In fact, in 1951, Henrici went to the United States and had a position at the American University in Washington, D.C. He was also a visitor at the National Bureau of Standards through a contract with his university. In 1956, he moved to the University of California at Los Angeles as an associate professor. Later he was promoted to full professor of mathematics and he remained there until 1962 when he returned to the ETH as a professor of mathematics, a position he held until the end of his life.

Henrici continued to hold his position at UCLA for the rest of his life. In 1985, he took on a part-time appointment as the William Kenan Distinguished Professor of Mathematics at the

[70] *Ueber die Lösung von ebenen Potentialproblemen mit scharfen und abgerundeten Ecken durch konforme Abbildung.*

[71] *Zur Funktionentheorie der Wellengleichung : mit Anwendungen auf spezielle Reihen und Integrale mit Besselschen, Whittakerschen und Mathieuschen Funktionen.* This can be translated as: On the complex analysis of wave equation with applications to specific series and integrals with Bessel, Whittaker, and Mathieu functions.

University of North Carolina at Chapel Hill. Throughout his career, he held a number of visiting professorships at Harvard, Stanford, and several other universities in the United States.

He married Elenore Jacottet in 1951. They divorced in 1972 and he married Marie-Louise Kaufmann in 1973. He had three children, Christoph, Katherine, and Andreas.

Peter Henrici died on March 13, 1987, in Zürich after a nine-month illness. He was a cultured and warm person with a sense of humor and modesty, an outstanding lecturer, and a gifted pianist.

Work

Henrici was an internationally recognized numerical analyst and complex analyst. He authored 11 books and over 80 research papers, and he was the advisor of 28 doctoral students. In his famous book *Discrete Variable Methods in Ordinary Differential Equations* [1644] published in 1962, he gave the first complete treatment of the notions of consistency, stability, and convergence for the numerical integration of ordinary differential equations. In 1964, his book *Elements of Numerical Analysis* [1645] contains a derivative free generalization of Steffensen's method with quadratic convergence for solving a system of nonlinear equations. His three volumes *Applied and Computational Complex Analysis* [1646] form a masterpiece with chapters on power series, integration, conformal mapping, location of zeros, special functions, integral transforms, asymptotics, continued fractions, discrete Fourier analysis, Cauchy integrals, construction of conformal maps, and univalent functions.

10.35 ▪ Karl Hessenberg

Karl Hessenberg

Life

Karl Adolf Hessenberg was born on September 8, 1904, in Frankfurt am Main, Germany. His father, Eduard Hessenberg (1873-1933), was a lawyer, and his mother was Emma Kugler (1878-1970). Eduard was the grandson of Heinrich Hoffmann (1809-1894), a psychiatrist and author of a famous book for children *Der Struwwelpeter*. Karl had two sisters and a brother Kurt (1908-1994) who became a music composer and professor at the Hochschule für Musik und Darstellende Kunst in Frankfurt.

Karl went to school in Frankfurt am Main between 1911 and 1923, and then joined Hartmann & Braun AG between April 1923 and April 1925. He studied electrical engineering at the Technische Hochschule in Darmstadt from 1925 to 1930.

From 1931 to 1932, he was an assistant to Alwin Oswal Walther (1898-1967), a pioneer of mechanical computing technology in Germany, at the Technische Hochschule Darmstadt. Afterwards, Hessenberg worked for the Elektrizitätswerk Rheinhessen in Worms from February 1932 to September 1936. Starting in October 1936, he was employed by A.E.G. (Allgemeine Elektricitäts-Gesellschaft AG, an electricity company) as an engineer, first in Berlin, and later in Frankfurt am Main.

He finished his dissertation under the supervision of Walther with the degree of Dr.-Ing. (Doktor-Ingenieur). It turned out that the dissertation contained some material already known, which delayed the submission to May 3, 1940, and the day of the defense to July 23, 1940. The dissertation was approved on February 11, 1942.

Karl married Gertrud Pflug (1908-2003). They had two daughters, Brigitte and Renate. According to Brigitte Bossert, the daughter of Karl,

> *He was a very modest man. During the war years he had earned a doctorate, but he never attached importance to this title.*

Karl Hessenberg died of cancer on February 22, 1959.[72]

Apparently, the Japanese professor Seiji Fujino was the first to be interested in the life and papers of Karl Hessenberg. Hessenberg's daughter wrote

> *Almost 40 years after his death, I received a letter in a roundabout way from a Japanese professor who had been trying in vain for a while to find out who Hessenberg was. Prof. Seiji Fujino had repeatedly come across the Hessenberg matrix in the literature during his research and he was interested in how this name had come about. [...] We entered into lively correspondence, and after lengthy research I was finally able to send him a copy of my father's doctoral thesis. Several visits to Frankfurt rounded off his efforts. In the meantime, he has written several publications about my father and his work, and a detailed report appeared in the Japanese press.*

Work

Karl was a near relative of the mathematician Gerhard Hessenberg (1874-1925). The *Hessenberg sum and product of ordinals* are named after G. Hessenberg, but Hessenberg's matrices are also often wrongly attributed to him.

Many authors cite Hessenberg's thesis with the title *Auflösung linearer Eigenwertaufgaben mit Hilfe der Hamilton-Cayleyschen Gleichung* (Solution of linear eigenvalue problems using the Hamilton-Cayley equation) as the origin of the Hessenberg matrices. This dissertation does not contain any Hessenberg matrices, even though a tridiagonal matrix is used as an example.

On the second page of the preamble of the dissertation there is the following comment:

> *One can remark that the method described in Section V can be enhanced slightly, which will be reported in a separate paper.*

This paper is a technical report of the IPM (Institute of Practical Mathematics) Darmstadt. It appeared on July 23, 1940, the day of Hessenberg's defense, with the title *Behandlung linearer*

[72]This biography was written with information taken from a page of the website http://www.hessenberg.de/karl1.html (accessed January 17, 2022) by Jens-Peter M. Zemke from the Technische Universität Hamburg-Harburg, Hamburg, Germany. Links to Hessenberg's documents which were scanned by Brigitte Bossert can be found at that url address.

Eigenwertaufgaben mit Hilfe der Hamilton-Cayleyschen Gleichung (Treatment of linear eigenvalue problems using the Hamilton-Cayley equation). In this report, Hessenberg used for the first time what are nowadays called Hessenberg matrices; see page 23, Equation (58).

Hessenberg described two methods for obtaining the characteristic polynomial. Hessenberg's method for computing eigenvalues was explained in the book [3343] by Rudolf Zurmühl (1904-1966) in 1950. The words "Hessenberg matrix" appeared in a paper [2676] by Edward Aaron Saibel (1903-1989) and W.J. Berger in 1953. Hessenberg was also cited in the book by Ewald Konrad Bodewig (1901-?) [353] in 1959 and in the book by James Hardy Wilkinson (1919-1986) [3248] in 1965.

10.36 ▪ Magnus R. Hestenes

Magnus Rudolph Hestenes

Life

Magnus Rudolph Hestenes was born February 13, 1906, in Brycelyn, Minnesota, USA. His father was Mons Elias Mathison Hestenes (1867-1912) and his mother was Anna Rodina Didrikson (1873-1946). Magnus had four sisters and one brother.

He graduated from the University of Wisconsin for his master's thesis in 1928 and the University of Chicago from which he obtained his Ph.D. thesis in 1932 under the supervision of Gilbert Ames Bliss (1876-1951). The title of his thesis was *Sufficient conditions for the general problem of Mayer with variable end-points*.

From 1932 to 1937, he was an instructor at various universities. After a year in Chicago, he left for Harvard as a National Research Fellow to work with Marston Morse (1892-1977). According to his own words in [1662]

> *In 1936 I developed an algorithm for constructing a set of mutually conjugate directions in Euclidean space for the purpose of studying quadric surfaces. I showed my results to Prof. Graustein, a geometer at Harvard University. His reaction was that it was too obvious to merit publication.*

In 1937, he obtained a position of assistant professor at the University of Chicago, later associate professor. During the latter years of World War II, he was a member of the Applied Mathematics Group at Columbia University concerned with aerial gunnery.

He left Chicago in 1947 to join the University of California at Los Angeles (UCLA) where he stayed until his retirement in 1973. From 1948 to 1952, he was consultant at the Rand Corporation, from 1949 to 1954 consultant at the National Bureau of Standards (NBS). He was working at the Institute of Numerical Analysis (INA) on the UCLA campus; see Chapter 5. He was Assistant Director of Research from mid-1950 and UCLA Liaison Officer up to the end of the INA in the second quarter of 1954.

Hestenes was chairman of the UCLA Mathematics Department from 1950 to 1958 and director of the university's computing facility from 1961 to 1963. During the sixties and seventies he was also consultant at the Institute for Defense Analyses and at the IBM Watson Research Laboratory. He became professor emeritus in 1973.

During his career he had a total of 34 Ph.D. students. He received the Guggenheim and Fulbright awards.

Hestenes was married to Susan Madelena Eastvold (1906-1985). They had three children. The eldest son, David Orlin Hestenes, is a theoretical physicist and science educator. In an interview he said

> *My father didn't try to teach me mathematics directly, but he created an atmosphere. I often saw my father sitting down and working with a piece of paper. In fact, he had amazing powers of concentration. My brother and I could run around making all sort of noise, and he would just sit there working away. Later on I asked him how he got such powers of concentration, because I myself am easily distracted by noise or anything. And he said it is because when he went to college he studied at a desk in the hallway of the dormitory, so he had to learn how to concentrate in the midst of all the noise of the other dorm students.*

Magnus R. Hestenes died of a heart attack on May 31, 1991, in Laguna Hill, California. He is buried in the Pacific View Memorial Park in Corona del Mar, California.

Work

For more about the introduction of the conjugate gradient algorithm at the INA, see Chapter 5. Hestenes' seminal paper with Eduard Ludwig Stiefel (1909-1978) was published in 1952 [1664].

Hestenes was much involved in variational theory and optimization. He published several books, *Calculus of Variations and Optimal Control Theory* in 1966, *Optimization Theory: The Finite Dimensional Case* in 1975, and *Conjugate Direction Methods in Optimization* in 1980.

10.37 ▪ Alston S. Householder

Life

Alston Scott Householder was born on May 5, 1904, in Rockford, Illinois, USA. But shortly after he was born, his family moved to Alabama where he spent his childhood.

After school, Householder went to Northwestern University where he received a BA (Bachelor of Arts) degree in philosophy in 1925. Then, he went to Cornell University, receiving an MA (Master of Arts) in 1927. He married Belle (who passed away in 1975) in 1926. They had two children, John and Jackie.

About his switch from philosophy to mathematics, Householder said [2206]

> *As a matter of fact, my original idea was to become a Methodist minister, and majoring in Philosophy seemed to be a reasonable thing to do. [...] I guess I suddenly discovered more or less by accident that I was interested in mathematics. Not only*

Alston Scott Householder

> *that, but that I could make a living teaching mathematics, and I didn't discover this really until I had already graduated and was well along toward a master's degree in philosophy. [...] I was interested in metaphysics and logic and actually, although my master's degree was in philosophy at Cornell, I took more mathematics courses than I did philosophy courses. I then began to discover that my real interest was in mathematics, rather than in philosophy. [...] I took off fall semester in 1935, and went back to the University of Chicago, and essentially finished up my thesis at that time.*

He taught mathematics at several places including Washburn College in Topeka, Kansas, up to 1937. He received a Ph.D. in mathematics in 1937 from the University of Chicago. The title of his thesis was *The dependence of a focal point upon curvature in the calculus of variations* and his advisor was Gilbert Ames Bliss (1876-1951).

However, Householder was more interested in mathematical biology than in the calculus of variations. In 1937, he joined N. Rashevsky's Committee of Mathematical Biology at the University of Chicago and stayed there until 1944. Nicolas Rashevsky (1899-1972) was one of the pioneers of mathematical biology and mathematical biophysics. Nevertheless, in 1938 Householder already wrote a paper [1755] about eigenvalues.

In 1944, Householder went to Washington to work with a group of psychologists and for the Navy, under the Applied Psychology Panel. Then, he went over to the Naval Research Laboratory. He joined the Mathematics Division of Oak Ridge National Laboratory (ORNL) in Tennessee in 1946. He remembered his arrival at ORNL:

> *I didn't develop an interest in numerical analysis until quite some time later. And that was after in fact I went to Oak Ridge, and became interested in computers. [...] They had a small computing group, a group of girls mainly [...] who had a section in the Physics Division that did computing, hand computing, mainly, for the physicists and, it seemed as though it would make sense to expand that somewhat, and so I was put in charge of that group, and it was made into a mathematical and computing section of the Physics Division. And then a year after that, since a number of our customers came from other divisions besides just the Physics Division, we were set*

> *up as what was called the Mathematics Panel. It was really an independent division, separate division, but being very small, only about fifteen people, "division" seemed like a rather imposing title, for such a small group, although Panel was not a very good name either really. But anyway, by that time I had become convinced that the Laboratory needed to have its own computing machine.*

A computer named *ORACLE* (Oak Ridge Automatic Computer and Logical Engine) was built at the Argonne National Laboratory and shipped to Oak Ridge in early 1954. In the 1960s this machine was replaced by commercial computers.

Householder was at the roots of the organization of the *Gatlinburg meeting* which had been renamed the *Householder Symposium* in his honor. He described the beginnings as follows [1751]:

> *In the summer of 1960 there happened to be assembled a group of people interested in matrices during the Ann Arbor summer session. Among them were the Todds, Wallace Givens, George Forsythe, Dick Varga, Fritz Bauer, Dave Young, Peter Henrici, Jim Wilkinson and possibly others. One evening at the Old German's, after a few pitchers of Michelob (dark), someone suggested having, perhaps, a third conference on matrix computations.* [...] *Naturally there has been some criticism, especially of the limited attendance. Admittedly, no committee, however constituted, can hope that its selections will be the best possible. I have tried to indicate the reasons for restricting attendance. Certainly no criticism of open meeting such as this one is intended. Open meetings serve certain purposes, the Gatlinburgs others.*

For more about the Gatlinburg meetings, see [2905] by Josef Stoer.

Householder left Oak Ridge National Laboratory in 1969 and became professor of mathematics at the University of Tennessee. He was president of the ACM (1954-1956) and SIAM (1963-1964) and vice-president of the MAA (1960-1961).

After the death of his first wife, he was remarried in 1984 to Heidi Vogg, the sister-in-law of Friedrich Ludwig Bauer (1924-2015). After his retirement in 1974 he settled in Malibu, California, where he died of a stroke on July 4, 1993.

Work

During his stay at ORNL Householder became an expert in numerical linear algebra. He was most influential through the publication of his books, *Principles of Numerical Analysis* in 1953 and *The Theory of Matrices in Numerical Analysis* in 1964. He is, of course, well known for the Householder reflectors or matrices. For real data, they are orthogonal matrices of the form

$$I - \frac{2}{v^T v} v v^T,$$

where v is a given vector. These transformations are used to zero some components of a vector. Householder did not invented these matrices. They were described in the book by Herbert Westren Turnbull (1885-1961) and Alexander Craig Aitken (1895-1967), *An Introduction to the Theory of Canonical Matrices* [3081] published in 1932. These matrices are described on page 103 of the 1944 edition. However, Householder showed how these transformations can be used to efficiently compute the QR factorization of a matrix [1744]. He also promoted the use of norms in numerical linear algebra. In some papers with Bauer, he showed how norms could be used to derive localization theorems for eigenvalues.

He is also remembered for proving the necessity part of the Householder-John theorem in a report in 1955 and a paper [1740] published in 1956. This result is used to study convergence of iterative methods based on matrix splittings for solving systems of linear equations.

Householder was very influential in the development of numerical linear algebra after World War II. In fact, most of the notation that we still use today was first introduced by him.

10.38 ▪ Carl G.J. Jacobi

Carl Gustav Jacob Jacobi

Life

Carl Gustav Jacob Jacobi was born in a prosperous Ashkenazi Jewish family on December 10, 1804, in Postdam, at that time in the Duchy of Magdeburg, Kingdom of Prussia.

The Duchy was abolished during the Napoleonic wars in 1807, and then made part of the Province of Saxony in 1815. The young Carl was first educated by his uncle Lehman, who taught him classical languages and elements of mathematics. In 1816, he entered the Potsdam Gymnasium. While in his first year, thanks to his abilities and the lessons of his uncle, he was moved to the final year class. Thus, by the end of the academic year 1816-17, at the age of 12, he had reached the necessary level to enter university. However, as the University of Berlin was not accepting students younger than 16 years old, he had to remain in the senior class until 1821. During this period, he studied Latin, Greek, philosophy, history, and mathematics by himself and received the highest awards. When he left high school he had already read Euler's *Introductio in Analysin Infinitorum* and had made attempts to solve quintic equations by radicals.

In 1821, Jacobi entered the University of Berlin. For two years, he followed lectures in philosophy and mathematics before deciding to choose mathematics. At that time, the level of mathematics at the University was too low for him and he had to study by himself the works of Euler, Lagrange, Laplace, and others.

In 1823, taught by Friedrich Theodor Poselger (1771-1838), he succeeded in the exams necessary for teaching mathematics, Greek, and Latin in secondary schools. Despite the fact that he was Jewish, he was offered a position at the Joachimsthal Gymnasium in Berlin. However, he decided to stay at the university to obtain a position there. In 1825, Jacobi obtained the degree of Doctor of Philosophy with a dissertation on the partial fraction expansion of rational fractions (*Disquisitiones analyticae de fractionibus simplicibus*) under the guidance of Enno Heeren Dirksen (1788-1850).

As explained by Leo Königsberger (1837-1921) in his speech at the celebration of the hundredth anniversary of Jacobi's birth delivered at the International Congress of Mathematicians in

Heidelberg on August 9, 1904,[73]

> *By introducing infinite series to determine the coefficients of a general partial fraction decomposition of rational functions, he brings a function-theoretical element into algebraic studies, which later proved its fertility in the development of the theory of functions, and at the same time uses the principles applied to interesting transformations of infinite series.*

Immediately after his dissertation, Jacobi defended his Habilitation and also converted to Christianity. Now qualified for teaching at the University as a *Privatdozent*, he lectured on the theory of curves and surfaces at the University of Berlin in 1825-1826. But the prospects in Berlin were not so good. In May 1826, Jacobi moved to the University of Königsberg where he joined Franz Ernst Neumann (1798-1895), who worked on crystallography and other physical applications of mathematics, and the astronomer Friedrich Wilhelm Bessel (1784-1846). At that time, he had already found important results in number theory and he was corresponding with Carl Friedrich Gauss (1777-1855) about the results on cubic residues he had obtained, inspired by Gauss's contributions on quadratic and biquadratic residues. Gauss was so impressed that he wrote to Bessel to obtain more information about Jacobi.

Jacobi also had remarkable new ideas about elliptic functions (independently and quite simultaneously as Niels Abel (1802-1829)) and their relation to the elliptic theta function. On August 5, 1827, he began a correspondence with Adrien Marie Legendre (1752-1833), the leading expert on the topic, who immediately realized that he had been surpassed by his young colleague. Thanks to the praise of Legendre, Jacobi was promoted to associate professor on December 28, 1827. This work culminated in his great treatise *Fundamenta Nova Theoriae Functionum Ellipticarum* (The foundations of a new theory of elliptic functions) in 1829. That same year, he became a tenured professor of mathematics at Königsberg University, a chair he held until 1842. Jacobi made a visit to Paris that summer and met French mathematicians. On his return, he visited Gauss in Göttingen.

On September 11, 1831, Jacobi married Marie Schwinck. They had five sons and three daughters. In May 1832, he was promoted to full professor after a four-hour disputation in Latin with the title *Mathesis est scientia eorum, quae per se clara sunt* (Mathematics is the science of the things that are clear by themselves). Jacobi had a high reputation as a teacher, and he attracted many students. He also organized research seminars gathering advanced students and also attracting his nearest colleagues; this was then a completely new thing in mathematics.

In 1834, Jacobi received some work from Ernst Eduard Kummer (1810-1893) who was at that time a teacher in a gymnasium in Liegnitz. He immediately recognized that Kummer had made advances beyond what he himself had achieved on third-order differential equations.

In July 1842, Bessel and Jacobi were sent by the king of Prussia to the annual meeting of the British Association for the Advancement of Science in Manchester to represent their country. They returned via Paris, where Jacobi gave a lecture at the Academy of Sciences.

Early in 1843, Jacobi became seriously ill with diabetes. Although he was from a wealthy family, he lost his money in the severe business depression of 1837 in Europe. Johann Peter Gustav Lejeune Dirichlet (1805-1859), with the help of Alexander von Humboldt (1769-1859), who was a friend of Jacobi, obtained support from Friedrich Willhelm IV, the king of Prussia, which enabled Jacobi to spend some months in Italy, as his doctor had advised.

He travelled with Carl Wilhelm Borchardt (1817-1880) and Dirichlet, lectured at a science meeting in Lucca, and arrived in Rome on November 16, 1843. After some time, his health greatly improved, and he started to work on the manuscripts of Diophantus' Arithmetica in the Vatican Library.

[73]This speech contains many interesting details on Jacobi's life; see https://mathshistory.st-andrews.ac.uk/Extras/Jacobi_memorial/

In June 1844, he had to return to Königsberg. But the climate there was too extreme for him and Friedrich Wilhelm IV allowed him to transfer to Berlin. He was even given a supplement to his salary to help offset the higher costs of living in Berlin, and also to help him with his medical expenses.

As a member of the Prussian Academy of Sciences, he was entitled, but not obliged, to lecture at the University of Berlin. But due to his poor health, he did not lecture too often.

During the revolutionary year of 1848, Jacobi became involved in a political discussion in the Constitutional Club. During an impromptu speech he made some remarks which brought him under fire from monarchists and republicans alike. As a consequence, his request to become officially associated with the University of Berlin was dismissed by the ministry of education. Moreover, in June 1849, the bonus on his salary was removed, and he had to give up his house in Berlin. He moved into an inn and his wife and children took up residence in the small town of Gotha, where life was less expensive.

Toward the end of 1849, Jacobi was offered a professorship in Vienna. Only after he had accepted it did the Prussian government realize the severe blow to its reputation that would result from his departure. Special concessions from the ministry finally led him to change his decision. His family, however, remained in Gotha for another year, until his eldest son graduated from the gymnasium. Jacobi, who lectured on number theory in the summer term of 1850, joined his family during vacations and worked on a paper about astronomy with his friend Peter Andreas Hansen (1795-1874).

Early in 1851, after another visit to his family, Jacobi contracted influenza. Hardly recovered, he fell ill with smallpox and died on February 18.

Work

Jacobi made fundamental contributions to elliptic functions, dynamics, differential equations, determinants, number theory, and numerical analysis. It is therefore impossible to describe all his achievements in this short section.

One of his greatest accomplishments was his theory of elliptic functions and their relation to the elliptic theta function. He also made fundamental contributions in the study of differential equations and to classical mechanics, in particular to the Hamilton-Jacobi theory, which is an alternative formulation of classical mechanics, equivalent to other formulations such as Newton's laws of motion, Lagrangian and Hamiltonian mechanics. The Hamilton-Jacobi equation is also the only formulation of mechanics in which the motion of a particle can be represented as a wave. In this sense, it is considered as the closest approach of classical mechanics to quantum mechanics.

Among Jacobi's work in mathematical physics is the attraction of ellipsoids. Colin Maclaurin (1698-1746) had shown that a homogeneous liquid mass may be rotated uniformly about a fixed axis without change of shape if this shape is an ellipsoid of revolution. Jacobi discovered that an ellipsoid with three different axes may satisfy the conditions of equilibrium. He was also interested in planetary theory and other particular dynamical problems. While contributing to celestial mechanics, he introduced the Jacobi integral in 1836, which is the only known conserved quantity for the circular restricted three-body problem.

In 1834, he proved that if a single-valued function of one variable is doubly periodic then the ratio of the periods is non-real. This result prompted much further work in this area, in particular by Liouville and Cauchy.

Jacobi also worked on determinants and studied the functional determinant now called the Jacobian. But he was not the first in this study since Cauchy had already written a paper on this topic in 1815. However, Jacobi published a long memoir *De determinantibus functionalibus* (On functional determinants) in 1841 devoted to this determinant in which he proved that if a set of

n functions in n variables are functionally related, then the Jacobian is identically zero, while if the functions are independent, the Jacobian cannot be identically zero [1792].

His name is related to the iterative solution of a system of linear equations, rotations, and the computation of eigenvalues. He was one of the first to introduce and study the symmetric polynomials that are now known as Schur polynomials, which form a linear basis for the space of all symmetric polynomials [1793]. The orthogonal polynomials named after Jacobi were introduced by him, in connection with the solution of the hypergeometric equation, in notes found by Eduard Heine (1821-1881) and published posthumously [1798]. His name is also attached to the Jacobi matrix which is a symmetric tridiagonal matrix with positive non-diagonal entries, even though Jacobi did not use matrices.

Jacobi was also interested in the history of mathematics. In January 1846, he gave a public lecture about René Descartes (1596-1650).

10.39 ▪ Camille Jordan

Camille Jordan

Life

Marie Ennemond Camille Jordan was born in Lyon, France, on January 5, 1838. He is mostly known for his fundamental work in algebra, group theory, and his quite influential *Cours d'Analyse*.

His father Esprit-Alexandre Jordan (1800-1888) had been a student at the École Polytechnique and was an engineer and a member of the French parliament in 1871. Jordan's mother Joséphine (1816-1882) was the sister of the famous painter Pierre Puvis de Chavannes (1824-1898) who realized many large wall decorations such as in the public library of Boston, in the Panthéon and in the great amphitheater of the Sorbonne in Paris. Other members of the family were also quite well known. Jordan was educated in an intellectual environment.

Jordan studied at the Collège d'Oullins and then at the Lycée de Lyon. He entered in the first rank at the École Polytechnique in 1855. After that, he was a student at École des Mines in 1861 and then, as many other French mathematicians of his time, he worked as an engineer until 1885, first in Privas in the south of France, then in Chalon-sur-Saône, and finally in Paris, in charge of monitoring stone quarries.

On January 14, 1861, he defended his thesis for the rank of *Docteur ès Sciences*. As was usual for this diploma until quite recently, the work had two parts: the research topic of the candidate that was *Sur le nombre des valeurs des fonctions* (On the number of values of functions), and another one chosen by the committee *Sur les périodes des fonctions inverses des intégrales des différentielles algébriques* (On the periods of inverse functions of integrals of algebraic differentials). The members of the committee were Jean-Marie Duhamel (1797-1872), Joseph Alfred Serret (1819-1885), and Victor Puiseux (1820-1883) who suggested the second topic.

In 1862, Jordan married Marie-Isabelle Munet (1843-1918), the daughter of the deputy mayor of Lyon. They had eight children, two daughters and six sons.

From 1873 on, Jordan was an examiner at the École Polytechnique, and he became Professor of Analysis on November 25, 1876, replacing Charles Hermite (1822-1901), who wanted to devote himself entirely to lectures at the Sorbonne. From 1875, he was the substitute of Serret at the Collège de France, and, in 1883, he became the successor of Joseph Liouville (1809-1882). Jordan had a reputation for eccentric choices of notation.

The *Journal de Mathématiques Pures et Appliquées* was a leading mathematical journal and played a very significant part in the development of mathematics throughout the 19th century. It was founded in 1836 by Joseph Liouville. In 1885, after Liouville's death, Jordan became editor of the journal, a role he kept for over 35 years until he passed away.

In 1912, Jordan retired from his positions. The final years of his life were saddened. He lost one of his daughters in 1912, and, between 1914 and 1916, three of his sons and the elder of his grandsons were killed during World War I. His wife died in 1918. Of his three remaining sons, Camille was a government minister, Alexandre Édouard was a professor of history at the Sorbonne, and the third son, Joseph Louis, was an engineer.

Among the honors received by Jordan was his election to the Académie des Sciences on April 4, 1881. He became its president in 1916. On July 12, 1890, he became an officer of the *Légion d'Honneur*. He was elected a foreign member of the Royal Society in 1919, and was the Honorary President of the International Congress of Mathematicians in Strasbourg in September 1920.

Jordan liked classical literature and museums. He did not go to theaters or concerts, but was much interested in playing chess. He traveled to many countries, Algeria, Egypt, Palestine, and the United States, and visited all Europe several times. But his passion were the Alps, where he climbed, for pleasure, many not too hazardous summits. Other personal reminiscences about Jordan were given by Henri Lebesgue (1875-1941) in [2019].

Jordan died in Paris on January 22, 1922.

Work

Beside his professional work as an engineer, Jordan dedicated much time to mathematical research. His work on group theory done between 1860 and 1870 was described in a major book *Traité des Substitutions et des Équations Algébriques* (Treatise on substitutions and algebraic equations) [1840] that was published in 1870. This treatise gave a comprehensive study of Galois' theory. It was the first ever group theory book. For 30 years, it remained the bible of all experts in group theory. For this work, Jordan was awarded the Poncelet Prize of the Académie des Sciences. The treatise contains the *Jordan normal form* theorem for matrices, not over the complex numbers but over a finite field.

It appears that Jordan was not aware of earlier results of this type by Karl Theodor Wilhelm Weierstrass (1815-1897) [3202], and it was the beginning of a vigourous feud with Leopold Kronecker (1823-1891) on the organization of the theory of bilinear forms; for details, see [416, 419, 420, 421] and Section 4.6.

After the publication of that treatise, Jordan's fame spread beyond France, and foreign students were willing to attend his lectures. Felix Klein (1849-1825) and Sophus Lie (1842-1899) came to Paris in 1870 to study with Jordan, who, at that time, was already developing his research in a new direction, that is, the determination of all groups of movements in the three-dimensional space. This may have influenced Lie's theory of "continuous groups" and Klein's idea of "discontinuous groups."

In 1874, Jordan published a paper [1842] whose title can be translated as *Memoir on bilinear forms*. Within this framework, he described what can be considered the Singular Value Decomposition (SVD) of the corresponding matrix; for the early history of the SVD, see [2876] and Section 4.3.

In 1875, Jordan published a paper [1844] on geometry in n dimensions with the aim of generalizing results obtained for two and three dimensions. This paper was partly translated to English by Gilbert Wright Stewart, who wrote

> *It is no exaggeration to call this work a foray by a great mathematician into largely unexplored territory, and Jordan's geometric intuition is something to behold. One of the high points of his work is his unearthing of what today we call the canonical angles between subspaces.* [...] *Moreover, he describes a general orthogonalization procedure that includes the widely used Gram-Schmidt method.*

Between 1882 and 1887, Jordan published his three volumes *Cours d'Analyse de l'École Polytechnique*. As explained by Jean Dieudonné (1906-1992) in [897],

> *In analysis his conception of what a rigorous proof should be was far more exacting than that of most of his contemporaries; and his Cours d'Analyse, which was first published in the early 1880's and had a very widespread influence, set standards which remained unsurpassed for many years. Jordan took an active part in the movement which started modern analysis in the last decades of the nineteenth century: independently of Peano, he introduced a notion of exterior measure for arbitrary sets in a plane or in n-dimensional space. The concept of a function of bounded variation originated with him, and he proved that such a function is the difference of two increasing functions, a result which enabled him to extend the definition of the length of a curve* [...]

10.40 ▪ Stefan Kaczmarz

Life

Stefan Marian Kaczmarz was born on March 20, 1895, in Sambor, in the district of Lwów, Poland, at that time in Austria-Hungary, now Sambir, Ukraine.

He had a sister and two brothers. He spent his early life in Kety, in the Silesian foothills east of Bielsko-Biala, where he attended primary school between 1901 and 1905. Then he went to Wadowice, about 20 km east of Kety and 35 km south west of Kraków, to study at the National Junior High School from 1905 to 1907. Then, the family moved to Tarnów, about 35 km east of Kraków, where he joined the Second National Gymnasium from 1907 to 1913. He graduated with honors on June 17, 1913.

In September 1913, Kaczmarz registered at the Jagiellonian University in Kraków and began to study mathematics, physics, and chemistry. At that time the Polish school of mathematics included, among others, the famous mathematicians Stanisław Zaremba (1863-1942) and Ivan Vladislavovich Śleszyński (1854-1931).

Stefan Kaczmarz

But World War I began in August 1914 and Kaczmarz enlisted in the Polish Regiment on September 1, and went to the western front in the spring of 1915. He was involved in the Carpathian campaign until March 1917, was then transferred to the artillery and sent to the front later in 1917. Affected at the School of Artillery, he refused to swear an oath of allegiance to the emperor and was interned by the Austrian authorities. He escaped, but was captured and assigned to another internment camp.

Released at the end of March 1918, he returned to the Jagiellonian University where he also followed the lectures of Tadeusz Banachiewicz (1882-1954) and Franciszek Leja (1885-1979). However, he was not free from his military duties and he continued to serve until February 1919.

After completing his seven semesters at the University, he volunteered to join the Polish Army in July 1920, and was sent to the School of Artillery in Poznań where he graduated on November 5, 1920. Then, he returned to the Jagiellonian University to complete the qualifications for becoming a schoolteacher. On October 1, 1921, he was appointed as an assistant in the Department of Mathematics at the Academy of Mining in Kraków. He also taught some classes at a girls' school in Kraków.

Poland became, de facto, an independent nation after the armistice of November 11, 1918. The University of Lwów was renamed the Jan Kazimierz University in 1919. Stefan Banach (1892-1945) obtained his Habilitation there in 1922, and was appointed an extraordinary professor of mathematics in July. Hugo Steinhaus (1887-1872) was also teaching at this place. They wanted to build a major research center in mathematics and they offered a position to Kaczmarz. He arrived in Lwów in October 1923 where he stayed until September 1939. He undertook research with Stanisław Ruziewicz (1889-1941) for obtaining a doctorate that he defended on October 13, 1924. The title was *The relationships between certain functional and differential equations*. A paper where he answered a question raised by Banach followed. Then, he published several papers related to series of orthogonal functions and Fourier series.

At the end of 1931, Kaczmarz was awarded a scholarship by the National Culture Fund to visit Cambridge in England, and Göttingen in Germany. He spent the first half of 1932 with Godfrey Harold Hardy (1877-1947) and Raymond Edward Alan Christopher Paley (1907-1933) in Cambridge, where he attended lectures by Norbert Wiener (1994-1964) on Fourier transforms and their applications. He then spent two months in Göttingen before returning to Poland. He taught many different theoretical courses in Lwów as well as some in applied mathematics.

When World War II started, Kaczmarz was working on the curvature of railway tracks after a train derailed near Lwów due to a too high speed in a curve. On September 1, 1939, German troops invaded Poland. On the following day Kaczmarz was assigned to Warsaw as a lieutenant in the reserve. He left Lwów by train to Warsaw on September 3, in his military uniform, sent a postcard to his wife the next day, and disappeared. The circumstances of his death remain unknown.

Kaczmarz loved to walk in the Bieszczady mountains in the oriental part of the Karpathians. He was described as a calm, quiet, and modest person with rather moderate scientific ambitions. He received the Independence Medal on April 29, 1929, the Independence Cross in 1933, the Golden Cross of Merit for his contribution to science in 1937, and the Bronze Medal for Long Service to the University of Lwów in 1938.

Work

In 1935, Kaczmarz co-authored a book with Steinhaus on the theory of orthogonal series.

It was in 1937 that his paper on his iterative method for solving systems of linear equations appeared [1858]. At this time, the method did not obtain much recognition. It was rediscovered in 1970 by researchers working in imaging and named the *Algebraic Reconstruction Technique* (ART). The method is well suited for parallel computations and large-scale problems since each step only requires one row of the matrix (or several rows simultaneously in its block version) and no matrix-vector products. The convergence of Kaczmarz's original method was studied in [1291] and [2993]. For an impressive list of publications on this method, see [598].

10.41 ▪ Leopold Kronecker

Leopold Kronecker

Life

Leopold Kronecker was born on December 7, 1823, in Liegnitz, Prussia (now Legnica, Poland) into a wealthy Jewish family. He had private tutors until he entered the Gymnasium at Liegnitz, where he was taught, in particular, by Ernst Eduard Kummer (1810-1893), who immediately recognized his talent for mathematics and took him well beyond what would be expected at school.

Kronecker entered the University of Berlin in 1841. His interest was not immediately focused on mathematics, but rather spread over other matters including astronomy and philosophy. He spent the summer of 1843 at the University of Bonn studying astronomy and the winter 1843-1844 at the University of Breslau (now Wrocław in Poland) following Kummer who had been appointed there as a full professor in 1842. Back to Berlin for the winter semester of 1844-1845, he studied mathematics with Johann Peter Gustav Lejeune Dirichlet (1805-1859) under whose supervision he defended his dissertation on algebraic number theory in 1845.

After his degree, Kronecker went back to his hometown to manage a large farming estate built up by his mother's uncle, a former banker. In 1848, he married his cousin Fanny Prausnitzer. They had six children. For several years Kronecker focused on business and, although he continued to study mathematics as a hobby and corresponded with Kummer, he did not publish any mathematical results. In 1853, he wrote a memoir on the algebraic solvability of equations extending the work of Évariste Galois (1811-1832) on the theory of equations.

In 1855, Kronecker returned to Berlin where Dirichlet, whose wife Rebecka came from the wealthy Mendelssohn family, introduced him to the Berlin elite. Kronecker did not look for a university position, but he rather wanted to take part in the mathematical life of the university, interact with the other mathematicians, and undertake research. That same year, Kummer came back to Berlin to fill up the vacancy which occurred when Dirichlet left for Göttingen. Carl Wilhelm Borchardt (1817-1880) was also lecturing at the University since 1848, and, in late 1855, he took over the editorship of *Crelle's Journal* after the death of its founder August Leopold Crelle (1780-1855). Karl Theodor Wilhelm Weierstrass (1815-1897) joined them in June 1856 and Kronecker became his close friend.

Kronecker rapidly became quite active in research, publishing a large number of works in a short time. They were on number theory, elliptic functions, and algebra, but, more importantly, he explored the interconnections between these topics. Kummer proposed Kronecker for election to the Berlin Academy in 1860, and the proposal was supported by Borchardt and Weierstrass. On January 23, 1861, Kronecker was elected to the Academy. Since members of the Berlin Academy had the right to lecture at the University, Kummer suggested that Kronecker exercise this right. This is what he did beginning in October 1862. His lectures were very much related to his research, that is, number theory, the theory of equations, the theory of determinants, and the theory of integrals. Apparently Kronecker's lectures were very difficult. It is reported by Helmut Hasse (1898-1979) in [1596] that Dmitry Fyodorovich Selivanov (1855-1932), one of the circle of young mathematicians in Berlin, recounted *And when the lecture was over, everyone shouts "Wonderful" but had not understood a thing*; see also [1636].

After the death of Georg Friedrich Bernhard Riemann (1826-1866) Kronecker was offered the chair of mathematics at the University of Göttingen, but he refused, preferring to keep his position at the Academy. He did accept honors such as election to the Paris Academy in that year. For many years he enjoyed good relations with his colleagues in Berlin and elsewhere. However, these relations began to deteriorate in the 1870s. Kronecker believed in the reduction of all mathematics to arguments involving only the integers and a finite number of steps. He is well known for his remark *God created the integers, everything else is the work of man.* He was the first to doubt the significance of non-constructive existence proofs. It appears that, from the early 1870s, Kronecker was opposed to the use of irrational numbers, upper and lower limits, and the Bolzano-Weierstrass theorem, because of their non-constructive nature. Another consequence of his philosophy of mathematics was that, for him, transcendental numbers could not exist. For these reasons, he tried to persuade Heinrich Eduard Heine (1821-1881) to withdraw his 1870 paper on trigonometric series and he was also opposed to the publication of the work of Georg Ferdinand Ludwig Philipp Cantor (1845-1918).

Although Kronecker's views about mathematics were well known to his colleagues throughout the 1870s and 1880s, it was not until 1886 that he made these views public, arguing against the theory of irrational numbers. After the proof of the transcendence of π by Carl Louis Ferdinand von Lindemann (1852-1939) in 1882, Kronecker complimented him but claimed that nothing was proved since transcendental numbers did not exist. Another feature of Kronecker's personality was that he tended to personally quarrel with those with whom he mathematically disagreed.

When Kummer retired from the university, Kronecker succeeded him as an ordinary professor in 1883. He also became a co-director of the seminar. This position increased his influence in Berlin and his international fame. He was elected a foreign member of the Royal Society of London on January 31, 1884. He was also a very influential figure within German mathematics. By 1888, Weierstrass felt that he could no longer work with Kronecker in Berlin and decided to go to Switzerland. However, realizing that Kronecker would be in a stronger position to influence the choice of his own successor, he finally decided to stay in Berlin.

The *Deutsche Mathematiker-Vereinigung* (German Association of Mathematicians) was set up in 1890 and the first meeting of the society was organised in Halle in September 1891. Despite the antagonism between Cantor and Kronecker, Cantor invited him to give a talk at this first meeting as a sign of respect for one of the senior and most eminent figures in German mathematics. However, Kronecker never addressed the meeting since his wife had been seriously injured in a climbing accident in the summer and died on August 23, 1891. Some months later, Kronecker died on December 29, 1891, in Berlin. In the last year of his life, he converted to Christianity.

Kronecker was the supervisor of Leopold Bernhard Gegenbauer (1849-1903), Kurt Wilhelm Sebastian Hensel (1861-1941), Adolf Kneser (1862-1930), Mathias Lerch (1860-1922), Franz Mertens (1840-1927), and Conrad Frédéric Jules Molk (1857-1914), among others. Molk was a French mathematician born in 1857 in Strasbourg, at that time in France, which, after the French-German war of 1870 and the defeat of France, became a German town. He studied at the ETH in Zürich, at the Sorbonne in Paris, and in Berlin. On July 24, 1884, he defended his thesis based on certain ideas of Kronecker in Paris (and not in Berlin, and not under the supervision of Kronecker, as often erroneously mentioned, but only dedicated to him). He then held several academic positions in France and, in 1890, he was appointed to the chair of applied mathematics at the University of Nancy. From 1902 until his death in 1914, Molk was the editor-in-chief of the *Encyclopédie des Sciences Mathématiques Pures et Appliquées* in 22 volumes. It was a French translation with re-writing and additions by many mathematicians and theoretical physicists from France, Germany, and several other European countries of the monumental encyclopedia *Encyklopädie der mathematischen Wissenschaften mit Einschluss ihrer Anwendungen* due to Felix Christian Klein (1849-1925) and Wilhelm Franz Meyer (1856-1934) and published in six volumes (20,000 pages) from 1898 to 1933. This publication had considerable success and brought honors to Molk. Despite its wide coverage, it does not contain material on linear algebra.

Work

Kronecker's research mainly focused on number theory and algebra. In an 1853 paper on the theory of equations and Galois theory, he formulated the Kronecker-Weber theorem, but without giving a complete proof. In 1886, Heinrich Martin Weber (1842-1913) published a proof, but with some gaps and errors. The first complete proof was given by David Hilbert (1862-1943) in 1896.

Kronecker introduced the structure theorem for finitely generated Abelian groups and studied elliptic functions. In a 1850 paper, *On the solution of the general equation of the fifth degree*, he studied the quintic equation by applying group theory. In algebraic number theory, he introduced the theory of divisors as an alternative to theory of ideals introduced by Julius Wilhelm Richard

Dedekind (1831-1916), which he did not find acceptable for philosophical reasons. Kronecker also contributed to the concept of continuity, reconstructing the form of irrational numbers in real numbers. In analysis, he rejected the formulation by Weierstrass of a continuous, nowhere differentiable function.

In linear algebra, during his work on the evaluation of determinants, he introduced the symbol δ named after him, and that he denoted by δ_{rs}, for which $\delta_{rs} = 0$ when $r \neq s$, and $\delta_{rs} = 1$ when $r = s$, r and s indicating a row and a column of the determinant. This notation was popularized by Hensel who edited and published Kronecker's work between 1895 and 1931.

The matrix product $\otimes$ was attributed to Kronecker by Hensel, even though it seems that it was defined by Johann Georg Zehfuss (1832-1901) in 1858 in a paper about determinants; see [1636].

Kronecker's feud with Camille Jordan (1838-1922) about the transformation of couples of bilinear forms is described in Chapter 4.

10.42 ▪ Alexei N. Krylov

Alexei Nikolaevich Krylov

Life

Alexei Nikolaevich Krylov was born on August 15, 1863. According to his grandson Andrey Pretrovich Kapitsa (1931-2011), he was borm in Visyaga, a hamlet near the town of Alatyr of Ardatov district in the Simbirsk (now Ulyanovsk) province of Russia. From his memoirs [1956], we know that he was baptized seven days after his birth in the village of Lipovka, which is wrongly indicated by some sources as his birthplace.

His father was Nikolai Alexandrovich Krylov (1830-1911), a retired artillery officer who participated in the Crimean war (1853-1856), and his mother was Sofia Viktorovna Lyapunova (1842-1913), a member of the noble family Lyapunov. The now famous mathematician Aleksandr Mikhailovich Lyapunov (1857-1918), a first cousin of his mother, was Aleksei's first teacher in mathematics; see [3267]. Krylov's father was a native of the Alatyr District and his mother was from Kazan.

Krylov's father was a landowner of an estate and of 2,000 acres of land, who loved hunting and traveling a lot. He was elected chairman of the Alatyr District Council in 1866 and the family moved to Alatyr. Krylov was the only child of his parents but we will see later in which circumstances he had a half-brother. His mother's sisters Elizaveta and Alexandra lived with the family. According to Krylov, the latter taught him reading, writing, prayers, sacred history, and French. In his memoirs [1956], Krylov gave interesting recollections of life in rural Russia at the end of the 19th century.

In 1872, his father sold the house and the estate and the family moved to Marseilles in France. In his memoirs, Krylov wrote that this was done because his father was ill and the doctor advised him to move to the south of France. However, since 2003, we know the true story from Andrey Kapitsa, the grandson of Krylov. It turns out that Krylov's father had an affair with his sister-in-law Alexandra who became pregnant. The child would have been an illegitimate one but this was badly regarded in Russia in those times. In Marseilles, the baby boy was declared from unknown parents and given the name Victor Henri (1872-1940). Nikolai and his wife adopted the child. He later became a famous physical chemist and physiologist. His third marriage was with his cousin Vera Vasylevna Lyapunova with whom he had four children.

So Krylov went to school in Marseilles, 14 Cours Julien, not far away from the *Vieux Port* (the old harbor), where he improved his French. In 1873, being ill, he traveled to Algeria with his father. In the spring of 1874, the family moved back to Russia, first to Taganrog and then to Sevastopol, where he was admitted to the second class of the Sevastopol District School where he learned Russian, Latin, and German. In 1875, Krylov's father moved to Libau and then to Riga where Krylov was admitted as a full boarder in a private German school. In 1877, he entered the German classical gymnasium in Riga. At that time, interested in the Navy, he asked his father to be sent to the Naval School. In September 1878, he was admitted to the junior preparatory class of the Naval School in Saint Petersburg. It was the sixth school in which he was educated. It is there that he became interested in mathematics.

In October 1884, Krylov became a michman which was a junior officer rank. He was assigned to the 8th Navy Crew and then joined the Compass Section of the Chief Hydrographic Office where he met Ivan Petrovich de Collong, also spelled as Kollong (Jean Alexander Heinrich Clapier de Colongue (1839-1901), who was born in Dünaburg now in Latvia).

Krylov worked under the supervision of de Collong on the compass deviation problem. Ships' armors were made of iron which amplified the deviation of the compasses. Previous studies of this problem were made by Gauss and Poisson. Krylov studied Gauss' work (in this, his knowledge of Latin was quite useful) and based his analysis of the problem on Gauss' general equations rather than on Poisson's work. For details on this problem, see [3267]. *Calculation of gradation forces in the compass deflector* published in *Hydrography Transactions* was Krylov's first published paper. In 1886, he published his second paper *On the arrangement of needles on the compass card* in the *Naval Transactions*.

During those years Krylov also improved the dromoscope of the French naval engineer François Ernest Fournier (1842-1934). A dromoscope is a mechanical device reproducing the relationship between the changing magnetic field acting on the compass and its deviation with respect to the ship's course using coefficients computed by Archibald Smith (1813-1872), a Scottish mathematician, but improved by Krylov. His work on the dromoscope was published in 1887 in the French journal *Revue Maritime et Coloniale*.

In the spring of 1885, Krylov was transferred to the 4th Naval Crew in Kronstadt where he followed a course on mine matters. At the beginning of 1886, he moved to Saint Petersburg and joined the commission led by de Collong in charge of recalculating the Naval Department pension fund. He stayed there until September 1887 and studied probabilities. Then, he was assigned to the Franco-Russian Works shipyard for practical studies in shipbuilding.

In September 1888, Krylov began his postgraduate education at the Shipbuilding Section of the Nikolaev Naval Academy where he had Alexandr Nikolayevich Korkin (1837-1901), a former student of Pafnuty Lvovich Chebyshev (1821-1894), as a professor of differential and integral calculus. He finished his studies there in 1890 and, on Korkin's recommandation, received an appointment as staff instructor in mathematics of the Naval School and a docent of the Naval Academy. During these times he also attended lectures of mathematics at Saint Petersburg University. At the Naval School, Krylov taught ship theory.

In November 1891, Krylov married his second cousin Elizaveta Dmitrievna Dranitsyna (1868-1948) with whom he had five children. The first two died in childhood. Then they had two sons, Nikolai and Aleksei, and a girl Anna Alekseevna (1903-1996). His daughter Anna wrote about him

> *In my childhood memories, father is a tall, broad-shouldered man, with thick black hair and a bushy black beard. Creativity absorbed him and was part of his life. Aleksei Krylov was never idle. To distract himself from one work, he found another, but he was not an armchair scientist. He was always among people. A brilliant storyteller and a wit, he liked humorous, risqué anecdotes and jokes. In his youth he played tennis, rode a bicycle and was fond of target shooting.*

In 1898, Krylov wrote a report for the Naval Technical Committee about the Shipbuiding Department of the Berlin Technical School that he visited. This led to the creation of the Saint Petersburg Polytechnical Institute in 1899, in which Krylov gave lectures later on.

In 1899, Krylov became the first foreigner to receive a gold medal from the Royal Institution of Naval Architects in London.

On January 1, 1900, by order of the Department of the Navy, Krylov was assigned to manage the towing tank and model basin in Saint Petersburg. On April 3, 1902, Czar Nicolas II visited the tank and *deigned to be entirely pleased with both the performed experiments and the observed order of things, and expressed his Imperial gratitude to the Superintendent of the tank* as it is written in Krylov's service record.

In 1902, he travelled to Algeria and France on the cruiser Askold to do measurements of the vibrations of the ship. Then, he was working on the question of ship unsinkability. In 1904, he traveled to Italy.

Starting January 1, 1908, Krylov was appointed chief inspector of shipbuilding and, in September, chairman of the Naval Technical Committee. He resigned from the Committee in 1910 after disagreements with the minister Vice Admiral Voyevodsky (1839-1937). Then, Krylov continued as a full professor at the Naval Academy teaching ship theory and differential and integral calculus. In 1910, he traveled to Belgium, France, and Germany to study the design of diesel engines. Up to 1912, he gave lectures on theoretical mechanics at the Institute of Railway Engineers as a full professor. He also became a consultant for several companies. In 1912, he wrote a report for the State Duma (Parliament) about rebuilding the Russian Navy. He also joined the State Council of the Russian Society of Steam Navigation and Trade and was later elected to the Society Board.

After Ivan Konstantinovich Grigorovich (1853-1930) became Naval Minister, he appointed Krylov as "general at the disposal of the Naval Minister." In 1913, at the request of the minister, Krylov conducted studies of ship rolling on a rented German boat that sailed from Hamburg to England, the Azores, and Portugal. From Hamburg, Krylov went to Wiesbaden and Paris to visit some factories.

In 1913 and the beginning of 1914, Krylov studied Newton's *Principia*, which he intended to translate from Latin to Russian.

In 1914, World War I began. Since there were no more lectures at the Naval Academy, Krylov completed his translation of Newton's work which was published in 1915 and 1916 with his commentaries.

In 1914, Krylov and his wife Elizaveta separated because he had an affair with Anna Bogdanova Feringer, the manager of the library of the Main Physics Observatory. During the war,

Elizaveta became a nurse, and in 1919, she left the country with her daughter. After living in several countries they finally settled in Paris. In 1927, Krylov's daughter Anna married Pyotr Kapitsa (1894-1984), a Russian physicist who, at that time, was working in Cambridge with Ernest Rutherford (1871-1937). They had two sons, Sergey (1928-2012) and Andrey (1931-2011). Kapitsa won the Nobel Prize in 1978 for his work on low-temperature physics. In 1935, the Kapitsa family returned to the USSR. Sergey had three children.

In 1915, Krylov was appointed chairman of the board of Putilov Works, a company building ships and manufacturing ammunitions. The company was under the state's sequestration. Then, on March 1916, Krylov was elected an ordinary member of the Academy of Sciences and he resigned from Putilov Works.

In May 1916, Krylov was chosen as the new director of the Main Physics Observatory and as the head of the Chief Military Meteorological Office. He was released from the observatory in January 1917 and became director of the Institute of Physics.

Then, the revolution began. It is amazing that being a high-ranked officer of the Czar's Navy, Krylov did not have too many problems with the new government. Moreover, his sons, Aleksei and Nikolay, who were officers in the White Army of General Anton Ivanovich Denikin (1872-1947), were killed in 1918. There is nothing about that in Krylov's memoirs. But his daughter Anna provided some explanations. She wrote

> *My father was always situated outside of political events. For his class he was an extraordinarily strange person, accepting any government and especially not paying any attention. Now, I understand that Aleksey Nikolaevich looked at our government like at an earthquake, a flood, a thunderstorm. That is something which exists, but it is necessary to continue one's work. As a matter of fact, after the October Revolution my father remained completely calm as before in his same position, teaching at the same Naval Academy. And eventually, he was offered to head the Academy, which he accepted. Certainly, this was strange to the highest degree: it was 1918, Papa was a full tsarist general, and notwithstanding that fact, he quite composedly became the head of the Academy. And here, he had to give lectures in advanced mathematics to such a contingent of students, who according to my opinion basically did not know any mathematics. This was a young composition, and they were not officers. But he was a completely brilliant lecturer, and he surpassed that all, and more importantly, his students surpassed it all. He, strictly speaking, brought up these people. Aleksey Nikolaevich considered that on him was given the responsibility for the fate of the Russian fleet, and he had to do his job. For many years he worked abroad; he could have stayed there, but that thought never entered his mind.*

In 1919, Krylov became commanding officer of the Naval Academy for a year and a half. In 1921, the Academy of Sciences decided to send a commission abroad to resume the scientific contacts with foreign countries. They first went to Germany and then moved to England. Then, Krylov was assigned to buy ships for sending by boat to Russia 750 steam locomotives and tenders that had been bought in Germany by the government. In January 1922, he was appointed chief of the maritime department of the Russian Railway Mission in Berlin. He also had to go to Sweden for the same kind of business related to the transportation of locomotives.

After the transportation of locomotives was completed Krylov was appointed to the board of the Russo-Norwegian Society, located in London, which was set up to transport wood from Archangelsk to London. Krylov, who was in charge of building and purchasing steamers, had to travel quite often to Bergen in Norway where the timber ships were built.

Later on in 1926-1927, Krylov had to deal with the building of oil tankers in French shipyards in Nantes and Caen. For this, Krylov had to stay for three years in France. So, from 1921 to 1927 Krylov was abroad most of the time.

In January 1928, back in the USSR, Krylov resumed giving lectures at the Naval Academy. He was still the director of the Institute of Physics and Mathematics. In 1938, he received the Order of Lenin which was the highest award in the USSR. In 1941, he received the Stalin Prize for his work on compass deviation.

When World War II began, he evacuated to Kazan where he wrote his memoirs from August 20 to September 15, 1941; see [1956]. He had a stroke in 1942. After the war, he lived in Moscow. On his 80th birthday he received the title of Hero of the Socialist Labor. Until 1945, he lived in an apartment of the Institute of Physics. In 1945, he moved to Leningrad (formerly and now Saint Petersburg). He passed away on October 26, 1945. He is buried in the Volkovskoye Cemetary in Saint Petersburg.

Here are two interesting quotes from Krylov, taken from his memoirs and based on experience:

> *After many years of practice, I became convinced that if any nonsense had become a routine, the more absurd the nonsense was, the harder it was to get rid of it.*

> *It is well known that commission productivity is in inverse proportion to the number of its members.*

Work

Krylov developed the first general and complete theory for compass deviation. With the encouragement by Professor de Collong, the young midshipman published his first technical papers.

The main goal of Krylov was to apply mathematics to the problems related to shipbuilding. All along his life, Krylov was interested in studying the vibration of ships and also unsinkability. In fluid mechanics, he made significant contributions to the theory of ships moving in shallow water. He also improved Fourier's method for solving boundary value problems. In 1912, he studied the acceleration of the convergence of Fourier series. He wrote more than 300 papers and several books, for instance, *Lectures on Approximate Calculations* in 1911, *On some Differential Equations of Mathematical Physics Having Application to Technical Problems* in 1913, and *Vibration of Ships* in 1936. His *Collected Works* in 12 volumes were finished to be published by the Academy of Sciences of the USSR in 1956.

For our purposes in this book, we are interested in Krylov's 1931 paper [1955]. In this paper, whose title can be translated as *On the numerical solution of the equation by which, in technical matters, frequencies of small oscillations of material systems are determined*, Krylov introduced what are now called Krylov vectors. He was interested in the coefficients of the characteristic polynomial of a matrix. Rather than developing $\det(\lambda I - A)$, he reviewed the methods that were published in previous years and proposed his own.

Let us assume that we have a vector v of grade n with respect to A of order n and that we are looking for the coefficients of the characteristic polynomial p_c. We can write

$$\begin{pmatrix} v & Av & \cdots & A^{n-1}v & A^n v \\ 1 & \lambda & \cdots & \lambda^{n-1} & \lambda^n - p_c(\lambda) \end{pmatrix} \begin{pmatrix} -\alpha_0 \\ \vdots \\ -\alpha_{n-1} \\ 1 \end{pmatrix} = \begin{pmatrix} 0 \\ \vdots \\ 0 \\ 0 \end{pmatrix},$$

where the α_j's are the coefficients of the polynomial. The first row of this matrix relation comes from the Cayley-Hamilton theorem and the second row from the definition of the characteristic polynomial. To obtain a nonzero solution the determinant of the matrix on the left must be zero.

With a few manipulations we get

$$p_c(\lambda) = \frac{\det\begin{pmatrix} v & Av & \cdots & A^{n-1}v & A^n v \\ 1 & \lambda & \cdots & \lambda^{n-1} & \lambda^n \end{pmatrix}}{\det\,(\,v \quad Av \quad \cdots \quad A^{n-1}v\,)}.$$

The interest of this formulation is that λ appears only in the last row of the matrix in the numerator rather than in every row as in $\lambda I - A$. However, numerically, this method can only work for small values of n but Krylov was only interested in small problems. This method was described in matrix term by Felix Ruvimowich Gantmacher (1908-1964) [1281] in 1934; see his book [1282]. For a description of Krylov's method, see also Alston Scott Householder's book [1745], page 148.

Krylov was not the only mathematician to use a sequence $v, Av, A^2v, \ldots$; see, for instance, Turnbull and Aitken [3081], pages 43, 46-51. They used these vectors for theoretical purposes to obtain the Frobenius canonical form and the Jordan canonical form of a matrix.

The term *Krylov sequence* appeared in a paper by Householder and Friedrich Ludwig Bauer [1752, p. 33] in 1959.

10.43 ▪ Vera Kublanovskaya

Vera Nikolaevna Kublanovskaya

Life

Vera Nikolaevna Kublanovskaya (née Totubalina) was born on November 21, 1920 in Krokhono, a small village in Vologda Oblast (USSR), on the bank of lake Beloye, where the river Sheskna had its source. She was one of the nine siblings of a farming and fishing family. Her native village was destroyed in the 1960s when the level of the lake was raised by the construction of a dam.

She went to school in Belozersk, 18 km from her birthplace, about 430 km east of Leningrad (before and now Saint Petersburg) and 470 km north of Moscow. After secondary school, she took a course for becoming a primary school teacher.

In 1939, after graduating with honors, she went to the Gerzen Pedagogical Institute in Leningrad. She was encouraged to pursue a career in mathematics by Dmitry Konstantinovich Faddeev (1907-1989). But World War II interrupted her studies and she had to go back to Krokhono because her mother was seriously ill. She worked there as a teacher of physics and physical culture.

In 1945, after a longer stay in her village due the siege of Leningrad, she wrote to Faddeev, who was the dean of the Faculty of Mathematics and Mechanics, and, thanks to his recommendation, she was accepted at Leningrad State University without having to pass the entrance exam. Kublanovskaya graduated in 1948 and joined the Steklov Mathematical Institute of the USSR Academy of Sciences (LOMI) in Leningrad as a junior researcher. She stayed there for her entire life.

She had contacts with Leonid Vitalyevich Kantorovich (1912-1986) who, after being one of the pioneers of linear programming and economics (for which he obtained a Nobel Prize in 1975), shifted back to mathematics, became interested in computational problems, and was involved in secret research on the Soviet nuclear weapons. Vera worked with him on this project in a secret Leningrad laboratory until 1955. For this reason, she was not allowed to travel abroad for a long time.

She got married and had two sons. In addition to her family and professional duties, Kublanovskaya undertook research for a candidate's degree (equivalent to a Ph.D.). She received it in 1955 for a work entitled *The application of analytic continuation in numerical analysis.* Following this, she was promoted to a senior researcher position at the Steklov Mathematical Institute.

Then, after the end of these secret works, Kantorovich organized a research group developing PRORAB, a universal computer language for BESM (*Bolshaya Elektronno-Schetnaya Mashina*), the first electronic computer constructed in the country in 1954. PRORAB could input an encoded computational task, decode it, and then execute it. Kublanovskaya worked in the group headed by Vera Nikolaevna Faddeeva (1906-1983), Faddeev's wife. Kublanovskaya was involved in the implementation of linear algebra algorithms in PRORAB.

In 1974, Kublanovskaya defended her Habilitation entitled *The use of orthogonal transformations in solving problems of algebra.* From 1968, she was teaching at the Leningrad Shipbuilding Institute (now the Saint Petersburg State Marine Technical University), where she was promoted to full professor in 1976. In October 1985, she was awarded an honorary doctorate from the University of Umeå in Sweden.

Vera Nikolaevna Kublanovskaya died on February 21, 2012.

Until her last days, she could not imagine life without mathematics. She devoted much time to her students, and her lessons inspired many of them. She supervised many undergraduate and graduate theses. She was one of the editors of *Zapiski Nauchnykh Seminarov LOMI*, and she supervised the publication of the series *Numerical Methods and Algorithms.* She authored around 120 papers, and when she was allowed to travel, she attended many international meetings and gave talks in the United Kingdom, Czechoslovakia, Hungary, the USA, Switzerland, and Sweden. Although she was a recognized scientist, she always remained modest, and enjoyed the esteem and affection of her colleagues.

More details on her life and works can be found in [59, 1856, 1935].

Work

An extended version of Kublanovskaya's doctoral work was published in 1959 [1958]. It contained applications of the method of analytic continuation to the solution of various problems in numerical analysis: solution of systems of linear equations, determination of the eigenpairs of a matrix, integration of differential equations by series expansion, finite differences for solving the Dirichlet problem, and solution of integral equations. She made numerical experiments, analyzed them, and discussed the accuracy that can be achieved.

Kublanovskaya's most famous work was published in three papers in 1960-1962; see [1959, 1960]. Inspired by the LR algorithm of Heinz Rutishauser (1918-1970), she developed, almost simultaneously with John Guy Figgis Francis, the QR algorithm (named *the method of one-side rotations* in the Russian literature) for computing the eigenvalues of a general matrix. In fact, she was using an LQ factorization, with L lower triangular and Q orthonormal. She gave a proof of convergence based on a determinantal theory [1393]. She did some computations with mechanical calculators but, apparently, her method was never used on a computer; see [1393].

She never came back to the QR algorithm later, but worked on numerical methods for spectral problems, such as generalized eigenvalue problems, matrix pencils, and, in particular, spectral problems for matrices with polynomial and rational occurrences of the parameters which led to a 202-page survey paper in 1997.

She applied these methods to inverse eigenvalue problems, nonlinear systems of algebraic equations in several variables, construction of irreducible factorizations, and methods for solving ill-conditioned rectangular linear systems. In 1966, she presented a method for computing the Jordan structure of a multiple eigenvalue by unitary transformations.

10.44 ▪ Joseph-Louis Lagrange

Joseph-Louis Lagrange
Collections École Polytechnique

Life

The names of three cities are attached to the life of Joseph-Louis Lagrange: Turin, Berlin, and Paris. He was born in Turin, the first born of 11 children, at number 29 of the street that today bears his name, on January 5, 1736, as Giuseppe Luigi Lagrangia or Giuseppe Lodovico De la Grange Tournier (and not Ludovico as it can often be found). Let us mention that Lagrange himself used several spellings of his name.

At that time, Turin was the capital of the kingdom of Piedmont-Sardinia, and thus, Lagrange was an Italian citizen, but later he obtained the French citizenship. His mother was Italian but his father was of French origin. He wanted his son to become a lawyer. Lagrange's father was wealthy but he lost most of his money in some hazardous business dealings.

Lagrange first studied at the College of Turin and his favourite subject was classical Latin. He read Cicero and Virgil. He was not much attracted by mathematics, but he found Greek geometry rather exciting. His interest in mathematics began when he read a copy of the work by the English astronomer Edmund Halley (1656-1742) [1551] published in 1693 on the use of algebra in optics. He was also attracted to physics by the excellent teaching of Giovanni Battista Beccaria (1716-1781), and he decided to devote himself to mathematics.

Charles Emmanuel III (1701-1773), the Duke of Savoy and King of Sardinia, appointed Lagrange as the *Sostituto del Maestro di Matematica* (mathematics assistant professor) at the Royal Military Academy of the Theory and Practice of Artillery on September 28, 1755. There, he taught courses in calculus and mechanics, and was the first to teach calculus in an engineering school. According to the Academy's military commander, Lagrange unfortunately was a problematic professor with his careless teaching style and his abstract reasoning.

Lagrange wrote several letters to Leonhard Euler (1707-1783) between 1754 and 1756 describing his mathematical results, in particular those obtained by applying the calculus of variations to mechanics. These results generalized those of Euler who immediately contacted Pierre Louis Moreau de Maupertuis (1698-1759), the president of the Berlin Academy and the discoverer of the *least action principle*. They offered Lagrange a position in Prussia but he declined the proposal. However, he was elected to the Berlin Academy on September 2, 1756.

In 1758, with the help of the chemist Giuseppe Saluzzo di Monesiglio (1734-1810) and the anatomist Gian Francesco Cigna (1734-1790), Lagrange founded a scientific society, which later became the Turin Academy of Sciences [545, 1593]. Most of his early writings were published in the five volumes of its transactions, named *Actes de la Société privée*, now usually known as the *Miscellanea Taurinensia*. In these papers, Lagrange made significant contributions to physical problems, analysis, number theory, and celestial mechanics. In 1759, Euler succeeded in having him nominated as an associate to the Berlin Academy.

In November 1763, Lagrange made a six-week visit to Paris where he met Jean Le Rond d'Alembert (1717-1783), Alexis Clairaut (1713-1765), Nicolas de Condorcet (1743-1794), and others but, unfortunately, he was seriously ill and had to go back to Turin. In 1765, d'Alembert, the editor with Denis Diderot (1713-1784) of the *Encyclopédie ou Dictionnaire raisonné des sciences, des arts et des métiers*, arranged for Lagrange to be offered a position at the Berlin Academy. But he again refused, writing

> *It seems to me that Berlin would not be at all suitable for me while M. Euler is there.*

By March 1766, d'Alembert knew that Euler was leaving Berlin for Saint Petersburg, and he wrote again to Lagrange to encourage him to accept a position in Berlin. After Euler left, Frederick II of Prussia (Frederick the Great) himself wrote to Lagrange expressing the wish of "the greatest king in Europe" to have "the greatest mathematician in Europe" resident at his court. Lagrange finally accepted the offer.

In 1767, he married his cousin Vittoria Conti but he did not wish to have children. As he wrote to d'Alembert, he had no taste for marriage, and would have avoided it altogether if not for his need of a nurse and housekeeper to provide him with the quiet conveniences necessary for a life of research [2686].

Lagrange's health was rather poor, and moreover, his wife died in 1783 after years of illness. His friend, d'Alembert, died that same year. Lagrange was very depressed by the events. In 1786, Frederick II died, and Lagrange decided to leave Berlin. Many offers came to him and he accepted the offer of the French King Louis XVI. On May 18, 1787, he left Berlin.

In France, he was received with many marks of distinction, he became a member of the French Académie des Sciences, and special apartments were prepared for him in the Louvre palace. But his enthusiasm was extinguished. It seems that he had lost his taste for mathematical research, and he became interested by metaphysics, the history of religions, general language theory, and botanics. However, soon after his arrival, the French revolution began. He was much stressed by the Reign of Terror and the arrest and even the execution of some scientists (in particular, the chemist Antoine Laurent Lavoisier (1743-1794) who intervened in his favor).

On March 26, 1791, the *meter* was instituted as the unit measure by the Assemblée Nationale (French Parliament), and in April the Commission des poids et des mesures (Commission of Weights and Measures) was created with Lagrange as a member. He promoted the official adoption by the French government of the metric system, including the use of decimals.

It was about 1792 that Lagrange met Renée-Françoise-Adélaïde Le Monnier (1767-1833), the daughter of his friend, the astronomer Pierre Charles Le Monnier (1715-1799). She insisted on marrying him and was a devoted wife to whom he became warmly attached. The École Centrale des Travaux Publics was founded on September 28, 1794, and, one year later, became École Polytechnique. Lagrange was a professor there [255].

After the Revolution, Lagrange fell into the favor of Napoléon Bonaparte, who enjoyed sharing geometric puzzles with him and Laplace. Appointed senator in 1799, he proposed and was the first signatory of the *Sénatus-consulte*, which, in 1802, annexed his fatherland Piedmont to France. As a consequence, he was given French citizenship [2479]. Napoléon named him to the *Légion d'Honneur* and Count of the Empire in 1808.

In 1810, he began a thorough revision of his *Mécanique analytique*, but he was able to complete only about two-thirds of it before his death in Paris on April 10, 1813, at 128 rue du Faubourg Saint-Honoré. Napoléon honored Lagrange with the *Grand Croix de l'Ordre Impérial de la Réunion* just two days before he died. He was buried that same year in the Panthéon in Paris.

According to Jean-Baptiste Joseph Delambre (1749-1822) in a note written for the first volume of Lagrange *Complete Works* [860],

> [He was] *Gentle, and even timid in conversation, he particularly liked questioning, either to make others stand out or to add their thoughts to his vast knowledge. When he spoke, he was always doubtful, and his first sentence usually started with "I don't know". He respected all opinions, was far from giving his own for rules; it was not that it would be easy to change them, and that he did not sometimes defend them with a warmth which increased until he noticed some alteration in himself; then he returned to his usual tranquility.*

The journal *Lettera Matematica*,[74] volume 2, issue 1-2, June 2014, is entirely devoted to Lagrange. It contains 15 papers; see [255, 545, 546, 2479].

Contemporary testimonies were written by Pietro Cossali (1748-1815) [748] and Delambre [860]. For more about Lagrange, see also [1928].

Work

On July 23, 1754, Lagrange published, under the name Luigi De la Grange Tournier, his first mathematical work as a letter written in Italian to Giulio Carlo da Fagnano (1682-1766). This work showed that Lagrange was working alone without the advice of a mathematical supervisor. The paper drew an analogy between the binomial theorem and the successive derivatives of the product of functions. Before this publication, Lagrange had sent the results to Euler. The month after the paper appeared, Lagrange found similar results in the correspondence between Jean Bernoulli (1667-1748) and Gottfried Wilhelm Leibniz (1646-1716). He was greatly upset by this discovery since he feared being branded a cheat who copies the results of others.

Wanting to produce interesting new results, he began to work on the tautochrone, which is a curve located in a vertical plane, where the time taken by a particle sliding along it under the uniform influence of gravity to its lowest point is independent of its starting point. His

[74] https://link.springer.com/journal/40329/volumes-and-issues/2-1 (accessed March 2022).

results contained his method of maxima and minima, which would contribute substantially to the new subject of the calculus of variations after Euler called it so in 1766. On August 12, 1755, Lagrange sent to Euler his results on the tautochrone and Euler replied on September 6, saying he was impressed by these new ideas.

In 1759, Lagrange [1968] was concerned with determining sufficient conditions for a stationary point of a function of several variables to be a minimum or a maximum. He first considered the case of one and two variables and derived the usual conditions. Then, he reduced the problem for three variables to the two-variable problem he solved before. He observed that it can be generalized to any number of variables and his method amounts to a recursive derivation of Gaussian elimination.

Lagrange was most productive when he was in Berlin. He worked on astronomy, mechanics, dynamics, fluid mechanics, probability, number theory, and the foundations of the calculus. He obtained an interesting result on the solution of Pell's equation. In 1770, he published his treatise *Réflexions sur la Résolution Algébrique des Équations* (Reflections on the algebraic solution of equations) in which he investigated why equations of degrees up to four could be solved by radicals. It was a first step towards group theory.

In 1773, Lagrange considered the problem of the rotation of a rigid body. The relations given by Lagrange amounts to writing $\det(A^T A) = [\det(A)]^2$ for a matrix A of order 3. In this work, we can see what we would consider today a dot product of vectors.

In 1776, he published a paper on the use of continued fractions in integral calculus in which he proposed a general method for obtaining the expansion into a continued fraction of the solution of a differential equation in one variable. He wrote that, since the form of these continued fractions does not allow to manipulate them easily, he reduced them into ordinary fractions, and he explained that these rational functions are exact up to the degree of the variable that is the sum of the degrees of the numerator and the denominator inclusively. This is the birth certificate of Padé approximants. Let us mention that the direct approach to Padé approximants was proposed by Johan Heinrich Lambert (1728-1777) in 1758 [433]. Each year, Lagrange sent the volume of the Memoirs of the Berlin Academy to d'Alembert. On December 12, 1778, when sending the 1776 volume, he wrote (our translation from French) *as usual, there is something from me, but nothing that can deserve your attention*. He was too modest.

Lagrange also decided to collect his works on mechanics, which would lead him to his *Traité de Méchanique Analitique* [sic] (Treatise on analytical mechanics). It was in this book that he introduced what is now known as the *Lagrange multipliers*. In France, Lagrange's treatise on mechanics was published in 1788. It summarized all the work done since the time of Isaac Newton (1643-1727), and promoted the use of the theory of differential equations. With this work, Lagrange transformed mechanics into a branch of mathematical analysis. After completing this work, Lagrange was exhausted.

The famous Lagrange formula for the interpolation polynomial can be found in his *Leçons Élémentaires sur les Mathématiques* given in the École Polytechnique (Œuvres, vol. 7, p. 286). In 1797, he published the first theory of functions of a real variable in *Théorie des Fonctions Analytiques* (Theory of analytic functions) although he failed to give enough attention to matters of convergence.

10.45 ▪ Edmond Laguerre

Life

Edmond Nicolas Laguerre was born on rue Rousseau in Bar-le-Duc, a historical city of the department of Meuse, France, on April 9, 1834. He was the son of Jacques Nicolas Laguerre (1796-1873), a hardware merchant, and his wife Christine Werly (1805-1885).

Edmond Nicolas Laguerre

He did his studies in various public establishments, his parents having successively placed him at the Stanislas college, at the *lycée* of Metz and at the Barbet institution so that he always had a comrade with him to watch over his already precarious health. He displayed a rare intelligence with a pronounced taste for languages and mathematics. His first work on the use of imaginary numbers in geometry dates back to 1851 and 1852. His first paper appeared in 1853 in the *Nouvelles Annales de Mathématiques* directed by Olry Terquem (1782-1862) who then noted

> *Profound investigator in geometry and analysis, the young Laguerre possessed an excessively rare spirit of abstraction, and it will not be too much to encourage the work of this man of the future.*

In this work, Laguerre gave the complete solution of the problem of the homogeneous transformation of angular relations, thus complementing and improving the work of Jean-Victor Poncelet (1788-1867) and Michel Chasles (1793-1880).

On November 1, 1853, he entered the 4th out of 110 at École Polytechnique (simply denoted below as Polytechnique). According to his description, he was 1.685 m tall, had light brown hair and eyebrows, a high forehead, a medium nose, gray blue eyes, a wide mouth, a round chin, a long face. He was nearsighted and has a mark near his left ear.

His professors were Jean-Marie Constant Duhamel (1797-1872) and Jacques Charles François Sturm (1803-1855) for analysis and Jules de La Gournerie (1814-1883) for geometry. During the school year 1853-1854, when occupying the position of *sergent-fourrier*, his teachers made the following observations on his work:

> *Diligent work but could be better regulated. Special question marks: consistently good or very good in analysis; at first very good but constantly decreasing since the beginning of the semester in descriptive geometry; too variable in physics; very good at chemistry. Notes of general questions: mediocre in analysis; very good in descriptive geometry.*

For the second semester, we find: *Good or fairly good results in all parts, but less satisfactory in general than those of the first semester.* In fact, he was second in the ranking for the first semester and 24th in the second one. As for his behavior, the reviews were less favorable: *Rather good behavior. Bad outfit. Weak and noisy student.* He received several punishments for bad demeanor, chatter, and singing during the study. He passed in his second year 59th out of 106. In 1854-1855, he was judged as follows: *Sustained work. Generally good or very good marks in analysis, mechanics and physics; very mediocre in chemistry*. His behavior and demeanor were passable. On the other hand he was always *very talkative and very careless* and, obviously he *could have done much better*. He was punished by two days in the police room for having *lit a fire in the study room*. He left Polytechnique 46th out of 94 with the following appreciations:

> *This very intelligent student could have stayed in the top of his class, but he did not work, Extremely dissipated. Must and can very well do at the application School.*

His output ranking prevented him from entering into civil careers. He entered 7th out of 41 in the military school École Impériale d'Application de l'Artillerie et du Génie in Metz on May 1, 1855. He did not seem to be more attentive than at Polytechnique:

> *Good behavior but has often been punished for delays in his work. Good outfit, but not very military. Has resources for mathematics, but has no taste for graphical works, draws poorly and slowly. Gave too much attention to subjects extraneous to school studies. He is the officer who is most late in his work. Speaks a little Italian.*

He left the school the 32th out of 40 and the Inspector General noted

> *Has lost a lot of ranks because, without being lazy, he took care of things extraneous to the work of the School. It is a fault which he can correct himself.*

On leaving the artillery school, Laguerre began a military career. He was lieutenant en second (second lieutenant) at the 3rd foot artillery regiment on December 6, 1856, then lieutenant on May 1, 1857. On March 13, 1863, he was appointed Capitaine (captain) and was employed as an assistant, at a weapons factory in Mutzig. On June 18, 1864, he gave up this job to become a teaching assistant for the course of descriptive geometry at Polytechnique.

On August 17, 1869, he married Marie Hermine Albrecht (1845-1934), daughter of Julie Caroline Durant de Mareuil, widow of Leopold Just Albrecht, deceased, owner, residing at chateau d'Aÿ in the department of Marne. His wife received a dowry of 24,000 francs in registered shares producing 1,200 francs of income. From this marriage two daughters were born.

At that time, he lived 3 rue Corneille in Paris, and later he moved to 61 boulevard Saint Michel. In November 1869, he was authorized to teach a course in superior geometry at the Sorbonne. During the siege of Paris in 1870, he was first appointed, on August 28, by Géneral Juste-Frédéric Riffault (1814-1895), to command in second the rampart battery, known as the École Polytechnique one. On November 12, he was assigned to the head of the 13th battery of the artillery regiment and took part, in that capacity, in the two battles at Champigny on November 30 and December 2, 1870. For his conduct, he was made a Chevalier de la Légion d'Honneur (Knight of the Legion of Honor). During the Paris insurrection, he

> *retained until March 27 the command of the men who remained in the battery, partially dismissed on March 14. After the forced dissolution of the battery, he joined in Tours the École Polytechnique where he had been reinstalled.*

After these events, he resumed his teaching at Polytechnique as well as his scientific work. On November 25, 1873, he was appointed lecturer for the course of analysis at Polytechnique and admission examiner on May 4, 1874, positions he held until his death. On May 31, 1877, he was raised to the rank of Chef d'Escadron (Commander). He was *very much loved and esteemed* at Polytechnique. In 1880, the Inspector General noted in his file

> *An excellent repeater of analysis, Commander Laguerre occupies a distinguished rank among our young geometers and he has a bright future ahead of him as a scholar.*

He had already published 114 articles by then. On July 5, 1882, he was made an Officier (Officer) of the Légion d'Honneur. In order to be able to devote himself entirely to his work, he took an early retirement on June 2, 1883.

On May 2, 1885, he was elected to the Academy of Sciences thanks to the action of Marie Ennemond Camille Jordan (1838-1922), who he had known when they were both students at Polytechnique. Shortly after Joseph Louis François Bertrand (1822-1900) entrusted him as deputy of the Chair of Mathematical Physics at Collège de France. He gave a very remarkable course there on the attraction of ellipsoids.

His already weak health and a continual fever forced him to give up all his occupations. He returned to Bar-le-Duc at the end of February 1886. Laguerre died on August 14, 1886, at 52 rue Tribel. Georges Henri Halphen (1844-1889) represented the Academy at his funeral and said a few words after reading a speech by the mathematician Joseph Louis François Bertrand (1822-1900) who was the brother-in-law of Charles Hermite (1822-1901).

This biography was written (in French) at the occasion of the International Congress *Polynômes Orthogonaux et Applications* held in Bar-le-Duc, October 15-18, 1984, on the occasion of the 150th anniversary of Laguerre's birth [431].

Work

The works of Laguerre cover various topics, which were analyzed by Eugène Rouché (1832-1910) [2609] in 1887, and by Jean Dieudonné (1906-1992) [896] in 1985 who gave a more contemporary and deeper review of them. It is interesting to compare these two texts together and also with that of Henri Poincaré (1854-1912) [2503] in 1887 and the notice [1979] Laguerre himself wrote in 1884 when he applied for the Academy of Sciences. These four documents are in French.

Laguerre was one of the founders of modern geometry. He first represented in a concrete way the imaginary points of the plane and the space. Then, he understood the important role of the area of the spherical triangle in spherical geometry, and extended the theory of foci to all algebraic curves. He pointed out several new properties of curves and anallagmatic surfaces (that is, whose form does not change by inversion), studied geodesic lines and the curvature of anallagmatic surfaces, extended Poncelet's closure theorem on the inscription of polygons into conics to hyperelliptic functions, and Joachimstahl's theorem on conics to second-order surfaces.

He gave an interpretation of homogeneous forms, and imagined two new systems of coordinates, one of which highlighting the tangents that can be drawn to a curve from an external point. He studied steering geometry, and Monge's method for representing three-dimensional objects in two dimensions.

Besides geometry, Laguerre developed, in a memoir [1976] published in 1867 in the *Journal de l'École Polytechnique*, all the essential points of the theory of linear substitutions. This made him one of the fathers of matrix theory; see Chapter 1.

In 1879, he examined a particular definite integral and expressed it as an asymptotic series. Then, he transformed it into a continued fraction the convergents of which involved the so-called *Laguerre polynomials*, and he proved their orthogonality property. This is the work he is best known for. Let us mention that other scholars were also led to the discovery of orthogonal polynomials via continued fractions. This is not surprising since, in both topics, a three-term recurrence relationship is involved.

In 1880, Laguerre proposed a numerical method for computing the zeros of a polynomial. It is almost guaranteed to always converge to some zero, whatever the initial guess is. Finally, let us cite his application of the principle of the last multiplier, which accomplishes, in all cases where the integration of a system of differential equations of motion is reduced to a first order differential equation of two variables, the integration of this last equation by giving its multipliers.

10.46 ▪ Cornelius Lanczos

Cornelius Lanczos
Courtesy of University of Manchester

Life

Cornelius Lanczos was born Kornél Löwy on February 2, 1893, in Székesfehérvár (Hungary), a city located about 60 miles southwest of Budapest. He was the eldest of the five children of a Jewish lawyer, Károly Löwy (1853-1939), who married Adél Hahn (1868-1944) in 1886. His father was for 23 years the president of the Székesfehérvár Bar as well as the president of the city's Neolog Jewish community. Lanczos had two brothers, Andrew and George, and two sisters, Anna and Gizi.

Lanczos' father changed the name of his children in 1906 in the process of Hungarization of the surnames. Lanczos attended a Jewish elementary school where the courses were taught in German but he also learned Hungarian and Hebrew. Then, he went to a Catholic secondary school (Cistercian Catholic gymnasium) where he learned Latin, Greek, and mathematics and from which he graduated in 1910.

The next step was the University of Budapest in the Faculty of Arts starting in September 1911. His physics teacher there was Baron Lorànd von Eötvös (1848-1919) who interested him in the theory of relativity and his mathematics teacher was Lipót Fejér (1880-1959) who was already a well known mathematician.

Lanczos graduated in 1915 and was then assistant to Károly Tangl (1869-1940), a professor of experimental physics at the Technical University in Budapest. His Ph.D. thesis that he obtained in 1921 under Prof. Rudolf Ortvay (1885-1945), at the university of Szeged, was about the theory of relativity with the title *The function theoretical relation to the Maxwellian aetherequation*. He sent his manuscript, written in German, to Albert Einstein (1879-1955) in 1919 from whom he received encouragement. In his thesis Lanczos made use of the quaternions of Sir William Rowan Hamilton (1805-1865); see Section 10.31.

Then, in 1922, the political situation in Hungary led him to move to Germany at the University of Freiburg where he spent three years continuing working in physics. He moved to Frankfurt in 1924 where he became the assistant to Erwin Madelung (1881-1972). During this period he married a German woman, Maria Elisabeth Rump (1895-1939), in 1927, and he spent a year (1928-1929) as an assistant to Einstein in Berlin. However, it seems there were some disagreements between Einstein and Lanczos concerning the problems he was supposed to work on. After a year Lanczos left, but he maintained correspondence and friendship with Einstein.

In the fall of 1929, Lanczos returned to Frankfurt. In 1931, he spent a year as a visiting professor at Purdue University in the USA at the invitation of Karl Lark-Horovitz (1892-1958). But Lanczos was not very happy with his life in West Lafayette. His wife, because of her illness, remained with her husband's parents in Székesfehérvár. In 1932, because of the economic and political troubles in Germany, he returned to the United States. He got a professorship at Purdue University where he stayed until 1946. Unfortunately, he had to travel without his wife who contracted tuberculosis soon after their wedding and could not get a visa. Lanczos traveled back to Europe every six months. In 1938, Lanczos obtained U.S. citizenship. But the political situation was getting worse in Europe. In May 1939, he went to Hungary again and he wrote to some friends

> *At the last minute great difficulties have arisen against the contemplated journey of my family. My wife is so sick again that her departure had to be postponed. I am in very bad shape* [...] *My wife has finally consented that the little boy shall come with me to America even if she is not able to join. I consider that quite imperative in view of the desperate future that awaits Europe.*

Lanczos brought his son Elmar (1933-1980) to the USA in August 1939 and his wife died from tuberculosis in October 1939.

Until 1937, Lanczos' papers were on physics and written in German. After that, his papers were written in English. In 1938, he published his first paper in numerical analysis about trigonometric interpolation of functions. However, he had already published a small note in the Bulletin of the AMS in 1935 about a new approximation method in solving linear differential equations with non-oscillating coefficients.

Most of the members of Lanczos' family died in concentration camps during World War II; only his aunt and a nephew survived. During this period he worked on Einstein's field equations, Dirac's equation, but began to be more and more interested in mathematical techniques, particularly for computations with mechanical calculators. In 1940, he published with Gordon Charles Danielson (1912-1983) a method for quickly evaluating Fourier coefficients that can be considered one of the ancestors of the FFT algorithm of James William Cooley (1926-2016) and John Wilder Tukey (1915-2000), published in 1965 [738]. In 1942, Lanczos was working in the aeronautical section of the Mechanical Department at Purdue University where he was in relation with Boeing.

During 1943-1944, Lanczos was associated with the National Bureau of Standards (NBS) for the Mathematical Tables Project. In 1945, he was offered a permanent position at Boeing Aircraft Company in Seattle where he stayed from 1946 to 1949. His work was to apply mathematics to airplane design. This is when he became interested in the computation of eigenvalues of matrices. In 1947, he also gave lectures at the University of Washington in Seattle on Fourier series. In January 1949, he joined the NBS which had founded in 1947 an Institute for Numerical Analysis (INA) on the UCLA campus in Los Angeles; see Chapter 5. There Lanczos turned his attention to the solution of linear systems and matrix eigenvalue problems in which he had been interested when working at Boeing. He investigated his method of "minimized iterations," a continuation of some work he began at Boeing. In the INA quarterly reports of 1953 Lanczos is listed as "on leave of absence during fiscal year 1953."

In 1952, Lanczos was invited to the Dublin Institute for Advanced Studies by the physicist Erwin Schrödinger (1887-1961). Then, he returned to the USA and was, for half a year in 1953, working for the North American Aviation company in Los Angeles. Around that time, during the McCarthyism era, Lanczos came under investigation for allegedly being a communist sympathiser. In a letter (see [1318]) Lanczos wrote

> *Yesterday two FBI agents visited me for a sort of interrogation. It is strange, after all, that because one formerly belonged to the Wallace party and made some pacifist*

> *and (previously regarded to be) Russian sympathizing statements, one remains a somewhat suspicious character until the end of one's life. Otherwise the people were very friendly, and I have no objection against the organization. However, this "file" can be terribly abused.*

In 1954, he decided to move permanently to the Dublin Institute for Advanced Studies in the School of Theoretical Physics. He resumed his physics research including the geometry of space time and married another German woman, Ilse Hildebrand (1901-1974), in 1955. His son Elmar decided to stay in the USA. He married, moved to Seattle, and had two sons. In Dublin, Lanczos was visiting professor in 1952-1953, full professor in 1954-1968, and emeritus professor afterwards. He received an honorary degree from Trinity College in Dublin in 1962. During his stay in Dublin, Lanczos rediscovered the Singular Value Decomposition (SVD). He also published a paper *Linear systems in self-adjoint form* for which he was awarded the AMS Chauvenet Prize in 1960. After his retirement Lanczos traveled to many places to give lectures. It is also worth mentioning that Lanczos was a gifted pianist.

He died of a heart attack in Budapest on his second visit back to Hungary, on June 25, 1974. He is buried in the Farkasrét Jewish Cemetary in Budapest. A six-volume edition of his collected works was published by North Carolina State University in 1998. Lanczos published over 120 papers and 8 books during his career:

The Variational Principles of Mechanics (1949, dedicated to Albert Einstein), *Applied Analysis* (1956), *Linear Differential Operators* (1961), *Albert Einstein and the Cosmic World Order* (1965), *Discourse on Fourier Series* (1966), *Numbers without End* (1968), *Space through the Ages* (1970), and *Einstein Decade: 1905-1915* (1974).

A special issue of *Computers and Mathematics with Applications* in memory of Lanczos was published in 1975 in which R. Butler from the University of Manchester wrote

> *He had a quite unique ability to present mathematical ideas according to the audience (he always enquired of the level) and to hold them throughout, always avoiding algebraic manipulation as far as possible. He told me that he always prepared every lecture in great detail and had preserved all his notes. This was essential he said, and also he added that a good lecturer must have the right mixture of introvert and extrovert personality. Certainly he had both, and it was especially noticeable how he enjoyed lecturing and how he became an actor from the moment of his introduction to an audience. His knowledge over most areas of mathematics was profound and this played no small role in his lecturing technique especially at question time.*

He also reported an interesting quote from Lanczos: *Retirement merely means one carries on, on half pay.*

For these biographical notes we used a paper about Lanczos' life prepared by Dianne Prost O'Leary [2370]. We also used the nice biographical essay by Barbara Gellai [1317] in [473] and her book [1318].

Videos of interviews with Lanczos are available at the University of Manchester.[75]

Work

Many Lanczos' papers are devoted to physics and, particularly, to the theory of relativity. Here we are only interested in his work related to numerical linear algebra.

Lanczos worked on what is now known as the Lanczos method for computing approximations of eigenvalues when he was with Boeing and with the INA in Los Angeles. During his stay at the INA, Lanczos worked on different projects whose titles changed a little bit over the years. The

[75] https://www.youtube.com/user/ManchesterMaths/videos (accessed August 2021)

details of those projects given below came from the quarterly reports of the National Applied Mathematics Laboratories of the NBS.

In 1949, he was responsible for project 11.1/1-49-AE1, whose aim was to develop a practical and economical method of evaluating the characteristic values (that is, eigenvalues) of arbitrary complex matrices, and project 11.1/1-49-AE2 for developing a practical and economical method for solving simultaneous algebraic equations by matrix iteration. Another project was 11.1/1-49-0DE4 to develop a practical and economical method of obtaining the characteristic values and functions of arbitrary linear differential operators. These projects were extended into project 11.1/1-50-3 on which Lanczos was joined by William Feller (1906-1970), George Elmer Forsythe (1917-1972), and Mark Kac (1914-1984). Its goal was the calculation of eigenvalues, eigenvectors, and eigenfunctions of linear operators. The comments on the project about Lanczos' method at that time were

> *The advantages claimed for the method are as follows: (a) The iterations are used in the most economical fashion, obtaining an arbitrary number of eigenvalues and eigensolutions by one single set of iterations, without reducing the order of the matrix. (b) The rapid accumulation of fatal rounding errors, common to all iteration processes if applied to matrices of high dispersion (large "spread" of the eigenvalues), is effectively counteracted by the method of "minimized iterations." (c) The method is directly translatable into analytical terms by replacing summation by integration. We then get a rapidly convergent analytical iteration process by which the eigenvalues and eigensolutions of linear differential and integral equations may be obtained.*

In September 1949, Lanczos gave a talk at the *Symposium on Large-Scale Digital Computing Machinery* organized in Harvard University. Walter Edwin Arnoldi (1917-1995), who later developed his own method for computing eigenvalues, was an attendee at this meeting. In the last quarter of 1949 the managers of project 11.1/1-50-3 were Lanczos, Magnus Rudolph Hestenes (1906-1991), William Karush (1917-1997), and Forsythe. At that time Lanczos was also working on Chebyshev polynomials, Fredholm integral equations, and the inverse of the Laplace transform.

These projects were continued in 1950. In the April-June quarterly report on project 11.1/1-50-3 we can read

> *An 8×8 symmetric matrix was specially constructed by Dr. Rosser so that the eigenvalues would be difficult to compute. Three teams were asked to tackle it using three different methods. Dr. Lanczos used the method described in publication (l) below. Dr. Karush used the gradient method described in publications (3) and (4). Dr. Forsythe was asked to use classical methods, but found them to be effectively impossible and used common sense together with ad hoc numerical experimentation. All three obtained correct results. Dr. Rosser plans to write a report comparing the solutions.*

In the paper published in 1950, he proposed constructing an orthogonal basis of what is now called a Krylov subspace. He demonstrated the algorithm's effectiveness on practical problems: the lateral vibration of a bar, the vibrations of a membrane and a string through numerical computations; see Chapter 6. In 1952, he discussed the solution of linear systems using the recurrence of minimized iterations and recognized that it was equivalent to the method of Conjugate Gradients of Hestenes and Eduard Stiefel (1909-1978) also published in 1952; see Chapter 5.

In 1950, Lanczos published his paper *An iteration method for the solution of the eigenvalue problem of linear differential and integral operators* [1984] in the *Journal of the National Bureau of Standards*. In 1952, he published his paper *Solution of systems of linear equations by minimized iterations* in the same journal.

When reading these two papers, we observe that in his own papers, Lanczos did not use a matrix framework (except for the multiplication by A at each step). Lanczos was thinking in terms of vectors and characteristic polynomials. He was not interested in the tridiagonal matrix he was implicitly computing. Let us formulate what he did for symmetric problems in more modern terms using matrices. Let $P_{k+1} = (p_0 \quad \cdots \quad p_{k+1})$ and $Q_{k+1} = (q_0 \quad \cdots \quad q_{k+1})$ with $p_0 = q_0 = b_0$. These two matrices are $n \times (k+2)$. We have the two matrix relations

$$AQ_k = P_{k+1}\underline{L}_{k+1}, \quad P_{k+1} = Q_{k+1}U_{k+1},$$

where $\underline{L}_{k+1}$ is a $(k+2) \times (k+1)$ lower triangular matrix with the ρ_i's on the diagonal and -1 on the subdiagonal and U_{k+1} is an upper triangular matrix of order $k+2$ with 1's on the diagonal and $-\sigma_i$'s on the first upper diagonal. The matrices P_k are orthogonal and the matrices $Q_k^T AQ_k$ are diagonal. If we eliminate Q_{k+1} we obtain

$$AP_k = P_{k+1}\underline{T}_{k+1},$$

where $\underline{T}_{k+1}$ is a $(k+2) \times (k+1)$ nonsymmetric tridiagonal matrix. This is the familiar relation that we use in the modern formulation of the Lanczos algorithm, except that the vectors are not normalized. The matrix $\underline{T}_{k+1}$ is the product $\underline{L}_{k+1}U_{k+1}$, that is, an LU factorization of the tridiagonal matrix. Note that this factorization is different from the LU factorization that can be used to derive the conjugate gradient method from the Lanczos algorithm. Lanczos was using the columns of the matrices Q_k as basis vectors and those of P_k are what we now use.

In 1958, Lanczos published the paper *Iterative solution of large-scale linear systems*. In this paper one can read

> *It is not advisable to assume that here is only one method by which a certain numerical problem can be solved. It is more desirable to have a collection of programming procedures to our disposal which can be adjusted to the specific demands of a given situation. A certain mathematical procedure may be inadequate to a certain type of problem but very well suited to another. It may likewise happen that certain procedure, discarded for practical reasons, reappears on the platform because certain changes in the technical design of the machine make the method practically feasible, while before it commanded only theoretical interest.*

The last sentence was certainly prophetic concerning the methods he developed at the beginning of the 1950s.

In this paper Lanczos scaled the symmetric positive definite matrix and the right-hand side dividing by an approximation of the largest eigenvalue, obtained by considering four moments. Then, he used polynomials constructed from the Chebyshev polynomials (of the first kind). This gives him a recurrence relation for the vectors he wanted to compute.

He concluded by writing

> *The paper discusses an iterative scheme for the solution of large scale systems of linear algebraic equations which does not demand the inversion of the matrix but obtains the solution in a series of successive approximations. The matrix is used as a whole and the basic operation is the multiplication of a fixed matrix with a constantly changing vector. The recurrence relation of the Chebyshev polynomials furnishes the recurrence scheme which has to be coded for the machine. The simplicity of this recurrence scheme is the attractive feature of the method. Moreover, preliminary experiments with small-scale matrices seem to indicate that the "signal-to-noise ratio" does not decrease with the number of iterations, even after several hundreds*

of iterations. Hence it seems that the method is in a favorable position relative to an accumulation of rounding errors. However, experiences with large matrices are not yet available.

In 1960, the Chauvenet prize was awarded to Lanczos for this paper. In an interview given in 1974 [1873], Lanczos said

I believe my most important contribution was in the fields of mathematics, to be precise, in numerical analysis–my discovery of a method now known as the Lanczos method. It is very little used today, because there are now a number of other methods, but it was particularly interesting in that the analysis of the matrix could be carried out, that is, all the eigenvectors could be obtained by a simple procedure.

10.47 ▪ Pierre-Simon Laplace

Pierre-Simon Laplace

Life

Pierre-Simon Laplace was born on March 23, 1749, in Beaumont-en-Auge, a small village in Normandy, France. On the village square, opposite each other, two houses bear a plaque indicating his birthplace.

His father, Pierre Laplace, was a cider producer and trader. His mother was Marie-Anne Sochon (1723-1768), from a farmers' family. Pierre-Simon attended the Benedictine priory school of his village between the ages of 7 and 16. His father wanted him to become a priest of the Roman Catholic church, and thus he went to the University of Caen to study theology in 1766. During his two years there, he discovered his mathematical talents and his love for the subject.

However, he left Caen without a degree but with an introduction letter to Jean Le Rond d'Alembert (1717-1783) from Pierre Le Canu, one of his professors. Tradition says that d'Alembert gave him a problem to solve and told him to come back in a week. Laplace brought the solution the next day. Second problem, same behavior. Much impressed, d'Alembert recommended him for a teaching position at the École Militaire.

On March 31, 1773, Laplace was elected corresponding member of the Académie des Sciences in Paris. His reputation steadily increased during the 1770s, a period he mostly dedicated to probability and celestial mechanics.

The 1780s was the period in which Laplace produced the results that made him one of the most important and influential scientists in the world. But Laplace was not modest about his abilities and achievements, and his relationships with his colleagues greatly suffered from his attitude.

In 1784, Laplace was appointed an examiner of the cadets of the École Militaire in Paris. In September 1785, he examined the young Napoléon Bonaparte, when he was 16 years old, in mathematics. He became a full member of the Académie des Sciences. The most productive period of his life began.

He deepened all the topics he had previously studied, and became interested in physics. Along his life, Laplace made important contributions to mathematical physics. He was interested in all topics that helped to interpret nature. He worked on hydrodynamics, the propagation of sound and tides, on the liquid state of matter, on the tension in the surface layer of water, on the rise of liquids in a capillarity tube, and on the cohesive forces in liquids. However, the main part of his works was devoted to celestial mechanics. In particular, he determined the attraction of a spheroid on an outside particle. On this occasion, he introduced the potential function, the Laplace equation, the Laplace coefficients (spherical harmonics), and he explained the angular velocity of the Moon about the Earth.

In 1787, Joseph-Louis Lagrange (1736-1823) arrived in Paris from Berlin; see Section 10.44. Despite a rivalry between them, each was to benefit greatly from the ideas of the other.

On March 15, 1788, Laplace married Marie-Charlotte de Courty de Romanges (1769-1862). They had a son and a daughter. They lived 50 km southeast of Paris, near Melun, during the most violent part of the French Revolution. Laplace was appointed, in May 1790, to the Commission des Poids et des Mesures (Commission of Weights and Measures) to work on the metric system. But the Reign of Terror began in 1793, and the Académie des Sciences and the other learned societies were abolished on August 8. In 1795, the École Normale was founded (but survived for only four months), and he lectured there on probability. He published these lectures in 1814 under the title of *Essai Philosophique sur les Probabilités* (Philosophical essay on probabilities). That same year saw the creation of the Bureau des Longitudes (Longitude Office), and he went on to lead it and the Paris Observatory.

In November 1799, immediately after seizing power in the coup of 18 Brumaire (November 9, 1799), Napoléon appointed Laplace to the duty of Minister of the Interior, but the appointment lasted only six weeks since he had no talent as an administrator. However, he was made a member of the Senate, then its chancellor, he received the *Légion d'Honneur*, and gained the title of *Comte de l'Empire*. This is why he was, from then on, known as "de Laplace."

In 1806, Laplace bought a house in Arcueil, at that time a village near Paris. The chemist Claude-Louis Berthollet (1748-1822) was his neighbor. Together, they founded the *Société d'Arcueil*, which gathered many scientists around them. Because of their closeness to Napoléon, Laplace and Berthollet had their hands on the scientific establishment and positions in the more prestigious offices.

The regular meetings almost completely ended around 1813 when François Arago (1786-1853) began to favor the wave theory of light proposed by Augustin Fresnel (1788-1827), which was opposed to the corpuscular theory supported by Laplace. Many of Laplace's other physical theories were contested, in particular in the works of Jean-Baptiste Joseph Fourier (1768-1830) and Sophie Germain (1776-1831), but he never accepted that his ideas could be wrong.

During his life, Laplace had always changed his views according to the political climate of the day. This behavior served him in the 1790s and 1800s, but deteriorated his personal relations with his colleagues who considered these changes of views as attempts to win favors. For instance on the relations between Laplace and Sylvestre-François Lacroix (1765-1843), see [3002].

In 1814, Napoléon's empire was falling, and Laplace hastened to tender his services to the Bourbons, and he became unpopular in political circles. During the Restoration (of the kingdom), he was rewarded with the title of marquis in 1817. In 1816, he was also elected to the Académie Française, but in 1826, he refused to sign the declaration issued by the Academy in favor of the freedom of the press, and he lost the last friends he had in politics.

Laplace died in Paris on March 5, 1827, the same day as Alessandro Volta (1745-1827). He was first buried in Paris, but in 1888 his remains were moved to Saint-Julien-de-Mailloc, a small village in Normandy.

For more details on Laplace, see the book by Charles Coulston Gillispie (1918-2015), Robert Fox, and Ivor Grattan-Guinness (1941-2014) [1354] in 1997.

Work

On March 28, 1770, Laplace presented at the Académie des Sciences in Paris a first paper on the maxima and minima of curves where he improved methods given by Lagrange. Another paper on difference equations followed less than four months later. In three years, he produced thirteen papers. The topics were extreme value problems, application of integral calculus to the solution of difference equations and their expansion into recurrent sequences, application of these techniques to the theory of games of chances, singular solutions of differential equations, probability theory, and various problems in mathematical astronomy. In particular, his 1771 paper *Recherches sur le calcul intégral aux différences infiniment petites, et aux différences finies* has to be mentioned since he proposed a method for the solution of difference equations. The paper was published in the *Miscellanea Taurinensia*, a journal founded by Lagrange. This was the second time their paths crossed.

Laplace gave a general method for expanding a determinant in terms of its complementary minors in 1772 [1993] (pp. 267-376; also Œuvres, vol. 8, pp. 369-477). More specifically, he showed that any determinant is equal to the sum of the co-factors of any row or column of the array multiplied by the elements that generated them.

In the 1870s, in order to explain the apparent instability of the orbit of Jupiter, which appeared to be shrinking, while that of Saturn was expanding, Euler and Lagrange had made approximations by ignoring the small terms in the equations of motion. Laplace noticed that although the terms were small, they could become important when integrated over a long period of time. Thus, he carried his analysis up to the cubic terms, and concluded that the two planets and the Sun must be in mutual equilibrium. This result was the starting point of his work on the stability of the solar system. Any scientific problem was, for him, the spark for inventing the necessary mathematical analysis for its solution.

In 1780-1784, Laplace collaborated with the chemist Antoine Laurent Lavoisier (1743-1794, executed by guillotine) on several experimental investigations. They designed their own equipment, and in 1783, they published their joint paper *Mémoire sur la chaleur* in which they discussed the kinetic theory of molecular motion. They measured the specific heat of various bodies, the expansion of metals with increasing temperature, and they also obtained the boiling points of ethanol and ether under pressure.

Laplace published his monumental *Exposition du Système du Monde* (Exposition of the system of the world) [1996] in 1796. It gave a general explanation of his nebular hypothesis, which considers the solar system as originating from the contracting and cooling of a large, flattened, and slowly rotating cloud of incandescent gas. Details were omitted. It also contained a summary of the history of astronomy. It is, in fact, a non-mathematical introduction to his most important *Traité de Mécanique Céleste* (Treatise on celestial mechanics) in five volumes, whose first volume appeared three years later. It contains the equation named after him although it was, in fact, known before his time. It is reported that when Napoléon congratulated Laplace for this work, he asked him why God was never mentioned. Laplace is said to have replied that he had not needed this assumption.

In two important papers in 1810 and 1811, Laplace developed the concept of characteristic function, and proved the first general central limit theorem. In 1812, he published his *Théorie Analytique des Probabilités* (Analytic theory of probabilities) [1998] in which he laid down many fundamental results in statistics. His work on the method of least squares is contained in the fourth chapter. He showed that the central limit theorem implies that the least squares estimates of a large number of independent observations not only maximize the likelihood function, but also minimize the expected a posteriori error. In this book, he also used a technique which is what we now call the Gram-Schmidt orthogonalization method; see [1990] and Chapter 4.

The book contained supplements about applications of probability to the determination of the masses of Jupiter, Saturn, and Uranus, triangulation methods in surveying, and problems of geodesy. Another supplement on generating functions was added in the 1825 edition. This theory was developed for the purpose of solving difference equations, and led Laplace to the integral transform bearing his name. His *Essai Philosophique sur les Probabilités* (Philosophical essay on probabilities) appeared in 1814. He argued that man's dependence on probability is a simple consequence of imperfect knowledge. He was thinking that whoever could follow every particle in the universe, and had unbounded powers of computation, would be able to know the past, and to predict the future with perfect certainty; see [1863] for an analysis.

10.48 ▪ Charles L. Lawson

The information given in this biography is issued from *Cleve's Corner: Cleve Moler on Mathematics and Computing*,[76] and from [1537].

Charles Lawrence Lawson

Courtesy of Cleve Moler (MATLAB blog: Cleve's Corner: Cleve Moler on Mathematics and Computing)

Life

Charles Lawrence (Chuck) Lawson was born on October 19, 1931, in the small town of Weiser, Idaho, USA, which is on the Snake River where Idaho and Oregon share a boundary. His mother was an optometrist. His father grew up on a farm in Wilder, Idaho. He went to the University of Idaho and then came to Weiser to teach mathematics and business in the high school.

[76]https://blogs.mathworks.com/cleve/2015/09/24/charles-lawson-1931-2015/ (accessed March 2022)

The young Chuck went to schools in Weiser, and then for his last year of high school he lived with his aunt in Hollywood, California, and attended and graduated from Hollywood High School. He entered the University of California at Berkeley in the fall of 1948. He obtained a BS in optometry in June 1952 and a certificate of completion in optometry in June 1953, a subject he was not particularly interested in. He wanted to do mathematics, and he was aware that computers were coming on the scene. Lawson took and passed state board exams in optometry in California, Oregon, and Idaho during the summer of 1953.

After he earned his degrees, he was immediately drafted into the Army. After one year as a private, he spent the next years doing eye exams as a second lieutenant. During these two years, he took correspondence courses in junior and senior level mathematics from UCLA. When he got out of the service in December 1956, he moved to the Los Angeles area and enrolled at UCLA in January of 1957 to study mathematics after the demise of the Institute for Numerical Analysis. But the SWAC, the *Standards Western Automatic Computer*, was still in operation. UCLA kept it, with modifications, for over 15 years. Lawson spent a lot of time using the machine. He wrote an assembler for it.

Lawson obtained a master's degree in 1959 with Peter Karl Henrici (1923-1987) that *was a little bit of an exploration of a somewhat obscure method for finding roots of a polynomial associated with the name of Routh.* In [1537] Lawson said about Henrici

> *He's a well-respected person in those years of Numerical Analysis. I unfortunately had little contact with him, even though I did my Master's work under him. I think he was busy with many other things, and having a Master's student wasn't a high priority with him. One fairly distinct recollection I have is stopping in the hall one day and saying that I'd like to make an appointment to see him, I wanted to talk over some things that I was working on with him whenever it was convenient for him. I remember him thinking for a little bit and saying, "I think eleven o'clock next Wednesday I'd have time that you could come and see me, and we could make an appointment for a time for you to come and talk with me." So I never had a feeling that I was really in a lot of close communication with him [...] With Henrici, you could probably add up the total amount of time I spent with him, it wouldn't have added up to more than a couple of hours the whole time. Maybe that's an exaggeration, but it wasn't much.*

Then, for his Ph.D., Lawson started to do some special reading in approximation theory under the direction of Theodore Samuel Motzkin (1908-1970). At the end of the third graduate year, he completed all the requirements and the examinations for the Ph.D., except the dissertation. So, in the fall of 1960, after three full years at UCLA, he took a position at the Jet Propulsion Laboratory (JPL), managed by Caltech for NASA in Pasadena, where he stayed for 36 years. He continued to work on his dissertation and weekly went back to UCLA to discuss things with Motzkin. His access to computers at that time was the *IBM 7094* at JPL, which was quite adequate to do his numerical experiments. Lawson obtained his degree in 1961. One of the parts of his dissertation *Contributions to the theory of linear least maximum approximation* was minimax approximation via weighted least squares.

Five years later, Lawson became the head of the Applied Mathematics Group at JPL. Over the years the group usually consisted of about half a dozen people. They consulted with the rest of the JPL on their mathematical computations, developed an extensive library of mathematical software, and engaged in a modest amount of numerical analysis research. Cleve Moler worked in the Math Group in the summer of 1961, just after graduating from Caltech, and then again the next year, after his first year of grad school at Stanford. Lawson was his supervisor.

Charles Lawson retired from JPL in 1996. He passed away on July 2, 2015, in his home in Laguna Woods, California. His passions included singing, playing the ukulele, and dancing.

Work

JPL was (and still is) responsible for ephemerides, descriptions of the position and motion of the planets and the Moon. One aspect of producing an ephemeris involves "lunar theory," semi-analytic Fourier-like series that attempt to describe the motion of the Moon. Lawson wanted to do symbolic manipulation of the trigonometric series involved in lunar theories. John McCarthy (1927-2011) had recently invented LISP and symbolic computing with lists. Lawson thought those ideas could be applied to lunar theories. Thus, with Moler, they began to write what today would be called a symbolic algebra system. They soon could represent trigonometric polynomials, and add, multiply, differentiate, and integrate them. They were working in Fortran and Moler even succeeded in hacking the Fortran compiler so that a routine was allowed to call itself such that they could implement recursion. But many questions remained open and the project was stopped.

In the late 1960s, Richard (Dick) Joseph Hanson (1938-2016) and Frederick (Fred) Thomas Krogh jointed Lawson's group. Hanson wanted to get a little more involved in applications than the group was, and he formed some close relationships with people in the Navigation Group that was involved in the control of spacecraft. A part of the techniques used was least squares fitting of computed trajectories, abstractly or in general, the same kind of thing Lawson's group was doing with ephemerides in 1960. Hanson and Lawson worked together on software for matrix computations, and published their important book *Solving Least Squares Problems* [2004] in 1974. Their software for least squares and splines was instrumental in the discovery, published in 1968, of irregularities in the Moon's gravitational field.

With Hanson, Lawson also worked on what they called "the French Curve Program" in 1965-1970, a name that obviously came from the fact that they used ideas from the French engineer Pierre Bézier (1910-1999). They wrote a program that, starting from a bunch of points, drew a curve through them. They took spline curves as their basis and allowed them to be of different orders with constraints on the values or the derivatives. As Lawson wrote,

> *So visually, it just meant if you had tried to do a curve fit with some conventional computer program and you didn't like the looks of it, but if you could say what was wrong with it, if you could say, "Well I really want this to be higher here," or, "I don't want this to turn down here," you could express that to the program, and it would get you a fit that would satisfy those constraints. So that was an interesting thing, and it combined ideas from approximation theory involving spline curves and linear algebra* [...]

The best known project of Lawson's group was the BLAS, the Basic Linear Algebra Subroutines. The project began modestly in 1972 when Krogh wrote assembly language routines for JPL's *UNIVAC 1108*; see Chapter 7.

10.49 ▪ Adrien-Marie Legendre

Life

Adrien-Marie Legendre was born on September 18, 1752. Not many details are known about his childhood and even his birthplace is not certain. According to some sources, it was Paris, while others mention that he was born in Toulouse and that the family moved to Paris when he was young.

Legendre was not very talkative about himself. We only know that he came from a wealthy family. He was educated at the Collège des Quatre-Nations (also known as Collège Mazarin, named after its founder) in Paris, where one of his professors was the abbot Joseph François Marie (1738-1801) who later used several of his pupil's results in his treatise on mechanics; see [838].

Legendre defended a thesis in physics and mathematics there in 1770. This was not a real research work, but mainly a plan for research, and a list of the literature he would study and the results he would try to prove. With no need for financial support, Legendre stayed in Paris and devoted himself to research.

At the request of J.F. Marie and thanks to Jean Le Rond d'Alembert (1717-1783), Legendre was appointed professor at the École Militaire where he taught from 1775 to 1780. In 1782, the Berlin Academy of Sciences decided to offer a prize on the topic *Determine the curve described by cannonballs and bombs, taking into consideration the resistance of the air; give rules for obtaining the ranges corresponding to different initial velocities and to different angles of projection.* Legendre won the prize with his essay *Recherches sur la trajectoire des projectiles dans les milieux résistants* (Research on the trajectories of projectiles in resistant media). Lagrange, who was the Director of Mathematics at the Academy, wrote to Laplace asking for more information about this young mathematician.

On March 30, 1783, Legendre was appointed an adjoint in the Académie des Sciences, and an associate two years later.

For several years, France and Great Britain had independently developed topographical measurements. In 1787, Legendre was appointed to the Anglo-French committee whose task was to work with the Royal Observatory of Greenwich in order to join the triangulation networks of both nations, and to calculate the precise distance between the Paris and the Greenwich observatories by means of trigonometry. He visited Dover and London together with Jean-Dominique Cassini de Thury (1748-1845) and Pierre Méchain (1744-1804), and the three also met the astronomer Frederick William Herschel (1738-1822), the discoverer of the planet Uranus. This work resulted in Legendre's election to the Royal Society of London, and to the publication of his *Mémoire sur les opérations trigonométriques dont les résultats dépendent de la figure de la terre* (Memoir on the trigonometric operations whose results depend on the shape of the earth) which contains his formulas for spherical triangles. Let us mention that due to the imprecision of the instruments used, these measurements had to be done again in 1817 by François Arago (1786-1853) and Claude-Louis Mathieu (1783-1875). Legendre was associated, in some sense, to the committee of the Assemblée Nationale that was nominated to define the length of the meter.

Legendre lost his private fortune in 1793 during the French Revolution. That year, he married Marguerite-Claudine Couhin (1772-1856), who was of great help to him for the rest of his life.

In 1793, the Académie des Sciences was closed but reopened in 1795 as the Institut National des Sciences et des Arts. Legendre became one of its six members in the mathematics section. When Napoléon reorganized the institute in 1803, Legendre was retained in the new geometry section.

From 1799 to 1812, Legendre served as an examiner for mathematics for graduating artillery students at the École Militaire, and, from 1799 to 1815, he was also an examiner at the École Polytechnique. Legendre was knighted by Napoléon in 1811. In 1824, he refused to vote for the government candidate at the Institut, and his pension from the École Militaire was stopped. It was then partially reinstated with the change in government in 1828.

Legendre was a fellow of the Royal Society of Edinburgh since 1820. In 1831, he was made an officer of the *Légion d'Honneur*, and a foreign honorary member of the American Academy of Arts and Sciences in 1832.

Legendre died in Paris on January 9, 1833, after a long and painful illness. He is buried in the Auteuil cemetery in the 16th district of Paris.

Curiously enough, there is no known portrait of Legendre, except a caricature by Julien Léopold Boilly (1796-1874). An engraving which was, in some publications (for instance, in [2930]), presented as a portrait of Legendre is, in fact, a portrayal of Louis Legendre (1752-1797), a politician; see [1032, 1033]. There is a statue by Alfred-Désiré Lanson (1851-1898) in the north courtyard of the Paris City Hall which is thought to be Legendre, but the first name is not mentioned. There is a difference of two years in the birthdate above the statue, but it seems likely that it is "our" Legendre.

Legendre (left) and Fourier (by J.L. Boilly)

Work

In the 1780s, Legendre turned his attention to the attraction of ellipsoids. He presented his results to the Académie des Sciences in Paris on January 22 and February 19, 1783. He gave a proof of a result due to Colin Maclaurin (1698-1748) stating that the attraction of two confocal ellipsoids at an external point lying on their principal axis is proportional to their masses. He also introduced the so-called Legendre functions and used them to determine, by means of power series, the attraction of an ellipsoid at any outside point. Legendre also presented several memoirs on the solution of indeterminate equations of the second degree, on continued fractions, on probability, and on the rotation of bodies which are not driven by any accelerating force.

On July 4, 1784, Legendre read to the Academy his memoir *Recherches sur la figure des planètes* (Research on the shape of planets). It is in this work that he introduced the orthogonal polynomials named after him by expanding their generating function $(1 - 2xz + z^2)^{-1/2}$ into a series of polynomials. As noticed in [838], there was a strong emulation between d'Alembert, Lagrange, Laplace, and Legendre on problems in celestial mechanics.

Over the next few years, Legendre published works in various areas. In number theory, in *Recherches d'analyse indéterminée* (Research on indefinite analysis) of 1785, he conjectured (but not proved since his proof was unsatisfactory) the law of quadratic reciprocity, subsequently proved by Gauss in 1801. He was also interested in the calculus of variations and, in 1786, he published his *Mémoire sur la manière de distinguer les maxima des minima dans le calcul des variations* (Memoir on the way to distinguish maxima from minima in the calculus of variations) where he considered the second derivative as a method for distinguishing between a maximum and a minimum. From 1786, he spent several decades developing a systematic study of the integrals he called elliptic functions since they were connected to the problem of finding the arc length of an ellipse. He showed that by adequate changes of variables, they can be reduced to three canonical forms [2512, pp. 19-20].

In 1794, he published the first edition of his best known book *Éléments de Géométrie* (Elements of geometry), which remained the leading elementary text on geometry for around 100 years. It replaced Euclid's *Elements* as a textbook in most of Europe, and was even more influential in the United States, where it had numerous translations starting in 1819, one of them even going to some 33 editions. In its following editions, he also gave a simple proof that π is irrational, the first proof that π^2 is irrational, and he conjectured that π is not the zero of any algebraic equation of finite degree with rational coefficients, which means that it is a transcendental number. Let us mention that the irrationality of π was first proved in 1761 by Johan Heinrich Lambert (1728-1777), and that, in 1882, Carl Louis Ferdinand Lindemann (1852-1939) proved its transcendence, thus ending negatively a problem open for two thousand years.

In his book *Éssai sur la Théorie des Nombres* (Essay on number theory), published in 1797-1798, Legendre conjectured the prime number theorem, which was rigorously proved by Jacques Hadamard (1865-1963) and Charles de la Vallée Poussin (1866-1962) in 1896. A much improved edition, in which he gave a new proof of the law of quadratic reciprocity, is dated 1806, after Gauss published its own in 1801. However, Gauss did not state that he was improving Legendre's result, but claimed the result for himself since his was the first completely rigorous proof. He also declared that he had obtained the law for the asymptotic distribution of primes before Legendre. One can easily understand that Legendre was hurt.

It was in an appendix to his *Nouvelles Méthodes pour la Détermination des Orbites des Comètes* (New methods for the determination of comet orbits), published in 1806, that Legendre introduced the least squares method. Gauss published his version of the least squares method in 1809, but, although he acknowledged Legendre's book, he claimed priority for himself. This attitude greatly hurt Legendre again, and he had to fight for many years to have his priority recognized.

Legendre's book *Exercices de Calcul Intégral sur Divers Ordres de Transcendantes et sur les Quadratures* (Exercises in integral calculus on various orders of transcendents and on quadratures) appeared in three volumes in 1811, 1817, and 1819. In the first volume, he introduced the basic properties of elliptic integrals, and also of beta and gamma functions. More results on the beta and gamma functions are given in the second volume, together with applications to mechanics, the rotation of the earth, the attraction of ellipsoids and other problems. The third volume is mostly devoted to tables of elliptic integrals.

Legendre came back to elliptic functions with the publication, from 1825 to 1837, of his *Traité des Fonctions Elliptiques* (Treatise on elliptic functions), where he gathered all his results on the topic. He reduced elliptic integrals to three standard forms, gave tables of the values of his elliptic integrals, and showed how to use them for solving important problems in mechanics and dynamics. But this work was rapidly outdated by the independent results of Niels Henrik Abel (1802-1829) and Carl Gustav Jacob Jacobi (1804-1851). In 1830, he gave a proof of Fermat's last theorem for $n = 5$, a result also obtained by Johann Peter Gustav Lejeune Dirichlet (1805-1859) in 1828. In 1832, using the Euclidean axioms except the parallel axiom, he failed to show that the sum of the angles of a triangle cannot be less than two right angles.

10.50 ▪ Gottfried W. von Leibniz

The life of Leibniz is so rich and so many biographies have already been written on him, that it is inappropriate to produce another extensive one, and difficult to bring new facts to light. Thus, we will restrict ourselves to the most important moments of his life, and to the works concerned with the topics relevant for this book.

Gottfried Wilhelm von Leibniz
Portrait by Christoph Bernhard Francke

Life

Gottfried Wilhelm (von) Leibniz was born on July 1, 1646, in Leipzig, a town of Saxony, a state of the Holy Roman Empire. His father, Friedrich Leibniz (1597-1652), a professor of moral philosophy at the University of Leipzig, died when he was six years old, and he was mostly raised by his mother Catharina Schmuck.

He followed a usual classical cursus at the Nicolai School in Leipzig, and had access to his father's library where he read philosophical and ecclesiastical books, mostly in Latin. In April 1661, he registered at the University of Leipzig where he studied philosophy, some mathematics, but also rhetoric, Latin, Greek, and Hebrew.

He obtained a bachelor's degree in December 1662. He defended his *Disputatio Metaphysica de Principio Individui* (Metaphysical disputation on the principle of individuation) on June 9, 1663, then he went to Jena to spend the summer term of 1663. Back in Leipzig, he obtained a master's degree in philosophy on February 9, 1664.

In 1666, at the age of 19, he published his first book *De Arte Combinatoria* (On the combinatorial art) in which he tried to prove the existence of God based on arguments from motion. That same year, Leibniz left Leipzig since the university refused to grant him a Doctorate in Law because of his age, and he went to the University of Altdorf bei Nürnberg, a small town outside the Free Imperial City of Nuremberg. There, in December 1666, he obtained his Doctorate in Law but refused the position he was offered because his interests turned in an entirely different direction. At this time, even though no official document of nobility has ever been found, he included *von* in his name.

After working for some time as a secretary in an alchemical society in Nuremberg, Leibniz met Baron Johann Christian von Boyneburg (1622-1672), chief minister of the Elector of Mainz, Johann Philipp von Schönborn (1605-1673). The Elector enrolled him for various legal, diplomatic, and even political duties. Among them, Leibniz proposed to the Elector a plan to protect German-speaking Europe by suggesting Louis XIV, the King of France, to conquest Egypt, and then, eventually, the Dutch East Indies. In return, France would agree to leave Germany and the Netherlands undisturbed. The Elector agreed, and in 1672, Leibniz was invited to Paris for discussions. But the plan was soon overtaken by the outbreak of the Franco-Dutch war.

In Paris, Leibniz met the Dutch physicist and mathematician Christiaan Huygens (1629-1695), the French oratorian priest and rationalist philosopher Nicolas Malebranche (1638-1715),

Antoine Arnauld (1612-1694), a leading French philosopher and mathematician, and he became lifelong friends with the German mathematician Ehrenfried Walther von Tschirnhaus (1651-1708).

Leibniz soon realized that his knowledge in mathematics and physics was quite poor and, under the guidance of Huygens, he began to self-study the writings of René Descartes (1596-1650) and Blaise Pascal (1623-1662). According to his notebooks, he had a critical breakthrough on November 11, 1675, when using integral calculus for obtaining the area under the graph of a function. He rapidly made a link between between the problem of tangent, which depends on the difference of ordinates, and the quadrature of a curve based on the sums of its ordinates. He also introduced various terms and notations but he did not publish anything on calculus until 1684. It must be noticed that around the same time, he emitted the hypothesis of infinitely small components in the universe.

After the failure of the plan about Egypt, the Elector sent his nephew and Leibniz to London in 1673 with a related mission to the English government. At this occasion, Leibniz met Henry Oldenburg (c.1619-1677), the first Secretary of the Royal Society founded in 1660, and the mathematician John Collins (1625-1683). He was invited to present his calculating machine, the *stepped reckoner* (see Chapter 7), which could execute all four arithmetic operations, at the Royal Society that nominated him an external fellow on April 19, 1673. Leibniz gradually improved his machine over the years. But von Boyneburg died in December 1672, the Elector the next February, thus leaving Leibniz without any support. He returned to Paris.

In 1673, the Duke John Frederick of Brunswick (1625-1679), with whom Leibniz was in contact for several years, offered him the post of counselor. Since no other proposal was coming, he accepted two years later, left Paris in October 1676, and arrived in Hanover in December. In 1677, he was promoted, at his request, to Privy Counselor of Justice, a post he held for the rest of his life. Leibniz served three consecutive dukes of Brunswick as historian, political adviser, and librarian of the ducal library. In his position, he played an important political, economical and intellectual role. He was involved in a project for draining water from the mines in the Harz mountains, using wind and water powers to operate pumps. He designed many different types of windmills, pumps, and gears, but the project ended in failure.

The Elector Ernest Augustus (1629-1698) commissioned Leibniz to write a history of the House of Brunswick, going back to the time of Charlemagne or earlier. Leibniz traveled extensively in Germany, Austria, and Italy, seeking and finding archives for this project. However, years and even decades passed, and the work still remained a project. The material collected was finally published in the 19th century, and extended over three volumes. In 1691, Leibniz was appointed Librarian of the Herzog August Library in Wolfenbüttel in Lower Saxony.

The dispute between Leibniz and Isaac Newton (1642-1727) on the introduction of differential and integral calculus began in 1708 when John Keill (1671-1721), a Scottish mathematician, natural philosopher, cryptographer, and fellow of the Royal Society, accused Leibniz of plagiarism. Leibniz employed his variant of calculus in 1674, and, in October 1684, he published his first paper on this topic, *Nova Methodus pro Maximis et Minimis, itemque tangentibus, quae nec fractas nec irrationales quantitates moratur, et singulare pro illis calculi genus* (A new method for maxima and minima, and for tangents, that is not hindered by fractional or irrational quantities, and a singular kind of calculus for the above mentioned). Newton claimed he was already working on a topic he named *the method of fluxions and fluents* in 1666 at the age of 23, but he did not publish his ideas.

The French mathematician Guillaume François Antoine, Marquis de l'Hôpital (1661-1704), wrote a text on Leibniz's calculus in 1696 in which he recognized that Newton's *Philosophiæ Naturalis Principia Mathematica* (Mathematical Principles of Natural Philosophy) of 1687 had *nearly all about this calculus*. Though Newton explained his geometrical form of calculus in Section I of Book I of the *Principia*, he only published his fluxional notation in part in 1693, and in full in 1704. A formal investigation by the Royal Society, undertaken in response to Leibniz's demand for a retraction, confirmed Keill's charge.

Historians of mathematics now tend to acquit Leibniz, pointing to important differences between the two approaches. This quarrel darkened the last years of his life. Anyway, finally, Leibniz's notation was retained and not Newton's.

In 1711, the Russian Czar Peter the Great (1672-1725) visited Hanover and met Leibniz, who became interested in Russian matters for the rest of his life. In 1712, Leibniz went to Vienna, where he was appointed Imperial Court Counsellor to the Habsburgs. On the death of Queen Anne of Great Britain and Ireland (1665-1714) in 1714, the Elector George Louis of Hanover (1660-1727) became King George I of Great Britain and Ireland. However, he did not include Leibniz in his London court since it would have been deemed insulting to Newton.

Leibniz died in Hanover on November 14, 1716. He was one of the greatest thinkers of the 17th and 18th centuries who made important contributions to philosophy, metaphysics, logic, religious ideas, history, physics, and mathematics.

Work

Leibniz worked on mathematics, logic, physics (a theory of force), philosophy, and calculus. He studied the binary system in arithmetic. He perfected his work on this system by 1679, but only published a paper about this in 1703 as Godefroy-Guillaume Leibnitz (notice the spelling of his name), *Explication de l'arithmétique binaire, qui se sert des seuls caractères 0 et 1 avec des remarques sur son utilité et sur ce qu'elle donne le sens des anciennes figures chinoises de Fohy*, in the *Mémoires de mathématique et de physique de l'Académie Royale des Sciences, 1703*.[77]

His most important mathematical papers were published between 1682 and 1692 in the very first scientific journal in Germany, *Acta Eruditorum*, founded in 1682 by the philosopher and scientist Otto Mencke (1644-1707). In 1684, Leibniz became interested in systems of linear equations. He arranged the coefficients into an array, and laid down the theory of determinants; see Chapter 3. He computed them using cofactors, a process now named the *Leibniz formula*. He also solved systems of linear equations using determinants, a procedure known as Cramer's rule, although Gabriel Cramer (1704-1752) obtained it only in 1750.

10.51 ▪ Ada Lovelace

Life

Augusta Ada Byron was the daughter of the English romantic poet and politician Lord George Gordon Byron (1788-1824) and his wife Anne Isabella Noel, née Milbanke (1792-1860), who was interested in mathematics (Byron called her "the Princess of Parallelograms") and later became involved in improving slum conditions and discussing rights for women.

Ada was born in Piccadilly, Middlesex (now London), England, on December 10, 1815. Her parents married on January 2, 1815, and separated on January 16, 1816, a month after her birth. Their separation was legalized in March 1816. Due to his scandalous life and ever-increasing debts, Byron was forced to leave England forever in April 1816, and Ada never saw him again. He died in Greece when Ada was eight years old.

Lady Byron tried to do everything possible in bringing up Ada so that she would not become a poet like her father. However, Ada always remained interested in him.

[77] https://hal.archives-ouvertes.fr/ads-00104781/document (accessed November 2021)

Ada King, Countess of Lovelace
Portrait by Alfred Edward Chalon

She was educated by a number of private tutors, some of them often for only a short period. Among them was William Frend (1757-1841), a clergyman, social reformer, and writer, who had tutored Ada's mother in mathematics, and William King (1786-1865), a physician and philanthropist from Brighton. Ada's favorite subject was geography but her mother promoted her interest in mathematics and logic, and she put a constant pressure on her daughter and even gave her punishments like solitary confinement.

Ada also showed a talent for foreign languages. During her childhood, she was often ill. At the age of eight, she had headaches that obscured her vision. In June 1829, she contracted measles from which she took a long while to recover. She had to stay in bed for nearly a year, and by 1831, she had to walk with crutches.

In 1833, Ada Byron was presented at Court. On June 5 of that year, she met Charles Babbage (1791-1871); see Section 10.3. Two weeks later Ada and her mother visited Babbage's London studio where the *Difference Engine* was on display. Ada was fascinated.

She attended mathematics and scientific demonstrations with her friend Mary Somerville (1780-1872), a Scottish scientist, writer, and polymath who wrote many works which influenced James Clerk Maxwell (1831-1879) and John Couch Adams (1819-1892). She was also in contact with Michael Faraday (1791-1967) and the writer Charles Dickens (1812-1870).

Ada married William, 8th Baron King (1805-1893), an English nobleman and scientist, on July 8, 1835. King was made Earl of Lovelace in 1838, Ada thus became Countess of Lovelace. They had three children.

In 1841, Ada began advanced study in mathematics with the mathematician and logician Augustus De Morgan (1806-1871). He taught her advanced calculus topics including Bernoulli's numbers.

In 1842-1843, Ada worked on Babbage's *Analytical Engine*; see below. After her work with Babbage, she continued to work on other projects. In 1844, she thought of creating a *calculus of the nervous system* but she never achieved it, although for that project, she visited the electrical engineer Andrew Crosse (1784-1855) in 1844 to learn how to do electrical experiments. We know that she had other scientific projects in mind but they were never realized. She lacked friends with whom to discuss them. She was also interested in metaphysics, magnetism, phrenology, and mesmerism.

Several scandals arose in her personal life with some male acquaintances, wine drinking, opium, and gambling on horses in which she lost a fortune. Around 1850, Lovelace fell out with her mother, almost certainly because she discovered that, for years, she was lying to her about her father Lord Byron.

By January 1852, Lovelace became seriously ill as the cancer which had been presumably the major cause of her previous health problems became more acute. During her illness that lasted several months, her mother excluded all of her friends and confidants, and she pushed Ada toward religion. She even lost contact with her husband after confessing to him something which caused him to abandon her bedside. It is not known what she told him. Up to the end, her mind remained as sharp as ever.

Ada Lovelace died on November 27, 1852, in Marylebone, London, from an uterine cancer, at the age of 37. She was buried, at her request, next to her father at the Church of St. Mary Magdalene in Hucknall, Nottinghamshire. According to many sites, a memorial plaque, written in Latin, to her and her father is in the chapel attached to Horsley Towers, the Lovelace estate in East Horsley, Surrey, but we were unable to find any trace of it.

The computer language Ada created on behalf of the United States Department of Defense was named after her. In 1981, the Association for Women in Computing inaugurated its Ada Lovelace Award. Ada Lovelace Day is an annual event celebrated on the second Tuesday of October. Its goal is to *raise the profile of women in science, technology, engineering, and math.*

Among the many writings on Ada Lovelace one can, for example, see the book [2857] by Dorothy Stein.

Work

In 1840, Babbage gave a seminar at the University of Turin about his *Analytical Engine*. Luigi Federico Menabrea (1809-1896), at that time a young Italian engineer and future Prime Minister of Italy, transcribed Babbage's lecture into French, and this transcript was published in volume 41 of the new series of the *Bibliothèque Universelle de Genève* in October 1842. The scientist and inventor Charles Wheatstone (1802-1875) asked Ada Lovelace to translate Menabrea's paper into English, work she did during a nine-month period in 1842-1843 but, most importantly, she complemented it with notes that were added to the translation and published in September 1843 under her initials AAL. The French title was *Notions sur la machine analytique de M. Charles Babbage* (English title: "Sketch of the Analytical Engine invented by Charles Babbage"... with notes by the translator).

Explaining the idea of the *Analytical Engine* and its differences with Babbage's first engine was a difficult task, since many scientists even did not grasp the concept, and the British establishment showed little interest in it. These notes, labeled from A to G, were around three times longer than Menabrea's paper itself. In note G, she described an algorithm for computing Bernoulli's numbers using the *Analytical Engine*; see Figure 10.3. It is considered the first ever published algorithm specifically designed for implementation on a calculator, even though this point is still a matter of debate among historians.

However, she was aware of the limitations of the Engine since she wrote

> *The Analytical Engine has no pretensions whatever to originate anything. It can do whatever we know how to order it to perform. It can follow analysis; but it has no power of anticipating any analytical relations or truths.*

The *Engine* was never completed, so her program was never tested. Lovelace and Babbage had a small quarrel about the publication since he wanted to leave the preface unsigned because of the criticism against the government, which could have been interpreted as a joint declaration. The publisher asked them to withdraw the paper but she refused. Their friendship recovered, and they continued to correspond.

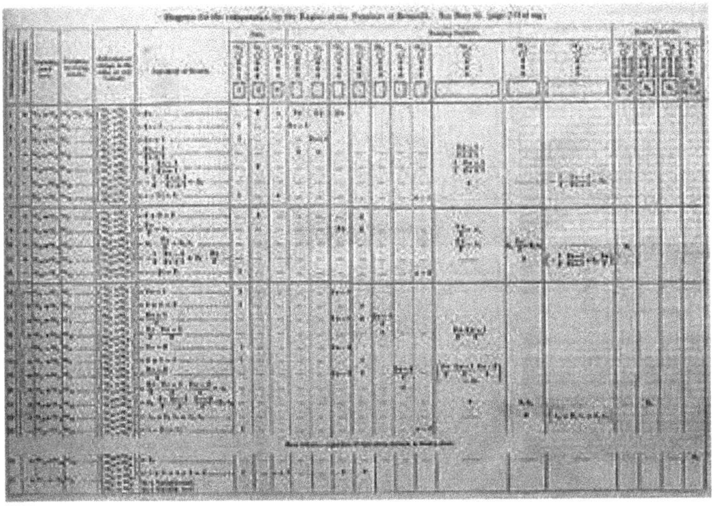

Figure 10.3. *Diagram from Note G*

10.52 ▪ John von Neumann

John von Neumann

Life

John von Neumann (Neumann János Lajos in Hungarian) was born on December 28, 1903, in Budapest, at that time in the Kingdom of Hungary. His father Maximilian Neumann von Margitta (1867-1929) was a banker and his mother was Margaret Kann (1881-1956). It was a wealthy, acculturated, and non-observant Jewish family.

John and his two brothers were educated at home until they were ten years old. They received lessons in English, French, German, Italian, and Ancient Greek. John was a child prodigy. By the age of eight, he was familiar with differential and integral calculus, but he was particularly interested in history.

He entered the Lutheran Fasori Evangélikus Gimnázium in 1911, one of the best schools in Budapest with a brilliant education system designed for the elite. The Nobel laureate in physics Eugene Wigner (1902-1992) was a year ahead of him, and they soon became friends. In addition to the Gymnasium, John had private lessons at home. In particular, at the age of 15, he began to study advanced calculus with Gábor Szegö (1895-1985).

On February 20, 1913, John's father was elevated to Hungarian nobility by the Emperor Franz Joseph for his service to the Austro-Hungarian Empire, but he did not modify his name. John, however, later used the German form with "von" to indicate the title. World War I had little effect on his family, and he completed his education at the Lutheran Gymnasium in 1921, winning the Eötvös Prize, a national prize for mathematics.

Von Neumann's father wanted his son to embrace a career able to bring him wealth. After asking the advice of Theodore von Karman (1881-1963), they chose chemistry, and John enrolled for a two-year, non-degree course in chemistry at the University of Berlin in 1921. He simultaneously registered at the Pázmány Péter University in Budapest for mathematics but he did not attend the lectures. However, together with Michael Fekete (1886-1957), the assistant at the University of Budapest who was tutoring him, he published a first paper on the location of the zeros of certain minimum polynomials in 1922. A year later, a paper giving a modern definition of ordinal numbers followed.

In September 1923, von Neumann passed the entrance exam for the prestigious Eidegenössische Technische Hochschule (ETH) in Zürich, and received his diploma in chemistry in 1926. In Zürich, he interacted with Hermann Weyl (1885-1955) and Georg Pólya (1887-1986) who were both there. That same year, he obtained his doctorate in mathematics from the University of Budapest with a thesis on the axiomatization of Cantor's set theory under the guidance (but did he need to be guided?) of Lipot (Leopold) Fejér (1880-1959).

Von Neumann obtained a Rockefeller Fellowship to study mathematics under David Hilbert (1862-1943) at the University of Göttingen during 1926-27. He also lectured at Berlin from 1926 to 1929, where he was appointed as a *Privatdozent*, the youngest ever elected in the university's history in any subject. Then, he went to Hamburg from 1929 to 1930.

By the end of 1929, von Neumann had already published 32 papers. They were mainly on two topics, far from set theory, but close to one another, quantum physics and operator theory, that he would later unify in his book *Mathematische Grundlagen der Quantunmechanik* (Mathematical foundations of quantum mechanics) published in 1932.

The American mathematician, geometer, and topologist Oswald Veblen (1880-1960) invited von Neumann to Princeton University to lecture on quantum theory in 1929. But before going there, he went back to Budapest to marry Marietta Kövesi (Kuper, 1909-1992) on New Year's Day in 1930. They had one daughter, born in 1935. Let us skip forward to say that they divorced in 1937, and that in October 1938, von Neumann married Klára Dán (1911-1963), also from Budapest, who he met during his last trip back to Budapest before World War II. Before marrying Marietta, von Neumann was baptized a Catholic in 1930. Von Neumann's father died in 1929. None of the family had converted to Christianity when he was alive, but all did afterwards.

In 1930, von Neumann became a visiting lecturer at Princeton University, and was appointed professor there in 1931. He was one of the original six mathematics professors at the newly founded Institute for Advanced Study at Princeton in 1933, a position he kept until his death. In 1933, he became co-editor of the *Annals of Mathematics* and two years later, of *Compositio Mathematica*, keeping both editorships for the rest of his life.

He and his wife Klára had an intensive social life in Princeton. They lived in one of the largest private houses of the town. John always wore formal suits, and he liked to eat and drink. He regularly played extremely loud German march music on his phonograph, and received complaints from his neighbors, including Albert Einstein (1879-1955). He was always working in noisy, chaotic environments. Despite being a notoriously bad driver, he enjoyed driving, occasioning a number of arrests and accidents. He had a lifelong passion for ancient history and was renowned for his historical knowledge, Byzantine history in particular. His closest friend in the United States was Stanisław Ulam (1909-1984), a former member of the famous mathematical school of Lwòw.

Von Neumann became an American citizen in 1937, and immediately tried to become a lieutenant in the United States Army's Officers Reserve Corps. He passed the exams easily but was not admitted because of his age. During the first years that he was in the United States, he continued to return to Europe during the summers. Until 1933, he still held academic positions in Germany but resigned when the Nazis came to power.

In 1938, von Neumann was awarded the Bôcher Prize by the American Mathematical Society for his memoir *Almost periodic functions and groups*.

During World War II, he was involved in the Manhattan Project, solving key problems in nuclear physics. He developed the mathematical models behind the explosive lenses used in the implosion-type nuclear weapon. From 1940, he was a member of the Scientific Advisory Committee at the Ballistic Research Laboratories at the Aberdeen Proving Ground in Maryland.

After the war, he consulted for a number of organizations, including the United States Air Force, the Army's Ballistic Research Laboratory, the Armed Forces Special Weapons Project, and the Lawrence Livermore National Laboratory. He was a member of the Navy Bureau of Ordnance from 1941 to 1955, and a consultant to the Los Alamos Scientific Laboratory from 1943 to 1955. From 1950 to 1955, he was also a member of the Armed Forces Special Weapons Project in Washington, D.C.

In 1955, President Eisenhower appointed him to the Atomic Energy Commission, and in 1956, he received its Enrico Fermi Award, knowing that he was incurably ill with cancer. But as a Hungarian émigré, he was much concerned that the Soviets could achieve nuclear superiority. Thus, he designed and promoted the policy of mutually assured destruction to limit the arms race.

It is impossible to list the numerous honors von Neumann received. He was elected to many academies, obtained several prizes and medals, and received two Presidential Awards. Details about his life and works can be found in the many testimonies about him; see, for instance, [93, 1555, 3089].

John von Neumann died on February 8, 1957, at the Walter Reed Hospital in Washington, D.C.

Work

Von Neumann made major contributions to many fields, including mathematics (foundations of mathematics, functional analysis, ergodic theory, representation theory, operator algebras, geometry, topology, numerical analysis), physics (quantum mechanics, hydrodynamics, quantum statistical mechanics), economics (game theory), computing (von Neumann architecture, linear programming, self-replicating machines, stochastic computing), and statistics. The works of von Neumann were analyzed in so many publications that it is impossible to quote them all. A quite complete reference is [93]. We will restrict ourselves mostly to his works related to numerical analysis.

Numerical methods existed for a long time before the advent of computers. But when they emerged, von Neumann was one of the first to realize their potential and to understand that they would induce a major change in calculations since they would be able to carry out long sequences of operations rapidly without human intervention, and to store a large amount of information on punched cards. Thus, more and more large and complex problems in mathematics, physics, economics, biology, and the earth sciences could be treated.

Herman Heine Goldstine (1913-2004) and von Neumann were convinced that a new barrier could be broken in the solution of differential and integral equations, and in the solution of systems of linear equations. However, they were conscious that the problem of numerical stability, due to the very high number of arithmetical operations that have to be carried out, might necessitate substantial changes in the numerical methods used for a long time. In particular, they advocated that direct methods for solving systems of linear equations have to be replaced by iterative ones which often need more multiplications but are more stable and, therefore, demand less precision. In 1946, they issued a document entitled *On the Principles of Large Scale Computing Machines* where they pointed out the future evolution of numerical analysis.

The question of the accuracy of matrix calculations was discussed by Harold Hotelling (1895-1973) as soon as 1943. He gave an example of a system of 11 linear equations which, solved by an elimination method, produced an error a million times greater than the error in the initial equations. And he supported the use of the Gauss-Seidel method.

In the fall of 1946 and early 1947, Goldstine and von Neumann realized that, for positive definite matrices, Gaussian elimination was quite stable. The numerical experiments conducted by Leslie Fox (1918-1993), Harry Douglas Huskey (1916-2017), and James Hardy Wilkinson (1919-1986) at the National Physical Laboratory in the UK led them to the same conclusion, but they claimed that extra digits could compensate round-off errors.

Alan Mathison Turing (1912-1954) also reassessed Hotelling's results. He visited Goldstine and von Neumann in Princeton in January 1947, and they had lengthy discussions on the topic; see Chapter 2.

The other works of von Neumann in numerical analysis concern partial differential equations, investigating the numerical stability of the methods used for their solution. He also studied the problem of shocks in fluid mechanics. He is credited for his invention, with Ulam, of Monte Carlo methods which were used at Los Alamos for the simulation of chain reactions in critical and supercritical systems. Von Neumann also took a strong interest in linear programming. He discovered the duality theorem, and suggested the equivalence between game theory and linear programming.

Von Neumann was, maybe, the last universal mathematician. He also was one of the pioneers of computer science, and he can be considered one of the initiators of modern numerical analysis.

10.53 ▪ Isaac Newton

Isaac Newton
Portrait by Godfrey Kneller

Life

According to the Gregorian calendar, Isaac Newton was born (prematurely) on January 4, 1643, in the manor house of Woolsthorpe, near Grantham in Lincolnshire, England. His father, who had the same first name and was a wealthy but illiterate farmer, died three months before his birth. Newton's mother remarried a clergyman two years later; she went to live with him, and they had three other children together. Isaac was left in the care of his grandmother. He did not have a happy childhood. He disliked his stepfather, he had some enmity towards his mother for her remarriage, and his grandfather left him nothing in his will.

From the age of about 12, Newton attended The King's School in Grantham, where he was taught Latin, Greek, and some basic mathematics, but he showed little promise in his studies.

His stepfather died in 1653. The young Isaac was removed from school in October 1659 and sent back to Woolsthorpe to live with his family. His mother wanted him to manage her estate, but he showed no talent for that. An uncle persuaded his mother that he had to return to school. Motivated by a desire for revenge, he became the best student, building sundials and models of windmills. Henry Stokes, the master of the school, insisted that Newton continue his studies. He entered Trinity College at Cambridge on June 5, 1661, as a subsizard (an undergraduate student who received a support for expenses in return for acting as a servant to other students) until he was awarded a scholarship in 1664. His aim was to obtain a law degree.

Instruction at Cambridge was based on the philosophy of Aristotle but Newton also studied René Descartes (1596-1650), Gassendi, Hobbes, and Boyle. The Copernican astronomy as explained by Galileo and Kepler's Optics attracted him. In 1665, he extended the binomial theorem, and began to think about what would later became calculus.

Soon after, he got his bachelor's degree in April 1665, but the university closed as a precaution against the Great Plague, and Newton returned to his home. There he laid the foundations for differential and integral calculus he named the *method of fluxions*, and which was based on the feeling that integration and differentiation of a function were inverse procedures. At the same time, he also developed his ideas on optics and the law of gravitation.

The University of Cambridge reopened in April 1667, and Newton was elected to a minor fellowship at Trinity College, and then to a major one in July 1668 after having obtained his master's degree. He is credited for the discovery of several interesting results such as Newton's identities and his iterative method for computing the square root of a number.

His research had impressed Isaac Barrow (1630-1677) who, in 1669, resigned from his professorship in favor of Newton who succeeded him as the Lucasian Professor at Cambridge, only one year after receiving his master's degree. At that time, any Fellow of a college at Cambridge had to take the holy orders and become an ordained Anglican priest. However, the terms of the Lucasian professorship required that the holder not be active in the church in order to have more time for science. King Charles II (1630-1685), whose permission was needed, agreed to the promotion, thus avoiding a conflict between Newton's religious ideas and the Anglican orthodoxy.

At Cambridge, Newton's first lectures were about optics. They began in January 1670. The chromatic aberration in a telescope using refracting lenses led him to the conclusion that the white light contained different types of rays. He decomposed the white light by using a prism and recomposed it with a lens and a second prism. Therefore, he constructed a reflecting telescope, which led to him being elected a Fellow of the Royal Society in 1672. That same year, he published a paper in which he tried to prove experimentally that light consisted in the motion of small particles rather than waves, a theory defended by Robert Hooke (1635-1703) and Christiaan Huygens (1629-1695). His theory of light issued in difficult relations with Hooke, and Newton delayed the publication of his book *Opticks* until 1704, after Hooke's death. Let us mention that the wave theory of light was unanimously accepted only after the works of Augustin Jean Fresnel (1788-1827) around 1819. It seems that Newton suffered a nervous breakdown in 1678.

The greatest achievement of Newton is his theory of universal gravitation. He had early versions of his three laws of motion as soon as 1666, when back in Woolsthorpe. The anecdote of the apple falling from a tree is well known. His main idea was that the Earth's gravity influenced the Moon, and counterbalanced its centrifugal force.

It took Newton several years to fully develop his ideas on celestial mechanics, which culminated in the publication of his *Philosophiae Naturalis Principia Mathematica* (Mathematical Principles of Natural Philosophy) on July 5, 1687. This work brought Newton an international reputation despite that continental scientists did not accept the idea of action at a distance, and continued to believe in the theory presented by Descartes in 1644 in which forces act through the aether filled with vortices.

James II (1633-1701) became king of Great Britain on February 6, 1685. He had converted to Catholicism in 1669. However, once on the throne, he was supported both by the Anglicans and the Catholics. But after some time, he favored the Catholics. Newton was a staunch Protestant and he firmly opposed the king who was filling vacant positions at Oxford and Cambridge with Roman Catholics. Rebellions arose, the Protestants asked the help of William III of Orange (1650-1702), and James, refusing to attack the invading army, escaped to France on December 23, 1688. The University of Cambridge elected Newton as one of its two members to the Convention Parliament on January 15, 1689. This Parliament declared that James had abdicated, and in February 1689, offered the crown to William and and his wife Mary, the true heiress to the English crown.

The dispute between Newton and Gottfried Wilhelm Leibniz (1646-1716) about the discovery of differential and integral calculus began in 1699; see Section 10.50. Newton claimed he was already working on this idea in 1666 when at home, but he did not publish his *fluxional notation for the calculus* until 1693 for its first part, and in full in 1704. Leibniz began to work on this problem in 1674, and his first paper on this topic appeared ten years later. The quarrel ended with Leibniz's death. Today, the consensus is that Leibniz and Newton independently invented the calculus.

Newton suffered a second breakdown in 1693 and retired from research. He decided to leave Cambridge and move to London to take up the lucrative post of warden of the Royal Mint in 1696, a position he held for the last 30 years of his life. In 1699, he became Master of the Mint, a more lucrative position. He took this work very seriously and he even successfully prosecuted 28 coiners.

Newton was elected President of the Royal Society in 1703, and was re-elected until his death. He was also an associate to the French Académie des Sciences. In April 1705, Queen Anne (1665-1714) knighted him during a royal visit to Trinity College in Cambridge.

In 1711, Newton published his *Methodus Differentialis* for the determination of a polynomial interpolating a certain number of points using *divided differences*. This study led him to a linear system with a Vandermonde matrix that he solved by elimination, and to the so-called Newton's interpolation formula.

Toward the end of his life, Newton lived at Cranbury Park, near Winchester, with his niece and her husband. He died in his sleep in London on March 20, 1727. Newton never married, and according to many testimonies, he was not a very easygoing man.

Work

For the work of Newton on elimination methods at the end of the 1680s, see Chapter 2.

10.54 ▪ Willgodt T. Odhner

Willgodt Theophil Odhner

Life

Willgodt Theophil Odhner was born in Westby, in the northern Wärmland province of Sweden, on August 10, 1845, the eldest of six children. His father was Theophil Dynamiel Odhner (1816-1863) and his mother was Fredrika Sofia Wall (1820-1874).

In 1854-1856, Odhner attended the school of Karlstad, where the family had moved in 1848. Then, he moved to Stockholm to work for his uncle. From 1864 to 1866, he followed courses at the Royal Institute of Technology in Stockholm, studying practical mechanics and mechanical technology, but he did not complete his studies.

At the end of 1868 or the beginning of 1869, he emigrated to Saint Petersburg in Russia. Initially, he found a job in a small mechanical workshop, but after a few months, he started to work in the factory of Ludvig Nobel (1831-1888), the brother of Alfred Nobel (1833-1896). This factory mainly manufactured weapons.

Odhner married Alma Skånberg (1853-1927) in 1871 with whom he had eight children.

There are several stories about how Odhner started being interested in calculating machines. The first one says that "he had an opportunity to repair a Thomas arithmometer and then became convinced that it is possible to solve the problem of mechanical calculation by a simpler and more appropriate way." The other story says that he had read an article about Thomas' arithmometer in *Dinglers Polytechnisches Journal* and thought it might be possible to construct a simpler calculating machine.

Aside from Nobel, Odhner had difficult relationships with the directors of the factory who thought the calculator project took too much time and he was, more or less, fired when Nobel was traveling abroad. Moreover, in 1877-1878, there was a war between Russia and Turkey and the Nobel factory received more military orders. In 1878, Odhner got a job at the Imperial Factory producing state papers. He stayed there for 14 years working on improving the printing machines.

In 1882, Odhner started his own business to produce paper cut in special forms which, apparently, was not very successful. In 1886, he founded his workshop which became the W.T. Odhner factory. In the beginning there were only members of his family as employees and only one machine. In 1892, Odhmer left the Imperial Factory and worked full time at his workshop. Short on money, he took as an associate Frank Hill, an Englishman. The company expanded, and in 1893, they built a new factory equipped to do mass production of calculators (Figure 10.4). In 1893, there were around 100 workers in the factory. The partnership with Hill probably ended in 1895. The factory produced not only calculators, but also, for instance, printing presses.

Willgodt Odhner died on September 2, 1903, in Saint Petersburg. After Odhner's death the production was continued by his sons Alexander and Georg and his son-in-law Karl Siewert. Around 29,000 machines were produced until the factory was forced to close down in 1918 and move to Sweden, to a new company Aktiebolaget Original-Odhner. In 1918, Axel Ehrenfrid Wibel in Stockholm got the rights to manufacture this type of machine under the name Facit.

Figure 10.4. *Odhner-Hill calculating machine*

Work

It is believed that Odhner started working on his calculating machine project in 1874. He did that in his spare time. Whether or not he was influenced by the machines of Baldwin and Staffel is not known for sure. His machine used a pinwheel instead of a stepped drum like almost all of the previous machines. He worked on his project for 15 years. It seems that a first prototype was finished in 1875. Odhner tried to convince Ludvig Nobel to start the production of the calculator. Nobel agreed to produce 14 machines. They were finished in 1877. Then, as we saw above, Odhner left the Nobel factory.

In 1878, he presented one of his machines for consideration by the Imperial Russian Technical Society. The report was mostly positive but also pinpointed some defects of the machine. Odhner got in relation with Karl Königsberger, a local businessman who tried to fill a patent for Odhner's 1877 machine in the USA. The answer of the patent office was that some claims infringed on a patent filed by Frank Stephen Baldwin (1838-1925). After some modifications the patent was accepted in October 1878. Then, Königsberger secured patents in Germany, Sweden, and Russia.

In 1890, the production of calculating machines started at Odhner's workshop. At this time he had an improved model (using a pinwheel with a variable number of teeth) as well as an advertising brochure and an instruction manual in Russian and German. He took all the rights for his machine from Königsberger and filed patents for his new model in Russia, France, Luxemburg, Belgium, Sweden, Norway, Austria-Hungary, England, Germany, Switzerland, and the USA. The production of the first year is estimated at 500 machines. Models with different numbers of decimal digits were manufactured. The German company Grimme, Natalis & Co. bought a license in 1892 and sold the machine under the name Brunsviga. This machine was a success, and in 1912, more than 20,000 had been sold. It was produced until 1958.

Most of the information in this section was obtained from Timo Leipälä, *The life and works of W.T. Odhner, parts I and II*, 12. Internationales Treffen der Rechenschiebersammler und 3. Symposium zur Entwicklung der Rechentechnik (2001).

10.55 ▪ Alexander M. Ostrowski

Alexander Markovich Ostrowski

Life

Alexander Markovich Ostrowski was born in Kiev, then in the Russian Empire and now in Ukraine, on September 25, 1893. He was the son of Mark Ostrowski, a merchant, and Vera Rashevskaya.

He first attended primary school in Kiev and a private school for a year. Then, he entered the Kiev School of Commerce where his teachers soon became aware of his extraordinary talents in mathematics. They recommended him to Dmitry Aleksandrovich Grave (1863-1939), a former student of Pafnuty Lvovich Chebyshev (1821-1894) and a professor of mathematics at the University of Kiev. Although he was only 15, Grave accepted him in his seminar on algebra. In 1911, Ostrowski obtained the title of Candidate of Commerce and he was awarded the Gold Medal. With Grave's help, he wrote his first memoir on Galois fields which appeared (in Ukrainian) in 1913.

Despite his successes, his entrance to the University of Kiev was rejected because he graduated from the School of Commerce and not from a high school, and also because there was a quota for Jewish students who were accepted by a lottery rather than on merit. Grave immediately wrote to Edmund Landau (1977-1938) and Kurt Hensel (1861-1941), and they invited Ostrowski to come to Germany. He accepted Hensel's offer at the Marburg University, and in 1912 he began to work there under his supervision.

World War I soon began and Ostrowski was considered as a civil prisoner with restricted movements. However, thanks to Hensel's intervention, he was allowed to use the university library. The four years of the war were not lost for him. He studied by himself and developed his famous theory of valuation on fields. He also studied foreign languages and music.

After the war, Ostrowski was free to travel and he went to Göttingen, the world center of mathematics at that time. Felix Klein (1849-1925) took him as one of his assistants. But for his doctorate, he worked with David Hilbert (1862-1943) and Landau. He obtained it with *summa cum laude* in 1920, solving in part Hilbert's 18th problem about lattices and sphere packing in the Euclidean space.

Then, Ostrowski went to Hamburg to work for his Habilitation as the assistant of Erich Hecke (1887-1947). There he studied a problem proposed by Hilbert on modules over polynomial rings. He was awarded his Habilitation in 1922. The following year, he accepted a lecturing position at Göttingen, teaching the recent developments of complex function theory.

After receiving a Rockefeller Research Fellowship, Ostrowski spent the academic year 1925-1926 at Oxford, Cambridge, and Edinburgh in the UK. But he was soon offered the Chair of

Mathematics at the University of Basel, Switzerland. He accepted and remained in Basel until his retirement in 1958.

In 1949, he married Margaret Sachs, a psychoanalyst who had been a student of the Swiss psychiatrist Carl Gustav Jung (1875-1961).

Alexander Ostrowski died on November 20, 1986, four years after his wife, in their villa Casa Almarost (ALexander MARgret OSTrowski) in Montagnola, overlooking the Lake of Lugano.

For details on Ostrowski, see the paper *Alexander M. Ostrowski (1893-1986): His life, work, and students*[78] by Walter Gautschi who was a student of Ostrowski in Basel.

Work

Although he was, at the origin, a pure mathematician, due to his numerous visits to the United States in the late forties and early fifties, Ostrowski became interested in applied mathematics and numerical analysis.

M-matrices and H-matrices were introduced by Ostrowski [2392] in 1937, even though he called them M-determinants and H-determinants.

After and even before his official retirement from Basel, Ostrowski widely traveled and lectured in several American universities. He was associated with the National Bureau of Standards, first at the Institute for Numerical Analysis at Los Angeles (see Chapter 5) and later in Washington. He became particularly interested in numerical methods for conformal mapping and problems, then at their beginning, related to the iterative solution of large systems of linear algebraic equations.

During his visits at the INA in the 1950s, he made several contributions to the Gerschgorin theory and he obtained bounds for eigenvalues of matrices.

In 1954, Ostrowski proved in [2397] that if $A = D - E - E^*$ is a Hermitian matrix with D Hermitian and positive definite, and $D - \omega E$ nonsingular for $0 \leq \omega \leq 2$, then the spectral radius of the SOR iteration matrix is strictly less than 1 if and only if A is positive definite and $0 < \omega < 2$. This result is often referred to as the Ostrowski-Reich theorem.

His monograph *Solution of Equations and Systems of Equations* was published in 1960. His work on algebraic equations involved methods for approximating their zeros, a study of the fundamental theorem of algebra, and Galois theory.

His results on the theory of norms of matrices and their applications to finding inequalities, to methods for solving linear systems, and locating and approximating eigenvalues of matrices have to be mentioned.

A few of Ostrowski's papers related to numerical linear algebra are [2392, 2394, 2395, 2396, 2397, 2398, 2399, 2400, 2407].

10.56 ▪ Oskar Perron

Life

Oskar Perron was born on May 7, 1880, in Frankenthal, Pfalz, a town in southwestern Germany, in the state of Rhineland-Palatinate. His father was a merchant and banker.

The young Oskar first went to the elementary school in 1886, to the Latin School for six years in the autumn of 1889, and then to the Humanist Gymnasium in Worms for two and a half years where he mainly studied classics. Despite the fact that his father wished him to pursue the family business, he studied mathematics in his spare time.

[78] https://www.cs.purdue.edu/homes/wxg/AMOengl.pdf (accessed August 2021)

Oskar Perron

In 1898, Perron obtained his Abitur and registered at the University of Munich to study mathematics and physics. As was usual at this time, he spent semesters at various universities, and, in particular, he studied at the University of Berlin. He began research in Munich under the guidance of Carl Louis Ferdinand von Lindemann (1852-1939). In 1902, he obtained his doctorate at the Ludwig-Maximilian University of Munich for his 43-page thesis whose title was *Über die Drehung eines starren Körpers um seinen Schwerpunkt bei Wirkung äusserer Kräfte* (On the rotation of a rigid body about its center of gravity by the action of external forces).

After his doctorate, Perron studied at Tübingen University and at Göttingen University where he worked with David Hilbert (1862-1943). In 1905, he published a paper entitled *Über eine Anwendung der Idealtheorie auf die Frage nach der Irreduzibilität algebraischer Gleichungen* (On an application of ideal theory to the question of the irreducibility of algebraic equations) in which he studied the algebraic dependence of polynomials. Perron completed his Habilitation *Grundlagen für eine Theorie des Jacobischen Kettenbruchalgorithmus* (Foundations for a theory of the Jacobian continued fraction algorithm) at Munich and was appointed a lecturer there in 1906. On July 28, 1906, he married Hermine Perron (1883-1961), who was connected to him via various family channels. They had three daughters.

In 1909, Hilbert took an active part in the selection of the successor of Hermann Minskowski (1864-1909) at the University of Göttingen. The last choice was between Perron and Edmund Georg Hermann Landau (1877-1938). There was considerable discussion on the respective merits of the two candidates. In the end, Felix Christian Klein (1849-1925) said

> *Oh, Perron is such a wonderful person. Everybody loves him. Landau is very disagreeable, very difficult to get along with. But we, being such a group as we are here, it is better that we have a man who is not so easy.*

Landau was chosen and Perron went to Tübingen as an extraordinary professor in 1910 [2552]. On December 13, 1913, Perron became an ordinary professor at Heidelberg, taking up the appointment in 1914. However, WW I stopped his career. From 1915 to 1918, he served in the Landsturm, a third-line reserve force consisting of older men. He was a lieutenant in a surveying unit until the end of the war. He received the Iron Cross. At the end of the war, he returned to Heidelberg where he taught until September 30, 1922, when he succeeded Alfred

Israel Pringsheim (1850-1941), the father-in-law of Thomas Mann (1875-1955), Nobel laureate in literature in 1929, in his chair at the Ludwig-Maximilian-University of Munich. There, he was a colleague of Constantin Carathéodory (1873-1950) and Heinrich Tietze (1880-1964) and the three became known as the "Munich mathematical triumvirate." Perron was elected to the Heidelberg Academy of Sciences in 1917, the German Academy of Scientists Leopoldina in 1919, the Bavarian Academy of Sciences in 1924, and the Göttingen Academy of Sciences in 1928.

The Nazi party came to power in early 1933. Perron was a strong opponent of it. In 1934, as president of the German Mathematical Society, he fought against the declarations of Ludwig Bieberbach (1886-1982) on Aryan and non-Aryan mathematics. Perron was in a difficult position since the Reich Ministry considered removing him from his position, but it seems that the decision was dropped. Later, despite many disagreements, he again manifested his opposition to the Nazis on several occasions.

Perron retired in 1951, but he continued to teach some courses at Munich until 1960, up to the age of 80. Moreover, he was still working and he published 18 papers between 1964 and 1973. The most remarkable of his books was *Nichteuklidische Elementargeometrie der Ebene* (Non-Euclidean elementary geometry of the plane) which he published in 1962. He had about 34 doctoral students.

Perron was awarded an honorary doctorate from the University of Tübingen in 1956, an honorary doctorate from the University of Mainz in 1960, and the Bavarian Order of Merit in 1959.

Perron loved mountains and their surroundings. His holidays were never complete without a stay in the mountains. He climbed the Totenkirchl (2,200 meters) in the Wilder Kaiser, Austria, more than 20 times, the last time when he was 74.

He died on February 22, 1975, in Munich.

Work

Perron had a total of 218 publications in a bibliography that he compiled himself. They cover a wide range of mathematical topics: differential equations, matrices and other topics in algebra, continued fractions, geometry, number theory, asymptotic expansions, and series. His work in analysis is remembered through the Perron integral (around 1910) that generalizes the notion of anti-derivative due to Isaac Newton (1643-1727) and is independent of the theory of measure of Henri-Léon Lebesgue (1875-1941). He also introduced subharmonic functions for solving the Dirichlet problem.

When in Munich around 1900, Perron followed the lectures on continued fractions by Pringsheim. Notice that, in 1882, Lindemann had became famous for his proof of the transcendental character of π, thus ending by a negative result a problem open for more than 2,000 years. His proof, using a variant of continued fractions and Padé approximants, was in the steps of the proof by Charles Hermite (1822-1901) for the transcendence of the number e in 1873. Thus, it is not surprising that Perron was interested in continued fractions and, in 1905, he published the paper *Über die Konvergenz von Kettenbrüchen mit positiven Gliedern* (Note on the convergence of continued fractions with positive terms) thus continuing 1898 Pringsheim's criterion for the convergence of a continued fraction. His work on this topic culminated in 1913 with his quite influential book *Die Lehre von den Kettenbrüchen* (Lessons on continued fractions), a second edition of which appeared in 1929, and a third one in 1954-1957. During the 19th century, continued fractions were a most important subject and a chapter was devoted to them in any textbook, and they were used in many applications [433]. However, before Perron's book, only one book had been entirely dedicated to them. It had been written in French by the

Russian mathematician Konstantin Alexandrovich Posse (1847-1928) under the name of Constantin Possé in 1886 [2513]. Neither Posse's nor Perron's books were translated into English.

For several decades, Perron also worked on Diophantine equations, a related topic. He is also known for the Jacobi-Perron algorithm studied in his Habilitation in 1906 (published in 1907 in *Mathematische Annalen*), which generalizes continued fractions to several dimensions in a natural way. It leads to recursively obtaining a sequence of simultaneous rational approximations to a set of real numbers [295].

The Perron-Frobenius theorem, proved by Perron in 1907 and Ferdinand Georg Frobenius (1849-1917) in 1912, describes the properties of the leading eigenvalue and of the corresponding eigenvectors of a non-negative real square matrix. It has important applications to probability theory (ergodicity of Markov chains), to the theory of dynamical systems, to economics (Okishio's theorem, Hawkins-Simon condition), to demography (Leslie population age distribution model), to social networks (DeGroot learning process), to Internet search engines (the PageRank algorithm), and even to ranking of football teams. Several proofs of this result can be found in the literature and various extensions as well.

In his Habilitation thesis on continued fractions, Perron had to solve some technical problems on matrices and their characteristic equations. A few months later he isolated these matrix problems in his paper *Zur Theorie der Matrices*, published in the *Mathematische Annalen*, and he gave a complete proof of the theorem [2468]. The entire filiation between Perron's work on continued fractions, this theorem, and its role in inspiring a slight generalization and the simplification of its proof by Frobenius in 1912 is explained in detail in [1602]. Roughly at the same time as Perron, Andrei Andreyevitch Markov (1856-1922), who was studying a type of probabilistic model, became interested in the existence of an eigenvalue with dominance properties of a stochastic non-negative matrix. He presented his results to the Saint Petersburg Academy of Sciences on December 7, 1907, but the paper was only published in 1908 [2141]. In his paper, Markov anticipated Frobenius' notion of an irreducible matrix, although restricted to stochastic matrices. A German translation was appended to the German edition of his lectures on probability theory in 1912, the same year as Frobenius' paper [2142]. There is no evidence that Frobenius knew it. The theory of Markov chains only became widespread in the 1930s and, then, rigorously developed through the Perron-Frobenius theorem.

Perron published an important book on irrational numbers in 1921. It only required a school level in mathematics as a prerequisite. It had several editions and Perron still revised it in 1960 when he was 80 years old.

For illustrating the danger of assuming that the solution of an optimization problem exists, Perron introduced the so-called Perron's paradox. Let N be the largest integer, if $N > 1$, then $N^2 > N$, which contradicts the definition of N. Hence $N = 1$.

As stated in [1602] *Perron was a talented and creative mathematician, whose work has not been accorded the historical attention it deserves.*

10.57 ▪ Lewis F. Richardson

Life

Lewis Fry Richardson was born on October 11, 1881 in Newcastle upon Tyne, England. His father David Richardson (1835-1913) operated a tanning and leather-manufacturing business. Fry was the maiden name of his mother Catherine (1838-1919). Lewis was the youngest of seven children.

Early on he showed an independent mind and an empirical approach. In 1898, he entered the Durham College of Science where he took courses in mathematics, physics, chemistry, botany, and zoology. Then, in 1900, he went to King's College in Cambridge and graduated with a

Lewis Fry Richardson

first-class degree in 1903. He spent the next ten years holding a series of positions in various academic and industrial laboratories.

In 1909, he married Dorothy Garnett (1885-1956). They adopted three children in the 1920s.

In 1910, Richardson submitted the paper [2566] for a D.Sc. and a fellowship at Cambridge, but it was rejected. The ideas contained in this work were too new, and the mathematics were considered approximate mathematics!

In 1913, Richardson became Superintendent of the Eskdalemuir Observatory in southern Scotland. During World War I, on May 16, 1916, he resigned and joined the Friends' Ambulance Unit (a Quaker organization) in France. He began to think about the causes of wars and how to prevent them. He suggested that the animosity between two countries could be measured, and that some differential equations are involved in the process. He published a book with these ideas, and returned to weather prediction.

In 1920, he became a lecturer in mathematics and physics at Westminster Training College, an institution training prospective schoolteachers up to a bachelor's degree. In 1926, he was elected a Fellow of the Royal Society of London.

Richardson left Westminster Training College in 1929 for the position of principal at the Technical College in Paisley, an industrial city near Glasgow. Although he had to teach 16 hours a week, he continued his research but came back to the study of the causes of wars and their prevention. Richardson wanted to *see whether there is any statistical connection between war, riot, and murder*. He began to accumulate such data, and decided to search for a relation between the probability of two countries going to war and the length of their common border. To his surprise, the lengths of the borders varied from one source to another. So, he investigated how to measure the length of a border, and he realized that it highly depends on the length of the ruler. Using a small ruler allows to follow more wiggles, more irregularities, than a long one which cuts the details. Thus, the smaller the ruler, the larger the result. The relation between the length of the border and that of the ruler leads to a new mathematical measure of rippling. At that time, Richardson's results on this problem were ignored by the scientific community, and they were only published posthumously. Today, they are considered to be at the origin of fractals.

In 1943, Richardson and his wife moved to their last home at Kilmun, 25 miles from Glasgow. He returned to his research on differential equations, and solved the associated system of linear equations by the so-called Richardson method.

Richardson died on September 30, 1953, in Kilmun. See [89] for a full-length biography.

Work

When serving as a chemist with the National Peat Industry Ltd., Richardson had to study the percolation of water. The process was described by the Laplace equation on an irregular domain. Richardson used finite differences and extrapolation to solve it [2566]. His iterative method for solving systems of linear equations is also described in this lengthy paper which was only accepted after much deliberation. While at Eskdalemuir Observatory, he again used finite differences for the problem of weather forecasting using a mathematical model. This was quite unusual at that time. Although he was certainly aware of the difficulty of the problem, since he estimated at 60,000 the number of people that have to be involved in the computations in order to obtain the prediction of tomorrow's weather before the day actually began, it seems that he did not realize that the problem was ill-conditioned. He also began to write a book on this topic [2567], published in 1922.

Over the years, Richardson made important contributions to fluid dynamics, in particular eddy-diffusion in the atmosphere. The so-called Richardson number is a fundamental quantity involving gradients of temperature and wind velocity.

The second paper where Richardson used extrapolation was published in 1927. It consists of two parts. The first one is due to him [2568], while the author of the second one is John Arthur Gaunt (1904-1944) [1296], one of his students.

In the mid-1930s, Richard Vynne Southwell (1888-1970) (see Section 10.64) published a series of papers in which he described his *relaxation method* for solving systems of linear equations. Richardson considered that the method was similar to his own finite difference method and suggested that Southwell should have acknowledged his paper. Southwell did not agree, arguing that the two methods were distinct. In later publications, Richardson referred, in polite terms, to this fact. In his later writings, Southwell however paid tribute to Richardson's *notable examples* and to *his ingenious device whereby approximate solutions may be improved.* In a letter Richardson wrote in June 1952 to Jule Gregory Charney (1917-1981), an American meteorologist who played an important role in developing numerical weather prediction, he commented about his method:

> *I think it is very like Southwell "relaxation method": he maintains that they are quite different. My method was first published in 1910: his 25 years later.*

10.58 ▪ Reuben L. Rosenberg

Reuben Louis Rosenberg

Life

It was difficult to find information about Rosenberg's life since snippets of information are disseminated on many websites.

Reuben Louis Rosenberg was born in Johannesburg, South Africa, on November 13, 1909. In 1929, he obtained a M.A. (Magister Artium) from the University of Cape Town in South Africa with a distinction in Applied Mathematics, and, in 1932, a Ph.D. at Berlin (magna cum laude). Then, he received a Beit Fellowship for Scientific Research from the Imperial College of Science, Technology and Medicine in London to undertake theoretical investigations in topics connected with quantum mechanics and obtain a D.I.C. (Diploma of Imperial College).

In 1934, he published, as a research fellow of Imperial College, a paper on the kinetic theory of the chromosphere. Then, it seems that he had taught at the University of the Witwatersrand in Johannesburg before obtaining an appointment, as a senior lecturer in Applied Mathematics, at the Natal University College (NUC), now University Kwazulu-Natal from 2004, in the town of Pietermaritzburg.

Rosenberg was involved in the Dramatic Society that staged, in 1940 (not for the first time), an outstanding performance of Oscar Wilde's *The Importance of Being Earnest*. At this occasion he was described "superb" for his "stage presence" and "diction." He himself financed the award of an annual "Oscar" for the best stage performance. Since, for some years, he was warden of the Men's Residence, or "Lord Warden" as he called himself with ironic reference to the Lady Warden on campus. When he donated a cup to be presented annually to the best student actor in a University Dramatic Society production it was appropriately engraved "The Lord Warden's Trophy."

In 1940, Rosenberg joined the armed forces as the Council of the College was specifically requested to release certain staff members from their academic responsibilities so that they could assist in military training and in other areas of expertise required for the war effort. In 1944, he is mentioned as belonging to the Department of Mathematics of the NUC at the occasion of a meeting of the Durban Library Committee regarding the safety of the collections. Duplication of library materials between the three Durban libraries was not encouraged but practical difficulties were encountered in the prevention of duplication between the two centers. Via the Pietermaritzburg Library Committee, Rosenberg requested that "a complete catalogue of the books and journals of the N.U.C. Library in Durban be made available in the Pietermaritzburg Library, pointing out that this will avoid excessive duplication and increase our library facilities and also that current numbers of certain journals, which by arrangement are obtained in Durban and not in Pietermaritzburg, be circulated in the Pietermaritzburg Library one month after their arrival in Durban."

In 1947-1948, he spent time continuing his research in atomic physics in the Physics Department of Bristol University, England. It is at this time that his paper with Philip Bernard Stein (1890-1974) on the solution of linear equations by iteration (the famous Stein-Rosenberg theorem, see Chapter 5) appeared (submitted on June 16, 1947) [2862]. Rosenberg resigned from NUC in 1949.

Then, Rosenberg spent two years at Bristol University followed by six years at Chelsea Polytechnic in London. In 1958, he joined the Mathematics Department of the University of New Brunswick in Canada, and rapidly became its head. He complained privately and publicly that students coming to the university were not well prepared. He was followed by other educators and his complaint led to a reform of secondary education.

In 1960-1962, Rosenberg was affiliated with the Division of Pure Physics, National Research Council in Ottawa, on sabbatical leave from University of New Brunswick. During this period, he wrote four papers on nuclear physics. In 1963, he was the Vice-President of the Canadian Mathematical Society, and, from 1964 to 1966, the President of the Association of University of

New Brunswick Teachers. He published several books for students, and a thesis in physics was defended under him at New Brunswick in 1965.

Then, from 1966, he was at Carleton University in Ottawa as a professor in the Department of Mathematics and later its chairman, member of the Senate of the University (1966-1968), chairman of the Interdepartmental Committees in Mathematics and Physics (1971-1972), and member of the General Science Committee for mathematics (1972-1973). From 1976, he was sessional lecturer in mathematics of the Department of Mathematics.

Rosenberg died on August 25, 1986, in Ottawa, where he is buried.

Endowed in 1986 by the daughters, friends, and academic colleagues of Rosenberg, the R.L. Rosenberg Memorial Scholarship in Mathematics is still awarded to an outstanding student entering a first-year honors program in the School of Mathematics and Statistics at Carleton University.

Work

Since he often changed places, it was difficult to find the papers published by Rosenberg. Let us comment on some of them.

In his paper on the kinetic theory of the chromosphere (received March 15, 1934) when he was at Imperial College, Rosenberg thanked the astronomer and mathematician William Hunter McCrea (1904-1999), who was then a Reader at Imperial College in London, for his many suggestions and discussions, and the Norwegian astrophysicist Svein Rosseland (1894-1985) for his interest in his work in which he was discussing the validity of one of his assumptions.

After his paper with Stein (received June 16, 1947; read October 16, 1947), it seems that Rosenberg never returned to numerical linear algebra.

In 1962-1964, Rosenberg, on sabbatical leave from the University of New Brunswick, published four papers with Ta-You Wu (1907-2000), a Chinese physicist who worked in the United States, Canada, mainland China, and Taiwan, and has been called the Father of Chinese Physics. The first three papers present a formulation of the theory of irreversible processes in ionized gases on the basis of the theory of Bogoliubov for neutral gases. In their last paper, the kinetic equation of Guernsey-Balescu for spatially homogeneous plasmas was solved as an initial value problem in the linearized approximation. The distribution functions for the electrons and the positive ions were expanded in series of associated Laguerre polynomials in the momentum, with coefficients which are functions of the time. The solution of the (infinite) systems of linear equations for these coefficients leads to the "spectrum of relaxation times," but the method used is not explained.

10.59 ▪ Heinz Rutishauser

Life

Heinz Rutishauser was born on January 30, 1918, in Weinfelden, Switzerland, 40 km south of Constance. His father Emil (1891-1931), who died when Heinz was 13 years old, was the headmaster of the Thurgauische Kantonsschule, a school in the nearby town of Frauenfeld. His mother Emma (1892-1934) passed away three years after his father. Thus, together with his younger brother and sister, he went to live in his uncle's house.

Rutishauser obtained his *Matura* (a secondary school exit exam) in 1936. From 1936 to 1942, he studied mathematics at the Eidgenössische Technische Hochschule (ETH, Federal Institute of Technology) in Zürich. His studies were interrupted by his military service that he spent in the artillery in the Gotthard region.

Heinz Rutishauser

From 1942 to 1945, he was an assistant of Walter Saxer (1896-1974), a mathematician specialist in insurance mathematics at ETH, and then a gymnasium teacher in Glarisegg and Trogen in 1947-1949. In 1948, he married Margrit Wirz. They had a daughter, Hanna (born 1950), who became a writer.

In 1950, Rutishauser defended his doctoral thesis *Über Folgen und Scharen von analytischen und meromorphen Funktionen mehrerer Variabeln, sowie von analytischen Abbildungen* (On the consequences and results of analytic and meromorphic functions of several variables, as well as of analytical maps) under the guidance of Saxer and Albert Pfluger (1907-1993), a specialist of complex function theory.

In 1948, Eduard Ludwig Stiefel (1909-1978, see Section 10.66) founded the Institut für Angewandte Mathematik (Institute for Applied Mathematics) at ETH Zürich to study the mathematical implications of computers. He recruited two assistants, Rutishauser and Ambrosius Paul Speiser (1922-2003), for working on the development of the first Swiss computer *ERMETH*. From 1949 to 1955, Rutishauser was employed as a research associate there. In 1948-1949, he and Speiser went to the USA and visited Howard Hathaway Aiken (1900-1973) at Harvard, John von Neumann (1903-1957) at Princeton, and other computing laboratories, to study the state of the art in computing. When they came back to Zürich, they learned that Stiefel had changed his plans and that ETH had rented the *Z4* computer of Konrad Zuze (1910-1995); see Section 10.78. Nevertheless, the *ERMETH* was finally built in 1955.

In 1951, Rutishauser defended his Habilitation *Automatische Rechenplanfertigung* (Automatic construction of computation plans) and became *Privatdozent.* As stated in [1494, 1495], he was a pioneer in *the automatic compilation of a suitably formulated algorithm and thus introduced the concept of what is now known as a compiler.* Rutishauser's ideas had a decisive impact on the creation of an internationally based programming language for scientific computing. He introduced several basic syntactic features to computer programming. In particular, he was involved in the definition and the design of the programming language Algol; see Chapter 8.

In 1955, he was appointed associate professor, and in 1968, he became full professor and the head of the Group for Computer Science, which later became the Computer Science Institute and ultimately in 1981 the Division of Computer Science at ETH Zürich.

At least from 1955, Rutishauser suffered from heart problems. In 1964, he had a first heart attack from which he recovered, but he became seriously ill in 1969. On November 10, 1970, he died in his office from acute heart failure.

For a paper on Rutishauser, see [226].

Work

In numerical analysis, Rutishauser first worked on the instability of numerical methods for solving ordinary differential equations. He also applied Romberg's method to numerical differentiation. However, besides his work in computer science, he is mostly known for the derivation of the qd-algorithm and the LR-algorithm.

The qd-algorithm is a method for determining, via continued fractions, the poles of a rational function expressed as a series in inverse powers of its variable. One step of this algorithm is equivalent to the product of two triangular matrices, a lower one and an upper one, each of them having only two nonzero diagonals, the main one and the diagonal next to it. Hence, the product is tridiagonal.

Rutishauser showed that multiplying these matrices together in the reverse order produced the next step of the qd-algorithm. Thus, Rutishauser had the idea of the LR method; see [1504]. For computing the eigenvalues of a general matrix A, he decomposed it into to the product $A = L_0 R_0$, where L_0 is lower triangular and R_0 is upper triangular. Then he constructed the new matrix $A_1 = R_0 L_0$, which is, in turn, decomposed into the product $A_1 = L_1 R_1$, with L_1 lower triangular and R_1 upper triangular, and so on.

Under some assumptions on A, if the sequence of matrices (A_k) converges, its limit is an upper triangular matrix whose diagonal contains the eigenvalues of A in decreasing order of magnitude. This method was later superseded by the QR algorithm; see Chapter 6. Rutishauser also contributed to gradient methods for systems of linear equations [1500, 1504].

10.60 ▪ Issai Schur

Issai Schur

Life

Issai (Schaia or Isaiah as on his grave) Schur was born on January 10, 1875, into a Jewish family, in Mogilev, Russian Empire (now Belarus). His father Moses Schur (?-c. 1893) was a businessman. Schur's mother was Golde Schur, née Landau (1835-1915).

In 1888, he went to Libau, now called Liepāja (Courland, now in Latvia), to live with his sister and brother-in-law. He attended the German-speaking Nicolai Gymnasium in Libau from 1888 to 1894, reaching the top grade in his final examination, and receiving a gold medal.

In October 1894, Schur entered the University of Berlin to study mathematics and physics. On November 27, 1901, he defended his doctoral dissertation *Über eine Klasse von Matrizen, die sich einer gegebenen Matrix zu ordnen lassen* (On a class of matrices that can be mapped to a given matrix) under the supervision of Ferdinand Georg Frobenius (1849-1917) and Lazarus Immanuel Fuchs (1833-1902). His work led the foundation of a general theory of the representation of linear groups.

From 1903 to 1913, Schur was a lecturer (*Privatdozent*) at the University of Berlin. Then, he became an associate professor at the University of Bonn from April 21, 1913 until April 1, 1916, as the successor to Felix Hausdorff (1868-1942).

In the following years, Frobenius tried several times to get Schur back to Berlin. The efforts of Frobenius were finally successful in 1916, when Schur succeeded Johannes Knoblauch (1855-1915) as associate professor. Frobenius died a year later, on August 3, 1917. Schur and Constantin Carathéodory (1873-1950) were candidates to his succession. Carathéodory was nominated.

In 1919, Schur was given a personal professorship. In 1921, he obtained the chair of Friedrich Hermann Schottky (1851-1935) who was retiring. In 1922, he was elected to the Prussian Academy of Sciences. Until 1933, Schur's school at the University of Berlin was, without any doubt, the most influential group of mathematicians in Berlin and among the most important in all of Germany. He was a charismatic leader and a good teacher.

In 1933, the events in Germany made Schur's life difficult. He was suspended and excluded from the university. His colleague Erhard Schmidt (1876-1959) fought for his reinstatement, and Schur was allowed to continue his lectures until the end of September 1935. During a trip to Switzerland to visit his daughter in Bern, he received a letter from Ludwig Bieberbach (1886-1882), on the behalf of the Rector, asking him to come back immediately to Berlin. It was to announce to him that he was dismissed from his professorship on September 30, 1935. Moreover, due to an intervention of Bieberbach, he was forced to resign from the commissions of the Academy on April 7, 1938, and, half a year later, from the Academy itself.

Schur received several invitations from abroad but he declined all of them because of the language barrier, and also because he could not understand that a German was not welcome in his own country.

Schur left Germany in early 1939, after the final humiliation of being forced to find a sponsor to pay the "Reichs flight tax" to allow him to leave Germany. He was not in good health. He first went to Bern where his wife, Regina Malke Frumkin (1881-c. 1965), joined him a few days later. After some weeks, they emigrated to Palestine. Without sufficient funds to live, Schur was forced to sell his mathematical books to the Institute for Advanced Study in Princeton.

Schur died from a heart attack on January 10, 1941, the day of his 66th birthday, in Tel Aviv, Palestine (now Israel).

Work

Schur is mainly known for his work on the representation theory of groups. Concerning linear algebra, in 1909, Schur introduced a factorization of a matrix $A = QLQ^*$ with L lower triangular and Q unitary in a paper [2717] whose title can be translated as *About the characteristic roots of a linear substitution with an application to the theory of integral equations*. Replacing L by R, an upper triangular matrix, this is now known as the *Schur factorization* or Schur decomposition.

In his 1917 paper [2718] (submitted in September 1916), whose title in English is *About power series that are bounded inside the unit circle*, Schur introduced the notion of what is now named the *Schur complement*, as well as the *Schur determinantal identity*. The name was introduced by Emilie Virginia Haynsworth (1916-1985) in June 1968; see Section 10.33. Everything you always wanted to know about the Schur complement can be found in the book [3324] edited by Fuzhen Zhang in 2005.

For Schur's contributions to analysis, see [1043].

10.61 ▪ Hermann Schwarz

Hermann Amandus Schwarz

Life

The German mathematician Karl Hermann Amandus Schwarz was born on January 25, 1843, in Hermsdorf am Kynast, Silesia (now Sobieszów, Jelenia Góra, Poland). His father Guido Wilhelm Schwarz was an architect.

He studied at the Gymnasium in Dortmund where his favorite subject was chemistry. Then, he entered the Gewerbeinstitut in Berlin (later the Technical University of Berlin) having in mind to study that topic. There, he followed the lectures of Karl Wilhelm Pohlke (1810-1876), a painter who had proposed in 1860 a fundamental result in axonometry, a graphical procedure that generates a planar image of a three-dimensional object. Through him, Schwarz became interested in geometry, and, in 1863, he published the first proof of Pohlke's result; see [2788].

Ernst Eduard Kummer (1810-1893) and Karl Theodor Wilhelm Weierstrass (1815-1897) rapidly persuaded Schwarz to turn to mathematics. He attended Weierstrass's lectures on integral calculus in 1861, and he combined his interest in geometry with Weierstrass's ideas in analysis. Kummer and Weierstrass became his advisors for his doctorate *De superficiebus in planum explicabilibus primorum septem ordinum*, defended in 1864 under the latinized name *Carolus Arminius Amandus Schwarz*.

In Berlin, Schwarz completed his exams to be qualified as a teacher, and in 1867, he was appointed as a *Privatdozent* to the University of Halle. In 1869, he was nominated as a full professor of mathematics at the Eidgenössische Technische Hochschule in Zürich, Switzerland. Then, in 1875, he obtained the chair of mathematics at Göttingen University. Finally, in 1892, he succeeded Weierstrass at the University of Berlin where he lectured until 1918. There, due to his teaching duties and the care for his students, his research activity lowered. He also had several interests outside mathematics.

In 1892, Schwarz became a member of the Berlin Academy of Science, and he was made an Honorary member of the London Mathematical Society. Among his 22 doctoral students one can find Leopold Fejér (1880-1959), Richard Fuchs (1873-1944), Gerhard Hessenberg (1874-1925), Herman (Chaim) Müntz (1884-1956), Theodor Vahlen (1869-1945), and Ernst Zermelo (1871-1953).

In 1912, he married Marie Kummer (1842-1921), a daughter of Ernst Kummer. They had six children. At some point, Schwarz became the captain of the local Voluntary Fire Brigade and, more surprisingly, he helped the stationmaster at the local railway station by closing the doors of the trains.

Schwarz died in Berlin on November 30, 1921.

Work

While in Berlin, Schwarz worked on surfaces of least area (a problem in the calculus of variations), and, in 1864, he discovered what is now known as the *Schwarz minimal surface*. In 1866, the Belgian mathematician Joseph Antoine Ferdinand Plateau (1801-1883) published a famous memoir on this topic, and, that same year, Weierstrass found a link between the theory of minimal surfaces and that of analytic functions.

The Riemann mapping theorem states that any simply connected region of the complex plane can be mapped conformally onto a disk. However, the proof given by Bernhard Riemann (1826-1866) used the Dirichlet problem, that is, the problem of finding the solution of the partial differential equation $\Delta u = 0$ in the interior of a given region that takes prescribed values on its boundary. In 1870, Schwarz [2720] first solved the problem for the square and the triangle, and then for general polygons. He gave a method to conformally map polygonal regions to the circle, and, by approximating an arbitrary simply connected region by polygons, he obtained a rigorous proof of the Riemann mapping theorem. The Schwarz-Christoffel mapping was discovered independently by Elwin Bruno Christoffel (1829-1900) [678] in 1867 and Schwarz [2719] in 1869.

This work by Schwarz is the basis of what is now called the *Schwarz alternating method* for computing the solution of an elliptic boundary value problem on a domain which is the union of two overlapping subdomains; see Figure 10.5 and Chapter 5. This is considered the first ever domain decomposition method.

In 1873, Schwarz studied the cases when Gauss' hypergeometric function is an algebraic function. More precisely, he established a list of 15 cases of parameters in which the hypergeometric equation has two independent solutions that are algebraic functions. In this work, he defined a conformal mapping of a triangle with arcs of circles as sides onto the unit disk. This function is an early example of an automorphic function, a topic later developed by Felix Christian Klein (1849-1925) and Henri Poincaré (1854-1912).

In 1885, on the occasion of Weierstrass' 70th birthday, Schwarz completely solved the question of whether a given minimal surface really yields a minimal area. For the proof, he constructed a function using successive approximations, an idea that led Émile Picard (1856-1941) to his existence proof of the solution of a differential equation (the Picard iterates).

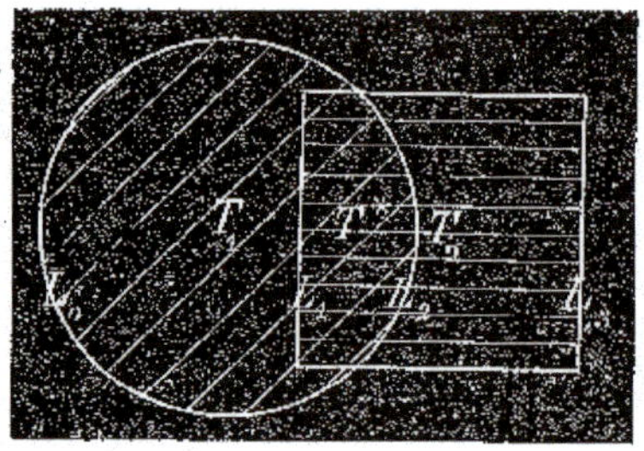

Figure 10.5. *Figure in Schwarz' 1870 paper*

The name Schwarz is also attached to the Cauchy-Schwarz inequality, sometimes also cited with the name of Bunyakovsky. This inequality for sums was first published by Augustin-Louis Cauchy (1789-1857) in 1821 in his *Cours d'Analyse de l'École Royale Polytechnique*, while the corresponding inequality for integrals was first proved in 1859 by the Russian mathematician Viktor Yakovlevich Bunyakovsky (1804-1889), who had studied in Paris with Cauchy. However, his work was written in French, and it does not seem to have widely circulated in Western Europe. For his work on minimal surfaces, Schwarz needed an analog of Cauchy's inequality for a two-dimensional integral, and he gave a modern proof of it in 1888 [2855].

10.62 ▪ Franz Schweins

Franz Ferdinand Schweins
Universitätsbibliothek Heidelberg

Life

Franz Ferdinand Schweins was born on March 24, 1780, in Fürstenberg, Germany. He was the son of the master carpenter Franz Schweins (1745-1813) and his mother was Gertrud Weitekamp (1753-1824).

He attended the high school in Paderborn and had to study theology at the request of his father. But early on he discovered a preference for mathematical studies, and was allowed to follow his inclination. From 1801 to 1802, he first studied at the Academy of Fine Arts in Kassel, and from 1802, he studied mathematics at the University of Göttingen.

In 1808, Schweins received his doctorate from the University of Göttingen with the dissertation *Circa problemeta aliqvot ad geometriam practicam spectantes*. He was nominated the same year as a *Privatdozent*. He also lectured on mathematics in Darmstadt.

In 1810, he became a private lecturer at the University of Heidelberg. In 1811, he was appointed associate professor. After he had rejected a call from the University of Greifswald, Schweins received in 1816 a full professorship of mathematics at the University of Heidelberg. In 1843, he became vice rector and was already in 1817, and 1832-1833, 1838-1839, and 1848-1849, a member of the Engeren Senate.

In 1819, 1825, 1832, 1838, and 1846, Schweins was elected Dean of the Philosophical Faculty of the University of Heidelberg. He worked there a little more than 46 years.

He was married to Catharina Barbara Bellosa (1790-1869) since 1826. They had a daughter. For his services he received in 1820 the title of Hofrat (Councilor) and in 1844 Secret Hofrat. Already in 1843, he received the Knight's Cross of the Order of the Zähringer Löwen. In October 1851, on the occasion of his 40th anniversary at the University of Heidelberg, Schweins was appointed to the Grand Duchy of Baden Secret Council.

Franz Ferdinand Schweins died on July 15, 1856, in Heidelberg.

Work

Combinatorial notions such as permutation and combination had been introduced by Blaise Pascal (1623-1662) and by Jacob Bernoulli (1654-1705). In the footsteps of Gottfried Wilhelm Leibniz (1646-1716), Carl Friedrich Hindenburg (1741-1808), a professor of physics and philosophy in Leipzig, founded the first modern school of combinatorics with the intention that this subject should occupy a major position in mathematics. Hindenburg used certain complicated notations for binomial coefficients and powers; one can feel the influence of the Rosenkreuzer secret codes here. The main advantages of the Hindenburg combinatorial school was the use of combinatorics in power series and the partial transition from the Latin language of Euler. The disadvantages were the limitation to formal computations and the old-fashioned notation.

Hindenburg had high hopes for his combinatorial school, and as a result Heinrich August Rothe (1773-1842) and Schweins formulated the q-binomial theorem, but without proof. Rothe introduced a sign for sums, which was used by Christoph Gudermann (1798-1852). In 1793, Rothe found a formula for the inversion of a formal power series, improving on a formula found without proof by Hieronymus Eschenbach (1764-1797) in 1789. This invention gave the combinatorial school a rise in Germany.

After beginning his studies in France, Edmund Külp (1801-1862) came back to Heidelberg, his native town, in order to obtain a doctorate. In his letters, he explains that Schweins

> *absolutely wants me to take care seriously combinatorial calculus, which I hate.*

He ends up abandoning the idea after finding that *the sickness has seized the minds too much powerfully*.

Schweins published many important discoveries in a book in 1825 [2726]. They were lost, however, and the fact that he had ever written on determinants was not brought to light until 1884 by Carl Gustav Jacob Jacobi (1804-1851) [2518].

Schweins found a determinantal identity, named after him; see, for instance, [18].

10.63 ▪ Philipp von Seidel

Philipp Ludwig von Seidel

Life

Philipp Ludwig von Seidel was born on October 24, 1821, in Zweibrücken, Germany. His father Justus Christian Felix (1783-1848) was working for the German post office and he had to move frequently. His mother was Juliane Philippine Reinhold (1793-1867). He had five brothers and sisters.

Due to his father's job, the young Philipp attended several different schools, first in Nördlingen, then in Nuremberg, and finally in Hof. He completed his school studies in the autumn of 1839, and then he received private lectures by Ludwig Christoph Schnürlein (1792-1852), a mathematics teacher at the Gymnasium in Hof who had studied under Gauss.

Seidel registered at the University of Berlin in 1840 and studied under Johann Peter Gustav Lejeune Dirichlet (1805-1859) and the astronomer Johann Franz Encke (1791-1865) for whom he carried out calculations at the observatory.

The custom among German students at this time was to spend time at various universities. Thus, in 1842, he moved to the Albertina in Königsberg where he studied under Carl Gustav Jacob Jacobi (1804-1851), Franz Ernst Neumann (1798-1895), and Friedrich Wilhelm Bessel (1784-1846) who, since Jacobi was leaving Königsberg for health reasons, advised him in 1843 to go to the Ludwig-Maximilians University in Munich for his doctorate.

With the help of the physicist Carl August von Steinheil (1801-1870), a former student of Bessel, Seidel obtained his doctorate from Munich in 1846 with the thesis *Über die beste Form der Spiegel in Teleskopen* (On the best form of mirrors in telescopes).

He defended his Habilitation at the University of Munich in 1847. The title was *Untersuchungen über die Konvergenz und Divergenz der Kettenbrüche* (Studies on the convergence and divergence of continued fractions). In this work he gave the first precise definitions of the convergence and divergence of an arithmetical continued fraction, see [433] by C.B. Seidel continued to work of these two topics throughout in his life.

In 1847, Seidel was appointed *Privatdozent* in Munich, then he became associate professor in 1851, and full professor in 1855. His lectures covered mathematics, including probability theory and the method of least squares, astronomy, and dioptrics. It is interesting to note that he never accepted Riemannian geometry. The famous physicist Max Planck (1858-1947) was among his students at the University of Munich.

In 1851, Seidel was elected an extraordinary member of the Bavarian Academy of Sciences, becoming a full member in 1861. In 1854, he was elected a corresponding member of the Göttingen Academy of Sciences. Since 1863 he was a corresponding member of the Prussian Academy of Sciences. In 1864, he was elected a member of the German Academy of Sciences Leopoldina. He also was a member of the Commission for the European Measurement of a Degree and of a group observing a transit of Venus.

From 1879 to 1882, Seidel was the successor of Johann von Lamont (1805-1879) and the director of the Bogenhausen observatory. Seidel was knighted (Bayerischer Ritter) in 1882.

Problems with his eyesight forced Seidel into early retirement. In the last decade of his life he became completely blind. He had never been married and had no immediate family to help him when he became ill. It was his unmarried sister Lucie Philippine Johanne (1819-1889) who looked after him. After her death, he certainly could not care for himself, and in his last seven years, he was looked after by the widow of a clergyman named Langhans.

Seidel died in Munich on August 13, 1896.

Work

Seidel was a mathematician, optician, and astronomer. In 1847, he found [2734] an incorrect lemma of Augustin-Louis Cauchy (1789-1857) [565, p. 131] stating that the limit of a convergent series of continuous functions is itself continuous. At first sight, Cauchy did not notice the distinction between simple and uniform convergence [1980]. Thus, the concept and the term of uniform convergence (*Gleichmässigen Konvergenz*) have to be credited to Seidel.

In 1855, he introduced the operations of contraction and extension of a continued fraction, and showed that a convergent continued fraction can be transformed into a divergent one and vice versa [2736].

In 1856, Seidel extended the Gaussian theory of rotationally symmetric lens systems to include monochromatic ray aberrations up to the third order. He mathematically identified five coefficients, now named after him, describing such aberrations [2073]. One year later, he published his widely acclaimed book *Theorie der Fehler, mit welchen die durch optische Instrumente gesehenen Bilder behaftet sind, und über die mathematischen Bedingungen ihrer Aufhebung* (Theory of errors affecting images seen through optical instruments, and on the mathematical conditions for their elimination), which was the standard work in the field for a long time [2737]. As stated by John William Strutt, also known as Lord Rayleigh (1842-1919), in 1908 [2932]

> *The theoretical investigation of this kind of aberration was one of Seidel's most important contributions to the subject, inasmuch as neither Airy nor Coddington appears to have contemplated it.*

Together with Steinheil, Seidel carried out the first photometric measurements on stars. He introduced probability theory into these measurements when considering atmospheric extinction, and derived trigonometric formulas for points lateral to the axis of an optical system. His work from 1856 soon became important for the production of improved telescopes by the optical-astronomical company *C.A. Steinheil & Söhne* founded by Steinheil in 1854.

Besides the application of probability theory to astronomy, Seidel also investigated the relation between the frequency of certain diseases and the climatic conditions in Munich. In particular, in 1865-1866, he published statistical studies which proved that the fluctuations in the groundwater level are related to the epidemic occurrence of typhus and cholera; see [1248].

After the Polytechnic School (today TU München) was founded by the King Ludwig II, Seidel received a teaching assignment for the method of least squares. It was in this context that he developed, in 1874, his method on the iterative solution of systems of linear equations, a method now known as the Gauss-Seidel method in numerical linear algebra; see Chapter 5. The title of his 33-page paper [2740] was *Über ein Verfahren, die Gleichungen, auf welche die Methode der kleinsten Quadrate führt, sowie lineare Gleichungen überhaupt, durch successive Annaherung aufzulösen* (About a procedure to solve the equations to which the method of least squares leads, as well as linear equations in general, by successive approximation).

10.64 ▪ Richard V. Southwell

Sir Richard Vynne Southwell
Courtesy of the Royal Society

Life

Richard Vynne Southwell was born in Norwich, Norfolk, UK, on July 2, 1888, the only son and the second of the three children of Edwin Batterbee Southwell (1854-1933), and his wife, Annie.

He was a delicate child and did not go to school until the age of nine. From 1898 to 1907, he was educated at King Edward VII's School in Norwich. His studies were almost entirely classical up to the age of sixteen, when he turned to mathematics.

He entered Trinity College in Cambridge in October 1907, and graduated in 1910 with first class honors in the Mechanical Sciences tripos. Immediately, he began to work on elastic stability, an almost unexplored topic at that time. He was coached by H.A. Webb, a fellow of Trinity College, who was interested in the theory of structures. He failed to secure a fellowship to Trinity College in 1911, but succeeded in 1912.

In October 1914, he was offered a position of College Lecturer at Trinity College, but he immediately volunteered when the war began and was sent to France as a Temporary Second Lieutenant in November 1914. The authorities having understood the importance of applied science in the conflict, Southwell was attached as a Lieutenant to the Royal Naval Air Service in charge of the development of non-rigid airships in May 1915. Promoted to Lieutenant-Commander in 1917, he was transferred to the R.A.F. with the rank of Major on April 1, 1918. Since he was assigned office work at the Admiralty, Bertram Hopkinson (1874-1918), a professor of mechanics and applied mathematics at Cambridge, managed to have him transferred to the staff of the

Royal Aircraft Establishment in Farnborough where he was in charge of the Aerodynamics and Structural Departments until his demobilization in March 1919. This experience brought him in touch with the leading aeronautical engineers of the time, and started his interest in the problems of aircraft structures.

In 1918, he married Isabella Wilhelmina Warburton Wingate (1896-1985), the daughter of a medical practitioner. They had four daughters.

After the war, Southwell returned to Trinity College, but in 1920, he accepted the position of Superintendent of the Aeronautics Department of the National Physical Laboratory where he began to work on space frames.

Although much of his works during the war had remained necessarily unpublished, they were well enough known for him to be elected to a Fellowship of the Royal Society in 1925. That same year, he returned to Trinity College as a Fellow and faculty lecturer in Mathematics. He became interested in vibration and hydrodynamic problems, and was a member of many governmental technical committees, including for the Air Ministry when the airships R.100 and R.101 were being devised. It was also the beginning of a long series of golfing weekends and holidays with, in particular, his friend Ernest Rutherford (1871-1937), who won the Nobel Prize in Chemistry in 1908 and was the director of the Cavendish Laboratory at Cambridge.

In 1929, Southwell became professor of Engineering Science at Oxford University. He took a large part in the teaching on the theory of structures and applied mechanics. His textbook, *An Introduction to the Theory of Elasticity for Engineers and Physicists*, appeared in January 1936. He established a research group, including his research assistant Derman Guy Christopherson (1915-2000). It is there that he developed his relaxation method.

In 1942, Southwell was invited to become the Rector of the Imperial College in London. His time in this position was not an easy one. Because of the war, educational work was more or less interrupted, and many members of the staff were involved in war work. When the war ended, a large number of students enrolled but there was little extra accommodation, limited finances, and few academic positions. Southwell was involved in the opening of a new student residence, Selkirk Hall. Despite his administrative duties, he also succeeded in maintaining his research activity, and published several papers describing applications of the relaxation method. Among his students, Olgierd Cecil Zienkiewicz (1921-2009), one of the early pioneers of the finite element method, defended his thesis in 1945. This is not surprising since in a paper dated 1945 [2837], Southwell explained

> *As my slides have shown, we present our solutions in the form of numerical values of the wanted solution (w) at nodal points of a uniform lattice or "net". The meshes (in theory) may be either hexagonal, or square, or triangular* [...]*; the values satisfy, not the governing equation (2) as it stands, but the approximation to it which results when its differentials are replaced by finite-difference approximations. Such replacement, of course, is no new device: indeed, it is hard to suggest an alternative, if the aim is to evolve a method applicable to any shape of boundary. I believe that triangular nets (which have some advantages in respect of accuracy) have not been employed before* [...]

In a later footnote, Southwell mentioned, without details, coarse meshes and finer ones, a remark that can be interpreted as related to what is now mesh refinement or to the multigrid method.

In 1948, at the age of 60, Southwell retired from his position at Imperial College. He was knighted that year. He continued to work and to publish. In the summer of 1959, he was involved in a street accident in Glasgow, and several operations were necessary before he could recover. But his memory became more and more unreliable, and the last years of his life were spent in seclusion. He died in St Andrew's Hospital, Northampton, on December 9, 1970, after a long illness.

Throughout his life, his colleagues found him a man of great charm and wit. A personal recollection of him is given in [679] together with a list of the honors he received and his bibliography.

Work

Southwell's first paper, published in 1912, was on the strength of struts. He then published several papers in theoretical applied mechanics.

After World War I, he wrote six papers on stress determination in space frames, and some others on tension coefficients, strain energy, and St. Venant's principle.

When in Oxford, Southwell developed his famous relaxation iterative method. In his very detailed obituary of Southwell [679] published in 1972, Christopherson explained the genesis of the method:

> *He was becoming increasingly preoccupied with the limitations of conventional mathematics in dealing with the problems with which the design engineer had to cope in practice. The fundamental differential equations could be derived, and in cases in which shapes and boundary conditions were simple - circles, rectangles and the like - could often be solved, but all too many elements in a practical engineering design are not circular or rectangular, and their most interesting features, from the point of stress determination, for example, are precisely those in which they differ from circles or rectangles. Not dissimilar difficulties arose in the apparently straightforward problem of the design of framed structures. Here the basic equations within the elastic range are linear, and there is in theory no difficulty in solving them. But even a moderately complicated framework may require the solution of some hundreds of simultaneous equations, and before the invention of computers this was in practice by orthodox methods almost an impossibility. It was these problems which led Southwell to the work which was to form the bulk of his research for the rest of his life. The technique originally described by the phrase "the systematic relaxation of constraints", which became generally known as the "Relaxation method", was devised by him first as a way by which engineers could understand and analyse the effect of any system of loading on a structural framework. The original idea was an engineering rather than a mathematical one. Suppose that the framework is in the first instance unloaded and unstressed. Round it, and supporting all the joints through a system of jacks, is erected in imagination a second perfectly rigid framework. The loads are then applied. Initially, of course, the imaginary rigid framework carries them all. The jacking system is then adjusted stage by stage, so that the loads are gradually transferred from the imaginary rigid framework to the real deformable one. When this process of "relaxation" has been carried on until the jacks are transmitting nothing, the whole system of loads has been transferred to the real framework and the calculated stresses in it are the true ones.*
>
> *Of course, such a calculation, performed in an automatic way by changing one variable at a time, would converge to the final solution only very slowly. But the basic idea can be easily developed so that the rate of convergence depends to a great extent on the intelligence and insight of the operator. In the case of a multistory framed building, for example, a single operation can be devised not just to change the displacement of one joint, but to adjust the position or the orientation of a whole floor, or of a "block" of many floors. When the resultant forces and moments on the various portions of the structure as a whole have reached the required values, the detail of displacements and rotations of individual joints and members are easily adjusted.*

These were the basic ideas presented by Southwell in two papers published in 1935, but he very soon saw that they could be applied over a much wider field than simply the analysis of frameworks. In mathematical terms the framework presents a set of linear equations of a particular kind. If the equations are written in the form of a determinant, the coefficients tend to become smaller as one moves away from the main diagonal. Such sets of equations occur not only in elastic structures, but in almost any other physical problem in which the quantities are linearly related. In particular, any linear differential equation, if replaced by a finite-difference analogue, results in just the type of problem which is most readily tractable by the relaxation technique. Accordingly, the idea could clearly be applied to any of the very wide range of problems in applied mechanics and related fields, in which what is wanted is a function given by a linear differential equation applying within a specified boundary on which the boundary condition is known.

The relaxation method was much developed in the decade 1932-1942 by Southwell and his group of students, among them, Leslie Fox (1918-1992), who defended his thesis on its use for solving the biharmonic equation in 1942; see Section 10.23

After having written several papers on this method since 1935, Southwell published his first monograph containing that topic in 1940 [2836]. A second volume appeared some years later [2838] in 1946 and a third one [2839] in 1956.

The rapid development of computers in the 1950s and 1960s rendered the method largely unnecessary. Other automatic relaxation methods were developed for computers, notably, by David Monaghan Young (1923-2008); see Section 10.77. During a visit to Harvard, where Young was working on his Ph.D. thesis, Southwell told David that *Any attempts to mechanize relaxation methods would be a waste of time*!

10.65 ▪ Philip B. Stein

Philip Bernard Stein

Life

Philip Bernard Stein was presumably born in Švėkšna, Šilutė District Municipality, Klaipėda County, Lithuania, on the estate of a Polish nobleman called Geminsitz. As Stein wrote[79] his official birthday was January 25, 1890, but, in fact, he did not know his actual birthday within a range of six months. He was the youngest child of a Jewish family. His father was Solomon Zebulin Stein and his mother Soroh Leah Malnik. He had four sisters.

In 1897, the family emigrated to Cape Colony in South Africa. Although the Colony had been British, it was independent since 1872. Stein attended the Normal College Boy's High School in Cape Town. He was awarded a minor bursary of £10 tenable for one year in the School Higher Examination of 1905. After graduating in 1906, he studied at the South African College in Cape Town, now the University of Cape Town. However, at this time, the College, although teaching at university level, did not have university status and, thus, could not deliver degrees. Stein sat the matriculation examination for the South African College in 1906 and he obtained a minor scholarship of £20 for one year. In 1909, while at the South African College, he was awarded the BA degree with honors in applied mathematics by the University of the Cape of Good Hope, and he also won the Ebden Scholarship, at the value of £200 per year, which allowed him to study abroad for three years. He entered Caius College in Cambridge and spent three years there, financed by the scholarship, studying for the Mathematical Tripos.

After his return to South Africa, Stein worked in the Audit Department of the South African Railway Offices in Johannesburg, but in about 1917, he joined the staff of the South African School of Mines and Technology in Johannesburg (from 1923 the University of the Witwatersrand). By this time he was a member of the South African Geographical Society and was scheduled to address its members in 1918 on "Cosmic theories." He also became a member of the South African Association for the Advancement of Science[80] in 1917.

Probably in 1918, he moved to the Natal University College in Pietermaritzburg (from 1950 the University of Natal) and was appointed there as a lecturer in 1920. That same year, he married Lily Rollnick, born in South Africa on February 15, 1920, with whom he had three children.

Up to that time, Stein was not a researcher and he had never studied for a doctorate. William Nicholas Roseveare (1864-1948), who had been the first professor of pure and applied mathematics at this college since 1910,[81] encouraged him to undertake research and, in 1926, he went on leave of absence with his family to the University of Cambridge in England. At Cambridge, John Edensor Littlewood (1885-1977) was his thesis advisor and Stein undertook research on the theory of functions. He was awarded a doctorate in 1931 for two dissertations: *On equalities for certain integrals in the theory of Picard functions* and *On the asymptotic distribution of the values of an integral function*.

Back home in 1931, Stein was appointed professor of mathematics and applied mathematics at the new college established in Durban by the University of Natal. He held this position until 1955.

Philip Bernard Stein died on January 7, 1974, in London.

Work

During his years in Durban Stein produced several papers. In his paper *On a theorem of M. Riesz* published in 1933, he gave a proof of the fundamental result of Marcel Riesz (1886-1969) on which the theory of conjugate functions in L_p spaces depends. The proof involves an ingenious use of Green's formula and is still the most elegant proof of the result [745]. This paper has been

[79] http://kehilalinks.jewishgen.org/sveksna/Stein_Family.pdf (accessed December 2021)
[80] http://www.s2a3.org.za/bio/Biograph_final.php?serial=2703 (accessed December 2021)
[81] https://www.s2a3.org.za/bio/Biograph_final.php?serial=2399 (accessed December 2021)

frequently cited and has served as a model for arguments generalizing the Riesz inequality. Stein also contributed to a problem concerning the real zeros of certain trigonometric functions.

In the late 1940s, Stein began to collaborate on numerical analysis with Reuben Louis Rosenberg (1909-1986), a long-standing lecturer in applied mathematics who resigned in 1949. Their collaboration led to their joint paper *On the solution of linear simultaneous equations by iteration* [2862] in 1948. The authors wrote that they *are indebted to Dr. Olga Taussky for advice on some points*. As it is written in [1665],

> *In the early post war years Philip Stein, together with a pupil, R.L. Rosenberg, submitted a paper to the London Mathematical Society. This paper was concerned with comparison of the two classical iterative methods for the solution of linear systems associated with the names of Jacobi and Gauss-Seidel. Olga Todd, as referee, noted its importance and novelty and gave them detailed advice and made sure that it was published. This paper* [2862] *has become a classic and the Stein-Rosenberg Theory is a standard chapter in courses on iterative matrix analysis.*

As a result of that paper, Stein visited the National Bureau of Standards in the USA and there wrote three papers *The convergence of Seidel iterants of nearly symmetric matrices* in 1951, *A note on bounds of multiple characteristic roots of a matrix* in 1952, and *Some general theorems on iterants* also in 1952. The latter paper was a report on a study of the properties of square matrices, performed under a contract between the National Bureau of Standards and the University of California at Los Angeles. While still at the Durban Campus of the University of Natal in those times, Stein was also affiliated with the University of California.

As mentioned by Jacob Lionel Bakst Cooper (1915-1977), who attended the South African College School in Cape Town from 1924 to 1931, Olga Taussky-Todd (1906-1995) suggested further problems to Stein; these are not completely solved yet, but Stein gave a partial solution of one of them in his paper *On the ranges of two functions of positive definite matrices* published in 1965, and wrote another much cited paper concerned with the problem of Lyapunov stability of matrices *On the ranges of two functions of positive definite matrices II* in 1967, with Allen Michael Pfeffer, a former student of John Todd (1911-2007).

Cooper concluded in [745]

> *His* [Stein's] *active mathematical life continued for many years after his retirement: he taught for some years in the University of Makerere and also for a period in University College, Cardiff. Stein was an excellent and conscientious teacher and a force stimulating mathematics at all levels in South Africa for the many years of his active life. His quiet humour, his liberal outlook and his general reasonableness will be remembered by his many friends.*

10.66 ▪ Eduard L. Stiefel

Life

Eduard Ludwig Stiefel was born April 21, 1909, in Zürich, Switzerland. He was the son of Eduard Stiefel (1875-1967), a well-known Swiss painter.

He did his studies in mathematics and physics at the Swiss Federal Institute of Technology (ETHZ) in Zürich. In 1932-1933, he visited the universities of Hamburg and Göttingen in Germany.

From 1933 to 1935, he was assistant in mathematics at ETH. He obtained his Ph.D. thesis from ETH in 1935 in topology, the title of his thesis being *Richtungsfelder und Fernparallelismus*

Eduard Ludwig Stiefel

in n-dimensionalen Mannigfaltigkeiten (Direction fields and teleparallelism in n-dimensional manifolds) under the supervision of Heinz Hopf (1894-1971). This laid the basis for the theory of vector fields on manifolds. He received his Habilitation in 1942. In 1939, he married Jeannette Beltrami.

From 1936 to 1943, he was lecturer in mathematics at ETH. In 1943, he became full professor of mathematics at ETH. From 1946 to 1948, he was head of the department of mathematics and physics and director of the Institute of Applied Mathematics (IAM) from 1948 to 1978.

During World War II, Stiefel, as an officer of the Swiss Army, had to some extent to work on computational problems. In the end, he reached the rank of colonel, commanding the artillery weather services.

Stiefel started his career as a pure mathematician, but after the war, he completely changed topics and shifted his interest to numerical analysis and scientific computing.

When starting the IAM, Stiefel hired some good collaborators. As assistants he chose the mathematician Heinz Rutishauser (1918-1970) and the electrical engineer Ambrosius Paul Speiser (1922-2003), who were former students at the ETH. Rutishauser had left the ETH three years before and was working as a high school teacher, while working on his dissertation in complex analysis in his spare time. Speiser was just getting his diploma in electrical engineering with a thesis related to computers.

It was decided that Stiefel and his two assistants should visit the USA to acquire some of the American knowledge about computers. His first trip was from October 18, 1948, until March 12, 1949. Rutishauser and Speiser stayed for longer, until the end of 1949. On their way they stopped in Amsterdam. Then, Stiefel visited New York at the IBM Watson Laboratory for Scientific Computation, the Institute for Mathematics and Mechanics at New York University, and the Computation Laboratory of the National Bureau of Standards. In Washington, he visited the Office of Naval Research and the National Bureau of Standards. In Boston, he visited Harvard University where Howard Hathaway Aiken (1900-1973) was designing his *Mark III* computer.

Finally, Stiefel visited the Institute of Advanced Study in Princeton, where John von Neumann (1903-1957) was working. Rutishauser stayed in Princeton to follow the work on von Neumann's computer project and Speiser stayed in Boston. After these visits, Stiefel was convinced that it was important to develop the use of numerical methods and computers even though not that many had been built yet. A first tentative design of a Swiss computer called *ERMETH* (Elektronische Rechenmaschine der ETH) was worked out by Speiser in Boston and, after his return, in Zürich.

When Stiefel came back from his trip to the USA, he anticipated that the design and the construction of the *ERMETH* would take several years. In 1949, he rented and moved to Zürich the *Z4* relay computing engine of the German designer Konrad Zuse (1910-1995); see Chapter 7 and Section 10.78. The machine was rented for five years for 30,000 SFr. Therefore, ETH became the first European university with an electronic computer in August 1950. On this machine, a linear system with 106 unknowns arising from a plate problem was solved by the conjugate gradient method in 1952; see [2891].

While working on the *Z4*, Stiefel's group was also working on the design and the construction of the *ERMETH*. Speiser was the technical director leading a group of five engineers and three mechanics. Rutishauser worked on the logical organization. This took some time and money and the machine was only running in July 1956. It had 16-digit decimal words, each of which contained two instructions, one 14-digit fixed point number, or one floating-point number with a 11-digit mantissa. A floating-point addition took 4 ms, a multiplication 18 ms. The magnetic drum could store 10,000 words. The *ERMETH* was in use at ETH until 1963. The machine is now at the Musée de la Communication in Bern.

In 1956-1957, Stiefel was president of the Swiss mathematical society. He was president of the Gesellschaft für Angewandte Mathematik und Mechanik (GAMM) from 1970 to 1974. He was also involved in local politics, being a community councilman of the city of Zürich from 1958 to 1966.

In 1959, Stiefel was one of the founding editors of the journal *Numerische Mathematik*. Later on, Stiefel became interested in celestial mechanics, particularly numerical methods for computing satellite orbits. He studied a method known as the Kustaanheimo-Stiefel regularization. Paul Kustaanheimo (1924-1997) was an astronomer and mathematician from Finland. Together they published a paper *Perturbation theory of Kepler motion based on spinor regularization* in the *J. reine angew. Math.* in 1965.

Eduard Stiefel died in November 25, 1978. A large part of the information in this biography comes from [1495, 1500] by Martin Hermann Gutknecht.

Work

Stiefel started to work in algebraic topology and differential geometry, particularly on Lie groups. His name is attached to the *Stiefel-Whitney classes* which are a set of topological invariants of a real vector bundle. Hassler Whitney (1907-1989) was an American mathematician. Stiefel's name is also attached to the *Stiefel manifold*.

One of the main achievements of Stiefel was the introduction of the conjugate gradient (CG) algorithm for solving symmetric positive definite systems of linear equations in 1951; see Sections 5.6 and 5.7 for the genesis of that method.

Stiefel first called it the n-step method since, mathematically, it delivers the exact solution of a linear system of order n in at most n steps. The method was first described in Section 5 of a paper submitted in July 1951 [2891]. In the same month, Stiefel traveled to the USA for a second visit, from July 1951 to February 1952. At the invitation of Olga Taussky-Todd (1906-1995), he went to the Institute for Numerical Analysis (INA) of the National Bureau of Standards, which was located on the UCLA campus in Los Angeles. There he attended the *Symposium on Simultaneous Linear Equations and the Determination of Eigenvalues*, held at the INA on August 23-25, 1951. His meeting with Magnus Rudolph Hestenes (1906-1991), who independently proposed the same method under the name conjugate gradient method, is described in Hestenes' biography; see Section 10.36. This led to the seminal joint paper [1664] published in 1952. Hestenes was thinking of the method from the point of view of the calculus of variations and Stiefel was thinking of it as an improvement (with finite termination) of what he called "relaxation." In his mind, this was meaning a step-by-step method; see also the booklet [1095] written in 1959 with his collaborators, Max E. Engeli, Theo Ginsburg (1926-1993), and Rutishauser.

Stiefel published several books including *Einführung in die Numerische Mathematik* (Introduction to numerical mathematics) in 1961 which was later translated to English and French.

10.67 ▪ James J. Sylvester

James Joseph Sylvester

Life

James Joseph Sylvester was born on September 3, 1814, in London, the sixth and youngest son of Abraham Joseph, one of seven children. The family originally resided in Liverpool. The name Sylvester was adopted by the eldest brother of James' father when he established himself in the United States. His example was followed by all the brothers.

Until 1829, Sylvester's education was done in private schools in London. Then, for his secondary education, he was sent to the Royal Institution in Liverpool where he remained for less than two years. In February 1830, he was awarded the first prize in the Mathematical School. He entered St John's College, Cambridge, in 1831. He was seriously ill in 1834-1835. In January 1837, he was second Wrangler in the Mathematical Tripos (see the note about Arthur Cayley for an explanation of what this was, Section 10.8). According to Percy Alexander MacMahon (1854-1929) [2112],

> *Sylvester was keenly disappointed at his failure to be senior of the year. He was always of an excitable disposition, and it is currently reported that, on hearing the result of the examination, he was much agitated.*

Unfortunately, he could not get his degree because, being a Jew, he could not sign the 39 articles of the Church of England. He also could not get a fellowship.

Sylvester published his first paper (on physics) whose title was *Analytical development of Fresnel's theory of crystals* in 1837 in the *Philosophical Magazine*. However, according to [71], he published a pamphlet in 1836.

In 1838, he was elected to the professorship of natural philosophy at University College in London (known in those times as London University), becoming a colleague of Augustus De Morgan (1806-1871), a famous British mathematician. In April 1839, he was elected Fellow of the Royal Society.

Sylvester crossed the Atlantic and became professor of mathematics at the University of Virginia in 1841-1842. On his way to America, he visited William Rowan Hamilton (1805-1865) in Dublin. Sylvester obtained a degree from Trinity College, Dublin, in 1841. He did not published any paper during his stay in the USA.

According to Alexander MacFarlane (1851-1913) [2108] the reason for his sudden departure from Virginia was the following:

> *Among his students were two brothers, fully imbued with the Southern ideas about honor. One day Sylvester criticised the recitation of the younger brother in a wealth of diction which offended the young man's sense of honor; he sent word to the professor that he must apologize or be chastised. Sylvester did not apologize, but provided himself with a sword-cane; the young man provided himself with a heavy walking-stick. The brothers lay in wait for the professor; and when he came along the younger brother demanded an apology, almost immediately knocked off Sylvester's hat, and struck him a blow on the bare head with his heavy stick. Sylvester drew his sword-cane, and pierced the young man just over the heart; who fell back into his brother's arms, calling out "I am killed". A spectator, coming up, urged Sylvester away from the spot. Without waiting to pack his books the professor left for New York, and took the earliest possible passage for England. The student was not seriously hurt; fortunately the point of the sword had struck fair against a rib.*

However, this story was considered untrue by Raymond Clare Archibald (1875-1955) who nevertheless believed in a dispute with a student and a lack of support from the board of the university; see [71] pp. 98–100. It seems that after living Virginia, Sylvester stayed for a while in New York before going back to his home country. So he did not take the earliest possible passage for England!

After his return to England, Sylvester turned to some other activities. In 1844, he became an actuary to the Legal and Equitable Life Assurance Company. He entered at the Inner Temple in 1846 and was called to the Bar in 1850. It is around those times that Sylvester met Arthur Cayley (1821-1895) who was studying law in Lincoln's Inn. In a paper published in 1851, Sylvester wrote

> *The theorem above enunciated was in part suggested in the course of a conversation with Mr Cayley (to whom I am indebted for my restoration to the enjoyment of mathematical life).*

There are many references to Cayley in Sylvester's papers.

In 1851, Sylvester published a paper on canonical forms about which he wrote (see *Collected Mathematical Papers* vol. II p. 714)

> *I discovered and developed the whole theory of canonical binary forms for odd degrees, and, as far as yet made out, for even degrees too, at one evening sitting, with a decanter of port wine to sustain nature's flagging energies, in a back office in Lincoln's Inn Fields. The work was done, and well done, but at the usual cost of racking thought-a brain on fire, and feet feeling, or feelingless, as if plunged in an ice-pail. That night we slept no more.*

In 1854, he was unsuccessfully a candidate for some professorships. However, in 1855, he was elected as a professor in the Royal Military College, Woolwich, where he stayed until 1870.

In 1861, he was awarded the Royal Medal from the Royal Society. In 1863, he was chosen corresponding member in the mathematics section by the French Academy of Sciences. He was elected member of the newly founded London Mathematical Society in 1865, was vice-president in 1866, and president in November 1866.

In 1870, Sylvester retired from Woolwich. According to [169],

> *For the next few years Sylvester resided near the Athenaeum Club, apparently somewhat undecided as to his course in life. We hear of him as reciting and singing at Penny Readings*[82] *and as being a candidate for the London School Board.*

In 1870, he also published the *Laws of Verse* [2965]. In this book there are short poems, sonnets, and translations from Horace and some German poets, like Schiller and Goethe.

In 1875, Johns Hopkins University was founded in Baltimore, Maryland, USA. Sylvester was professor of mathematics at that university from 1876 to 1883. According to the editor of Sylvester's *Collected Mathematical Papers*, Henry Frederick Baker (1866-1956) [169],

> *It was an experiment in educational method; Sylvester was free to teach whatever he wished in the way he thought best.*

Sylvester was instrumental in the publication of the *American Journal of Mathematics*, whose first volume appeared in 1878 with an article by Sylvester entitled *On an application of the new atomic theory to the graphical representation of the invariants and covariants of binary quantics - with three appendices*, pp. 64–125.

In 1880, Sylvester was awarded the Copley medal by the Royal Society. This medal was created after a donation by Sir Godfrey Copley (c. 1653-1709), an English landowner, art collector, and Tory politician. It is given for "outstanding achievements in research in any branch of science." Sylvester received the medal for "his long continued investigations and discoveries in mathematics." That same year he was elected Honorary Fellow of St John's College at Cambridge. In 1887, he was awarded the De Morgan gold medal. He was the second recipient after Cayley in 1884.

In 1883, Sylvester received an honorary degree from the University of Oxford. In December 1883, he was elected chair of Savilian Professor of Geometry in Oxford. This professorship had been founded in 1619 by Sir Henry Savile (1549-1622), an English scholar and mathematician, Warden of Merton College, Oxford, and Provost of Eton. A farewell meeting was organized on December 20, 1883, at Johns Hopkins University. In his inaugural lecture in 1885, he said

> *It is now two years and seven days since a message by the Atlantic cable containing the single word "Elected" reached me in Baltimore informing me that I had been appointed Savilian Professor of Geometry in Oxford, so that for three weeks I was in the unique position of filling the post and drawing the pay of Professor of Mathematics in each of two Universities: one, the oldest and most renowned, the other -an infant Hercules- the most active and prolific in the world, and which realises what only existed as a dream in the mind of Bacon - the House of Solomon in the New Atlantis.*

In 1890, Sylvester was awarded an honorary degree by the University of Cambridge. During his stay at Oxford he founded the Oxford Mathematical Society.

[82] Penny Readings consisted of readings and other performances, for which the admission charge was one penny.

According to [169],

> *In 1893 his eyesight began to be a serious trouble to him, and in 1894 he applied for leave to resign the active duties of his chair. After that he lived mainly in London or at Tunbridge Wells, sad and dejected because his mathematical power was failing.* [...] *At the beginning of March* [1897], *he had a paralytic seizure while working in his rooms at Hertford Street, Mayfair. He never spoke again, and died 15 March 1897. He was buried with simple ceremonial at the Jewish Cemetery at Dalston on March 19.*

In 1889 Cayley [592] wrote

> *He is a Fellow of New College, Oxford; Foreign Associate of the United States National Academy of Sciences; Foreign Member of the Royal Academy of Sciences, Gottingen, of the Royal Academy of Sciences of Naples, and of the Academy of Sciences of Boston ; Corresponding Member of the Institute of France, of the Imperial Academy of Science of St. Petersburg, of the Royal Academy of Science of Berlin, of the Lyncei of Rome, of the Istituto Lombardo, and of the Société Philomathique. He has been long connected with the editorial staff of the Quarterly Journal of Mathematics (under one or another of its titles), and was the first editor of, and is a considerable contributor to, the American Journal of Mathematics.*

In his obituary notice, MacMahon [2112] wrote

> *His handwriting was bad, and a trouble to his printers. His papers were finished with difficulty. No sooner was the manuscript in the editor's hands than alterations, corrections, ameliorations and generalisations would suggest themselves to his mind, and every post would carry further directions to the editors and printers* [...] *He was fond of billiards, whist and chess. He liked occasionally going into the society of ladies, but was never married.*

Examples of Sylvester's handwriting can be seen in [71].

More information about Sylvester's life and work can be found in [592, 2112, 169, 2108, 71, 1155, 770, 2463, 2464, 1682, 1685] (ordered by date).

Work

Sylvester (probably, as we have seen above) published his first paper in 1837. He worked on many topics in mathematics but his main interest was *Invariant Theory* as well as number theory in the end of his life. Here, we consider only some papers related to our topic, that is, matrices and linear algebra.

His *Collected Mathematical Papers* (referred to as CMP below), edited by H.F. Baker, were published by Cambridge University Press, vol. I 1904, vol. II 1908, vol. III 1909, vol. IV 1912. These volumes contain 342 papers. There is an interesting biographical notice [169] in volume IV.

In the paper titled *Additions to the articles, "On a new class of theorems", and "On Pascal's theorem"* [2955] in 1850, Sylvester coined the term "matrix." He wrote

> *For this purpose we must commence, not with a square, but with an oblong arrangement of terms consisting, suppose, of m lines and n columns. This will not in itself represent a determinant, but is, as it were, a Matrix out of which we may form various systems of determinants by fixing upon a number p, and selecting at will p lines and p columns, the squares corresponding to which may be termed determinants of the pth order.*

As we can see, at this time, for him a matrix represented only an array of numbers from which determinants can be formed. The word "matrix" appears in several papers published in the next years, for instance, in

- *An essay on canonical forms, supplement to a sketch of a memoir* on elimination, transformation and canonical forms* [2957] in 1851.
- *An enumeration of the contacts of lines and surfaces of the second order* [2956] in 1851.
- *On the relation between the minor determinants of linearly equivalent quadratic functions* [2959] in 1851.

  In that paper, on page 247 of CMP vol. I, we can read an explanation of why Sylvester chose the word "matrix":

  > *I have in previous papers defined a "Matrix" as a rectangular array of terms, out of which different systems of determinants may be engendered, as from the womb of a common parent; these cognate determinants being by no means isolated in their relations to one another, but subject to certain simple laws of mutual dependence and simultaneous deperition. The condensed representation of any such Matrix, according to my improved Vandermondian notation, will be*

  $$\left\{ \begin{matrix} a_1, & a_2, & \dots & a_n \\ \alpha_1, & \alpha_2, & \dots & \alpha_n \end{matrix} \right\}.$$

- *A demonstration of the theorem that every homogeneous quadratic polynomial is reducible by real orthogonal substitutions to the form of a sum of positive and negative squares* [2960] in 1852.
- *On a theory of the syzygetic* relations of two rational integral functions, comprising an application to the theory of Sturm's functions, and that of the greatest algebraical common measure* [2963] in 1853.

  In this long paper, starting on page 580 in CMP vol. I, there is a glossary of terms in which we can read on page 583,

  > *Inverse. -The inverse to a given square matrix is formed by selecting in its turn each component of the given matrix, substituting unity in its place, making all the other components in the same line and column therewith zero, and finally writing the value of the determinant corresponding to the matrix thus modified in lieu of the selected component. If the determinant to the matrix be equal to unity, its second inverse, that is the inverse to its inverse, will be identical, term for term, with the original matrix.*
  >
  > *Matrix. -A square or rectangular arrangement of terms in lines and columns.*

Note that the rule for the inverse was also given in Cayley's memoir in 1858.

Sylvester returned to matrix theory much later. In the paper titled *Lectures on the principles of universal algebra* [2966] in 1883, we can read

> *Much as I owe in the way of fruitful suggestion to Cayley's immortal memoir, the idea of subjecting matrices to the additive process and of their consequent amenability to the laws of functional operation was not taken from it, but occurred to me independently before I had seen the memoir or was acquainted with its contents; and indeed*

> *forced itself upon my attention as a means of giving simplicity and generality to my formula for the powers or roots of matrices, published in the Comptes Rendus of the Institute for 1882 (Vol. 94, pp. 55, 396).*

He also wrote

> *Cauchy has taught us what is to be understood by the product of one rectangular array or matrix by another of the same length and breadth, and we have only to consider the case of rectangles degenerating each to a single line and column respectively, to understand what is meant by the product of the multiplication of the two parallels spoken of above.*

Sylvester also coined the term "latent roots" for what we now call "eigenvalues." In [2966], he wrote

> *My memoir on Tchebycheff's method concerning the totality of prime numbers within certain limits, was the indirect cause of turning my attention to the subject, as (through the systems of difference-equations therein employed to contract Tchebycheff's limits) I was led to the discovery of the properties of the latent roots of matrices, and had made considerable progress in developing the theory of matrices considered as quantities, when on writing to Prof. Cayley upon the subject he referred me to the memoir in question: all this only proves how far the discovery of the quantitative nature of matrices is removed from being artificial or factitious, but, on the contrary, was bound to be evolved, in the fulness of time, as a necessary sequel to previously acquired cognitions.*

In [2967] in 1883, Sylvester explained why he chose the word "latent" and what the "latent roots" are:

> *It will be convenient to introduce here a notion (which plays a conspicuous part in my new theory of multiple algebra), namely that of the latent roots of a matrix - latent in a somewhat similar sense as vapour may be said to be latent in water or smoke in a tobacco-leaf. If from each term in the diagonal of a given matrix, λ be subtracted, the determinant to the matrix so modified will be a rational integer function of λ; the roots of that function are the latent roots of the matrix.*

Another contribution to linear algebra is Sylvester's determinantal formula in the paper *On the relation between the minor determinants of linearly equivalent quadratic functions* [2959] in 1851.

Sylvester is probably mostly known by students of mathematics for the *Law of inertia*, a proof of which he gave using quadratic forms in the paper *A demonstration of the theorem that every homogeneous quadratic polynomial is reducible by real orthogonal substitutions to the form of a sum of positive and negative squares* [2960] in 1852. This law states that the number of negative, zero, and positive eigenvalues of a Hermitian matrix A is invariant under a congruence transformation, that is, X^*AX, where X is a nonsingular matrix.

In 1883, Sylvester gave a formula for a general function of a matrix whose eigenvalues are distinct; see [2967]. This formula uses what is now known as the Lagrange interpolation polynomial.

Sylvester also considered equations whose unknowns are matrices. In the paper *Equations in matrices* [2968] in 1884, he considered the equation $px + q = 0$, where p, x, q are matrices.

In the paper *Sur l'équation en matrices* $px = xq$ [2973] in 1884, he solved the homogeneous case of what we know now as a *Sylvester equation*. He wrote the equations for the entries for $n = 2$.

In the paper *Sur la résolution générale de l'équation linéaire en matrices d'un ordre quelconque* [2971] in 1884, polynomial matrix equations are considered. In that paper, he wrote the following about the Cayley-Hamilton theorem[83] (our translation):

> *It is in the Lectures, published in 1844, that for the first time appeared the beautiful conception of the identical equation applied to the matrices of the third order, wrapped in a language peculiar to Hamilton, after him laid bare by Mr. Cayley in a very important Memoir on matrices in the Philosophical Transactions for 1857 or 1858, and extended by him to matrices of any order, but without proof; this proof was given later by the late Mr. Clifford (see his posthumous works), by Mr. Buchheim in the Mathematical Messenger (walking, as he confesses, in the footsteps of Mr. Tait, of Edinburgh), by Mr. Ed. Weyr, by ourselves, and probably by others; but the four methods cited above appear to be quite distinct from one another.*

Note that the general case of the theorem was proved by Ferdinand Georg Frobenius (1849-1917) in 1878.

10.68 ▪ Olga Taussky-Todd

Olga Taussky and John Todd

Life

Olga Taussky was born on August 30, 1906, in Olmütz, Austro-Hungarian Empire (now Olomouc, Czech Republic), into a Jewish family.

Her father, Julius David Taussky (1860-1925), was an industrial chemist who also wrote articles for a newspaper. Her mother, Ida Pollach (1875-1951), was a housewife. Olga had two sisters, Ilona and Herta.

The family moved to Vienna when Olga was three years old. She did very well in the primary school, in particular in arithmetic, but her favorite subjects were essay writing and grammar. She also had music lessons.

It was World War I, and the family was almost starving. In 1916, they moved to Linz. When she was 14 years old, Olga entered the high school where she spent a year before studying at the Gymnasium. She suddenly became interested in mathematics. She entered Vienna University

[83] *C'est dans les Lectures, publiées en 1844, que pour la première fois a paru la belle conception de l'équation identique appliquée aux matrices du troisième ordre, enveloppée dans un langage propre à Hamilton, après lui mise à nu par M. Cayley dans un très important Mémoire sur les matrices dans les Philosophical Transactions pour 1857 ou 1858, et étendue par lui aux matrices d'un ordre quelconque, mais sans démonstration; cette démonstration a été donnée plus tard par feu M. Clifford (voir ses ceuvres posthumes), par M. Buchheim dans le Mathematical Messenger (marchant, comme il l'avoue, sur les traces de M. Tait, d'Edimbourg), par M. Ed. Weyr, par nous-même, et probablement par d'autres; mais les quatre méthodes citées plus haut paraissent être tout à fait distinctes l'une de l'autre.*

and began to study chemistry under the pressure of her family, but soon turned back to mathematics.

She undertook a thesis under the supervision of Philipp Furtwängler (1869-1940) on algebraic number fields just as class field theory was being developed. She was a fellow student of Kurt Gödel (1906-1978); see [3013]. She defended her thesis in 1930, and continued to work on the same topic. At a meeting of the German Mathematical Society, Hans Hahn (1879-1934) recommended her to Richard Courant (1888-1972) and, in 1931, she became his assistant in Göttingen. There, she worked with Emil Artin (1898-1962) and Emmy Noether (1882-1935). But during the summer of 1932, she was advised by Courant not to return to Göttingen after her holidays because of the political situation.

In 1932-1933, Taussky tutored in Vienna, then she spent a year at Bryn Mawr College, a women's liberal arts college, in Pennsylvania, USA. Then, she took up a research fellowship from Girton College at the University of Cambridge, UK, in 1935. Godfrey Harold Hardy (1877-1947) helped her obtain a teaching post in a London college in 1937 where she soon met John (Jack) Todd (1911-2007); see Section 10.70. They got married in 1938.

During World War II, Olga and Jack moved from place to place. She also worked at the National Physical Laboratory in Teddington between 1943 and 1946 where, for war purposes, she studied aerodynamics.

Olga went into applied work for the Ministry of Aircraft Production during the war. The problems included analysis of aircraft designs for their stability properties. The tools were the localization of eigenvalues, stability analysis (testing whether the real parts of all eigenvalues are less than 0, or anyway not too far above 0), and numerical computation.

After the war, the Todds emigrated to the United States where they occupied various positions, in particular at the National Bureau of Standards; see Section 5.6. In 1957, Olga and Jack both joined the California Institute of Technology (Caltech) in Pasadena, California, where they stayed until their retirements, which was in 1976 for Olga. But she continued doing mathematics.

In 1958, she was the first women since Emmy Noether in 1934 invited to give a one-hour lecture at an AMS meeting. She received many honors from several universities and was vice-president of the American Mathematical Society in 1985.

Olga Taussky-Todd died on October 7, 1995, in Pasadena from the consequences of a broken hip from which she had not fully recovered.

For papers on Taussky's life and work, see [225, 1408, 2705, 2706, 3084].

Work

Olga Taussky had many interests in pure and applied mathematics. She worked in number theory including class field theory, group theory, topological algebra, and differential equations. She wrote many papers on her own, and also two papers on matrices with her husband John.

Before World War II, she was concerned with pure mathematics; however, after the war, she continued to write papers on matrix theory, group theory, and algebraic number theory, but also papers on numerical analysis.

In 1948, she derived bounds for eigenvalues of matrices in [3003]. In 1949, she published a paper [3004] titled *A recurring theorem on determinants* in which she proved that a strictly diagonally dominant matrix is nonsingular, a result which goes back to the French mathematician Lucien Lévy (1853-1912) in 1881. This result has been rediscovered many times. However, Olga also gave a simple proof of the fact that if the matrix A is just diagonally dominant with equality holding in at most $n-1$ rows and if A is irreducible, then A is nonsingular.

A consequence of the diagonal dominance theorem is that the eigenvalues are contained in what we now call *Gerschgorin's circles*, even though this result was known long before Gerschgorin's paper [1341] in 1931. She strengthened the theorem and promoted the method together with Alfred Theodor Brauer (1894-1985). She wrote about its applications, transmissions, and independent rediscoveries over the decades.

During her stay at NBS in the 1950s, she edited proceedings of several conferences on the solution of linear equations and eigenvalue problems [2432, 3006].

In the paper [1720] co-authored with Alan Jerome Hoffman (1924-2021) in 1954, she gave a characterization of normal matrices.

In a joint paper [3012] with Hans Julius Zassenhaus (1912-1991) in 1959, she proved that for every square matrix A there exists a nonsingular symmetric matrix S transforming A into its transpose A^T, that is, $A^T = S^{-1}AS$. Another result was that every nonsingular matrix transforming A into its transpose is symmetric if and only if A is similar to a companion matrix.

In 1968, she wrote a paper [3008] on the role of positive definite matrices in the study of the eigenvalues of general matrices. On the same topic, in 1972, she published a paper [3009] on the role of symmetric matrices in the study of general matrices.

For more about Taussky's work on matrices, see her paper *How I became a torchbearer for matrix theory* [3010]. At the end of this paper, she wrote

> *Some advice. When you observe an interesting property of numbers, ask if perhaps you are not seeing, in the 1x1 case, an interesting property of matrices. Think of $GL(n, F)$ or $SL(n, F)$, $GL(n, Z)$ or $SL(n, Z)$. When you have a pair of interesting matrices study the pencil that they generate, or even the algebra. When the determinant of a certain matrix turns out to be important, ask about the matrix as a whole, for instance as in the case of the discriminant matrix, as suggested by the discriminant of an algebraic number field. When a polynomial in one variable interests you, ask about the matrices of which it is the characteristic polynomial. When people look down on matrices, remind them of great mathematicians such as Frobenius, Schur, C.L. Siegel, Ostrowski, Motzkin, Kac, etc., who made important contributions to the subject. I am proud to have been a torchbearer for matrix theory, and I am happy to see that there are many others to whom the torch can be passed.*

10.69 ▪ Charles-Xavier Thomas de Colmar

Charles-Xavier Thomas de Colmar

Life

Charles-Xavier Thomas (known as Thomas de Colmar) was born May 5, 1785, in Colmar in the east of France. His father Joseph Antoine Thomas (1758-1831) was a medical doctor and his mother was Françoise Xavière Entzlen (1759-1817).

After his studies, he held a position in the administration in charge of collecting taxes. In 1809, he joined the Napoléonic army in Portugal and Spain. He was the general manager for the food supplies of the army. In Spain, he met his wife Francisca de Paula Garcia de Ampudia Alvarez (1794-1874). They had ten children. At some point he was imprisoned by the enemy but he managed to escape.

He left the army at the beginning of 1814 and started to be interested in insurance companies. After a trip to England, in 1819 he co-founded the "Compagnie du Phénix." After disagreements with the board of the company, he left and in 1829 founded another insurance company, "La Compagnie du Soleil" with which he became very successful and wealthy. In 1843, he founded yet another insurance company, "La Compagnie de l'Aigle."

In 1850, he bought the Maisons-Laffitte castle, a very large and nice castle 20 kilometers west of Paris. He passed away in March 1870. He is buried in the Père-Lachaise cemetery in Paris.

Work

Even though he was a successful businessman, Thomas de Colmar is best known for his invention of a calculating machine, the *Arithmomètre* (arithmometer), which was the first commercially successful calculating machine capable of performing addition, subtraction, multiplication, and division. He probably got interested in automating computations when he was in Spain because there he had to do some accounting. Another hypothesis is that this was motivated by the computations that had to be done in his insurance companies.

Thomas used some techniques that were used before by Blaise Pascal (1623-1662) and Gottfried Wilhelm Leibniz (1646-1716). Like Pascal, he used the complement of a number to be able to do subtractions only with additions, and like Leibniz, he used a stepped wheel. The arithmometer had three parts, concerned with setting, counting, and recording. In the first models, the mechanism was activated by pulling a silk ribbon. This was later replaced by a crank.

In 1820, Thomas filed a patent in France for the arithmometer. The first prototype of the machine has been lost. But in 1822, he asked a skilled watchmaker named Devrine to construct a more robust machine which is now at the Smithsonian Institute in Washington. The invention of the arithmometer was praised in the February 1822 issue of the *Bulletin de la Société d'Encouragement pour l'Industrie Nationale* (Bulletin for the strengthening of the national industry) [1214] in which one can find a detailed description of the machine with beautiful engravings; see Figures 10.6 and 10.7.

Apparently nothing new happened until 1846 when Thomas hired a new worker, Pierre-Hippolyte Piolaine, who was the son of a watchmaker, to build a new machine which was finished in 1848. In 1850, Thomas had a reliable machine and he filed new patents in France, England, and Belgium. The arithmometer was shown at exhibitions in London in 1851 and in Paris in 1855. To promote his machine, Thomas offered models with richly decorated boxes to several VIPs of the time.

Several improved models and different variations of the machine were produced over the years. It is estimated that when Thomas passed away, 900 arithmometers had been produced. The manufacture of these machines was continued under the management of Thomas' son, Louis André Nicolas Thomas de Bojano (1818-1881) from 1870 to 1881, and then Thomas' grandson up to 1887. In 1888, the company was sold to Louis Payen and machines were manufactured up until World War I. A tentative attempt to restart the production in 1920 was a failure.

The arithmometers were produced and sold in Europe and the United States up until the beginning of the 20th century and were widely imitated and marketed by many manufacturers. For instance, Arthur Burkhardt (1857-1918) of Glashutte was the first maker in Germany to build an arithmometer of the Thomas de Colmar type, delivering his machine in 1878.

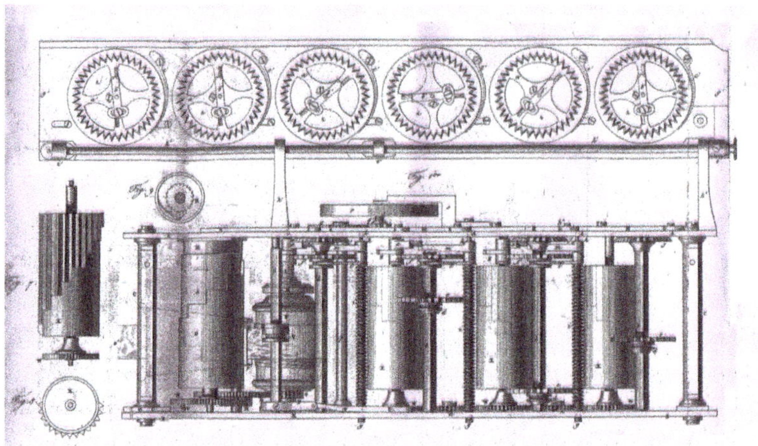

Figure 10.6. *Arithmometer, 1822*

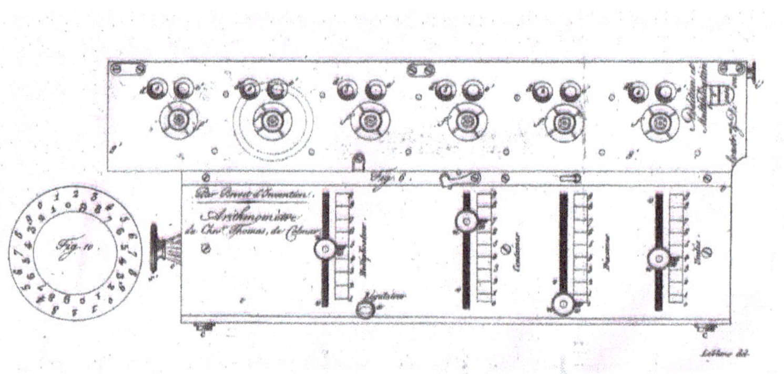

Figure 10.7. *Arithmometer, 1822*

Another machine which was later very successful was invented by Willgodt Theophil Odhner (1845-1905); see Section 10.54.

There are more than 350 arithmometers that are identified in museums and private collections.

10.70 ▪ John Todd

Life

John Todd, known among friends and colleagues as Jack, was born in Carnacally, Ireland, on May 16, 1911. He was the eldest of three brothers and he had a sister.

His parents were primary school teachers. When he was six years old, his parents moved to Belfast. There he attended primary and secondary schools. When he was 11, he entered the Methodist College in Belfast after winning a scholarship. He studied only mathematics since he wanted to become an engineer. After graduating from this college, he went to Queen's University Belfast in 1928, and received his B.Sc. in 1931.

Then, he entered St. John's College at Cambridge University as a postgraduate student since he had a first class degree from Queen's. But there was no analyst at St. John's College to supervise Todd's studies. He was advised to go to Godfrey Harold Hardy's office, but he was out watching cricket. So he went to John Edensor Littlewood (1885-1977), who advised him against getting a doctorate and told him just to do research.

John Todd

After two years of research on transfinite superpositions of absolutely continuous functions, Todd was appointed to Queen's University in Belfast where John Semple (1904-1985), whose most important work was in algebraic geometry, was a professor. In 1936, Semple moved to King's College in London, and he invited Todd to join him one year later. Initially, Todd taught aspects of real analysis, in particular measure theory. When another professor became ill, he was asked to take over the course on group theory. He developed an interest in this area, and fought a challenging research problem for which he asked the advice of Olga Taussky (1906-1995), a matrix and number theorist, who was at Westfield College. In 1936, he (and also Olga) was an invited speaker at the International Congress of Mathematicians in Oslo, Norway, where he gave a talk on *Transfinite superpositions of absolutely continuous functions*. Olga and John were married in 1938.

At the beginning of World War II, King's College was evacuated to Bristol, and Todd was told that he was not needed to teach any longer and should find another job. Thus, he returned to Belfast to teach at Methodist College in 1940-1941. As part of the war effort, he worked for the British Admiralty from 1941 to 1945, initially on ways of counteracting acoustic mines. Todd rapidly persuaded his superiors to establish what became the Admiralty Computing Service, centralizing much of numerical computations for the naval service. He accompanied John von Neumann (1903-1957) during his visit with the Admiralty, and he introduced him to their computing facility.

In 1945, Todd was part of a group that visited Germany for investigating mathematics and computers that could be interesting for the Navy. During this mission, they discovered an old hunting lodge, *Lorenzenhof*, in Oberwolfach in the Black Forest, which had been used as a research center for mathematics since the fall of 1944. Todd wrote [3054]

> *We were exhausted after being on the road for over a month and asked if we could rest there for the weekend. For safety, we posted notices on the main entrance to the effect that the building was under the protection of the British Navy. The next day Reuter* [Gerd Edzard Harry Reuter (1921-1992), a German-born mathematician who emigrated to Great Britain] *went off to Heidelberg to fill our gas tank and get rations, and I was left alone* [...]
>
> *We were having a discussion on the patio when they arose a commotion among the servants. It was caused by a foraging party of Moroccan troups who wanted to occupy the building. I quickly got into proper dress with hat and in my best French persuaded them to leave the mathematicians and* même les poules *undisturbed. The very distinguished sergeant asked if it would be permitted to shake the hand of a British naval officer. Of course I said, "Yes", and they left to try their luck elsewhere* [...]
>
> *This incident kept "Lorenzenhof" intact until the local government was set up.*

In September 1947, the Todds emigrated to the United States, and helped to establish the Institute for Numerical Analysis (INA) at the National Bureau of Standards (NBS) in Los Angeles [3055]; see Section 5.6. They also spent three months visiting von Neumann at the Institute for Advanced Study in Princeton before going to UCLA in Los Angeles in April 1948. At the NBS, Todd was chief of the Computation Laboratory and Olga was a mathematics consultant. They moved to Washington a year later where they stayed for 10 years. In 1957, they joined the California Institute of Technology in Pasadena, California.

Todd was one of the founders of the journal *Numerische Mathematik.* He served on the board for 49 years.

John Todd died at his home in Pasadena, California, on June 21, 2007.

Work

Even before setting up the Admiralty Computing Service, Todd was interested in computing. In 1946, he had to teach a numerical analysis course at King's College in London. In the last joint paper he wrote with Olga Taussky [3011], one can read

> *In 1946 one of us (J.T.) offered a course at Kings' College, London (KCL) on Numerical Mathematics. While we had some wartime experience in numerical mathematics, including characteristic values of matrices, we had had little to do with the solution of systems of linear equations. In order to see how this topic should be presented, we made a survey of Math. Rev. (at that time easy!) and found a review (MR7 (1944), 488), of a paper by Henry Jensen* [1816], *written by E. Bodewig. Jensen stated* Cholesky's method seems to possess all advantages. *So, it was decided to follow Cholesky and, since the method was clearly explained, we did not try to find the original paper. Leslie Fox, then in the newly formed Mathematics Division of the (British) National Physical Laboratory (NPL), audited the course and apparently found the Cholesky Method attractive, for he took it back to NPL, where he and his colleagues studied it deeply* [1206, 3079]. *From these papers, the Cholesky (or sometimes Choleski) Method made its way into the tool boxes of numerical linear algebraists via the textbooks of the 1950s.*

Here is the review by Henry Jensen (1915-1974) published in *Mathematical Reviews*:

> MR0015921 (7,488d) 65.0X
>
> Jensen, Henry
> An attempt at a systematic classification of some methods for the solution of normal equations.
> Geodætisk Institut, København, Meddelelse 1944, (1944) No. 18, 45 pp.
>
> *Explaining and comparing the usual methods for a direct solution (that is, noniterative) of linear equations, the author arrives at the following conclusions. The determinantal solution by Chiò's rule and the method of equal coefficients are only practicable in the case of few equations. Gauss's method gives a very plain solution for all systems. Banachiewicz's method of Cracovians is not superior to the Gaussian algorithm. Boltz's and Krüger's methods are useful in special systems, while the best of all methods seems to be Cholesky's. The explanations are illustrated by two examples which are computed by all methods or a part of them.*
>
> Reviewed by E. Bodewig
>
>

Thus, after these lectures, Cholesky's method became to be known to numerical analysts. Let us quote Todd again [3055, p. 257]:

> *In early postwar years, local and national centers for applied mathematics, and in particular for computing, were organized in many countries.*
>
> *I returned to King's College, London, and gave my first course in numerical mathematics in 1946. We had two Marchand ACT 10M machines for that class. In this course I introduced the Cholesky method as the preferred one for the solution of positive definite systems $Ax = b$. This was taken up by L. Fox, who analyzed it deeply with his National Physical Laboratory colleagues H.D. Huskey, J.H. Wilkinson, and Turing. It has indeed become one of the workhorses of numerical linear algebraists.*

In fact, it was proved by Otto Toeplitz (1881-1940) in 1907 [3056] that if a matrix H is Hermitian, there exists a lower triangular matrix P such that $H = PP^T$, but he did not give a method for its construction. The method was found later by André-Louis Cholesky (1875-1918). For more about this method, and the discovery of the original manuscript describing his method, see Section 10.12.

Todd's first papers in numerical linear algebra were concerned about ill-conditioning; see [3048, 3049]. He studied the condition numbers of matrices arising from finite difference methods for second-order elliptic problems. Todd was instrumental in the adoption of condition numbers of matrices.

He was much interested in computations; see the paper [2337] published in 1958 with Morris Newman (1924-2007) on the evaluation of matrix inversion programs.

Later in his life, he wrote several papers on the history of computational mathematics. Todd edited or wrote several books including the two-volume *Basic Numerical Mathematics* [3053, 3052].

The contribution of John Todd to numerical analysis is described in [2113]; see also [2207]. Let us quote what he wrote in 1955 [3050]:

> *The profession of numerical analysis is not yet so desirable that it is taken up by choice; indeed, although it is one of the oldest professions, it is only now becoming respectable. Most of those who are now working in this field have been, more or less, drafted into it, either in World War I or in World War II, or more recently. The question at issue is why have we stayed in this field and not returned to our earlier interests.*
>
> *Our answer is that numerical analysis is an attractive subject where mathematics of practically all sorts can be used significantly, and from which, on the other hand, many of its branches can benefit.*

10.71 ▪ Otto Toeplitz

Life

Otto Toeplitz was born in Breslau, Germany (now Wrocław, Poland) on August 1, 1881, into a Jewish family of several teachers of mathematics. He was the son of Emil Abraham Toeplitz (1852-1917), a mathematician, and Pauline Ernestine Lubliner (1857-1892).

Toeplitz grew up in Breslau where he attended the Gymnasium. Then, as it was natural for him to study mathematics, he entered the University of Breslau, and was awarded a doctorate

Otto Toeplitz

in algebraic geometry in 1905 whose title was *Über Systeme von Formen, deren Funktionaldeterminante identisch verschwindet* (About systems of forms, whose functional determinant vanishes identically) under the supervision of Jacob Rosanes (1842-1922) and Friedrich Otto Rudolf Sturm (1841-1919), not to be confused with Jacques Charles François Sturm (1803-1855), who introduced Sturm sequences.

In 1906, Toeplitz went to the University of Göttingen, the world's leading center for mathematics at that time, and stayed there for seven years. The Faculty of Mathematics included David Hilbert (1862-1943), Felix Klein (1849-1925), and Hermann Minkowski (1864-1909).

Toeplitz joined the group working with Hilbert, including Max Born (1882-1970), who had studied in Breslau one year after Toeplitz and would be awarded the Nobel Prize in Physics in 1954, Richard Courant (1888-1972), and Ernst Hellinger (1883-1950), who became a close friend of Toeplitz and a co-author. In 1907, Toeplitz defended his Habilitation *Zur Transformation der Scharen bilinearer Formen von unendlichvielen Veränderlichen* (On the transformation of families of bilinear forms of infinitely many variables), and he was nominated *Privatdozent.*

Hilbert was then completing his theory of integral equations, and Toeplitz began to extend the classical theories of infinite linear and quadratic forms on n-dimensional spaces to infinite dimensional ones. He published five papers directly related to Hilbert's spectral theory of operators. At the same time, he also published a paper on the necessary and sufficient conditions for a linear summation process to converge to the same limit as the original sequence (the Toeplitz theorem), and he found the basic ideas of the so-called Toeplitz operators and matrices.

In 1913, Toeplitz obtained an extraordinary professorship at the University of Kiel. He was promoted to a professorship in 1920. With Hellinger, they began the project to write a major encyclopedic article on integral equations. They worked on it for many years, and it finally appeared in 1927. In 1928, Toeplitz accepted an offer of a chair at the University of Bonn.

Toeplitz was deeply interested in the history of mathematics, and as soon as 1920, he advocated a "genetic method" in teaching mathematics. It consisted in a presentation of each topic by introducing only the ideas that led to it. He wrote

> *The historian - the mathematical historian as well - must record all that has been, whether good or bad. I, on the contrary, want to select and utilize from mathematical history only the origins of those ideas which came to prove their value.*

He began to write a book using these ideas, but it was left unfinished. It was edited by Gottfried Maria Hugo Köthe (1905-1989), and posthumously published in German in 1946, with an English translation in 1963 with the title *The Calculus: A Genetic Approach.* The book introduced the subjects by showing how they developed from classical problems of Greek mathematics. Toeplitz was able to read Greek texts and knew Plato quite well.

In 1929, with the historian of sciences Otto Eduard Neugebauer (1899-1990) and the philologist and philosopher Julius Stenzel (1883-1935), he founded the journal *Quellen und Studien zur Geschichte der Mathematik, Astronomie und Physik* (Sources and studies on the history of mathematics, astronomy and physics).

Toeplitz was also greatly interested in school mathematics. With the German-born American mathematician Hans Adolph Rademacher (1892-1969), he wrote a classic of popular mathematics *Von Zahlen und Figuren*, first published in 1930, and later translated into English as *Enjoyment of Mathematics*.

In the 1930s, Toeplitz developed a general theory of infinite dimensional spaces and criticized Stefan Banach's work as being too abstract. In a joint paper with Köthe in 1934, he introduced, in the context of linear sequence spaces, important new concepts and theorems.

The Law for the Restoration of the Professional Civil Service was passed by the Nazi regime on April 7, 1933. Professors of Jewish origin were to be removed from teaching. However, Toeplitz was able to keep his position since he had been appointed before 1914. However, he was dismissed after the decisions made at the Nuremberg party congress in the autumn of 1935. Then, he started to work for the Jewish community in Bonn and its vicinity, and he opened a Jewish school for children. On the national level, he selected gifted students for scholarships which allowed them to study abroad.

In 1939, Toeplitz finally emigrated to Mandatory Palestine, where he became the scientific advisor to the rector of the Hebrew University of Jerusalem.

Toeplitz was married to Erna Henschel (1896-1976). They had three children. He died from tuberculosis in Jerusalem (under British Mandate at that time) on February 15, 1940.

As explained by Born in [376], the work of Toeplitz was quite influential in the development of quantum mechanics, in particular for its matrix and operator formalism.

Work

Toeplitz is, of course, known to people working in linear algebra for the matrices with constant diagonals named after him.

It is less well known that, in a paper on the Jacobian transformation of quadratic forms of infinitely many variables [3056], published in 1907, he showed that a symmetric positive definite matrix can be factorized into the product of a lower triangular matrix by its transpose, a result that would later be put into an algorithmic form by André Louis Cholesky (1875-1918) who was, most certainly, not aware of Toeplitz's result.

10.72 ▪ Alan M. Turing

Life

Alan Mathison Turing was born June 23, 1912, in London. His father was Julius Mathison Turing (1873-1947) who was working for the Indian Civil Service and his mother was Ethel Sara Stoney (1881-1976). It seems that his father's family was of Scottish origin. Alan had an older brother, John. When Alan was one year old, his mother joined Alan's father in India, leaving Alan in England with some friends of the family.

After preparatory school, he went to Sherborne School in Dorset from 1926 to 1931. He entered King's College in Cambridge University in 1931. This is when he became interested

Alan Mathison Turing

in mathematical logic. He was a second class in the Tripos part I and a wrangler in Part II with a "b." He graduated in 1935 and won a Smith's Prize in 1936. He was elected Fellow of King's College in 1935 with a dissertation on the central limit theorem of probability that he rediscovered, unaware that it had already been proved.

Working on a problem posed by David Hilbert (1862-1943), in 1936 he published *On computable numbers, with an application to the Entscheidungsproblem* [the decision problem] in which he introduced what is now known as the *Turing machine*. Then, he spent two years in Princeton, USA, working with Alonzo Church (1903-1995), who had been solving the same problem with his "lambda-calculus," a few months before Turing. They both had shown that a general solution to the Entscheidungsproblem is impossible. In Princeton, he met the already famous John von Neumann (1903-1957). Herman Heine Goldstine (1913-2004), who was a collaborator of von Neumann, later wrote the following about the relations of von Neumann and Turing:

> *I should mention that as a graduate student, Turing got his Ph.D. at Princeton under Princeton logician Alonzo Church. Turing's thesis was essentially on a paper computer. Johnny von Neumann's office was just a few doors away from Turing's, and Johnny followed everything that Turing did. There was a meeting of minds there from 1937 on. Von Neumann was so impressed with Turing that he wanted him to stay as his assistant. But Turing wanted to go back because he had a call from the Foreign Office to work at Bletchley.*

In 1938, Turing returned to Cambridge. The war started in 1939 and for the next six years Turing worked for the British government at the Government Code and Cypher School at Bletchley Park located in Buckinghamshire. Turing was recruited in 1938 as the government was preparing for the possibility of a war with Germany. He came to Bletchley Park on September 4, 1939.

Using some previous work by Polish codebreakers, Turing designed an electro-mechanical machine known as "the Bombe" which, by the autumn of 1940, was used to break Enigma-enciphered messages. Enigma was a German enciphering machine. Later, Turing was the leader of the naval Enigma decryption team. After receiving captured material, his team succeeded in breaking the U-boat code in the summer of 1941. In December 1942, Turing left Bletchley Park for the USA to help with the US high-speed "Bombes," and with voice decryption research.

During his stay in the USA, Turing met again von Neumann and Goldstine who were working on their paper about rounding errors in the computation of the inverse of a matrix [3158].

Goldstine remembered that

> *He was a very stubborn man. Von Neumann and I had the idea that Gauss's method of elimination - which is a way of solving a system of linear equations- was probably the correct way to use on a computer. We had a mathematical proof, and Turing came over for a few weeks and worked with us. He had a somewhat different idea, and we never were able to get him to agree with what we said.*

About the relations between von Neumann and Turing and their works on computing, it is probably fair to say that von Neumann was influenced by Turing's work on the Turing machine and that, even though they disagreed at that time, Turing was influenced by von Neumann's work on matrices.

Turing returned to England in August 1943, not to Bletchley Park but to nearby Hanslope Park, where he worked on secured voice communications systems. Turing was appointed Officer of the Most Excellent Order of the British Empire (OBE) in 1946 for his wartime service.

Turing joined the National Physical Laboratory (NPL) in London in 1945 to design and build an automatic computing machine named ACE (*Automatic Computing Engine*). James Hardy Wilkinson (1919-1986) was Turing's assistant; see Section 10.76. However, after a while, Turing's design was judged too ambitious and it was decided to build a smaller machine, the *Pilot ACE*. Due to these problems, Turing took a year's sabbatical at Cambridge. He returned for a short time in the middle of 1948, disliked what he saw, and went off permanently to Manchester.

In his 1970 address for the Turing award, Wilkinson [3253] wrote

> *From 1946 to 1948 I had the privilege of working with the great man himself at the National Physical Laboratory. I use the term "great man" advisedly because he was indeed a remarkable genius.* [...] *My career was certainly profoundly influenced by the association and, without it, it is unlikely that I would have remained in the computer field.* [...] *In 1946 I joined the newly formed Mathematics Division at the National Physical Laboratory. It was there that I first met Alan Turing, though he was, of course, known to me before by reputation, but mainly as an eccentric.* [...] *Working with Turing was tremendously stimulating, perhaps at times to the point of exhaustion.* [...] *Turing occasionally had days when he was "unapproachable" and at such times it was advisable to exercise discretion. I soon learned to recognize the symptoms and would exercise my right (or, as I usually put it, "meet my obligations") of working in the Computing Section until the mood passed, which it usually did quite quickly. Turing had a strong predilection for working things out from first principles, usually in the first instance without consulting any previous work on the subject, and no doubt it was this habit which gave his work that characteristically original flavor.* [...] *However, I feel bound to say that his published work fails to give an adequate impression of his remarkable versatility as a mathematician. His knowledge ranged widely over the whole field of pure and applied mathematics and seemed, as it were, not merely something he had learned from books, but to form an integral part of the man himself.*

In 1948, Turing published *Rounding-off errors in matrix processes*, which we will discuss below. That same year he was offered a readership at the University of Manchester by Maxwell Herman Newman (1897-1984) who he knew from his years at Bletchley Park. In Manchester he worked on different topics including the Manchester Computer project. In 1950, he published *Computing machinery and intelligence* in which he studied problems related to what we now call artificial intelligence. He also worked on applications of mathematics to biology, publishing *The chemical basis of morphogenesis* in 1952.

Turing was elected a Fellow of the Royal Society of London in 1951. He was arrested for homosexuality in March 1952 and was put on trial. He was given a choice between going to jail or having hormonal injections for a year, which he accepted.

He died on June 7, 1954, allegedly committing suicide with cyanide poisoning. However, the exact circumstances of his death have not been clearly established so far.

For more about Turing, see [2338].

Work

We have seen above that Turing worked on many different topics, mathematical logic, computer science, and mathematical biology. What is of concern to us here is his 1948 paper *Rounding-off errors in matrix processes* [3079]. Wilkinson, who was Turing's colleague at NPL, wrote in [3253]

> *When I joined NPL in 1946 the mood of pessimism about the stability of elimination methods for solving linear systems was at its height and was a major talking point. Bounds had been produced which purported to show that the error in the solution would be proportional to 4^n and this suggested that it would be impractical to solve systems even of quite modest order.* [...] *However, it happened that some time after my arrival, a system of 18 equations arrived in Mathematics Division and after talking around it for some time we finally decided to abandon theorizing and to solve it. A system of 18 is surprisingly formidable, even when one has had previous experience with 12, and we accordingly decided on a joint effort. The operation was manned by Fox, Goodwin, Turing, and me, and we decided on Gaussian elimination with complete pivoting. Turing was not particularly enthusiastic, partly because he was not an experienced performer on a desk machine and partly because he was convinced that it would be a failure.* [...] *I'm sure that this experience made quite an impression on him and set him thinking afresh on the problem of rounding errors in elimination processes. About a year later he produced his famous paper "Rounding-off errors in matrix processes" which together with the paper of J. von Neumann and H. Goldstine did a great deal to dispel the gloom.*

At the beginning of his paper, Turing wrote

> *This paper contains descriptions of a number of methods for solving sets of linear simultaneous equations and for inverting matrices, but its main concern is with the theoretical limits of accuracy that may be obtained in the application of these methods, due to rounding-off errors.* [...] *The writer was prompted to carry out this research largely by the practical work of L. Fox in applying the elimination method. Fox found that no exponential build-up of errors such as that envisaged by Hotelling actually occurred. In the meantime another theoretical investigation was being carried out by J. v. Neumann, who reached conclusions similar to those of this paper for the case of positive definite matrices, and communicated them to the writer at Princeton in January 1947 before the proofs given here were complete.*

In that paper he defined measures of work as the number of elementary operations $(+, -, \times, \div)$ and of reads or writes. He chose to record only multiplications and writes, a division being set equal to two multiplications. Strangely enough, considering that Turing was a visionary man, he wrote some statements which were proved to be wrong later:

> *If, however, we wish to solve a number of sets of equations with the same matrix A it is more convenient to work out the inverse and apply it to each of the vectors b*

as well as

> *It seems probable that with the advent of electronic computers it will become standard practice to find the inverse.*

He considered triangular solutions and described a matrix which has zeros above the diagonal as "lower triangular" and one which has zeros below as "upper triangular." Then, he stated a theorem on the LDU factorization of a matrix and showed that the elimination method produced the LU factorization through elementary matrices describing the combinations of rows:

> *The process of replacing the rows of a matrix by linear combinations of other rows may be regarded as left-multiplication of the matrix by another matrix, this second matrix having coefficients which describe the linear combinations required.*

The pivoting for stability is seen as permutations of rows, and he wrote

> *There seems to be a definite advantage in using the largest pivot in the column.*

This was known later as partial pivoting.

Turing also considered what is now known as the Gauss-Jordan method. This was proposed by Wilhem Jordan (1842-1899), a German geodesist. In this method the elimination is done both in the lower and upper parts of the matrix to end up with a diagonal matrix. Turing was thinking that

> *It may be the best method for use with electronic computing machinery.*

However, today this is not the method of choice to compute the inverse of a matrix, if it is needed anyway, something which does not happen too often.

Turing discussed the Cholesky method (that he wrote Choleski, despite what is written in his reference list) for symmetric systems and he generalized the same recipe for nonsymmetric systems, which gives another implementation of the elimination method. Other methods were briefly described, like the QR factorization.

For the study of rounding errors, Turing introduced several norms, $N(A)$ which is now known as the Frobenius norm, $B(A) = \max_x \|Ax\|/\|x\|$ which is the spectral or Euclidean norm, and $M(A) = \max_{i,j} |a_{i,j}|$ which is the max norm (note that this is different from the infinity norm). Then, he defined the N-condition number (and coined the term) of a matrix A of order n as

$$\frac{1}{n}N(A)N(A^{-1}),$$

and the M-condition number as $nM(A)M(A^{-1})$. The choice of $N(A)$ as a measure of the degree of ill-conditioning was justified by statistical arguments. He also compared these numbers to the determinant of the matrix as a measure of ill-conditioning, writing

> *However, the determinant may differ very greatly from the above-defined condition numbers as a measure of conditioning.*

He also remarked that

> *The best conditioned matrices are the orthogonal ones.*

In passing, he also defined the residual vector and what is iterative refinement.

Then, using these definitions, Turing studied the rounding-off errors in Jordan's method and in the Gaussian elimination process. He assumed that in the elementary operation,

$$a_{i,j}^{(r-1)} - \frac{a_{r,j}^{(r-1)} a_{i,r}^{(r-1)}}{a_{r,r}^{(r-1)}},$$

there is an error at most ϵ, and that in

$$x_{i,j}^{(r-1)} - \frac{x_{r,j}^{(r-1)} a_{i,r}^{(r-1)}}{a_{r,r}^{(r-1)}},$$

there is an error at most ϵ', where $X_r^{(r)} = J_r \cdots J_1$, the matrices J_i being the elementary matrices doing the combination of rows.

He denoted error matrices S_i, $i = 1, \ldots, n$, and error vectors s_i, $i = 1, \ldots, n$. Without error in the back substitution and ignoring second-order terms, he obtained

$$\text{error in } x \approx U^{-1} X_n \sum_{r=1}^{n} X_r^{-1}(s_r' - S_r U^{-1} X_n b),$$

where, for the sake of simplicity, $X_r = X_r^{(r)}$. Notice that in his formulas s_r' is not defined. Then, using partial pivoting, $M(X_r^{-1}) = 1$ with $X_r^{-1} = I - \sum_{s \le r}(I - J_s)$ and

$$|\text{error in } x_m| = (A^{-1} \sum_{r=1}^{n} X_r^{-1}(s_r' - S_r A^{-1} b))_m,$$

$$= \frac{n^2(n+1)}{2} M(A^{-1})\epsilon' + \frac{n^4(n+1)}{2} [M(A^{-1})]^2 M(b)\epsilon.$$

However, according to Nicholas John Higham [1681], page 185, the square in $[M(A^{-1})]^2$ can be easily eliminated. One has to add to this result the errors from the backward substitution. Turing estimated that this was of a smaller order.

Knowing what has been done later, we may think that this paper, despite the reputation of his author, is not really rigorous. But nevertheless, it has been very influential in rehabilitating Gaussian elimination as well as the paper of von Neumann and Goldstine [3158] who considered Gaussian elimination for symmetric positive definite matrices. They did a fixed point arithmetic analysis and derived an a priori bound on the residual of inverting a symmetric positive definite matrix. Their paper did not consider solving equations, even though they studied the factorization of the matrix for the sake of computing the inverse. We observe that the condition number appeared unnamed in that paper as the ratio of the largest eigenvalue to the smallest one.

A thorough and rigorous analysis was published by Wilkinson [3244] in 1961, introducing the growth factor of the intermediate quantities $a_{i,j}^{(r)}$ and using floating point arithmetic.

10.73 ▪ Alexandre-Théophile Vandermonde

Alexandre-Théophile Vandermonde

Life

Alexandre-Théophile Vandermonde was born on February 28, 1735, in Paris, France. The Vandermonde family was of Flemish origin. His father Jacques-François (1692-1746) was a surgeon who already had a son Charles-Augustin (1727-1762) from a first marriage when he was in China. Vandermonde's mother was Jeanne Dailly (1689-1777).

Vandermonde graduated in law in 1755 and obtained a "licence" in 1757 but his first interest was in music and playing the violin. He started to be interested in mathematics a little before 1770. He met Denis Diderot (1713-1784), Jean Le Rond d'Alembert (1717-1783) who was publishing *L'Encyclopédie*, and Alexis Fontaine des Bertins (1704-1771) who was a mathematician.

Vandermonde's first mathematical work *Mémoire sur la résolution des équations* (Memoir on the solution of equations) opened for him the gates of the Académie des Sciences as "adjoint géomètre" in 1771. In that paper he developed the theory of symmetric functions with the aim of understanding solvability of algebraic equations.

In 1771, he presented to the Académie a memoir titled *Mémoire sur l'élimination* (Memoir on elimination) [3121] that was published in 1776. His two other mathematical works were *Remarques sur les problèmes de situation* (Remarks on situation problems) and *Mémoire sur les irrationnelles des différents ordres avec une application au cercle* (Memoir on irrationals of different orders with application to the circle) in 1772. And that was the end of the mathematical career of Vandermonde.

Then, he became interested in physics, engineering, chemistry, and music. He collaborated with Jacques de Vaucanson (1709-1782), an inventor who created sophisticated automatas like "the flute player" and "the digestive duck." After his death, Vaucanson's collection of machines became the Cabinet des machines du Roi of which Vandermonde became the curator in 1783.

Vandermonde participated actively in the works of the Académie and wrote many reports, for instance, in 1786 with Claude-Louis Berthollet (1748-1822) and Gaspard Monge (1746-1818) about the manufacturing of iron and steel. Still interested in music, he published in 1778 a system of harmony *Système d'harmonie applicable à l'état actuel de la musique* that started some controversies. He wrote another memoir on music in 1780.

Vandermonde was an enthusiastic supporter of the French revolution in 1789. He participated in several committees and joined a few political societies that were created in those times, in particular the Club des Jacobins.

In 1794, Vandermonde obtained a position at the newly created Conservatoire des Arts et Métiers. In February 1795, he was appointed professor of political economy at the École Normale. He became interested in economy during his missions for the government of the republic.

Unfortunately, Vandermonde had health problems and he died from tuberculosis on January 1, 1796, in Paris. For more details on Vandermonde as an economist, see [1125, 1618].

Work

It is generally believed that the most important mathematical work of Vandermonde is his first memoir in 1771; see, for instance, [2020]. Positive appraisals on this memoir were given by Étienne Bézout (1730-1783) and Nicolas de Condorcet (1743-1794). According to [2020], Leopold Kronecker (1823-1891) said that

> *the development of algebra started with Vandermonde's memoir in 1770.*

Vandermonde can be considered a predecessor of Évariste Galois (1811-1832) and Niels Henrik Abel (1802-1829).

Of more importance for us is the memoir on elimination. In this work, Vandermonde introduced objects that we can recognize as determinants. He was probably the first to consider them as functions of interest and he established some of their properties. Of course, Gottfried Wilhelm Leibniz (1646-1716) had previously been working on determinants but he communicated his results only in letters to some other mathematicians and therefore had not much influence on his followers. On the contrary, Vandermonde's memoir was cited by Augustin-Louis Cauchy (1789-1857) and Carl Gustav Jacob Jacobi (1804-1851); see Chapter 3.

For a discussion of whether or not a Vandermonde determinant appears in disguise in his paper [3121], see [3293].

10.74 ▪ Richard S. Varga

Richard Steven Varga
Courtesy of Claude Brezinski

Life

Richard Steven Varga was born on October 9, 1928, in Cleveland, Ohio, USA.

His parents were both born in Hungary; they emigrated separately to the USA where they met and married in Cleveland. Richard's father was an expert tool and die maker, and his mother was an expert in passementerie sewing.

Varga attended public schools, and went to the West Technical High School in Cleveland since his father wanted him to become a draftsman. Then, he registered at Case Institute of Technology (now Case Western Reserve University), originally in mechanical engineering, but he rapidly switched to mathematics. It is perhaps his passage in this university which gave him the taste for mechanics since he always repaired his cars himself. He dedicated some time for sports, and was a member of the college wrestling team. He also won some table tennis tournaments.

He graduated with a B.S. in mathematics in June 1950, and was looking for a job as an actuary when an older professor at Case, Max Morris, persuaded him to apply to Harvard University for graduate studies in mathematics. He went there, completed his master's degree in this first year, and was awarded an A.M. in June 1951.

That same year he married Esther Marie Pfister (1926-2015). They had one daughter, Gretchen. Then, Varga undertook research at Harvard with the American analyst Joseph Leonard Walsh (1895-1973) as his thesis advisor. In June 1954, he completed his Ph.D. thesis in complex approximation theory [3127].

From 1954 to 1960, he worked for Westinghouse Electric Corporation, Bettis Atomic Power Laboratory in Pittsburgh. There, he had to solve two- and three-dimensional multigroup diffusion equations. The method of choice for solving the systems of linear equations that were involved was the new SOR method of David Monaghan Young (1923-2008). The famous Garrett Birkhoff (1911-1996) was a professor at Harvard and a consultant at Bettis, and they worked together on several research topics there. In [3135], Varga wrote

> *I had completely changed, in a short time, from an analyst in complex approximation theory to a numerical analyst!*

In 1960, Varga became a professor of mathematics at Case Institute of Technology. In 1969, Varga moved to Kent State University, as a professor of mathematics. He was Director of the Institute for Numerical Mathematics from 1980 to 1988, and from 1988 to 2006 he was Research Director. He stayed there until his retirement in 2006, directing 25 Ph.D. theses.

Varga was editor-in-chief of *Numerische Mathematik* for 14 years. He also was one of the founding editors of the online electronic journal *Electronic Transactions on Numerical Analysis*, and organized several international conferences.

Varga received many honours including the Hans Schneider Prize from the International Linear Algebra Society in 2005, the President's Medal from Kent State University in 1981, and the von Humboldt Prize in 1982. The University of Karlsruhe (Germany) awarded him an honorary degree in 1991, and he received a similar honor from the University of Lille (France) in 1993.

Moreover, he was always kind to young researchers. According to [537],

> *In addition to his contributions as a researcher, Richard Varga has also promoted Numerical Analysis and its interplay with analysis, matrix theory and approximation theory by encouraging friends and colleagues to work in these areas and by organizing conferences that promote the interplay between these fields.*

Richard Varga passed away on February 25, 2022.

Work

The first edition of Varga's influential book *Matrix Iterative Analysis* [3132] was published in 1962. There, as he wrote in [3135],

> *newer things like M-matrices and the Perron-Frobenius theory of nonnegative matrices, graph theory, matrix Padé approximations, regular splittings of matrices, and finite difference and finite element approximations of second-order elliptic p.d.e.s were treated.*

A revised and expanded editiion was published in 2000.

Varga published about 240 papers, and wrote several books on various topics such as functional analysis and approximation theory, polynomial and rational interpolation and approximation, and Gerschgorin's circles [3136].

Varga was also very much interested in the Riemann hypothesis, and he thought that Padé approximation could be a way to prove it.

As it was written in [274] in 2006,

> *His research contributions span several areas of numerical mathematics, matrix theory, approximation theory, complex analysis, and computational number theory* [...] *Richard Varga, however, is not only a great scientist but also a wonderful human being. He is even known to be a serious table tennis player, he is a former wrestler, he likes to repair his car by himself, and anybody who has heard him sing can tell you about his beautiful and powerful voice. His knowledge of languages and flair for telling jokes and anecdotes all contribute to make Richard Varga a very charming man.*

10.75 ▪ Karl Weierstrass

Karl Weierstrass

Life

Karl Theodor Wilhelm Weierstrass was born into a Roman Catholic family in Ostenfelde, a village near Ennigerloh, in the Province of Westphalia, Kingdom of Prussia, on October 31, 1815. At that time, his father Wilhelm was the secretary to the mayor of Ostenfelde. He was a well educated man who had a broad knowledge of arts and sciences. His mother was Theodora Vonderforst. Karl was the eldest of four children, none of whom married.

When Karl was eight years old, his father became a tax inspector, a job needing to move from place to place within short periods of time. Thus, Karl frequently changed school as the family moved around Prussia. Karl's mother died in 1827 and his father Wilhelm remarried one year later. In 1829, Wilhelm Weierstrass became an assistant at the main tax office in Paderborn, and Karl entered the Catholic Gymnasium Theodorianum there. He excelled at this place despite having to take on a part-time job as a bookkeeper to help the family income.

At the Gymnasium, Weierstrass reached a level of mathematical competence far beyond what was usually expected. He regularly read *Crelle's Journal* and gave mathematical lessons to one of his brothers. Since his father wanted him to make a career in the Prussian administration, after graduating from the Gymnasium in 1834, Karl entered the University of Bonn to study law, finance, and economics. Weierstrass suffered from the conflict of either obeying his father or studying mathematics. As a result, he did not attend either the mathematics lectures or the lectures of his planned course. He continued to study mathematics by himself reading the book *Mécanique Céleste* by Pierre Simon de Laplace (1749-1827) and then a work by Carl Gustav Jacob Jacobi (1804-1851) on elliptic functions, a topic he learned through transcripts of lectures by Christoph Gudermann (1798-1822). Weierstrass also spent four years engaging in intensive fencing and drinking. The outcome was that he left the university in 1838 without a degree.

Weierstrass's father was upset but he allowed Karl to study at the Theological and Philosophical Academy of Münster in order to obtain the necessary exams to become a secondary school teacher. He enrolled there on May 22, 1839. Gudermann was lecturing in Münster and Weierstrass attended his lectures on elliptic functions. Leaving Münster in the autumn of 1839, he studied for the teacher's examination which he registered for in March 1840. But by this time, his father had changed jobs again, becoming director of a salt works in January 1840, and the family moved to Westernkotten near Lippstadt on the Lippe River, west of Paderborn. By April 1841, after qualifying as a teacher, Weierstrass began a one-year probation at the Gymnasium in Münster. Although he did not publish them at this time, he wrote three short papers in 1841 and 1842 presenting the concepts on which he would later base his theory of functions of a complex variable.

In 1843, he taught in Deutsch Krone in West Prussia (now Poland) and, from 1848, at the Lyceum Hosianum in Braunsberg where besides mathematics he also taught physics, botany, geography, history, German, calligraphy, and even gymnastics. Due to the problems he had as a student, his demanding teaching job, and the stress of working on mathematics in his free time, Weierstrass, from around 1850 and over about a period of 12 years, began to suffer from quite severe attacks of dizziness. During his stay in Braunsberg, he published papers on Abelian functions in the school prospectus but they remained unnoticed by mathematicians.

In 1854, Weierstrass published *Zur Theorie der Abelschen Functionen* (On the theory of Abelian functions) in *Crelle's Journal*. This paper did not give the full theory of inversion of hyperelliptic integrals that he had developed, but rather a preliminary description of his methods involving the representation of Abelian functions as converging power series. With this paper Weierstrass gained some fame. The University of Königsberg awarded him an honorary doctoral degree on March 1, 1854. In 1855, he applied for the chair at the University of Breslau left vacant after Ernst Eduard Kummer (1810-1893) left for Berlin. Kummer tried to influence things so that he could go to Berlin, but Weierstrass was not appointed. However, he was promoted to senior lecturer at Braunsberg and he obtained a one-year leave of absence to devote himself to advanced mathematical study. He had already decided that he would never return to schoolteaching.

Weierstrass published a full version of his theory of inversion of hyperelliptic integrals in his paper *Theorie der Abelschen Functionen* in 1856. Several universities offered him a chair, among them the Gewerbeinstitut (Industry Institute, later the Technische Hochschule) in Berlin. He accepted the offer on June 14, 1856. Offers continued to arrive. While attending a conference in Vienna in September 1856, he was offered a chair at any Austrian university of his choice. Before he had decided how to answer, the University of Berlin also offered him a professorship in October. He quickly accepted, but having already accepted the offer from the Industry Institute, he was unable to formally hold this chair for some years.

Weierstrass's lectures attracted students from all over the world. The topics of his lectures included the application of Fourier series and integrals to mathematical physics (1856/57), an introduction to the theory of analytic functions, the theory of elliptic functions, and applications to problems in geometry and mechanics. In 1859-1860, he lectured on an introduction to analysis where he tackled the foundations of the subject for the first time. In 1860-1861, he taught integral calculus. But following his previous problems, his health deteriorated in December 1861 when he completely collapsed. It took him about a year to recover sufficiently, but not completely, to lecture again. From this time on, he lectured sitting down while a student wrote for him on the blackboard. Then, the attacks he had since 1850 stopped and were replaced by chest problems. In his 1863-1864 course on the general theory of analytic functions, Weierstrass began to formulate his theory of the real numbers and proved that the complex numbers are the only commutative algebraic extension of the real numbers.

Finally, in 1864, Weierstrass took his professorship at the Friedrich-Wilhelms-Universität Berlin, which later became the Humboldt Universität zu Berlin. He continued to lecture a four-semester course until 1890. The four courses were Introduction to the theory of analytic functions, Elliptic functions, Abelian functions, and Calculus of variations or applications of elliptic functions. Weierstrass and his two colleagues Kummer and Leopold Kronecker (1823-1891) gave Berlin a reputation as a leading university for studying mathematics. Kronecker was a close friend of Weierstrass for many years but in 1877 Kronecker's opposition to the work of Georg Ferdinand Ludwig Philipp Cantor (1845-1918) ended their friendship. This became so bad that Weierstrass decided to leave Berlin and go to Switzerland. However, realizing that Kronecker would be in a stronger position to influence the choice of his own successor, he finally decided to stay in Berlin.

In 1870, at the age of fifty-five, Weierstrass met Sofia Vasilyevna Kovalevskaya (1850-1891) from Russia, who he tutored privately after failing to secure her admission to the University. They had an intense relationship that was never entirely clarified. It was through his efforts that she received an honorary doctorate from Göttingen, and he also used his influence to help her obtain a position of *Privatedocent* at the University of Stockholm in 1883. They corresponded for 20 years between 1871 to 1890. More than 160 letters were exchanged, but Weierstrass burnt Kovalevskaya's letters after her death.

Weierstrass decided to supervise himself the publication of his own complete works since he had a large amount of unpublished material from his lecture courses that, without his help, would be difficult to gather. The first two volumes appeared in 1894 and 1895. The remaining volumes appeared slowly: volume 3 in 1903, volume 4 in 1902, volumes 5 and 6 in 1915, and volume 7 in 1927.

Weierstrass was confined to a wheelchair, immobile, and dependent for the last three years of his life. He died in Berlin from pneumonia on February 19, 1897.

Work

Weierstrass's approach still dominates the theory of functions of a complex variable. This is clearly seen from the contents and style of his lectures, particularly the introduction course on analysis he gave in 1859-1860. Its contents were numbers with his power series approach to irrational numbers as limits of convergent series, entire functions, the notion of uniform convergence and functions defined by infinite product, analytic continuation, points of singularity, analytic functions of several variables, and contour integrals.

In 1872, Weierstrass found a continuous function that had no derivative at any point. It is an example of fractal curve.

The Casorati-Weierstrass theorem describes the behavior of holomorphic functions near their essential singularities. It was published by Weierstrass in 1876 (in German) and by Julian-Karl Vasilievich Sokhotski (1842-1927) in 1868 in his master's thesis (in Russian). So it was called Sokhotski's theorem in the Russian literature and Weierstrass' theorem in the Western literature. The same theorem was published by Felice Casorati (1835-1890) in 1868, and by Charles Auguste Briot (1817-1882) and Jean-Claude Bouquet (1819-1885) in the first edition of their book *Théorie des fonctions doublement périodiques et, en particulier, des fonctions elliptiques* in 1859. However, Briot and Bouquet removed this theorem from the second edition in 1875.

The Weierstrass approximation theorem states that every continuous function defined on a closed interval can be uniformly approximated as closely as desired by a polynomial. The original version of this result was established by Weierstrass in 1885 using the Weierstrass transform.

Weierstrass gave an axiomatic definition of determinants in the lessons he delivered in 1886-1887; see Chapter 3. For his work on bilinear forms, see Chapter 4.

10.76 ▪ James H. Wilkinson

James Hardy Wilkinson
Stanford University, 1982. Courtesy of Gérard Meurant

Life

James (Jim) Hardy Wilkinson was born on September 27, 1919, in Strood, Kent, UK, near the town of Rochester, not far from the Medway river, southeast of London. He was the third children of five, having a brother and three sisters. His father, James William Wilkinson, was the 12th child of a family of 13, and his mother was Kathleen Charlotte Hardy.

After working in his brother's dairy, Jim's father set up his own dairy business. The family was quite poor with the children helping in the dairy business. It failed during the 1930s economic depression and Jim's father had no employment until World War II. However, despite this poverty, the children received a good education.

At the age of 11, Jim won a scholarship to Sir Joseph Williamson's Mathematical School in Rochester whose headmaster was Mr. E.D. Clarke, who recognized Jim's abilities very early. Jim won a scholarship in mathematics to Trinity College, Cambridge, on his first attempt when he was 16 years old. There he studied mathematics under Godfrey Harold Hardy (1877-1947), John Edensor Littlewood (1885-1977), and Abram Samoilovitch Besicovitch (1891-1970).

He completed the three parts of the Mathematical Tripos and was Senior Wrangler in 1939.[84] However, World War II began and Jim never got a Ph.D. from Cambridge.

In January 1940, he was drafted into military work at the Cambridge Mathematical Laboratory of the Ordnance Board of the Ministry of Supply. He had to compute trajectories using hand-desk calculators. There he met E.T. (Chas) Goodwin, who was later Superintendent of the Mathematics Division at the National Physical Laboratory (NPL). In mid-1943, Jim was transferred to the Ministry of Supply Armament Research Laboratory at Fort Halstead, close to Sevenoaks (Kent). There, he had to solve problems in the thermodynamics of explosions.

Importantly for the rest of his life, it is in Fort Halstead that Jim met his wife, Heather Norah Ware (1924-2011). She had a first class honors degree in mathematics from King's College, London, in 1943. They had two children, Jenny and David.

In 1945, Jim asked Goodwin for a transfer from Fort Halstead. At this time Goodwin was in charge of the Desk Machine Computing Section in the NPL Mathematics Division. Jim joined the NPL group in May 1946. This is where he met Alan Mathison Turing (1912-1954). Turing needed an assistant for his plans to build an automatic computing machine named ACE (*Automatic Computing Engine*). It was decided that Wilkinson would work half-time for Goodwin and half-time for Turing. In practice he worked almost full-time for Turing. In Wilkinson's words (see [2324]),

> *I thought quite seriously of going back to Cambridge, but it was at this stage that I began to hear rumours about electronic computers and these appeared to me to provide facilities for solving partial differential equations in a reasonable way. There was clearly no future in solving them on hand computers. Shortly afterwards I discovered that a Mathematics Division was being set up at NPL. I got in touch with E.T. Goodwin who had been a colleague of mine at Cambridge in the Maths Lab. He was one of the first to join this new division. He invited me to have a chat with him at NPL and there I met Turing who I knew already by reputation as something of an eccentric. Turing and I had a long discussion and I was very impressed with him.*

Harry Douglas Huskey (1916-2017) came for a one-year sabbatical, on leave from California and at NPL invitation, to help with the machine building. Huskey suggested strongly to build a smaller machine (the *Pilot ACE*) rather than the original ACE project. At that time it was also not clear which NPL division should build the machine. Due to these problems, Turing took a year's sabbatical at Cambridge, leaving Jim in charge of the ACE with a team of five or six members. Turing returned for a short time in the middle of 1948, disliked what he saw, and went off permanently to Manchester.

The detailed design of the *Pilot ACE* started in the beginning of 1949 and the machine started to work in May 1950; see [3238]. There was a three-day demonstration for the press in November 1950 and Jim said (see [1205])

> *The Pilot ACE worked virtually perfectly for the whole three days, a level of performance which it had not been within striking distance of achieving before and which it did not achieve again for some considerable time.*

Anyway, the machine worked for the next five years. With the addition of a magnetic drum store and some improvements it was marketed by the English Electric Company under the name *DEUCE*.

With the early use of the *Pilot ACE*, Jim became interested in rounding error analysis, starting with the evaluation of polynomials. Beresford Neill Parlett [2447] wrote that

[84] For an explanation, see the biography of Arthur Cayley, Section 10.8

One could say that the decade 1947-1957 was the exciting learning period in which Wilkinson, and his colleagues at NPL, discovered how automatic computation differed from human computation assisted by desk top calculating machines. By dint of trying every method that they could think of and watching the progress of their computations on punched cards, paper tape, or even lights on the control console, these pioneers won an invaluable practical understanding of how algorithms behave when implemented on computers.

Wilkinson worked at the NPL from 1946 up until his retirement in 1980. He was promoted to Chief Scientific Officer, a rare distinction, in 1974. He was living in Teddington not too far from the NPL.

Wilkinson was a frequent visitor to the United States. According to Gene Howard Golub (1932-2007) and Clive Barry Moler [1387] [1387], all in all, he took almost 200 trips across the Atlantic. In 1958, he began giving short courses at the University of Michigan Summer College of Engineering. The notes from these courses were the preliminary versions of his two books. The following anecdote is reported in [1387]:

During the early days of the University of Michigan Summer Conferences in Computing and Numerical Analysis, Wilkinson and the other lecturers and participants would retire to Ann Arbor's Pretzel Bell restaurant and pub to try to escape the rigors of the Midwest in July. Olga Tausky-Todd observed these late evening sessions, but did not participate as fully as the others. One day she wanted to take a visitor to lunch and so asked Wilkinson if it "was possible to order beer in more modest quantities" at the Pretzel Bell. Jim replied, "Yes, I assume it is, but it has never been one of my ambitions."

Jim was also a frequent visitor to the Argonne National Laboratory close to Chicago. In 1962, he received a D.Sc. from Cambridge University in recognition of his contributions. In the 1960s, he published two books summarizing his work and experience about rounding errors in numerical linear algebra and the behavior of the algorithms known so far for solving linear systems and eigenvalue problems. The book *Rounding Errors in Algebraic Processes* appeared in 1964 and *The Algebraic Eigenvalue Problem* in 1965. He was one of the founders of the journal *Linear Algebra and its Applications* and become a main editor in 1967.

He was elected to the Royal Society in 1969. Wilkinson was the first professional numerical analyst to be elected to the Fellowship of the Royal Society. In 1970, he received the ACM Turing Award and the SIAM von Neumann Award.

In 1971, the *Handbook for Automatic Computation* volume II appeared that was edited by Jim and Christian Reinsch. It was a collection of programs, written in Algol 60, for solving a variety of linear algebra problems. This book was very influential on the development of numerical software for linear algebra.

In 1973, Wilkinson became a Distinguished Fellow of the British Computing Society, and in 1977, an honorary fellow of the Institute for Mathematics and its Applications (UK). In 1977, after being discharged of his duties for the Council of the Royal Society, he accepted a professorship in the Computer Science department at Stanford University, California, that lasted until 1984. Usually he taught there in the Winter quarter. It is there that G.M., being a visiting scientist in that department, met Wilkinson in 1982 and took the photo you have seen above.

In 1983, Jim was elected a Fellow of the Japan Society for the Promotion of Science and in 1987 he received the Chauvenet Prize of the Mathematical Association of America for his paper on *The perfidious polynomial* [3254].

Jim Wilkinson passed away on October 5, 1986, having a heart attack while he was working in his garden in Teddington.

The Argonne National Laboratory has a J.H. Wilkinson Fellowship in Computational Science and SIAM has two prizes in his honor: the James H. Wilkinson Prize in Numerical Analysis and Scientific Computing and the James H. Wilkinson Prize for Numerical Software.

Interesting articles about Wilkinson's life and work are [3253, 1387, 230, 2447, 1565] (ordered by date). The most detailed article is the bibliographical memoir [1205] written by Leslie Fox (1918-1990).

Work

Despite his early work on the *Pilot ACE* (see [3237]), Wilkinson is most well known for his contributions to numerical linear algebra, particularly for his use of backward error analysis of algorithms. He was not the first to use backward error analysis, but he promoted it as an efficient technique to study algorithms in finite precision arithmetic and he applied it to most of the numerical linear algebra algorithms that were known at that time.

Backward analysis consists in showing that the computed solution in finite precision arithmetic is the exact solution of a "nearby problem." It is hoped that the perturbation is small in some sense.

One of Jim's first works in backward error analysis was concerned with the computation of the polynomial $p_n(x) = a_0x^n+a_1x^{n-1}+\cdots+a_n$ for a given value of x; see [1205, 3241, 3242]. He wrote a routine for assembling the coefficients of the polynomial from the factored form $p_{20}(x) = (x-1)(x-2)\cdots(x-20)$. A second routine used Horner's rule to compute $p_{20}(x)$ and a third used a Newton's iteration to compute the roots. He wanted to compute the largest zero starting from $x_0 = 21$. He was expecting monotonic quadratic convergence to the zero $x = 20$ but this did not happen. His successive estimates were around $x = 20$ with random movements in both directions. His analysis suggested that the largest zeros were very sensitive to small relative changes in the coefficients a_i's. The theory of fixed-point arithmetic was established at that time, but Jim published in his first book a complete theory of rounding errors in floating-point arithmetic that was done by software in the *Pilot ACE* computer. Backward and forward error analysis for Horner's rule was given in [3247] written in 1963.

Then, Wilkinson applied backward error analysis to Gaussian elimination for solving linear systems $Ax = b$. He had already some experience with rounding errors in Gaussian elimination since, during World War II, he solved by hand with a desk calculator a linear system of order 12. In his 1970 Turing Award Lecture [3253], he wrote

> *I used ten-decimal computation more as a safety precaution.* [...] *I slowly lost figures until the final reduced equation was of the form, say,* $.0000376235\,x_{12} = .0000216312$. *At this stage I can remember thinking to myself that the computed* x_{12} *derived from this relation could scarcely have more than six correct figures. I substituted my solution in the original equations to see how they checked.* [...] *To my astonishment the left-hand side agreed with the given right-hand side to ten figures.*

Another experiment with Gaussian elimination for solving (still without an electronic computer) a linear system of order 18 took place at NPL. In a subsequent paper, Fox, Huskey, and Wilkinson [1206] showed empirical evidence in support of Gaussian elimination, writing that

> *in our practical experience on matrices of orders up to the twentieth, some of them very ill-conditioned, the errors were in fact quite small.*

To simplify, the computed solution $\hat{x}$ satisfies $(A+E)\hat{x} = b$ with $\|E\|_\infty\| \le \gamma_{3n}gn^2\|A\|_\infty$ and $\gamma_k = cku/(1-cku)$, where c is a small constant and u is the unit round-off. Moreover, the norm of the residual can be shown to be bounded by a quantity proportional to u. There is no floating point error analysis of the Cholesky factorization in Wilkinson's books, but he gave a detailed analysis later.

Wilkinson's ideas about backward error analysis are summarized in the following sentences:

> *There is still a tendency to attach too much importance to the precise error bounds obtained by an a priori error analysis. In my opinion, the bound itself is usually the least important part of it. The main object of such an analysis is to expose the potential instabilities, if any, of an algorithm so that, hopefully, from the insight thus obtained one might be led to improved algorithms.*

A great deal of Wilkinson's works is devoted to the eigenvalue problem. He analyzed many methods for computing the eigenvalues and eigenvectors of matrices; see his 1965 book [3248]. He considered the QR iterations and studied what is now known as Wilkinson's shift to accelerate convergence.

In 1970, he wrote a paper [1754] with Richard Steven Varga (1928-2022) and Alston Scott Householder (1904-1993).

Together with Golub, he studied the Jordan canonical form, explaining the shortcomings of this form in practical computations. Later he became interested in the Kronecker form of a pencil of matrices (A, B).

Wilkinson was also very influential on the software development for linear algebra problems. The *Handbook for Automatic Computation* [3256] was a landmark in this area and the translation of some of its routines from Algol to Fortran led to the development of the package EISPACK for solving eigenproblems at Argonne National Laboratory. Jim was also in contact with Argonne during the development of LINPACK, a package for solving linear systems; see Chapter 8. Jim's knowledge was also important for the developers of the library written at the Numerical Algorithms Group (NAG) in the UK who also started from translations of the *Handbook*.

10.77 ▪ David M. Young

David Young

Life

David Monaghan Young Jr. was born October 20, 1923, in Quincy, Massachusetts, USA, a small town south of Boston. He was the son of David Monaghan Young (1885-1967), originally from Scotland, and Madge Colby Tooker Young (1895-1983).

He went to North Quincy High School and very early on, became interested in mathematics. However, the family was not wealthy since David's father lost his accountant job during the economy depression. There was not enough money for college education. However, David took the entrance exam to the Webb Institute of Naval Architecture which was giving scholarships covering the whole tuition. He started his studies at the Webb Institute in September 1941. For the USA the war started in December 1941, but the students at Webb were allowed to finish their degrees. David joined the U.S. Navy in December 1942, becoming an officer in March 1945. He graduated with a bachelor of science degree in 1944 and left the Navy in March 1946, but stayed in the U.S. Naval Reserves for many years.

Then, Young went to Harvard University to study mathematics. He received a master of arts degree in 1947 and started working for a Ph.D. under the supervision of Garrett Birkhoff (1911-1996), the son of George David Birkhoff (1884-1944). The selected topic was relaxation methods for solving linear systems. He also took a course on computer programming and was able to use the *Mark I* computer at the Harvard Computation Laboratory. During a visit to Harvard, Sir Richard Vynne Southwell (1888-1970), who was "the" expert in relaxation methods, told David that *any attempts to mechanize relaxation methods would be a waste of time*. Fortunately, he was wrong.

David married Mildred Victoria Acker (1924-2017) on October 9, 1949. They had three children: William David, Arthur Earle, and Carolyn Ellen.

David Young obtained a Ph.D. thesis from Harvard University in May 1950. Its title was *Iterative methods for solving partial difference equations of elliptic type* [3304]. David and his wife wrote all the equations and the symbols in ink in the three copies of the thesis (151 pages). Gene Howard Golub (1932-2007) wrote later that *David Young's thesis is one of the monumental works of modern numerical analysis*. It was extremely difficult for David to publish a paper based on what he had done in his thesis. It took him four years to have the paper published after several revisions [3306].

In the fall of 1950, David became an instructor of mathematics at Harvard. He met Richard Steven Varga (1928-2022) who was a first-year graduate student in the Mathematics Department. They even became roommates.

In 1950-1951, Young worked in the Computing Laboratory, Aberdeen Proving Ground in Maryland. There he used the *ORDVAC* computer. In the fall of 1952, he became assistant professor of mathematics at the University of Maryland and was promoted to associate professor in 1953.

The family moved to Los Angeles in 1954 where David worked for a while in the industry as a manager of the Mathematical Analysis Department at Ramo-Wooldbridge Corporation (a space technology company). He visited several California universities and met Gene Golub for the first time.

Young moved to Austin (Texas) in the summer of 1958 and stayed there for the rest of his career. He joined the Department of Mathematics of the University of Texas as professor of mathematics and director of the newly founded Computation Center. There he expanded the Computation Center, being able to install a Control Data Corporation *CDC 6600* (which was a supercomputer in those times) in 1966 with a grant from the National Science Foundation.

In 1966, he took part in the foundation of the Department of Computer Science. In 1970, he became director of the Center for Numerical Analysis which participated in NATS (National Activity for Testing Software) for the testing and certification of EISPACK, jointly with Argonne National Laboratory and Stanford University.

David Young retired in 1999. He published around 200 papers and 4 books including [3307, 1533]. He died December 21, 2008 in Austin, Texas and is buried in the Austin Memorial Park Cemetery.

More details on his life and work can be found in [1901].

Work

Relaxation methods are iterative methods for solving linear systems $Ax = b$; see Chapter 5. Starting from a given initial iterate x^0, the residual vector $r^0 = b - Ax^0$, a component r_i^0 is chosen and x_i^1 is computed to zero the ith residual, the other components of x^1 being identical to those of x^0. This process is repeated until convergence and parameters can be used to under or over "relax." These techniques were used for computations done with hand-desk calculation machines. The problem is, of course, the choice of the component to be "relaxed." One can choose, for instance, the index of the largest component in magnitude of the residual. Some other choices could be made by the person doing the computation. As we have seen above, Southwell was thinking that relaxation methods cannot be used efficiently on digital computers.

The point Successive OverRelaxation method (SOR) is defined by

$$x_i^{k+1} = \frac{\omega}{a_{i,i}}(b_i - \sum_{j=1}^{i-1} a_{i,j}x_j^{k+1} - \sum_{j=i+1}^{n} a_{i,j}x_j^k) + (1-\omega)x_i^k,$$

where ω is a real parameter, $\omega > 0$. In matrix form we have

$$(D + \omega L)x^{k+1} = \omega b - \omega U x^k + (1-\omega)Dx^k,$$

where the matrix A with entries $a_{i,j}$ is written as $A = D + L + U$, where D is diagonal and L (resp., U) is the strictly lower (resp., upper) triangular part of A. The iteration matrix, denoted by $\mathcal{L}_\omega$, is given by

$$\mathcal{L}_\omega = \left(\frac{1}{\omega}D + L\right)^{-1}\left[\frac{1-\omega}{\omega}D - U\right].$$

Young considered symmetric matrices. He proved that, under some hypothesis, there exists an optimal value ω_b,

$$\omega_b = \frac{2}{1 + (1 - \rho(J(A))^2)^{\frac{1}{2}}},$$

where $J(A)$ is the Jacobi iteration matrix, that is, $J(A) = -D^{-1}(L+U) = I - D^{-1}A$ and ρ is the spectral radius. Moreover, $\rho(\mathcal{L}_{\omega_b}) = \omega_b - 1$. Young's theory relies on the fact that for matrices *with a consistent ordering* (see Chapter 5), there is a relationship between the eigenvalues λ of $\mathcal{L}_\omega$ and μ of $J(A)$, namely,

$$(\lambda + \omega - 1)^2 = \omega^2\mu^2\lambda.$$

For the symmetric linear system arising from 5-point finite difference method in a square, the spectral radius of the optimal SOR method is $1 - 2\pi h + O(h^2)$, where h is the mesh size, compared to $1 - 2\pi^2 h^2 + O(h^4)$ for the Gauss-Seidel method (which corresponds to $\omega = 1$). It gives a large improvement of the convergence rate when $h \to 0$.

Unfortunately, for many problems, the spectral radius of the Jacobi matrix is unknown. Some techniques to compute approximations were proposed; see [1533, 3307].

Young worked also on the Richardson method and, in the 1980s, on generalizations of the conjugate gradient method to nonsymmetric linear systems, the Orthodir and Orthores iterative methods which are mathematically equivalent respectively to GMRES and to the Full Orthogonalization method (FOM); see [1809, 3310] and Chapter 5.

Since 1970, Young collaborated with David Ronald Kincaid on the ITPACK and NSPCG projects. These are software Fortran packages providing access to iterative methods for solving linear systems.

10.78 ▪ Konrad Zuse

Konrad Zuse

By Wolfgang Hunscher, Dortmund - Own work, CC BY-SA 3.0

Life

Konrad Ernst Otto Zuse was born in Berlin (Germany) on June 22, 1910. In 1912, the family moved to Braunsberg in Prussia (now in Poland) where Zuse's father worked in the post office. Zuse had one sister. He went to elementary school in Braunsberg. In 1923, the family moved to Hoyerswerda in Saxony where he finished his studies in 1928.

He went to what is now the Berlin Technical University. At first he hesitated between different areas, but finally chose civil engineering in which he graduated in 1935. He obtained a position with the Henschel Aviation Company near Berlin in 1935.

In 1934, he became interested in computing machines since he found quite boring the many calculations a civil engineer had to do by hand. After studying the available mechanical calculating machines of the time, from 1936 to 1938, Zuse built his first mechanical computer, the *Z1*, in his parents' apartment. On January 30, 1944, the *Z1* was destroyed by a British air raid.

In 1939, Zuse was called to military service. Unsatisfied with the *Z1*, Zuse built the *Z2* that was shown to the DVL (Deutsche Versuchsanstalt für Luftfahrt), which is now the German Aerospace Center, in 1940. He replaced binary switching metal sheets by old phone relays in the arithmetic and control units. The *Z2* was also destroyed by an allied air raid during World War II.

He obtained some funding from the DVL and founded a company called the Zuse Apparatebau in Berlin on April 1, 1940. The company was located in the Methfesselstrasse 7 and 10 in Berlin-Kreuzberg where he built the *Z3* in 1941. The *Z3* was the first fully operational electromechanical computer. By 1945 the company had about 20 employees, but the building and the machine were destroyed by an allied air raid in 1943.

Zuse married Gisela Brandes in January 1945. They had five children, their first son, Horst, being born in November 1945.

Zuse built the *Z4* computer at another location at Oranienstrasse 6. It was almost completed in 1944. To escape the air raids and the Soviet army, the *Z4* was packed and moved from Berlin on February 14, 1945, to Göttingen where the machine was completed. Then, the computer was transported to Hinterstein in Bad Hindelang in southern Bavaria where it was hidden. Unable

Figure 10.8. *Zuse's Z4 computer*

to work on hardware, Zuse developed a programming language that he called *Plankalkül*; see Chapter 8 and [232, 1923].

Some people claimed that there was a meeting between some British scientists, including Alan Mathison Turing (1912-1954), and Zuse in Göttingen in 1947, but so far there is no evidence of that. What is known is that Zuse visited England for three weeks in February 1948.

In 1948, the *Z4* was built in a former flour mill in Hopferau (Allgäu, Bavaria), near the Austrian border. In 1949, Eduard Ludwig Stiefel (1909-1978) from ETH in Switzerland visited Zuse in Hopferau to see and evaluate the *Z4*; see Section 10.66. ETH rented the machine for five years. The *Z4* was delivered to the ETH Zürich on July 11, 1950. In 1954, it was transferred to the Institut Franco-Allemand des Recherches de St. Louis in France, where it was used until 1959. Today, the *Z4* is on display in the Deutsche Museum in Munich.

In 1949, Zuse founded the company Zuse KG with five employees in a small village called Neukirchen which is 120 km north of Frankfurt. It was the first computer company in Germany. In 1957, Zuse KG was moved to Bad Hersfeld. The first thing done was to put the *Z4* back to work and to enhance it at the request of ETH, for instance, adding a conditional branch capability.

Zuse had some contacts with IBM but not for too long, and then, with Remington Rand. During a trip to the USA he met with Howard Hathaway Aiken (1900-1973) and John Adam Presper Eckert (1919-1995). Then, Zuse KG set up a collaboration with Remington Rand Switzerland, but at some point, this was ended by the Remington Rand headquarters in the USA.

In 1955, Zuse KG built the *Z11*, a relay computer. It was followed by the *Z22*, using vacuum tubes, and the *Z23*, a transistorized version in 1961. Around 100 *Z23* machines were sold. The *Z31* appeared in 1963 but only 7 machines were sold.

Zuse rebuilt the *Z3* at his Zuse KG company between 1960 and 1961 to demonstrate the performance of his machine and to justify his patents for which there was still a pending justice decision. In 1967, a decision was made by the German patent court and Zuse lost his 26-year fight about the invention of the *Z3*.

After 1964, Zuse KG was no longer owned and controlled by Zuse. It was bought by Rheinstahl, then by the Brown Boveri Company, and in 1967, by Siemens. Zuse had a consultant contract with Siemens AG until 1969 when he left his former company.

After leaving Zuse KG, Konrad Zuse wrote his autobiography. The first edition was published in 1970 by a small publisher and reprinted in 1983 by Springer [3344]. In 1993, Springer published an English translation.

In 1987 and 1989, helped by two students, Zuse rebuilt his first mechanical computer, the *Z1*. It is now on exhibition at the German Museum of Technology in Berlin.

Konrad Zuse passed away on December 18, 1995, in Hünfeld, Hesse, from heart problems.

Work

Zuse's first computer was a binary machine because he wanted to use binary switching elements. He designed a binary floating point unit. He developed a mechanical memory of 64 32-bit words in which each memory cell could be addressed by the punch tape (a perforated 35-mm film) which was used to input the data. He also constructed a control unit and implemented input and output devices for conversion from the binary to the decimal number system and vice versa. The movement of the plates for the memory was driven by an electrical motor which produced a clock frequency of one Hertz.

This machine was a programmable mechanical binary floating point computer. But the programs could not be stored since the memory was too small for that. Zuse submitted two patents for this machine which never worked satisfactorily because of the lack of reliability of the binary switching metal sheets.

The *Z2* used the mechanical memory of the *Z1*, but, for the arithmetic and control unit, Zuse used 800 electrical phone relays. The arithmetic unit consisted of a 16-bit fixed-point engine. The next machine, the *Z3*, was built completely with relays, 600 in the arithmetic unit and about 2,800 for the memory and control unit.

The *Z4* contained about 2,200 relays and a mechanical memory of 64 words made with metal plates. The frequency was 30 Hertz. It had a floating point arithmetic unit. The word length was 32 bits (1 sign bit, 24-bit mantissa, 7-bit exponent) and the machine did 25-30 operations per second which led to about 11 multiplications per second. The input was done with a decimal keyboard or a punched tape. The output device was a Mercedes typewriter.

Much later, in 1961, the *Z23* had a 40-bit word length and used a drum memory with 8,192 words as main storage, with 256 words of rapid access ferrite memory. The *Z23* used about 2,700 transistors and 7,700 diodes. It was a transistor binary machine.

Bibliography

[1] A. Abdelfattah, H. Anzt, A. Bouteiller, A. Danalis, J. J. Dongarra, M. Gates, A. Haidar, J. Kurzak, P. Luszczek, S. Tomov, S. Wood, P. Wu, I. Yamazaki, and A. YarKhan. SLATE Working Note 1: Roadmap for the development of a linear algebra library for exascale computing. Technical Report ICL-UT-17-02, Innovative Computing Laboratory, University of Tennessee, 2017. (Cited on page 377.)

[2] K. Abe and G. L. G. Sleijpen. BiCR variants of the hybrid BiCG methods for solving linear systems with nonsymmetric matrices. *J. Comput. Appl. Math.*, 234:985–994, 2010. (Cited on page 242.)

[3] K. Abe and G. L. G. Sleijpen. Hybrid Bi-CG methods with a Bi-CG formulation closer to the IDR approach. *Appl. Math. Comput.*, 218(22):10889–10899, 2012. (Cited on page 243.)

[4] F. F. Abeles. Chiò's and Dodgson's determinantal identities. *Linear Algebra Appl.*, 454:130–137, 2014. (Cited on pages 114 and 405.)

[5] O. S. Adams. *Application of the Theory of Least Squares to the Adjustment of Triangulation.* Government Printing Office, Washington, DC, 1915. Special publication no. 28. (Cited on pages 66 and 448.)

[6] W. A. Adcock. An automatic simultaneous equation computer and its use in solving secular equations. *Review of Scientific Instruments*, 19(3):181–187, 1948. (Cited on page 332.)

[7] R. Adrain. Research concerning the probabilities of the errors which happen in making observations. *The Analyst or Mathematical Museum*, 1:93–108, 1808. (Cited on page 57.)

[8] M. Afanasjew, M. Eiermann, O. G. Ernst, and S. Güttel. Implementation of a restarted Krylov subspace method for the evaluation of matrix functions. *Linear Algebra Appl.*, 429(10):229–314, 2008. (Cited on page 390.)

[9] S. N. Afriat. Analytic functions of finite dimensional linear transformations. *Proc. Cambridge Philos. Soc.*, 55(1):51–61, 1959. (Cited on page 387.)

[10] K. Aihara, K. Abe, and E. Ishiwata. An alternative implementation of the IDRstab method saving vector updates. *JSIAM Letters*, 3:69–72, 2011. (Cited on page 243.)

[11] T. J. Aird. The IMSL library. In W. R. Cowell, editor, *Sources and Development of Mathematical Software*, pages 264–301. Prentice Hall, 1984. (Cited on page 375.)

[12] A. C. Aitken. On Bernoulli's numerical solution of algebraic equations. *Proc. Roy. Soc. Edinburgh*, 46:289–305, 1927. (Cited on pages 290, 294, 308, and 405.)

[13] A. C. Aitken. On the latent roots of certain matrices. *Proc. Edinburgh Math. Soc.*, 1(2):135–138, 1928. (Cited on page 405.)

[14] A. C. Aitken. Further numerical studies in algebraic equations and matrices. *Proc. Roy. Soc. Edinburgh*, 51:80–90, 1930. (Cited on pages 290, 291, and 405.)

[15] A. C. Aitken. On the evaluation of determinants, the formation of their adjugates, and the practical solution of simultaneous linear equations. *Proc. Edinburgh Math. Soc.*, 3(3):207–219, 1933. (Cited on pages 16 and 405.)

[16] A. C. Aitken. The evaluation of the latent roots and the latent vectors of a matrix. *Proc. Roy. Soc. Edinburgh*, 57:269–304, 1936. (Cited on pages 290, 291, and 405.)

[17] A. C. Aitken. XII.-Studies in practical mathematics. I. The evaluation, with applications, of a certain triple product matrix. *Proc. Roy. Soc. Edinburgh*, 57:172–181, 1938. (Cited on page 73.)

[18] A. C. Aitken. *Determinants and Matrices*. Oliver and Boyd, Edinburgh and London, 1939. (Cited on pages 12, 117, 381, 405, and 557.)

[19] A. C. Aitken. On the iterative solution of a system of linear equations. *Proc. Roy. Soc. Edinburgh*, 63(1):52–60, 1950. (Cited on pages 174 and 405.)

[20] A. C. Aitken. *Gallipoli to the Somme: Recollections of a New Zealand Infantryman*. Oxford University Press, Oxford, 1963. (Cited on page 404.)

[21] H. Akaike. Block Toeplitz matrix inversion. *SIAM J. Appl. Math.*, 24(2):234–241, 1973. (Cited on page 386.)

[22] E. L. Akim et al. In memory of Radii Petrovich Fedorenko (1930-2010). *Comput. Math. Math. Phys.*, 50(8):1459–1463, 2010. (Cited on page 453.)

[23] A. G. Aksoy. Al-Khwarizmi and the hermeneutic circle: Reflections on a trip to Samarkand. *J. Humanist. Math.*, 6(2):114–127, 2016. (Cited on page 42.)

[24] R. Al Homsi. Equation solving in Indian mathematics. Technical Report U.U.D.M. 2018:27, University of Uppsala, 2018. (Cited on page 42.)

[25] G. Alefeld. Zur konvergenz des Peaceman-Rachford verfahrens. *Numer. Math.*, 26:409–419, 1976. (Cited on page 179.)

[26] G. Alefeld and R. S. Varga. Zur konvergenz des symmetrischen relaxationsverfahrens. *Numer. Math.*, 25(3):291–295, 1975. (Cited on page 178.)

[27] L. Alfonsi. *Étienne Bézout (1730-1783): mathématicien, académicien et professeur au siècle des Lumiéres*. PhD thesis, Université Paris VI, France, 2005. (Cited on pages 54 and 105.)

[28] L. Alfonsi. Étienne Bézout: Analyse algébrique au siècle des Lumiéres. *Rev. Hist. Math.*, 14(2): 211–287, 2008. (Cited on pages 54 and 105.)

[29] L. Alfonsi. *Étienne Bézout (1730-1783) Mathématicien des Lumiéres*. L'Harmattan, Paris, France, 2011. (Cited on pages 54 and 105.)

[30] J. L. Aliaga, D. Boley, R. Freund, and V. Hernández. A Lanczos-type method for multiple starting vectors. *Math. Comput.*, 69(232):1577–1601, 2000. (Cited on pages 235 and 242.)

[31] D. M. Allen. The relationship between variable selection and data augmentation and a method for prediction. *Technometrics*, 16(1):125–127, 1974. (Cited on page 392.)

[32] F. Alouges and P. Loreaux. Massively parallel preconditioners. *Numer. Algorithms*, 14(4):361–375, 1997. (Cited on page 264.)

[33] S. C. Althoen and R. McLaughlin. Gauss-Jordan reduction: A brief history. *Am. Math. Mon.*, 94(2):130–142, 1987. (Cited on page 61.)

[34] M. Altman. On the approximate solution of linear algebraic equations. *Bull. Pol. Acad. Sci. Math.*, 5:365–370, 1957. (Cited on page 189.)

[35] M. Altman. Approximation methods in functional analysis. Lecture Notes, California Institute of Technology, 1959. (Cited on page 207.)

[36] M. Altman. A general method of steepest ortho-descent. *Bull. Pol. Acad. Sci. Math.*, 9:645–651, 1961. (Cited on pages 192, 193, and 406.)

[37] M. Altman. A general method with minimum ortho-residual. *Bull. Pol. Acad. Sci. Math.*, 9:739–744, 1961. (Cited on pages 192, 193, and 406.)

[38] M. Altman. A generalized method of steepest descent. *Bull. Pol. Acad. Sci. Math.*, 9:733–737, 1961. (Cited on pages 192, 193, and 406.)

[39] M. Altman. A gradient relaxation method for linear algebraic equations. *Bull. Pol. Acad. Sci. Math.*, 9:653–657, 1961. (Cited on pages 192, 193, and 406.)

[40] M. Altman. A linear iterative method with a vector parameter. *Bull. Pol. Acad. Sci. Math.*, 9:169–6174, 1961. (Cited on pages 192, 193, and 406.)

[41] M. Altman. A method of steepest ortho-descent. *Bull. Pol. Acad. Sci. Math.*, 9:575–580, 1961. (Cited on pages 192, 193, and 406.)

[42] G. G. Alway and D. W. Martin. An algorithm for reducing the bandwidth of a matrix of symmetrical configuration. *Comput. J.*, 8(3):264–272, 1965. (Cited on page 88.)

[43] U. Amaldi. Sulle sostituzioni lineari commutabili. *Rendiconti dell'Istituto Lombardo, serie II*, 45:433–445, 1912. (Cited on page 396.)

[44] P. R. Amestoy, A. Buttari, I. S. Duff, A. Guermouche, J.-Y. L'Excellent, and B. Uçar. MUMPS. In D. Padua, editor, *Encyclopedia of Parallel Computing*, pages 1232–1238. Springer, 2011. (Cited on pages 91 and 378.)

[45] P. R. Amestoy, A. Buttari, J.-Y. L'Excellent, and T. Mary. Bridging the gap between flat and hierarchical low-rank matrix formats: The multilevel block low-rank format. *SIAM J. Sci. Comput.*, 41(3):A1414–A1442, 2019. (Cited on pages 91 and 378.)

[46] P. R. Amestoy, A. Buttari, J.-Y. L'Excellent, and T. Mary. Performance and scalability of the block low-rank multifrontal factorization on multicore architectures. *ACM Trans. Math. Soft. (TOMS)*, 45(1):article 2, 2019. (Cited on pages 91 and 378.)

[47] P. R. Amestoy, T. A. Davis, and I. S. Duff. An approximate minimum degree ordering algorithm. *SIAM J. Matrix Anal. Appl.*, 17(4):886–905, 1996. (Cited on page 87.)

[48] P. R. Amestoy, I. S. Duff, and J.-Y. L'Excellent. Multifrontal parallel distributed symmetric and unsymmetric solvers. *Comput. Methods Appl. Mech. Engrg.*, 184(2-4):501–520, 2000. (Cited on pages 91 and 378.)

[49] P. R. Amestoy, I. S. Duff, J.-Y. L'Excellent, and J. Koster. A fully asynchronous multifrontal solver using distributed dynamic scheduling. *SIAM J. Matrix Anal. Appl.*, 23(1):15–41, 2001. (Cited on pages 91 and 378.)

[50] P. R. Amestoy, I. S. Duff, J.-Y. L'Excellent, and F.-H. Rouet. Parallel computation of entries of A^{-1}. *SIAM J. Sci. Comput.*, 37(2):C268–C284, 2015. (Cited on page 91.)

[51] P. R. Amestoy, I. S. Duff, and C. Puglisi. Multifrontal QR factorization in a multiprocessor environment. *Numer. Linear Algebra Appl.*, 3(4):275–300, 1996. (Cited on pages 133 and 378.)

[52] P. R. Amestoy, X. S. Li, and E. G. Ng. Diagonal Markowitz scheme with local symmetrization. *SIAM J. Matrix Anal. Appl.*, 29(1):228–244, 2007. (Cited on page 378.)

[53] P. R. Amestoy, X. S. Li, and S. Pralet. Unsymmetric ordering using a constrained Markowitz scheme. *SIAM J. Matrix Anal. Appl.*, 29(1):302–327, 2007. (Cited on page 378.)

[54] P. R. Amestoy and C. Puglisi. An unsymmetrized multifrontal LU factorization. *SIAM J. Matrix Anal. Appl.*, 24(2):553–569, 2002. (Cited on page 378.)

[55] B. Amiot. Discussion des valeurs générales fournies par la résolution de trois équations du premier degré entre trois inconnues. *Nouvelles Annales de Mathématiques 1re série*, 6:407–421, 1847. (Cited on page 56.)

[56] E. Anderson, Z. Bai, C. Bischof, L. S. Blackford, J. Demmel, J. Dongarra, J. Du Croz, A. Greenbaum, S. Hammarling, A. McKenney, and D. Sorensen. *LAPACK Users' Guide*. SIAM, Philadelphia, 1st edition, 1992. (Cited on page 369.)

[57] R. S. Anderssen and G. H. Golub. Richardson's non-stationary matrix iterative procedure. Technical Report STAN-CS-72-304, Computer Science Dept., Stanford University, 1972. (Cited on pages 163 and 181.)

[58] Anonymous. Lamberto Cesari: 1910-1990. *Notices Amer. Math. Soc.*, 37(6):677, 1990. (Cited on page 429.)

[59] Anonymous. To the memory of V.N. Kublanoskaya. J. Math. Sci., 191 (2013) 1-3; Translated from Zapiski Nauchnykh Seminarov POMI, 405 (2012) 7-12, 2013. (Cited on page 507.)

[60] R. Ansorge. Connections between the Cimmino-method and the Kaczmarz-method for the solution of singular and regular systems of equations. *Computing*, 33:367–375, 1984. (Cited on page 192.)

[61] H. Anzt, E. Chow, and J. Dongarra. ParILUT—a new parallel threshold ILU factorization. *SIAM J. Sci. Comput.*, 40(4):C503–C519, 2018. (Cited on page 261.)

[62] H. Anzt, T. K. Huckle, J. Bräckle, and J. Dongarra. Incomplete sparse approximate inverses for parallel preconditioning. *Parallel Comput.*, 71:1–22, 2018. (Cited on page 263.)

[63] H. Anzt, S. Tomov, J. Dongarra, and V. Heuveline. A block-asynchronous relaxation method for graphics processing units. *J. Parallel Dist. Com.*, 73(12):1613–1626, 2013. (Cited on page 197.)

[64] J. R. Appleyard and I. M. Cheshire. Nested factorisation. In *SPE 1983 Reservoir Simulation Symposium, San Francisco*, 1983. paper SPE 12264. (Cited on page 259.)

[65] J. R. Appleyard, I. M. Cheshire, and R. K. Pollard. Special techniques for fully implicit simulators. Technical Report CSS 112, AERE Harwell, UK, 1981. (Cited on page 259.)

[66] I. Arany. Another method for finding pseudo-peripheral nodes. *Annales Universitatis Scientiarum Budapestinensis de Rolando Eötvös nominatae*, 4:39–49, 1983. (Cited on page 88.)

[67] I. Arany. The method of Gibbs-Poole-Stockmeyer is non-heuristic. *Annales Univ. Sci. Budapest., Sect. Comp.*, 4:29–37, 1983. (Cited on page 88.)

[68] I. Arany, L. Szoda, and W. F. Smyth. An improved method for reducing the bandwidth of sparse symmetric matrices. In *Proceedings IFIP Congress*, pages 6–10. IFIP, 1971. (Cited on page 88.)

[69] I. Arany, L. Szoda, and W. F. Smyth. Minimizing the bandwidth of sparse matrices. *Annales Univ. Sci. Budapest., Sect. Comp.*, 9:129–151, 1973. (Cited on page 88.)

[70] P. Arbenz. Divide and conquer algorithms for the bandsymmetric eigenvalue problem. *Parallel Comput.*, 18(10):1105–1128, 1992. (Cited on page 316.)

[71] R. A. Archibald. Unpublished letters of James Joseph Sylvester and other new information concerning his life and work. *Osiris*, 1:85–154, 1936. Published by The University of Chicago Press on behalf of The History of Science Society. (Cited on pages 569 and 571.)

[72] M. E. Argentati, A. V. Knyazev, K. Neymeyr, E. E. Ovtchinnikov, and M. Zhou. Convergence theory for preconditioned eigenvalue solvers in a nutshell. *Found. Comput. Math.*, 17(3):713–727, 2015. (Cited on page 319.)

[73] J. H. Argyris. Energy theorems and structural analysis. *Aircraft Engineering and Aerospace Technology*, 26(11):383–394, 1954. (Cited on page 84.)

[74] M. Arioli. An analysis of GMRES worst case convergence. Unpublished report, 2009. (Cited on page 240.)

[75] M. Arioli and I. S. Duff. Using FGMRES to obtain backward stability in mixed precision. *Electron. Trans. Numer. Anal.*, 33:31–44, 2009. (Cited on page 242.)

[76] M. Arioli, I. S. Duff, J. Noailles, and D. Ruiz. A block projection method for sparse matrices. *SIAM J. Sci. Statist. Comput.*, 13(1):47–70, 1992. (Cited on page 192.)

[77] M. Arioli, I. S. Duff, D. Ruiz, and M. Sadkane. Block Lanczos techniques for accelerating the Block Cimmino method. *SIAM J. Sci. Comput.*, 16(6):1478–1511, 1995. (Cited on page 192.)

[78] M. Arioli and C. Fassino. Roundoff error analysis of algorithms based on Krylov subspace methods. *BIT Numerical Mathematics*, 36(2):189–205, 1996. (Cited on page 234.)

[79] M. Arioli, V. Pták, and Z. Strakoš. Krylov sequences of maximal length and convergence of GMRES. *BIT Numerical Mathematics*, 38(4):636–643, 1998. (Cited on pages 231, 244, and 305.)

[80] W. E. Arnoldi. The principle of minimized iterations in the solution of the matrix eigenvalue problem. *Quart. Appl. Math.*, 9:17–29, 1951. (Cited on pages 206, 219, 220, 300, and 303.)

[81] S. F. Ashby. ChebyCode: a FORTRAN, implementation of Manteuffel's adaptive Chebyshev algorithm. Technical Report 1203, Department of Computer Science, University of Illinois at Urbana-Champaign, 1985. (Cited on page 182.)

[82] S. F. Ashby. *Polynomial preconditioning for conjugate gradient methods.* PhD thesis, Dept. of Computer Science, University of Illinois, 1987. (Cited on page 262.)

[83] S. F. Ashby. Minimax polynomial preconditioning for Hermitian linear systems. *SIAM J. Matrix Anal. Appl.*, 12(4):766–789, 1991. (Cited on page 262.)

[84] S. F. Ashby, T. A. Manteuffel, and J. S. Otto. A comparison of adaptive Chebyshev and least squares polynomial preconditioning for Hermitian positive definite linear systems. *SIAM J. Sci. Statist. Comput.*, 13(1):1–29, 1992. (Cited on page 262.)

[85] C. C. Ashcraft and R. G. Grimes. On vectorizing incomplete factorizations and SSOR preconditioners. *SIAM J. Sci. Statist. Comput.*, 9(1):122–151, 1988. (Cited on page 260.)

[86] C. C. Ashcraft and R. G. Grimes. SPOOLES: An object-oriented sparse matrix library. In *Proc. 1999 SIAM Conf. Parallel Processing for Scientific Computing*, 1999. (Cited on page 371.)

[87] C. C. Ashcraft, R. G. Grimes, and J. G. Lewis. Accurate symmetric indefinite linear equation solvers. *SIAM J. Matrix Anal. Appl.*, 20(2):513–561, 1998. (Cited on page 78.)

[88] C. C. Ashcraft, R. G. Grimes, J. G. Lewis, B. W. Peyton, and H. D. Simon. Progress in sparse matrix methods for large linear systems on vector supercomputers. *Intl. J. Supercomp. Appl.*, 1(4):10–30, 1987. (Cited on page 91.)

[89] O. M. Ashford. *Prophet or Professor? The Life and Work of Lewis Fry Richardson.* Adam Hilger Ltd., Bristol and London, 1985. (Cited on page 547.)

[90] E. Asplund. Inverses of matrices $\{a_{ij}\}$ which satisfy $a_{ij} = 0$ for $j > i+p$. *Math. Scand.*, 7:57–60, 1959. (Cited on page 383.)

[91] S. O. Asplund. Finite boundary value problems solved by Green's matrix. *Math. Scand.*, 7:49–56, 1959. (Cited on page 383.)

[92] W. Aspray. An interview with Friedrich L. Bauer. Charles Babbage Institute, 1987. (Cited on page 414.)

[93] W. Aspray. *John von Neumann and the Origins of Modern Computing.* The MIT Press, Cambridge, Massachussets, 1990. (Cited on page 536.)

[94] G. P. Astrakhantsev. An iterative method of solving elliptic net problems. *Zh. Vychisl. Mat. Mat. Fiz.*, 11:439–448, 1971. (Cited on page 269.)

[95] R. Astudillo and M. B. van Gijzen. The induced dimension reduction method applied to convection-diffusion-reaction problems. In B. Karasözen, M. Manguoglu, M. Tezer-Sezgin, S. Goktepe, and O. Ugur, editors, *Numerical Mathematics and Advanced Applications ENUMATH 2015, Lecture Notes in Computational Science and Engineering*, volume 112, pages 295–303, 2016. (Cited on page 319.)

[96] R. Astudillo and M. B. van Gijzen. A restarted induced dimension reduction method to approximate eigenpairs of large unsymmetric matrices. *J. Comput. Appl. Math.*, 296:24–36, 2016. (Cited on page 319.)

[97] J. V. Atanasoff. Computing machine for the solution of large systems of linear algebraic equations. In B. Randell, editor, *The Origins of Digital Computers*, pages 315–335. Springer, 1982. Originally, Internal Rep., Iowa State University, Ames (1940). (Cited on page 335.)

[98] J. V. Atanasoff. Advent of electronic digital computing. *Ann. Hist. Comput.*, 6(3):229–282, 1984. (Cited on page 335.)

[99] G. Auchmuty. A posteriori error estimates for linear equations. *Numer. Math.*, 61(1):1–6, 1992. (Cited on page 244.)

[100] J. L. Aurentz, A. P. Austin, M. Benzi, and V. Kalantzis. Stable computation of generalized matrix functions via polynomial interpolation. *SIAM J. Matrix Anal. Appl.*, 40(1):210–234, 2019. (Cited on page 390.)

[101] J. L. Aurentz, T. Mach, L. Robol, R. Vandebril, and D. S. Watkins. *Core-Chasing Algorithms for the Eigenvalue Problem.* SIAM, Philadelphia, 2018. (Cited on page 319.)

[102] J. L. Aurentz, T. Mach, L. Robol, R. Vandebril, and D. S. Watkins. Fast and backward stable computation of roots of polynomials, part II: Backward error analysis; companion matrix and companion pencil. *SIAM J. Matrix Anal. Appl.*, 39(3):1245–1269, 2018. (Cited on page 314.)

[103] J. L. Aurentz, T. Mach, R. Vandebril, and D. S. Watkins. Fast and backward stable computation of roots of polynomials. *SIAM J. Matrix Anal. Appl.*, 36(3):942–973, 2015. (Cited on pages 293 and 314.)

[104] J. L. Aurentz, T. Mach, R. Vandebril, and D. S. Watkins. Fast and stable unitary QR algorithm. *Electron. Trans. Numer. Anal.*, 44:327–341, 2015. (Cited on page 293.)

[105] L. Autonne. Sur les groupes linéaires, réels et orthogonaux. *Bull. S.M.F.*, 30:121–134, 1902. (Cited on pages 8, 123, 136, and 142.)

[106] L. Autonne. Sur l'Hermitien. *Rendiconti del Circolo Matematico di Palermo*, 16(1):104–128, 1902. (Cited on pages 8 and 136.)

[107] L. Autonne. Sur la canonisation des formes bilinéaires. *Nouvelles Annales de Mathématiques, 4ème série*, 3:57–64, 1903. (Cited on page 136.)

[108] L. Autonne. Sur l'hypohermitien. *Bull. S.M.F.*, 31:140–155, 1903. (Cited on pages 8 and 136.)

[109] L. Autonne. Sur quelques propriétés des matrices hypohermitiennes n-aires. *Bull. S.M.F.*, 31:268–271, 1903. (Cited on pages 8 and 136.)

[110] L. Autonne. *Sur les Formes Mixtes*. Gauthier-Villars, Paris, 1905. Annales de l'Université de Lyon, fascicule 16. (Cited on pages 13, 136, and 149.)

[111] L. Autonne. *Sur les Matrices Linéaires Échangeables à une Matrice Donnée*. Gauthier-Villars, 1910. 14e cahier, 2e sér. du Journal de l'Ecole Polytechnique. (Cited on pages 136, 149, and 396.)

[112] L. Autonne. Sur les matrices hypohermitiennes et sur les matrices unitaires. *C.R. Acad. Sci. Paris, Sér 1*, 156:858–860, 1913. (Cited on pages 13 and 136.)

[113] L. Autonne. Sur les matrices hypohermitiennes et sur les matrices unitaires. *Annales de l'Université de Lyon*, 38:1–77, 1915. (Cited on pages 13 and 136.)

[114] H. Avron, A. Druinsky, and A. Gupta. Revisiting asynchronous linear solvers: Provable convergence rate through randomization. *J. ACM*, 62(6):1–27, 2015. (Cited on page 197.)

[115] O. Axelsson. A generalized SSOR method. *BIT*, 12:443–467, 1972. (Cited on page 255.)

[116] O. Axelsson. On preconditioning and convergence acceleration for sparse matrix problems. Technical Report Report 74-10, CERN, Geneva, Switzerland, 1974. (Cited on pages 255 and 259.)

[117] O. Axelsson. A class of iterative methods for finite element equations. *Comput. Methods Appl. Mech. Eng.*, 9(2):123–137, 1976. (Cited on pages 209, 255, and 259.)

[118] O. Axelsson. Conjugate gradient type methods for unsymmetric and inconsistent systems of linear equations. *Linear Algebra Appl.*, 29:1–16, 1980. (Cited on page 222.)

[119] O. Axelsson. Incomplete block matrix factorization preconditioning methods. The ultimate answer? *J. Comput. Appl. Math.*, 12:3–18, 1985. (Cited on page 259.)

[120] O. Axelsson. A general incomplete block matrix factorization method. *Linear Algebra Appl.*, 74:179–190, 1986. (Cited on page 259.)

[121] O. Axelsson. A generalized conjugate gradient, least square method. *Numer. Math.*, 51(2):209–227, 1987. (Cited on page 222.)

[122] O. Axelsson. *Iterative Solution Methods*. Cambridge University Press, 1994. (Cited on pages 230, 257, and 259.)

[123] O. Axelsson. A survey of algebraic multilevel iteration (AMLI) methods. *BIT Numerical Mathematics*, 43(5):863–879, 2003. (Cited on page 271.)

[124] O. Axelsson. Milestones in the development of iterative solution methods. *J. Elec. Comput. Engrg.*, 2010(ID 972794), 2010. (Cited on page 246.)

[125] O. Axelsson, S. Brinkkemper, and V. P. Il'in. On some versions of incomplete block matrix factorization iterative methods. *Linear Algebra Appl.*, 58:3–15, 1984. (Cited on page 259.)

[126] O. Axelsson and V. Eijkhout. Vectorizable preconditioners for elliptic difference equations in three space dimensions. *J. Comput. Appl. Math.*, 27:299–321, 1989. (Cited on page 260.)

[127] O. Axelsson and I. Gustafsson. On the use of preconditioned conjugate gradient methods for Red/Black ordered five point difference schemes. *J. Comp. Phys.*, 35:284–289, 1980. (Cited on page 256.)

[128] O. Axelsson and J. Karátson. Mesh independent superlinear PCG rates via compact-equivalent operators. *SIAM J. Numer. Anal.*, 45(4):1495–1516, 2007. (Cited on page 264.)

[129] O. Axelsson and J. Karátson. Equivalent operator preconditioning for elliptic problems. *Numer. Algorithms*, 50(3):297–380, 2009. (Cited on page 264.)

[130] O. Axelsson and J. Karátson. Superior properties of the PRESB preconditioner for operators on two-by-two block form with square blocks. *Numer. Math.*, 146(2):335–368, 2020. (Cited on page 264.)

[131] O. Axelsson and G. Lindskog. On the rate of convergence of the preconditioned conjugate gradient method. *Numer. Math.*, 48:499–523, 1986. (Cited on page 210.)

[132] O. Axelsson and S. Margenov. On multilevel preconditioners which are optimal with respect to both problem and discretization parameters. *Comput. Methods Appl. Math.*, 3(1):6–22, 2003. (Cited on page 271.)

[133] O. Axelsson and B. Polman. On approximate factorization methods for block matrices suitable for vector and parallel processors. *Linear Algebra Appl.*, 77:3–26, 1986. (Cited on pages 259 and 260.)

[134] O. Axelsson and B. Polman. Block preconditioning and domain decomposition methods, II. *J. Comput. Appl. Math.*, 24:55–72, 1988. (Cited on page 267.)

[135] O. Axelsson and P.S. Vassilevski. Algebraic multilevel preconditioning methods, I. *Numer. Math.*, 56:157–177, 1989. (Cited on page 270.)

[136] O. Axelsson and P.S. Vassilevski. A survey of multilevel preconditioned iterative methods. *BIT Numerical Mathematics*, 29(4):769–793, 1989. (Cited on page 270.)

[137] O. Axelsson and P.S. Vassilevski. Algebraic multilevel preconditioning methods, II. *SIAM J. Numer. Anal.*, 27(6):1569–1590, 1990. (Cited on page 270.)

[138] E. H. Ayachour. *Application de la biorthogonalité aux méthodes de projection.* PhD thesis, Université de Lille, France, 1998. (Cited on page 236.)

[139] E. H. Ayachour. Avoiding look-ahead in the Lanczos method and Padé approximation. *Appl. Math.*, 26(1):33–62, 1999. (Cited on page 236.)

[140] E. H. Ayachour. A fast implementation for GMRES method. *J. Comput. Appl. Math.*, 159(2,):269–283, 2003. (Cited on page 240.)

[141] K. I. Babenko. Toeplitz and Hankel matrices. *Russian Math. Surveys*, 41(1):209–218, 1986. (Cited on page 385.)

[142] M. Baboulin, J. Dongarra, J. Herrmann, and S. Tomov. Accelerating linear system solutions using randomization techniques. *ACM Trans. Math. Soft. (TOMS)*, 39(2):1–13, 2013. (Cited on page 197.)

[143] A. K. Bag. Mathematics and mathematical researches in India during fifth to twentieth centuries - Profiles and prospects. *Indian Journal of History of Science*, 47(3):473–512, 2012. (Cited on page 42.)

[144] J. Baglama, D. Calvetti, and L. Reichel. Fast Leja points. *Electron. Trans. Numer. Anal.*, 7(124-140):119–120, 1998. (Cited on pages 185 and 229.)

[145] J. Baglama, D. Calvetti, and L. Reichel. Algorithm 827: irbleigs: A MATLAB program for computing a few eigenpairs of a large sparse Hermitian matrix. *ACM Trans. Math. Soft. (TOMS)*, 29(3):337–348, 2003. (Cited on page 306.)

[146] J. Baglama, D. Calvetti, and L. Reichel. IRBL: An implicitly restarted block-Lanczos method for large-scale Hermitian eigenproblems. *SIAM J. Sci. Comput.*, 24(5):1650–1677, 2003. (Cited on page 306.)

[147] J. Baglama and L. Reichel. Augmented implicitly restarted Lanczos bidiagonalization methods. *SIAM J. Sci. Comput.*, 27(1):19–42, 2005. (Cited on page 139.)

[148] J. Baglama and L. Reichel. Restarted block Lanczos bidiagonalization methods. *Numer. Algorithms*, 43:251–272, 2006. (Cited on page 140.)

[149] J. Baglama and L. Reichel. Augmented GMRES-type methods. *Numer. Linear Algebra Appl.*, 14(4):337–350, 2007. (Cited on page 234.)

[150] C. Baheux. *Algorithmes d'implémentation de la méthode de Lanczos*. PhD thesis, Université de Lille, France, 1994. (in French). (Cited on page 228.)

[151] C. Baheux. New implementations of Lanczos method. *J. Comput. Appl. Math.*, 57:3–15, 1995. (Cited on page 228.)

[152] J. Bahi, S. Contassot-Vivier, and R. Couturier. *Parallel Iterative Algorithms: From Sequential to Grid computing*. Chapman and Hall/CRC, 2007. (Cited on page 197.)

[153] J. Bahi and J.-C. Miellou. Contractive mappings with maximum norms: comparison of constants of contraction and application to asynchronous iterations. *Parallel Comput.*, 19(5):511–523, 1993. (Cited on page 196.)

[154] Z. Bai. Error analysis of the Lanczos algorithm for the nonsymmetric eigenvalue problem. *Math. Comput.*, 62(205):209–226, 1994. (Cited on page 303.)

[155] Z. Bai, D. Day, and Q. Ye. ABLE: An adaptive block Lanczos method for non-Hermitian eigenvalue problems. *SIAM J. Matrix Anal. Appl.*, 20(4):1060–1082, 1999. (Cited on page 303.)

[156] Z. Bai and J. W. Demmel. On a block implementation of Hessenberg multishift QR iteration. *Intl. J. High Speed Computing*, 1(1):97–112, 1989. (Cited on page 314.)

[157] Z. Bai and J. W. Demmel. Computing the generalized singular value decomposition. *SIAM J. Sci. Comput.*, 14(6):1464–1486, 1993. (Cited on page 141.)

[158] Z. Bai, J. Demmel, J. Dongarra, A. Ruhe, and H. van der Vorst. *Templates for the Solution of Algebraic Eigenvalue Problems*. SIAM, Philadelphia, 2000. (Cited on page 307.)

[159] Z. Bai, D. Hu, and L. Reichel. Implementation of GMRES method using QR factorisation. In J. Dongarra, K. Kennedy, P. Messina, D. C. Sorensen, and R. G. Voigt, editors, *Proc. Fifth SIAM Conference on Parallel Processing for Scientific Computing*, pages 84–91. SIAM, 1992. (Cited on page 229.)

[160] Z. Bai, D. Hu, and L. Reichel. A Newton basis GMRES implementation. *IMA J. Numer. Anal.*, 14:563–581, 1994. (Cited on page 229.)

[161] Z.-Z. Bai, G. H. Golub, and M. K. Ng. Hermitian and skew-Hermitian splitting methods for non-Hermitian positive definite linear systems. *SIAM J. Matrix Anal. Appl.*, 24(3):603–626, 2003. (Cited on page 179.)

[162] A. H. Baker. *On improving the performance of the linear solver GMRES*. PhD thesis, University of Colorado, Boulder, 2003. (Cited on page 241.)

[163] A. H. Baker, J. M. Dennis, and E. R. Jessup. An efficient block variant of GMRES. Technical Report CU-CS-957-03, University of Colorado, Boulder, 2003. (Cited on page 241.)

[164] A. H. Baker, J. M. Dennis, and E. R. Jessup. On improving linear solver performance: A block variant of GMRES. *SIAM J. Sci. Comput.*, 27(5):1608–1626, 2006. (Cited on page 241.)

[165] A. H. Baker, R.D. Falgout, T. V. Kolev, and U. M. Yang. Scaling hypre's multigrid solvers to 100,000 cores. In M. Berry et al., editor, *High Performance Scientific Computing: Algorithms and Applications*. Springer, 2012. (Cited on pages 270 and 378.)

[166] A. H. Baker, E. R. Jessup, and T. V. Kolev. A simple strategy for varying the restart parameter in GMRES(m). *J. Comput. Appl. Math.*, 230(2):751–761, 2009. (Cited on page 232.)

[167] A. H. Baker, E. R. Jessup, and T. A. Manteuffel. A technique for accelerating the convergence of restarted GMRES. *SIAM J. Matrix Anal. Appl.*, 26(4):962–984, 2005. (Cited on page 241.)

[168] C. T. Baker, L. Fox, D. F. Mayers, and K. Wright. Numerical solution of Fredholm integral equations of first kind. *Comput. J.*, 7(2):141–148, 1964. (Cited on page 393.)

[169] H. F. Baker. Biographical notice. In *Collected Mathematical Papers of J. J. Sylvester, volume IV*, pages xv–xxxvii, 1912. (Cited on pages 570 and 571.)

[170] N. S. Bakhvalov. On the convergence of a relaxation method under natural constraints on an elliptic operator. *Zh. Vychisl. Mat. Mat. Fiz.*, 6:861–883, 1966. (Cited on page 269.)

[171] A. B. Bakushinskii. A general method of constructing regularizing algorithms for a linear incorrect equation in Hilbert space. *Zh. Vychisl. Mat. Mat. Fiz.*, 7:672–677, 1967. (Cited on page 392.)

[172] S. Balay, S. Abhyankar, M. F. Adams, J. Brown, P. Brune, K. Buschelman, L. Dalcin, A. Dener, V. Eijkhout, W. D. Gropp, D. Karpeyev, D. Kaushik, M. G. Knepley, D. A. May, L. Curfman McInnes, R. T. Mills, T. Munson, K. Rupp, P. Sanan, B. F. Smith, S. Zampini, and H. Zhang. PETSc users' manual. Technical Report ANL-95/11 - Revision 3.12, Argonne National Laboratory, 2019. (Cited on page 378.)

[173] R. S. Ball. *Great Astronomers*. Isbister, London, 1895. (Cited on pages 479 and 481.)

[174] G. Ballard, E. Carson, J. Demmel, M. Hoemmen, N. Knight, and O. Schwartz. Communication lower bounds and optimal algorithms for numerical linear algebra. *Acta Numerica*, 23:1–155, 2014. (Cited on page 243.)

[175] H. R. Baltzer. *Theorie und Anwendungen der Determinanten mit Beziehung auf die Orginalquellen*. Hirzel, Leipzig, 1857. French translation by J. Hoüel, Théorie et Applications des Déterminants avec l'indication des sources originales, Mallet-Bachelier Paris, 1861. (Cited on pages 115 and 116.)

[176] S. Banach. Sur les opérations dans les ensembles abstraits et leur application aux équations intégrales. *Fund. Math.*, 3(1):133–181, 1922. (Cited on page 15.)

[177] S. Banach. Théorie des Opérations Linéaires. Warszowa. Reprinted in S. Banach, Ouvres 2 (1963), 1932. (Cited on page 15.)

[178] T. Banachiewicz. Zur berechnung der determinanten, wie such der inversen, und zur darauf basierten auflösung der systeme lineare gleichungen. *Acta Astronom., Sér C.*, 3:41–67, 1937. (Cited on pages 25 and 412.)

[179] T. Banachiewicz. Méthode de résolution numérique des équations linéaires, du calcul des déterminants et des inverses, et de réduction des formes quadratiques. *Bulletin international de l'Académie polonaise des sciences et des lettres, Classe des sciences mathématiques et naturelles. Série A*, pages 393–404, 1938. (Cited on page 70.)

[180] T. Banachiewicz. Principes d'une nouvelle technique de la méthode des moindres carrés. Méthode de résolution numérique des équations linéaires, du calcul des déterminants et des inverses, et de réduction des formes quadratiques. *Bulletin international de l'Académie polonaise des sciences et des lettres, Classe des sciences mathématiques et naturelles, Série A*, pages 134–135, 1938. (Cited on pages 73 and 412.)

[181] T. Banachiewicz. An outline of the Cracovian algorithm of the method of least squares. *Astron. J.*, 50:38–41, 1942. (Cited on pages 73 and 412.)

[182] T. Banachiewicz. The role of cracovians in astronomy. *Vistas in Astronomy*, 1:200–206, 1955. (Cited on page 412.)

[183] R. E. Bank and T. F. Chan. An analysis of the composite step biconjugate gradient method. *Numer. Math.*, 66(1):295–319, 1993. (Cited on pages 225, 230, and 232.)

[184] R. E. Bank and T. F. Chan. A composite step bi-conjugate gradient algorithm for nonsymmetric linear systems. *Numer. Algorithms*, 7(1):1–16, 1994. (Cited on pages 225, 230, and 232.)

[185] R. E. Bank and C. C. Douglas. Sharp estimates for multigrid rates of convergence with general smoothing and acceleration. *SIAM J. Numer. Anal.*, 22(4):617–633, 1985. (Cited on page 270.)

[186] R. E. Bank and T. F. Dupont. An optimal order process for solving elliptic finite element equations. *Math. Comput.*, 36:35–51, 1981. (Cited on page 269.)

[187] R. E. Bank, T. F. Dupont, and H. Yserentant. The hierarchical basis multigrid method. *Numer. Math.*, 52(4):427–458, 1988. (Cited on page 270.)

[188] R. E. Bank and D. J. Rose. On the complexity of sparse Gaussian elimination via bordering. *SIAM J. Sci. Statist. Comput.*, 11(1):145–160, 1990. (Cited on page 93.)

[189] R. E. Bank and R. K. Smith. General sparse elimination requires no permanent integer storage. *SIAM J. Sci. Statist. Comput.*, 8(4):574–584, 1987. (Cited on pages 93 and 371.)

[190] I. Bar-On. Interlacing properties of tridiagonal symmetric matrices with applications to parallel computing. *SIAM J. Matrix Anal. Appl.*, 17(3):548–562, 1996. (Cited on page 280.)

[191] J. Baranger and M. Duc-Jacquet. Matrices tridiagonales symétriques et matrices factorisables. *RIRO*, 3:61–66, 1971. (Cited on page 383.)

[192] E. H. Bareiss. Numerical solution of linear equations with Toeplitz and vector Toeplitz matrices. *Numer. Math.*, 13:404–424, 1969. (Cited on page 386.)

[193] V. Bargmann, D. Montgomery, and J. von Neumann. Solution of linear systems of high order. In A. H. Taub, editor, *John von Neumann Collected Works, Vol. 5*, pages 421–477. Macmillan, New York, 1946. (Cited on page 75.)

[194] D. Barkai, K. J. M. Moriarty, and C. Rebbi. A modified conjugate gradient solver for very large systems. *Comput. Phys. Comm.*, 36(1):1–8, 1985. (Cited on page 214.)

[195] J. L. Barlow. More accurate bidiagonal reduction for computing the singular value decomposition. *SIAM J. Matrix Anal. Appl.*, 23(3):761–798, 2002. (Cited on page 138.)

[196] J. L. Barlow. Block Gram-Schmidt downdating. *Electron. Trans. Numer. Anal.*, 43:163–187, 2015. (Cited on page 130.)

[197] J. L. Barlow. Block modified Gram-Schmidt algorithms and their analysis. *SIAM J. Matrix Anal. Appl.*, 40(4):1257–1290, 2019. (Cited on page 130.)

[198] J. L. Barlow and E. H. Bareiss. Probabilistic error analysis of Gaussian elimination in floating point and logarithmic arithmetic. *Computing*, 34:349–364, 1985. (Cited on page 79.)

[199] J. L. Barlow and A. Smoktunowicz. Reorthogonalized block classical Gram-Schmidt. *Numer. Math.*, 123:395–423, 2013. (Cited on page 130.)

[200] J. L. Barlow, A. Smoktunowicz, and H. Erbay. Improved Gram-Schmidt type downdating methods. *BIT Numerical Mathematics*, 45(2):259–285, 2005. (Cited on page 130.)

[201] S. T. Barnard, A. Pothen, and H. Simon. A spectral algorithm for envelope reduction of sparse matrices. *Numer. Linear Algebra Appl.*, 2(4):317–334, 1995. (Cited on page 90.)

[202] S. Barnett. Leverrier's algorithm: A new proof and extensions. *SIAM J. Matrix Anal. Appl.*, 10(4):551–556, 1989. (Cited on page 289.)

[203] S. Barnett. Leverrier's algorithm for orthogonal bases. *Linear Algebra Appl.*, 236:245–263, 1996. (Cited on page 289.)

[204] R. Barrett, M. Berry, T. F. Chan, J. Demmel, J. Donato, J. Dongarra, V. Eijkhout, R. Pozo, C. Romine, and H. van der Vorst. *Templates for the Solution of Linear Systems: Building Blocks for Iterative Methods*. SIAM, Philadelphia, 1994. (Cited on page 246.)

[205] W. W. Barrett. A theorem on inverse of tridiagonal matrices. *Linear Algebra Appl.*, 27:211–217, 1979. (Cited on page 383.)

[206] W. W. Barrett and P. J. Feinsilver. Inverses of banded matrices. *Linear Algebra Appl.*, 41:111–130, 1981. (Cited on page 383.)

[207] W. W. Barrett and C. R. Johnson. Determinantal formulae for matrices with sparse inverses. *Linear Algebra Appl.*, 56:73–88, 1984. (Cited on page 383.)

[208] I. Barrodale and G. F. Stuart. A new variant of Gaussian elimination. *IMA J. Appl. Math.*, 19:39–47, 1977. (Cited on page 78.)

[209] R. H. Bartels and G. W. Stewart. Algorithm 432: Solution of the matrix equation $AX + XB = C$. *Comm. ACM*, 15:820–826, 1972. (Cited on pages 397 and 398.)

[210] W. Barth, R. S. Martin, and J. H. Wlkinson. Calculation of the eigenvalues of a symmetric tridiagonal matrix by the method of bisection. *Numer. Math.*, 9:386–393, 1967. (Cited on page 362.)

[211] D. P. Bartlett. *General Principles of the Method of Least Squares: With Applications.* MIT, 1915. (Cited on page 62.)

[212] M. S. Bartlett. An inverse matrix adjustment arising in discriminant analysis. *Ann. Math. Stat.*, 22:107–111, 1951. (Cited on page 83.)

[213] H. Bateman. A formula for the solving function of a certain integral equation of the second kind. *Messenger Math.*, 2(37):179–187, 1908. (Cited on page 136.)

[214] G. M. Baudet. Asynchronous iterative methods for multiprocessors. *J. ACM*, 25(2):226–244, 1978. (Cited on page 196.)

[215] F. L. Bauer. Beiträge zum Danilewski-verfahren. In *Aktuelle Problem der Rechentechnik. Bericht über das Internationale Mathematiker-Kolloquium, Dresden,22. bis 27. November 1955*. VEB Deutscher Verlag der Wissenschaften, Berlin, 1957. (Cited on pages 291 and 414.)

[216] F. L. Bauer. Das verfahren der treppeniteration und verwandte verfahren zur lösung algebraischer eigenwertprobleme. *J. Appl. Math. Phys.*, 8:214–235, 1957. (Cited on pages 295, 310, and 414.)

[217] F. L. Bauer. On modern matrix iteration processes of Bernoulli and Graeffe type. *J. ACM*, 5:246–257, 1958. (Cited on pages 310 and 414.)

[218] F. L. Bauer. The quotient-difference and the epsilon algorithms. In R. E. Langer, editor, *On Numerical Approximation*, pages 361–370, Madison, WI, 1959. University of Wisconsin Press. (Cited on pages 310 and 414.)

[219] F. L. Bauer. On the field of values subordinate to a norm. *Numer. Math.*, 4:103–113, 1962. (Cited on pages 16, 288, and 414.)

[220] F. L. Bauer. Optimally scaled matrices. *Numer. Math.*, 5:73–87, 1963. (Cited on pages 80, 251, and 414.)

[221] F. L. Bauer. Elimination with weighted row combinations for solving linear equations and least squares problems. *Numer. Math.*, 7:338–352, 1965. (Cited on pages 361 and 414.)

[222] F. L. Bauer. Genauigkeitsfragen bei der lösung linearer gleichungssysteme. *Zamm.-Z Angew. Math. Me.*, 46(7):409–421, 1966. (Cited on pages 23, 79, and 414.)

[223] F. L. Bauer. Theory of norms. Technical Report CS 75, Computer Science Dpt., Stanford University, CA, 1967. (Cited on pages 16 and 414.)

[224] F. L. Bauer. Remarks on optimally scaled matrices. *Numer. Math.*, 13(1):1–3, 1969. (Cited on pages 80 and 414.)

[225] F. L. Bauer. Olga Taussky-Todd 30.8.1906 - 7.10.1995. *Linear Algebra Appl.*, 280:9–12, 1998. (Cited on page 575.)

[226] F. L. Bauer. My years with Rutishauser. *Informatik-Spektrum*, 34(5), 2011. (Cited on page 552.)

[227] F. L. Bauer and C. T. Fike. Norms and exclusion theorems. *Numer. Math.*, 2(1):137–141, 1960. (Cited on pages 16 and 414.)

[228] F. L. Bauer and A. S. Householder. Moments and characteristic roots. *Numer. Math.*, 2(1):42–53, 1960. (Cited on page 414.)

[229] F. L. Bauer and A. S. Householder. Some inequalities involving the Euclidean condition of a matrix. *Numer. Math.*, 2(1):308–311, 1960. (Cited on pages 16, 22, and 414.)

[230] F. L. Bauer and A. S. Householder. In memoriam, J. H. Wilkinson (1919-1987). *Numer. Math.*, 51:1–2, 1987. (Cited on page 598.)

[231] F. L. Bauer, J. Stoer, and C. Witzgall. Absolute and monotonic norms. *Numer. Math.*, 3(1):257–264, 1961. (Cited on pages 16 and 414.)

[232] F. L. Bauer and H. Wössner. The "Plankalkül" of Konrad Zuse: A forerunner of today's programming languages. *Comm. ACM*, 15(7):678–685, 1972. (Cited on pages 358 and 603.)

[233] C. A. Beattie, M. Embree, and J. Rossi. Convergence of restarted Krylov subspaces to invariant subspaces. *SIAM J. Matrix Anal. Appl.*, 25(4):1074–1109, 2004. (Cited on page 306.)

[234] C. A. Beattie, M. Embree, and D. C. Sorensen. Convergence of polynomial restart Krylov methods for eigenvalue computations. *SIAM Rev.*, 47(3):492–515, 2005. (Cited on page 306.)

[235] C. A. Beattie and I. C. F. Ipsen. Inclusion regions for matrix eigenvalues. *Linear Algebra Appl.*, 358(1-3):281–291, 2003. (Cited on page 287.)

[236] R. Beauwens. Convergence analysis of some factorization iterative methods for M-matrices. Technical Report COO-2280-5, Northwestern University, Evanston, IL, 1973. (Cited on pages 253 and 254.)

[237] R. Beauwens. Factorization iterative methods, M-operators and H-operators. *Numer. Math.*, 31(4):335–357, 1978. Erratum, *Numer. Math.* 49 (1986), pp. 457-458. (Cited on page 255.)

[238] R. Beauwens. OBV methods. Technical Report COO-2280-38, Northwestern University, Evanston, IL, 1979. (Cited on page 255.)

[239] R. Beauwens. On Axelsson's perturbations. *Linear Algebra Appl.*, 68:221–242, 1985. (Cited on page 255.)

[240] R. Beauwens. Lower eigenvalue bounds for pencils of matrices. *Linear Algebra Appl.*, 85:101–119, 1987. (Cited on page 264.)

[241] R. Beauwens and M. Ben Bouzid. On sparse block factorization iterative methods. *SIAM J. Numer. Anal.*, 24(5):1066–1076, 1987. (Cited on page 259.)

[242] R. Beauwens and M. Ben Bouzid. Existence and conditioning properties of sparse approximate block factorizations. *SIAM J. Numer. Anal.*, 25(4):941–956, 1988. (Cited on page 259.)

[243] R. Beauwens and L. Quenon. Existence criteria for partial matrix factorizations in iterative methods. *SIAM J. Numer. Anal.*, 13(4):615–643, 1976. (Cited on page 254.)

[244] R. Beauwens and R. Wilmet. Conditioning analysis of positive definite matrices by approximate factorizations. *J. Comput. Appl. Math.*, 26(3):257–269, 1989. (Cited on page 255.)

[245] R. H. Beck. *Einführung in die Axiomatik der Algebra.* Walter de Gruyter, Berlin, Leipzig, 1926. (Cited on page 11.)

[246] B. Beckermann. Image numérique, GMRES et polynômes de Faber. *C.R. Acad. Sci. Paris, Sér 1*, 340:855–860, 2005. (Cited on page 241.)

[247] B. Beckermann, S. A. Goreinov, and E. E. Tyrtyshnikov. Some remarks on the Elman estimate for GMRES. *SIAM J. Matrix Anal. Appl.*, 27(3):772–778, 2006. (Cited on page 241.)

[248] B. Beckermann and S. Güttel. Superlinear convergence of the rational Arnoldi method for the approximation of matrix functions. *Numer. Math.*, 121(2):205–236, 2012. (Cited on page 390.)

[249] B. Beckermann and L. Reichel. The Arnoldi process and GMRES for nearly symmetric matrices. *SIAM J. Matrix Anal. Appl.*, 30(1):102–120, 2008. (Cited on page 242.)

[250] T. Beelen and P. Van Dooren. Computational aspects of the Jordan canonical form. In M. G. Cox and S. Hammarling, editors, *Reliable Numerical Computation*, pages 57–72, Oxford University Press, New York, 1990. (Cited on page 150.)

[251] G. G. Bejarano and B. R. Rosenblatt. A solution of simultaneous linear equations and matrix inversion with high speed computing devices. *M.T.A.C.*, 7(42):77–81, 1953. (Cited on page 77.)

[252] G. G. Belford and E. H. Kaufman Jr. An application of appproximation theory to an error estimate in linear algebra. *Math. Comput.*, 28(127):711–712, 1974. (Cited on page 208.)

[253] M. Belge, M. E. Kilmer, and E. L. Miller. Efficient determination of multiple regularization parameters in a generalized L-curve framework. *Inverse Probl.*, 18(4):1161–1183, 2002. (Cited on page 392.)

[254] B. Belhoste. *Augustin-Louis Cauchy. A Biography*. Springer, New York, 1991. (Cited on page 419.)

[255] B. Belhoste. When an academician becomes professor: the case of Joseph-Louis Lagrange. *Lett. Mat. Int.*, 2:25–34, 2014. (Cited on page 510.)

[256] M. Bellalij, G. Meurant, and H. Sadok. The distance of an eigenvector to a Krylov subspace and the convergence of the Arnoldi method for eigenvalue problems. *Linear Algebra Appl.*, 504:387–405, 2016. (Cited on page 304.)

[257] M. Bellalij, L. Reichel, and H. Sadok. Some properties of range restricted GMRES methods. *J. Comput. Appl. Math.*, 290:310–318, 2015. (Cited on page 245.)

[258] M. Bellalij, Y. Saad, and H. Sadok. On the convergence of the Arnoldi process for eigenvalue problems. Technical Report umsi-2007-12, Minnesota Supercomputer Institute, University of Minnesota, Minneapolis, MN, 2007. (Cited on page 304.)

[259] M. Bellalij, Y. Saad, and H. Sadok. Further analysis of the Arnoldi process for eigenvalue problems. *SIAM J. Numer. Anal.*, 48(2):393–407, 2010. (Cited on page 304.)

[260] M. Beller. Matrix theory before Schrödinger. Philosphy, problems, consequences. *Isis*, 74:469–491, 1983. (Cited on page 26.)

[261] R. Bellman. *Introduction to Matrix Analysis*. Mc Graw-Hill, 1960. (Cited on page 388.)

[262] E. Beltrami. Sulle funzioni bilineari. *Giornale di Matematiche ad Uso degli Studenti Delle Università Italiane*, 11:98–106, 1873. (Cited on pages 134 and 417.)

[263] J.-D. Benamou and B. Desprès. A domain decomposition method for the Helmholtz equation and related optimal control problems. *J. Comp. Phys.*, 136(1):68–82, 1997. (Cited on page 268.)

[264] I. Bendixson. Sur les racines d'une équation fondamentale. *Acta Math.*, 25:359–365, 1902. (Cited on page 286.)

[265] A. A. Bennett. Normalized geometric systems. *Proc. Nat. Acad. Sci. U.S.A.*, 7(3):84–89, 1921. (Cited on page 15.)

[266] Cdt. Benoît. Note sur une méthode de résolution des équations normales provenant de l'application de la méthode des moindres carrés à un système d'équations linéaires en nombre inférieur à celui des inconnues (Procédé du commandant Cholesky). *Bull. Géodésique*, 2:67–77, 1924. (Cited on pages 21, 68, and 436.)

[267] M. Benzi. Preconditioning techniques for large linear systems: a survey. *J. Comp. Phys.*, 182(2):418–477, 2002. (Cited on pages 265 and 430.)

[268] M. Benzi. Gianfranco Cimmino's contributions to numerical mathematics. Atti del Seminario di Analisi Matematica, Dipartimento di Matematica dell'Università di Bologna. Volume Speciale: Ciclo di Conferenze in Memoria di Gianfranco Cimmino, Marzo-Aprile 2004, Tecnoprint, Bologna, 2005. (Cited on pages 191, 192, 430, and 439.)

[269] M. Benzi. A generalization of the Hermitian and skew-Hermitian splitting iteration. *SIAM J. Matrix Anal. Appl.*, 31(2):360–374, 2009. (Cited on page 180.)

[270] M. Benzi. Some uses of the field of values in numerical analysis. *Bolletino dell'Unione Matematica Italiana*, 14:159–177, 2021. (Cited on page 287.)

[271] M. Benzi and P. Boito. Quadrature rule-based bounds for functions of adjacency matrices. *Linear Algebra Appl.*, 433:637–652, 2010. (Cited on pages 213 and 391.)

[272] M. Benzi and P. Boito. Matrix functions in network analysis. *GAMM Mitteilungen*, 43(3):article e202000012, 2020. (Cited on page 391.)

[273] M. Benzi, J. K. Cullum, and M. Tůma. Robust approximate inverse preconditioning for the conjugate gradient method. *SIAM J. Sci. Comput.*, 22(4):1318–1332, 2000. (Cited on page 263.)

[274] M. Benzi, L. Cvetković, and M. Neumann. Preface. *Numer. Algorithms*, 42:205–206, 2006. (Cited on page 592.)

[275] M. Benzi, E. Estrada, and C. Klymko. Ranking hubs and authorities using matrix functions. *Linear Algebra Appl.*, 431:2447–2474, 2013. (Cited on pages 213 and 391.)

[276] M. Benzi and G. H. Golub. A preconditioner for generalized saddle point problems. *SIAM J. Matrix Anal. Appl.*, 26(1):20–41, 2004. (Cited on page 265.)

[277] M. Benzi, G. H. Golub, and J. Liesen. Numerical solution of saddle point problems. *Acta Numerica*, 14:1–137, 2005. (Cited on page 265.)

[278] M. Benzi, J. C. Haws, and M. Tůma. Preconditioning highly indefinite and nonsymmetric matrices. *SIAM J. Sci. Comput.*, 22(4):1333–1353, 2000. (Cited on page 263.)

[279] M. Benzi, W. Joubert, and G. Mateescu. Numerical experiments with parallel orderings for ILU preconditioners. *Electron. Trans. Numer. Anal.*, 8:88–114, 1999. (Cited on page 261.)

[280] M. Benzi, R. Kouhia, and M. Tůma. Stabilized and block approximate inverse preconditioners for problems in solid and structural mechanics. *Comput. Methods Appl. Mech. Engrg.*, 190(49-50):6533–6554, 2001. (Cited on page 263.)

[281] M. Benzi, C. D. Meyer, and M. Tůma. A sparse approximate inverse preconditioner for the conjugate gradient method. *SIAM J. Sci. Comput.*, 17(5):1135–1149, 1996. (Cited on pages 82 and 263.)

[282] M. Benzi, D. B. Szyld, and A. Van Duin. Orderings for incomplete factorization preconditioning of nonsymmetric problems. *SIAM J. Sci. Comput.*, 20(5):1652–1670, 1999. (Cited on page 260.)

[283] M. Benzi and M. Tůma. Numerical experiments with two approximate inverse preconditioners. *BIT Numerical Mathematics*, 38(2):234–241, 1998. (Cited on page 263.)

[284] M. Benzi and M. Tůma. A sparse approximate inverse preconditioner for nonsymmetric linear systems. *SIAM J. Sci. Comput.*, 19(3):968–994, 1998. (Cited on page 263.)

[285] M. Benzi and M. Tůma. A comparative study of sparse approximate inverse preconditioners. *Appl. Numer. Math.*, 30(2-3):305–340, 1999. (Cited on page 263.)

[286] M. Benzi and M. Tůma. Orderings for factorized sparse approximate inverse preconditioners. *SIAM J. Sci. Comput.*, 21(5):1851–1868, 2000. (Cited on page 263.)

[287] M. Benzi and M. Tůma. A robust incomplete factorization preconditioner for positive definite matrices. *Numer. Linear Algebra Appl.*, 10:385–400, 2003. (Cited on page 263.)

[288] T. J. Bergin, editor. *50 years of army computing, from ENIAC to MSRC*, 2000. (Cited on page 337.)

[289] E. R. Berkelamp. *Algebraic Coding Theory*. McGraw-Hill, New York, 1968. (Cited on page 386.)

[290] A. Berman and R. J. Plemmons. *Nonnegative Matrices in the Mathematical Sciences*. Academic Press, 1979. Reprinted by SIAM, Philadelphia, 1994. (Cited on pages 80 and 177.)

[291] M. Bern, J. R. Gilbert, B. Hendrickson, N. Nguyen, and S. Toledo. Support-graph preconditioners. *SIAM J. Matrix Anal. Appl.*, 27(4):930–951, 2006. (Cited on page 264.)

[292] J. A. B. Bernardes and S. L. Gonzaga de Oliveira. A systematic review of heuristics for profile reduction of symmetric matrices. *Procedia Comput. Sci.*, 51:221–230, 2015. (Cited on page 88.)

[293] C. Bernardi, Y. Maday, and A. T. Patera. Domain decomposition by the mortar element method. In H. G. Kaper and M. Garbey, editors, *Asymptotic and Numerical Methods for Partial Differential Equations with Critical Parameters*, pages 269–286, Dordrecht, 1993. Springer. (Cited on page 268.)

[294] D. Bernoulli. Observationes de seriebus quae formantur ex additione vel subtractione quacunque terminorum se mutuo consequentium. *Commentarii Academiae Scientiarum Imperialis Petropolitanae*, 3:85–100, 1728. (Cited on page 290.)

[295] L. Bernstein. *The Jacobi-Perron algorithm - its theory and application*. Springer-Verlag, 1971. Lecture Notes Math., vol. 207. (Cited on page 546.)

[296] C. E. Berry. A criterion of convergence for the classical iterative method of solving linear simultaneous equations. *Ann. Math. Stat.*, 16(4):398–400, 1945. (Cited on page 172.)

[297] M. W. Berry, D. Mezher, B. Philippe, and A. Sameh. Parallel algorithms for the singular value decomposition. In E. J. Kontoghiorghes, editor, *Handbook of Parallel Computing and Statistics*, pages 117–164. Chapman and Hall/CRC, 2005. (Cited on page 138.)

[298] M. W. Berry and A. Sameh. An overview of parallel algorithms for the singular value and symmetric eigenvalue problems. *J. Comput. Appl. Math.*, 27:191–213, 1989. (Cited on page 138.)

[299] J. Bertrand. *Méthode des Moindres Carrés, Mémoires sur la Combinaison des Observations par C. F. Gauss*. Mallet-Bachelier, Paris, 1855. (Cited on pages 16 and 61.)

[300] D. Bertsekas. Distributed asynchronous computation of fixed points. *Math. Programming*, 27:107–120, 1983. (Cited on page 196.)

[301] D. Bertsekas and J. Tsitsiklis. *Parallel and Distributed Computation, Numerical Methods*. Prentice Hall, Englewood Cliffs, NJ, 1989. (Cited on pages 196 and 197.)

[302] V. A. Le Besgue. Thèses de mécanique et d'astronomie. *J. Math. Pures Appl.*, 1ère série, 2:337–365, 1837. (Cited on page 281.)

[303] F. W. Bessel. Uber die berechnung der geographischen längen und breiten aus geodätischen vermessungen. *Astronomische Nachrichten*, 4(86):241–254, 1825. (Cited on page 61.)

[304] F. W. Bessel and J. J. Baeyer. *Gradmessung in Ostpreußen und ihre Verbindung mit Preußischen und Russischen Dreieckskettten. Ausgeführt von F. W. Bessel, Director der Königsberger Sternwarte, und J. J. Baeyer, Major im Generalstabe*. Berlin: In Commission bei F. Dümmler, 1838. (Cited on page 61.)

[305] I. Bethune, J. M. Bull, N. J. Dingle, and N. J. Higham. Performance analysis of asynchronous Jacobi's method implemented in MPI, SHMEM and OpenMP. *Int. J. High Performance Computing Applications*, 28(1):97–111, 2014. (Cited on page 197.)

[306] M. Bečka, G. Okša, and M. Vajteršic. Dynamic ordering for a parallel block-Jacobi SVD algorithm. *Parallel Comput.*, 28(2):243–262, 2002. (Cited on page 139.)

[307] M. Bečka, G. Okša, and M. Vajteršic. Parallel block-Jacobi SVD methods. In M. W. Berry, K. A. Gallivan, E. Gallopoulos, A. Grama, B. Philippe, Y. Saad, and F. Saied, editors, *High-Performance Scientific Computing*, pages 185–197. Springer, London, 2012. (Cited on page 139.)

[308] M. Bečka, G. Okša, and M. Vajteršic. New dynamic orderings for the parallel one-sided block-Jacobi SVD algorithm. *Parallel Process. Lett.*, 25(2):1550003, 2015. (Cited on page 139.)

[309] J. Bečvář. *Z Historie Lineární Algebry*. Matfyzpress, Praha, 2007. (in Czech). (Cited on page 12.)

[310] E. Bézout. Recherches sur le degré des équations résultantes de l'évanouissement des inconnues et sur les moyens qu'il convient d'employer pour trouver ces équations. *Mémoires de l'Académie Royale des Sciences pour 1764*, pages 288–338, 1767. (Cited on pages 54 and 105.)

[311] E. Bézout. *Théorie Générale des Équations Algébriques*. Ph.D. Pierres, Paris, 1779. (Cited on pages 54 and 106.)

[312] E. Bézout. *Cours de Mathématiques, à l'usage des gardes du pavillon et de la marine, troisième partie*. Ph.D. Pierres, Paris, 1781. (Cited on page 54.)

[313] E. Bézout. *Cours de Mathématiques, à l'usage du corps de l'artillerie, seconde partie*. Ph.D. Pierres, Paris, 1788. (Cited on page 54.)

[314] R. Bhatia and P. Rosenthal. How and why to solve the operator equation $AX - XB = Y$. *Bull. London Math. Soc.*, 29(1):1–21, 1997. (Cited on page 394.)

[315] W. G. Bickley. Review of "The Escalator Method in Engineering Vibration Problems" by Joseph Morris. *Aeronaut. J.*, 51(442):865–866, 1947. (Cited on page 297.)

[316] W. G. Bickley and J. McNamee. Matrix and other direct methods for the solution of systems of linear difference equations. *Philos. Trans. R. Soc. Lond.*, 252(1005):69–131, 1960. (Cited on page 77.)

[317] J. Bienaymé. Sur la probabilité des erreurs d'après la méthode des moindres carrés. *J. Math. Pures Appl.*, 17:33–78, 1852. (Cited on page 61.)

[318] J. Bienaymé. Remarques sur les différences qui distinguent l'interpolation de M. Cauchy de la méthode des moindres carrés, et qui assurent la supériorité de cette méthode. *J. Math. Pures Appl.*, 18:299–308, 1853. (Cited on page 61.)

[319] J. Bienaymé. Considérations à l'appui de la découverte de Laplace sur la loi de probabilité dans la méthode des moindres carrés. *J. Math. Pures Appl.*, 2ème série 12:158–176, 1867. (Cited on page 62.)

[320] K. R. Biermann. *Die Mathematik und ihre Dozenten an der Berliner Universität 1810-1920*. Akademie-Verlag, Berlin, 1973. (Cited on page 463.)

[321] D. Bindel, J. Demmel, W. Kahan, and O. Marques. On computing Givens rotations reliably and efficiently. *ACM Trans. Math. Soft. (TOMS)*, 28(2):206–238, 2002. (Cited on page 131.)

[322] J. M. Binet. Mémoire contenant un système de formules analytiques, et leur application à des considerations géométriques. *Nouveau Bulletin des Sciences, par la Société Philomatique*, Tome III:243–246, 1812. (Cited on pages 3, 108, and 109.)

[323] J. M. Binet. Mémoire sur un système de formules analytiques, et leur application à des considerations géométriques. *J. Ecole Polytechnique*, 9:280–302, 1812. (Cited on pages 3 and 108.)

[324] D. A. Bini, P. Boito, Y. Eidelman, L. Gemignani, and I. Gohberg. A fast implicit QR eigenvalue algorithm for companion matrices. *Linear Algebra Appl.*, 432(8):2006–2031, 2010. (Cited on page 293.)

[325] D. A. Bini, F. Di Benedetto, E. Tyrtyshnikov, and M. Van Barel, eds. *Structured Matrices in Numerical Linear Algebra*. Springer INdAM Series 30, 2019. (Cited on page 387.)

[326] D. A. Bini, B. Iannazzo, and B. Meini. *Numerical Solution of Algebraic Riccati Equations*. SIAM, Philadelphia, 2012. (Cited on page 399.)

[327] G. Birkhoff and J. A. George. Elimination by nested dissection. In J. F. Traub, editor, *Complexity of Sequential and Parallel Numerical Algorithms*, pages 221–269. Academic Press, New York, 1973. (Cited on page 89.)

[328] G. D. Birkhoff. Boundary value and expansion problems of ordinary linear differential equations. *Trans. Amer. Math. Soc.*, 9(4):373–395, 1908. (Cited on page 288.)

[329] M. Sh. Birman. On a variant of the method of successive approximations. *Vestn. LGU*, 9:69–76, 1952. (Cited on page 163.)

[330] C. H. Bischof. Incremental condition estimation. *SIAM J. Matrix Anal. Appl.*, 11(2):312–322, 1990. (Cited on page 24.)

[331] C. H. Bischof, J. G. Lewis, and D. J. Pierce. Incremental condition estimation for sparse matrices. *SIAM J. Matrix Anal. Appl.*, 11(4):644–659, 1990. (Cited on page 24.)

[332] C. H. Bischof and C. F. Van Loan. The WY representation for products of Householder matrices. *SIAM J. Sci. Statist. Comput.*, 8(1):s2–s13, 1987. (Cited on page 133.)

[333] C. H. Bischof and P. T. P. Tang. Generalizing incremental condition estimation. *Numer. Linear Algebra Appl.*, 1:149–163, 1992. (Cited on page 24.)

[334] R. R. Bitmead and B. D. O. Anderson. Asymptotically fast solution of Toeplitz and related systems of linear equations. *Linear Algebra Appl.*, 34:103–116, 1980. (Cited on page 386.)

[335] Å. Björck. Solving linear least squares problems by Gram-Schmidt orthogonalization. *BIT Numerical Mathematics*, 7(1):1–21, 1967. (Cited on page 129.)

[336] Å. Björck. Numerics of Gram-Schmidt orthogonalization. *Linear Algebra Appl.*, 197:297–316, 1994. (Cited on page 129.)

[337] Å. Björck. *Numerical Methods for Least Squares Problems*. SIAM, Philadelphia, 1996. (Cited on page 129.)

[338] Å. Björck and C. Bowie. An iterative algorithm for computing the best estimate of an orthogonal matrix. *SIAM J. Numer. Anal.*, 8(2):358–364, 1971. (Cited on page 142.)

[339] Å. Björck and G. H. Golub. Numerical methods for computing angles between linear subspaces. *Math. Comput.*, 27(123):579–594, 1973. (Cited on page 144.)

[340] Å. Björck, E. Grimme, and P. Van Dooren. An implicit shift bidiagonalization algorithm for ill-posed systems. *BIT Numerical Mathematics*, 34(4):510–534, 1994. (Cited on page 139.)

[341] Å. Björck and S. Hammarling. A Schur method for the square root of a matrix. *Linear Algebra Appl.*, 52:127–140, 1983. (Cited on page 388.)

[342] Å. Björck and C. C. Paige. Loss and recapture of orthogonality in the modified Gram-Schmidt algorithm. *SIAM J. Matrix Anal. Appl.*, 13(1):176–190, 1992. (Cited on page 129.)

[343] Å. Björck, H. Park, and L. Eldén. Accurate downdating of least squares solutions. *SIAM J. Matrix Anal. Appl.*, 15(2):549–568, 1994. (Cited on page 130.)

[344] P. Bjørstad and O. B. Widlund. Iterative methods for the solution of elliptic problems on regions partitioned into substructures. *SIAM J. Numer. Anal.*, 23(6):1097–1120, 1986. (Cited on page 267.)

[345] A. N. Black and R. V. Southwell. Relaxation methods applied to engineering problems. II. Basic theory, with applications to surveying and to electrical networks, and an extension to gyrostatic systems. *Proc. R. Soc. Lond.*, 164:447–467, 1938. (Cited on page 169.)

[346] L. S. Blackford, J. Choi, A. Cleary, E. D'Azevedo, J. Demmel, I. Dhillon, J. Dongarra, S. Hammarling, G. Henry, A. Petitet, K. Stanley, D. Walker, and R. C. Whaley. *ScaLAPACK User's Guide*. SIAM, 1997. (Cited on page 376.)

[347] L. S. Blackford, J. Demmel, J. Dongarra, I. Duff, S. Hammarling, G. Henry, M. Heroux, L. Kaufman, A. Lumsdaine, A. Petitet, R. Pozo, K. Remington, and R. C. Whaley. An updated set of basic linear algebra subprograms (BLAS). *ACM Trans. Math. Soft. (TOMS)*, 28(2):135–151, 2002. (Cited on page 366.)

[348] M. Bôcher. *Introduction to Higher Algebra*. The MacMillan Company, New York, 1907. (Cited on pages 11, 13, 148, and 149.)

[349] M. Bôcher. *An Introduction to the Study of Integral Equations*. Cambridge University Press, London, Edinburgh, 1909. (Cited on page 126.)

[350] M. Bôcher and L. Brand. On linear equations with an infinite number of variables. *Ann. Math.*, 13(1/4):167–186, 1911. (Cited on page 15.)

[351] E. Bodewig. Bericht über die vershiedenen methoden zur lösung eines systems linear gleichungen mit reellen koffizienten. I through V. *Proc. Royal Acad. Sci. Amsterdam*, 50, 51:[50] 930–941, 1104–116, [51] 53–64, 211–219, 1947-1948. (Cited on pages 76 and 189.)

[352] E. Bodewig. A practical refutation of the iteration method for the algebraic eigenproblem. *Math. Comput.*, 8(48):237–240, 1954. (Cited on page 295.)

[353] E. Bodewig. *Matrix Calculus*. North-Holland, 2nd edition, 1959. First edition 1956. (Cited on pages 22, 76, 158, 192, 292, 295, and 487.)

[354] R. F. Boisvert. Mathematical software: past, present, and future. *Math. Comput. Simulation*, 54(4-5):227–241, 2000. (Cited on page 364.)

[355] R. F. Boisvert, M. J. Donahue, D. W. Lozier, R. McMichael, and B. W. Rust. Mathematics and measurement. *J. Res. Natl. Inst. Stand. Technol.*, 106(1):293–313, 2001. (Cited on page 201.)

[356] R. F. Boisvert, R. Pozo, K. Remington, R. F. Barrett, and J. J. Dongarra. Matrix market: A web resource for test matrix collections. In R. F. Boisvert, editor, *Quality of Numerical Software*, pages 125–137, Boston, 1997. Springer. (Cited on page 373.)

[357] P. Boldi, M. Santini, and S. Vigna. PageRank as a function of the damping factor. In *Proc. of the Fourteenth International World Wide Web Conference*, pages 557–566. ACM Press, 2005. (Cited on page 295.)

[358] D. L. Boley. Translation of "Sulle funzione bilineari" by E. Beltrami. In M. Moonen and B. De Moor, editors, *SVD and Signal Processing, III*, pages 5–18. Elsevier, 1995. (Cited on page 134.)

[359] D. L. Boley, S. Elhay, G. H. Golub, and M. H. Gutknecht. Nonsymmetric Lanczos and finding orthogonal polynomials associated with indefinite weights. *Numer. Algorithms*, 1(1):21–43, 1991. (Cited on page 229.)

[360] D. L. Boley and G. H. Golub. The Lanczos-Arnoldi algorithm and controllability. *Syst. Control Lett.*, 4(6):317–324, 1984. (Cited on page 223.)

[361] D. L. Boley and G. H. Golub. The nonsymmetric Lanczos algorithm and controllability. *Syst. Control Lett.*, 16(2):97–105, 1991. (Cited on page 223.)

[362] M. Bollhöfer. A robust ILU with pivoting based on monitoring the growth of the inverse factors. *Linear Algebra Appl.*, 338:201–218, 2001. (Cited on page 258.)

[363] M. Bollhöfer. A robust and efficient ILU that incorporates the growth of the inverse triangular factors. *SIAM J. Sci. Comput.*, 25(1):86–103, 2003. (Cited on page 258.)

[364] M. Bollhöfer and V. Mehrmann. Algebraic multilevel methods and sparse approximate inverses. *SIAM J. Matrix Anal. Appl.*, 24(1):191–218, 2002. (Cited on page 271.)

[365] M. Bollhöfer and Y. Notay. JADAMILU: a software code for computing selected eigenvalues of large sparse symmetric matrices. *Comput. Phys. Comm.*, 177(12):951–964, 2007. (Cited on page 317.)

[366] M. Bollhöfer and Y. Saad. Multilevel preconditioners constructed from inverse-based ILUs. *SIAM J. Sci. Comput.*, 27(5):1627–1650, 2006. (Cited on page 271.)

[367] Y. Bolshakov and B. Reichstein. Unitary equivalence in an indefinite scalar product: an analogue of singular-value decomposition. *Linear Algebra Appl.*, 222:155–226, 1995. (Cited on page 144.)

[368] Y. Bolshakov, C. V. van der Mee, A. C. M. Ran, B. Reichstein, and L. Rodman. Extension of isometries in finite-dimensional indefinite scalar product spaces and polar decompositions. *SIAM J. Matrix Anal. Appl.*, 18(3):752–774, 1997. (Cited on page 144.)

[369] Y. Bolshakov, C. V. van der Mee, A. M. Ran, B. Reichstein, and L. Rodman. Polar decompositions in finite dimensional indefinite scalar product spaces: General theory. *Linear Algebra Appl.*, 261(1-3):91–141, 1997. (Cited on page 144.)

[370] H. Boltz. Entwicklungsverfahren zur ausgleichung geodätischer netze nach der methode der kleinsten quadrate. Veröffentlichen des Preussischen Geodâtischen Instituts, N. F. no. 90, 1923. (Cited on page 70.)

[371] E. G. Boman and B. Hendrickson. Support theory for preconditioning. *SIAM J. Matrix Anal. Appl.*, 25(3):694–717, 2003. (Cited on page 264.)

[372] C. Bonfanti. Information technology in Italy: the origins and the early years (1954-1965). In A. Tatnall, editor, *Reflections on the History of Computing*, pages 320–347. Springer, Berlin, Heidelberg, 2012. (Cited on page 339.)

[373] M. Bonnet and G. Meurant. Résolution de systèmes d'équations linéaires par la méthode du gradient conjugué avec préconditionnement. Technical Report N-2159, CEA/DAM, 1980. (Cited on page 210.)

[374] G. Boole and J. F. Moulton. *A Treatise on the Calculus of Finite Differences*. Macmillan and Company, 1872. (Cited on page 162.)

[375] C. W. Borchardt. Neue eigenschaft der gleichung, mit deren hülfe man die seculären störungen der planeten bestimmt. *Journal für die reine und angewandte Mathematik (Crelle Journal)*, 30:38–45, 1846. (Cited on page 289.)

[376] M. Born. Otto Toeplitz. *Nature*, 145:617, 1940. (Cited on page 583.)

[377] H. Bosmans. Le "De Arte Magna" de Guillaume Gosselin. *Bibliotheca Mathematica*, III(7):44–46, 1906. (Cited on page 48.)

[378] N. Bosner and J. L. Barlow. Block and parallel versions of one-sided bidiagonalization. *SIAM J. Matrix Anal. Appl.*, 29(3):927–953, 2007. (Cited on page 138.)

[379] C. Bossut. *Essai sur l'histoire générale des mathématiques*. Louis, Libraire, Paris, 1802. Two volumes. (Cited on page 55.)

[380] C. Bossut et al. *Encyclopédie Méthodique, Mathématiques, Tome Premier*. Panckouke, Paris, 1784. (Cited on page 54.)

[381] M. A. Botchev and L. A. Knizhnerman. ART: Adaptive residual-time restarting for Krylov subspace matrix exponential evaluations. *J. Comput. Appl. Math.*, 364:article 112311, 2020. (Cited on page 390.)

[382] A. Böttcher and K. Rost. Topics in the numerical linear algebra of Toeplitz and Hankel matrices. *GAMM-Mitt.*, 27(2):174–188, 2004. (Cited on page 386.)

[383] A. Bouras and V. Frayssé. A relaxation strategy for inexact matrix-vector products for Krylov methods. Technical Report TR/PA/00/15, CERFACS, 2000. (Cited on page 239.)

[384] A. Bouras and V. Frayssé. Inexact matrix-vector products in Krylov methods for solving linear equations: A relaxation strategy. *SIAM J. Matrix Anal. Appl.*, 26(3):660–678, 2005. (Cited on page 239.)

[385] P. L. M. Bourdon. *Éléments d'Algèbre*. Ninth edition. Bachelier, Paris, 1843. (Cited on page 56.)

[386] R. Bouyouli, K. Jbilou, A. Messaoudi, and H. Sadok. On block minimal residual methods. *Appl. Math. Lett.*, 20(3):284–289, 2007. (Cited on page 243.)

[387] H. J. Bowdler, R. S. Martin, G. Peters, and J. H. Wilkinson. Solution of real and complex systems of linear equations. *Numer. Math.*, 8:217–234, 1966. (Cited on page 361.)

[388] H. J. Bowdler, R. S. Martin, C. Reinsch, and J. H. Wilkinson. The QR and QL algorithms for symmetric matrices. *Numer. Math.*, 11:293–306, 1968. (Cited on page 362.)

[389] A. H. Bowker. On the norm of a matrix. *The Annals of Mathematical Statistics*, 18(2):285–288, 1947. (Cited on page 15.)

[390] A. Bracha-Barak and P. E. Saylor. A symmetric factorization procedure for the solution of elliptic boundary value problems. *SIAM J. Numer. Anal.*, 10(1):190–206, 1973. (Cited on page 254.)

[391] D. Braess. The contraction number of a multigrid method for solving the Poisson equation. *Numer. Math.*, 37:387–404, 1981. (Cited on page 269.)

[392] D. Braess and W. Hackbusch. A new convergence proof for the multigrid method including the V-cycle. *SIAM J. Numer. Anal.*, 20(5):967–975, 1983. (Cited on page 270.)

[393] K. Braman, R. Byers, and R. Mathias. The multishift QR algorithm. Part I: Maintaining well-focused shifts and level 3 performance. *SIAM J. Matrix Anal. Appl.*, 23(4):929–947, 2002. (Cited on page 314.)

[394] K. Braman, R. Byers, and R. Mathias. The multishift QR algorithm. Part II: Aggressive early deflation. *SIAM J. Matrix Anal. Appl.*, 23(4):948–973, 2002. (Cited on page 314.)

[395] J. H. Bramble. *Multigrid Methods*. Pitman Research Notes in Mathematical Sciences Vol. 294. Longman Scientific & Technical, Essex, UK, 1993. (Cited on page 270.)

[396] J. H. Bramble and J. E. Pasciak. The analysis of smoothers for multigrid algorithms. *Math. Comput.*, 58:467–488, 1992. (Cited on page 270.)

[397] J. H. Bramble, J. E. Pasciak, and A. H. Schatz. The construction of preconditioners for elliptic problems by substructuring I. *Math. Comput.*, 47(175):103–134, 1986. (Cited on page 267.)

[398] J. H. Bramble, J. E. Pasciak, and A. H. Schatz. An iterative method for elliptic problems on regions partitioned into substructures. *Math. Comput.*, 46(174):361–369, 1986. (Cited on page 267.)

[399] J. H. Bramble, J. E. Pasciak, and A. H. Schatz. The construction of preconditioners for elliptic problems by substructuring II. *Math. Comput.*, 49(179):1–16, 1987. (Cited on page 267.)

[400] J. H. Bramble, J. E. Pasciak, and A. H. Schatz. The construction of preconditioners for elliptic problems by substructuring III. *Math. Comput.*, 51(184):415–430, 1988. (Cited on page 267.)

[401] J. H. Bramble, J. E. Pasciak, and A. H. Schatz. The construction of preconditioners for elliptic problems by substructuring IV. *Math. Comput.*, 53(187):1–24, 1989. (Cited on page 267.)

[402] J. H. Bramble, J. E. Pasciak, J. P. Wang, and J. Xu. Convergence estimates for multigrid algorithms without regularity assumptions. *Math. Comput.*, 57(195):23–45, 1991. (Cited on page 270.)

[403] J. H. Bramble, J. E. Pasciak, J. P. Wang, and J. Xu. Convergence estimates for product iterative methods with applications to domain decomposition. *Math. Comput.*, 57(195):1–21, 1991. (Cited on page 267.)

[404] C. W. Brand. An incomplete factorization preconditioning using repeated Red/Black ordering. *Numer. Math.*, 61:433–454, 1992. (Cited on page 257.)

[405] L. Brand. On infinite systems of linear integral equations. *The Annals of Mathematics*, 14(1/4):101–118, 1912. (Cited on page 126.)

[406] A. Brandt. Multi-level adaptive technique (MLAT) for fast numerical solution to boundary value problems. In H. Cabannes and R. Temam, editors, *Proceedings of the Third International Conference on Numerical Methods in Fluid Mechanics*, pages 82–89. Springer, 1973. Lecture Notes in Physics 18. (Cited on page 269.)

[407] A. Brandt. Multi-level adaptive solutions to boundary-value problems. *Math. Comput.*, 31:333–390, 1977. (Cited on page 269.)

[408] A. Brandt. Guide to multigrid development. In W. Hackbusch and U. Trottenberg, editors, *Multigrid Methods, Proceedings of the Koln-Porz Conference*, pages 220–312. Springer, 1982. Lecture Notes in Mathematics 960. (Cited on page 269.)

[409] A. Brandt. Algebraic multigrid theory: The symmetric case. *Appl. Math. Comput.*, 19:23–56, 1986. (Cited on page 269.)

[410] A. Brandt. Rigorous quantitative analysis of multigrid, I: Constant coefficients two-level cycle with L_2-norm. *SIAM J. Numer. Anal.*, 31(6):1695–1730, 1994. (Cited on page 269.)

[411] A. Brandt, S. McCormick, and J. Ruge. Multigrid methods for differential eigenproblems. *SIAM J. Sci. Statist. Comput.*, 4(2):244–260, 1983. (Cited on page 318.)

[412] A. Brauer. Limits for the characteristic roots of a matrix. *Duke Math. J.*, 13:387–395, 1946. (Cited on page 286.)

[413] A. Brauer. Limits for the characteristic roots of a matrix II. *Duke Math. J.*, 14:21–26, 1947. (Cited on pages 286 and 474.)

[414] R. Brauer and H. Weyl. Spinors in n dimensions. *Amer. J. Math.*, 57(2):425–449, 1935. (Cited on page 288.)

[415] R. K. Brayton, F. G. Gustavson, and R. A. Willoughby. Some results on sparse matrices. *Math. Comput.*, 24(112):937–954, 1970. (Cited on page 92.)

[416] F. Brechenmacher. *Histoire du théorème de Jordan de la décomposition matricielle (1870-1930)*. PhD thesis, École des hautes études en sciences sociales, 2005. (Cited on pages 12, 148, and 495.)

[417] F. Brechenmacher. Les matrices: formes de représentations et pratiques opératoires (1850-1930). hal-00142789, 2006. (Cited on page 12.)

[418] F. Brechenmacher. Regards croisés sur Camille Jordan. *Matapli (SMAI)*, 78:57–67, 2006. (Cited on page 12.)

[419] F. Brechenmacher. Algebraic generality vs arithmetic generality in the controversy between C. Jordan and L. Kronecker (1874). arXiv preprint arXiv:0712.2566, 2007. (Cited on pages 12, 148, and 495.)

[420] F. Brechenmacher. A controversy and the writing of a history: the discussion of "small oscillations" (1760-1860) from the standpoint of the controversy between Jordan and Kronecker (1874). *Bull. Belgian Math. Soc.*, 13(5):941–944, 2007. (Cited on pages 12, 148, and 495.)

[421] F. Brechenmacher. La controverse de 1874 entre Camille Jordan et Leopold Kronecker. *Rev. Hist. Math.*, 13(2):187–257, 2007. (Cited on pages xv, 12, 148, and 495.)

[422] F. Brechenmacher. L'identité algébrique d'une pratique portée par la discussion sur l'équation à l'aide de laquelle on détermine les inégalités séculaires des planètes (1766-1874). In *Sciences et Techniques en Perspective IIe série, fac. 1*, pages 5–85. Librairie Blanchard, Paris, 2007. (Cited on pages 12 and 278.)

[423] F. Brechenmacher. Une étude quantitative sur le temps long des occurrences du terme "matrice" en mathématiques. hal-00502403, 2008. (Cited on page 12.)

[424] F. Brechenmacher. Une histoire de l'universalité des matrices mathématiques. *Revue de Synthèse*, 131(4):569–603, 2010. (Cited on page 12.)

[425] J. L. Brenner and G. W. Reitwiesner. Remark on determination of characteristic roots by iteration. *M.T.A.C.*, 9(51):117–118, 1955. (Cited on page 295.)

[426] R. P. Brent, F. G. Gustavson, and D. Y. Y. Yun. Fast solution of Toeplitz systems of equations and computation of Padé approximants. *J. Algo.*, 1(3):259–295, 1980. (Cited on page 386.)

[427] R. P. Brent and F. T. Luk. The solution of singular-value and symmetric eigenvalue problems on multiprocessor arrays. *SIAM J. Sci. Statist. Comput.*, 6(1):69–84, 1985. (Cited on page 138.)

[428] C. Brezinski. Computation of the eigenelements of a matrix by the ε-algorithm. *Linear Algebra Appl.*, 11:7–20, 1975. (Cited on page 294.)

[429] C. Brezinski. Rational approximation to formal power series. *J. Approx. Theory*, 25:295–317, 1979. (Cited on page 228.)

[430] C. Brezinski. *Padé-type Approximation and General Orthogonal Polynomials*, volume ISNM 50. Birkhäuser, Basel, 1980. (Cited on pages 221, 223, 228, 308, and 385.)

[431] C. Brezinski. Edmond Nicolas Laguerre. In C. Brezinski, A. Draux, A. P. Magnus, P. Maroni, and A. Ronveaux, editors, *Polynômes Orthogonaux et Applications*, pages xxi–xxvi, Heidelberg, 1985. Springer-Verlag. Lecture Notes in Mathematics, vol. 1171. (Cited on page 514.)

[432] C. Brezinski. Other manifestations of the Schur complement. *Linear Algebra Appl.*, 111:231–247, 1988. (Cited on page 25.)

[433] C. Brezinski. *History of Continued Fractions and Padé Approximants*. Springer-Verlag, Berlin, 1991. (Cited on pages 511, 545, and 558.)

[434] C. Brezinski. The methods of Vorobyev and Lanczos. *Linear Algebra Appl.*, 234:21–41, 1996. (Cited on pages 247 and 249.)

[435] C. Brezinski. *Projection Methods for Systems of Equations*. North-Holland, Amsterdam, 1997. (Cited on pages 186, 247, and 249.)

[436] C. Brezinski. Error estimates for the solution of linear systems. *SIAM J. Sci. Comput.*, 21(2):764–781, 1999. (Cited on pages 213 and 244.)

[437] C. Brezinski. The block Lanczos and Vorobyev methods. *C.R. Acad. Sci. Paris, Sér 1*, 331:137–142, 2000. (Cited on page 249.)

[438] C. Brezinski. Formal orthogonal polynomials. In *Computational Aspects of Linear Control*, pages 73–85, Boston, MA, 2002. Springer. (Cited on page 228.)

[439] C. Brezinski. La méthode de Cholesky. *Rev. Hist. Math.*, 11:205–238, 2005. (Cited on pages 68 and 437.)

[440] C. Brezinski. The life and work of André Cholesky. *Numer. Algorithms*, 43:279–288, 2006. (Cited on page 68.)

[441] C. Brezinski and M. Gross-Cholesky. La vie et les travaux d'André-Louis Cholesky. *Bull. Soc. Amis. Bib. Éc. Polytech.*, 39:7–32, 2005. (Cited on page 68.)

[442] C. Brezinski and M. Redivo-Zaglia. Breakdowns in the computation of orthogonal polynomials. In A. Cuyt, editor, *Nonlinear Numerical Methods and Rational Approximation, II*, pages 49–59, Kluwer, Dordrecht, 1994. (Cited on page 225.)

[443] C. Brezinski and M. Redivo-Zaglia. Hybrid procedures for solving linear systems. *Numer. Math.*, 67:1–19, 1994. (Cited on page 232.)

[444] C. Brezinski and M. Redivo-Zaglia. Treatment of near-breakdown in the CGS algorithm. *Numer. Algorithms*, 7:33–73, 1994. (Cited on page 229.)

[445] C. Brezinski and M. Redivo-Zaglia. Look-ahead in BiCGstab and other methods for linear systems. *BIT*, 35:169–201, 1995. (Cited on page 229.)

[446] C. Brezinski and M. Redivo-Zaglia. Transpose-free Lanczos-type algorithms for nonsymmetric linear systems. *Numer. Algorithms*, 17:67–103, 1998. (Cited on page 229.)

[447] C. Brezinski and M. Redivo-Zaglia. A Schur complement approach to a general extrapolation algorithm. *Linear Algebra Appl.*, 368:279–301, 2003. (Cited on pages 25 and 26.)

[448] C. Brezinski and M. Redivo-Zaglia. The PageRank vector: Properties, computation, approximation, and acceleration. *SIAM J. Matrix Anal. Appl.*, 28(2):551–575, 2006. (Cited on page 295.)

[449] C. Brezinski and M. Redivo-Zaglia. Convergence acceleration of Kaczmarz's method. *J. Eng. Math.*, 93:3–19, 2015. (Cited on pages 189 and 190.)

[450] C. Brezinski and M. Redivo-Zaglia. The genesis and early developments of Aitken's process, Shanks transformation, the ε-algorithm, and related fixed point methods. *Numer. Algorithms*, 80:11–133, 2019. (Cited on page 311.)

[451] C. Brezinski and M. Redivo-Zaglia. *Extrapolation and Rational Approximation. The Works of the Main Contributors*. Springer Nature, Switzerland, 2020. (Cited on page 311.)

[452] C. Brezinski, M. Redivo-Zaglia, G. Rodriguez, and S. Seatzu. Extrapolation techniques for ill-conditioned linear systems. *Numer. Math.*, 81(1):1–29, 1998. (Cited on page 392.)

[453] C. Brezinski, M. Redivo-Zaglia, G. Rodriguez, and S. Seatzu. Multi-parameter regularization techniques for ill-conditioned linear systems. *Numer. Math.*, 94(2):203–228, 2003. (Cited on page 392.)

[454] C. Brezinski, M. Redivo-Zaglia, and H. Sadok. Avoiding breakdown and near-breakdown in Lanczos type algorithms. *Numer. Algorithms*, 1:261–284, 1991. (Cited on page 228.)

[455] C. Brezinski, M. Redivo-Zaglia, and H. Sadok. Addendum to "Avoiding breakdown and near-breakdown in Lanczos type algorithms". *Numer. Algorithms*, 2:133–136, 1992. (Cited on page 228.)

[456] C. Brezinski, M. Redivo-Zaglia, and H. Sadok. A breakdown-free Lanczos type algorithm for solving linear systems. *Numer. Math.*, 63:29–38, 1992. (Cited on page 228.)

[457] C. Brezinski, M. Redivo-Zaglia, and H. Sadok. Breakdowns in the implementation of the Lanczos method for solving linear systems. *Computers Math. Applic.*, 33(1):31–44, 1997. (Cited on pages 225 and 228.)

[458] C. Brezinski, M. Redivo-Zaglia, and H. Sadok. New look-ahead Lanczos-type algorithms for linear systems. *Numer. Math.*, 83(1):53–85, 1999. (Cited on page 228.)

[459] C. Brezinski, M. Redivo-Zaglia, and H. Sadok. A review of formal orthogonality in Lanczos-based methods. *J. Comput. Appl. Math.*, 140(1):81–98, 2002. (Cited on page 228.)

[460] C. Brezinski, G. Rodriguez, and S. Seatzu. Error estimates for linear systems with applications to regularization. *Numer. Algorithms*, 49(1-4):85–104, 2008. (Cited on pages 244 and 392.)

[461] C. Brezinski, G. Rodriguez, and S. Seatzu. Error estimates for the regularization of least squares problems. *Numer. Algorithms*, 51(1):61–76, 2009. (Cited on pages 244 and 392.)

[462] C. Brezinski and H. Sadok. Avoiding breakdown in the CGS algorithm. *Numer. Algorithms*, 1:199–206, 1991. (Cited on page 228.)

[463] C. Brezinski and H. Sadok. Lanczos-type algorithms for solving systems of linear equations. *Appl. Numer. Math.*, 11:443–473, 1993. (Cited on page 221.)

[464] C. Brezinski and D. Tournès. *André-Louis Cholesky (1875-1918), Mathematician, Topographer and Army Officer*. Birkhäuser, Basel, 2014. (Cited on pages 68 and 436.)

[465] C. Brezinski and L. Wuytack. Numerical analysis in the twentieth century. In C. Brezinski and L. Wuytack, editors, *Numerical analysis: Historical Developments in the 20th Century*, pages 1–40, North-Holland, Amsterdam, 2001. (Cited on page xv.)

[466] W. L. Briggs. *A Multigrid Tutorial.* SIAM, Philadelphia, 1987. (Cited on page 270.)

[467] W. L. Briggs, V. E. Henson, and S. F. McCormick. *A Multigrid Tutorial.* 2nd edition, SIAM, Philadelphia, 2000. (Cited on page 270.)

[468] S. Brin and L. Page. The anatomy of a large-scale hypertextual web search engine. *Comput. Networks ISDN Syst.*, 30:107–117, 1998. (Cited on page 295.)

[469] F. Brioschi. *La Teorica dei Determinanti e le sue Principali Applicazioni.* Bizzoni, 1854. (Cited on pages 115 and 289.)

[470] F. Brioschi. Note sur un théorème relatif aux déterminants gauches. *J. Math. Pures Appl.*, 19:253–256, 1854. (Cited on pages 115 and 289.)

[471] F. Brioschi. *Théorie des Déterminants et leurs Principales Applications.* Mallet-Bachelier, Paris, 1856. (Cited on page 381.)

[472] R. A. Brooker and F. H. Sumner. The method of Lanczos for calculating the characteristic roots and vectors of a real symmetric matrix. *Proc. Instn. Elect. Engrs B.*, 103:114–119, 1956. (Cited on page 301.)

[473] J. D. Brown, M. T. Chu, D. C. Ellison, and R. J. Plemmons, editors. *Proceedings of the Cornelius Lanczos International Centenary Conference, 1993.* SIAM, Philadelphia, 1994. (Cited on pages 517 and 667.)

[474] P. N. Brown. A theoretical comparison of the Arnoldi and GMRES algorithms. *SIAM J. Sci. Statist. Comput.*, 12(1):58–78, 1991. (Cited on pages 225 and 227.)

[475] P. N. Brown and A. C. Hindmarsh. Reduced storage matrix methods in stiff ODE systems. *Appl. Math. Comput.*, 31:40–91, 1989. (Cited on page 225.)

[476] P. N. Brown and H. F. Walker. GMRES on (nearly) singular systems. *SIAM J. Matrix Anal. Appl.*, 18(1):37–51, 1997. (Cited on page 235.)

[477] E. T. Browne. The characteristic roots of a matrix. *Bull. Amer. Math. Soc.*, 36(10):705–710, 1930. (Cited on pages 136 and 286.)

[478] E. T. Browne. Limits to the characteristic roots of a matrix. *Am. Math. Mon.*, 46(5):252–265, 1939. (Cited on page 286.)

[479] R. Bru, J. Cerdán, J. Marín, and J. Mas. Preconditioning sparse nonsymmetric linear systems with the Sherman-Morrison formula. *SIAM J. Sci. Comput.*, 25(2):701–715, 2003. (Cited on page 264.)

[480] R. Bru, L. Elsner, and M. Neumann. Models of parallel chaotic iteration methods. *Linear Algebra Appl.*, 103:175–192, 1988. (Cited on page 196.)

[481] R. Bru, J. Marín, J. Mas, and M. Tůma. Balanced incomplete factorization. *SIAM J. Sci. Comput.*, 30(5):2302–2318, 2008. (Cited on page 258.)

[482] R. A. Brualdi. Matrices, eigenvalues and directed graphs. *Linear Multilinear Algebra*, 11:143–165, 1982. (Cited on page 286.)

[483] R. A. Brualdi and S. Mellendorf. Regions in the complex plane containing the eigenvalues of a matrix. *Am. Math. Mon.*, 101(10):975–985, 1994. (Cited on page 286.)

[484] R. A. Brualdi and H. Schneider. Determinantal identities: Gauss, Schur, Cauchy, Sylvester, Kronecker, Jacobi, Binet, Laplace, Muir, and Cayley. *Linear Algebra Appl.*, 52/53:769–791, 1983. (Cited on page 25.)

[485] A. M. Bruaset. *A Survey of Preconditioned Iterative Methods.* Vol. 328. CRC Press, 1995. (Cited on page 257.)

[486] G. H. Bruce, D. W. Peaceman, H. A. Rachford, and J. D. Rice. Calculations of unsteady-state gas flow through porous media. *J. Pet. Technol.*, 5(3):79–92, 1953. (Cited on page 382.)

[487] M. Brunetti. Old and new proofs of Cramer's rule. *Appl. Math. Sci.*, 8(133):6689–6697, 2014. (Cited on page 105.)

[488] G.H. Bryan. Eugenio Beltrami. *Nature*, 61 (1589):568–569, 1900. (Cited on page 417.)

[489] A. Buchheim. Mathematical notes no. 3. *Messenger Math.*, 13:62–66, 1884. (Cited on pages 7 and 426.)

[490] A. Buchheim. On the theory of matrices. *Proc. Lond. Math. Soc.*, 16:63–82, 1884. (Cited on pages 7, 9, and 387.)

[491] A. Buchheim. XX. An extension of a theorem of Professor Sylvester's relating to matrices. *Philos. Mag.*, 22(135):173–174, 1886. (Cited on page 387.)

[492] H. M. Bücker and M. Sauren. A variant of the biconjugate gradient method suitable for massively parallel computing. In *International Symposium on Solving Irregularly Structured Problems in Parallel*, Springer, Berlin, Heidelberg, 1997. (Cited on page 235.)

[493] H. M. Bücker and M. Sauren. Reducing global synchronization in the biconjugate gradient method. In L. T. Yang, editor, *Parallel Numerical Computation with Applications*, pages 63–76, Springer, Boston, 1999. (Cited on page 235.)

[494] W. K. Bühler. *Gauss. A Biographical Study*. Springer-Verlag, Berlin, Heildelberg, New York, 1981. (Cited on page 469.)

[495] R. Bujakiewicz-Korońska and J. Koroński. The Krakowian calculus and other mathematical ideas of Tadeusz Banachiewicz. *Zeszyty Nauk. Politech. Slask. Mat.-Fiz*, 76:23–46, 1996. (in Polish). (Cited on page 412.)

[496] R. Bujakiewicz-Korońska and J. Koroński. The life of Tadeusz Banachiewicz and his scientific activity. *Studia Historiae Scientiarum*, 15:275–300, 2016. (Cited on page 412.)

[497] Z. Bujanović. On the permissible arrangements of Ritz values for normal matrices in the complex plane. *Linear Algebra Appl.*, 438:4606–4624, 2013. (Cited on page 307.)

[498] N. I. Buleev. A numerical method for the solution of two-dimensional and three-dimensional equations of diffusion. *Mat. Sb.*, 51:227–238, 1960. (in Russian). (Cited on page 252.)

[499] N. I. Buleev. The method of incomplete factorization for solving two-dimensional and three-dimensional equations of the diffusion type. *Zh. Vychisl. Mat. Mat. Fiz.*, 10(4):1042–1044, 1970. (in Russian), English translation: *U.S.S.R. Comput. Math. Math. Phys.*, 10:4 (1970), pp. 307-310. (Cited on page 252.)

[500] A. Bultheel and M. van Barel. *Linear Algebra: Rational Approximation and Orthogonal Polynomials*. North-Holland, Amsterdam, 1997. (Cited on page 385.)

[501] J. R. Bunch and L. Kaufman. Some stable methods for calculating inertia and solving symmetric linear systems. *Math. Comput.*, 31(137):63–179, 1977. (Cited on pages 78 and 217.)

[502] J. R. Bunch, L. Kaufman, and B. N. Parlett. Decomposition of a symmetric matrix. *Numer. Math.*, 27(1):95–109, 1976. (Cited on pages 78 and 368.)

[503] J. R. Bunch, C. P. Nielsen, and D. C. Sorensen. Rank-one modification of the symmetric eigenproblem. *Numer. Math.*, 31:31–48, 1978. (Cited on page 315.)

[504] J. R. Bunch and B. N. Parlett. Direct methods for solving symmetric indefinite systems of linear equations. *SIAM J. Numer. Anal.*, 8(4):639–655, 1971. (Cited on pages 78 and 368.)

[505] J. R. Bunch and D. J. Rose. *Sparse Matrix Computations*. Academic Press, 1976. (Cited on pages 86 and 364.)

[506] O. Buneman. A compact non-iterative Poisson solver. Technical Report 294, Stanford University Institute for Plasma Research, Stanford University, 1969. (Cited on page 385.)

[507] H. T. Burgess. A practical method of determining elementary divisors. *Ann. Math., second series*, 18(1):4–6, 1916. (Cited on page 149.)

[508] J. V. Burke and A. Greenbaum. Some equivalent characterizations of the polynomial numerical hull of degree k. Technical Report NA-04-29, The Mathematical Institute, University of Oxford, 2004. (Cited on page 240.)

[509] J. V. Burke and A. Greenbaum. Characterizations of the polynomial numerical hull of degree k. *Linear Algebra Appl.*, 419:37–47, 2006. (Cited on page 240.)

[510] A. R. Burks. *Who Invented the Computer?: The Legal Battle that Changed Computing History*. Prometheus Books, 2010. (Cited on page 336.)

[511] A. W. Burks, H. H. Goldstine, and J. von Neumann. Preliminary discussion of the logical design of an electronic computing instrument. Institute for Advanced Study, Princeton University, 1946. (Cited on page 338.)

[512] W. Burnside. On the reduction of a linear substitution to its canonical form. *Proc. Lond. Math. Soc.*, 1(1):180–194, 1898. (Cited on page 150.)

[513] P. A. Businger and G. H. Golub. Linear least squares solutions by Householder transformations. *Numer. Math.*, 7(3):269–276, 1965. (Cited on page 361.)

[514] P. A. Businger and G. H. Golub. Algorithm 358: Singular value decomposition of a complex matrix. *Comm. ACM*, 12(10):564–565, 1969. (Cited on page 137.)

[515] J. Buteo. *Logistica, quae & Arithmetica Vulgò Dicitur, in Libros Quinque Digesta*. Gulielmum Rouillium, Lyon, 1559. (Cited on pages 46 and 48.)

[516] A. Buttari, J. Langou, J. Kurzak, and J. Dongarra. Parallel tiled QR factorization for multicore architectures. *Concurr. Comp.-Pract. E.*, 20(13):1573–1590, 2008. (Cited on page 133.)

[517] B. L. Buzbee, F. W. Dorr, J. A. George, and G. H. Golub. The direct solution of the discrete Poisson equation on irregular regions. *SIAM J. Numer. Anal.*, 8(4):722–736, 1971. (Cited on page 266.)

[518] B. L. Buzbee, G. H. Golub, and C. W. Nielson. On direct methods for solving Poisson's equations. *SIAM J. Numer. Anal.*, 7(4):627–656, 1970. (Cited on pages 266 and 385.)

[519] R. Byers and H. Xu. A new scaling for Newton's iteration for the polar decomposition and its backward stability. *SIAM J. Matrix Anal. Appl.*, 30(2):822–843, 2008. (Cited on page 143.)

[520] B. Bylina and J. Bylina. Solving linear systems using the method of WZ factorization with BLAS1 and BLAS2 libraries. *Studia Informatica*, 25(1):57–67, 2004. (Cited on page 152.)

[521] C.-J. Lin and J. J. Moré. Incomplete Cholesky factorizations with limited memory. *SIAM J. Sci. Comput.*, 21(1):24–45, 1999. (Cited on page 258.)

[522] X.-C. Cai and M. Sarkis. A restricted additive Schwarz preconditioner for general sparse linear systems. *SIAM J. Sci. Comput.*, 21(2):792–797, 1999. (Cited on page 268.)

[523] F. Cajori. *A History of Mathematical Notation, volume I Notations in Elementary Mathematics*. The Open Court Publishing Company, 1928. (Cited on page 117.)

[524] F. Cajori. *A History of Mathematical Notation, volume II Notations Mainly in Higher Mathematics*. The Open Court Publishing Company, 1929. (Cited on page 117.)

[525] D. Calvetti, G. H. Golub, and L. Reichel. An adaptive Chebyshev iterative method for nonsymmetric linear systems based on modified moments. *Numer. Math.*, 67(1):21–40, 1994. (Cited on page 185.)

[526] D. Calvetti, G. H. Golub, and L. Reichel. A computable error bound for matrix functionals. *J. Comput. Appl. Math.*, 103:301–306, 1999. (Cited on page 213.)

[527] D. Calvetti, P. C. Hansen, and L. Reichel. L-curve curvature bounds via Lanczos bidiagonalization. *Electron. Trans. Numer. Anal.*, 14:20–35, 2002. (Cited on page 392.)

[528] D. Calvetti, S.-M. Kim, and L. Reichel. Quadrature rules based on the Arnoldi process. *SIAM J. Matrix Anal. Appl.*, 26(3):765–781, 2005. (Cited on page 390.)

[529] D. Calvetti, B. Lewis, and L. Reichel. GMRES-type methods for inconsistent systems. *Linear Algebra Appl.*, 316(1-3):157–169, 2000. (Cited on page 239.)

[530] D. Calvetti, B. Lewis, and L. Reichel. GMRES, L-curves, and discrete ill-posed problems. *BIT Numerical Mathematics*, 42(1):44–65, 2002. (Cited on page 239.)

[531] D. Calvetti, B. Lewis, and L. Reichel. On the regularizing properties of the GMRES method. *Numer. Math.*, 91(4):605–625, 2002. (Cited on pages 239 and 393.)

[532] D. Calvetti, S. Morigi, L. Reichel, and F. Sgallari. Computable error bounds and estimates for the conjugate gradient method. *Numer. Algorithms*, 25:79–88, 2000. (Cited on page 213.)

[533] D. Calvetti, S. Morigi, L. Reichel, and F. Sgallari. An iterative method with error estimators. *J. Comput. Appl. Math.*, 127:93–119, 2001. (Cited on page 213.)

[534] D. Calvetti, J. Petersen, and L. Reichel. A parallel implementation of the GMRES method. In L. Reichel, A. Ruttan, and R. S. Varga, editors, *Numerical Linear Algebra*, pages 31–45, De Gruyter, Berlin, 1993. (Cited on page 229.)

[535] D. Calvetti and L. Reichel. Adaptive Richardson iteration based on Leja points. *J. Comput. Appl. Math.*, 71(2):267–286, 1996. (Cited on page 185.)

[536] D. Calvetti and L. Reichel. An adaptive Richardson iteration method for indefinite linear systems. *Numer. Algorithms*, 12(1):125–149, 1996. (Cited on page 185.)

[537] D. Calvetti and L. Reichel. Preface. *Numer. Algorithms*, 25:xi–xiv, 2000. (Cited on page 591.)

[538] D. Calvetti and L. Reichel. Tikhonov regularization of large linear problems. *BIT Numerical Mathematics*, 43(2):263–283, 2003. (Cited on page 392.)

[539] D. Calvetti, L. Reichel, and F. Sgallari. Application of anti-Gauss quadrature rules in linear algebra. In W. Gautschi, G. H. Golub, and G. Opfer, editors, *Applications and Computation of Orthogonal Polynomials*, pages 41–56, Basel, 1999. Birkhäuser. (Cited on page 213.)

[540] D. Calvetti, L. Reichel, and A. Shuibi. Tikhonov regularization of large symmetric problems. *Numer. Linear Algebra Appl.*, 12(2-3):127–139, 2005. (Cited on page 392.)

[541] D. Calvetti, L. Reichel, and D. C. Sorensen. An implicitly restarted Lanczos method for large symmetric eigenvalue problems. *Electron. Trans. Numer. Anal.*, 2:1–21, 1994. (Cited on page 306.)

[542] D. Calvetti, L. Reichel, and H. Xu. A CS decomposition for orthogonal matrices with application to eigenvalue computation. *Linear Algebra Appl.*, 476:197–232, 2015. (Cited on page 146.)

[543] S. L. Campbell, I. C. F. Ipsen, C. T. Kelley, and C. D. Meyer. GMRES and the minimal polynomial. *BIT Numerical Mathematics*, 36(4):664–675, 1996. (Cited on page 234.)

[544] Z. H. Cao. Avoiding breakdown in variants of the BI-CGSTAB algorithm. *Linear Algebra Appl.*, 263:113–132, 1997. (Cited on page 235.)

[545] S. Caparrini. Lagrange and the Turin Academy of Sciences. *Lett. Mat. Int.*, 2:35–37, 2014. (Cited on pages 509 and 510.)

[546] S. Caparrini. The personality of Lagrange. *Lett. Mat. Int.*, 2:9–11, 2014. (Cited on page 510.)

[547] A. Capelli and G. Garbieri. *Corso di Analisi Algebrica.* F. Sachetto, Padua, 1886. (Cited on page 115.)

[548] R. Carden. A simple algorithm for the inverse field of values problem. *Inverse Probl.*, 25:1–9, 2009. (Cited on page 288.)

[549] R. Carden. *Ritz values and Arnoldi convergence for non-Hermitian matrices.* PhD thesis, Rice University, Houston, 2011. (Cited on page 307.)

[550] R. Carden and M. Embree. Ritz value localization for non-Hermitian matrices. *SIAM J. Matrix Anal. Appl.*, 33(4):1320–1338, 2012. (Cited on page 307.)

[551] R. Carden and D. J. Hansen. Ritz values of normal matrices and Ceva's theorem. *Linear Algebra Appl.*, 438(11):4114–4129, 2013. (Cited on page 307.)

[552] D. Carlson. What are Schur complements, anyway? *Linear Algebra Appl.*, 74:257–275, 1986. (Cited on page 25.)

[553] D. Carlson, E. Haynsworth, and T. Markham. A generalization of the Schur complement by means of the Moore-Penrose inverse. *SIAM J. Appl. Math.*, 26(1):169–175, 1974. (Cited on page 25.)

[554] D. Carlson, T. L. Markham, and F. Uhlig. Emilie Haynsworth, 1916-1985. *Linear Algebra Appl.*, 75:269–276, 1986. (Cited on page 483.)

[555] E. C. Carson. *Communication-avoiding Krylov subspace methods in theory and practice.* PhD thesis, University of California at Berkeley, 2015. Technical Report UCB/EECS-2015-179. (Cited on page 243.)

[556] E. C. Carson. The adaptive s-step conjugate gradient method. *SIAM J. Matrix Anal. Appl.*, 39(3):1318–1338, 2018. (Cited on page 215.)

[557] E. C. Carson and J. Demmel. A residual replacement strategy for improving the maximum attainable accuracy of s-step Krylov subspace methods. *SIAM J. Matrix Anal. Appl.*, 35(1):22–43, 2014. (Cited on page 243.)

[558] E. C. Carson and N. J. Higham. Accelerating the solution of linear systems by iterative refinement in three precisions. *SIAM J. Sci. Comput.*, 40(2):A817–A847, 2018. (Cited on page 355.)

[559] E. C. Carson, M. Rozložník, Z. Strakoš, P. Tichý, and M. Tůma. The numerical stability analysis of pipelined conjugate gradient methods: Historical context and methodology. *SIAM J. Sci. Comput.*, 40(5):A3549–A3580, 2018. (Cited on page 216.)

[560] L. M. Carvalho, S. Gratton, R. F. Lago, and X. Vasseur. A flexible generalized conjugate residual method with inner orthogonalization and deflated restarting. *SIAM J. Matrix Anal. Appl.*, 32(4):1212–1235, 2011. (Cited on page 244.)

[561] E. Carvallo. Sur les systèmes linéaires, le calcul des symboles différentiels et leur application à la physique mathématique. *Monatsh Math. Phys.*, 2:225–266, 1891. (Cited on page 387.)

[562] G. Cassinis. Risoluzione dei sistemi di equazioni algebriche lineari. *Rend. Sem. Math. Fis. Milano*, 17:62–78, 1946. (Cited on page 430.)

[563] A. L. Cauchy. Mémoire sur le nombre de valeurs qu'une fonction peut acquérir, lorsqu'on y permute de toutes les manières possibles les quantités qu'elle renferme. *Journal de l'École Polytechnique*, 10(17):1–27, 1815. (Cited on pages 109 and 419.)

[564] A. L. Cauchy. Mémoire sur les fonctions qui ne peuvent obtenir que deux valeurs égales et de signes contraires par suite des transpositions opérées entre les variables qu'elles renferment. *Journal de l'École Polytechnique*, 10(17):29–112, 1815. (Cited on pages 4, 109, and 420.)

[565] A. L. Cauchy. *Cours d'Analyse de l'École Royale Polytechnique, 1ère partie Analyse Algébrique.* Debure frères, libraires du Roi, 1821. (Cited on pages 111, 419, 464, and 559.)

[566] A. L. Cauchy. Sur les formules qui résultent de l'emploi du signe $>$ ou $<$, et sur les moyennes entre plusieurs quantités. In *Ouvres Complètes*, volume 2, tome III, pages 360–377. Gauthier-Villars et Fils, 1821. (Cited on pages 14 and 123.)

[567] A. L. Cauchy. Leçons sur les applications du calcul infinitésimal à la géométrie. In *Ouvres Complètes*, volume 2, tome V, pages 11–528. Gauthier-Villars et Fils, 1826. (Cited on page 280.)

[568] A. L. Cauchy. Sur les centres, les plans principaux et les axes principaux des surfaces du second degré. In *Exercices de Mathématiques*. De Bure Frères, 1828. (Cited on pages 123 and 280.)

[569] A. L. Cauchy. Sur l'équation à l'aide de laquelle on détermine les inégalités séculaires des mouvements des planètes. In *Ouvres Complètes*, volume 2, tome IX, pages 174–95. Gauthier-Villars et Fils, 1829. (Cited on pages 4, 111, 123, 280, and 420.)

[570] A. L. Cauchy. Mémoire sur l'équation qui a pour racines les moments d'inertie principaux d'un corps solide. In *Ouvres Complètes*, volume 1, tome II, pages 79–81. Gauthier-Villars et Fils, 1830. (Cited on page 280.)

[571] A. L. Cauchy. Mémoire sur l'interpolation. *J. Math. Pures Appl.*, 2:193–205, 1837. (Cited on page 61.)

[572] A. L. Cauchy. Mémoire sur l'élimination d'une variable entre deux équations algébriques. In *Exercices d'Analyse et de Physique Mathématique, tome I*. Bachelier, Paris, 1840. (Cited on page 111.)

[573] A. L. Cauchy. Mémoire sur les dilatations, les condensations et les rotations produits par un changement de forme dans un système de points matériels. *Ouvres Série (2)*, 12:343–377, 1841. (Cited on page 142.)

[574] A. L. Cauchy. Mémoire sur les fonctions alternées et sur les sommes alternées. In *Exercices d'Analyse et de Physique Mathématique, tome II*. Bachelier, Paris, 1841. (Cited on pages 111 and 420.)

[575] A. L. Cauchy. Note sur la formation des fonctions alternées qui servent à résoudre le problème de l'élimination. *C.R. Acad. Sci. Paris, Sér 1*, XII:414–426, 1841. (Cited on pages 111 and 420.)

[576] A. L. Cauchy. Méthode générale pour la résolution des systèmes d'équations simultanées. *C.R. Acad. Sci. Paris, Sér 1*, 25:536–538, 1847. (Cited on page 170.)

[577] R. L. Causey. Computing eigenvalues of non-Hermitian matrices by methods of Jacobi type. *J. Soc. Ind. Appl. Math.*, 6(2):172–181, 1958. (Cited on page 151.)

[578] R. L. Causey and R. T. Gregory. On Lanczos' algorithm for tridiagonalizing matrices. *SIAM Rev.*, 3(4):322–328, 1961. (Cited on page 301.)

[579] A. Cayley. On a theorem in the geometry of position. *Cambridge Math. J.*, 2:267–271, 1841. (Cited on pages 112 and 423.)

[580] A. Cayley. On the theory of determinants. *Trans. Cambridge Philos. Soc.*, 8:1–16, 1843. (Cited on page 112.)

[581] A. Cayley. On the theory of groups as depending on the symbolic equation $\theta^n = 1$. *Philos. Mag.*, 7:40–47, 1854. (Cited on page 423.)

[582] A. Cayley. Recherches sur les matrices dont les termes sont des fonctions linéaires d'une seule indéterminée. *Journal für die reine und angewandte Mathematik (Crelle Journal)*, 50:313–317, 1855. (Cited on pages 4 and 424.)

[583] A. Cayley. Remarques sur la notation des fonctions algébriques. *Journal für die reine und angewandte Mathematik (Crelle Journal)*, 50:282–285, 1855. (Cited on pages 4 and 424.)

[584] A. Cayley. Sur la transformation d'une fonction quadratique en elle-même par des substitutions linéaires. *Journal für die reine und angewandte Mathematik (Crelle Journal)*, 50:288–299, 1855. (Cited on pages 4 and 424.)

[585] A. Cayley. Note sur la méthode d'élimination de Bezout. *Journal für die reine und angewandte Mathematik (Crelle Journal)*, 53:366–367, 1857. (Cited on pages 4 and 424.)

[586] A. Cayley. On the determination of the value of a certain determinant. *Quart. J. Math.*, 2:163–166, 1857. (Cited on page 381.)

[587] A. Cayley. A memoir on the theory of matrices. *Philos. Trans. R. Soc. Lond.*, 148:17–37, 1858. (Cited on pages 3, 387, and 424.)

[588] A. Cayley. A supplementary memoir on the theory of matrices. *Philos. Trans. R. Soc. Lond.*, 156:25–35, 1866. (Cited on page 426.)

[589] A. Cayley. On the extraction of the square root of a matrix of the third order. *Proc. Roy. Soc. Edinburgh*, 7:675–682, 1872. (Cited on page 387.)

[590] A. Cayley. On the matrical equation $qQ - Qq' = 0$. *Messenger Math.*, 14:176–178, 1885. (Cited on page 395.)

[591] A. Cayley. On the quaternion equation $qQ - Qq' = 0$. *Messenger Math.*, 14:108–112, 1885. (Cited on page 395.)

[592] A. Cayley. Science Worthies, XXV- James Joseph Sylvester. *Nature*, XXXIX(1001):217–219, 1889. (Cited on page 571.)

[593] U. V. Çatalyürek and C. Aykanat. Hypergraph-partitioning-based decomposition for parallel sparse-matrix vector multiplication. *IEEE Trans. Parallel Distributed Systems*, 10(7):673–693, 1999. (Cited on page 90.)

[594] J. Cecconi. Lamberto Cesari. *Bollettino dell'Unione Matematica Italiana, Sezione A, Serie VII*, 6(1):125–151, 1992. (Cited on page 429.)

[595] F. Cecioni. Sulle equazioni fra matrici $AX = XB, X^m = A$. *Atti Acc. Naz. Lincei*, 18(5):566–571, 1909. (Cited on page 396.)

[596] F. Cecioni. Sopra alcune operazioni algebriche sulle matrici. *Annali della Scuola Normale Superiore di Pisa, Classe di Scienze, serie 1*, 11:1–141, 1910. (Cited on page 396.)

[597] A. Cegielski. Methods for finding fixed points of nonexpansive operators in a Hilbert space. Technical Report M 08/20, Technische Universität Ilmenau, Institut für Mathematik, 2008. (Cited on page 186.)

[598] A. Cegielski. A list of publications on the Kaczmarz method (up to 2010). http://staff.uz.zgora.pl/acegiels/Publications-Kaczmarz-method.pdf, 2010. (Cited on pages 188, 189, and 498.)

[599] A. Cegielski. *Iterative Methods for Fixed Point Problems in Hilbert Spaces*. Springer-Verlag, Berlin, Heidelberg, 2012. Lecture Notes in Mathematics, Vol. 2057. (Cited on page 186.)

[600] L. Cesari. Sulla risoluzione dei sistemi di equazioni lineari per approssimazioni successive. *La Ricerca Scientifica, Progr. Tecn. Econom. Naz., serie II, anno VIII*, 1(11-12):512–522, 1937. (Cited on pages 168, 430, and 439.)

[601] L. Cesari. Sulla risoluzione dei sistemi di equazioni lineari per approssimazioni successive. Technical Report 12, Istituto per le applicazioni del calcolo, Consiglio Nazionale delle Richerche, Roma, 1937. (Cited on pages 168, 250, 261, 430, and 439.)

[602] L. Cesari. Sulla risoluzione dei sistemi di equazioni lineari per approssimazioni successive. *Rass. Poste, dei Teleg. e dei Telef., anno IX*, 4:285–296, 1937. (Cited on page 430.)

[603] L. Cesari. Sulla risoluzione dei sistemi di equazioni lineari per approssimazioni successive. *Rendic. Reale Accademia Nazionale dei Lincei, Classe Scienze Fis., Mat., Natur., ser. 6a*, 25:422–428, 1937. (Cited on page 430.)

[604] L. Cesari. Leonida Tonelli (1885-1946). *Bollettino dell'Unione Matematica Italiana, Serie 3*, 1(1):64–67, 1946. (Cited on page 427.)

[605] J. L. Chabert. Gauss et la méthode des moindres carrés. *Rev. Hist. Sci.*, 42(1-2):5–26, 1989. (Cited on page 58.)

[606] J. L. Chabert. *A History of Algorithms. From the Pebble to the Microchip*. Springer-Verlag, Berlin, Heidelberg, 1999. (Cited on page 59.)

[607] A. B. Chace. *The Rhind Mathematical Papyrus, Volume I*. Mathematical Association of America, 1929. (Cited on page 34.)

[608] A. B. Chace, L. Bull, and H. P. Manning. *The Rhind Mathematical Papyrus, Volume II*. Mathematical Association of America, 1929. (Cited on page 34.)

[609] F. Chaitin-Chatelin and V. Frayssé. *Lectures on Finite Precision Computations*. SIAM, Philadelphia, 1996. (Cited on page 432.)

[610] F. Chaitin-Chatelin and E. Traviesas-Cassan. PRECISE and the quality of reliable numerical software. In *Accuracy and Reliability in Scientific Computing*, pages 95–108. SIAM, Philadelphia, 2005. (Cited on page 432.)

[611] R. H. Chan, C. Greif, and D. P. O' Leary. *Milestones in Matrix Computation: The Selected Works of Gene H. Golub with Commentaries*. Oxford University Press, Oxford, 2007. (Cited on pages 478 and 671.)

[612] T. F. Chan. An improved algorithm for computing the singular value decomposition. *ACM Trans. Math. Soft. (TOMS)*, 8(1):72–83, 1982. (Cited on page 137.)

[613] T. F. Chan. Rank revealing QR factorizations. *Linear Algebra Appl.*, 88-89:67–82, 1987. (Cited on page 13.)

[614] T. F. Chan. Fourier analysis of relaxed incomplete factorization preconditioners. *SIAM J. Sci. Statist. Comput.*, 12(3):668–680, 1991. (Cited on page 257.)

[615] T. F. Chan, L. De Pillis, and H. van der Vorst. A transpose-free squared Lanczos algorithm and application to solving nonsymmetric linear systems. Technical Report 91-17 CAM, Dept. of Mathematics, UCLA, 1991. (Cited on page 236.)

[616] T. F. Chan, L. De Pillis, and H. van der Vorst. Transpose-free formulations of Lanczos-type methods for nonsymmetric linear systems. *Numer. Algorithms*, 17(1):51–66, 1998. (Cited on page 236.)

[617] T. F. Chan and H. C. Elman. Fourier analysis of iterative methods for elliptic problems. *SIAM Rev.*, 31(1):20–49, 1989. (Cited on page 178.)

[618] T. F. Chan, E. Gallopoulos, V. Simoncini, T. Szeto, and C. H. Tong. A quasi-minimal residual variant of the Bi-CGStab algorithm for nonsymmetric systems. *SIAM J. Sci. Comput.*, 15(2):338–347, 1994. (Cited on page 232.)

[619] T. F. Chan, K. R. Jackson, and B. Zhu. Alternating-direction incomplete factorizations. *SIAM J. Numer. Anal.*, 20(2):239–257, 1983. (Cited on page 264.)

[620] T. F. Chan, S. D. Margenov, and P. S. Vassilevski. Performance of block-ILU factorization preconditioners based on block-size reduction for 2D elasticity systems. *SIAM J. Sci. Comput.*, 18(5):1355–1366, 1997. (Cited on page 259.)

[621] T. F. Chan and T. Mathew. Domain decomposition algorithms. *Acta Numerica*, 3:61–143, 1994. (Cited on page 268.)

[622] T. F. Chan and G. Meurant. Fourier analysis of block preconditioners. Technical Report CAM 90-04, University of California, Los Angeles, 1990. (Cited on page 259.)

[623] T. F. Chan and T. Szeto. A composite step conjugate gradients squared algorithm for solving nonsymmetric linear systems. *Numer. Algorithms*, 7(1):17–32, 1994. (Cited on page 232.)

[624] T. F. Chan and T. Szeto. Composite step product methods for solving nonsymmetric linear systems. *SIAM J. Sci. Comput.*, 17(6):1491–1508, 1996. (Cited on page 232.)

[625] T. F. Chan, W. P. Tang, and W. L. Wan. Wavelet sparse approximate inverse preconditioners. *BIT Numerical Mathematics*, 37(3):644–660, 1997. (Cited on page 264.)

[626] T. F. Chan and P. S. Vassilevski. A framework for block ILU factorizations using block-size reduction. *Math. Comput.*, 64(209):129–156, 1995. (Cited on page 259.)

[627] T. F. Chan and W. L. Wan. Analysis of projection methods for solving linear systems with multiple right-hand sides. *SIAM J. Sci. Comput.*, 18(6):1698–1721, 1997. (Cited on page 235.)

[628] T. F. Chan and Q. Ye. A mixed product Krylov subspace method for solving nonsymmetric linear systems. *Amer. J. Math.*, 1(3):422–434, 1997. (Cited on page 235.)

[629] W. M. Chan and A. George. A linear time implementation of the Reverse Cuthill-McKee algorithm. *BIT Numerical Mathematics*, 20(1):8–14, 1980. (Cited on page 88.)

[630] R. Chandra. *Conjugate gradient methods for partial differential equations*. PhD thesis, Yale University, 1978. (Cited on pages 217, 255, and 264.)

[631] R. Chandra, S. C. Eisenstat, and M. H. Schultz. The modified conjugate residual method for partial differential equations. In R. Vichnevetsky, editor, *Advances in Computer Methods for Partial Differential Equations: Proceedings of the AICA International Symposium on Computer Methods for Partial Differential Equations, held at Lehigh University, Bethlehem, Pennsylvania, June 17-19, 1975*, pages 13–19. AICA, 1977. (Cited on page 217.)

[632] S. Chandrasekaran, M. Gu, X. Sun, J. Xia, and J. Zhu. A superfast algorithm for Toeplitz systems of linear equations. *SIAM J. Matrix Anal. Appl.*, 29(4):1247–1266, 2007. (Cited on page 387.)

[633] S. Chandrasekaran, M. Gu, J. Xia, and J. Zhu. A fast QR algorithm for companion matrices. In *Recent Advances in Matrix and Operator Theory*, pages 111–143. Birkhäuser, Basel, 2007. (Cited on page 293.)

[634] S. Chandrasekaran and I. C. F. Ipsen. On rank-revealing factorisations. *SIAM J. Matrix Anal. Appl.*, 15(2):592–622, 1994. (Cited on pages 13 and 79.)

[635] S. Chandrasekaran and I. C. F. Ipsen. On the sensitivity of solution components in linear systems of equations. *SIAM J. Matrix Anal. Appl.*, 16(1):93–112, 1995. (Cited on page 79.)

[636] K. M. Chandy and L. Lamport. Distributed snapshots: determining global states of distributed systems. *ACM Trans. Comput. Syst.*, 3(1):63–75, 1985. (Cited on page 197.)

[637] A. Chang. Application of sparse matrix methods in electric power system analysis. In R. A. Willoughby, editor, *Proceedings of the Symposium on Sparse Matrices and their Applications*, pages 113–122. IBM, Thomas J. Watson Research Center, 1969. (Cited on page 87.)

[638] X.-W. Chang and C. C. Paige. Componentwise perturbation analyses for the QR factorization. *Numer. Math.*, 88(2):319–345, 2001. (Cited on page 133.)

[639] X.-W. Chang, C. C. Paige, and D. Titley-Peloquin. Structure in loss of orthogonality. *Linear Algebra Appl.*, 623:149–167, 2021. (Cited on page 302.)

[640] A. Chapman and Y. Saad. Deflated and augmented Krylov subspace techniques. *Numer. Linear Algebra Appl.*, 4:43–66, 1997. (Cited on page 234.)

[641] B. A. Chartres and J. C. Geuder. Computable error bounds for direct solution of linear equations. *J. ACM*, 14(1):63–71, 1967. (Cited on page 79.)

[642] G. C. Chase. History of mechanical computing machinery. In *Proceedings of the 1952 ACM National Meeting (Pittsburgh)*. ACM, 1952. Reprinted in *Annals of the History of Computing* v 3 n 2 (1980), pp. 198-226. (Cited on page 325.)

[643] A. Châtelet. Sur certains ensembles de tableaux et leur application à la théorie des nombres. *Annales scientifiques de l'É.N.S*, 28:105–202, 1911. (Cited on page 8.)

[644] F. Chatelin. *Spectral Approximation of Linear Operators*. Academic Press, New York, 1983. Reprinted by SIAM, Philadelphia, 2011. (Cited on pages 319 and 432.)

[645] F. Chatelin. *Eigenvalues of Matrices*. Wiley, Chichester, 1993. Reprinted by SIAM, Philadelphia, 2012. (Cited on pages 319 and 432.)

[646] F. Chatelin. *Qualitative Computing, A Computational Journey into Nonlinearity*. World Scientific Publishing Co., Singapore, 2012. (Cited on page 432.)

[647] W. Chauvenet. *A Treatise on the Method of Least Squares*. J. B. Lippincott, Philadelphia, 1868. (Cited on page 62.)

[648] Y. Chauvet and G. Meurant. Multitasking on the CRAY X-MP. *The Journal of Systems and Software*, 2:17–20, 1986. Proceedings of International Workshop on Modeling and Performance Evaluation of Parallel Systems, IMAG Grenoble (1984). (Cited on page 214.)

[649] D. Chazan and W. Miranker. Chaotic relaxation. *Linear Algebra Appl.*, 2(2):199–222, 1969. (Cited on pages 195 and 196.)

[650] K. Chemla. Résonances entre démonstration et procédure. Remarques sur le commentaire de Liu Hui (3ème siècle) aux Neuf Chapitres sur les Procédures Mathématiques (1er siècle). In K. Chemla, editor, *Regards Obliques sur l'Argumentation en Chine, Extrême-Orient, Extrême-Occident*, pages 91–129. Presses Universitaires de Vincennes, 1992. (Cited on pages 38 and 39.)

[651] K. Chemla. Different concepts of equations in the Nine Chapters on Mathematical Procedures and in the commentary on it by Liu Hui (3rd Century). *Historia Scientiarum*, 4(2):113–137, 1994. (Cited on pages 36, 38, and 39.)

[652] K. Chemla. Algebraic equations east and west until the middle ages. In K. Hashimoto, C. Jami, and L. Skar, editors, *East Asian Science: Tradition and Beyond, Seventh International Conference on the History of Science in East Asia (1993)*, pages 83–89, Osaka, 1995. Kansai University Press. (Cited on page 43.)

[653] K. Chemla. Aperçu sur l'histoire des mathématiques en Chine ancienne dans le contexte d'une histoire internationale. In D. Tournès, editor, *L'Océan Indien au Carrefour des Mathématiques Arabes, Chinoises, Européennes et Indiennes*, pages 71–90. I. U. F. M. de La Réunion, 1998. Actes du colloque à Saint-Denis de la Réunion (3-7 novembre 1997). (Cited on pages 35 and 39.)

[654] K. Chemla. Reading proofs in Chinese commentaries: algebraic proofs in an algorithmic context. In K. Chemla, editor, *The History of Mathematical Proof in Ancient Traditions*, pages 423–486. Cambridge University Press, 2012. (Cited on pages 38 and 39.)

[655] K. Chemla and G. Shuchun. *Les Neuf Chapitres. Le Classique Mathématique de la Chine Ancienne et ses Commentaires*. Dunod, 2004. (Cited on pages 35 and 39.)

[656] G. Chen and Z. Jia. An analogue of the results of Saad and Stewart for harmonic Ritz vectors. *J. Comput. Appl. Math.*, 167(2):493–498, 2004. (Cited on page 307.)

[657] K. Chen. *Matrix Preconditioning Techniques and Applications*. Cambridge University Press, UK, 2005. (Cited on page 265.)

[658] T. Chen and E. C. Carson. Predict-and-recompute conjugate gradient variants. *SIAM J. Sci. Comput.*, 42(5):A3084–A3108, 2020. (Cited on page 216.)

[659] Y. Chen, T. A. Davis, W. W. Hager, and S. Rajamanickam. Algorithm 887: CHOLMOD, supernodal sparse Cholesky factorization and update/downdate. *ACM Trans. Math. Soft. (TOMS)*, 35(3):22:1–22:14, 2008. (Cited on page 372.)

[660] Y. T. Chen and R. P. Tewarson. On the optimal choice of pivots for the Gaussian elimination. *Computing*, 9(3):245–250, 1972. (Cited on page 78.)

[661] J. M. Chesneaux and A. Matos. Breakdown and near-breakdown control in the CGS algorithm using stochastic arithmetic. *Numer. Algorithms*, 11:99–116, 1996. (Cited on page 228.)

[662] F. Chiò. *Mémoire sur les Fonctions Connues sous les Noms de Résultantes ou de Déterminans*. A. Pons et C., Torino, 1853. (Cited on pages 114 and 405.)

[663] D. Choi and A. Greenbaum. Roots of matrices in the study of GMRES convergence and Crouzeix's conjecture. *SIAM J. Matrix Anal. Appl.*, 36(1):289–301, 2015. (Cited on page 240.)

[664] J. Choi, J. J. Dongarra, S. Ostrouchov, A. Petitet, D. Walker, and R. C. Whaley. A proposal for a set of parallel basic linear algebra subprograms. In *International Workshop on Applied Parallel Computing*, pages 107–114, Berlin, Heidelberg, 1995. Springer. LNCS 1041. (Cited on page 376.)

[665] J. Choi, J. J. Dongarra, R. Pozo, and D. W. Walker. ScaLAPACK: A scalable linear algebra library for distributed memory concurrent computers. In *The Fourth Symposium on the Frontiers of Massively Parallel Computation*, pages 120–127. IEEE, 1992. (Cited on page 376.)

[666] J. Choi, J. J. Dongarra, and D. W. Walker. PB-BLAS: A set of parallel block basic linear algebra subprograms. *Concurrency-Pract. Exp.*, 8(7):517–535, 1996. (Cited on page 376.)

[667] A. Cholesky. *Cours de Topographie. 2ème Partie, Topographie Générale*. École Spéciale des Travaux Publics, Paris, 7ème édition, 1937. (Cited on page 436.)

[668] C. Chorianopoulos, P. Psarrakos, and F. Uhlig. A method for the inverse numerical range problem. *Electron. J. Linear Algebra*, 20:198–206, 2010. (Cited on page 288.)

[669] E. Chow, A. J. Cleary, and R. D. Falgout. Design of the hypre preconditioner library. In M. Henderson, C. Anderson, and S. Lyons, editors, *Proc. of the SIAM Workshop on Object Oriented Methods for Inter-operable Scientific and Engineering Computing*. SIAM, Philadelphia, 1998. (Cited on page 378.)

[670] E. Chow and A. Patel. Fine-grained parallel incomplete LU factorization. *SIAM J. Sci. Comput.*, 37(2):C169–C193, 2015. (Cited on page 261.)

[671] E. Chow and Y. Saad. Approximate inverse preconditioners for general sparse matrices. Technical Report UMSI 94-101, Supercomputer Institute, University of Minnesota, 1994. (Cited on page 263.)

[672] E. Chow and Y. Saad. Approximate inverse techniques for block-partitioned matrices. *SIAM J. Sci. Comput.*, 18(6):1657–1675, 1997. (Cited on page 263.)

[673] E. Chow and Y. Saad. Approximate inverse preconditioners via sparse-sparse iterations. *SIAM J. Sci. Comput.*, 19(3):995–1023, 1998. (Cited on page 263.)

[674] J. Christianidis. Une interprétation byzantine de Diophante. *Historia Math.*, 25(1):22–28, 1997. (Cited on page 41.)

[675] J. Christianidis. The way of Diophantus: Some clarifications on Diophantus' method of solution. *Historia Math.*, 34(3):289–305, 2007. (Cited on page 41.)

[676] J. Christianidis and A. Megremi. Tracing the early history of algebra: Testimonies on Diophantus in the Greek-speaking world (4th-7th century CE). *Historia Math.*, 47:16–38, 2019. (Cited on page 41.)

[677] J. Christianidis and J. Oaks. Practicing algebra in late antiquity: The problem-solving of Diophantus of Alexandria. *Historia Math.*, 40(2):127–163, 2013. (Cited on page 41.)

[678] E. B. Christoffel. Sopra un problema proposto da Dirichlet. *Ann. Mat. Pura Appl. Serie II*, 4:1–9, 1870. (Cited on page 555.)

[679] D. G. Christopherson. Richard Vynne Southwell, 1888-1970. *Biographical Memoirs of Fellows of the Royal Society*, 18:549–565, 1972. (Cited on pages 169 and 562.)

[680] D. G. Christopherson and R. V. Southwell. Relaxation methods applied to engineering problems. III. Problems involving two independent variables. *Proc. R. Soc. Lond.*, 168:317–350, 1938. Series A, Mathematical and Physical Sciences. (Cited on page 171.)

[681] A. T. Chronopoulos. *A class of parallel iterative methods implemented on multiprocessors*. PhD thesis, University of Illinois at Urbana-Champaign, 1987. (Cited on page 215.)

[682] A. T. Chronopoulos. S-step iterative methods for (non)symmetric indefinite linear systems. *SIAM J. Numer. Anal.*, 28(6):1776–1789, 1991. (Cited on page 227.)

[683] A. T. Chronopoulos and C. W. Gear. On the efficient implementation of preconditioned s-step conjugate gradient methods on multiprocessors with memory hierarchy. *Parallel Comput.*, 11(1):37–53, 1989. (Cited on page 215.)

[684] A. T. Chronopoulos and C. W. Gear. S-step iterative methods for symmetric linear systems. *J. Comput. Appl. Math.*, 25(2):153–168, 1989. (Cited on page 215.)

[685] A. T. Chronopoulos and A. B. Kucherov. Block s-step Krylov iterative methods. *Numer. Linear Algebra Appl.*, 17(1):3–15, 2010. (Cited on page 243.)

[686] A. T. Chronopoulos and S. Ma. On squaring Krylov subspace iterative methods for nonsymmetric linear systems. Technical Report 89-67, Dept. Computer Science, University of Minnesota, 1989. (Cited on page 226.)

[687] A. T. Chronopoulos and C. D. Swanson. Parallel iterative s-step methods for unsymmetric linear systems. *Parallel Comput.*, 22(5):623–641, 1996. (Cited on page 227.)

[688] D. Chu, L. De Lathauwer, and B. De Moor. On the computation of the restricted singular value decomposition via the cosine-sine decomposition. *SIAM J. Matrix Anal. Appl.*, 22(2):580–601, 2000. (Cited on page 141.)

[689] D. Chu, L. De Lathauwer, and B. De Moor. A QR-type reduction for computing the SVD of a general matrix product/quotient. *Numer. Math.*, 95(1):101–121, 2003. (Cited on page 141.)

[690] D. Chu and B. De Moor. On a variational formulation of the QSVD and the RSVD. *Linear Algebra Appl.*, 311(1-3):61–78, 2000. (Cited on page 141.)

[691] D. Chu and B. De Moor. On the nonuniqueness of the factorization factors in the product singular value decomposition. *Linear Algebra Appl.*, 314(1-3):191–203, 2000. (Cited on page 141.)

[692] H. Chu. A method of Samuelson and minimal polynomials. *J. Math. Anal. Appl.*, 21:396–404, 1968. (Cited on page 292.)

[693] M. T. Chu, R. E. Funderlic, and G. H. Golub. On a variational formulation of the generalized singular value decomposition. *SIAM J. Matrix Anal. Appl.*, 18(4):1082–1092, 1997. (Cited on page 141.)

[694] J. Chung, J. G. Nagy, and D. P. O'Leary. A weighted-GCV method for Lanczos-hybrid regularization. *Electron. Trans. Numer. Anal.*, 28:149–167, 2008. (Cited on page 392.)

[695] Nicolas Chuquet. *Le Triparty en la Science des Nombres.* With a note by M. Aristide Marre, Imprimerie des Sciences Mathématiques et Physiques, Rome, 1881. (Cited on page 44.)

[696] G. Ciaramella and M. J. Gander. Happy 25th Anniversary DDM!... But how fast can the Schwarz method solve your logo? In R. Haynes, S. MacLachlan, X.-C. Cai, L. Halpern, H. H. Kim, A. Klawonn, and O. Widlund, editors, *Domain Decomposition Methods in Science and Engineering XXV*, pages 92–99. Springer, 2020. Lecture Notes in Computational Science and Engineering, Vol. 138. (Cited on page 267.)

[697] G. Ciaramella and M. J. Gander. *Iterative Methods and Preconditioners for Systems of Linear Equations.* SIAM, Philadelphia, 2022. (Cited on page 268.)

[698] P. Ciarlet Jr. Repeated Red-Black ordering: a new approach. *Numer. Algorithms*, 7:295–324, 1994. (Cited on page 257.)

[699] G. C. Cifoletti. *Mathematics and rhetoric: Jacques Peletier, Guillaume Gosselin and the making of the French algebraic tradition.* PhD thesis, Princeton University, 1993. (Cited on page 46.)

[700] G. C. Cifoletti. La question de l'algèbre. mathématiques et rhétorique des hommes de droit dans la France du XVIe siècle. *Annales. Histoire, Sciences Sociales*, 50(6):1385–1416, 1995. (Cited on page 46.)

[701] G. Cimmino. Calcolo approssimato per le soluzioni dei sistemi di equazioni lineari. *La Ricerca Scientifica, Series II*, 9:326–333, 1938. (Cited on pages 191 and 439.)

[702] G. Cimmino. Un metodo Monte Carlo per la risoluzione numerica dei sistemi di equazioni lineari. *Rendiconti dell'Accademia delle Scienze dell'Istituto di Bologna*, XII(2):39–44, 1967. (Cited on pages 192 and 439.)

[703] G. Cimmino. An unusual way of solving linear systems. *Rendiconti della Classe di Scienze Fisiche, Matematiche e Naturali dell'Accademia Nazionale dei Lincei, VIII*, 80:6–7, 1986. Reprinted in [707], pp. 633-634. (Cited on page 439.)

[704] G. Cimmino. La regola di Cramer svincolata dalla nozione di determinante. *Rendiconti dell'Accademia delle Scienze dell'Istituto di Bologna, XIII*, 10:115–122, 1986/1987. Reprinted in [707], pp. 635-642. (Cited on page 439.)

[705] G. Cimmino. Dante e la Matematica. *Atti della Accademia Pontaniana, Nuova Serie*, 36:5–16, 1987/88. (Cited on page 438.)

[706] G. Cimmino. On some identities involving spherical means. *Rendiconti della Classe di Scienze Fisiche, Matematiche e Naturali dell'Accademia Nazionale dei Lincei, VIII*, 83:69–73, 1989. Reprinted in [707], pp. 649-652. (Cited on page 439.)

[707] G. Cimmino. *Opere Scelte.* C. Sbordone and G. Trombetti, eds., Accademia di Scienze Fisiche e Matematiche della Società Nazionale di Scienze Lettere e Arti in Napoli, Giannini, Napoli, 2002. (Cited on pages 439 and 638.)

[708] S. Cinquini. Ricordo di Lamberto Cesari. *Rendiconti dell'Istituto Lombardo di Scienze e Lettere, Parte Generale e Atti Ufficiali*, 125:219–236, 1992. (Cited on pages 428 and 429.)

[709] M. Cipolla. Sulle matrici espressioni analitiche di un'altra. *Rendiconti del Circolo Matematico di Palermo*, 56(1):144–154, 1932. (Cited on page 388.)

[710] D. W. Clark. *Elements of Algebra.* Harper and Brothers, New York, 1843. (Cited on page 56.)

[711] A. R. Clarke. *Geodesy.* Clarendon Press, Oxford, 1880. (Cited on page 62.)

[712] B. I. Clasen. Sur une nouvelle méthode de résolution des équations linéaires et sur l'application de cette méthode au calcul des déterminants. *Ann. Soc. Sci. Bruxelles*, 12:251–281, 1888. also in *Mathesis*, t. IX, 1889, pp. 320-350. (Cited on page 67.)

[713] A. K. Cline, A. R. Conn, and C. F. Van Loan. Generalizing the LINPACK condition estimator. In J. P. Hennart, editor, *Numerical Analysis, Lectures Notes in Mathematics 909*, pages 73–83, Berlin, Heidelberg, 1982. Springer. (Cited on page 23.)

[714] A. K. Cline, C. B. Moler, G. W. Stewart, and J. H. Wilkinson. An estimate for the condition number of a matrix. *SIAM J. Numer. Anal.*, 16(2):368–375, 1979. (Cited on pages 23 and 368.)

[715] A. K. Cline and R. K. Rew. A set of counter-examples to three condition number estimators. *SIAM J. Sci. Statist. Comput.*, 4(4):602–611, 1983. (Cited on page 24.)

[716] M. Clint and A. Jennings. The evaluation of eigenvalues and eigenvectors of real symmetric matrices by simultaneous iteration method. *Comput. J.*, 13(1):76–80, 1970. (Cited on page 296.)

[717] M. Clint and A. Jennings. A simultaneous iteration method for the unsymmetric eigenvalue problem. *IMA J. Appl. Math.*, 8(1):111–121, 1971. (Cited on page 296.)

[718] R. Clint Whaley, A. Petitet, and J. Dongarra. Automated empirical optimization of software and the ATLAS project. *Parallel Comput.*, 27(1-2):3–35, 2001. (Cited on page 366.)

[719] A. I. Cohen. Rate of convergence of several conjugate gradient algorithms. *SIAM J. Numer. Anal.*, 9(2):248–259, 1972. (Cited on page 208.)

[720] H. T. Colebrooke. *Miscellaneous Essays, vol. II*. Trubner and Co., London, 1873. Dissertations on the algebra of the Hindus. (Cited on page 42.)

[721] E. Coleman. *Resilience for asynchronous iterative methods for sparse linear systems*. PhD thesis, Old Dominion University, 2019. (Cited on page 197.)

[722] E. Coleman, E. J. Jensen, and M. Sosonkina. Fault recovery methods for asynchronous linear solvers. *Int. J. Parallel Program.*, 49(1):51–80, 2021. (Cited on page 197.)

[723] A. R. Collar. The first fifty years of aeroelasticity. *Aerospace (Royal Aeronautical Society Journal)*, 5:12–20, 1978. (Cited on page 71.)

[724] L. Collatz. Einschließungssatz für die charakteristischen zahlen von matrizen. *Mathematische Zeitschrift*, 48(1):221–226, 1942. (Cited on page 286.)

[725] L. Collatz. Fehlerabschätzung für das iterationsverfahren zur auflösung linearer gleichungssysteme. *Zamm.-Z Angew. Math. Me.*, 22(6):357–361, 1942. (Cited on page 170.)

[726] C.-J. Collet and J. Itard. Un mathématicien humaniste Claude-Gaspar Bachet de Méziriac (1581-1638). *Revue d'Histoire des Sciences et de leurs Applications*, 1(1):26–50, 1947. (Cited on page 39.)

[727] T. P. Collignon, G. L. G. Sleijpen, and M. B. van Gijzen. Interpreting IDR(s) as a deflation process. Technical Report 10-21, Delft University of Technology, 2010. (Cited on page 244.)

[728] T. P. Collignon and M. B. van Gijzen. Solving large sparse linear systems efficiently on grid computers using an asynchronous iterative method as a preconditioner. In G. Kreiss, P. Lötstedt, A. Målqvist, and M. Neytcheva, editors, *Numerical Mathematics and Advanced Applications 2009*, pages 261–268, Berlin, Heidelberg, 2010. Springer. Proceedings of ENUMATH 2009, the 8th European Conference on Numerical Mathematics and Advanced Applications, Uppsala, July 2009. (Cited on page 197.)

[729] T. P. Collignon and M. B. van Gijzen. Minimizing synchronization in IDR(s). *Numer. Linear Algebra Appl.*, 18(5):805–825, 2011. (Cited on page 244.)

[730] L. J. Comrie. On the construction of tables by interpolation. *Monthly Notices of the Royal Astronomical Society*, 88:506–523, 1928. (Cited on page 333.)

[731] L. J. Comrie. The application of the Hollerith tabulating machine to Brown's tables of the moon. *Monthly Notices of the Royal Astronomical Society*, 92:694, 1932. (Cited on page 333.)

[732] L. J. Comrie. The application of commercial calculating machines to scientific computing. *M.T.A.C.*, 2:149–159, 1946. (Cited on page 333.)

[733] P. Concus and G. H. Golub. Use of fast direct methods for the efficient numerical solution of nonseparable elliptic equations. *SIAM J. Numer. Anal.*, 10(6):1103–1120, 1973. (Cited on pages 252 and 263.)

[734] P. Concus, G. H. Golub, and D. P. O'Leary. A generalized conjugate gradient method for the numerical solution of elliptic partial differential equations. In J. R. Bunch and D. J. Rose, editors, *Sparse Matrix Computations*, pages 309–332. Academic Press, 1976. (Cited on page 209.)

[735] P. Concus, G.H. Golub, and G. Meurant. Block preconditioning for the conjugate gradient method. *SIAM J. Sci. Statist. Comput.*, 6(1):220–252, 1985. (Cited on pages 94 and 258.)

[736] P. Concus and G. Meurant. On computing INV block preconditionings for the conjugate gradient method. *BIT*, 26:493–504, 1986. (Cited on page 258.)

[737] J. W. Cooley. How the FFT gained acceptance. In S. G. Nash, editor, *A History of Scientific Computing*, pages 133–140. ACM Press, 1990. (Cited on page 384.)

[738] J. W. Cooley and J. W. Tukey. An algorithm for the machine calculation of complex Fourier series. *Math. Comput.*, 19(90):297–301, 1965. (Cited on pages 384 and 516.)

[739] S. Cools. Analyzing and improving maximal attainable accuracy in the communication hiding pipelined BiCGStab method. *Parallel Comput.*, 86:16–35, 2019. PMAA'18 Special Issue. (Cited on page 246.)

[740] S. Cools, J. Cornelis, P. Ghysels, and W. Vanroose. Improving strong scaling of the conjugate gradient method for solving large linear systems using global reduction pipelining. arXiv preprint arXiv:1905.06850, 2019. EuroMPI'19: Proceedings of the 26th European MPI Users' Group Meeting. (Cited on page 216.)

[741] S. Cools, J. Cornelis, and W. Vanroose. Numerically stable recurrence relations for the communication hiding pipelined conjugate gradient method. *IEEE Trans. Parallel Distrib. Syst.*, 30(11):2507–2522, 2019. (Cited on pages 216 and 246.)

[742] S. Cools and W. Vanroose. The communication-hiding pipelined BiCGStab method for the efficient parallel solution of large unsymmetric linear systems. *Parallel Comput.*, 65:1–20, 2017. (Cited on page 246.)

[743] S. Cools, W. Vanroose, E. F. Yetkin, E. Agullo, and L. Giraud. On rounding error resilience, maximal attainable accuracy and parallel performance of the pipelined conjugate gradients method for large-scale linear systems in PETSc. In *Proceedings of the EASC'16 Exascale Applications and Software Conference 2016*, pages 1–10. ACM, 2016. (Cited on page 216.)

[744] S. Cools, E. F. Yetkin, E. Agullo, L. Giraud, and W. Vanroose. Analyzing the effect of local rounding error propagation on the maximal attainable accuracy of the pipelined conjugate gradient method. *SIAM J. Matrix Anal. Appl.*, 39(1):426–450, 2018. (Cited on page 216.)

[745] J. L. B. Cooper. Philip Stein. *Bull. London Math. Soc.*, 7(3):321–322, 1975. (Cited on pages 564 and 565.)

[746] J. Cornelis, S. Cools, and W. Vanroose. The communication-hiding conjugate gradient method with deep pipelines. arXiv preprint arXiv:1801.04728, 2018. (Cited on page 216.)

[747] P. Cossali. *Origine, Trasporto in Italia, Primi Progressi in Essa dell'Algebra: Storia Critica di Nuove Dispuisizioni Analytiche e Metafisiche Arricchita. Vol. 2.* Reale Tipogr. Parmense, 1799. (Cited on page 56.)

[748] P. Cossali. *Elogio di Luigi Lagrange*. Bettoni, Padova, 1813. (Cited on page 510.)

[749] E. Cosserat and F. Cosserat. Sur la théorie de l'élasticité. Premier mémoire. *Annales de la Faculté des Sciences de Toulouse, 1ère série*, 10(3-4):1–116, 1896. (Cited on page 142.)

[750] S. Costiner and S. Ta'asan. Adaptive multigrid techniques for large-scale eigenvalue problems: Solutions of the Schrödinger problem in two and three dimensions. *Phys. Rev. E*, 51(4):3704, 1995. (Cited on page 318.)

[751] R. W. Cottle. Manifestations of the Schur complement. *Linear Algebra Appl.*, 8(3):189–211, 1974. (Cited on pages 25 and 117.)

[752] R. W. Cottle. William Karush and the KKT theorem. Doc. Math., Extra volume: Optimization stories, 2012. (Cited on page 199.)

[753] R. Courant and D. Hilbert. *Methoden der Mathematischen Physik, Volume I (Second edition)*. Springer, Berlin, 1931. (Cited on pages 126, 136, and 285.)

[754] W. Cowell. *Portability of Numerical Software*. Springer, 1977. Workshop, Oak Brook, Illinois, June 21-23, 1976, LNCS 57. (Cited on page 364.)

[755] D. E. Crabtree and E. V. Haynsworth. An identity for the Schur complement of a matrix. *Proc. Am. Math. Soc.*, 22:364–366, 1969. (Cited on page 25.)

[756] E. J. Craig. *Iteration procedures for simultaneous equations*. PhD thesis, Massachusetts Institute of Technology, 1954. (Cited on pages 207 and 220.)

[757] E. J. Craig. The N-step iteration procedures. *J. Math. Phys.*, 34(1-4):64–73, 1955. (Cited on pages 207 and 220.)

[758] G. Cramer. *Introduction à l'Analyse des Lignes Courbes Algébriques*. Frères Cramer et CL. Philibert, Genéve, 1750. (Cited on pages 53 and 104.)

[759] S. H. Crandall. Iterative procedures related to relaxation methods for eigenvalue problems. *Proc. R. Soc. Lond.*, 207:416–423, 1951. (Cited on page 295.)

[760] S. H. Crandall. *Engineering Analysis: A Survey of Numerical Procedures*. McGraw-Hill, New York, 1956. (Cited on page 208.)

[761] P. Craven and G. Wahba. Smoothing noisy data with spline functions. *Numer. Math.*, 31(4):377–403, 1979. (Cited on page 392.)

[762] A. L. Crelle. *Rechentafeln, Welche Alles Multipliciren und Dividiren mit Zahlen unter Tausend Ganz Ersparen, bei Grösseren Zahlen aber die Rechnungerleichtern und Sicherer Machen*. Maurerschen Buchhandlung, Berlin, 1820. (Cited on page 63.)

[763] L. Cremona. Eugenio Beltrami. *Rendiconti del Circolo Matematico di Palermo*, 14:275–289, 1900. (Cited on page 417.)

[764] T. Crilly. Cayley's anticipation of a generalised Cayley-Hamilton theorem. *Historia Math.*, 5:211–219, 1978. (Cited on pages 422 and 423.)

[765] T. Crilly. *The mathematics of Arthur Cayley with particular reference to linear algebra*. PhD thesis, Middlesex Polytechnic, 1981. Available from Middlesex University's Research Repository. (Cited on page 427.)

[766] T. Crilly. The rise of Cayley's invariant theory (1841-1862). *Historia Math.*, 13:241–254, 1986. (Cited on page 427.)

[767] T. Crilly. The decline of Cayley's invariant theory (1863-1895). *Historia Math.*, 15:332–347, 1988. (Cited on page 427.)

[768] T. Crilly. A gemstone in matrix algebra. *Math. Gazette*, 76(475):182–188, 1992. (Cited on page 427.)

[769] T. Crilly. A Victorian mathematician: Arthur Cayley (1821-1895). *Math. Gazette*, 79(485):259–262, 1995. (Cited on page 427.)

[770] T. Crilly. A mathematician extraordinary-James Joseph Sylvester (1814-1897). *Math. Gazette*, 81(490):7–11, 1997. (Cited on page 571.)

[771] T. Crilly. The young Arthur Cayley. *Notes and Records Roy. Soc. London*, 52(2):267–282, 1998. (Cited on page 427.)

[772] T. Crilly. Arthur Cayley as Sadleirian professor: A glimpse of mathematics teaching at 19th-century Cambridge. *Historia Math.*, 26:125–160, 1999. (Cited on page 427.)

[773] T. Crilly. *Arthur Cayley: Mathematician Laureate of the Victorian Age*. Johns Hopkins University Press, 2006. (Cited on page 427.)

[774] H. Cross. Analysis of continuous frames by distributing fixed-end moments. *Trans. Am. Soc. Civ. Eng.*, 96(1):1–10, 1932. (Cited on page 166.)

[775] P. D. Crout. A short method for evaluating determinants and solving systems of linear equations with real or complex coefficients. *Trans. Amer. Inst. Elect. Engrs*, 60(12):1235–1241, 1941. (Cited on pages 71 and 446.)

[776] M. Crouzeix. Bounds for analytical functions of matrices. *Integ. Equat. Oper. Th.*, 48(4):461–477, 2004. (Cited on page 240.)

[777] M. Crouzeix. Numerical range and functional calculus in Hilbert space. *J. Funct. Anal.*, 244(2):668–690, 2007. (Cited on page 241.)

[778] M. Crouzeix and C. Palencia. The numerical range is a $(1+\sqrt{2})$-spectral set. *SIAM J. Matrix Anal. Appl.*, 38(2):649–655, 2017. (Cited on page 241.)

[779] M. Crouzeix, B. Philippe, and M. Sadkane. The Davidson method. *SIAM J. Sci. Comput.*, 15(1):62–76, 1994. (Cited on page 317.)

[780] H. Crowder and P. Wolfe. Linear convergence of the conjugate gradient method. *IBM J. Res. Dev.*, 16(4):431–433, 1972. (Cited on page 210.)

[781] M. J. Crowe. *A History of Vector Analysis: The Evolution of the Idea of a Vectorial System.* University of Notre Dame Press, 1967. (Cited on pages 3 and 122.)

[782] F. E. Croxton and D. J. Cowden. *Applied General Statistics.* Prentice-Hall, New York, 1939. (Cited on page 66.)

[783] R. Cuevas, C. E. Schaerer, and A. Bhaya. A proportional-derivative control strategy for varying the restart parameter in GMRES(m). In *Anais do XXXIII CNMAC*, volume 3, pages 1000–1001, 2010. (Cited on page 232.)

[784] C. E. Cullis. *Matrices and Determinoids, volume I.* Cambridge University Press, 1913. (Cited on pages 11 and 13.)

[785] C. E. Cullis. *Matrices and Determinoids, volume II.* Cambridge University Press, 1918. (Cited on page 11.)

[786] J. Cullum. Peaks, plateaus, numerical instabilities in a Galerkin minimal residual pair of methods for solving $Ax = b$. *Appl. Numer. Math.*, 19(3):255–278, 1995. (Cited on page 227.)

[787] J. Cullum. Iterative methods for solving $Ax = b$, GMRES/FOM versus QMR/BiCG. *Adv. Comput. Math.*, 6:1–24, 1996. (Cited on pages 227 and 234.)

[788] J. Cullum and W. E. Donath. A block Lanczos algorithm for computing the q algebraically largest eigenvalues and a corresponding eigenspace of large, sparse, real symmetric matrices. In *1974 IEEE Conference on Decision and Control including the 13th Symposium on Adaptive Processes*, pages 505–509. IEEE, 1974. (Cited on page 303.)

[789] J. Cullum and A. Greenbaum. Relations between Galerkin and norm-minimizing iterative methods for solving linear systems. *SIAM J. Matrix Anal. Appl.*, 17(2):223–247, 1996. (Cited on page 227.)

[790] J. K. Cullum and R. A. Willoughby. *Lanczos Algorithms for Large Symmetric Eigenvalue Computations, Volume 1: Theory.* Birkhauser, 1985. (Cited on pages 302 and 372.)

[791] J. K. Cullum and R. A. Willoughby. *Lanczos Algorithms for Large Symmetric Eigenvalue Computations, Volume 2: Programs.* Birkhauser, 1985. (Cited on pages 302 and 372.)

[792] J. J. M. Cuppen. A divide and conquer method for the symmetric tridiagonal eigenproblem. *Numer. Math.*, 36(2):177–195, 1981. (Cited on page 315.)

[793] A. R. Curtis and J. K. Reid. The solution of large sparse unsymmetric systems of linear equations. *IMA J. Appl. Math.*, 8(3):344–353, 1971. (Cited on pages 86 and 92.)

[794] A. R. Curtis and J. K. Reid. On the automatic scaling of matrices for Gaussian elimination. *IMA J. Appl. Math.*, 10(l):118–124, 1972. (Cited on page 80.)

[795] J. H. Curtiss. A generalization of the method of conjugate gradients for solving systems of linear algebraic equations. *M.T.A.C.*, 8(48):189–193, 1954. (Cited on page 207.)

[796] E. H. Cuthill. Several strategies for reducing the bandwidth of matrices. In D. J. Rose and R. A. Willoughby, editors, *Sparse Matrices and their Applications*, pages 157–166, New York, 1972. Plenum Press. (Cited on page 88.)

[797] E. H. Cuthill and J. McKee. Reducing the bandwidth of sparse symmetric matrices. In *24th ACM National Conference, ACM Publ P-69,*, pages 157–172. ACM, 1969. (Cited on pages 88 and 370.)

[798] G. Cybenko. Round-off error propagation in Durbin's, Levinson's and Trench's algorithms. Proc. Int. Conf. on Acoust., Speech and Sig. Processing, Washington, DC, 1979. (Cited on page 386.)

[799] G. Cybenko. The numerical stability of the Levinson-Durbin algorithm for Toeplitz systems of equations. *SIAM J. Sci. Statist. Comput.*, 1(3):303–319, 1980. (Cited on page 386.)

[800] H. Ltaief, D. Sukkari, and D. Keyes. A high performance QDWH-SVD solver using hardware accelerators. *ACM Trans. Math. Soft. (TOMS)*, 43(1):1–25, 2016. (Cited on page 139.)

[801] A. Dahan Dalmenico. L'essor des mathématiques appliquées aux Etats-Unis : l'impact de la seconde guerre mondiale. *Rev. Hist. Math.*, 2:149–213, 1996. (Cited on page 337.)

[802] G. Dahlquist, S. C. Eisenstat, and G. H. Golub. Bounds for the error of linear systems of equations using the theory of moments. *J. Math. Anal. Appl.*, 37:151–166, 1972. (Cited on pages 212 and 244.)

[803] G. Dahlquist, G. H. Golub, and S. G. Nash. Bounds for the error in linear systems. In R. Hettich, editor, *Proceedings of the Workshop on Semi-infinite Programming*, pages 154–172, Berlin, 1978. Springer. (Cited on pages 212 and 244.)

[804] J. W. Daniel. The conjugate gradient method for linear and nonlinear operator equations. *SIAM J. Numer. Anal.*, 4(1):10–26, 1967. (Cited on page 208.)

[805] J. W. Daniel. Convergence of the conjugate gradient method with computationally convenient modifications. *Numer. Math.*, 10(2):125–131, 1967. (Cited on page 208.)

[806] J. W. Daniel. A correction concerning the convergence rate for the conjugate gradient method. *SIAM J. Numer. Anal.*, 7(2):277–280, 1970. (Cited on page 208.)

[807] J. W. Daniel, W. B. Gragg, L. Kaufman, and G. W. Stewart. Reorthogonalization and stable algorithms for updating the Gram-Schmidt QR factorization. *Math. Comput.*, 30(136):772–795, 1976. (Cited on page 130.)

[808] G. C. Danielson and C. Lanczos. Some improvements in practical Fourier analysis and their application to X-ray scattering from liquids. *J. Franklin Inst.*, 233(5):435–452, 1942. (Cited on page 384.)

[809] A. M. Danilevskii. On the numerical solution of the secular equation. *Mat. Sb.*, 2:169–172, 1937. (Cited on page 291.)

[810] G. B. Dantzig. *Linear Programming and Extensions*. Princeton University Press, 1965. (Cited on pages 74, 84, and 85.)

[811] G. B. Dantzig and W. Orchard-Hays. The product form for the inverse in the simplex method. *M.T.A.C.*, 8:64–67, 1954. (Cited on page 92.)

[812] G. B. Dantzig, A. Orden, and P. Wolfe. The generalized simplex method. Technical Report Rand P-392-1, Rand Corporation, 1953. (Cited on page 84.)

[813] J. Dauben. Suan Shu Shu (A book on numbers and computations): English translation with commentary. *Archive for History of Exact Sciences*, 6(2):91–178, 2008. (Cited on page 35.)

[814] E. R. Davidson. The iterative calculation of a few of the lowest eigenvalues and corresponding eigenvectors of large real-symmetric matrices. *J. Comp. Phys.*, 17:87–94, 1975. (Cited on page 316.)

[815] C. Davies. *Elements of Algebra, translated from the French of M. Bourdon*. Wiley and Long, New York, 1835. (Cited on page 56.)

[816] C. Davis. The rotation of eigenvectors by a perturbation. *J. Math. Anal. Appl.*, 6(2):159–173, 1963. (Cited on page 144.)

[817] C. Davis. The rotation of eigenvectors by a perturbation-II. *J. Math. Anal. Appl.*, 11:20–27, 1965. (Cited on page 144.)

[818] C. Davis and W. M. Kahan. Some new bounds on perturbation of subspaces. *Bull. Amer. Math. Soc.*, 75(4):863–868, 1969. (Cited on pages 144 and 145.)

[819] C. Davis and W. M. Kahan. The rotation of eigenvectors by a perturbation. III. *SIAM J. Numer. Anal.*, 7(1):1–46, 1970. (Cited on pages 144 and 285.)

[820] P. Davis and P. Rabinowitz. A multiple purpose orthonormalizing code and its uses. *J. ACM*, 1:183–191, 1954. (Cited on page 128.)

[821] P. Davis and P. Rabinowitz. Advances in orthonormalizing computation. In Franz L. Alt, editor, *Advances in Computers*, volume 2, pages 55–133. Elsevier, 1961. (Cited on page 128.)

[822] P. J. Davis. *The Education of a Mathematician*. A K Peters, Natick, Massachusetts, 2000. (Cited on page 483.)

[823] T. Davis. Forty two problems of first degree from Diophantus' Arithmetica. Master's thesis, Wichita State University, Wichita, KS, 2010. (Cited on pages 40 and 41.)

[824] T. A. Davis. Algorithm 832: UMFPACK v4.3-an unsymmetric-pattern multifrontal method. *ACM Trans. Math. Soft. (TOMS)*, 30(2):196–199, 2004. (Cited on pages 91 and 372.)

[825] T. A. Davis. A column pre-ordering strategy for the unsymmetric-pattern multifrontal method. *ACM Trans. Math. Soft. (TOMS)*, 30(2):165–195, 2004. (Cited on page 372.)

[826] T. A. Davis. *Direct Methods for Sparse Linear Systems*. SIAM, 2006. (Cited on pages 94 and 372.)

[827] T. A. Davis. Algorithm 915, SuiteSparseQR: Multifrontal multithreaded rank-revealing sparse QR factorization. *ACM Trans. Math. Soft. (TOMS)*, 38(1):8:1–8:22, 2011. (Cited on pages 133 and 372.)

[828] T. A. Davis and I. S. Duff. An unsymmetric-pattern multifrontal method for sparse LU factorization. *SIAM J. Matrix Anal. Appl.*, 18(1):140–158, 1997. (Cited on pages 91 and 371.)

[829] T. A. Davis and I. S. Duff. A combined unifrontal/multifrontal method for unsymmetric sparse matrices. *ACM Trans. Math. Soft. (TOMS)*, 25(1):1–20, 1999. (Cited on pages 91 and 372.)

[830] T. A. Davis and Y. Hu. The University of Florida sparse matrix collection. *ACM Trans. Math. Soft. (TOMS)*, 38(1):1–25, 2011. (Cited on pages 94 and 373.)

[831] T. A. Davis and E. Palamadai Natarajan. Algorithm 907: KLU, a direct sparse solver for circuit simulation problems. *ACM Trans. Math. Soft. (TOMS)*, 37(3):1–17, 2010. (Cited on pages 371 and 372.)

[832] T. A. Davis, S. Rajamanickam, and W. M. Sid-Lakhdar. A survey of direct methods for sparse linear systems. *Acta Numerica*, 25:383–566, 2016. (Cited on pages 91, 93, 94, and 370.)

[833] E. Davison and F. Man. The numerical solution of $A'Q + QA = -C$. *IEEE Trans. Automat. Contr.*, 13(4):448–449, 1968. (Cited on page 398.)

[834] E. F. D'Azevedo, V. Eijkhout, and C. H. Romine. Conjugate gradient algorithms with reduced synchronization overhead on distributed memory multiprocessors. Technical Report Lapack Working Notes 56, University of Tennessee, 1999. (Cited on page 215.)

[835] E. F. D'Azevedo, P. A. Forsyth, and W.-P. Tang. Ordering methods for preconditioned conjugate gradient methods applied to unstructured grid problems. *SIAM J. Matrix Anal. Appl.*, 13(3):944–961, 1992. (Cited on page 260.)

[836] E. F. D'Azevedo, P. A. Forsyth, and W.-P. Tang. Towards a cost-effective ILU preconditioner with high level fill. *BIT Numerical Mathematics*, 32(3):442–463, 1992. (Cited on page 260.)

[837] E. F. D'Azevedo and C. H. Romine. Reducing communication costs in the conjugate gradient algorithm on distributed memory multiprocessors. Technical Report ORNL/TM-12192, Oak Ridge National Laboratory, 1992. (Cited on page 215.)

[838] E. de Beaumont. Éloge historique d'Adrien-Marie Legendre. *Mémoires de l'Académie des Sciences*, 32:XXXVII–LXXXVII, 1864. (Cited on pages 526 and 527.)

[839] H. T. de Beer. The history of the Algol effort. Master's thesis, Technische Universiteit Eindhoven, The Netherlands, 2006. (Cited on page 360.)

[840] C. de Boor and G. H. Golub. The numerically stable reconstruction of a Jacobi matrix from spectral data. *Linear Algebra Appl.*, 21(3):245–260, 1978. (Cited on page 316.)

[841] C. de Boor and J. R. Rice. Extremal polynomials with application to Richardson iteration for indefinite linear systems. *SIAM J. Sci. Statist. Comput.*, 3(1):47–57, 1982. (Cited on pages 182 and 262.)

[842] F. de Dinechin, L. Forget, J. M. Muller, and Y. Uguen. Posits: The good, the bad and the ugly. In *Proceedings of the Conference for Next Generation Arithmetic 2019*, pages 1–10. ACM, 2019. (Cited on page 355.)

[843] P. J. de Jonge. A comparative study of algorithms for reducing the fill-in during Cholesky factorization. *Bull. Géodésique*, 66:296–305, 1992. (Cited on page 88.)

[844] O. De la Cruz Cabrera, M. Matar, and L. Reichel. Analysis of directed networks via the matrix exponential. *J. Comput. Appl. Math.*, 355:182–192, 2019. (Cited on page 391.)

[845] B. De Moor. Generalizations of the singular value and QR decompositions. *Signal Process.*, 25(2):135–146, 1991. (Cited on page 141.)

[846] B. De Moor. On the structure of generalized singular value and QR decompositions. *SIAM J. Matrix Anal. Appl.*, 15(1):347–358, 1994. (Cited on page 141.)

[847] B. De Moor and G. H. Golub. The restricted singular value decomposition: Properties and applications. *SIAM J. Matrix Anal. Appl.*, 12(3):401–425, 1991. (Cited on page 141.)

[848] B. De Moor and P. Van Dooren. Generalizations of the singular value and QR decompositions. *SIAM J. Matrix Anal. Appl.*, 13(4):993–1014, 1992. (Cited on page 141.)

[849] B. De Moor and H. Zha. A tree of generalizations of the ordinary singular value decomposition. *Linear Algebra Appl.*, 147:469–500, 1991. (Cited on page 141.)

[850] C. de Polignac. *Méthodes optimales de produits de matrices.* PhD thesis, Faculté des Sciences de Grenoble, France, 1970. (Cited on page 80.)

[851] J. A. De Séguier. Sur la théorie des matrices. *Bull. S.M.F.*, 36:20–40, 1908. (Cited on page 149.)

[852] E. de Sturler. Nested Krylov methods based on GCR. *J. Comput. Appl. Math.*, 67:15–41, 1996. (Cited on page 230.)

[853] E. de Sturler. Truncation strategies for optimal Krylov subspace methods. *SIAM J. Numer. Anal.*, 36(3):864–889, 1999. (Cited on page 231.)

[854] E. de Sturler and D. R. Fokkema. Nested Krylov methods and preserving the orthogonality. In D. Melson, T. Manteuffel, and S. F. Mc Cormick, editors, *Sixth Copper Mountain Conference on Multigrid Methods*, pages 111–126. NASA conference publication 3324, 1993. (Cited on page 230.)

[855] E. de Sturler and J. Liesen. Block-diagonal and constraint preconditioners for nonsymmetric indefinite linear systems. Part I: Theory. *SIAM J. Sci. Comput.*, 26(5):1598–1619, 2005. (Cited on page 265.)

[856] E. de Sturler and H. A. van der Vorst. Reducing the effect of global communication in GMRES(m) and CG on parallel distributed memory computers. *Appl. Numer. Math.*, 18(4):441–459, 1995. (Cited on page 229.)

[857] T. J. Dekker. Algol 60 procedures in numerical algebra. Technical Report NR 8, Mathematisch Centrum, Amsterdam, 1969. (Cited on page 364.)

[858] T. J. Dekker. Algol 60 procedures in numerical algebra, part 1. Technical Report MC Tracts 22, Mathematisch Centrum, Amsterdam, 1975. Third edition. (Cited on page 364.)

[859] T. J. Dekker and W. Hoffmann. Algol 60 procedures in numerical algebra part 2. Technical Report MC Tracts 23, Mathematisch Centrum, Amsterdam, 1968. (Cited on page 364.)

[860] J.-B. Delambre. Notice sur la vie et les ouvrages de M. Le Comte J.-L. Lagrange. in *Ouvres de Lagrange*, J.-A. Serret ed., Gauthier-Villars, Paris, Tome I, 1867. (Cited on page 510.)

[861] J. Della-Dora. Numerical linear algorithms and group theory. *Linear Algebra Appl.*, 10(3):267–283, 1975. (Cited on page 313.)

[862] J. Della-Dora and F. Robert. Noël Gastinel and computational mathematics. *Linear Algebra Appl.*, 78:237–247, 1986. (Cited on pages 465 and 466.)

[863] M. A. DeLong and J. M. Ortega. SOR as a parallel preconditioner. In L. Adams and J. L. Nazareth, editors, *Linear and Nonlinear Conjugate Gradient-related Methods*, pages 143–148. SIAM, Philadelphia, 1996. (Cited on page 259.)

[864] M. A. DeLong and J. M. Ortega. SOR as a preconditioner II. *Appl. Numer. Math.*, 26(4):465–482, 1998. (Cited on page 259.)

[865] J. W. Demmel. The condition number of equivalence transformations that block diagonalize matrix pencils. In B. Kågström and A. Ruhe, editors, *Matrix Pencils*, pages 2–16. Lecture Notes in Mathematics 973. Springer, 1982. (Cited on page 252.)

[866] J. W. Demmel. The condition number of equivalence transformations that block diagonalize matrix pencils. *SIAM J. Numer. Anal.*, 20(3):599–610, 1983. (Cited on page 252.)

[867] J. W. Demmel. On condition numbers and the distance to the nearest ill-posed problem. *Numer. Math.*, 51(3):251–289, 1987. (Cited on page 22.)

[868] J. W. Demmel. On floating point errors in Cholesky. Technical Report Lapack Working Note # 14, University of Tennessee, 1989. (Cited on page 80.)

[869] J. W. Demmel. *Applied Numerical Linear Algebra*. SIAM, Philadelphia, 1997. (Cited on page 234.)

[870] J. W. Demmel, J. J. Dongarra, J. Du Croz, A. Greenbaum, S. Hammarling, and D. Sorensen. Prospectus for the development of a linear algebra library for high-performance computers. Technical Report 97, Argonne National Laboratory, 1987. (Cited on page 369.)

[871] J. W. Demmel, S. C. Eisenstat, J. R. Gilbert, X. S. Li, and J. W. Liu. A supernodal approach to sparse partial pivoting. *SIAM J. Matrix Anal. Appl.*, 20(3):720–755, 1999. (Cited on pages 92 and 371.)

[872] J. W. Demmel, J. R. Gilbert, and X. S. Li. An asynchronous parallel supernodal algorithm for sparse Gaussian elimination. *SIAM J. Matrix Anal. Appl.*, 20(4):915–952, 1999. (Cited on pages 92 and 378.)

[873] J. W. Demmel, L. Grigori, M. Hoemmen, and J. Langou. Communication-optimal parallel and sequential QR and LU factorizations. *SIAM J. Sci. Comput.*, 34(1):A206–A239, 2012. (Cited on page 133.)

[874] J. W. Demmel, M. Gu, S. Eisenstat, I. Slapničar, K. Veselić, and Z. Drmač. Computing the singular value decomposition with high relative accuracy. *Linear Algebra Appl.*, 299(1-3):21–80, 1999. (Cited on page 137.)

[875] J. W. Demmel and N. J. Higham. Stability of block algorithms with fast level-3 BLAS. *ACM Trans. Math. Soft. (TOMS)*, 18(3):274–291, 1992. (Cited on page 94.)

[876] J. W. Demmel, N. J. Higham, and R. S. Schreiber. Stability of block LU factorization. *Numer. Linear Algebra Appl.*, 2(2):173–190, 1995. (Cited on page 94.)

[877] J. W. Demmel and W. Kahan. Accurate singular values of bidiagonal matrices. *SIAM J. Sci. Statist. Comput.*, 11(5):873–912, 1990. (Cited on pages 137 and 316.)

[878] J. W. Demmel and K. Veselić. Jacobi's method is more accurate than QR. *SIAM J. Matrix Anal. Appl.*, 13(4):1204–1245, 1992. (Cited on pages 281 and 313.)

[879] A. P. Dempster, M. Schatzoff, and N. Wermuth. A simulation study of alternatives to ordinary least squares. *J. Am. Stat. Assoc.*, 72(357):77–91, 1977. (Cited on page 394.)

[880] J. E. Dendy. Black box multigrid. *J. Comp. Phys.*, 48:366–386, 1982. (Cited on page 269.)

[881] E. D. Denman and Alex N. Beavers Jr. The matrix sign function and computations in systems. *Appl. Math. Comput.*, 2(1):63–94, 1976. (Cited on page 389.)

[882] J. Desplanques. Théorème d'algèbre. *Journal de Mathématiques Spéciales*, 9:12–13, 1887. (Cited on page 286.)

[883] M. Dessagnes. *La circulation des tablettes cunéiformes mathématiques et la constitution des collections de la fin du XIXe siècle à la première moitié du XXe siècle*. PhD thesis, Université Paris Sorbonne Cité, 2017. (Cited on page 34.)

[884] J.-L. Destouches. *La Mécanique Ondulatoire*. Presses Universitaires de France, Paris, 1971. Collection "Que sais-je", vol. 311. (Cited on page 26.)

[885] I. S. Dhillon. *A new $O(n^2)$ algorithm for the symmetric tridiagonal eigenvalue/eigenvector problem*. PhD thesis, EECS Department, University of California, Berkeley, 1997. (Cited on page 316.)

[886] I. S. Dhillon and B. N. Parlett. Relatively robust representations of symmetric tridiagonals. *Linear Algebra Appl.*, 309:121–151, 2000. (Cited on pages 311 and 316.)

[887] I. S. Dhillon and B. N. Parlett. Orthogonal eigenvectors and relative gaps. *SIAM J. Matrix Anal. Appl.*, 25(3):858–899, 2004. (Cited on page 316.)

[888] I. S. Dhillon and B. N. Parlett. Multiple representations to compute orthogonal eigenvectors of symmetric tridiagonal matrices. *Linear Algebra Appl.*, 387:1–28, 2004. (Cited on page 316.)

[889] I. S. Dhillon, B. N. Parlett, and C. Vömel. Glued matrices and the MRRR algorithm. *SIAM J. Sci. Comput.*, 27(2):496–510, 2005. (Cited on page 316.)

[890] I. S. Dhillon, B. N. Parlett, and C. Vömel. The design and implementation of the MRRR algorithm. *ACM Trans. Math. Soft. (TOMS)*, 32(4):533–560, 2006. (Cited on page 316.)

[891] E. Di Napoli, E. Polizzi, and Y. Saad. Efficient estimation of eigenvalue counts in an interval. *Numer. Linear Algebra Appl.*, 23(4):674–692, 2016. (Cited on page 318.)

[892] L. E. Dickson. Canonical form of a linear homogeneous substitution in a Galois field. *Amer. J. Math.*, 22(2):121–137, 1900. (Cited on page 147.)

[893] L. E. Dickson. Theory of linear groups in an arbitrary field. *Trans. Amer. Math. Soc.*, 2(4):363–394, 1901. (Cited on page 147.)

[894] L. E. Dickson. Canonical form of a linear homogeneous transformation in an arbitrary realm of rationality. *Amer. J. Math.*, 24(2):101–108, 1902. (Cited on page 147.)

[895] L. E. Dickson. *Modern Algebraic Theories*. B. H. Sanborn and Co., 1926. (Cited on pages 11 and 149.)

[896] J. Dieudonné. La vie et l'ouvre de Laguerre. In C. Brezinski, A. Draux, A. P. Magnus, P. Maroni, and A. Ronveaux, editors, *Polynômes Orthogonaux et Applications*, pages 1–15, Heidelberg, 1985. Springer-Verlag. Lecture Notes in Mathematics, vol. 1171. (Cited on page 514.)

[897] J. Dieudonné. Camille Jordan. In *Complete Dictionary of Scientific Biography*, New York, 2008. Charles Scribner's Sons. (Cited on page 496.)

[898] E. W. Dijkstra and C. S. Scholten. Termination detection for diffusing computations. *Inf. Process. Lett.*, 11(1):1–4, 1980. (Cited on page 197.)

[899] Q. V. Dinh, R. Glowinski, and J. Périaux. Application of domain decomposition technique to the numerical solution of the Navier-Stokes equations. *Numer. Meth. Engng.*, 1:383–404, 1980. (Cited on page 267.)

[900] U. Dini. Eugenio Beltrami. *Annali di matematica pura ed applicata*, 4(1):151–152, 1900. (Cited on page 417.)

[901] P. A. M. Dirac. On the theory of quantum mechanics. *Proc. R. Soc. Lond.*, 112(762):661–677, 1926. (Cited on page 288.)

[902] M. d'Ocagne. *Le Calcul Simplifié par les Procédés Mécaniques et Graphique*. Gauthier-Villars, Paris, 1905. (Cited on page 325.)

[903] C. L. Dodgson. Condensation of determinants, being a new and brief method for computing their arithmetical values. *Proc. R. Soc. Lond.*, 15:150–155, 1867. (Cited on page 114.)

[904] C. L. Dodgson. *An Elementary Treatise on Determinants: With their Application to Simultaneous Linear Equations and Algebraical Geometry*. Macmillan and Company, London, 1867. (Cited on pages 115 and 116.)

[905] D. S. Dodson, R. G. Grimes, and J. G. Lewis. Sparse extensions to the Fortran Basic Linear Algebra Subprograms. *ACM Trans. Math. Soft. (TOMS)*, 17(2):253–263, 1991. (Cited on page 366.)

[906] S. Doi. On parallelism and convergence of incomplete LU factorizations. *Appl. Numer. Math.*, 7(5):417–436, 1991. (Cited on page 260.)

[907] S. Doi and A. Hoshi. Large-numbered multicolor MILU preconditioning on SX-3/14. *Intl. J. Comput. Math.*, 44(1-4):143–152, 1992. (Cited on page 260.)

[908] S. Doi and A. Lichnewsky. Some parallel and vector implementations of preconditioned iterative methods on Cray-2. *Intl. J. High Speed Computing*, 2(02):143–179, 1990. (Cited on page 260.)

[909] S. Doi and A. Lichnewsky. A graph-theory approach for analyzing the effects of ordering on ILU preconditionning. Technical Report RR-1452, INRIA, Rocquencourt, France, 1991. (Cited on page 260.)

[910] S. Doi and T. Washio. Ordering strategies and related techniques to overcome the trade-off between parallelism and convergence in incomplete factorizations. *Parallel Comput.*, 25(13-14):1995–2014, 1999. (Cited on page 260.)

[911] V. Dolean, P. Jolivet, and F. Nataf. *An Introduction to Domain Decomposition Methods: Algorithms, Theory, and Parallel Implementation*. SIAM, Philadelphia, 2015. (Cited on page 268.)

[912] J. J. Dongarra. LINPACK working note 3, Fortran BLAS timing. Technical Report ANL-80-24. CM-P00068636, Argonne National Laboratory, 1980. (Cited on page 365.)

[913] J. J. Dongarra, J. R. Bunch, C. B. Moler, and G. W. Stewart. Preliminary Linpack user's guide. Technical Report ANL-AMD-TM-313, Argonne National Laboratory, 1977. Linpack working note # 9. (Cited on page 368.)

[914] J. J. Dongarra, J. R. Bunch, C. B. Moler, and G. W. Stewart. *LINPACK User's Guide*. SIAM, Philadelphia, 1979. (Cited on pages 23 and 368.)

[915] J. J. Dongarra, J. Du Croz, S. Hammarling, and I. S. Duff. Algorithm 679: A set of level 3 basic linear algebra subprograms: model implementation and test programs. *ACM Trans. Math. Soft. (TOMS)*, 16(1):18–28, 1990. (Cited on page 365.)

[916] J. J. Dongarra, J. Du Croz, S. Hammarling, and R. J. Hanson. Algorithm 656: An extended set of FORTRAN basic linear algebra subprograms. *ACM Trans. Math. Soft. (TOMS)*, 14(1):18–32, 1988. (Cited on page 365.)

[917] J. J. Dongarra, J. Du Croz, S. Hammarling, and R. J. Hanson. An extended set of FORTRAN basic linear algebra subprograms. *ACM Trans. Math. Soft. (TOMS)*, 14(1):1–17, 1988. (Cited on page 365.)

[918] J. J. Dongarra, I. S. Duff, M. Gates, A. Haidar, S. Hammarling, N. J. Higham, J. Hogg, P. Valero Lara, P. Luszczek, M. Zounon, S. D. Relton, S. Tomov, T. Costa, and S. Knepper. Batched BLAS (Basic Linear Algebra Subprograms) 2018 specification. Technical Report ICL-UTK-1170-2018, Innovative Computing Laboratory, University of Tennessee, 2018. (Cited on page 376.)

[919] J. J. Dongarra, G. H. Golub, E. Grosse, C. B. Moler, and K. Moore. Netlib and NA-Net: Building a scientific computing community. *IEEE Ann. Hist. Comput.*, 30(2):30–41, 2008. (Cited on page 379.)

[920] J. J. Dongarra and E. Grosse. Distribution of mathematical software via electronic mail. *ACM SIGNUM Newsletter*, 20(3):45–47, 1985. (Cited on page 379.)

[921] J. J. Dongarra and E. Grosse. Distribution of mathematical software via electronic mail. *Comm. ACM*, 30(5):403–407, 1987. (Cited on page 379.)

[922] J. J. Dongarra, F. G. Gustavson, and A. Karp. Implementing linear algebra algorithms for dense matrices on a vector pipeline machine. *SIAM Rev.*, 26(1):91–112, 1984. (Cited on page 84.)

[923] J. J. Dongarra, S. Hammarling, and A. Petitet. Case studies on the development of ScaLAPACK and the NAG numerical PVM library. In R. F. Boisvert, editor, *Quality of Numerical Software*, pages 236–248, Boston, MA, 1997. Springer. (Cited on page 376.)

[924] J. J. Dongarra, M. A. Heroux, and P. Luszczek. High-performance conjugate-gradient benchmark: A new metric for ranking high-performance computing systems. *The International Journal of High Performance Computing Applications*, 30(1):3–10, 2016. (Cited on pages 352 and 369.)

[925] J. J. Dongarra and A. R Hinds. Unrolling loops in Fortran. *Software- Practice and Experience*, 9(3):219–226, 1979. (Cited on page 365.)

[926] J. J. Dongarra, P. Luszczek, and A. Petitet. The LINPACK benchmark: past, present and future. *Concurr. Comp.-Pract. E.*, 15(9):803–820, 2003. (Cited on pages 352 and 369.)

[927] J. J. Dongarra, H. W. Meuer, and E. Strohmaier. TOP500 supercomputer sites. *Supercomputer*, 13: 89–111, 1997. (Cited on page 340.)

[928] J. J. Dongarra and C. B. Moler. Implementation guide for Linpack. Technical Report ANL-80-105, Argonne National Laboratory, 1980. Linpack working note # 13. (Cited on page 368.)

[929] J. J. Dongarra and C. B. Moler. Eispack: A package for solving matrix eigenvalue problems. In W. R. Cowell, editor, *Sources and Development of Mathematical Software*, pages 68–87, Englewood Cliffs, NJ, 1984. Prentice-Hall. (Cited on page 367.)

[930] J. J. Dongarra and D. C. Sorensen. A fully parallel algorithm for the symmetric eigenvalue problem. *SIAM J. Sci. Statist. Comput.*, 8(2):S139–S154, 1987. (Cited on page 315.)

[931] J. J. Dongarra and G. W. Stewart. Linpack - A package for solving linear systems. Technical Report ANL-82-30, Argonne National Laboratory, 1982. Linpack working note 15. (Cited on page 368.)

[932] J. J. Dongarra and D. Walker. MPI: A standard message passing interface. *Supercomputer*, 12(1): 56–68, 1996. (Cited on page 376.)

[933] J. J. Dongarra and J. Waśniewski. Lapack95 - high performance linear algebra package. *Math. Model. Anal.*, 5(1):44–54, 2000. (Cited on page 369.)

[934] J. D. P. Donnelly. Periodic chaotic relaxation. *Linear Algebra Appl.*, 4(2):117–128, 1971. (Cited on page 196.)

[935] M. H. Doolittle. Explanations and illustrations of method employed in the office in solution of normal equations and the adjustment of a triangulation. In *Report of the Superintendant of the U.S. Coast and Geodetic Survey showing the progress of the work during fiscal year ending June 1878*, pages 115–120. Government Publishing Office, Washington, 1881. (Cited on pages 62 and 448.)

[936] J. L. Dorier. Analyse historique de l'émergence des concepts élémentaires d'algèbre linéaire. hal-02138152, 1990. (Cited on page 102.)

[937] J. L. Dorier. Emergence du concept de rang dans l'étude des systèmes d'équations linéaires. *Cahiers du Séminaire d'Histoire des Mathématiques 2éme série*, 3:159–190, 1993. (Cited on page 116.)

[938] J. L. Dorier. A general outline of the genesis of vector space theory. *Historia Math.*, 22(3):227–261, 1995. (Cited on page 3.)

[939] F. W. Dorr. The direct solution of the discrete Poisson equation on a rectangle. *SIAM Rev.*, 12(2):248–263, 1970. (Cited on page 382.)

[940] Z. Dostál. Conjugate gradient method with preconditioning by projector. *Int. J. Comput. Math.*, 23(3-4):315–323, 1988. (Cited on page 217.)

[941] Z. Dostál. Projector preconditioning and domain decomposition methods. *Appl. Math. Comput.*, 37(2):75–81, 1990. (Cited on page 217.)

[942] C. C. Douglas. Multi-grid algorithms with applications to elliptic boundary-value problems. *SIAM J. Numer. Anal.*, 21(2):236–254, 1984. (Cited on page 270.)

[943] C. C. Douglas. Multigrid methods in science and engineering. *IEEE Comput. Sci. Eng.*, 3(4):55–68, 1996. (Cited on page 270.)

[944] C. C. Douglas. A review of numerous parallel multigrid methods. In G. Astfalk, editor, *Applications on Advanced Architecture Computers*, pages 187–202. SIAM, Philadelphia, 1996. (Cited on page 270.)

[945] C. C. Douglas and J. Douglas, Jr. A unified convergence theory for abstract multigrid or multilevel algorithms, serial and parallel. *SIAM J. Numer. Anal.*, 30(1):136–158, 1993. (Cited on page 270.)

[946] J. Douglas and H. H. Rachford. On the numerical solution of heat conduction problems in two and three space variables. *Trans. Amer. Math. Soc.*, 82(2):421–439, 1956. (Cited on page 179.)

[947] J. Douglas, Jr. On the numerical integration of $\frac{\partial^2 u}{\partial x^2} + \frac{\partial^2 u}{\partial y^2} = \frac{\partial u}{\partial t}$ by implicit methods. *J. Soc. Ind. Appl. Math.*, 3(1):42–65, 1955. (Cited on page 179.)

[948] J. Douglas Jr. Round-off error in the numerical solution of the heat equation. *J. ACM*, 6:48–58, 1959. (Cited on page 77.)

[949] G. Scorza Dragoni. Ricordo di Carlo Miranda. *Rend. Accad. Naz. Sci. XL Mem. Mat.*, (5) 8((1)): 275–277, 1984. (Cited on page 438.)

[950] G. Scorza Dragoni. Gianfranco Cimmino (12 marzo 1908 - 30 maggio 1989). *Atti della Accademia Nazionale dei Lincei, Classe di Scienze Fisiche, Matematiche e Naturali. Rendiconti Lincei. Supplemento, Serie IX*, 2(1):59–68, 1991. (Cited on pages 438, 439, and 440.)

[951] A. Draux. *Polynômes Orthogonaux Formels. Applications*, volume LNM 974. Springer-Verlag, 1983. (Cited on page 228.)

[952] P. Drineas and M. W. Mahoney. Lectures on randomized numerical linear algebra. In M. W. Mahoney, J. C. Duchi, and A. C. Gilbert, editors, *The Mathematics of Data 25*. American Mathematical Society/SIAM, 2018. (Cited on page 140.)

[953] J. Drkošová, A. Greenbaum, M. Rozložník, and Z. Strakoš. Numerical stability of GMRES. *BIT*, 35(3):309–330, 1995. (Cited on page 233.)

[954] Z. Drmač. Implementation of Jacobi rotations for accurate singular value computation in floating point arithmetic. *SIAM J. Sci. Comput.*, 18(4):1200–1222, 1997. (Cited on page 138.)

[955] Z. Drmač. Algorithm 977: A QR-preconditioned QR SVD method for computing the SVD with high accuracy. *ACM Trans. Math. Soft. (TOMS)*, 44(1):1–30, 2017. (Cited on page 138.)

[956] Z. Drmač and K. Veselić. New fast and accurate Jacobi SVD algorithm: I. *SIAM J. Matrix Anal. Appl.*, 29(4):1322–1342, 2008. (Cited on page 138.)

[957] Z. Drmač and K. Veselić. New fast and accurate Jacobi SVD algorithm: II. *SIAM J. Matrix Anal. Appl.*, 29(4):1343–1362, 2008. (Cited on page 138.)

[958] V. Druskin, A. Greenbaum, and L. Knizhnerman. Using nonorthogonal Lanczos vectors in the computation of matrix functions. *SIAM J. Sci. Comput.*, 19(1):38–54, 1998. (Cited on page 389.)

[959] V. Druskin and L. Knizhnerman. Two polynomial methods of calculating functions of symmetric matrices. *U.S.S.R. Comput. Math. Math. Phys.*, 29(6):112–121, 1989. (Cited on page 389.)

[960] V. Druskin and L. Knizhnerman. Krylov space approximations of eigenpairs and matrix functions in exact and computer arithmetic. *Numer. Linear Algebra Appl.*, 2(3):205–217, 1995. (Cited on page 389.)

[961] V. Druskin and L. Knizhnerman. Extended Krylov subspaces: Approximation of the matrix square root and related functions. *SIAM J. Matrix Anal. Appl.*, 19(3):755–771, 1998. (Cited on page 389.)

[962] V. Druskin and L. Knizhnerman. Error estimates for the simple Lanczos process when computing functions of symmetric matrices and eigenvalues. *J. Comput. Math. and Math. Phys.*, 31(7):970–983, 991. (Cited on page 389.)

[963] V. Druskin, L. Knizhnerman, and V. Simoncini. Analysis of the rational Krylov subspace and ADI methods for solving the Lyapunov equation. *SIAM J. Numer. Anal.*, 49(5):1875–1898, 2011. (Cited on page 398.)

[964] M. Dryja. A capacitance matrix method for Dirichlet problem on polygon region. *Numer. Math.*, 39(1):51–64, 1982. (Cited on page 266.)

[965] M. Dryja, W. Proskurowski, and O. Widlund. Method of domain decomposition with cross points for elliptic finite element problems. In *Proceedings of the International Symposium on Optimal Algorithms, Blagovegrad, Bulgaria*, pages 97–111. Bulgarian Academy of Sciences, Sofia, 1986. (Cited on page 267.)

[966] M. Dryja and O. Widlund. Some domain decomposition algorithms for elliptic problems. In L. Hayes and D. Kincaid, editors, *Iterative Methods for Large Linear Systems*, pages 273–291, San Diego, 1989. Academic Press. (Cited on page 267.)

[967] M. Dryja and O. Widlund. Towards a unified theory of domain decomposition algorithms for elliptic problems. In T. F. Chan, R. Glowinski, J. Périaux, and O. Widlund, editors, *Third International Symposium on Domain Decomposition Methods for Partial Differential Equations*, pages 3–21. SIAM, 1990. (Cited on page 267.)

[968] E. D'Sylva and G. A. Miles. The S.S.O.R. iteration scheme for equations with σ_1 ordering. *Comput. J.*, 6(4):366–367, 1964. (Cited on page 178.)

[969] K. Du, J. Duintjer Tebbens, and G. Meurant. Any admissible harmonic Ritz value set is possible for GMRES. *Electron. Trans. Numer. Anal.*, 47:36–56, 2017. (Cited on pages 244 and 305.)

[970] L. Du, T. Sogabe, B. Yu, Y. Yamamoto, and S. L. Zhang. A block IDR(s) method for nonsymmetric linear systems with multiple right-hand sides. *J. Comput. Appl. Math.*, 235(14):4095–4106, 2011. (Cited on page 244.)

[971] L. Du, T. Sogabe, and S-L. Zhang. Quasi-minimal residual smoothing technique for the IDR(s) method. *JSIAM Letters*, 3:13–16, 2011. (Cited on page 244.)

[972] P. F. Dubois, A. Greenbaum, and G. H. Rodrigue. Approximating the inverse of a matrix for use in iterative algorithms on vector processors. *Computing*, 22(3):257–268, 1979. (Cited on page 261.)

[973] P. F. Dubois and G. H. Rodrigue. Analysis of the recursive doubling algorithm. Technical Report UCRL-79071, Lawrence Livermore Laboratory, Livermore, CA, 1976. (Cited on page 255.)

[974] A. A. Dubrulle. An optimum iteration for the matrix polar decomposition. *Electron. Trans. Numer. Anal.*, 8:21–25, 1999. (Cited on page 143.)

[975] I. S. Duff. On permutations to block triangular form. *IMA J. Appl. Math.*, 19(3):339–342, 1977. (Cited on page 94.)

[976] I. S. Duff. A survey of sparse matrix research. *Proc. IEEE*, 65(4):500–535, 1977. (Cited on page 94.)

[977] I. S. Duff. MA32-A package for solving sparse unsymmetric systems using the frontal method. Technical Report AERE-R-10079, AERE Harwell, Oxfordshire, 1981. (Cited on page 89.)

[978] I. S. Duff. *Sparse Matrices and their Uses*. Academic Press, 1981. Proceedings of the IMA Numerical Analysis Group Conference held at the University of Reading, 9th-11th July, 1980. (Cited on page 86.)

[979] I. S. Duff. A survey of sparse matrix software. Technical Report AERE-R-10512, AERE Harwell, Oxfordshire, 1982. (Cited on pages 94 and 370.)

[980] I. S. Duff. Design features of a frontal code for solving sparse unsymmetric linear systems out-of-core. *SIAM J. Sci. Statist. Comput.*, 5(2):270–280, 1984. (Cited on page 89.)

[981] I. S. Duff. A survey of sparse matrix software. In W. R. Cowell, editor, *Sources and Development of Mathematical Software*, pages 165–199. Prentice-Hall, Englewood Cliffs, NJ, 1984. (Cited on page 94.)

[982] I. S. Duff. Data structures, algorithms and software for sparse matrices. In D J. Evans, editor, *Sparsity and its Applications*, pages 1–29. Cambridge University Press, Cambridge, UK, 1985. (Cited on page 92.)

[983] I. S. Duff. Parallel implementation of multifrontal schemes. *Parallel Comput.*, 3:193–204, 1986. (Cited on page 91.)

[984] I. S. Duff. Direct solvers. *Computer Physics Reports*, 11:1–20, 1989. (Cited on page 94.)

[985] I. S. Duff. Multiprocessing a sparse matrix code on the Alliant FX/8. *J. Comput. Appl. Math.*, 27:229–239, 1989. (Cited on page 91.)

[986] I. S. Duff. Matrix methods. Technical Report RAL-TR-1998-076, Rutherford Appleton Laboratory, Oxfordshire, 1998. (Cited on page 94.)

[987] I. S. Duff, A. M. Erisman, C. W. Gear, and J. K. Reid. Some remarks on inverses of sparse matrices. Technical Report ANL/MCS-TM-51, Argonne National Laboratory, 1985. (Cited on page 85.)

[988] I. S. Duff, A. M. Erisman, C. W. Gear, and J. K. Reid. Sparsity structure and Gaussian elimination. *ACM Signum newsletter*, 23(2):2–8, 1988. (Cited on page 85.)

[989] I. S. Duff, A. M. Erisman, and J. K. Reid. On George's nested dissection method. *SIAM J. Numer. Anal.*, 13(5):686–695, 1976. (Cited on page 90.)

[990] I. S. Duff, A. M. Erisman, and J. K. Reid. *Direct Methods for Sparse Matrices*. Oxford University Press, 2nd edition, 2017. First edition 1987. (Cited on pages 92 and 94.)

[991] I. S. Duff, N. I. M. Gould, J. K. Reid, J. A. Scott, and K. Turner. The factorization of sparse symmetric indefinite matrices. *IMA J. Numer. Anal.*, 11:181–204, 1991. (Cited on page 78.)

[992] I. S. Duff, R. G. Grimes, and J. G. Lewis. Sparse matrix test problems. *ACM Trans. Math. Soft. (TOMS)*, 15(1):1–14, 1989. (Cited on page 373.)

[993] I. S. Duff, R. G. Grimes, and J. G. Lewis. The Rutherford-Boeing sparse matrix collection. Technical Report RAL-TR-97-031, Rutherford Appleton Laboratory, 1997. (Cited on page 373.)

[994] I. S. Duff, R. Guivarch, D. Ruiz, and M. Zenadi. The augmented block Cimmino distributed method. *SIAM J. Sci. Comput.*, 37(3):A1248–A1269, 2015. (Cited on page 192.)

[995] I. S. Duff, M. A. Heroux, and R. Pozo. An overview of the sparse basic linear algebra subprograms: The new standard from the BLAS technical forum. *ACM Trans. Math. Soft. (TOMS)*, 28(2):239–267, 2002. (Cited on page 366.)

[996] I. S. Duff, M. Marrone, and G. Radicati. A proposal for user level sparse BLAS. Technical Report RAL-92-087, Rutherford Appleton Laboratory, 1992. (Cited on page 366.)

[997] I. S. Duff, M. Marrone, G. Radicati, and C. Vittoli. Level 3 basic linear algebra subprograms for sparse matrices: a user-level interface. *ACM Trans. Math. Soft. (TOMS)*, 23(3):379–401, 1997. (Cited on page 366.)

[998] I. S. Duff and G. Meurant. The effect of ordering on preconditioned conjugate gradients. *BIT*, 29:635–657, 1989. (Cited on pages 164 and 260.)

[999] I. S. Duff and J. K. Reid. A comparison of sparsity orderings for obtaining a pivotal sequence in Gaussian elimination. *J. Inst. Maths Applics*, 14:281–291, 1974. (Cited on pages 87 and 92.)

[1000] I. S. Duff and J. K. Reid. Algorithm 529: Permutations to block triangular form. *ACM Trans. Math. Soft. (TOMS)*, 4(2):189–192, 1978. (Cited on page 94.)

[1001] I. S. Duff and J. K. Reid. An implementation of Tarjan's algorithm for the block triangularization of a matrix. *ACM Trans. Math. Soft. (TOMS)*, 4(2):137–147, 1978. (Cited on page 94.)

[1002] I. S. Duff and J. K. Reid. Some design features of a sparse matrix code. *ACM Trans. Math. Soft. (TOMS)*, 5(1):18–35, 1979. (Cited on page 374.)

[1003] I. S. Duff and J. K. Reid. The multifrontal solution of indefinite sparse symmetric linear equations. *ACM Trans. Math. Soft. (TOMS)*, 9(3):302–325, 1983. (Cited on pages 91, 228, and 374.)

[1004] I. S. Duff and J. K. Reid. The multifrontal solution of unsymmetric sets of linear equations. *SIAM J. Sci. Statist. Comput.*, 5(3):633–641, 1984. (Cited on pages 91 and 374.)

[1005] I. S. Duff and J. K. Reid. The design of MA48: a code for the direct solution of sparse unsymmetric linear systems of equations. *ACM Trans. Math. Soft. (TOMS)*, 22(2):187–226, 1996. (Cited on page 92.)

[1006] I. S. Duff and J. A. Scott. The design of a new frontal code for solving sparse unsymmetric systems. *ACM Trans. Math. Soft. (TOMS)*, 22(1):30–45, 1996. (Cited on page 89.)

[1007] I. S. Duff and G. W. Stewart. *Sparse Matrix Proceedings, 1978*. SIAM, 1979. (Cited on pages 86 and 364.)

[1008] I. S. Duff and B. Uçar. On the block triangular form of symmetric matrices. *SIAM Rev.*, 52(3):455–470, 2010. (Cited on page 94.)

[1009] I. S. Duff and B. Uçar. Combinatorial problems in solving linear systems. In U. Naumann and O. Schenk, editors, *Combinatorial Scientific Computing*, pages 21–68. Chapman and Hall, CRC Press, 2012. (Cited on page 94.)

[1010] I. S. Duff and H. A. van der Vorst. Developments and trends in the parallel solution of linear systems. *Parallel Comput.*, 25:1931–1970, 1999. (Cited on page 93.)

[1011] I. S. Duff and C. Vömel. Algorithm 818: A reference model implementation of the Sparse BLAS in Fortran 95. *ACM Trans. Math. Soft. (TOMS)*, 28(2):268–283, 2002. (Cited on page 366.)

[1012] I. S. Duff and C. Vömel. Incremental norm estimation for dense and sparse matrices. *BIT*, 42:300–322, 2002. (Cited on page 24.)

[1013] H. M. Dufour. Résolution des systèmes linéaires par la méthode des résidus conjugués. *Bulletin Géodésique*, 71:65–87, 1964. (Cited on page 216.)

[1014] J. Duintjer Tebbens and G. Meurant. Any Ritz value behavior is possible for Arnoldi and for GMRES. *SIAM J. Matrix Anal. Appl.*, 33(3):958–978, 2012. (Cited on page 244.)

[1015] J. Duintjer Tebbens and G. Meurant. Prescribing the behavior of early terminating GMRES and Arnoldi iterations. *Numer. Algorithms*, 65(1):69–90, 2014. (Cited on pages 244 and 305.)

[1016] J. Duintjer Tebbens and G. Meurant. On the convergence of Q-OR and Q-MR Krylov methods for solving linear systems. *BIT Numerical Mathematics*, 56(1):77–97, 2016. (Cited on page 245.)

[1017] J. Duintjer Tebbens and G. Meurant. On the residual norms, the Ritz values and the harmonic Ritz values that can be generated by restarted GMRES. *Numer. Algorithms*, 84(4):1329–1352, 2020. (Cited on page 245.)

[1018] J. Duintjer Tebbens, G. Meurant, H. Sadok, and Z. Strakoš. On investigating GMRES convergence using unitary matrices. *Linear Algebra Appl.*, 450:83–107, 2014. (Cited on pages 117, 238, and 245.)

[1019] J. Duintjer Tebbens and M. Tůma. Efficient preconditioning of sequences of nonsymmetric linear systems. *SIAM J. Sci. Comput.*, 29(5):1918–1941, 2007. (Cited on page 264.)

[1020] J. Duintjer Tebbens and M. Tůma. Preconditioner updates for solving sequences of linear systems in matrix-free environment. *Numer. Linear Algebra Appl.*, 17(6):997–1019, 2010. (Cited on page 264.)

[1021] J. Duintjer Tebbens and M. Tůma. On incremental condition estimators in the 2-norm. *SIAM J. Matrix Anal. Appl.*, 35(1):174–197, 2014. (Cited on page 24.)

[1022] A. L. Dulmage and N. S. Mendelsohn. Coverings of bipartite graphs. *Canadian J. Math.*, 10:517–534, 1958. (Cited on page 94.)

[1023] A. L. Dulmage and N. S. Mendelsohn. On the inversion of sparse matrices. *Math. Comput.*, 16(80):494–496, 1962. (Cited on page 94.)

[1024] A. L. Dulmage and N. S. Mendelsohn. Two algorithms for bipartite graphs. *J. Soc. Ind. Appl. Math.*, 11(1):183–194, 1963. (Cited on page 94.)

[1025] D. B. Duncan and J. F. Kenney. *On the Solution of Normal Equations and Related Topics*. Edwards Brothers, Ann Arbor, MI, 1946. (Cited on page 73.)

[1026] W. J. Duncan. Some devices for the solution of large sets of simultaneous linear equations. *Philos. Mag. Ser. 7*, 35:660–670, 1944. (Cited on page 83.)

[1027] W. J. Duncan and A. R. Collar. On a method for the solution of oscillation problems by matrices. *Philos. Mag.*, 17:865–909, 1934. (Cited on page 296.)

[1028] T. F. Dupont. A factorization procedure for the solution of elliptic difference equations. *SIAM J. Numer. Anal.*, 5(4):753–782, 1968. (Cited on page 254.)

[1029] T. F. Dupont, R. P. Kendall, and H.H. Rachford. An approximate factorization procedure for solving self-adjoint elliptic difference equations. *SIAM J. Numer. Anal.*, 5(3):559–573, 1968. (Cited on pages 253 and 254.)

[1030] E. Durand. *Solutions Numériques des Équations Algébriques, I et II.* Masson, Paris, 1960-1961. (Cited on page 94.)

[1031] J. Durbin. The fitting of time-series models. *Revue de l'Institut International de Statistique*, 28(3):233–244, 1960. (Cited on page 386.)

[1032] P. Duren. Changing faces: the mistaken portrait of Legendre. *Notices Amer. Math. Soc.*, 56(11):1440–1443, 2009. (Cited on page 527.)

[1033] P. Duren. Epilogue on Legendre portraits. *Notices Amer. Math. Soc.*, 57(6):703, 2010. (Cited on page 527.)

[1034] P. S. Dwyer. Combined expansions of products of symmetric power sums and of sums of symmetric power products with applications to sampling. *Ann. Math. Stat.*, 9:1–47, 1938. (Cited on page 450.)

[1035] P. S. Dwyer. The Doolittle technique. *Ann. Math. Stat.*, 12:449–458, 1941. (Cited on pages 66, 73, and 450.)

[1036] P. S. Dwyer. The evaluation of determinants. *Psychometrika*, 6:191–204, 1941. (Cited on page 450.)

[1037] P. S. Dwyer. The evaluation of linear forms. *Psychometrika*, 6:355–365, 1941. (Cited on page 450.)

[1038] P. S. Dwyer. The solution of simultaneous equations. *Psychometrika*, 6:101–109, 1941. (Cited on pages 72 and 450.)

[1039] P. S. Dwyer. A matrix presentation of least squares and correlation theory with matrix justification of improved methods of solution. *Ann. Math. Stat.*, 15:82–89, 1944. (Cited on pages 73 and 450.)

[1040] P. S. Dwyer. The square root method and its use in correlation and regression. *J. Amer. Statist. Assoc.*, 40:493–503, 1945. (Cited on page 73.)

[1041] P. S. Dwyer. *Linear Computations.* John Wiley, New York, 1951. (Cited on pages 73 and 450.)

[1042] J. Dyer. *Acceleration of the convergence of the Kaczmarz method and iterated homogeneous transformations.* PhD thesis, University of California, Los Angeles, 1965. (Cited on page 190.)

[1043] H. Dym and V. Katsnelson. Contributions of Issai Schur to analysis. In A. Joseph, A. Melnikov, and R. Rentschler, editors, *Studies in Memory of Issai Schur*, pages xci–clxxxiii. Birkhäuser, 2003. (Cited on page 554.)

[1044] P. J. Eberlein. Solution to the complex eigenproblem by a norm reducing Jacobi type method. *Numer. Math.*, 14:232–245, 1970. (Cited on pages 281 and 362.)

[1045] P. J. Eberlein. On the Schur decomposition of a matrix for parallel computation. *IEEE Trans. Comput.*, 2:167–174, 1987. (Cited on page 151.)

[1046] P. J. Eberlein and J. Boothroyd. Solution to the eigenproblem by a norm reducing Jacobi type method. *Numer. Math.*, 11:1–12, 1968. (Cited on pages 281 and 362.)

[1047] C. Eckart and G. Young. The approximation of one matrix by another of lower rank. *Psychometrika*, 1(3):211–218, 1936. (Cited on page 136.)

[1048] C. Eckart and G. Young. A principal axis transformation for non-Hermitian matrices. *Bull. Amer. Math. Soc.*, 45(2):118–121, 1939. (Cited on page 136.)

[1049] A. Edelman and H. Murakami. Polynomial roots from companion matrix eigenvalues. *Math. Comput.*, 64(210):763–776, 1995. (Cited on page 293.)

[1050] A. Eden and G. Irzik. German mathematicians in exile in Turkey: Richard von Mises, William Prager, Hilda Geiringer, and their impact on Turkish mathematics. *Historia Math.*, 39:432–459, 2012. (Cited on page 472.)

[1051] A. Edrei. Sur les déterminants récurrents et les singularités d'une fonction donnée par son développement de Taylor. *Compositio mathematica*, 7:20–88, 1940. (Cited on page 185.)

[1052] J. Egerváry. On rank-diminishing operations and their applications to the solution of linear equations. *Z. Angew. Math. Phys.*, 11:376–386, 1960. (Cited on page 81.)

[1053] R. Ehrig and P. Deuflhard. GMERR-an error minimizing variant of GMRES. Technical Report SC 97-63, Konrad-Zuse-Zentrum für Informationstechnik Berlin, 1997. (Cited on page 233.)

[1054] L. W. Ehrlich. The block symmetric successive overrelaxation method. *J. Soc. Ind. Appl. Math.*, 12(4):807–826, 1964. (Cited on page 177.)

[1055] M. Eiermann. On semiiterative methods generated by Faber polynomials. *Numer. Math.*, 56:139–156, 1989. (Cited on page 183.)

[1056] M. Eiermann. Fields of values and iterative methods. *Linear Algebra Appl.*, 180:167–197, 1993. (Cited on pages 236 and 287.)

[1057] M. Eiermann and O. G. Ernst. Geometric aspects of the theory of Krylov subspace methods. *Acta Numerica*, 10:251–312, 2001. (Cited on pages 236 and 239.)

[1058] M. Eiermann and O. G. Ernst. A restarted Krylov subspace method for the evaluation of matrix functions. *SIAM J. Numer. Anal.*, 44(6):2481–2504, 2006. (Cited on page 390.)

[1059] M. Eiermann, O. G. Ernst, and O. Schneider. Analysis of acceleration strategies for restarted minimal residual methods. *J. Comput. Appl. Math.*, 123(1):261–292, 2000. (Cited on pages 232, 234, and 239.)

[1060] M. Eiermann, X. Li, and R. S. Varga. On hybrid semi-iterative methods. *SIAM J. Numer. Anal.*, 26(1):152–168, 1989. (Cited on page 184.)

[1061] M. Eiermann and W. Niethammer. On the construction of semiiterative methods. *SIAM J. Numer. Anal.*, 20(6):1153–1160, 1983. (Cited on page 183.)

[1062] M. Eiermann, W. Niethammer, and R. S. Varga. A study of semiiterative methods for nonsymmetric systems of linear equations. *Numer. Math.*, 47:505–533, 1985. (Cited on page 183.)

[1063] V. Eijkhout. Analysis of parallel incomplete point factorizations. *Linear Algebra Appl.*, 154:723–740, 1991. (Cited on page 260.)

[1064] V. Eijkhout. Computational variants of the CGS and BiCGstab methods. Technical Report UT-CS-94-241, Dept. of Computer Science, University of Tennessee, 1994. (Cited on page 232.)

[1065] S. C. Eisenstat. Efficient implementation of a class of preconditioned conjugate gradient methods. *SIAM J. Sci. Statist. Comput.*, 2(1):1–4, 1981. (Cited on page 256.)

[1066] S. C. Eisenstat, H. C. Elman, and M. H. Schultz. Variational iterative methods for nonsymmetric systems of linear equations. *SIAM J. Numer. Anal.*, 20(2):345–357, 1983. (Cited on pages 221 and 223.)

[1067] S. C. Eisenstat, H. C. Elman, M. H. Schultz, and A. H. Sherman. The (new) Yale sparse matrix package. In G. Birkhoff and A. Schoenstadt, editors, *Elliptic Problem Solvers II*, pages 45–52. Academic Press, 1984. (Cited on page 371.)

[1068] S. C. Eisenstat, M. C. Gursky, M. H. Schultz, and A. H. Sherman. Yale sparse matrix package II: The non-symmetric codes. Technical Report TR 114, Department of Computer Science, Yale University, 1977. (Cited on page 371.)

[1069] S. C. Eisenstat, M. C. Gursky, M. H. Schultz, and A. H. Sherman. Yale sparse matrix package I: The symmetric codes. *Internat. J. Numer. Methods Engrg.*, 18(8):1145–1151, 1982. (Cited on pages 87 and 370.)

[1070] S. C. Eisenstat, J. W. Lewis, and M. H. Schultz. Optimal block diagonal scaling of block 2-cyclic matrices. *Linear Algebra Appl.*, 44:181–186, 1982. (Cited on page 252.)

[1071] S. C. Eisenstat and J. W. H. Liu. Structural representations of Schur complements in sparse matrices. In J. A. George, J. R. Gilbert, and J. W. H. Liu, editors, *Graph Theory and Sparse Matrix Computation*, volume 56 of *IMA Volumes in Applied Mathematics*, pages 85–100, Springer-Verlag, New York, 1993. (Cited on page 93.)

[1072] S. C. Eisenstat, M. H. Schultz, and A. H. Sherman. Applications of an element model for Gaussian elimination. In J. R. Bunch and D. J. Rose, editors, *Sparse Matrix Computations*, pages 85–96. Academic Press, 1976. (Cited on page 87.)

[1073] S. C. Eisenstat, M. H. Schultz, and A. H. Sherman. Software for sparse Gaussian elimination with limited core storage. In I. S. Duff and G. W. Stewart, editors, *Sparse Matrix Proceedings 1978*, pages 135–153. SIAM, Philadelphia, 1979. (Cited on page 371.)

[1074] S. C. Eisenstat, M. H. Schultz, and A. H. Sherman. Algorithms and data structures for sparse symmetric Gaussian elimination. *SIAM J. Sci. Statist. Comput.*, 2(2):225–237, 1981. (Cited on page 87.)

[1075] F. G. Eisenstein. Untersuchungen über die cubischen formen mit zwei variabeln. *Journal für die reine und angewandte Mathematik (Crelle Journal)*, 27:89–104, 1844. (Cited on page 4.)

[1076] F. G. Eisenstein. Allgemeine untersuchungen über die formen dritten grades mit drei variabeln, welche der kreistheilung ihre entstehung verdanken. *Journal für die reine und angewandte Mathematik (Crelle Journal)*, 28 and 29:289–374, 19–53, 1844-1845. (Cited on page 4.)

[1077] D. El Baz, P. Spiteri, J.-C. Miellou, and M. Jarraya. Mathematical study of perturbed asynchronous iterations designed for distributed termination. *Int. Math. J.*, 1(5):491–503, 2002. (Cited on page 196.)

[1078] M. N. El Tarazi. Some convergence results for asynchronous algorithms. *Numer. Math.*, 39:325–240, 1982. (Cited on page 196.)

[1079] L. Elbouyahyaoui, A. Messaoudi, and H. Sadok. Algebraic properties of the block GMRES and block Arnoldi methods. *Electron. Trans. Numer. Anal.*, 33:207–220, 2009. (Cited on page 243.)

[1080] L. Eldén and V. Simoncini. Solving ill-posed linear systems with GMRES and a singular preconditioner. *SIAM J. Matrix Anal. Appl.*, 33(4):1369–1394, 2012. (Cited on page 393.)

[1081] T. Elfving. Block-iterative methods for consistent and inconsistent linear equations. *Numer. Math.*, 35:1–12, 1980. (Cited on page 191.)

[1082] H. C. Elman. *Iterative methods for large sparse nonsymmetric systems of linear equations*. PhD thesis, Computer Science Dept., Yale University, 1982. Research Report 229. (Cited on pages 221, 223, and 224.)

[1083] H. C. Elman. A stability analysis of incomplete LU factorizations. *Math. Comput.*, 47(175):191–217, 1986. (Cited on page 257.)

[1084] H. C. Elman. Approximate Schur complement preconditioners on serial and parallel computers. *SIAM J. Sci. Comput.*, 10(3):581–605, 1989. (Cited on page 261.)

[1085] H. C. Elman, Y. Saad, and P. E. Saylor. A hybrid Chebyshev Krylov subspace algorithm for solving nonsymmetric systems of linear equations. *SIAM J. Sci. Statist. Comput.*, 7(3):840–855, 1986. (Cited on pages 182 and 262.)

[1086] H. C. Elman, D. J. Silvester, and A. J. Wathen. *Finite Elements and Fast Iterative Solvers: with Applications in Incompressible Fluid Dynamics*. Oxford University Press, 2014. (Cited on page 265.)

[1087] H. C. Elman and R. L. Streit. Polynomial iteration for nonsymmetric indefinite linear systems. In *Numerical Analysis*, pages 103–117, Springer, Berlin, Heidelberg, 1986. (Cited on pages 182 and 262.)

[1088] L. Elsner. A note on optimal block-scaling of matrices. *Numer. Math.*, 44:127–128, 1984. (Cited on page 252.)

[1089] M. Embree. How descriptive are GMRES convergence bounds? Technical Report 99/08, Mathematical Institute, University of Oxford, UK, 1999. (Cited on page 236.)

[1090] M. Embree. The tortoise and the hare restart GMRES. *SIAM Rev.*, 45(2):259–266, 2003. (Cited on page 232.)

[1091] M. Embree. The Arnoldi eigenvalue iteration with exact shifts can fail. *SIAM J. Matrix Anal. Appl.*, 31(1):1–10, 2009. (Cited on page 306.)

[1092] M. Embree, R. B. Morgan, and H. V. Nguyen. Weighted inner products for GMRES and GMRES-DR. *SIAM J. Sci. Comput.*, 39(5):S610–S632, 2017. (Cited on page 246.)

[1093] M. Embree, J. A. Sifuentes, K. M. Soodhalter, D. B. Szyld, and F. Xue. Short-term recurrence Krylov subspace methods for nearly Hermitian matrices. *SIAM J. Matrix Anal. Appl.*, 33(2):480–500, 2012. (Cited on page 242.)

[1094] H. W. Emmons. The numerical solution of partial differential equations. *Quart. Appl. Math.*, 2(3):173–195, 1944. (Cited on page 171.)

[1095] M. Engeli, T. Ginsburg, H. Rutishauser, and E. Stiefel. *Refined Iterative Methods for Computations of the Solution and the Eigenvalues of Self-adjoint Boundary Value Problems.* Birkhäuser Verlag, Basel/Stuttgart, 1959. (Cited on pages 207, 208, 301, and 568.)

[1096] H. W. Engl. Regularization methods for the stable solution of inverse problems. *Surv. Math. Ind.*, 3(4):71–143, 1993. (Cited on page 391.)

[1097] H. W. Engl, M. Hanke, and A. Neubauer. *Regularization of Inverse Problems.* Kluwer Academic, 1996. (Cited on page 391.)

[1098] J. Erhel. A parallel GMRES version for general sparse matrices. *Electron. Trans. Numer. Anal.*, 3(12):160–176, 1995. (Cited on pages 133 and 229.)

[1099] J. Erhel and F. Guyomarc'h. An augmented conjugate gradient method for solving consecutive symmetric positive definite linear systems. *SIAM J. Matrix Anal. Appl.*, 21(4):1279–1299, 2000. (Cited on page 217.)

[1100] T. Ericsson. Computing functions of matrices using Krylov subspace methods. Technical Report TR, Chalmers University, Sweden, 1990. (Cited on page 389.)

[1101] A. M. Erisman, R. G. Grimes, J. G. Lewis, W. G. Poole, Jr., and H. D. Simon. Evaluation of orderings for unsymmetric sparse matrices. *SIAM J. Sci. Statist. Comput.*, 8(4):600–624, 1987. (Cited on page 88.)

[1102] O. G. Ernst. Minimal and orthogonal residual methods and their generalizations for solving linear operator equations. *Habilitation thesis, TU Freiberg, Germany*, 2000. (Cited on page 236.)

[1103] O. G. Ernst. Residual-minimizing Krylov subspace methods for stabilized discretizations of convection-diffusion equations. *SIAM J. Matrix Anal. Appl.*, 21(4):1079–1101, 2000. (Cited on page 238.)

[1104] R. Escalante and M. Raydan. *Alternating Projection Methods.* SIAM, Philadelphia, 2011. (Cited on page 188.)

[1105] R. Estrin, D. Orban, and M. A. Saunders. Euclidean-norm error bounds for SYMMLQ and CG. *SIAM J. Matrix Anal. Appl.*, 40(1):235–253, 2019. (Cited on page 213.)

[1106] R. Estrin, D. Orban, and M. A. Saunders. LNLQ: An iterative method for least-norm problems with an error minimization property. *SIAM J. Matrix Anal. Appl.*, 40(3):1102–1124, 2019. (Cited on page 213.)

[1107] R. Estrin, D. Orban, and M. A. Saunders. LSLQ: An iterative method for linear least-squares with an error minimization property. *SIAM J. Matrix Anal. Appl.*, 40(1):254–275, 2019. (Cited on page 373.)

[1108] L. Euler. *Intoductio in Analysin Infinitorum, tomus secundus.* Marcus-Michaelem Bousquet, Lausanne, 1748. (Cited on page 123.)

[1109] L. Euler. Du mouvement de rotation des corps solides autour d'un axe variable. In *Mémoires de l'Académie des Sciences de Berlin*, pages 154–193. Académie des Sciences de Berlin, 1765. (Cited on page 279.)

[1110] L. Euler. Recherches sur la connoissance mécanique des corps. In *Mémoires de l'Académie des Sciences de Berlin*, pages 131–153. Académie des Sciences de Berlin, 1765. (Cited on page 279.)

[1111] L. Euler. Formulae generales pro translatione quacunque corporum rigidorum. *Novi Commentarii Acad. Petrop.*, 20:189–207, 1775. (Cited on page 123.)

[1112] D. J. Evans. The use of preconditioning in iterative methods for solving linear equations with symmetric positive definite matrices. *J. Inst. Maths Applics*, 4(3):295–314, 1968. (Cited on page 259.)

[1113] D. J. Evans. *Software for Numerical Mathematics*. Academic Press, 1974. Proceedings of the Loughborough University of Technology Conference of the Institute of Mathematics and its Applications held in April 1973. (Cited on page 364.)

[1114] D. J. Evans. *Preconditioning Methods: Theory and Applications*. Gordon and Breach Science Publishers, Inc., 1983. (Cited on page 259.)

[1115] D. J. Evans and M. Hatzopoulos. The solution of certain banded systems of linear equations using the folding algorithm. *Comput. J.*, 19(2):184–187, 1976. (Cited on page 152.)

[1116] D. J. Evans and M. Hatzopoulos. A parallel linear system solver. *Int. J. Comput. Math.*, 7(3):227–238, 1979. (Cited on page 152.)

[1117] J.-C. Evard and F. Uhlig. On the matrix equation $f(X) = A$. *Linear Algebra Appl.*, 162:447–519, 1992. (Cited on page 394.)

[1118] G. C. Everstine. A comparison of three resequencing algorithms for the reduction of matrix profile and wavefront. *Internat. J. Numer. Methods Engrg.*, 14(6):837–853, 1979. (Cited on page 88.)

[1119] V. Faber, A. Greenbaum, and D. E. Marshall. The polynomial numerical hulls of Jordan blocks and related matrices. *Linear Algebra Appl.*, 374:231–246, 2003. (Cited on page 240.)

[1120] V. Faber, W. Joubert, E. Knill, and T. Manteuffel. Minimal residual method stronger than polynomial preconditioning. *SIAM J. Matrix Anal. Appl.*, 17(4):707–729, 1996. (Cited on page 234.)

[1121] V. Faber, J. Liesen, and P. Tichý. The Faber-Manteuffel theorem for linear operators. *SIAM J. Numer. Anal.*, 46(3):1323–1337, 2008. (Cited on page 224.)

[1122] V. Faber, J. Liesen, and P. Tichý. Properties of worst-case GMRES. *SIAM J. Matrix Anal. Appl.*, 34(4):1500–1519, 2013. (Cited on page 240.)

[1123] V. Faber and T. Manteuffel. Necessary and sufficient conditions for the existence of a conjugate gradient method. *SIAM J. Numer. Anal.*, 21(2):352–362, 1984. (Cited on page 224.)

[1124] V. Faber and T. Manteuffel. Orthogonal error methods. *SIAM J. Numer. Anal.*, 24(1):170–187, 1987. (Cited on page 224.)

[1125] G. Faccarello. Du Conservatoire à l'Ecole Normale. *Les cahiers d'histoire du CNAM*, 2-3:17–57, 1993. (Cited on page 590.)

[1126] D. K. Faddeev. Properties of the inverse of a Hessenberg matrix. *J. Sov. Math.*, 24(1):118–120, 1984. in Numerical Methods and Computational Issues, volume 5, V. P. Ilin and V. N. Kublanoskaya eds., 1981 (in Russian). (Cited on pages 218 and 383.)

[1127] D. K. Faddeev and V. N. Faddeeva. *Computational Methods of Linear Algebra*. Freeman & Co, San Francisco and London, 1963. Original Moscow (1960). (Cited on pages 94 and 297.)

[1128] D. K. Faddeev and I. S. Sominsky. *Problems in Higher Algebra*. Mir, Moskow and Leningrad, 1949. (Cited on page 289.)

[1129] R. D. Falgout, J. E. Jones, and U. M. Yang. The design and implementation of hypre, a library of parallel high performance preconditioners. In A.M. Bruaset and A. Tveito, editors, *Numerical Solution of Partial Differential Equations on Parallel Computers*, pages 267–294. Springer, 2006. (Cited on page 378.)

[1130] R. D. Falgout and U. M. Yang. Hypre: A library of high performance preconditioners. In P. M. A. Sloot, C. J. K. Tan, J. J. Dongarra, and A. G. Hoekstra, editors, *Computational Science - ICCS 2002 Part III*, pages 632–641. Springer, 2002. Lecture Notes in Computer Science 2331. (Cited on page 378.)

[1131] K. Fan. Note on circular disks containing the eigenvalues of a matrix. *Duke Math. J.*, 25:441–445, 1958. (Cited on page 286.)

[1132] K. Fan. Topological proofs for certain theorems on matrices with non-negative elements. *Monatsh Math. Phys.*, 62(3):219–237, 1958. (Cited on page 177.)

[1133] K. Fan. Note on M-matrices. *The Quarterly Journal of Mathematics*, 11(1):43–49, 1960. (Cited on page 254.)

[1134] K. Fan and A. J. Hoffman. Lower bounds for the rank and location of the eigenvalues of a matrix. In O. Taussky, editor, *Contributions to the solution of systems of linear equations and the determination of eigenvalues*, pages 117–130. National Bureau of Standards, Washington, DC, 1954. Applied Mathematics Series, Vol. 39. (Cited on page 286.)

[1135] K. Fan and A. J. Hoffman. Some metric inequalities in the space of matrices. *Proc. Am. Math. Soc.*, 6(1):111–116, 1955. (Cited on page 142.)

[1136] H.-R. Fang and Y. Saad. A filtered Lanczos procedure for extreme and interior eigenvalue problems. *SIAM J. Sci. Comput.*, 34(4):A2220–A2246, 2012. (Cited on page 306.)

[1137] L. Fantappié. Le calcul des matrices. *C.R. Acad. Sci. Paris, Sér 1*, 186:619–621, 1928. (Cited on page 388.)

[1138] R. W. Farebrother. A memoir of the life of M.H. Doolittle. *Bull. Inst. Math. Appl.*, 23:102, 1987. (Cited on page 448.)

[1139] R. W. Farebrother. A. C. Aitken and the consolidation of matrix theory. *Linear Algebra Appl.*, 264:3–12, 1997. (Cited on page 405.)

[1140] R. W. Farebrother. Henry Agard Wallace and machine calculation. *Image*, 40:21–23, 2008. (Cited on page 449.)

[1141] R.W. Farebrother. *Fitting Linear Relationships: A History of the Calculus of Observations 1750-1900*. Springer-Verlag, New York, 1999. (Cited on page 25.)

[1142] C. Farhat, J. Mandel, and F. X. Roux. Optimal convergence properties of the FETI domain decomposition method. *Comput. Methods Appl. Mech. Engrg.*, 115(3-4):365–385, 1994. (Cited on page 268.)

[1143] C. Farhat and F. X. Roux. A method of finite element tearing and interconnecting and its parallel solution algorithm. *Internat. J. Numer. Methods Engrg.*, 32:1205–1227, 1991. (Cited on page 267.)

[1144] A. B. Farnell. Limits for the field of values of a matrix. *Am. Math. Mon.*, 52(9):488–493, 1945. (Cited on page 287.)

[1145] H. Fassbender, M. Griebel, O. Holtz, G. W. Stewart, and C. Zenger. Curriculum vitae of Friedrich Ludwig Bauer. *Linear Algebra Appl.*, 417:299–300, 2006. (Cited on page 414.)

[1146] D. Fearnley-Sander. Hermann Grassmann and the creation of linear algebra. *Am. Math. Mon.*, 86(10):809–817, 1979. (Cited on page 3.)

[1147] R. P. Fedorenko. A relaxation method for solving elliptic difference equations. *Zh. Vychisl. Mat. Mat. Fiz.*, 1:922–927, 1961. Also in *U.S.S.R. Comput. Math. and Math. Phys.*, 1 (1962), pp. 1092-1096. (Cited on pages 269 and 454.)

[1148] R. P. Fedorenko. The speed of convergence of one iterative process. *Zh. Vychisl. Mat. Mat. Fiz.*, 4:559–563, 1964. Also in *U.S.S.R. Comput. Math. and Math. Phys.*, 4 (1964), pp. 227-235. (Cited on pages 269 and 454.)

[1149] R. P. Fedorenko. Iterative methods for elliptic difference equations. *Russian Math. Surveys*, 28(2):129–195, 1973. (Cited on page 269.)

[1150] D. G. Feingold and R. S. Varga. Block diagonally dominant matrices and generalizations of the Gerschgorin circle theorem. *Pacific J. Math.*, 12(4):1241–1250, 1962. (Cited on pages 94 and 287.)

[1151] W. Feller and G. E. Forsythe. New matrix transformations for obtaining characteristic vectors. *Quart. Appl. Math.*, 8(4):325–331, 1951. (Cited on page 132.)

[1152] C. Fenu, D. Martin, L. Reichel, and G. Rodriguez. Block Gauss and anti-Gauss quadrature with application to networks. *SIAM J. Matrix Anal. Appl.*, 34(4):1655–1684, 2013. (Cited on pages 213 and 391.)

[1153] K. V. Fernando and B. N. Parlett. Accurate singular values and differential qd algorithms. *Numer. Math.*, 67:191–229, 1994. (Cited on pages 311 and 316.)

[1154] K. V. Fernando, B. N. Parlett, and I. S. Dhillon. A way to find the most redundant equation in a tridiagonal system. Technical Report PAM-635, Center for Pure and Applied Mathematics, University of California, Berkeley, 1995. (Cited on page 316.)

[1155] L. S. Feuer. America's first Jewish professor: James Joseph Sylvester at the University of Virginia. *Amer. Jewish Arch. J.*, 36(2):152–201, 1984. (Cited on page 571.)

[1156] C. M. Fiduccia and R. M. Mattheyses. A linear-time heuristic for improving network partitions. In *19th Design Automation Conference*, pages 175–181. IEEE, 1982. (Cited on page 90.)

[1157] M. Fiedler. Some estimates of spectra of matrices. In *Symposium on the Numerical Treatment of Ordinary Differential Equations, Integral and Integro-Differential Equations, Rome, 1960*, pages 33–36. Birkhauser Verlag, 1961. (Cited on page 287.)

[1158] M. Fiedler. Algebraic connectivity of graphs. *Czechoslovak Math. J.*, 23(98):298–305, 1973. (Cited on page 90.)

[1159] M. Fiedler. Geometry of the nnumerical range of matrices. *Linear Algebra Appl.*, 37:81–96, 1981. (Cited on page 287.)

[1160] M. Fiedler. *Special Matrices and their Applications in Numerical Mathematics*. Martinus Nijhoff Publishers, Dordrecht, The Netherlands, 1st edition, 1986. reprinted by Dover (2008). (Cited on page 80.)

[1161] M. Fiedler. Laplacian of graphs and algebraic connectivity. In *Combinatorics and Graph Theory (Warsaw, 1987)*, pages 57–70. Banach Center Publications 25(1), 1989. (Cited on page 90.)

[1162] M. Fiedler. Vlastimil Pták (8 November 1925-9 May 1999). *Linear Algebra Appl.*, 310(1-3):21–22, 2000. (Cited on pages 16 and 231.)

[1163] M. Fiedler and V. Muller. Professor Vlastimil Pták died. *Czechoslovak Math. J.*, 50(2):445–447, 2000. (Cited on pages 16 and 231.)

[1164] M. Fiedler and V. Pták. Generalized norms of matrices and the location of the spectrum. *Czechoslovak Math. J.*, 12(4):558–571, 1962. (Cited on page 16.)

[1165] M. Fiedler and V. Pták. On matrices with non-positive off-diagonal elements and positive principal minors. *Czechoslovak Math. J.*, 12(3):382–400, 1962. (Cited on pages 16, 80, and 177.)

[1166] G. A. Fierheller. *Do Not Fold, Spindle or Mutilate: The "Hole" Story of Punched Cards*. Stewart Publishing and Printing, 2014. (Cited on page 333.)

[1167] P. J. E. Finck. *Eléments d'Algèbre*. Derivaux, Strasbourg, 1846. (Cited on page 101.)

[1168] J. Finger. Über die gegenseitigen beziehungen gewisser in der mechanik mit vortheil anwendbaren flächen zweiter ordnung nebst anwendungen auf probleme der astatik. *Wien Sitzgsber. 2a*, 101:1105–1142, 1892. (Cited on page 142.)

[1169] J. W. Fischbach. Some applications of gradient method. Technical Report 828, Ballistic Research Laboratories, Aberdeen Proving Grounds, Maryland, 1954. (Cited on page 207.)

[1170] B. Fischer. *Polynomial Based Iteration Methods for Symmetric Linear Systems*. Wiley Teubner, 1996. (Cited on pages 217 and 262.)

[1171] B. Fischer and G. H. Golub. On the error computation for polynomial based iteration methods. In A. Greenbaum and M. Luskin, editors, *Recent Advances in Iterative Methods*, pages 59–67, Springer, New York, 1994. (Cited on pages 212 and 244.)

[1172] B. Fischer and L. Reichel. A stable Richardson iteration method for complex linear systems. *Numer. Math.*, 54(2):225–242, 1989. (Cited on page 184.)

[1173] C. F. Fischer and R. A. Usmani. Properties of some tridiagonal matrices and their application to boundary value problems. *SIAM J. Numer. Anal.*, 6(1):127–142, 1969. (Cited on page 383.)

[1174] E. Fischer. Über quadratische formen mit reelen koeffizienten. *Monatsh Math. Phys.*, 16:234–249, 1905. (Cited on page 285.)

[1175] M. Flagg. Solving a system of linear equations using ancient Chinese methods. Linear Algebra. 1., https://digitalcommons.ursinus.edu/triumphs_linear/1, 2017. (Cited on pages 35 and 39.)

[1176] D. A. Flanders and G. H. Shortley. Numerical determination of fundamental modes. *J. Appl. Phys.*, 21(12):1326–1332, 1950. (Cited on pages 180 and 300.)

[1177] R. Fletcher. Conjugate gradient methods for indefinite systems. In G. A. Watson, editor, *Numerical Analysis. Lecture Notes in Mathematics, vol 506*, pages 73–89, Springer, Berlin, Heidelberg, 1976. (Cited on pages 217, 221, and 456.)

[1178] R. Fletcher and M. J. D. Powell. A rapidly convergent descent method for minimization. *Comput. J.*, 6(2):163–168, 1963. (Cited on page 456.)

[1179] R. Fletcher and C. M. Reeves. Function minimization by conjugate gradients. *Comput. J.*, 7:149–154, 1964. (Cited on pages 208 and 456.)

[1180] D. R. Fokkema, G. L. G. Sleijpen, and H. van der Vorst. Generalized conjugate gradient squared. *J. Comput. Appl. Math.*, 71(1):125–146, 1996. (Cited on page 234.)

[1181] D. R. Fokkema, G. L. G. Sleijpen, and H. van der Vorst. Jacobi-Davidson style QR and QZ algorithms for the reduction of matrix pencils. *SIAM J. Sci. Comput.*, 20(1):94–125, 1998. (Cited on page 317.)

[1182] D. C.-L. Fong and M. Saunders. LSMR: An iterative algorithm for sparse least-squares problems. *SIAM J. Sci. Comput.*, 33(5):2950–2971, 2011. (Cited on page 373.)

[1183] A. R. Forsyth. Proof of a theorem by Cayley in regard to matrices. *Messenger Math.*, 13:139–142, 1884. (Cited on page 7.)

[1184] A. I. Forsythe and G. E. Forsythe. IBM experiments with accelerated gradient methods for linear equations. Technical Report 1643, National Bureau of Standards, 1952. (Cited on page 461.)

[1185] G. E. Forsythe. Gauss to Gerling on relaxation. *M.T.A.C.*, 5:255–258, 1951. (Cited on pages 155 and 469.)

[1186] G. E. Forsythe. Bibliographical survey of Russian mathematical monographs, 1930 to 1951. Technical Report 1628, National Bureau of Standards, 1952. (Cited on page 461.)

[1187] G. E. Forsythe. A numerical analyst's fifteen-foot shelf. *M.T.A.C.*, 7(44):221–228, 1953. (Cited on page 461.)

[1188] G. E. Forsythe. Solving linear algebraic equations can be interesting. *Bull. Amer. Math. Soc.*, 59(4):299–329, 1953. (Cited on pages 70, 162, 189, and 460.)

[1189] G. E. Forsythe. Relaxation methods. Technical report, Dept. of Mathematics, Univesity of California, Los Angeles, 1955. (Cited on page 461.)

[1190] G. E. Forsythe. Today's computational methods of linear algebra. *SIAM Rev.*, 9(3):489–515, 1967. (Cited on page 460.)

[1191] G. E. Forsythe. What to do till the computer scientist comes. *Am. Math. Mon.*, 75(5):454–462, 1968. (Cited on page 461.)

[1192] G. E. Forsythe. Pitfalls in computation, or why a math book isn't enough. *Am. Math. Mon.*, 77(9):931–956, 1970. (Cited on page 461.)

[1193] G. E. Forsythe and P. Henrici. The cyclic Jacobi method for computing the principal values of a complex matrix. *Trans. Amer. Math. Soc.*, 94:1–23, 1960. (Cited on page 136.)

[1194] G. E. Forsythe, M. R. Hestenes, and J. B. Rosser. Iterative methods for solving linear equations. *Bull. Amer. Math. Soc.*, 57(6):480, 1951. (Cited on page 461.)

[1195] G. E. Forsythe, M. A. Malcolm, and C. B. Moler. *Computer Methods for Mathematical Computations*. Prentice-Hall, Inc., Englewood Cliffs, NJ, 1977. (Cited on page 94.)

[1196] G. E. Forsythe and C. B. Moler. *Computer Solution of Linear Algebraic Systems*. Prentice-Hall, Inc., Englewood Cliffs, NJ, 1967. (Cited on pages 77, 80, 94, 364, and 459.)

[1197] G. E. Forsythe and T. S. Motzkin. An extension of Gauss' transformation for improving the condition of systems of linear equations. *M.T.A.C.*, 6(37):9–17, 1952. (Cited on pages 156 and 461.)

[1198] G. E. Forsythe and E. G. Straus. On best conditioned matrices. *Proc. Am. Math. Soc.*, 6:340–345, 1955. (Cited on pages 80 and 251.)

[1199] G. E. Forsythe and W. R. Wasow. *Finite-difference Methods for Partial Differential Equations*. John Wiley and Sons, 1960. (Cited on pages 84 and 459.)

[1200] L. V. Foster and T. A. Davis. Algorithm 933: Reliable calculation of numerical rank, null space bases, pseudoinverse solutions, and basic solutions using SuiteSparseQR. *ACM Trans. Math. Soft. (TOMS)*, 40(1):1–23, 2013. (Cited on page 14.)

[1201] L. Fox. A short account of relaxation methods. *Quart. J. Mech. Appl. Math.*, 1:253–280, 1948. (Cited on page 172.)

[1202] L. Fox. Practical solution of linear equations and inversion of matrices. In O. Taussky, editor, *Contributions to the Solution of Systems of Linear Equations and the Determination of Eigenvalues*, pages 1–54. National Bureau of Standards, Washington, DC, 1954. Applied Mathematics Series, Vol. 39. (Cited on page 75.)

[1203] L. Fox. *The Numerical Solution of Two-point Boundary Problems in Ordinary Differential Equations*. Oxford University Press, 1957. (Cited on page 462.)

[1204] L. Fox. *An Introduction to Numerical Linear Algebra*. Oxford University Press, 1964. (Cited on pages 77 and 462.)

[1205] L. Fox. James Hardy Wilkinson. *Biogr. Mem. Fellows Royal Soc.*, 33:671–708, 1987. The Royal Society, London. (Cited on pages 596 and 598.)

[1206] L. Fox, H. D. Huskey, and J. H. Wilkinson. Notes on the solution of linear algebraic equations. *Quart. J. Mech. Appl. Math.*, 1:149–173, 1948. (Cited on pages 74, 75, 204, 462, 580, and 598.)

[1207] L. Fox, H. D. Huskey, and J. H. Wilkinson. The solution of algebraic linear simultaneous equations by punched card methods. Technical Report Ma/23/106, National Physical Laboratory, United Kingdom, 1948. (Cited on pages 75, 333, and 462.)

[1208] L. Fox and D. F. Mayers. *Computing Methods for Scientists and Engineers*. Oxford University Press, 1968. (Cited on page 462.)

[1209] J. S. Frame. Machines for solving algebraic equations. *M.T.A.C.*, 1(9):337–353, 1945. (Cited on page 332.)

[1210] J. S. Frame. A simple recursion formula for inverting a matrix (abstract). *Bull. Amer. Math. Soc.*, 55:1045, 1949. (Cited on page 289.)

[1211] R. Franci. The history of algebra in Italy in the 14th and 15th centuries. Some remarks on recent historiography. *Actes d'Història de la Cìencia i de la Tècnica, nova època*, 3(2):175–194, 2010. (Cited on page 44.)

[1212] J. G. F. Francis. The QR transformation: a unitary analogue to the LR transformation-Part 1. *Comput. J.*, 4(3):265–271, 1961. (Cited on pages 122, 311, and 312.)

[1213] J. G. F. Francis. The QR transformation-Part 2. *Comput. J.*, 4(4):332–345, 1962. (Cited on pages 311, 312, and 313.)

[1214] M. Francoeur and al. Rapports sur l'arithmomètre de M. Thomas. *Bulletin de la Société d'Encouragement pour l'Industrie Nationale*, CCXII, 1822. (Cited on page 577.)

[1215] W. L. Frank. Computing eigenvalues of complex matrices by determinant evaluation and by methods of Danilewski and Wielandt. *J. Soc. Ind. Appl. Math.*, 6(4):378–392, 1958. (Cited on pages 292 and 295.)

[1216] W. L. Frank. Solution of linear systems by Richardson's method. *J. ACM*, 7(3):274–286, 1960. (Cited on page 208.)

[1217] S. P. Frankel. Convergence rates of iterative treatments of partial differential equations. *M.T.A.C.*, 4(30):65–75, 1950. (Cited on pages 165 and 175.)

[1218] V. Frayssé, L. Giraud, and S. Gratton. A set of flexible-GMRES routines for real and complex arithmetics. Technical Report TR/PA/98/20, CERFACS, Toulouse, France, 1998. (Cited on page 236.)

[1219] V. Frayssé, L. Giraud, S. Gratton, and J. Langou. Algorithm 842: A set of GMRES routines for real and complex arithmetics on high performance computers. *ACM Trans. Math. Soft. (TOMS)*, 31(2):228–238, 2005. (Cited on page 241.)

[1220] V. Frayssé, L. Giraud, and H. Kharraz-Aroussi. On the influence of the orthogonalization scheme on the parallel performance of GMRES. In *Euro-Par'98 Parallel Processing*, Springer, Berlin, Heidelberg, 1998. (Cited on page 229.)

[1221] R. A. Frazer. Note on the escalator method. *Philos. Mag.*, 38:287–289, 1947. (Cited on page 297.)

[1222] R. A. Frazer, W. J. Duncan, and A. R. Collar. *Elementary Matrices and some Applications to Dynamics and Differential Equations*. Cambridge University Press, 1938. (Cited on pages 71, 219, and 388.)

[1223] T. Freeth, D. Higgon, A. Dacanalis, L. MacDonald, M. Georgakopoulou, and A. Wojcik. A model of the cosmos in the ancient Greek Antikythera mechanism. *Nature Scientific Reports*, 11(1):1–15, 2021. See the corrections in *Scientific Reports 11*, 17361 (2021). (Cited on page 324.)

[1224] M. Freitag. *Inner-outer iterative methods for eigenvalue problems - convergence and preconditioning*. PhD thesis, University of Bath, UK, 2007. (Cited on page 317.)

[1225] R. W. Freund. Quasi-kernel polynomials and their use in non-Hermitian matrix iterations. *J. Comput. Appl. Math.*, 43(1-2):135–158, 1992. (Cited on page 237.)

[1226] R. W. Freund. A transpose-free quasi-minimal residual algorithm for non-Hermitian linear systems. *SIAM J. Sci. Comput.*, 14(2):470–482, 1993. (Cited on page 230.)

[1227] R. W. Freund, G. H. Golub, and N. M. Nachtigal. Iterative solution of linear systems. *Acta Numerica*, 1:57–100, 1992. (Cited on page 246.)

[1228] R. W. Freund, M. H. Gutknecht, and N. M. Nachtigal. An implementation of the look-ahead Lanczos algorithm for non-Hermitian matrices, part I. Technical Report RIACS 90.45, NASA Ames Research Center, Mountain View, CA, 1990. (Cited on page 228.)

[1229] R. W. Freund, M. H. Gutknecht, and N. M. Nachtigal. An implementation of the look-ahead Lanczos algorithm for non-Hermitian matrices. Technical Report RIACS 91.09, NASA Ames Research Center, Mountain View, CA, 1991. (Cited on page 228.)

[1230] R. W. Freund, M. H. Gutknecht, and N. M. Nachtigal. An implementation of the look-ahead Lanczos algorithm for non-Hermitian matrices. *SIAM J. Sci. Comput.*, 14(1):137–158, 1993. (Cited on page 228.)

[1231] R. W. Freund and M. Hochbruck. A biconjugate gradient type algorithm on massively parallel architectures. In R. Vichnevetsky and J. J. H. Miller, editors, *IMACS'91*, pages 720–721, Dublin, 1991. Criterion Press. (Cited on page 229.)

[1232] R. W. Freund and M. Hochbruck. A biconjugate gradient-type algorithm for the iterative solution of non-Hermitian linear systems on massively parallel architectures. In C. Brezinski and U. Kulisch, editors, *Computational and Applied Mathematics, I: Algorithms and Theory*, pages 169–178, North Holland, Amsterdam, 1992. (Cited on page 229.)

[1233] R. W. Freund and M. Hochbruck. On the use of two QMR algorithms for solving singular systems and applications in Markov chain modeling. *Numer. Linear Algebra Appl.*, 1(4):403–420, 1994. (Cited on page 232.)

[1234] R. W. Freund and M. Malhotra. A block QMR algorithm for non-Hermitian linear systems with multiple right-hand sides. *Linear Algebra Appl.*, 254(1-3):119–157, 1997. (Cited on page 235.)

[1235] R. W. Freund and N. M. Nachtigal. An implementation of the look-ahead Lanczos algorithm for non-Hermitian matrices, part II. Technical Report RIACS 90.46, NASA Ames Research Center, Mountain View, CA, 1990. (Cited on page 228.)

[1236] R. W. Freund and N. M. Nachtigal. QMR: a quasi-minimal residual method for non-Hermitian linear systems. *Numer. Math.*, 60(3):315–339, 1991. (Cited on pages 225, 227, and 228.)

[1237] R. W. Freund and N. M. Nachtigal. Implementation details of the coupled QMR algorithm. Technical Report RIACS 92.19, NASA Ames Research Center, Mountain View, CA, 1992. (Cited on page 228.)

[1238] R. W. Freund and N. M. Nachtigal. An implementation of the QMR method based on coupled two-term recurrences. Technical Report RIACS 92.15, NASA Ames Research Center, Mountain View, CA, 1992. (Cited on page 228.)

[1239] R. W. Freund and N. M. Nachtigal. An implementation of the QMR method based on coupled two-term recurrences. *SIAM J. Sci. Comput.*, 15(2):313–337, 1994. (Cited on page 228.)

[1240] R. W. Freund and T. Szeto. A quasi-minimal residual squared algorithm for non-Hermitian linear systems. Technical Report 91.26, RIACS, NASA Ames Research Center, Mountain View, CA. (Cited on page 229.)

[1241] R. W. Freund and T. Szeto. A quasi-minimal residual squared algorithm for non-Hermitian linear systems. Technical Report CAM Report 92-19, Dept. of Mathematics, UCLA, 1992. (Cited on page 229.)

[1242] R .W. Freund and T. Szeto. A transpose-free quasi-minimal residual squared algorithm for non-Hermitian linear systems. In R. Vichnevetsky, D. Knight, and G. Richter, editors, *Advances in Computer Methods for Partial Differential Equations - VII, IMACS*, pages 258–264. IMACS, 1992. (Cited on page 229.)

[1243] J. Friberg. *A Remarkable Collection of Babylonian Mathematical Texts. Manuscripts in the Schøyen Collection, Cuneiform Texts I.* Springer, 2007. (Cited on page 33.)

[1244] J. Friberg and F. N. H. Al-Rawi. *New Mathematical Cuneiform Texts.* Springer, 2016. (Cited on pages 33 and 34.)

[1245] V. M. Fridman. The method of minimum iterations with minimum errors for a system of linear algebraic equations with a symmetrical matrix. *Zh. Vychisl. Mat. Mat. Fiz.*, 2(2):341–342, 1962. (In Russian), English translation: *U.S.S.R. Comput. Math. Math. Phys.*, 2:2 (1963), 362-363. (Cited on pages 208 and 216.)

[1246] M. Frigo and S. G. Johnson. FFTW: An adaptive software architecture for the FFT. In *Proc. 1998 IEEE Intl. Conf. Acoustics Speech and Signal Processing*, pages 1381–1384. IEEE, 1998. (Cited on page 384.)

[1247] M. Frigo and S. G. Johnson. The design and implementation of FFTW3. *Proceedings of the IEEE*, 93(2):216–231, 2005. Special issue on "Program Generation, Optimization, and Platform Adaptation". (Cited on page 384.)

[1248] R. Fritsch. Seidel, Ludwig. *Neue Deutsche Biographie*, 24:176–178, 2010. (Cited on page 559.)

[1249] F. G. Frobenius. Über das Pfaffsche problem. *Journal für die reine und angewandte Mathematik (Crelle Journal)*, 82:230–315, 1877. (Cited on page 116.)

[1250] F. G. Frobenius. Über lineare substitutionen und bilineare formen. *Journal für die reine und angewandte Mathematik (Crelle Journal)*, 84:1–63, 1878. (Cited on pages 6, 123, 148, 150, 288, 396, and 464.)

[1251] F. G. Frobenius. Theorie der linearen formen mit ganzen coefficienten. *Journal für die reine und angewandte Mathematik (Crelle Journal)*, 86:482–544, 1879. (Cited on pages 6, 12, 148, 150, 290, and 464.)

[1252] F. G. Frobenius. Über relationen zwischen den näherungsbruchen von potenzreihen. *Journal für die reine und angewandte Mathematik (Crelle Journal)*, 90:1–17, 1881. (Cited on page 464.)

[1253] F. G. Frobenius. Über die cogredienten transformationen der bilinearen formen. *S'ber. Akad. der Wiss. Berlin*, 2 of Gesammelte Abhundlungen:695–704, 1896. (Cited on page 388.)

[1254] F. G. Frobenius. Über vertauschbare matrizen. *S'ber. Akad. der Wiss. Berlin*, 2 of Gesammelte Abhundlungen:601–614, 1896. (Cited on pages 6 and 290.)

[1255] F. G. Frobenius. Über matrizen aus positiven elementen. *S'ber. Akad. der Wiss. Berlin*, pp. 471-476, 1908. Gesammelte Abhundlungen, Springer, Berlin, 1968, vol. 3, pp. 404-409. (Cited on page 117.)

[1256] F. G. Frobenius. Über die mit einer matrix vertauschbaren matrizen. Sitzungsber. Preuss. Akad. Wiss., 1910. (Cited on page 396.)

[1257] F. G. Frobenius. Über den rang einer matrix. *S'ber. Akad. der Wiss. Berlin*, 12:20–29, 1911. (Cited on page 13.)

[1258] F. G. Frobenius. Über matrizen aus nicht negativen elementen. *S.-B. Preuss. Akad. Wiss. (Berlin)*, 26:456–477, 1912. (Cited on page 464.)

[1259] A. Frommer and U. Glässner. Restarted GMRES for shifted linear systems. *SIAM J. Sci. Comput.*, 19(1):15–26, 1998. (Cited on page 236.)

[1260] A. Frommer, S. Güttel, and M. Schweitzer. Efficient and stable Arnoldi restarts for matrix functions based on quadrature. *SIAM J. Matrix Anal. Appl.*, 35(2):661–683, 2014. (Cited on page 390.)

[1261] A. Frommer, K. Kahl, Th. Lippert, and H. Rittich. 2-norm error bounds and estimates for Lanczos approximations to linear systems and rational matrix functions. *SIAM J. Matrix Anal. Appl.*, 34(3):1046–1065, 2013. (Cited on page 213.)

[1262] A. Frommer, K. Lund, M. Schweitzer, and D. B. Szyld. The Radau-Lanczos method for matrix functions. *SIAM J. Matrix Anal. Appl.*, 38(3):710–732, 2017. (Cited on page 390.)

[1263] A. Frommer, H. Schwandt, and D. B. Szyld. Asynchronous weighted additive Schwarz methods. *Electron. Trans. Numer. Anal.*, 5:48–61, 1997. (Cited on page 196.)

[1264] A. Frommer and V. Simoncini. Matrix functions. In W. H. A. Schilders, H. A. van der Vorst, and J. Rommes, editors, *Model Order Reduction: Theory, Research Aspects and Applications*, pages 275–304. Volume 13 of Mathematics in Industry. Springer, 2008. (Cited on page 388.)

[1265] A. Frommer and D. B. Szyld. Asynchronous two-stage iterative methods. *Numer. Math.*, 69(2):141–153, 1994. (Cited on page 196.)

[1266] A. Frommer and D. B. Szyld. On asynchronous iterations. *J. Comput. Appl. Math.*, 123(1-2):201–216, 2000. (Cited on page 196.)

[1267] S. Fujino. GPBiCG(m, ℓ): a hybrid of BiCGSTAB and GPBiCG methods with efficiency and robustness. *Appl. Numer. Math.*, 41(1):107–117, 2002. (Cited on page 239.)

[1268] S. Fujino and S. Doi. Optimizing multicolor ICCG methods on some vector computers. In R. Beauwens, editor, *Proceedings IMACS Int. Symp. Iterative Methods in Linear Algebra*. North Holland, 1991. (Cited on page 260.)

[1269] S. Fujino, M. Fujiwara, and M. Yoshida. BiCGSafe method based on minimization of associate residual. *Transactions of JSCES*, 2005. (Cited on page 240.)

[1270] L. E. Fuller and J. D. Logan. On the evaluation of determinants by Chio's method. *Two-year College Math. J.*, 6(1):8–10, 1975. (Cited on page 114.)

[1271] R. E. Funderlic, M. Neumann, and R. J. Plemmons. LU decompositions of generalized diagonally dominant matrices. *Numer. Math.*, 40:57–69, 1982. (Cited on page 80.)

[1272] R. E. Funderlic and R. J. Plemmons. LU decomposition of M-matrices by elimination without pivoting. *Linear Algebra Appl.*, 41:41–99, 1981. (Cited on page 80.)

[1273] A. Galántai. *Projectors and Projection Methods*. Kluwer Academic Publishers, Boston, 2004. (Cited on page 186.)

[1274] A. Galántai. The rank reduction procedure of Egerváry. *Cent. Eur. J. Oper. Res.*, 18(1):5–24, 2009. (Cited on page 81.)

[1275] E. Gallopoulos and Y. Saad. Efficient solution of parabolic equations by Krylov approximation methods. *SIAM J. Sci. Statist. Comput.*, 13(5):1236–1264, 1992. (Cited on page 390.)

[1276] M. J. Gander. Schwarz methods over the course of time. *Electron. Trans. Numer. Anal.*, 31(5):228–255, 2008. (Cited on page 267.)

[1277] M. J. Gander and G. Wanner. From Euler, Ritz, and Galerkin to modern computing. *SIAM Rev.*, 54(4):627–666, 2012. (Cited on page 284.)

[1278] M. J. Gander and G. Wanner. The origins of the alternating Schwarz method. In J. Erhel, M. J. Gander, L. Halpern, G. Pichot, T. Sassi, and O. Widlund, editors, *Domain Decomposition Methods in Science and Engineering XXI*, pages 487–495. Springer, 2014. (Cited on page 265.)

[1279] W. Gander. Algorithms for the polar decomposition. *SIAM J. Sci. Statist. Comput.*, 11(6):1102–1115, 1990. (Cited on page 143.)

[1280] W. Gander and G. H. Golub. Cyclic reduction-history and applications. In *Scientific Computing (Hong Kong, 1997)*, pages 73–85, 1997. (Cited on page 384.)

[1281] F. R. Gantmacher. On the algebraic analysis of Krylov's method of transforming the secular equation. In *Trans. Second Math. Congress, II*, pages 45–48, 1934. (In Russian). (Cited on page 506.)

[1282] F. R. Gantmacher. *The Theory of Matrices, Vol. 1*. AMS Chelsea Publishing, Providence, RI, 1959. (Cited on pages 147 and 506.)

[1283] F. R. Gantmacher and M. G. Krein. *Oscillation Matrices and Kernels and Small Vibrations of Mechanical Systems*. State Publishing House, Moscow, 1950. First edition 1941 (in Russian). (Cited on page 382.)

[1284] P. R. Garabedian. Estimation of the relaxation factor for small mesh size. *M.T.A.C.*, 10(56):183–185, 1956. (Cited on page 176.)

[1285] B. S. Garbow, J. M. Boyle, J. J. Dongarra, and C. B. Moler. *Matrix Eigensystem Routines - EISPACK Guide Extension*. Springer, 1977. vol. 51 Lecture Notes in Computer Science. (Cited on page 367.)

[1286] N. Gastinel. Procédé itératif pour la résolution numérique d'un système d'équations linéaires. *C.R. Acad. Sci. Paris, Sér 1*, 246:2571–2574, 1958. (Cited on pages 193 and 194.)

[1287] N. Gastinel. *Matrices du second degré et normes générales en analyse numérique linéaire*. PhD thesis, Université de Grenoble, France, 1960. (Cited on pages 16 and 194.)

[1288] N. Gastinel. Sur-décomposition de normes générales et procédés itératifs. *Numer. Math.*, 5:142–151, 1963. (Cited on pages 16 and 195.)

[1289] N. Gastinel. Propriétés de certains ensembles normés de matrices. *Numer. Math.*, 7:255–260, 1965. (Cited on page 16.)

[1290] N. Gastinel. *Analyse Numérique Linéaire*. Hermann, Paris, 1966. Translated to English as "Numerical Linear Algebra", Academic Press, New York, 1970. (Cited on pages 94, 189, 190, 466, and 666.)

[1291] N. Gastinel. *Numerical Linear Algebra*. Academic Press, New York, 1970. English translation of [1290]. (Cited on pages 94 and 498.)

[1292] N. Gastinel. Sur l'extension de normes sur des algèbres de matrices. *Numer. Math.*, 17:71–83, 1971. (Cited on page 16.)

[1293] N. Gastinel and J. L. Joly. Condition numbers and general projection method. *Linear Algebra Appl.*, 3:185–224, 1970. (Cited on page 195.)

[1294] M. Gates, S. Tomov, and J. J. Dongarra. Accelerating the SVD two stage bidiagonal reduction and divide and conquer using GPUs. *Parallel Comput.*, 74:3–18, 2018. (Cited on page 138.)

[1295] A. Gaul, M. H. Gutknecht, J. Liesen, and R. Nabben. A framework for deflated and augmented Krylov subspace methods. *SIAM J. Matrix Anal. Appl.*, 34(2):495–518, 2013. (Cited on page 244.)

[1296] J. A. Gaunt. The deferred approach to the limit. Part II: Interpenetrating lattices. *Philos. Trans. R. Soc. Lond.*, 226:350–361, 1927. (Cited on page 548.)

[1297] C. F. Gauss. *Disquisitiones Arithmeticae*. Gerhard Fleischer, Leipzig, 1801. (Cited on page 468.)

[1298] C. F. Gauss. *Theoria Motus Corporum Coelestium in Sectionibus Solem Ambientium*. Perthes & Besser, Hamburg, 1809. (Cited on page 58.)

[1299] C. F. Gauss. Disquisitio de elementis ellipticis palladis. In *Carl Friedrich Gauss Werke, Vol VI*, pages 1–24. Königlichen Gesellschaft der Wissenschaften zu Göttingen, 1810. (Cited on pages 16, 17, 59, 60, and 469.)

[1300] C. F. Gauss. Methodus nova integralium valores per approximationem inveniendi. *Commentationes Societatis Regiae Scientarium Gottingensis Recentiores*, 3:39–76, 1814. (Cited on page 469.)

[1301] C. F. Gauss. *Determinatio Attracionis, quam in Punctum Quodvis Positionis Datae Exerceret Planeta, si eius Massa per Totam Orbitam, Ratione Temporis, quo Singulae Partes Discribuntur, Uniformiter Esset Dispertita*. Henricum Dieterich, Gottingen, 1818. (Cited on page 123.)

[1302] C. F. Gauss and C. L. Gerling. [letters of Gauss to and from Gerling]. In C. Schaefer, editor, *Briefwechsel zwischen Carl Friedrich Gauss und Christian Ludwig Gerling*. Otto Elsner Verlagsgesellschaft, Berlin, 1927. (Cited on page 155.)

[1303] W. Gautschi. Leonhard Euler: His life, the man, and his works. *SIAM Rev.*, 50(1):3–33, 2008. (Cited on pages 275 and 451.)

[1304] We. Gautschi. The asymptotic behaviour of powers of matrices. *Duke Math. J.*, 20(1):127–140, 1953. (Cited on pages 16 and 170.)

[1305] We. Gautschi. The asymptotic behaviour of powers of matrices II. *Duke Math. J.*, 20(3):375–379, 1953. (Cited on pages 16 and 170.)

[1306] B. Gavin, A. Międlar, and E. Polizzi. FEAST eigensolver for nonlinear eigenvalue problems. *J. Comput. Sci.*, 27:107–117, 2018. (Cited on page 318.)

[1307] B. Gavin and E. Polizzi. Krylov eigenvalue strategy using the FEAST algorithm with inexact system solves. *Numer. Linear Algebra Appl.*, 25(5):e2188, 2018. (Cited on page 318.)

[1308] J. Gavin and A. Schärlig. *Sur les Doigts, jusqu'à 9999: la Numération Digitale, des Anciens à la Renaissance*. Presses Polytechniques et Universitaires Romandes, 2014. (Cited on page 323.)

[1309] M. K. Gavurin. The use of polynomials of best approximation for the improvement of the convergence of iteration processes. *Uspekhi Mat. Nauk*, 5(3):156–160, 1950. (Cited on page 163.)

[1310] E. S. Gawlik, Y. Nakatsukasa, and B. D. Sutton. A backward stable algorithm for computing the CS decomposition via the polar decomposition. *SIAM J. Matrix Anal. Appl.*, 39(3):1448–1469, 2018. (Cited on page 146.)

[1311] S. Gazzola, P. C. Hansen, and J. G. Nagy. IR Tools: A MATLAB package of iterative regularization methods and large-scale test problems. *Numer. Algorithms*, 81(3):773–811, 2019. (Cited on page 393.)

[1312] S. Gazzola, S. Noschese, P. Novati, and L. Reichel. Arnoldi decomposition, GMRES, and preconditioning for linear discrete ill-posed problems. *Appl. Numer. Math.*, 142:102–121, 2019. (Cited on page 393.)

[1313] S. Gazzola and L. Reichel. A new framework for multi-parameter regularization. *BIT Numerical Mathematics*, 56(3):919–949, 2016. (Cited on page 392.)

[1314] H. Geiringer. On the solution of systems of linear equations by certain iteration methods. In *Reissner Anniversary Volume*, pages 365–393. J. W. Edwards, Ann Arbor, Michigan, 1949. (Cited on pages 173 and 473.)

[1315] A. Geist, A. Beguelin, J. J. Dongarra, W. Jiang, R. Manchek, and V. S. Sunderam. *Using PVM*. MIT Press, 1994. (Cited on page 376.)

[1316] I. Gelfand. Normierte ringe. *Mat. Sb.*, 9(51):3–24, 1941. (Cited on page 170.)

[1317] B. Gellai. Cornelius Lanczos, a biographical essay. In *[473]*, pages xxi–xlviii, 1994. (Cited on page 517.)

[1318] B. Gellai. *The Intrinsic Nature of Things: The Life and Science of Cornelius Lanczos*. American Mathematical Society, 2010. (Cited on pages 332, 516, and 517.)

[1319] W. M. Gentleman. Least squares computations by Givens transformations without square roots. *IMA J. Appl. Math.*, 12(3):329–336, 1973. (Cited on page 131.)

[1320] W. M. Gentleman. Error analysis of QR decompositions by Givens transformations. *Linear Algebra Appl.*, 10(3):189–197, 1975. (Cited on page 131.)

[1321] A. George and J. Friberg. Six more mathematical cuneiform texts in the Schøyen collection. In D. Minuto and R. Pintaudi, editors, *Papyri Graecae Schøyen. Essays and Texts in Honour of Martin Schøyen*, pages 123–195, Florence, Italy, 2010. Edizioni Gonnelli. (Cited on page 323.)

[1322] J. A. George. *Computer implementation of the finite element method*. PhD thesis, Dept. of Computer Science, Stanford University, 1971. (Cited on page 88.)

[1323] J. A. George. Nested dissection of a regular finite element mesh. *SIAM J. Numer. Anal.*, 10(2):345–363, 1973. (Cited on page 89.)

[1324] J. A. George and Kh. Ikramov. Is the polar decomposition finitely computable? *SIAM J. Matrix Anal. Appl.*, 17(2):348–354, 1996. (Cited on page 143.)

[1325] J. A. George and Kh. Ikramov. Addendum: Is the polar decomposition finitely computable? *SIAM J. Matrix Anal. Appl.*, 18(1):264, 1997. (Cited on page 143.)

[1326] J. A. George and Kh. Ikramov. Block LU factorization is stable for block matrices whose inverses are block diagonally dominant. *J. Math.Sci.*, 127(3):1962–1968, 2005. (Cited on page 94.)

[1327] J. A. George and W.-H. Liu. A note on fill for sparse matrices. *SIAM J. Numer. Anal.*, 12(3):452–455, 1975. (Cited on page 86.)

[1328] J. A. George and J. W. H. Liu. An implementtion of a pseudo-peripheral node finder. *ACM Trans. Math. Soft. (TOMS)*, 5:286–295, 1979. (Cited on page 88.)

[1329] J. A. George and J. W. H. Liu. A fast implementation of the minimum degree algorithm using quotient graphs. *ACM Trans. Math. Soft. (TOMS)*, 6:337–358, 1980. (Cited on page 87.)

[1330] J. A. George and J. W. H. Liu. A minimal storage implementation of the miniraum degree algorithm. *SIAM J. Numer. Anal.*, 17(2):282–299, 1980. (Cited on page 87.)

[1331] J. A. George and J. W. H. Liu. *Computer Solution of Large Sparse Positive Definite Systems*. Prentice Hall, 1981. (Cited on pages 89, 90, 94, and 371.)

[1332] J. A. George and J. W. H. Liu. The evolution of the minimum degree ordering algorithm. *SIAM Rev.*, 31(1):1–19, 1989. (Cited on pages 87 and 370.)

[1333] J. A. George, J. W. H. Liu, and E. Ng. *Computer Solution of Sparse Linear Systems*. Prentice Hall, 1994. (Cited on page 371.)

[1334] J. A. George and E. G. Ng. An implementation of Gaussian elimination with partial pivoting for sparse systems. *SIAM J. Sci. Comput.*, 6(2):390–409, 1985. (Cited on page 93.)

[1335] J. A. George and E. G. Ng. Symbolic factorization for sparse Gaussian elimination with partial pivoting. *SIAM J. Sci. Statist. Comput.*, 8(6):877–898, 1987. (Cited on page 93.)

[1336] J. A. George and E. G. Ng. On the complexity of sparse *QR* and *LU* factorization of finite-element matrices. *SIAM J. Sci. Statist. Comput.*, 9(5):849–861, 1988. (Cited on page 90.)

[1337] J. A. George and E. G. Ng. Parallel sparse Gaussian elimination with partial pivoting. *Annals of Oper. Res.*, 22(1):219–240, 1990. (Cited on page 93.)

[1338] T. Gergelits, K.-A. Mardal, B. F. Nielsen, and Z. Strakoš. Laplacian preconditioning of elliptic PDEs: Localization of the eigenvalues of the discretized operator. *SIAM J. Numer. Anal.*, 57(3):1369–1394, 2019. (Cited on page 263.)

[1339] T. Gergelits and Z. Strakoš. Composite convergence bounds based on Chebyshev polynomials and finite precision conjugate gradient computations. *Numer. Algorithms*, 65(4):759–782, 2014. (Cited on page 210.)

[1340] C. L. Gerling. *Die Ausgleichungs-Rechnungen der Practischen Geometrie oder die Methode der Kleinsten Quadrate mit ihren Anwendungen für Geodätische Aufgaben*. Friedrich und Andreas Perthes, Hamburg, 1843. (Cited on page 156.)

[1341] S. A. Gerschgorin. Über die abgrenzung der eigenwerte einer matrix. *Izv. Akad. Nauk SSSR, Ser. Fiz.-Mat.*, 6:749–754, 1931. (Cited on pages 286, 474, and 575.)

[1342] H. Gfrerer. An a posteriori parameter choice for ordinary and iterated Tikhonov regularization of ill-posed problems leading to optimal convergence rates. *Math. Comput.*, 49:507–522, 1987. (Cited on page 392.)

[1343] P. Ghysels, T. J. Ashby, K. Meerbergen, and W. Vanroose. Hiding global communication latency in the GMRES algorithm on massively parallel machines. *SIAM J. Sci. Comput.*, 35(1):C48–C71, 2013. (Cited on page 244.)

[1344] P. Ghysels, X. S. Li, F.-H. Rouet, S. Williams, and A. Napov. An efficient multicore implementation of a novel HSS-structured multifrontal solver using randomized sampling. *SIAM J. Sci. Comput.*, 38(5):S358–S384, 2016. (Cited on page 377.)

[1345] P. Ghysels and W. Vanroose. Hiding global synchronization latency in the preconditioned conjugate gradient algorithm. *Parallel Comput.*, 40(7):224–238, 2014. (Cited on page 215.)

[1346] J. W. Gibbs and E. B. Wilson. *Vector Analysis*. Yale University Press, 1901. (Cited on pages 3 and 122.)

[1347] N. E. Gibbs, W. G. Poole, Jr., and P. K. Stockmeyer. An algorithm for reducing the bandwidth and profile of a sparse matrix. *SIAM J. Numer. Anal.*, 13(2):236–250, 1976. (Cited on page 88.)

[1348] N. E. Gibbs, W. G. Poole, Jr., and P. K. Stockmeyer. A comparison of several bandwidth and profile reduction algorithms. *ACM Trans. Math. Soft. (TOMS)*, 2(4):322–330, 1976. (Cited on page 88.)

[1349] J. R. Gilbert. Predicting structure in sparse matrix computations. *SIAM J. Matrix Anal. Appl.*, 15(1):62–79, 1994. (Cited on page 93.)

[1350] J. R. Gilbert, X. S. Li, E. G. Ng, and B. W. Peyton. Computing row and column counts for sparse QR and LU factorization. *BIT*, 41(4):693–710, 2001. (Cited on page 93.)

[1351] J. R. Gilbert and J. W. H. Liu. Elimination structures for unsymmetric sparse *LU* factors. *SIAM J. Matrix Anal. Appl.*, 14(2):334–352, 1993. (Cited on page 93.)

[1352] J. R. Gilbert and E. G. Ng. Structural representations of Schur complements in sparse matrices. In J. A. George, J. R. Gilbert, and J. W. H. Liu, editors, *Graph Theory and Sparse Matrix Computation*, volume 56 of *IMA Volumes in Applied Mathematics*, pages 107–139, New York, 1993. Springer Verlag. (Cited on page 93.)

[1353] J. R. Gilbert and T. Peierls. Sparse partial pivoting in time proportional to arithmetic operations. *SIAM J. Sci. Statist. Comput.*, 9(5):862–874, 1988. (Cited on pages 93 and 371.)

[1354] C. C. Gillispie, R. Fox, and I. Grattan-Guinness. *Pierre-Simon Laplace, 1749-1827: A life in Exact Science*. Princeton University Press, Princeton, 1997. (Cited on page 522.)

[1355] T. Ginsburg. The conjugate gradient method. *Numer. Math.*, 5:191–200, 1963. (Cited on pages 208 and 361.)

[1356] G. Giorgi. Nuove osservazioni sulle funzioni di matrici. *Rendiconti della Reale Accademia dei Lincei*, 8:1–8, 1928. (Cited on page 388.)

[1357] L. Giraud, S. Gratton, and J. Langou. Convergence in backward error of relaxed GMRES. *SIAM J. Sci. Comput.*, 29(2):710–728, 2007. (Cited on page 242.)

[1358] L. Giraud, S. Gratton, X. Pinel, and X. Vasseur. Flexible GMRES with deflated restarting. *SIAM J. Sci. Comput.*, 32(4):1858–1878, 2010. (Cited on page 243.)

[1359] L. Giraud and J. Langou. When modified Gram-Schmidt generates a well-conditioned set of vectors. *IMA J. Numer. Anal.*, 22(4):521–528, 2002. (Cited on page 129.)

[1360] L. Giraud and J. Langou. A robust criterion for the modified Gram-Schmidt algorithm with selective reorthogonalization. *SIAM J. Sci. Comput.*, 25(2):417–441, 2003. (Cited on pages 129 and 130.)

[1361] L. Giraud, J. Langou, and M. Rozložník. On the loss of orthogonality in the Gram-Schmidt orthogonalization process. *Comp. Math. App.*, 50:1069–1075, 2005. (Cited on page 129.)

[1362] L. Giraud, J. Langou, M. Rozložník, and J. van den Eshof. Rounding error analysis of the classical Gram-Schmidt orthogonalization process. *Numer. Math.*, 101:87–100, 2005. (Cited on page 129.)

[1363] W. Givens. Fields of values of a matrix. *Proc. Am. Math. Soc.*, 3(2):206–209, 1952. (Cited on page 287.)

[1364] W. Givens. A method of computing eigenvalues and eigenvectors suggested by classical results on symmetric matrices. In L. J. Paige and O. Taussky, editors, *Proceedings of the 1951 Symposium "Simultaneous Linear Equations and the Determination of Eigenvalues", NBS 29*, pages 117–122. National Bureau of Standards, 1953. (Cited on pages 131 and 476.)

[1365] W. Givens. Numerical computation of the characteristic values of a real symmetric matrix. Technical Report ORNL-1574, Oak Ridge National Laboratory, 1954. (Cited on pages 131, 308, and 476.)

[1366] W. Givens. The characteristic value-vector problem. *J. ACM*, 4:298–307, 1957. (Cited on page 308.)

[1367] W. Givens. Computation of plane unitary rotations transforming a general matrix to triangular form. *J. Soc. Ind. Appl. Math.*, 6(1):26–50, 1958. (Cited on pages 131, 151, 308, and 476.)

[1368] A. Gloden. La méthode du Luxembourgeois B.I. Clasen pour la résolution d'un système d'équations linéaires. *Revue d'histoire des sciences et de leurs applications*, 6(2):168–170, 1953. (Cited on page 67.)

[1369] H. H. Goldstine. *A History of Numerical Analysis from the 16th through the 19th Century*. Springer, 1977. (Cited on page xv.)

[1370] H. H. Goldstine. Remembrance of things past. In S.G. Nash, editor, *A History of Scientific Computing*, pages 5–16. ACM Press, 1990. (Cited on page 76.)

[1371] H. H. Goldstine, F. J. Murray, and J. von Neumann. The Jacobi method for real symmetric matrices. *J. ACM*, 6(1):59–96, 1959. (Cited on pages 131 and 281.)

[1372] H. H. Goldstine and J. von Neumann. Planning and coding of problems for an electronic computing instrument, Parts I, II, III. Institute for Advanced Study, Princeton, New Jersey, 1947-1948. (Cited on page 358.)

[1373] H. H. Goldstine and J. von Neumann. Numerical inverting of matrices of high order. II. *Proc. Am. Math. Soc.*, 2(2):188–202, 1951. (Cited on pages 20 and 79.)

[1374] H. H. Goldstine and J. von Neumann. On the principles of large scale computing machines. In A. H. Taub, editor, *John von Neumann Collected Works, Vol. 5*, pages 1–33, New York, 1963. Macmillan. (Cited on page 19.)

[1375] G. H. Golub. *The use of Chebyshev matrix polynomials in the iterative solution of linear equations compared to the method of successive overrelaxation*. PhD thesis, Dept. of Computer Science, University of Illinois at Urbana-Champaign, 1959. (Cited on page 181.)

[1376] G. H. Golub. Bounds for the round-off errors in the Richardson second order method. *BIT Numerical Mathematics*, 2(4):212–223, 1962. (Cited on page 181.)

[1377] G. H. Golub. Numerical methods for solving linear least squares problems. *Numer. Math.*, 7:206–216, 1965. (Cited on page 13.)

[1378] G. H. Golub. Least squares, singular values and matrix approximations. *Aplikace matematiky*, 13(1):44–51, 1968. (Cited on page 137.)

[1379] G. H. Golub. Some modified matrix eigenvalue problems. *SIAM Rev.*, 15(2):318–334, 1973. (Cited on page 212.)

[1380] G. H. Golub, M. Heath, and G. Wahba. Generalized cross-validation as a method to choosing a good ridge parameter. *Technometrics*, 21(2):215–223, 1979. (Cited on page 392.)

[1381] G. H. Golub, A. Hoffman, and G. W. Stewart. A generalization of the Eckart-Young-Mirsky matrix approximation theorem. *Linear Algebra Appl.*, 88:317–327, 1987. (Cited on page 136.)

[1382] G. H. Golub and W. Kahan. Calculating the singular values and pseudo-inverse of a matrix. *SIAM J. Numer. Anal.*, 2(2):205–224, 1965. (Cited on page 137.)

[1383] G. H. Golub and D. Mayers. The use of preconditioning over irregular regions. In R. Glowinski and J. L. Lions, editors, *Computing Methods in Applied Science and Engineering VI*, pages 3–14. North-Holland, 1984. (Cited on page 267.)

[1384] G. H. Golub and G. Meurant. Matrices, moments and quadrature. In D. F. Griffiths and G. A. Watson, editors, *Numerical Analysis 1993*, volume 303 of *Pitman Research Notes in Mathematics*, pages 105–156. Longman Sci. Tech., 1994. Reprinted in [611]. (Cited on pages 212, 244, 390, and 478.)

[1385] G. H. Golub and G. Meurant. Matrices, moments and quadrature II or how to compute the norm of the error in iterative methods. *BIT*, 37(3):687–705, 1997. (Cited on page 212.)

[1386] G. H. Golub and G. Meurant. *Matrices, Moments and Quadrature with Applications*. Princeton University Press, 2010. (Cited on pages 117, 205, 213, 297, 390, 392, and 478.)

[1387] G. H. Golub and C. B. Moler. Dedication to J. H. Wilkinson. *Linear Algebra Appl.*, 88/89(1-3):1–3, 1987. (Cited on pages 597 and 598.)

[1388] G. H. Golub, S. Nash, and C. Van Loan. A Hessenberg-Schur method for the problem $AX+XB = C$. *IEEE Trans. Automat. Contr.*, 24(6):909–913, 1979. (Cited on page 397.)

[1389] G. H. Golub and D. P. O'Leary. Some history of the conjugate gradient and Lanczos algorithms: 1948-1976. *SIAM Rev.*, 31(1):50–102, 1989. (Cited on pages 201 and 216.)

[1390] G. H. Golub and M. L. Overton. The convergence of inexact Chebyshev and Richardson iterative methods for solving linear systems. *Numer. Math.*, 53(5):571–593, 1988. (Cited on page 184.)

[1391] G. H. Golub and C. Reinsch. Singular value decomposition and least squares solutions. *Numer. Math.*, 14:403–420, 1970. (Cited on pages 137 and 362.)

[1392] G. H. Golub and Z. Strakoš. Estimates in quadratic formulas. *Numer. Algorithms*, 8(2):241–268, 1994. (Cited on pages 212 and 244.)

[1393] G. H. Golub and F. Uhlig. The QR algorithm: 50 years later its genesis by John Francis and Vera Kublanovskaya and subsequent developments. *IMA J. Numer. Anal.*, 29(3):467–485, 2009. (Cited on pages 311 and 508.)

[1394] G. H. Golub and R. Underwood. The block Lanczos method for computing eigenvalues. In J. Rice, editor, *Mathematical Software III, Proceedings of a Symposium conducted by the Mathematics Research Center, the University of Wisconsin-Madison, March 28-30, 1977*, pages 361–377. Academic Press, 1977. (Cited on page 303.)

[1395] G. H. Golub and H. A. van der Vorst. 150 years old and still alive: Eigenproblems. In I. S. Duff and G. A. Watson, editors, *The State of the Art in Numerical Analysis*, pages 93–119. Oxford University Press, 1997. (Cited on page 320.)

[1396] G. H. Golub and H. A. van der Vorst. Closer to the solutions: Iterative linear solvers. In I. S. Duff and G. A. Watson, editors, *The State of the Art in Numerical Analysis*, pages 63–92. Oxford University Press, 1997. (Cited on page 246.)

[1397] G. H. Golub and H. A. van der Vorst. Eigenvalue computation in the 20th century. *J. Comput. Appl. Math.*, 123(1-2):35–65, 2000. (Cited on page 320.)

[1398] G. H. Golub and C. F. Van Loan. An analysis of the total least squares problem. *SIAM J. Numer. Anal.*, 17(6):883–893, 1980. (Cited on page 213.)

[1399] G. H. Golub and C. F. Van Loan. *Matrix Computations*. 3rd edition. Johns Hopkins University Press, Baltimore, 1996. (Cited on pages 319 and 478.)

[1400] G. H. Golub and J. M. Varah. On a characterization of the best ℓ_2-scaling of a matrix. *SIAM J. Numer. Anal.*, 11(3):472–479, 1974. (Cited on pages 80 and 252.)

[1401] G. H. Golub and R. S. Varga. Chebyshev semi-iterative methods, successive overrelaxation iterative methods, and second order Richardson iterative methods. I. *Numer. Math.*, 3(1):147–156, 1961. (Cited on page 181.)

[1402] G. H. Golub and R. S. Varga. Chebyshev semi-iterative methods, successive overrelaxation iterative methods, and second order Richardson iterative methods. II. *Numer. Math.*, 3(1):157–168, 1961. (Cited on page 181.)

[1403] G. H. Golub and U. von Matt. Generalized cross-validation for large scale problems. In S. van Huffel, editor, *Recent Advances in Total Least Squares Techniques and Errors-in-variable Modelling*, pages 139–148, Philadelphia, 1997. SIAM. (Cited on page 392.)

[1404] G. H. Golub and U. von Matt. Tikhonov regularization for large scale problems. In G. H. Golub, S. H. Lui, F. Luk, and R. Plemmons, editors, *Scientific Computing*, pages 3–26, Berlin, 1997. Springer. (Cited on page 392.)

[1405] G. H. Golub and J. H. Welsch. Calculation of Gauss quadrature rules. *Math. Comput.*, 23(106):221–230, 1969. (Cited on page 212.)

[1406] G. H. Golub and J. H. Wilkinson. Ill-conditioned eigensystems and the computation of the Jordan canonical form. *SIAM Rev.*, 18(4):578–619, 1976. (Cited on page 149.)

[1407] G. H. Golub and Q. Ye. Inexact inverse iteration for generalized eigenvalue problems. *BIT*, 40(4):671–684, 2000. (Cited on page 317.)

[1408] J. R. Goodstein. Olga Taussky-Todd. *Notices Amer. Math. Soc.*, 67(3):345–353, 2020. (Cited on page 575.)

[1409] S. Goossens and D. Roose. Ritz and harmonic Ritz values and the convergence of FOM and GMRES. *Numer. Linear Algebra Appl.*, 6(4):281–293, 1999. (Cited on page 237.)

[1410] D. Gordon. Parallel ART for image reconstruction in CT using processor arrays. *Int. J. Parallel Emergent Distrib. Syst.*, 21:365–380, 2005. (Cited on page 187.)

[1411] D. Gordon. The Cimmino-Kaczmarz equivalence and related results. *Appl. Anal. Optim.*, 2:253–270, 2018. (Cited on page 439.)

[1412] R. Gordon, R. Bender, and G. T. Herman. Algebraic reconstruction techniques (ART) for three-dimensional electron microscopy and x-ray photography. *J. Theor. Biol.*, 29:471–481, 1970. (Cited on page 187.)

[1413] G. Gosselin. *De Arte Magna, seu de Occulta Parte Numerorum quae et Algebra et Almucabala Vulgo Dicitur*. Gilles Bets, Paris, 1577. (Cited on page 47.)

[1414] D. Gough. The success story of the transfer and development of methods from geophysics to helioseismology. In B. H. Jacobsen, K. Mosegaard, and P. Sibani, editors, *Inverse Methods*, pages 1–31. Springer, Berlin, Heidelberg, 1996. (Cited on page 394.)

[1415] N. I. Gould and J. A. Scott. On approximate inverse preconditioners. Technical Report RAL-TR-95-026, Rutherford Appleton Laboratory, UK, 1995. (Cited on page 263.)

[1416] N. I. Gould and J. A. Scott. Sparse approximate-inverse preconditioners using norm-minimization techniques. *SIAM J. Sci. Comput.*, 19(2):605–625, 1998. (Cited on page 263.)

[1417] R. Gower and P. Richtárik. Randomized iterative methods for linear systems. *SIAM J. Matrix Anal. Appl.*, 36(4):1660–1690, 2015. (Cited on page 191.)

[1418] T. Gowers. Is Mathematics discoverd or invented? In J. Polkinghorne, editor, *Meaning in Mathematics*, pages 3–12. Oxford University Press, 2011. (Cited on page xvii.)

[1419] W. B. Gragg. The Padé table and its relation to certain algorithms of numerical analysis. *SIAM Rev.*, 14(1):1–62, 1972. (Cited on page 385.)

[1420] W. B. Gragg. Matrix interpretations and applications of the continued fraction algorithm. *Rocky Mountain J. Math.*, 4(2):213–225, 1974. (Cited on page 221.)

[1421] W. B. Gragg and W. J. Harrod. The numerically stable reconstruction of Jacobi matrices from spectral data. *Numer. Math.*, 44(3):317–335, 1984. (Cited on page 316.)

[1422] W. B. Gragg and L. Reichel. On the application of orthogonal polynomials to the iterative solution of linear systems of equations with indefinite or non-Hermitian matrices. *Linear Algebra Appl.*, 88:349–371, 1987. (Cited on page 183.)

[1423] J. P. Gram. Ueber die entwickelung reeller functionen in reihen mittelst der methode der kleinsten quadrate. *Journal für die reine und angewandte Mathematik (Crelle Journal)*, 94:41–73, 1883. (Cited on page 124.)

[1424] H. Grassmann. *Die Lineale Ausdehnungslehre*. Wiegand, Leipzig, 1844. (Cited on pages 3 and 122.)

[1425] H. Grassmann. *Die Ausdehnungslehre*. Enslin, Berlin, 1862. (Cited on pages 3 and 122.)

[1426] I. Grattan-Guinness. The mathematics of the past: Distinguishing its history from our heritage. *Historia Math.*, 31(2):163–185, 2004. (Cited on page xvi.)

[1427] I. Grattan-Guinness and W. Ledermann. Matrix theory. In I. Grattan-Guinness, editor, *Companion Encyclopedia of the History and Philosophy of the Mathematical Sciences*, volume 1, pages 775–786. Routledge, 1994. (Cited on page 10.)

[1428] R. P. Graves. *Life of Sir William Rowan Hamilton*. Hodges, Higgis & Co, Dublin, 1882. vol. I 1882, vol. II 1885, vol. III 1889. (Cited on page 481.)

[1429] P. R. Graves-Morris. A "Look-around Lanczos" algorithm for solving a system of linear equations. *Numer. Algorithms*, 3-4:247–274, 1997. (Cited on page 235.)

[1430] P. R. Graves-Morris and A. Salam. Avoiding breakdown in van der Vorst's method. *Numer. Algorithms*, 21:205–223, 1999. (Cited on page 237.)

[1431] J. Gray. The history of the concept of a finite-dimensional vector space. *Historia Math.*, 7:65–70, 1980. (Cited on page 3.)

[1432] J. F. Grcar. *Analysis of the Lanczos algorithm and of the approximation problem in Richardson's method*. PhD thesis, University of Illinois at Urbana-Champaign, 1981. (Cited on page 302.)

[1433] J. F. Grcar. How ordinary elimination became Gaussian elimination. *Historia Math.*, 38(2):163–218, 2011. (Cited on pages 39, 70, and 76.)

[1434] J. F. Grcar. John von Neumann's analysis of Gaussian elimination and the origins of modern numerical analysis. *SIAM Rev.*, 53(4):607–682, 2011. (Cited on page 19.)

[1435] J. F. Grcar. Mathematicians of Gaussian elimination. *Notices Amer. Math. Soc.*, 58(6):782–792, 2011. (Cited on pages 33, 39, and 59.)

[1436] J. F. Grcar. Review of The Chinese Roots of Linear Algebra by Roger Hart. arXiv preprint arXiv:1104.2282, 2011. (Cited on pages 33 and 39.)

[1437] A. Greenbaum. *Convergence properties of the conjugate gradient algorithm in exact and finite precision arithmetic*. PhD thesis, University of California, Berkeley, 1981. (Cited on pages 211 and 302.)

[1438] A. Greenbaum. Analysis of a multigrid method as an iterative technique for solving linear systems. *SIAM J. Numer. Anal.*, 21(3):473–485, 1984. (Cited on page 270.)

[1439] A. Greenbaum. Behavior of slightly perturbed Lanczos and conjugate gradient recurrences. *Linear Algebra Appl.*, 113:7–63, 1989. (Cited on pages 211 and 302.)

[1440] A. Greenbaum. The Lanczos and conjugate gradient algorithms in finite precision arithmetic. In J. D. Brown, M. T. Chu, D. C. Ellison, and R.J. Plemmons, editors, *Proceedings of the Cornelius Lanczos International Centenary Conference (1993)*, pages 49–60. SIAM, Philadelphia, 1994. (Cited on pages 211 and 302.)

[1441] A. Greenbaum. Estimating the attainable accuracy of recursively computed residual methods. *SIAM J. Matrix Anal. Appl.*, 18(3):535–551, 1997. (Cited on pages 211 and 233.)

[1442] A. Greenbaum. *Iterative Methods for Solving Linear Systems*. SIAM, Philadelphia, 1997. (Cited on pages 211, 235, and 302.)

[1443] A. Greenbaum. Generalizations of the field of values useful in the study of polynomial functions of a matrix. *Linear Algebra Appl.*, 347:233–249, 2002. (Cited on pages 185, 236, and 240.)

[1444] A. Greenbaum. Some theoretical results derived from polynomial numerical hulls of Jordan blocks. *Electron. Trans. Numer. Anal.*, 18:81–90, 2004. (Cited on pages 236 and 240.)

[1445] A. Greenbaum, H. Liu, and T. Chen. On the convergence rate of variants of the conjugate gradient algorithm in finite precision arithmetic. *SIAM J. Sci. Comput.*, 43(5):S496–S515, 2021. (Cited on page 216.)

[1446] A. Greenbaum and M. L. Overton. Numerical investigation of Crouzeix's conjecture. *Linear Algebra Appl.*, 542:225–245, 2018. (Cited on page 241.)

[1447] A. Greenbaum, V. Pták, and Z. Strakoš. Any nonincreasing convergence curve is possible for GMRES. *SIAM J. Matrix Anal. Appl.*, 17(3):465–469, 1996. (Cited on page 231.)

[1448] A. Greenbaum and G. H. Rodrigue. Incomplete Cholesky conjugate gradient method for the STAR (5-point operator). Technical Report UCID-17574, Lawrence Livermore Laboratory, Livermore, 1977. (Cited on page 255.)

[1449] A. Greenbaum and G. H. Rodrigue. Optimal preconditioners of a given sparsity pattern. *BIT Numerical Mathematics*, 29(4):610–634, 1989. (Cited on page 252.)

[1450] A. Greenbaum, M. Rozložník, and Z. Strakoš. Numerical behaviour of the modified Gram-Schmidt GMRES implementation. *BIT*, 37(3):706–719, 1997. (Cited on page 234.)

[1451] A. Greenbaum and Z. Strakoš. Predicting the behavior of finite precision Lanczos and conjugate gradient computations. *SIAM J. Matrix Anal. Appl.*, 13(1):121–137, 1992. (Cited on pages 211 and 303.)

[1452] A. Greenbaum and Z. Strakoš. Matrices that generate the same Krylov residual spaces. In G. H. Golub, A. Greenbaum, and M. Luskin, editors, *Recent Advances in Iterative Methods*, pages 95–118. Springer, 1994. (Cited on pages 231 and 238.)

[1453] A. Greenbaum and L. N. Trefethen. GMRES/CR and Arnoldi/Lanczos as matrix approximation problems. *SIAM J. Sci. Comput.*, 15(2):359–368, 1994. (Cited on page 231.)

[1454] D. Greenspan. Methods of matrix inversion. *Am. Math. Mon.*, 62(5):303–318, 1955. (Cited on page 77.)

[1455] J. Greenstadt. A method for finding roots of arbitrary matrices. *M.T.A.C.*, 9(50):47–52, 1955. (Cited on pages 151 and 281.)

[1456] J. Greenstadt. Some numerical experiments in triangularizing matrices. *Numer. Math.*, 4(1):187–195, 1962. (Cited on pages 151 and 281.)

[1457] R. T. Gregory. Computing eigenvalues and eigenvectors of a symmetric matrix on the ILLIAC. *M.T.A.C.*, 7(44):215–220, 1953. (Cited on page 281.)

[1458] R. T. Gregory. Results using Lanczos' method for finding eigenvalues of arbitrary matrices. *J. Soc. Ind. Appl. Math.*, 6(2):182–188, 1958. (Cited on page 301.)

[1459] R. T. Gregory and D. L. Karney. *A Collection of Matrices for Testing Computational Algorithms.* Wiley-Interscience, New York, 1969. (Cited on page 373.)

[1460] M. S. Grewal and A. P. Andrews. *Kalman Filtering: Theory and Practice Using MATLAB.* John Wiley & Sons, New York, 3rd edition, 2008. (Cited on page 399.)

[1461] D. A. Grier. *When Computers Were Human.* Princeton University Press, Princeton and Oxford, 2005. (Cited on page 449.)

[1462] L. Grigori, E. Boman, S. Donfack, and T. A. Davis. Hypergraph-based unsymmetric nested dissection ordering for sparse LU factorization. *SIAM J. Sci. Comput.*, 32(6):3426–3446, 2010. (Cited on page 90.)

[1463] L. Grigori, M. Cosnard, and E. G. Ng. On the row merge tree for sparse LU factorization with partial pivoting. *BIT Numerical Mathematics*, 47(1):45–76, 2007. (Cited on page 93.)

[1464] L. Grigori, J. R. Gilbert, and M. Cosnard. Symbolic and exact structure prediction for sparse Gaussian elimination with partial pivoting. *SIAM J. Matrix Anal. Appl.*, 30(4):1520–1545, 2008. (Cited on page 93.)

[1465] L. Grigori, S. Moufawad, and F. Nataf. Enlarged Krylov subspace conjugate gradient methods for reducing communication. *SIAM J. Matrix Anal. Appl.*, 37(2):744–773, 2016. (Cited on page 216.)

[1466] R. Grimes, D. Kincaid, and D. Young. ITPACK 2A: A FORTRAN implementation of adaptive accelerated iterative methods for solving large sparse linear systems. Technical Report CNA-164, Center for Numerical Analysis, University of Texas, Austin, 1980. (Cited on page 372.)

[1467] G. Grioli. Una proprietà di minimo nella cinematica delle deformazioni finite. *Boll. Un. Math. Ital.*, 2:252–255, 1940. (Cited on page 142.)

[1468] M. J. Grote and T. Huckle. Parallel preconditioning with sparse approximate inverses. *SIAM J. Sci. Comput.*, 18(3):838–853, 1997. (Cited on page 263.)

[1469] M. J. Grote and H. D. Simon. Parallel preconditioning and approximate inverses on the Connection Machine. In R. Sincovec, editor, *Proceedings of the Sixth SIAM Conference on Parallel Processing for Scientific Computing*, pages 519–523. SIAM, Philadelphia, 1993. (Cited on page 262.)

[1470] M. Gu and S. C. Eisenstat. A stable and efficient algorithm for the rank-one modification of the symmetric eigenproblem. *SIAM J. Matrix Anal. Appl.*, 15(4):1266–1276, 1994. (Cited on page 316.)

[1471] M. Gu and S. C. Eisenstat. A divide-and-conquer algorithm for the bidiagonal SVD. *SIAM J. Matrix Anal. Appl.*, 16(1):79–92, 1995. (Cited on page 138.)

[1472] M. Gu and S. C. Eisenstat. A divide-and-conquer algorithm for the symmetric tridiagonal eigenproblem. *SIAM J. Matrix Anal. Appl.*, 16(1):172–191, 1995. (Cited on page 315.)

[1473] M. Gu and S. C. Eisenstat. Efficient algorithms for computing a strong rank-revealing QR factorization. *SIAM J. Sci. Comput.*, 17(4):848–869, 1996. (Cited on page 13.)

[1474] A. El Guennouni, K. Jbilou, and H. Sadok. A block version of BiCGStab for linear systems with multiple right-hand sides. *Electron. Trans. Numer. Anal.*, 16:129–142, 2003. (Cited on page 240.)

[1475] A. El Guennouni, K. Jbilou, and H. Sadok. The block-Lanczos method for linear systems with multiple right-hand sides. *Appl. Numer. Math.*, 51:243–256, 2004. (Cited on page 240.)

[1476] N. Guglielmi, M. L. Overton, and G. W. Stewart. An efficient algorithm for computing the generalized null space decomposition. *SIAM J. Matrix Anal. Appl.*, 36(1):38–54, 2015. (Cited on page 150.)

[1477] A. Guilbaud and G. Jouve. La résolution des équations aux dérivées partielles dans les Opuscules mathématiques de D'Alembert (1761-1783). *Rev. Hist. Math.*, 15(1):59–122, 2009. (Cited on page 277.)

[1478] R. Guivarch and P. Spiteri. Implantation de méthodes de sous-domaines asynchrones avec PVM et MPI sur IBM-SP2. *Calculateurs Parallèles*, 10(4):431–438, 1998. (Cited on page 196.)

[1479] B. C. Gunter and R. A. van de Geijn. Parallel out-of-core computation and updating of the QR factorization. *ACM Trans. Math. Soft. (TOMS)*, 31(1):60–78, 2005. (Cited on page 133.)

[1480] C.-H. Guo. On a quadratic matrix equation associated with an M-matrix. *IMA J. Numer. Anal.*, 23(1):11–27, 2003. (Cited on page 394.)

[1481] S. Guo. *Jiuzhang Suanshu (Nine Chapters on Mathematics)*. Liaoning Educational Press, 1990. (Cited on page 35.)

[1482] A. Gupta. Improved symbolic and numerical factorization algorithms for unsymmetric sparse matrices. *SIAM J. Matrix Anal. Appl.*, 24(2):529–552, 2002. (Cited on page 377.)

[1483] A. Gupta. Recent advances in direct methods for solving unsymmetric sparse systems of linear equations. *ACM Trans. Math. Soft. (TOMS)*, 28(3):301–324, 2002. (Cited on page 377.)

[1484] A. Gupta. A shared and distributed memory parallel general sparse direct solver. *Applic. Algebra in Eng. Comm. and Comput.*, 18(3):263–277, 2007. (Cited on page 377.)

[1485] J. Gustafson. First electronic digital calculating machine forerunner to Cornell's FPS-164/MAX. *Forefronts*, 1(6), 1985. (Cited on page 335.)

[1486] J. Gustafson. Reconstruction of the Atanasoff-Berry computer. In *The First Computers: History and Architectures*, pages 91–106, Cambridge, MA, 2000. The MIT Press. (Cited on page 335.)

[1487] J. Gustafson. *The End of Error, Unum Computing*. CRC Press, 2015. (Cited on page 355.)

[1488] I. Gustafsson. A class of first order factorization methods. *BIT*, 18:142–156, 1978. (Cited on page 255.)

[1489] I. Gustafsson. On modified incomplete factorization methods. In J. Hinze, editor, *Numerical Integration of Differential Equations and Large Linear Systems*, pages 334–351, Berlin, Heidelberg, 1982. Springer. Proceedings of two workshops held at the University of Bielefeld, Spring 1980. (Cited on page 255.)

[1490] F. G. Gustavson and J. Wasniewski. Gauss's, Cholesky's and Banachiewicz's contributions to least squares. Technical Report IMM-TR-2011, No. 16, Technical University of Denmark, 2011. (Cited on page 59.)

[1491] F. V. Gustavson. Some basic techniques for solving sparse systems of linear equations. In D. J. Rose and R. A. Willoughby, editors, *Sparse Matrices and their Applications*, pages 41–52. Plenum Press, 1972. (Cited on page 86.)

[1492] F. V. Gustavson. Finding the block lower triangular form of a sparse matrix. In J. R. Bunch and D. J. Rose, editors, *Sparse Matrix Computations*, pages 275–290. Academic Press, 1976. (Cited on page 94.)

[1493] M. H. Gutknecht. Stationary and almost stationary iterative (k, l)-step methods for linear and nonlinear systems of equations. *Numer. Math.*, 56:179–213, 1989. (Cited on page 184.)

[1494] M. H. Gutknecht. The pionner days of scientific computing in Switzerland. In S. G. Nash, editor, *A History of Scientific Computing*, pages 301–313. ACM Press, 1990. (Cited on page 551.)

[1495] M. H. Gutknecht. The pioneer days of scientific computing in Switzerland. *IPS Windows*, 2:23–27, 1991. And *Crosscuts*, v 2 n 3 (1993), pp. 14-17. (Cited on pages 551 and 567.)

[1496] M. H. Gutknecht. A completed theory of the unsymmetric Lanczos process and related algorithms. Part I. *SIAM J. Matrix Anal. Appl.*, 13(2):594–639, 1992. (Cited on page 228.)

[1497] M. H. Gutknecht. Variants of BICGSTAB for matrices with complex spectrum. *SIAM J. Sci. Comput.*, 14(5):1020–1033, 1993. (Cited on page 230.)

[1498] M. H. Gutknecht. A completed theory of the unsymmetric Lanczos process and related algorithms. Part II. *SIAM J. Matrix Anal. Appl.*, 15(1):15–58, 1994. (Cited on page 228.)

[1499] M. H. Gutknecht. Lanczos-type solvers for nonsymmetric linear systems of equations. *Acta Numerica*, 6:271–397, 1997. (Cited on page 235.)

[1500] M. H. Gutknecht. Numerical analysis in Zürich - 50 years ago. *Zürich Intelligencer*, 10-15, 2007. (Cited on pages 552 and 567.)

[1501] M. H. Gutknecht. IDR explained. *Electron. Trans. Numer. Anal.*, 36(3):126–148, 2010. (Cited on page 243.)

[1502] M. H. Gutknecht. Spectral deflation in Krylov solvers: A theory of coordinate space based methods. *Electron. Trans. Numer. Anal.*, 39:156–185, 2012. (Cited on pages 239 and 244.)

[1503] M. H. Gutknecht. Revisiting (k, ℓ)-step methods. *Numer. Algorithms*, 69(2):455–469, 2015. (Cited on page 184.)

[1504] M. H. Gutknecht and B. N. Parlett. From qd to LR, or, how were the qd and LR algorithms discovered? *IMA J. Numer. Anal.*, 31(3):741–754, 2010. (Cited on pages 309, 310, and 552.)

[1505] M. H. Gutknecht and K. J. Ressel. Look-ahead procedures for Lanczos-type product methods based on three-term Lanczos recurrences. *SIAM J. Matrix Anal. Appl.*, 21(4):1051–1078, 2000. (Cited on page 238.)

[1506] M. H. Gutknecht and S. Röllin. The Chebyshev iteration revisited. *Parallel Comput.*, 28(2):263–283, 2002. (Cited on page 185.)

[1507] M. H. Gutknecht and M. Rozložník. By how much can residual minimization accelerate the convergence of orthogonal residual methods? *Numer. Algorithms*, 27:189–213, 2001. (Cited on page 239.)

[1508] M. H. Gutknecht and M. Rozložník. Residual smoothing techniques: Do they improve the limiting accuracy of iterative solvers? *BIT Numerical Mathematics*, 41(1):86–114, 2001. (Cited on page 239.)

[1509] M. H. Gutknecht and Z. Strakoš. Accuracy of two three-term and three two-term recurrences for Krylov space solvers. *SIAM J. Matrix Anal. Appl.*, 22(1):213–229, 2000. (Cited on page 211.)

[1510] M. H. Gutknecht and J.-P. M. Zemke. Eigenvalue computations based on IDR. *SIAM J. Matrix Anal. Appl.*, 34(2):283–311, 2013. (Cited on page 319.)

[1511] S. Güttel. *Rational Krylov methods for operator functions.* PhD thesis, Technische Universität Bergakademi Freiberg, Freiberg, Germany, 2010. (Cited on page 390.)

[1512] S. Güttel. Rational Krylov approximation of matrix functions: Numerical methods and optimal pole selection. *GAMM Mitteilungen*, 36(1):8–31, 2013. (Cited on page 390.)

[1513] S. Güttel, D. Kressner, and K. Lund. Limited-memory polynomial methods for large-scale matrix functions. *GAMM Mitteilungen*, 43(3):article e202000019, 2020. (Cited on page 390.)

[1514] S. Güttel, E. Polizzi, P. T. P. Tang, and G. Viaud. Zolotarev quadrature rules and load balancing for the FEAST eigenvalue solver. *SIAM J. Sci. Comput.*, 37(4):2100–2122, 2015. (Cited on page 318.)

[1515] S. Güttel and F. Tisseur. The nonlinear eigenvalue problem. *Acta Numerica*, 26:1–94, 2017. (Cited on page 318.)

[1516] L. Guttman. Enlargement methods for computing the inverse matrix. *Ann. Math. Stat.*, 17:336–343, 1946. (Cited on page 83.)

[1517] G. J. Habetler and E. L. Wachspress. Symmetric successive overrelaxation in solving diffusion difference equations. *Math. Comput.*, 15(76):356–362, 1961. (Cited on pages 176 and 177.)

[1518] K. Habgood and I. Arel. A condensation-based application of Cramer's rule for solving large-scale linear systems. *J. Discrete Algorithms*, 10:98–109, 2012. (Cited on pages 114 and 117.)

[1519] M. Habu and T. Nodera. GMRES(m) algorithm with changing the restart cycle adaptively. In *Proceedings of Algorithmy 2000 Conference on Scientific Computing*, pages 354–263, 2000. (Cited on page 232.)

[1520] G. D. Hachtel. The sparse tableau approach to finite element assembly. In J. R. Bunch and D. J. Rose, editors, *Sparse Matrix Computations*, pages 349–363. Academic Press, 1976. (Cited on page 86.)

[1521] G. D. Hachtel, R. K. Brayton, and F. G. Gustavson. The sparse tableau approach to network analysis and design. *IEEE Trans. Circuit Theory*, 18(1):101–113, 1971. (Cited on page 86.)

[1522] W. Hackbusch. Convergence of multigrid iterations applied to difference equations. *Math. Comput.*, 34(150):425–440, 1980. (Cited on page 269.)

[1523] W. Hackbusch. *Multigrid Methods and Applications*. Springer, 1985. (Cited on page 270.)

[1524] W. Hackbusch. *Iterative Solution of Large Sparse Systems of Equations*. Springer, 1994. (Cited on page 230.)

[1525] J. Hadamard. Essai sur l'étude des fonctions données par leur développement de Taylor. *J. Math. Pures Appl.*, 4ème série, 8:101–186, 1892. (Cited on page 308.)

[1526] J. Hadamard. *Leçons sur la Propagation des Ondes*. Hermann et fils, Paris, 1903. (Cited on page 286.)

[1527] S. M. Hadfield and T. A. Davis. Lost pivot recovery for an unsymmetric-pattern multifrontal method. Technical Report TR-94-029, CISE Dept., Univ. of Florida, 1994. (Cited on page 91.)

[1528] S. M. Hadfield and T. A. Davis. The use of graph theory in a parallel multifrontal method for sequences of unsymmetric pattern sparse matrices. *Cong. Numer.*, 108:43–52, 1995. (Cited on page 91.)

[1529] A. Hadjidimos. Accelerated overrelaxation method. *Math. Comput.*, 32(141):149–157, 1978. (Cited on page 178.)

[1530] A. Hadjidimos. A survey of the iterative methods for the solution of linear systems by extrapolation, relaxation and other techniques. *J. Comput. Appl. Math.*, 20:37–51, 1987. (Cited on page 178.)

[1531] A. Hadjidimos. Successive overrelaxation (SOR) and related methods. *J. Comput. Appl. Math.*, 123(1-2):177–199, 2000. (Cited on page 178.)

[1532] P. Hagander. Numerical solution of $A^tS + SA + Q = 0$. Technical Report TFRT-7008, Report 6920, Department of Automatic Control, Lund Institute of Technology (LTH), 1969. (Cited on page 398.)

[1533] L. A. Hageman and D. M. Young. *Applied Iterative Methods*. Academic Press, 1981. Reprinted by Dover, 2003. (Cited on pages 175, 600, and 601.)

[1534] W. W. Hager. Condition estimates. *SIAM J. Sci. Statist. Comput.*, 5(2):311–316, 1984. (Cited on page 23.)

[1535] W. W. Hager. Updating the inverse of a matrix. *SIAM Rev.*, 31(2):221–239, 1989. (Cited on page 83.)

[1536] T. Haigh. An interview with Brian Ford, Steve Hague and Sven Hammarling. SIAM, Philadelphia, 2004. (Cited on page 375.)

[1537] T. Haigh. An interview with Charles L. Lawson. SIAM, Philadelphia, 2004. (Cited on pages 523 and 524.)

[1538] T. Haigh. An interview with Charles W. Johnson. SIAM, Philadelphia, 2004. (Cited on page 375.)

[1539] T. Haigh. An interview with Cleve Moler. SIAM, Philadelphia, 2004. (Cited on page 367.)

[1540] T. Haigh. An interview with Jack J. Dongarra. SIAM, Philadelphia, 2004. (Cited on page 367.)

[1541] T. Haigh. An interview with James C.T. Pool. SIAM, Philadelphia, 2004. (Cited on page 367.)

[1542] T. Haigh. An interview with Tom Aird. SIAM, Philadelphia, 2004. (Cited on page 375.)

[1543] T. Haigh. An interview with Gene Golub. SIAM, Philadelphia, 2005. (Cited on pages 181 and 477.)

[1544] T. Haigh. An interview with Iain S. Duff. SIAM, Philadelphia, 2005. (Cited on page 94.)

[1545] T. Haigh. An interview with William M. Kahan. SIAM, Philadelphia, 2005. (Cited on page 344.)

[1546] T. Haigh. Cleve Moler: Mathematical software pioneer and creator of Matlab. *IEEE Annals of the history of computing*, 30(1):87–91, 2008. (Cited on page 379.)

[1547] J. K. Hale. In memoriam Lamberto Cesari (1910-1990). *J. Diff. Equ.*, 96:201–202, 1992. (Cited on page 429.)

[1548] N. Halko, P. G. Martinsson, and J. A. Tropp. Finding structure with randomness: Probabilistic algorithms for constructing approximate matrix decompositions. *SIAM Rev.*, 53(2):217–288, 2011. (Cited on page 140.)

[1549] A. Hall. On Gauss' method of elimination. *Annals of Mathematics*, 8(1/6):64, 1893-1894). (Cited on page 62.)

[1550] F. J. Hall, T. L. Markham, M. Rozložník, and J. Stuart. Miroslav Fiedler (7. 4. 1926-20. 11. 2015). *Czechoslovak Math. J.*, 66(3):585–590, 2015. (Cited on pages 16 and 177.)

[1551] E. Halley. An instance of the excellence of the modern algebra, in the resolution of the problem of finding the foci of optick glasses universally. *Philos. Trans. R. Soc. Lond.*, 17:960–969, 1693. (Cited on page 508.)

[1552] E. Hallman. Sharp 2-norm error bounds for LSQR and the conjugate gradient method. *SIAM J. Matrix Anal. Appl.*, 41(3):1183–1207, 2020. (Cited on page 213.)

[1553] P. R. Halmos. *Finite-dimensional Vector Spaces*. Van Nostrand, 2nd edition, 1958. (Cited on page 288.)

[1554] P. R. Halmos. *A Hilbert Space Problem Book*. Springer, 1967. (Cited on page 288.)

[1555] P. R. Halmos. The legend of John von Neumann. *Am. Math. Mon.*, 80:382–394, 1973. (Cited on page 536.)

[1556] G. B. Halsted. Eugenio Beltrami. *Am. Math. Mon.*, 9(3):59–63, 1902. (Cited on page 417.)

[1557] H. L. Hamburger. Hermitian transformations of deficiency-index (1,1), Jacobi matrices and undetermined moment problems. *Amer. J. Math.*, 66(4):489–522, 1944. (Cited on page 382.)

[1558] W. R. Hamilton. On a new species of imaginary quantities connected with a theory of quaternions. *Proceedings of the Royal Irish Academy (1836-1869)*, 2:424–434, 1840. (Cited on pages 3 and 481.)

[1559] W. R. Hamilton. *Lectures on Quaternions: Containing a Systematic Statement of a New Mathematical Method; of which the Principles Were Communicated in 1843 to the Royal Irish Academy;* Hodges and Smith, 1853. (Cited on page 481.)

[1560] W. R. Hamilton. On the existence of a symbolic and biquadratic equation which is satisfied by the symbol of linear or distributive operation on a quaternion. *The London, Edinburgh and Dublin Philosophical Magazine and Journal of Science*, 24:127–128, 1862. (Cited on page 481.)

[1561] W. R. Hamilton. On a new and general method of inverting a linear and quaternion function of a quaternion. *Proceedings of the Royal Irish Academy*, 8:182–183, 1864. (Cited on page 481.)

[1562] W. R. Hamilton. On the existence of a symbolic and biquadratic equation, which is satisfied by the symbol of linear operation in quaternions. *Proceedings of the Royal Irish Academy*, 8:190–191, 1864. (Cited on page 481.)

[1563] S. Hammarling. A note on modifications to the Givens plane rotation. *IMA J. Appl. Math.*, 13:215–218, 1974. (Cited on page 131.)

[1564] S. Hammarling. Numerical solution of the stable, non-negative definite Lyapunov equation. *IMA J. Numer. Anal.*, 2(3):303–323, 1982. (Cited on page 398.)

[1565] S. Hammarling. James Hardy Wilkinson 27 September 1919 - 5 October 1986. Technical Report MIMS Eprint 2019-10, Manchester Institute for Mathematical Sciences, University of Manchester, 2019. (Cited on page 598.)

[1566] M. Hanke. *Conjugate Gradient Type Methods for Ill-posed Problems.* Longman, Harlow, 1995. (Cited on page 391.)

[1567] M. Hanke. Limitations of the L-curve method in ill-posed problems. *BIT Numerical Mathematics*, 36(2):287–301, 1996. (Cited on page 392.)

[1568] M. Hanke. On Lanczos based methods for the regularization of discrete ill-posed problems. *BIT Numerical Mathematics*, 41(5):1008–1018, 2001. (Cited on page 393.)

[1569] M. Hanke and P. C. Hansen. Regularization methods for large scale problems. *Surv. Math. Ind.*, 3(4):253–315, 1993. (Cited on page 391.)

[1570] M. Hanke and T. Raus. A general heuristic for choosing the regularization parameter in ill-posed problems. *SIAM J. Sci. Comput.*, 17(4):956–972, 1996. (Cited on page 392.)

[1571] H. Hankel. Über eine besondere classe des symmetrischen determinanten. Inaugural Dissertation, Universität Göttingen, 1861. (Cited on page 385.)

[1572] T. L. Hankins. In defence of biographies: The use of biography in the history of science. *Hist. Sci.*, 17:1–16, 1979. (Cited on page xix.)

[1573] T. L. Hankins. *Sir William Rowan Hamilton.* The Johns Hopkins University Press, Baltimore, 1980. (Cited on page 481.)

[1574] J. Hannah. False position in Leonardo of Pisa's Liber Abbaci. *Historia Math.*, 34:306–332, 2007. (Cited on page 43.)

[1575] E. R. Hansen. On the Danilewski method. *J. ACM*, 10(1):102–109, 1963. (Cited on page 291.)

[1576] P. C. Hansen. The truncated SVD as a method for regularization. *BIT Numerical Mathematics*, 27(4):534–553, 1987. (Cited on page 394.)

[1577] P. C. Hansen. Analysis of discrete ill-posed problems by means of the L-curve. *SIAM Rev.*, 34(4):561–580, 1992. (Cited on page 392.)

[1578] P. C. Hansen. Regularization tools: a Matlab package for analysis and solution of discrete ill-posed problems. *Numer. Algorithms*, 6(1):1–35, 1994. (Cited on page 393.)

[1579] P. C. Hansen. *Rank-deficient and Discrete Ill-posed Problems: Numerical Aspects of Linear Inversion*. SIAM, Philadelphia, 1998. (Cited on page 391.)

[1580] P. C. Hansen. Regularization Tools version 4.0 for Matlab 7.3. *Numer. Algorithms*, 46(2):189–194, 2007. (Cited on page 393.)

[1581] P. C. Hansen. *Discrete Inverse Problems: Insight and Algorithms*. SIAM, Philadelphia, 2010. (Cited on page 391.)

[1582] P. C. Hansen and T. K. Jensen. Noise propagation in regularizing iterations for image deblurring. *Electron. Trans. Numer. Anal.*, 31:204–220, 2008. (Cited on page 393.)

[1583] P. C. Hansen, T. K. Jensen, and G. Rodriguez. An adaptive pruning algorithm for the discrete L-curve criterion. *J. Comput. Appl. Math.*, 198(2):483–492, 2007. (Cited on page 392.)

[1584] P. C. Hansen, J. G. Nagy, and D. P. O'Leary. *Deblurring Images: Matrices, Spectra, and Filtering*. SIAM, Philadelphia, 2006. (Cited on page 391.)

[1585] P. C. Hansen and D. P. O'Leary. The use of the L-curve in the regularization of discrete ill-posed problems. *SIAM J. Sci. Comput.*, 14(6):1487–1503, 1993. (Cited on page 392.)

[1586] P. C. Hansen, T. Sekii, and H. Shibahashi. The modified truncated SVD method for regularization in general form. *SIAM J. Sci. Statist. Comput.*, 13(5):1142–1150, 1992. (Cited on page 394.)

[1587] R. J. Hanson. A numerical method for solving Fredholm integral equations of the first kind using singular values. *SIAM J. Numer. Anal.*, 8(3):616–622, 1971. (Cited on page 393.)

[1588] R. J. Hanson, F. T. Krogh, and C. L. Lawson. A proposal for standard linear algebra subprograms. Technical Report Memorandum 33-660, Jet Propulsion Laboratory, California, 1973. (Cited on page 364.)

[1589] F. Harary. A graph theoretic method for the complete reduction of a matrix with a view toward finding its eigenvalues. *J. Math. Phys.*, 38(1-4):104–111, 1959. (Cited on page 86.)

[1590] F. Harary. A graph theoretic approach to matrix inversion by partitioning. *Numer. Math.*, 4(1):128–135, 1962. (Cited on page 86.)

[1591] F. Harary. Sparse matrices and graph theory. In J. K. Reid, editor, *Large Sparse Sets of Linear Equations*, pages 139–150, Academic Press, London and New York, 1971. (Cited on page 86.)

[1592] T. Harriot. *Artis Analyticæ Praxis*. Robert Barker, London, 1631. (Cited on page 50.)

[1593] B. Harris. Lagrange: A well-behaved function. *The Montana Mathematics Enthusiast*, 4:128–137, 2007. (Cited on page 509.)

[1594] R. Hart. *The Chinese Roots of Linear Algebra*. Johns Hopkins University Press, Baltimore, 2011. (Cited on page 39.)

[1595] U. Hashagen. An interview with Fritz Bauer. SIAM, Philadelphia, 2004. (Cited on page 414.)

[1596] H. Hasse. Kurt Hensel zum Gedächtnis. *Journal für die reine und angewandte Mathematik (Crelle Journal)*, 187:1–13, 1950. (Cited on page 499.)

[1597] F. Hausdorff. Der wertevorrat einer bilinearform. *Math. Z.*, 3:314–316, 1919. (Cited on page 287.)

[1598] H. E. Hawkes. The reduction of families of bilinear forms. *Amer. J. Math.*, 32(2):101–114, 1910. (Cited on page 149.)

[1599] T. Hawkins. The theory of matrices in the 19th century. In *Proc. International Congress of Mathematicians, Vancouver*, pages 561–570, 1974. (Cited on pages 6, 12, and 424.)

[1600] T. Hawkins. Cauchy and the spectral theory of matrices. *Historia Math.*, 2:1–29, 1975. (Cited on pages 12, 275, 280, 281, and 420.)

[1601] T. Hawkins. Weierstrass and the theory of matrices. *Archive for History of Exact Sciences*, 17(2):119–163, 1977. (Cited on pages 10 and 12.)

[1602] T. Hawkins. Continued fractions and the origins of the Perron-Frobenius theorem. *Archive for History of Exact Sciences*, 62(6):655–717, 2008. (Cited on page 546.)

[1603] T. Hawkins. *The Mathematics of Frobenius in Context*. Springer, 2013. (Cited on page 12.)

[1604] K. Hayami and M. Sugihara. A geometric view of Krylov subspace methods on singular systems. *Numer. Linear Algebra Appl.*, 18(3):449–469, 2011. (Cited on page 244.)

[1605] K. Hayami and M. Sugihara. Corrigendum to: A geometric view of Krylov subspace methods on singular systems. *Numer. Linear Algebra Appl.*, 21(5):701–702, 2014. (Cited on page 244.)

[1606] K. Hayami, J.-F. Yin, and T. Ito. GMRES methods for least squares problems. *SIAM J. Matrix Anal. Appl.*, 31(5):2400–2430, 2010. (Cited on page 243.)

[1607] Ta. Hayashi. *The Bakhshālī Manuscript: an Ancient Indian Mathematical Treatise*. Forsten, Groningen, 1995. Groningen Oriental Studies 11. (Cited on page 41.)

[1608] Ts. Hayashi. The "Fukudai" and determinants in Japanese mathematics. *Proc. Tokyo Mathematico-Physical Soc., Ser. 2*, 5(16):254–271, 1910. (Cited on pages 99 and 100.)

[1609] R. M. Hayes. Iterative methods of solving linear problems on Hilbert space. In O. Taussky, editor, *Contributions to the Solution of Systems of Linear Equations and the Determination of Eigenvalues*, pages 71–104. National Bureau of Standards, Washington, DC, 1954. (Cited on page 207.)

[1610] E. V. Haynsworth. Determination of the inertia of a partitioned Hermitian matrix. *Linear Algebra Appl.*, 1:73–81, 1968. (Cited on pages 117 and 484.)

[1611] E. V. Haynsworth. On the Schur complement. Basel Mathematical Notes, BMN 20, 1968. (Cited on pages 25 and 484.)

[1612] M. T. Heath. Some extensions of an algorithm for sparse linear least squares problems. *SIAM J. Sci. Statist. Comput.*, 3(2):223–237, 1982. (Cited on page 13.)

[1613] M. T. Heath, A. J. Laub, C. C. Paige, and R. C. Ward. Computing the singular value decomposition of a product of two matrices. *SIAM J. Sci. Statist. Comput.*, 7(4):1147–1159, 1986. (Cited on page 141.)

[1614] M. T. Heath, E. G. Ng, and B. W. Peyton. Parallel algorithms for sparse linear systems. *SIAM Rev.*, 33(3):420–460, 1991. (Cited on page 93.)

[1615] M. T. Heath and P. Raghavan. A Cartesian parallel nested dissection algorithm. *SIAM J. Matrix Anal. Appl.*, 16(1):235–253, 1995. (Cited on page 371.)

[1616] M. T. Heath and P. Raghavan. Performance of a fully parallel sparse solver. *Intl. J. Supercomp. Appl. High Perf. Comput.*, 11(1):49–64, 1997. (Cited on page 371.)

[1617] T. L. Heath. *Diophantos of Alexandria, a Study in the History of Greek Algebra*. Cambridge University Press, 1885. (Cited on page 40.)

[1618] J. Hecht. Un exemple de multidisciplinarité: Alexandre Vandermonde (1735-1796). *Population*, 26(4):641–676, 1971. (Cited on page 590.)

[1619] B. A. Hedman. An earlier date for "Cramer's rule". *Historia Math.*, 26:365–368, 1999. (Cited on pages 53, 104, and 105.)

[1620] A. Heeffer. The tacit appropriation of Hindu algebra in renaissance practical arithmetic. *Ganita Bhārāti*, 29(1-2):1–60, 2007. (Cited on pages 41 and 42.)

[1621] A. Heeffer. A conceptual analysis of early Arabic algebra. In S. Rahman, T. Street, and H.Tahiri, editors, *The Unity of Science in the Arabic Tradition*, pages 89–128, Springer, Dordrecht, 2008. (Cited on page 43.)

[1622] A. Heeffer. The emergence of symbolic algebra as a shift in predominant models. *Foundations of science*, 13(2):149–161, 2008. (Cited on page 48.)

[1623] A. Heeffer. The reception of ancient Indian mathematics by western historians. In *Ancient Indian Leaps into Mathematics*, pages 135–152. Birkhäuser, Boston, 2009. (Cited on page 42.)

[1624] A. Heeffer. From the second unknown to the symbolic equation. In *Philosophical Aspects of Symbolic Reasoning in Early Modern Mathematics*, volume 26, pages 57–101. College Publications, 2010. (Cited on page 44.)

[1625] A. Heeffer. The rhetoric of problems in algebra textbooks from Pacioli to Euler. Centre for Logic and Philosophy of Science, Universiteit Gent, Belgium, 2011. (Cited on page 53.)

[1626] A. Heeffer. The rule of quantity by Chuquet and de la Roche and its influence on German cossic algebra. In S. Rommevaux, M. Spiesser, and M. R. Massa Esteve, editors, *Pluralité de l'Algébre à la Renaissance*, pages 127–1147. Editions Honoré Champion, 2012. (Cited on page 44.)

[1627] A. Heeffer and T. Rothman. On remembering Cardano anew. *The Mathematical Intelligencer*, 36(4):53–66, 2014. (Cited on page 45.)

[1628] C. J. Hegedüs. Purcell's method, Egerváry and related results. *Cent. Eur. J. Oper. Res.*, 18(1):25–35, 2010. (Cited on page 81.)

[1629] I. Heger. Über die auflosung eines systemes von mehreren unbestimmten gleichungen des ersten grades in ganzen zahlen. *Denkschr. Akad. Wiss. (Wien)*, 14(2):1–121, 1858. (Cited on page 13.)

[1630] M. T. Heideman, D. H. Johnson, and C. S. Burrus. Gauss and the history of the fast Fourier transform. *Arch. Hist. Exact Sci.*, 34(3):265–277, 1985. (Cited on page 384.)

[1631] G. Heinig and K. Rost. Fast algorithms for Toeplitz and Hankel matrices. *Linear Algebra Appl.*, 435(1):1–59, 2011. (Cited on page 387.)

[1632] E. Hellinger and O. Toeplitz. Zur einordnung der kettenbruchtheorie in die theorie der quadratischen formen von unendlichvielen veränderlichen. *Journal für die reine und angewandte Mathematik (Crelle Journal)*, 144:212–238, 1914. (Cited on page 382.)

[1633] G. Helmberg, P. Wagner, and G. Veltkamp. On Faddeev-Leverrier's method for the computation of the characteristic polynomial of a matrix and of eigenvectors. *Linear Algebra Appl.*, 185:219–233, 1993. (Cited on page 289.)

[1634] P. W. Hemker. Introduction to multigrid methods. *Nieuw Archief voor Wiskunke*, 3(21):71–101, 1981. (Cited on page 269.)

[1635] P. W. Hemker. On the order of prolongations and restrictions in multigrid procedures. *J. Comput. Appl. Math.*, 32(3):423–429, 1990. (Cited on page 270.)

[1636] H. V. Henderson, F. Pukelsheim, and S. R. Searle. On the history of the Kronecker product. *Linear Multilinear Algebra*, 14:113–120, 1983. (Cited on pages 499 and 501.)

[1637] H. V. Henderson and S. R. Searle. On deriving the inverse of a sum of matrices. *SIAM Rev.*, 23(1):53–60, 1981. (Cited on page 83.)

[1638] B. Hendrickson and R. Leland. A multi-level algorithm for partitioning graphs. In *Supercomputing '95: Proc. 1995 ACM/IEEE Conf. on Supercomputing*, pages 1–14. IEEE Computer Society, 1995. (Cited on page 90.)

[1639] P. Hénon, P. Ramet, and J. Roman. PaStiX: A high-performance parallel direct solver for sparse symmetric definite systems. *Parallel Comput.*, 28(2):301–321, 2002. (Cited on pages 92 and 377.)

[1640] P. Hénon and Y. Saad. A parallel multistage ILU factorization based on a hierarchical graph decomposition. *SIAM J. Sci. Comput.*, 28(6):2266–2293, 2006. (Cited on page 271.)

[1641] P. Henrici. On the speed of convergence of cyclic and quasicyclic Jacobi methods for computing eigenvalues of Hermitian matrices. *SIAM J. Soc. Ind. Appl. Math.*, 6(2):144–162, 1958. (Cited on page 281.)

[1642] P. Henrici. The quotient difference algorithm. In *NBS Appl. Math. Series, 49*, pages 23–46. National Bureau of Standards, 1958. (Cited on page 310.)

[1643] P. Henrici. Bounds for iterates, inverses, spectral variation and fields of values of non-normal matrices. *Numer. Math.*, 4:24–40, 1962. (Cited on page 287.)

[1644] P. Henrici. *Discrete Variable Methods in Ordinary Differential Equations*. John Wiley, New York, 1962. (Cited on page 485.)

[1645] P. Henrici. *Elements of Numerical Analysis*. John Wiley, 1964. (Cited on page 485.)

[1646] P. Henrici. *Applied and Computational Complex Analysis, 3 volumes*. John Wiley, 1974-1986. (Cited on page 485.)

[1647] K. Hensel. Über potenzreihen von matrizen. *Journal für die reine und angewandte Mathematik (Crelle Journal)*, 155:107–110, 1826. (Cited on page 387.)

[1648] K. Hensel. Theorie der körper von matrizen. *Journal für die reine und angewandte Mathematik (Crelle Journal)*, 127:116–166, 1904. (Cited on pages 8, 148, 396, and 464.)

[1649] C. Hermite. Sur une question relative à la théorie des nombres. *J. Math. Pures Appl.*, 14:21–30, 1849. (Cited on page 113.)

[1650] C. Hermite. Sur la théorie des formes quadratiques ternaires indéfinies. *Journal für die reine und angewandte Mathematik (Crelle Journal)*, 47:307–312, 1854. (Cited on page 4.)

[1651] C. Hermite. Remarque sur un théorème de M. Cauchy. *C.R. Acad. Sci. Paris, Sér 1*, 41:181–183, 1855. (Cited on page 4.)

[1652] M. A. Heroux and J. M. Willenbring. A new overview of the Trilinos project. *Scientific Programming*, 20:83–88, 2012. (Cited on page 378.)

[1653] J. G. Herriot. In memory of George E. Forsythe. *Comm. ACM*, 15(8):719–720, 1972. (Cited on page 458.)

[1654] M. Herzberger. The normal equations of the method of least squares and their solution. *Quart. Appl. Math.*, 7:217–223, 1949. (Cited on page 74.)

[1655] K. Hessenberg. Behandlung linearer eigenwertaufgaben mit hilfe der Hamilton-Cayleyschen gleichung. Technical Report 1) Bericht der Reihe Numerische Verfahren, Institut für Praktische Mathematik, Technische Hochschule Darmstadt, 1940. (Cited on pages 218 and 291.)

[1656] K. Hessenberg. *Die berechnung der eigenwerte und eigenlösungen linearer gleichungesysteme.* PhD thesis, Technische Hochschule Darmstadt, Germany, 1940. (Cited on page 292.)

[1657] D. Hestenes. Grassmann's legacy. In H. J. Petsche, A. Lewis, J. Liesen, and S. Russ, editors, *From Past to Future: Graßmann's Work in Context*, Basel, 2011. Springer. (Cited on page 3.)

[1658] M. R. Hestenes. Iterative computational methods. *Comm. Pure Appl. Math.*, 8(1):85–95, 1955. (Cited on page 207.)

[1659] M. R. Hestenes. The conjugate gradient method for solving linear systems. In J. H. Curtiss, editor, *Numerical Analysis, Proc. Symp. Appl. Math VI,*, pages 83–102. American Mathematical Society, 1956. (Cited on page 216.)

[1660] M. R. Hestenes. Inversion of matrices by biorthogonalization and related results. *J. Soc. Ind. Appl. Math.*, 6(1):51–90, 1958. (Cited on page 136.)

[1661] M. R. Hestenes. Iterative methods for solving linear equations. *J. Optim. Theory Appl.*, 11(4):323–334, 1973. (Cited on page 207.)

[1662] M. R. Hestenes. Conjugacy and gradients. In S.G. Nash, editor, *A History of Scientific Computing*, pages 167–179. ACM Press, 1990. (Cited on pages 202 and 487.)

[1663] M. R. Hestenes and W. Karush. A method of gradients for the calculation of the characteristic roots and vectors of a real symmetric matrix. *J. Res. Natl. Bur. Stand.*, 47(1):45–61, 1951. (Cited on page 300.)

[1664] M. R. Hestenes and E. Stiefel. Methods of conjugate gradients for solving linear systems. *J. Res. Natl. Bur. Stand.*, 49(6):409–436, 1952. (Cited on pages 16, 201, 202, 216, 251, 488, and 568.)

[1665] M. R. Hestenes and J. Todd. NBS-INA. the Institute for Numerical Analysis, UCLA 1947-1954. Technical Report Special Publication 730, NIST, 1991. (Cited on pages 200, 201, and 565.)

[1666] C. Hettle. The symbolic and mathematical influence of Diophantus' Arithmetica. *Journal of Humanistic Mathematics*, 5(1):139–166, 2015. (Cited on page 41.)

[1667] K. J. Heuvers and D. Moak. Matrix solutions of the functional equation of the gamma function. *Aequationes Mathematicae*, 33(1):1–17, 1987. (Cited on page 387.)

[1668] M. Heyouni. The global Hessenberg and CMRH methods for linear systems with multiple right-hand sides. *Numer. Algorithms*, 26(4):317–332, 2001. (Cited on page 237.)

[1669] M. Heyouni and H. Sadok. On a variable smoothing procedure for Krylov subspace methods. *Linear Algebra Appl.*, 268:131–149, 1998. (Cited on pages 236 and 237.)

[1670] J. E. Hicken and D. W. Zingg. A simplified and flexible variant of GCROT for solving nonsymmetric linear systems. *SIAM J. Sci. Comput.*, 32(3):1672–1694, 2010. (Cited on page 231.)

[1671] D. J. Higham. Condition numbers and their condition numbers. *Linear Algebra Appl.*, 214:193–213, 1995. (Cited on page 24.)

[1672] N. J. Higham. Computing the polar decomposition—with applications. *SIAM J. Sci. Statist. Comput.*, 7(4):1160–1174, 1986. (Cited on page 142.)

[1673] N. J. Higham. A survey of condition number estimation for triangular matrices. *SIAM Rev.*, 29(4):575–596, 1987. (Cited on page 24.)

[1674] N. J. Higham. Algorithm 674: FORTRAN codes for estimating the one-norm of a real or complex matrix, with applications to condition estimation. *ACM Trans. Math. Soft. (TOMS)*, 14(4):381–396, 1988. (Cited on page 24.)

[1675] N. J. Higham. Experience with a matrix norm estimator. *SIAM J. Sci. Statist. Comput.*, 11(4):804–809, 1990. (Cited on page 24.)

[1676] N. J. Higham. Algorithm 694: A collection of test matrices in MATLAB. *ACM Trans. Math. Soft. (TOMS)*, 17(3):289–305, 1991. (Cited on page 373.)

[1677] N. J. Higham. Perturbation theory and backward error for $AX - XB = C$. *BIT*, 33:124–136, 1993. (Cited on page 397.)

[1678] N. J. Higham. The matrix sign decomposition and its relation to the polar decomposition. *Linear Algebra Appl.*, 212:3–20, 1994. (Cited on page 143.)

[1679] N. J. Higham. *Accuracy and Stability of Numerical Algorithms*. SIAM, Philadelphia, 1996. (Cited on page 80.)

[1680] N. J. Higham. QR factorization with complete pivoting and accurate computation of the SVD. *Linear Algebra Appl.*, 309:153–174, 2000. (Cited on page 138.)

[1681] N. J. Higham. *Accuracy and Stability of Numerical Algorithms*. 2nd edition. SIAM, Philadelphia, 2002. (Cited on pages 77, 80, 133, 292, 344, and 588.)

[1682] N. J. Higham. Cayley, Sylvester, and early matrix theory. *Linear Algebra Appl.*, 428:39–43, 2008. (Cited on page 571.)

[1683] N. J. Higham. *Functions of Matrices: Theory and Computation*. SIAM, Philadelphia, 2008. (Cited on pages 144, 387, 388, and 389.)

[1684] N. J. Higham. Gaussian elimination. *Interdisciplinary Reviews: Computational Statistics*, 3(3):230–238, 2011. (Cited on page 80.)

[1685] N. J. Higham. Sylvester's influence on applied mathematics. *Math. Today*, 50(4):202–206, 2014. (Cited on page 571.)

[1686] N. J. Higham. Numerical linear algebra and matrix analysis. Technical Report MIMS EPrint 2015-103, The University of Manchester, 2015. (Cited on page 80.)

[1687] N. J. Higham and A. Al-Mohy. The complex step approximation to the Fréchet derivative of a matrix function. *Numer. Algorithms*, 53(1):133–148, 2010. (Cited on page 389.)

[1688] N. J. Higham and T. Mary. A new approach to probabilistic rounding error analysis. *SIAM J. Sci. Comput.*, 41(5):A2815–A2835, 2019. (Cited on page 79.)

[1689] N. J. Higham and T. Mary. Sharper probabilistic backward error analysis for basic linear algebra kernels with random data. *SIAM J. Sci. Comput.*, 42(5):A3427–A3446, 2020. (Cited on page 79.)

[1690] N. J. Higham, C. Mehl, and F. Tisseur. The canonical generalized polar decomposition. *SIAM J. Matrix Anal. Appl.*, 31(4):2163–2180, 2010. (Cited on page 144.)

[1691] N. J. Higham and P. Papadimitriou. A new parallel algorithm for computing the singular value decomposition. In *Proceedings of the Fifth SIAM Conference on Applied Linear Algebra*. SIAM, Philadelphia, 1994. (Cited on pages 138 and 143.)

[1692] N. J. Higham and P. Papadimitriou. A parallel algorithm for computing the polar decomposition. *Parallel Comput.*, 20(8):1161–1173, 1994. (Cited on pages 138 and 143.)

[1693] N. J. Higham and S. D. Relton. Estimating the condition number of the Fréchet derivative of a matrix function. *SIAM J. Sci. Comput.*, 36(6):C617–C634, 2014. (Cited on page 389.)

[1694] N. J. Higham and R. S. Schreiber. Fast polar decomposition of an arbitrary matrix. *SIAM J. Sci. Statist. Comput.*, 11(4):648–655, 1990. (Cited on page 142.)

[1695] D. Hilbert. Grundzüge einer allgemeinen theorie der linearen integralgleichungen. Nachrichten von d. Königl. Ges. d. Wissensch. zu Göttingen, 1904. (Cited on page 288.)

[1696] F. B. Hildebrand. *Introduction to Numerical Analysis*. McGraw-Hill Book Company, 2nd edition, 1974. First edition 1956. (Cited on pages 94 and 446.)

[1697] R. O. Hill, Jr. and B. N. Parlett. Refined interlacing properties. *SIAM J. Matrix Anal. Appl.*, 13(1):239–247, 1992. (Cited on page 280.)

[1698] W. D. Hillis. *The Connnection Machine*. MIT Press, 1986. (Cited on page 346.)

[1699] H. Hilton. Some properties of symmetric and orthogonal substitutions. *Proc. Lond. Math. Soc.*, 2(1):94–99, 1913. (Cited on page 288.)

[1700] J. B. Hiriart-Urruty and H. Caussinus. Sarrus, Borel, Deltheil, le Rouergue et ses mathématiciens. *Gazette de la SMF*, 104:89–97, 2005. (Cited on page 101.)

[1701] A. Hirsch. Sur les racines d'une équation fondamentale: extrait d'une lettre de M. A. Hirsch à M. I. Bendixson. *Acta Math.*, 25:367–370, 1902. (Cited on page 286.)

[1702] F .L. Hitchcock. On the solution of the linear matrix equation by double multiplication. *Proc. Natl. Acad. Sci. USA*, 8:78–83, 1922. (Cited on page 395.)

[1703] D. Ho, F. Chatelin, and M. Bennani. Arnoldi-Tchebychev procedure for large scale nonsymmetric matrices. *Rairo - Math. Model. Numer. Anal.*, 24(1):53–65, 1990. (Cited on page 305.)

[1704] M. Hochbruck and C. Lubich. On Krylov subspace approximations to the matrix exponential operator. *SIAM J. Numer. Anal.*, 34(5):1911–1925, 1997. (Cited on page 390.)

[1705] M. Hochbruck and C. Lubich. Error analysis of Krylov methods in a nutshell. *SIAM J. Sci. Comput.*, 19(2):695–701, 1998. (Cited on page 235.)

[1706] M. Hochbruck, C. Lubich, and H. Selhofer. Exponential integrators for large systems of differential equations. *SIAM J. Sci. Comput.*, 19(5):1552–1574, 1998. (Cited on page 390.)

[1707] M. E. Hochstenbach. A Jacobi-Davidson type SVD method. *SIAM J. Sci. Comput.*, 23(2):606–628, 2001. (Cited on page 139.)

[1708] M. E. Hochstenbach. A Jacobi-Davidson type method for the generalized singular value problem. *Linear Algebra Appl.*, 431(3-4):471–487, 2009. (Cited on page 141.)

[1709] M. E. Hochstenbach and Y. Notay. The Jacobi-Davidson method. *GAMM Mitteilungen*, 29(2):368–382, 2006. (Cited on page 317.)

[1710] M. E. Hochstenbach and Y. Notay. Homogeneous Jacobi-Davidson. *Electron. Trans. Numer. Anal.*, 29:19–30, 2007. (Cited on page 317.)

[1711] M. E. Hochstenbach and Y. Notay. Controlling inner iterations in the Jacobi-Davidson method. *SIAM J. Matrix Anal. Appl.*, 31(2):460–477, 2009. (Cited on page 317.)

[1712] M. E. Hochstenbach and L. Reichel. An iterative method for Tikhonov regularization with a general linear regularization operator. *J. Integral Equations Appl.*, 22(3):463–480, 2010. (Cited on page 393.)

[1713] U. Hochstrasser. *Die anwendung der methode der konjugierten gradienten und ihrer modifikationen auf die lösung linearer randwertprobleme*. PhD thesis, ETH Zürich, 1954. (Cited on page 207.)

[1714] R. W. Hockney. A fast direct solution of Poisson's equation using Fourier analysis. *J. ACM*, 12:95–113, 1965. (Cited on page 384.)

[1715] R. W. Hockney. The potential calculation and some applications. *Methods Comput. Phys.*, 9:135–211, 1970. (Cited on page 384.)

[1716] R. W. Hockney and C. R. Jesshope. *Parallel Computers 2*. Adam Hilger, Bristol, 1988. (Cited on page 346.)

[1717] P. G. Hoel. On methods of solving normal equations. *The Annals of Mathematical Statistics*, 12(3):354–359, 1941. (Cited on page 66.)

[1718] M. F. Hoemmen. *Communication-avoiding Krylov subspace methods*. PhD thesis, University of California at Berkeley, 2010. (Cited on page 243.)

[1719] M. F. Hoemmen. A communication-avoiding, hybrid-parallel, rank-revealing orthogonalization method. In *2011 IEEE International Parallel & Distributed Processing Symposium*, 2011. (Cited on page 133.)

[1720] A. J. Hoffman and O. Taussky. A characterization of normal matrices. *J. Res. Natl. Bur. Stand.*, 52:17–19, 1954. (Cited on page 576.)

[1721] W. Hoffmann. Iterative algorithms for Gram-Schmidt orthogonalization. *Computing*, 41:335–348, 1989. (Cited on page 129.)

[1722] W. Hoffmann and B. N. Parlett. A new proof of global convergence for the tridiagonal *QL* algorithm. *SIAM J. Numer. Anal.*, 15(5):929–937, 1978. (Cited on page 312.)

[1723] J. D. Hogg, J. K. Reid, and J. A. Scott. Design of a multicore sparse Cholesky factorization using DAGs. *SIAM J. Sci. Comput.*, 32(6):3627–3649, 2010. (Cited on page 92.)

[1724] O. Hölder. Über einen mittelwertsatz. *Nachrichten von der Königl. Gesellschaft der Wissenschaften und der Georg-Augusts-Universität zu Göttingen*, 2:38–47, 1889. (Cited on page 14.)

[1725] M. P. Holmes, J. Isbell, C. Lee, and A. G. Gray. QUIC-SVD: Fast SVD using cosine trees. In *Advances in Neural Information Processing Systems*, pages 673–680, 2009. (Cited on page 140.)

[1726] A. Hood. *Localizing the eigenvalues of matrix-valued functions: analysis and applications*. PhD thesis, Cornell University, 2017. (Cited on page 285.)

[1727] P. Hood. Frontal solution program for unsymmetric matrices. *Internat. J. Numer. Methods Engrg.*, 10(2):379–399, 1976. (Cited on pages 89 and 374.)

[1728] J. Hook and N. Dingle. Performance analysis of asynchronous parallel Jacobi. *Numer. Algorithms*, 77(1-3):831–866, 2018. (Cited on page 197.)

[1729] R. A. Horn and C. R. Johnson. *Matrix Analysis*. Cambridge University Press, 1987. 2nd edition (2013). (Cited on page 148.)

[1730] P. Horst. A method of determining the coefficients of a characteristic equation. *Ann. Math. Stat.*, 6(2):83–84, 1935. (Cited on page 289.)

[1731] H. Hotelling. Analysis of a complex of statistical variables into principal components. *J. Educ. Psychol.*, 24:417–441, 1933. (Cited on page 295.)

[1732] H. Hotelling. Some new methods in matrix calculation. *Ann. Math. Stat.*, 14(1):1–34, 1943. (Cited on pages 15, 19, 74, and 172.)

[1733] H. Hotelling. Practical problems of matrix calculation. In J. Neyman, editor, *Proceedings of the [First] Berkeley Symposium on Mathematical Statistics and Probability, 1946*, pages 275–293. The Regents of the University of California, 1949. (Cited on pages 172 and 174.)

[1734] S.-H. Hou. A simple proof of the Leverrier-Faddeev characteristic polynomial algorithm. *SIAM Rev.*, 40(3):706–709, 1998. (Cited on page 289.)

[1735] A. S. Householder. Some numerical methods for solving systems of linear equations. *Am. Math. Mon.*, 57:453–459, 1950. (Cited on pages 74, 174, and 192.)

[1736] A. S. Householder. *Principles of Numerical Analysis*. McGraw-Hill Book Company, New York, 1953. (Cited on pages 301 and 430.)

[1737] A. S. Householder. On norms of vectors and matrices. Technical Report ORNL-1756, Oak Ridge National Laboratory, 1954. (Cited on page 16.)

[1738] A. S. Householder. The generation of error in digital computation. Technical Report ORNL-1983, Oak Ridge National Laboratory, 1955. (Cited on page 77.)

[1739] A. S. Householder. Terminating and nonterminating iterations for solving linear systems. *J. Soc. Ind. Appl. Math.*, 3(2):67–72, 1955. (Cited on page 16.)

[1740] A. S. Householder. On the convergence of matrix iterations. *J. ACM*, 3(4):314–324, 1956. (Cited on pages 16, 176, and 490.)

[1741] A. S. Householder. A survey of some closed methods for inverting matrices. *J. Soc. Ind. Appl. Math.*, 5(3):155–169, 1957. (Cited on page 132.)

[1742] A. S. Householder. A class of methods for inverting matrices. *J. Soc. Ind. Appl. Math.*, 6(2):189–195, 1958. (Cited on pages 22 and 132.)

[1743] A. S. Householder. On matrices with non-negative elements. *Monatsh Math. Phys.*, 62(3):238–242, 1958. (Cited on page 177.)

[1744] A. S. Householder. Unitary triangularization of a nonsymmetric matrix. *J. ACM*, 5(4):339–342, 1958. (Cited on pages 132, 308, 399, and 490.)

[1745] A. S. Householder. *The Theory of Matrices in Numerical Analysis*. Blaisdell, New York, 1964. Reprinted by Dover, New York, (1975). (Cited on pages 15, 16, and 506.)

[1746] A. S. Householder. The Kantorovich and some related inequalities. *SIAM Rev.*, 7(4):463–473, 1965. (Cited on page 208.)

[1747] A. S. Householder. Moments and characteristic roots. II. *Numer. Math.*, 11:126–128, 1968. (Cited on page 293.)

[1748] A. S. Householder. Norms and the localization of roots of matrices. *Bull. Amer. Math. Soc.*, 74(5):816–830, 1968. (Cited on pages 15 and 16.)

[1749] A. S. Householder. The Padé table, the Frobenius identities, and the qd algorithm. *Linear Algebra Appl.*, 4(2):161–174, 1971. (Cited on page 310.)

[1750] A. S. Householder. George E. Forsythe. *SIAM J. Numer. Anal.*, 10(2):viii–xi, 1973. (Cited on page 459.)

[1751] A. S. Householder. The Gatlinburgs. *SIAM Rev.*, 16(3):340–343, 1974. (Cited on page 490.)

[1752] A. S. Householder and F. L. Bauer. On certain methods for expanding the characteristic polynomials. *Numer. Math.*, 1:29–37, 1959. (Cited on pages 218, 291, 292, 293, 301, and 506.)

[1753] A. S. Householder and F. L. Bauer. On certain iterative methods for solving linear systems. *Numer. Math.*, 2:55–59, 1960. (Cited on page 414.)

[1754] A. S. Householder, R. S. Varga, and J. H. Wilkinson. A note on Gerschgorin's inclusion theorem for eigenvalues of matrices. *Numer. Math.*, 16(2):141–144, 1970. (Cited on page 599.)

[1755] A. S. Householder and G. Young. Matrix approximation and latent roots. *Am. Math. Mon.*, 45(3):165–171, 1938. (Cited on page 489.)

[1756] J. L. Howland and J. A. Senez. A constructive method for the solution of the stability problem. *Numer. Math.*, 16(1):1–7, 1970. (Cited on page 398.)

[1757] J. Høyrup. Al-Khwârizmî, Ibn Turk, and the Liber Mensurationum: On the origins of Islamic algebra. *Erdem*, 2(5):445–484, 1986. (Cited on page 42.)

[1758] J. Høyrup. *Lengths, Widths, Surfaces: A Portrait of Old Babylonian Algebra and its Kin*. Springer, 2002. (Cited on pages 33 and 34.)

[1759] J. Høyrup. The tortuous ways toward a new understanding of algebra in the Italian abbacus school (14th-16th centuries). In O. Figueras, J. L. Cortina, S. Alatorre, T. Rojano, and A. Sepulveda, editors, *Proceedings of the Joint Meeting PME32-PME NA XXX*, pages 1–15. Cinvestav-UMSNH, 2008. (Cited on page 33.)

[1760] J. Høyrup. Old babylonian "algebra", and what it teaches us about possible kinds of mathematics. *Ganita Bhāratī*, 32:87–110, 2012. (Cited on pages 33 and 34.)

[1761] J. Høyrup. Was Babylonian mathematics created by "Babylonian mathematicians"? In H. Neumann and S. Paulus, editors, *Wissenskultur im Alten Orient: Weltanschauung, Wissenschaften, Techniken, Technologien*, pages 105–119. Harrassowitz Verlag, 2012. (Cited on page 33.)

[1762] D. Y. Hu and L. Reichel. Krylov-subspace methods for the Sylvester matrix equation. *Linear Algebra Appl.*, 172:283–313, 1992. (Cited on page 397.)

[1763] Y. F. Hu and J. A. Scott. A multilevel algorithm for wavefront reduction. *SIAM J. Sci. Comput.*, 23(4):1352–1375, 2001. (Cited on page 89.)

[1764] T. Huckle. Low-rank modification of the unsymmetric Lanczos algorithm. *Math. Comput.*, 64(212):1577–1588, 1995. (Cited on pages 228 and 263.)

[1765] T. Huckle. Efficient computation of sparse approximate inverses. *Numer. Linear Algebra Appl.*, 5(1):57–71, 1998. (Cited on page 263.)

[1766] T. Huckle. Approximate sparsity patterns for the inverse of a matrix and preconditioning. *Appl. Numer. Math.*, 30(2-3):291–303, 1999. (Cited on page 263.)

[1767] V. Hug and T. Steiner. Une lettre d'Euler à d'Alembert retrouvée. *Historia Math.*, 42:84–94, 2015. (Cited on page 277.)

[1768] T. J. R. Hughes, I. Levit, and J.Winget. An element-by-element solution algorithm for problems of structural and solid mechanics. *Comput. Methods Appl. Mech. Engrg.*, 36(2):241–254, 1983. (Cited on page 264.)

[1769] H. D. Huskey. Characteristics of the Institute for Numerical Analysis computer. *M.T.A.C.*, 4(30):103–108, 1950. (Cited on page 198.)

[1770] D. Hysom and A. Pothen. A scalable parallel algorithm for incomplete factor preconditioning. *SIAM J. Sci. Comput.*, 22(6):2194–2215, 2001. (Cited on page 261.)

[1771] B. Iannazzo. A note on computing the matrix square root. *Calcolo*, 40(4):273–283, 2003. (Cited on page 389.)

[1772] Y. Ikebe. On inverses of Hessenberg matrices. *Linear Algebra Appl.*, 24:93–97, 1979. (Cited on pages 218 and 383.)

[1773] V. P. Il'in. *Iterative Incomplete Factorization Methods*. World Scientific, 1992. (Cited on page 252.)

[1774] A. Imakura, R. C. Li, and S. L. Zhang. Locally optimal and heavy ball GMRES methods. *Japan J. Indust. Appl. Math.*, 33(2):471–499, 2016. (Cited on page 241.)

[1775] A. Imakura, T. Sogabe, and S. L. Zhang. An efficient variant of the GMRES(m) method based on the error equations. *E. Asian J. Appl. Math.*, 2:1–22, 2012. (Cited on page 241.)

[1776] A. Imakura, T. Sogabe, and S. L. Zhang. A look-back-type restart for the restarted Krylov subspace methods for solving non-Hermitian linear systems. *Japan J. Indust. Appl. Math.*, 35(2):835–859, 2018. (Cited on page 241.)

[1777] M. H. Ingraham and H. C. Trimble. On the matric equation $TA = BT + C$. *Amer. J. Math.*, 63(1):9–28, 1941. (Cited on page 396.)

[1778] I. S. Iohvidov. *Hankel and Toeplitz Matrices and Forms, Algebraic Theory*. Birkhäuser, Boston, 1982. (Cited on page 386.)

[1779] I. C. F. Ipsen. A different approach to bounding the minimal residual norm in Krylov methods. Technical Report CRSC-TR98-19, Dept. of Mathematics, North Carolina State University, 1998. (Cited on page 235.)

[1780] I. C. F. Ipsen. Expressions and bounds for the GMRES residual. *BIT Numerical Mathematics*, 40(3):524–535, 2000. (Cited on pages 236, 238, and 240.)

[1781] I. C. F. Ipsen and C. D. Meyer. The idea behind Krylov methods. *Appl. Numer. Math.*, pages 889–899, 1998. (Cited on page 236.)

[1782] I. C. F. Ipsen and R. Rehman. Perturbation bounds for determinants and characteristic polynomials. *SIAM J. Matrix Anal. Appl.*, 30(2):762–776, 2008. (Cited on page 293.)

[1783] I. C. F. Ipsen. A history of inverse iteration. In B. Huppert and H. Schneider, editors, *Helmut Wielandt, Mathematische Werke, Mathematical Works, vol. 2: Linear Algebra and Analysis*, pages 464–72, Berlin, 1996. Walter de Gruyter. (Cited on page 295.)

[1784] B. M. Irons. A frontal solution program for finite element analysis. *Internat. J. Numer. Methods Engrg.*, 2(1):5–32, 1970. (Cited on pages 89 and 374.)

[1785] E. Isaacson and H. B. Keller. *Analysis of Numerical Methods*. Wiley, New York, 1966. Reprinted by Dover, New York, 1994. (Cited on page 94.)

[1786] T. Ito and K. Hayami. Preconditioned GMRES methods for least squares problems. *Japan J. Indust. Appl. Math.*, 25(2):185, 2008. (Cited on page 242.)

[1787] V. K. Ivanov. On linear problems which are not well-posed. *Soviet Math. Dokl.*, 3:981–983, 1962. (Cited on page 392.)

[1788] N. Bosner J. L. Barlow and Z. Drmač. A new stable bidiagonal reduction algorithm. *Linear Algebra Appl.*, 397:35–84, 2005. (Cited on page 138.)

[1789] C. G. J. Jacobi. Über Gauss' neue methode, die werthe der integrale näherungsweise zu finden. *Journal für die reine und angewandte Mathematik (Crelle Journal)*, 1:301–308, 1826. (Cited on page 469.)

[1790] C. G. J. Jacobi. Euleri formulae de transformatione coordinatarum. *Journal für die reine und angewandte Mathematik (Crelle Journal)*, 2:188–189, 1827. (Cited on page 123.)

[1791] C. G. J. Jacobi. De binis quibuslibet functionibus homogeneis secundi ordinis per substitutiones lineares in alias binas transformandis, quas solis quadratis variabilum constant .... *Journal für die reine und angewandte Mathematik (Crelle Journal)*, 12:1–69, 1834. (Cited on page 281.)

[1792] C. G. J. Jacobi. De determinantibus functionalibus. *Journal für die reine und angewandte Mathematik (Crelle Journal)*, 22:319–359, 1841. (Cited on page 494.)

[1793] C. G. J. Jacobi. De fonctionibus alternatibus earumque divisione per productum et differentiis elementorum conflatum. *Journal für die reine und angewandte Mathematik (Crelle Journal)*, 28:360–371, 1841. (Cited on pages 112 and 494.)

[1794] C. G. J. Jacobi. De formatione et proprietatibus determinantium. *Journal für die reine und angewandte Mathematik (Crelle Journal)*, 22:285–318, 1841. (Cited on page 112.)

[1795] C. G. J. Jacobi. Ueber eine neue Auflösungsart der bei der Methode der kleinsten Quadrate vorkommenden lineären Gleichungen. *Astronomische Nachrichten*, 22(20):297–306, 1845. (Cited on pages 130 and 157.)

[1796] C. G. J. Jacobi. Über die darstellung einer reihe gegebener werthe durch eine gebrochene rationale function. *Journal für die reine und angewandte Mathematik (Crelle Journal)*, 30:127–156, 1846. (Cited on pages 281 and 464.)

[1797] C. G. J. Jacobi. Über ein leichtes verfahren die in der theorie der säcularstörungen vorkommenden gleichungen numerisch aufzulösen. *Journal für die reine und angewandte Mathematik (Crelle Journal)*, 30:51–94, 1846. (Cited on pages 158, 281, and 317.)

[1798] C. G. J. Jacobi. Untersuchungen über die differentialgleichung der hypergeometrischen reihe. *Journal für die reine und angewandte Mathematik (Crelle Journal)*, 56:149–165, 1859. (Cited on page 494.)

[1799] D. A. H. Jacobs. *Numerical Software: Needs and Availability*. Academic Press, 1978. Proceedings of IMA Conference, Sussex 1977. (Cited on page 364.)

[1800] C. Jagels and L. Reichel. The extended Krylov subspace method and orthogonal Laurent polynomials. *Linear Algebra Appl.*, 431(3-4):441–458, 2009. (Cited on page 390.)

[1801] C. Jagels and L. Reichel. Recursion relations for the extended Krylov subspace method. *Linear Algebra Appl.*, 434(7):1716–1732, 2011. (Cited on page 390.)

[1802] J. R. Jain. An efficient algorithm for a large Toeplitz set of linear equations. *IEEE Trans. Acoust. Speech Signal Process.*, 27(6):612–615, 1979. (Cited on page 386.)

[1803] W. Jalby and B. Philippe. Stability analysis and improvement of the block Gram-Schmidt algorithm. *SIAM J. Sci. Statist. Comput.*, 12(5):1058–1073, 1991. (Cited on page 130.)

[1804] K. R. James and W. Riha. Convergence criteria for successive overrelaxation. *SIAM J. Numer. Anal.*, 12(2):137–143, 1975. (Cited on page 178.)

[1805] A. Jameson. Solution of the equation $AX + XB = C$ by inversion of an $M \times M$ or $N \times N$ matrix. *SIAM J. Appl. Math.*, 16(5):1020–1023, 1968. (Cited on page 397.)

[1806] K. Jbilou. Low rank approximate solutions to large Sylvester matrix equations. *Appl. Math. Comput.*, 177:365–376, 2006. (Cited on page 398.)

[1807] K. Jbilou, A. Messaoudi, and H. Sadok. Global GMRES algorithm for solving nonsymmetric linear systems of equations with multiple right-hand sides. *Appl. Numer. Math.*, 31:49–63, 1999. (Cited on page 237.)

[1808] K. C. Jea. *Generalized conjugate gradient acceleration of iterative methods*. PhD thesis, Center for Numerical Analysis, University of Texas at Austin, 1982. Report CNA-176. (Cited on page 222.)

[1809] K. C. Jea and D. M. Young. On the simplification of generalized conjugate-gradient methods for nonsymmetrizable linear systems. *Linear Algebra Appl.*, 52:399–417, 1983. (Cited on pages 222, 223, and 601.)

[1810] A. Jennings. A compact storage scheme for the solution of symmetric linear simultaneous equations. *Comput. J.*, 9(3):281–285, 1966. (Cited on page 86.)

[1811] A. Jennings. Influence of the eigenvalue spectrum on the convergence rate of the conjugate gradient method. *IMA J. Appl. Math.*, 20(1):61–72, 1977. (Cited on page 210.)

[1812] A. Jennings. *Matrix Computations for Engineers and Scientists*. John Wiley, 1977. (Cited on page 94.)

[1813] A. Jennings and G. M. Malik. Partial elimination. *IMA J. Appl. Math.*, 20(3):307–316, 1977. (Cited on page 254.)

[1814] A. Jennings and G. M. Malik. The solution of sparse linear equations by the conjugate gradient method. *Internat. J. Numer. Methods Engrg.*, 12(1):141–158, 1978. (Cited on page 210.)

[1815] E. J. Jensen, E. Coleman, and M. Sosonkina. Predictive modeling of the performance of asynchronous iterative methods. *J. Supercomput.*, 75(8):5084–5105, 2019. (Cited on page 197.)

[1816] H. Jensen. *An Attempt at a Systematic Classification of some Methods for the Solution of Normal Equations*. Geodaetisk Institut Kobenhan, Meddelelse no. 18. Bianco Lunos Bogtrykkeri A/S, København, 1944. (Cited on pages 75 and 580.)

[1817] T. K. Jensen and P. C. Hansen. Iterative regularization with minimal residual methods. *BIT Numerical Mathematics*, 47(1):103–120, 2007. (Cited on page 393.)

[1818] J. A. G. Jess and H. G. M. Kees. A data structure for parallel LU decomposition. *IEEE Trans. Comput.*, C-31(3):231–239, 1982. (Cited on page 93.)

[1819] E. R. Jessup and D. C. Sorensen. A parallel algorithm for computing the singular value decomposition of a matrix. *SIAM J. Matrix Anal. Appl.*, 15(2):530–548, 1994. (Cited on page 138.)

[1820] F. Jézéquel and J. M. Chesneaux. CADNA: a library for estimating round-off error propagation. *Comput. Phys. Comm.*, 178(12):933–955, 2008. (Cited on page 238.)

[1821] Z. Jia. The convergence of generalized Lanczos methods for large unsymmetric eigenproblems. *SIAM J. Matrix Anal. Appl.*, 16(3):843–862, 1995. (Cited on page 304.)

[1822] Z. Jia. On IOM(q): The incomplete orthogonalization method for large unsymmetric linear systems. *Numer. Linear Algebra Appl.*, 3(6):491–512, 1996. (Cited on page 234.)

[1823] Z. Jia. Refined iterative algorithms based on Arnoldi's process for large unsymmetric eigenproblems. *Linear Algebra Appl.*, 259:1–23, 1997. (Cited on page 306.)

[1824] Z. Jia. On IGMRES: An incomplete generalized minimal residual method for large unsymmetric linear systems. *Sci. China Ser. A*, 41(12):1278–1288, 1998. (Cited on page 234.)

[1825] Z. Jia. The convergence of harmonic Ritz values, harmonic Ritz vectors, and refined harmonic Ritz vectors. *Math. Comput.*, 74(251):1441–1456, 2004. (Cited on page 307.)

[1826] Z. Jia and D. Niu. An implicitly restarted refined bidiagonalization Lanczos method for computing a partial singular value decomposition. *SIAM J. Matrix Anal. Appl.*, 25(1):246–265, 2003. (Cited on page 139.)

[1827] Z. Jia and D. Niu. A refined harmonic Lanczos bidiagonalization method and an implicitly restarted algorithm for computing the smallest singular triplets of large matrices. *SIAM J. Sci. Comput.*, 32(2):714–744, 2010. (Cited on page 140.)

[1828] P. Jiránek. *Limiting accuracy of iterative methods*. PhD thesis, Technical University Liberec, Czech Republic, 2007. (Cited on page 242.)

[1829] P. Jiránek and M. Rozložník. Adaptive version of Simpler GMRES. *Numer. Algorithms*, 53(1):93–112, 2010. (Cited on page 242.)

[1830] P. Jiránek, M. Rozložník, and M. H. Gutknecht. How to make simpler GMRES and GCR more stable. *SIAM J. Matrix Anal. Appl.*, 30(4):1483–1499, 2008. (Cited on page 242.)

[1831] S. Jochi. *The influence of Chinese mathematical arts on Seki Kowa*. PhD thesis, School of Oriental and African Studies, University of London, UK, 1993. (Cited on page 100.)

[1832] T. Joffredo. *Approches biographiques de l' "Introduction à l'analyse des lignes courbes algébriques" de Gabriel Cramer*. PhD thesis, Université de Lorraine, France, 2017. (Cited on page 441.)

[1833] F. John. Advanced numerical analysis. Lecture Notes, Department of Mathematics, New York University, 1956. (Cited on page 176.)

[1834] C. R. Johnson. Numerical determination of the field of values of a general complex matrix. *SIAM J. Numer. Anal.*, 15(3):595–602, 1978. (Cited on page 287.)

[1835] O. G. Johnson, C. A. Micchelli, and G. Paul. Polynomial preconditioners for conjugate gradient calculations. *SIAM J. Numer. Anal.*, 20(2):362–376, 1983. (Cited on page 261.)

[1836] W. W. Johnson. On Gauss' method of substitution. *Annals of Mathematics*, 7(1/5):65–69, 1892-1893. (Cited on page 62.)

[1837] L. Johnsson. Highly concurrent algorithms for solving linear systems of equations. Technical Report 5079-TR-83, California Institute of Technology, 1983. (Cited on page 214.)

[1838] L. Johnsson. Highly concurrent algorithms for solving linear systems of equations. In G. Birkhoff and A. Schoenstadt, editors, *Elliptic Problem Solvers, II*, pages 105–126. Academic Press, 1984. (Monterey, California, 1983). (Cited on page 214.)

[1839] M. T. Jones and P. E. Plassmann. An improved incomplete Cholesky factorization. *ACM Trans. Math. Soft. (TOMS)*, 21(1):5–17, 1995. (Cited on page 257.)

[1840] C. Jordan. *Traité des Substitutions et des Équations Algébriques*. Gauthier-Villars, Paris, 1870. (Cited on pages 147 and 495.)

[1841] C. Jordan. Sur les polynômes bilinéaires. *C.R. Acad. Sci. Paris, Sér 1*, 74:1395–1399, 1873. (Cited on page 147.)

[1842] C. Jordan. Mémoire sur les formes bilinéaires. *J. Math. Pures Appl.*, 19:35–54, 1874. (Cited on pages 134 and 496.)

[1843] C. Jordan. Sur la réduction des formes bilinéaires. *C.R. Acad. Sci. Paris, Sér 1*, 78:614–617, 1874. (Cited on page 134.)

[1844] C. Jordan. Essai sur la géométrie à n dimensions. *Bull. S.M.F.*, 3:103–174, 1875. (Cited on pages 144 and 496.)

[1845] W. Jordan. *Handbuch der Vermessungskunde, vol. 1*. Metzler, 1888. (Cited on pages 61 and 67.)

[1846] W. Joubert. *Generalized conjugate gradient and Lanczos methods for the solution of nonsymmetric systems of linear equations*. PhD thesis, Center for Numerical Analysis, University of Texas at Austin, 1990. (Cited on page 227.)

[1847] W. Joubert. Lanczos methods for the solution of nonsymmetric systems of linear equations. *SIAM J. Matrix Anal. Appl.*, 13(3):926–943, 1992. (Cited on pages 225 and 227.)

[1848] W. Joubert. On the convergence behavior of the restarted GMRES algorithm for solving nonsymmetric linear systems. *Numer. Linear Algebra Appl.*, 1(5):427–447, 1994. (Cited on page 232.)

[1849] G. Jouve. The first works of D'Alembert and Euler about the problem of vibrating strings from the perspective of their correspondence. *Centaurus*, 59(4):300–307, 2017. (Cited on page 277.)

[1850] I. V. Jovović and B. J. Malešsević. A note on solutions of the matrix equation $AXB = C$. *Scientific publications of the state University of Novi Pazar, Ser. A: Appl. Math. Inform. Mech.*, 6(1):45–55, 2014. (Cited on page 394.)

[1851] C. Jugé. *Jacques Peletier du Mans, Essai sur sa Vie, son Ouvre, son Influence*. A. Lemerre, Paris, 1907. (Cited on page 46.)

[1852] A. Juhel. Le Verrier et la première détermination des valeurs propres d'une matrice. Bibnum. Textes fondateurs de la science, 2011. (Cited on page 289.)

[1853] E. Jürgens. Numerische berechnung von determinanten. *Jahresbericht der Deutschen Mathematiker-Vereinigung*, 9:131–136, 1901. (Cited on page 21.)

[1854] J. H. Justice. An algorithm for inverting positive definite Toeplitz matrices. *SIAM J. Appl. Math.*, 23(3):289–291, 1972. (Cited on page 386.)

[1855] J. H. Justice. The Szegö recursion relation and inverses of positive definite Toeplitz matrices. *SIAM J. Math. Anal.*, 5(3):503–508, 1974. (Cited on page 386.)

[1856] B. Kågström, V. Khazanov, F. Uhlig, and A. Ruhe. Obituaries; Vera Kublanoskaya. *SIAM News*, 45(6), 2012. (Cited on page 507.)

[1857] B. Kågström and A. Ruhe. An algorithm for numerical computation of the Jordan normal form of a complex matrix. *ACM Trans. Math. Soft. (TOMS)*, 6(3):398–419, 1980. (Cited on page 150.)

[1858] S. Kaczmarz. Angenäherte auflösung von systemen linearer gleichungen. *Bulletin International de l'Académie Polonaise des Sciences et des Lettres*, A 35:355–357, 1937. Translated to English, "Approximate solution of systems of linear equations", *Intern. J. Control*, 57.6 (1993), pp. 1269-1271. (Cited on pages 188 and 498.)

[1859] W. Kahan. The rate of convergence of the extrapolated Gauss-Seidel iterations. *J. ACM*, 4:521–522, 1957. (Cited on page 175.)

[1860] W. Kahan. *Gauss-Seidel methods of solving large systems of linear equations*. PhD thesis, University of Toronto, 1958. (Cited on page 175.)

[1861] W. Kahan. Numerical linear algebra. *Canad. Math. Bull.*, 9(5):757–801, 1966. (Cited on page 22.)

[1862] W. Kahan. How Cray's arithmetic hurts scientific computation. Cray Users Group meeting in Toronto, 1990. (Cited on page 342.)

[1863] J.-P. Kahane. Hasard et déterminisme chez Laplace. arXiv preprint arXiv:0709.4382, 2007. (Cited on page 523.)

[1864] D. Kahaner, C. B. Moler, and S. Nash. *Numerical Methods and Software*. Prentice-Hall, Englewood Cliffs, NJ, 1989. (Cited on page 364.)

[1865] C. Kamath and A. Sameh. A projection method for solving nonsymmetric linear systems on multiprocessors. *Parallel Comput.*, 9(3):291–312, 1989. (Cited on page 192.)

[1866] E. Kamgnia and B. Philippe. Counting eigenvalues in domains of the complex field. *Electron. Trans. Numer. Anal.*, 40:1–16, 2013. (Cited on page 287.)

[1867] S. Kamvar, T. Haveliwala, and G. H. Golub. Adaptive methods for the computation of PageRank. *Linear Algebra Appl.*, 386:51–65, 2004. (Cited on page 295.)

[1868] K. Kang, W. Lothian, J. Sears, and B. D. Sutton. Simultaneous multidiagonalization for the CS decomposition. *Numer. Algorithms*, 66(3):479–493, 2014. (Cited on page 146.)

[1869] S. Kaniel. Estimates for some computational techniques in linear algebra. *Math. Comput.*, 20:369–378, 1966. (Cited on pages 208 and 301.)

[1870] L. V. Kantorovich. The method of successive approximation for functional equations. *Acta Math.*, 71(1):63–97, 1939. (Cited on page 15.)

[1871] L. V. Kantorovich. Functional analysis and applied mathematics. *Uspekhi Mat. Nauk*, 3:89–185, 1948. (In Russian). (Cited on page 208.)

[1872] I. E. Kaporin. High quality preconditioning of a general symmetric positive definite matrix based on its $U^TU + U^TR + R^TU$-decomposition. *Numer. Linear Algebra Appl.*, 5(6):483–509, 1998. (Cited on page 258.)

[1873] I. Kardos. *Scientists Face to Face*. Corvina Kidó, Budapest, 1978. (Cited on page 520.)

[1874] R. Karlson. A study of some roundoff effects of the GMRES method. Technical Report Li-TH-MAT-R-1990-11, Department of Mathematics, Linköping University, Sweden, 1991. (Cited on page 227.)

[1875] W. Karush. An iterative method for finding characteristic vectors of a symmetric matrix. *Pacific J. Math.*, 1(2):233–248, 1951. (Cited on page 300.)

[1876] G. Karypis and V. Kumar. A fast and high quality multilevel scheme for partitioning irregular graphs. *SIAM J. Sci. Comput.*, 20(1):359–392, 1998. (Cited on page 90.)

[1877] G. Karypis and V. Kumar. A parallel algorithm for multilevel graph partitioning and sparse matrix ordering. *J. Parallel Dist. Com.*, 48(1):71–95, 1998. (Cited on page 90.)

[1878] T. Kato. On the upper and lower bounds of eigenvalues. *J. Phys. Soc. Japan*, 4(4-6):334–339, 1949. (Cited on page 285.)

[1879] T. Kato. Estimation of iterated matrices, with application to the von Neumann condition. *Numer. Math.*, 2(1):22–29, 1960. (Cited on page 170.)

[1880] V. J. Katz and K. H. Parshall. *Taming the Unknown: A History of Algebra from Antiquity to the Early Twentieth Century*. Princeton University Press, 2014. (Cited on page xvi.)

[1881] G. R. Kaye. *The Bakhshālī Manuscript: A Study in Medieval Mathematics*. Delhi Manager of Publications, 1933. (Cited on page 41.)

[1882] R. Kehl, R. Nabben, and D. B. Szyld. Adaptive multilevel Krylov methods. *Electron. Trans. Numer. Anal.*, 51:512–528, 2017. (Cited on page 246.)

[1883] H. B. Keller. On the solution of singular and semidefinite linear system by iterations. *SIAM J. Numer. Anal.*, 2(2):281–290, 1965. (Cited on page 176.)

[1884] C. Kenney and A. J. Laub. Condition estimates for matrix functions. *SIAM J. Matrix Anal. Appl.*, 10(2):191–209, 1989. (Cited on page 389.)

[1885] B. W. Kernighan and S. Lin. An efficient heuristic procedure for partitioning graphs. *Bell System Tech. J.*, 49(2):291–307, 1970. (Cited on page 90.)

[1886] D. S. Kershaw. The incomplete Cholesky-conjugate gradient method for the iterative solution of systems of linear equations. *J. Comp. Phys.*, 26(1):43–65, 1978. (Cited on page 255.)

[1887] D S. Kershaw. Problem of unstable pivots in the incomplete LU-conjugate gradient method. Technical Report UCRL-81103, Lawrence Livermore Laboratory, Livermore, 1978. (Cited on page 255.)

[1888] D. S. Kershaw. On the problem of unstable pivots in the incomplete LU conjugate gradient method. *J. Comp. Phys.*, 38:114–123, 1980. (Cited on page 255.)

[1889] J. Kestyn, E. Polizzi, and P. T. P. Tang. FEAST eigensolver for non-Hermitian problems. *SIAM J. Sci. Comput.*, 38(5):S772–S799, 2016. (Cited on page 318.)

[1890] I. M. Khabaza. An iterative least-square method suitable for solving large sparse matrices. *Comput. J.*, 6(2):202–206, 1963. (Cited on page 220.)

[1891] I. N. Khlodovskii. On the general case of the transformation of the secular equation by the method of A. Krylov. *Izvestiya Rossiiskoi Akademii Nauk. Seriya Matematicheskaya*, 8:1077–1102, 1933. (Cited on page 291.)

[1892] L. Kieffer. Le mathématicien Luxembourgeois Bernard Isidore Clasen (1829-1902). *Scientiarum Historia*, 3(3):117–123, 1961. (Cited on page 67.)

[1893] A. Kiełbasiński and K. Ziętak. Numerical behaviour of Higham's scaled method for polar decomposition. *Numer. Algorithms*, 32(2-4):105–140, 2003. (Cited on page 143.)

[1894] A. Kiełbasiński and K. Ziętak. Note on "A new scaling for Newton's iteration for the polar decomposition and its backward stability" by R. Byers and H. Xu. *SIAM J. Matrix Anal. Appl.*, 31(3):1538–1539, 2010. (Cited on page 143.)

[1895] M. E. Kilmer and D. P. O'Leary. Choosing regularization parameters in iterative methods for ill-posed problems. *SIAM J. Matrix Anal. Appl.*, 22(4):1204–1221, 2001. (Cited on page 393.)

[1896] M. E. Kilmer and G. W. Stewart. Iterative regularization and MINRES. *SIAM J. Matrix Anal. Appl.*, 21(2):613–628, 1999. (Cited on page 393.)

[1897] S. K. Kim and A. T. Chronopoulos. A class of Lanczos-like algorithms implemented on parallel computers. *Parallel Comput.*, 17:763–778, 1991. (Cited on page 229.)

[1898] S. K. Kim and A. T. Chronopoulos. An efficient nonsymmetric Lanczos method on parallel vector computers. *J. Comput. Appl. Math.*, 42(3):357–374, 1992. (Cited on page 229.)

[1899] D. R. Kincaid. On complex second-degree iterative methods. *SIAM J. Numer. Anal.*, 11(2):211–218, 1974. (Cited on page 182.)

[1900] D. R. Kincaid, T. C. Oppe, and W. D. Joubert. An introduction to the NSPCG software package. *Internat. J. Numer. Methods Engrg.*, 27(3):589–608, 1989. (Cited on page 372.)

[1901] D. R. Kincaid, R. S. Varga, and C. H. Warlick. The life and times of Dr David M. Young, Jr. *Numer. Linear Algebra Appl.*, 17:743–757, 2010. (Cited on page 600.)

[1902] I. P. King. An automatic reordering scheme for simultaneous equations derived from network analysis. *Internat. J. Numer. Methods Engrg.*, 2:523–533, 1970. (Cited on page 89.)

[1903] N. G. Kingsbury and P. J. W. Rayner. Digital filtering using logarithmic arithmetic. *Electronics Letters*, 7(2):56–58, 1971. (Cited on page 355.)

[1904] R. Kippenhahn. Uber den wertevorrat einer matrix. *Mathematische Nachrichten*, 6(3-4):193–228, 1951. Translated into English by P. F. Zachlin and M. E. Hochstenbach, "On the numerical range of a matrix", *Linear Multilinear Algebra*, 56 (1-2), 185-225, 2008. (Cited on page 287.)

[1905] G. Kitagawa. An algorithm for solving the matrix equation $X = FXF^T + S$. *Int. J. Control*, 25(5):745–753, 1977. (Cited on page 398.)

[1906] T. H. Kjeldsen. A contextualized historical analysis of the Kuhn-Tucker theorem in nonlinear programming: the impact of World War II. *Historia Math.*, 27(4):331–361, 2000. (Cited on page 199.)

[1907] G. Kjellberg. On the convergence of the successive over-relaxation applied to a class of linear systems of equations with complex eigenvalues. *Ericsson Technics*, 2:245–258, 1958. (Cited on page 176.)

[1908] D. L. Kleinman. On an iterative technique for Riccati equation computations. *IEEE Trans. Automat. Contr.*, 13(1):114–115, 1968. (Cited on page 399.)

[1909] D. A. Klempt. *Lehrbuch zur Einführung in die Moderne Algebra*. Teubner, 1880. (Cited on page 56.)

[1910] V. V. Klyuyev and N. I. Kokovkin-Shcherbak. On the minimization of the number of arithmetic operations for the solution of linear algebraic systems of equations. *J. Comput. Math. and Math. Phys.*, 5:21–33, 1965. (In Russian). A translation, by G.J. Tee, is available as Tech. Report CS24, Computer Sci. Dept. Stanford University. Another translation was published in U.S.S.R. Comput. Math. Math. Phys., v. 5: n. 1 (1965), pp. 25-43. (Cited on page 81.)

[1911] L. Knizhnerman. Calculation of functions of unsymmetric matrices using Arnoldi's method. *Comput. Math. Math. Phys.*, 31(1):1–9, 1992. (Cited on page 389.)

[1912] L. Knizhnerman. On GMRES-equivalent bounded operators. *SIAM J. Matrix Anal. Appl.*, 22(1):195–212, 2000. (Cited on page 238.)

[1913] L. Knizhnerman and V. Simoncini. A new investigation of the extended Krylov subspace method for matrix function evaluations. *Numer. Linear Algebra Appl.*, 17(4):615–638, 2010. (Cited on page 390.)

[1914] E. Knobloch. Der beginn der determinantentheorie. In *Leibnizens Nachgelassene Studien zum Determinantenkalkul*, Hildesheim, 1980. Gerstenberg. (Cited on page 102.)

[1915] E. Knobloch. Déterminants et élimination chez Leibniz. *Rev. Hist. Sci.*, 54(2):143–164, 2001. (Cited on pages 102 and 103.)

[1916] E. Knobloch. Review of the book "Roger Hart, The Chinese Roots of Linear Algebra". *East Asian Science, Technology, and Medicine*, 34:225–227, 2012. (Cited on page 39.)

[1917] E. Knobloch. Determinant theory, symmetric functions, and dyadic. In M. R. Antognazza, editor, *The Oxford Handbook of Leibniz*, pages 225–246. Oxford University Press, 2018. (Cited on pages 102 and 326.)

[1918] W. R. Knorr. Arithmêtikê stoicheiôsis: On Diophantus and Hero of Alexandria. *Historia Math.*, 20(2):180–192, 1993. (Cited on page 39.)

[1919] C. G. Knott. *Life and Scientific Work of Peter Guthrie Tait*. Cambridge University Press, 1911. (Cited on pages 4 and 5.)

[1920] D. E. Knuth. *The Art of Computer Programming, volume 1*. Addison-Wesley, 1968. (Cited on page 86.)

[1921] D. E. Knuth. Ancient babylonian algorithms. *Comm. ACM*, 15(7):671–677, 1972. (Cited on pages 34 and 323.)

[1922] D. E. Knuth. George Forsythe and the development of computer science. *Comm. ACM*, 15(8):721–726, 1972. (Cited on page 459.)

[1923] D. E. Knuth and L. T. Pardo. The early development of programming languages. In N. Metropolis, J. Howlett, and G. C. Rota, editors, *A History of Computing in the Twentieth Century*, pages 197–273. Academic Press, 1980. (Cited on pages 358 and 603.)

[1924] A. V. Knyazev. Preconditioned eigensolvers - an oxymoron? *Electron. Trans. Numer. Anal.*, 7:104–123, 1998. (Cited on page 319.)

[1925] A. V. Knyazev. Toward the optimal preconditioned eigensolver: Locally optimal block preconditioned conjugate gradient method. *SIAM J. Sci. Comput.*, 23(2):517–541, 2001. (Cited on page 319.)

[1926] A. V. Knyazev, M. E. Argentati, I. Lashuk, and E. E. Ovtchinnikov. Block Locally Optimal Preconditioned Eigenvalue Xolvers (BLOPEX) in Hypre and PETSc. *SIAM J. Sci. Comput.*, 29(5):2224–2239, 2007. (Cited on page 319.)

[1927] J. Kociński. Groups with involution, and quasigroups with Cracovian representations. arXiv preprint, arXiv: 1509.08740, 2015. (Cited on page 70.)

[1928] T. Koetsier. Joseph Louis Lagrange (1736-1813): His life, his work and his personality. *Nieuw Archief Voor Wiskunde*, 4:191–205, 1986. (Cited on page 510.)

[1929] E. G. Kogbetliantz. Solution of linear equations by diagonalization of coefficients matrix. *Quart. Appl. Math.*, 13(2):123–132, 1955. (Cited on page 136.)

[1930] W. Kohn. A variational iteration method for solving secular equations. *J. Chem. Phys.*, 17(7):670, 1949. (Cited on page 295.)

[1931] E. Kokiopoulou, C. Bekas, and E. Gallopoulos. Computing smallest singular triplets with implicitly restarted Lanczos bidiagonalization. *Appl. Numer. Math.*, 49:39–61, 2004. (Cited on page 139.)

[1932] G. Kollias, A. Grama, and Z. Li. Asynchronous iterative algorithms. In D. Padua, editor, *Encyclopedia of Parallel Computing*, pages 87–95. Springer, Boston, 2011. (Cited on page 197.)

[1933] L. Yu. Kolotilina and A. Yu. Yeremin. Factorized sparse approximate inverse preconditionings I. Theory. *SIAM J. Matrix Anal. Appl.*, 14(1):45–58, 1993. (Cited on page 262.)

[1934] L. Yu. Kolotilina and A. Yu Yeremin. Factorized sparse approximate inverse preconditioning II: Solution of 3D FE systems on massively parallel computers. *Intl. J. High Speed Computing*, 7(2):191–215, 1995. (Cited on page 262.)

[1935] Ya. Kon'kova, V. N. Simonova, and V. B. Khazanov. Vera Nikolaevna Kublanovskaya. A short biography. *J. Math. Sci.*, 114:1755–1756, 2003. Translated from *Zapiski Nauchnykh Seminarov POMI*, 268 (2000), pp. 5-8. (Cited on page 507.)

[1936] J. Kopal, M. Rozložník, and M. Tůma. Factorized approximate inverses with adaptive dropping. *SIAM J. Sci. Comput.*, 38(3):A1807–A1820, 2016. (Cited on page 263.)

[1937] A. Korganoff and M. Pavel-Parvu. *Méthodes de Calcul Numérique, Tome 2: Eléments de Théorie des Matrices Carrées et Rectangles en Analyse Numérique*. Dunod, Paris, 1967. (Cited on page 94.)

[1938] M. Kormes. Numerical solution of the boundary value problem for the potential equation by means of punched cards. *Review of Scientific Instruments*, 14(8):248–250, 1943. (Cited on page 170.)

[1939] A. A. Kosinski. Cramer's rule is due to Cramer. *Math. Mag.*, 74(4):310–312, 2001. (Cited on page 105.)

[1940] O. Kouteynikoff. Guillaume Gosselin, algébriste de la Renaissance. *Mnémosyne*, Num. Spécial 2.:4–84, 2004. (Cited on page 48.)

[1941] Z. Kovarik. Some iterative methods for improving orthonormality. *SIAM J. Numer. Anal.*, 7(3):386–389, 1970. (Cited on page 142.)

[1942] G. Kowalewski. *Einfurhung in die Determinantentheorie*. Walter de Gruyter and Co., Berlin, Leipzig, 1924. First edition 1909. (Cited on pages 126 and 294.)

[1943] B. Krasnopolsky. The reordered BiCGStab method for distributed memory computer systems. *Procedia Comput. Sci.*, 1:213–218, 2012. (Cited on page 244.)

[1944] C. Krattenthaler. Advanced determinant calculus. In *The Andrews Festschrift*, pages 349–426, Springer, Berlin, Heidelberg, 2001. (Cited on page 117.)

[1945] C. Krattenthaler. Advanced determinant calculus: a complement. *Linear Algebra Appl.*, 411:68–166, 2005. (Cited on page 117.)

[1946] M. Krein. Über das spektrum des Jakobischen form in verbindung mit der theorie der torsionsschwingungen von walzen. *Mat. Sb.*, 40(4):455–466, 1933. (Cited on page 382.)

[1947] H. Kreis. *Contribution à la théorie des systèmes linéaires*. PhD thesis, Universität Zürich, 1906. (Cited on page 396.)

[1948] D. Kressner. *Numerical Methods for General and Structured Eigenvalue Problems*. Springer, 2005. (Cited on page 319.)

[1949] D. Kressner. On the use of larger bulges in the QR algorithm. *Electron. Trans. Numer. Anal.*, 20:50–63, 2005. (Cited on page 314.)

[1950] D. Kressner. The effect of aggressive early deflation on the convergence of the QR algorithm. *SIAM J. Matrix Anal. Appl.*, 30(2):805–821, 2008. (Cited on page 314.)

[1951] U. Kristiansen. Use of the Rayleigh quotient in string, chain, and plate vibration. 40th Scandinavian Symposium on Physical Acoustics, Geilo, Norway, Jan. 29 - Feb. 1, 2017, 2017. (Cited on page 282.)

[1952] F. T. Krogh. On the use of assembly code for heavily used modules in linear algebra. Technical Report JPL CM-303, Jet Propulsion Laboratory, California, 1972. (Cited on page 364.)

[1953] L. Kronecker. Über sturmsche functionen. *Monatsber. d. König. Preuss. Akad. Wiss. Berlin*, pages 95–121, 1878. (Cited on page 464.)

[1954] L. Kronecker. *Vorlesungen uber die Theorie der Determinanten, Erste band. Erste bis Einundzwanzigste Vorlesungen*. B. G. Teubner, 1903. (Cited on pages 11 and 116.)

[1955] A. N. Krylov. O čislemnon rešenii uravnenija, kotorym v techničeskih voprasah opredeljajutsja častoy malyh kolebaniĭ material'nyh. *Izv. Adad. Nauk SSSR, ser Fiz.-Mat.*, 4:491–539, 1931. Possible translation: "On the numerical solution of the equation by which the frequency of small oscillations is determined in technical problems". (Cited on pages 217, 291, and 505.)

[1956] A. N. Krylov. *Moi Vospominaniya*. Izdatel'stvo Akademii Nauk SSSR, Moscow and Leningrad, 1942. Translated to English by Laura N. Meyerovich with the title *Professor Krylov's Navy, Memoir of Alexey N. Krylov*, Magnet Publishing (2014). (Cited on pages 501, 502, and 505.)

[1957] M. Kubínová and K. M. Soodhalter. Admissible and attainable convergence behavior of block Arnoldi and GMRES. *SIAM J. Matrix Anal. Appl.*, 41(2):464–486, 2020. (Cited on page 244.)

[1958] V. N. Kublanovskaya. Application of analytic continuation by means of change of variables in numerical analysis. In *The Works on the Approximate Analysis, Trudy Mat. Inst. Steklov., 53*, pages 145–185. Acad. Sci. USSR, Moscow, Leningrad, 1959. (Cited on page 507.)

[1959] V. N. Kublanovskaya. On some algorithms for solving of full problem of eigenvalues. *Zh. Vychisl. Mat. Mat. Fiz.*, 1:555–570, 1961. (Cited on pages 136, 312, and 507.)

[1960] V. N. Kublanovskaya. On some algorithms for the solution of the complete eigenvalue problem. *USSR Comput. Math. and Math. Phys.*, 1:637–657, 1962. (Cited on pages 136, 312, and 507.)

[1961] V. N. Kublanovskaya. On some algorithms for the solution of the complete eigenvalue problem for a degenerate matrix. *USSR Comput. Math. and Math. Phys.*, 6(4):1–14, 1966. (Cited on page 149.)

[1962] G. Kumfert and A. Pothen. Two improved algorithms for envelope and wavefront reduction. *BIT Numerical Mathematics*, 37(3):559–590, 1997. (Cited on page 88.)

[1963] J. Kurzak, M. Gates, A. YarKhan, I. Yamazaki, P. Wu, P. Luszczek, J. Finney, and J. J. Dongarra. SLATE Working Note 5: Parallel BLAS performance report. Technical Report ICL-UT-18-01, Innovative Computing Laboratory, University of Tennessee, 2018. (Cited on page 376.)

[1964] Y. Kuznetsov. New algorithms for approximate realization of implicit difference schemes. *Sov. J. Numer. Anal. Math. Modelling*, 3(2):99–114, 1988. (Cited on page 267.)

[1965] P. Laasonen. On the iterative solution of the matrix equation $AX^2 - I = 0$. *M.T.A.C.*, 12:109–116, 1958. (Cited on pages 389 and 394.)

[1966] J. Laderman. The square root method for solving simultaneous linear equations. *M.T.A.C.*, 3:13–16, 1948. (Cited on page 73.)

[1967] G. López Lagomasino, L. Reichel, and L. Wunderlich. Matrices, moments, and rational quadrature. *Linear Algebra Appl.*, 429(10):2540–2554, 2008. (Cited on page 391.)

[1968] J. L. Lagrange. Recherches sur la méthode des maximis et minimis. *Miscellanea philosophico-mathematica societatis privatæ Taurinensis*, 1:18–32, 1759. in Ouvres Complètes, tome 1, pages 1-20. (Cited on pages 57 and 511.)

[1969] J. L. Lagrange. Solution de différents problèmes de calcul intégral. *Miscellanea Taurinensia*, III, 1766. in Ouvres Complètes, tome 1, pages 471-668. (Cited on page 277.)

[1970] J. L. Lagrange. Nouvelle solution du problème du mouvement de rotation d'un corps de figure quelconque qui n'est animé par aucune force accélératrice. *Nouveaux Mémoires de l'Académie des Sciences et Belles-Lettres de Berlin*, pages 85–128, 1773. in Ouvres Complètes, tome 3, pages 577-616. (Cited on pages 108, 123, and 279.)

[1971] J. L. Lagrange. Solution analytique de quelques problèmes sur les pyramides triangulaires. *Nouveaux Mémoires de l'Académie des Sciences et Belles-Lettres de Berlin*, pages 149–173, 1773. in Ouvres Complètes, tome 3, pages 661-692. (Cited on page 122.)

[1972] J. L. Lagrange. Recherches sur les équations séculaires des mouvements des nouds et des inclinaisons des orbites des planètes. *Mémoires de l'Académie Royale des Sciences de Paris*, pages 97–174, 1774. in Ouvres Complètes, tome 6, pages 635-709. (Cited on page 278.)

[1973] J. L. Lagrange. Théorie des variations séculaires des éléments des planètes: Première partie. *Nouveaux Mémoires de l'Académie des Sciences et des Belles-Lettres de Berlin*, pages ?–?, 1781. in Ouvres Complètes, tome 5, pages 125-207. (Cited on page 278.)

[1974] J. L. Lagrange. Théorie des variations séculaires des éléments des planètes: Seconde partie. *Nouveaux Mémoires de l'Académie des Sciences et des Belles-Lettres de Berlin*, pages 169–292, 1782. in Ouvres Complètes, tome 5, pages 211-344. (Cited on page 278.)

[1975] J. L. Lagrange. *Méchanique Analitique*. chez la Veuve Desaint, Paris, 1788. (Cited on page 279.)

[1976] E. Laguerre. Sur le calcul des systèmes linéaires. *J. École Impériale Polytechnique*, Tome XXV, Cahier 62:215–264, 1867. Also Ouvres Complètes, Tome I, pp. 221-267. (Cited on pages 5, 6, 387, and 514.)

[1977] E. Laguerre. *Notes sur la Résolution des Équations Numériques*. Gauthier-Villars, Paris, 1880. (Cited on page 289.)

[1978] E. Laguerre. Sur une méthode pour obtenir par approximation les racines d'une équation algébrique qui a toutes ses racines réelles. *Nouvelles Annales de Mathématiques*, 2ème série, 19:161–171, 1880. Also Ouvres Complètes, Tome I, pp. 87-103. (Cited on page 289.)

[1979] E. Laguerre. Notice sur les travaux mathématiques de M. Laguerre (pour servir d'appui à sa candidature à l'Académie des sciences). Gauthier-Villars, Paris, 1884. (Cited on page 514.)

[1980] I. Lakatos. *Proofs and Refutations*. Cambridge University Press, Cambridge, 1976. (Cited on page 559.)

[1981] C. Lallemand. *Compensation d'un Réseau de Nivellements par la Méthode des Coefficients Indéterminés*. Béranger, Paris and Liège, 1912. (Cited on pages 66 and 448.)

[1982] P. Lancaster and L. Rodman. *Algebraic Riccati Equations*. Oxford University Press, 1995. (Cited on page 399.)

[1983] C. Lanczos. Trigonometric interpolation of empirical and analytical functions. *J. Math. Phys.*, 17(1-4):123–199, 1938. (Cited on page 180.)

[1984] C. Lanczos. An iteration method for the solution of the eigenvalue problem of linear differential and integral operators. *J. Res. Natl. Bur. Stand.*, 45:255–282, 1950. (Cited on pages 126, 205, 218, 219, 297, 308, and 518.)

[1985] C. Lanczos. Solution of systems of linear equations by minimized iterations. *J. Res. Natl. Bur. Stand.*, 49:33–53, 1952. (Cited on pages 180, 205, 218, 300, and 303.)

[1986] C. Lanczos. Iterative solution of large-scale linear systems. *J. Soc. Ind. Appl. Math.*, 6(1):91–109, 1958. (Cited on page 207.)

[1987] C. Lanczos. *Linear Differential Operators*. D. Van Nostrand Company, London, 1961. Reprinted by SIAM, Philadelphia, 1996. (Cited on page 3.)

[1988] G. Landsberg. Über fundamentalsysteme und bilineare formen. *Journal für die reine und angewandte Mathematik (Crelle Journal)*, 116:331–349, 1896. (Cited on pages 150, 395, and 396.)

[1989] P. Langevin. La valeur éducative de l'histoire des sciences. *Bulletin de la Société Française de Pédagogie*, 22:692–700, 1926. (Cited on page xviii.)

[1990] J. Langou. Translation and modern interpretation of Laplace's Théorie Analytique des Probabilités, pages 505-512, 516-520. arXiv preprint, arXiv:0907.4695v1, 2009. (Cited on pages 124, 218, and 523.)

[1991] J. Lankeit, P. Neff, and Y. Nakatukasa. The minimization of matrix logarithms: On a fundamental property of the unitary polar factor. *Linear Algebra Appl.*, 449:28–42, 2014. (Cited on page 143.)

[1992] P. S. Laplace. Mémoire sur la probabilité des causes par les événements. *Mémoires de l'Académie Royale des Sciences de Paris (Savants étrangers)*, 6:621–656, 1774. Translated by Stephen M. Stigler in *Statistical Science*, 1 (3), 1986. (Cited on page 59.)

[1993] P. S. Laplace. Recherches sur le calcul intégral et sur le système du monde. *Histoire de l'Académie Royale des Sciences avec les Mémoires de Mathématique et de Physique pour l'année 1772*, pages 267–376, 1776. (Cited on pages 107 and 522.)

[1994] P. S. Laplace. Mémoire sur les approximations des formules qui sont fonctions de très grands nombres. *Mémoires de l'Académie royale des sciences de Paris, année 1782*, pages 209–291, 1785. (Cited on page 59.)

[1995] P. S. Laplace. Mémoire sur les approximations des formules qui sont fonctions de très grands nombres (suite). *Mémoires de l'Académie royale des sciences de Paris, année 1783*, pages 295–338, 1786. (Cited on page 59.)

[1996] P. S. Laplace. *Exposition du Système du Monde*. Imprimerie de Crapelet, 1796. (Cited on page 522.)

[1997] P. S. Laplace. Mémoires sur les approximations de formules qui sont fonctions de très grands nombres et sur leurs applications aux probabilités. *Mémoires de l'Académie des Sciences de Paris*, pages 353–415, 559–565, 1809. (Cited on page 59.)

[1998] P. S. Laplace. *Théorie Analytique des Probabilités*. Courcier, Paris, 3rd edition, 1820. (Cited on pages 25, 124, and 523.)

[1999] A. J. Laub. A Schur method for solving algebraic Riccati equations. *IEEE Trans. Automat. Contr.*, 24:913–921, 1979. (Cited on page 399.)

[2000] H. Laurent. Exposé d'une théorie nouvelle des substitutions linéaires. *Nouvelles Annales de Mathématiques 3e série*, 15:345–365, 1896. (Cited on page 8.)

[2001] D. P. Laurie. Anti-Gaussian quadrature formulas. *Math. Comput.*, 65(214):739–747, 1996. (Cited on page 213.)

[2002] D. P. Laurie. Accurate recovery of recursion coefficients from Gaussian quadrature formulas. *J. Comput. Appl. Math.*, 112:165–180, 1999. (Cited on pages 311 and 316.)

[2003] F. F. Lawless and P. Wang. A simulation study of ridge and other regression estimators. *Commun. Statist. Theor. Meth.*, 5:307–323, 1976. (Cited on page 394.)

[2004] C. L. Lawson and R. J. Hanson. *Solving Least Squares Problems*. SIAM, Philadelphia, 1995. First edition, Prentice-Hall (1974). (Cited on pages 137, 392, 482, and 525.)

[2005] C. L. Lawson, R. J. Hanson, F. T. Krogh, and D. R. Kincaid. Basic linear algebra subprograms for Fortran usage. Technical Report SAND77-0898, Sandia Laboratories, Albuquerque, New Mexico, 1977. (Cited on pages 364 and 482.)

[2006] C. L. Lawson, R. J. Hanson, F. T. Krogh, and D. R. Kincaid. Algorithm 539: Basic linear algebra subprograms for Fortran usage. *ACM Trans. Math. Soft. (TOMS)*, 5(3):308–325, 1979. (Cited on pages 364 and 365.)

[2007] C. Le Calvez and B. Molina. Implicitly restarted and deflated GMRES. *Numer. Algorithms*, 21(1-4):261–285, 1999. (Cited on page 233.)

[2008] Jean Le Rond d'Alembert. *Traité de Dynamique*. David l'aîné, Paris, 1743. First edition. (Cited on page 275.)

[2009] Jean Le Rond d'Alembert. Recherches sur le calcul intégral. In *Histoire de l'Académie Royale des Sciences et Belles Lettres, 1746*, pages 182–224. Ambroise Haude, Berlin, 1748. (Cited on page 277.)

[2010] Jean Le Rond d'Alembert. Suite des recherches sur le calcul intégral. In *Histoire de l'Académie Royale des Sciences et Belles Lettres, 1748*, pages 249–290. Haude and Stener, Berlin, 1750. (Cited on page 277.)

[2011] Jean Le Rond d'Alembert. *Traité de Dynamique*. David, Paris, 1758. Second edition. (Cited on page 276.)

[2012] P. Le Tallec, J. Mandel, and M. Vidrascu. A Neumann-Neumann domain decomposition algorithm for solving plate and shell problems. *SIAM J. Numer. Anal.*, 35(2):836–867, 1998. (Cited on page 268.)

[2013] U. Le Verrier. Sur les variations séculaires des éléments des orbites pour les sept planètes principales. *J. Math. Pures Appl.*, 1(5):220–254, 1840. (Cited on page 289.)

[2014] C. Lebaud. On the QR algorithm with shift. *Int. J. Comput. Math.*, 2(1-4):343–358, 1968. (Cited on page 313.)

[2015] C. Lebaud. Remarques sur la convergence de la méthode QR. *Revue Française d'Informatique et de Recherche Opérationnelle*, 2(13):103–112, 1968. (Cited on page 313.)

[2016] C. Lebaud. L'algorithme double QR avec "shift". *Numer. Math.*, 16(2):163–180, 1970. (Cited on page 313.)

[2017] V. I. Lebedev. Iterative methods for the solution of operator equations with their spectrum lying on several intervals. *Zh. Vychisl. Mat. Mat. Fiz.*, 9(6):1247–1252, 1969. (Cited on page 163.)

[2018] V. I. Lebedev and S. A. Finogenov. The order of choice of the iteration parameters in the cyclic Chebyshev iteration method. *Zh. Vychisl. Mat. Mat. Fiz.*, 11(2):425–438, 1971. English translation: *USSR Comput. Math. Math. Phys.* 11 (2), (1971), pp. 155-170. (Cited on page 163.)

[2019] H. Lebesgue. Notice sur la vie et les travaux de M. Camille Jordan. *Mémoires de l'Académie des Sciences de l'Institut de France*, 58:29–66, 1923. (Cited on page 495.)

[2020] H. Lebesgue. L'oeuvre mathematique de Vandermonde. *Thalès*, 4:28–42, 1937. (Cited on pages 107 and 590.)

[2021] A.-M. Legendre. *Nouvelles Méthodes pour la Détermination des Orbites des Comètes, Appendice*. Didot, Paris, 1805. Second edition, Courcier, 1806. Reproduced by L. Puissant in *Traité de Géodésie*, Courcier, 1805, pages 137-141. (Cited on page 57.)

[2022] N. J. Lehmann. Calculation of eigenvalue bounds in linear problems. *Arch. Math.*, 2:139–147, 1949. (Cited on page 287.)

[2023] N. J. Lehmann. Optimale eigenwerteinschließungen. *Numer. Math.*, 5(1):246–272, 1963. (Cited on page 287.)

[2024] N. J. Lehmann. Zur verwendung optimaler eigenwerteingrenzungen bei der lösung symmetrischer matrizenaufgaben. *Numer. Math.*, 8(1):42–55, 1966. (Cited on page 287.)

[2025] R. B. Lehoucq. *Analysis and implementation of an implicitly restarted Arnoldi iteration*. PhD thesis, Rice University, Houston, 1995. (Cited on page 306.)

[2026] R. B. Lehoucq. The computation of elementary unitary matrices. *ACM Trans. Math. Soft. (TOMS)*, 22(4):393–400, 1996. (Cited on page 133.)

[2027] R. B. Lehoucq. Implicitly restarted Arnoldi methods and subspace iteration. *SIAM J. Matrix Anal. Appl.*, 23(2):551–562, 2001. (Cited on page 306.)

[2028] R. B. Lehoucq and D. C. Sorensen. Deflation techniques for an implicitly restarted Arnoldi iteration. *SIAM J. Matrix Anal. Appl.*, 17(4):789–821, 1996. (Cited on page 306.)

[2029] R. B. Lehoucq, D. C. Sorensen, and C. Yang. *ARPACK Users' Guide: Solution of Large-Scale Eigenvalue Problems*. SIAM, Philadelphia, 1998. (Cited on pages 306, 372, and 378.)

[2030] G. W. Leibniz. *Leibnizens Gesammelte Werke, Sér. 3, t. 2*. Handschriften der königlichen Bibliothek zu Hannover, 1850. published by G. H. Pertz. (Cited on page 102.)

[2031] A. W. Leissa. The historical basis of the Rayleigh and Ritz methods. *J. Sound Vib.*, 287:961–978, 2005. (Cited on page 285.)

[2032] F. Leja. Propriétés des points extrémaux des ensembles plans et leur application à la représentation conforme. *Annales Polonici Mathematici*, 3(2):317–342, 1957. (Cited on page 185.)

[2033] F. Leja. Sur certaines suites liées aux ensembles plans et leur application à la représentation conforme. *Annales Polonici Mathematici*, 1(4):8–13, 1957. (Cited on page 185.)

[2034] O. M. Leland. *Practical Least Squares*. McGraw-Hill Book Company, New York, 1921. (Cited on page 66.)

[2035] J.-G. Lemoine. Les anciens procédés de calcul sur les doigts en orient et en occident. *Revue des études islamiques*, 6, 1932. (Cited on page 323.)

[2036] S. J. Leon, Å. Björck, and W. Gander. Gram-Schmidt orthogonalization: 100 years and more. *Numer. Linear Algebra Appl.*, 20(3):492–532, 2013. (Cited on page 129.)

[2037] A. S. Leonov. On the choice of regularization parameters by means of the quasi-optimality and ratio criteria. *Doklady Akademii Nauk SSSR*, 240(1):18–20, 1978. (Cited on page 392.)

[2038] A. O. Leuschner. Celestial mechanics: A survey of the status of the determination of the general perturbations of the minor planets. Vol. 25. National research council of the National Academy of Sciences, 1922. (Cited on page 60.)

[2039] R. J. LeVeque and L. N. Trefethen. Fourier analysis of the SOR iteration. *IMA J. Numer. Anal.*, 8(3):273–279, 1988. (Cited on page 178.)

[2040] N. Levinson. The Wiener (root mean square) error criterion in filter design and prediction. *J. Math. Phys.*, 25(1-4):261–278, 1946. (Cited on page 386.)

[2041] L. Lévy. Sur la possibilité de l'équilibre électrique. *C.R. Acad. Sci. Paris, Sér 1*, XCIII:706–708, 1881. (Cited on page 286.)

[2042] R. Levy. Restructuring the structural stiffness matrix to improve computational efficiency. *Jet Propulsion Laboratory Quarterly Technical Review*, 1:91–70, 1971. (Cited on page 89.)

[2043] S. Levy, D. Goldenberg, and G. Zibritosky. Simply supported long rectangular plate under combined axial load and normal pressure. Technical Report NACA-949, National Advisory Committee for Aeronautics, 1944. (Cited on page 72.)

[2044] D. I. Lewin. An interview with Cleve Moler. *Computers in Physics*, 9(1):8–9, 1995. (Cited on page 379.)

[2045] M. Lewin. On the coefficients of the characteristic polynomial of a matrix. *Discrete Math.*, 125(1-3):255–262, 1994. (Cited on page 289.)

[2046] C. K. Li and Y. T. Poon. Convexity of the joint numerical range. *SIAM J. Matrix Anal. Appl.*, 21(2):668–678, 2000. (Cited on page 288.)

[2047] H. B. Li, T. Z. Huang, T. X. Gu, and X. P. Liu. From Sylvester's determinant identity to Cramer's rule. arXiv preprint arXiv:1407.1412, 2014. (Cited on pages 114 and 117.)

[2048] J. Li. *Jiuzhang Suanshu Jiaozheng (Critical edition of The Nine Chapters on Mathematical Procedures).* Xi'an: Shaanxi kexue jishu chubanshe, 1993. (in Chinese). (Cited on page 35.)

[2049] N. Li and Y. Saad. Crout versions of ILU factorization with pivoting for sparse symmetric matrices. *Electron. Trans. Numer. Anal.*, 20:75–85, 2005. (Cited on page 258.)

[2050] N. Li, Y. Saad, and E. Chow. Crout versions of ILU for general sparse matrices. *SIAM J. Sci. Comput.*, 25(2):716–728, 2003. (Cited on page 258.)

[2051] R. Li, Y. Xi, L. Erlandson, and Y. Saad. The eigenvalues slicing library (EVSL): Algorithms, implementation, and software. *SIAM J. Sci. Comput.*, 41(4):C393–C415, 2019. (Cited on pages 306, 318, and 372.)

[2052] R. Li, Y. Xi, E. Vecharynski, C. Yang, and Y. Saad. A thick-restart Lanczos algorithm with polynomial filtering for Hermitian eigenvalue problems. *SIAM J. Sci. Comput.*, 38(4):A2512–A2534, 2016. (Cited on pages 306 and 318.)

[2053] R. C. Li. Convergence of CG and GMRES on a tridiagonal Toeplitz linear system. *BIT Numerical Mathematics*, 47(3):577–599, 2007. (Cited on page 241.)

[2054] R. C. Li and W. Zhang. The rate of convergence of GMRES on a tridiagonal Toeplitz linear system. *Numer. Math.*, 112:267–293, 2009. (Cited on page 241.)

[2055] R. C. Li and W. Zhang. The rate of convergence of GMRES on a tridiagonal Toeplitz linear system, II. *Linear Algebra Appl.*, 431:2425–2436, 2009. (Cited on page 241.)

[2056] X. S. Li. *Sparse Gaussian elimination on high performance computers*. PhD thesis, University of California at Berkeley, 1996. (Cited on page 371.)

[2057] X. S. Li. An overview of SuperLU: Algorithms, implementation, and user interface. *ACM Trans. Math. Soft. (TOMS)*, 31(3):302–325, 2005. (Cited on pages 92 and 371.)

[2058] X. S. Li and J. W. Demmel. SuperLU_DIST: A scalable distributed-memory sparse direct solver for unsymmetric linear systems. *ACM Trans. Math. Soft. (TOMS)*, 29(2):110–140, 2003. (Cited on pages 92 and 378.)

[2059] X. S. Li, J. W. Demmel, J. R. Gilbert, L. Grigori, M. Shao, and I. Yamazaki. SuperLU Users' Guide. Technical Report LBNL-44289, Lawrence Berkeley National Laboratory, 1999. (Cited on pages 92 and 371.)

[2060] X. S. Li and M. Shao. A supernodal approach to incomplete LU factorization with partial pivoting. *ACM Trans. Math. Soft. (TOMS)*, 37(4):1–20, 2011. (Cited on page 258.)

[2061] Z. Li, Y. Saad, and M. Sosonkina. pARMS: A parallel version of the algebraic recursive multilevel solver. *Numer. Linear Algebra Appl.*, 10(5-6):485–509, 2003. (Cited on pages 271 and 372.)

[2062] A. Lichnewsky. Some vector and parallel implementations for preconditioned conjugate gradient algorithms. In J. S. Kowalik, editor, *Proceedings of the NATO Workshop on High-Speed Computations*, pages 343–359. NATO ASI Series, 1984. (Cited on page 260.)

[2063] H. Liebmann. Die angengäherte ermittelung harmonischer funktionen und konformer abbildungen. *Bayerischen Akad. Wiss. München*, 47:385–416, 1918. (Cited on pages 163 and 164.)

[2064] J. Liesen. Computable convergence bounds for GMRES. *SIAM J. Matrix Anal. Appl.*, 21(3):882–903, 2000. (Cited on pages 237 and 238.)

[2065] J. Liesen, M. Rozložník, and Z. Strakoš. Least squares residuals and minimal residual methods. *SIAM J. Sci. Comput.*, 23(5):1503–1525, 2002. (Cited on page 239.)

[2066] J. Liesen and Z. Strakoš. GMRES convergence analysis for a convection-diffusion model problem. *SIAM J. Sci. Comput.*, 26(6):1989–2009, 2005. (Cited on page 238.)

[2067] J. Liesen and Z. Strakoš. *Krylov Subspace Methods, Principles and Analysis.* Oxford University Press, 2013. (Cited on pages 224, 231, 245, and 246.)

[2068] J. Liesen and P. Tichý. Convergence analysis of Krylov subspace methods. *GAMM Mitteilungen*, Band 27(Heft 2), 2004. (Cited on page 238.)

[2069] J. Liesen and P. Tichý. The worst-case GMRES for normal matrices. *BIT Numerical Mathematics*, 44:79–98, 2004. (Cited on pages 240 and 241.)

[2070] J. Liesen and P. Tichý. On best approximations of polynomials in matrices in the matrix 2-norm. *SIAM J. Matrix Anal. Appl.*, 31(2):853–863, 2009. (Cited on page 231.)

[2071] J. Liesen and P. Tichý. The field of values bound on ideal GMRES. arXiv preprint, arXiv:1211.5969, 2012. (Cited on pages 236 and 287.)

[2072] I. C. Liggett. The Gauss-Seidel method of solution of simultaneous linear equations. In *Industrial Computation Seminar*, pages 62–65, New York, 1950. IBM. (Cited on page 174.)

[2073] P. D. Lin and R. B. Johnson. Seidel aberration coefficients: An alternative computational method. *Optics Express*, 27:19712–19725, 2019. (Cited on page 559.)

[2074] P.-L. Lions. On the Schwarz alternating method. I. In R. Glowinski, G. H. Golub, G. A. Meurant, and J. Périaux, editors, *First International Symposium on Domain Decomposition Methods for Partial Differential Equations*, pages 1–42. SIAM, Philadelphia, 1988. (Cited on page 267.)

[2075] P.-L. Lions. On the Schwarz alternating method. II: Stochastic interpretation and orders properties. In T. F. Chan, R. Glowinski, J. Périaux, and O. Widlund, editors, *Proceedings of the Second International Symposium on Domain Decomposition Methods for Partial Differential Equations*, pages 47–70. SIAM, Philadelphia, 1989. (Cited on page 267.)

[2076] P.-L. Lions. On the Schwarz alternating method. III: A variant for nonoverlapping subdomains. In T. F. Chan, R. Glowinski, J. Périaux, and O. Widlund, editors, *Third International Symposium on Domain Decomposition Methods for Partial Differential Equations*, pages 202–223. SIAM, Philadelphia, 1990. (Cited on page 267.)

[2077] R. J. Lipton, D. J. Rose, and R. E. Tarjan. Generalized nested dissection. *SIAM J. Numer. Anal.*, 16(2):346–358, 1979. (Cited on page 90.)

[2078] R. J. Lipton and R. E. Tarjan. A separator theorem for planar graphs. *SIAM J. Appl. Math.*, 36(2):177–189, 1979. (Cited on page 90.)

[2079] J. W. H. Liu. A compact row storage scheme for Cholesky factors using elimination trees. *ACM Trans. Math. Soft. (TOMS)*, 12(2):127–148, 1986. (Cited on page 87.)

[2080] J. W. H. Liu. A partial pivoting strategy for sparse symmetric matrix decomposition. *ACM Trans. Math. Soft. (TOMS)*, 13(2):173–182, 1987. (Cited on page 78.)

[2081] J. W. H. Liu. The role of elimination trees in sparse factorization. *SIAM J. Matrix Anal. Appl.*, 11(1):134–172, 1990. (Cited on page 87.)

[2082] J. W. H. Liu. A generalized envelope method for sparse factorization by rows. *ACM Trans. Math. Soft. (TOMS)*, 17(1):112–129, 1991. (Cited on page 88.)

[2083] J. W. H. Liu. The multifrontal method for sparse matrix solution: Theory and practice. *SIAM Rev.*, 34(1):82–109, 1992. (Cited on page 91.)

[2084] J. W. H. Liu, E. G. Ng, and B. W. Peyton. On finding supernodes for sparse matrix computations. *SIAM J. Matrix Anal. Appl.*, 14(1):242–252, 1993. (Cited on page 91.)

[2085] J. W. H. Liu and A. H. Sherman. Comparative analysis of the Cuthill-McKee and the reverse Cuthill-McKee ordering algorithms for sparse matrices. *SIAM J. Numer. Anal.*, 13(2):198–213, 1976. (Cited on page 88.)

[2086] R. K. Livesley. The analysis of large structural systems. *Comput. J.*, 3(1):34–39, 1960. (Cited on pages 85 and 208.)

[2087] A. Loewy. Über einen fundamentalsatz für matrizen oder lineare homogene differential-syssteme. Vol. 5, Sitzungsberichte der Heidelberger Akademie der Wissenschaften, Mathematisch-naturwissenschaftliche Klasse : Abt. A, 1918. (Cited on page 290.)

[2088] S. Loisel and P. Maxwell. Path-following method to determine the field of values of a matrix with high accuracy. *SIAM J. Matrix Anal. Appl.*, 39(4):1726–1749, 2018. (Cited on page 287.)

[2089] A. T. Lonseth. Systems of linear equations with coefficients subject to error. *Ann. Math. Stat.*, 13(3):332–337, 1942. (Cited on pages 19 and 75.)

[2090] A. T. Lonseth. On relative errors in systems of linear equations. *Ann. Math. Stat.*, 15(3):323–325, 1944. (Cited on pages 19 and 75.)

[2091] A. T. Lonseth. The propagation of error in linear problems. *Trans. Amer. Math. Soc.*, 62(2):193–212, 1947. (Cited on pages 15, 19, and 75.)

[2092] G. Loria. Eugenio Beltrami e le sue opere matematiche. *Bibliotheca Mathematica*, 3:392–440, 1901. (Cited on page 417.)

[2093] M. Lotkin. Characteristic values of arbitrary matrices. *Quart. Appl. Math.*, 14(3):267–275, 1956. (Cited on page 308.)

[2094] A. N. Lowan and J. Laderman. Table of Fourier coefficients. *J. Math. Phys.*, 22(3):136–147, 1943. (Cited on page 74.)

[2095] H. Ltaief, D. Sukkari, A. Esposito, Y. Nakatsukasa, and D. Keyes. Massively parallel polar decomposition on distributed-memory systems. *ACM Trans. Parallel Comput. (TOPC)*, 6(1):1–15, 2019. (Cited on page 139.)

[2096] G. Ludyk. *Quantum Mechanics in Matrix Form.* Springer Nature, Cham, Switzerland, 2018. (Cited on page 26.)

[2097] D. G. Luenberger. Hyperbolic pairs in the method of conjugate gradients. *SIAM J. Appl. Math.*, 17(6):1263–1267, 1969. (Cited on page 217.)

[2098] D. G. Luenberger. The conjugate residual method for constrained minimization problems. *SIAM J. Numer. Anal.*, 7(3):390–398, 1970. (Cited on page 217.)

[2099] F. T. Luk. Computing the singular-value decomposition on the ILLIAC IV. *ACM Trans. Math. Soft. (TOMS)*, 6(4):524–539, 1980. (Cited on page 138.)

[2100] F. T. Luk and S. Qiao. Computing the CS-decomposition on systolic arrays. *SIAM J. Sci. Statist. Comput.*, 7(4):1121–1125, 1986. (Cited on page 145.)

[2101] N. N. Luzin. On academician A.N. Krylov's method of generating the secular equation. *Izvestiya Rossiiskoi Akademii Nauk. Seriya Matematicheskaya*, 7:903–958, 1931. (Cited on page 291.)

[2102] N. N. Luzin. On some properties on the inversing multiplier in the process of M. A. Krylov. *Izvestiya Rossiiskoi Akademii Nauk. Seriya Matematicheskaya*, 5:595–638, 1932. (Cited on page 291.)

[2103] A.M. Lyapunov. *On the stability of ellipsoidal forms of equilibrium of a rotating liquid.* PhD thesis, St Petersburg University, 1884. in Collection of Works (Izd. Akad. Nauk SSSR, Moscow, 1959), Vol. 3, pp. 5-113. (Cited on pages 394 and 398.)

[2104] M. S. Lynn. On the equivalence of SOR, SSOR and USSOR as applied to σ_1-ordered systems of linear equations. *Comput. J.*, 7:72–75, 1964. (Cited on page 178.)

[2105] E.-C. Ma. A finite series solution of the matrix equation $AX - XB = C$. *SIAM J. Appl. Math.*, 14(3):490–495, 1966. (Cited on page 395.)

[2106] C. C. MacDuffee. *The Theory of Matrices.* Springer, Berlin, Heidelberg, 1933. (Cited on pages 12, 13, 68, 136, 149, 290, 388, and 394.)

[2107] C. C. MacDuffee. *Vectors and Matrices.* The Mathematical Association of America, 1943. (Cited on pages 9 and 15.)

[2108] A. MacFarlane. *Lectures on Ten British Mathematicians.* Chapman and Hall, 1916. (Cited on pages 569 and 571.)

[2109] T. Mach, M. S. Pranić, and R. Vandebril. Computing approximate extended Krylov subspaces without explicit inversion. *Electron. Trans. Numer. Anal.*, 40:414–435, 2013. (Cited on page 390.)

[2110] T. Mach, M. S. Pranić, and R. Vandebril. Computing approximate (block) rational Krylov subspaces without explicit inversion with extensions to symmetric matrices. *Electron. Trans. Numer. Anal.*, 43:100–124, 2014. (Cited on page 390.)

[2111] C. Maclaurin. *A Treatise of Algebra.* A. Millar and J. Nourse, London, 1748. (Cited on pages 53 and 104.)

[2112] P. A. MacMahon. James Joseph Sylvester. *Nature*, 55(1430):492–494, 1897. (Cited on pages 568 and 571.)

[2113] N. Madden. John Todd and the development of modern numerical analysis. *Irish Math. Soc. Bulletin*, 69:11–23, 2012. (Cited on page 581.)

[2114] L. O. Mafteiu-Scai. The bandwidths of a matrix. A survey of algorithms. *Annals of West University of Timisoara, Mathematics and Computer Science*, 52(2):183–223, 2014. (Cited on page 88.)

[2115] G .A. Maggi. *Eugenio Beltrami.* Annuario della R. Università di Pisa, 1900-1901. (Cited on page 417.)

[2116] F. Magoulès and C. Venet. Asynchronous iterative sub-structuring methods. *Math. Comput. Simul.*, 145:34–49, 2018. (Cited on page 197.)

[2117] M. W. Mahoney. Lecture notes on randomized linear algebra. ArXiv Preprint ArXiv:1608.04481, 2016. (Cited on page 140.)

[2118] J. F. Maitre. *Sur certaines normes et fonctionnelles dans les espaces de matrices et d'opérateurs.* PhD thesis, Institut National Polytechnique de Grenoble, Université Joseph Fourier, Grenoble, France, 1974. (Cited on page 16.)

[2119] J.-F. Maitre and F. Musy. Multigrid methods: Convergence theory in a variational framework. *SIAM J. Numer. Anal.*, 21(4):657–671, 1984. (Cited on page 270.)

[2120] H. Majima. Seki Takakazu, his life and bibliography. In E. Knobloch, H. Komatsu, and D. Liu, editors, *Seki, Founder of Modern Mathematics in Japan*, pages 3–20. Springer, 2013. (Cited on page 99.)

[2121] J. Màlek and Z. Strakoš. *Preconditioning and the Conjugate Gradient Method in the Context of Solving PDEs*. SIAM, Philadelphia, 2015. (Cited on pages 216 and 265.)

[2122] L. Maligranda. Why Hölder's inequality should be called Roger's inequality. *Math. Inequalities Appl.*, 1(1):69–83, 1998. (Cited on page 14.)

[2123] R. R. M. Mallock. An electrical calculating machine. *Proc. R. Soc. Lond.*, 140(841):457–483, 1933. (Cited on page 332.)

[2124] S. Man-Keung. An excursion in ancient Chinese mathematics. *Paleontological Society Papers*, 6:159–166, 2000. (Cited on page 39.)

[2125] J. Mandel. Algebraic study of multigrid methods for symmetric, definite problems. *Appl. Math. Comput.*, 25(1):39–56, 1988. (Cited on page 270.)

[2126] J. Mandel and C. R. Dohrmann. Convergence of a balancing domain decomposition by constraints and energy minimization. *Numer. Linear Algebra Appl.*, 10:639–659, 2003. (Cited on page 268.)

[2127] J. Mandel, S. F. McCormick, and J. Ruge. An algebraic theory for multigrid methods for variational problems. *SIAM J. Numer. Anal.*, 25(1):91–110, 1988. (Cited on page 270.)

[2128] J. Mandel and S.V. Parter. On the multigrid F-cycle. *Appl. Math. Comput.*, 37(1):19–36, 1990. (Cited on page 270.)

[2129] M. Manguoglu, A. H. Sameh, and O. Schenk. PSPIKE: A parallel hybrid sparse linear system solver. In H. Sips, D. Epema, and H.-X. Lin, editors, *Euro-Par 2009 Parallel Processing*, pages 797–808, Berlin, Heidelberg, 2009. Springer. (Cited on page 377.)

[2130] L. Mansfield. On the use of deflation to improve the convergence of conjugate gradient iteration. *Com. Appl. Num. Meth.*, 4(2):151–156, 1988. (Cited on page 217.)

[2131] P. Mansion. Rapport sur ce Mémoire. *Mathesis*, IX:351–359, 1889. (Cited on page 67.)

[2132] T. A. Manteuffel. The Tchebychev iteration for nonsymmetric linear systems. *Numer. Math.*, 28(3):307–327, 1977. (Cited on pages 181, 182, and 262.)

[2133] T. A. Manteuffel. Adaptive procedure for estimating parameters for the nonsymmetric Tchebychev iteration. *Numer. Math.*, 31(2):183–208, 1978. (Cited on pages 181, 182, and 262.)

[2134] T. A. Manteuffel. An incomplete factorization technique for positive definite linear systems. *Math. Comput.*, 34(150):473–497, 1980. (Cited on page 255.)

[2135] G. I. Marchuk and Y. A. Kuznetsov. On optimal iteration processes. *Soviet Math. Dokl.*, 9(4):1041–1045, 1968. (Cited on page 220.)

[2136] G. I. Marchuk and Y. A. Kuznetsov. Méthodes itératives et fonctionnelles quadratiques. In J. L. Lions and G. I. Marchuk, editors, *Sur les Méthodes Numériques en Sciences Physiques et Économiques*, pages 1–117. Dunod, 1974. (Cited on page 220.)

[2137] M. Marcus. An inequality connecting the P-condition number and the determinant. *Numer. Math.*, 4(1):350–353, 1962. (Cited on page 22.)

[2138] J. Mařík and V. Pták. Norms, spectra and combinatorial properties of matrices. *Czechoslovak Math. J.*, 10(2):181–196, 1960. (Cited on page 16.)

[2139] P. Maritz. Sir Thomas Muir, 1844-1934. *Linear Algebra Appl.*, 411:3–67, 2005. (Cited on page 107.)

[2140] J. D. Markel and A. H. Gray. On autocorrelation equations, as applied to speech analysis. *IEEE Trans. Audio Electroacoust.*, AU-21:69–79, 1973. (Cited on page 386.)

[2141] A. A. Markov. Rasprostranenie predel'nyh teorem ischisleniya veroyatnostej na summu velichin svyazannyh v cep'. *Zapiski Akademii Nauk po Fiziko-matematicheskomu otdeleniyu, VIII seriya*, 25(3), 1908. Translated into German, *Ausdehnung der Satze uber die Grenzwerte in der Wahrscheinlichkeitsrechnung auf eine Summe verketteter Grossen*, in: A.A. Markoff (Ed.), Wahrscheinlichkeitsrechnung, B.G. Teuber, Leipzig, (1912), pp. 272-298. Translated into English, *Extension of the limit theorems of probability theory to a sum of variables connected in a chain* in: R.A. Howard (Ed.), Dynamic Probabilities Systems, vol. 1, Wiley, New York, 1971, pp. 552-576. (Cited on page 546.)

[2142] A. A. Markov. *Wahrscheinlichkeitsrechnung*. B.G. Teubner, Leipzig, 1912. (Cited on page 546.)

[2143] H. M. Markowitz. The elimination form of the inverse and its application to linear programming. Technical Report P-680, The Rand Corporation, Santa Monica, California, 1955. (Cited on pages 85 and 92.)

[2144] H. M. Markowitz. The elimination form of the inverse and its application to linear programming. *Management Science*, 3(3):255–269, 1957. (Cited on page 85.)

[2145] O. A. Marques, C. Vömel, J. W. Demmel, and B. N. Parlett. Algorithm 880: A testing infrastructure for symmetric tridiagonal eigensolvers. *ACM Trans. Math. Soft. (TOMS)*, 35(1):1–13, 2008. (Cited on page 373.)

[2146] A. W. Marshall and I. Olkin. Norms and inequalities for condition numbers. *Pacific J. Math.*, 15(1):241–247, 1965. (Cited on page 22.)

[2147] A. W. Marshall and I. Olkin. Norms and inequalities for condition numbers, II. *Linear Algebra Appl.*, 2:167–172, 1969. (Cited on page 23.)

[2148] D. W. Martin and G. J. Tee. Iterative methods for linear equations with symmetric positive definite matrix. *Comput. J.*, 4(3):242–254, 1961. (Cited on page 208.)

[2149] E. Martin. *Die Rechenmaschinen und ihre Entwicklungsgeschichte*. Johannes Meyer, 1925. English translation, *The Calculation Machines, their History and Development*, The MIT Press, 1992. (Cited on page 325.)

[2150] R. S. Martin, G. Peters, and J. H. Wilkinson. Symmetric decomposition of a positive definite matrix. *Numer. Math.*, 7:362–383, 1965. (Cited on page 361.)

[2151] R. S. Martin, G. Peters, and J. H. Wilkinson. Iterative refinement of the solution of a positive definite system of equations. *Numer. Math.*, 8:203–216, 1966. (Cited on page 361.)

[2152] R. S. Martin, G. Peters, and J. H. Wilkinson. The QR algorithm for real Hessenberg matrices. *Numer. Math.*, 14:219–231, 1970. (Cited on pages 313 and 362.)

[2153] R. S. Martin, C. Reinsch, and J. H. Wilkinson. Householder's tridiagonalization of a symmetric matrix. *Numer. Math.*, 11:181–195, 1968. (Cited on page 362.)

[2154] R. S. Martin, C. Reinsch, and J. H. Wilkinson. The QR algorithm for band symmetric matrices. *Numer. Math.*, 16:85–92, 1970. (Cited on page 362.)

[2155] R. S. Martin and J. H. Wilkinson. Symmetric decomposition of positive definite band matrices. *Numer. Math.*, 7:355–361, 1965. (Cited on pages 85 and 361.)

[2156] R. S. Martin and J. H. Wilkinson. Solution of symmetric and unsymmetric band equations and the calculations of eigenvectors of band matrices. *Numer. Math.*, 9:279–301, 1967. (Cited on pages 85 and 362.)

[2157] R. S. Martin and J. H. Wilkinson. The implicit QL algorithm. *Numer. Math.*, 12:377–383, 1968. (Cited on page 362.)

[2158] R. S. Martin and J. H. Wilkinson. The modified LR algorithm for complex Hessenberg matrices. *Numer. Math.*, 12:369–376, 1968. (Cited on page 362.)

[2159] R. S. Martin and J. H. Wilkinson. Reduction of the symmetric eigenproblem $Ax = \lambda Bx$ and related problems to standard form. *Numer. Math.*, 11:99–110, 1968. (Cited on page 362.)

[2160] R. S. Martin and J. H. Wilkinson. Similarity reduction of a general matrix to Hessenberg form. *Numer. Math.*, 12:349–368, 1968. (Cited on page 362.)

[2161] P.-G. Martinsson and J. Tropp. Randomized numerical linear algebra: Foundations & algorithms. arXiv preprint arXiv:2002.01387, 2020. (Cited on page 140.)

[2162] J.-C. Martzloff. *A History of Chinese Mathematics*. Springer, Berlin, 1997. (Cited on page 39.)

[2163] T. Mary. *Block low-rank multifrontal solvers: Complexity, performance, and scalability*. PhD thesis, Université Toulouse 3 Paul Sabatier, France, 2017. (Cited on pages 91 and 378.)

[2164] K. J. Maschhoff and D. C. Sorensen. P ARPACK: An efficient portable large scale eigenvalue package for distributed memory parallel architectures. In *International Workshop on Applied Parallel Computing*, Berlin, Heidelberg, 1996. Springer. (Cited on page 378.)

[2165] J. L. Massey. Shift-register synthesis and BCH decoding. *IEEE Trans. Information Theory*, 15:122–127, 1969. (Cited on page 386.)

[2166] N. Mastronardi and P. Van Dooren. Computing the Jordan structure of an eigenvalue. *SIAM J. Matrix Anal. Appl.*, 38(3):949–966, 2017. (Cited on page 150.)

[2167] A. M. Matsokin and S. V. Nepomnyaschikh. A Schwarz alternating method in a subspace. *Soviet Math. (Iz. VUZ)*, 29:78–84, 1985. (Cited on page 267.)

[2168] R. M. M. Mattheij. Stability of block *LU*-decompositions of matrices arising from BVP. *SIAM J. Alg. Disc. Meth.*, 5(3):314–331, 1984. (Cited on page 94.)

[2169] R. M. M. Mattheij. The stability of LU-decompositions of block tridiagonal matrices. *Bull. Austral. Math. Soc.*, 29:177–205, 1984. (Cited on page 94.)

[2170] L. Maurer. Zur theorie der linearen substitutionen. Inauguraldissertation, Kaiser-Wilhelms-Universität Strassburg, 1887. (Cited on page 396.)

[2171] J. E. Maxfield and H. Minc. On the matrix equation $X'X = A$. *Proc. Edinburgh Math. Soc.*, 13(2):125–129, 1962. (Cited on page 394.)

[2172] K. O. May. Historical vices, I. Logical attribution. *Historia Math.*, 2:185–187, 1975. (Cited on page xvi.)

[2173] K. O. May. Historical vices, II. Priority chasing. *Historia Math.*, 2:315–317, 1975. (Cited on page xvi.)

[2174] K. O. May. What is good history and who should do it? *Historia Math.*, 2:449–455, 1975. (Cited on page xvi.)

[2175] C. McCarthy and G. Strang. Optimal conditioning of matrices. *SIAM J. Numer. Anal.*, 10(2):370–388, 1973. (Cited on pages 80 and 252.)

[2176] P. McCorduck. An interview with Albert Bowker. OH 6, Charles Babbage Institute, University of Minnesota, 1979. (Cited on page 458.)

[2177] P. McCorduck. An interview with Alexandra Forsythe. OH 17, Charles Babbage Institute, University of Minnesota, 1979. (Cited on page 458.)

[2178] P. McCorduck. An interview with Gene Golub. OH 20, Charles Babbage Institute, University of Minnesota, 1979. (Cited on page 459.)

[2179] P. McCorduck. An interview with John Herriot. OH 21, Charles Babbage Institute, University of Minnesota, 1979. (Cited on page 459.)

[2180] S. F. McCormick. The methods of Kaczmarz and row orthogonalization for solving linear equations and least squares problems in Hilbert space. *Indiana Univ. Math. J.*, 26:1137–1150, 1977. (Cited on page 190.)

[2181] S. F. McCormick. Multigrid methods for variational problems: General theory for the *V*-cycle. *SIAM J. Numer. Anal.*, 22(4):634–643, 1985. (Cited on page 270.)

[2182] S. F. McCormick. *Multigrid Methods*. SIAM, Philadelphia, 1987. (Cited on page 270.)

[2183] K. McManus, S. Johnson, and M. Cross. Communication latency hiding in a parallel conjugate gradient method. In *Eleventh International Conference on Domain Decomposition Methods (Greenwich, England, 1998)*, pages 306–313. DDM.org, 1999. (electronic). (Cited on page 215.)

[2184] J. McNamee. Algorithm 408: A sparse matrix package (part I). *Comm. ACM*, 14(4):265–273, 1971. (Cited on page 86.)

[2185] J. McNamee. Algorithm 408: A sparse matrix package (part II): Special cases. *ACM Trans. Math. Soft. (TOMS)*, 9(3):340–345, 1983. (Cited on page 86.)

[2186] K. Meerbergen, R. B. Morgan, and A. V. Knyazev. Preconditioning techniques. In Z. Bai, J. Demmel, J. Dongarra, A. Ruhe, and H. van der Vorst, editors, *Templates for the Solution of Algebraic Eigenvalue Problems: A Practical Guide*, pages 337–368. SIAM, Philadelphia, 2000. (Cited on page 319.)

[2187] R. Mehmke. On Seidel's method to solve linear equations with a very large number of unknowns by successive approximations. *Math. Sb*, 16(2):342–345, 1892. (Cited on pages 160 and 161.)

[2188] R. Mehmke. Über die zweckmäßigste art, lineare gleichungen durch elimination aufzulösen. *Zamm.-Z Angew. Math. Me.*, 10(5):508–514, 1930. (Cited on page 67.)

[2189] R. Mehmke and P. A. Nekrasov. Solutions of linear systems of equations by means of successive approximations. *Mat. Sb.*, 16(3):437–459, 1892. (Cited on page 160.)

[2190] V. Mehrmann. Ralph Byers 1955-2007. *Linear Algebra Appl.*, 428:2410–2414, 2008. (Cited on page 143.)

[2191] J. A. Meijerink and H. A. van der Vorst. An iterative solution method for linear systems of which the coefficient matrix is a symmetric M-matrix. *Math. Comput.*, 31(137):148–162, 1977. (Cited on page 254.)

[2192] J. A. Meijerink and H. A. van der Vorst. Guidelines for the usage of incomplete decompositions in solving sets of linear equations as they occur in practical problems. *J. Comp. Phys.*, 44:134–155, 1981. (Cited on page 254.)

[2193] G. Meinardus. Über eine verallgemeinerung einer ungleichung von LV Kantorowitsch. *Numer. Math.*, 5(1):14–23, 1963. (Cited on page 208.)

[2194] J. Meinguet. Refined error analyses of Cholesky factorization. *SIAM J. Numer. Anal.*, 20(6):1243–1250, 1983. (Cited on page 80.)

[2195] B. Meini. The matrix square root from a new functional perspective: Theoretical results and computational issues. *SIAM J. Matrix Anal. Appl.*, 26(2):362–376, 2004. (Cited on page 389.)

[2196] A. Melman. A numerical comparison of methods for solving secular equations. *J. Comput. Appl. Math.*, 86:237–249, 1997. (Cited on page 315.)

[2197] A. Melman. A unifying convergence analysis of second-order methods for secular equation. *Math. Comput.*, 66(217):333–344, 1997. (Cited on page 315.)

[2198] A. Melman. Generalizations of Gershgorin disks and polynomial zeros. *Proc. Am. Math. Soc.*, 138(7):2349–2364, 2010. (Cited on pages 285 and 474.)

[2199] A. Melman. Modified Gershgorin disks for companion matrices. *SIAM Rev.*, 54(2):355–373, 2012. (Cited on pages 285 and 474.)

[2200] A. Melman. Generalization and variations of Pellet's theorem for matrix polynomials. *Linear Algebra Appl.*, 439:1550–1567, 2013. (Cited on page 285.)

[2201] A. Melman. Cauchy-type inclusion and exclusion regions for polynomial zeros. *The Teaching of Mathematics*, XVII(1):39–50, 2014. (Cited on pages 285 and 474.)

[2202] X. Meng, M. A. Saunders, and M. W. Mahoney. LSRN: A parallel iterative solver for strongly over- or underdetermined systems. *SIAM J. Sci. Comput.*, 36(2):C95–C118, 2014. (Cited on page 373.)

[2203] G. Merchant and T. Parks. Efficient solution of a Toeplitz-plus-Hankel coefficient matrix system of equations. *IEEE Trans. Acoust. Speech Signal Process.*, 30(1):40–44, 1982. (Cited on page 387.)

[2204] M. Merriman. *Elements of the Method of Least Squares*. MacMillan and Co., London, 1877. (Cited on page 62.)

[2205] M. Merriman. A list of writings relating to the method of least squares with historical and critical notes. *Trans. Connecticut Academy of Arts and Sciences*, 4:151–232, 1877. Reprinted in [2900, Vol. 1]. (Cited on page 449.)

[2206] R. R. Mertz. Interview with Alston S. Householder. Smithsonian National Museum of American History, Computer Oral History Collection, 1970. (Cited on page 488.)

[2207] R. R. Mertz. Interview with John Todd. Smithsonian National Museum of American History, Computer Oral History Collection, 1971. (Cited on page 581.)

[2208] W. H. Metzler. On the roots of matrices. *Amer. J. Math.*, 14(4):326–377, 1892. (Cited on pages 148 and 387.)

[2209] G. Meurant. The block preconditioned conjugate gradient method on vector computers. *BIT*, 24:623–633, 1984. (Cited on page 258.)

[2210] G. Meurant. Numerical experiments for the preconditioned conjugate gradient method on the CRAY X-MP/2. Technical Report LBL-18023, Lawrence Berkeley Laboratory, 1984. (Cited on pages 214, 215, and 216.)

[2211] G. Meurant. Multitasking the conjugate gradient on the CRAY X-MP/48. Technical Report NA-85-33, Computer Science Department, Stanford University, 1985. (Cited on page 214.)

[2212] G. Meurant. Multitasking the conjugate gradient method on the CRAY X-MP/48. *Parallel Comput.*, 5:267–280, 1987. (Cited on page 214.)

[2213] G. Meurant. Practical use of the conjugate gradient method on parallel supercomputers. *Comput. Phys. Comm.*, 53:467–477, 1989. (Cited on page 261.)

[2214] G. Meurant. The conjugate gradient method on vector and parallel supercomputers. In W. L. Hogarth and B. J. Noye, editors, *Computational Techniques and Applications*. Hemisphere Publishing Corp., 1990. (Cited on pages 214 and 261.)

[2215] G. Meurant. A review on the inverse of symmetric tridiagonal and block tridiagonal matrices. *SIAM J. Matrix Anal. Appl.*, 13(3):707–728, 1992. (Cited on page 383.)

[2216] G. Meurant. The computation of bounds for the norm of the error in the conjugate gradient algorithm. *Numer. Algorithms*, 16:77–87, 1997. (Cited on pages 212 and 244.)

[2217] G. Meurant. *Computer Solution of Large Linear Systems*. North-Holland, Amsterdam, 1999. (Cited on pages 94, 186, 214, 237, and 260.)

[2218] G. Meurant. Numerical experiments in computing bounds for the norm of the error in the preconditioned conjugate gradient algorithm. *Numer. Algorithms*, 22:353–365, 1999. (Cited on pages 212 and 244.)

[2219] G. Meurant. Gaussian elimination for the solution of linear systems of equations. In P. G. Ciarlet and J. L. Lions, editors, *Handbook of Numerical Analysis, vol. 7*, pages 3–170. North-Holland, 2000. (Cited on page 94.)

[2220] G. Meurant. A multilevel AINV preconditioner. *Numer. Algorithms*, 29:107–129, 2002. (Cited on page 271.)

[2221] G. Meurant. Estimates of the ℓ_2 norm of the error in the conjugate gradient algorithm. *Numer. Algorithms*, 40(2):157–169, 2005. (Cited on pages 213 and 244.)

[2222] G. Meurant. *The Lanczos and Conjugate Gradient Algorithms, from Theory to Finite Precision Computations*. SIAM, Philadelphia, 2006. (Cited on pages 211 and 301.)

[2223] G. Meurant. Estimates of the norm of the error in solving linear systems with FOM and GMRES. *SIAM J. Sci. Comput.*, 33(5):2686–2705, 2011. (Cited on page 243.)

[2224] G. Meurant. On the residual norm in FOM and GMRES. *SIAM J. Matrix Anal. Appl.*, 32(2):394–411, 2011. (Cited on page 243.)

[2225] G. Meurant. The computation of isotropic vectors. *Numer. Algorithms*, 60(2):193–204, 2012. (Cited on page 288.)

[2226] G. Meurant. GMRES and the Arioli, Pták and Strakoš parametrization. *BIT Numerical Mathematics*, 52(3):687–702, 2012. (Cited on page 231.)

[2227] G. Meurant. Necessary and sufficient conditions for GMRES complete and partial stagnation. *Appl. Numer. Math.*, 75:100–107, 2014. (Cited on page 240.)

[2228] G. Meurant. On the location of the Ritz values in the Arnoldi process. *Electron. Trans. Numer. Anal.*, 43:188–212, 2015. (Cited on page 307.)

[2229] G. Meurant. The coefficients of the FOM and GMRES residual polynomials. *SIAM J. Matrix Anal. Appl.*, 38(1):96–117, 2017. (Cited on page 245.)

[2230] G. Meurant. An optimal Q-OR Krylov subspace method for solving linear systems. *Electron. Trans. Numer. Anal.*, 47:127–152, 2017. (Cited on page 245.)

[2231] G. Meurant. On prescribing the convergence behavior of the conjugate gradient algorithm. *Numer. Algorithms*, 84(4):1353–1380, 2020. (Cited on page 211.)

[2232] G. Meurant and J. Duintjer Tebbens. The role eigenvalues play in forming GMRES residual norms with non-normal matrices. *Numer. Algorithms*, 68(1):143–165, 2015. (Cited on pages 244 and 245.)

[2233] G. Meurant and J. Duintjer Tebbens. *Krylov Methods for Nonsymmetric Linear Systems*. Springer International Publishing, 2020. Springer Series in Computational Mathematics, Vol. 57. (Cited on pages 186 and 246.)

[2234] G. Meurant, J. Papež, and P. Tichý. Accurate error estimation in CG. *Numer. Algorithms*, 88(3):1337–1359, 2021. (Cited on page 213.)

[2235] G. Meurant and Z. Strakoš. The Lanczos and conjugate gradient algorithms in finite precision arithmetic. *Acta Numerica*, 15:471–542, 2006. (Cited on pages 211 and 301.)

[2236] G. Meurant and P. Tichý. On computing quadrature-based bounds for the A-norm of the error in conjugate gradients. *Numer. Algorithms*, 62(2):163–191, 2012. (Cited on page 213.)

[2237] G. Meurant and P. Tichý. Erratum to: On computing quadrature-based bounds for the A-norm of the error in conjugate gradients. *Numer. Algorithms*, 66(3):679–680, 2014. (Cited on page 213.)

[2238] G. Meurant and P. Tichý. Approximating the extreme Ritz values and upper bounds for the A-norm of the error in CG. *Numer. Algorithms*, 82(3):937–968, 2019. (Cited on page 213.)

[2239] C. D. Meyer. *Matrix Analysis and Applied Linear Algebra*. SIAM, Philadelphia, 2000. (Cited on page 94.)

[2240] F. Meyer. *Elemente der Arithmetik und Algebra*. H. W. Schmidt, 1885. (Cited on page 56.)

[2241] R. Meyer-Spasche. A constructive method of solving the Liapounov equation for complex matrices. *Numer. Math.*, 19(5):433–438, 1972. (Cited on page 398.)

[2242] R. Meyer-Spasche. Round-trip of an algorithm. In *Österreichisches Symposium zur Geschichte der Mathematik*, pages 195–204. Österreichische Gesellschaft für Wissenschaftsgeschichte, 2017. (Cited on page 295.)

[2243] J.-C. Miellou. Algorithmes de relaxation chaotique à retards. *RAIRO Analyse Numérique*, 9(R1):55–82, 1975. (Cited on page 196.)

[2244] J.-C. Miellou, P. Spiteri, and D. El Baz. A new stopping criterion for linear perturbed asynchronous iterations. *J. Comput. Appl. Math.*, 219(2):471–483, 2008. (Cited on page 197.)

[2245] J.-C. Miellou, P. Spiteri, and G.R. Perrin. An inexpensive method of performance evaluation for subdomain decomposition parallel algorithms of three dimensional elliptic problems. *J. Syst. Softw.*, 6(1-2):169–173, 1986. (Cited on page 196.)

[2246] Y. Mikami. On the Japanese theory of determinants. *Isis*, 2(1):9–36, 1914. (Cited on pages 99 and 101.)

[2247] S. G. Mikhlin. On the Schwarz algorithm. *Doklady Akademii Nauk SSSR*, 77:569–571, 1951. (Cited on page 265.)

[2248] D. F. Miller. The iterative solution of the matrix equation $XA + BX + C = 0$. *Linear Algebra Appl.*, 105:131–137, 1988. (Cited on page 398.)

[2249] G. A. Miller. On the history of determinants. *Am. Math. Mon.*, 37(5):216–219, 1930. (Cited on page 12.)

[2250] K. Miller. Numerical analogs to the Schwarz alternating procedure. *Numer. Math.*, 7(2):91–103, 1965. (Cited on page 265.)

[2251] H. Minkowski. Zur theorie der einheiten in den algebraischen zahlkörpern. Nachrichten von der Gesellschaft der Wissenschaften zu Göttingen, Mathematisch-Physikalische Klasse, 1900, 1900. (Cited on page 286.)

[2252] W. L. Miranker. A survey of parallelism in numerical analysis. *SIAM Rev.*, 13(4):524–547, 1971. (Cited on page 196.)

[2253] L. Mirsky. *An Introduction to Linear Algebra.* Clarendon Press, Oxford, 1955. (Cited on page 117.)

[2254] L. Mirsky. Symmetric gauge functions and unitarily invariant norms. *Quart. Appl. Math.*, 11(1):50–59, 1960. (Cited on pages 22 and 136.)

[2255] P. Misra, E. S. Quintana, and P. M. Van Dooren. Numerically reliable computation of characteristic polynomials. In *Proceedings of 1995 American Control Conference - ACC'95. Vol. 6*, pages 4025–4029. IEEE, 1995. (Cited on page 292.)

[2256] H. F. Mitchell. Inversion of a matrix of order 38. *M.T.A.C.*, 3(23):161–166, 1948. (Cited on page 76.)

[2257] D. Mitra. Asynchronous relaxations for the numerical solution of differential equations by parallel processors. *SIAM J. Sci. Statist. Comput.*, 8(1):s43–s58, 1987. (Cited on page 196.)

[2258] D. Mitrović. Étude théorique d'un principe nouveau de construction des machines électriques servant à résoudre les systèmes d'équations algébriques linéaires. *Annales de la faculté des Sciences de Toulouse*, 4ème série tome 16:169–203, 1952. (Cited on page 332.)

[2259] C. B. Moler. Iterative refinement in floating point. *J. ACM*, 14(2):316–321, 1967. (Cited on page 77.)

[2260] C. B. Moler. *Numerical Computing with MATLAB: Revised Reprint.* SIAM, Philadelphia, 2004. (Cited on page 380.)

[2261] C. B. Moler. EISPACK, matrix eigensystem routines. Cleve's corner, The MathWorks, 2018. (Cited on page 367.)

[2262] C. B. Moler. The historic MATLAB user's guide. Cleve's corner, The MathWorks, 2018. (Cited on page 380.)

[2263] C. B. Moler. LINPACK, linear equation package. Cleve's corner, The MathWorks, 2018. (Cited on page 368.)

[2264] C. B. Moler. MATLAB history, modern MATLAB Part 1 and Part 2. Cleve's corner, The MathWorks, 2018. (Cited on page 380.)

[2265] C. B. Moler. The MATLAB technical computing environment. Cleve's corner, The MathWorks, 2018. (Cited on page 380.)

[2266] C. B. Moler and G. W. Stewart. An algorithm for generalized matrix eigenvalue problems. *SIAM J. Numer. Anal.*, 10(2):241–256, 1973. (Cited on pages 314 and 367.)

[2267] C. B. Moler and C. Van Loan. Nineteen dubious ways to compute the exponential of a matrix. *SIAM Rev.*, 20(4):801–836, 1978. (Cited on page 389.)

[2268] C. B. Moler and C. Van Loan. Nineteen dubious ways to compute the exponential of a matrix, twenty-five years later. *SIAM Rev.*, 45(1):3–49, 2003. (Cited on page 389.)

[2269] M. Montagnac and J. M. Chesneaux. Dynamic control of a bicgstab algorithm. *Appl. Numer. Math.*, 32(1):103–117, 2000. (Cited on page 238.)

[2270] G. H. Moore. The axiomatization of linear algebra: 1875-1940. *Historia Math.*, 22:262–303, 1995. (Cited on page 3.)

[2271] I. Moret and P. Novati. An interpolatory approximation of the matrix exponential based on Faber polynomials. *J. Comput. Appl. Math.*, 131(1-2):361–380, 2001. (Cited on page 389.)

[2272] R. B. Morgan. Computing interior eigenvalues of large matrices. *Linear Algebra Appl.*, 154-156:289–309, 1991. (Cited on page 307.)

[2273] R. B. Morgan. Generalizations of Davidson's method for computing eigenvalues of large nonsymmetric matrices. *J. Comp. Phys.*, 101:287–291, 1992. (Cited on page 317.)

[2274] R. B. Morgan. A restarted GMRES method augmented with eigenvectors. *SIAM J. Matrix Anal. Appl.*, 16(4):1154–1171, 1995. (Cited on page 233.)

[2275] R. B. Morgan. On restarting the Arnoldi method for large nonsymmetric eigenvalue problems,. *Math. Comput.*, 65:1213–1230, 1996. (Cited on page 306.)

[2276] R. B. Morgan. Implicitly restarted GMRES and Arnoldi methods for nonsymmetric systems of equations. *SIAM J. Matrix Anal. Appl.*, 21(4):1112–1135, 2000. (Cited on page 233.)

[2277] R. B. Morgan. GMRES with deflated restarting. *SIAM J. Sci. Comput.*, 24(1):20–37, 2002. (Cited on pages 233 and 306.)

[2278] R. B. Morgan and D. S. Scott. Generalizations of Davidson's method for computing eigenvalues of sparse symmetric matrices. *SIAM J. Sci. Statist. Comput.*, 7(3):817–825, 1986. (Cited on page 317.)

[2279] R. B. Morgan and M. Zeng. A harmonic restarted Arnoldi algorithm for calculating eigenvalues and determining multiplicity. *Linear Algebra Appl.*, 415(1):96–113, 2006. (Cited on page 306.)

[2280] K. Mori. Georg von Charasoff and anticipation of von Mises iteration in economic analysis. *Cahiers d'Économie Politique/Papers in Political Economy*, 2:65–89, 2016. (Cited on page 294.)

[2281] S. Morigi, L. Reichel, F. Sgallari, and F. Zama. Iterative methods for ill-posed problems and semiconvergent sequences. *J. Comput. Appl. Math.*, 193(1):157–167, 2006. (Cited on page 393.)

[2282] K. Morikuni and K. Hayami. Inner-iteration Krylov subspace methods for least squares problems. *SIAM J. Matrix Anal. Appl.*, 34(1):1–22, 2013. (Cited on page 245.)

[2283] K. Morikuni and K. Hayami. Convergence of inner-iteration GMRES methods for rank-deficient least squares problems. *SIAM J. Matrix Anal. Appl.*, 36(1):225–250, 2015. (Cited on page 245.)

[2284] K. Morikuni, L. Reichel, and K. Hayami. FGMRES for linear discrete ill-posed problems. *Appl. Numer. Math.*, 75:175–187, 2014. (Cited on page 393.)

[2285] K. Moriya and T. Nodera. Usage of the convergence test of the residual norm in the Tsuno-Nodera version of the GMRES algorithm. *ANZIAM J.*, 49:293–308, 2007. (Cited on page 232.)

[2286] V. A. Morozov. On the solution of functional equations by the method of regularization. *Doklady Akademii Nauk SSSR*, 167(3):510–512, 1966. (Cited on page 392.)

[2287] V. A. Morozov. *Regularization Methods for Ill-posed Problems*. CRC, 1993. (Cited on page 392.)

[2288] J. Morris. On a simple method for solving simultaneous linear equations by a successive approximation process. *J. R. Aeronaut. Soc.*, 39(292):349–351, 1935. (Cited on page 296.)

[2289] J. Morris. A successive approximation process for solving simultaneous linear equations. Technical Report A.R.C.R. & M., 1711, Aeronautical Research Council, 1936. (Cited on page 296.)

[2290] J. Morris. Relaxation and iteration. *J. R. Aeronaut. Soc.*, 42(331):613–616, 1938. (Cited on page 296.)

[2291] J. Morris. Simultaneous equations. an approximate and successive approximation method based on iteration. *Aircr. Eng. Aerosp. Technol.*, 11(5):199–200, 1939. (Cited on page 296.)

[2292] J. Morris. XVI. An escalator process for the solution of linear simultaneous equations. *Philos. Mag.*, 37(265):106–120, 1946. (Cited on page 296.)

[2293] J. Morris. *The Escalator Method in Engineering Vibration Problems*. Chapman and Hall, London, 1947. (Cited on pages 296 and 297.)

[2294] J. Morris and J. W. Head. Lagrangian frequency equation, an "escalator" method for numerical solution. *14*, pages 312–316, 1942. (Cited on page 296.)

[2295] J. Morris and J. W. Head. LXXXVI. the "Escalator" process for the solution of Lagrangian frequency equations. *Philos. Mag.*, 35(250):735–759, 1944. (Cited on page 296.)

[2296] R. Morris. Tapered floating point: A new floating-point representation. *IEEE Transactions on Computers*, 100(12):1578–1579, 1971. (Cited on page 355.)

[2297] D. Moskovitz. The numerical solution of Laplace's and Poisson's equations. *Quart. Appl. Math.*, 2(2):148–163, 1944. (Cited on page 382.)

[2298] F. R. Moulton. On the solutions of linear equations having small determinants. *Am. Math. Mon.*, 20(8):242–249, 1913. (Cited on page 18.)

[2299] T. Muir. A reinvestigation of the problem of the automorphic linear transformation of a bipartite quadric. *Amer. J. Math.*, 20(3):215–228, 1898. (Cited on page 7.)

[2300] T. Muir. *The Theory of Determinants; in the Historical Order of Development*. Macmillan and Co., London, 2nd edition, 1906. First edition 1890, reprinted by Dover in 1960. (Cited on pages 107, 109, 111, 112, and 381.)

[2301] T. Muir. The literature of Cayleyan matrices. *Trans. Royal Society of South Africa*, 18(3):219–227, 1929. (Cited on page 7.)

[2302] T. Muir. *Contributions to the History of Determinants*. Blackie and Son, London and Glasgow, 1930. (Cited on page 107.)

[2303] H. P. Mulholland. On the distribution of a convex even function of several independent rounding-off errors. *Proc. Am. Math. Soc.*, 3(2):310–321, 1952. (Cited on page 20.)

[2304] C. R. Muntz. Solution directe de l'équation séculaire et des problémes analogues transcendents. *C.R. Acad. Sci. Paris, Sér 1*, pages 43–46, 1913. (Cited on pages 126 and 294.)

[2305] C. R. Muntz. Sur la solution des équations séculaires et des équations intégrales. *C.R. Acad. Sci. Paris, Sér 1*, pages 860–862, 1913. (Cited on page 294.)

[2306] F. D. Murnaghan. On the field of values of a square matrix. *Proceedings of the National Academy of Sciences of the United States of America*, 18(3):246–248, 1932. (Cited on page 287.)

[2307] F. D. Murnaghan and A. Wintner. A canonical form for real matrices under orthogonal transformations. *Proceedings of the National Academy of Sciences of the United States of America*, 17(7):417–420, 1931. (Cited on page 151.)

[2308] M. F. Murphy, G. H. Golub, and A. J. Wathen. A note on preconditioning for indefinite linear systems. *SIAM J. Sci. Comput.*, 21(6):1969–1972, 2000. (Cited on page 264.)

[2309] C. J. Murray. *The Supermen, the Story of Seymour Cray and the Technical Wizards behind the Supercomputer*. John Wiley and Sons, 1997. (Cited on page 445.)

[2310] F. J. Murray. Linear transformations between Hilbert spaces and the application of this theory to linear partial differential equations. *Trans. Amer. Math. Soc.*, 37(2):301–338, 1935. (Cited on page 128.)

[2311] F. J. Murray. Bilinear transformations in Hilbert space. *Trans. Amer. Math. Soc.*, 45(3):474–507, 1939. (Cited on page 128.)

[2312] C. Musco and C. Musco. Randomized block Krylov methods for stronger and faster approximate singular value decomposition. In C. Cortes, N. Lawrence, D. Lee, M. Sugiyama, and R. Garnett, editors, *Advances in Neural Information Processing Systems 28 (NIPS 2015)*, pages 14243–14253. NIPS, 2015. (Cited on page 319.)

[2313] F. Musy and M. Charnay. Sur le théorème de Stein-Rosenberg. *RAIRO*, 8(R2):95–107, 1974. (Cited on page 196.)

[2314] R. Nabben. Decay rates of the inverse of nonsymmetric tridiagonal and band matrices. *SIAM J. Matrix Anal. Appl.*, 20(3):820–837, 1999. (Cited on page 383.)

[2315] N. M. Nachtigal. *A look-ahead variant of the Lanczos algorithm and its application to the quasi-minimal residual method for non Hermitian linear systems*. PhD thesis, MIT, 1991. RIACS Tech. Report 91.19, NASA Ames. (Cited on page 228.)

[2316] N. M. Nachtigal, S. C. Reddy, and L. N. Trefethen. How fast are nonsymmetric matrix iterations? *SIAM J. Matrix Anal. Appl.*, 13(3):778–795, 1992. (Cited on page 225.)

[2317] N. M. Nachtigal, L. Reichel, and L. N. Trefethen. A hybrid GMRES algorithm for nonsymmetric linear systems. *SIAM J. Matrix Anal. Appl.*, 13(3):796–825, 1992. (Cited on page 229.)

[2318] A. Nagel. Astronomisch-Geodätische Arbeiten für die Europäische Gradmessung im Königreich Sachsen. II. Abtheilung. Das Trigonometrische Netz I. Ordnung. Berlin, 1890. (Cited on page 61.)

[2319] Y. Nakatsukasa, Z. Bai, and F. Gygi. Optimizing Halley's iteration for computing the matrix polar decomposition. *SIAM J. Matrix Anal. Appl.*, 31(5):2700–2720, 2010. (Cited on pages 138 and 143.)

[2320] Y. Nakatsukasa and R. W. Freund. Computing fundamental matrix decompositions accurately via the matrix sign function in two iterations: The power of Zolotarev's functions. *SIAM Rev.*, 58(3):461–493, 2016. (Cited on pages 139 and 143.)

[2321] Y. Nakatsukasa and N. J. Higham. Backward stability of iterations for computing the polar decomposition. *SIAM J. Matrix Anal. Appl.*, 33(2):460–479, 2012. (Cited on page 138.)

[2322] Y. Nakatsukasa and N. J. Higham. Stable and efficient spectral divide and conquer algorithms for the symmetric eigenvalue decomposition and the SVD. *SIAM J. Sci. Comput.*, 35(3):A1325–A1349, 2013. (Cited on page 138.)

[2323] Y. Nakatsukasa and J. A. Tropp. Fast and accurate randomized algorithms for linear systems and eigenvalue problems. arXiv:2111.00113, 2021. (Cited on page 319.)

[2324] J. C. Nash. An interview with J. H. Wilkinson on 13 July, 1984. IAM, 2004. (Cited on page 596.)

[2325] National Physical Laboratory. *Automatic Digital Computation.* Her Majesty's Stationary Office, London, 1954. Proceedings of a symposium held at the National Physical Laboratory, March 25-28, 1953. (Cited on page 77.)

[2326] NBS. *Further Contributions to the Solution of Systems of Linear Equations and the Determination of Eigenvalues, NBS 49.* National Bureau of Standards, USA, 1956. (Cited on page 200.)

[2327] P. Neff, J. Lankeit, and A. Madeo. On Grioli's minimum property and its relation to Cauchy's polar decomposition. *Int. J. Eng. Sci.*, 80:209–217, 2014. (Cited on page 142.)

[2328] P. Neff, Y. Nakatsukasa, and A. Fischle. A logarithmic minimization property of the unitary polar factor in the spectral and Frobenius norms. *SIAM J. Matrix Anal. Appl.*, 35(3):1132–1154, 2014. (Cited on page 144.)

[2329] P. A. Nekrasov. Determination of the unknowns by the method of least squares with a very large number of unknowns. *Mat. Sb.*, 12(1):189–204, 1885. (in Russian). (Cited on page 159.)

[2330] P. A. Nekrasov. On the problem of solving systems of linear equations with a large number of unknowns by successive approximations. *Apppendix to the Notes of the Imperial Academy of Sciences of St. Petersburg*, LXIX(5):1–18, 1892. (in Russian). (Cited on page 160.)

[2331] E. Netto. *Vorlesungen über Algebra, volume I.* B. G. Teubner, Leipzig, 1896. (Cited on pages 11 and 13.)

[2332] E. Netto. *Elementare Algebra: Akademische Vorlesungen für Studierende der Ersten Semester.* B. G. Teubner, 1904. (Cited on page 11.)

[2333] O. Neugebauer. *The Exact Sciences in Antiquity.* 2nd edition. Brown University Press, 1957. (Cited on page 323.)

[2334] A. Neumaier and R. S. Varga. Exact convergence and divergence domains for the symmetric successive overrelaxation (SSOR) methods applied to H-matrices. *Linear Algebra Appl.*, 58:261–272, 1984. (Cited on page 178.)

[2335] A. Neuman, L. Reichel, and H. Sadok. Implementations of range restricted iterative methods for linear discrete ill-posed problems. *Linear Algebra Appl.*, 436(10):3974–3990, 2012. (Cited on page 393.)

[2336] O. Nevanlinna. *Convergence of Iterations for Linear Equations.* Lectures in Mathematics ETH Zürich. Birkhäuser Verlag, Basel, 1993. (Cited on page 236.)

[2337] M. Newman and J. Todd. The evaluation of matrix inversion programs. *J. Soc. Ind. Appl. Math.*, 6(4):466–476, 1958. (Cited on page 581.)

[2338] M. H. Newman. Alan Mathison Turing. *Biogr. Mem. Fellows Royal Soc.*, 1:253–263, 1955. The Royal Society, London. (Cited on page 586.)

[2339] I. Newton. *Universal Arithmetick: or a Treatise of Arithmetical Composition and Resolution*. J. Senex and al., London, 1720. Translated by the late Mr. Raphson and revised and corrected by Mr. Cunn, 1st edition 1720, 2nd corrected edition 1728, 3rd edition, with notes by T. Wilder, W. Johnston ed., London, 1769. (Cited on pages 51 and 52.)

[2340] I. Newton. *Arithmetica Universalis: Sive de Compositione et Resolutione Arithmetica Liber*. B. and S. Tooke, London, 2nd edition, 1722. First edition 1707. (Cited on pages 51 and 52.)

[2341] E. G. Ng. Sparse matrix methods. In L. Hogben, editor, *Handbook of Linear Algebra, second edition*, Chapter 53, pages 931–951. Chapman and Hall, CRC Press, 2013. (Cited on page 94.)

[2342] E. G. Ng and B. W. Peyton. A supernodal Cholesky factorization algorithm for shared-memory multiprocessors. *SIAM J. Sci. Comput.*, 14(4):761–769, 1993. (Cited on page 91.)

[2343] R. A. Nicolaides. On multigrid convergence in the indefinite case. *Math. Comput.*, 32(144):1082–1086, 1978. (Cited on page 269.)

[2344] R. A. Nicolaides. On some theoretical and practical aspects of multigrid methods. *Math. Comput.*, 33(147):933–952, 1979. (Cited on page 269.)

[2345] R. A. Nicolaides. Deflation of conjugate gradients with applications to boundary value problems. *SIAM J. Numer. Anal.*, 24(2):355–365, 1987. (Cited on pages 209 and 217.)

[2346] W. Niethammer. Iterationsverfahren und allgemeine Euler-Verfahren. *Mathematische Zeitschrift*, 102(4):288–317, 1967. (Cited on page 183.)

[2347] W. Niethammer. On different splittings and the associated iteration methods. *SIAM J. Numer. Anal.*, 16(2):186–200, 1979. (Cited on page 178.)

[2348] W. Niethammer and R. S. Varga. The analysis of k-step iterative methods for linear systems from smmability theory. *Numer. Math.*, 41:177–206, 1983. (Cited on pages 182 and 183.)

[2349] Y. Nievergelt. A tutorial history of least squares with applications to astronomy and geodesy. *J. Comput. Appl. Math.*, 121:37–72, 2000. (Cited on page 61.)

[2350] E. S. Nikolaev and A. A. Samarskii. Choice of the iteration parameters in the Richardson method. *Zh. Vychisl. Mat. Mat. Fiz.*, 12(4):900–973, 1972. English translation: USSR Comput. Math. Math. Phys., 12 (4) (1972), pp. 141-158. (Cited on page 163.)

[2351] N. Nikolski. *Toeplitz Matrices and Operators*. Cambridge University Press, 2020. (Cited on page 385.)

[2352] Q. Niu and L. Z. Lu. Restarted GMRES method augmented with the combination of harmonic Ritz vectors and error approximations. *Int. J. Math. Comput. Phys. Elec. Comput. Engrg.*, 6(8):857–864, 2012. (Cited on page 233.)

[2353] Y. Notay. On the convergence rate of the conjugate gradients in presence of rounding errors. *Numer. Math.*, 65(1):301–317, 1993. (Cited on page 211.)

[2354] Y. Notay. A multilevel block incomplete factorization preconditioning. *Appl. Numer. Math.*, 31(2):209–225, 1999. (Cited on page 270.)

[2355] Y. Notay. Robust parameter-free algebraic multilevel preconditioning. *Numer. Linear Algebra Appl.*, 9(6-7):409–428, 2002. (Cited on page 271.)

[2356] Y. Notay. Is Jacobi-Davidson faster than Davidson? *SIAM J. Matrix Anal. Appl.*, 26(2):522–543, 2005. (Cited on page 317.)

[2357] Y. Notay. Aggregation-based algebraic multilevel preconditioning. *SIAM J. Matrix Anal. Appl.*, 27(4):998–1018, 2006. (Cited on page 271.)

[2358] P. Novati. Some properties of the Arnoldi-based methods for linear ill-posed problems. *SIAM J. Numer. Anal.*, 55(3):1437–1455, 2017. (Cited on page 393.)

[2359] J. A. Oaks. Polynomials and equations in medieval Italian algebra. *Bollettino di Storia delle Scienze Matematiche*, 30:23–60, 2010. (Cited on page 44.)

[2360] J. A. Oaks. François Viète's revolution in algebra. *Archive for History of Exact Sciences*, 72(3):245–302, 2018. (Cited on page 48.)

[2361] W. Oettli and W. Prager. Compatibility of approximate solution of linear equations with given error bounds for coefficients and right-hand sides. *Numer. Math.*, 6(1):405–409, 1964. (Cited on page 79.)

[2362] W. Oettli, W. Prager, and J. H. Wilkinson. Admissible solutions of linear systems with not sharply defined coefficients. *J. Soc. Ind. Appl. Math., Ser. B Numer. Anal.*, 2(2):291–299, 1965. (Cited on page 79.)

[2363] T. Ogawa. A review of the history of Japanese mathematics. *Rev. Hist. Math.*, 7(1):137–155, 2001. (Cited on page 101.)

[2364] E. C. Ogbuobiri, W. F. Tinney, and J. W. Walker. Sparsity-directed decomposition for Gaussian elimination on matrices. *IEEE Trans. Power App. Syst.*, 1:141–150, 1970. (Cited on page 87.)

[2365] R. Oldenburger. Infinite powers of matrices and characteristic roots. *Duke Math. J.*, 6(2):357–361, 1940. (Cited on page 170.)

[2366] D. P. O'Leary. The block conjugate gradient algorithm and related methods. *Linear Algebra Appl.*, 29:293–322, 1980. (Cited on pages 209 and 303.)

[2367] D. P. O'Leary. Estimating matrix condition numbers. *SIAM J. Sci. Statist. Comput.*, 1(2):205–209, 1980. (Cited on pages 23 and 223.)

[2368] D. P. O'Leary. Yet another polynomial precondtioner for the conjugate gradient method. *Linear Algebra Appl.*, 154-156:377–388, 1991. (Cited on page 262.)

[2369] D. P. O'Leary. Conjugate gradients and related KMP algorithms: The beginnings. In L. Adams and J. L. Nazareth, editors, *Linear and Nonlinear Conjugate Gradient-related Methods*, pages 1–8, SIAM, Philadelphia, 1996. (Cited on pages 201, 207, and 208.)

[2370] D. P. O'Leary. Commentary on an iteration method for the solution of the eigenvalue problem of linear differential and integral operators by Cornelius Lanczos. In D. R. Lide, editor, *A Century of Excellence in Measurements, Standards, and Technology - A Chronicle of Selected NBS/NIST Publications 1901-2000*, pages 77–80. NIST Special Publication 958, 2001. (Cited on pages 201 and 517.)

[2371] D. P. O'Leary. Commentary on methods of conjugate gradients for solving linear systems by Magnus R. Hestenes and Eduard Stiefel. In D. R. Lide, editor, *A Century of Excellence in Measurements, Standards, and Technology - A Chronicle of Selected NBS/NIST Publications 1901-2000*, pages 81–85. NIST Special Publication 958, 2001. (Cited on page 201.)

[2372] D. P. O'Leary. Gene Howard Golub, 1932-2007. *Linear Algebra Appl.*, 428:2405–2409, 2008. (Cited on page 478.)

[2373] D. P. O'Leary and J. A. Simmons. A bidiagonalization-regularization procedure for large scale discretizations of ill-posed problems. *SIAM J. Sci. Statist. Comput.*, 2(4):474–489, 1981. (Cited on page 393.)

[2374] D. P. O'Leary and P. Whitman. Parallel QR factorization by Householder and modified Gram-Schmidt algorithms. *Parallel Comput.*, 16(1):99–112, 1990. (Cited on page 133.)

[2375] T. A. Oliphant. An extrapolation procedure for solving linear systems. *Quart. Appl. Math.*, 20(3):257–265, 1962. (Cited on page 253.)

[2376] T. A. Oliphant. A note on extrapolation procedures for solving linear systems. *Quart. Appl. Math.*, 21(1):72–75, 1963. (Cited on page 253.)

[2377] K. C. O'Meara, J. Clark, and C. Vinsonhaler. *Advanced Topics in Linear Algebra: Weaving Matrix Problems through the Weyr Form*. Oxford University Press, 2011. (Cited on page 148.)

[2378] Y. Onoue, S. Fujino, and N. Nakashima. A difference between easy and profound preconditionings of IDR(s) method. *Transactions of JSCES*, Paper 20080023, 2008. (in Japanese). (Cited on page 242.)

[2379] Y. Onoue, S. Fujino, and N. Nakashima. Improved IDR(s) method for gaining very accurate solutions. *World Academy of Science, Engineering and Technology*, 55, 2009. (Cited on page 242.)

[2380] Y. Onoue, S. Fujino, and N. Nakashima. An overview of a family of new iterative methods based on IDR theorem and its estimation. In *Proceedings of the International MultiConference of Engineers and Computer Scientists 2009, Vol II IMECS*, 2009. (Cited on page 242.)

[2381] G. Opfer and G. Schober. Richardson's iteration for nonsymmetric matrices. *Linear Algebra Appl.*, 58:343–361, 1984. (Cited on page 182.)

[2382] T. C. Oppe, W. D. Joubert, and D. R. Kincaid. NSPCG user's guide. Technical Report CNA-216, Center for Numerical Analysis, The University of Texas at Austin, 1988. (Cited on page 372.)

[2383] W. Orchard-Hays. An efficient form of inverse for sparse matrices. In *Proceedings of the 1956 11th ACM National Meeting*, pages 154–157, New York, 1956. Association for Computing Machinery. (Cited on page 85.)

[2384] ORNL. *Sparse Matrix Symposium, Fairfield Glade 1982. Software Catalog*. Oak Ridge National Laboratory, 1982. M. Heath Ed. (Cited on page 370.)

[2385] J. M. Ortega. An error analysis of Householder's method for the symmetric eigenproblem. *Numer. Math.*, 5:211–225, 1963. (Cited on page 308.)

[2386] J. M. Ortega and H. F. Kaiser. The LL^T and QR methods for symmetric tridiagonal matrices. *Comput. J.*, 6(1):99–101, 1963. (Cited on page 312.)

[2387] J. M. Ortega and R. G. Voigt. Solution of partial differential equations on vector and parallel computers. *SIAM Rev.*, 27(2):149–240, 1985. (Cited on page 178.)

[2388] E. L. Ortiz and A. Pinkus. Herman Müntz: a mathematician's odyssey. *Math. Intell.*, 27:22–31, 2005. (Cited on page 294.)

[2389] N. Osada. Determinants by Seki Takakazu from the group-theoretic viewpoint. In *Mathematics of Takebe Katahiro and History of Mathematics in East Asia*, pages 83–99. World Scientific, 2018. Advanced Studies in Pure Mathematics 79. (Cited on page 100.)

[2390] E. E. Osborne. On preconditioning of matrices. *J. ACM*, 7(4):338–345, 1960. (Cited on page 251.)

[2391] O. Østerby and Z. Zlatev. *Direct Methods for Sparse Matrices*. Springer, 1983. (Cited on page 371.)

[2392] A. M. Ostrowski. Über die determinanten mit überwiegender hauptdiagonale. *Commentarii Mathematici Helvetici*, 10(1):69–96, 1937. (Cited on pages 177 and 543.)

[2393] A. M. Ostrowski. Sur les conditions générales pour la régularité des matrices. *Univ. Roma. Ist. Naz. Alta Mat. Rend. Mat. e. Appl.*, 5(10):156–168, 1951. (Cited on page 286.)

[2394] A. M. Ostrowski. On absolute convergence of linear iteration processes. Technical Report 1728, National Bureau of Standards, 1952. (Cited on page 543.)

[2395] A. M. Ostrowski. On Gauss' speeding up device in the theory of single step iteration. Technical Report 1728, National Bureau of Standards, 1952. (Cited on page 543.)

[2396] A. M. Ostrowski. On over and under relaxation in the theory of the cyclic single step iteration. *M.T.A.C.*, 7(43):152–159, 1953. (Cited on page 543.)

[2397] A. M. Ostrowski. On the linear iteration procedures for symmetric matrices. *Rend. Mat. Appl.*, 14:140–163, 1954. (Cited on pages 176 and 543.)

[2398] A. M. Ostrowski. Uber normen von matrizen. *Mathematische Zeitschrift*, 63(1):2–18, 1955. (Cited on pages 16 and 543.)

[2399] A. M. Ostrowski. Determinanten mit überwiegender hauptdiagonale und die absolute konvergenz von linearen iterationsprozessen. *Commentarii Mathematici Helvetici*, 30(1):175–210, 1956. (Cited on pages 195 and 543.)

[2400] A. M. Ostrowski. On Gauss' speeding up device in the theory of single step iteration. *M.T.A.C.*, 12(62):116–132, 1958. (Cited on pages 173 and 543.)

[2401] A. M. Ostrowski. On the convergence of the Rayleigh quotient iteration for the computation of the characteristic roots and vectors. I. *Arch. Ration. Mech. Anal.*, 1(1):233–241, 1958. (Cited on page 295.)

[2402] A. M. Ostrowski. On the convergence of the Rayleigh quotient iteration for the computation of the characteristic roots and vectors. II. *Arch. Ration. Mech. Anal.*, 2(1):423–428, 1958. (Cited on page 295.)

[2403] A. M. Ostrowski. On the convergence of the Rayleigh quotient iteration for the computation of the characteristic roots and vectors. III. *Arch. Ration. Mech. Anal.*, 3(1):325–340, 1959. (Cited on page 295.)

[2404] A. M. Ostrowski. On the convergence of the Rayleigh quotient iteration for the computation of the characteristic roots and vectors. IV. *Arch. Ration. Mech. Anal.*, 3(1):341–347, 1959. (Cited on page 295.)

[2405] A. M. Ostrowski. On the convergence of the Rayleigh quotient iteration for the computation of the characteristic roots and vectors. V. *Arch. Ration. Mech. Anal.*, 3(1):472–481, 1959. (Cited on page 295.)

[2406] A. M. Ostrowski. On the convergence of the Rayleigh quotient iteration for the computation of the characteristic roots and vectors. VI. *Arch. Ration. Mech. Anal.*, 4(1):153–165, 1959. (Cited on page 295.)

[2407] A. M. Ostrowski. A new proof of Haynsworth's quotient formula for Schur complements. *Linear Algebra Appl.*, 4:389–392, 1971. (Cited on pages 25 and 543.)

[2408] D. V. Ouellette. Schur complement and statistics. *Linear Algebra Appl.*, 36:187–295, 1981. (Cited on pages 25 and 117.)

[2409] Oxford University. Leslie Fox, 1918-1993. Booklet edited by the Computing Laboratory, 1993. (Cited on page 461.)

[2410] M. H. C. Paardekooper. An eigenvalue algorithm for skew symmetric matrices. *Numer. Math.*, 17(3):189–202, 1971. (Cited on page 281.)

[2411] C. C. Paige. Practical use of the symmetric Lanczos process with reorthogonalization. *BIT*, 10:183–195, 1970. (Cited on page 301.)

[2412] C. C. Paige. *The computation of eigenvalues and eigenvectors of very large sparse matrices*. PhD thesis, University of London, 1971. (Cited on pages 208, 209, 301, and 303.)

[2413] C. C. Paige. Computational variants of the Lanczos method for the eigenproblem. *J. Inst. Maths Applics*, 10:373–381, 1972. (Cited on pages 209 and 302.)

[2414] C. C. Paige. Error analysis of the Lanczos algorithm for tridiagonalizing a symmetric matrix. *J. Inst. Maths Applics*, 18:341–349, 1976. (Cited on pages 209 and 302.)

[2415] C. C. Paige. Accuracy and effectiveness of the Lanczos algorithm for the symmetric eigenproblem. *Linear Algebra Appl.*, 34:235–258, 1980. (Cited on pages 209 and 302.)

[2416] C. C. Paige. A note on a result of Sun Ji-Guang: Sensitivity of the CS and GSV decompositions. *SIAM J. Numer. Anal.*, 21(1):186–191, 1984. (Cited on page 145.)

[2417] C. C. Paige. Computing the generalized singular value decomposition. *SIAM J. Sci. Statist. Comput.*, 7(4):1126–1146, 1986. (Cited on page 141.)

[2418] C. C. Paige. A useful form of unitary matrix obtained from any sequence of unit 2-norm n-vectors. *SIAM J. Matrix Anal. Appl.*, 31(2):565–583, 2009. (Cited on pages 129 and 302.)

[2419] C. C. Paige. An augmented stability result for the Lanczos Hermitian matrix tridiagonalization process. *SIAM J. Matrix Anal. Appl.*, 31(5):2347–2359, 2010. (Cited on page 302.)

[2420] C. C. Paige. Accuracy of the Lanczos process for the eigenproblem and solution of equations. *SIAM J. Matrix Anal. Appl.*, 40(4):1371–1398, 2019. (Cited on page 302.)

[2421] C. C. Paige and I. Panayotov. Hessenberg matrix properties and Ritz vectors in the finite-precision Lanczos tridiagonalization process. *SIAM J. Matrix Anal. Appl.*, 32(4):1079–1094, 2011. (Cited on page 302.)

[2422] C. C. Paige, I. Panayotov, and J.-P.M. Zemke. An augmented analysis of the perturbed two-sided Lanczos tridiagonalization process. *Linear Algebra Appl.*, 447:119–132, 2014. (Cited on pages 245 and 302.)

[2423] C. C. Paige, B. N. Parlett, and H. A. van der Vorst. Approximate solutions and eigenvalue bounds from Krylov subspaces. *Numer. Linear Algebra Appl.*, 2(2):115–133, 1995. (Cited on page 307.)

[2424] C. C. Paige, M. Rozložník, and Z. Strakoš. Modified Gram-Schmidt (MGS), least squares, and backward stability of MGS-GMRES. *SIAM J. Matrix Anal. Appl.*, 28(1):264–284, 2006. (Cited on pages 130 and 241.)

[2425] C. C. Paige and M. A. Saunders. Solution of sparse indefinite systems of linear equations. *SIAM J. Numer. Anal.*, 12(4):617–629, 1975. (Cited on pages 208, 217, 221, and 224.)

[2426] C. C. Paige and M. A. Saunders. Towards a generalized singular value decomposition. *SIAM J. Numer. Anal.*, 18(3):398–405, 1981. (Cited on pages 141 and 145.)

[2427] C. C. Paige and M. A. Saunders. Algorithm 583; LSQR: Sparse linear equations and least-squares problems. *ACM Trans. Math. Soft. (TOMS)*, 8(2):195–209, 1982. (Cited on page 373.)

[2428] C. C. Paige and M. A. Saunders. LSQR: An algorithm for sparse linear equations and sparse least squares. *ACM Trans. Math. Soft. (TOMS)*, 8(1):43–71, 1982. (Cited on pages 373 and 393.)

[2429] C. C. Paige and P. Van Dooren. On the quadratic convergence of Kogbetliantz's algorithm for computing the singular value decomposition. *Linear Algebra Appl.*, 77:301–313, 1986. (Cited on page 136.)

[2430] C. C. Paige and M. Wei. History and generality of the CS decomposition. *Linear Algebra Appl.*, 208-209:303–326, 1994. (Cited on page 144.)

[2431] C. C. Paige and W. Wülling. Properties of a unitary matrix obtained from a sequence of normalized vectors. *SIAM J. Matrix Anal. Appl.*, 35(2):526–545, 2014. (Cited on page 302.)

[2432] L. J. Paige and O. Taussky. *Proceedings of the 1951 Symposium "Simultaneous Linear Equations and the Determination of Eigenvalues", NBS 29*. National Bureau of Standards, USA, 1953. (Cited on pages 70, 200, 202, 203, and 576.)

[2433] L. Painvin. Sur un certain système d'équations linéaires. *J. Math. Pures Appl.*, 2:41–46, 1858. (Cited on page 381.)

[2434] V. Pan. *Structured Matrices and Polynomials: Unified Superfast Algorithms*. Birkäuser, Basel, 2001. (Cited on page 387.)

[2435] T. J. Park and J. C. Light. Unitary quantum time evolution by iterative Lanczos reduction. *J. Chem. Phys.*, 85(10):5870–5876, 1986. (Cited on page 389.)

[2436] W. V. Parker. The matrix equation $AX = XB$. *Duke Math. J.*, 17(1):43–51, 1950. (Cited on page 396.)

[2437] W. V. Parker. Characteristic roots and field of values of a matrix. *Bull. Amer. Math. Soc.*, 57(2):103–108, 1951. (Cited on page 287.)

[2438] B. N. Parlett. The development and use of methods of LR type. *SIAM Rev.*, 6(3):275–295, 1964. (Cited on page 310.)

[2439] B. N. Parlett. Laguerre's method applied to the matrix eigenvalue problem. *Math. Comput.*, 18(87):464–485, 1964. (Cited on page 289.)

[2440] B. N. Parlett. Convergence of the QR algorithm. *Numer. Math.*, 7(2):187–193, 1965. (Cited on page 313.)

[2441] B. N. Parlett. Correction to convergence of the QR algorithm. *Numer. Math.*, 10(2):163–164, 1967. (Cited on page 313.)

[2442] B. N. Parlett. Global convergence of the basic QR algorithm on Hessenberg matrices. *Math. Comput.*, 22(104):803–817, 1968. (Cited on page 313.)

[2443] B. N. Parlett. Normal Hessenberg and moment matrices. *Linear Algebra Appl.*, 6:37–43, 1973. (Cited on page 313.)

[2444] B. N. Parlett. The Rayleigh quotient iteration and some generalizations for nonnormal matrices. *Math. Comput.*, 28:679–693, 1974. (Cited on page 282.)

[2445] B. N. Parlett. A recurrence among the elements of functions of triangular matrices. *Linear Algebra Appl.*, 14(2):117–121, 1976. (Cited on page 388.)

[2446] B. N. Parlett. *The Symmetric Eigenvalue Problem.* Prentice Hall, 1980. Reprinted by SIAM, Philadelphia, 1998. (Cited on pages 287, 302, 315, and 319.)

[2447] B. N. Parlett. The contribution of J. H. Wilkinson to numerical analysis. In S. G. Nash, editor, *A History of Scientific Computing*, pages 17–30. ACM Press, 1990. (Cited on pages 596 and 598.)

[2448] B. N. Parlett. Reduction to tridiagonal form and minimal realizations. *SIAM J. Matrix Anal. Appl.*, 13(2):567–593, 1992. (Cited on page 228.)

[2449] B. N. Parlett. The new qd algorithms. *Acta Numerica*, 4:459–491, 1995. (Cited on page 311.)

[2450] B. N. Parlett. For tridiagonals T replace T with LDL^T. *J. Comput. Appl. Math.*, 123(1-2):117–130, 2000. (Cited on page 316.)

[2451] B. N. Parlett. The QR algorithm. *Comput. Sci. Eng.*, 2(1):38–42, 2000. (Cited on page 311.)

[2452] B. N. Parlett. Very early days of matrix computations. *SIAM News*, 36(9), 2003. (Cited on page 12.)

[2453] B. N. Parlett. A review of books by D. S. Watkins and D. Kressner. *SIAM Rev.*, 52(4):771–774, 2010. (Cited on page 319.)

[2454] B. N. Parlett and I. S. Dhillon. Fernando's solution to Wilkinson's problem: An application of double factorization. *Linear Algebra Appl.*, 267:247–279, 1997. (Cited on page 316.)

[2455] B. N. Parlett, F. M. Dopico, and C. Ferreira. The inverse eigenvector problem for real tridiagonal matrices. *SIAM J. Matrix Anal. Appl.*, 37(2):577–597, 2016. (Cited on page 316.)

[2456] B. N. Parlett and W. G. Poole, Jr. A geometric theory for the *QR*, *LU* and power iterations. *SIAM J. Numer. Anal.*, 10(2):389–412, 1973. (Cited on page 313.)

[2457] B. N. Parlett and J. Le. Forward instability of tridiagonal QR. *SIAM J. Matrix Anal. Appl.*, 14(1):279–316, 1993. (Cited on page 313.)

[2458] B. N. Parlett and C. Reinsch. Balancing a matrix for calculation of eigenvalues and eigenvectors. *Numer. Math.*, 13:293–304, 1969. (Cited on page 362.)

[2459] B. N. Parlett and D. S. Scott. The Lanczos algorithm with selective orthogonalization. *Math. Comput.*, 33(145):217–238, 1979. (Cited on page 302.)

[2460] B. N. Parlett and G. Strang. Matrices with prescribed Ritz values. *Linear Algebra Appl.*, 428:1725–1739, 2008. (Cited on page 305.)

[2461] B. N. Parlett, D. R. Taylor, and Z. A. Liu. A look-ahead Lanczos algorithm for unsymmetric matrices. *Math. Comput.*, 44(169):105–124, 1985. (Cited on page 225.)

[2462] M. Parodi. *Sur quelques propriétés des valeurs caractéristiques des matrices carrées.* Gauthier-Villars, Paris, 1952. Mémor. Sci. Math. 118. (Cited on page 287.)

[2463] K. H. Parshall. Telling the life of a mathematician: The case of J. J. Sylvester. *Rev. Hist. Math.*, 5:285–302, 1999. (Cited on page 571.)

[2464] K. H. Parshall. *James Joseph Sylvester. Jewish Mathematician in a Victorian World.* Johns Hopkins University Press, 2006. (Cited on page 571.)

[2465] B. Parsons. General k-part stationary iterative solutions to linear systems. *SIAM J. Numer. Anal.*, 24(1):188–198, 1987. (Cited on page 183.)

[2466] S. V. Parter. On the eigenvalues and eigenvectors of a class of matrices. *J. Soc. Ind. Appl. Math.*, 8(2):376–388, 1960. (Cited on page 86.)

[2467] S. V. Parter. The use of linear graphs in Gauss elimination. *SIAM Rev.*, 3(2):119–130, 1961. (Cited on pages 86 and 87.)

[2468] W. Parys. Why didn't Charasoff and Remak use Perron-Frobenius mathematics? *The European Journal of the History of Economic Thought*, 21(6):991–1014, 2014. (Cited on page 546.)

[2469] E. Pascal. *Determinanti. Teoria ed Applicazioni.* Ulrico Hoepli, Milano, 1897. (Cited on page 11.)

[2470] E. Pascal. *Repertorio di Matematiche Superiori: 1: Analisi.* Ulrico Hoepli, Milano, 1898. (Cited on page 11.)

[2471] E. Pascal. Eugenio Beltrami. *Rendiconti del R. Istituto Lombardo di Scienze e Lettere*, 34, ser. 2:57–108, 1901. (Cited on page 417.)

[2472] D. W. Peaceman and H. H. Rachford, Jr. The numerical solution of parabolic and elliptic differential equations. *J. Soc. Ind. Appl. Math.*, 3(1):28–41, 1955. (Cited on page 179.)

[2473] G. Peacock. *A Treatise on Algebra. Vol. 2.* J. and J. J. Deighton, 1830. (Cited on page 56.)

[2474] G. Peano. Integrazione per serie delle equazioni differenziali lineari. *Atti della Reale Accademia delle Scienze di Torino*, 22:437–446, 1887. (Cited on page 14.)

[2475] G. Peano. Intégration par séries des équations différentielles linéaires. *Math. Ann.*, 32(3):450–456, 1888. (Cited on pages 14 and 387.)

[2476] J. Pejlare and K. Bråting. Writing the history of mathematics: Interpretations of the mathematics of the past and its relation to the mathematics of today. In B. Sriraman, editor, *Handbook of the Mathematics of the Arts and Sciences*, pages 1–26. Springer, 2019. (Cited on page xvi.)

[2477] A.-E. Pellet. Sur un mode de séparation des racines des équations et la formule de Lagrange. *Bull. Sci. Math.*, 5:393–395, 1881. (Cited on page 285.)

[2478] E. Penchèvre. *Histoire de la théorie de l'élimination.* PhD thesis, Université Paris VII, 2006. (Cited on page 54.)

[2479] L. Pepe. Lagrange (1736-1813): a life in mathematics. *Lett. Mat. Int.*, 2:3–8, 2014. (Cited on page 510.)

[2480] D. V. Perevozkin and G. A. Omarova. Preconditioning methods based on spanning tree algorithms. *Journal of Physics*, 1715(1):1–5, 2021. (Cited on page 264.)

[2481] O. Perlot. *Préconditionnements de systèmes linéaires sur machines massivement parallèles CM-2 et CM-5.* PhD thesis, Université Paris VI, 1995. (Cited on page 262.)

[2482] O. Perron. Zur theorie der matrizen. *Math. Ann.*, 64:248–263, 1907. (Cited on page 464.)

[2483] O. Perron. *Algebra, Volumes I Die Grundlagen and II Theorie der Algebraischen Gleichung.* Walter de Gruyter, Berlin, Leipzig, 1927. (Cited on page 11.)

[2484] J. Pestana and A. J. Wathen. Natural preconditioning and iterative methods for saddle point systems. *SIAM Rev.*, 57(1):71–91, 2015. (Cited on page 265.)

[2485] G. Peters and J. H. Wilkinson. Eigenvectors of real and complex matrices by LR and QR triangularizations. *Numer. Math.*, 16:181–204, 1970. (Cited on page 362.)

[2486] S. S. Petrova and A. D. Solov'ev. The origin of the method of steepest descent. *Historia Math.*, 24:361–375, 1997. (Cited on page 170.)

[2487] H. Petzold. Wilhelm Cauer and his mathematical device. In B. Finn, editor, *Exposing Electronics, Artefacts Series. Studies in the History of Science and Technology*, pages 45–74. CRC Press, 2000. (Cited on page 332.)

[2488] R. Pexton. UCRL lectures on numerical analysis and applied mathematics. Technical Report UCRL-2119, University of California Radiation Laboratory, 1952. (Cited on page 72.)

[2489] B. Philippe and L. Reichel. On the generation of Krylov subspace bases. *Appl. Numer. Math.*, 62(9):1171–1186, 2012. (Cited on page 229.)

[2490] D. L. Phillips. A technique for the numerical solution of certain integral equations of the first kind. *J. ACM*, 9(1):84–97, 1962. (Cited on page 391.)

[2491] H. B. Phillips. Functions of matrices. *Amer. J. Math.*, 41(4):266–278, 1918. (Cited on page 388.)

[2492] J. L. Phillips. The triangular decomposition of Hankel matrices. *Math. Comput.*, 25(115):599–602, 1971. (Cited on page 386.)

[2493] E. Picard. Sur les équations intégrales de première espèce et sur certains problèmes de Physique mathématique. *C.R. Acad. Sci. Paris, Sér 1*, 148:1563–1568, 1909. (Cited on page 136.)

[2494] D. J. Pierce and R. J. Plemmons. Fast adaptive condition estimation. *SIAM J. Matrix Anal. Appl.*, 13(1):274–291, 1992. (Cited on page 24.)

[2495] S. Pincherle. *Algebra Complementare: Teoria delle Equazioni. Vol. 141*. U. Hoepli, 1893. (Cited on page 56.)

[2496] S. Pincherle. *Lezioni di Algebra Complementare Dettate nella R. Università di Bologna e Redatte per Uso degli Studenti. Vol. 1*. N. Zanichelli, 1906. (Cited on pages 11 and 56.)

[2497] B. Pini. Gianfranco Cimmino. *Bollettino dell'Unione Matematica Italiana, Sezione A, Serie VII*, 18(2):116–123, 1991. (Cited on pages 438 and 439.)

[2498] S. Pissanetsky. *Sparse Matrix Technology*. Academic Press, 1984. (Cited on page 94.)

[2499] P. Pizzetti. Sulla compensazione delle osservazioni secondo il metodo dei minimi quadrati, nota I, II. *Rendiconti della Reale Accademia Nazionale dei Lincei*, 3:230–235 and 288–293, 1887. (Cited on page 161.)

[2500] R. J. Plemmons. M-matrix characterizations. I- Nonsingular M-matrices. *Linear Algebra Appl.*, 18:175–188, 1977. (Cited on page 177.)

[2501] K. Plofker. *Mathematics in India*. Princeton University Press, 2008. (Cited on page 42.)

[2502] K. Plofker, A. Keller, T. Hayashi, C. Montelle, and D. Wujastyk. The Bakhshālī manuscript: A response to the Bodleian Library's radiocarbon dating. *History of Science in South Asia*, 5(1):134–150, 2017. (Cited on page 41.)

[2503] H. Poincaré. Notice sur la vie et les travaux de M. Laguerre, membre de la section de géometrie. *C.R. Acad. Sci. Paris, Sér 1*, 104:1643–1650, 1887. (Cited on page 514.)

[2504] H. Poincaré. Sur les groupes continus. Memoirs presented to the Cambridge Philosophical Society on the occasion of the Jubilee of Sir George Gabriel Stokes, Bart. Hon. LL. D., Hon. Sc. D., Lucasian Professor, Cambridge University Press, 1900. (Cited on page 388.)

[2505] E. Polizzi. Density-matrix-based algorithm for solving eigenvalue problems. *Phys. Rev. B*, 79(11):115112, 2009. (Cited on page 318.)

[2506] E. Polizzi and A. H. Sameh. A parallel hybrid banded system solver: The SPIKE algorithm. *Parallel Comput.*, 32(2):177–194, 2006. (Cited on page 377.)

[2507] B. T. Polyak. Random algorithms for solving convex inequalities. In D. Butnariu, Y. Censor, and S. Reich, editors, *Inherently Parallel Algorithms in Feasibility and Optimization and their Application*, pages 409–422. Elsevier Science Publishers, Amsterdam, 2001. Studies in Computational Mathematics, vol. 8. (Cited on page 191.)

[2508] J. C. T. Pool. James Wallace Givens, Jr. 1910-1993. *Notices Amer. Math. Soc.*, 41:29–30, 1994). (Cited on page 476.)

[2509] E. L. Poole and J. M. Ortega. Multicolor ICCG methods for vector computers. *SIAM J. Numer. Anal.*, 24(6):1394–1418, 1987. (Cited on page 260.)

[2510] G. Poole and T. Boullion. A survey on M-matrices. *SIAM Rev.*, 16(4):419–427, 1974. (Cited on page 177.)

[2511] G. Poole and L. Neal. Gaussian elimination: when is scaling beneficial? *Linear Algebra Appl.*, 162:309–324, 1992. (Cited on page 78.)

[2512] P. Popescu-Pampu. *What Is the Genus?* Springer International Publishing, 2016. (Cited on page 528.)

[2513] C. Possé. Sur quelques applications des fractions continues algébriques. Imprimerie de l'Académie Impériale des Sciences, St. Petersburg, 1886. (Cited on page 546.)

[2514] A. Pothen, H. D. Simon, and K.-P. Liou. Partitioning sparse matrices with eigenvectors of graphs. *SIAM J. Matrix Anal. Appl.*, 11(3):430–452, 1990. (Cited on page 90.)

[2515] A. Pothen and S. Toledo. Elimination structures in scientific computing. In D. Mehta and S. Sahni, editors, *Handbook on Data Structures and Applications*, Chapter 59. Chapman and Hall, CRC Press, 2004. (Cited on page 93.)

[2516] J. E. Potter. Matrix quadratic solutions. *SIAM J. Appl. Math.*, 14(3):496–501, 1966. (Cited on page 399.)

[2517] J. Poulson, B. Marker, R. A. van de Geijn, J. R. Hammond, and N. A. Romero. Elemental: A new framework for distributed memory dense matrix computations. *ACM Trans. Math. Soft. (TOMS)*, 39(2):13:1–13:24, 2013. (Cited on page 377.)

[2518] G. B. Price. Some identities in the theory of determinants. *Am. Math. Mon.*, 54:75–90, 1947. (Cited on page 557.)

[2519] W. Proskurowski and O. Widlund. On the numerical solution of Helmholtz's equation by the capacitance matrix method. *Math. Comput.*, 30(135):433–468, 1976. (Cited on page 266.)

[2520] J. S. Przemieniecki. Matrix structural analysis of substructures. *Am. Inst. Aero. Astro. J.*, 1(1):138–147, 1963. (Cited on pages 89 and 266.)

[2521] J. S. Przemieniecki. *Theory of Matrix Structural Analysis*. McGraw-Hill, 1968. (Cited on page 266.)

[2522] V. Pták. Norms and the spectral radius of matrices. *Czechoslovak Math. J.*, 12(4):555–557, 1962. (Cited on page 16.)

[2523] E. W. Purcell. The vector method of solving simultaneous linear equations. *J. Math. Phys.*, 32(1-4):180–183, 1953. (Cited on page 81.)

[2524] L. Pursell and S. Y. Trimble. Gram-Schmidt orthogonalization by Gauss elimination. *Am. Math. Mon.*, 98(6):544–549, 1991. (Cited on page 127.)

[2525] L. D. Pustyl'nikov. Toeplitz and Hankel matrices and their applications. *Russian Math. Surveys*, 39:63–98, 1984. (Cited on page 385.)

[2526] T. Püttmann. Kelvin: A simultaneoous calculator. http://www.math-meets-machines.de/ (Accessed November 2021), 2019. (Cited on page 332.)

[2527] A. Quarteroni and A. Valli. *Domain Decomposition Methods for Partial Differential Equations*. Oxford Science Publications, 1999. (Cited on page 268.)

[2528] P. Raghavan. DSCPACK: Domain-separator codes for the parallel solution of sparse linear systems. Technical Report CSE-02-004, Dept. Computer Science and Engineering, The Pennsylvania State University, 2002. (Cited on page 371.)

[2529] H. F. Rainsford. *Survey Adjustments and Least Squares*. Constable and Co. Ltd., London, 1957. (Cited on page 75.)

[2530] R. Ralha. One-sided reduction to bidiagonal form. *Linear Algebra Appl.*, 358:219–238, 2003. (Cited on page 138.)

[2531] B. Randell. On Alan Turing and the origins of digital computers. Technical report, University of Newcastle upon Tyne, Computing Laboratory, 1972. (Cited on page 336.)

[2532] B. Randell. Colossus: Godfather of the computer. In B. Randell, editor, *The Origins of Digital Computers*, pages 349–354, Springer, Berlin, Heidelberg, 1982. (Cited on page 336.)

[2533] B. Randell. From Analytical Engine to electronic digital computer: The contributions of Ludgate, Torres and Bush. *Ann. Hist. Comput.*, 4(4):327–341, 1982. (Cited on pages 330, 331, and 334.)

[2534] B. Randell. A Turing enigma. Technical Report CS-TR-1342, University of Newcastle upon Tyne, Computing Laboratory, 2012. Also International Conference on Concurrency Theory. Springer, Berlin, Heidelberg, 2012. (Cited on pages 336 and 337.)

[2535] S C. S. Rao. Existence and uniqueness of WZ factorization. *Parallel Comput.*, 23(8):1129–1139, 1997. (Cited on page 152.)

[2536] T. Rapcsák. The life and the works of Jenő Egerváry (1891-1958). *Cent. Eur. J. Oper. Res.*, 18(1):59–71, 2010. (Cited on page 81.)

[2537] J. Raphson. *Analysis Aequationum Universalis: seu ad aequationes algebraicas resolvendas methodus generalis, & expedita, ex nova infinitarum serierum methodo, deducta ac demonstrata.* typis TB prostant venales apud A. & I. Churchill (second edition 1702), London, 1690. (Cited on page 155.)

[2538] R. Rashed. Les travaux perdus de Diophante (I). *Rev. Hist. Sci.*, 27(2):97–122, 1974. (Cited on pages 39 and 41.)

[2539] R. Rashed. Les travaux perdus de Diophante (II). *Rev. Hist. Sci.*, 28(1):3–30, 1975. (Cited on pages 39 and 41.)

[2540] A. S. Rasulov, M. T. Bakoev, and D. R. Akabirhodjaeva. Probabilistic approach to the asynchronous iteration. *J. Appl. Math. Phys.*, 2(1):32–40, 2014. (Cited on page 197.)

[2541] Raytheon Co. Research on linear systems of a very large size. Technical Report N67-16086 and N67-25495, National Aeronautics and Space Administration, Washington, DC, 1966. (Cited on page 193.)

[2542] M. Redivo-Zaglia. Pseudo-Schur complements and their properties. *Appl. Numer. Math.*, 50:511–519, 2004. (Cited on page 26.)

[2543] T. Regińska. A regularization parameter in discrete ill-posed problems. *SIAM J. Sci. Comput.*, 17(3):740–749, 1996. (Cited on page 392.)

[2544] R. Rehman. *Numerical computation of the characteristic polynomial of a complex matrix.* PhD thesis, North Carolina State University, Raleigh, NC, 2010. (Cited on page 293.)

[2545] R. Rehman and I. C. F. Ipsen. Computing characteristic polynomials from eigenvalues. *SIAM J. Matrix Anal. Appl.*, 32(1):90–114, 2011. (Cited on page 293.)

[2546] E. Reich. On the convergence of the classical iterative method of solving linear simultaneous equations. *Ann. Math. Stat.*, 20(3):448–451, 1949. (Cited on page 173.)

[2547] L. Reichel. Polynomials by conformal mapping for the Richardson iteration method for complex linear systems. *SIAM J. Numer. Anal.*, 25(6):1359–1368, 1988. (Cited on page 184.)

[2548] L. Reichel. The application of Leja points to Richardson iteration and polynomial preconditioning. *Linear Algebra Appl.*, 154-156:389–414, 1991. (Cited on pages 163 and 184.)

[2549] L. Reichel and M. M. Spalević. A new representation of generalized averaged Gauss quadrature rules. *Appl. Numer. Math.*, 165:614–619, 2021. (Cited on page 391.)

[2550] L. Reichel, M. M. Spalević, and T. Tang. Generalized averaged Gauss quadrature rules for the approximation of matrix functionals. *BIT Numerical Mathematics*, 56(3):1045–1067, 2016. (Cited on page 391.)

[2551] L. Reichel and Q. Ye. Breakdown-free GMRES for singular systems. *SIAM J. Matrix Anal. Appl.*, 26(4):1001–1021, 2005. (Cited on page 241.)

[2552] C. Reid. *Hilbert.* Springer-Verlag, Heidelberg, 1970. (Cited on page 544.)

[2553] J. K. Reid. A note on the stability of Gaussian elimination. *IMA J. Appl. Math.*, 8(3):374–375, 1971. (Cited on pages 77 and 79.)

[2554] J. K. Reid. On the method of conjugate gradients for the solution of large sparse systems of linear equations. In J. K. Reid, editor, *Proc. Conference on Large Sparse Sets of Linear Equations.* Academic Press, New York, 1971. (Cited on page 209.)

[2555] J. K. Reid. *Proc. Conference on Large Sparse Sets of Linear Equations.* Academic Press, New York, 1971. (Cited on page 86.)

[2556] J. K. Reid. The use of conjugate gradients for systems of linear equations possessing "Property A". *SIAM J. Numer. Anal.*, 9(2):325–332, 1972. (Cited on page 209.)

[2557] C. Reinsch and F. L. Bauer. Rational QR transformation with Newton shift for symmetric tridiagonal matrices. *Numer. Math.*, 11:264–272, 1968. (Cited on page 362.)

[2558] E. Remes. Sur un procédé convergent d'approximations successives pour déterminer les polynômes d'approximation. *C.R. Acad. Sci. Paris, Sér 1*, 198:2063–2065, 1934. (Cited on page 182.)

[2559] O. Rendel, A. Rizvanolli, and J. P. M. Zemke. IDR: A new generation of Krylov subspace methods? *Linear Algebra Appl.*, 439(4):1040–1061, 2013. (Cited on page 245.)

[2560] K. J. Ressel and M. H. Gutknecht. QMR smoothing for Lanczos-type product methods based on three-term recurrences. *SIAM J. Sci. Comput.*, 19(1):55–73, 1998. (Cited on page 236.)

[2561] A. Reusken. Convergence analysis of the Gauss-Seidel preconditioner for discretized one dimensional Euler equations. *SIAM J. Numer. Anal.*, 41(4):1388–1405, 2003. (Cited on page 259.)

[2562] J. F. Riccati. Animadversationes in aequationes diffentiales secundi gradus. *Acta Eruditorum Lipsiae*, 8:67–73, 1724. (Cited on page 394.)

[2563] A. C. Rice, R. J. Wilson, and J. H. Garner. From student club to national society: The founding of the London Mathematical Society in 1865. *Historia Math.*, 22(4):402–421, 1995. (Cited on page xvi.)

[2564] J. R. Rice. Experiments on Gram-Schmidt orthogonalization. *Math. Comput.*, 20(94):325–328, 1966. (Cited on page 128.)

[2565] J. R. Rice. *Mathematical Software*. Proceedings of the Mathematical Software Symposium 1970. Academic Press, 1971. (Cited on page 364.)

[2566] L. F. Richardson. The approximate arithmetical solution by finite difference of physical problems involving differential equations, with an application to the stress in a masonry dam. *Philos. Trans. R. Soc. Lond.*, 210:307–357, 1910. ser A. (Cited on pages 70, 162, 294, 547, and 548.)

[2567] L. F. Richardson. *Weather Prediction by Numerical Process*. Cambridge University Press, Cambridge, UK, 1922. (Cited on pages 70 and 548.)

[2568] L. F. Richardson. The deferred approach to the limit. Part I: Single lattice. *Philos. Trans. R. Soc. Lond.*, 226:299–349, 1927. ser A. (Cited on page 548.)

[2569] H. Richter. Über matrixfunktionen. *Math. Ann.*, 122(1):16–34, 1950. (Cited on page 388.)

[2570] R. D. Richtmyer. *Difference Methods for Initial-value Problems*. Wiley-Interscience, New York, 1957. (Cited on page 84.)

[2571] J. L. Rigal and J. Gaches. On the compatibility of a given solution with the data of a linear system. *J. ACM*, 14(3):543–548, 1967. (Cited on page 78.)

[2572] J. D. Riley. Iteration procedures for the Dirichlet difference problem. *M.T.A.C.*, 8(47):125–131, 1954. (Cited on page 176.)

[2573] J. D. Riley. Solving systems of linear equations with a positive definite, symmetric, but possibly ill-conditioned matrix. *M.T.A.C.*, 9:96–101, 1955. (Cited on page 176.)

[2574] R. F. Rinehart. The equivalence of definitions of a matric function. *Am. Math. Mon.*, 62(6):395–414, 1955. (Cited on page 388.)

[2575] J. Rissanen. Recursive identification of linear systems. *SIAM J. Contr.*, 9(3):420–430, 1971. (Cited on page 386.)

[2576] J. Rissanen. Algorithms for the triangular decomposition of block Hankel and Toeplitz matrices with application to factoring positive matrix polynomials. *Math. Comput.*, 27:147–154, 1973. (Cited on page 386.)

[2577] J. Rissanen. Solution of linear equations with Hankel and Toeplitz matrices. *Numer. Math.*, 22:361–366, 1974. (Cited on page 386.)

[2578] W. Ritz. Über eine neue methode zur lösung gewisser variationsprobleme der mathematischen physik. *Journal für die reine und angewandte Mathematik (Crelle Journal)*, 135:1–61, 1908. (Cited on page 284.)

[2579] W. Ritz. Theorie der transversalschwingungen einer quadratischen platte mit freien rändern. *Annalen der Physik*, 28:737–786, 1909. (Cited on page 284.)

[2580] F. Robert. *Etude et utilisation de normes vectorielles en analyse numérique linéaire*. PhD thesis, Université de Grenoble, France, 1968. (Cited on page 16.)

[2581] F. Robert, M. Charnay, and F. Musy. Itérations chaotiques série-parallèle pour des équations nonlinéaires de point fixe. *Aplikace Matematiky*, 20(1):1–38, 1975. (Cited on page 196.)

[2582] Y. Robert. Regular incomplete factorizations of real positive definite matrices. *Linear Algebra Appl.*, 48:105–117, 1982. (Cited on page 256.)

[2583] J. D. Roberts. Linear model reduction and solution of the algebraic Riccati equation by use of the sign function. *Internat. J. Control*, 32(4):677–687, 1980. (Cited on page 398.)

[2584] G. H. Rodrigue and D. Wolitzer. Preconditioning by incomplete block cyclic reduction. *Math. Comput.*, 42(166):549–565, 1984. (Cited on page 260.)

[2585] L. J. Rogers. An extension of a certain theorem in inequalities. *Messenger Math.*, 17(10):145–150, 1888. (Cited on page 14.)

[2586] H. Rohrbach. Bemerkungen zu einem determinantensatz von Minkowski. *Jahresber. Deutsch. Math. Verein.*, 40:49–53, 1931. (Cited on page 286.)

[2587] C. Roland, B. Beckermann, and C. Brezinski. Altman's methods revisited. *Appl. Math. (Warsaw)*, 31:353–368, 2004. (Cited on pages 193 and 405.)

[2588] S. Röllin and M. H. Gutknecht. Variations of Zhang's Lanczos-type product method. *Appl. Numer. Math.*, 41(1):119–133, 2002. (Cited on page 239.)

[2589] D. J. Rose. *Symmetric elimination on sparse positive definite systems and potential flow network problem*. PhD thesis, Harvard University, 1970. (Cited on page 87.)

[2590] D. J. Rose. Triangulated graphs and the elimination process. *J. Math. Anal. Appl.*, 32:597–609, 1970. (Cited on page 87.)

[2591] D. J. Rose. A graph-theoretic study of the numerical solution of sparse positive definite systems of linear equations. In R. C. Read, editor, *Graph Theory and Computing*, pages 183–217. Academic Press, 1972. (Cited on page 87.)

[2592] D. J. Rose and R. E. Tarjan. Algorithmic aspects of vertex elimination on directed graphs. *SIAM J. Appl. Math.*, 34(1):176–197, 1978. (Cited on page 92.)

[2593] D. J. Rose, R. E. Tarjan, and G. S. Lueker. Algorithmic aspects of vertex elimination on graphs. *SIAM J. Comput.*, 5(2):266–283, 1976. (Cited on page 89.)

[2594] D. J. Rose and R. A. Willoughby. *Sparse Matrices and their Applications*. Springer, Boston, 1972. (Cited on pages 86 and 364.)

[2595] F. Rosen. *The Algebra of Mohammed Ben Musa*. Oriental Translation Fund, London, 1831. (Cited on pages 42 and 43.)

[2596] R. Rosen. Matrix bandwidth minimization. In *Proceedings 23rd National Conference ACM*, pages 585–595. ACM, 1968. (Cited on page 88.)

[2597] D. Rosenblatt. On linear models and the graphs of Minkowski-Leontief matrices. *Econometrica*, 25(2):25–338, 1957. (Cited on page 86.)

[2598] P. C. Rosenbloom. The difference equation method for solving the Dirichlet problem. In E. F. Beckenbach, editor, *Construction and Applications of Conformal Maps*, pages 231–237. National Bureau of Standards, 1952. (Cited on page 382.)

[2599] J. L. Rosenfeld. A case study on programming for parallel processors. Technical Report RC-1864, IBM, Thomas J. Watson Research Center, 1967. (Cited on page 195.)

[2600] E. C. Ross. *Elements of Algebra, translated from the French of M. Bourdon for the use of the cadets*. E. B. Clayton, New York, 1831. (Cited on page 56.)

[2601] J. B. Rosser, C. Lanczos, M. R. Hestenes, and W. Karush. Separation of close eigenvalues of a real symmetric matrix. *J. Res. Natl. Bur. Stand.*, 47(4):291–297, 1951. (Cited on pages 299 and 300.)

[2602] W. E. Roth. The equations $AX - YB = C$ and $AX - XB = C$ in matrices. *Proc. Am. Math. Soc.*, 3:392–396, 1952. (Cited on page 396.)

[2603] E. Rothberg and A. Gupta. Efficient sparse matrix factorization on high-performance workstations - Exploiting the memory hierarchy. *ACM Trans. Math. Soft. (TOMS)*, 17(3):313–334, 1991. (Cited on page 91.)

[2604] E. Rothberg and A. Gupta. An evaluation of left-looking, right-looking, and multifrontal approaches to sparse Cholesky factorization on hierarchical-memory machines. *Intl. J. High Speed Computing*, 5(4):537–593, 1993. (Cited on page 91.)

[2605] E. Rothberg and A. Gupta. An efficient block-oriented approach to parallel sparse Cholesky factorization. *SIAM J. Sci. Comput.*, 15(6):1413–1439, 1994. (Cited on page 92.)

[2606] V. Rotkin and S. Toledo. The design and implementation of a new out-of-core sparse Cholesky factorization method. *ACM Trans. Math. Soft. (TOMS)*, 30(1):19–46, 2004. (Cited on page 371.)

[2607] E. Rouché. Mémoire sur la série de Lagrange. *J. de l'École Impériale Polytechnique*, 22:193–224, 1862. (Cited on page 285.)

[2608] E. Rouché. Note sur les équations linéaires. *J. de l'École Polytechnique*, 81:221–228, 1880. (Cited on page 116.)

[2609] E. Rouché. Edmond Laguerre, sa vie et ses travaux. *Nouvelles Annales de Mathématiques, 3e série*, 6:105–173, 1887. (Cited on pages 6 and 514.)

[2610] M. Rozložník. *Numerical stability of the GMRES method*. PhD thesis, Czech Technical University, Prague, 1996. (Cited on page 233.)

[2611] M. Rozložník. *Saddle-point Problems and their Iterative Solution*. Birkhäuser, Basel, 2018. (Cited on page 265.)

[2612] M. Rozložník and V. Simoncini. Krylov subspace methods for saddle point problems with indefinite preconditioning. *SIAM J. Matrix Anal. Appl.*, 24(2):368–391, 2002. (Cited on page 265.)

[2613] M. Rozložník, Z. Strakoš, and M. Tůma. On the role of orthogonality in the GMRES method. In *SOFSEM'96: Theory and Practice of Informatics*, pages 409–416, Springer, Berlin, Heidelberg, 1996. (Cited on page 234.)

[2614] M. Rozložník and R. Weiss. On the stable implementation of the generalized minimal error method. *J. Comput. Appl. Math.*, 98(1):49–62, 1998. (Cited on page 233.)

[2615] S. T. Rubin. *Ett nytt sätt att lösa normalekvationer*. Svensk Lantmäteritidskrift, 1926. (Cited on page 70.)

[2616] J. W. Ruge and K. Stüben. Efficient solution of finite difference and finite element equations by algebraic multigrid (AMG). In D. J. Paddon and H. Holstein, editors, *Multigrid Methods for Integral and Differential Equations*, pages 169–212. Clarendon Press, 1985. (Cited on page 270.)

[2617] J. W. Ruge and K. Stüben. Algebraic multigrid. In S. F. Mc Cormick, editor, *Multigrid Methods*, pages 73–130. SIAM, Philadelphia, 1987. (Cited on page 270.)

[2618] A. Ruhe. On the quadratic convergence of a generalization of the Jacobi method to arbitrary matrices. *BIT Numerical Mathematics*, 8(3):210–231, 1968. (Cited on page 281.)

[2619] A. Ruhe. An algorithm for numerical determination of the structure of a general matrix. *BIT Numerical Mathematics*, 10(2):196–216, 1970. (Cited on page 149.)

[2620] A. Ruhe. Numerical aspects of Gram-Schmidt orthogonalization of vectors. *Linear Algebra Appl.*, 52:591–601, 1983. (Cited on page 129.)

[2621] A. Ruhe. Rational Krylov sequence methods for eigenvalue computation. *Linear Algebra Appl.*, 58:391–405, 1984. (Cited on page 307.)

[2622] A. Ruhe. The rational Krylov algorithm for nonsymmetric eigenvalue problems. III: Complex shifts for real matrices. *BIT Numerical Mathematics*, 34(1):165–176, 1994. (Cited on page 307.)

[2623] A. Ruhe. Rational Krylov algorithms for nonsymmetric eigenvalue problems. II. Matrix pairs. *Linear Algebra Appl.*, 197:283–295, 1994. (Cited on page 307.)

[2624] A. Ruhe. Rational Krylov: A practical algorithm for large sparse nonsymmetric matrix pencils. *SIAM J. Sci. Comput.*, 19(5):1535–1551, 1998. (Cited on page 307.)

[2625] A. Ruhe and T. Wiberg. The method of conjugate gradients used in inverse iteration. *BIT*, 12(4):543–554, 1972. (Cited on page 317.)

[2626] S. M. Rump. Bounds for the componentwise distance to the nearest singular matrix. *SIAM J. Matrix Anal. Appl.*, 18(1):83–103, 1997. (Cited on page 22.)

[2627] S. M. Rump. Structured perturbations and symmetric matrices. *Linear Algebra Appl.*, 278:121–132, 1998. (Cited on page 22.)

[2628] S. M. Rump. Ill-conditioned matrices are componentwise near to singularity. *SIAM Rev.*, 41(1):102–112, 1999. (Cited on page 22.)

[2629] S. M. Rump. Structured perturbations part II: Componentwise distances. *SIAM J. Matrix Anal. Appl.*, 25(1):31–56, 2003. (Cited on page 22.)

[2630] S. M. Rump and C.-P. Jeannerod. Improved backward error bounds for LU and Cholesky factorizations. *SIAM J. Matrix Anal. Appl.*, 35(2):684–698, 2014. (Cited on page 79.)

[2631] C. Runge. *Praxis der Gleichungen.* De Gruyter and Co., Berlin, Leipzig, 1921. (Cited on page 165.)

[2632] C. Runge and H. König. *Vorlesungen über Numerisches Rechnen.* Springer, Berlin, 1924. (Cited on page 170.)

[2633] D. E. Rutherford. On the solution of the matrix equation $AX + XB = C$. *Nederl. Akad. Wetensch. Proc. Ser. A*, 35:54–59, 1932. (Cited on page 396.)

[2634] H. Rutishauser. Beiträge zur kenntnis des biorthogonalisierungs-algorithmus von Lanczos. *Z. Angew. Math. Phys.*, 4(1):35–56, 1953. (Cited on pages 301 and 308.)

[2635] H. Rutishauser. Anwendungen des Quotienten-Differenzen-Algorithmus. *Z. Angew. Math. Phys.*, 5(6):496–508, 1954. (Cited on page 309.)

[2636] H. Rutishauser. Der Quotienten-Differenzen-Algorithmus. *Z. Angew. Math. Phys.*, 5(3):233–251, 1954. (Cited on page 309.)

[2637] H. Rutishauser. Bestimmung der eigenwerte und eigenvektoren einer matrix mit hilfe des Quotienten-Differenzen-Algorithmus. *Z. Angew. Math. Phys.*, 6(5):387–401, 1955. (Cited on pages 309 and 310.)

[2638] H. Rutishauser. Une méthode pour la détermination des valeurs propres d'une matrice. *C.R. Acad. Sci. Paris, Sér 1*, 240:34–36, 1955. (Cited on pages 310 and 311.)

[2639] H. Rutishauser. *Der Quotienten-Differenzen-Algorithmus.* Birkhäuser Verlag, Basel, 1957. (Cited on page 310.)

[2640] H. Rutishauser. Solution of eigenvalue problems with the LR-transformation. National Bureau of Standards, Applied Mathematics Series 49, 1958. (Cited on page 310.)

[2641] H. Rutishauser. The Jacobi method for real symmetric matrices, 1966. (Cited on pages 281 and 362.)

[2642] H. Rutishauser. *Handbook for Automatic Computation: A Description of Algol 60.* Springer-Verlag, 1967. (Cited on pages 129 and 360.)

[2643] H. Rutishauser. Computational aspects of F.L. Bauer's simultaneous iteration method. *Numer. Math.*, 13:4–13, 1969. (Cited on page 296.)

[2644] H. Rutishauser. Simultaneous iteration method for symmetric matrices. *Numer. Math.*, 16:205–223, 1970. (Cited on page 362.)

[2645] H. Rutishauser. *Lectures on Numerical Mathematics.* Birkhäuser, Boston, 1990. (Cited on page 310.)

[2646] H. Rutishauser and F. L. Bauer. Détermination des vecteurs propres d'une matrice par une méthode itérative avec convergence quadratique. *C.R. Acad. Sci. Paris, Sér 1*, 240:1680–1681, 1955. (Cited on page 310.)

[2647] H. Rutishauser and H. R. Schwarz. The LR transformation method for symmetric matrices. *Numer. Math.*, 5:273–289, 1963. (Cited on page 310.)

[2648] Y. Saad. On the rates of convergence of the Lanczos and the block-Lanczos methods. *SIAM J. Numer. Anal.*, 17(5):687–706, 1980. (Cited on page 301.)

[2649] Y. Saad. Variations on Arnoldi's method for computing eigenelements of large unsymmetric matrices. *Linear Algebra Appl.*, 34:269–295, 1980. (Cited on pages 222, 303, and 305.)

[2650] Y. Saad. Krylov subspace methods for solving large nonsymmetric linear systems. *Math. Comput.*, 37:105–126, 1981. (Cited on page 223.)

[2651] Y. Saad. The Lanczos biorthogonalization algorithm and other oblique projection methods for solving large unsymmetric systems. *SIAM J. Numer. Anal.*, 19(3):485–506, 1982. (Cited on page 223.)

[2652] Y. Saad. Chebyshev acceleration techniques for solving nonsymmetric eigenvalue problems. *Math. Comput.*, 42(166):567–588, 1984. (Cited on page 305.)

[2653] Y. Saad. Practical use of some Krylov subspace methods for solving indefinite and nonsymmetric linear systems. *SIAM J. Sci. Statist. Comput.*, 5(1):203–228, 1984. (Cited on page 223.)

[2654] Y. Saad. Practical use of polynomial preconditionings for the conjugate gradient method. *SIAM J. Sci. Statist. Comput.*, 6(4):865–881, 1985. (Cited on pages 214 and 262.)

[2655] Y. Saad. Least squares polynomials in the complex plane and their use for solving nonsymmetric linear systems. *SIAM J. Numer. Anal.*, 24(1):155–169, 1987. (Cited on pages 182 and 262.)

[2656] Y. Saad. Krylov subspace methods on supercomputers. *SIAM J. Sci. Statist. Comput.*, 10(6):1200–1232, 1989. (Cited on page 215.)

[2657] Y. Saad. Analysis of some Krylov subspace approximations to the matrix exponential operator. *SIAM J. Numer. Anal.*, 29(1):209–228, 1992. (Cited on page 390.)

[2658] Y. Saad. *Numerical Methods for Large Eigenvalue Problems*. Manchester University Press, 1992. Revised edition, Classics in Applied Mathematics 66, SIAM, Philadelphia, 2011. (Cited on pages 304, 305, and 319.)

[2659] Y. Saad. A flexible inner-outer preconditioned GMRES algorithm. *J. Sci. Comput.*, 14(2):461–469, 1993. (Cited on page 230.)

[2660] Y. Saad. ILUT: A dual threshold incomplete ILU preconditioner. *Numer. Linear Algebra Appl.*, 1:387–402, 1994. (Cited on page 257.)

[2661] Y. Saad. Analysis of augmented Krylov subspace methods. *SIAM J. Matrix Anal. Appl.*, 18(2):435–449, 1997. (Cited on page 234.)

[2662] Y. Saad. *Iterative Methods for Sparse Linear Systems*. 2nd edition, SIAM, Philadelphia, 2003. (Cited on page 186.)

[2663] Y. Saad. Multilevel ILU with reorderings for diagonal dominance. *SIAM J. Sci. Comput.*, 27(3):1032–1057, 2005. (Cited on page 271.)

[2664] Y. Saad. Iterative methods for linear systems of equations: A brief historical journey. In S. C. Brenner, I. Shparlinski, C.-W. Shu, and D.B. Szyld, editors, *75 years of Mathematics of Computation*, pages 197–216. American Mathematical Society, 2019. (Cited on page 246.)

[2665] Y. Saad and M. H. Schultz. Conjugate gradient-like algorithms for solving nonsymmetric linear systems. *Math. Comput.*, 44(170):417–424, 1985. (Cited on pages 223 and 224.)

[2666] Y. Saad and M. H. Schultz. GMRES: a generalized minimum residual algorithm for solving nonsymmetric linear systems. Technical Report YALEU/DCS/RR-254, Yale University, 1985. (Cited on page 224.)

[2667] Y. Saad and M. H. Schultz. GMRES: A generalized minimum residual algorithm for solving nonsymmetric linear systems. *SIAM J. Sci. Statist. Comput.*, 7(3):856–869, 1986. (Cited on pages 131, 224, and 239.)

[2668] Y. Saad and B. Suchomel. ARMS: An algebraic recursive multilevel solver for general sparse linear systems. *Numer. Linear Algebra Appl.*, 9(5):359–378, 2002. (Cited on page 271.)

[2669] Y. Saad and H. A. van der Vorst. Iterative solution of linear systems in the 20th century. *J. Comput. Appl. Math.*, 123(1):1–33, 2000. (Cited on pages 246 and 430.)

[2670] Y. Saad and K. Wu. DQGMRES: A direct quasi-minimal residual algorithm based on incomplete orthogonalization. *Numer. Linear Algebra Appl.*, 3(4):329–343, 1996. (Cited on page 225.)

[2671] Y. Saad, M. Yeung, J. Erhel, and F. Guyomarc'h. A deflated version of the conjugate gradient algorithm. *SIAM J. Sci. Comput.*, 21(5):1909–1926, 2000. (Cited on page 217.)

[2672] Y. Saad and J. Zhang. BILUTM: A domain-based multilevel block ILUT preconditioner for general sparse matrices. *SIAM J. Matrix Anal. Appl.*, 21(1):279–299, 1999. (Cited on page 270.)

[2673] H. Sadok. CMRH: a new method for solving nonsymmetric linear systems based on the Hessenberg reduction algorithm. *Numer. Algorithms*, 20(4):303–321, 1999. (Cited on page 237.)

[2674] H. Sadok. Analysis of the convergence of the minimal and the orthogonal residual methods. *Numer. Algorithms*, 40:201–216, 2005. (Cited on page 241.)

[2675] H. Sadok and D. B. Szyld. A new look at CMRH and its relation to GMRES. *BIT Numerical Mathematics*, 52(2):485–501, 2012. (Cited on page 244.)

[2676] E. Saibel and W. J. Berger. On finding the characteristic equation of a square matrix. *M.T.A.C.*, 7(44):228–236, 1953. (Cited on pages 218, 292, and 487.)

[2677] T. Sakurai and H. Sugiura. A projection method for generalized eigenvalue problems using numerical integration. *J. Comput. Appl. Math.*, 159(1):119–128, 2003. (Cited on page 318.)

[2678] G. Salmon. *Lessons Introductory to the Modern Higher Algebra.* Hodges, Smith and Co, Dublin, 1859. (Cited on page 115.)

[2679] G. Salmon. Science Worthies, XXII- Arthur Cayley. *Nature*, XXVIII(725):481–485, 1883. (Cited on page 421.)

[2680] A. Sameh. On Jacobi and Jacobi-like algorithms for a parallel computer. *Math. Comput.*, 25(115):579–590, 1971. (Cited on page 281.)

[2681] A. H. Sameh and Z. Tong. The trace minimization method for the symmetric generalized eigenvalue problem. *J. Comput. Appl. Math.*, 123(1-2):155–175, 2000. (Cited on page 318.)

[2682] A. H. Sameh and J. A. Wisniewski. A trace minimization algorithm for the generalized eigenvalue problem. *SIAM J. Numer. Anal.*, 19(6):1243–1259, 1982. (Cited on page 317.)

[2683] P. A. Samuelson. A method of determining explicitly the coefficients of the characteristic equation. *Ann. Math. Stat.*, 13(4):424–429, 1942. (Cited on page 292.)

[2684] J. M. Sanz-Serna. La cuadratura gaussiana según Gauss. *La Gaceta dela RSME*, 22(1):101–116, 2019. Also arXiv:1807-03506. (Cited on page 469.)

[2685] S. Sarasvati and U. Jyotishmati. *The Bakhshali Manuscript, an Ancient Treatise of Indina Arithmetic.* Arvind Printers, 1919. (Cited on page 42.)

[2686] G. Sarton. Lagrange's personality (1736-1813). *Proc. Amer. Phil. Soc.*, 88:457–496, 1944. (Cited on page 509.)

[2687] H. Sassenfeld. Ein hinreichendes konvergenzkriterium und eine fehlerabschätzung für die iteration in einzelschritten bei linearen gleichungen. *Zamm.-Z Angew. Math. Me.*, 31(3):92–94, 1951. (Cited on page 176.)

[2688] N. Sato and W. F. Tinney. Techniques for exploiting the sparsity or the network admittance matrix. *IEEE Transactions on Power Apparatus and Systems*, 82(69):944–950, 1963. (Cited on page 87.)

[2689] F. E. Satterthwaite. Error control in matrix calculation. *Ann. Math. Stat.*, 15(4):373–387, 1944. (Cited on pages 19 and 75.)

[2690] M. A. Saunders, H. D. Simon, and E. L. Yip. Two conjugate-gradient-type methods for unsymmetric linear equations. *SIAM J. Numer. Anal.*, 25(4):927–940, 1988. (Cited on page 236.)

[2691] S. A. Savari and D. Bertsekas. Finite termination of asynchronous iterative algorithms. *Parallel Comput.*, 22(1):39–56, 1996. (Cited on page 197.)

[2692] N. Schappacher. Diophantus of Alexandria: A text and its history. http://irma.math.unistra.fr/~schappa/NSch/Publications.html (Accessed November 2021), 1998. (Cited on pages 39 and 41.)

[2693] A. Schärlig. *Un Portrait de Gerbert d'Aurillac: Inventeur d'un Abaque, Utilisateur Précoce des Chiffres Arabes, et Pape de l'An Mil.* Presses Polytechniques et Universitaires Romandes, Lausanne, Switzerland, 2011. (Cited on page 324.)

[2694] S. Schechter. Relaxation methods for linear equations. *Comm. Pure Appl. Math.*, 12:313–335, 1959. (Cited on page 195.)

[2695] O. Schenk and K. Gärtner. Solving unsymmetric sparse systems of linear equations with PARDISO. *Future Gener. Comp. Sy.*, 20(3):475–487, 2004. (Cited on page 377.)

[2696] O. Schenk, K. Gärtner, W. Fichtner, and A. Stricker. PARDISO: A high-performance serial and parallel sparse linear solver in semiconductor device simulation. *Future Gener. Comp. Sy.*, 18(1):69–78, 2001. (Cited on page 377.)

[2697] W. Schmeidler. *Vorträge über Determinanten und Matrizen mit Anwendungen in Physik und Technik.* Akademieverlag, Berlin, 1949. (Cited on page 173.)

[2698] E. Schmidt. Zur theorie der linearen und nichtlinearen integralgleichungen I. *Math. Ann.*, 63:433–467, 1907. (Cited on pages 125, 126, and 135.)

[2699] E. Schmidt. Zur theorie der linearen und nichtlinearen integralgleichungen II. *Math. Ann.*, 64:161–174, 1907. (Cited on page 125.)

[2700] E. Schmidt. Uber die auflösung linearer gleichungen mit unendlich vielen ubbekannten. *Rend. Circ. Mat. Palermo. Ser. 1*, 25:53–77, 1908. (Cited on pages 14 and 125.)

[2701] E. Schmidt. Zur theorie der linearen und nichtlinearen integralgleichungen III. *Math. Ann.*, 65(3):370–399, 1908. (Cited on page 125.)

[2702] R. J. Schmidt. XXXII. On the numerical solution of linear simultaneous equations by an iterative method. *The London, Edinburgh, and Dublin Philosophical Magazine and Journal of Science*, 32(214):369–383, 1941. (Cited on page 170.)

[2703] H. Schneider. Regions of exclusion for the latent roots of a matrix. *Proc. Am. Math. Soc.*, 5:320–322, 1954. (Cited on page 287.)

[2704] H. Schneider. The concepts of irreducibility and full indecomposability of a matrix in the works of Frobenius, Köning and Markov. *Linear Algebra Appl.*, 18:139–162, 1977. (Cited on page 6.)

[2705] H. Schneider. Olga Taussky-Todd's influence on matrix theory and matrix theorists. *Linear Multilinear Algebra*, 5:197–224, 1977. (Cited on pages xvi, 10, and 575.)

[2706] H. Schneider. Some personal reminiscences of Olga Taussky-Todd. *Linear Algebra Appl.*, 280:15–19, 1998. (Cited on page 575.)

[2707] H. Schneider and G. Strang. Comparison theorems for supremum norms. *Numer. Math.*, 4:15–20, 1962. (Cited on page 16.)

[2708] O. Schneider. *Krylov subspace methods and their generalizations for solving singular linear operator equations with applications to continuous time Markov chains.* PhD thesis, Technischen Universität Bergakademie Freiberg, Freiberg, Germany, 2006. (Cited on page 241.)

[2709] W. Schönauer. *Scientific Computing on Vector Computers*, volume 2 of *Special Topics in Supercomputing*. North-Holland, 1987. (Cited on page 225.)

[2710] A. Schönhage. Zur konvergenz des Jacobi-verfahrens. *Numer. Math.*, 3(1):374–380, 1961. (Cited on page 281.)

[2711] R. Schreiber. A new implementation of sparse Gaussian elimination. *ACM Trans. Math. Soft. (TOMS)*, 8(3):256–276, 1982. (Cited on page 87.)

[2712] R. Schreiber and C. Van Loan. A storage-efficient *WY* representation for products of Householder transformations. *SIAM J. Sci. Statist. Comput.*, 10(1):53–57, 1989. (Cited on page 133.)

[2713] R. Schreiber and B. Parlett. Block reflectors: Theory and computation. *SIAM J. Numer. Anal.*, 25(1):189–205, 1988. (Cited on page 133.)

[2714] G. Schubring. *Conflicts Between Generalization, Rigor, and Intuition.* Springer, New York, 2005. (Cited on page 419.)

[2715] E. G. Schultchen. *Iterative solution of linear equation in finite element analysis.* PhD thesis, Lehigh University, Bethlehem, Pennsylvania, 1973. (Cited on page 209.)

[2716] E. G. Schultchen and C. N. Kostem. Solution of linear equations by iterative methods in finite element analysis. Technical Report 400.10, Lehigh University, Bethlehem, Pennsylvania, 1973. (Cited on page 209.)

[2717] I. Schur. Ueber die charakteristischen wurzeln einer linearen substitution mit einer anwendung auf die theorie der integralgleichungen. *Math. Ann.*, 66:488–510, 1909. (Cited on pages 68, 151, and 554.)

[2718] I. Schur. Uber potenzreihen, die im innem des einheitskreises beschrankt sind. *Journal für die reine und angewandte Mathematik (Crelle Journal)*, 147:205–232, 1917. (Cited on pages 24, 117, and 554.)

[2719] H. A. Schwarz. Conforme abbildung der oberflache eines tetraeders auf die oberflache einer kugel. *Journal für die reine und angewandte Mathematik (Crelle Journal)*, 20:70–105, 1869. (Cited on page 555.)

[2720] H. A. Schwarz. Über einen grenzübergang durch alternierendes verfahren. *Vierteljahrsschrift Naturf. Ges. Zürich*, 15(3):272–286, 1870. (Cited on pages 265 and 555.)

[2721] H. R. Schwarz. Critère de stabilité pour des systèmes d'équations différentielles à coefficients constants. *C.R. Acad. Sci. Paris, Sér 1*, 241(1):15–16, 1955. (Cited on page 292.)

[2722] H. R. Schwarz. Critère de stabilité pour des systèmes d'équations différentielles à coefficients constants complexes. *C.R. Acad. Sci. Paris, Sér 1*, 242(3):325–327, 1956. (Cited on page 292.)

[2723] H. R. Schwarz. *Ein verfahren zur stabilitätsfrage bei matrizen-eigenwertproblemen*. PhD thesis, Eidgenössischen Technischen Hochschule, Zürich, 1956. (Cited on page 292.)

[2724] H. R. Schwarz. Tridiagonalization of a symmetric band matrix. *Numer. Math.*, 12:231–241, 1968. (Cited on page 362.)

[2725] A. Schwarzenberg-Czerny. On matrix factorization and efficient least squares solution. *Astronomy and Astrophysics Supplement*, 110:405–410, 1995. (Cited on page 70.)

[2726] F. Schweins. Theorie der Differenzen und Differentiale. u. s. w. Heidelberg, Theorie der Producte mit Versetzungen, 1825. (Cited on pages 111 and 557.)

[2727] M. Schweitzer. Any finite convergence curve is possible in the initial iterations of restarted FOM. *Electron. Trans. Numer. Anal.*, 45:133–145, 2016. (Cited on page 245.)

[2728] H. Schwerdtfeger. *Les Fonctions de Matrices. I. Les Fonctions Univalentes*. Actualités Scientifiques et Industrielles, 649. Hermann, Paris, 1938. (Cited on page 388.)

[2729] D. S. Scott. *Analysis of the symmetric Lanczos algorithm*. PhD thesis, University of California, Berkeley, USA, 1978. (Cited on page 302.)

[2730] D. S. Scott. How to make the Lanczos algorithm converge slowly. *Math. Comput.*, 33(145):239–247, 1979. (Cited on page 302.)

[2731] J. A. Scott. A frontal solver for the 21st century. *Commun. Numer. Meth. Engng.*, 22(10):1015–1029, 2006. (Cited on page 89.)

[2732] R. F. Scott. *A Treatise on the Theory of Determinants*. Cambridge University Press, 1880. (Cited on page 7.)

[2733] M. K. Seager. Parallelizing conjugate gradient for the CRAY X-MP. *Parallel Comput.*, 3(1):35–47, 1986. (Cited on page 215.)

[2734] P. L. Seidel. Note über eine Eigenschaft der Reihen, welche discontinuirliche Functionen darstellen. *Abhandlungen der Mathem.-Physikalische, Classe der Königlich Bayerischen Akademie der Wissenschaften*, 5:381–394, 1847. (Cited on page 559.)

[2735] P. L. Seidel. Untersuchungen über die Konvergenz und Divergenz der Kettenbrüche. Habilitationsschrift, Ludwig-Maximilians-Universität München, 1847. (Cited on page 158.)

[2736] P. L. Seidel. Bemerkungen über den Zusammenhang zwischen dem Bildungsgesetze eines Kettenbruches und der Art des Fortgangs seiner Näherungsbrüche. *Abhandlungen der Bayerischen Akademie der Wissenschaften, Mathematisch-naturwissenschaftliche Klasse*, 7:559–602, 1855. (Cited on page 559.)

[2737] P. L. Seidel. *Theorie der Fehler, mit Welchen die durch Optische Instrumente Gesehenen Bilder behaftet sind, und über die Mathematischen Bedingungen ihrer Aufhebung*. Cotta, München, 1857. (Cited on page 559.)

[2738] P. L. Seidel. Resultate photometrischer Messungen an 208 Fixsternen. *Astronomische Nachrichten*, 60:307–320, 1863. (Cited on page 158.)

[2739] P. L. Seidel. Über die Berechnung der wahrscheinlichsten Werthe solcher Unbeckannten zwischen welchen Bedingungs-Gleichungen bestehen. *Astronomische Nachrichten*, 84:193–210, 1874. (Cited on page 158.)

[2740] P. L. Seidel. Über ein Verfahren, die Gleichungen, auf welche die Methode der kleinsten Quadrate führt, sowie lineare Gleichungen überhaupt, durch successive Annaherung aufzulösen. *Abhandlungen der Bayerischen Akademie der Wissenschaften. Mathematisch-Naturwissenschaftliche Abteilung*, 11(3):81–108, 1874. (Cited on pages 158 and 560.)

[2741] T. Sekii. Two-dimensional inversion for solar internal rotation. *Publications of the Astronomical Society of Japan*, 43:381–411, 1991. (Cited on page 394.)

[2742] T. Sekimoto and S. Fujino. A proposal of variants of BiCGSafe method for solving linear systems in realistic problems. In *Proceedings of the World Congress on Engineering 2012 Vol I WCE*, 2012. (Cited on page 240.)

[2743] M. Serfati. *La constitution de l'écriture symbolique mathématique*. PhD thesis, Université Paris I, 1997. (In French). (Cited on pages 49 and 102.)

[2744] M. Serfati. Mathématiques et pensée symbolique chez Leibniz. *Rev. Hist. Sci.*, 54(2):165–222, 2001. (Cited on page 102.)

[2745] D. Sesana and V. Simoncini. Spectral analysis of inexact constraint preconditioning for symmetric saddle point matrices. *Linear Algebra Appl.*, 438(6):2683–2700, 2013. (Cited on page 265.)

[2746] J. Sesiano. *Books IV to VII of Diophantus' Arithmetica, in the Arabic Translation Attributed to Qustā ibn Lūqā*. Springer, 1982. (Cited on pages 39 and 41.)

[2747] L. F. Shampine, H. A. Watts, and J. R. Rice. *Mathematical Software III*. Academic Press, 1977. (Cited on page 364.)

[2748] D. Shanks. An analogy between transient and mathematical sequences and some nonlinear sequence-to-sequence transforms suggested by it. Part I. Technical Report Memorandum 9994, Naval Ordnance Laboratory, White Oak, Maryland, 1949. (Cited on page 310.)

[2749] D. Shanks. Non-linear transformations of divergent and slowly convergent sequences. *J. Math. and Phys.*, 34:1–42, 1955. (Cited on page 310.)

[2750] H. Shapiro. The Weyr characteristic. *Am. Math. Mon.*, 106(10):919–929, 1999. (Cited on page 148.)

[2751] J. W. Sheldon. On the numerical solution of elliptic difference equations. *M.T.A.C.*, 9:101–112, 1955. (Cited on page 176.)

[2752] K. Shen, J. N. Crossley, and A. W.-C. Lun. *The Nine Chapters on the Mathematical Art: Companion and Commentary*. Oxford University Press, 1999. (Cited on pages 35 and 39.)

[2753] W. F. Sheppard. Central-difference formulae. *Proc. Lond. Math. Soc.*, 1(1):449–488, 1899. (Cited on page 162.)

[2754] A. H. Sherman. Algorithms for sparse Gaussian elimination with partial pivoting. *ACM Trans. Math. Soft. (TOMS)*, 4(4):330–338, 1978. (Cited on pages 92 and 371.)

[2755] J. Sherman. Computations of inverse matrices by means of IBM machines. *IBM Technical Newsletter*, 3:22–25, 1951. (Cited on page 77.)

[2756] J. Sherman and W. J. Morrison. Adjustment of an inverse matrix corresponding to changes in the elements of a given column or a given row ofthe original matrix. *Ann. Math. Stat.*, 20:621, 1949. (Cited on page 83.)

[2757] J. Sherman and W. J. Morrison. Adjustment of an inverse matrix corresponding in one element of a given matrix. *Ann. Math. Stat.*, 21(1):124–127, 1950. (Cited on page 83.)

[2758] O. Sheynin. C.F. Gauss and the method of least squares. *Śląski Przegląd Statystyczny*, 12:9–38, 2014. (Cited on page 59.)

[2759] G. H. Shortley and R. Weller. The numerical solution of Laplace's equation. *J. Applied Physics*, 9(5):334–348, 1938. (Cited on page 165.)

[2760] G. H. Shortley, R. Weller, and B. Fried. Numerical solution of Laplace's and Poisson's equations. *Ohio State University Studies, Engineering Series, Bull.* 107, 1942. (Cited on page 165.)

[2761] R. B. Sidje. Alternatives for parallel Krylov subspace basis computation. *Numer. Linear Algebra Appl.*, 4(4):305–331, 1997. (Cited on page 229.)

[2762] R. B. Sidje. Expokit: Software package for computing matrix exponentials. *ACM Trans. Math. Soft. (TOMS)*, 24(1):130–156, 1998. (Cited on page 390.)

[2763] R. B. Sidje and B. Philippe. Parallel Krylov subspace basis computation. In J. Tankoano, editor, *Proceedings of the Second African Conference on Research in Computer Science, CARI'94*, pages 421–440. Orstom, 1994. (Cited on pages 133 and 229.)

[2764] R. Siegmund-Schultze. Hilda Geiringer-von Mises, Charlier series, ideology, and the human side of the emancipation of applied mathematics at the University of Berlin during the 1920s. *Historia Math.*, 20:364–381, 1993. (Cited on page 472.)

[2765] L. E. Sigler. *Fibonacci's Liber Abaci, A Translation into Modern English of Leonardo Pisano Book of Calculation*. Springer, New York, 2002. (Cited on page 43.)

[2766] H. D. Simon. *The Lanczos algorithm for solving symmetric linear systems*. PhD thesis, University of California, Berkeley, 1982. (Cited on page 302.)

[2767] H. D. Simon. Analysis of the symmetric Lanczos algorithm with reorthogonalization methods. *Linear Algebra Appl.*, 61:101–131, 1984. (Cited on page 302.)

[2768] H. D. Simon. The Lanczos algorithm with partial reorthogonalization. *Math. Comput.*, 42(165): 115–142, 1984. (Cited on page 302.)

[2769] H. D. Simon. Partitioning of unstructured problems for parallel processing. *Computing systems in engineering*, 2(2-3):135–148, 1991. (Cited on page 90.)

[2770] V. Simoncini. A stabilized QMR version of block BICG. *SIAM J. Matrix Anal. Appl.*, 18(2):419–434, 1997. (Cited on page 235.)

[2771] V. Simoncini. On the convergence of restarted Krylov subspace methods. *SIAM J. Matrix Anal. Appl.*, 22(2):430–452, 2000. (Cited on page 237.)

[2772] V. Simoncini. Block triangular preconditioners for symmetric saddle-point problems. *Appl. Numer. Math.*, 49(1):63–80, 2004. (Cited on page 265.)

[2773] V. Simoncini. On a non-stagnation condition for GMRES and application to saddle point matrices. *Electron. Trans. Numer. Anal.*, 37:202–213, 2010. (Cited on page 240.)

[2774] V. Simoncini. Computational methods for linear matrix equations. *SIAM Rev.*, 58(3):377–441, 2016. (Cited on page 398.)

[2775] V. Simoncini and M. Benzi. Spectral properties of the Hermitian and skew-Hermitian splitting preconditioner for saddle point problems. *SIAM J. Matrix Anal. Appl.*, 26(2):377–389, 2004. (Cited on page 265.)

[2776] V. Simoncini and E. Gallopoulos. An iterative method for nonsymmetric systems with multiple right-hand sides. *SIAM J. Sci. Comput.*, 16(4):917–933, 1994. (Cited on page 233.)

[2777] V. Simoncini and E. Gallopoulos. Convergence properties of block GMRES and matrix polynomials. *Linear Algebra Appl.*, 247:97–119, 1996. (Cited on page 233.)

[2778] V. Simoncini and E. Gallopoulos. A hybrid block GMRES method for nonsymmetric systems with multiple right-hand sides. *J. Comput. Appl. Math.*, 66(1-2):457–469, 1996. (Cited on page 233.)

[2779] V. Simoncini and D. B. Szyld. Flexible inner-outer Krylov subspace methods. *SIAM J. Numer. Anal.*, 40(6):2219–2239, 2003. (Cited on page 239.)

[2780] V. Simoncini and D. B. Szyld. Theory of inexact Krylov subspace methods and applications to scientific computing. *SIAM J. Sci. Comput.*, 25(2):454–477, 2003. (Cited on page 239.)

[2781] V. Simoncini and D. B. Szyld. On the occurence of superlinear convergence of exact and inexact Krylov subspace methods. *SIAM Rev.*, 47(2):247–272, 2005. (Cited on pages 239 and 241.)

[2782] V. Simoncini and D. B. Szyld. Recent computational developments in Krylov subspace methods for linear systems. *Numer. Linear Algebra Appl.*, 14(1):1–59, 2007. (Cited on pages 234 and 239.)

[2783] V. Simoncini and D. B. Szyld. New conditions for non-stagnation of minimal residual methods. *Numer. Math.*, 109:477–487, 2008. (Cited on page 240.)

[2784] V. Simoncini and D. B. Szyld. Interpreting IDR as a Petrov-Galerkin method. *SIAM J. Sci. Comput.*, 32(4):1898–1912, 2010. (Cited on page 243.)

[2785] M.-K. Siu. Equations in China: Two millenia of innovation, transmission and re-transmission. Preprint, The University of Hong Kong, Department of Mathematics, 2018. (Cited on page 39.)

[2786] R. D. Skeel. Scaling for numerical stability in Gaussian elimination. *J. ACM*, 26(3):494–526, 1979. (Cited on pages 23, 79, and 80.)

[2787] R. D. Skeel. Iterative refinement implies numerical stability for Gaussian elimination. *Math. Comput.*, 35(151):817–832, 1980. (Cited on page 79.)

[2788] Z. Sklenáriková and M. Pémová. The Pohlke-Schwarz theorem and its relevancy in the didactics of mathematics. Quaderni di Ricerca in Didattica, G.R.I.M. (Department of Mathematics, University of Palermo, Italy), n17, 2007. (Cited on page 554.)

[2789] G. L. G. Sleijpen, A. G. L. Booten, D. R. Fokkema, and H. A. van der Vorst. Jacobi-Davidson type methods for generalized eigenproblems and polynomial eigenproblems. *BIT Numerical Mathematics*, 36:595–633, 1996. (Cited on page 317.)

[2790] G. L. G. Sleijpen and D. R. Fokkema. BiCGstab(ℓ) for linear equations involving unsymmetric matrices with complex spectrum. *Electron. Trans. Numer. Anal.*, 11:11–32, 1993. (Cited on page 230.)

[2791] G. L. G. Sleijpen, P. Sonneveld, and M. B. van Gijzen. Bi-CGSTAB as an induced dimension reduction method. *Appl. Numer. Math.*, 60(11):1100–1114, 2010. (Cited on page 243.)

[2792] G. L. G. Sleijpen, J. van den Eshof, and M. B. van Gijzen. Restarted GMRES with inexact matrix-vector products. In *International Conference on Numerical Analysis and its Applications*, pages 494–502, Springer, Berlin, Heidelberg, 2004. (Cited on page 239.)

[2793] G. L. G. Sleijpen and H. van der Vorst. Maintaining convergence properties of BiCGstab methods in finite precision arithmetic. *Numer. Algorithms*, 10:203–223, 1995. (Cited on page 233.)

[2794] G. L. G. Sleijpen and H. van der Vorst. An overview of approaches for the stable computation of hybrid BiCG methods. *Appl. Numer. Math.*, 19(3):235–254, 1995. (Cited on page 233.)

[2795] G. L. G. Sleijpen and H. A. van der Vorst. A Jacobi-Davidson iteration method for linear eigenvalue problems. *SIAM J. Matrix Anal. Appl.*, 17(2):401–425, 1996. (Cited on pages 233 and 317.)

[2796] G. L. G. Sleijpen and H. A. van der Vorst. A Jacobi-Davidson iteration method for linear eigenvalue problems. *SIAM Rev.*, 42(2):267–293, 2000. (Cited on page 317.)

[2797] G. L. G. Sleijpen, H. A. van der Vorst, and E. Meijerink. Efficient expansion of subspaces in the Jacobi-Davidson method for standard and generalized eigenproblems. *Electron. Trans. Numer. Anal.*, 7:75–89, 1998. (Cited on page 317.)

[2798] G. L. G. Sleijpen and M. B. van Gijzen. Exploiting BiCGSTAB(ℓ) strategies to induce dimension reduction. *SIAM J. Sci. Comput.*, 32(5):2687–2709, 2010. (Cited on page 243.)

[2799] S. W. Sloan. An algorithm for profile and wavefront reduction of sparse matrices. *Internat. J. Numer. Methods Engrg.*, 23(2):239–251, 1986. (Cited on page 88.)

[2800] P. Smit. Generating identical Ritz values. Technical report, Tilburg University, The Netherlands, 1995. (Cited on page 305.)

[2801] P. Smit and M. H. C. Paardekooper. The effects of inexact solvers in algorithms for symmetric eigenvalue problems. *Linear Algebra Appl.*, 287(1-3):337–357, 1999. (Cited on page 317.)

[2802] B. Smith, J. Boyle, J. Dongarra, B. Garbow, Y. Ikebe, V. Klema, and C. Moler. *Matrix Eigensystem Routines, EISPACK Guide*. Springer Verlag, 2nd edition, 1976. Lecture Notes in Computer Science, Volume 6. (Cited on page 367.)

[2803] B. Smith, J. Boyle, B. Garbow, Y. Ikebe, V. Klema, and C. Moler. *Matrix Eigensystem Routines, EISPACK Guide*. Springer Verlag, 1st edition, 1974. Lecture Notes in Computer Science, Volume 6. (Cited on page 367.)

[2804] B. F. Smith, P. E. Bjørstad, and W. D. Gropp. *Domain Decomposition: Parallel Multilevel Methods for Elliptic Partial Differential Equations*. Cambridge University Press, 1996. (Cited on page 268.)

[2805] D. E. Smith. *A Source Book in Mathematics*. McGraw-Hill Book Company, 1929. (Cited on pages 102 and 103.)

[2806] D. E. Smith and Y. Mikami. *A History of Japanese Mathematics*. The Open Court Publishing Company, Chicago, 1914. (Cited on page 101.)

[2807] H. J. S. Smith. On systems of linear indeterminate equations and congruences. *Philos. Trans. R. Soc. Lond.*, 151:293–326, 1861. (Cited on page 116.)

[2808] R. A. Smith. Matrix equation $XA + BX = C$. *SIAM J. Appl. Math.*, 16(1):198–201, 1968. (Cited on page 397.)

[2809] F. Smithies. *Integral Equations*. Cambridge University Press, Cambridge, 1958. (Cited on page 393.)

[2810] L. Smoch. *Application des méthodes de Krylov à la résolution de systèmes singuliers*. PhD thesis, Université du Littoral, Calais, France, 1999. (in French). (Cited on page 237.)

[2811] L. Smoch. Some results about GMRES in the singular case. *Numer. Algorithms*, 22(2):193–212, 1999. (Cited on page 237.)

[2812] L. Smoch. Spectral behaviour of GMRES applied to singular systems. *Adv. Comput. Math.*, 27(2):151–166, 2007. (Cited on pages 237 and 241.)

[2813] A. Smoktunowicz and I. Wröbel. Numerical stability of the Richardson second order method. *Demonstratio Mathematica*, 38(1):255–264, 2005. (Cited on page 186.)

[2814] D. C. Smolarski and P. E. Saylor. Implementation of an adaptive algorithm for Richardson's method. *Linear Algebra Appl.*, 154:615–646, 1991. (Cited on pages 182 and 262.)

[2815] W. F. Smyth. Algorithms for the reduction of the matrix bandwidth and profile. *J. Comput. Appl. Math.*, 12-13:551–561, 1985. (Cited on page 88.)

[2816] R. A. Snay. Reducing the profile of sparse symmetric matrices. *Bull. Géodésique*, 50(4):341–352, 1976. (Cited on page 88.)

[2817] M. Snir, W. Gropp, S. Otto, S. Huss-Lederman, J. J. Dongarra, and D. Walker. *MPI–the Complete Reference: the MPI Core*. MIT Press, 1998. (Cited on page 376.)

[2818] F. E. Snyder and H. M. Livingston. Coding of a Laplace boundary value problem for the UNIVAC. *M.T.A.C.*, 3(25):341–350, 1949. (Cited on page 336.)

[2819] S. L. Sobolev. Schwarz algorithm in the theory of elasticity. *Doklady Akademii Nauk SSSR*, IV:243–246, 1936. (Cited on page 265.)

[2820] T. Sogabe, M. Sugihara, and S. L. Zhang. An extension of the conjugate residual method to non-symmetric linear systems. *J. Comput. Appl. Math.*, 226:103–113, 2009. (Cited on page 242.)

[2821] W. Somers. *Encyclopedia of Union College History*. Union College Press, Schenectady, NY, 2003. (Cited on page 448.)

[2822] C. Somigliana. Inaugurazione del Monumento a Eugenio Beltrami nella R. Università di Pavia. Pavia, Italy, 1909. (Cited on page 417.)

[2823] P. Sonneveld. CGS, a fast Lanczos-type solver for nonsymmetric linear systems. Technical Report 84-16, Department of Mathematics and Informatics, Delft University of Technology, The Netherlands, 1984. (Cited on page 223.)

[2824] P. Sonneveld. CGS, a fast Lanczos-type solver for nonsymmetric linear systems. *SIAM J. Sci. Statist. Comput.*, 10(1):36–52, 1989. (Cited on page 223.)

[2825] P. Sonneveld. AGS-IDR-CGS-BiCGSTAB-IDR(s): The circle closed. A case of serendipity. In *Proceedings of the International Kyoto Forum 2008 on Krylov Subspace Methods*, pages 1–14, 2008. (Cited on page 242.)

[2826] P. Sonneveld. On the convergence behavior of IDR(s) and related methods. *SIAM J. Sci. Comput.*, 34(5):A2576–A2598, 2012. (Cited on page 244.)

[2827] P. Sonneveld. A history of Krylov product methods - A case of serendipity. Technical Report 17-05, Insitute of Applied Mathematics, Delft University of Technology, The Netherlands, 2017. (Cited on pages 221, 226, and 242.)

[2828] P. Sonneveld and M. B. van Gijzen. IDR(s): A family of simple and fast algorithms for solving large nonsymmetric systems of linear equations. *SIAM J. Sci. Comput.*, 31(2):1035–1062, 2008. (Cited on page 242.)

[2829] K. M. Soodhalter. Stagnation of block GMRES and its relationship to block FOM. *Electron. Trans. Numer. Anal.*, 46:162–189, 2017. (Cited on page 240.)

[2830] D. C. Sorensen. Implicit application of polynomial filters in a k-step Arnoldi method. *SIAM J. Matrix Anal. Appl.*, 13(1):357–385, 1992. (Cited on page 305.)

[2831] M. Sosonkina, L. T. Watson, R. K. Kapania, and H. F. Walker. A new adaptive GMRES algorithm for achieving high accuracy. *Numer. Linear Algebra Appl.*, 5:275–297, 1998. (Cited on page 232.)

[2832] J. -M. Souriau. *Calcul Linéaire, Tome I.* Presses Universitaires de France, Paris, 1959. (Cited on page 289.)

[2833] J.-M. Souriau. Une méthode pour la décomposition spectrale et l'inversion d'une matrice. *C.R. Acad. Sci. Paris, Sér 1*, 220:1010–1011, 1948. (Cited on page 289.)

[2834] R. V. Southwell. Stress-calculation in frameworks by the method of "systematic relaxation of constraints". I and II. *Proc. R. Soc. Lond.*, 151:56–95, 1935. (Cited on page 168.)

[2835] R. V. Southwell. Stress-calculation in frameworks by the method of "systematic relaxation of constraints". III. *Proc. R. Soc. Lond.*, 153:41–76, 1935. (Cited on page 169.)

[2836] R. V. Southwell. *Relaxation Methods in Engineering Science.* Oxford University Press, Oxford, 1940. (Cited on pages 169 and 563.)

[2837] R. V. Southwell. On relaxation methods: A mathematics for engineering science. *Proc. R. Soc. Lond.*, 184:258–288, 1945. (Cited on page 561.)

[2838] R. V. Southwell. *Relaxation Methods in Theoretical Physics.* Clarendon Press, Oxford, 1946. (Cited on page 563.)

[2839] R. V. Southwell. *Relaxation Methods in Theoretical Physics II.* Oxford University Press, Oxford, 1956. (Cited on page 563.)

[2840] R. Speek. Robot mathematician solves nine simultaneous equations. Faculty of Science and Engineering, University of Groningen, 2018. (Cited on page 332.)

[2841] B. Speelpenning. The generalized element method. Technical Report UIUCDCS-R-78-946, University of Illinois at Urbana-Champaign, 1978. (Cited on page 89.)

[2842] D. A. Spielman and S.-H. Teng. Nearly linear time algorithms for preconditioning and solving symmetric, diagonally dominant linear systems. *SIAM J. Matrix Anal. Appl.*, 35(3):835–885, 2014. (Cited on page 264.)

[2843] M. Spiesser. Questions sur la diffusion du Liber Abbaci en France au XVème siècle à travers l'étude des traités commerciaux. *Bollettino di Storia delle Scienze Matematiche*, 24(1):114–135, 2004. (Cited on page 44.)

[2844] M. Spiesser. L'ouvre de Nicolas Chuquet dans le contexte des savoirs mathématiques de la fin du XVe siècle. Histoire littéraire de la France, tome 43-1, 2005. (Cited on page 44.)

[2845] M. Spiesser. L'algèbre de Nicolas Chuquet dans le contexte français de l'arithmétique commerciale. *Rev. Hist. Math.*, 12:7–33, 2006. (Cited on page 44.)

[2846] W. Spottiswoode. *Elementary Theorems Relating to Determinants*. Longman, Brown, Green, and Longman, London, 1851. (Cited on page 113.)

[2847] W. Spottiswoode. Elementary theorems relating to determinants, second edition. *Journal für die reine und angewandte Mathematik (Crelle Journal)*, 51:209–271, 328–381, 1856. (Cited on page 381.)

[2848] G. Starke. Field-of-values analysis of preconditioned iterative methods for nonsymmetric elliptic problems. *Numer. Math.*, 78(1):103–117, 1997. (Cited on pages 236 and 287.)

[2849] G. Starke and R. S. Varga. A hybrid Arnoldi-Faber iterative method for nonsymmetric systems of linear equations. *Numer. Math.*, 64(1):213–240, 1993. (Cited on pages 185 and 229.)

[2850] A. Stathopoulos. Nearly optimal preconditioned methods for Hermitian eigenproblems under limited memory. Part I: Seeking one eigenvalue. *SIAM J. Sci. Comput.*, 29(2):481–514, 2007. (Cited on page 317.)

[2851] A. Stathopoulos and J. R. McCombs. Nearly optimal preconditioned methods for Hermitian eigenproblems under limited memory. Part II: Seeking many eigenvalues. *SIAM J. Sci. Comput.*, 29(5):2162–2188, 2007. (Cited on page 317.)

[2852] A. Stathopoulos and Y. Saad. Restarting techniques for the (Jacobi-) Davidson symmetric eigenvalue methods. *Electron. Trans. Numer. Anal.*, 7:163–181, 1998. (Cited on page 317.)

[2853] A. Stathopoulos, Y. Saad, and K. Wu. Dynamic thick restarting of the Davidson, and the implicitly restarted Arnoldi methods. *SIAM J. Sci. Comput.*, 19(1):227–245, 1998. (Cited on pages 306 and 317.)

[2854] J. Stedall. *A large discourse concerning algebra: John Wallis' 1685 Treatise of Algebra*. PhD thesis, The Open University, 2000. (Cited on page 50.)

[2855] J. M. Steele. *The Cauchy-Schwarz Master Class. An Introduction to the Art of Mathematical Inequalities*. Cambridge University Press, Cambridge, 2004. (Cited on page 556.)

[2856] L. A. Steen. Highlights in the history of spectral theory. *Am. Math. Mon.*, 80(4):359–381, 1973. (Cited on page 275.)

[2857] D. Stein. *Ada: A Life and a Legacy*. The MIT Press, Cambridge, Massachusetts, 1985. (Cited on page 533.)

[2858] P. Stein. The convergence of Seidel iterants of nearly symmetric matrices. *M.T.A.C.*, 5(36):237–240, 1951. (Cited on page 173.)

[2859] P. Stein. A note on bounds of multiple characteristic roots. *J. Res. Natl. Bur. Stand.*, 48:59, 1952. (Cited on page 173.)

[2860] P. Stein. A note on the bounds of the real parts of the characteristic roots of a matrix. *J. Res. Natl. Bur. Stand.*, 48:106–110, 1952. (Cited on page 173.)

[2861] P. Stein. Some general theorems on iterants. *J. Res. Natl. Bur. Stand.*, 48(1):82–83, 1952. (Cited on page 173.)

[2862] P. Stein and R. L. Rosenberg. On the solution of linear simultaneous equations by iteration. *J. Lond. Math. Soc.*, 1(2):111–118, 1948. (Cited on pages 173, 549, and 565.)

[2863] R. G. Steinke. An approximate inversion method for five point difference matrix equations. Technical Report LA-UR-79-162, Los Alamos Scientific Laboratory, 1979. (Cited on page 255.)

[2864] M. Stepànovà. From the algorithm Fang Cheng to the matrix theory. In *WDS'10 Proceedings of Contributed Papers, Part I*, pages 127–132, Matfyzpress, Prague, 2010. (Cited on page 36.)

[2865] M. Stepànovà. The change of paradigm in the matrix theory. In *WDS'11 Proceedings of Contributed Papers*, pages 85–90, 2011. (Cited on page 8.)

[2866] D. E. Stewart and T. S. Leyk. Error estimates for Krylov subspace approximations of matrix exponentials. *J. Comput. Appl. Math.*, 72(2):359–369, 1996. (Cited on page 390.)

[2867] G. W. Stewart. Modifying pivot elements in Gaussian elimination. *Math. Comput.*, 28(126):537–542, 1974. (Cited on page 78.)

[2868] G. W. Stewart. The convergence of the method of conjugate gradients at isolated extreme points of the spectrum. *Numer. Math.*, 24:85–93, 1975. (Cited on page 209.)

[2869] G. W. Stewart. Simultaneous iteration for computing invariant subspaces of non-Hermitian matrices. *Numer. Math.*, 25:123–136, 1976. (Cited on page 296.)

[2870] G. W. Stewart. On the perturbation of pseudo-inverses, projections and linear least squares problems. *SIAM Rev.*, 19(4):634–662, 1977. (Cited on page 145.)

[2871] G. W. Stewart. Research, development, and LINPACK. In J. R. Rice, editor, *Mathematical Software III*, pages 1–14, Academic Press, New York, 1977. (Cited on pages 80 and 145.)

[2872] G. W. Stewart. Computing the CS decomposition of a partitioned orthonormal matrix. *Numer. Math.*, 40(3):297–306, 1982. (Cited on page 145.)

[2873] G. W. Stewart. A method for computing the generalized singular value decomposition. In B. Kågström and A. Ruhe, editors, *Matrix Pencils*, pages 207–220, Springer, Berlin, Heidelberg, 1983. (Cited on page 141.)

[2874] G. W. Stewart. Rank degeneracy. *SIAM J. Sci. Statist. Comput.*, 5(2):403–413, 1984. (Cited on page 13.)

[2875] G. W. Stewart. A Jacobi-like algorithm for computing the Schur decomposition of a nonhermitian matrix. *SIAM J. Sci. Statist. Comput.*, 6(4):853–864, 1985. (Cited on page 151.)

[2876] G. W. Stewart. On the early history of the singular value decomposition. *SIAM Rev.*, 35(4):551–566, 1993. (Cited on pages 134 and 496.)

[2877] G. W. Stewart. Gauss, statistics, and Gaussian elimination. *J. Comput. Graph. Statist.*, 4(1):1–11, 1995. (Cited on page 58.)

[2878] G. W. Stewart. *Matrix Algorithms, Volume I: Basic Decompositions*. SIAM, Philadelphia, 1998. (Cited on page 121.)

[2879] G. W. Stewart. The decompositional approach to matrix computation. *Comput. Sci. Eng.*, 2(1):50–59, 2000. (Cited on page 121.)

[2880] G. W. Stewart. A generalization of Saad's theorem on Rayleigh-Ritz approximations. *Linear Algebra Appl.*, 327:115–119, 2001. (Cited on page 305.)

[2881] G. W. Stewart. *Matrix Algorithms, Volume II: Eigensystems*. SIAM, Philadelphia, 2001. (Cited on page 314.)

[2882] G. W. Stewart. A Krylov-Schur algorithm for large eigenproblems. *SIAM J. Matrix Anal. Appl.*, 23(3):601–614, 2001. (Cited on page 306.)

[2883] G. W. Stewart. The Gram-Schmidt algorithm and its variations. Technical Report TR-2004-84, Dept. of Mathematics, University of Maryland, 2004. (Cited on page 129.)

[2884] G. W. Stewart. Block Gram-Schmidt orthogonalization. *SIAM J. Sci. Comput.*, 31(1):761–775, 2008. (Cited on page 130.)

[2885] G. W. Stewart. On downdating QR factorizations. Technical Report TR-2010-01, Dept. of Mathematics, University of Maryland, 2010. (Cited on page 130.)

[2886] G. W. Stewart. Camille Jordan 1875, "Essay on the geometry of n dimensions". Technical report, University of Maryland, 2016. (Cited on page 144.)

[2887] G. W. Stewart. A canonical CS representation of a pair of subspaces. *SIAM J. Matrix Anal. Appl.*, 37(2):539–549, 2016. (Cited on page 146.)

[2888] G. W. Stewart and J.-G. Sun. *Matrix Perturbation Theory*. Academic Press, New York, 1990. (Cited on page 79.)

[2889] M. Stewart and P. Van Dooren. On the factorization of hyperbolic and unitary transformations into rotations. *SIAM J. Matrix Anal. Appl.*, 27(3):876–890, 2006. (Cited on page 145.)

[2890] L. Stickelberger. Über reelle orbhogonale substitutionen. *Programm der Polyt. Schule, Zurich*, 1877. (Cited on page 68.)

[2891] E. Stiefel. Über einige methoden der relaxationsrechnung. *Z. Angew. Math. Phys.*, 3(1):1–33, 1952. (Cited on pages 202, 207, and 567.)

[2892] E. Stiefel. Some special methods of relaxation technique. In L. J. Paige and O. Taussky, editors, *Simultaneoous Linear Equations and the Determination of Eigenvalues*, pages 43–48. National Bureau of Standards, Applied Math. Ser. 29, 1953. (Cited on page 207.)

[2893] E. Stiefel. Zur interpolation von tabellierten funktionen durch exponentialsummen und zur berechnung von eigenwerten aus den schwarzschen konstanten. *Zamm.-Z Angew. Math. Me.*, 33(8-9):260–262, 1953. (Cited on page 309.)

[2894] E. Stiefel. Relaxationsmethoden bester strategie zur lösung linearer gleichungssysteme. *Commentarii Mathematici Helvetici*, 29:157–179, 1955. (Cited on pages 207, 216, and 220.)

[2895] E. Stiefel. On solving Fredholm integral equations. Applications to conformal mapping and variational problems of potential theory. *J. Soc. Ind. Appl. Math.*, 4(2):63–85, 1956. (Cited on page 180.)

[2896] E. Stiefel. Recent developments in relaxation techniques. In *Proc. International Congress of Mathematicians 1954, Vol.1*, pages 384–391. North-Holland, Amsterdam, 1957. (Cited on page 207.)

[2897] E. Stiefel. Kernel polynomial in linear algebra and their numerical applications. In *Further Contributions to the Solution of Simultaneoous Linear Equations and the Determination of Eigenvalues*, pages 1–22. National Bureau of Standards, Applied Math. Ser. 49, 1958. (Cited on pages 181 and 207.)

[2898] T. J. Stieltjes. Quelques recherches sur la théorie des quadratures dites mécaniques. *Ann. Sci. Ecole Norm. Sup. Paris*, Sér. 3 tome 1:409–426, 1884. Also in Ouvres complètes, vol. 1, Noordhoff, Groningen, 1914, pp. 377-396. (Cited on pages 205 and 297.)

[2899] T. J. Stieltjes. Sur les racines de l'equation $X_n = 0$. *Acta Math.*, 9:385–400, 1887. (Cited on page 177.)

[2900] S. M. Stigler. *American Contributions to Mathematical Statistics in the Nineteenth Century*. Arno Press, New York, 1980. (Cited on page 708.)

[2901] S. M. Stigler. Gauss and the invention of least squares. *The Annals of Statistics*, 9:465–474, 1981. (Cited on page 58.)

[2902] J. Stoer. On the characterization of least upper bound norms in matrix space. *Numer. Math.*, 6:302–314, 1964. (Cited on page 16.)

[2903] J. Stoer. *Einführung in die Numerische Mathematik I*. Springer Verlag, Berlin, 1972. (Cited on page 79.)

[2904] J. Stoer. Solution of large linear systems of equations by conjugate gradient type methods. In A. Bachem, M. Götschel, and B. Korte, editors, *Mathematical Programming, The State of the Art*, pages 540–565. Springer, Berlin, Heidelberg, 1983. (Cited on page 208.)

[2905] J. Stoer. The Gatlinburg Symposia and their influence on the development of numerical linear algebra. *Image*, 46:13–25, 2011. (Cited on page 490.)

[2906] J. Stoer and R. Bulirsch. *Introduction to Numerical Analysis*. Springer Verlag, New York, 1980. (Cited on page 79.)

[2907] J. Stoer and R. Freund. On the solution of large indefinite systems of linear equations by conjugate gradient algorithms. In R. Glowinski and J. L. Lions, editors, *Computing Methods in Applied Sciences and Engineering V*. North-Holland, 1982. (Cited on page 208.)

[2908] M. Stoll. A Krylov-Schur approach to the truncated SVD. *Linear Algebra Appl.*, 436(8):2795–2806, 2012. (Cited on page 140.)

[2909] M. Stoll. A literature survey of matrix functions for data science. *GAMM Mitteilungen*, 43(3):article e202000013, 2020. (Cited on page 390.)

[2910] B. J. Stone. Best possible ratios of certain matrix norms. *Numer. Math.*, 4(1):114–116, 1962. (Cited on page 16.)

[2911] H. L. Stone. Iteraive solution of implicit approximations of multidimensional partial differential equations. *SIAM J. Numer. Anal.*, 5(3):530–558, 1968. (Cited on page 254.)

[2912] H. S. Stone. An efficient parallel algorithm for the solution of a tridiagonal linear system of equations. *J. ACM*, 20(1):27–38, 1973. (Cited on page 255.)

[2913] M. H. Stone. *Linear Transformations in Hilbert Space and their Applications to Analysis*. American Mathematical Soc., New York, 1932. (Cited on page 382.)

[2914] A. Storjohann. An $o(n^3)$ algorithm for the Frobenius normal form. In *Proceedings of the 1998 International Symposium on Symbolic and Algebraic Computation*, pages 101–105. ACM, 1998. (Cited on page 150.)

[2915] A. Storjohann. *Algorithms for matrix canonical forms*. PhD thesis, Department of Computer Science, Swiss Federal Institute of Technology (ETH), 2000. (Cited on page 150.)

[2916] N. Stoyko. Le développement de l'utilisation de Cracoviens en France. *Acta Astronomica*, 9:203–205, 1953. (Cited on page 412.)

[2917] P. D. Straffin. Liu Hui and the first golden age of Chinese mathematics. *Math. Mag.*, 71(3):163–181, 1998. (Cited on page 35.)

[2918] Z. Strakoš. Performance of the E 2345 array processor. *Computers and Artificial Intelligence*, 4(3):273–284, 1985. (Cited on page 214.)

[2919] Z. Strakoš. Effectivity and optimizing of algorithms and programs on the host-computer/array-processor system. *Parallel Comput.*, 4(2):189–207, 1987. (Cited on page 214.)

[2920] Z. Strakoš. On the real convergence rate of the conjugate gradient method. *Linear Algebra Appl.*, 154-156:535–549, 1991. (Cited on page 211.)

[2921] Z. Strakoš. Model reduction using the Vorobyev moment problem. *Numer. Algorithms*, 51:363–379, 2009. (Cited on page 249.)

[2922] Z. Strakoš and A. Greenbaum. Open questions in the convergence analysis of the Lanczos process for the real symmetric eigenvalue problem. IMA preprint 934, 1992. (Cited on page 211.)

[2923] Z. Strakoš and P. Tichý. On error estimation in the conjugate gradient method and why it works in finite precision computations. *Electron. Trans. Numer. Anal.*, 13:56–80, 2002. (Cited on page 212.)

[2924] Z. Strakoš and P. Tichý. Error estimation in preconditioned conjugate gradients. *BIT Numerical Mathematics*, 45:789–817, 2005. (Cited on page 212.)

[2925] Z. Strakoš and P. Tichý. On efficient numerical approximation of the bilinear form $c^* A^{-1} b$. *SIAM J. Sci. Comput.*, 33(2):565–587, 2011. (Cited on pages 221 and 390.)

[2926] G. Strang and S. MacNamara. Functions of difference matrices are Toeplitz plus Hankel. *SIAM Rev.*, 56(3):525–546, 2014. (Cited on page 387.)

[2927] V. Strassen. Gaussian elimination is not optimal. *Numer. Math.*, 13(4):354–356, 1969. (Cited on page 80.)

[2928] J. C. Strikwerda. A probabilistic analysis of asynchronous iteration. *Linear Algebra Appl.*, 349(1-3):125–154, 2002. (Cited on page 196.)

[2929] T. Strohmer and R. Vershynin. A randomized Kaczmarz algorithm with exponential convergence. *J. Fourier Anal. Appl.*, 15:262–278, 2009. (Cited on page 191.)

[2930] D. J. Struik. *A Concise History of Mathematics*. G. Bell and Sons, London, 1954. (Cited on pages xvi and 527.)

[2931] J. W. Strutt. *The Theory of Sound*. MacMillan and Co., London, 1877. Second edition 1894, J.W. Strutt is also known as Lord Rayleigh. (Cited on page 282.)

[2932] J. W. Strutt. LXII. Hamilton's principle and the five aberrations of von Seidel. *The London, Edinburgh, and Dublin Philosophical Magazine and Journal of Science*, 15(90):677–687, 1908. (Cited on page 559.)

[2933] J. W. Strutt. On the calculation of Chladni's figures for a square plate. *Philos. Mag.*, 22:225–229, 1911. (Cited on page 285.)

[2934] K. Stüben. Algebraic multigrid (AMG): experiences and comparisons. *Appl. Math. Comput.*, 13:419–452, 1983. (Cited on page 270.)

[2935] K. Stüben and U. Trottenberg. Multigrid methods: Fundamental algorithms, model problem analysis and applications. In Springer, editor, *Multigrid Methods, Proceedings of the Koln-Porz Conference*, pages 1–176, 1982. Lecture Notes in Mathematics 960. (Cited on page 269.)

[2936] J. C. F. Sturm. Analyse d'un mémoire sur la résolution des équations numériques. *Bulletin des Sciences Mathématiques, Physiques et Chimiques*, 12:419–422, 1829. See also "Extrait d'un mémoire de M. Sturm", same volume, pp. 422-425. (Cited on page 281.)

[2937] J. C. F. Sturm. Compte-rendu des 40ème, 41ème et 42ème livraisons des Exercices de Mathématiques de A.-L. Cauchy. *Bulletin des Sciences Mathématiques, Physiques et Chimiques*, 12:301–303, 1829. (Cited on page 281.)

[2938] J. C. F. Sturm. Extrait d'un mémoire sur l'intégration d'un système d'équations différentielles linéaires. *Bulletin des Sciences Mathématiques, Physiques et Chimiques*, 12:313–322, 1829. (Cited on page 281.)

[2939] Y. Su, A. Bhaya, E. Kaszkurewicz, and V. S. Kozyakin. Further results on convergence of asynchronous linear iterations. *Linear Algebra Appl.*, 281(1-3):11–24, 1998. (Cited on page 196.)

[2940] D. Sukkari, H. Ltaief, A. Esposito, and D. Keyes. A QDWH-based SVD software framework on distributed-memory manycore systems. *ACM Trans. Math. Soft. (TOMS)*, 45(2):1–21, 2019. (Cited on page 139.)

[2941] J.-G. Sun. Rounding-error and perturbation bounds for the Cholesky and LDL^T factorizations. *Linear Algebra Appl.*, 173:77–97, 1992. (Cited on page 80.)

[2942] B. D. Sutton. Computing the complete CS decomposition. *Numer. Algorithms*, 50(1):33–65, 2009. (Cited on pages 145 and 146.)

[2943] B. D. Sutton. Stable computation of the CS decomposition: Simultaneous bidiagonalization. *SIAM J. Matrix Anal. Appl.*, 33(1):1–21, 2012. (Cited on page 146.)

[2944] B. D. Sutton. Divide and conquer the CS decomposition. *SIAM J. Matrix Anal. Appl.*, 34(2):417–444, 2013. (Cited on page 146.)

[2945] C. D. Swanson and A. T. Chronopoulos. Orthogonal s-step methods for nonsymmetric linear systems of equations. In *Proceedings of the 6th International Conference on Supercomputing*, pages 456–462. ACM, 1992. (Cited on page 227.)

[2946] P. N. Swarztrauber. A direct method for the discrete solution of separable elliptic equations. *SIAM J. Numer. Anal.*, 11(6):1136–1150, 1974. (Cited on page 385.)

[2947] P. N. Swarztrauber. The methods of cyclic reduction, Fourier analysis and the FACR algorithm for the discrete solution of Poisson's equation on a rectangle. *SIAM Rev.*, 19(3):490–501, 1977. (Cited on page 385.)

[2948] P. N. Swarztrauber and R. A. Sweet. The direct solution of the discrete Poisson equation on a disk. *SIAM J. Numer. Anal.*, 10(5):900–907, 1973. (Cited on page 385.)

[2949] P. N. Swarztrauber and R. A. Sweet. Algorithm 541, efficient Fortran subprograms for the solution of separable elliptic partial differential equations. *ACM Trans. Math. Soft. (TOMS)*, 5(3):352–364, 1979. (Cited on page 385.)

[2950] D. R. Sweet. *Numerical methods for Toeplitz matrices.* PhD thesis, Department of Computing Science, University of Adelaide, Australia, 1982. (Cited on pages 386 and 387.)

[2951] R. A. Sweet. A generalized cyclic reduction algorithm. *SIAM J. Numer. Anal.*, 11(3):506–520, 1974. (Cited on page 385.)

[2952] R. A. Sweet. A cyclic reduction algorithm for solving block tridiagonal systems of arbitrary dimension. *SIAM J. Numer. Anal.*, 14(4):706–720, 1977. (Cited on page 385.)

[2953] F. J. Swetz and Ang Tian Se. A brief chronological and bibliographic guide to the history of Chinese mathematics. *Historia Math.*, 11(1):39–56, 1984. (Cited on page 39.)

[2954] J. J. Sylvester. On derivation of coexistence: Part II. Being the theory of simultaneous simple homogeneous equations. *Philos. Mag.*, xvi:37–43, 1840. (Cited on page 111.)

[2955] J. J. Sylvester. Additions to the articles, "on a new class of theorems", and "on Pascal's theorem". *Philos. Mag.*, xxxvil:363–370, 1850. CMP vol. I p. 145. (Cited on pages 4, 113, and 571.)

[2956] J. J. Sylvester. An enumeration of the contacts of lines and surfaces of the second order. *Philos. Mag.*, 1:119–140, 1851. CMP vol. I p. 219. (Cited on pages 13 and 572.)

[2957] J. J. Sylvester. An essay on canonical forms, supplement to a sketch of a memoir on elimination, transformation and canonical forms. *George Bell, Fleet Street*, 1851. CMP vol. I p. 203. (Cited on page 572.)

[2958] J. J. Sylvester. On a certain fundamental theorem of determinants. *Philos. Mag.*, 2:142–145, 1851. CMP vol. I p. 252-255. (Cited on page 113.)

[2959] J. J. Sylvester. On the relation between the minor determinants of linearly equivalent quadratic functions. *Philos. Mag.*, 1:295–305, 1851. CMP vol. I p. 241. (Cited on pages 3, 4, 25, 572, and 573.)

[2960] J. J. Sylvester. A demonstration of the theorem that every homogeneous quadratic polynomial is reducible by real orthogonal substitutions to the form of a sum of positive and negative squares. *Philos. Mag.*, 4:138–142, 1852. CMP vol. I p. 378. (Cited on pages 123, 290, 572, and 573.)

[2961] J. J. Sylvester. On a fundamental rule in the algorithm of continued fractions. *Philos. Mag.*, 6:297–299, 1853. CMP vol I pp. 641-644. (Cited on page 381.)

[2962] J. J. Sylvester. On a remarkable modification of Sturm's theorem. *Philos. Mag.*, 4:446–456, 1853. CMP vol I pp. 609-619. (Cited on page 381.)

[2963] J. J. Sylvester. On a theory of the syzygetic relations of two rational integral functions, comprising an application to the theory of Sturm's functions, and that of the greatest algebraical common measure. *Philos. Trans. R. Soc. Lond.*, CXLIII:407–548, 1853. CMP vol I p. 427. (Cited on page 572.)

[2964] J. J. Sylvester. Théorème sur les déterminants. *Nouvelles Annales de Mathématiques*, 13:305, 1854. CMP vol II pp. 28. (Cited on page 381.)

[2965] J. J. Sylvester. *The Laws of Verse: Or principles of versification exemplified in metrical translations, together with an annotated reprint of the inaugural presidential address to the mathematical and physical section of the British Association at Exeter.* Longmans, Green, 1870. (Cited on page 570.)

[2966] J. J. Sylvester. Lectures on the principles of universal algebra. *Amer. J. Math.*, 6(1):270–286, 1883. CMP vol. IV pp. 208-224. (Cited on pages 288, 290, 572, and 573.)

[2967] J. J. Sylvester. On the equation to the secular inequalities in the planetary theory. *Philos. Mag.*, xvi:267–269, 1883. CMP vol. IV pp. 110-111. (Cited on pages 288, 290, 387, and 573.)

[2968] J. J. Sylvester. Equations in matrices. *Johns Hopkins University Circulars*, III:122, 1884. (Cited on page 573.)

[2969] J. J. Sylvester. The genesis of an idea, or story of a discovery relating to equations in multiple quantity. *Nature*, 31:35–36, 1884. (Cited on page 7.)

[2970] J. J. Sylvester. On involuants and other allied species of invariant to matrix systems. *Johns Hopkins University Circulars*, III:9–12, 1884. (Cited on page 13.)

[2971] J. J. Sylvester. Sur la résolution générale de l'équation linéaire en matrices d'un ordre quelconque. *C.R. Acad. Sci. Paris, Sér 1*, XCIX:409–412, 1884. CMP vol. IV pp. 199-205. (Cited on pages 395 and 574.)

[2972] J. J. Sylvester. Sur la solution du cas le plus général des équations linéaires en quantités binaires, c'est-à-dire en quaternions ou en matrices du second ordre. *C.R. Acad. Sci. Paris, Sér 1*, XCIX:117–118, 1884. CMP vol. IV pp. 181-182. (Cited on page 395.)

[2973] J. J. Sylvester. Sur l'équation en matrices $px = xq$. *C.R. Acad. Sci. Paris, Sér 1*, XCIX:67–71, 1884. CMP vol. IV pp. 176-180. (Cited on pages 395 and 573.)

[2974] J. J. Sylvester. Sur l'équation linéaire trinôme en matrices d'un ordre quelconque. *C.R. Acad. Sci. Paris, Sér 1*, XCIX:527–529, 1884. CMP vol. IV pp. 206-207. (Cited on page 395.)

[2975] J. J. Sylvester. Sur la réduction biorthogonale d'une forme linéo-linéaire à sa forme canonique. *C.R. Acad. Sci. Paris, Sér 1*, 108:651–653, 1889. (Cited on page 135.)

[2976] J. J. Sylvester. A new proof that a general quadric may be reduced to its canonical form (that is, a linear function of squares) by means of a real orthogonal substitution. *Messenger Math.*, 19:1–5, 1889-1890. (Cited on pages 135 and 136.)

[2977] J. J. Sylvester. On the reduction of a bilinear quantic of the nth order to the form of a sum of n products by a double orthogonal substitution,. *Messenger Math.*, 19:42–46, 1889-1890. (Cited on page 135.)

[2978] O. Szász. Az Hadamard-féle determinanstétel egy elemi bebizonyitása. *Mathematikai és Physikai Lapok*, 19:221–227, 1910. (Cited on pages 126 and 218.)

[2979] O. Szász. Ein elementarer beweis des Hadamardschen determinantensatzes. *Monatsh Math. Phys.*, 28(1):253–257, 1917. (Cited on page 126.)

[2980] O. Szász. *Collected Mathematical Papers*. Stechert-Hafner, University of Cincinnati, 1955. Edited by H.D. Lipsich. (Cited on page 126.)

[2981] G. Szegö. Beitrag zur theorie der thetafunktionen. *Sitz Preuss. Akad. Wiss. Phys. Math. Ki.*, XIX:242–252, 1926. (Cited on page 386.)

[2982] G. Szegö. *Orthogonal Polynomials*. American Mathematical Society, Providence, RI, 1939. (Cited on page 180.)

[2983] R. Sznajder. Kaczmarz algorithm revisited. *Czasopismo Techniczne 2015, Nauki Podstawowe*, 2-NP/2015:247–254, 2015. (Cited on page 189.)

[2984] D. B. Szyld. Different models of parallel asynchronous iterations with overlapping blocks. *Comput. Appl. Math.*, 17:101–115, 1998. (Cited on page 196.)

[2985] D. B. Szyld, E. Vecharynski, and F. Xue. Preconditioned eigensolvers for large-scale nonlinear Hermitian eigenproblems with variational characterizations. II. Interior eigenvalues. *SIAM J. Sci. Comput.*, 37(6):A2969–A2997, 2015. (Cited on page 319.)

[2986] D. B. Szyld and J. A. Vogel. FQMR: A flexible quasi-minimal residual method with inexact preconditioning. *SIAM J. Sci. Comput.*, 23(2):363–380, 2001. (Cited on page 239.)

[2987] D. B. Szyld and F. Xue. Preconditioned eigensolvers for large-scale nonlinear Hermitian eigenproblems with variational characterizations. I. Extreme eigenvalues. *Math. Comput.*, 85(302):2887–2918, 2016. (Cited on page 319.)

[2988] H. Taber. On the theory of matrices. *Amer. J. Math.*, 12(4):337–396, 1890. (Cited on pages 8 and 387.)

[2989] H. Taber. On the matrix equation $\varphi\Omega = \Omega\varphi$. *Proc. Amer. Acad. Arts. Sci.*, 26:64–66, 1890-1891. (Cited on page 395.)

[2990] H. Taber. On certain identities in the theory of matrices. *Amer. J. Math.*, 13(2):159–172, 1891. (Cited on page 387.)

[2991] S. Takenaka. On linear integral equations in infinite number of unknown functions. *Tohoku Mathematical Journal*, First Series 23:167–196, 1924. (Cited on page 126.)

[2992] H. Takeno. A theorem concerning the characteristic equation of the matrix of a tensor of the second order. *Tensor NS*, 3:119–122, 1954. (Cited on page 289.)

[2993] K. Tanabe. Projection method for solving a singular system of linear equations and its applications. *Numer. Math.*, 17:203–214, 1971. (Cited on pages 190 and 498.)

[2994] P. T. P. Tang and E. Polizzi. FEAST as a subspace iteration eigensolver accelerated by approximate spectral projection. *SIAM J. Matrix Anal. Appl.*, 35(2):354–390, 2014. (Cited on page 318.)

[2995] W. P. Tang. Generalized Schwarz splittings. *SIAM J. Sci. Statist. Comput.*, 13(2):573–595, 1992. (Cited on page 268.)

[2996] M. Tanio and M. Sugihara. GBi-CGSTAB(s,L): IDR(s) with higher-order stabilization polynomials. *J. Comput. Appl. Math.*, 235(3):765–784, 2010. (Cited on pages 240 and 243.)

[2997] P. Tannery. A quelle époque vivait Diophante? *Bulletin des Sciences Mathématiques et Astronomiques, 2ème série*, 3(1):261–269, 1879. (Cited on pages 39 and 41.)

[2998] P. Tannery. La perte de sept livres de Diophante. *Bulletin des Sciences Mathématiques et Astronomiques, 2ème série*, 8(1):192–206, 1884. (Cited on page 41.)

[2999] P. Tannery. Sur une épigramme attribuée à Diophante. *Revue des Études Grecques*, 4(16):377–382, 1891. (Cited on page 41.)

[3000] R. A. Tapia, J. E. Dennis, Jr., and J. P. Schäfermeyer. Inverse, shifted inverse, and Rayleigh quotient iteration as Newton's method. *SIAM Rev.*, 60(1):3–55, 2018. (Cited on page 295.)

[3001] R. E. Tarjan. Graph theory and Gaussian elimination. In J. R. Bunch and D. J. Rose, editors, *Sparse Matrix Computations*, pages 3–22. Academic Press, 1976. (Cited on page 92.)

[3002] R. Taton. Laplace et Sylvestre-François Lacroix. *Revue d'histoire des sciences et de leurs applications*, 6(4):350–360, 1953. (Cited on page 521.)

[3003] O. Taussky. Bounds for characteristic roots of matrices. *Duke Math. J.*, 15(4):1043–1044, 1948. (Cited on page 575.)

[3004] O. Taussky. A recurring theorem on determinants. *Am. Math. Mon.*, 56(10):672–676, 1949. (Cited on page 575.)

[3005] O. Taussky. Note on the condition of matrices. *Math. Comput.*, 4(30):111–112, 1950. (Cited on page 21.)

[3006] O. Taussky. *Contributions to the Solution of Systems of Linear Equations and the Determination of Eigenvalues, NBS 39*. National Bureau of Standards, USA, 1954. (Cited on pages 200 and 576.)

[3007] O. Taussky. Some topics concerning bounds for eigenvalues of finite matrices. In J. Todd, editor, *Survey of Numerical Analysis*, pages 279–297. McGraw-Hill, 1962. (Cited on page 286.)

[3008] O. Taussky. Positive-definite matrices and their role in the study of the characteristic roots of general matrices. *Adv. Math.*, 2(2):175–186, 1968. (Cited on page 576.)

[3009] O. Taussky. The role of symmetric matrices in the study of general matrices. *Linear Algebra Appl.*, 5(2):147–154, 1972. (Cited on page 576.)

[3010] O. Taussky. How I became a torchbearer for matrix theory. *Am. Math. Mon.*, 95(9):801–812, 1988. (Cited on page 576.)

[3011] O. Taussky and J. Todd. Cholesky, Toeplitz and the triangular factorization of symmetric matrices. *Numer. Algorithms*, 41:197–202, 2006. (Cited on pages 68 and 580.)

[3012] O. Taussky and H. Zassenhaus. On the similarity transformation between a matrix and its transpose. *Pacific J. Math.*, 9(3):893–896, 1959. (Cited on page 576.)

[3013] O. Taussky-Todd. Remembrances of Kurt Gödel. *Engineering and Science*, 51(2):24–28, 1987. (Cited on page 575.)

[3014] D. R. Taylor. *Analysis of the look ahead Lanczos algorithm*. PhD thesis, University of California, Berkeley, 1982. (Cited on page 225.)

[3015] G. J. Tee. Semyon Aronovich Gerschgorin. *Image*, 32:2–3, and 5, 2004. (Cited on page 474.)

[3016] D. Teets and K.Whitehead. The discovery of Ceres: How Gauss became famous. *Math. Mag.*, 72(2):83–93, 1999. (Cited on page 60.)

[3017] C. Temperton. Direct methods for the solution of the discrete Poisson equation: Some comparisons. *J. Comp. Phys.*, 31(1):1–20, 1979. (Cited on page 385.)

[3018] C. Temperton. On the FACR(l) algorithm for the discrete Poisson equation. *J. Comp. Phys.*, 34(3):314–329, 1980. (Cited on page 385.)

[3019] C. Temperton. A note on prime factor FFT algorithms. *J. Comp. Phys.*, 52:198–204, 1983. (Cited on page 384.)

[3020] C. Temperton. Implementation of a self sorting in place prime factor FFT algorithm. *J. Comp. Phys.*, 58:283–299, 1985. (Cited on page 384.)

[3021] C. Temperton. A generalized prime factor FFT algorithm for any $N = 2^p 3^q 5^r$. *SIAM J. Sci. Comput.*, 13(3):676–686, 1992. (Cited on page 384.)

[3022] G. Temple. The general theory of relaxation methods applied to linear systems. *Proc. Roy. Soc. A*, 169:476–500, 1939. (Cited on page 169.)

[3023] G. F. J. Temple. The accuracy of Rayleigh's method of calculating the natural frequencies of vibrating systems. *Proc. Roy. Soc. A*, 211:204–224, 1952. (Cited on page 282.)

[3024] G. F. J. Temple and W. G. Bickley. *Rayleigh's Principle and Its Applications to Engineering: The Theory and Practice of the Energy Method for the Approximate Determination of Critical Loads and Speeds*. Dover, New York, 1956. First edition, Oxford University Press, 1933. (Cited on pages 282 and 285.)

[3025] M. Terquem. Notice sur l'élimination, formules de Cramer. *Nouvelles Annales de Mathématiques : journal des candidats aux Écoles Polytechnique et Normale*, 1e série, 5:153–162, 1846. (Cited on page 56.)

[3026] R. P. Tewarson. On the product form of inverses of sparse matrices. *SIAM Rev.*, 8(3):336–342, 1966. (Cited on page 92.)

[3027] R. P. Tewarson. The product form of inverses of sparse matrices and graph theory. *SIAM Rev.*, 9(1):91–99, 1967. (Cited on page 92.)

[3028] R. P. Tewarson. Row-column permutation of sparse matrices. *Comput. J.*, 10(3):300–305, 1967. (Cited on page 92.)

[3029] R. P. Tewarson. The Crout reduction for sparse matrices. *Comput. J.*, 12(2):158–159, 1969. (Cited on page 92.)

[3030] R. P. Tewarson. Projection methods for solving sparse linear systems. *Comput. J.*, 12:77–80, 1969. (Cited on pages 192 and 193.)

[3031] R. P. Tewarson. Computations with sparse matrices. *SIAM Rev.*, 12(4):527–543, 1970. (Cited on pages 92 and 94.)

[3032] D. J. Thomas and J. M. Smith. Joseph Raphson, F.R.S. *Notes and Records Roy. Soc. London*, 44(2):151–167, 1990. (Cited on page 52.)

[3033] L. H. Thomas. Elliptic problems in linear difference equations over a network. Watson Sc. Comp. Lab. Rep., Columbia University, New York, 1949. (Cited on page 382.)

[3034] M. Thomas. *La règle à calcul, instrument de l'ère industrielle : le rôle de la France*. PhD thesis, Université de Nantes, 2014. (Cited on page 325.)

[3035] W. Thomson. On a machine for the solution of simultaneous linear equations. *Proc. R. Soc. Lond.*, 28(190-195):111–113, 1879. W. Thomson is also known as Lord Kelvin. (Cited on page 331.)

[3036] W. Thomson and P. G. Tait. *Treatise on Natural Philosophy, Part I.* Cambridge University press, 1888. (Cited on page 331.)

[3037] P. Tichý, J. Liesen, and V. Faber. On worst-case GMRES, ideal GMRES, and the polynomial numerical hull of a Jordan block. *Electron. Trans. Numer. Anal.*, 26:453–473, 2007. (Cited on page 240.)

[3038] P. Tichý and J. Zítko. Derivation of BiCG from the conditions defining Lanczos method for solving a system of linear equations. *Appl. Math.*, 43(5):381–388, 1998. (Cited on page 221.)

[3039] A. N. Tikhonov. On the stability of inverse problems. *Doklady Akademii Nauk SSSR*, 39(5):195–198, 1943. (Cited on page 391.)

[3040] A. N. Tikhonov. Solution of incorrectly formulated problems and the regularization method. *Soviet Math. Dokl.*, 4:1035–1038, 1963. (Cited on page 391.)

[3041] A. N. Tikhonov. *Ill-posed Problems in Natural Sciences, Proceedings of the International Conference Moscow 1991*. TVP Science, Moscow, 1992. Editor. (Cited on page 391.)

[3042] A. N. Tikhonov and V. Y. Arsenin. *Solutions of Ill-posed Problems*. Mir, Moscow, 1976. Translation to English, Wiley, 1977. (Cited on page 391.)

[3043] A. N. Tikhonov and A. Goncharsky. *Ill-posed Problems in the Natural Sciences*. Mir, Moscow, 1987. (Cited on page 392.)

[3044] A. N. Tikhonov, A. V. Goncharsky, V. V. Stepanov, and A. G. Yagola. *Numerical Methods for the Solution of Ill-posed Problems*. Nauka, Moscow, 1990. Translation into English, Springer Science, 1995. (Cited on page 392.)

[3045] W. F. Tinney and J. W. Walker. Direct solutions of sparse network equations by optimally ordered triangular factorization. *Proc. IEEE*, 55(11):1801–1809, 1967. (Cited on page 87.)

[3046] M. Tismenetsky. A new preconditioning technique for solving large sparse linear systems. *Linear Algebra Appl.*, 154:331–353, 1991. (Cited on page 257.)

[3047] D. Titley-Peloquin, J. Pestana, and A. J. Wathen. GMRES convergence bounds that depend on the right-hand-side vector. *IMA J. Numer. Anal.*, 34(2):462–479, 2014. (Cited on page 245.)

[3048] J. Todd. The condition of certain matrices. I. *Quart. J. Mech. Appl. Math.*, 2(4):469–472, 1949. (Cited on pages 21 and 581.)

[3049] J. Todd. The condition of a certain matrix. *Math. Proc. Cambridge Phil. Soc.*, 46(1):116–118, 1950. (Cited on pages 21 and 581.)

[3050] J. Todd. Motivation for working in numerical analysis. *Comm. Pure Appl. Math.*, 8:97–116, 1955. (Cited on page 581.)

[3051] J. Todd. Numerical analysis at the National Bureau of Standards. *SIAM Rev.*, 17(2):361–370, 1975. (Cited on page 201.)

[3052] J. Todd. *Basic Numerical Mathematics: Vol. 1: Numerical Analysis*. Birkhäuser Verlag, Basel, Stuttgart, 1979. (Cited on page 581.)

[3053] J. Todd. *Basic Numerical Mathematics: Vol. 2: Numerical Algebra*. Birkhäuser Verlag, Basel, Stuttgart, 1979. (Cited on page 581.)

[3054] J. Todd. Oberwolfach - 1945. In E. F. Beckenbach and W. Walter, editors, *General Inequalities 3, 3rd International Conference on General Inequalities, Oberwolfach, April 26- May 2, 1981*, pages 19–22, Basel, 1983. Springer. (Cited on page 579.)

[3055] J. Todd. The prehistory and early history of computation at the U.S. National Bureau of Standards. In S. G. Nash, editor, *A History of Scientific Computing*, pages 251–268, ACM Press, New York, 1990. (Cited on pages 201, 580, and 581.)

[3056] O. Toeplitz. Die Jacobische transformation der quadratischen formen von unendlich vielen veränderlichen. *Nach. Akad. Wiss. Göttingen, Math. Phys. Kl., II*, pages 101–110, 1907. (Cited on pages 68, 381, 581, and 583.)

[3057] O. Toeplitz. Das algebraische analogon zu einem satze von Fejér. *Math. Z.*, 2:187–197, 1918. (Cited on page 287.)

[3058] K.-C. Toh. GMRES vs. ideal GMRES. *SIAM J. Matrix Anal. Appl.*, 18(1):30–36, 1997. (Cited on page 234.)

[3059] K.-C. Toh and L. N. Trefethen. Pseudozeros of polynomials and pseudospectra of companion matrices. *Numer. Math.*, 68(3):403–425, 1994. (Cited on page 293.)

[3060] K.-C. Toh and L. N. Trefethen. The Chebyshev polynomials of a matrix. *SIAM J. Matrix Anal. Appl.*, 20(2):400–419, 1998. (Cited on page 234.)

[3061] C. B. Tompkins. Projection methods in calculation of some linear problems. multilithed typescript, 52 pp., part of Engineering Research Associates, Logistics papers, issue no. IV, Appendix I to Bimonthly Progress Report No. 17, Contract N6onr-240, Task Order I, Office of Naval Research, Project NR 047 010, 1949. (Cited on page 189.)

[3062] C. H. Tong. A family of quasi-minimal residual methods for nonsymmetric linear systems. *SIAM J. Sci. Comput.*, 15(1):89–105, 1994. (Cited on page 232.)

[3063] C. H. Tong and Q. Ye. Analysis of the finite precision bi-conjugate gradient algorithm for nonsymmetric linear systems. *Math. Comput.*, 69(232):1559–1575, 2000. (Cited on page 238.)

[3064] L. Torres y Quevedo. Essais sur l'automatique; sa définition, étendue théorique de ses applications. *Revue Générale des Sciences Pures et Appliquées*, 21:601–611, 1915. (Cited on page 331.)

[3065] A. Toselli and O. Widlund. *Domain Decompposition Methods - Algorithms and Theory*. Springer Series in Computational Mathematics, 2004. Vol. 34. (Cited on page 268.)

[3066] D. S. Tracy. In Memoriam: Paul Sumner Dwyer 1901-1982. *Amer. Statist.*, 37:303–304, 1983. (Cited on page 74.)

[3067] L. N. Trefethen. Approximation theory and numerical linear algebra. In J. C. Mason and M. G. Cox, editors, *Algorithms for Approximation II*, Chapman and Hall, London, 1990. (Cited on page 226.)

[3068] L. N. Trefethen. Computation of pseudospectra. *Acta Numerica*, pages 247–295, 1999. (Cited on page 226.)

[3069] L. N. Trefethen. Gene H. Golub (1932-2007): Mathematician and godfather of numerical analysis. *Nature*, 450(7172):962, 2007. (Cited on page 478.)

[3070] L. N. Trefethen and M. Embree. *Spectra and Pseudospectra: The Behavior of Nonnormal Matrices and Operators*. Princeton University Press, 2005. (Cited on page 226.)

[3071] W. F. Trench. An algorithm for the inversion of finite Toeplitz matrices. *J. Soc. Ind. Appl. Math.*, 12(3):515–522, 1964. (Cited on page 386.)

[3072] W. F. Trench. An algorithm for the inversion of finite Hankel matrices. *J. Soc. Ind. Appl. Math.*, 13(4):1102–1107, 1965. (Cited on page 386.)

[3073] U. Trottenberg, C. W. Oosterlee, and A. Schuller. *Multigrid*. Elsevier, 2000. (Cited on page 270.)

[3074] C. A. Truesdell. Cauchy and the modern mechanics of continua. *Rev. Hist. Sci.*, 45(1):5–24, 1992. (Cited on page xv.)

[3075] J. N. Tsitsiklis. A comparison of Jacobi and Gauss-Seidel parallel iterations. *Appl. Math. Letters*, 2(2):167–170, 1989. (Cited on page 196.)

[3076] N. Tsuno and T. Nodera. The speedup of the GMRES(m) method using the early restarting procedure. *Trans. Inform. Process. Soc. Japan*, 40:1760–1773, 1999. (Cited on page 232.)

[3077] L. B. Tuckerman. On the mathematically significant figures in the solution of simultaneous linear equations. *Ann. Math. Stat.*, 12(3):307–316, 1941. (Cited on page 19.)

[3078] A. D. Tuff and A. Jennings. An iterative method for large systems of linear structural equations. *Internat. J. Numer. Methods Engrg.*, 7(2):175–183, 1973. (Cited on page 254.)

[3079] A. M. Turing. Rounding-off errors in matrix processes. *Quart. J. Mech. Appl. Math.*, 1:287–308, 1948. (Cited on pages 15, 20, 74, 76, 251, 580, and 586.)

[3080] H. W. Turnbull. *The Theory of Determinants, Matrices and Invariants*. Blackie and Son, 1929. (Cited on page 12.)

[3081] H. W. Turnbull and A. C. Aitken. *An Introduction to the Theory of Canonical Matrices*. Blackie and Son, 1932. Reprinted by Dover (2004). (Cited on pages 8, 12, 15, 131, 149, 150, 218, 388, 396, 405, 490, and 506.)

[3082] E. E. Tyrtyshnikov. New fast algorithms for systems with Hankel and Toeplitz matrices. *USSR Comput. Math. and Math. Phys.*, 29(3):1–6, 1989. (Cited on page 387.)

[3083] S. Ubaru and Y.Saad. Fast methods for estimating the numerical rank of large matrices. In M. F. Balcan and K. Q. Weinberger, editors, *Proceedings of the 33rd International Conference on Machine Learning*, pages 468–477. JMLR, 2016. (Cited on page 14.)

[3084] F. Uhlig. Olga Taussky, a torchbearer for mathematics and teacher of mathematicians. *PAMM Proc. Appl. Math. Mech.*, 7(1):1170501–1170503, 2007. (Cited on page 575.)

[3085] F. Uhlig. An inverse field of values problem. *Inverse Probl.*, 24:1–19, 2008. (Cited on page 288.)

[3086] F. Uhlig. Finding John Francis who found QR fifty years ago. *Image*, 43:19–21, 2009. (Cited on page 311.)

[3087] F. Uhlig. An honorary doctorate for John Francis. *Image*, 56:11–13, 2016. (Cited on page 312.)

[3088] F. Uhlig. Zhang Neural Networks for fast and accurate computations of the field of values. *Linear Multilinear Algebra*, 68(9):1894–1910, 2020. (Cited on page 287.)

[3089] S. Ulam. John von Neumann 1903-1957. *Bull. Amer. Math. Soc.*, 64:1–49, 1958. (Cited on page 536.)

[3090] R. Underwood. An iterative block Lanczos method for the solution of large sparse symmetric eigenproblems. Technical Report STAN-CS-75-496, Computer Science Dept., Stanford University, California, 1975. (Cited on page 303.)

[3091] R. Underwood. An approximate factorization procedure based on the block Cholesky decomposition and its use with the conjugate gradient method. Technical Report NEDO-11386, General Electric, San Jose, CA, 1976. (Cited on page 254.)

[3092] A. Üresin and M. Dubois. Sufficient conditions for the convergence of asynchronous iterations. *Parallel Comput.*, 10(1):83–92, 1989. (Cited on page 196.)

[3093] A. Üresin and M. Dubois. Effects of asynchronism on the convergence rate of iterative algorithms. *J. Parallel Dist. Com.*, 34(1):66–81, 1996. (Cited on page 196.)

[3094] A. van der Sluis. Condition numbers and equilibration of matrices. *Numer. Math.*, 14(1):14–23, 1969. (Cited on pages 80 and 251.)

[3095] A. van der Sluis. Condition, equilibration and pivoting in linear algebraic systems. *Numer. Math.*, 15(1):74–86, 1970. (Cited on pages 80 and 251.)

[3096] A. van der Sluis and H. A. van der Vorst. The rate of convergence of conjugate gradients. *Numer. Math.*, 48:543–560, 1986. (Cited on page 210.)

[3097] H. van der Vorst. A vectorizable variant of some ICCG methods. *SIAM J. Sci. Statist. Comput.*, 3(3):350–356, 1982. (Cited on page 260.)

[3098] H. van der Vorst. An iterative solution method for solving $f(A)x = b$, using Krylov subspace information obtained for the symmetric positive definite matrix A. *J. Comput. Appl. Math.*, 18(2):249–263, 1987. (Cited on page 389.)

[3099] H. van der Vorst. Bi-CGSTAB: A fast and smoothly converging variant of Bi-CG for the solution of nonsymmetric linear systems. Technical report, Dept. of Math., University of Utrecht, The Netherlands, 1990. (Cited on page 226.)

[3100] H. van der Vorst. Bi-CGSTAB: A fast and smoothly converging variant of Bi-CG for the solution of nonsymmetric linear systems. *SIAM J. Sci. Statist. Comput.*, 13(2):631–644, 1992. (Cited on page 227.)

[3101] H. van der Vorst. *Iterative Krylov Methods for Large Linear Systems*. Cambridge University Press, 2003. (Cited on pages 240 and 389.)

[3102] H. van der Vorst and P. Sonneveld. CGSTAB, a more smoothly converging variant of CG-S. Technical report, Department of Mathematics and Informatics, Delft University of Technology, 1990. (Cited on page 226.)

[3103] H. van der Vorst and C. Vuik. The superlinear convergence behavior of GMRES. *J. Comput. Appl. Math.*, 48:327–341, 1993. (Cited on page 230.)

[3104] H. van der Vorst and C. Vuik. GMRESR: a family of nested GMRES methods. *Numer. Linear Algebra Appl.*, 1(4):369–386, 1994. (Cited on page 230.)

[3105] H. van der Vorst and Q. Ye. Residual replacement strategies for Krylov subspace iterative methods for the convergence of true residuals. *SIAM J. Sci. Comput.*, 22(3):835–852, 2000. (Cited on page 233.)

[3106] M. B. van Gijzen, G. L. G. Sleijpen, and J. P. M. Zemke. Flexible and multi-shift induced dimension reduction algorithms for solving large sparse linear systems. *Numer. Linear Algebra Appl.*, 22(1):1–25, 2015. (Cited on page 245.)

[3107] M. B. van Gijzen and P. Sonneveld. Algorithm 913: An elegant IDR(s) variant that efficiently exploits biorthogonality properties. *ACM Trans. Math. Soft. (TOMS)*, 38(1):5.1–5.19, 2011. (Cited on page 244.)

[3108] H. P. M. van Kempen. On quadratic convergence of the special cyclic Jacobi method. *Numer. Math.*, 9(1):19–22, 1966. (Cited on page 281.)

[3109] H. P. M. van Kempen. On the convergence of the classical Jacobi method for real symmetric matrices with non-distinct eigenvalues. *Numer. Math.*, 9(1):11–18, 1966. (Cited on page 281.)

[3110] C. F. Van Loan. Generalizing the singular value decomposition. *SIAM J. Numer. Anal.*, 13(1):76–83, 1976. (Cited on page 141.)

[3111] C. F. Van Loan. On Stewart's singular value decomposition for partitioned orthogonal matrices. Technical Report STAN-CS-79-767, Department of Computer Science, Stanford University, 1979. (Cited on page 145.)

[3112] C. F. Van Loan. Computing the CS and the generalized singular value decompositions. *Numer. Math.*, 46(4):479–491, 1985. (Cited on pages 141 and 145.)

[3113] C. F. Van Loan. *Computational Frameworks for the Fast Fourier Transform*. SIAM, Philadelphia, 1992. (Cited on page 384.)

[3114] J. Van Rosendale. Minimizing inner product data dependencies in conjugate gradient iteration. Technical Report TR 172178, ICASE, NASA Langley Research Center, 1983. (Cited on page 214.)

[3115] H. van Rossum. *A Theory of Orthogonal Polynomials Based on the Padé Table*, volume 4. Van Gorcum, 1953. (Cited on page 228.)

[3116] A. van Weerden. *A Victorian Marriage, Sir William Rowan Hamilton*. BoekenGilde, The Netherlands, 2017. (Cited on page 481.)

[3117] R. Vandebril. Chasing bulges or rotations? A metamorphosis of the QR-algorithm. *SIAM J. Matrix Anal. Appl.*, 32(1):217–247, 2011. (Cited on page 314.)

[3118] R. Vandebril, M. Van Barel, and N. Mastronardi. *Matrix Computations and Semiseparable Matrices, Volume I: Linear Systems*. Johns Hopkins University Press, Baltimore, Maryland, 2008. (Cited on page 387.)

[3119] R. Vandebril, M. Van Barel, and N. Mastronardi. *Matrix Computations and Semiseparable Matrices, Volume II: Eigenvalue and Singular Value Methods*. Johns Hopkins University Press, Baltimore, Maryland, 2008. (Cited on page 387.)

[3120] R. Vandebril and D. S. Watkins. A generalization of the multishift QR algorithm. *SIAM J. Matrix Anal. Appl.*, 33(3):759–779, 2012. (Cited on page 314.)

[3121] A.-T. Vandermonde. Mémoire sur l'élimination. *Histoire de l'Académie Royale des Sciences avec les Mémoires de Mathématique et de Physique pour l'année 1772*, pages 516–532, 1776. (Cited on pages 106, 589, and 590.)

[3122] D. Vanderstraeten. An accurate parallel block Gram-Schmidt algorithm without reorthogonalization. *Numer. Linear Algebra Appl.*, 7(4):219–236, 2000. (Cited on page 130.)

[3123] J. M. Varah. Computing invariant subspaces of a general matrix when the eigensystem is poorly conditioned. *Math. Comput.*, 24(109):137–149, 1970. (Cited on page 149.)

[3124] J. M. Varah. On the solution of block-tridiagonal systems arising from certain finite-difference equations. *Math. Comput.*, 26(120):859–868, 1972. (Cited on page 94.)

[3125] J. M. Varah. A practical examination of some numerical methods for linear discrete ill-posed problems. *SIAM Rev.*, 21(1):100–111, 1979. (Cited on page 394.)

[3126] J. M. Varah. Pitfalls in the numerical solution of linear ill-posed problems. *SIAM J. Sci. Statist. Comput.*, 4(2):164–176, 1983. (Cited on pages 391 and 394.)

[3127] R. S. Varga. *Properties of a special set of entire functions and their respective partial sums.* PhD thesis, Harvard University, 1954. (Cited on page 591.)

[3128] R. S. Varga. A comparison of the successive overrelaxation method and semi-iterative methods using Chebyshev polynomials. *J. Soc. Ind. Appl. Math.*, 5(2):39–46, 1957. (Cited on page 180.)

[3129] R. S. Varga. Orderings of the successive overrelaxation scheme. *Pacific J. Math.*, 9:925–939, 1959. (Cited on page 175.)

[3130] R. S. Varga. P-cyclic matrices: a generalization of the Young-Frankel successive overrelaxation scheme. *Pacific J. Math.*, 9:617–628, 1959. (Cited on page 175.)

[3131] R. S. Varga. Factorization and normalized iterative methods. In R. E. Langer, editor, *Boundary Problems in Differential Equations*, pages 121–142. University of Wisconsin Press, Madison, 1960. Proceedings of a Symposium conducted by the Mathematics Research Center at the University of Wisconsin, Madison, April 20-22, 1959. (Cited on pages 177 and 253.)

[3132] R. S. Varga. *Matrix Iterative Analysis.* Prentice-Hall, Englewood Cliffs, N.J., 1962. A revised and expanded edition was published by Springer in 2000. (Cited on pages 165, 177, 178, 181, 208, and 592.)

[3133] R. S. Varga. Minimal Gerschgorin sets. *Pacific J. Math.*, 15(2):719–729, 1965. (Cited on page 287.)

[3134] R. S. Varga. On recurring theorems on diagonal dominance. *Linear Algebra Appl.*, 13:1–9, 1976. (Cited on page 178.)

[3135] R. S. Varga. Reminiscences of Richard S. Varga. *Numer. Algorithms*, 25:xv–xvi, 2000. (Cited on pages 591 and 592.)

[3136] R. S. Varga. *Geršgorin and his Circles.* Springer-Verlag, Berlin, 2004. (Cited on pages 285, 287, 474, and 592.)

[3137] R. S. Varga, L. Cvetković, and Vladimir Kostić. Approximation of the minimal Geršgorin set of a square complex matrix. *Electron. Trans. Numer. Anal.*, 30:398–405, 2008. (Cited on page 287.)

[3138] R. S. Varga, E. B. Saff, and V. Mehrmann. Incomplete factorizations of matrices and connections with H-matrices. *SIAM J. Numer. Anal.*, 17(6):787–793, 1980. (Cited on page 256.)

[3139] Z. Vavřín. Miroslav Fiedler and Vlastimil Ptâk: life and work. *Linear Algebra Appl.*, 223:3–29, 1995. (Cited on page 16.)

[3140] E. Vecharynski and J. Langou. Any admissible cycle-convergence behavior is possible for restarted GMRES at its initial cycles. *Numer. Linear Algebra Appl.*, 18(3):499–511, 2011. (Cited on page 245.)

[3141] H. J. Veenstra. Veltmann's device for solving a system of linear equations. Bachelor thesis, Faculty of Science and Engineering, University of Groningen, 2012. (Cited on page 332.)

[3142] W. Veltmann. Auflösung linearer gleichungen. *Zeitschrift für Mathematik und Physik*, XXXI:257–272, 1886. (Cited on page 332.)

[3143] S. Venturini. On Cimmino integrals as residues of zeta function. In S. Coen, editor, *Mathematicians in Bologna 1861-1960*, pages 519–538. Birkhäuser, Basel, 2012. (Cited on page 439.)

[3144] F. M. Verzuh. The solution of simultaneous linear equations with the aid of the 602 calculating punch. *M.T.A.C.*, 3(27):453–462, 1949. (Cited on page 333.)

[3145] F. Viète. *In Artem Analyticem Isagoge.* J. Mettayer, Tours, 1591. (Cited on page 48.)

[3146] M. B. Villarino. Gauss on Gaussian quadrature. *Am. Math. Mon.*, 127:125–138, 2020. (Cited on page 469.)

[3147] P. K. W. Vinsome. Orthomin, an iterative method for solving sparse sets of simultaneous linear equations. In *Proceedings of the Fourth SPE Symposium on Numerical Simulation of Reservoir Performance*. Society of Petroleum Engineers of AIME, 1976. (Cited on page 221.)

[3148] C. Vinti. Lamberto Cesari. *Atti Acc. Naz. Lincei, Rend. Lincei, Suppl., Serie IX*, Supplement 3:59–74, 1992. (Cited on page 429.)

[3149] B. Vital. *Etude de quelques méthodes de résolution de problèmes linéaires de grande taille sur multiprocesseur*. PhD thesis, Université de Rennes I, Rennes, France, 1990. (in French). (Cited on page 227.)

[3150] V. V. Voevodin and E. E. Tyrtyshnikov. On generalizations of conjugate direction methods. In *Numerical Methods of Algebra*, pages 3–9. Moscow State University Press, Moscow, 1981. (Cited on page 224.)

[3151] V. V. Voevodin, E. E. Tyrtyshnikov, O. B. Arushnian, M. K. Samarin, J. M. Boyle, W. R. Cowell, K. W. Dritz, and B. S. Garbow. The Toeplitz package users' guide - preliminary version, Argonne National Lab., 1979. (Cited on page 386.)

[3152] C. R. Vogel. Non-convergence of the L-curve regularization parameter selection method. *Inverse Probl.*, 12(4):535–547, 1996. (Cited on page 392.)

[3153] J. A. Vogel. Flexible BiCG and flexible Bi-CGSTAB for nonsymmetric linear systems. *Appl. Math. Comput.*, 188(1):226–233, 2007. (Cited on page 239.)

[3154] G. von Charasoff. *Das System des Marxismus. Darstellung und Kritik*. Hans Bondy, Berlin, 1910. (Cited on page 294.)

[3155] R. von Mises and H. Pollaczek-Geiringer. Praktische verfahren der gleichungsauflösung. *Zamm.-Z Angew. Math. Me.*, 9(1):58–77, 1929. (Cited on pages 166, 172, and 473.)

[3156] R. von Mises and H. Pollaczek-Geiringer. Praktische verfahren der gleichungsauflösung. *Zamm.-Z Angew. Math. Me.*, 9(2):152–164, 1929. (Cited on pages 166, 294, and 473.)

[3157] J. von Neumann. First draft of a report on the EDVAC. Technical report, University of Pennsylvania, Moore School of Electrical Engineering, 1945. (Cited on page 338.)

[3158] J. von Neumann and H. H. Goldstine. Numerical inverting of matrices of high order. *Bull. Amer. Math. Soc.*, 53:1021–1099, 1947. (Cited on pages 15, 19, 75, 132, 584, and 588.)

[3159] W. Sartorius von Waltershausen. *Carl Friedrich Gauss, a Memorial*. S. Hirzel, Leipzig, 1856. Translated to English by Helen Wirthington Gauss, Colorado Springs, 1966. (Cited on page 470.)

[3160] Y. V. Vorobyev. *Methods of Moments in Applied Mathematics*. Gordon and Breach Science Publishers, New York, 1965. (Cited on page 246.)

[3161] A. Voss. Über die mit einer bilinearen form vertauschbaren bilinearen formen. *Sitzb. Akad München*, 19:283–300, 1889. (Cited on pages 395 and 396.)

[3162] C. Vuik. New insights in GMRES-like methods with variable preconditioners. *J. Comput. Appl. Math.*, 61(2):189–204, 1995. (Cited on page 230.)

[3163] C. Vuik, A. Segal, and J. A. Meijerink. An efficient preconditioned CG method for the solution of a class of layered problems with extreme contrasts in the coefficients. *J. Comp. Phys.*, 152(1):385–403, 1999. (Cited on page 217.)

[3164] E. L. Wachspress. *Iterative Solution of Elliptic Systems*. Prentice-Hall, Engelwood Cliffs, N.J., 1962. (Cited on page 182.)

[3165] E. L. Wachspress. Iterative solution of the Lyapunov matrix equation. *Appl. Math. Lett.*, 1:87–90, 1988. (Cited on page 398.)

[3166] D. N. Wakam and J. Erhel. Parallelism and robustness in GMRES with a Newton basis and deflated restarting. *Electron. Trans. Numer. Anal.*, 40:381–406, 2013. (Cited on pages 229 and 244.)

[3167] D. N. Wakam, J. Erhel, and W. D. Gropp. Parallel adaptive deflated GMRES. In *Domain Decomposition Methods in Science and Engineering XX*, pages 631–638, Springer, Berlin, Heidelberg, 2013. (Cited on page 244.)

[3168] H. F. Walker. Implementation of the GMRES and Arnoldi methods using Householder transformations. Technical Report UCRL-93589, Lawrence Livermore Laboratory, 1985. (Cited on pages 225 and 233.)

[3169] H. F. Walker. Implementation of the GMRES method using Householder transformations. *SIAM J. Sci. Statist. Comput.*, 9(1):152–163, 1988. (Cited on page 225.)

[3170] H. F. Walker. Implementations of the GMRES method. *Comput. Phys. Comm.*, 53:311–320, 1989. (Cited on page 225.)

[3171] H. F. Walker. Residual smoothing and peak/plateau behavior in Krylov subspace methods. *Appl. Numer. Math.*, 19(3):279–286, 1995. (Cited on page 232.)

[3172] H. F. Walker and L. Zhou. A simpler GMRES. *Numer. Linear Algebra Appl.*, 1(6):571–581, 1994. (Cited on pages 231 and 238.)

[3173] H. A. Wallace and G. W. Snedecor. Correlation and machine calculation. *Iowa State College Bulletin*, 35, 1925. (Cited on page 449.)

[3174] J. R. Wallis, R. P. Kendall, and T. E. Little. Constrained residual acceleration of conjugate residual methods. In *SPE 13536, Reservoir Simulation Symposium, Austin Texas*. Society of Petroleum Engineers, 1985. (Cited on page 225.)

[3175] J. L. Walsh. Approximation by polynomials in the complex domain. *Mémorial des Sciences Mathématiques*, 73:1–72, 1935. (Cited on page 127.)

[3176] P. J. Walsh. Algorithm 127: Ortho. *Comm. ACM*, 5(10):511–513, 1962. (Cited on page 128.)

[3177] R. E. Walters. Comparison of experimental with calculated results for the lifting effectiveness of a flexible 45° sweptback wing of aspect ratio 6.0 at Mach numbers from 0.8 to 1.3. Technical Report RM L54B16, NACA Langley Aeronautical Laboratory, 1954. (Cited on page 72.)

[3178] L. Wantzel. Recherches sur les moyens de reconnaître si un problème de géométrie peut se résoudre avec la règle et le compas. *J. Math. Pures Appl.*, 2:366–372, 1837. (Cited on page 468.)

[3179] W. H. Ware. The history and development of the electronic computer project at the Institute for Advanced Study. Technical Report P-377, The Rand Corporation, 1953. (Cited on page 338.)

[3180] A. Warusfel. *L'ouvre mathématique de Descartes dans La Géométrie*. PhD thesis, Université Paris 4, 2010. (Cited on page 49.)

[3181] D. S. Watkins. Understanding the *QR* algorithm. *SIAM Rev.*, 24(4):427–440, 1982. (Cited on page 313.)

[3182] D. S. Watkins. Some perspectives on the eigenvalue problem. *SIAM Rev.*, 35(3):430–471, 1993. (Cited on page 314.)

[3183] D. S. Watkins. Shifting strategies for the parallel *QR* algorithm. *SIAM J. Sci. Comput.*, 15(4):953–958, 1994. (Cited on page 314.)

[3184] D. S. Watkins. Forward stability and transmission of shifts in the *QR* algorithm. *SIAM J. Matrix Anal. Appl.*, 16(2):469–487, 1995. (Cited on page 314.)

[3185] D. S. Watkins. The transmission of shifts and shift blurring in the QR algorithm. *Linear Algebra Appl.*, 241-243:877–896, 1996. (Cited on page 314.)

[3186] D. S. Watkins. Performance of the *QZ* algorithm in the presence of infinite eigenvalues. *SIAM J. Matrix Anal. Appl.*, 22(2):364–375, 2000. (Cited on page 314.)

[3187] D. S. Watkins. QR-like algorithms for eigenvalue problems. *J. Comput. Appl. Math.*, 123(1-2):67–83, 2000. (Cited on page 314.)

[3188] D. S. Watkins. *The Matrix Eigenvalue Problem: GR and Krylov Subspace Methods*. SIAM, Philadelphia, 2007. (Cited on pages 314 and 319.)

[3189] D. S. Watkins. The *QR* algorithm revisited. *SIAM Rev.*, 50(1):133–145, 2008. (Cited on page 314.)

[3190] D. S. Watkins. Francis's algorithm. *Am. Math. Mon.*, 118(5):387–403, 2011. (Cited on page 314.)

[3191] D. S. Watkins and L. Elsner. Chasing algorithms for the eigenvalue problem. *SIAM J. Matrix Anal. Appl.*, 12(2):374–384, 1991. (Cited on page 313.)

[3192] D. S. Watkins and L. Elsner. Convergence of algorithms of decomposition type for the eigenvalue problem. *Linear Algebra Appl.*, 143:19–47, 1991. (Cited on page 313.)

[3193] J. W. Watts. A method for improving line successive overrelaxation in anisotropic problems-a theoretical analysis. *Soc. Pet. Eng. J.*, 13(2):105–118, 1973. (Cited on page 225.)

[3194] F. V. Waugh. A simplified method of determining multiple regression constants. *Journal of the American Statistical Association*, 30(192):694–700, 1935. (Cited on page 66.)

[3195] F. V. Waugh and P. S. Dwyer. Compact computation of the inverse of a matrix. *The Annals of Mathematical Statistics*, 16(3):259–271, 1945. (Cited on page 73.)

[3196] H. Wayland. Expansion of determinantal equations into polynomial form. *Quart. Appl. Math.*, 2(4):277–306, 1945. (Cited on page 292.)

[3197] J. H. M. Wedderburn. Note on the linear matrix equation. *Proc. Edinburgh Math. Soc.*, 22:49–53, 1904. (Cited on page 396.)

[3198] J. H. M. Wedderburn. The absolute value of the product of two matrices. *Bull. Amer. Math. Soc.*, 31(7):304–308, 1925. (Cited on page 15.)

[3199] J. H. M. Wedderburn. *Lectures on Matrices*. American Mathematical Society, 1934. (Cited on pages 9, 12, 15, 149, and 388.)

[3200] U. Wegner. Bemerkungen zu den iterationsverfahren für lineare gleichungsysteme. *Zamm.-Z Angew. Math. Me.*, 31(8-9):243–245, 1951. (Cited on page 176.)

[3201] U. Wegner. Bemerkungen zur matrizentheorie. *Zamm.-Z Angew. Math. Me.*, 33(8-9):262–264, 1953. (Cited on page 289.)

[3202] K. Weierstrass. Zur theorie der quadratischen und bilinearen formen. *Monatsberichte der Königlich Preussischen Akademie der Wissenschaften zu Berlin*, pages 310–338, 1868. (Cited on pages 147 and 495.)

[3203] H. F. Weinberger. Error bounds in the Rayleigh-Ritz approximation of eigenvectors. *J. Res. Natl. Bur. Stand.*, 64(5):217–225, 1960. (Cited on page 285.)

[3204] C. Weisbecker. *Improving multifrontal solvers by means of algebraic Block Low-Rank representations*. PhD thesis, Institut National Polytechnique de Toulouse, France, 2013. (Cited on page 378.)

[3205] R. Weiss. *Convergence behavior of generalized conjugate gradient methods*. PhD thesis, University of Karlsruhe, Germany, 1990. (Cited on page 232.)

[3206] R. Weiss. Error-minimizing Krylov subspace methods. *SIAM J. Sci. Comput.*, 15(3):511–527, 1994. (Cited on page 232.)

[3207] R. Weiss. Properties of generalized conjugate gradient methods. *Numer. Linear Algebra Appl.*, 1(1):45–63, 1994. (Cited on page 232.)

[3208] R. Weiss. A theoretical overview of Krylov subspace methods. *Appl. Numer. Math.*, 19(3):207–233, 1995. (Cited on page 233.)

[3209] R. Weiss. *Parameter-free Iterative Linear Solvers*. Wiley VCH, 1996. (Cited on page 233.)

[3210] J. Weissinger. Zur theorie und anwendung des iterationsverfahrens. *Mathematische Nachrichten*, 8(1):193–212, 1952. (Cited on page 176.)

[3211] J. Weissinger. Verallgemeinerungen des Seidelschen iterationsverfahrens. *Zamm.-Z Angew. Math. Me.*, 33(4):155–163, 1953. (Cited on page 176.)

[3212] R. Weitzenböck. Über die matrixgleichung $AX + XB = C$. *Proc. Akad. Wetensch. Amsterdam*, 35:60–61, 1932. (Cited on page 396.)

[3213] R. Weitzenböck. Über die matrixgleichung $X^2 = A$. *Proc. Akad. Wetensch. Amsterdam*, 35:157–161, 1932. (Cited on page 396.)

[3214] J. Wellstein. Über symmetrische, alternierende und orthogonale normalformen von matrizen. *Journal für die reine und angewandte Mathematik (Crelle Journal)*, 163:166–182, 1930. (Cited on page 148.)

[3215] P. Wesseling. A robust and efficient multigrid method. In W. Hackbusch and U. Trottenberg, editors, *Multigrid methods, Proceedings of the Koln-Porz Conference*, pages 614–630. Springer, 1982. Lecture Notes in Mathematics 960. (Cited on page 269.)

[3216] P. Wesseling. Theoretical and practical aspects of a multigrid method. *SIAM J. Sci. Statist. Comput.*, 3(4):387–407, 1982. (Cited on page 269.)

[3217] P. Wesseling. *An Introduction to Multigrid Methods*. John Wiley, 1992. (Cited on page 270.)

[3218] P. Wesseling and P. Sonneveld. Numerical experiments with a multiple grid and a preconditioned Lanczos type method. In R. Rautmann, editor, *Approximation Methods for Navier-Stokes Problems, Paderborn, Germany 1979, Lecture Notes in Mathematics*. Springer, 1980. (Cited on page 222.)

[3219] H. Weyl. Das asymptotische verteilungsgesetz der eigenwerte linearer partieller differentialgleichungen (mit einer anwendung auf die theorie der hohlraumstrahlung). *Math. Ann.*, 71(4):441–479, 1912. (Cited on page 136.)

[3220] H. Weyl. Inequalities between the two kinds of eigenvalues of a linear transformation. *Proc. Nat. Acad. Sci.*, 35(7):408–411, 1949. (Cited on page 136.)

[3221] E. Weyr. Répartition des matrices en espèces et formation de toutes les espèces. *C.R. Acad. Sci. Paris, Sér 1*, 100:966–969, 1885. (Cited on pages 7 and 148.)

[3222] E. Weyr. Note sur la théorie de quantités complexes formées avec n unités principales. *Bull. Sci. Math. II*, 11:205–215, 1887. (Cited on page 387.)

[3223] E. Weyr. O theorii forem bilinearných. Nákl. Jubilejního Fondu Král. České Společnosti Nauk, 1889. (Cited on pages 7, 116, and 148.)

[3224] E. Weyr. Zur theorie der bilinearen formen. *Monatsh Math. Phys.*, 1(1):163–236, 1890. (Cited on pages 7, 116, and 148.)

[3225] G. F. Wheeler and W. P. Crummett. The vibrating string controversy. *Am. J. Phys.*, 55(1):33–37, 1987. (Cited on page 277.)

[3226] A. N. Whitehead. *A Treatise on Universal Algebra with Applications*. Cambridge University Press, 1898. (Cited on pages 11 and 13.)

[3227] E. T. Whittaker and G. Robinson. *The Calculus of Observations*. Blackie and Son Limited, London and Glasgow, 1924. (Cited on pages 70 and 165.)

[3228] O. Widlund and M. Dryja. An additive variant of the Schwarz alternating method for the case of many subregions. Technical Report Technical Report 339, Ultracomputer Note 131, Department of Computer Science, Courant Institute, 1987. (Cited on page 267.)

[3229] H. Wielandt. Das iterationsverfahren bei nicht selbstadjugierten linearen eigenwertprobleme. *Math. Zeitschrift*, 50:93–143, 1944. First appeared as: Beiträge zur mathematischen Behandlung komplexer Eigenwertprobleme, Teil II, Bericht B 43/J/21, Aerodynamische Versuchsanstalt Göttingen, Germany, 1943. (Cited on page 295.)

[3230] H. Wielandt. Inclusion theorems for eigenvalues. In L. J. Paige and O. Taussy, editors, *Simultaneous Linear Equations and the Determination of Eigenvalues*, pages 75–78. National Bureau of Standards Applied Mathematics Series 29, 1953. (Cited on page 295.)

[3231] R. A. Wiggins and E. A. Robinson. Recursive solution to the multi-channel filtering problem. *J. Geophys. Res.*, 70:1885–1891, 1965. (Cited on page 386.)

[3232] J. B. Wilbur. The mechanical solution of simultaneous equations. *J. Franklin Inst.*, 222(6):715–724, 1936. (Cited on pages 331 and 332.)

[3233] H. S. Wilf. Matrix inversion by the annihilation of rank. *J. Soc. Ind. Appl. Math.*, 7(2):149–151, 1959. (Cited on page 83.)

[3234] M. V. Wilkes. Computers then and now. *J. ACM*, 15(1):1–7, 1968. (Cited on page 337.)

[3235] M. V. Wilkes. Computers then and now - Part 2. In R. E. Beck and M. L. Soffa, editors, *Proceedings of the 1996 ACM 24th Annual Conference on Computer Science*, pages 115–119. ACM, 1996. (Cited on page 337.)

[3236] M. V. Wilkes, D. J. Wheeler, and S. Gill. *The Preparation of Programs for an Electronic Digital Computer.* Addison-Wesley Press, 1951. (Cited on page 358.)

[3237] J. H. Wilkinson. Linear algebra on the Pilot ACE. In National Physical Laboratory, editor, *Automatic Digital Computation*, pages 129–136, London, 1954. Her Majesty's Stationary Office. Proceedings of a symposium held at the National Physical Laboratory, March 25-28, 1953. (Cited on pages 77, 295, and 598.)

[3238] J. H. Wilkinson. The Pilot ACE. In National Physical Laboratory, editor, *Automatic Digital Computation*, pages 5–16, London, 1954. Her Majesty's Stationary Office. Proceedings of a symposium held at the National Physical Laboratory, March 25-28, 1953. (Cited on pages 338 and 596.)

[3239] J. H. Wilkinson. The calculation of the eigenvectors of codiagonal matrices. *Comput. J.*, 1(2):90–96, 1958. (Cited on page 295.)

[3240] J. H. Wilkinson. The calculation of eigenvectors by the method of Lanczos. *Comput. J.*, 1:148–152, 1959. (Cited on page 301.)

[3241] J. H. Wilkinson. The evaluation of zeros of ill-conditioned polynomials. Part I. *Numer. Math.*, 1:150–166, 1959. (Cited on page 598.)

[3242] J. H. Wilkinson. The evaluation of zeros of ill-conditioned polynomials. Part II. *Numer. Math.*, 1:167–180, 1959. (Cited on page 598.)

[3243] J. H. Wilkinson. Householder's method for the solution of the algebraic eigenvalue problem. *Comput. J.*, 3:23–28, 1960. (Cited on page 308.)

[3244] J. H. Wilkinson. Error analysis of direct methods of matrix inversion. *J. ACM*, 8:281–330, 1961. (Cited on pages 23, 77, 80, and 588.)

[3245] J. H. Wilkinson. Note on the quadratic convergence of the cyclic Jacobi process. *Numer. Math.*, 4(1):296–300, 1962. (Cited on page 281.)

[3246] J. H. Wilkinson. Plane-rotations in floating-point arithmetic. In *Experimental Arithmetic, High Speed Computing and Mathematics*, pages 185–198, Providence, RI, 1963. American Mathematical Society. volume 15 of Proceedings of Symposia in Applied Mathematics. (Cited on page 131.)

[3247] J. H. Wilkinson. *Rounding Errors in Algebraic Processes.* Prentice-Hall, Englewood Cliffs, NJ, 1964. (Cited on pages 77, 94, and 598.)

[3248] J. H. Wilkinson. *The Algebraic Eigenvalue Problem.* Oxford University Press, Oxford, UK, 1965. (Cited on pages 23, 77, 79, 94, 131, 225, 281, 291, 292, 295, 300, 301, 303, 311, 313, 319, 487, and 599.)

[3249] J. H. Wilkinson. Convergence of the LR, QR, and related algorithms. *Comput. J.*, 8(1):77–84, 1965. (Cited on pages 308, 310, and 311.)

[3250] J. H. Wilkinson. Error analysis of transformations based on the use of matrices of the form $I - 2ww^H$. In L. B. Rall, editor, *Error in Digital Computation, volume 22*, pages 77–101. John Wiley and Sons, 1965. Proceedings of a symposium conducted by the Mathematics Research Center, United States Army, at the University of Wisconsin. (Cited on page 133.)

[3251] J. H. Wilkinson. A priori error analysis of algebraic processes. In I. G. Petrovsky, editor, *Proc. International Congress of Mathematicians, Moscow 1966*, pages 629–640, Mir, Moscow, 1968. (Cited on page 80.)

[3252] J. H. Wilkinson. Modern error analysis. *SIAM Rev.*, 13(4):548–568, 1971. (Cited on pages 20 and 76.)

[3253] J. H. Wilkinson. Some comments from a numerical analyst. 1970 Turing Lecture. *J. ACM*, 18(2):137–147, 1971. (Cited on pages 585, 586, and 598.)

[3254] J. H. Wilkinson. The perfidious polynomial. In G. H. Golub, editor, *Studies in Numerical Analysis*, volume 24 of *Studies in Mathematics*, pages 1–28, Washington, DC, 1984. Mathematical Association of America. (Cited on pages 293 and 597.)

[3255] J. H. Wilkinson. The Pilot ACE at the National Physical Laboratory. In B. J. Copeland, editor, *Alan Turing's Computing Engine*, pages 93–104. Oxford University Press, 2005. Reprinted from *The Radio and Electronic Engineer*, vol. 45 (1975). (Cited on page 338.)

[3256] J. H. Wilkinson and C. Reinsch. *Handbook for Automatic Computation, Volume II, Linear Algebra*. Springer-Verlag, 1971. (Cited on pages 85, 94, 133, 281, 363, and 599.)

[3257] P R. Willems and B. Lang. Twisted factorizations and qd-type transformations for the MR^3 algorithm—new representations and analysis. *SIAM J. Matrix Anal. Appl.*, 33(2):523–553, 2012. (Cited on page 316.)

[3258] R. A. Willoughby. *Proceedings of the Symposium on Sparse Matrices and their Applications, held at the IBM Watson Center, September 9-10, 1968*. International Business Machines Corporation, Thomas J. Watson Research Center, 1969. (Cited on page 86.)

[3259] R. A. Willoughby. A survey of sparse matrix technology. In R. F. Hartung, editor, *Computer Oriented Analysis of Shell Structures*, pages 65–166. Wright-Patterson Air Force Base, 1971. (Cited on page 94.)

[3260] R. Wilson. The equation $px = xq$ in a linear associative algebra. *Proc. Lond. Math. Soc.*, 30:359–366, 1930. (Cited on page 396.)

[3261] R. Wilson. Eliminants of characteristic equations. *Proc. Lond. Math. Soc.*, 33:517–524, 1932. (Cited on page 396.)

[3262] A. Wintner and F. D. Murnaghan. On a polar representation of non-singular square matrices. *Proceedings of the National Academy of Sciences of the United States of America*, 17(12):676–678, 1931. (Cited on page 142.)

[3263] N. Wirth. *Algorithms Plus Data Equals Programs*. Prentice-Hall, 1976. (Cited on page 380.)

[3264] H. Wittmeyer. *Einfluß der Änderung einer Matrix auf die Lösung des zugehörigen Gleichungssystems, sowie auf die charakteristischen Zahlen und die Eigenvektoren*. PhD thesis, Technische Universität Darmstadt, Germany, 1935. (Cited on page 167.)

[3265] H. Wittmeyer. Einfluß der änderung einer matrix auf die lösung des zugehörigen gleichungssystems, sowie auf die charakteristischen zahlen und die eigenvektoren. *Zamm.-Z Angew. Math. Me.*, 16(5):287–300, 1936. (Cited on page 19.)

[3266] H. Wittmeyer. Über die lösung von linearen gleichungssystemen durch iteration. *Zamm.-Z Angew. Math. Me.*, 16(5):301–310, 1936. (Cited on page 167.)

[3267] P. Wlodkowski. Taming Poseidon's beast of incertainty: The auspicious debut of marine engineer Aleksej Nikolaevich Krylov (1863-1945). *The International Journal for the History of Engineering & Technology*, 85(1):140–158, 2015. (Cited on pages 501 and 502.)

[3268] J. Wolfson-Pou and E. Chow. Reducing communication in distributed asynchronous iterative methods. *Procedia Comput. Sci.*, 80:1906–1916, 2016. (Cited on page 197.)

[3269] J. Wolfson-Pou and E. Chow. Modeling the asynchronous Jacobi method without communication delays. *J. Parallel Dist. Com.*, 128:84–98, 2019. (Cited on page 197.)

[3270] Y. K. Wong. An application of orthogonalization process to the theory of least squares. *Ann. Math. Statist.*, 6:53–75, 1935. (Cited on pages 71, 127, and 128.)

[3271] M. A. Woodbury. Inverting modified matrices. Technical Report Memorandum Rept. 42, Statistical Research Group, Princeton University, Princeton, NJ, 1950. (Cited on page 83.)

[3272] H. Woźniakowski. Numerical stability of the Chebyshev method for the solution of large linear systems. *Numer. Math.*, 28(2):191–209, 1977. (Cited on page 181.)

[3273] H. Woźniakowski. Roundoff-error analysis of a new class of conjugate-gradient algorithms. *Linear Algebra Appl.*, 29:507–529, 1980. (Cited on page 210.)

[3274] Z. Woźnicki. *Two-sweep iterative methods for solving large linear systems and their application to the numerical solution of multi-group multi-dimensional neutron diffusion equation*. PhD thesis, Institute of Nuclear Research, Swierk, Poland, 1973. (Cited on page 253.)

[3275] T. W. Wright. *A Treatise on the Adjustment of Observations: With Applications to Geodetic Work and Other Measures of Precision*. D. Van Nostrand, New York, 1884. (Cited on pages 66 and 448.)

[3276] T. W. Wright and J. F. Hayford. *A Treatise on the Adjustment of Observations: With Applications to Geodetic Work*. D. Van Nostrand, New York, 2nd edition, 1906. (Cited on pages 66 and 448.)

[3277] H. E. Wrigley. Accelerating the Jacobi method for solving simultaneous equations by Chebyshev extrapolation when the eigenvalues of the iteration matrix are complex. *Comput. J.*, 6(2):169–176, 1963. (Cited on page 181.)

[3278] K. Wu. Restarted variants of DQGMRES. In *Proceedings of 1997 International Workshop on Computational Science and Engineering*, 1997. (Cited on page 225.)

[3279] K. Wu and H. Simon. Thick-restart Lanczos method for large symmetric eigenvalue problems. *SIAM J. Matrix Anal. Appl.*, 22(2):602–616, 2000. (Cited on pages 233 and 306.)

[3280] W. Wülling. The stabilization of weights in the Lanczos and conjugate gradient methods. *BIT Numerical Mathematics*, 45(2):395–414, 2005. (Cited on page 303.)

[3281] W. Wülling. On stabilization and convergence of clustered Ritz values in the Lanczos method. *SIAM J. Matrix Anal. Appl.*, 27(3):891–908, 2006. (Cited on page 303.)

[3282] P. Wynn. On a device for computing the $e_m(s_n)$ transformation. *M.T.A.C.*, 10:91–96, 1956. (Cited on pages 294 and 310.)

[3283] P. Wynn. Upon systems of recursions which obtain among the quotients of the Padé table. *Numer. Math.*, 8:264–269, 1966. (Cited on page 464.)

[3284] P. Wynn. A general system of orthogonal polynomials. *Quart. J. Math. Oxford*, 2(18):81–96, 1967. (Cited on page 228.)

[3285] H. Xu. Functions of a matrix and Krylov matrices. *Linear Algebra Appl.*, 434(1):185–200, 2011. (Cited on page 389.)

[3286] Y. Xu. Chinese-US mathematical relations, 1859-1949. In M. C. Barr and K. H. Parshall, editors, *Mathematics Unbound: The Evolution of an International Mathematical Research Community, 1800-1945*, pages 287–310. American Mathematical Society, 2002. (Cited on page 128.)

[3287] P. Yalamov and D. J. Evans. Round-off analysis of the WZ matrix factorisation method. *Int. J. Comput. Math.*, 53(1-2):61–81, 1994. (Cited on page 152.)

[3288] P. Yalamov and D. J. Evans. The WZ matrix factorisation method. *Parallel Comput.*, 21(7):1111–1120, 1995. (Cited on page 152.)

[3289] T. Yamamoto. On Lanczos' algorithm for tri-diagonalization. *J. Sci. Hiroshima Univ. Ser. A-I*, 32:259–284, 1968. (Cited on page 301.)

[3290] L. T. Yang and R. P. Brent. The improved Krylov subspace methods for large and sparse linear systems on bulk synchronous parallel architectures. In *Proceedings of the International Parallel and Distributed Processing Symposium (IPDPS' 03)*. IEEE, 2003. (Cited on page 235.)

[3291] L. T. Yang and R. P. Brent. The improved parallel BiCG method for large and sparse unsymmetric linear systems on distributed memory architectures. In *Proceedings of the 16th International Parallel and Distributed Processing Symposium (IPDPS 2002)*, pages 349–360. IEEE, 2003. (Cited on page 235.)

[3292] M. Yannakakis. Computing the minimum fill-in is NP-complete. *SIAM J. Alg. Disc. Meth.*, 2(1):77–79, 1981. (Cited on page 88.)

[3293] B. Ycart. A case of mathematical eponymy: the Vandermonde determinant. *Rev. Hist. Math.*, 9(1):43–77, 2013. (Cited on pages 107 and 590.)

[3294] K. Ye and L.-H. Lim. Every matrix is a product of Toeplitz matrices. *Found. Comput. Math.*, 16:577–598, 2016. (Cited on page 386.)

[3295] Q. Ye. A breakdown-free variation of the nonsymmetric Lanczos algorithms. *Math. Comput.*, 62(205):179–207, 1994. (Cited on page 231.)

[3296] A. Yu. Yeremin, L. Y. Kolotilina, and A. A. Nikishin. Factorized sparse approximate inverse preconditionings III. Iterative construction of preconditioners. *Zapiski Nauchnykh Seminarov POMI*, 248:17–48, 1998. Translated in *Journal of Mathematical Sciences*, 101.4 (2000), pp. 3237-3254. (Cited on page 262.)

[3297] A. Yu. Yeremin and A. A. Nikishin. Factorized-sparse-approximate-inverse preconditionings of linear systems with unsymmetric matrices. *Journal of Mathematical Sciences*, 121(4):2448–2457, 2004. (Cited on page 262.)

[3298] M. C. Yeung. On the solution of singular systems by Krylov subspace methods. In *Ninth International Symposium on Distributed Computing and Applications to Business Engineering and Science (DCABES)*. IEEE, 2010. (Cited on page 243.)

[3299] M. C. Yeung. ML(n) BiCGSTAB: reformulation, analysis and implementation. *Numer. Math. Theor. Meth. Appl.*, 5(3):447–492, 2012. (Cited on page 237.)

[3300] M. C. Yeung. ML(n)BiCGStabt: a ML(n)BiCGStab variant with A-transpose. *J. Comput. Appl. Math.*, 272:57–69, 2014. (Cited on page 237.)

[3301] M. C. Yeung and T. F. Chan. ML(k)BiCGSTAB: A BiCGSTAB variant based on multiple Lanczos starting vectors. *SIAM J. Sci. Comput.*, 21(4):1263–1290, 1999. (Cited on pages 237 and 242.)

[3302] K. Yoo and H. Park. Accurate downdating of a modified Gram-Schmidt QR decomposition. *BIT Numerical Mathematics*, 36(1):166–181, 1996. (Cited on page 130.)

[3303] C. N. Yood. *Argonne National Laboratory and the emergence of computer and computational science, 1946-1992*. PhD thesis, The Pennsylvania State University, 2005. (Cited on page 476.)

[3304] D. M. Young. *Iterative methods for solving partial difference equations of elliptic type*. PhD thesis, Harvard University, Cambridge, Massachusetts, 1950. (Cited on pages 165, 171, 174, and 600.)

[3305] D. M. Young. On Richardson's method for solving linear equations with positive definite matrices. *J. Math, and Physics*, 32:243–255, 1953. (Cited on page 163.)

[3306] D. M. Young. Iterative methods for solving partial difference equations of elliptic type. *Trans. Amer. Math. Soc.*, 76:92–111, 1954. (Cited on pages 174 and 600.)

[3307] D. M. Young. *Iterative Solution of Large Linear Systems*. Academic Press, 1971. Reprinted by Dover (2003). (Cited on pages 175, 600, and 601.)

[3308] D. M. Young. Second-degree iterative methods for the solution of large linear systems. *J. Approx. Theory*, 5:137–148, 1972. (Cited on page 181.)

[3309] D. M. Young. A historical review of iterative methods. In S. G. Nash, editor, *A History of Scientific Computing*, pages 180–194. ACM Press, 1990. (Cited on page 175.)

[3310] D. M. Young and K. C. Jea. Generalized conjugate-gradient acceleration of nonsymmetrizable iterative methods. *Linear Algebra Appl.*, 34:159–194, 1980. (Cited on pages 222, 223, 224, and 601.)

[3311] D. M. Young and D. Kincaid. The ITPACK package for large sparse linear systems. In M. H. Schultz, editor, *Elliptic Problem Solvers*, pages 163–185. Academic Press, 1981. (Cited on page 372.)

[3312] D. M. Young and D. R. Kincaid. Linear stationary second-degree methods for the solution of large linear systems. In T. M. Rassias, H. M. Srivastava, and A. Yanushauskas, editors, *Topics in Polynomials of One and Several Variables and their Applications: Volume dedicated to the memory of P.L. Chebyshev (1821-1894)*, pages 609–629. World Scientific Company, 1993. (Cited on page 185.)

[3313] H. Yserentant. Old and new convergence proofs for multigrid methods. *Acta Numerica*, 2:285–326, 1993. (Cited on page 270.)

[3314] Y. X. Yuan. Jiu Zhang Suan Shu and the Gauss algorithm for linear equations. *Documenta Math.*, Extra Volume ISMP:9–14, 2012. (Cited on pages 35 and 39.)

[3315] S. Zadoo. *Critical study of the Bakhsāli manuscript.* PhD thesis, University of Kashmir, 1992. (Cited on page 42.)

[3316] I. Zavorin. *Analysis of GMRES convergence by spectral factorization of the Krylov matrix.* PhD thesis, University of Maryland, 2001. (Cited on page 240.)

[3317] I. Zavorin, D. P. O'Leary, and H. C. Elman. Complete stagnation of GMRES. *Linear Algebra Appl.*, 367:165–183, 2003. (Cited on page 240.)

[3318] J. P. M. Zemke. *Krylov subspace methods in finite precision: A unified approach.* PhD thesis, Technical University of Hamburg, Germany, 2003. (Cited on page 303.)

[3319] J. P. M. Zemke. (Hessenberg) eigenvalue-eigenmatrix relations. *Linear Algebra Appl.*, 414(2-3):589–606, 2006. (Cited on page 303.)

[3320] J. P. M. Zemke. Abstract perturbed Krylov methods. *Linear Algebra Appl.*, 424(2-3):405–434, 2007. (Cited on page 303.)

[3321] J. P. M. Zemke. Variants of IDR with partial orthonormalization. *Electron. Trans. Numer. Anal.*, 46:245–272, 2017. (Cited on page 246.)

[3322] N. Zettili. *Quantum Mechanics, Concepts and Application.* Wiley, Chichester, UK, 2nd edition, 2009. (Cited on page 26.)

[3323] H. Zha. A numerical algorithm for computing restricted singular value decomposition of matrix triplets. *Linear Algebra Appl.*, 168:1–26, 1992. (Cited on page 141.)

[3324] F. Zhang. *The Schur Complement and its Applications.* Springer Science, New York, 2005. (editor). (Cited on pages 26 and 554.)

[3325] J. Zhang. On preconditioning Schur complement and Schur complement preconditioning. *Electron. Trans. Numer. Anal.*, 10:115–130, 2000. (Cited on page 271.)

[3326] J. Zhang. Sparse approximate inverse and multilevel block ILU preconditioning techniques for general sparse matrices. *Appl. Numer. Math.*, 35:67–86, 2000. (Cited on page 270.)

[3327] L. Zhang and T. Nodera. A new adaptive restart for GMRES(m) method. *ANZIAM J.*, 46:409–425, 2005. (Cited on page 232.)

[3328] S.-L. Zhang. GPBi-CG: Generalized product-type methods based on Bi-CG for solving nonsymmetric linear systems. *SIAM J. Sci. Comput.*, 18(2):537–551, 1997. (Cited on pages 235 and 239.)

[3329] S.-L. Zhang. A class of product-type Krylov-subspace methods for solving nonsymmetric linear systems. *J. Comput. Appl. Math.*, 149(1):297–305, 2002. (Cited on page 235.)

[3330] W. Zhang. *GMRES on a tridiagonal Toeplitz linear system.* PhD thesis, University of Kentucky, 2007. (Cited on page 241.)

[3331] L. Zhou and H. F. Walker. Residual smoothing techniques for iterative methods. *SIAM J. Sci. Comput.*, 15(2):297–312, 1994. (Cited on pages 232 and 236.)

[3332] J. Zítko. Generalization of convergence conditions for a restarted GMRES. *Numer. Linear Algebra Appl.*, 7:117–131, 2000. (Cited on page 237.)

[3333] J. Zítko. Convergence conditions for a restarted GMRES method augmented with eigenspaces. *Numer. Linear Algebra Appl.*, 12(4):373–390, 2005. (Cited on page 237.)

[3334] J. Zítko. How residual bounds for restarted GMRES describe the real behaviour. In *PAMM: Proceedings in Applied Mathematics and Mechanics*, volume 8. Wiley-VCH Verlag, 2008. (Cited on page 237.)

[3335] J. Zítko. Some remarks on the restarted and augmented GMRES method. *Electron. Trans. Numer. Anal.*, 31:221–227, 2008. (Cited on page 237.)

[3336] J. Zítko and D. Nádhera. Behaviour of the augmented GMRES method. In *SNA'10, Nové Hrady*, Prague, Czech Republic, 2010. Charles University. (Cited on page 237.)

[3337] Z. Zlatev. On some pivotal strategies in Gaussian elimination by sparse technique. *SIAM J. Numer. Anal.*, 17(1):18–30, 1980. (Cited on page 92.)

[3338] Z. Zlatev. *Computational Methods for General Sparse Matrices*. Springer, 1991. (Cited on page 94.)

[3339] Z. Zlatev, J. Wasniewski, and K. Schaumburg. *Y12M: Solution of Large and Sparse Systems of Linear Algebraic Equations*. Springer, Berlin, Heidelberg, 1981. Lecture Notes in Computer Science 121. (Cited on page 371.)

[3340] S. Zohar. Toeplitz matrix inversion: The algorithm of W.F. Trench. *J. ACM*, 16(4):592–601, 1969. (Cited on page 386.)

[3341] S. Zohar. The solution of a Toeplitz set of linear equations. *J. ACM*, 21(2):272–276, 1974. (Cited on page 386.)

[3342] K. Zorychta. Comparison of Purcell's method with the elimination method of Gauss. *Computatio*, 2:65–74, 1964. (Cited on page 81.)

[3343] R. Zurmühl. *Matrizen, eine Darstellung für Ingenieure*. Springer-Verlag, Berlin, Heidelberg, 1950. (Cited on pages 218, 292, and 487.)

[3344] K. Zuse. *Der Computer - Mein Lebenswerk.* Springer-Verlag, 1983. English translation, *The Computer - My Life*, Springer, 1993. (Cited on page 603.)

Name index

Subject index